Tumors of the Central Nervous System

AFIP Atlas of Tumor Pathology

Editorial Director: Kelley S. Hahn
Production Editor: Dian S. Thomas
Editorial/Scanning Assistant: Mirlinda Q. Caton
Copyeditor: Audrey Kahn
Scanning Technician: Kenneth Stringfellow

Available from the American Registry of Pathology
Armed Forces Institute of Pathology
Washington, DC 20306-6000
www.afip.org
ISBN 1-933477-01-6
978-1-933477-01-5

AFIP ATLAS OF TUMOR PATHOLOGY

Fourth Series
Fascicle 7

TUMORS OF THE CENTRAL NERVOUS SYSTEM

by

Peter C. Burger, MD
Professor of Pathology
The Johns Hopkins University School of Medicine
Baltimore, Maryland

Bernd W. Scheithauer, MD
Consultant in Pathology
Mayo Clinic
Professor of Pathology
Mayo Graduate School of Medicine
Rochester, Minnesota

Published by the
American Registry of Pathology
Washington, DC
in collaboration with the
Armed Forces Institute of Pathology
Washington, DC
2007

AFIP ATLAS OF TUMOR PATHOLOGY

EDITORS' NOTE

The Atlas of Tumor Pathology has a long and distinguished history. It was first conceived at a Cancer Research Meeting held in St. Louis in September 1947 as an attempt to standardize the nomenclature of neoplastic diseases. The first series was sponsored by the National Academy of Sciences-National Research Council. The organization of this Sisyphean effort was entrusted to the Subcommittee on Oncology of the Committee on Pathology, and Dr. Arthur Purdy Stout was the first editor-in-chief. Many of the illustrations were provided by the Medical Illustration Service of the Armed Forces Institute of Pathology (AFIP), the type was set by the Government Printing Office, and the final printing was done at the Armed Forces Institute of Pathology (hence the colloquial appellation "AFIP Fascicles"). The American Registry of Pathology (ARP) purchased the Fascicles from the Government Printing Office and sold them virtually at cost. Over a period of 20 years, approximately 15,000 copies each of nearly 40 Fascicles were produced. The worldwide impact of these publications over the years has largely surpassed the original goal. They quickly became among the most influential publications on tumor pathology, primarily because of their overall high quality, but also because their low cost made them easily accessible the world over to pathologists and other students of oncology.

Upon completion of the first series, the National Academy of Sciences-National Research Council handed further pursuit of the project over to the newly created Universities Associated for Research and Education in Pathology (UAREP). A second series was started, generously supported by grants from the AFIP, the National Cancer Institute, and the American Cancer Society. Dr. Harlan I. Firminger became the editor-in-chief and was succeeded by Dr. William H. Hartmann. The second series' Fascicles were produced as bound volumes instead of loose leaflets. They featured a more comprehensive coverage of the subjects, to the extent that the Fascicles could no longer be regarded as "atlases" but rather as monographs describing and illustrating in detail the tumors and tumor-like conditions of the various organs and systems.

Once the second series was completed, with a success that matched that of the first, ARP, UAREP, and AFIP decided to embark on a third series. Dr. Juan Rosai was appointed as editor-in-chief, and Dr. Leslie H. Sobin became associate editor. A distinguished Editorial Advisory Board was also convened, and these outstanding pathologists and educators played a major role in the success of this series, the first publication of which appeared in 1991 and the last (number 32) in 2003.

The same organizational framework will apply to the current fourth series, but with UAREP no longer in existence, ARP will play the major role. New features will include a hardbound cover, illustrations almost exclusively in color, and an accompanying electronic version of each Fascicle. There will also be increased emphasis (wherever appropriate) on the cytopathologic (intraoperative, exfoliative, and/or fine needle aspiration) and molecular features that are important in diagnosis and prognosis. What will not change from the three previous series, however, is the goal of providing the practicing pathologist with thorough, concise, and up-to-date information on the nomenclature and classification; epidemiologic, clinical, and pathogenetic features; and, most importantly, guidance in the diagnosis of the tumors and tumorlike lesions of all major organ systems and body sites.

As in the third series, a continuous attempt will be made to correlate, whenever possible, the nomenclature used in the Fascicles with that proposed by the World Health Organization's Classification of Tumours, as well as to ensure a consistency of style throughout. Close cooperation between the various authors and their respective liaisons from the Editorial Board will continue to be emphasized in order to minimize unnecessary repetition and discrepancies in the text and illustrations.

Particular thanks are due to the members of the Editorial Advisory Board, the reviewers (at least two for each Fascicle), the editorial and production staff, and—first and foremost—the individual Fascicle authors for their ongoing efforts to ensure that this series is a worthy successor to the previous three.

Steven G. Silverberg, MD
Leslie H. Sobin, MD

ACKNOWLEDGEMENTS

No complex project like this could be done without the help of others. We are pleased to acknowledge our debt to the photographic and digital imaging support from both of our departments. In Baltimore, profound gratitude goes to Mr. Norman Barker, Associate Professor and Director of Pathology Photography, Digital Imaging, and Computer Graphics for his uncompromising attention to detail in photomicrography and digital imaging of whole mount tissue sections. Mrs. Ellen Winslow was meticulous also in digital imaging of radiologic studies and gross specimens. In Rochester, Mr. Eric M. Sheahan of Media Support Services was an invaluable contributor as well.

At Johns Hopkins, Pat Goldthwaite, with a magic touch in retrieving slides, assisted also by obtaining histories and copies of references. Mrs. Rosemary Rogers stood guard, protecting professional time for the long project. Dr. Ziya Gokaslan and his associate Mr. Ian Suk in the Department of Neurosurgery could not have been more helpful in obtaining intraoperative images of spinal neoplasms.

At Mayo Clinic, Denise Chase was of invaluable assistance with her excellent and always cheerful assistance in transcription services.

Contributors of material are acknowledged in figure captions throughout the book. Special thanks go to Drs. Arie Perry of Washington University in St. Louis, Constance Griffin at Johns Hopkins, and Jaclyn Biegel of the Children's Hospital of Philadelphia for supplying images of fluorescence in situ hybridization studies.

Finally, we appreciate the helpful suggestions of the two reviewers Drs. Richard A. Prayson and Marc K. Rosenblum.

Peter C. Burger, MD

Bernd W. Scheithauer, MD

DEDICATIONS

To my wife Paula and daughter Elizabeth for their love and support during a long, labor-intensive project

Peter C. Burger, MD

To Nany . . .

Bernd W. Scheithauer, MD

CONTENTS

1 OVERVIEW AND CLASSIFICATION

In the decade since the publication of the previous Fascicle on tumors of the central nervous system (CNS), many new entities have been described and the prognostic significance of certain tumor subtypes has been established. Grading criteria have been refined and molecular features have been correlated with tumor type and grade. While important to affected patients, the newly described lesions are usually rarities, unlikely to be seen by a single pathologist, whereas the refined grading criteria of common lesions, such as meningiomas, are both practical and globally relevant. An updated international consensus, or at least general agreement, on CNS tumor classification was published in 2007 by the World Health Organization (WHO) (1) and is outlined in Table 1-1.

In light of these developments, the present volume has been revised extensively. The classifications of both existing entities and newly described lesions have been brought into close alignment with the WHO system. Among the new lesions/variants are chordoid glioma of the third ventricle, cerebellar liponeurocytoma, large cell/anaplastic medulloblastoma, and papillary glioneuronal tumor. Entities described in the new millennium, such as papillary tumor of the pineal region, have also been included. The new WHO classification/grading system for meningiomas is described in detail.

The present book retains the outline of its predecessor and differs only slightly from the organization of the WHO book (1). For example, hemangiopericytoma and melanocytic tumors are discussed here in separate chapters, Tumors of Mesenchymal Tissue and Melanocytic Neoplasms, respectively, whereas the WHO publication combines them in a single chapter on meningeal tumors. These, and other minor variations, are only cosmetic distinctions that do not detract from the similarity of our classification scheme and that of the WHO.

Not withstanding widespread pessimism, CNS tumors are often surgically curable, thanks to advances in both neuroimaging and surgical techniques. Such cures are usually found among patients with relatively well-circumscribed tumors, but sometimes achieved even in those with diffuse, infiltrating lesions. It is essential to recognize the better-delimited tumors since they often lend themselves to resection, or at least debulking, that can permit a period of observation until the need for additional therapy, if any, becomes apparent. The significance of tumor architecture as it pertains to diagnosis and treatment is thus emphasized in the descriptions of individual entities throughout the book. Our first Appendix subdivides CNS tumors on the basis of demarcation; the large section on astrocytic neoplasms in chapter 3 is organized around this feature.

As three-dimensional, multiplanar forms of macroscopic examination, neuroradiologic techniques are important to pathologists for establishing the degree of circumscription, chronicity, vascularity, mineralization, and metabolic activity of various tumors. We discuss these techniques in regard to both neoplastic and non-neoplastic lesions. Chapter 2 reviews normal radiologic anatomy.

Despite the sophistication of neuroimaging, reactive lesions, such as demyelinating diseases, occasionally are biopsied in expectation of a neoplasm, with the very real potential for pathologic misdiagnosis and overtreatment. The admonition that pathologists remain alert to the possibility of a non-neoplastic lesion is repeated throughout this book. It is stressed in chapter 19 on reactive and inflammatory masses, and in two appendices that present the histologic features of non-neoplastic and low-grade lesions. A third appendix is an algorithm that ensures that non-neoplastic and low-grade entities are given full consideration in the differential diagnosis of a suspected CNS neoplasm.

Grading criteria, always artificial and arbitrary, are discussed at length throughout the text. For example, the prognostic implications of mitotic counts and MIB-1 indices are recurring themes. The latter is an increasingly appreciated prognostic factor, with results often calculated to tenths of a percent, although there is surprisingly little information regarding interinstitutional reproducibility.

Techniques of the burgeoning field of molecular diagnostics are increasingly employed in tumor classification and grading. Examples relevant to CNS tumors include the status of chromosomes 1p and 19q in infiltrating gliomas; gain of chromosome 7, amplification and mutation of *EGFR*, and loss of heterozygosity on chromosomes 10, 17p, and 9p in diffuse astrocytomas; amplification or overexpression of *c-myc* in medulloblastomas; chromosomal imbalances in grades II and III meningiomas; mutations of *INI1* in atypical teratoid/rhabdoid tumor; and an ever growing list of gene expression profiles in multiple tumor types. We illustrate approaches of practical value, while assuming that morphologic findings will endure as a context in which to evaluate molecular findings.

Table 1-1

WORLD HEALTH ORGANIZATION CLASSIFICATION OF TUMORS OF THE NERVOUS SYSTEM[a]

Tumor	Code
TUMOURS OF NEUROEPITHELIAL TISSUE	
Astrocytic Tumours	
Pilocytic astrocytoma	9421/1
Pilomyxoid astrocytoma	*9425/3*[b]
Subependymal giant cell astrocytoma	9384/1
Pleomorphic xanthoastrocytoma	9424/3
Diffuse astrocytoma	9400/3
Fibrillary astrocytoma	9420/3
Protoplasmic astrocytoma	9410/3
Gemistocytic astrocytoma	9411/3
Anaplastic astrocytoma	9401/3
Glioblastoma	9440/3
Giant cell glioblastoma	9441/3
Gliosarcoma	9442/3
Gliomatosis cerebri	9381/3
Oligodendroglial Tumours	
Oligodendroglioma	9450/3
Anaplastic oligodendroglioma	9451/3
Oligoastrocytic Tumours	
Oligoastrocytoma	9382/3
Anaplastic oligoastrocytoma	9382/3
Ependymal Tumours	
Subependymoma	9383/1
Myxopapillary ependymoma	9394/1
Ependymoma	9391/3
Cellular	9391/3
Papillary	9393/3
Clear cell	9391/3
Tanycytic	9391/3
Anaplastic ependymoma	9392/3
Choroid Plexus Tumours	
Choroid plexus papilloma	9390/0
Atypical choroid plexus papilloma	*9390/1*
Choroid plexus carcinoma	9390/3
Other Neuroepithial Tumours	
Astroblastoma	9430/3
Chordoid glioma of the 3rd ventricle	9444/1
Angiocentric glioma	*9431/1*
Neuronal and Mixed Neuronal-Glial Tumours	
Dysplastic gangliocytoma of cerebellum (Lhermitte-Duclos)	9493/0
Desmoplastic infantile astrocytoma/ganglioglioma	9412/1
Dysembryoplastic neuroepithelial tumour	9413/0
Gangliocytoma	9492/0
Ganglioglioma	9505/1
Anaplastic ganglioglioma	9505/3
Central neurocytoma	9506/1
Extraventricular neurocytoma	*9506/1*
Cerebellar liponeurocytoma	*9506/1*
Papillary glioneuronal tumour	*9509/1*
Rosette-forming glioneuronal tumour of the 4th ventricle	*9509/1*
Paraganglioma of the filum terminale	8680/1
Tumours of the Pineal Region	
Pineal parenchymal tumours	
Pineocytoma	9361/1
Pineal parenchymal tumour of intermediate differentiation	9362/3
Pineoblastoma	9362/3
Papillary tumour of the pineal region	*9395/3*

[a]Louis DN, Ohgaki H, Wiestler OD, Cavenee WK, eds. WHO Classification of Tumours of the Central Nervous System. Lyon: IARC Press; 2007.

[b]The italicized numbers are provisional codes proposed for the 4th edition of ICD-O. While they are expected to be incorporated into the next ICD-O edition, they currently remain subject to change.

Table 1-1 (Continued)

Tumour	Code
Embryonal Tumours	
Medulloblastoma	9470/3
Desmoplastic/nodular medulloblastoma	*9471/3*
Medulloblastoma with extensive nodularity	*9471/3*
Anaplastic medulloblastoma	*9474/3*
Large cell medulloblastoma	9474/3
CNS primitive neuroectodermal tumour (PNET)	9473/3
Medulloepithelioma	9501/3
Neuroblastoma	9500/3
Ganglioneuroblastoma	9490/3
Ependymoblastoma	9392/3
Atypical teratoid/rhabdoid tumor	9508/3
TUMOURS OF CRANIAL AND PARASPINAL NERVES	
Schwannoma (Neurinoma)	9560/0
Cellular	9560/0
Plexiform	9560/0
Melanotic	9560/0
Neurofibroma	9540/0
Plexiform	9550/0
Perineurioma	9571/0
Intraneural perineurioma	9571/0
Soft tissue perineurioma	9571/0
Malignant Peripheral Nerve Sheath Tumour (MPNST)	9540/3
Epithelioid	9540/3
MPNST with divergent meschymal and/or epithelial differentiation	9540/3
Melanotic	9540/3
TUMOURS OF THE MENINGES	
Tumours of Meningothelial Cells	
Meningioma	9530/0
Meningothelial	9531/0
Fibrous (fibroblastic)	9532/0
Transitional (mixed)	9537/0
Psammomatous	9533/0
Angiomatous	9534/0
Microcystic	9530/0
Secretory	9530/0
Lymphoplasmacyte-rich	9530/0
Metaplastic	9530/0
Clear Cell	9538/1
Chordoid	9538/1
Atypical	9539/1
Papillary	9538/3
Rhabdoid	9538/3
Anaplastic	9530/3
Mesenchymal Tumours	
Lipoma	8850/0
Angiolipoma	8861/0
Hibernoma	8880/0
Liposarcoma (intracranial)	8850/3
Solitary fibrous tumour	8815/0
Fibrosarcoma	8810/3
Malignant fibrous histiocytoma	8830/3
Leiomyoma	8890/0
Leiomyosarcoma	8890/3
Rhabdomyoma	8900/0
Rhabdomyosarcoma	8900/3
Chondroma	9220/3
Chondrosarcoma	9220/3
Osteoma	9180/0
Osteosarcoma	9180/3
Osteochondroma	9210/0
Haemangioma	9120/0
Epithelioid haemangioendothelioma	9133/1
Haemangiopericytoma	9150/1
Angiosarcoma	9120/3
Kaposi sarcoma	9140/3
Primary Melanocytic Lesions	
Diffuse melanocytosis	8728/0
Melanocytoma	8728/1
Malignant melanoma	8720/3
Meningeal melanomatosis	8728/3
Other Neoplasms Related to the Meninges	
Haemangioblastoma	9161/1
LYMPHOMAS AND HAEMOPOIETIC NEOPLASMS	
Malignant lymphomas	9590/3
Plasmacytoma	9731/3
Granulocytic sarcoma	9930/3
GERM CELL TUMOURS	
Germinoma	9064/3
Embryonal carcinoma	9070/3
Yolk sac tumour	9071/3
Choriocarcinoma	9100/3
Teratoma	9080/1
Mature	9080/0
Immature	9080/3
Teratoma with malignant transformation	9084/3
Mixed germ cell tumours	9085/3
TUMOURS OF THE SELLAR REGION	
Craniopharyngioma	9350/1
Adamantinomatous	9351/1
Papillary	9352/1
Granular cell tumour	9582/0
Pituicytoma	*9432/1*
Spindle cell oncocytoma	*8291/0*
METASTATIC TUMOURS	

REFERENCES

1. Louis DN, Ohgaki H, Wiestler OD, Cavenee WK, eds. WHO Classification of Tumours of the Central Nervous System. Lyon: IARC Press; 2007. (In press)

2 EMBRYOLOGY AND NORMAL ANATOMY

EMBRYOLOGY

The developing central nervous system (CNS) first emerges as a plate of embryonic epithelium which, when uplifted, delimits a midline longitudinal trench, the primitive neural groove. In tectonic fashion, by a process of uplifting and subvention, the opposing edges of the groove approach the midline, make contact in the mid-dorsal region, and undergo rostral to caudal fusion. The result is a cylinder that remains open at both ends until the time of closure of the anterior and posterior neuropores. Early in embryogenesis, the cylinder becomes constricted to form four segments: the prosencephalon, mesencephalon, rhombencephalon, and a caudal elongation that remains undivided as the forerunner of the spinal cord. In accordance with the C-shape of the early embryo, the neural tube is angulated or flexed in the cervical and cephalic regions. A subsequent posterior shift of the cerebrum necessitates an additional kink in the region of the future pons.

The crowning achievement in the development of the human brain is the appearance and enlargement of the telencephalic vesicles. These voluminous, aneurysmal expansions bud from the prosencephalon and expand dramatically as they begin to assert their role as the brain's dominant division, the cerebrum. Their lumens, the future lateral ventricles, retain their connection with the third ventricle via the foramina of Monro. These narrow apertures are vulnerable to obstruction by such regional lesions as colloid cyst, neurocytoma, and a wide spectrum of gliomas.

The second prosencephalic division, the diencephalon, gives rise to the thalami and hypothalamus. The mesencephalon, or midbrain, retains a somewhat tubular profile and a short simple lumen, the aqueduct of Sylvius, which persists as a narrow but functionally important channel for the passage of cerebrospinal fluid into the fourth ventricle. In subsequent development, the rhombencephalon, or hindbrain, evolves into the metencephalon (pons and cerebellum) and myelencephalon (medulla).

As a primitive reflex and relay structure, the spinal cord lacks the complicated structure of the more rostral segments subserving cognition, intellect, and coordination. As a result, the cord retains the basic tubular architecture of the early developing nervous system, as well as the dorsal and ventral segregation of sensory and motor functions, respectively. Motor functions initiated ventrally in the anterior gray columns are conducted to the periphery through the anterior roots, whereas sensory impulses are received dorsally through the posterior roots. Although the central canal is patent along its length at birth, it becomes discontinuous by puberty, leaving only clusters and small tubules of ependymal cells (see fig. 2-40). The most common intramedullary cord tumor (ependymoma) is derived from these cells.

NORMAL ANATOMY

Only a brief review of normal anatomy can be presented here. The reader can consult other sources for more details, particularly how they relate to microscopic study (1).

Skull

As a densely calcified structure, the skull is highly absorptive of an X-ray beam and is therefore bright in computerized tomographic (CT) images (fig. 2-1) (2,3). By magnetic resonance imaging (MRI), on the other hand, it is dark in both T1- and T2-weighted images, with the exception of the fine linear band of brightness due to diploic adipose tissue visualized in T1-weighted images (fig. 2-2).

Several localizing anatomic terms for cranial structures used frequently by neurosurgeons are tuberculum sellae, dorsum sellae, and Meckel's cave. The tuberculum sellae is the top of the anterior wall of the sella turcica. Meningiomas are

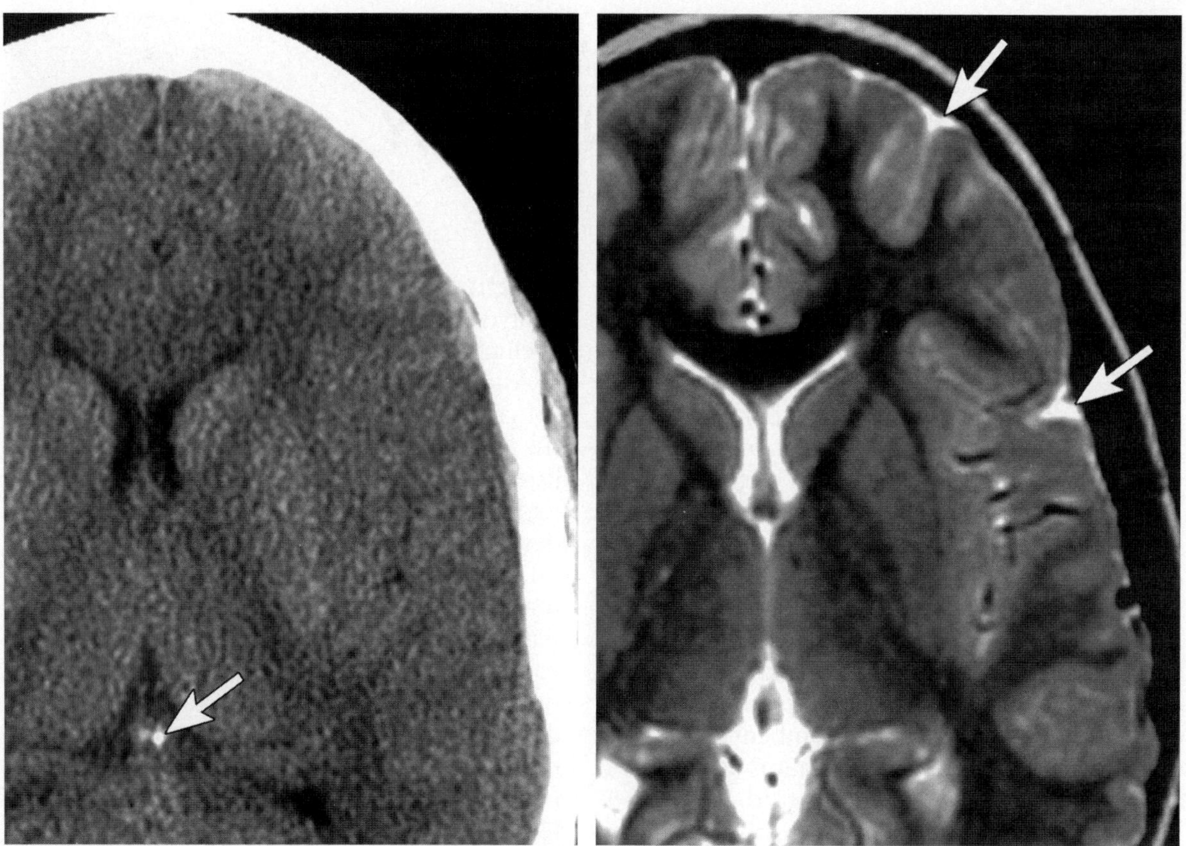

Figure 2-1

COMPARISON OF COMPUTERIZED TOMOGRAPHY AND MAGNETIC RESONANCE IMAGING: NORMAL SKULL AND BRAIN

Representative images illustrate the relative sensitivities and resolution of computerized tomography (CT) and magnetic resonance imaging (MRI). CT (left) is sensitive to the presence of calcium, as is evident in the highly absorptive skull and normal pineal calcifications (arrow), but insensitive to differing densities of brain parenchyma. In contrast, MRI (right), as here in a T2-weighted image, resolves details of the basal ganglia, thalamus, and cerebral cortex. Cerebrospinal fluid, in either the subarachnoid space (arrows) or the ventricular system, is hyperintense, i.e., bright. The skull is black.

common in this region. The dorsum sellae is the sella's sloping posterior wall, which bears the posterior clinoid processes. Within the middle cranial fossa is Meckel's cave, a medial recess housing the trigeminal ganglion. Meningiomas and melanocytomas occur in this region.

The configuration of the skull base permits subdivision of supratentorial structures into those of the anterior and middle cranial fossae; the floors of these fossae support the frontal and the temporal lobes, respectively. The dural venous sinuses described below merge at the torcula, a point of confluence of the sagittal and transverse sinuses, which underlie a major bony landmark, the inion.

Meninges

The brain is invested by a tough, fibrous and nearly acellular layer, the dura. This layer generates a dark MRI signal, which radiologically is added to the dark lamina provided by the inner table of the skull (figs. 2-1, 2-2). The intracranial dura is reduplicated to form an arching midline fold, the falx cerebri, which partially separates the two cerebral hemispheres. A similar membrane forms a peaked septation, the tentorium cerebelli, which serves as a landmark that conveniently places most intracranial structures into either the supratentorial or infratentorial compartment. The cerebrum, basal ganglia, thalamus, and hypothalamus are supratentorial,

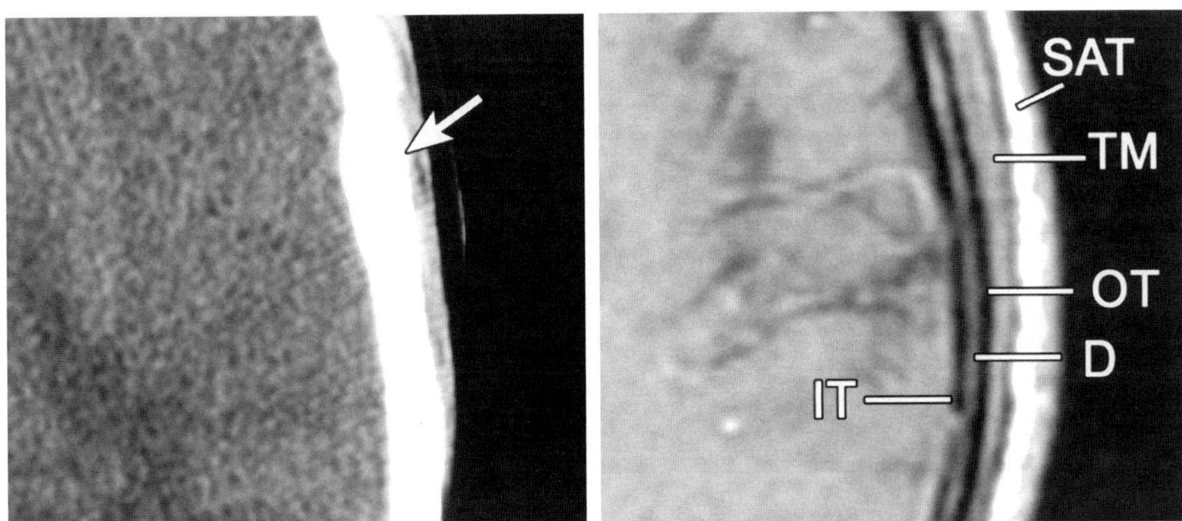

Figure 2-2

COMPARISON OF COMPUTERIZED TOMOGRAPHY AND MAGNETIC RESONANCE IMAGING: NORMAL SKULL AND BRAIN

Left: By CT, the densely calcified skull is a hyperattenuating, or white, lamina (arrow) closely apposed to the surface of the brain.

Right: By MRI, on the other hand, as here in a postcontrast T1-weighted image, the cranial vault is a band of blackness interrupted only by a thin central layer of bright signal corresponding to the diploic adipose tissue. SAT = subcutaneous adipose tissue, TM = temporalis muscle, OT = outer table, D = diploe, IT = inner table.

whereas the cerebellum and most of the brain stem are infratentorial. A hiatus in the tentorium accommodates the midbrain and permits passage of cranial nerves and vessels.

The arachnoid membrane covers the subarachnoid space, which is filled with cerebrospinal fluid that has a high or "bright" signal intensity in T2-weighted images (fig. 2-1). A number of MRI sequences, such as fluid-attenuated inversion recovery (FLAIR), suppress the signal of this and other compartmentalized liquids, to help the observer focus upon the lesion at hand. Interstitial water, i.e., intralesional or perilesional edema, remains bright.

The arachnoid membrane is delicate and transparent in the young, but, as life proceeds, succumbs gradually to clinically insignificant fibrosis, particularly over the convexities of the cerebral hemispheres. This physiologic process should be distinguished from pathologic opacifications due to neoplastic or inflammatory cell infiltrates. The arachnoid membrane normally strips readily from the brain.

Arachnoidal cells play an active role in fluid transport since they line the arachnoid villi and abut the lumens of venous sinuses (fig. 2-3). These same cells fan out diffusely over the surface of the arachnoid membrane as its inconspicuous covering (fig. 2-4), but with age aggregate and can become sufficiently prominent to be grossly apparent as white flecks. Whorls, psammoma bodies, and intranuclear pseudoinclusions identical to those seen in meningiomas are common histologic features both in arachnoidal villi and in diffusely distributed cell clusters (fig. 2-5). Normal meningothelial cells are uniformly immunoreactive for vimentin and epithelial membrane antigen (4). Their epithelial ultrastructural features, well-formed desmosomes and tonofilaments (5,6), are evident in meningiomas, the neoplasm derived from these cells.

The arachnoid membrane is composed, in large part, of arachnoid cells, but also of specialized fibroblasts that form trabeculae within the subarachnoid space. The outer surface of this space consists of horizontally oriented arachnoidal cells with tight junctions similar to those of the brain's vasculature. This arachnoid barrier layer is joined to the innermost layer of the dura, the dural border layer. Despite the commonly observed separation of the arachnoid membrane from the dura, both at

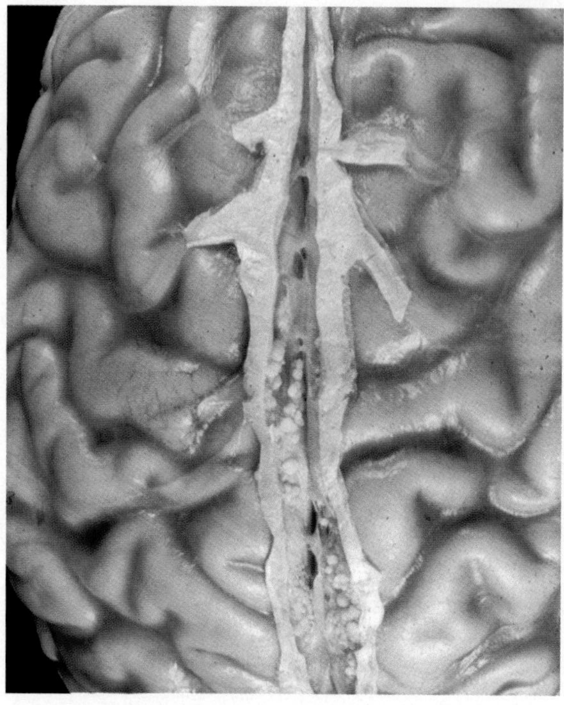

Figure 2-3

NORMAL BRAIN AND MENINGES

The dura has been dissected leaving only the portions enclosing the superior sagittal sinus and overlying the terminal portions of the superior cortical veins as they enter the sinus. The diaphanous arachnoid membrane wraps the brain. Arachnoid villi protrude into the sinus.

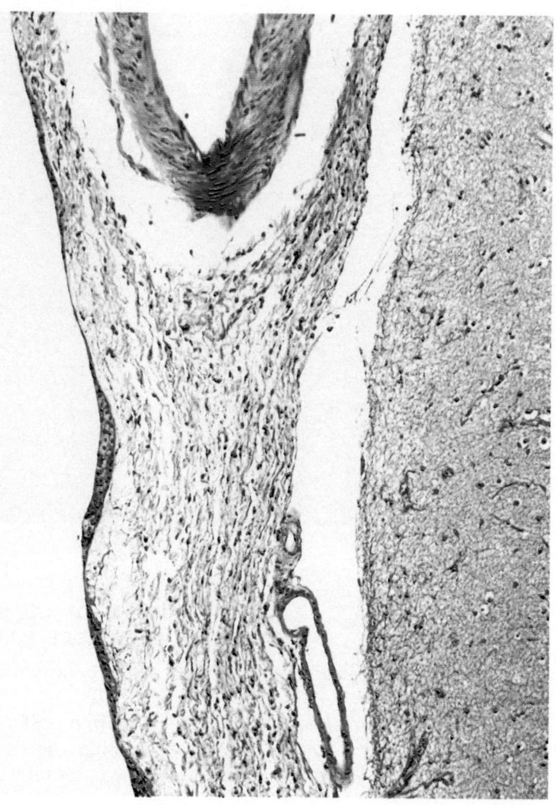

Figure 2-4

NORMAL MENINGES AND SUBARACHNOID SPACE

The arachnoid is a delicate membrane covered by a continuous layer of meningothelial cells that, with age, form small nests. The large vessel with a clearly defined muscularis is typical of arteries in the subarachnoid space, but not those in brain parenchyma.

surgery and at autopsy, the subdural space is an artifactual cleft produced by dehiscence within the dural border layer (7,8).

The subarachnoid space extends into the brain as a perivascular sleeve known as the Virchow-Robin space. Its depth corresponds approximately to the midportion of the cortical ribbon. Although microscopic studies suggest that this space does not normally communicate directly with the subarachnoid space, in practice it is patent to neoplastic cells, microorganisms, and inflammatory infiltrates (9).

Melanocytes are inconspicuous leptomeningeal cells that are concentrated at the base of the brain and ventral surface of the cervical spinal cord (figs. 2-6, 2-7). The MRI signal characteristics of melanin, i.e., bright in precontrast T1-weighted images and dark in T2-weighted images, are evident in the basal meninges of heavily pigmented patients (10). Rare melanocytic neoplasms arise from these cells.

The delicate pia mater abuts the glia limitans of the brain, a surface layer formed by the expanded process of subpial astrocytes that fan out over the surface of the brain. The cells of this specialized astrocytic layer may be related to the histogenesis of a special form of superficial glioma, pleomorphic xanthoastrocytoma. Basal laminae are common to both the normal cells and to this neoplasm. The pia is also reflected to form a cuff around vessels. Although often considered subarachnoid in location, such vessels are actually separated from the pia by this layer (9,11).

Vasculature

On the dura are terminal extensions of the middle meningeal artery, which is a terminal extension of the external carotid system. These

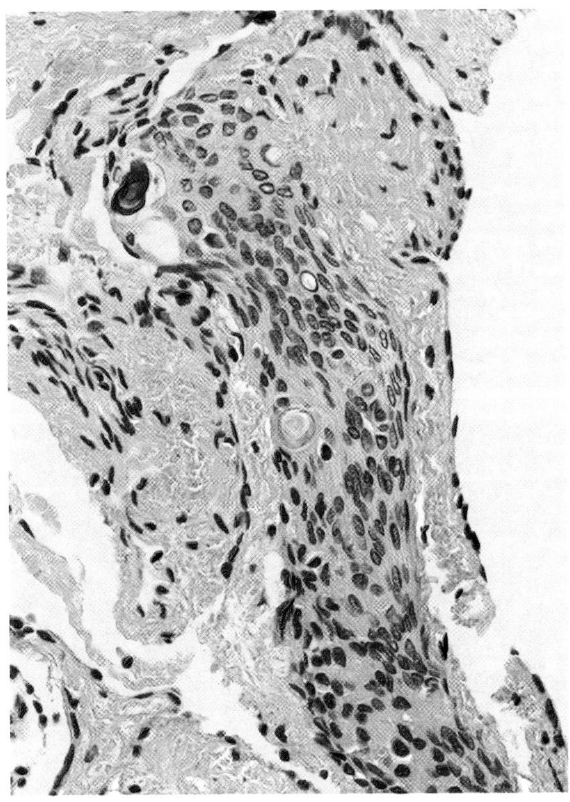

Figure 2-5

NORMAL MENINGOTHELIAL CELLS

With their intranuclear pseudoinclusions, whorls, and psammoma bodies, small clusters of meningothelial cells are similar to those of meningioma.

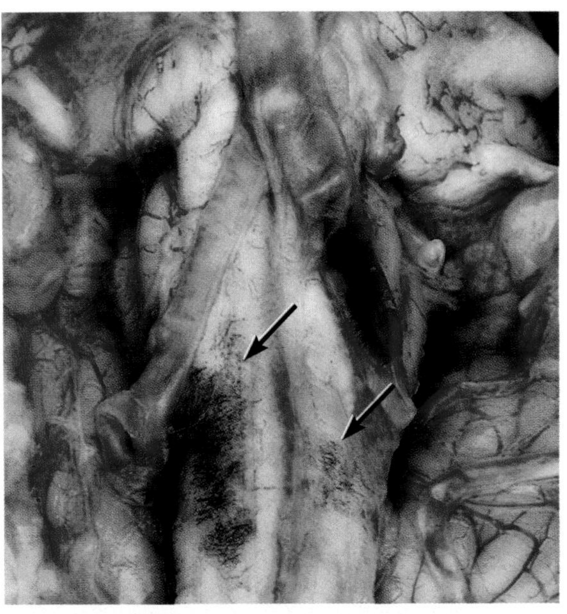

Figure 2-6

NORMAL MENINGEAL MELANOCYTES

Lacey patches of pigmentation (arrows) are common in the leptomeninges of darkly pigmented individuals. The anterior surfaces of the medulla and upper spinal cord are most noticeably affected.

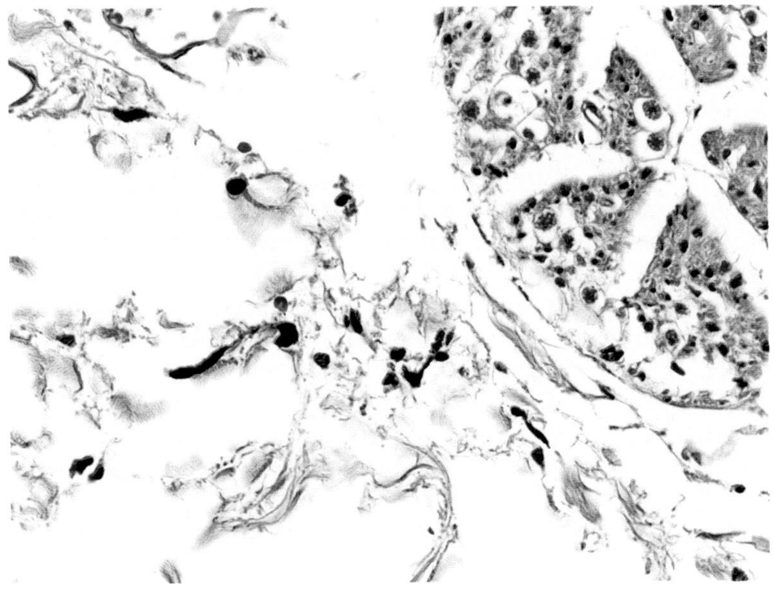

Figure 2-7

NORMAL MENINGEAL MELANOCYTES

If not for their pigment, meningeal melanocytes would be difficult to identify in hematoxylin and eosin (H&E)-stained sections.

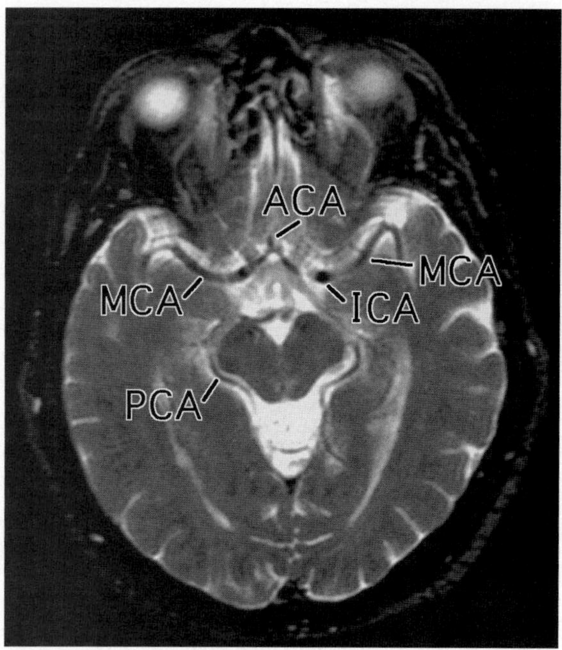

Figure 2-8

NORMAL CEREBRAL VESSELS

By MRI, especially in T2-weighted images, flowing blood leaves behind a dark signal, or "flow void." In this axial section near the base of the brain, the profiles of the major arteries (except the basilar) are readily evident. ICA = internal carotid artery, MCA = middle cerebral artery, ACA = anterior cerebral artery, and PCA = posterior cerebral artery.

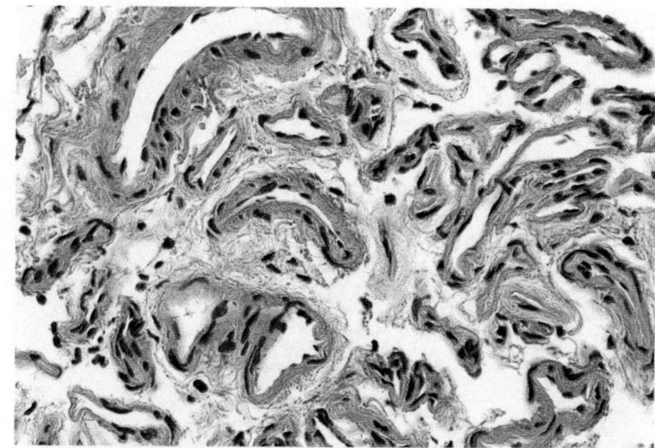

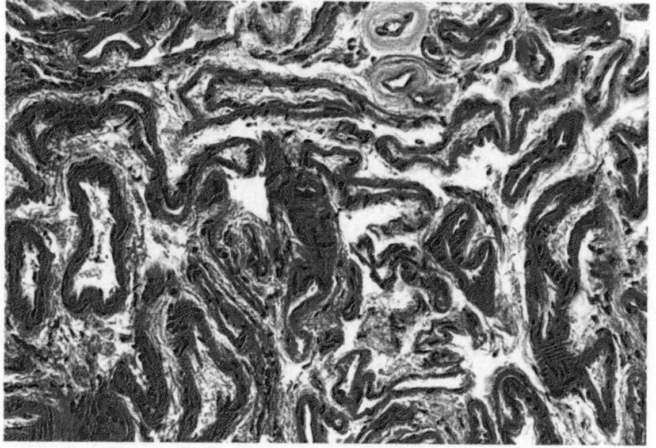

Figure 2-9

ARTIFACTUALLY COMPACTED NORMAL CEREBRAL BLOOD VESSELS SIMULATING VASCULAR MALFORMATION

Top: When artifactually apposed, as they often are in surgical specimens, otherwise normal vessels in the subarachnoid space can be misinterpreted as those of a vascular malformation.

Bottom: The uniformly and normally muscularized media of the compacted vessels helps rule out a vascular malformation (Masson trichrome stain).

branches are commonly recruited to supply dura-based tumors such as meningioma and meningeal hemangiopericytoma, but are largely unaffected by intraparenchymal neoplasms such as gliomas, which derive their vasculature principally from the internal carotid or vertebrobasilar system. Determination of a lesion's vascular supply thus serves as a helpful neuroradiologic clue in the preoperative differential diagnosis of CNS neoplasms.

The richly vascular brain is supplied by both the internal carotid ("anterior") and vertebrobasilar ("posterior") circulations, which are joined by the posterior communicating arteries. MRI, especially in T2-weighted images, records flowing blood as a dark signal ("flow void") (fig. 2-8). Magnetic resonance angiography (MRA) is a noninvasive technique that, in some cases, obviates traditional angiography.

After entering the intracranial cavity, the arteries traverse the subarachnoid space to penetrate the substance of the nervous system. In contrast to other body sites, the media of intraparenchymal vessels is so thin as to be scarcely discernible. In superficial cortical biopsies, the normally rich vasculature of the subarachnoid space may be artifactually compressed to simulate a vascular malformation (fig. 2-9, top). In contrast to malformed vessels, these juxtaposed vessels have a normal complement of smooth muscle (fig. 2-9, bottom). Arteries with a well-defined muscular coat are limited, almost by definition, to the subarachnoid space, a feature that helps identify this compartment when overrun by neoplastic or inflammatory infiltrates (fig. 2-4).

The dural venous sinuses are low pressure, avalvular channels that direct blood from cerebral veins to the internal jugular system. Blood from the cerebral cortex is channeled to the superior sagittal sinus, whereas the effluent from deep cerebral structures is passed on to the straight sinus. Thereafter, blood flow proceeds sequentially from the torcula, or confluence of the sinuses, through the transverse and sigmoid sinuses, to exit the cranium and return to the heart via the internal jugular veins and superior vena cava. Venous blood from the temporal lobe cortex proceeds directly into the adjacent transverse and sigmoid sinuses. Two large veins on the brain convexity, those of Labbé and Trolard, represent both landmarks to the surgeon and potential obstacles to en bloc resection. The former passes from the sylvian fissure inferiorly to the transverse sinus, while the latter courses superiorly along the general region of the central sulcus to the superior sagittal sinus.

Protruding into dural venous sinuses are small botryoid masses known as arachnoid villi; these convey cerebrospinal fluid from the subarachnoid space to the venous circulation (fig. 2-3). Also termed pacchionian granulations, these unique structures become prominent with age. When large, they induce small pits, foveolae granulares, in the inner table of the skull, and occasionally are large enough to mimic a pathologic condition due to the resultant lucency in the skull (12).

Ultrastructurally, the presence of tight junctions between endothelial cells and the apposition of astrocytic foot processes to vessel walls are distinctive features of the brain vasculature. These intimate endothelial-endothelial tight junctions are the basis of the blood-brain barrier, which requires metabolites, drugs, and toxins to pass through, rather than between, endothelial cells. At the light microscopic level, the space occupied by the astrocytic processes abutting vessels appears as an artifactual clear zone, the result of autolytic cellular imbibition of water.

Cerebrum

Cerebral Cortex. In T2-weighted MRIs, cortical gray matter is readily distinguished from underlying white matter since the latter has a lower, i.e., darker, signal (fig. 2-1) (2,3). The opposite is true in T1-weighted images, although the distinction between gray and white matter is not as distinct (fig. 2-2). The density of cortical gray matter in T2-weighted images is brighter than white matter. The signal characteristics of cortical gray matter are often used as points of reference. Largely because of their content of water, most lesions are brighter than gray or white matter, but there are some notable exceptions, and meningiomas in particular are usually isointense or even hypointense to gray matter in T2-weighted images.

To accommodate its expanse, the brain surface is convoluted, forming gyri in the cerebrum and folia in the cerebellum. Certain prominent clefts in the cerebral cortex serve as important anatomic landmarks. The generally horizontal sylvian fissure delimits the superior border of the temporal lobe, while the central sulcus (rolandic fissure) angles somewhat posteriorly in the midportion of the cerebral hemisphere to separate the precentral, or motor cortex, of the frontal lobe from the postcentral, or sensory cortex, of the parietal lobe. The parietal and occipital lobes are partially demarcated by the short parieto-occipital fissure, a prominent feature of the medial surfaces of the hemispheres.

Generous surgical resections of brain tissue are possible, particularly in relatively "silent" areas of the cerebrum. These include the frontal lobes anterior to the motor cortex, the anterior nonspeech-dominant temporal lobe, and the cerebellar hemispheres. Although homonymous hemianopsia may result, large resections of the occipital lobe may be performed. On the other hand, some regions are approached with trepidation for fear of significant postoperative neurologic deficits. These include the parietal lobe, especially of the dominant hemisphere, the extraparietal speech areas of the dominant hemispheres, and the precentral (motor) region of the frontal lobe, all of which are charged with essential functions. Pathologists can expect small specimens from these areas. Scant specimens are also the rule from lesions in the thalamus, basal ganglia, pineal region, and spinal cord. The request for a frozen section diagnosis from a small specimen from this, or any, site, should be weighed carefully against the inevitable frozen section artifact that often complicates biopsy interpretation, since little or no nonfrozen tissue may remain.

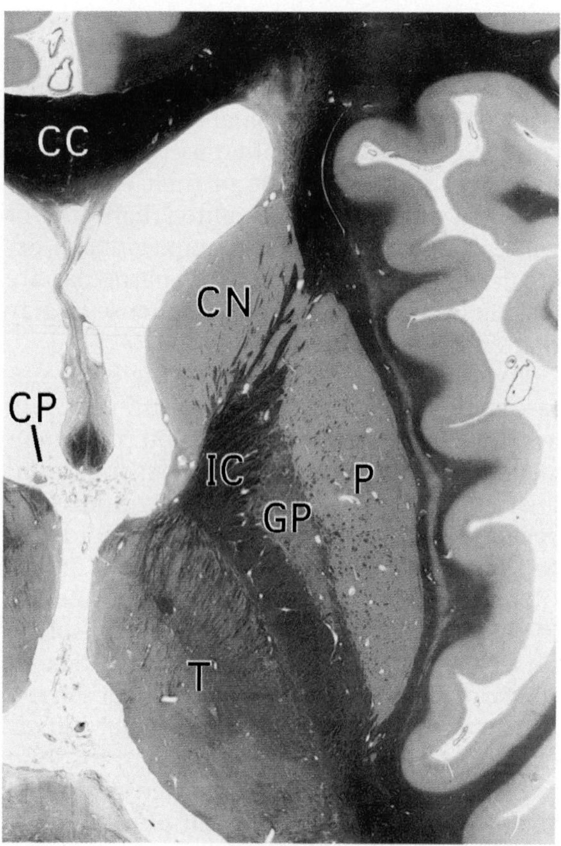

Figure 2-10

NORMAL BRAIN

The section, taken in the same plane as the CT and MRI images in figure 2-1, is stained with H&E/Luxol fast blue. There is a sharp distinction between the white matter and both cerebral cortex and deep gray matter. Distinct small bundles of myelinated fibers ("pencil bundles of Wilson") are present in the basal ganglia, especially in the caudate nucleus. CN = caudate nucleus, P = putamen, GP = globus pallidus, IC = internal capsule, CC = corpus callosum, CP = choroid plexus in foramen of Monro.

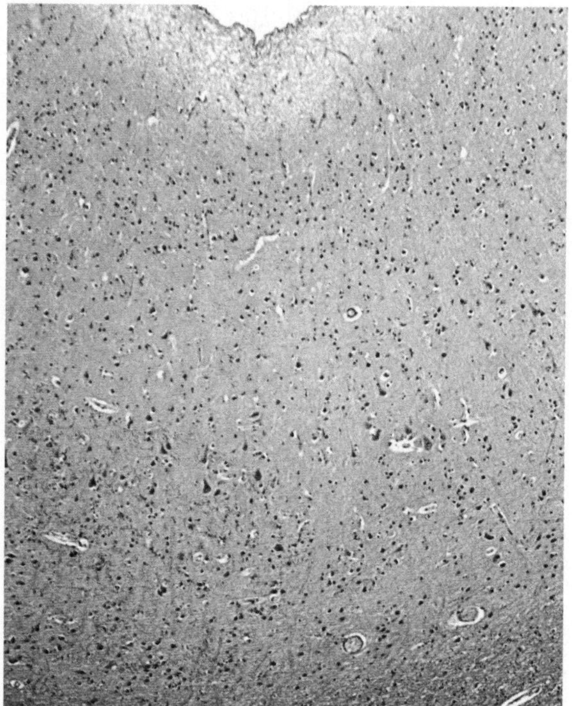

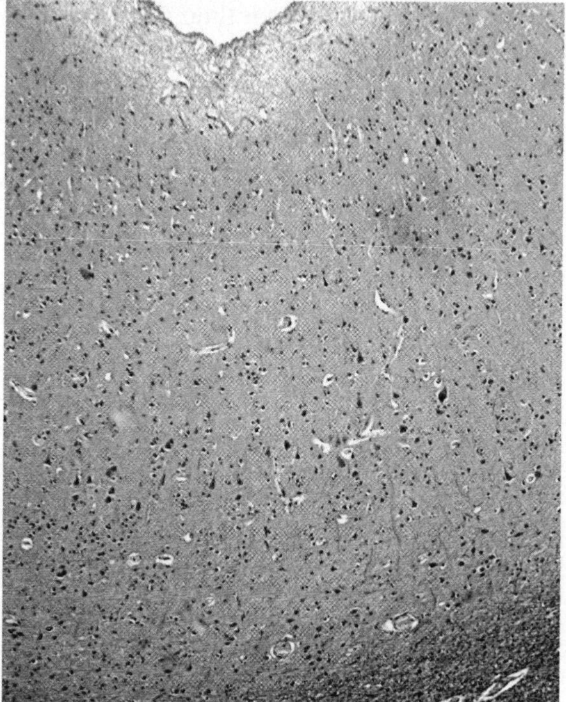

Figure 2-11

NORMAL BRAIN

The gray-white junction is well seen with an H&E stain (top) but is more apparent when the latter is combined with a method for staining myelin, such as Luxol fast blue (bottom).

Histologically, the cerebral cortex is well demarcated from underlying white matter along the gray-white junction, which is seen best after staining for myelin (figs. 2-10, 2-11). Although not always obvious in fragmented surgical specimens, the cortex has both horizontal and vertical patterns of organization. Such lamination is well seen in the phylogenetically newer neocortex, such as the primary sensory and motor cortices. Horizontal layering results in the development of cortical laminae that vary in number with site. Best defined is the paucicellular molecular layer that lies immediately beneath the

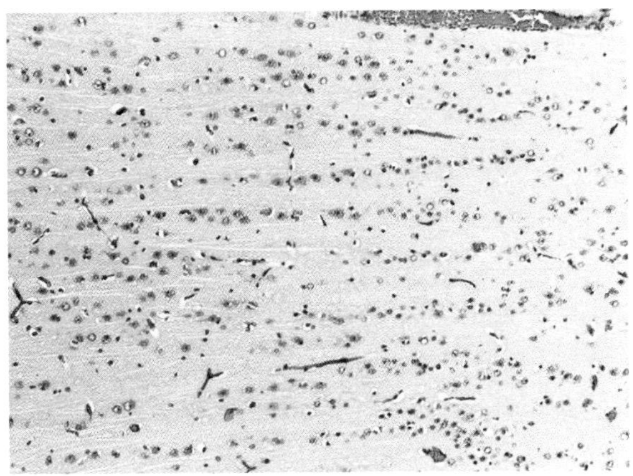

Figure 2-12

NORMAL CEREBRAL CORTEX

Parallel columns of large neurons are a normal finding, not evidence of cortical dysplasia or gangliocytoma.

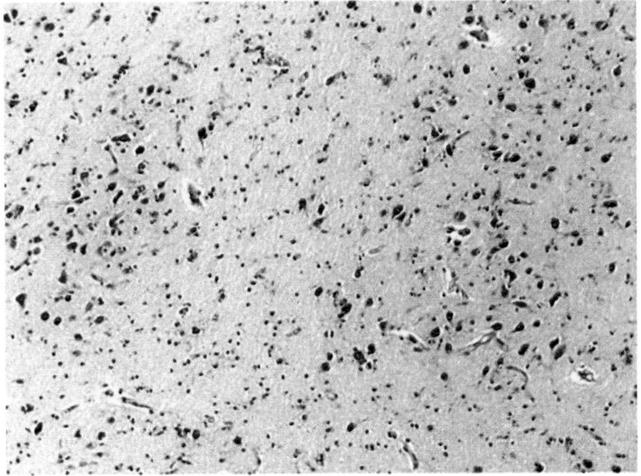

Figure 2-13

NORMAL NEURONAL CLUSTERING IN THE AMYGDALA

The normal clustering of neurons in the amygdala should not be misinterpreted as evidence of a ganglion cell tumor (cresyl fast violet stain).

pial surface. Composed in part of astrocytic processes, its most superficial portion is intensely immunoreactive for glial fibrillary acidic protein (GFAP). This layer is enlarged and even more GFAP positive in the seizure-associated, but nonspecific, change known as Chaslin gliosis.

The vertical organization of cerebral cortex is apparent as the columns of neurons seen fortuitously in some sections (fig. 2-12). Large pyramidal neurons within these columns are distinctly polar and project conspicuous apical dendrites toward the pial surface. Axons are simpler in structure and usually emerge from the basal portions of the cells.

Neurons are not as layered at supratentorial sites other than the cerebral neocortex. This is especially true in the amygdala where clustering is normal and should not be interpreted as a neoplastic or hamartomatous process (fig. 2-13).

Two general classes of normal cortical neurons are recognized: large pyramidal cells and small, round, nonpyramidal cells. Dendrites of the pyramidal cells branch extensively and generally extend horizontally and toward the cortical surface (see fig. 2-18). This orientation is evident even when tissue is overrun by infiltrative gliomas, such as oligodendroglioma or astrocytoma, and is, in concert with lamination, useful in distinguishing normal neurons from those of a ganglion cell tumor. The term ganglion cell is often applied to such large neurons, some of which contain stacks of rough endoplasmic reticulum, the ultrastructural counterpart of the basophilic patches known as Nissl substance. Although ganglion cells appear triangular in routine microsections, their angularity is to some extent artifactual; the cells' normal rotund configuration is observed in rapidly fixed tissues (fig. 2-14). In typical biopsy specimens, some cortical neurons are dark and contracted as an artifact of the trauma of excision (fig. 2-15). Such cells should not be equated with

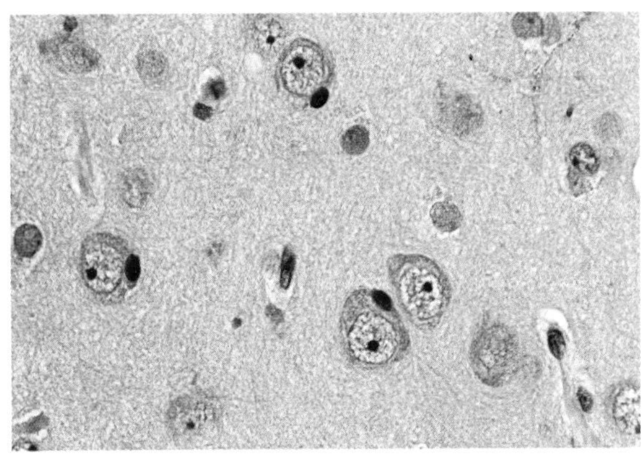

Figure 2-14

NORMAL CORTICAL GANGLION CELLS WITH PERINEURONAL ("SATELLITE") OLIGODENDROGLIA

Ganglion cells in rapidly fixed specimens are rotund, not shrunken and triangular. Small oligodendrocytes attend large cortical neurons.

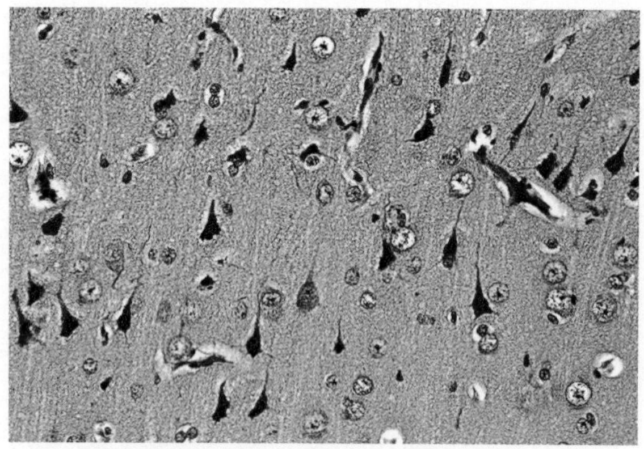

Figure 2-15

NORMAL CORTICAL NEURONS WITH CONDENSATION PRODUCED BY SURGICAL TRAUMA

Cytoplasmic condensation of cortical neurons, a common artifactual consequence of trauma inherent in a biopsy, should not be interpreted as evidence of disease, such as ischemia.

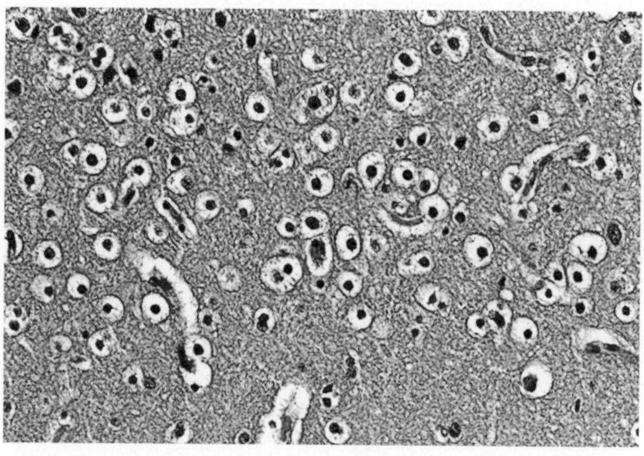

Figure 2-16

NORMAL CORTICAL NEURONS WITH AUTOLYTIC CHANGE SIMULATING OLIGODENDROGLIOMA

After autolytically imbibing water, small neurons in superficial cortical layers can, in aggregate, be misinterpreted as oligodendroglioma.

hypereosinophilic "red neurons" with karyolysis present in infarcts and in some viral infections (13).

Small neurons in the superficial cortical lamina are prone to autolytic imbibition of water and so mimic oligodendrocytes (fig. 2-16). In a large specimen, the affected layer can usually be followed horizontally to a point where cytoplasmic vacuolation diminishes and the neuronal nature of the cells is more readily apparent. Neuronal binucleation, a common diagnostic feature of ganglion cell tumors, is exceedingly rare in normal cortex.

Selective staining of neurons and their processes may be accomplished by routine histochemical or immunohistochemical methods. Classic histochemical procedures include the cresyl fast violet stain for cytoplasmic Nissl substance and the Bodian or Bielschowsky silver impregnation method for the demonstration of axons. The cresyl violet method highlights not only rough endoplasmic reticulum (Nissl substance), but chromatin and nucleoli as well (fig. 2-13). In Bodian or Bielschowsky preparations, axons appear as long dark threads of uniform thickness due to silver deposition on their neurofilaments. It is mainly axons, rather than dendrites, that are stained. The noncommittal term neurite is used by some to designate a neuronal process, either axonal or dendritic, although the term should be applied only to axons.

Immunohistochemical reagents are useful for identifying neurons. Immunopositivity for synaptophysin, a synaptic vesicle–associated protein, is considered neuron specific. It is reliably exhibited throughout gray matter where it appears as fine, diffuse background staining of the neuropil (fig. 2-17). To a lesser extent, it labels perikarya or axons. Granular surface staining is prominent in, but not specific to, ganglion cell tumors. Punctate and cytoplasmic staining for chromogranin can be present in some normal cortical neurons. Neurofilament protein reactivity, using antibodies to phosphorylated epitopes, is generally restricted to neurofilament-rich structures such as axons rather than perikarya or dendrites. The cell body is better stained with antibodies to the nonphosphorylated protein. Reactivity to neurofilament proteins is taken as prima facie evidence of neuronal differentiation. Other neuron-labeling antibodies are microtubule-associated protein-2 (MAP-2) and CD99.

An additional antibody, NeuN, stains the nuclei of normal neurons, both large and small, and those of the better differentiated or "mature" ganglion cell tumors (fig. 2-18). Staining is most dramatic in the cerebellum, where internal granular cells are positive but Purkinje cells are not. Staining for neuron-specific enolase (NSE) provides far less than conclusive evidence that a cell is a neuron, since staining is often

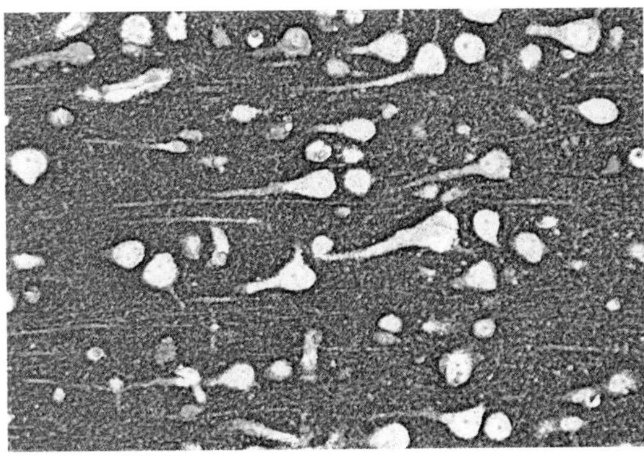

Figure 2-17

NORMAL CEREBRAL CORTEX

The fibrillar background, or neuropil, is intensely immunoreactive for synaptophysin. Internally, individual neurons are negative.

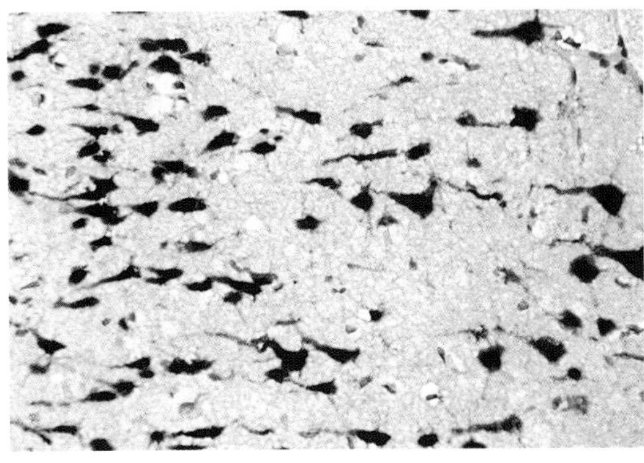

Figure 2-18

NORMAL CEREBRAL CORTEX

Mature neurons can be demonstrated with a number of immunohistochemical reagents including the NeuN antibody applied here. As is typical of normal cortex, but not of gangliocytoma, the regimented cells uniformly direct their apical dendrites to the pial surface. Ganglion cell tumors are unpredictably, and often only sparsely, reactive for NeuN.

nonspecific. Nonspecificity aside, immunoreactivity for NSE is reliably present in gray matter in which staining is diffuse.

Although the myelin content of gray matter is minimal when compared to that of white matter, widely separated myelinated fibers are seen within the cerebral cortex. A prominent horizontal layer of myelinated fibers, the outer stripe of Baillarger, or stripe of Gennari, serves as a macroscopic marker of the primary visual or "striate" cortex of the medial occipital lobes.

Non-neuronal constituents of cortical gray matter include astrocytes, oligodendrocytes, and microglia. As throughout most of the brain, with the notable exception of the choroid plexus stroma, fibrous tissue is scant and confined to the adventitia of blood vessels.

Nuclei of astrocytes within the gray matter are round and, although considerably smaller than those of large cortical neurons, approximate those of smaller neurons in both size and chromatin density. Such astrocytes are referred to as protoplasmic. Their extensions, branching more frequently than those of the fibrillary astrocytes of white matter, create a finer, "woolier" domain of processes after staining for GFAP. Given their paucity of cytoplasmic intermediate filaments, normal cortical astrocytes are only weakly reactive for GFAP. Reactive astrocytes, on the other hand, are strongly labeled, highlighting their symmetric radial array of elongate, often nonbranching processes.

Eye-catching astrocytic products, in both gray and white matter, include polyglucosan bodies, or corpora amylacea. They are readily identified when aggregated about vessels or aligned linearly in the subpial or subependymal regions. These slightly basophilic, strongly periodic acid–Schiff (PAS)- or Gomori methenamine silver (GMS)-positive, faintly laminated structures are contained within the terminations of astrocyte processes and tend to accumulate with age. Corpora amylacea should not be misinterpreted as fungi such as *Cryptococcus* (fig. 2-19).

Oligodendrocytes are most numerous in white matter where their multiple plate-like processes wrap about axons to form myelin sheaths. In small number, oligodendrocytes are also normal inhabitants of gray matter. Here they are relatively restricted in distribution, several hovering about each large neuron, particularly in deep cortical layers (fig. 2-14). The presence of such "satellite oligodendroglia" should not be interpreted as perineuronal satellitosis of an infiltrating glioma. Smaller size, greater nuclear uniformity, delicate chromatin, and inconspicuous nucleoli distinguish normal satellite oligodendrocytes from their neoplastic counterpart.

Present occasionally in normal tissues, such as the pineal gland (see fig. 2-48), certain neoplasms, and reactive states, are intracytoplasmic, hyaline, often corkscrew-shaped structures

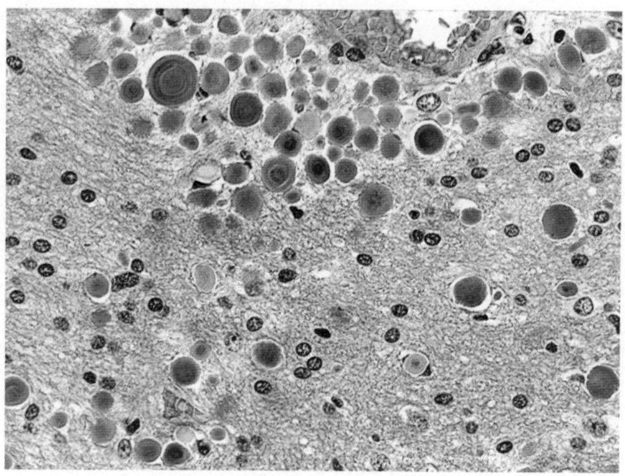

Figure 2-19

CORPORA AMYLACEA

Accumulating with age, corpora amylacea congregate in subpial, perivascular, and subependymal regions. They should not be confused with infectious organisms such as *Cryptococcus*.

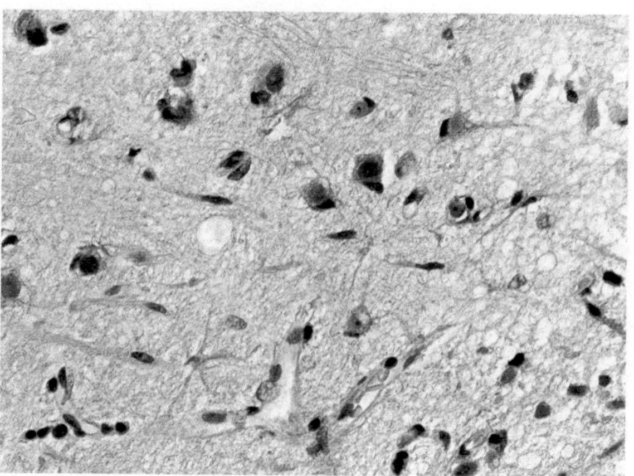

Figure 2-20

REACTIVE MICROGLIA

Difficult to identify in their native state, reactive microglia acquire enough cytoplasm, bipolar, to become apparent in H&E-stained sections.

termed Rosenthal fibers. Although they are found in some neoplasms, most notably pilocytic astrocytomas, they are not evidence per se of neoplasia. In reactive states, Rosenthal fibers appear most often in chronic gliosis in the region of the third ventricle (particularly in the settings of craniopharyngioma and pineal cyst), cerebellum (around hemangioblastoma), brain stem, and spinal cord (any longstanding lesion), the same general regions favored by pilocytic astrocytomas. This congruence in distribution raises the possibility that these regions harbor a subset of astrocytes to which both this form of gliosis and pilocytic neoplasms can be attributed. Rosenthal fibers are not unique to these lesions, however, and may be seen in other areas such as the subpial zone, and, occasionally, in infiltrating astrocytomas. Rosenthal fibers reside in GFAP-positive processes but are themselves negative, being reactive instead for alphaB-crystallin.

Microglia are mesenchymal cells of monocyte/macrophage lineage that, while ubiquitous, are inconspicuous and morphologically nonspecific. Their small thin nuclei are about as long as those of endothelial cells, but considerably thinner. The cytoplasm is normally not discernible from the surrounding neuropil, but becomes thick enough to be seen when the cell is responding, or "activated" (fig. 2-20). Although microglia occur in both gray and white matter, they are discussed here since it is in gray matter that they are most apparent in disease states, such as encephalitis. Reactivity of these cells by immunohistochemical (lysozyme) or lectin (Ricinus communis agglutinin-1) staining leaves no doubt about their presence. Lectin staining visualizes the cytoplasm as bipolar extensions that branch, often at almost right angles, to permeate the adjacent brain. In pathologic states associated with tissue destruction, microglia round up to assume the conventional portly profile of phagocytic cells. During this transition, they become immunoreactive for macrophage markers such as CD68, as well as for common leukocyte antigen (CD45) (fig. 2-21). In large destructive lesions, the cells are joined and greatly outnumbered by hematogenous macrophages that have entered the scene. Phagocytes engorged with debris are known as "gitter cells." Such lipid-rich cells are particularly abundant in demyelinating disease and infarcts. The macrophage nature of the former is readily demonstrated by immunohistochemistry with antisera to CD68. The rare occurrence of neoplasms composed of microglial cells is discussed in chapter 12.

Very few lymphocytes are seen in the normal CNS. The finding of even a few, generally about vessels, is evidence of disease.

The gray matter is frequently seen in smears prepared during intraoperative evaluation of CNS lesions. Normal large neurons, often

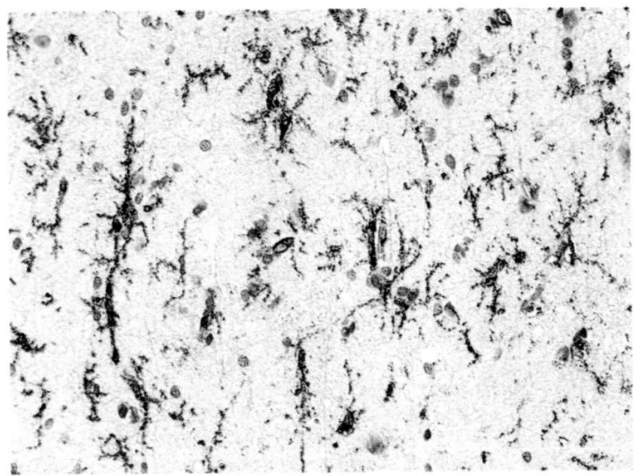

Figure 2-21

REACTIVE MICROGLIA

While negative in their resting state, activated microglia react with antibodies to macrophage markers such as HAM-56 (illustrated here) and CD68. The bipolarity and small "wooly" extensions are typical. The cells were responding to an infarct not seen in this illustration.

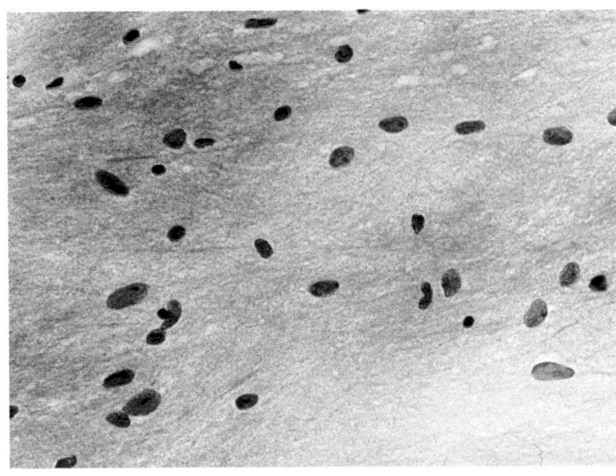

Figure 2-22

NORMAL CEREBRAL CORTEX: SMEAR PREPARATION

Devoid of their cytoplasm as they typically are in smear preparations, nuclei of normal large cortical neurons should not be confused with the "naked nuclei" of an infiltrating glioma.

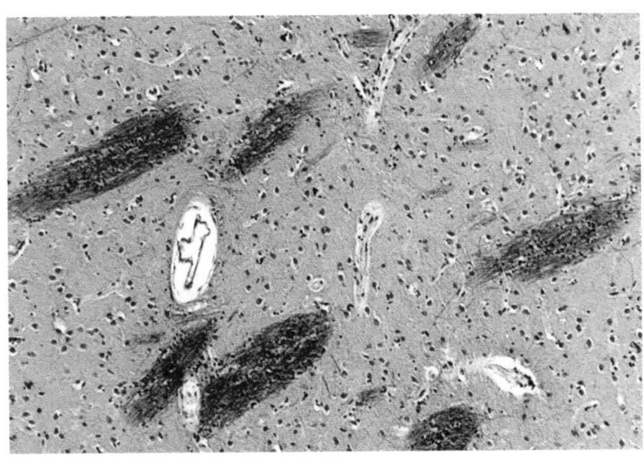

Figure 2-23

"PENCIL BUNDLES" IN CORPUS STRIATUM

Small bundles of myelinated axons known as "pencil bundles of Wilson" are a distinctive feature of the basal ganglia and thalamus, especially the corpus striatum (caudate nucleus, internal capsule, and putamen). They are seen best after staining for myelin, as here with the H&E/Luxol fast blue stain.

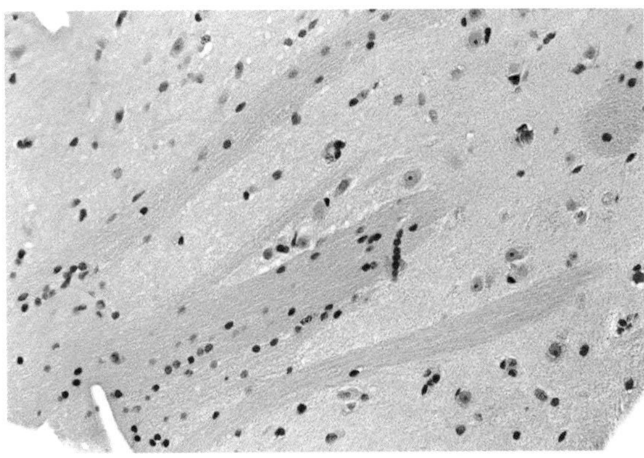

Figure 2-24

"PENCIL BUNDLES" IN CORPUS STRIATUM

Often seen in specimens obtained stereotactically, "pencil bundles" are reliable evidence of the origin of the specimen.

stripped of their cytoplasm, can be misinterpreted as neoplastic elements (fig. 2-22).

Basal Ganglia and Thalamus. Referred to collectively as "deep gray matter," these deep-seated structures are known for their general lack of lamination and the presence of myelinated axons in excess of those encountered in the cerebral cortex. In the corpus striatum (an anatomic group formed of caudate nucleus, putamen, and the intervening internal capsule), and to a lesser extent the thalamus and basal ganglia, myelinated axons are compacted into small white matter bundles known as the "pencil bundles of Wilson" (fig. 2-23). These distinctive formations are even evident in stereotactic biopsy specimens and readily identify the tissue's site of origin (fig. 2-24).

Hippocampus. Because of its complicated architecture, particularly when viewed in tangential or oblique sections, normal hippocampus

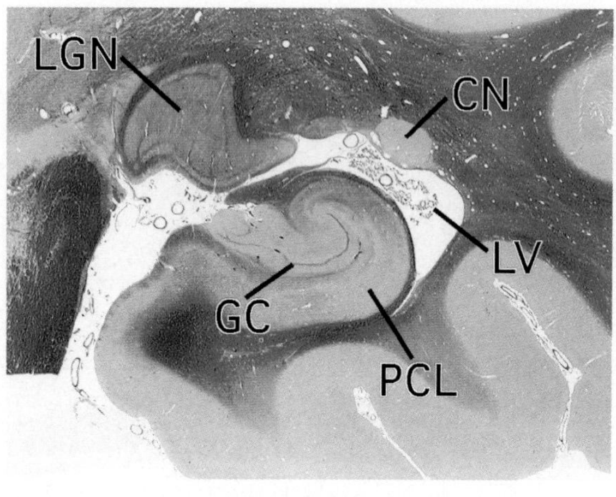

Figure 2-25

NORMAL HIPPOCAMPUS

The normal anatomy of the hippocampus is best seen in the coronal plane. PCL = pyramidal cells of Ammon's horn, GC = granule cells of the dentate gyrus, LV = lateral ventricle with choroid plexus, LGN = lateral geniculate nucleus, CN = caudate nucleus (tail).

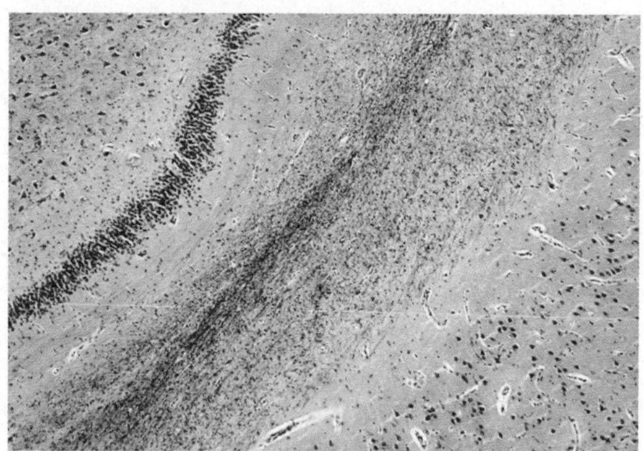

Figure 2-26

NORMAL HIPPOCAMPUS

The distinctive lamina of small granular cells at the top left is a conspicuous feature of the hippocampus.

may be misinterpreted as lesional tissue. In the optimal coronal plane, the formation consists of a C-shaped lamina of small granular cell neurons, a molecular layer, and a spiral of large pyramidal neurons in Ammon's horn (figs. 2-25, 2-26). The anatomy is readily recognized in such sections, but the difference between normality and disease (neuronal depopulation and gliosis) may be difficult to assess in randomly oriented surgical material (fig. 2-27). In the setting of chronic epilepsy, either of the layers becomes

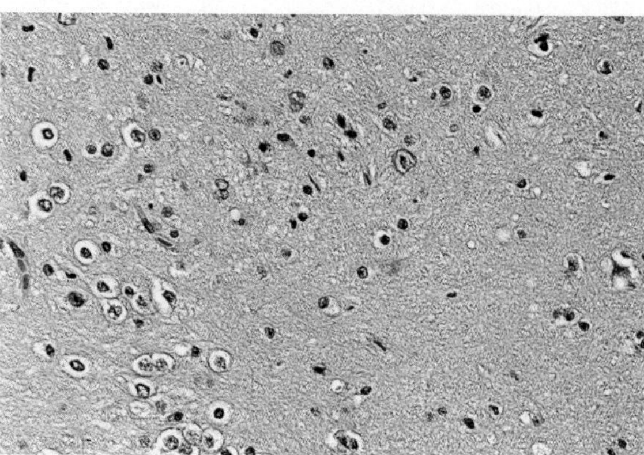

Figure 2-27

NORMAL HIPPOCAMPUS

Tangential sections obscure the normal lamination and can create the impression of a neuronal neoplasm or hamartoma. As with small neurons at any site, the granule cells can acquire oligodendrocyte-like perinuclear halos.

depopulated and gliotic (hippocampal, or mesial, sclerosis). On neuroimaging, this is a contractile process unlikely to be confused with a neoplasm. Microscopically, however, the gliosis can resemble an infiltrating glioma.

White Matter. The radiologic features of normal cerebral white matter are illustrated in figures 2-1 (CT) and 2-2 (MRI). In the latter, it appears darker than gray matter in a T2-weighted image, but is lighter in a T1-weighted image. Macroscopically, the deep white matter of the cerebral hemispheres (centrum semiovale) and compact pathways (including the corpus callosum, fornices, anterior commissure, internal capsules, cerebral peduncles, and medullary pyramids) are glistening white and sharply demarcated from adjacent gray matter, just as they are in histologic sections stained for myelin (figs. 2-10, 2-25). Edema and infiltrating neoplasms rob white matter, especially the centrum semiovale, of its normal sheen, and supplant it with a dull yellow tinge or fine granularity. Compact white matter pathways, such as the corpus callosum, are not as susceptible to vasogenic edema.

Since it only transmits, rather than initiates, chemoelectric impulses, white matter is simpler than gray matter in its cellular composition and organization. It is composed largely of axonal processes and their accompanying myelin sheaths. Although occasional neurons are encountered in normal cerebral white matter,

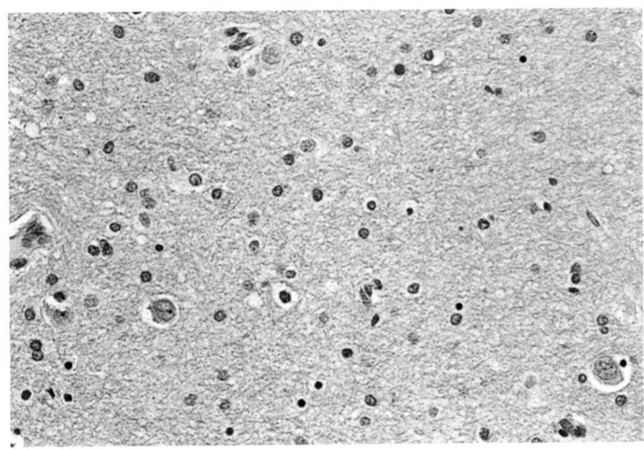

Figure 2-28

NORMAL NEURONS IN WHITE MATTER

Scattered neurons are common in normal white matter, particularly in the temporal lobe.

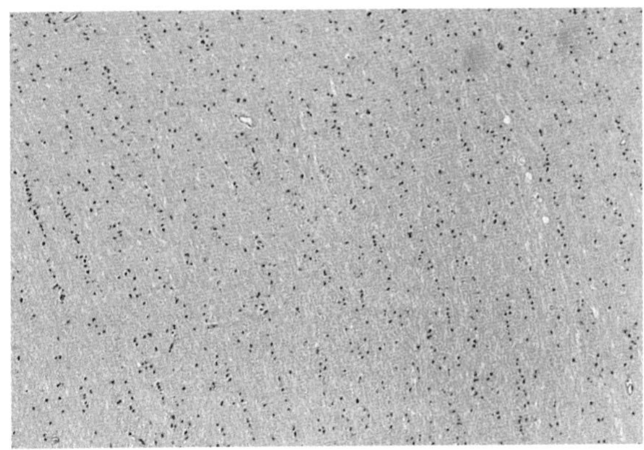

Figure 2-29

NORMAL CEREBRAL WHITE MATTER

Fortuitous sections capture short columns of neurons that should not be confused with rows of infiltrating tumor cells.

particularly in the temporal lobe, they are few and are largely concentrated beneath the cortical ribbon (fig. 2-28). Such cells should not be interpreted necessarily as evidence of hamartoma, microdysgenesis, or ganglion cell tumor.

Oligodendrocytes, with their round nuclei, are abundant in white matter. In fortuitous sections, these cells orient themselves along fiber pathways in rows or fascicles (fig. 2-29). Normal oligodendrocytes are prone to autolysis, during which they imbibe fluid to produce a "fried egg" appearance similar to, although usually less prominent than, that seen in oligodendrogliomas. Autolytic change in astrocytes is limited largely to the perivascular terminations of their processes, which create a clear space about small vessels.

In some cases, an unexplained phenomenon is the hypercellularity produced by oligodendrocyte-like cells populating the white matter. Perivascular aggregation is prominent (fig. 2-30), as is the "infiltration" of deeper cortical layers where the cells hover about neurons. Although such formations are seen most often in temporal lobes resected to control seizures, they appear incidentally in other specimens as well. This finding resembles a very, very well-differentiated oligodendroglioma.

Myelin, the lipid rich "insulation" applied to axons by the wrapping of oligodendrocyte processes, is highlighted by a variety of special stains. The best known, the Luxol fast blue method, is generally combined either with hematoxylin and eosin (H&E) (figs. 2-10, 2-11, 2-25) or with the PAS method. Immunohistochemical staining for galactocerebroside, myelin basic protein, and leu-7 can also be used, but there is no dependable consensus marker of oligodendrocytes in clinical material. In tissue sections, particularly those of edematous white matter, axons with their myelin sheaths have irregular tubular profiles that may superficially resemble hyphae (fig. 2-31). Axons are best

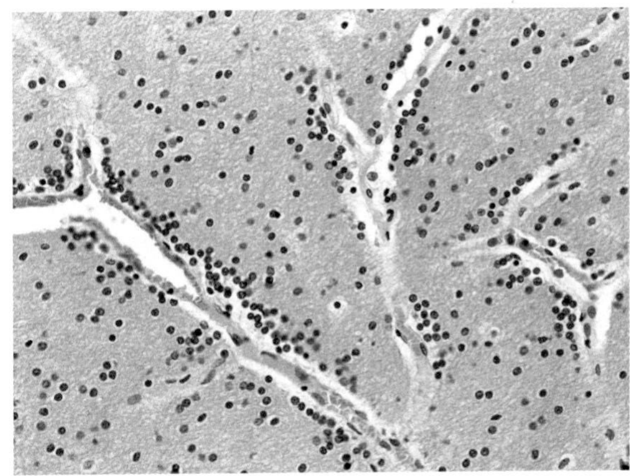

Figure 2-30

NORMAL CEREBRAL WHITE MATTER

A distinctive form of apparent hypercellularity is produced by the presence of small uniform cells in perivascular and perineuronal arrangements. While this phenomenon closely resembles well-differentiated oligodendroglioma, it has no prognostic significance and is usually seen in temporal lobes removed for seizure control.

immunolabeled with antibodies to phosphorylated neurofilament protein.

Astrocytes in the white matter, referred to as fibrillary or fibrous types, are putative precursors of the common infiltrative, or diffuse, astrocytomas discussed in the chapter that follows. Although the nuclei of these cells are somewhat larger and less hyperchromatic than those of oligodendrocytes, there is a degree of overlap in their nuclear characteristics, and it is often difficult to assign individual cells to the astrocytic or oligodendroglial category on the basis of H&E-stained sections alone. Diagnostically, the identity of normal cell types is usually not questioned, but only whether an infiltrating astrocytic or oligodendroglioma is present. This depends, in part, on retaining a mental image of the tissue's normal cellularity. Unfortunately, apparent cellularity varies considerably from specimen to specimen, depending in large part upon section thickness (fig. 2-32).

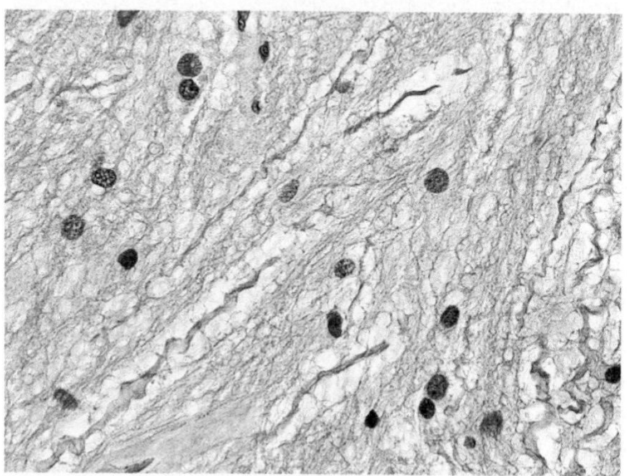

Figure 2-31

NORMAL AXONS

Separated by edema fluid, large axons can resemble hyphae.

In smear preparations, normal tissue may appear hypercellular when cells are heaped up. As a consequence, thick smears can easily be misinterpreted as evidence of a glioma (fig. 2-33). Smears often suggest a degree of hypercellularity that is not confirmed in tissue sections.

Ventricles and Choroid Plexus. The choroid plexus, the source of cerebrospinal fluid, is a contrast-enhancing, intraventricular garland present in the third, lateral, and fourth ventricles, but not the aqueduct of Sylvius (figs. 2-34, 2-35). Two parallel strands course in the roof of the third ventricle, and diverge as they pass through their respective foramen of Monro to reach the lateral ventricles (fig. 2-36), including bodies, trigones (atria), and temporal horns (fig. 2-37). In the trigone, the juncture of the body with temporal and occipital horns, the plexus

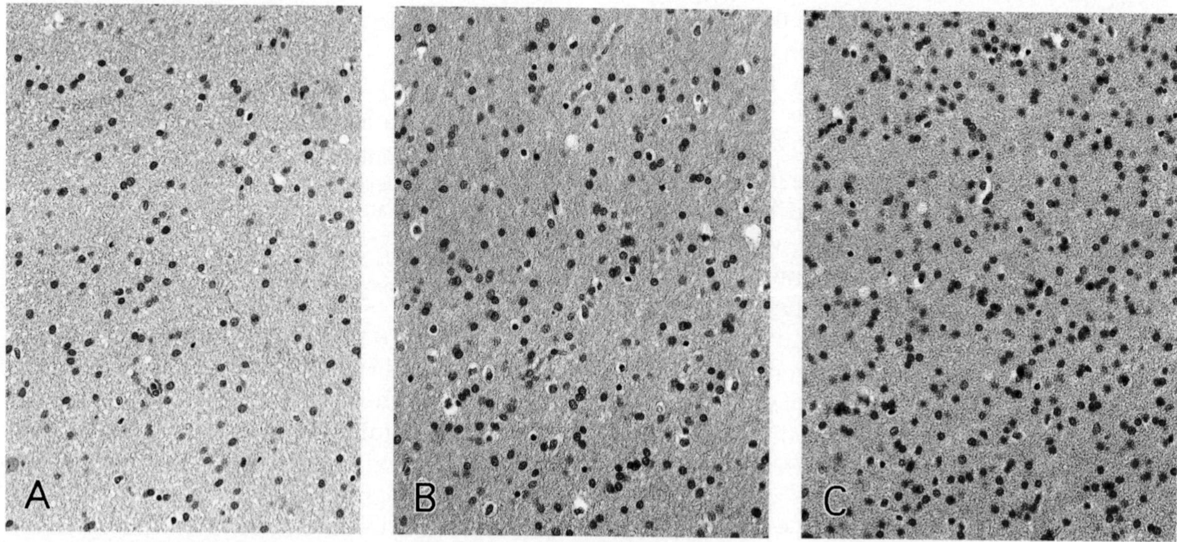

Figure 2-32

NORMAL WHITE MATTER

An increase in cellularity is simulated by increments in section thickness from 5 (A) to 10 (B) to 15 µm (C).

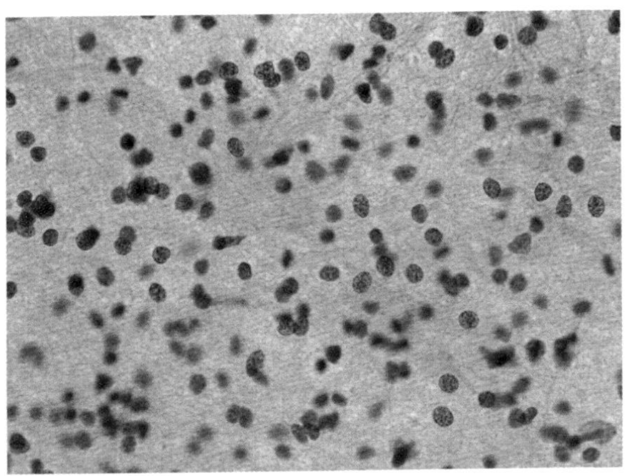

Figure 2-33

NORMAL CEREBRAL WHITE MATTER: SMEAR PREPARATION

The apparent cellularity of normal and neoplastic tissue can be exaggerated in smear preparations with heaped up areas several cells deep. In contrast to a glioma, there is only minimal variation of cell size and shape.

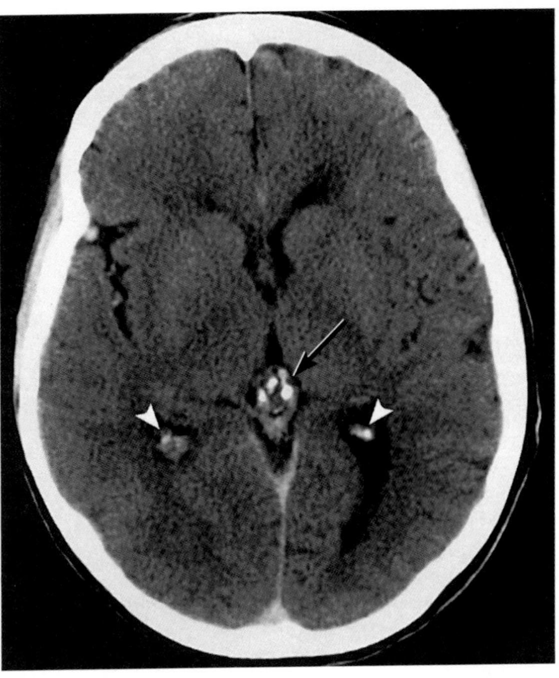

Figure 2-34

NORMAL CALCIFICATION OF CHOROID PLEXUS AND PINEAL GLAND

As seen by CT, the choroid plexus glomus (arrowheads) and the pineal gland (arrow) are normally calcified after childhood.

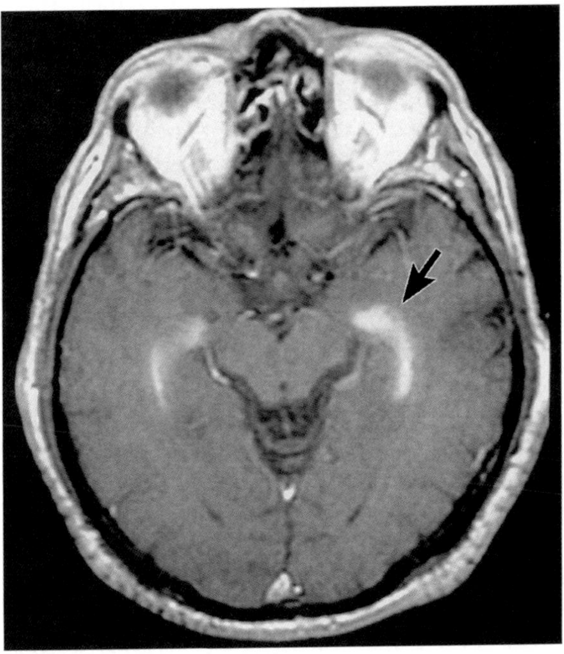

Figure 2-35

NORMAL CHOROID PLEXUS

The choroid plexus (arrow), here within the temporal horn, is normally contrast enhancing.

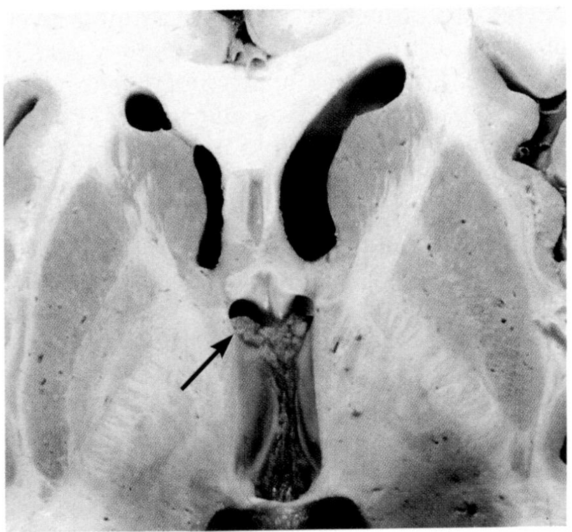

Figure 2-36

NORMAL BRAIN AND CHOROID PLEXUS

As viewed from below in an axial horizontal section, the finely papillary choroid plexus extends along the roof of the third ventricle and passes through the foramina of Monro (arrow).

often forms an especially prominent mass known as the glomus. Xanthomas and xanthogranulomatous lesions occur in this region.

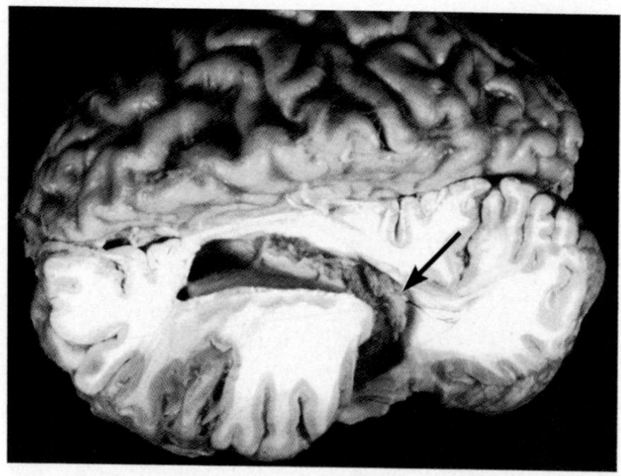

Figure 2-37

NORMAL BRAIN AND CHOROID PLEXUS

Emerging from the third ventricle through the foramen of Monro, the choroid plexus courses posteriorly to the trigone or atrium (arrow) and recurves anteriorly into the temporal horn.

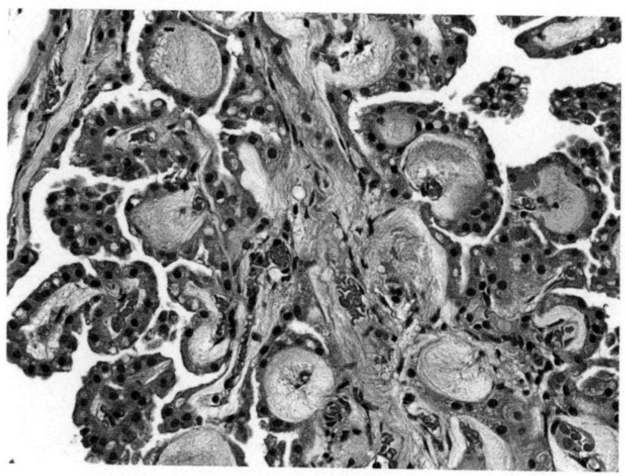

Figure 2-38

NORMAL CHOROID PLEXUS

The normal epithelium has a "hobnail" or "cobblestone" profile that differs from the flat surface of a papilloma.

From the roof of the fourth ventricle, the plexus extends laterally through the foramina of Luschka to reach the subarachnoid space at the cerebellopontine angle, explaining the occasional papilloma that occurs lateral to the brain stem.

Histogenetically related to ependyma, choroid plexus epithelium consists of a simple layer of cuboidal to columnar cells squarely seated upon a basement membrane underlain by well-defined papillary fibrovascular stroma (14,15). Unlike ependyma, choroid plexus displays an apical "cobblestone" profile (fig. 2-38), and, by light microscopy, lacks cilia. Normal choroid plexus is immunoreactive for cytokeratins, particularly low molecular weight forms, vimentin, S-100 protein, prealbumin (transthyretin), and synaptophysin (16–19). With age, circumnuclear masses of argyrophilic filaments, so-called Biondi's rings, become prominent within choroid plexus cells. These become even more numerous in the setting of Alzheimer's disease (20).

As the choroid plexus forms from invaginating meningeal fibrovascular tissue along the choroidal fissure, meningothelial cells are carried into its stroma, providing the substrate for intraventricular meningiomas (fig. 2-39).

Ependyma. Among the various forms of glia, ependymal cells are unique since they form an epithelial layer that lines the ventricular system and comprises the discontinuous remnants of the central canal of the postpubertal spinal cord. When encountered in the small, fragmented biopsy specimens so typical of the spinal cord, such rests can be misinterpreted as a neoplasm (fig. 2-40). The same is true of disorganized ependymal cell remnants near the lateral ventricle (fig. 2-41). Although in adults scattered cilia are found on ultrastructural examination, few if any persist at the light microscopic level. Unlike choroid plexus, no basal lamina underlies the ependymal layer. Instead, the latter rests directly upon subependymal glia at the anteroposterior extremes of the ventricular cavities.

Normal ependyma is immunoreactive for S-100 protein and, to a variable degree, for GFAP and EMA (21). A special variant of the ependymal cell, the tanycyte, is an elongated and tapered cell that, unlike ordinary columnar ependymal cells, projects long basilar processes into the subependymal layer. Occasionally, these processes extend to the pial surface, particularly in the region of the third ventricle and hypothalamus.

Pineal Gland (Epiphysis). The pineal is an approximately 1-cm appendage to the posterior wall of the third ventricle (figs. 2-42, 2-43). During embryogenesis, the organ evolves from biphasic tissue that consist of dark, melanin-containing, S-100–positive cells and pale synaptophysin-immunoreactive neurosecretory cells (fig. 2-44) (22). In adults, the gland is a cytologically more homogeneous, lobulated,

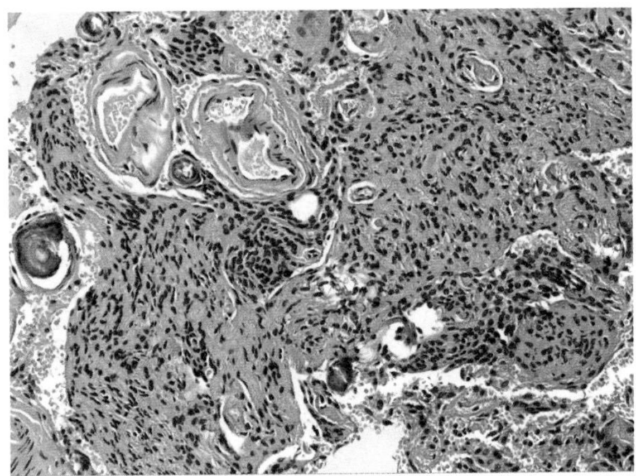

Figure 2-39

NORMAL MENINGOTHELIAL CELLS IN CHOROID PLEXUS

Nested meningothelial cells in the stroma of the choroid plexus are logical sources of intraventricular meningiomas.

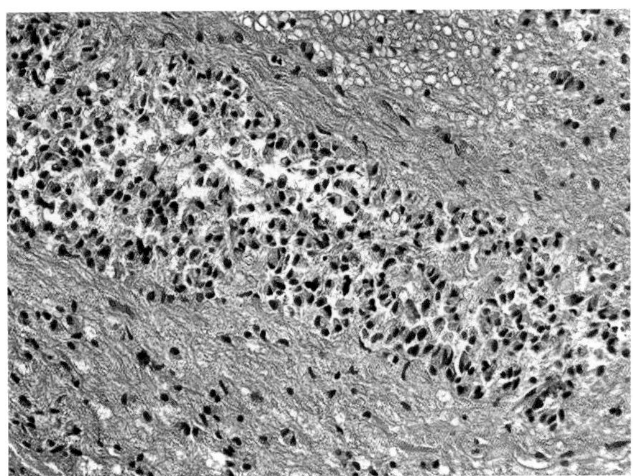

Figure 2-40

NORMAL REMNANTS OF THE SPINAL CORD CENTRAL CANAL

As compact, unstructured clusters or lobules of dark cells, remnants of the spinal canal can be misinterpreted as a neoplasm.

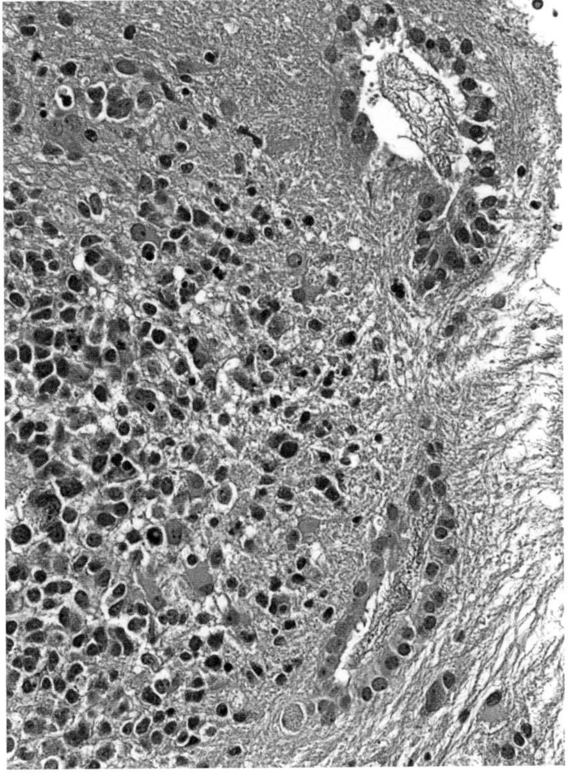

Figure 2-41

PARAVENTRICULAR EPENDYMAL TUBULES (ADJACENT TO GLIOBLASTOMA)

Small paraventicular aggregates or tubules of ependymal cells should not be misinterpreted as ependymoma or ependymal differentiation in a periventricular neoplasm such as this glioblastoma.

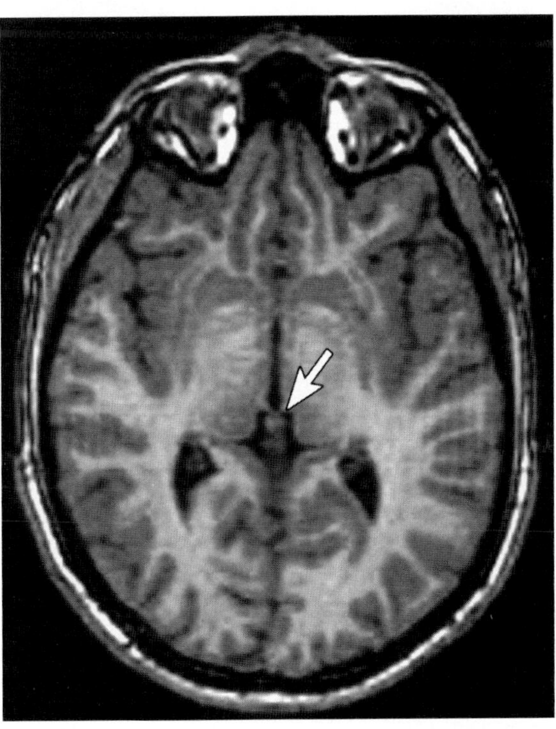

Figure 2-42

NORMAL PINEAL GLAND

As seen in a precontrast T1-weighted MRI, the pineal gland (arrow) is a small midline structure in the immediate environs of the posterior third ventricle.

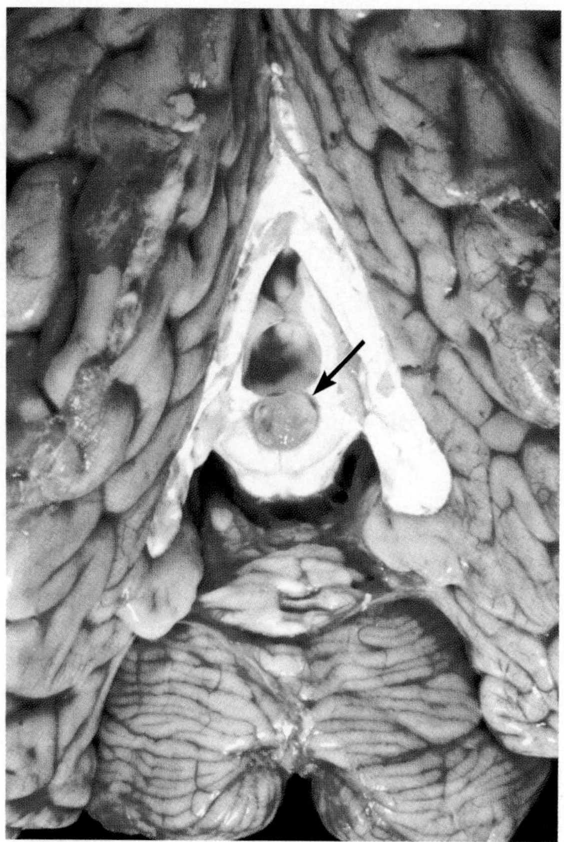

Figure 2-43

NORMAL PINEAL GLAND

Viewed from above, the pineal gland (arrow) is a well-circumscribed mass related to the posterior third ventricle. Note the gland's proximity to the quadrigeminal plate.

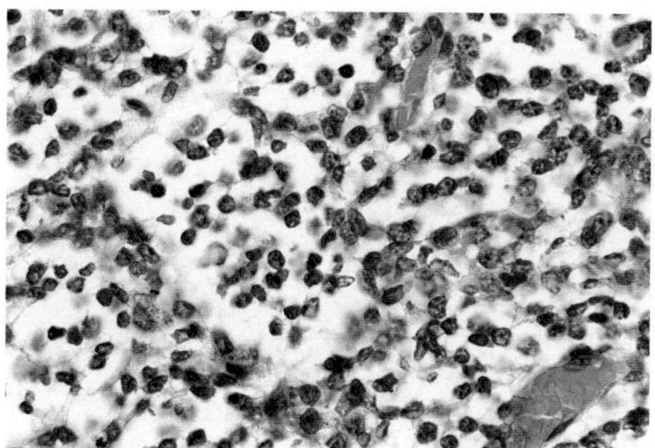

Figure 2-44

NORMAL PINEAL GLAND OF A NEWBORN

At birth, the pineal gland is a cellular organ with both dark, often pigmented, type I cells and nonpigmented type II cells with more cytoplasm.

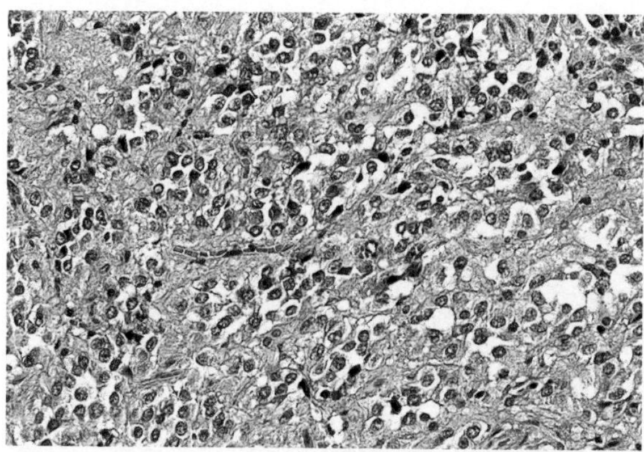

Figure 2-45

NORMAL PINEAL GLAND OF AN ADULT

The normal adult gland, as here in the wall of a pineal cyst, can be mistaken for a neoplasm such as pineocytoma. The gland's lobularity is a helpful diagnostic feature.

synaptophysin-positive structure composed of neurosecretory cells in which melanin is only occasionally seen at the ultrastructural level (figs. 2-45, 2-46). As is discussed in chapter 6, the gland, at any stage of development, may resemble a pineal parenchymal neoplasm or even a glioma, particularly when the pineal's characteristic lobulation is inapparent, and the sharp interface between the parenchyma and the glial stroma is not present in the specimen. Normal pineal astrocytes are well seen with GFAP immunostains as they circumscribe lobules of pineocytes (fig. 2-47). The supportive nature of these glia is readily evident in such situations. When dispersed more randomly, however, the cells may appear neoplastic. Pilocytic astrocytes accompanied by Rosenthal fibers are common in normal pineal stroma and increase greatly in number in the walls of small incidental or large symptomatic glial cysts of the pineal gland (fig. 2-48).

Infundibulum. The normal infundibulum varies somewhat in cellularity, and can simulate a glioma, as is discussed in chapter 3 (see fig. 3-190).

Cerebellum

The juxtaposition of Purkinje cells and diminutive internal granule cells illustrates the extremes of neuronal size (fig. 2-49). Particularly in specimens studied at frozen section, care must be taken not to misinterpret the darkly chromatinic granular cells as medulloblastoma

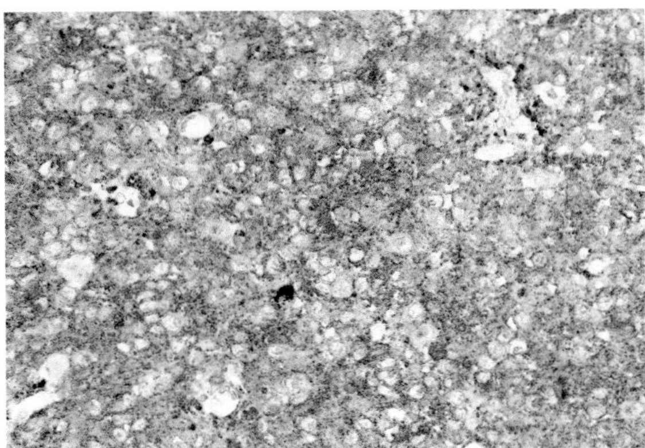

Figure 2-46

NORMAL PINEAL GLAND

Normal pineocytes are dependably immunoreactive for synaptophysin.

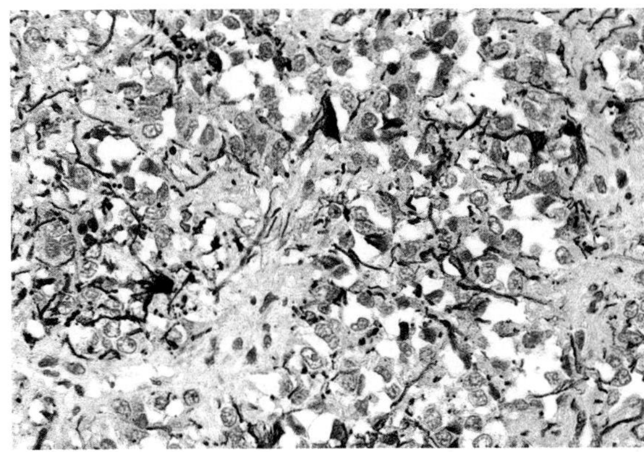

Figure 2-47

NORMAL PINEAL GLAND

Large glial fibrillary acidic protein (GFAP)-positive, stellate astrocytes are normal stromal constituents that should not be interpreted as evidence of a glioma.

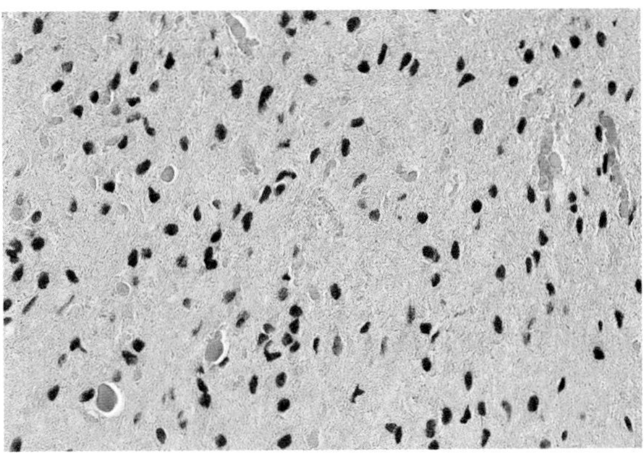

Figure 2-48

NORMAL PINEAL GLAND

Rosenthal fibers, small eosinophilic hyaline structures, are often prominent in the gland's glial stroma, particularly about pineal cysts.

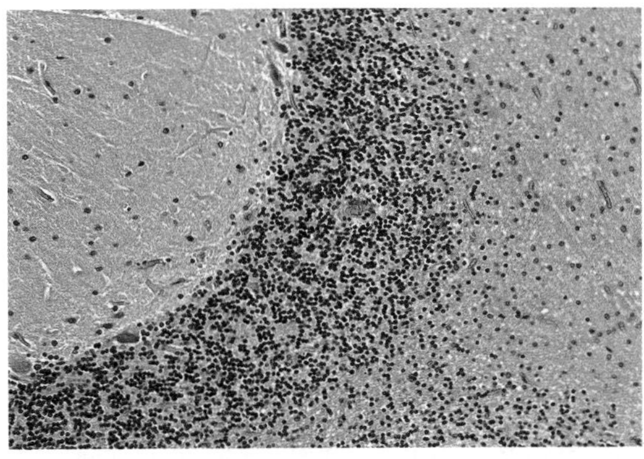

Figure 2-49

NORMAL CEREBELLUM

The normal cerebellum contains a paucicellular superficial molecular layer, a Purkinje cell layer, an internal granule cell layer, and white matter. Deep nuclei such as the dentate are not illustrated.

cells. Contributing to the similitude are small, synaptophysin-positive patches of neuropil that resemble Homer-Wright rosettes (figs. 2-50, 2-51). Normal granule cells are smaller than those of small cell neoplasms, either primary or metastatic. In addition, they are uniformly round, have distinct central nucleoli, and are free of mitotic activity. They are, unlike both Purkinje cells and most cells in medulloblastomas, immunoreactive for NeuN.

Until the 10th or 11th month of postnatal life, the molecular layer is overlain by small cells of the external granular cell layer. The external granular cell layer dwindles as its cells are eliminated by apoptosis or emigration along the processes of Bergmann glia into the internal granular cell layer (fig. 2-52) (23). These small superficial cells continue to proliferate in the postnatal period, and thus have long been suspected to be precursors of some medulloblastomas, as is discussed in chapter 5.

Elongated astrocytes, known as Bergmann glia, are confined to the layer that includes two types of neurons, Purkinje cells and basket cells.

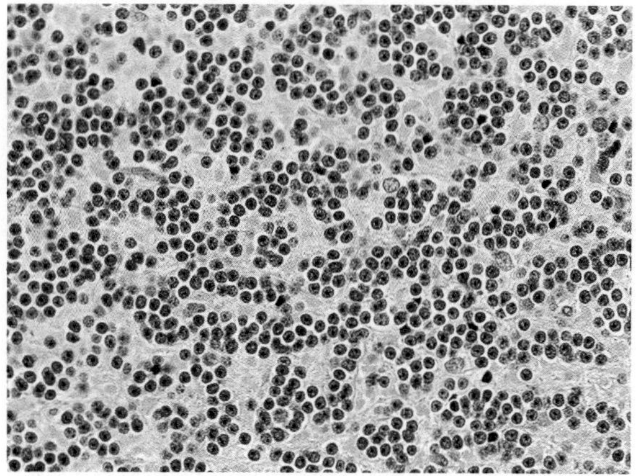

Figure 2-50

NORMAL CEREBELLAR INTERNAL GRANULAR CELL LAYER

The dense cellularity of the internal granule cell layer is interrupted by small fibrillar zones that superficially resemble neuroblastic (Homer-Wright) rosettes.

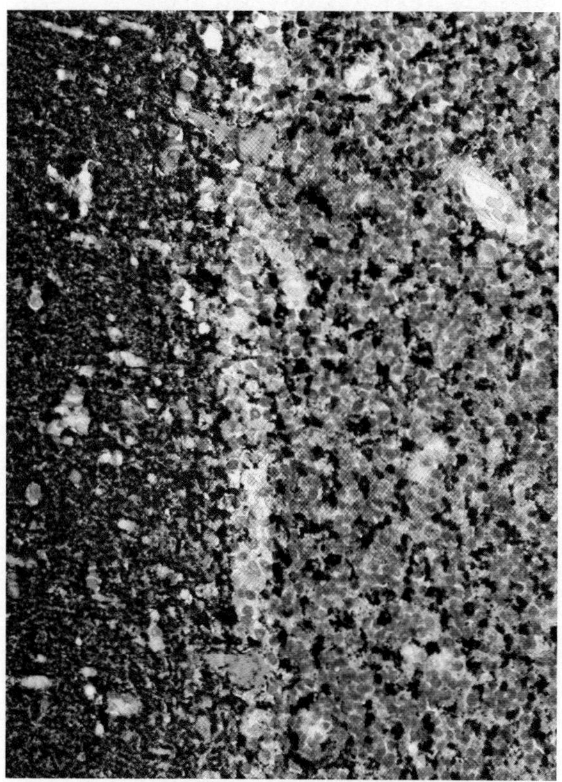

Figure 2-51

NORMAL CEREBELLUM

The molecular layer, left, and the cores of the fibrillar zones of the internal granular cell layer, right, are intensely immunoreactive for the neuronal marker synaptophysin.

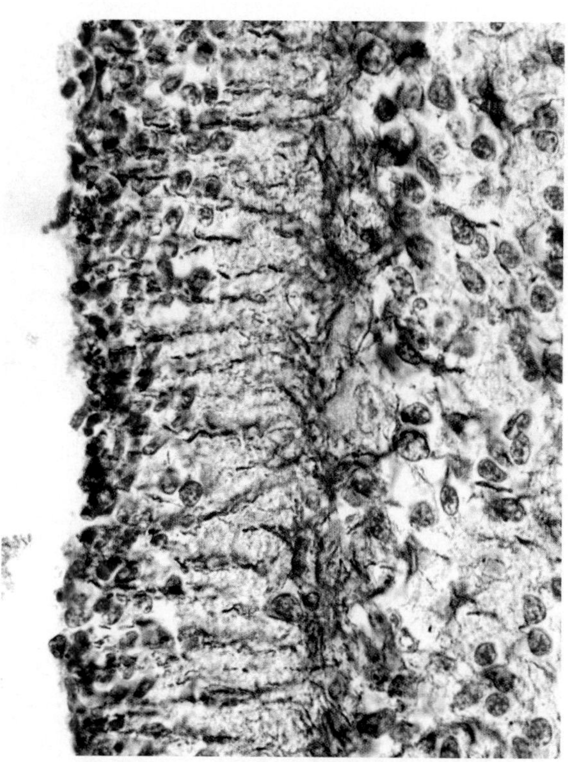

Figure 2-52

NORMAL DEVELOPING CEREBELLUM

Seen after immunostaining for GFAP, the normal cerebellum of a 1-month-old is covered by a superficial layer of external granular cells (left) that may be the source of some medulloblastomas. This layer is depleted postnatally within the first year as the cells migrate into the internal granular cell layer along GFAP-positive processes of Bergmann glia.

The nuclei of these glial cells are larger and less dense than those of the subsequent internal granule cells, but smaller than those of Purkinje cells (figs. 2-52, 2-53). Bergmann glia respond to injury, especially to the loss of Purkinje cells, by hypertrophy and hyperplasia.

Aside from Purkinje cells, other large neurons are concentrated in deep nuclei, including the dentate, fastigium, and globiform nuclei. Incorporated into a neoplasm such as glioma or medulloblastoma, these cells can erroneously suggest neuronal (ganglion cell) differentiation. Their advanced differentiation and linear, anatomic orientation are helpful diagnostic features to the contrary.

In smear preparations, the small, densely packed internal granule cells are easily misinterpreted as a neoplasm. The presence of Purkinje

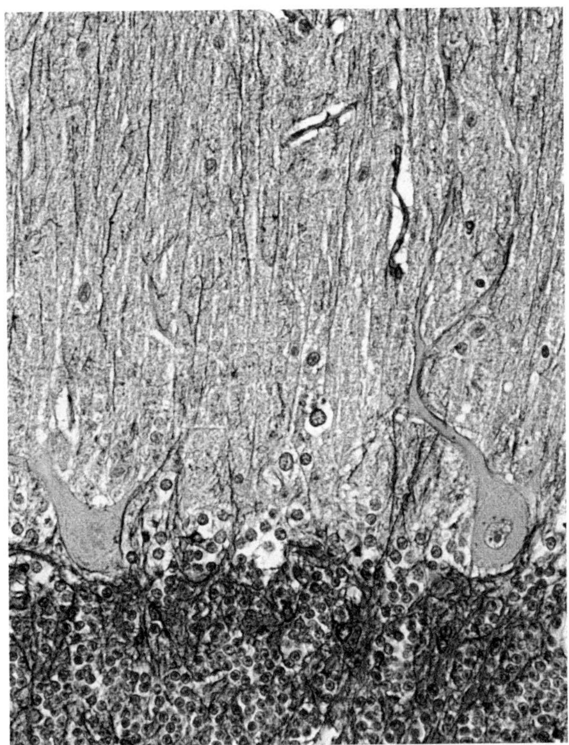

Figure 2-53

NORMAL CEREBELLUM OF AN ADULT

Staining for GFAP highlights the processes of Bergmann glia (astrocytes) that extend to the pial surface.

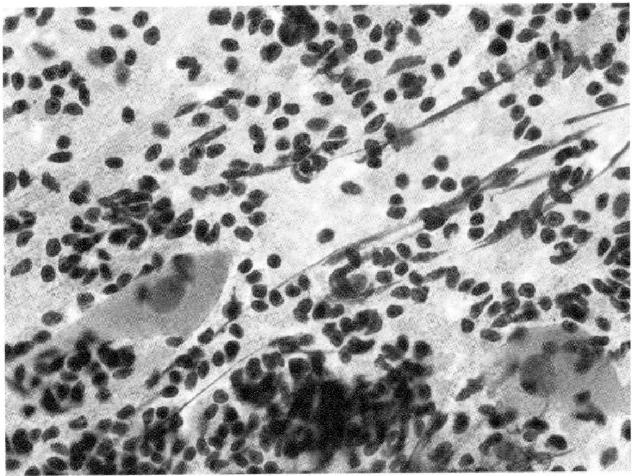

Figure 2-54

NORMAL CEREBELLUM: SMEAR PREPARATION

The number, density, and small size of internal granular cells can be misinterpreted as evidence of a neoplasm. A Purkinje cell is a helpful reminder that the small cells may be normal.

cells, the marked cytologic uniformity of the granule cells, and an absence of mitotic figures and apoptotic bodies are reassuring evidence that they are a normal tissue component (fig. 2-54).

Spine and Spinal Cord

The anatomic compartments of the spinal column, meningeal spaces, spinal cord, and nerve roots are seen well by neuroimaging techniques (fig. 2-55), which permit radiologists, clinicians, and pathologists alike to localize a lesion. From an anatomic and diagnostic standpoint, intraspinal structures lie within one of three compartments: extradural (outside the dura but within the bony spinal canal), intradural-extramedullary (within the dura but outside the spinal cord), and intramedullary (within the spinal cord or its terminal extension, filum terminale). The intradural-extramedullary compartment can be further subdivided into the subdural and subarachnoid spaces. Lesions in these three compartments are summarized in Appendix B.

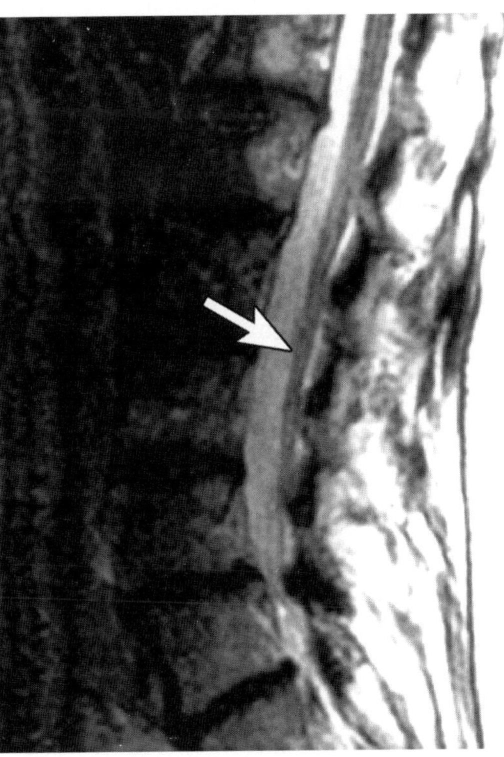

Figure 2-55

NORMAL CAUDA EQUINA

Seen in the midsagittal plane, the nerve roots in the cauda equina (arrow) are concentrated in the posterior portion of the dural sac, as seen here in a T2-weighted MRI. The bright areas in two lumbar vertebrae are metastases from a carcinoma of the lung.

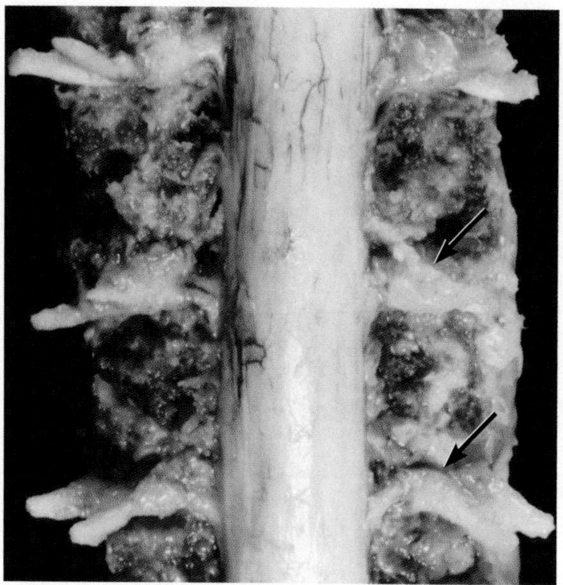

Figure 2-56

NORMAL SPINAL ANATOMY

The closely apposed spine and dura leave little intervening epidural space. Arrows indicate dorsal root ganglia whose microscopic appearance is illustrated in figures 2-60 and 2-61.

The dura is largely unattached to vertebrae, except anteriorly where it is bonded to the posterior surfaces of vertebral bodies (fig. 2-56). Regional epidural structures include vertebrae, intervertebral discs, segments of spinal nerves distant to the point at which they enter the dural sac, and scant adipose tissue. It has been debated whether the epidural space normally contains lymphoid tissue, an understandable question given the occurrence of seemingly primary epidural lymphomas.

Structures assigned to the intradural-extramedullary compartment include leptomeninges and the proximal portions of the spinal nerve roots. Drop metastases and inflammatory states often involve this compartment.

The term intramedullary refers to structural lesions within the spinal cord, including its caudal extension, the filum terminale (figs. 2-57, 2-58). The latter, a string-like terminal extension of the cord, is composed largely of fibrous tissue, astroglia, and nests of ependymal cells. In about 10 percent of normal individuals, small lobules of adipose tissue are also found in the filum (24). In a limited surgical exposure, the

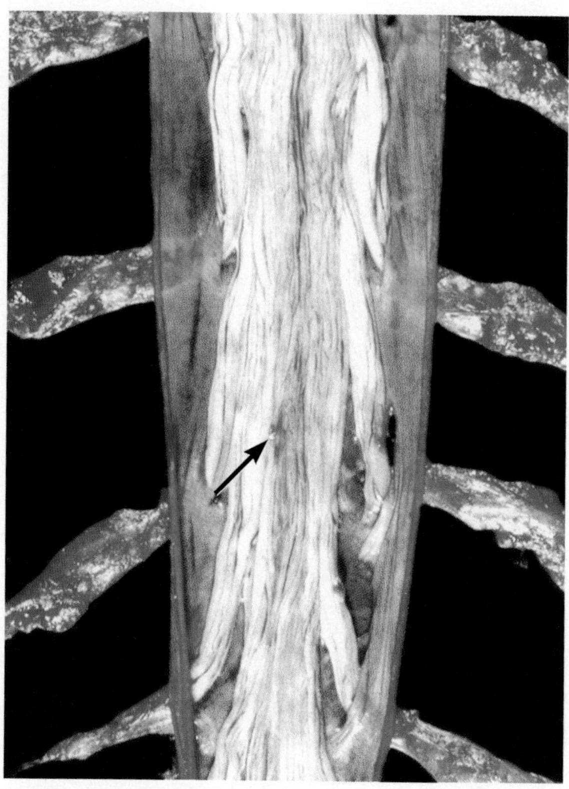

Figure 2-57

NORMAL CAUDA EQUINA

The dura has been opened to expose the cauda equina with its sheaf of nerve roots and the caudal extension of the spinal cord (filum terminale). Prominent vessels on the surface of the latter are helpful in identification. The small hemorrhagic focus (arrow) is the site of an antemortem lumbar puncture.

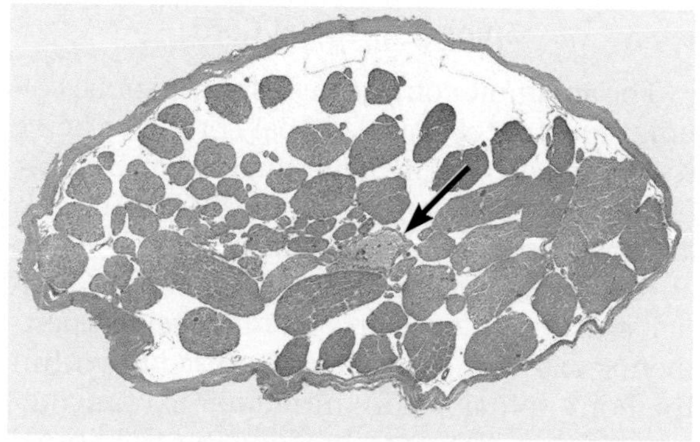

Figure 2-58

NORMAL CAUDA EQUINA

As seen in a transverse whole mount section, the cauda equina region, delimited by the dura, contains both spinal nerve roots and the centrally placed conus medullaris (arrow) or, at lower levels, the filum terminale.

distinction of filum from nerve root may be difficult. Caudal peripheral nerves often delicately adhere to its outer surface. The filum is the usual locus of myxopapillary ependymoma and paraganglioma, although either can arise, exceptionally, in a nerve root. Hemangioblastomas, especially in von Hippel-Lindau disease, can also arise from a nerve root.

As distinguished from the brain, the spinal cord is organized in a reverse manner, with white matter without and a gray matter within. At the center lies the central canal that after adolescence is no longer a continuous lumen, but becomes discontinuous clusters of plump ependymal cells or spaces that can be misconstrued as an abnormality (see fig. 2-40).

Peripheral-Central Nervous System Junction

The contact of the CNS with the outside world is mediated through cranial and spinal nerves. Of the former, the optic and olfactory "nerves" are really tracts of the CNS in which oligodendrocytes, rather than Schwann cells, are the myelinating element. The 10 remaining pairs of cranial nerves, or 11 if the vestigial "cranial nerve zero" on the undersurface of the frontal lobe is included (25), are potential sites of schwannomas. Sensory nerves or nerve roots are usually affected. Only the eighth nerve is involved with any frequency, and its vestibular, rather than acoustic, division is typically the site of origin. The reason for this is unclear, since the density of normal Schwann cells is the same in both divisions (26).

With the exception of the eighth, or vestibulo-acoustic, nerve, the zone of transition between the brain (with its oligodendrocytes) and the peripheral nervous system (PNS) (with its Schwann cells) is positioned within a millimeter or two of its external or pial surface. In the eighth nerve, the transition zone is situated 1 cm peripherally, placing it well within the internal auditory meatus. Vestibular schwannomas, therefore, begin in the auditory canal, and only after a period of enlargement arrive at the cerebellopontine angle. Given the resolution of MRI scans, many of these are now detected at the intracanalicular stage. Among the cranial nerves, and for unexplained reasons, the eighth nerve is the favorite site for rare hamartomatous "lipomas," as is discussed in chapter 13.

In spinal nerve roots, the CNS-PNS transition zone is situated close to the pial surface (fig. 2-59, left), with a short unmyelinated segment, the Obersteiner-Redlich zone, interposed between the terminal domains of CNS and the PNS (fig. 2-59, right). Intradural spinal nerve roots are, therefore, components of the peripheral, rather than central, nervous system.

Occasionally submitted as surgical specimens from the intervertebral spaces are small nubbins of tan tissue representing dorsal root ganglion. These are recognized histologically by the mixture of well-formed ganglion cell and peripheral nerve fibers (fig. 2-60). Unlike gangliocytomas, cells of the dorsal root ganglion contain finely granular brown pigment and are surrounded by S-100–positive satellite cells (fig. 2-61).

REFERENCES

1. Fuller GN, Burger PC. Central nervous system. In: Sternberg S, ed. Histopathology for pathologists. New York: Lippincott-Raven; 1997:243–82.
2. Burger PC, Nelson JS, Boyko OB. Diagnostic synergy in radiology and surgical neuropathology: neuroimaging techniques and general interpretive guidelines. Arch Pathol Lab Med 1998;122: 609–19.
3. Burger PC, Nelson JS, Boyko OB. Diagnostic synergy in radiology and surgical neuropathology: radiographic findings of specific pathologic entities. Arch Pathol Lab Med 1998;122:620–32.
4. Winek RR, Scheithauer BW, Wick MR. Meningioma, meningeal hemangiopericytoma (angioblastic meningioma), peripheral hemangiopericytoma, and acoustic schwannoma. A comparative immunohistochemical study. Am J Surg Pathol 1989;13:251–61.
5. Kida S, Yamashima T, Kubota T, Ito H, Yamamoto S. A light and electron microscopic and immunohistochemical study of human arachnoid villi. J Neurosurg 1988;69:429–35.

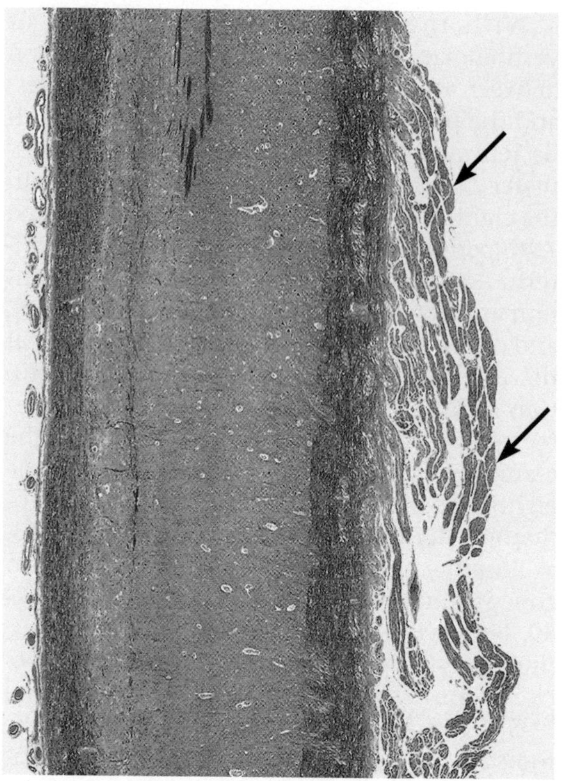

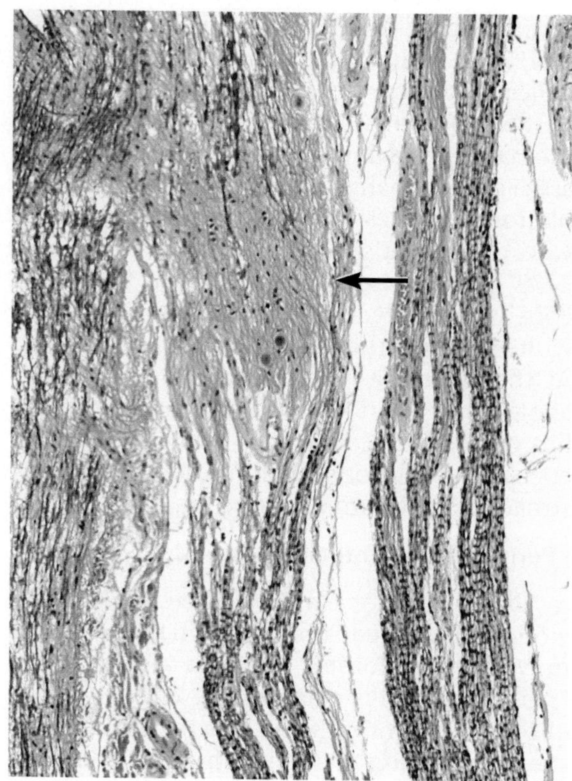

Figure 2-59

NORMAL SPINAL CORD AND NERVE ROOTS

Left: As illustrated in a whole mount sagittal section, sensory roots (arrows) angle superiorly to enter the posterior aspect of the spinal cord. Only a few anterior roots are seen.

Right: At higher magnification, a short unmyelinated segment (Obersteiner-Redlich zone) (arrow) is shown.

6. Yamashima T, Kida S, Yamamoto S. Ultrastructural comparison of arachnoid villi and meningiomas in man. Mod Pathol 1988;1:224–34.
7. Haines DE, Harkey HL, al-Mefty O. The "subdural" space: a new look at an outdated concept. Neurosurgery 1993;32:111–20.
8. O'Rahilly R, Muller F. The meninges in human development. J Neuropathol Exp Neurol 1986;45:588–608.
9. Hutchings M, Weller RO. Anatomical relationships of the pia mater to cerebral blood vessels in man. J Neurosurg 1986;65:316–25.
10. Gebarski SS, Blaivas MA. Imaging of normal leptomeningeal melanin. AJNR Am J Neuroradiol 1996;17:55–60.
11. Alcolado R, Weller RO, Parrish EP, Garrod D. The cranial arachnoid and pia mater in man: anatomical and ultrastructural observations. Neuropathol Appl Neurobiol 1988;14:1–17.
12. Rosenberg AE, O'Connell JX, Ojemann RG, Plata MJ, Palmer WE. Giant cystic arachnoid granulations: a rare cause of lytic skull lesions. Hum Pathol 1993;24:438–41.
13. Queiroz LS, Lopes de Faria J. Evolution of dark neurons in experimental brain stab wounds. Virchows Arch B Cell Pathol 1978;28:361–70.
14. Dohrmann GJ, Bucy PC. Human choroid plexus: a light and electron microscopic study. J Neurosurg 1970;33:506–16.
15. Strazielle N, Ghersi-Egea JF. Choroid plexus in the central nervous system: biology and physiopathology. J Neuropathol Exp Neurol 2000;59:561–74.
16. Herbert J, Cavallaro T, Dwork AJ. A marker for primary choroid plexus neoplasms. Am J Pathol 1990;136:1317–25.
17. Kepes JJ, Collins J. Choroid plexus epithelium (normal and neoplastic) expresses synaptophysin. A potentially useful aid in differentiating carcinoma of the choroid plexus from metastatic papillary carcinomas. J Neuropathol Exp Neurol 1999;58:398–401.

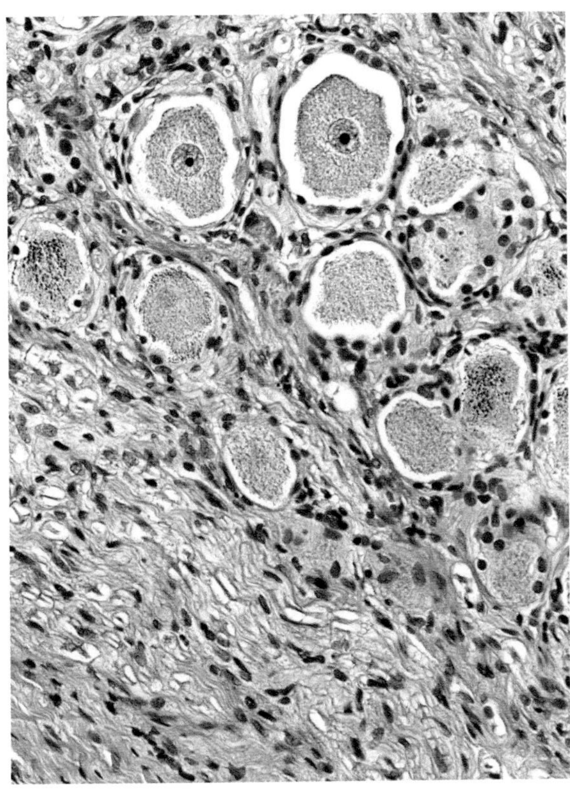

Figure 2-60

NORMAL DORSAL ROOT GANGLION

The organized mixture of large ganglion cells and small wavy bundles of peripheral nerves are typical of this structure that is sometimes submitted to pathology as a suspected neoplasm.

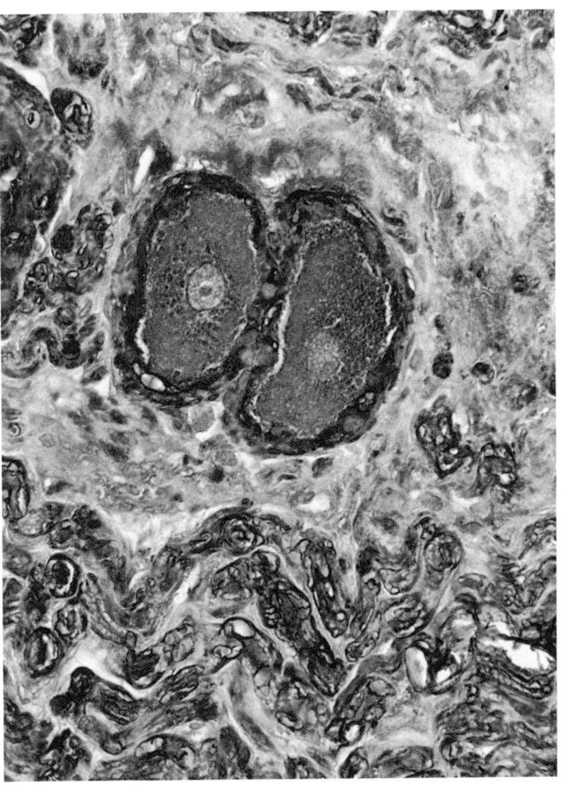

Figure 2-61

NORMAL DORSAL ROOT GANGLION

S-100-protein–positive perineuronal satellite cells are characteristic of sensory ganglia, not of gangliocytoma. Schwann cell–derived myelin sheaths are also intensely S-100-protein positive.

18. Lach B, Scheithauer BW. Colloid cyst of the third ventricle: a comparative ultrastructural study of neuraxis cysts and choroid plexus epithelium. Ultrastruct Pathol 1992;16:331–49.
19. Miettinen M, Clark R, Virtanen I. Intermediate filament proteins in choroid plexus and ependyma and their tumors. Am J Pathol 1986;123:231–40.
20. Wen GY, Wisniewski HM, Kascsak RJ. Biondi ring tangles in the choroid plexus of Alzheimer's disease and normal aging brains: a quantitative study. Brain Res 1999;832:40–6.
21. Uematsu Y, Rojas-Corona RR, Llena JF, Hirano A. Distribution of epithelial membrane antigen in normal and neoplastic human ependyma. Acta Neuropathol (Berl) 1989;78:325–8.
22. Min KW, Seo IS, Song J. Postnatal evolution of the human pineal gland. An immunohistochemical study. Lab Invest 1987;57:724–8.
23. Friede RL. Dating the development of human cerebellum. Acta Neuropathol (Berl) 1973;23:48–58.
24. McLendon RE, Oakes WJ, Heinz ER, Yeates AE, Burger PC. Adipose tissue in the filum terminale: a computed tomographic finding that may indicate tethering of the spinal cord. Neurosurgery 1988;22:873–76.
25. Fuller GN, Burger PC. Nervus terminalis (cranial nerve zero) in the adult human. Clin Neuropathol 1990;9:279–83.
26. Tallan EM, Harner SG, Beatty CW, Scheithauer BW, Parisi JE. Does the distribution of Schwann cells correlate with the observed occurrence of acoustic neuromas? Am J Otol 1993;14:131–4.

3 TUMORS OF NEUROGLIA AND CHOROID PLEXUS

ASTROCYTIC NEOPLASMS

Although nearly all astrocytomas are infiltrative to some degree, nevertheless, they can be usefully divided into infiltrative and localized types. Those of the astrocytoma-anaplastic astrocytoma-glioblastoma continuum are infiltrative, or "diffuse," by definition, particularly the gliomatosis cerebri variant. More circumscribed tumors include pilocytic astrocytoma, pleomorphic xanthoastrocytoma, and subependymal giant cell astrocytoma. A classification of central nervous system (CNS) tumors by the degree of circumscription is given in Appendix A.

DIFFUSELY INFILTRATING ASTROCYTIC TUMORS

For practical purposes, the most frequent astrocytomas are best defined as infiltrating gliomas that cannot be classified as oligodendroglial. Implicit in this negative definition is the fact that astrocytomas vary considerably in terms of their histologic and cytologic appearance, and they thus may not all represent the same entity. In any case, infiltrating astrocytic tumors comprise a significant proportion of all primary brain tumors and exhibit a seemingly continuous spectrum of differentiation and tumor grade.

Although these biologically malignant tumors occur in patients of all ages and arise at all levels of the neuraxis, they do not do so randomly. In adults, most occur in the cerebral hemispheres, whereas in children, the same neoplasms typically occur in the brain stem (1–4) or thalamus (5–8). Less commonly affected sites in both children and adults include the spinal cord (9–12) and cerebellum (13–15). Astrocytic tumors in the latter site are almost always of the pilocytic type. The same is true of the optic nerve where only a rare astrocytoma, generally high grade, is infiltrative (or diffuse) (16–18).

Among diffusely infiltrative astrocytomas of the cerebral hemispheres, a close correlation is observed between histologic grade and four clinical variables: patient age, duration of symptoms, neurologic performance status, and length of postoperative survival. With occasional exceptions, lesions in older patients are more anaplastic, biologically aggressive, recently symptomatic, and destructive of neurologic function.

A cardinal property of infiltrating astrocytomas is a propensity to undergo anaplastic change. This process is common, and relates somewhat to patient age, since it occurs more often and rapidly in older individuals. It is also especially frequent in gemistocytic astrocytomas (19,20). Anaplastic progression is readily seen when initial specimens are compared with ones taken later in the course of the disease, or when a single specimen has tumor of differing grades (21,22). Thus, tumors that are largely grade II astrocytoma often include foci of grade III anaplastic astrocytoma and even glioblastoma. Conversely, many glioblastomas consist, in part, of preexisting grade II and/or III tumor. When incomplete, this progression produces borderline lesions that are difficult to grade. In spite of its importance, it is not clear how often anaplastic change occurs in a grade II astrocytoma.

The grading of diffusely infiltrating astrocytoma necessarily applies artificially abrupt points of division to what is a continuum. Since the process by which the tumor evolves toward glioblastoma proceeds by accumulation of molecular events, such milestones are detectable and are becoming part of routine pathologic analysis. Inasmuch as some are of prognostic significance, these events also represent incremental increases in tumor grade at the cytologic level. It may be possible to recognize the effects of these molecular events at an earlier, focal, cytologic level than on the basis of classic histologic criteria.

Despite exciting developments relevant to tumor classification and grading, the discussion of infiltrative astrocytic neoplasms focuses largely on classic morphologic criteria. Comments regarding newer methods are added when they contribute to diagnosis and prognosis.

The grading system in this Fascicle is the three-tiered scheme of the 2007 World Health Organization (WHO) formulation (23–25), which recognizes diffuse astrocytoma (grade II), anaplastic astrocytoma (grade III), and glioblastoma (grade IV). There are no grade I infiltrating astrocytomas since this grade is reserved for the heterogeneous group of unrelated, localized, slowly growing lesions mentioned above. The widely used term "low-grade astrocytoma" encompasses both grade II infiltrating astrocytomas and the localized, grade I lesions. While grade I and grade II lesions have the negative property of not being "malignant," i.e., neither grade III nor grade IV, they have little else in common and the imprecise term misleadingly suggests that grade I astrocytomas are simply better differentiated versions of grade II diffuse astrocytomas.

The 2007 WHO system of astrocytoma grading is a modification of the St. Anne-Mayo grading scheme that was originally designed for application to astrocytomas of the diffusely infiltrative type, specifically as seen in needle biopsy specimens (26). The latter employs a four-tiered format based on the presence or absence of four variables: nuclear atypia, mitoses, microvascular proliferation, and necrosis. Rare tumors devoid of these features are designated grade 1, those with one variable are grade 2, those with two variables are grade 3, and tumors with three or four features are grade 4. This appealingly simple system was applied retrospectively to a large series of patients with long follow-up, with reproducible and prognostically useful results (26,27). Since the four parameters essentially appear in temporal sequence (nuclear abnormalities followed by mitotic activity and then by microvascular proliferation and/or necrosis), the scheme actually lends itself to further simplification. In practice, grade 2 tumors show atypia alone, grade 3 tumors add mitoses, and grade 4 lesions acquire microvascular proliferation, necrosis, or both. Grade 1 tumors are so rare among infiltrative astrocytomas (1 percent) as to be "nonexistent."

This volume and the 2007 WHO system recognize only three lesions: diffuse astrocytoma (grade II), anaplastic astrocytoma (grade III), and glioblastoma (grade IV). In an additional departure from the original St. Anne-Mayo system, the WHO approach does not accept a solitary mitosis in a sizable tumor as necessarily placing the lesion in the grade III category. The justification for this departure is discussed below (28). The nonspecific term "microvascular proliferation" has replaced the more strictly defined "endothelial proliferation" of both the St. Anne-Mayo and the previous 2000 WHO grading schemes.

The etiology of infiltrating astrocytomas is unknown; genetic factors, such as Turcot's syndrome (29), or ionizing radiation (30–33) can be incriminated only rarely. Most radiation-induced gliomas are grade IV at the time of diagnosis. Rare infiltrating astrocytomas also occur in patients with Ollier's disease and Maffucci's syndrome (34–38) and Li-Fraumeni syndrome (39). Infiltrative astrocytic neoplasms are relatively uncommon among blacks (40).

Diffuse Astrocytoma

Definition. The WHO grade II *diffuse astrocytoma* is a well-differentiated, diffusely infiltrative lesion.

Clinical Features. Diffuse astrocytomas most frequently affect the cerebral hemispheres of young to middle-aged adults (figs. 3-1–3-3), and the brain stem (figs. 3-4, 3-5) and thalamus of children. Occasional tumors also occur in the cerebellum (see figs. 3-148, 3-149) or spinal cord (fig. 3-6). In any site, diffuse astrocytoma should be distinguished from other astrocytomas, particularly those that are better circumscribed. The critical distinction between infiltrating and better circumscribed lesions is summarized in Appendix A. With the rare exception of meningeal gliomatosis, discussed separately, astrocytomas are intraparenchymal. The differential diagnosis of spinal lesions by anatomic compartment is given in Appendix B.

As with any expanding intracranial mass, diffuse astrocytomas may produce nonspecific symptoms of mass effect, seizures, and neurologic deficits in accord with the lesion's size, location, and rate of growth. Seizures are more common than functional deficits due to parenchymal destruction, which occurs more often in higher-grade tumors. Astrocytomas in the brain stem produce neurologic signs caused by dysfunction of cranial nerve nuclei and compression of the sensorimotor tracts that traverse the pons and medulla. In combination with radiographic evidence of brain stem enlargement

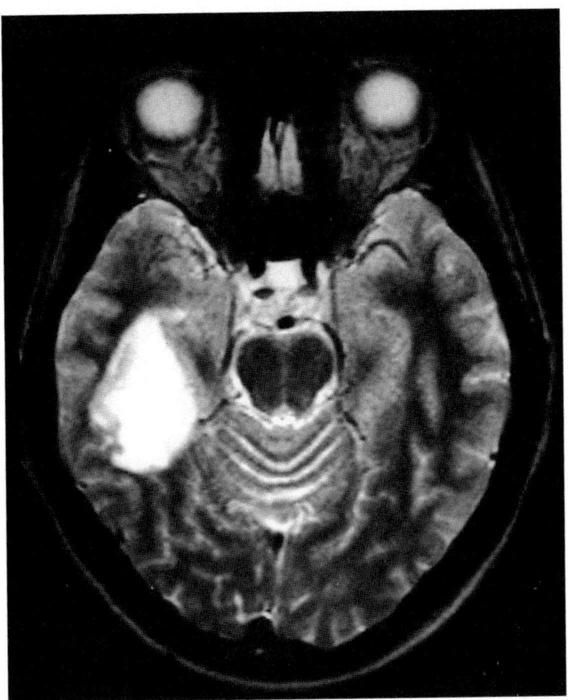

Figure 3-1

GRADE II DIFFUSE ASTROCYTOMA

As in this example in the temporal lobe, grade II diffuse astrocytomas create an area of high signal intensity on T2-weighted magnetic resonance imaging (MRI). (Fig. 3-5, right from Fascicle 10, 3rd Series.)

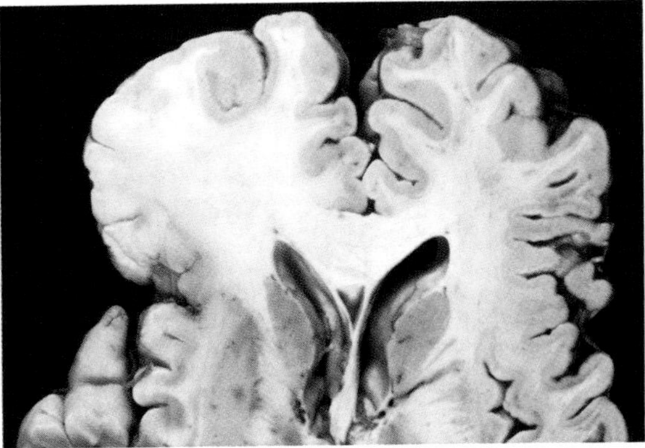

Figure 3-2

GRADE II DIFFUSE ASTROCYTOMA

Superficial infiltrating gliomas, in this case astrocytic, typically obscure the gray-white junction.

or hypertrophy, these symptoms are often considered sufficiently specific to justify radiotherapy without histologic confirmation.

Radiologic Findings. By magnetic resonance imaging (MRI), diffuse astrocytomas (grade II) present as ill-defined areas of low signal intensity on T1-weighted images. Due to their content of bright edema fluid, they are even more obvious on T2-weighted (figs. 3-1, 3-4) or fluid-attenuated inversion recovery (FLAIR) images. While some extend in finger-like fashion into white matter pathways, such as the external capsule and extreme capsule, other, more superficially situated tumors often appear well defined.

Thalamic astrocytomas can be bilateral, sometimes strikingly so, and may extend into adjacent lobes or the brain stem. Diffuse astrocytomas primary to the brain stem overrun the tissue, usually the pons, which is expanded until it partially engulfs the basilar artery (fig. 3-4). Pilocytic astrocytomas, in contrast, are more likely to arise in the medulla or midbrain, and are

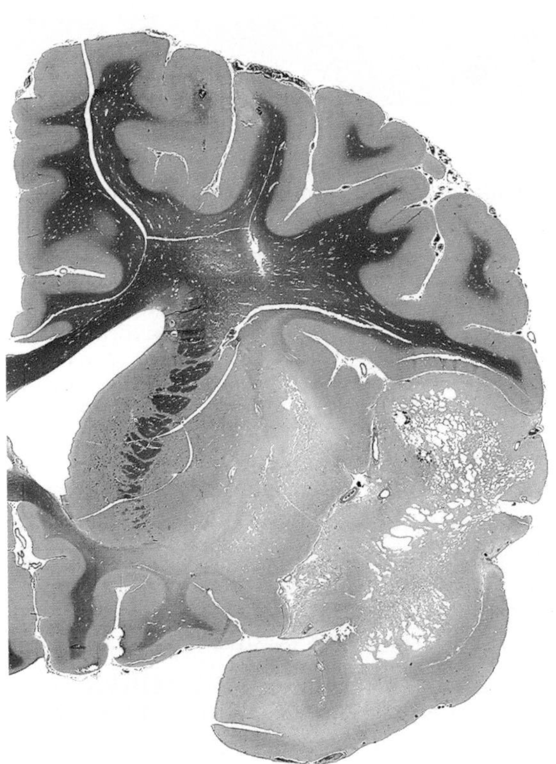

Figure 3-3

GRADE II DIFFUSE ASTROCYTOMA

Grade II astrocytomas are usually centered in the white matter and can, as in this case, be extensively microcystic. There was 6-year history of seizures with this temporal lobe lesion in a 36-year-old man.

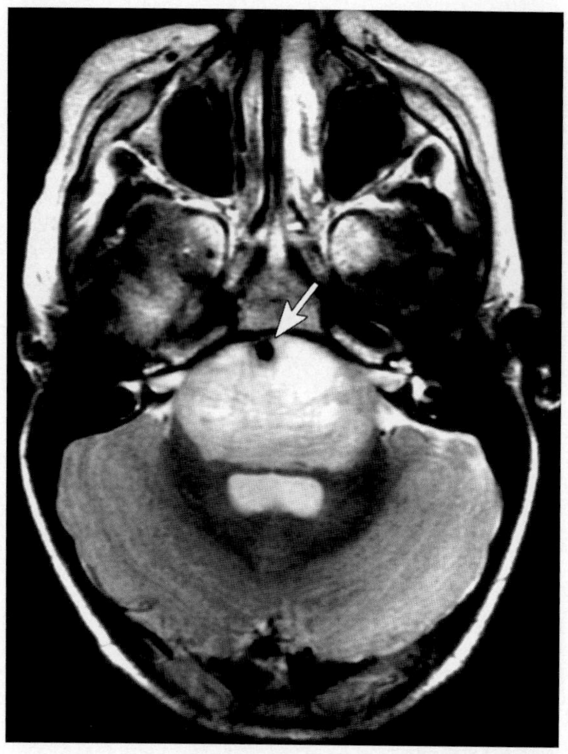

Figure 3-4

GRADE II DIFFUSE ASTROCYTOMA OF THE PONS

Diffuse astrocytomas of the brain stem, usually in the pons, generate a hyperintense signal on T2-weighted images that extends throughout the expanded structure. As is typical, the clinical and radiologic findings were so specific that radiotherapy and chemotherapy proceeded without histologic confirmation. The 3-year-old boy survived 16 months. Characteristically, the expanded pons partially surrounds the basilar artery (arrow).

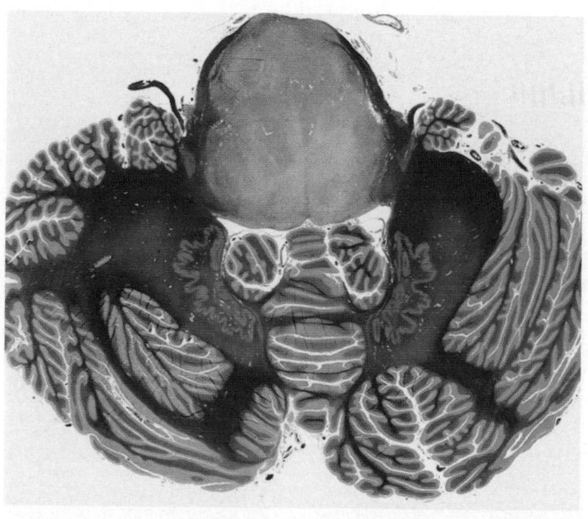

Figure 3-5

GRADE II DIFFUSE ASTROCYTOMA OF THE PONS

Diffuse astrocytomas of the brain stem are intrinsic, infiltrating lesions that efface architectural features (hematoxylin and eosin [H&E]/Luxol fast blue stain).

well circumscribed, often exophytic, and almost always contrast enhancing (see fig. 3-121) (4). In any site, diffuse enhancement almost obviates consideration of a grade II diffuse astrocytoma. Instead, attention should be directed to the possibility of a localized lesion such as pilocytic astrocytoma or pleomorphic xanthoastrocytoma. Irregular enhancement in diffuse astrocytoma generally signifies regions of progression to anaplastic astrocytoma or glioblastoma. Most diffuse astrocytomas of the cerebral hemispheres are solid, but gemistocytic variants are occasional cystic exceptions.

Astrocytomas in the spinal cord can span one or many spinal segments (fig. 3-6). Although some are associated with a syrinx, the latter, especially when large, is more typical of discrete intramedullary tumors such as pilocytic astrocytoma, ependymoma, and hemangioblastoma (see Appendix D for a list of cystic CNS tumors).

Gross Findings. Early in their evolution, deep-seated astrocytomas may be too subtle to be macroscopically apparent. Later, they expand the white matter and even involve the overlying cerebral cortex (figs. 3-2, 3-3). Since they may be largely restricted to white matter, a small biopsy from the cortex may not yield diagnostic tissue.

Astrocytomas vary in texture. Some are firm, while others are soft and nearly gelatinous. The intratumoral clear fluid-filled cysts seen in some cases help distinguish astrocytoma from glioblastoma, which has a necrotic center with potentially hemorrhagic, semiliquid contents. At autopsy or in gross sections of lobectomy specimens, grade II astrocytomas expand the white matter and blur the ordinarily well-demarcated gray-white junction.

Astrocytomas of the brain stem generally have their epicenter within the pons, a structure that enlarges as it accommodates the proliferating cells (fig. 3-5). Grade IV astrocytomas, in contrast, are understandably necrotic. The expanding tumor encroaches posteriorly and superiorly upon the fourth ventricle, often to the point of functional obstruction, while protruding anteroinferiorly to displace and engulf

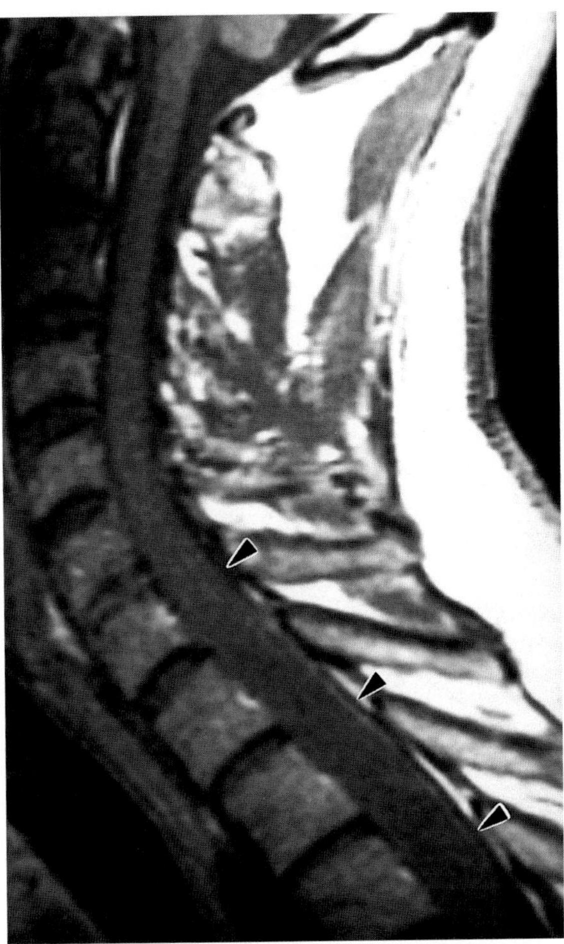

Figure 3-6

GRADE II DIFFUSE ASTROCYTOMA OF THE SPINAL CORD

Diffuse astrocytomas, grade II, of the spinal cord are noncontrast-enhancing lesions (arrowheads) that enlarge the cord over multiple segments. This might become a contrast-enhancing lesion with anaplastic transformation. (Fig. 3-9 from Fascicle 10, 3rd Series.)

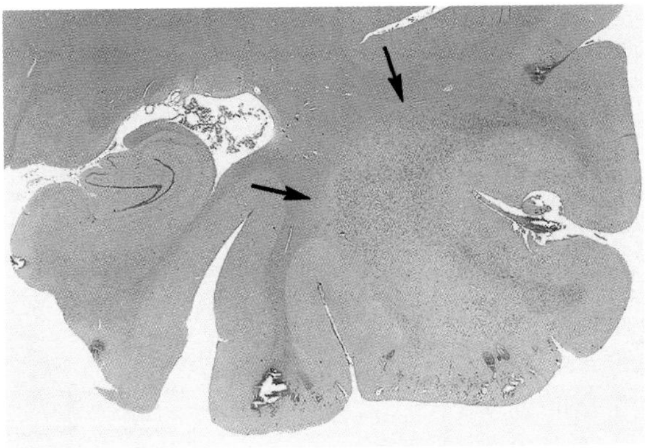

Figure 3-7

GRADE II DIFFUSE ASTROCYTOMA

An incidental postmortem finding in an accident victim, the pale staining grade II astrocytoma (arrows) is centered in the white matter, where it incorporates preexisting parenchyma. Although the lesion is diffusely infiltrative, it has a macroscopically abrupt interface with the surrounding white matter. Hemorrhagic cortical contusions are present inferiorly.

the basilar artery. Collectively, these features are virtually diagnostic.

Astrocytoma of the spinal cord produces an ill-defined fusiform enlargement. A fluid-filled cyst, or syrinx, may either lie within the tumor or extend from its rostral or caudal poles into uninvolved spinal cord.

Microscopic Findings. At low-power magnification, most well-differentiated astrocytomas are regions of hypercellularity that are generally situated predominantly within the white matter (fig. 3-7). The cellularity is approximately two or three times normal. The tumor infiltrates widely here and in gray matter to incorporate preexisting structures, including neurons, axons, oligodendrocytes, and astrocytes (figs. 3-8, 3-9). Nonetheless, there may be little disturbance of architectural features in underlying affected tissue. The margin of such tumors is indistinct since tumor cells lie individually dispersed within intact parenchyma and often follow fiber pathways for some distance. Such spread of tumor cells is often even more apparent in anaplastic astrocytoma or glioblastoma. Astrocytomas often enter the overlying cortex or deep hemispheric structures. Proliferation in subpial, perivascular, perineuronal, and subependymal zones produces what are termed "secondary structures of Scherer" (41,42). Such characteristic patterns of tumor cell accumulation are usually more prominent in oligodendrogliomas, astrocytomas undergoing anaplastic transformation, or in so-called gliomatosis cerebri.

In all but the best-differentiated neoplasms, astrocytoma cells are irregularly distributed as compared to the more uniform cellularity of normal white matter (see fig. 2-32). In addition to hypercellularity, which is often subtle, nearly all astrocytomas possess cells with enlarged, cigar-shaped, irregular, hyperchromatic nuclei.

The cytoplasm of astrocytoma cells varies considerably both in amount and configuration

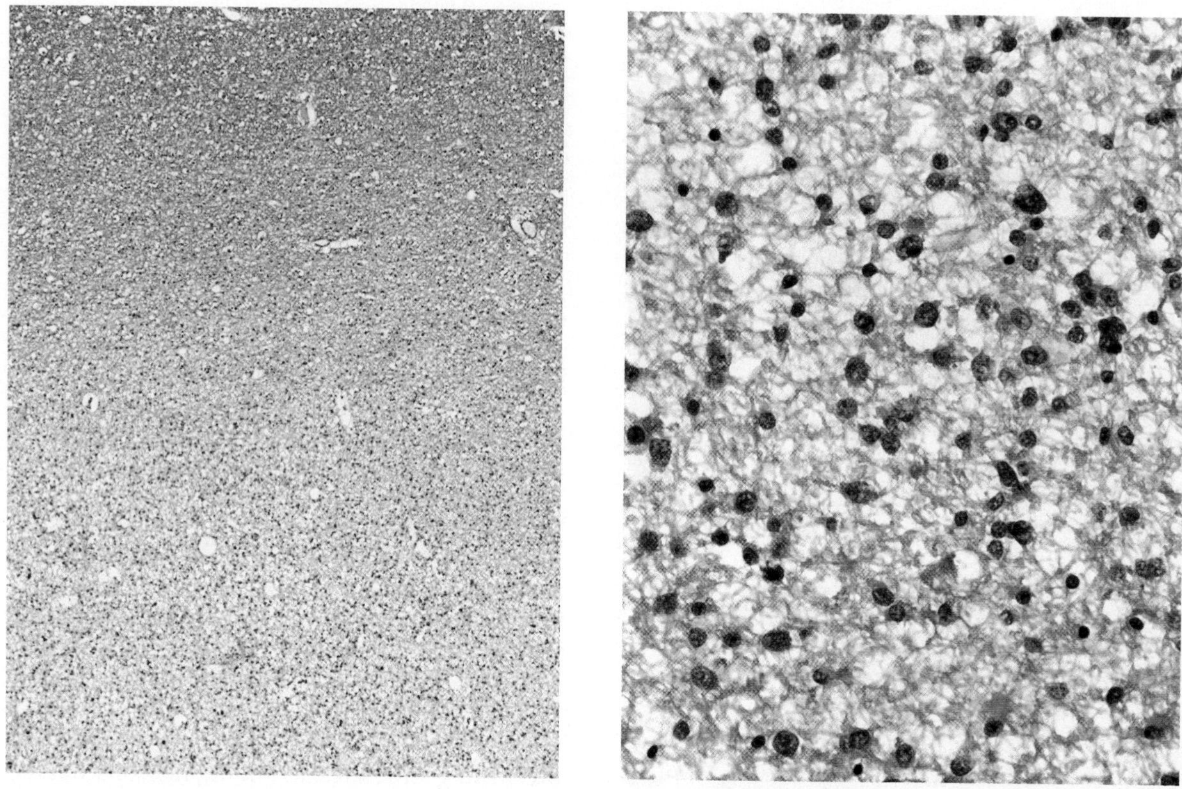

Figure 3-8

GRADE II DIFFUSE ASTROCYTOMA

Left: Higher magnification of the lesion seen in figure 3-7 illustrates the "diffuseness" by which the infiltrating lesion incorporates preexisting parenchymal elements.

Right: The slight to moderate degree of nuclear pleomorphism is typical of grade II diffuse astrocytomas.

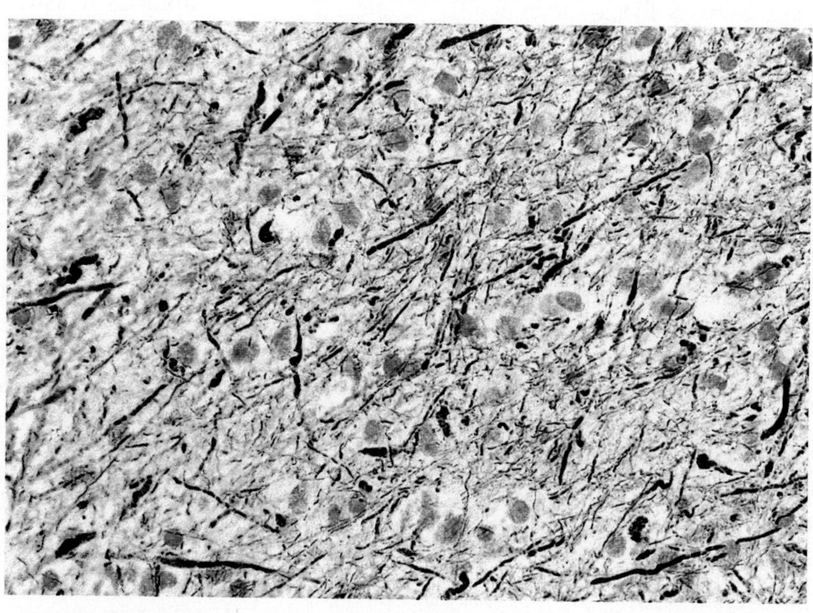

Figure 3-9

GRADE II DIFFUSE ASTROCYTOMA

Diffuse astrocytomas are infiltrating processes, as is clearly seen after staining for axons with antibodies to neurofilament protein. Tumor cells are intercalated among the immunohistochemically positive axons.

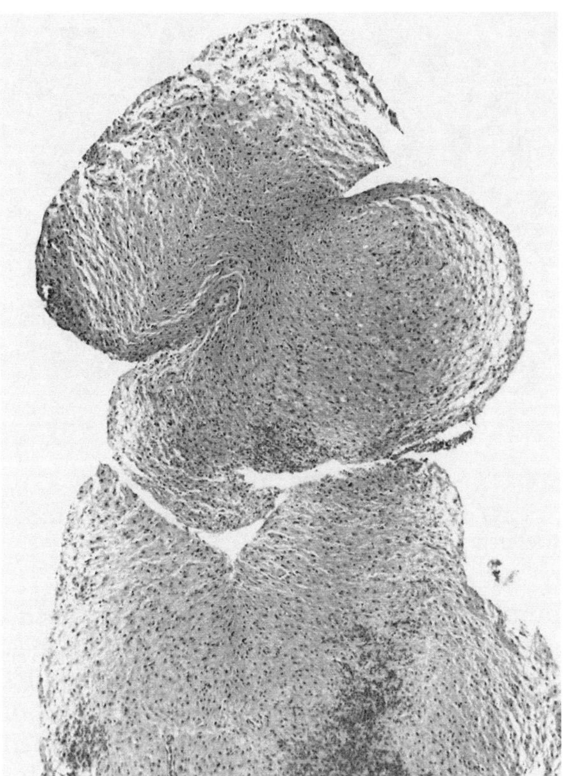

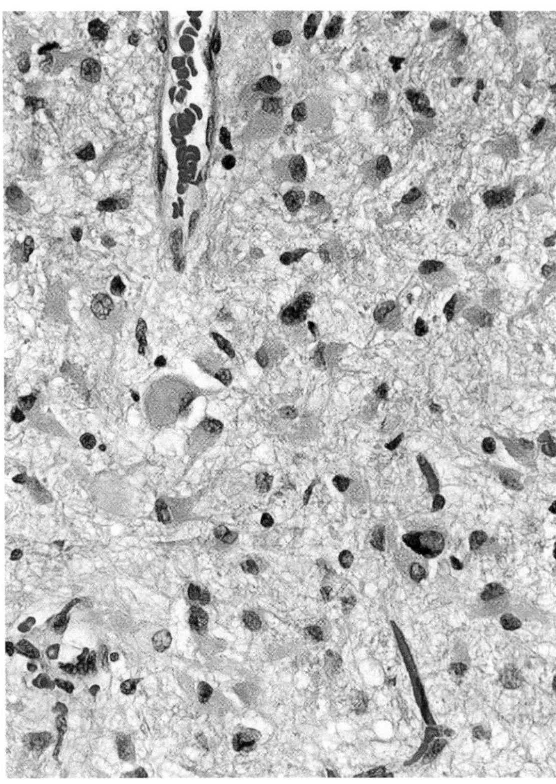

Figure 3-10

GRADE II DIFFUSE ASTROCYTOMA: NEEDLE BIOPSY SPECIMEN

Left: There is a modest increase in cellularity in this stereotactically obtained needle biopsy specimen of diffuse astrocytoma, grade II.

Right: Hypercellularity, cytologic atypia, and coarse processes are typical of many grade II diffuse astrocytomas.

(figs. 3-10, 3-11). In some paucicellular lesions, the tumor cells lie isolated in virtually intact brain tissue; their fibril-free cytoplasm is scant and devoid of processes ("naked nuclei") (fig. 3-11). Such tumors are recognized more on the basis of their increased, irregularly distributed cellularity and nuclear atypia than by the presence of "astrocytic" cytologic features. In more cellular lesions, the cells have varying quantities of fibril-containing cytoplasm and short, often asymmetric processes (fig. 3-10). The latter are the basis of the overtly astrocytic quality of such tumors. Many lesions have both cells with naked nuclei and those with the smallest amounts of cytoplasm, but enough to suggest an astrocytic nature of the lesion (fig. 3-11). Perinuclear halos are not uncommon, and would lend an oligodendroglioma appearance were it not for the nuclear hyperchromatism and polymorphism

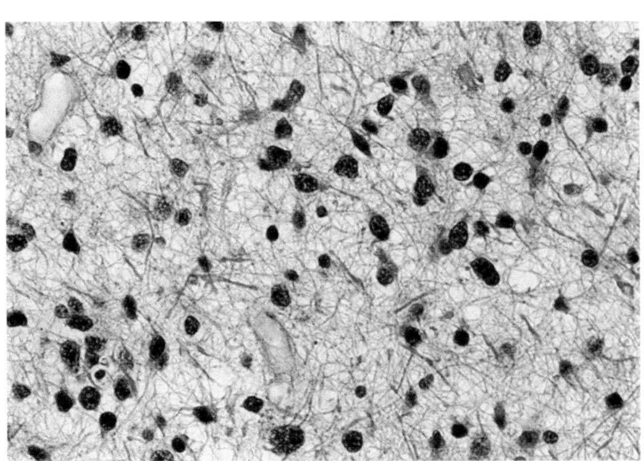

Figure 3-11

GRADE II DIFFUSE ASTROCYTOMA

"Naked," mildly atypical nuclei are common in grade II diffuse astrocytomas.

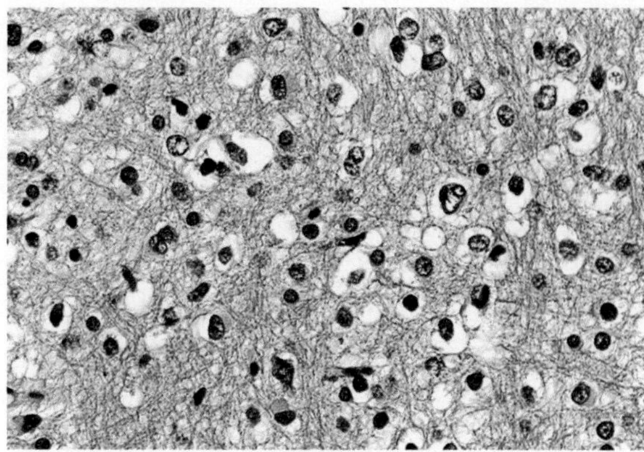

Figure 3-12

GRADE II DIFFUSE ASTROCYTOMA RESEMBLING OLIGODENDROGLIOMA

Perinuclear halos always raise the issue of oligodendroglioma or mixed glioma. The dark, somewhat angulated nuclei and absence of nucleoli are more characteristic of astrocytoma than oligodendroglioma.

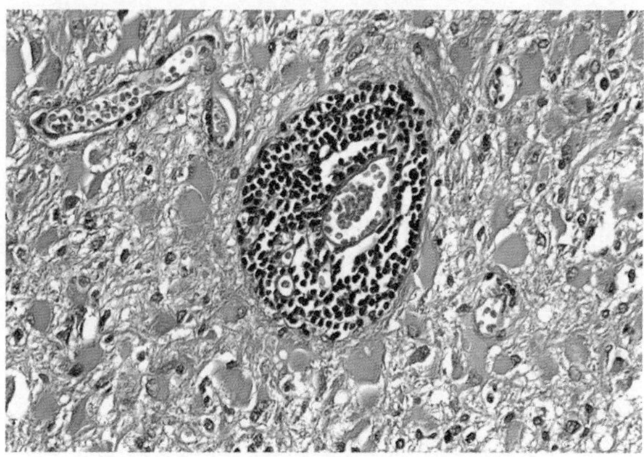

Figure 3-13

GEMISTOCYTIC ASTROCYTOMA

A perivascular cuff of lymphocytes or regions of diffuse astrocytoma in which gemistocytes predominate are common in gemistocytic astrocytoma.

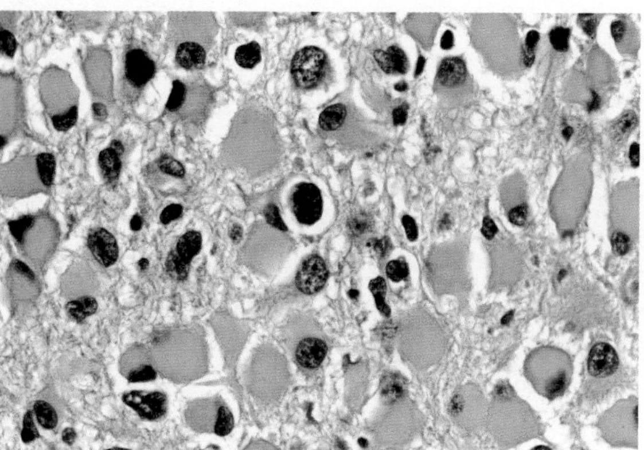

Figure 3-14

GEMISTOCYTIC ASTROCYTOMA

Cells with dense nuclei and little cytoplasm are widely scattered throughout the lesion.

(albeit subtle in some cases), fibrillar background, and absence of nucleoli (fig. 3-12).

Gemistocytes occur in varying numbers, and only a few astrocytomas have enough glassy cytoplasm to justify the term *gemistocytic astrocytoma* (figs. 3-13, 3-14). By definition, most cells in the gemistocytic variant of diffuse astrocytoma exhibit a plump, glassy cell body and an eccentric corona of short, stout-to-delicate processes. Aside from their population density, such cells, when scattered and well differentiated, can mimic reactive astrocytes. The principal distinguishing features of reactive astrocytes are their uniform distribution, minimal increase in number, and symmetric stellate configuration of long-radiating processes. Such cells are best seen in smear preparations or on immunostains for glial fibrillary acidic protein (GFAP).

Characteristically, the gemistocytic lesion contains scattered cells with larger, more hyperchromatic nuclei and scant, if any, visible cytoplasm (fig. 3-14). As is discussed below in Immunohistochemical Findings, this population appears to be the principal proliferating component of the tumor in light of its high MIB-1 index relative to that of gemistocytes (fig. 3-15). The appearance in many gemistocytic astrocytomas of small, poorly differentiated cells with scant cytoplasm, in concert with an increased mitotic rate or MIB-1 labeling index, indicates malignant transformation (see fig. 3-35). Most gemistocytic tumors, therefore, fall into the category of anaplastic astrocytoma, or into the highest grade, glioblastoma, when microvascular proliferation and/or necrosis are present. Although we recognize the potential of gemistocytic astrocytomas to behave aggressively (19,20), we do not automatically classify them as grade III (anaplastic) astrocytoma, although a comment about the biologic behavior of such lesions would then be appropriate.

The presence of microcystic spaces is a distinctive and diagnostically helpful feature of gliomas (fig. 3-16). These often round, fluid-filled

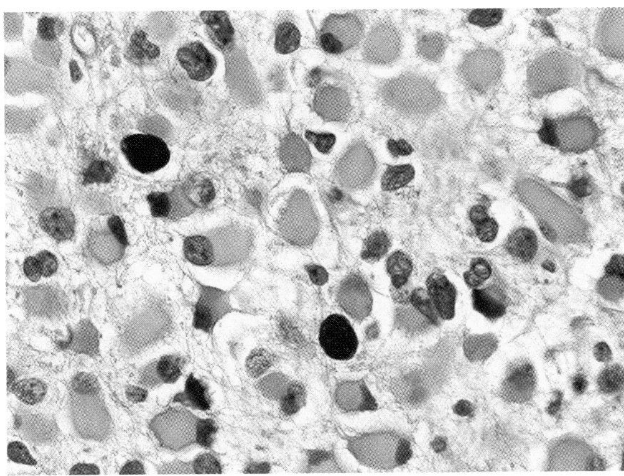

Figure 3-15

GEMISTOCYTIC ASTROCYTOMA

As inferred from the high MIB-1 index, cells with dark nuclei and scant cytoplasm are the lesion's principal proliferating component. Gemistocytes themselves have a low index.

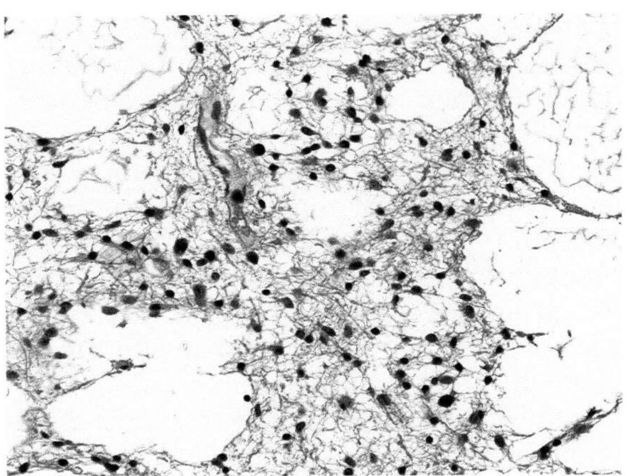

Figure 3-16

GRADE II DIFFUSE ASTROCYTOMA

Microcystic change is common in astrocytoma, but rare in gliosis.

microcavities characterize well-differentiated gliomas of both astrocytic or oligodendroglial type, and are only rarely found in reactive gliosis. Varying in size and contour, and frequently filled with proteinaceous material, they differ from the linear clefts produced by ice crystals formed during frozen section processing. Microcysts are seen typically in younger patients and, unless they are accompanied by clear evidence of malignant transformation, generally indicate a well-differentiated astrocytoma or oligodendroglioma. Microcysts may occur in either gray or white matter, but are more prominent in the former.

Small numbers of perivascular lymphocytes are commonly observed in astrocytomas, particularly gemistocytic variants (fig. 3-13).

The nuclei of astrocytoma cells vary in both type and degree of differentiation. Those in small, very well-differentiated tumors are uniform, nearly round to oval, and have little hyperchromasia or nuclear prominence. Aside from gemistocytic tumors, in which nuclei are often round, those in more typical astrocytomas are elongate, irregular in contour, and hyperchromatic. Prominent nucleoli are uncommon in diffuse astrocytomas, but small nucleoli are often present in gemistocytic tumors.

Almost by definition, mitoses are rare in grade II astrocytoma and, by the strict definition of the original St. Anne-Mayo system, none are permitted. There has been, however, reluctance to elevate the grade of an astrocytoma based on the finding of a solitary mitosis, especially when the specimen is sizable and otherwise typical of grade II in terms of cellularity, degree of cytologic atypia, and abundance of microcysts. Several large studies have confirmed that grade III astrocytomas, diagnosed as such by the original St. Anne-Mayo system on the basis of a solitary mitosis, behave much like grade II astrocytomas without mitoses (28). Another study found a relationship between the outcome and the number of microscopic fields required to find a mitosis (43).

The WHO system is more tolerant of mitoses, accepting "marked mitotic activity" for grade III lesions. What is considered marked is not specified, however, thus leaving considerable uncertainty as to how grade III (anaplastic) astrocytomas are defined (23,24). At present, some pathologists lean toward the original St. Anne-Mayo system and see any mitotic activity as evidence of grade III, a lesion that, almost by definition, is treated with radiotherapy. In a sizable specimen, we prefer to see several mitoses. On the other hand, a solitary mitosis in a small stereotactic biopsy specimen can be used to justify a grade III designation in the appropriate cytologic context.

Since well-differentiated astrocytomas occasionally have coagulative necrosis, the presence of necrosis alone, in the absence of other features of atypia or anaplasia, is not diagnostic of glioblastoma. The issue is especially problematic, however, when necrosis, but without pseudopalisading or other features of a high-grade lesion, occurs in a gemistocytic astrocytoma. The presence of necrosis in a well-differentiated neoplasm, particularly one devoid of mitoses, should prompt reconsideration of the diagnosis, exclusion of other forms of glioma, and review of the patient's history in an effort to explain the finding. For example, prior radiotherapy will produce multiple small foci of parenchymal necrosis. Tumoral necrosis may also be seen in pilocytic astrocytoma and low-grade ependymoma, in which cases it is ischemic, "infarct-like" necrosis of no prognostic significance. It even occurs, occasionally, in an otherwise typical grade II astrocytoma.

At any point in tumor progression, anaplastic transformation of a grade II astrocytoma results in the emergence of clones of cells more prone to infiltration of both cortex and white matter fiber pathways. At this juncture, although cellularity may be low and mitoses hard to find, nuclei tend to be smaller, denser, and elongate. Labeling indices are typically high in these diagnostically challenging lesions.

Cartilage is a rare feature of astrocytoma (44). An occasional diffuse astrocytoma, usually a gemistocytic variant, may feature Rosenthal fibers in some number. These eosinophilic hyaline structures are much more commonly associated with pilocytic astrocytomas and are, accordingly, discussed in detail in that section.

Rare diffuse astrocytomas of grade II or III contain small, circular, synaptophysin-positive regions of "neuropil" surrounded by neurocytes or small ganglion cells. This is described in chapter 4.

Frozen Section. Although the typical features of diffuse astrocytoma grade II are usually obvious in both permanent and frozen section specimens, variations in section thickness and staining characteristics often complicate interpretation in the latter. The presence of ice crystals in edematous white matter compounds the problem, since these clefts compress intervening tissue and produce artifactual increases in cell density (fig. 3-17). The problem is especially acute in small specimens of the sort obtained by stereotactic needle biopsy. In such cases, adequate samples of nonfrozen tissue must be reserved for optimal processing. Every effort should be made to avoid freezing an entire specimen since the process distorts the nuclei of normal astrocytes and oligodendrocytes. In both frozen and permanent sections, the nuclei have contorted, hyperchromatic profiles, resembling those of astrocytoma. Furthermore, even if an astrocytoma seems apparent, it is nonetheless wise to make only a more general diagnosis of "low-grade infiltrating glioma" since it is difficult, and rarely if ever necessary, to distinguish astrocytoma from oligodendroglioma in frozen sections alone.

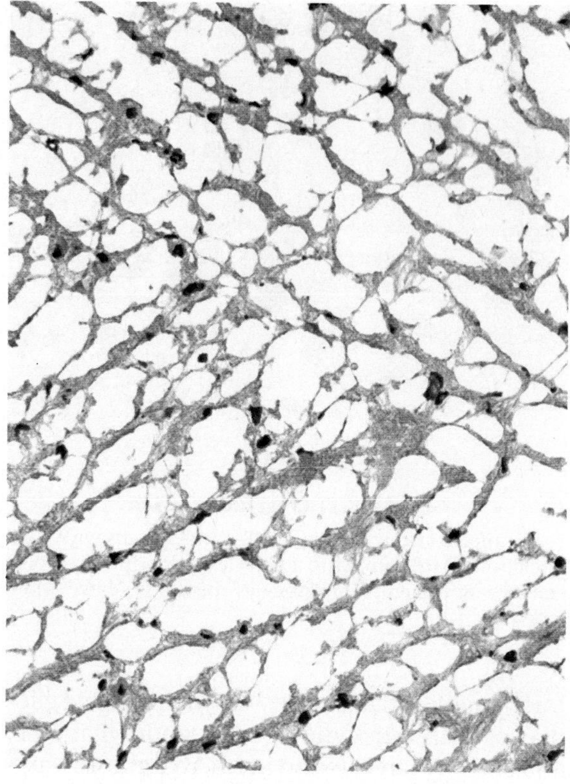

Figure 3-17

GRADE II DIFFUSE ASTROCYTOMA: FROZEN SECTION

Extensive ice crystals often confound intraoperative interpretation of astrocytoma infiltrating edematous white matter.

A firm intraoperative diagnosis of infiltrative astrocytoma requires an unequivocal increase in cellularity, irregular distribution of glial cells, nuclear atypia, and, in some instances, the presence of microcysts. If only slight hypercellularity and vague nuclear abnormalities are

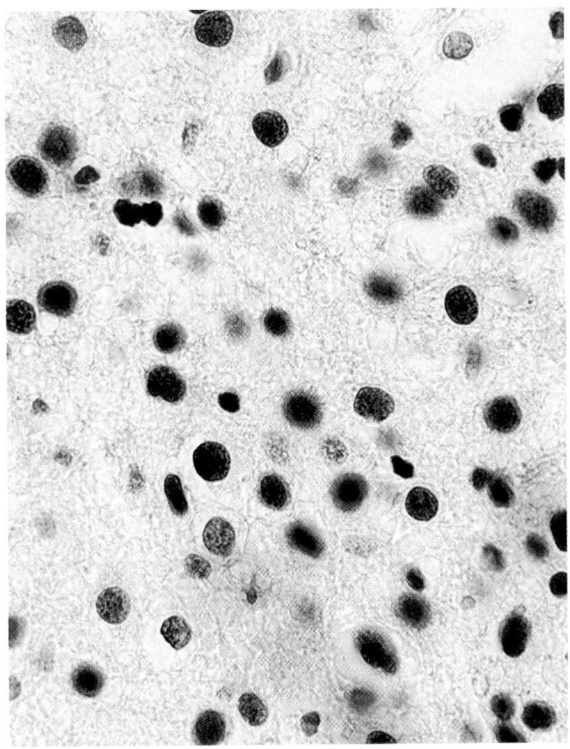

Figure 3-18

GRADE II DIFFUSE ASTROCYTOMA

Some grade II diffuse astrocytomas are strongly and diffusely immunoreactive for p53 protein. Such staining is rare in oligodendrogliomas, where mutations of *p53* are uncommon.

seen, additional tissue should be requested and the diagnosis deferred to permanent sections.

Needle Biopsy Specimens. Since needle biopsy specimens of astrocytoma can be exceedingly difficult to interpret, a definitive diagnosis may have to await permanent sections. Trochar biopsies 1 cm in length and 2 cm in diameter are often preferable to smaller, partly crushed specimens obtained from cup style forceps or the fragments aspirated through thin needles (fig. 3-10).

Immunohistochemical Findings. Virtually all diffuse astrocytomas are S-100 protein reactive, with staining that is cytoplasmic and often also nuclear. The cytoplasm often is immunopositive for GFAP, particularly in areas of medium cellularity, whereas a negative reaction is the rule in hypocellular regions containing naked nuclei. In gemistocytic tumors, GFAP reactivity is usually prominent at the periphery of the cells whereas the central region is largely negative. Reactivity for vimentin is almost always seen as it is in many other cell types. We use it mainly for confirming the antigenic viability of a specimen. Wide-spectrum keratin reactivity is common in tumor cells, and particularly in reactive astrocytes (45). Little if any reactivity occurs in the latter with antibodies such as CAM5.2. Staining for p53 protein is variable in diffuse astrocytomas; nuclei are diffusely immunoreactive in about a third (fig. 3-18). There are only a few, if any, cells positive for p53 in most cases. A higher percentage is present in gemistocyte-rich lesions (20).

Reported MIB-1 labeling indices of astrocytoma grade II have generally been less than 2 percent, often even 1 percent or less, but there is considerable case-to-case and region-to-region variation and little information about interinstitutional variation in the sensitivity of the technique (28,46–48). One study found the mean, median, and upper range of MIB-1 as 2.3, 2.0, and 7.6 percent, respectively (28). A labeling index greater than 1.5 percent was a poor prognostic sign in one study (46), whereas another investigation found a separation at 2 percent (47). In gemistocytic variants, the oft-present cells with dark nuclei and little surrounding cytoplasm have a high MIB-1 index; since this index is low in the predominant gemistocytes, the overall rate is often well within the grade II range.

Ultrastructural Findings. The cells of diffuse astrocytoma vary greatly in size, configuration, content of organelles, and fibrillarity. Most show evidence of differentiation, particularly by the presence of intermediate filaments both in perinuclear cytoplasm and in processes. Such filaments vary in distribution from parallel bundles to skeins or tight whorls. Gemistocytic astrocytes are distinctive in their relative abundance of subplasmalemmal filaments, as well as filaments enmeshed among organelles at the center of the cell body (49–51). In addition to processes, filopodia may be present. Randomly distributed, often poorly formed intercellular junctions may be encountered, but well-formed desmosomes are lacking.

Cytologic Findings. Cytologic preparations, preferably of smear type, are invaluable diagnostic aids. Astrocytomas vary considerably in appearance, depending not only upon their histologic subtype but also upon cellularity. When isolated cells are present in otherwise intact parenchyma, the nuclei, even in smears, usually appear naked, in that they lack apparent cytoplasm. In contrast, when smears are made of cellular tumors, the often-aggregated cells have elongate cytoplasmic processes with variable fibrillarity (fig. 3-19). The feltwork of glial processes that characterizes cellular tumors is more readily seen in cytologic preparations than in frozen or routine histologic sections. These cellular extensions are often delicate and most cannot be related to a particular cell. On the other hand, cytologic preparations of gliotic tissue exhibit classic reactive astrocytes with long symmetrical processes radiating from their cell bodies (see fig. 19-9).

The nuclei of astrocytoma cells are generally more angulated and hyperchromatic than those of oligodendroglioma cells, although differentiating between these two gliomas may be no easier in cytologic smears than in tissue sections. A smear of normal white matter is illustrated in figure 2-33.

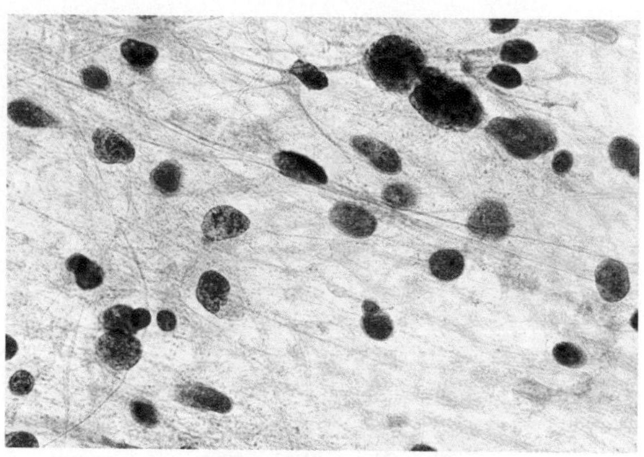

Figure 3-19

GRADE II DIFFUSE ASTROCYTOMA: SMEAR PREPARATION

Mild nuclear pleomorphism and fine cytoplasmic processes are typical of diffuse astrocytomas. Nuclei are more irregular and nucleoli less prominent than in oligodendrogliomas.

Molecular and Cytogenetic Findings. There is no specific molecular profile for grade II astrocytoma, although *p53* mutations are not uncommon, particularly in gemistocytic lesions (20,52,53). The relationship of *p53* mutations and immunostaining to prognosis is discussed in the section Prognosis, below. One study found mutations of *p53* in all of 11 astrocytomas with more than 5 percent gemistocytes that progressed to grade III astrocytoma but in 61 percent of astrocytomas with less than 5 percent gemistocytes (20). While gemistocytes have a low MIB-1 index, as discussed above, they are neoplastic as evidenced by *p53* mutations identical to those of nongemistocytic cells (54).

Cells in some astrocytomas gain a portion of chromosome 7, usually all or part of the long arm (55,56). This may be an event in tumor progression since patients harboring this gain, often as trisomy, do less well than those that do not (56). Loss of chromosome 19q is not uncommon; a focal loss at 19.q13.3 is sometimes present (57).

Differential Diagnosis. The diagnosis of grade II astrocytoma can be difficult and all available information should be used. This begins with awareness of clinical data and the results of neuroradiologic imaging. If the radiographic appearance of a lesion differs significantly from the typical astrocytoma described under Radiologic Findings, above, the diagnosis should be questioned. The diagnosis would be suspect, for example, in the face of contrast enhancement as is outlined in Appendix N. Appendix L summarizes non-neoplastic CNS lesions that can be misinterpreted as neoplasms, and Appendix K lists CNS tumors that can be overgraded. Since nearly all astrocytomas, to some extent, displace surrounding structures, caution is also in order in the absence of mass effect, although some sizable astrocytomas have little or no mass effect. The diagnosis should be questioned if the lesion is contracting, no matter what the histologic or cytologic appearance. The presence of diffuse contrast enhancement or a cyst-mural nodule architecture is also atypical of diffuse astrocytoma, and should generate skepticism regarding the diagnosis. A contrast-enhancing ring virtually rules out the diagnosis of grade II diffuse astrocytoma, since the pattern is more characteristic of lesions such as glioblastoma, some pilocytic astrocytomas, abscesses, metastases, infarcts, or demyelinating disease. Even focal contrast enhancement in an infiltrative astrocytoma is strong presumptive evidence of malignant transformation, unless it can be explained by prior radiotherapy.

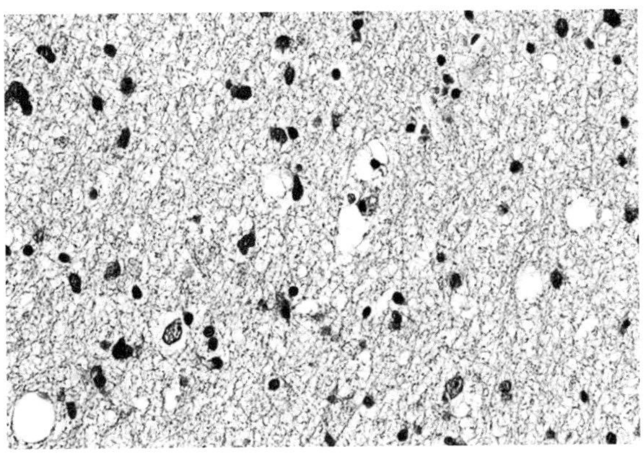

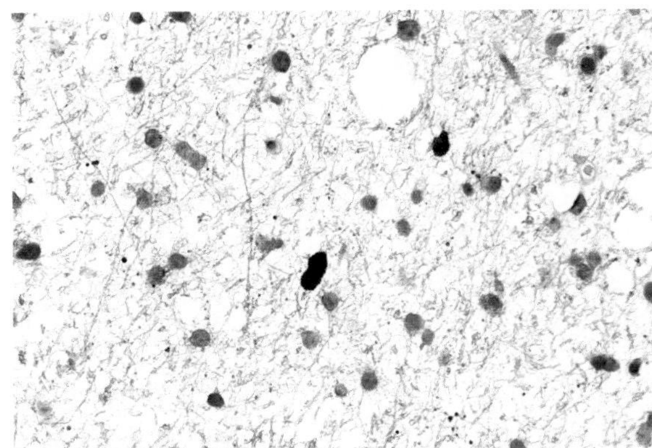

Figure 3-20

POSSIBLE GRADE II DIFFUSE ASTROCYTOMA

Left: There may be only a subtle increase in cellularity produced by atypical cells with scant cytoplasm.
Right: While not diagnostic, immunopositivity for MIB-1 in diffuse atypical cells supports the diagnosis of astrocytoma.

Radiologic findings are extremely important in the assessment of spinal cord lesions where specimens are notoriously small. The typical diffuse astrocytoma is bright on T2-weighted images and lacks contrast enhancement. Enhancement in an intramedullary astrocytic tumor of the diffuse type almost invariably indicates that it is of high grade (anaplastic astrocytoma or glioblastoma). It does not have this association in a discrete neoplasm such as pilocytic astrocytoma, ependymoma, hemangioblastoma, and ganglioglioma, where contrast enhancement is routine.

With radiologic data in hand, histologic examination can immediately focus on cellularity as an indicator of the presence of an abnormality. Caution is appropriate, however, when the sole abnormality is uniform hypercellularity in the absence of atypia or prominent cytoplasm, and consideration must be given to the possibility that the slide simply shows a thickly cut section of normal brain (see fig. 2-32). In some cases, even in sections of normal thickness, oligodendrocytes seem increased in number along vessels and about neurons (see fig. 2-30). These mimic the secondary structures so common in infiltrative gliomas of astrocytic and oligodendroglial type.

One of the most vexing challenges is the evaluation of specimens, usually small, in which there are widely scattered, mildly atypical cells, a few of which are MIB-1 positive (fig. 3-20). Such tissue is often seen on MRI as a T2-weighted or FLAIR bright image that is sometimes expansile and for which there is no logical alternative to infiltrating glioma. Although this scenario is highly suspicious for a glioma, we require a diagnosis based on histologic and cytologic features, not MIB-1 staining alone. Mitoses strengthen the case for a glioma, but are not conclusive. Molecular studies are likely to contribute to this issue. In situ hybridization can be used in tissue sections to detect gains of chromosome 7 (58), although the finding may suggest tumor progression to grade III since patients with such gains do less well than those without (56).

The specificity of immunohistochemistry for p53 protein is not clear. One study found frequent staining of normal and reactive cells (59), but another found reactive cells to be negative, outside of progressive multifocal leukoencephalopathy in which virally infected cells are strongly positive (60). Immunostaining for p53 is now widely employed, with the feeling that diffuse, strong staining of cytologically suspicious infiltrating cells lends support to the diagnosis of astrocytoma.

If the lesion is clearly so abnormal that the possibility of astrocytoma comes immediately to mind, the search can proceed for nuclear atypia, microcysts, and secondary structures such as cortical involvement with satellitosis and subpial tumor cell accumulation. The presence of any one of these features, even in the setting of minimal hypercellularity, clearly indicates an infiltrating glioma.

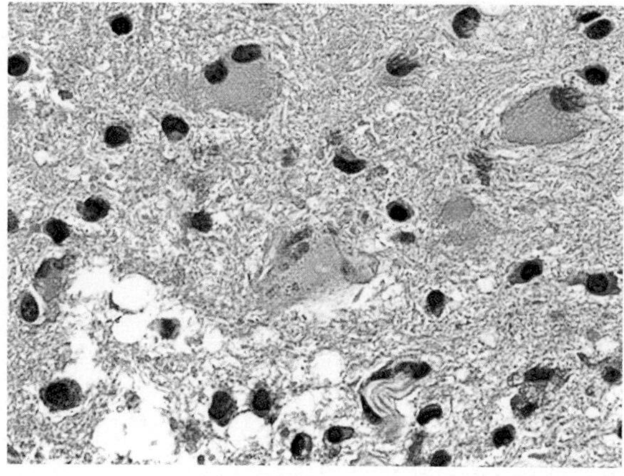

Figure 3-21

FIBRILLARY GLIOSIS

The amount of cytoplasm, but not notably the number of cells, increases in active gliosis. The multinucleated Creutzfeldt cell near the center of the illustration is commonly seen in reactive lesions.

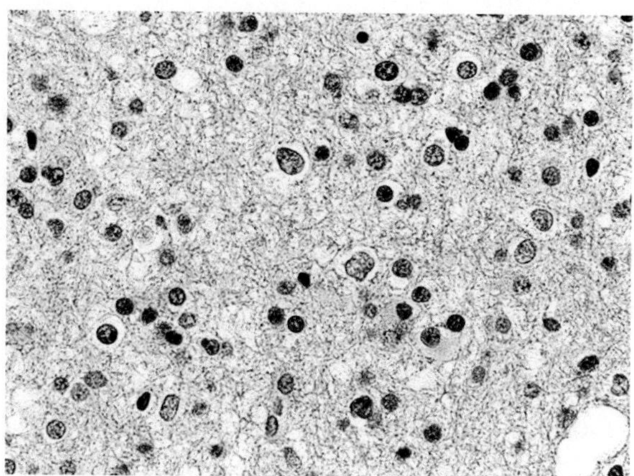

Figure 3-22

GRADE II DIFFUSE ASTROCYTOMA

Diffuse astrocytomas without glassy cytoplasm are recognized by an increase in cell density and the degree of cytologic atypia.

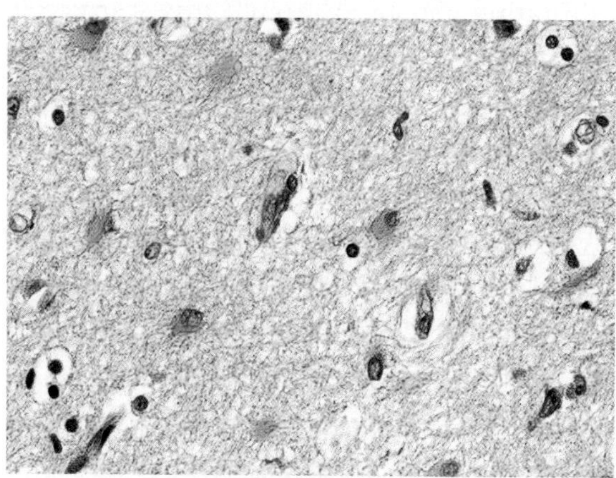

Figure 3-23

FIBRILLARY GLIOSIS

Reactive astrocytes, here responding to a neighboring focus of toxoplasmosis, are hypertrophic, but only marginally hyperplastic.

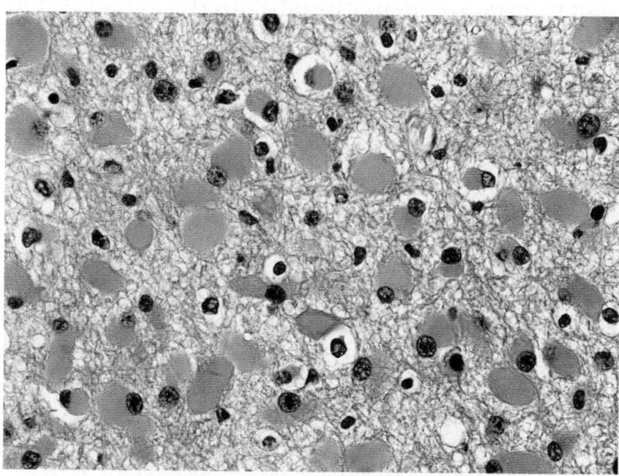

Figure 3-24

GEMISTOCYTIC ASTROCYTOMA

Neoplastic gemistocytes are increased in number and have more nuclear atypia than reactive gemistocytes.

Since reactive gliosis does not produce a substantial increase in cellularity (figs. 3-21, 3-22), its histologic distinction from well-differentiated astrocytoma generally poses little difficulty, although the distinction between paucicellular gemistocytic astrocytoma and gliosis can be difficult, if not impossible (figs. 3-23, 3-24). In all but the most florid examples, reactive cells are noted more for nuclear uniformity and cellular hypertrophy than for hyperplasia. Uniform cell distribution, prominence of cytoplasm, and stellate symmetry of elongated processes are characteristic. The cells are thus individually conspicuous, sometimes even alarming, but are not substantially increased in number. Furthermore, the cytoplasm of reactive astrocytes is prominent in contrast to the usually limited cytoplasm of all but gemistocytic neoplastic astrocytes. Most astrocytomas, even if not patently "astrocytic" in appearance, are too cellular to suggest gliosis;

gliosis, although obviously astrocytic, is not particularly cellular. An apparent exception is found near vascular malformations, where, in addition to prominent astrocytes, the presence of histiocytes and compaction of tissue results in definite hypercellularity.

The possibility of non-neoplastic conditions must always be considered in the differential diagnosis of diffuse astrocytoma. Of special importance are macrophage-rich lesions, such as demyelinating diseases (fig. 3-25), which can histologically mimic astrocytoma, oligodendroglioma, and mixed oligoastrocytoma, sometimes very closely. Demyelinating disease is discussed in detail in chapter 19. The algorithm for evaluation of intracranial and intraspinal lesions given in Appendix O encourages pathologists to give full consideration to reactive lesions listed in Appendix L.

At low-power magnification, the possibility of demyelinating disease is suggested by the presence of perivascular cuffs of lymphocytes, a prominent feature of the acute and subacute phases of demyelination. Essential to the diagnosis of demyelinating disease is recognizing numerous foamy macrophages. These cytologically bland cells abound in the parenchyma and, as they exit with their burden of phagocytized myelin debris, accumulate two to three deep in perivascular spaces where they wait to enter the vasculature. Such round phagocytes have discrete cell borders, uniform spherical nuclei, vacuolated cytoplasmic content of periodic acid–Schiff (PAS)-positive myelin debris, and immunoreactivity for macrophage markers. Fully developed macrophages, i.e., engorged foam cells, are infrequent in brain tumors, even malignant examples, although surprisingly many activated microglia, and lesser numbers of histiocytes, may be seen with macrophage markers. Perivascular cuffing by foamy macrophages is unusual in an astrocytoma and should prompt consideration of a reactive process.

In the clinical setting of immunosuppression, most commonly in patients with acquired immunodeficiency syndrome (AIDS), progressive multifocal leukoencephalopathy must be included in the differential diagnosis. This demyelinating process of viral etiology is characterized by the presence of pleomorphic astrocytes that appear neoplastic due to their bizarre,

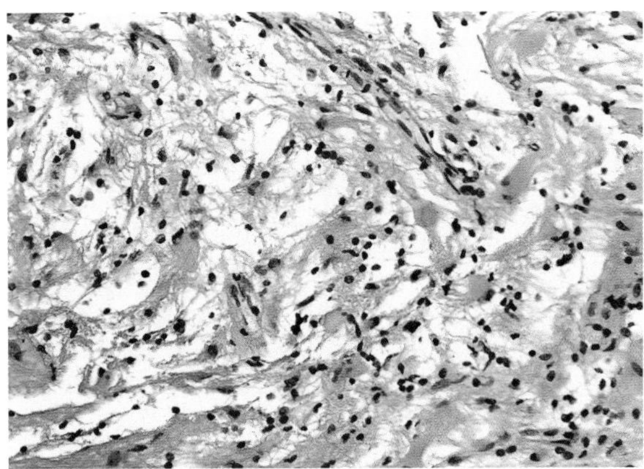

Figure 3-25

DEMYELINATING DISEASE MIMICKING ASTROCYTOMA

Demyelinating disease can closely resemble astrocytoma, when macrophages are inconspicuous.

markedly hyperchromatic nuclei (see figs. 19-23, 19-24). The reactive nature of the lesion is recognized by the abundance of foamy macrophages, or "gitter cells," and the purple, "ground-glass" textured viral inclusions within the nuclei of oligodendrocytes. These enlarged cells are most evident at the margin of the lesion. Cerebral infarct, an additional macrophage-rich lesion potentially misinterpreted as astrocytoma, is discussed in chapter 19.

When reactive conditions are excluded and the presence of neoplasm is established, the two principal diagnostic considerations become oligodendroglioma and diffusely infiltrating astrocytic tumors of higher grade. Oligodendroglioma deserves principal consideration since, like astrocytoma, it is infiltrating, usually supratentorial, and favors patients in the third and fourth decades. It poses no diagnostic difficulty when histologically classic, but when less cellular, less haloed, or more fibril-rich (see below) may be difficult to distinguish from a well-differentiated astrocytoma (fig. 3-26). In recent years, formerly strict diagnostic criteria for oligodendroglioma have become lax, and many astrocytomas with little more than focal round nuclei are classified as oligodendroglioma or "mixed glioma" (61). In general, even low-grade, well-differentiated astrocytomas have greater nuclear pleomorphism and more prominent cytoplasm and process formation than oligodendrogliomas. Astrocytomas usually lack

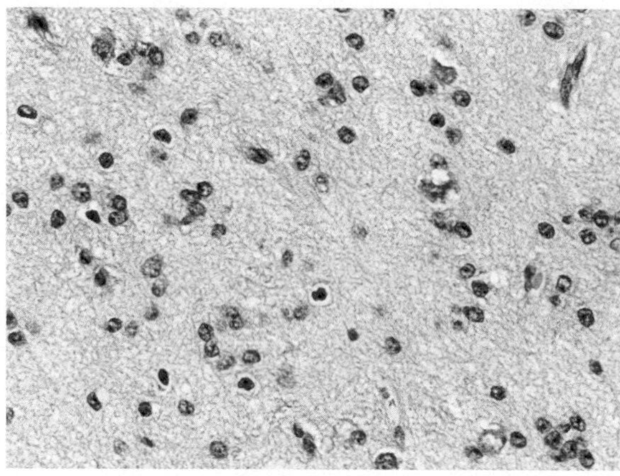

Figure 3-26

OLIGODENDROGLIOMA RESEMBLING DIFFUSE ASTROCYTOMA

The distinction between oligodendroglioma and astrocytoma can be difficult when perinuclear halos are lacking. Elsewhere, this lesion had the classic histologic features of oligodendroglioma. The small, but clearly defined nucleoli are more consistent with oligodendroglioma than astrocytoma. Codeletion of chromosomes 1p and 19q was present.

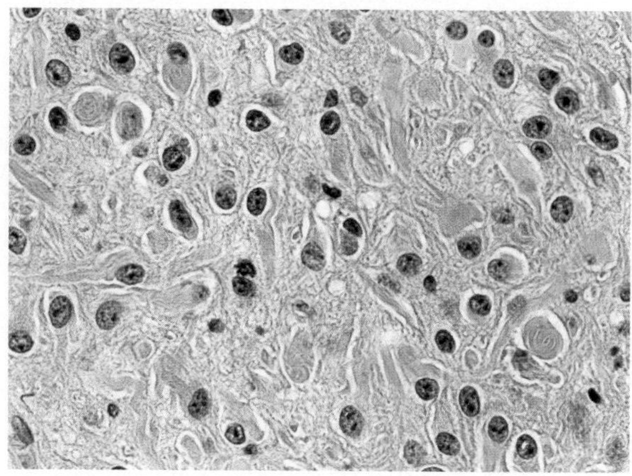

Figure 3-27

OLIGODENDROGLIOMA RESEMBLING DIFFUSE ASTROCYTOMA

Highly fibrillated cells in oligodendrogliomas can have "astrocytic" cytoplasmic features. The round nuclei with defined nucleoli, however, are more characteristic of oligodendrocytes, as are intracytoplasmic whorls of filaments.

the calcification and the degree of cortical involvement and perineuronal satellitosis that characterize oligodendrogliomas.

As is discussed in the sections on oligodendroglioma and mixed gliomas, some oligodendrogliomas include cells with eccentric, eosinophilic, GFAP-positive cell bodies ("minigemistocytes") or those with round, oligodendroglial-like nuclei and apical caps of fibrillated, GFAP-positive cytoplasm ("gliofibrillary oligodendrocytes") that lend a decidedly astrocytic appearance (fig. 3-27). Such cells vary in number and often merge imperceptively with ones more obviously astrocytic, thus causing great uncertainty as to which tumors are astrocytic, oligodendroglial, or "mixed."

Some well-differentiated gliomas consist of cells that, by straddling morphologic boundaries, resemble astrocytic-oligodendroglial hybrids. Such tumors feature rather uniform cell size and contain both cells with round nuclei and those that have the angulation and hyperchromatism more typical of astrocytoma. Perinuclear halos are present but restrained in degree. The increasingly employed molecular genetic profiling so useful to the identification oligodendrogliomas and their distinction from astrocytoma is appropriate in such cases.

It is extremely important to distinguish diffuse astrocytomas from pilocytic astrocytomas. The former, even when well differentiated, are diffusely invasive and potentially aggressive, with the ability to infiltrate tissue and to transform into a grade III or IV lesion. The latter, on the other hand, are relatively discrete, and often amenable to excision. Even if not, they are slowly growing and are associated with a far more favorable prognosis in adults, although in children there may not be such a dramatic difference in outcome (28,62).

Pilocytic lesions should be suspected on clinical and radiographic grounds alone, given their general but not invariable occurrence in young patients, radiologic circumscription and contrast enhancement, and frequent cyst-mural nodule architecture. A salient microscopic feature is pattern variation, which is often biphasic, with densely compacted, elongated, or piloid cells associated with Rosenthal fibers alternating with loose knit, microcyst-rich tissue accompanied by eosinophilic granular bodies (EGBs). The significance of Rosenthal fibers and EGBs cannot be overemphasized since they are rare in diffuse astrocytomas. These and other histologic features that suggest low-grade or non-neoplastic lesions are summarized in Appendix M. Compared to diffuse astrocytoma,

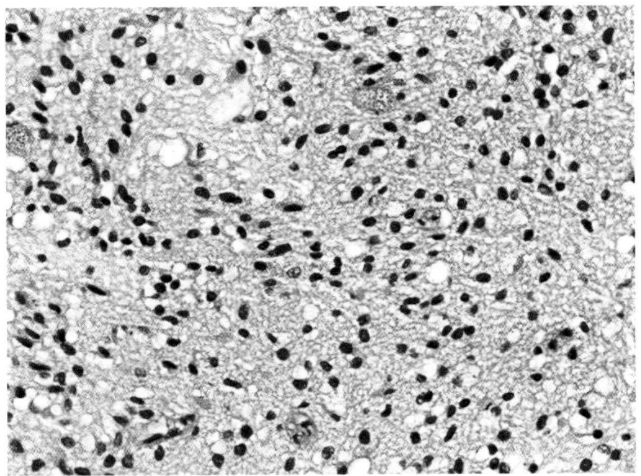

Figure 3-28

"DIFFUSE" PILOCYTIC ASTROCYTOMA RESEMBLING DIFFUSE ASTROCYTOMA

Some pilocytic astrocytomas, particularly of the cerebellum, have an infiltrative quality that, at least focally, resembles the growth pattern of diffuse astrocytoma. The uniformly spaced, round, cytologically bland nuclei are more consistent with pilocytic astrocytoma.

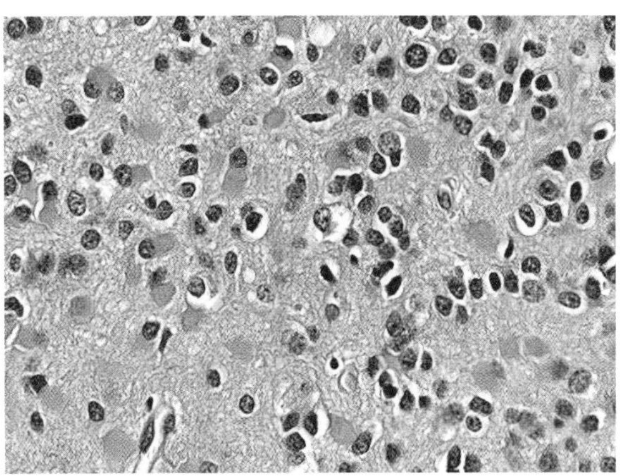

Figure 3-29

EPENDYMOMA RESEMBLING DIFFUSE ASTROCYTOMA

Out of context, highly fibrillar areas with small gemistocytes can be misinterpreted as evidence of diffuse astrocytoma.

pilocytic lesions are more solid and considerably less invasive of brain parenchyma, as seen at high magnification. Nevertheless, the two tumors can exhibit considerable morphologic overlap (fig. 3-28), particularly in the cerebellum where some pilocytic lesions, at least at the periphery, incorporate host tissue.

Occasional ependymomas consist in large part of fibrillar cells that phenotypically resemble those of diffuse astrocytoma (fig. 3-29, see figs. 3-252, 3-253). The presence of vague perivascular pseudorosettes or of obvious ependymoma in other parts of such lesions is sufficient justification for a diagnosis of ependymoma.

The distinction of diffuse astrocytoma (grade II) from anaplastic astrocytoma (grade III) may pose a problem, depending largely upon the grading criteria applied and the size of the specimen. Increases in cellularity, nuclear hyperchromasia, pleomorphism, and mitotic activity, and elevated MIB-1 labeling indices all characterize anaplastic astrocytoma.

Some astrocytomas contain many cells with large, hyperchromatic, obviously hyperploid nuclei that suggest anaplasia, but are paucicellular and devoid of mitotic activity. Such tumors may occur in younger patients as low-density, noncontrast-enhancing lesions. Despite the worrisome cytologic features, the lesions are generally considered grade II.

At times, there is such a disparity between low cellularity and bizarre cytologic features that it is appropriate to communicate the quandary to clinicians, who must then make therapeutic decisions based on integrated clinical, radiologic, and pathologic data. Such worrisome lesions can undergo rapid evolution or exhibit foci of radiographic contrast enhancement, features which, when coupled with histologic data, strongly suggest a diagnosis of high-grade glioma. If these cells are recognized during surgery, the surgeon can take additional samples.

Treatment and Prognosis. Although well-differentiated diffuse astrocytomas appear indolent when compared to the overtly malignant glioblastoma and the somewhat less aggressive anaplastic astrocytoma, most patients with such astrocytomas of the cerebrum are dead within 10 years. The median survival times in two sizable series were approximately 8 years (63) and 7 years (64), but was lower, 4 years, in another (65). The median 5-year survival period was 5.6 years in one large population-based study (66, 67). In this same study, the median survival period for patients with gemistocytic lesions was significantly shorter than that for those with fibrillary astrocytomas, 3.8 years versus 5.9 years. At present, the precise frequency of progression of astrocytoma to a tumor of higher grade is unknown, but there is at least a strong tendency for this transformation, particularly in gemistocytic variants (63–65).

As might be expected, given the close association between age and prognosis in patients with diffuse astrocytic neoplasms, younger individuals with well-differentiated astrocytomas do considerably better than older patients (64). Progression to high-grade malignancy is also more likely in the latter, and occurs more precipitously. Complicating therapeutic decision making is the fact that astrocytomas occurring in early life (through the second decade) may remain stable for years, showing neither clinical nor radiographic evidence of progression. In two studies, tumor-related deaths occurred only in association with high-grade progression, not among patients with only persistent grade II disease (64,63).

The precise role of surgery in the management of cerebral hemispheric astrocytomas occurring in both children and adults remains to be established. Some have a radiologic and macroscopic (fig. 3-7) appearance suggesting surgery may be possible. Rejecting the generally accepted notion that grade II astrocytomas are so "diffuse" as to preclude resection, some surgeons attempt the feat, if possible with a peritumoral margin of radiologically normal brain on T2-weighted MRIs. Longer follow-up periods will be necessary to test the validity of this approach in adults (68). In children, complete resection appears to increase progression-free and overall survival rates, and in one series such patients had only slightly inferior overall survival rates compared to those with pilocytic astrocytomas (69). Gain of chromosome 7 appears to be associated with a shorter survival period (56).

The role of postoperative radiotherapy in the treatment of diffuse astrocytoma of the cerebral hemisphere is much debated. When the tumor has been totally excised by radiologic criteria, treatment is often held in reserve, being administered upon recurrence or at such time as malignant transformation is apparent as foci of contrast enhancement. Radiation is administered with less hesitation when residual disease is apparent radiologically, particularly if the patient remains symptomatic. The benefit of radiotherapy is not clear, but one randomized study found that irradiation did not increase overall survival (estimated 5-year survival rate, 66 percent irradiated versus 63 percent nonirradiated) but did prolong time to progression (70). Short, 18 months or less, survival time attends diffuse astrocytoma of the brain stem, even those that do not contrast enhance, despite aggressive radiotherapy and chemotherapy.

There are few long-term studies of spinal cord astrocytomas in which a clear distinction is made between diffuse and pilocytic lesions. Some indicate a 5-year survival rate of less than 50 percent for patients with the diffuse variety, whereas a more favorable outlook is suggested in other studies (71,72).

The relationship of survival to molecular status is unclear, but two factors have been associated with a poorer outcome: *p53* gene mutation (almost always associated with immunoreactivity for p53 protein and especially common in gemistocytic variants) (53) and gain of chromosome 7 (56). Another study found unfavorable prognostic factors to be age greater than 50 years, gemistocytic subtype, and mutation in *p53* (73). Immunopositivity for p53 was less significant ($p = 0.064$).

Anaplastic Astrocytoma

Definition. *Anaplastic astrocytoma* is a grade III infiltrating astrocytoma, intermediate between diffuse astrocytoma grade II and glioblastoma grade IV.

Clinical Features. In accord with the statistical relationship between grade and patient age with the diffuse astrocytomas, anaplastic astrocytomas generally occur a decade later than grade II tumors and a decade earlier than glioblastomas. While cerebral hemispheric examples occur most often in the fifth decade, they also appear occasionally in children.

Radiologic Findings. Given its varying histologic definition, it is no surprise that a precise radiographic characterization of anaplastic astrocytoma has not been achieved. Some lesions partially enhance following the intravenous administration of contrast agents, but not with the perinecrotic ring or "rim" pattern that typifies glioblastoma (fig. 3-30). Use of double-dose contrast increases the frequency with which regions of enhancement are detected.

Gross Findings. Intraoperatively, anaplastic astrocytomas may appear as somewhat firm,

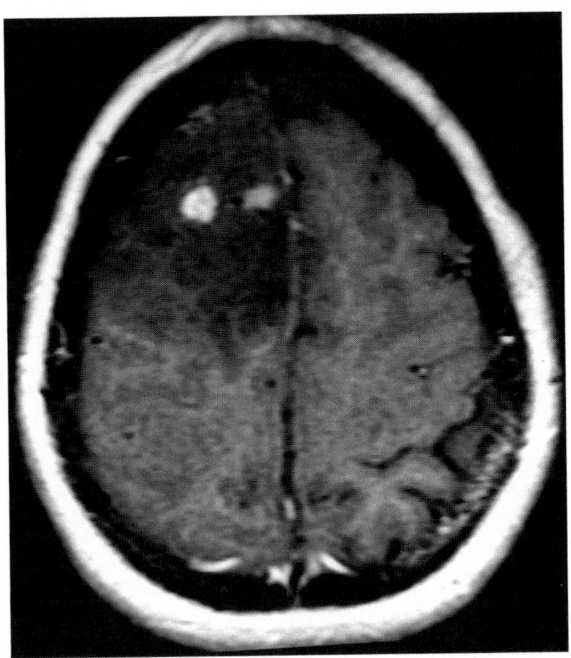

Figure 3-30

ANAPLASTIC ASTROCYTOMA

Anaplastic astrocytoma may or may not contrast enhance. A large area of dark signal representing grade II astrocytoma is present in the frontal lobe on the left. Higher-grade foci are apparent as the two white contrast-enhancing foci within this darker zone. (Fig. 3-1 from Fascicle 10, 3rd Series.)

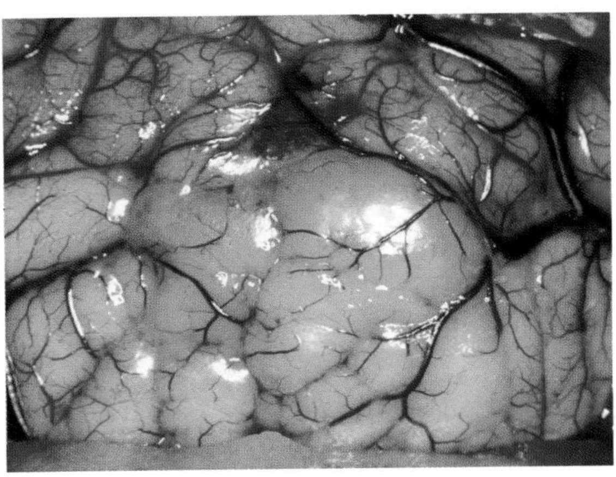

Figure 3-31

ANAPLASTIC ASTROCYTOMA

Anaplastic (grade III) astrocytomas are more likely to be expansile masses than their grade II counterparts. This lesion, in a 40-year-old man with a recent onset of seizures, infiltrates and expands the cerebral cortex.

discernible masses, but some are no different in appearance than grade II counterparts. When sufficiently cellular, their texture is soft. Cortical involvement results in gyral expansion and pallor (fig. 3-31).

Microscopic Findings. Although usually clearly astrocytic in nature, anaplastic astrocytoma surpasses grade II diffuse astrocytoma in terms of cellularity, nuclear pleomorphism, and hyperchromasia (figs. 3-32, 3-33), but lacks the marked cellularity, vascular proliferation, and necrosis of glioblastoma. Vessels may be prominent and tangles of well-formed capillaries lined by a delicate single layer of endothelium may be seen; multilayered vascular proliferation is not observed. Mitotic figures are encountered, albeit sometimes in small number. Microcysts may be seen in mitotically active anaplastic astrocytomas, although they are more common in grade II diffuse astrocytoma, pilocytic astrocytoma, and oligodendroglioma.

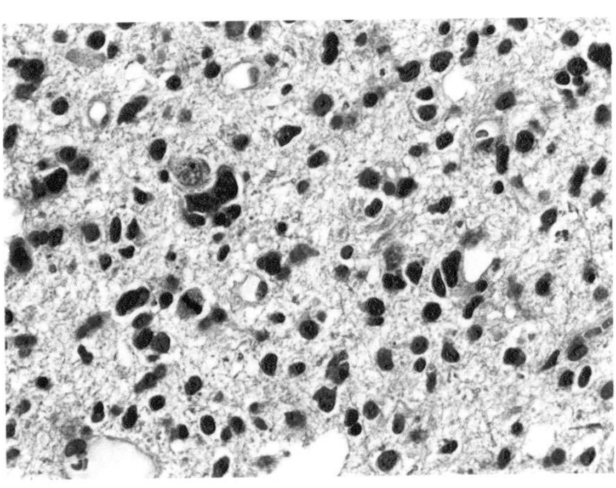

Figure 3-32

ANAPLASTIC ASTROCYTOMA

This degree of nuclear pleomorphism, even in a paucicellular tumor, suggests the possibility of a grade III process. Mitoses were present.

Some anaplastic astrocytomas have a gemistocytic component that is accompanied by dispersed or clustered small hyperchromatic cells. Although gemistocytic astrocytomas are not, to us, always grade III, a mitotically active, small cell component (fig. 3-34), or more overt anaplastic transformation (fig. 3-35), is clear evidence of a grade III designation. Gemistocytic astrocytomas of either grade are especially prone to anaplastic transformation.

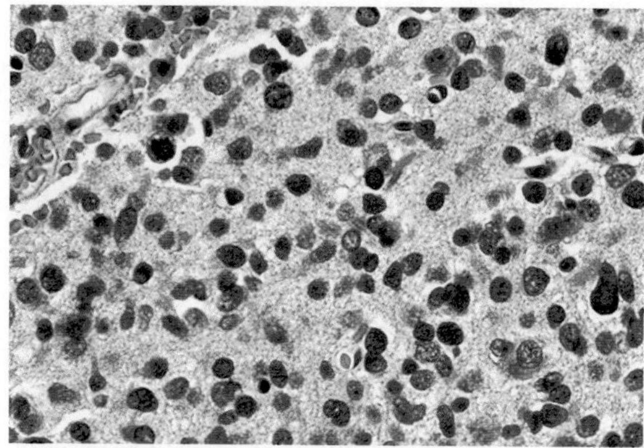

Figure 3-33

ANAPLASTIC ASTROCYTOMA

The illustrated medium cell density, degree of nuclear pleomorphism, and presence of mitotic activity are typical of grade III astrocytomas. Cytologic features typical of diffuse astrocytoma are nuclear pleomorphism, hyperchromasia, and only inconspicuous nucleoli. The lesion lacks the monotony of oligodendroglioma.

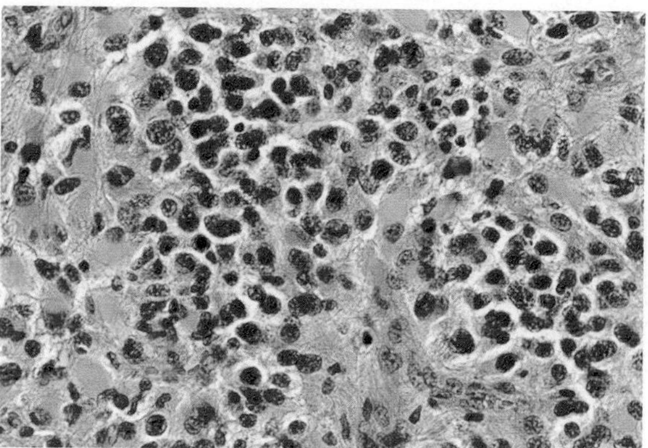

Figure 3-35

ANAPLASTIC GEMISTOCYTIC ASTROCYTOMA

Small undifferentiated cells show anaplastic transformation.

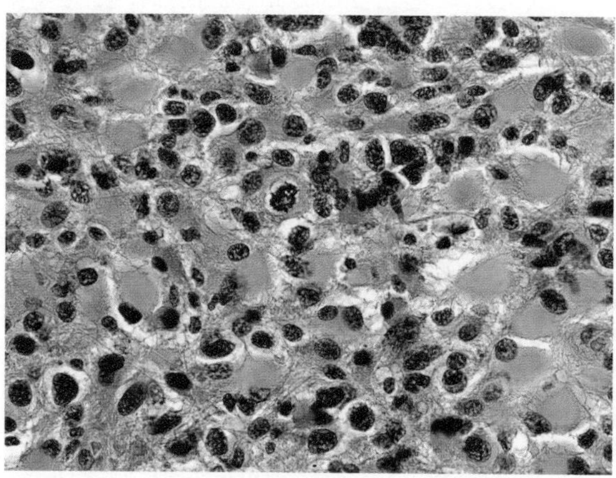

Figure 3-34

ANAPLASTIC GEMISTOCYTIC ASTROCYTOMA

Gemistocytic astrocytomas are often grade III on the basis of mitotic activity and a prominent component of undifferentiated cells.

The diagnosis of anaplastic astrocytoma is usually not difficult since most qualify on the basis of nuclear atypia and mitotic activity. In practice, particularly with small specimens, the diagnosis is sometimes reached, albeit cautiously, in the absence of mitoses, but in the face of high cellularity, overt nuclear atypia, and a high MIB-1 labeling index. This approach, although not a part of any published grading scheme, seems to us a practical solution to a common problem. Also problematic, and not uncommon in stereotactically obtained specimens, are paucicellular astrocytomas with a high MIB-1 index but no mitoses (fig. 3-36). A descriptive diagnosis is appropriate, but aggressive behavior should be anticipated. The extent to which molecular features can contribute is discussed in the section Differential Diagnosis.

Some cellular specimens with mitotic activity represent "near misses" of a glioblastoma, but others, even when studied in large specimens, are highly cellular, unequivocally anaplastic, mitotically active, and "small cell," yet lack necrosis both radiologically and histologically. Some are amplified for epidermal growth factor receptor (*EGFR*) and have loss on chromosome 10 as well, thus qualifying in molecular terms as glioblastoma (74).

Frozen Section. The appearance of anaplastic astrocytoma in smears is highly variable. In tumors of low or moderate cellularity, only nuclear atypia may be evident. Mitoses in such tumors are often hard to find. On the other hand, in cellular lesions, the cells appear loosely cohesive, their processes entwined in a fibrillar background, the sine qua non of astrocytic tumors. Nuclei are typically hyperchromatic, but their degree of pleomorphism is highly variable. Frozen sections of edematous white matter with infiltrating tumor cells are difficult to interpret because of the ice crystals that compress and distort the parenchyma. Cytologic features of tumors cells are difficult enough to evaluate in

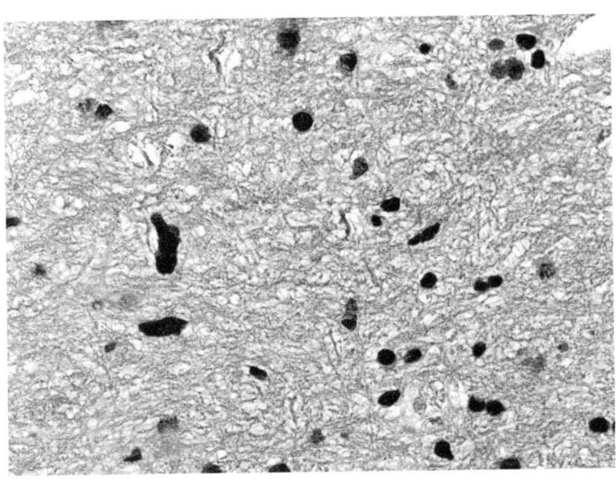

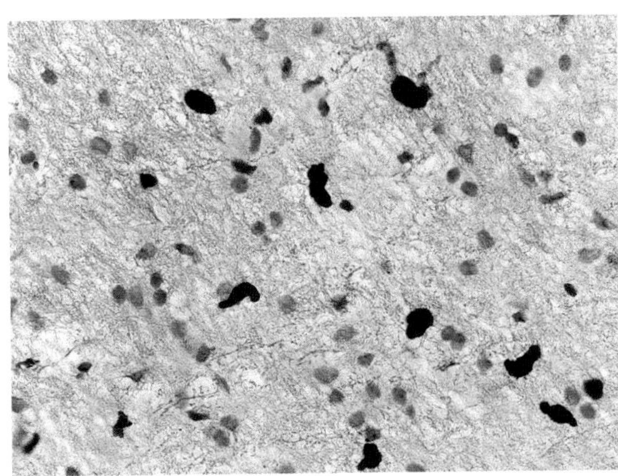

Figure 3-36

DIFFUSE ASTROCYTOMA POSSIBLY GRADE III

Left: Diffuse astrocytomas can be difficult to grade when composed of only small, widely spaced cells with substantial atypia.

Right: A high MIB-1 rate suggests that the lesion is, or will behave like, a grade III neoplasm.

frozen section specimens, but become even more obscure in this setting. They are much better preserved in gray matter.

Immunohistochemical Findings. The immunohistochemical features of anaplastic astrocytoma are the same as those of glioblastoma and grade II diffuse astrocytoma and include positivity for GFAP and S-100 protein. Reactivity for vimentin, sometimes impressive in degree, is of little diagnostic significance, but gives assurance that the specimen's antigenic viability has not been compromised by fixation and tissue processing. Although more of a pitfall in glioblastoma and gliosarcoma, reactivity for wide-spectrum keratin is regularly seen.

Not surprisingly, reported ranges of proliferation indices are quite variable and overlap those of glioblastoma, and at the low end, with grade II lesions. In one study, the mean, median, and range of the MIB-1 index were 6.0, 4.4, and 0.1 to 25.7 percent, respectively (28).

Cytologic Findings. Cytologic features vary with histologic grade. At the higher end are small elongate cells with bipolar processes that create a fibrillar background. In better-differentiated lesions, nuclei are often larger and more irregular in size and shape than those in grade II astrocytomas. Although present almost by definition in grade III lesions, mitoses may be difficult to find in smear preparations.

Differential Diagnosis. In most cases, anaplastic astrocytomas are sufficiently cellular to form a macroscopically and microscopically obvious abnormality, and neither normal brain nor reactive gliosis is a realistic diagnostic alternative. Although the possibility of reactive gliosis must always be considered, for completeness sake if nothing else, it is easily dismissed based on the finding of unacceptable cellularity, pleomorphism, and mitotic activity. Exceptions are paucicellular lesions, as illustrated in figure 3-36. where it is not always easy to exclude a reactive lesion. Considerable nuclear pleomorphism may be seen in gliosis, particularly in the subacute or chronic reaction of demyelinating diseases, including the distinctive variant, progressive multifocal leukoencephalopathy. The presence of multinucleated astrocytes (Creutzfeldt cells) should always suggest the possibility of a reactive, and often macrophage-rich, lesion. Non-neoplastic lesions potentially misinterpreted as neoplasms are listed in Appendix L.

The principal neoplasms in the differential diagnosis include other infiltrative astrocytic neoplasms of lower and high grade (astrocytoma and glioblastoma), oligodendroglial tumors, as well as pilocytic astrocytoma. Gliomatosis cerebri, of both grade II and III types, is discussed later in this chapter.

Given the progressive transformation that characterizes diffuse astrocytic tumors, it is not surprising that anaplastic astrocytoma merges imperceptibly with astrocytoma grade II and glioblastoma, and that, short of arbitrary definitions, no sharp histologic or cytologic line of demarcation distinguishes these lesions. Only necrosis and microvascular proliferation serve as points of distinction from glioblastoma. A key factor underlying a meaningful diagnosis of anaplastic astrocytoma is often the extent to which the specimen is representative. Well-selected tissue obtained by stereotactic biopsy, especially serial biopsies along a trajectory, is likely to be more informative than tissue timidly obtained by open biopsy, or a generous specimen missing a tumor's most malignant element.

If amplification of *EGFR*, especially mutant forms, occurs only in glioblastoma, its detection could provide a means to distinguish anaplastic astrocytoma from glioblastoma. Unfortunately, the genetic alteration occurs in less than half of glioblastomas, and can be focal. Other molecular findings typical of glioblastoma can be sought indirectly by immunohistochemistry or directly by fluorescent in situ hybridization (FISH)- or polymerase chain reaction (PCR)-based techniques. For example, immunohistochemistry for mutant *EGFRvIII* could identify glioblastoma cells when tissue features of glioblastoma are not evident (see fig. 3-78) (75). Certainly, anaplastic astrocytomas with glioblastoma features, such as mutations in *PTEN*, are more aggressive than those without such features (76). Expression of the mutant EGFRvIII protein is also prognostically unfavorable in anaplastic astrocytomas (75).

Such molecular approaches help distinguish anaplastic astrocytoma from glioblastoma in molecular terms, but it is not yet entirely clear how this will be incorporated in clinical practice, short of those situations where the molecular findings confirm a strong suspicion derived from the histologic sections and MIB-1 staining. Some densely cellular anaplastic astrocytomas have cytologic and molecular features of glioblastoma, but without necrosis or microvascular proliferation. A comment may be appended to the diagnosis of anaplastic astrocytoma indicating that cellularity and proliferative activity are so marked as to suggest the tumor may be, or is about to become, glioblastoma.

Anaplastic astrocytoma shares a number of histologic features with oligodendroglioma, not only with the anaplastic variant but also with better-differentiated examples previously subjected to freezing. The characteristically round nuclei of neoplastic oligodendrocytes become angulated and hyperchromatic in the process. Perinuclear clearing, a diagnostically helpful artifact of delayed fixation, also may be lost to freezing. In tissue not previously frozen, distinctive oligodendroglial qualities are preserved, namely, nuclear uniformity, delicate chromatin, and roundness. This contrasts with greater nuclear pleomorphism and chromatin coarseness in most anaplastic astrocytomas. Oligodendrogliomas are more prone to dense cortical involvement with perineuronal satellitosis and subpial accumulation of tumor cells.

Contrary to expectation, the presence or absence of GFAP positivity generally does not help distinguish an anaplastic astrocytoma from an oligodendroglial tumor. GFAP positivity is often observed in cellular oligodendrogliomas, particularly in higher-grade lesions. Such positive cells should therefore not necessarily be interpreted as astrocytes. GFAP reactivity usually occurs in the form of "minigemistocytes," cells with plump, paranuclear cytoplasm that is either hyaline or fibrillar in quality. Such cells can be as reactive for GFAP as gemistocytic astrocytes, if not more so. We consider these and the "gliofibrillary oligodendrocytes" described above to be oligodendrocyte variants when they possess nuclear characteristics of oligodendroglioma and merge imperceptibly with conventional oligodendrocytes. As a rule, these cell variants are focal in an otherwise typical oligodendroglioma, usually grade III, but can be dominant or even exclusive. The differential diagnosis in the last instance focuses on gemistocytic astrocytoma.

As previously discussed with diffuse astrocytomas, limited or nonrepresentative sections of a pilocytic astrocytoma can easily be mistaken for anaplastic astrocytoma. Hypercellularity, nuclear atypia, occasional mitoses, glomeruloid vascular proliferation, and even necrosis in some pilocytic tumors create the confusion. Attention to clinical factors (young patient age), neuroimaging data (typical locations, cyst-mural nodule architecture, conspicuous contrast enhancement), and histologic scrutiny for telltale features (PAS-positive eosinophilic granular bodies, Rosenthal fibers, low MIB-1 labeling indices) usually resolve the issue. There are times, however, particularly if the specimen is small, where this issue is not resolved pathologically.

Treatment and Prognosis. Anaplastic astrocytoma is a highly malignant tumor: patients have a median survival period, following both surgery and radiation therapy, of approximately 2 to 3 years. There is considerable variation in survival time given the diagnostic criteria and patient selection. In a population-based study, the median survival time was only 1.6 years (66, 67). Occasional lesions seed the cerebrospinal fluid (CSF) spaces (77). Survival is favorably affected by presentation in early life and maximal resection. The use of labeling indices, such as MIB-1, in the assessment of prognosis is unclear. The presence of molecular changes associated with glioblastoma, such as mutation of *PTEN*, has, not surprisingly, been associated with a shorter survival period (76). The role for chemotherapy remains unsettled, but may be more efficacious for anaplastic astrocytoma than glioblastoma.

Glioblastoma

Definition. *Glioblastoma* is a highly malignant astrocytic glioma that appears to arise either de novo or in transition from diffuse astrocytoma and anaplastic astrocytoma.

General Features. Glioblastoma is synonymous with *astrocytoma grade IV*, since most are cytologically astrocytic, at least in part, or are temporally or spatially associated with a lower-grade (II or III) astrocytoma. Without question, some glioblastomas are so poorly differentiated at the time of presentation that they provide little or no histologic evidence of an astrocytic precursor lesion.

Glioblastomas that arise in transition from an often sizable, better-differentiated astrocytic tumor have been referred to as *secondary glioblastomas* (see figs. 3-42, 3-47) (21). More often, the well-differentiated and anaplastic cells lie admixed, obscuring any evolutionary milestones (22). Other glioblastomas are densely cellular and homogeneously anaplastic, and exhibit none of the less cellular and better-differentiated components seen in secondary tumors. It remains a matter of debate whether these "primary" variants are truly malignant de novo or have overrun and obscured a precursor lesion. The molecular definitions of primary and secondary glioblastoma and their implications for gliomagenesis are discussed below.

Largely because individual lesions may be histologically monomorphic, the term "multiforme" has been dropped by the WHO. Many lesions, however, retain the multiforme modifier, particularly in the widely used acronymic designation, "GBM."

Clinical Features. Affecting primarily the cerebral hemispheres of adults (figs. 3-37–3-41) and the thalami and brain stem of children, glioblastoma is the most common malignant glioma. In the optic nerve (16–18) and cerebellum (78–80), however, it is entirely overshadowed by pilocytic astrocytoma. Glioblastomas are uncommon in the spinal cord (fig. 3-42) (11,81).

In the cerebral hemispheres, glioblastomas appear at any age, but are most frequent after the fifth decade. Most are solitary, but occasional examples appear radiologically separate and warrant the designation *multicentric glioblastoma* (82). In such cases, the diagnosis of glioblastoma may come as a surprise to clinicians, although individually, the gliomas are deep-seated and typical of the entity. Metastases are usually more superficial. True multicentricity is difficult to establish, however, even at autopsy.

The presenting symptoms of patients with glioblastoma are similar to those of patients with better-differentiated diffuse astrocytomas, although accompanying neurologic deficits are more frequent, more abrupt in onset, and more rapid in evolution. Unlike lower-grade lesions whose infiltrating and insinuating qualities carry the cells unobtrusively into intact parenchyma with little resultant mass effect, at least initially, glioblastomas are often expansile and edema generating. As a result, they are more likely to produce frank neurologic deficits and signs of increased intracranial pressure. Acute hemorrhage precipitates symptoms in occasional cases. Glioblastomas can evolve within months or even weeks from a lesion with little or no MRI abnormality to a large, threatening, contrast-enhancing mass.

Radiologic Findings. On MRI, glioblastomas typically have an enhancing ring or rim in postcontrast T1-weighted images, and a generally broad zone of surrounding edema evident in

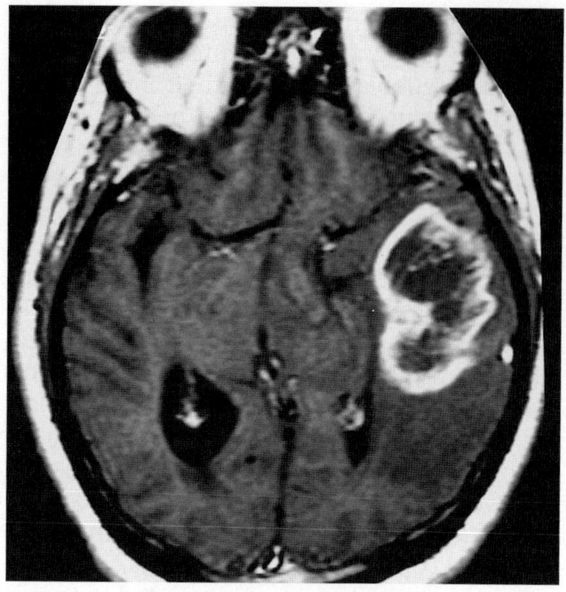

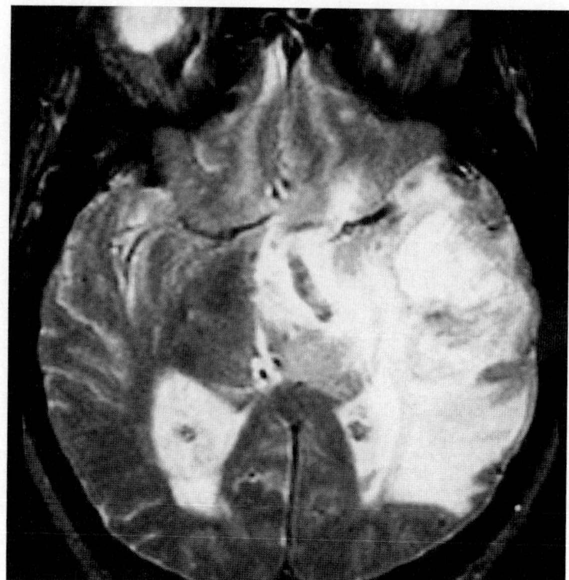

Figure 3-37

GLIOBLASTOMA

Left: By MRI, glioblastomas almost always have a "ring" or "rim" of contrast enhancement that circumscribes a dark area of necrosis.

Right: A broad, bright zone of cerebral edema is seen on T2-weighted images. Infiltrating tumor cells typically surround the contrast-enhancing region. (Fig. 3-28 from Fascicle 10, 3rd Series.)

T2-weighted or FLAIR images (fig. 3-37). The central, hypodense core of the lesion represents tumoral necrosis, whereas the contrast-enhancing rim is highly cellular neoplasm with abnormal vessels that are permeable to contrast agents. The peripheral zone of low attenuation indicates vasogenic edema containing varying numbers of isolated infiltrative tumor cells (83–85). Although glioblastomas often appear cystic given their low-density centers, untreated tumors are rarely fluid-filled. Indeed, a diagnosis of glioblastoma should be carefully reconsidered if a cyst of significant size is observed. Primary brain tumors with prominent, fluid-filled cavities are often well-differentiated, prognostically favorable lesions, such as pilocytic astrocytoma, pleomorphic xanthoastrocytoma, or ganglion cell tumor. Appendix D tabulates cystic CNS tumors.

An unusual variant, *giant cell glioblastoma,* is more discrete, and may lack the large central necrotic core so typical of conventional glioblastoma (fig. 3-40).

Gross Findings. Tumors that infiltrate the cortex produce irregularity, firmness, and abnormalities of the surface vasculature. Small thrombosed "black veins" are a distinctive and highly suggestive feature. Some tumors become attached to, and at least partially derive their blood supply from, the dura. Glioblastomas that reach the meninges often incite marked desmoplasia, with resulting induration and discreteness that can simulate a meningioma or a metastasis. Symmetric, necrotic lesions bridging the corpus callosum create the classic "butterfly" pattern (figs. 3-38, 3-39).

At operation, a gray fleshy rim, corresponding to the region of contrast enhancement, may be encountered. With further dissection, the presence of a necrotic core provides strong presumptive evidence of the diagnosis. In transected lobectomy specimens, glioblastomas appear to pathologists as largely necrotic masses with peripheral rinds of viable, fleshy, gray tissue. The latter varies from case to case, being substantial in some and almost absent in others. Surrounding white matter varies in texture. It may be firm and granular, indicating gross neoplastic involvement, or simply wet and dull as a consequence of edema and diffusely infiltrating tumor cells. In such cases, the ordinarily

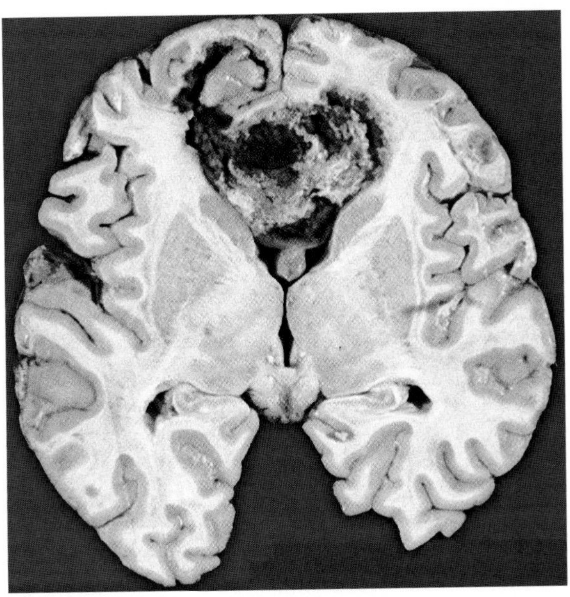

Figure 3-38

GLIOBLASTOMA

The untreated glioblastoma, here in the classic "butterfly" distribution, is necrotic and hemorrhagic. (Courtesy of Dr. Rodney D. McComb, Omaha, NE.)

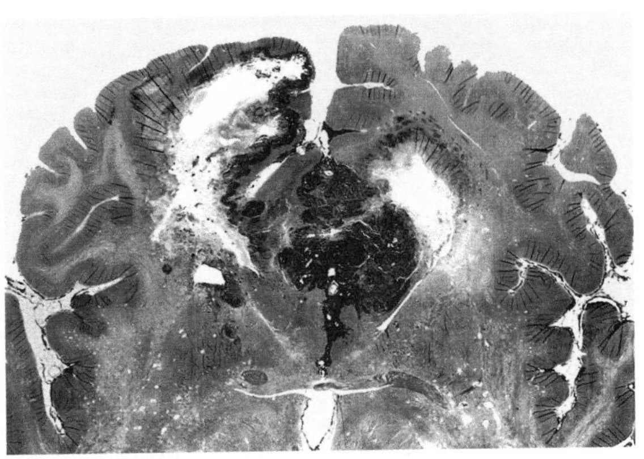

Figure 3-39

GLIOBLASTOMA

As seen in a whole mount histologic section, the glioblastoma of figure 3-38 has a highly cellular, dark rind of neoplasm about a central area of necrosis (cresyl fast violet stain).

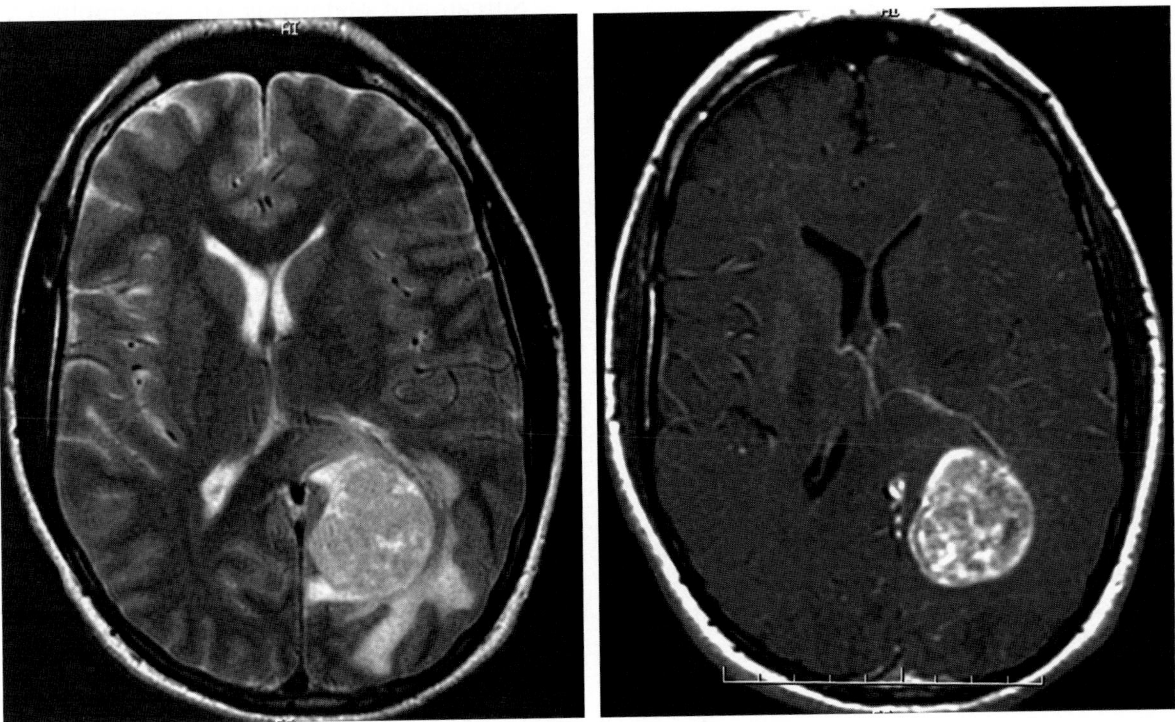

Figure 3-40

GIANT CELL GLIOBLASTOMA

Unlike the classic glioblastoma, the giant cell type may enhance solidly, i.e., homogeneously, without a core of necrosis.

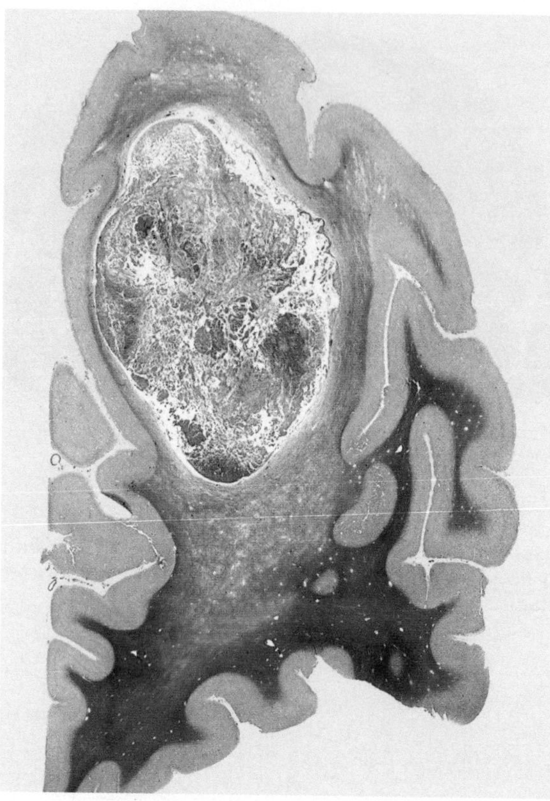

Figure 3-41

GIANT CELL GLIOBLASTOMA

The discreteness of some giant cell glioblastomas is illustrated in a whole mount histologic section (H&E/Luxol fast blue stain).

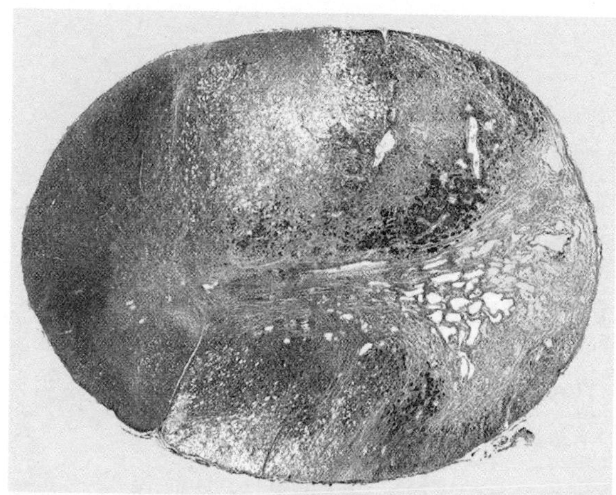

Figure 3-42

SECONDARY GLIOBLASTOMA

Glioblastomas are uncommon in the spinal cord. This darkly staining example arose from a microcystic gemistocytic astrocytoma.

glistening, ivory white matter acquires a faintly yellow hue and contributes to a loss of demarcation between gray and white matter.

Glioblastomas usually infiltrate brain tissue diffusely, but some are so circumscribed as to mimic metastatic carcinoma. The similarity is especially evident in the giant cell variant or gliosarcoma, tumors so well demarcated that the disbelieving surgeon may dismiss a frozen section diagnosis of glioma (figs. 3-40, 3-41). Leptomeningeal involvement sometimes takes the form of diffuse sheet-like infiltrates or discrete spinal deposits.

Like their supratentorial counterparts, glioblastomas of the brain stem are hemorrhagic and necrotic. Anterior displacement and partial engulfment of the basilar artery is characteristic, as is extension of tumor along fiber pathways such as the middle cerebellar peduncle.

Glioblastomas of the spinal cord often have spread within the local, or even distant, subarachnoid space by the time of diagnosis.

Spread and Metastasis. The geographic extent of a glioblastoma is of great relevance when planning radiation therapy. It is fortunate for pathologists that a detailed, or even partial, study of surgical margins is only infrequently requested. It is almost impossible to determine tumor extent by routine histopathology alone, although immunohistochemistry or FISH may identify tumor-specific proteins or genetic changes. For therapeutic purposes, tumor topography is inferred from radiographic images.

Thus, it is neither practical nor possible to stage glioblastomas histopathologically, but the distribution of their cells is not entirely random for they exhibit a well-known tendency to spread along compact fiber pathways, such as the corpus callosum, optic radiation, anterior commissure, fornix, and subependymal zones (figs. 3-43, 3-44). Such dissemination patterns underlie the poor prognosis associated with tumors situated near these "anatomic siphons." Such extension produces tumors whose three-dimensional configuration belies their generally smooth contours in contrast-enhanced scans. Cells extending along these pathways are typically small, undifferentiated, and somewhat polar (fig. 3-45). Within cerebral cortex, these highly infiltrative

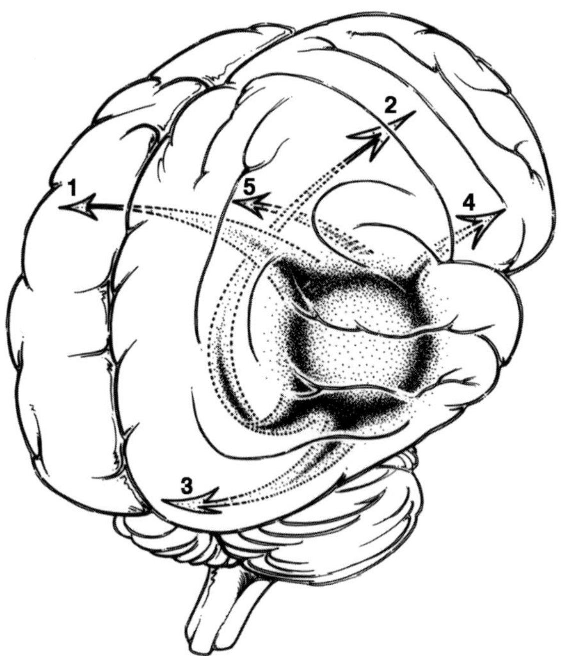

Figure 3-43

GLIOBLASTOMA

Common routes of tumor extension include: 1) corpus callosum, 2) fornix, 3) optic radiation, 4) association pathway, and 5) anterior commissure. (Modified from fig. 4 from Burger PC. Classification, grading, and patterns of malignant gliomas. In: Apuzzo ML, ed. Malignant cerebral gliomas. Park Ridge, IL: American Association of Neurological Surgeons; 1990:3–17.)

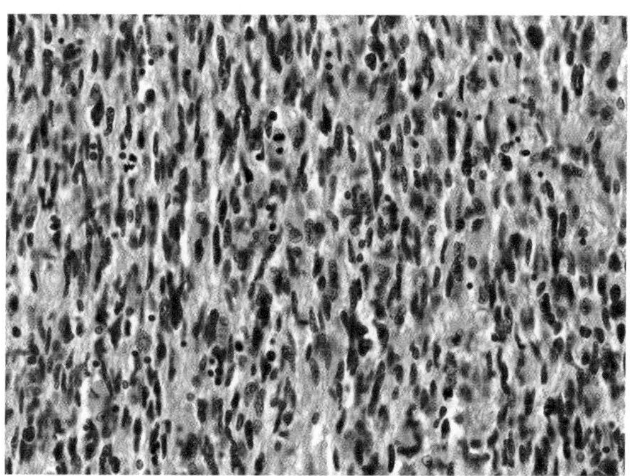

Figure 3-45

GLIOBLASTOMA

Glioblastomas infiltrate freely along compact white matter pathways, as here where tumor cells squeeze between the closely packed axons of the corpus callosum.

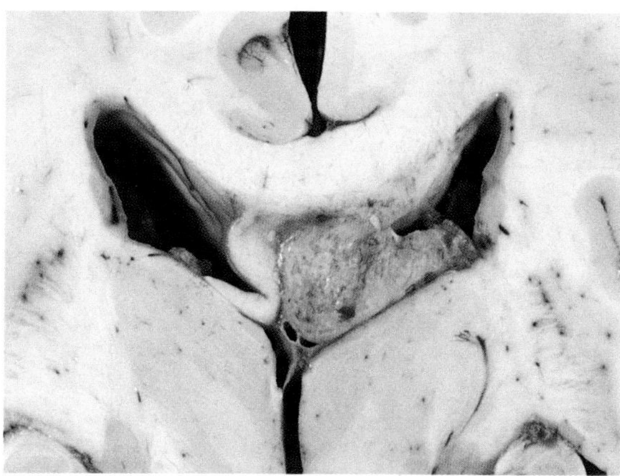

Figure 3-44

GLIOBLASTOMA

As a compact white matter pathway, the fornix is especially susceptible to tumor infiltration, often, as in this case, from a lesion in a temporal lobe.

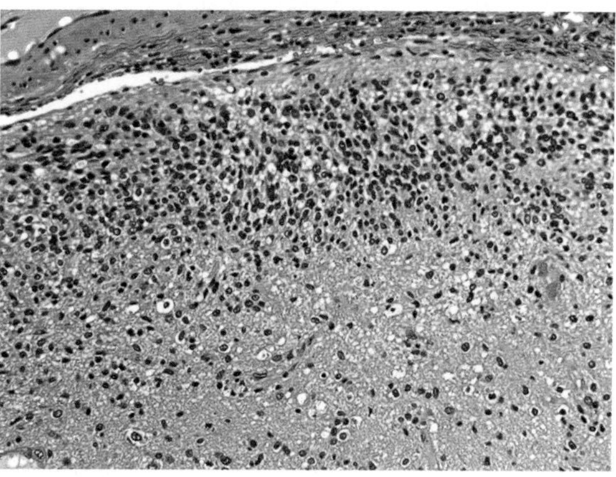

Figure 3-46

GLIOBLASTOMA

Subpial concentration is common in glioblastomas as it is in other infiltrating astrocytic and oligodendroglial neoplasms.

lesions produce "secondary structures," such as perineuronal satellitosis and aggregation in subpial and perivascular spaces (fig. 3-46).

Several studies have established that much of the edematous zone surrounding glioblastomas, as defined by its hyperintensity on T2-weighted MRI, contains infiltrating tumor cells (85–87), adversely affecting resectability. Only an occasional glioblastoma is sufficiently localized, and detected early enough, that therapies are curative.

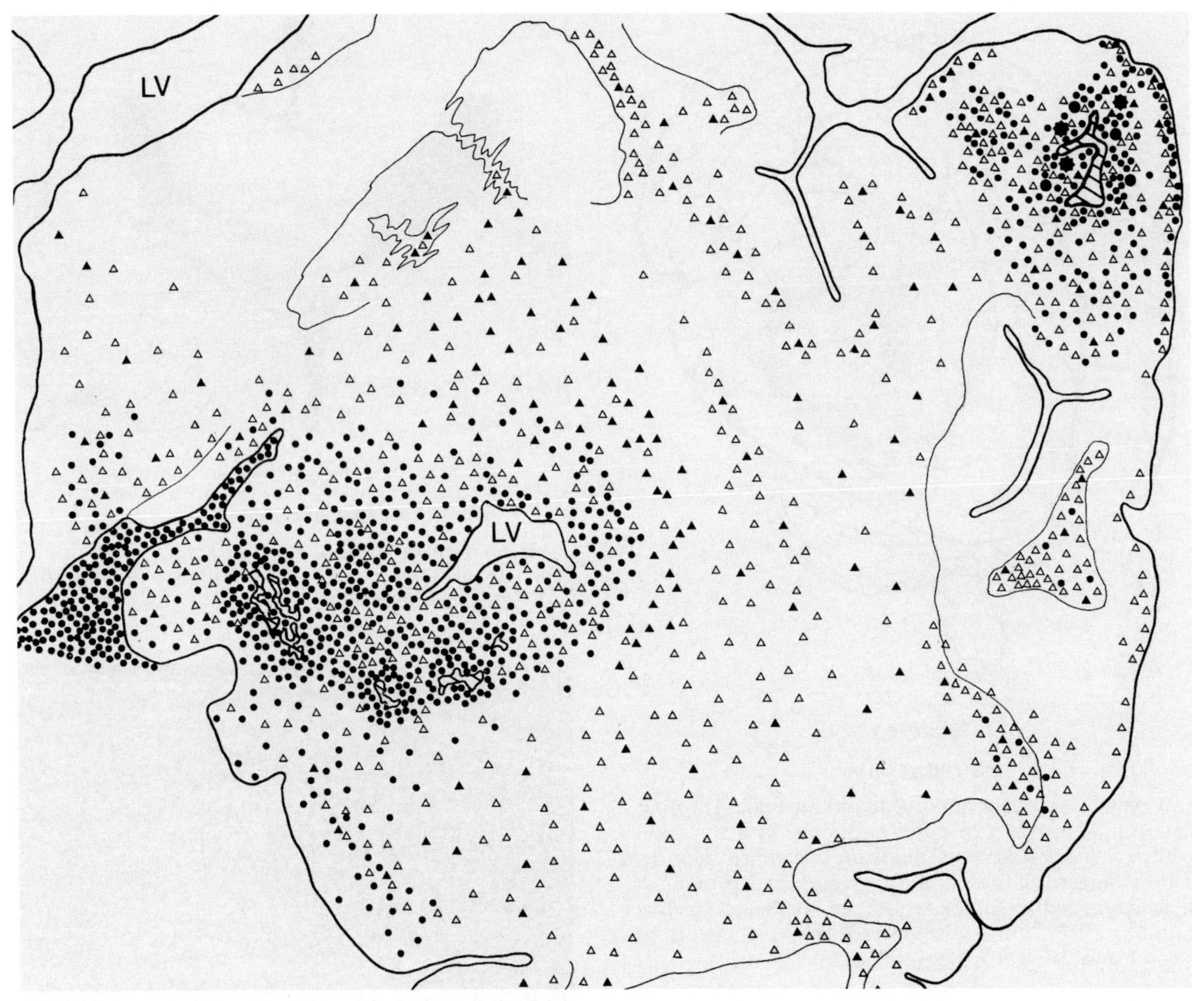

Figure 3-47

SECONDARY GLIOBLASTOMA

The diagram illustrates the astrocytoma component as low density regions of solid and open triangles. The two foci of glioblastoma are the densely populated regions in the superior temporal gyrus and periventricular area. The better-differentiated component had been apparent clinically and radiologically for 6 years. (LV = lateral ventricle). (Fig. 1 from Burger PC, Kleihues P. Cytologic composition of the untreated glioblastoma with implications for the evaluation of needle biopsies. Cancer 1989;63:2014–23.)

Glioblastomas of the spinal cord often disseminate freely in the spinal subarachnoid space.

Microscopic Findings. In light of a seemingly endless intratumoral and intertumoral heterogeneity, it is impossible to characterize in these pages all the variations of this polymorphic glioma. Nonetheless, certain tissue patterns are sufficiently common or diagnostically important to warrant description. Some rare variants deserve space out of proportion to their incidence since they mimic other neoplasms, and thus present problems in differential diagnosis.

Some glioblastomas develop in transition from a better-differentiated diffuse astrocytoma of grade II or III, as determined either by history with prior biopsy, or study of large sections wherein the presence of tumor progression is inescapable (figs. 3-42, 3-47, 3-48). Cytologically, the cells of such secondary tumors vary greatly in size, configuration, content of glial fibrils, and extent of process formation, but share a common "astrocytic quality" attributable largely to their pink cytoplasm. Lesions that are also likely to be secondary but in which the evidence is less

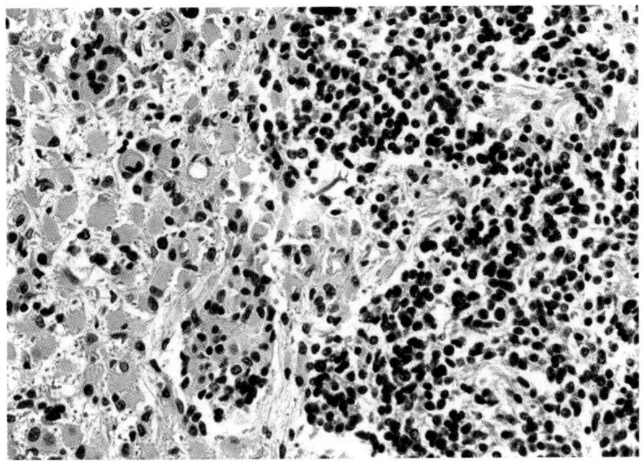

Figure 3-48

SECONDARY GLIOBLASTOMA

Secondary glioblastomas arise within a preexisting astrocytoma of lower grade. In this lesion from the spinal cord illustrated in figure 3-42, small, dark, undifferentiated cells are intermixed with gemistocytes.

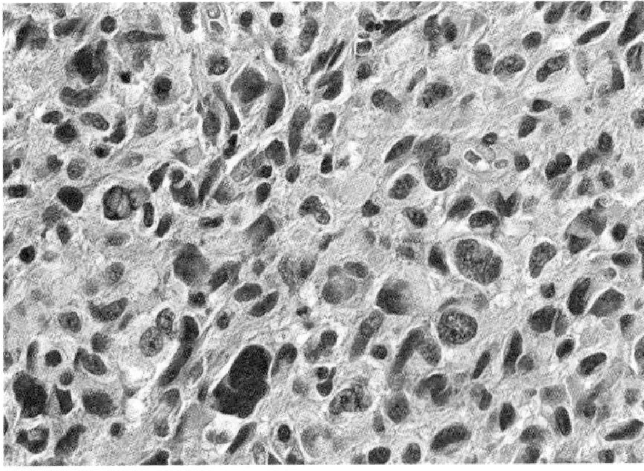

Figure 3-49

GLIOBLASTOMA

Glioblastomas often have an astrocytic quality lent by cells with prominent eosinophilic cytoplasm. A substantial degree of nuclear polymorphism is common in such lesions.

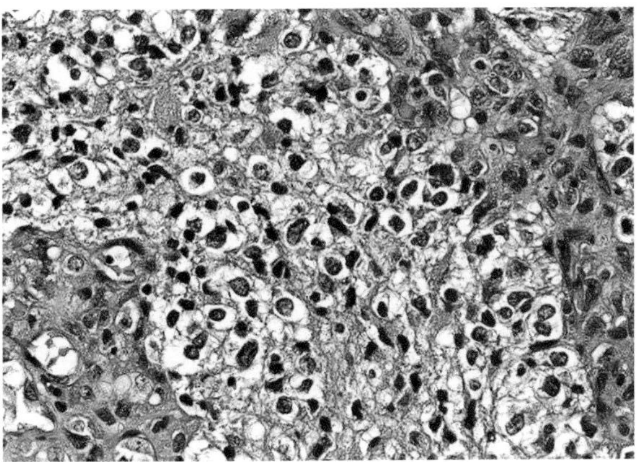

Figure 3-50

GLIOBLASTOMA

Perinuclear halos are common in glioblastomas and should not be used, in themselves, as a criterion for oligodendroglioma. Nuclear roundness and overall cytologic monotony are criteria for oligodendroglioma, and this lesion has neither.

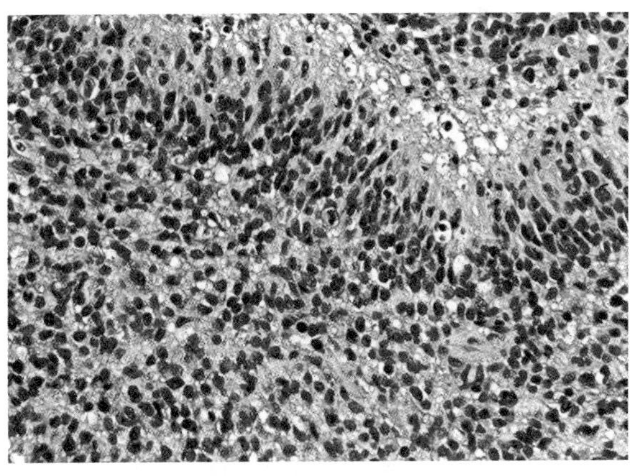

Figure 3-51

SMALL CELL GLIOBLASTOMA

Small cell glioblastomas are uniformly composed of small cells with little cytoplasm. Necrosis with prominent pseudopalisading is common.

clear, have a mixture of underdifferentiated or poorly differentiated cells with fibrillary or gemistocytic elements (fig. 3-49). An element of polymorphism is the rule in most of these tumors. As previously noted, gemistocytic tumors undergoing anaplastic transformation often acquire small, hyperchromatic, undifferentiated-appearing cells, a cardinal feature of the shift toward glioblastoma (see fig. 3-35).

Often, glioblastomas have an astrocytic quality because of small amounts of pink cytoplasm, but no other compelling evidence that the lesion has arisen by progression from a lower-grade astrocytoma. Some glioblastomas have an oligodendroglioma-like quality due to prominent perinuclear halos, and a degree of nuclear roundness (fig. 3-50). Monotonous small cell glioblastomas with or without halos also resemble anaplastic oligodendroglioma (see fig. 3-53).

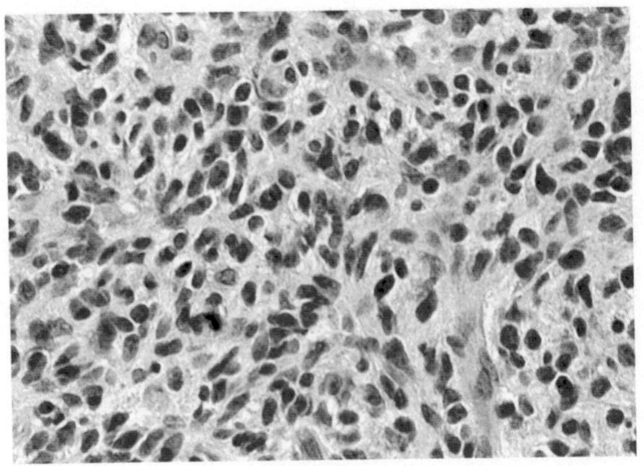

Figure 3-52

SMALL CELL GLIOBLASTOMA

Cells of small cell glioblastoma are uniform in size, shape, and chromatin conformation. Mitoses are frequent.

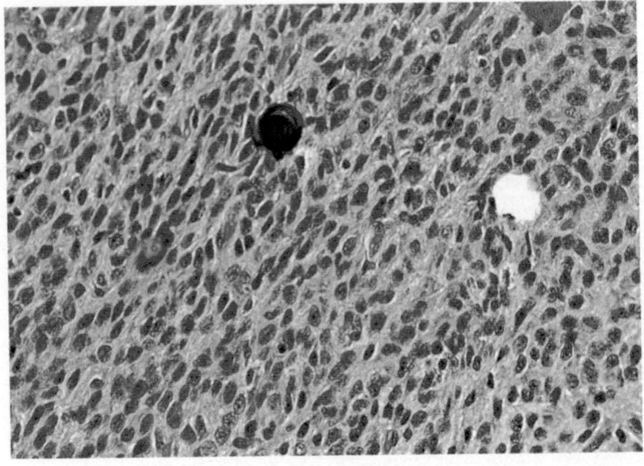

Figure 3-53

SMALL CELL GLIOBLASTOMA

As a consequence of the cellular monomorphism, small cell glioblastoma can resemble oligodendroglioma. Scattered calcospherites add to the illusion. The nuclei of glioblastoma are subtly elongate, however, not round.

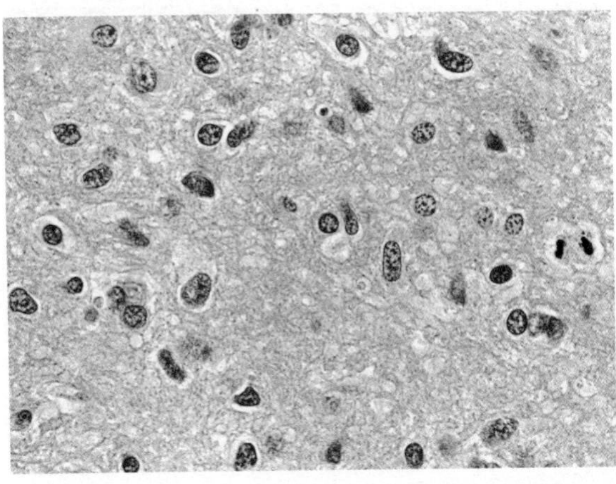

Figure 3-54

SMALL CELL GLIOBLASTOMA

Infiltrating small tumor cells can be difficult to identify when dispersed in the zone of infiltration. Note the mitosis.

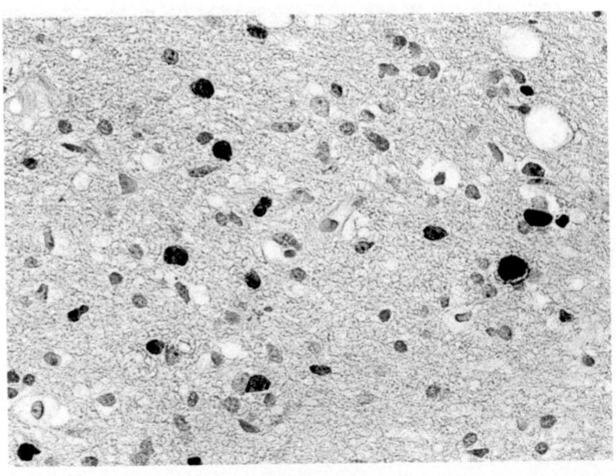

Figure 3-55

SMALL CELL GLIOBLASTOMA

Isolated small tumor cells have a high MIB-1 index.

Other glioblastomas are composed of monotonous, small, largely GFAP-negative cells which congregate in high density (figs. 3-51, 3-52) (74,88). These highly infiltrative cells have slightly elongate nuclei that are remarkably uniform in size, and bland in cytologic qualities. While decidedly monomorphous, these tumors lack the uniform nuclear roundness of oligodendroglioma, a differential diagnosis that often arises, especially with common small cell glioblastomas with microcalcifications (fig. 3-53). The perivascular orientation of tumor cells in this same variant may closely resemble the pseudorosettes of ependymoma (see fig. 3-292). Necrosis with pseudopalisading is often prominent and mitotic activity is high (fig. 3-51). Given the cells' size and cytologic blandness, they are difficult to distinguish from reactive, or even normal, cells when present in low density at the lesion's infiltrating margin (fig. 3-54). The lesion's high MIB-1 index is useful in identification (fig. 3-55), as is amplification of *EGFR,* which is

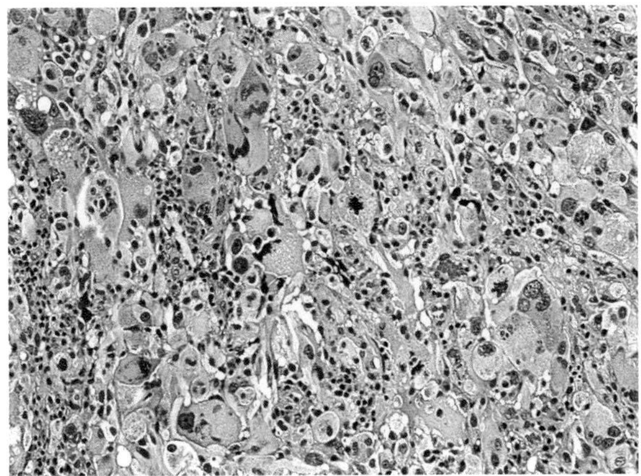

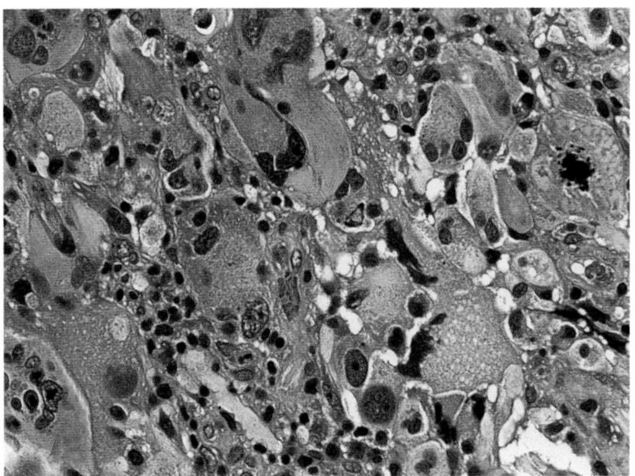

Figure 3-56

GIANT CELL GLIOBLASTOMA

Left: Giant cell glioblastomas are markedly polymorphic.
Right: Lymphocytes are often intermixed with pleomorphic cells.

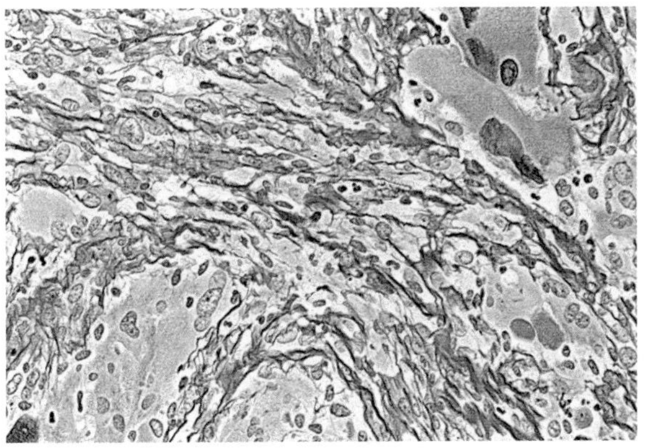

Figure 3-57

GIANT CELL GLIOBLASTOMA

Reticulin is often prominent in giant cell glioblastomas.

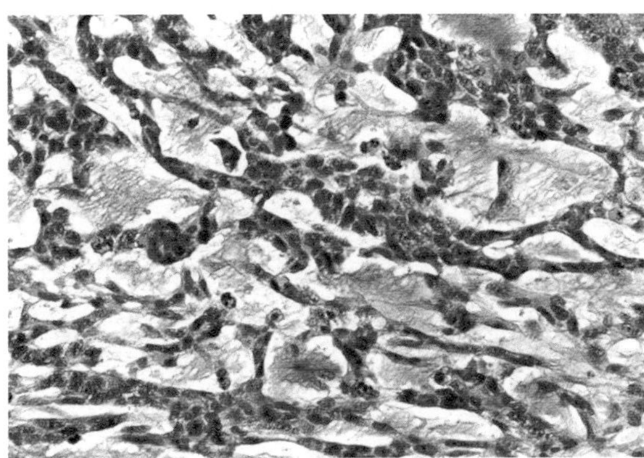

Figure 3-58

GLIOBLASTOMA

Strands of epithelioid cells in a mucopolysaccharide background create a resemblance to metastatic carcinoma or chordoma in some glioblastomas.

common in such tumors (74,88). A somewhat better-differentiated small cell variant of glioblastoma is composed of GFAP-positive cells with short, delicate, often bipolar processes.

At the opposite end of the spectrum from small cell lesions are those composed primarily of pleomorphic tumor giant cells (fig. 3-56) (89). Since the latter are present in many glioblastomas, no clear distinction can be drawn between truly giant cell variants and tumors containing such cells in more limited numbers. Nonetheless, these cells are so monstrous and so dominant in some lesions that legitimate claim to the giant cell variant of glioblastoma can be made. Intercellular reticulin and even collagen deposition, unusual features in glioblastoma, produce a firm, remarkably localized lesion in some cases, and also help to individualize the giant cell variant (fig. 3-57). Scattered lymphocytes are also common.

Another variant is sarcoma-like glioblastoma, which consists either of fascicles of spindle cells or contains stellate cells in a myxoid stroma (fig. 3-58). It is not always clear whether such lesions are simply glioblastoma variants or represent distinct clinicopathologic entities. Other variants

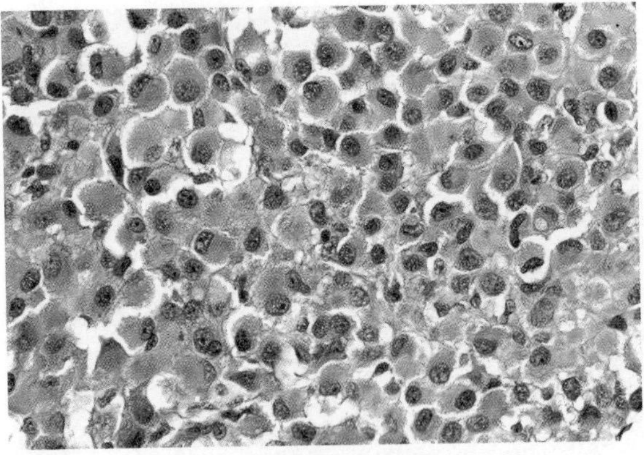

Figure 3-59

GLIOBLASTOMA

Glioblastomas with epithelioid features such as distinct cell borders, round nuclei, and prominent nucleoli can simulate metastatic carcinoma or melanoma.

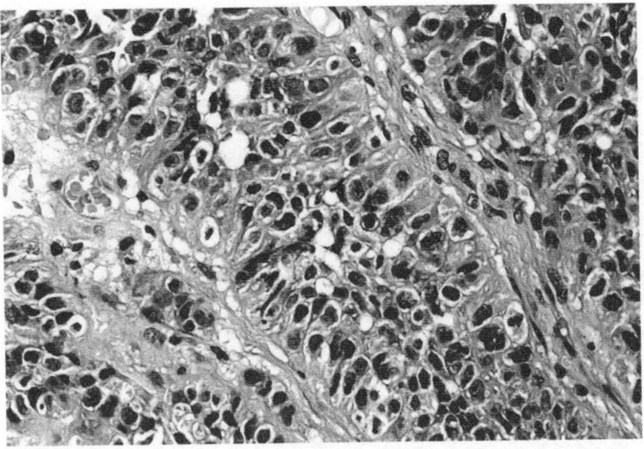

Figure 3-60

GLIOBLASTOMA

Columnar cells with distinct cell borders give some glioblastomas an epithelial quality. Elsewhere, this was a typical, infiltrating glioblastoma. The tumor's immunoreactivity for cytokeratins is illustrated in figure 3-74.

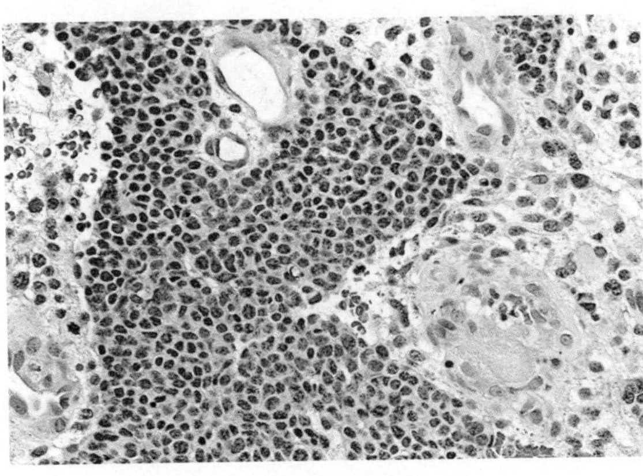

Figure 3-61

GLIOBLASTOMA

Compact lobules of small cells suggest small cell carcinoma, an impression supported by immunopositivity for cytokeratins. Similar foci of small dark cells in glioblastomas can be positive for neuronal makers.

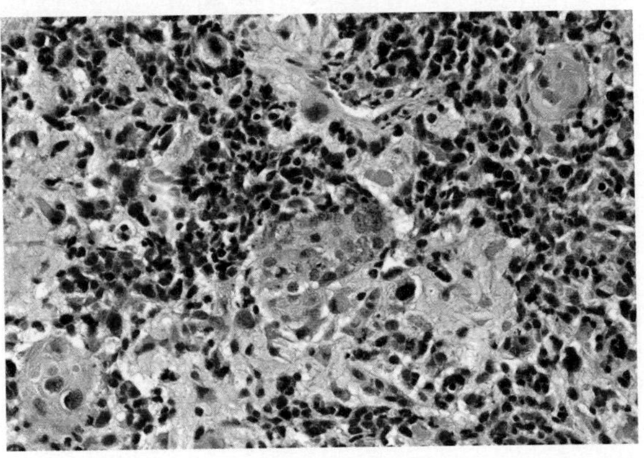

Figure 3-62

GLIOBLASTOMA

Overtly epithelial differentiation in the form of squamous pearls does not exclude glioblastoma.

are epithelioid (fig. 3-59), lipid-rich epithelioid (90), rhabdoid (91), and granular cell. The granular cell variant of diffuse astrocytic neoplasms is discussed separately later in this chapter.

Rare glioblastomas have a carcinoma-like appearance that is produced by cohesive astrocytic cells with discrete, epithelial-like cell borders (fig. 3-60) or lobules of small cohesive cells (fig. 3-61); adenoid and papillary formations; and even squamous pearls (fig. 3-62) (92,93). Such unexpected tissue components may be seen in conventional glioblastomas but are more often a component of gliosarcoma. Immunoreactivity for GFAP, even when focal, attests to their glial nature. Identical *p53* mutations in the epithelial component and the obviously glial component in two cases established the clonal nature of the lesion (94).

Not surprisingly, epithelial cells in glioblastoma are often immunoreactive for cytokeratins. Keratin immunostains must be cautiously interpreted to distinguish glioblastoma from

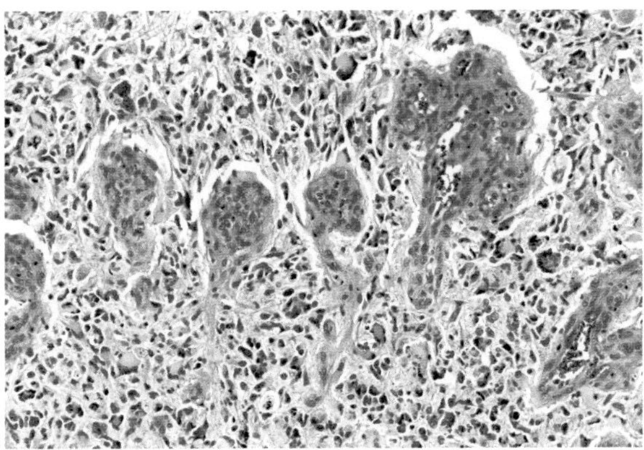

Figure 3-63

GLIOBLASTOMA

Glomeruloid vascular proliferation is usually directed toward a region of necrosis, which in this case was off the top of the illustration.

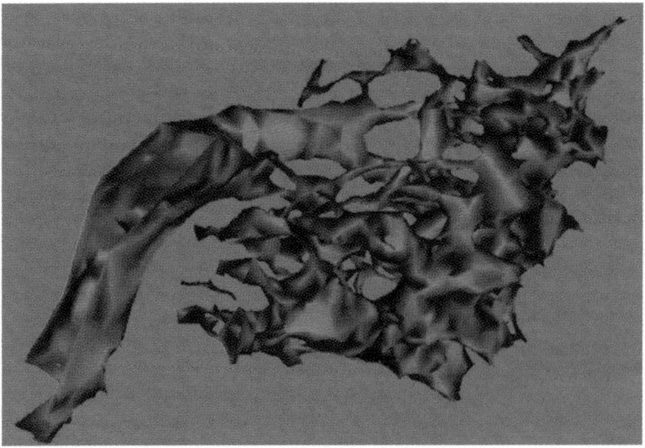

Figure 3-64

GLIOBLASTOMA

A reconstruction from serial histologic sections illustrates the glomeruloid architecture of this common form of microvascular proliferation. (Courtesy of Dr. Pieter Wesseling, Nijmegen, Netherlands.)

metastatic carcinoma, as is discussed in the sections Immunohistochemical Findings and Differential Diagnosis.

Other metaplastic tissues occasionally seen in glioblastoma include bone, cartilage, or skeletal muscle. Occasional glioblastomas have a sometimes prominent synaptophysin-immunoreactive "PNET" component that may have features closely resembling large cell/anaplastic change as seen in medulloblastoma.

From the above discussion, it is clear that the diagnosis of glioblastoma is not based purely on cytologic qualities, but depends also upon identification of what are epiphenomena in an infiltrative, mitotically active astrocytic neoplasm. These features are vascular proliferation and necrosis, and the presence of at least one is essential for the diagnosis.

Vascular proliferation assumes two forms. Most common is a well-known variant that forms globular masses resembling the glomerular tufts of the kidney (figs. 3-63, 3-64). This is often conspicuous about foci of necrosis and may exhibit directional growth or tropism, as illustrated by arcades that are spatially related to zones of necrosis. The proliferation is a response to tissue factors such as vascular endothelial growth factor released from ischemic tumor cells (95,96). Although it is often assumed that this proliferation, now referred to as "microvascular proliferation," is endothelial in nature, immunohistochemical studies have clearly shown a subset of cells to be reactive for smooth muscle antigens (97,98). Indeed, only the flattened cells immediately lining the glomeruloid lumens are positive for factor VIII–related antigen. Thus, pericytes or smooth muscle cells also proliferate.

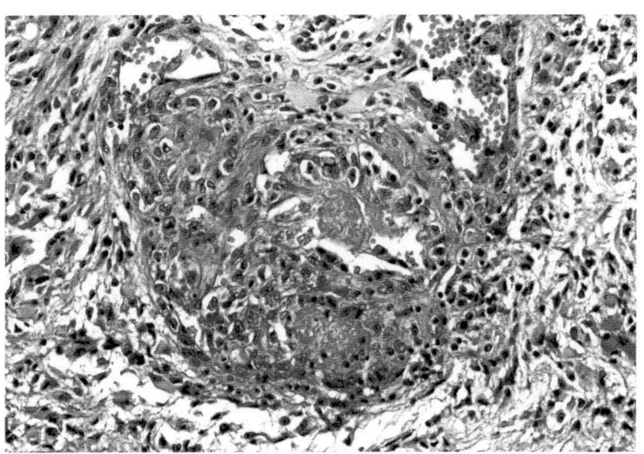

Figure 3-65

GLIOBLASTOMA

Proliferation of endothelial cells within small and medium-sized vessels is a less common, but prognostically more significant, form of vascular hyperplasia than the classic glomeruloid type.

The second form of vascular hyperplasia has a more legitimate claim to the term "endothelial proliferation" since it is intraluminal and consists largely of endothelial cells within small to medium-sized vessels (fig. 3-65). There is thus a single lumen, or one that has been obliterated

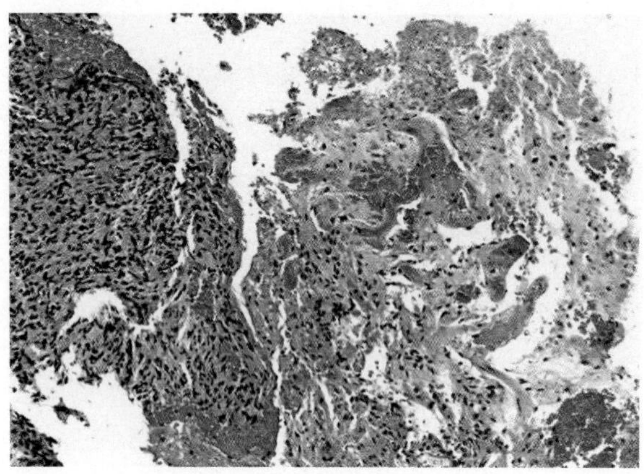

Figure 3-66

GLIOBLASTOMA: NEEDLE BIOPSY SPECIMEN

As illustrated in a needle biopsy specimen, necrosis in glioblastomas does not always have peripheral pseudopalisading.

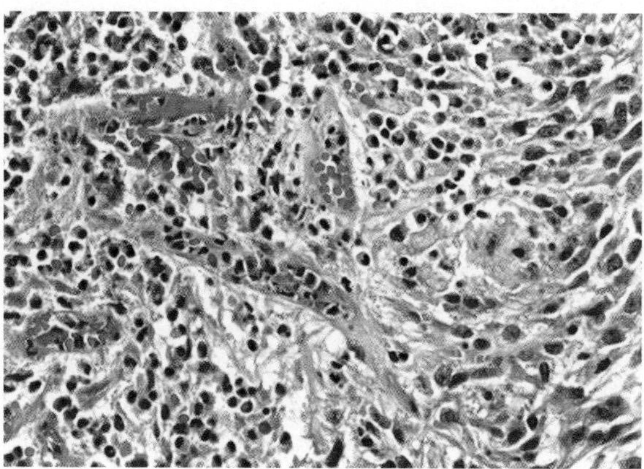

Figure 3-67

GLIOBLASTOMA

Dominated by acute inflammation, some glioblastomas can be difficult to distinguish from an abscess.

by the proliferating endothelial cells. Since larger vessels may be affected, the proliferation is not necessarily "microvascular." Endothelial proliferation is less common than glomeruloid microvascular proliferation and appears to have a more constant correlation with high-grade gliomas and a poor prognosis. Microvascularity of the glomeruloid type is a common feature of low-grade gliomas such as pilocytic astrocytoma, especially in the wall of accompanying cysts, whereas endothelial proliferation within larger vessels is not.

Simple enlargement, or hypertrophy, of vascular cells is common in malignant gliomas but does not qualify as vascular proliferation. It should nevertheless prompt a search for vessels with more prognostically significant proliferation. Telangiectatic vessels without multilayered endothelial proliferations are relatively common in malignant gliomas, and are also seen in indolent neoplasms, such as pilocytic astrocytoma, and are of no prognostic significance. Other innocuous vascular changes include gaping vessels, capillaries in a connective tissue–rich stroma (granulation tissue), closely packed abnormal channels of varying size suggesting a vascular malformation, and collagenous mural thickening.

Necrosis, the second cardinal feature of glioblastoma, takes the form of either large confluent areas of parenchymal destruction, inclusive of vasculature, or small, often multiple, serpiginous foci (95). On MRI scan, it is the large confluent zones of necrosis that comprise the tumor's hypodense center. Necrotic areas, particularly smaller foci, often feature peripheral accumulations of somewhat radially oriented cells (see fig. 3-51) (30,95). Such "pseudopalisading" occurs almost exclusively in high-grade astrocytomas, in some malignant oligodendrogliomas or ependymomas, and rarely, in other tumors, such as medulloblastoma. This concentration, while dramatic, is not always present and is not required for diagnosis (fig. 3-66).

Since the brain is generally fastidious about evacuating necrotic debris, it is somewhat surprising that the extensive necrosis seen in many glioblastomas does not attract larger numbers of macrophages. Nonetheless, activated microglia and fully formed macrophages are readily identified by specific immunohistochemical markers and can, in some instances, complicate interpretation of the biopsy (99,100). Perivascular inflammatory cells, primarily T lymphocytes, are a common feature in gemistocytic tumors (101). Only rarely are lymphoid follicles present.

Acute inflammation confounds the interpretation of some glioblastomas when, in the rare lesion, it obscures the neoplasm (fig. 3-67). Multiple sections may be required to find convincing tumor, and in some cases one is never identified. Abscess becomes an attractive diagnosis to the pathologist, but not to the surgeon who sees a lesion with all of the features of glioblastoma. A similar situation in which neoplastic tissue is overshadowed by reactive changes occurs in occasional glioblastomas that are extensively infarcted.

Histologic Effects of Therapy. In many instances, radiotherapy checks the rapid enlargement of glioblastoma and results a transient phase of remission or quiescence. The interlude is marked by stability, if not regression, of neurologic deficits and diminution in the size of the contrast-enhancing mass. In other patients, the tumor enlarges inexorably, seemingly unaffected by the treatment. Unfortunately, any period of response is short-lived since the glioblastoma usually reasserts itself within a year in the form of clinical deterioration and/or the appearance of one or more expanding new foci of contrast enhancement. Reoperation may be contemplated to relieve pressure and to provide space for future tumor enlargement. Positron emission tomography (PET) using fluorodeoxyglucose has proven useful in the important distinction between a recurrence and radionecrosis. The former may be metabolically active or "hot"; the latter is usually not.

Histologic specimens taken during the quiescent, post-treatment phase typically differ dramatically from those of the pretreatment period (102,103). Treated tumors are often paucicellular and feature conspicuously pleomorphic astrocytes. Of the latter, some are radiation-affected tumor cells, and others are reactive astrocytes with radiation-induced atypia (fig. 3-68). Small cells with markedly hyperchromatic nuclei and inconspicuous cytoplasm may represent new tumor cells, but their rarity makes that interpretation difficult. Such lesions are, of course, glioblastomas, but are best characterized as "persistent" or "quiescent" rather than "recurrent." The last term should be reserved for actively proliferating tumors in which small cells without abundant cytoplasm or bizarre nuclei are well represented. Since radiotherapy suppresses proliferative activity, the presence of mitotic activity and/or significant MIB-1 labeling help to justify a diagnosis of recurrence. The finding of only widely scattered MIB-1–positive cells in irradiated tumors poses a problem, since not all such cells may be neoplastic.

Brain parenchyma surrounding previously irradiated glioblastomas is often punctuated by foci of necrosis that represent the brain's delayed response to radiation. Additional changes attributable to radiation include fibrinoid necrosis and vascular thickening. When occurring in the tumor bed, these changes represent desirable consequences of radiotherapy and, as such, are more appropriately designated "radiation effect" or "tumoral radionecrosis" rather than the pejorative term "radionecrosis." It is histologically similar, however. In most instances in which a previously irradiated tumor is biopsied and radiation effect is evident, viable tumor is also apparent; only in a few instances does the contrast-enhancing mass consist solely of radionecrosis.

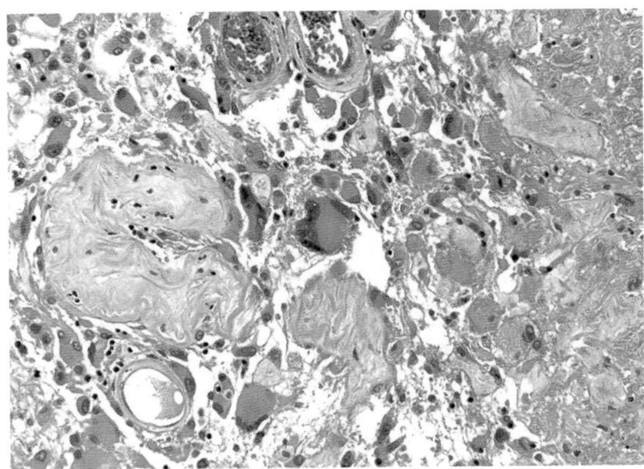

Figure 3-68

PREVIOUSLY IRRADIATED GLIOBLASTOMA

The postirradiation phase of "quiescence" is paucicellular, with marked cellular atypia and foci of coagulation necrosis.

Since small foci of actively growing tumor, irradiated quiescent neoplasm, radiation effect, radiation necrosis, and gliosis can coexist in a specimen, the pathologist is often not in a position to define which single factor, or combination of factors, underlies the clinical and radiographic findings that prompted the resection (fig. 3-69). Radiation-induced changes notwithstanding, the finding of even a focal, small cell, mitotically active lesion, particularly one with necrosis with pseudopalisading, is evidence that the lesion is "active" or truly recurrent. In the absence of such a finding, as in a specimen with only pleomorphism and paucicellularity, we do not make a diagnosis of active tumor recurrence. Our diagnosis for the lesion shown in figure 3-68 would be glioblastoma multiforme, with a

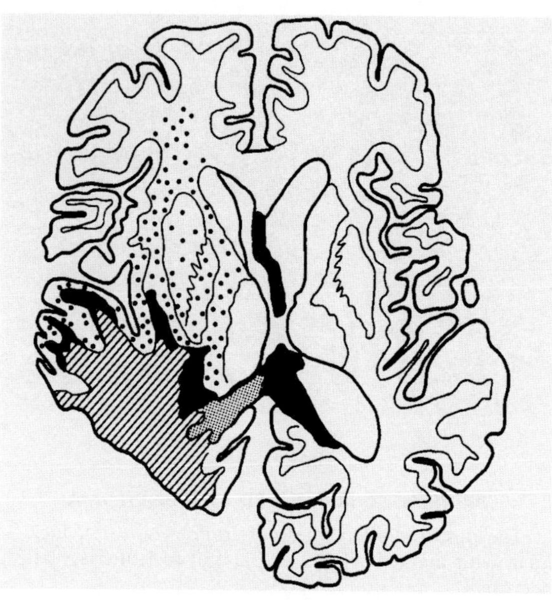

Figure 3-69

PREVIOUSLY IRRADIATED GLIOBLASTOMA

The diagram illustrates common tissue reactions encountered in irradiated tumors at the time of recurrence. The solid black areas represent highly cellular malignant neoplasm that has crossed the corpus callosum and extended along the ipsilateral fornix. Isolated solid circles are zones of infiltrating neoplasm within largely intact white matter, a classic feature of diffuse astrocytoma. The finely stippled region represents paucicellular tumor with the cellular pleomorphism typical of the "quiescent" phase of glioblastoma. Areas of radiation effect are diagonally hatched. (Fig. 3-46 from Fascicle 10, 3rd Series.)

comment that the tumor is typical of a radiated glioblastoma in a phase of quiescence and not one of active recurrence. Radionecrosis is most clearly an entity when it occurs following treatment of extracranial neoplasms (see chapter 19).

Characterization of the lesion along the above guidelines has significant prognostic significance for patients with recurrent lesions. In one study, factors that affected prognosis were: 1) tissue that was pure recurrent tumor versus that which contained both recurrent tumor and radiation effect, and 2) the relative amounts of tissue with radiation effect and recurrent tumor (103).

The effects of chemotherapy upon the cytologic and histologic appearance of glioblastoma cells are poorly understood. BCNU-impregnated wafers placed in the tumor bed can produce necrosis of both parenchyma and vasculature.

Frozen Section. A diagnosis of glioblastoma is most easily made on specimens from the contrast-enhanced cellular area and, if present, the necrotic core. One can be certain that the contrast-enhancing area, whether of a glioblastoma or other enhancing lesion, has not been sampled if the specimen is so hypocellular as to suggest only a low-grade infiltrating glioma or reactive gliosis. Specimens from the contrast-enhancing zone typically consist of markedly cellular tumor with a proportionate increase in vascularity.

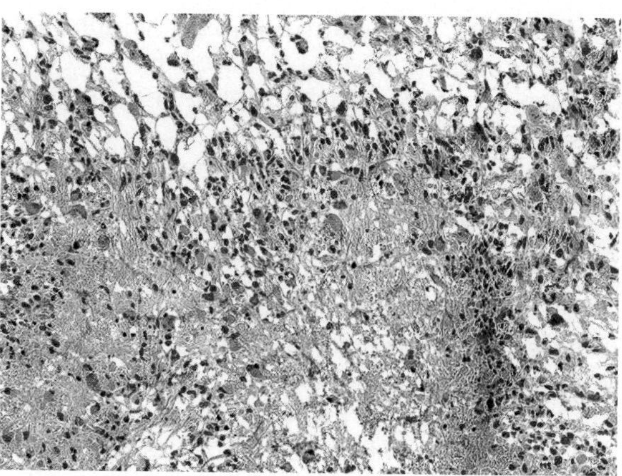

Figure 3-70

GLIOBLASTOMA: FROZEN SECTION

Glioblastomas often appear less cellular in frozen sections but necrosis is seen well. Tumor cell processes are usually somewhat separated and thus better resolved than they are in paraffin-embedded tissue.

Prognostically important features, such as mitosis, vascular proliferation, and necrosis, must be carefully sought (fig. 3-70). Vascular proliferation is readily identified, either as glomeruloid microvascular hyperplasia or larger caliber vessels filled with layers of endothelial cells. Necrosis may be difficult to identify in smears as well as in frozen section specimens, where it can be readily confused with normal, albeit artifactually altered, white or gray matter. The alignment of nuclei so typical of pseudopalisading is a helpful low-power feature, since it demarcates zones of necrosis that may not be obvious in frozen tissue. In the setting of an infiltrating glial neoplasm, palisading necrosis is virtually diagnostic of glioblastoma, except for occasional malignant oligodendrogliomas.

Frozen sections of previously treated lesions often disclose extensive necrosis without palisading, hyalinized vessels, and a few pleomorphic tumor cells (fig. 3-71). Such findings are not evidence of active disease.

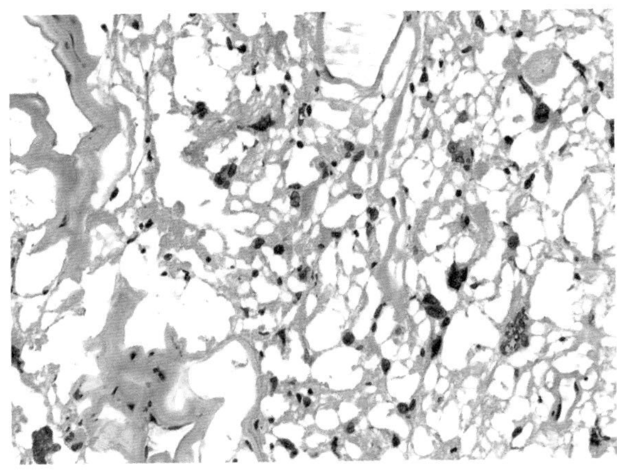

Figure 3-71

PREVIOUSLY IRRADIATED GLIOBLASTOMA: FROZEN SECTION

Necrosis and pleomorphic tumor cells in low density are typical of previously irradiated glioblastomas.

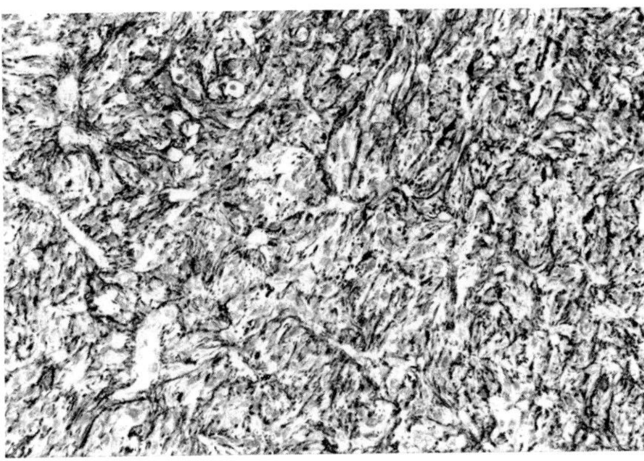

Figure 3-72

GLIOBLASTOMA

At least some cells are glial fibrillary acidic protein (GFAP) positive in most glioblastomas. Almost all are immunoreactive in this case.

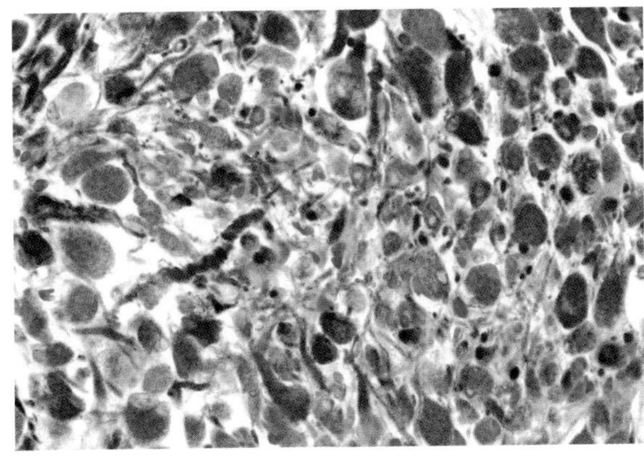

Figure 3-73

GLIOBLASTOMA

Glioblastomas can be immunoreactive for broad spectrum cytokeratins. Note the cytoplasmic processes typical of glioblastoma, not metastasis.

Immunohistochemical Findings. The immunophenotype of glioblastoma includes reactivity for GFAP (fig. 3-72), and, of course, vimentin and S-100 protein. While vimentin reactivity is entirely nonspecific and of little diagnostic significance, it does indicate the immunoviability of the specimen. On the other hand, immunoreactivity for GFAP in neoplastic cells clearly establishes the glial nature of the tumor, although not necessarily branding it one of astrocytic type. As might be expected, the degree and geographic extent of GFAP reactivity in glioblastoma vary greatly. Generally, astrocyte-like cells are strongly positive, while small undifferentiated and giant cells are often negative or weakly stained. Between these extremes, generalization is not possible. Some very "astrocytic" lesions may be largely immunonegative, whereas small cell tumors, particularly those with cells extending into fine bipolar processes, may be strongly and diffusely positive.

Reactive astrocytes are strongly GFAP positive, symmetrically disposed, and anchored by long, radially arranged, tapering processes that grasp nearby vessels. Neoplastic cells are often less reactive for GFAP, irregularly distributed, lack the star-like symmetry of their processes, and do not terminate extensively upon blood vessels. It is unfortunate that in largely undifferentiated glioblastomas, wherein immunoreactivity is most desperately needed, it may be difficult to relate GFAP-positive processes or even cytoplasm to neoplastic nuclei. One must be conservative in this effort, accepting as diagnostic only unequivocally GFAP-reactive cells with nuclear features of malignancy. In such instances, finding S-100 protein staining alone is suggestive but not diagnostic of glioblastoma.

Although immunoreactivity for epithelial markers such as cytokeratins seems alien to resting astrocytes, it is common in both neoplastic (fig. 3-73) and reactive astroglia, but particularly the latter. Immunopositivity is especially likely with antibodies AE1/AE3; CAM5.2 is

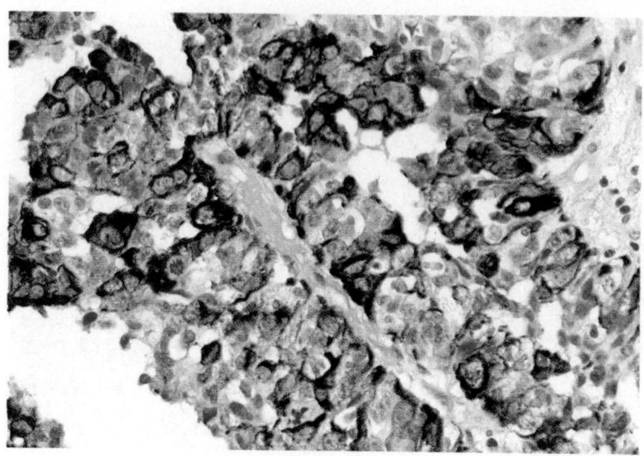

Figure 3-74

GLIOBLASTOMA

The epithelial phenotype extends to the protein level in glioblastomas that are immunoreactive for cytokeratins. The appearance of the lesion using the H&E stain is illustrated in figure 3-60.

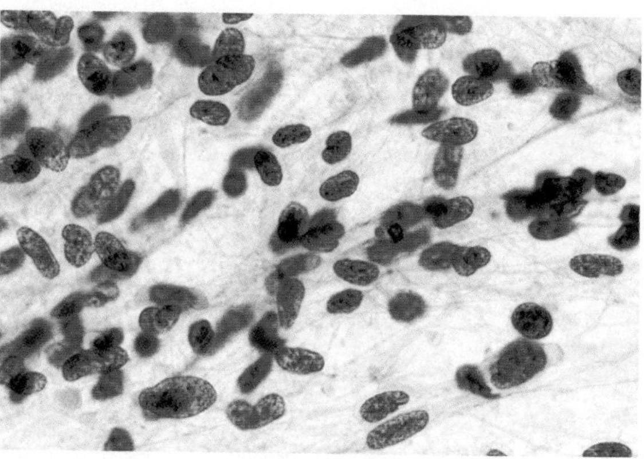

Figure 3-75

GLIOBLASTOMA: SMEAR PREPARATION

Small cells with bipolar processes differ fundamentally from the smooth-contoured, sharply defined cells in lymphomas and metastatic neoplasms.

more likely to be negative (104). Such reactivity obviously compromises the diagnostic value of the antibody for distinguishing glioblastoma from metastatic carcinoma, and staining should be done in concert with that for GFAP and epithelial membrane antigen (EMA) or other epithelial markers. Positivity for cytokeratins is most misleading in glioblastomas with epithelioid cytologic features or an "adenoid" growth pattern (fig. 3-74) (93,105). In those rare metaplastic glioblastomas and gliosarcomas exhibiting true squamous or glandular differentiation, full spectrum keratin staining can be expected; such tumors are also immunopositive for EMA. The immunoreactivity of some glioblastomas for EGFR is discussed under Differential Diagnosis (see fig. 3-78).

Giant cell glioblastomas frequently react for class III beta-tubulin and GFAP, but not as frequently for neuronal markers such as synaptophysin and neurofilament protein as does pleomorphic xanthoastrocytoma (106). The incidence of immunoreactivity for p53 protein is considerably higher in the giant cell glioblastoma (106) in accord with the high incidence of *p53* mutations (107).

Ultrastructural Findings. The ultrastructural findings in glioblastomas are in accordance with the light microscopic degree of differentiation (50,51). Intermediate filaments are abundant in obviously astrocytic cells, but are often sparse or absent in cells that appear undifferentiated. In contrast to most metastatic carcinomas, with their well-formed desmosomes, intercellular contacts in glioblastoma and other astrocytomas are often rudimentary. In poorly differentiated lesions, junctions consist of little more than ill-defined, apposed submembrane densities ("intermediate junctions"). Tonofilament bundles, either cytoplasmic or desmosome-associated, are also lacking in astrocytoma and glioblastoma. Although filopodia may be seen, well-formed microvilli are not.

Cytologic Findings. In all but the most epithelioid variants, glioblastomas disclose their glial origin in the form of fine processes that surround clumps of cells and extend, often in a bipolar fashion, from individual cells (fig. 3-75). Most glioblastomas have a predominant population of nuclei that are, in comparison to carcinoma, rather small. Chromatin conformation is also typical, being more evenly dispersed than in the typical carcinoma. Nucleoli are less obvious.

Molecular and Cytogenetic Findings. In terms of cytogenetic findings, many glioblastomas are "unstable," as evidenced by the multitude of abnormalities seen in these tumors as a group and in individual cases. As summarized in figure 3-76, losses on chromosomes 6, 10, 15, and 18, and the Y chromosome is especially common (108,109). Of these, chromosome 10 appears to be the most closely associated with the tumor (109). Gains are less frequent and often

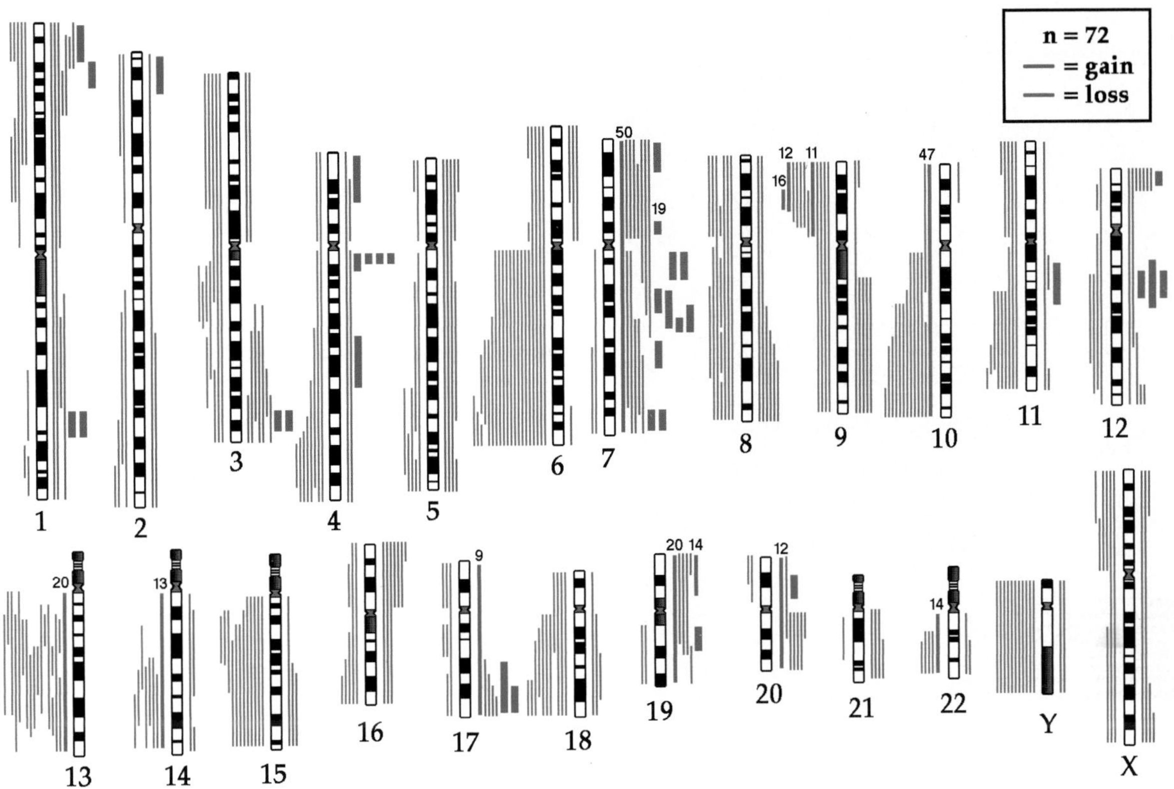

Figure 3-76

GLIOBLASTOMA: COMPARATIVE GENOMIC HYBRIDIZATION

Chromosome imbalances are frequent in glioblastomas. Losses (red) and gains (green) illustrate the genomic instability in a series of glioblastomas. Losses on chromosome 6, 10, 13, 15, and Y are especially common. Gains are most common on chromosome 7. (Courtesy of Dr. Burt G. Feurstein, San Francisco, CA.)

involve chromosome 7. From the vantage point of the hematoxylin and eosin (H&E)-stained section at least, small cell lesions appear remarkably stable and uniform in DNA content.

Molecular abnormalities in glioblastoma are numerous and include deletions and mutations of *p16* and *PTEN*, loss of heterozygosity of chromosome 17p, mutations in *p53*, and amplification of *MDM2* and *EGFR* (fig. 3-77). One large population-based study found loss of heterozygosity (LOH) on 10q in 69 percent of glioblastomas, *EGFR* amplification in 34 percent, *p53* mutations in 31 percent, $p16^{INK4a}$ homozygous deletions in 31 percent, and *PTEN* mutations in 24 percent (66,66). Some of these abnormalities, such as loss of chromosome 10, *PTEN* mutations, *p16* deletions, and *EGFR* amplification, appear to distinguish glioblastoma from grade III astrocytoma, and can, in concept, be used to identify isolated invading tumor cells. One study suggested that the histologic phenotype of glioblastoma is associated with LOH on chromosome 10q-ter (66,67,110).

In attempts to make a coherent story out of this complexity, genetic changes have been correlated with each other in order to define specific molecular pathways that correlate, in turn, with histopathologic and clinical findings such as patient age and the duration of preoperative symptoms (111). From this has emerged a clinical/molecular concept of "primary" and "secondary" glioblastomas (66,67,112), although it is unclear how distinct this differentiation is, or the extent to which it is therapeutically and prognostically relevant. On the whole, but with many exceptions, secondary types present in younger patients, more often women who have a longer duration of symptoms, and usually lie

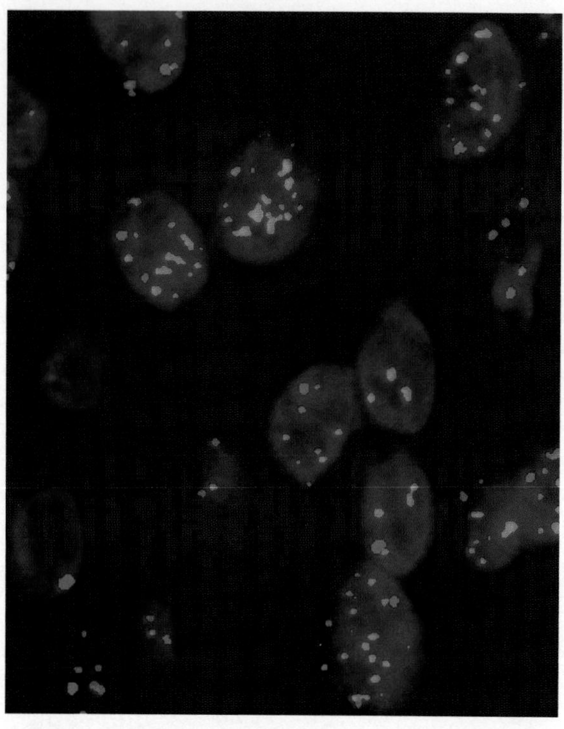

Figure 3-77

GLIOBLASTOMA: EPIDERMAL GROWTH FACTOR RECEPTOR *(EGFR)* AMPLIFICATION/FLUORESCENCE IN SITU HYBRIDIZATION (FISH)

Occurring in approximately one third of glioblastomas, amplification of *EGFR* is visualized by FISH as multiple coarse signals (red) distributed randomly in a "double minute" pattern. The green signals are those of centromere enumeration probe 7. (Courtesy of Dr. Arie Perry, St. Louis, MO.)

in the cerebral hemispheres. In only approximately 5 percent of patients in one study was there clinical and histopathologic evidence of tumor progression from a lower-grade precursor astrocytoma (66,67). Diffuse astrocytomas of the brain stem in children may be of this type. The lesion corresponds to what Scherer (21) also described as secondary glioblastoma (see fig. 3-47). When these patients are defined clinically by a short duration of preoperative symptoms, the tumors have a high frequency of mutations of *p53* but infrequent mutations in *PTEN* or amplification of *EGFR* (66,67,113, 114). Loss of chromosome 19q in the region of a presumed tumor suppressor gene(s) as yet unidentified is more common in secondary tumors. One study suggests that chromosome 10 loss preferentially affects the long arm (115).

The considerably more frequent primary form of glioblastoma appears more abruptly and generally later in life than the secondary type, and more often in men. Genetically, this variant less often has mutations of *p53*, but exhibits complex genetic abnormalities including deletions, often homozygous, of *p16* and *DMBT* (deleted in malignant brain tumors), as well as mutations in *PTEN*. Amplification of *EGFR* occurs in 30 to 40 percent of such lesions, and is commonly associated with homozygous deletions in $p16^{INK4a}$ (66,67). Correlation with histologic features is incomplete, although large areas of necrosis may be typical of primary glioblastomas (113). Glioblastomas uniformly exhibiting the small cell phenotype have a high rate of *EGFR* amplification and thus often fall into the primary category (74,88,116).

The giant cell glioblastoma is known for a high incidence in *p53* mutations and low incidence of *EGFR* amplification (107,117,118).

Pediatric glioblastomas may evolve along different lines than those in adults, with a greater utilization of the *p53* pathway and fewer cases with amplification of *EGFR* (119,120,120a). Microsatellite instability also may be more common (119,121).

Differential Diagnosis. Not infrequently, a stereotactically obtained specimen is either from a lesion's periphery, so that few if any tumor cells are present, or comes from a point so deep in the lesion's core as to be largely necrotic. In either situation, the presence or absence of a neoplasm is the principal issue. In the case of the former, the diagnosis rests upon recognizing the increase in number and atypia of what are usually small infiltrating tumor cells. There may be little nuclear pleomorphism or hyperchromasia, and the diagnosis may not be possible at this point, although deeper sections may reveal diagnostic tissue. A high MIB-1 index is consistent with a neoplasm, but is only indirect evidence since a firm diagnosis rests on cytologic features. Strong immunoreactivity for p53 protein in somewhat atypical cells is also suggestive of neoplasia, but is not a diagnostic finding. Procedures such as FISH for *EGFR* amplification (fig. 3-77), immunohistochemistry for mutant *EGFRvIII* expression (fig. 3-78), or analysis for specific chromosomal losses and gains are now being used to distinguish tumor cells from gliosis (58).

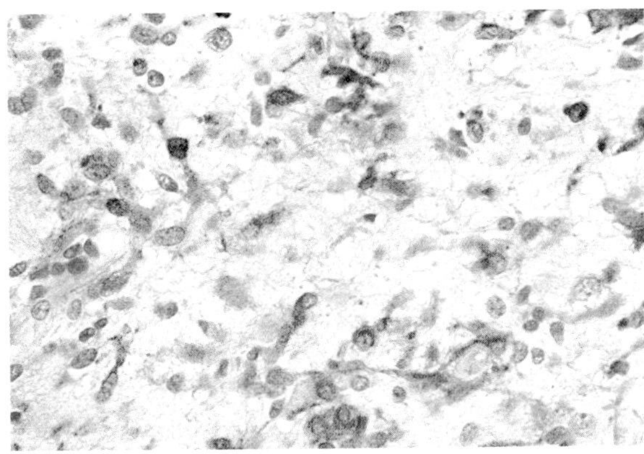

Figure 3-78

GLIOBLASTOMA

Immunoreactivity for the mutated *EGFRvIII* can, in concept, be used to identify cells as neoplastic. (Courtesy of Dr. Kenneth D. Aldape, Houston, TX.)

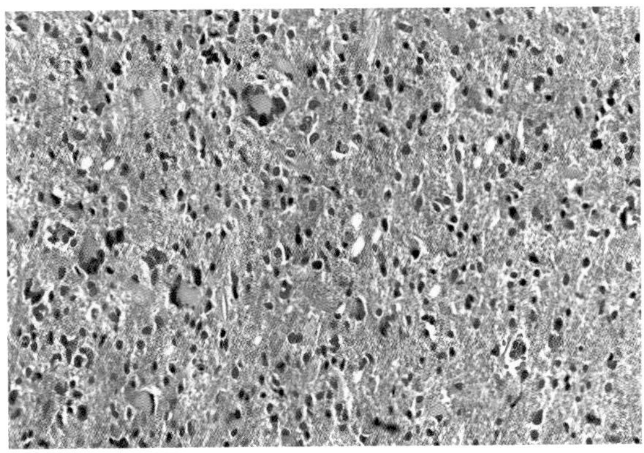

Figure 3-79

GLIOBLASTOMA

Diffuse infiltration, a cardinal finding in glioblastomas, aids in the distinction from both metastasis and well-circumscribed glioma.

If the tissue is clearly neoplastic, establishing the diagnosis of glioblastoma begins with the determination that the tumor is primary rather than metastatic. The lesion must then be identified as a malignant astrocytic neoplasm and distinguished from other gliomas. Primary CNS lymphoma must also be considered in the face of an anaplastic lesion.

That a tumor is a primary rather than a metastatic neoplasm is generally based on a glioblastoma's infiltrative nature, which is most evident at the periphery (see figs. 3-54, 3-55, 3-79). The glioma's distinctive cytologic features in smear preparations are also helpful. Infiltration is manifest by the overrunning of brain parenchyma, both gray and white matter. Staining for phosphorylated neurofilament protein or by the Bielschowsky silver method assesses the lesion's infiltrative quality, and can be used to demonstrate axons in more cellular areas where preexisting parenchyma is not apparent in H&E-stained sections. Metastatic neoplasms evolve from the tumor microembolus stage into large, cohesive masses that displace rather than diffusely infiltrate brain tissue. By tracking along blood vessels, however, some do infiltrate tissues and simulate a glioma (see fig. 20-12) in that isolated neoplastic cells are occasionally seen in brain parenchyma around metastases, e.g., melanoma, small cell carcinoma, and lymphoma. Nonetheless, the extent of such infiltration is generally limited. Reactive astrocytes are much less often incorporated into the substance of metastases than of gliomas, but when they are they occur in small number about vessels, particularly at the periphery of the lesion. Unlike glioblastomas, necrosis in metastases usually spares blood vessels and a collar of neoplastic cells.

Even if the infiltrating nature of a tumor is difficult to establish, the cytologic features of glioblastoma aid in the distinction from metastatic neoplasms. Those features particular to glioblastoma include small poorly differentiated cells, cells distinctly astrocytic in appearance with little or no tendency to nuclear molding, inconspicuous nucleoli and, in some cases, remarkable cellular and nuclear pleomorphism. The small cells of the glioblastoma are often even smaller than those of small cell carcinoma and lie within a fine fibrillar background composed of neoplastic cell processes and/or remnants of normal neuropil or white matter. The astrocytic nature of some cells is recognized by their fibrillated or glassy, eosinophilic cytoplasm. Abundance of the latter is the hallmark of gemistocytic astrocytoma, a tumor in which transitions from plump unequivocal astrocytes to small, hyperchromatic, undifferentiated cells are frequently noted.

There are, nevertheless, monomorphous, highly cellular, small cell glioblastomas that, due to their compact architecture, prompt consideration of a metastasis. Carcinoma becomes even more of a diagnostic possibility when glioblastomas contain cohesive epithelioid cells with discrete borders (see fig. 3-59) or when true epithelial differentiation is present (see figs. 3-60–3-62). When single cell infiltration is not present, the distinction between glioblastoma and metastatic carcinoma requires immunohistochemical or ultrastructural analysis.

In theory, immunohistochemistry should distinguish metastatic carcinoma from largely undifferentiated glioblastomas that show GFAP reactivity at least focally. Such immunopositive neoplastic cells are usually best seen in perivascular regions where they exhibit a degree of differentiation more advanced than the majority population of anaplastic-appearing tumor cells with barely discernible cytoplasm. The presence of GFAP-reactive cells deep within the interior of the lesion also provides presumptive evidence of its primary glial nature. In any case, glioblastomas are S-100 protein immunoreactive, although this is not specific for glial differentiation but also seen in melanoma.

Immunohistochemical markers of epithelial differentiation, although useful for eliminating metastatic carcinoma from the differential diagnosis, should be interpreted with caution and in concert with staining for GFAP. The problem of nonspecific keratin reactivity in glial cells is best illustrated in brain tissue adjacent to metastatic carcinoma, wherein reactive astrocytes can be strongly cytokeratin positive, particularly when antisera such as for AE1/AE3 are used. Not infrequently, astrocytes stain more intensely for cytokeratins than they do for GFAP. In one study, cytokeratins AE3, AE5, and KS-1A3 stained variable numbers of normal and reactive astrocytes, whereas AE1, M20, CK-E3, and KS-B17.2 were negative (122). Another study found only rare focal immunoreactivity in glioblastomas with CAM5.2, CK7, and CK20, and no positivity with BerEp4 (104). The ever-enlarging repertoire of "tissue specific" antibodies useful in identifying the source of metastases is discussed in chapter 20.

The relationship of glioblastoma and oligodendroglioma is a difficult issue and, given the spectrum of oligodendroglioma-like changes seen focally in many glioblastomas, cannot be resolved in all instances short of molecular characterization, and even the issue of mixed glioma remains. Small cell glioblastomas, frequently calcified, are particularly likely to be confused with anaplastic oligodendroglioma (116). When highly anaplastic, some oligodendroglial tumors have many features in common with glioblastoma, namely, high cellularity, brisk mitotic rate, vascular proliferation, and necrosis with pseudopalisading. If the neoplastic cells have obvious, unequivocal oligodendroglial features, we classify such tumors as anaplastic oligodendroglioma, grade III. The issue of grade IV oligodendroglioma is discussed later. Perivascular halos alone should not be over interpreted as sufficient criteria for the diagnosis. The oligodendroglioma quality may be noted in the pathology report.

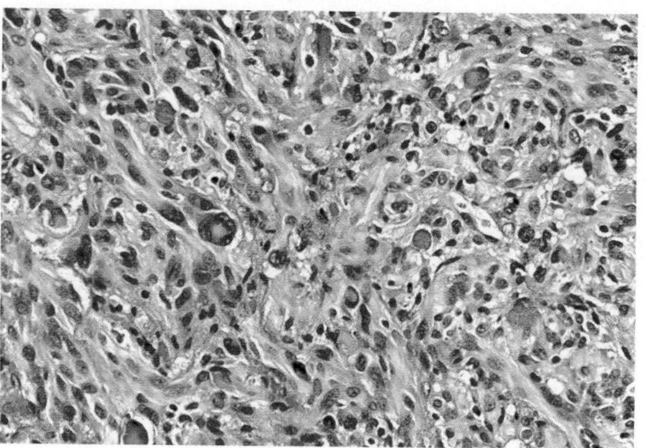

Figure 3-80

PLEOMORPHIC XANTHOASTROCYTOMA SIMULATING GLIOBLASTOMA

Cellular pleomorphism and high cellularity are incomplete and unreliable evidence of anaplasia. Note the eosinophilic granular bodies.

Occasional small cell glioblastomas feature perivascular pseudorosettes resembling those of ependymoma (see fig. 3-292). Distinguishing such lesions from ependymoma is based primarily upon the often diffusely infiltrative quality and lack of necrosis with pseudopalisading of the former. In addition, ependymomas may exhibit EMA-positive microlumens.

The often difficult distinction of giant cell glioblastoma from pleomorphic xanthoastrocytoma (fig. 3-80) is discussed with the latter tumor. The presence of glomeruloid vascular proliferation, and sometimes necrosis, may create the impression that a pilocytic astrocytoma is a high-grade glioma, even glioblastoma (fig. 3-81). This issue is discussed with pilocytic astrocytoma.

Primary central nervous system lymphoma (PCNSL) can enter into the differential

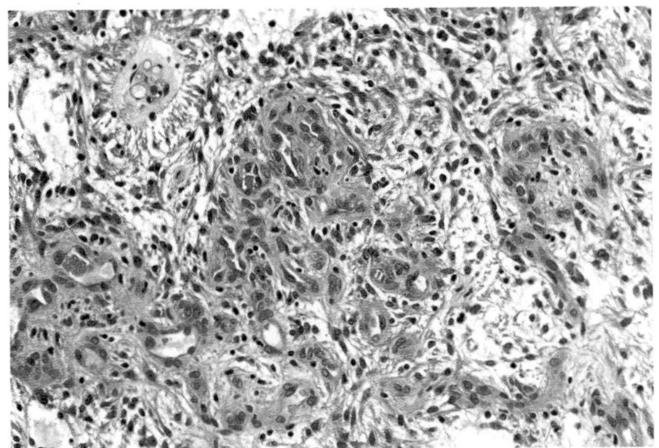

Figure 3-81

PILOCYTIC ASTROCYTOMA RESEMBLING GLIOBLASTOMA

Florid glomeruloid vascular proliferation is present in some pilocytic astrocytomas.

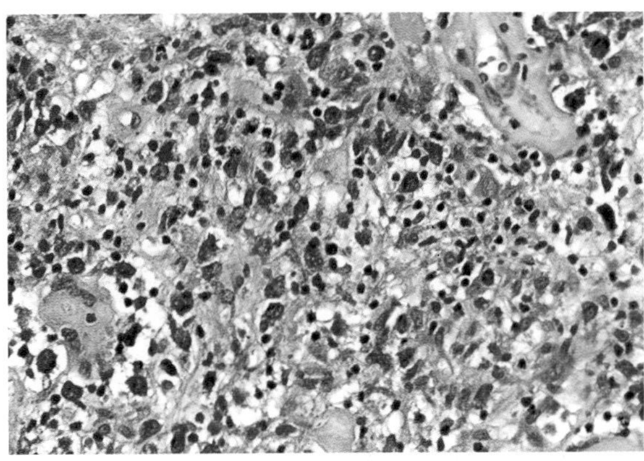

Figure 3-82

PRIMARY CENTRAL NERVOUS SYSTEM LYMPHOMA (PCNSL) SIMULATING GLIOBLASTOMA

There is enough cytologic polymorphism in some PCNSLs to simulate glioblastoma. Both tumors diffusely infiltrate the brain.

diagnosis, most commonly when an older patient presents with a contrast-enhancing intracerebral mass. PCNSL becomes even more suspect with a multicentric or subependymal process, particularly when it occurs in sites unusual for glioblastoma such as the fornix, septum pellucidum, brain stem, and cerebellum. Sporadic lymphomas typically are homogeneously enhancing without the ring configuration so typical of glioblastomas. Central necrosis is infrequently seen in lymphomas except those associated with immunosuppression, particularly in patients with AIDS.

In frozen sections, lymphoma is characterized by patchy hypercellularity due largely to tumor angiocentricity, conspicuous reactive gliosis, frequent apoptosis, histiocytic reaction, and general lack of microvascular proliferation. Lymphomas can, however, appear similar to glioblastomas if intrinsically pleomorphic (fig. 3-82), or when general nuclear roundness is artifactually altered by the freezing process. In brief, the cells of lymphoma lack cohesion, detach individually, possess scant cytoplasm, and feature generally round or reniform, vesicular nuclei with coarse chromatin and sizable nucleoli. Antibodies used to identify lymphoid neoplasms are discussed with PCNSL, later in the chapter. PCNSLs are GFAP negative.

Treatment. Glioblastomas have frustrated virtually every attempt at curative therapy. Not only are they beyond the reach of local control when first detected, but resistant clones emerge to reverse any initial radiotherapeutic or chemotherapeutic success. Thus, it is not surprising that many treatment modalities and combinations thereof have been applied. At present, baseline, or "standard," treatment after the initial diagnosis consists of radiotherapy, with most patients treated with chemotherapy as well. Radiotherapy is usually administered in a focused, conformal fashion, which spares as much normal brain tissue as possible. The maximal dose is directed to the contrast-enhancing component of the tumor, and a smaller dose to the abnormal area and a peripheral margin as seen on T2-weighted MRI. Chemotherapy may be given as an adjuvant, i.e., after radiotherapy, or may be held in reserve until the all too predictable time of recurrence.

An unknown, but not trivial, percentage of patients die with, but not directly of, the tumor (102,123,124). Pulmonary embolism is not uncommon.

Patterns of Recurrence. At the time of death, densely cellular tumor is usually seen at the original tumor site (fig. 3-83), frequently involving adjacent fiber tracts such as the corpus callosum, fornix, or anterior commissure (see figs. 3-43, 3-44). Contralateral recurrence is not uncommon. On occasion, usually in cases in which prolonged local control is achieved, tumors recur far from the original site. Extensive subependymal

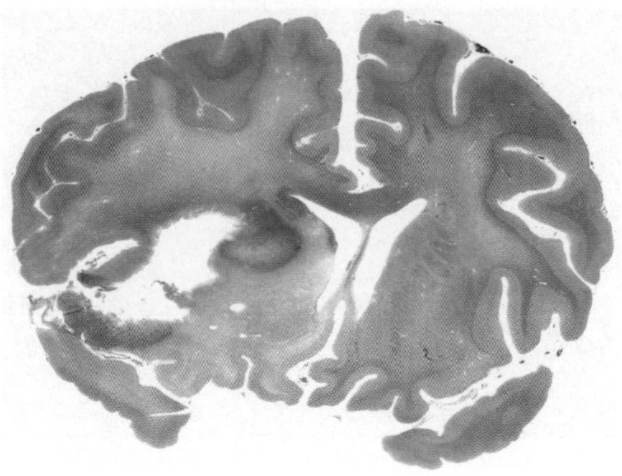

Figure 3-83

RECURRENT GLIOBLASTOMA

Seen as the darkly stained region in a section stained with cresyl fast violet, glioblastomas usually recur in and around the original tumor bed.

spread is common. Cerebrospinal dissemination occurs in about 10 percent of cases and appears to be more frequent in patients with long postoperative survival periods (125). Only a rare lesion metastasizes systemically to such sites as bone, lymph node, liver, or lung (126–135).

Prognosis. Despite radiotherapy, the median survival period of patients with cerebral glioblastoma is approximately 12 months. Survival rates depend in part on the age and neurological status of the patient, since those that are older or have a poorer neurological performance status do less well (66,67). In one population-based study where all patients were considered, and a substantial proportion, 30 percent, of the patients were over the age of 70 years, survival rates at 6 months, 1 year, 2 years, and 3 years were only 42 percent, 18 percent, 3 percent, and 1 percent, respectively (66,67). Noteworthy in this study was an operation rate for partial or complete resection of only 54 percent.

Longer survival periods are more likely in younger patients and, in the opinions of some observers, in cases in which a better differentiated astrocytoma was the precursor lesion (secondary glioblastoma) (136), but this may be largely the effect of the younger age of such patients (66,67). The especially poor outlook for elderly patients with glioblastomas is well known (137). Although difficult to confirm, the extent of resection may also affect survival (138,139). Patients with tumors of giant cell type appear to fare somewhat better (140,141), perhaps in part as a consequence of the degree of tumor circumscription. In one study, patients with giant cell glioblastoma, defined as tumors with greater than 20 percent multinucleated giant cells, were younger than other GBM patients and survived longer, although the survival difference was not significant (120a). An oligodendroglioma appearance was more common in glioblastomas with long-term survival in once study, but there was no overall prognostic significance of an oligodendroglioma component when corrected for age and gender (120a). The issue of grade IV mixed glioma is discussed in the section on oligodendrogliomas. Occasional supratentorial glioblastomas seed CSF pathways (77).

Understandably, the prognosis associated with glioblastoma of the brain stem is especially poor (142). Subarachnoid dissemination is common. Short survival periods are also expected for patients with spinal cord glioblastomas, in which spinal CSF dissemination is especially likely (10–12,71).

Due in part to the overall short survival periods and the strong (negative) prognostic impact of advancing patient age, it has been difficult to identify independent molecular prognostic factors, or therapeutic targets. Loss on chromosome 10 was a negative prognostic factor in one study (66,67). In regard to specific abnormalities, *EGFR* has received the most attention, with variable opinions as to the prognostic significance of amplification. One study found such amplification to be a strong prognostic predictor, but, paradoxically, one that was associated with longer survival periods in patients older than 60 years (76). It has also been suggested that *EGFR* overexpression is a negative prognostic factor only when associated with wild type *p53* (143). One study found that coexpression of EGFRvIII and PTEN is associated with responsiveness to inhibitors of EGFR kinase (143a). Silencing of the *MGMT* gene by promoter methylation, thus compromising repair of drug-induced DNA damage in tumor cells, was a prognostic factor in GBM patients with the alkylating agent temazolamide (143b).

Other studies, while confirming the close association between *EGFR* amplification and

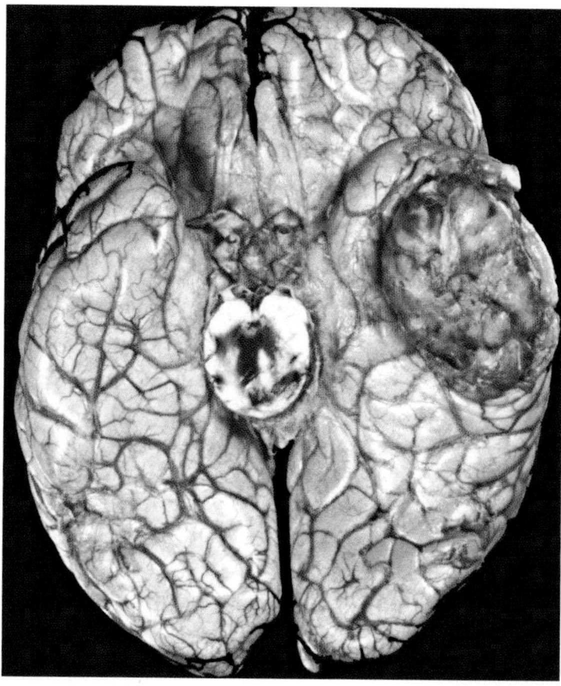

Figure 3-84

GLIOSARCOMA

Gliosarcoma is usually a well-circumscribed mass. This lesion produced terminal transtentorial herniation with secondary brain hemorrhages.

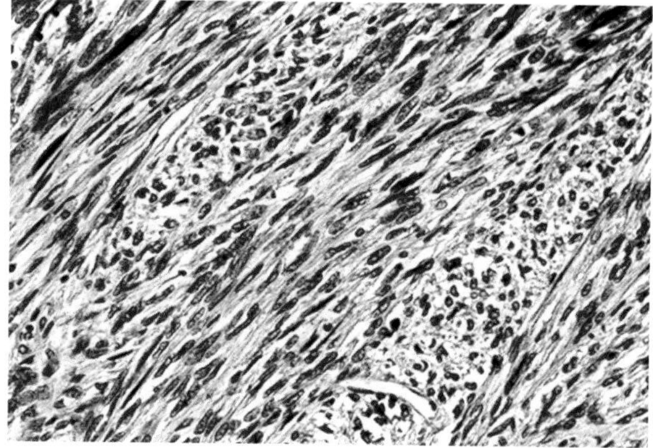

Figure 3-85

GLIOSARCOMA

The sarcoma often has the classic herringbone pattern of fibrosarcoma.

glioblastoma, and its negative prognostic impact in histologically diagnosed grade III astrocytomas, find *EGFR* not predictive of outcome in glioblastoma patients (144,145). The significance of mutations in *p53* is unclear. No prognostic effect has been noted in some studies (144), but others have found longer survival periods in the presence of a mutation (146). Patients with tumors in which the *MGMT* (O^6-methylguanine-DNA methyltransferase) promoter was methylated did significantly better than those with tumors in which this gene promoter was not methylated (146a).

Gliosarcoma

Definition. *Gliosarcoma* is a malignant glioma that has coexisting glial and mesenchymal components. Largely as a reflection of the molecular findings discussed below, gliosarcoma is now considered a variant of glioblastoma rather than a distinct nosological entity.

Clinical Features. Approximately 2 percent of glioblastomas have an eye-catching, although prognostically insignificant, sarcomatous component. These tumors occur at the same sites as do glioblastomas (147,148).

Gross Findings. Unlike the more infiltrative and hence macroscopically ill-defined glioblastoma, collagen-rich gliosarcoma is typically firm and well circumscribed (fig. 3-84).

Microscopic Findings. The sarcomatous element is a spindle cell proliferation, exhibiting either the precise herringbone architecture so typical of well-differentiated fibrosarcoma (fig. 3-85) or the disorganized fascicles and cellular pleomorphism of malignant fibrous histiocytoma (149). In practice, most lesions fall between these extremes. In some tumors, the mesenchymal element is topographically related to blood vessels, and appears to merge with proliferating vascular or adventitial tissue. In others, vasocentricity is less conspicuous, the pattern being a complex, marbled admixture of obvious glioblastoma and sarcomatous tissue (fig. 3-86). In still others, the distinction between glial and mesenchymal components is not readily apparent (fig. 3-87). Reticulin, rich in the mesenchymal component, circumscribes but does not penetrate islands of glial tissue (fig. 3-88).

Lines of mesenchymal differentiation other than fibroblastic include epithelial (150), myofibroblastic, cartilage (fig. 3-89) (150), bone (fig. 3-90) (151), angiosarcoma (152), smooth muscle (153), and, on rare occasion, striated muscle (154,155). Multiple forms of mesenchyme may coexist in the same lesion (149). As is discussed

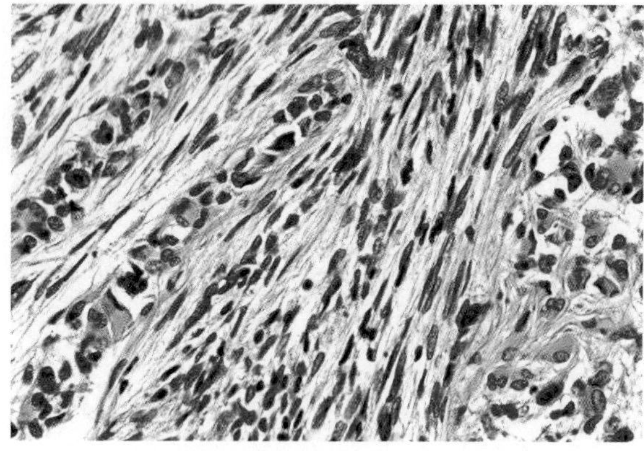

Figure 3-86

GLIOSARCOMA

Mesenchymal and glial components are clearly distinguishable in some cases. Glassy cytoplasm with processes helps identify the glial component.

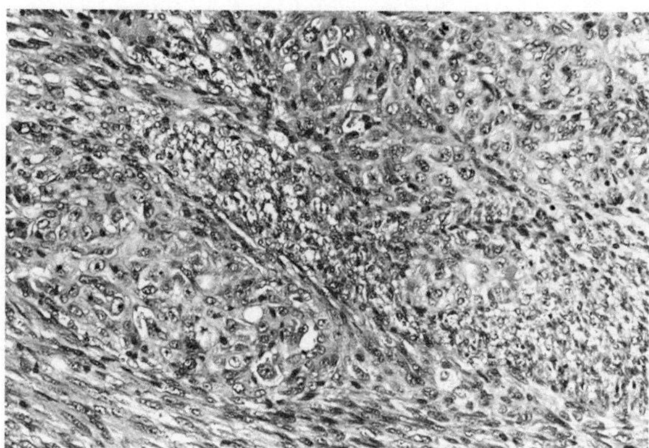

Figure 3-87

GLIOSARCOMA

Glial and sarcomatous components can be difficult to distinguish in H&E-stained sections alone. Here, the segregated glial cells are plumper and have more prominent nucleoli than those of the surrounding sarcoma.

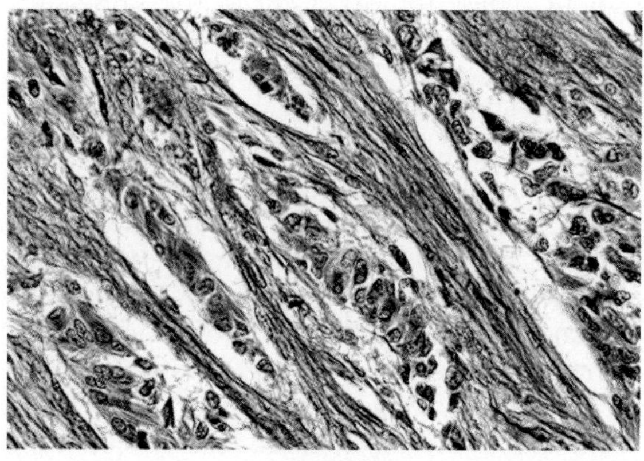

Figure 3-88

GLIOSARCOMA

Pericellular reticulin is abundant in mesenchymal regions, but absent in the glial tissue.

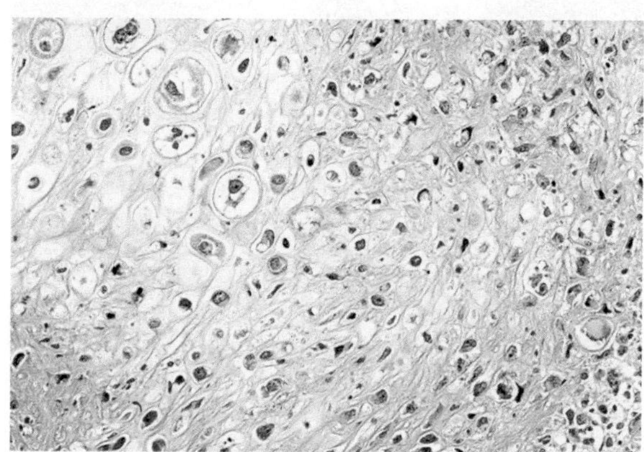

Figure 3-89

GLIOSARCOMA

Cartilage is a common form of metaplasia in gliosarcoma.

below, some gliosarcomas are such an intimate admixture of mesenchymal and glial elements that the sarcoma appears to result from metaplasia of the glial component. The simple presence of intense desmoplasia in a malignant glioma is not the equivalent of gliosarcoma.

In most gliosarcomas, the glial component is astrocytic and in every way resembles glioblastoma. It can be inconspicuous in advanced lesions where the glia are dispersed in the sarcoma as clusters or even as single cells whose detection requires immunohistochemical confirmation of GFAP (see fig. 3-92). With time, the sarcoma may so overshadow the parent glioma that the whole lesion appears solid and demarcated, and the only obvious glial component is that which clings to the surface of the excised mass.

Some cases have striking epithelial features, with tubules, canals, glands, and even squames (150,156,157). These features are also present in some glioblastomas without a mesenchymal component.

Although exuberant, the mesenchymal component in desmoplastic malignant glioma is not always clearly neoplastic, and regional gradations in the degree of cytologic atypia are

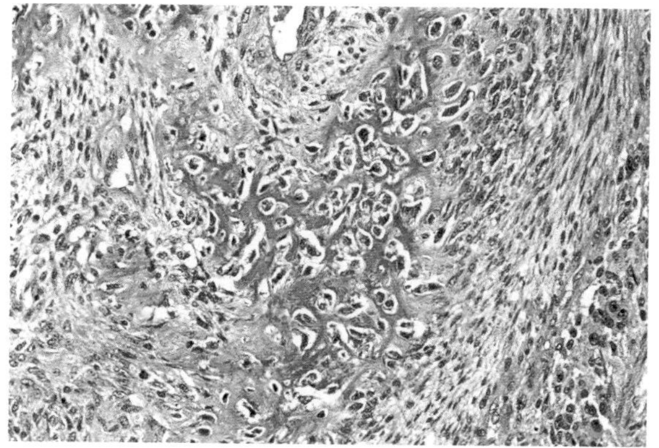

Figure 3-90

GLIOSARCOMA

Osseous metaplasia is an unusual transformation in gliosarcoma.

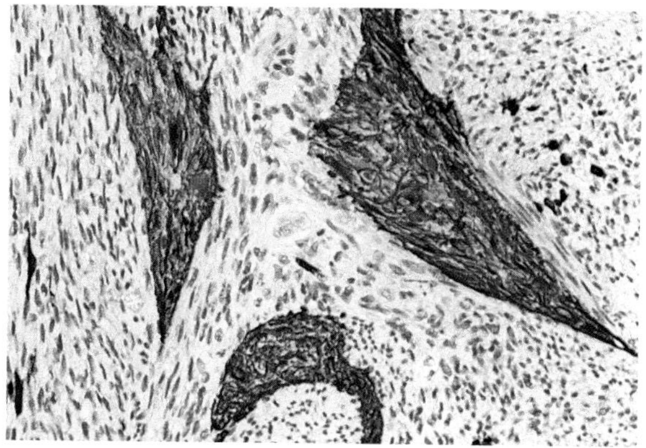

Figure 3-91

GLIOSARCOMA

Islands of glial tissue are strongly immunoreactive for GFAP.

common. Often, it seems that tissue that is, in areas, overtly atypical has matured into tissue that is cicatricial rather than neoplastic. Such lesions may demonstrate active microvascular proliferation with perivascular zones of plump adventitial pericytes and fibroblasts. Away from the immediate perivascular zones, however, proliferation appears to stabilize or even regress, as nuclear to cytoplasmic ratios, mitotic activity, and nuclear atypia all diminish as collagen deposition proportionally increases. The same is true of gliomas, nearly all malignant, that secondarily involve the dura to incite intense desmoplasia and mimic sarcoma.

In practical terms, the identical prognoses of glioblastoma and gliosarcoma relieve the pathologist of the difficult task of making an arbitrary distinction between gliosarcoma and glioblastoma with a robust mesenchymal reaction.

Immunohistochemical Findings. Whether disposed in large lobules (fig. 3-91) or as isolated cells (fig. 3-92), the glial component is identified by its reactivity for GFAP. The finding of S-100 protein and even GFAP reactivity in the sarcomatous element, as well as the molecular data described below, support the notion that the sarcoma results by metaplasia from glial cells. Epithelial components are immunoreactive for cytokeratins (150). Immunoreactivity for p53 is found in some cases, in both glial and mesenchymal areas (158). The immunophenotype of the sarcoma component varies considerably, ranging from nonspecific vimentin positivity to tumors in which smooth muscle actin suggests myofibroblastic or myoid differentiation (153,155).

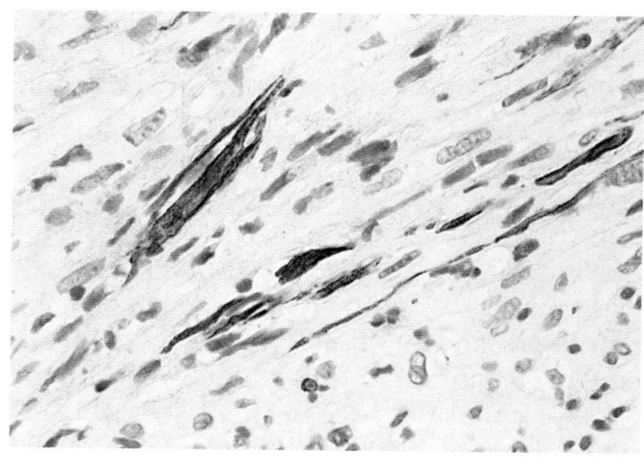

Figure 3-92

GLIOSARCOMA

GFAP-positive glia can be widely dispersed in mesenchymal regions.

Ultrastructural Findings. In addition to obvious astrocytes, poorly differentiated elements and cells resembling fibroblasts, histiocytes, and myofibroblasts have been noted (159).

Cytologic Findings. The glial components are astrocytes, often including gemistocytes, typical of glioblastoma. Sarcomatous elements are variably undifferentiated spindle-shaped, rhabdoid, chondroid (fig. 3-93), and osteoblastic (160).

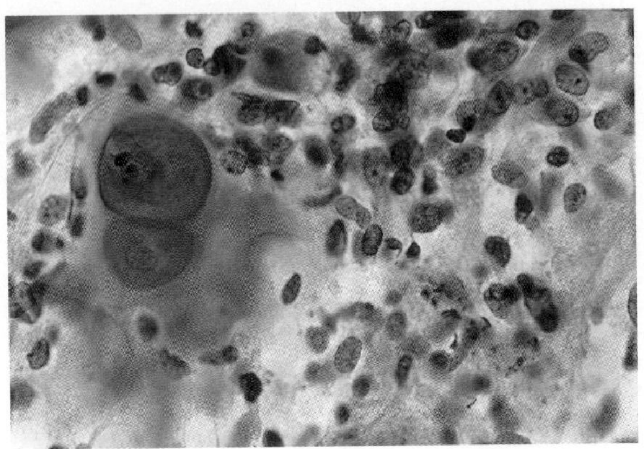

Figure 3-93

GLIOSARCOMA: SMEAR PREPARATION

Chondroid metaplasia is not uncommon in gliosarcoma.

Molecular Findings. The finding of identical mutations in *p53* and similar comparative genomic hybridization profiles in glial and mesenchymal components has been interpreted as evidence that gliosarcomas are clonal, albeit phenotypically biphasic, neoplasms (161–164). This is in accord with the suggestion that gliosarcomas are only intensely desmoplastic glioblastomas (165), although in some cases, there is clearly a phenotypically mesenchymal component, whatever its origin. One study found the molecular changes consistent with primary glioblastoma except for the absence of *EGFR* amplification (163).

Treatment and Prognosis. Gliosarcomas are no more or less malignant than glioblastomas (148,166,167). Occasional gliosarcomas erode through the skull (168).

Granular Cell Astrocytoma

Definition. *Granular cell astrocytoma* is an infiltrative neoplasm that has a prominent component of lysosome-rich granular cells.

General Features. Infrequently, large numbers of granular cells appear within an otherwise typical diffuse astrocytic neoplasm. When granular cells are the predominant or even exclusive element, the resemblance to the parent lesion is obscured and the lesion can be confused readily with non-neoplastic processes such as cerebral infarcts and demyelinating disease, with granular cell neoplasms of the infundibulum (a presumably pituicyte-derived neoplasm), or with very rare peripheral nerve tumors in intimate association with brain or spinal cord.

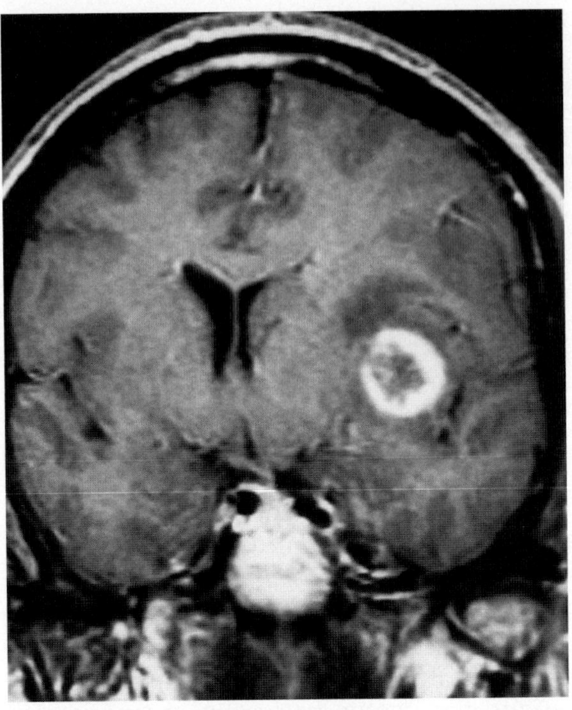

Figure 3-94

GRANULAR CELL ASTROCYTOMA

Appropriately since most are glioblastoma variants, granular cell astrocytomas often contrast enhance in a rim configuration.

Clinical and Radiologic Findings. Clinical presentation, patient profile, and neuroradiologic aspects of these tumors are nonspecific, and simply those of any underlying diffuse astrocytic neoplasm. The vast majority arise in the cerebral hemispheres (169–171). Since granular cell change usually occurs in high-grade diffuse astrocytomas, most are contrast enhancing (fig. 3-94).

Gross Findings. Diffuse astrocytomas with granular cell components are sufficiently rare that no general statements can be made about macroscopic findings. As a rule, the tumors are infiltrative and ill-defined. Grade IV examples resemble, and in fact are, glioblastomas.

Microscopic Findings. Granular cell astrocytomas consist in large part of distinctive cells with discrete borders and granular cytoplasm (figs. 3-95, 3-96). These often form homogeneous sheets, but in some cases combine intermediate, less granulated forms with cell processes (fig. 3-

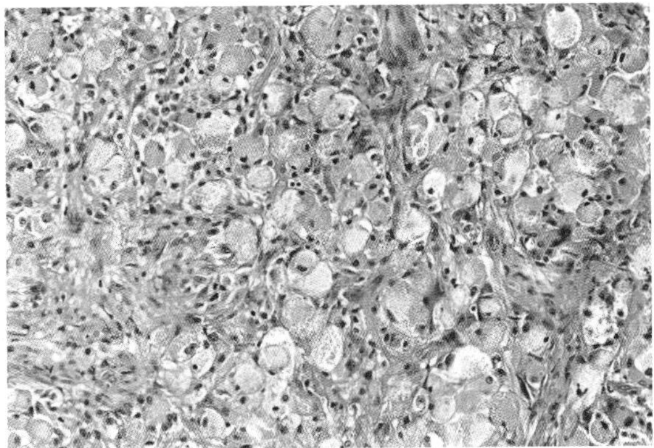

Figure 3-95

GRANULAR CELL ASTROCYTOMA

The compact population of granular cells is distinctive even at low magnification.

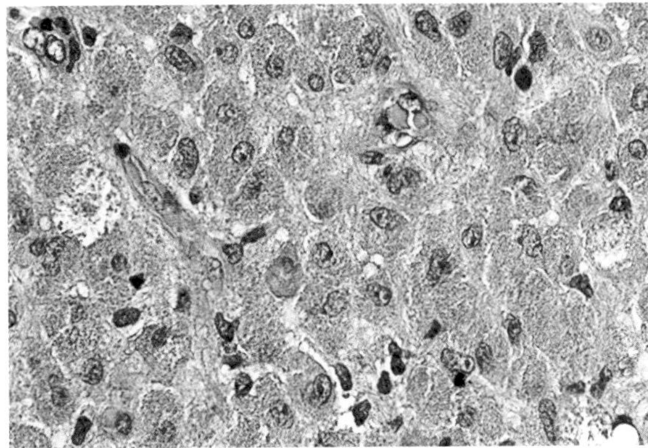

Figure 3-96

GRANULAR CELL ASTROCYTOMA

In spite of the aggressive nature of the lesion, granular cells themselves are cytologically bland and mitoses few.

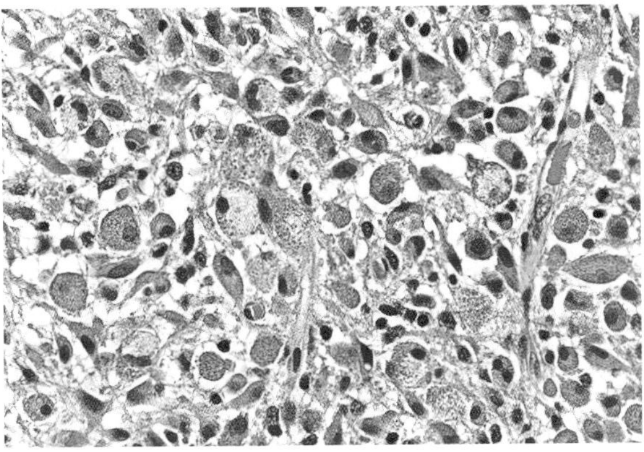

Figure 3-97

GRANULAR CELL ASTROCYTOMA

Short tumor cell processes suggest a glial, and not macrophage, origin of the granular cells.

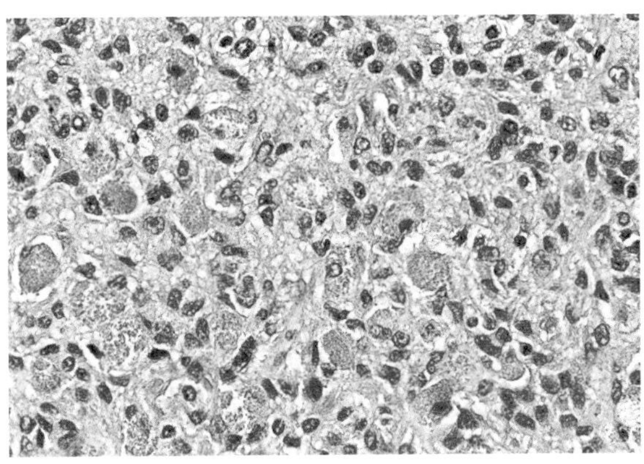

Figure 3-98

GRANULAR CELL ASTROCYTOMA

Transition to conventional diffuse astrocytoma, here grade III, is common, but is not present in all specimens.

97) and regions of typical, diffuse high-grade astrocytoma (fig. 3-98). The cytologic transition may be abrupt or gradual. Areas of infiltrative growth, while the rule, are not always present.

The distinctive cytoplasmic granularity is due to the accumulation of PAS-positive granules that often displace the nucleus (fig. 3-99). Perivascular lymphocytes mark some cases. As expected, nuclear cytologic features vary somewhat with tumor grade, and as a rule, the nuclei are larger and coarser than those of normal astrocytes. Conspicuous nucleoli are apparent in some cases, however, they lack the marked hyperchromasia and pleomorphism of glioblastoma. To a limited extent, reticulin may invest individual cells (fig. 3-100).

Immunohistochemical Findings. Although the granular cells are generally negative for GFAP, faint staining may be seen in some (fig. 3-101) (169,170). Often, especially in areas of advanced granular cell change, the granular cells are positive only along their circumference. As a result, it may be difficult to determine whether reactivity resides in the tumor cell or belongs to a nearby, compressed reactive astrocyte. Immunoreactivity for macrophage markers such as CD68 varies in proportion to the granule content, but there is often faint

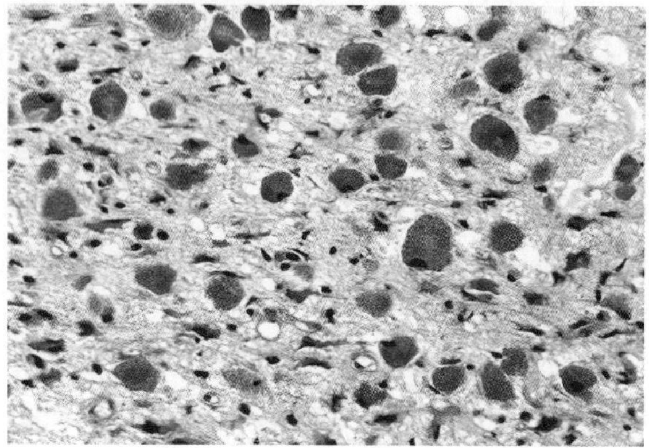

Figure 3-99

GRANULAR CELL ASTROCYTOMA

The granular cells are periodic acid–Schiff (PAS) positive.

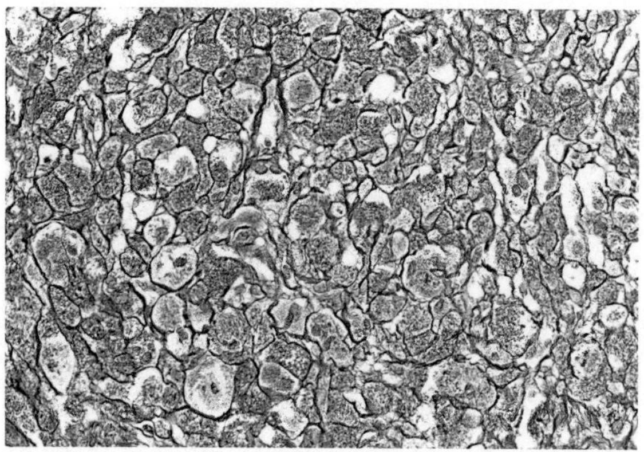

Figure 3-100

GRANULAR CELL ASTROCYTOMA

Abundant reticulin, in some cases pericellular, is present in a minority of cases.

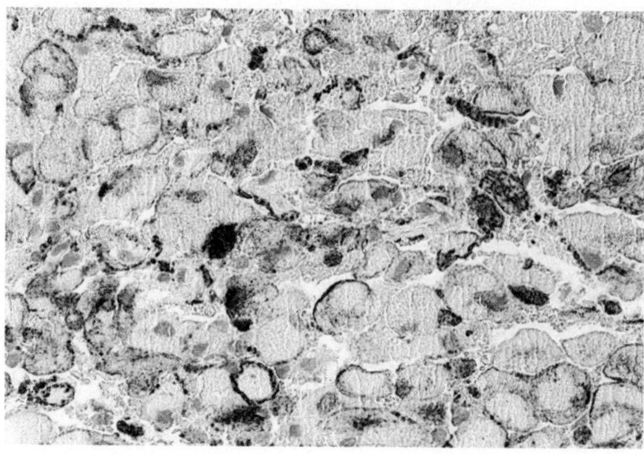

Figure 3-101

GRANULAR CELL ASTROCYTOMA

A superificial rim of cytoplasmic staining for GFAP establishes the nature of the granular cells. In many cases, however, it is difficult to determine if immunoreactivity for GFAP is present in tumor cells or only in processes of adjacent reactive astrocytes.

staining of the granular cells in accord with their lysosomal content (fig. 3-102, left). In our experience, another macrophage marker, HAM56, may be negative (fig. 3-102, right). Whereas the granular cells often have a low MIB-1 labeling index (169,172), the index may be high in any nongranular neoplastic cells in the same lesion (169).

Ultrastructural Findings. The cytoplasm of the granulated cells is filled with secondary lysosomes (171,173). The number of intermediate filaments varies considerably.

Cytologic Findings. The cytoplasmic granularity is clearly seen. Some cells have processes in accord with the astrocytic nature (fig. 3-103).

Molecular Findings. As would be expected given the apparent relation of the lesion to diffuse astrocytoma, allelic losses on chromosomes 1p, 9p, 10q, 17p, and 19q are common (174).

Differential Diagnosis. The cytologically bland cells resemble macrophages, and lesions such as cerebral infarct and demyelinating disease must be considered. Granular cells are considerably larger than macrophages and their cytoplasm is coarsely granular, whereas that of macrophages is finely granular or foamy (fig. 3-104). Nuclei of granular cells are more hyperchromatic, and transition to more conventional astrocytoma occurs in some cases. Nevertheless, these are subjective or semiquantitative distinctions, and it is often difficult to dispel the feeling that the lesion is reactive even when all of the above features of granular cell astrocytoma are present. A decision may await immunohistochemistry, which unfortunately may also be inconclusive since the lysosome-rich cells may be immunoreactive for CD68 and negative for GFAP. Thus, it can be difficult to finalize the diagnosis of what is, in essence, a malignant glioma on the basis of cells that are cytologically bland and resemble macrophages.

Treatment and Prognosis. Although the lesions may not appear cytologically or histologically anaplastic, granular cell astrocytomas, overall, behave as grade III or IV infiltrating

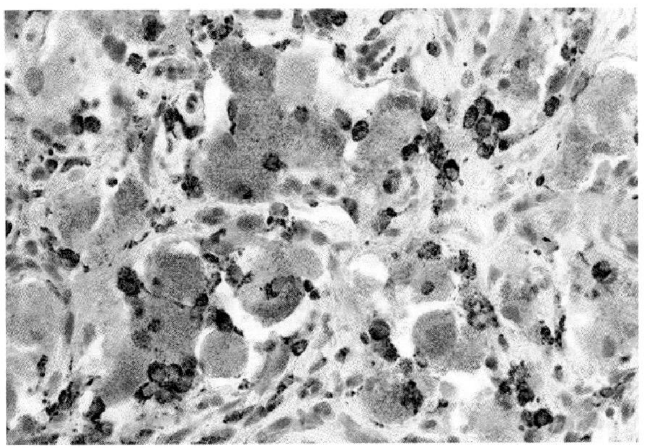

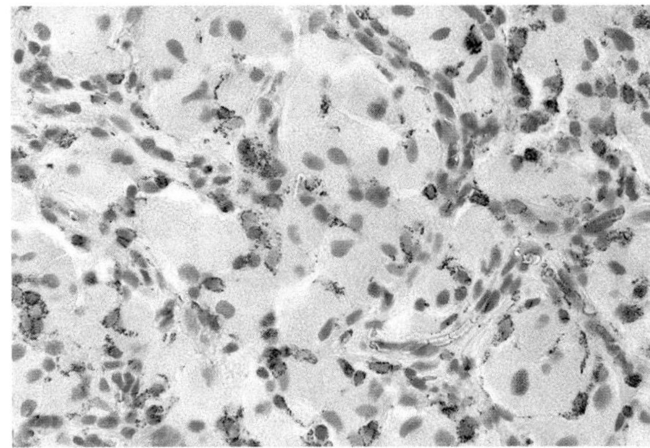

Figure 3-102

GRANULAR CELL ASTROCYTOMA

Left: The granular cells may be immunoreactive for CD68.
Right: Staining for another macrophage marker, HAM56, may be negative.

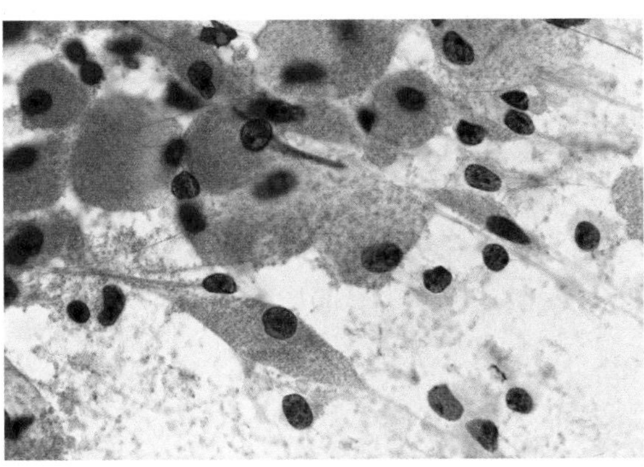

Figure 3-103

GRANULAR CELL ASTROCYTOMA: SMEAR PREPARATION

While the cells have distinct cell borders that resemble those of macrophages, they also have short processes typical of astrocytes.

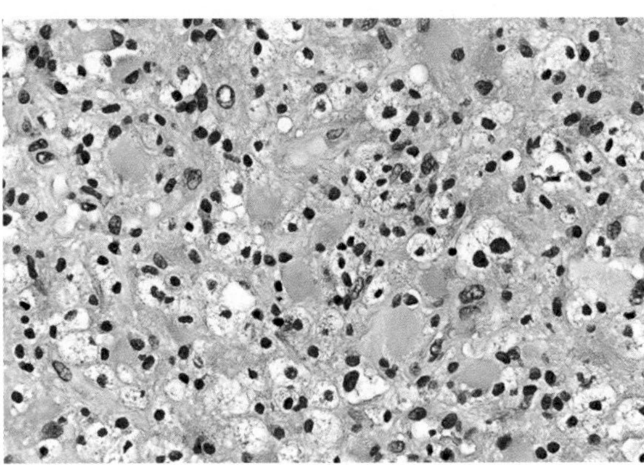

Figure 3-104

DEMYELINATING DISEASE RESEMBLING GRANULAR CELL ASTROCYTOMA

Macrophages closely resemble neoplastic granular astrocytes, although the phagocytes are smaller, more uniform, and finely rather than coarsely granular.

astrocytomas. Despite radiation therapy, most patients with anaplastic or GBM-like examples die within 2 years of surgery (169–172).

Gliomatosis Cerebri

Definition. *Gliomatosis cerebri* is a remarkably extensive, diffusely infiltrating glioma of astrocytic, or rarely oligodendroglial, type. It is often grade III, but some cases are histologically and biologically grade II.

General Features. The term "gliomatosis" is generally reserved for lesions in which the infiltrative capability is seemingly out of proportion to the degree of anaplasia observed and often divorced from any tendency to form zones of concentrated cellularity. Although the occasional oligodendroglioma is sufficiently diffuse to be a "-tosis" (175,176), the entity is more closely associated with infiltrating astrocytomas. By convention, sizable infiltrating tumors with a solid, often necrotic, epicenter present at the time of initial presentation are not included in the gliomatosis category.

Clinical Features. The lesion usually occurs in adults (175,177,178), but children may be affected. Symptoms depend upon a tumor's location

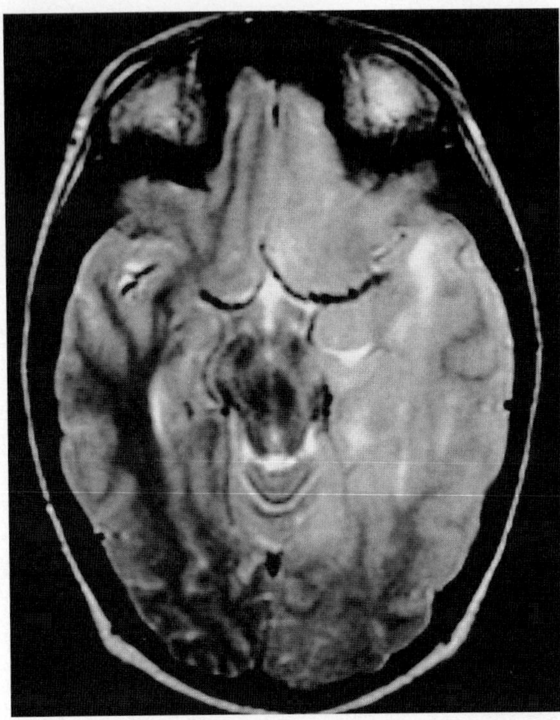

Figure 3-105

GLIOMATOSIS CEREBRI

As is evident in this T2-weighted image, gliomatosis cerebri permeates large regions of brain. As here, bilaterality is not uncommon. (Fig. 3-55 from Fascicle 10, 3rd Series.)

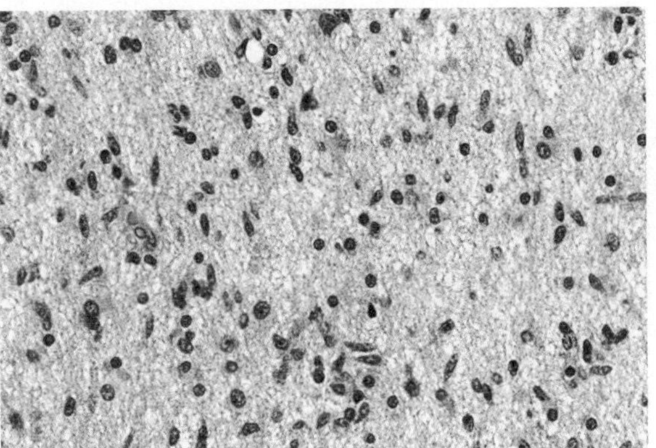

Figure 3-106

GLIOMATOSIS CEREBRI

The cells are diffusely infiltrative within the cortex. While there is no "typical" appearance since the lesion is defined by extent rather than cytologic features, the small, elongated, bland nuclei are most common.

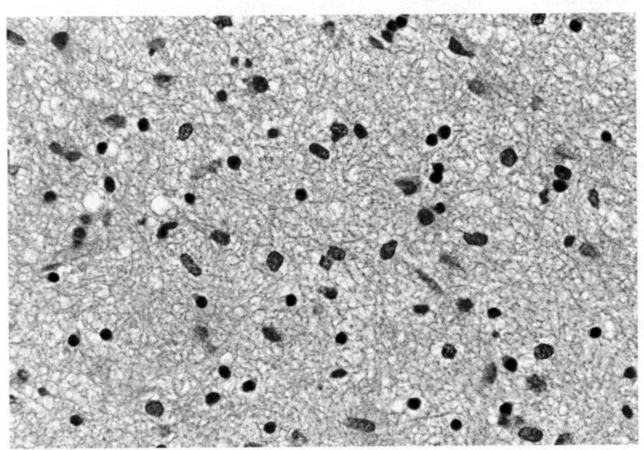

Figure 3-107

GLIOMATOSIS CEREBRI

Distinguishing gliomatosis from gliosis depends largely on the degree of cytologic atypia. The distinction can be difficult, and in some cases impossible, as in this case.

or extent. Seizures, personality changes, aphasia, and hemiplegia are common (175,178,179)

Radiologic Findings. By definition, gliomatosis cerebri is radiologically widespread, and the lesion often occupies the greater part of a cerebral hemisphere, if not both sides of the brain. In contiguity, it may thus involve the supratentorial compartment, posterior fossa, and even the spinal cord. At a minimum, three lobes should be involved by some definitions. The basal ganglia and thalamus are often involved. As expected, based upon the histologic features of typical cases, the lesion is an extensive area of high signal intensity in T2-weighted or FLAIR MRI scans in many cases (fig. 3-105) (175,177–181), with surprisingly little mass effect given the size of the lesion. Contrast-enhancing foci, which may appear late in the course of the disease, correspond to zones of anaplastic transformation.

Gross Findings. Gliomatosis cerebri is not only widely invasive and ill-defined, but may not be evident at all, even in a generous lobectomy specimen. Often, however, the gray-white junction is effaced as it is with any infiltrating glioma.

Microscopic Findings. A wide variety of histologic and cytologic features is encountered, including considerable variation in cellularity. The latter is usually slight to moderate (figs. 3-106, 3-107), and in some cases so minimally increased that the diagnosis is difficult if not impossible to make. Some grade III cases are cytologically atypical and mitotically active (fig. 3-108). "Secondary structures," i.e., the accumulation of neoplastic cells around vessels, neurons,

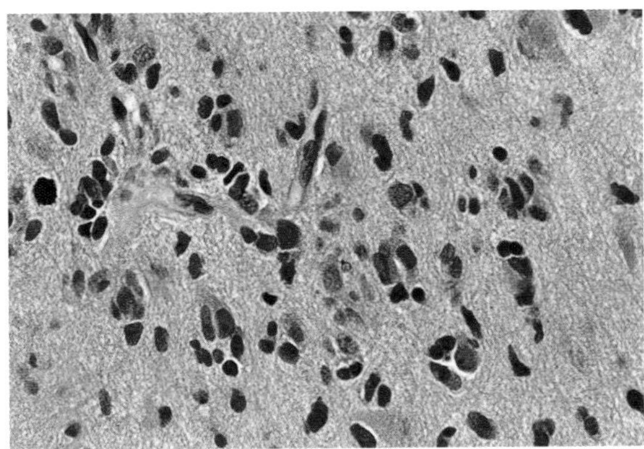

Figure 3-108

GLIOMATOSIS CEREBRI

While the classic lesion is well differentiated, some are grade III, with considerable nuclear hyperchromatism, atypia, and mitoses.

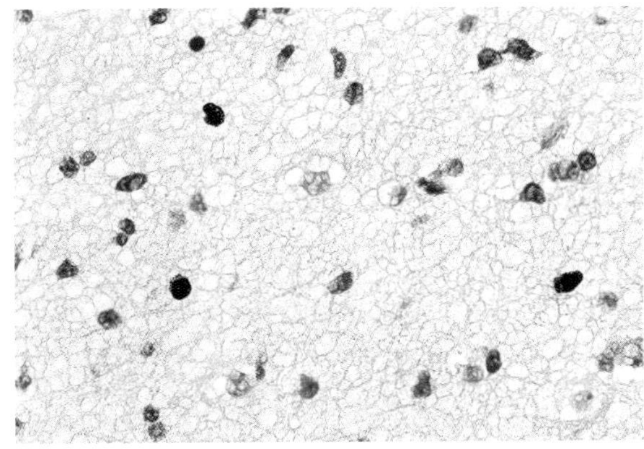

Figure 3-109

GLIOMATOSIS CEREBRI

Some cases of gliomatosis are grade III based on cytologic atypia and mitotic activity. The MIB-1 index is high in such cases, although difficult to estimate given the lesion's low cellularity.

and subpial and subependymal regions, are common. Small foci of anaplastic transformation are present in some cases. Unlike ordinary, highly cellular malignant gliomas, gliomatosis cerebri produces little parenchymal destruction. Vascular proliferation and necrosis are absent or sparse.

The degree of cytologic atypia also varies, ranging from bland cells difficult to distinguish from normal or reactive glia, to those with overt nuclear atypia. As a rule, the number of mitoses is low per unit area, but the number at a given tumor cell density may be high in grade III lesions. While nuclear elongation is held by some to be particularly characteristic, nuclear conformation varies greatly, from round, through oval, to elongate.

In many cases, the cytoplasm is hardly discernible, particularly in H&E-stained sections. In other lesions there are cells with more cytoplasm or obvious processes that are overtly astrocytic. In most cases, the cytologic features are not those associated with a high-grade lesion, but those of a low- to medium-grade process.

Frozen Section. The features of diffuse astrocytoma are discussed above. A diagnosis of "infiltrating glioma" or "primary neoplasm, most consistent with glioma" suffices.

Immunohistochemical Findings. The elongated cell processes so inconspicuous in H&E-stained sections are only sometimes positive for GFAP. As a result, immunostaining for this marker is of little aid in diagnosis. Cells obviously more astrocytic in phenotype are of course GFAP positive. This is particularly true in areas of secondary structure formation wherein the cells are more numerous and likely to acquire visible cytoplasm and processes.

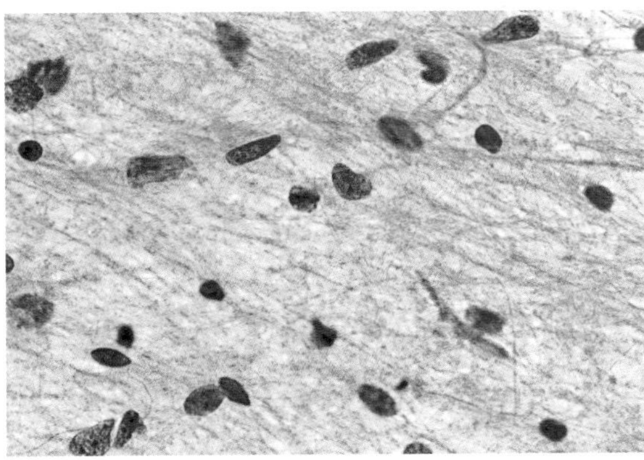

Figure 3-110

GLIOMATOSIS CEREBRI

The elongated, bland cells of the "typical case" are seen well in smear preparations.

The MIB-1 rate varies considerably, from values so low as to make it difficult to find a positive cell, to the levels usually associated with grade III astrocytoma. The presence of positive cells is supportive of the diagnosis, but not in itself diagnostic (fig. 3-109).

Cytologic Findings. The findings are those of diffuse astrocytoma (fig. 3-110).

Molecular and Cytogenetic Findings. No unique molecular profile has emerged although the number of studied cases is yet small. As suspected, given the phenotypical similarity of gliomatosis cerebri to diffuse astrocytoma, genetic findings such as mutations in *p53* (3 of 7 tumors) and combined *PTEN* mutation and *EGFR* overexpression in one case, have been reported (175). An additional case studied postmortem with sampling of multiple areas revealed widespread similar changes consistent with a clonal nature (182). A mutation in codon 234, in exon 7, a "hotspot" in diffuse astrocytomas, was found in almost all samples.

Differential Diagnosis. Gliomatosis cerebri can be difficult to distinguish from reactive gliosis or even normal brain, the extent and intensity of T2-weighted or FLAIR signal notwithstanding. These sometimes difficult issues are discussed in the sections on grade II and grade III diffuse astrocytomas.

Microgliomatosis, an exceedingly rare, poorly understood lesion, is difficult to exclude solely on histologic and cytologic grounds. Instead it is identified on the basis of immunostaining for macrophage markers (CD68, HAM56, Ricinis communis agglutinin). It is unclear just how many previously reported cases of gliomatosis cerebri actually were cases of microgliomatosis. One should be particularly suspicious in cases of "gliomatosis" featuring marked nuclear/cellular elongation, a characteristic of microglia or "rod cells" in reactive processes such as viral encephalitis.

Treatment and Prognosis. Given the problems inherent in defining the entity, it is difficult to generalize regarding the biologic behavior of gliomatosis cerebri. It is also no surprise that its clinical course varies considerably, with survival periods that range from months to many years (175,178). There was a median survival period of 14 months in one study (175). Examples featuring cytologic malignancy and high mitotic indices might be expected to behave more aggressively than those with cytologically bland nuclei and no mitoses, but this has been difficult to confirm given the paucity of patients (175).

Meningeal Gliomatosis

Definition. *Meningeal gliomatosis* is a glioma residing primarily within the leptomeninges.

General Features. The diagnosis of meningeal gliomatosis is applied rigorously only to leptomeningeal gliomas for which, despite careful search, no intraparenchymal source can be found. In practice, however, the term is used for neoplasms that, although not entirely leptomeningeal, have a parenchymal component that is so small as to be considered an ingrowth of the meningeal lesion. Without a careful autopsy it is obviously impossible to determine whether any glioma is truly primary in the leptomeninges. Even then skeptics would assume that an intraparenchymal primary neoplasm had escaped detection.

Primary diffuse meningeal gliomas may well be derived from small heterotopic nodules or from surface protrusions of glial tissue into the subarachnoid space (183,184). Favored sites include the region of the medulla near the foramina of Luschka and at the caudal end of the spinal cord. Small discrete gliomas in the subarachnoid space can arise from such foci (185,186).

We discuss meningeal gliomatosis in this section on astrocytic tumors since most are variants of diffuse astrocytomas. A rare, generally pediatric neoplasm in which oligodendrocyte-like cells fill the subarachnoid space has been referred to as *diffuse leptomeningeal oligodendrogliomatosis* (187–191). The nature of these latter lesions, which have not been proved to be oligodendroglial (or even glial), is not clear. Their biologic behavior is poorly understood.

Clinical Features. As might be expected given the diffuse nature of the process, there is no typical clinicopathologic setting. Patients of all ages are affected. Tumors are often predominantly intraspinal (192–195), but may diffusely affect both the intraspinal and intracranial compartments (196–198). Presenting symptoms, understandably variable, include headache, confusion, and cranial or spinal nerve deficits, among others. At presentation, many cases are assumed on clinical or radiologic grounds to be infectious processes such as tuberculous meningitis.

Radiologic Findings. Initially, the lesion may be an ill-defined, nonenhancing leptomeningeal process, but with evolution, it often becomes enhancing (fig. 3-111).

Gross Findings. The tumors thicken and opacify the leptomeninges, while encasing and infiltrating nerve roots and surrounding vessels (fig. 3-112).

Microscopic Findings. The tumors range from well-differentiated astrocytoma (fig. 3-113) to anaplastic astrocytoma (fig. 3-114, top) or even glioblastoma. Tumor cells often assume a spindle configuration and are flattened between reticulin fibers. At least some cells are immunoreactive for GFAP (fig. 3-114, bottom). Pericellular laminin immunoreactivity may also be present.

Differential Diagnosis. Since most meningeal gliomas are drop metastases, a primary spinal or intracranial intraparenchymal glioma must be sought. When densely packed and reticulin rich, the process may resemble meningeal sarcoma. The general reactivity of meningeal gliomatosis for GFAP confirms its glial nature, and excludes consideration of a mesenchymal, e.g., fibrohistiocytic, tumor.

Treatment and Prognosis. The prognosis is poor for patients with these progressive, diffuse lesions.

Protoplasmic Astrocytoma

Definition. *Protoplasmic astrocytoma* is a relatively circumscribed lesion composed of cells resembling the protoplasmic astrocytes of gray matter.

Clinical and Radiologic Findings. Little is known about this uncommon, rather poorly defined lesion about which there are no

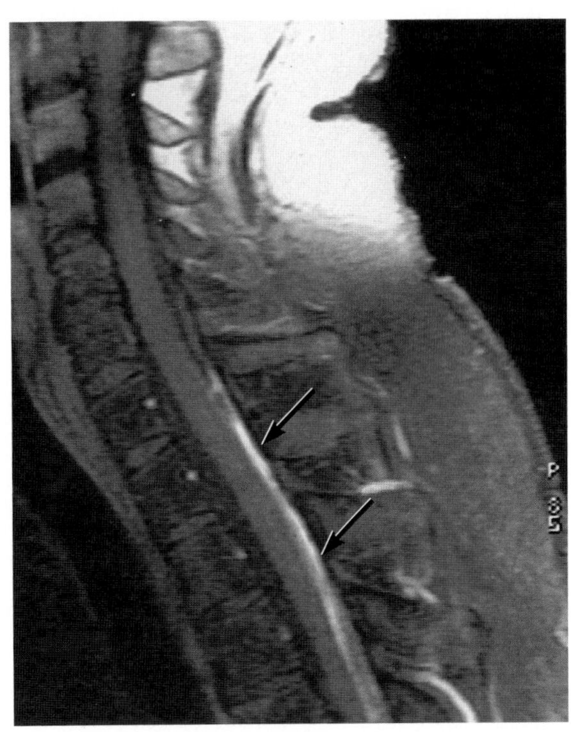

Figure 3-111

MENINGEAL GLIOMATOSIS

Leptomeningeal sheets of contrast enhancement (arrows) are present in a 36-year-old man with leg and arm pain. The lesion's histologic features are illustrated in figure 3-114.

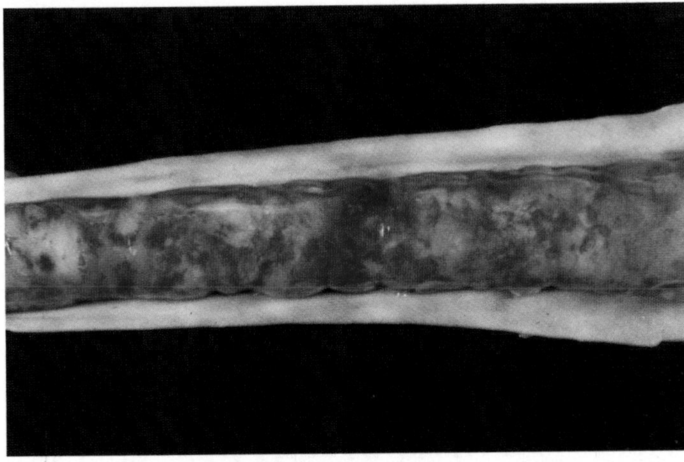

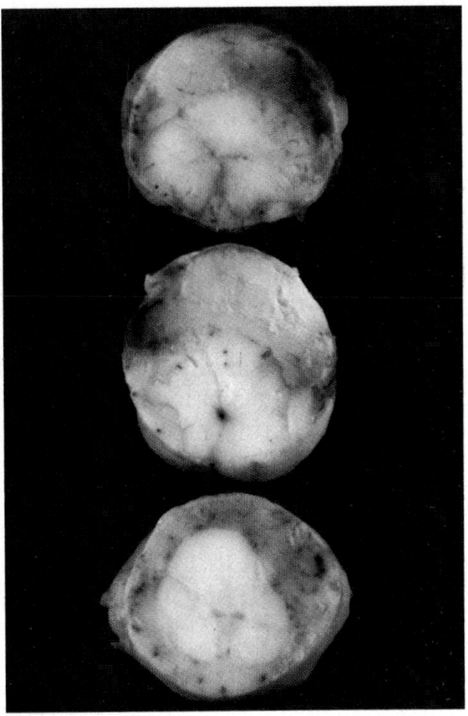

Figure 3-112

MENINGEAL GLIOMATOSIS

Above: In this spinal example, the tumor fills the subarachnoid space as a soft, gray, focally hemorrhagic infiltrate.

Right: On cut section, the tumor surrounds the spinal cord, thickens the subarachnoid space, and encases nerve roots. No intraparenchymal primary lesion was discovered despite histologic sampling at multiple levels.

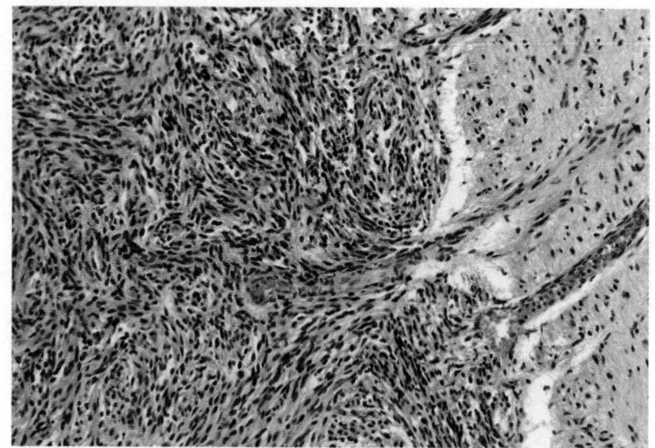

Figure 3-113

MENINGEAL GLIOMATOSIS

Neoplastic astrocytes in the meninges are often fusiform cells in a fascicular arrangement.

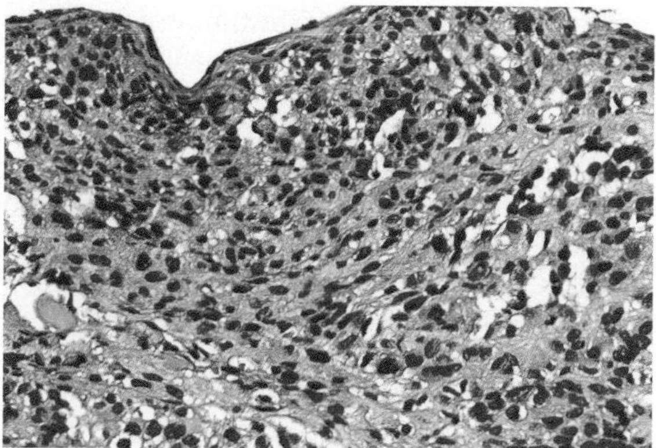

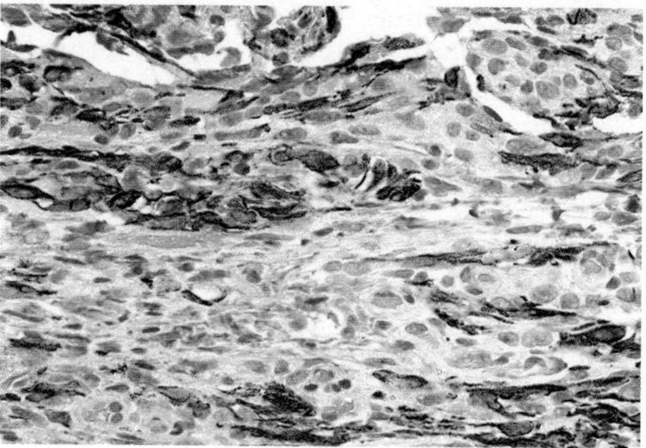

Figure 3-114

MENINGEAL GLIOMATOSIS

Top: This grade III example is densely cellular and mitotically active.

Bottom: GFAP immunoreactivity can be used to identify the lesion's glial nature.

consensus diagnostic criteria. Indeed, it may not exist as an entity, and may be only a nonspecific pattern and component of other tumors, specifically oligodendroglioma and pilocytic astrocytoma. Most appear in children or young adults as superficial, sometimes cortical-based masses (199).

Gross Findings. The lesion often appears gelatinous and, unlike diffuse astrocytoma, relatively well circumscribed.

Microscopic Findings. Protoplasmic astrocytomas exhibit low to moderate cellularity, and are noted for their "cobweb" cytoarchitectural pattern, which results from small, poorly fibrillated cells with "limp," radiating processes enmeshed in a mucoid matrix. The mucoid matrix contributes to the formation of often prominent microcysts (fig. 3-115). Nuclei are uniform in size, round to oval, and show little if any mitotic activity. As in pilocytic astrocytomas, protoplasmic astrocytomas may exhibit glomeruloid vessel proliferation, particularly within cyst walls.

Immunohistochemical Findings. Limited GFAP reactivity is confined to the paranuclear cytoplasm. As a rule, processes are negative.

Differential Diagnosis. Unlike pilocytic astrocytomas, protoplasmic tumors lack a biphasic, compact-microcystic pattern; Rosenthal fibers and granular bodies; and hyalinized vessels. Oligodendrogliomas frequently exhibit similar loose textured components, microcysts, and scant GFAP immunoreactivity. Classic oligodendroglioma features, such as monotonous

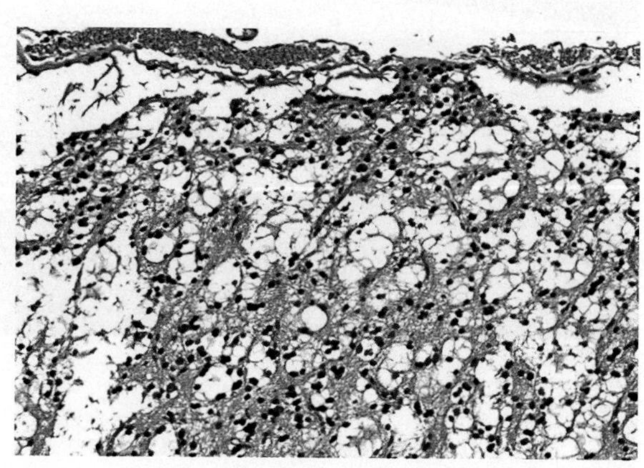

Figure 3-115

PROTOPLASMIC ASTROCYTOMA

Microcysts create conspicuous spongiosis.

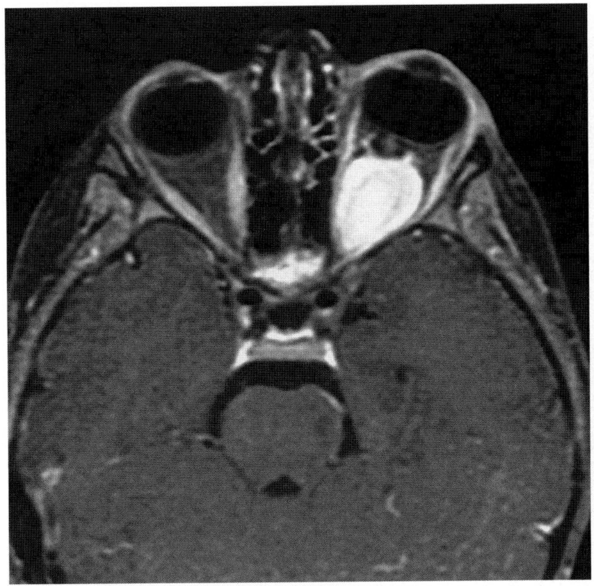

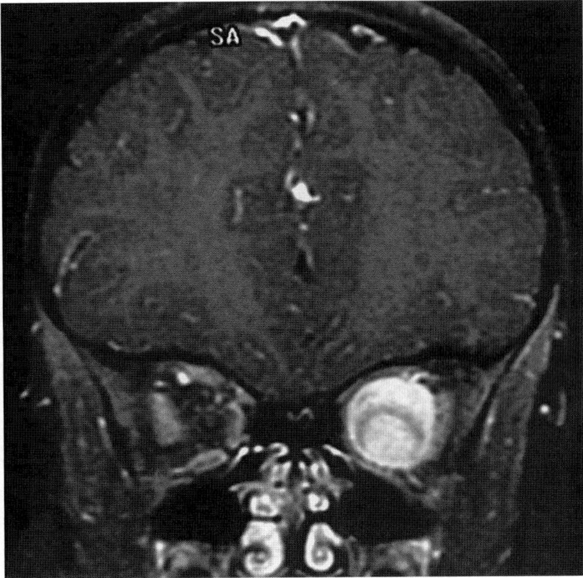

Figure 3-116

PILOCYTIC ASTROCYTOMA OF THE OPTIC NERVE

As here in a 4-year-old boy with proptosis, pilocytic astrocytomas of the optic nerve often have a distinctive neuroradiologic profile: an enlarged nerve on the right, surrounded by abnormal tissue in the leptomeninges. The complex is seen axially (left) and coronally (right) in postcontrast T1-weighted images.

nuclear roundness and prominent perineuronal satellitosis help make the distinction.

Treatment and Prognosis. The behavior of clinical protoplasmic astrocytoma is not well characterized, but anaplastic transformation is seen (199).

WELL-CIRCUMSCRIBED ASTROCYTIC NEOPLASMS

Pilocytic Astrocytoma

Definition. *Pilocytic astrocytoma* is a relatively circumscribed and often cystic neoplasm composed of compacted fibril-rich and loose textured fibril-poor astrocytes.

General Features. Although both WHO grade I pilocytic astrocytomas and grade II diffuse astrocytomas are often conjoined in the category "low-grade astrocytoma," they are distinct, unrelated entities that differ markedly in their capacity for tissue infiltration and malignant degeneration. Pilocytic astrocytoma, WHO grade I, is not a lower-grade version of the well-differentiated, grade II diffuse astrocytoma. As is discussed below, a pilocytic astrocytoma variant, pilomyxoid astrocytoma, is considered WHO grade II. Appendix A lists CNS tumors by the degree of circumscription.

Clinical and Radiologic Findings. Pilocytic astrocytomas arise throughout the neuraxis and are particularly common in children and young adults. Some tumors are associated with neurofibromatosis type 1, as is discussed in chapter 18. Although the general histologic appearance of pilocytic astrocytoma is similar at any location, the clinical and radiographic features are presented here from a topographic perspective. Dissemination into the subarachnoid space is apparent at diagnosis in a small percentage of cases (200–202). Pilocytic astrocytoma is the most common form of astrocytoma in the visual system and cerebellum.

Pilocytic astrocytomas of the visual system arise either in the optic nerve proper (figs. 3-116, 3-117), where they produce a lesion loosely known as *optic nerve glioma*, or in the optic chiasm, where they overlap clinically and radiologically with pilocytic astrocytomas of the hypothalamus (fig. 3-118). Distinguishing between chiasmatic and hypothalamic types is not always possible.

Pilocytic tumors of the optic nerve cause loss of vision and, in the case of intraorbital lesions,

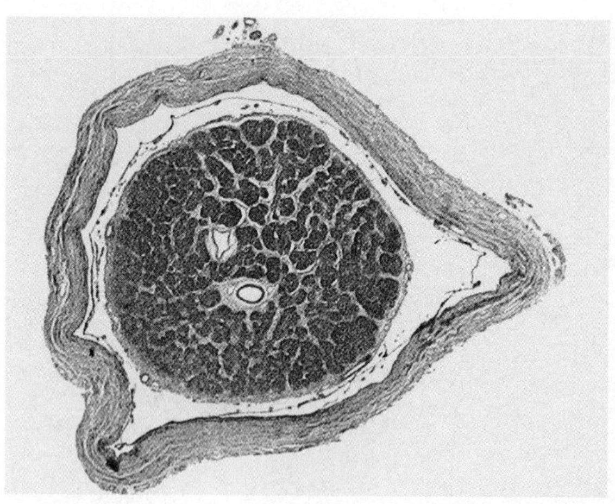

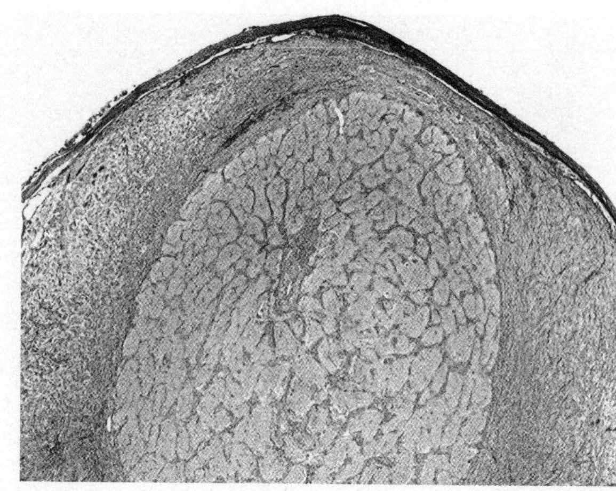

Figure 3-117

NORMAL OPTIC NERVE AND PILOCYTIC ASTROCYTOMA OF THE OPTIC NERVE

Illustrated in the macrosection on the left is a cross section of normal optic nerve. From outside in are the nerve sheath (dura), delicate arachnoid, and the septated nerve with ophthalmic artery and vein. Reproduced at the same magnification, on the right, is a pilocytic astrocytoma that enlarges the still septated nerve while expanding within the leptomeninges and stretching the nerve sheath (H&E/Luxol fast blue stain).

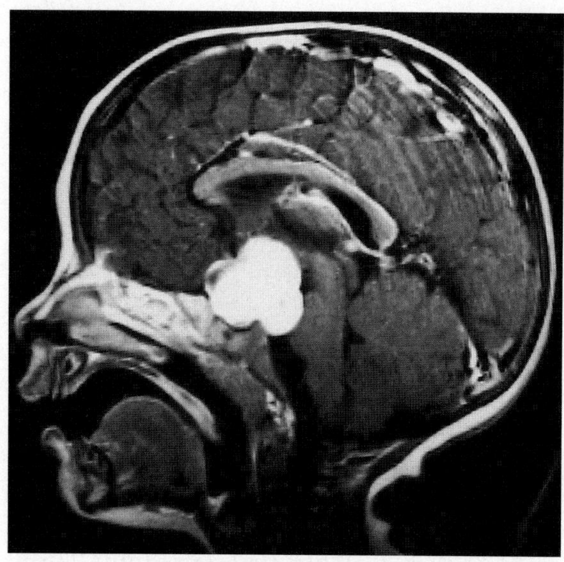

Figure 3-118

PILOCYTIC ASTROCYTOMA OF THE HYPOTHALAMIC REGION

At this commonly affected site, the contrast-enhancing neoplasm lies largely within the third ventricle. (Fig. 3-63 from Fascicle 10, 3rd Series.)

proptosis (203). Tumors more centrally placed, as in the chiasm, compromise visual acuity, often bilaterally, and/or interfere with hypothalamic-pituitary function. Optic nerve lesions may be unilateral or bilateral, with the latter particularly common in patients with neurofibromatosis type 1. Although the optic nerve is an extension of the CNS, such syndrome-associated gliomas arise in the peripheral, or type 1, form of the disease.

Radiologically, pilocytic astrocytomas of the optic nerve are contrast-enhancing, fusiform enlargements that extend a centimeter or more in length (fig. 3-116, left). The rigidity of the optic nerve sheath (dura) somewhat restricts the tumor's growth, and promotes expansion along the long axis of the nerve. A fusiform rather than globular mass is formed. As the lesion occupies and permeates the tissue, the nerve becomes large or "hypertrophic" as each compartment is expanded and gradually replaced by astrocytoma. The tumor often extends into the subarachnoid space, where it forms a thick, circumferential sleeve around the variably enlarged nerve; this is seen clearly with MRI (fig. 3-116, right). As a presumed result of tumor infiltration, a chiasmatic or hypothalamic lesion commonly produces signal abnormalities posteriorly along the optic tract to the level of the lateral geniculate bodies, sometimes bilaterally.

Pilocytic astrocytomas of the hypothalamus and third ventricular region primarily affect children and present as seemingly discrete tumors. When they extend into the third ventricle, they become largely intraventricular (figs. 3-118, 3-119). (See

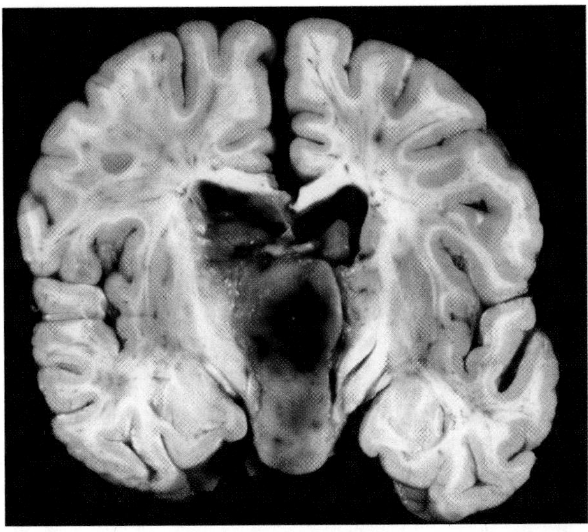
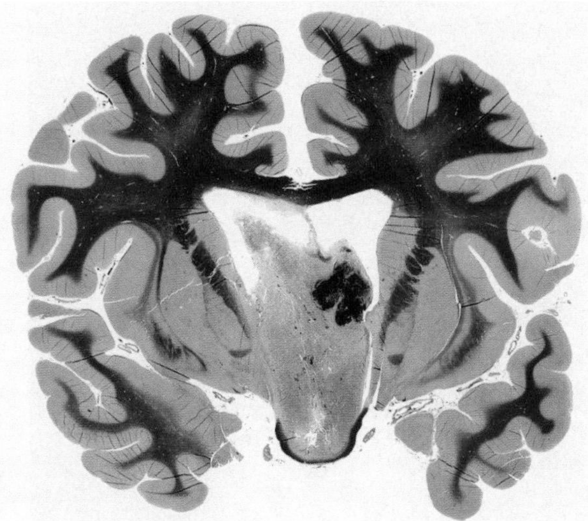

Figure 3-119

PILOCYTIC ASTROCYTOMA OF THE HYPOTHALAMUS

Left: Hypothalamic pilocytic astrocytomas can be largely intraventricular.
Right: The overall discreteness of pilocytic astrocytoma is well seen in this whole mount histologic section.

Appendix C for a listing of intraventricular tumors.) Since many tumors show an element of chiasmal involvement, their distinction from optic pathway tumors is somewhat artificial.

As at other sites, radiographic contrast enhancement generally distinguishes pilocytic tumors from well-differentiated diffuse astrocytomas, which generally lack this feature.

Pilocytic astrocytomas of the cerebral hemispheres generally occur in patients somewhat older than those with visual system or hypothalamic involvement (204,205). Since young adults are most often affected, hemispheric pilocytic tumors sometimes arise in the same patient population as the more common diffuse astrocytoma, precisely the lesion with which pilocytic tumors must not be confused. Pilocytic astrocytomas are usually circumscribed and enhancing masses, often taking the form of a mural nodule within a cyst. (Appendix D lists cystic CNS tumors.) Such an image is highly atypical for diffuse astrocytoma, an ill-defined tumor that almost by definition is not contrast enhancing unless modified by anaplastic transformation or radiation therapy.

Pilocytic astrocytoma of the cerebellum, long referred to by the nonspecific term *cerebellar astrocytoma*, resembles cerebral examples in its somewhat discrete appearance and frequent cyst/mural nodule architecture (fig. 3-120) (206, 207). Cerebellar pilocytic astrocytomas usually present during the second decade of life and produce symptoms referable to either obstruction of CSF flow or to cerebellar dysfunction. Despite gross circumscription, such tumors often permeate surrounding tissue, particularly

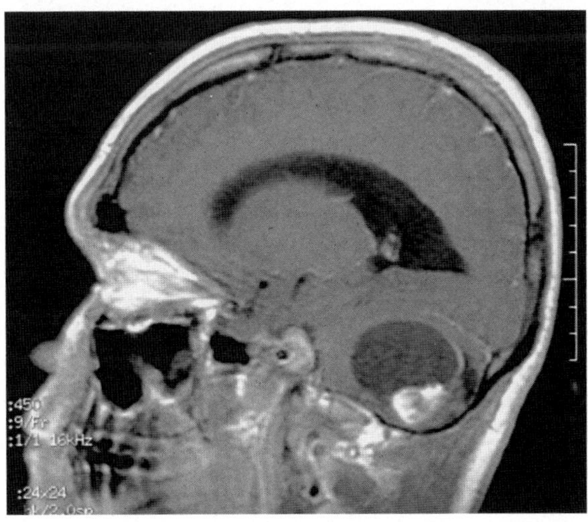

Figure 3-120

PILOCYTIC ASTROCYTOMA OF THE CEREBELLUM

Pilocytic astrocytomas of the cerebellum often form a cyst/mural nodule complex.

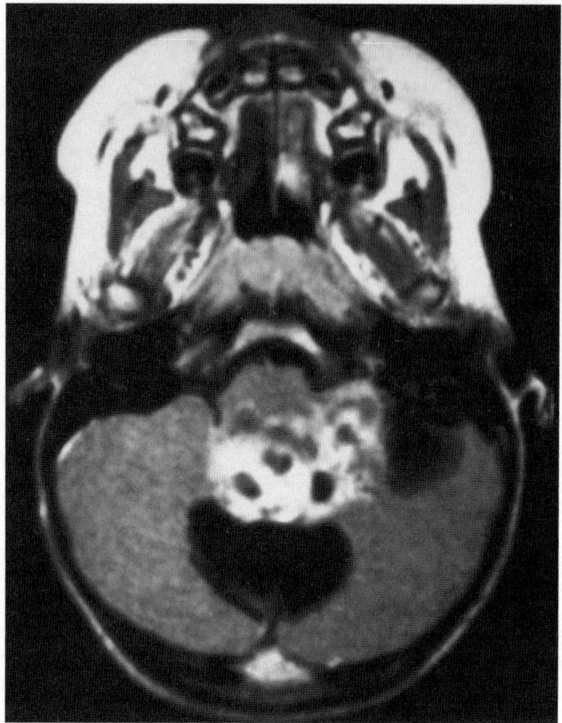

Figure 3-121

PILOCYTIC ASTROCYTOMA OF THE BRAIN STEM

Pilocytic astrocytomas of the brain stem are contrast enhancing, often multicystic, and usually exophytic. Grade II diffuse astrocytomas, in contrast, are intrinsic and enhance only when high grade, and then usually in a rim pattern typical of glioblastoma at any site. (Fig. 3-68 from Fascicle 10, 3rd Series.)

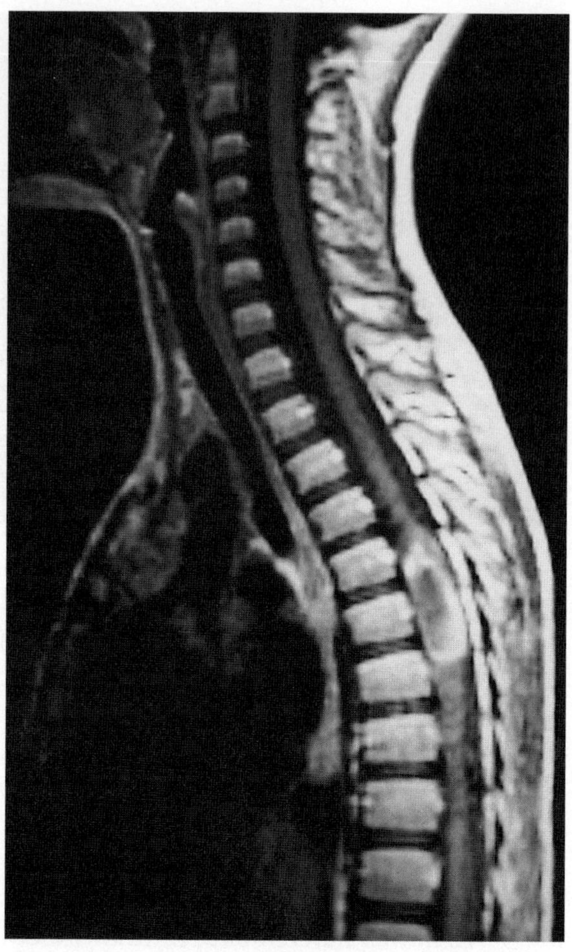

Figure 3-122

PILOCYTIC ASTROCYTOMA OF THE SPINAL CORD

Pilocytic astrocytomas of the spinal cord are discrete and usually contrast enhancing. This radiologic appearance contrasts with the infiltrating, and essentially unresectable, diffuse astrocytoma illustrated in figure 3-6. (Fig. 3-69 from Fascicle 10, 3rd Series.)

white matter, for some distance (208). In addition, they frequently involve the overlying leptomeninges. Eccentric growth may carry the tumor into the cerebellopontine angle, thus mimicking an ependymoma or vestibular schwannoma. Midline examples may protrude into the fourth ventricle, suggesting a diagnosis of ependymoma or medulloblastoma.

Pilocytic astrocytomas of the brain stem are considerably less common than the classic brain stem glioma of the diffuse type, and are also considerably more favorable prognostically. Described radiologically as well circumscribed or dorsally exophytic (209–211), these sometimes bulky lesions are discrete, contrast enhancing, and often multicystic (fig. 3-121). Some examples are largely intraventricular; others are largely intraparenchymal, but even then they are often subject to at least partial resection.

The clinical and radiographic features of *pilocytic astrocytomas of the spinal cord* are not well established, and the ratio of pilocytic to diffuse lesions remains to be defined. Pilocytic lesions are more common in young patients and are usually associated with a long history of symptoms. Circumscribed and contrast enhancing, they appear in profile like a link of sausages, often with a cyst (syrinx) at one or both poles (fig. 3-122). Known as *holocord astrocytoma*, some impressive long, solid or systic lesions pan many spinal segments. At the time of surgery the lesion presents as a fusiform

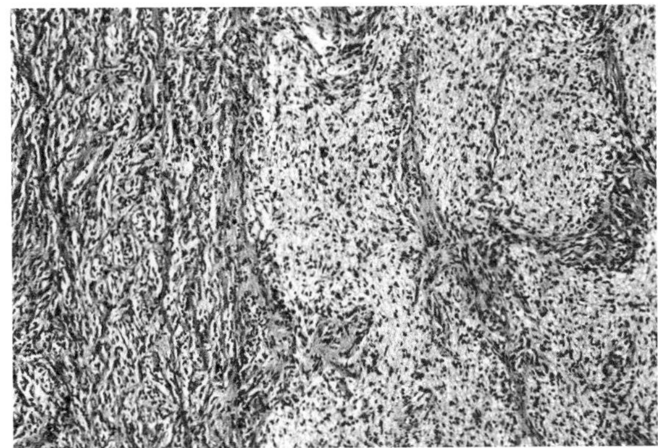

Figure 3-123

PILOCYTIC ASTROCYTOMA OF THE OPTIC NERVE

The classic pilocytic astrocytoma consists of an intraparenchymal component that enlarges the nerve, and a meningeal element that expands in the subarachnoid compartment.

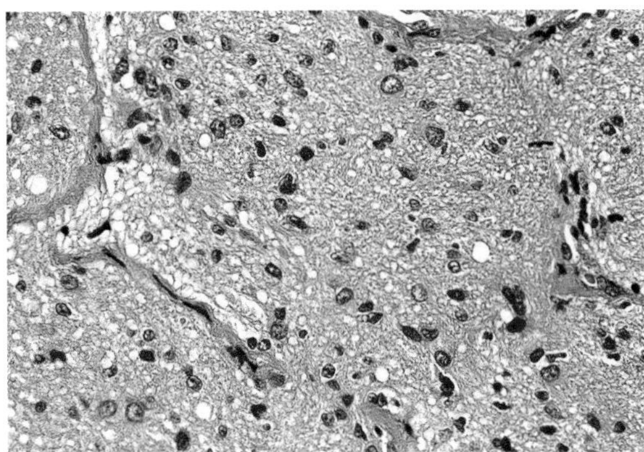

Figure 3-124

PILOCYTIC ASTROCYTOMA OF THE OPTIC NERVE

Pilocytic astrocytomas may infiltrate visual pathways, as is demonstrated here at the surgical margin near the optic chiasm. Only widely scattered, mildly pleomorphic tumor cells are present.

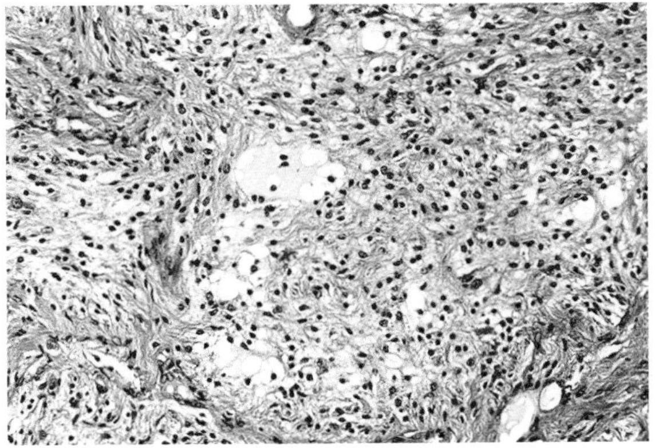

Figure 3-125

PILOCYTIC ASTROCYTOMA OF THE OPTIC NERVE

Classic spongy, microcystic tissue such as this may be only focal in pilocytic astrocytomas of the optic nerve.

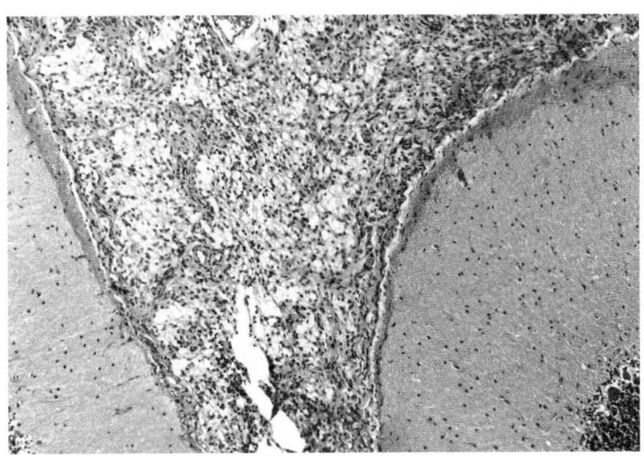

Figure 3-126

PILOCYTIC ASTROCYTOMA

Pilocytic astrocytomas of the cerebellum often fill the regional subarachnoid space.

enlargement similar to that of other intramedullary tumors, including diffuse astrocytoma with which pilocytic lesions are most likely to be confused (see fig. 3-6).

Microscopic Findings. Pilocytic astrocytomas of the optic nerve may exhibit the classic features described below but, in part because of location, are histologically somewhat idiosyncratic. Thus, while they may have compact and microcystic patterns, they are unusual in their intrinsic tendency to enlarge the optic nerve from within (figs. 3-117, 3-123). The degree of cellularity ranges from the obvious to the subtle, the latter best seen at the surgical margin near the optic chiasm (fig. 3-124). Within compartments in advanced lesions there may be the same microcystic tissue seen in pilocytic lesions at other sites (fig. 3-125). At other loci, pilocytic astrocytomas are bulky masses that, although they incorporate preexisting tissue, are macroscopically more displacing than infiltrating.

Pilocytic astrocytomas in any location are prone to break through the pia and expand within the overlying subarachnoid space. Especially frequent in the optic nerve (figs. 3-117, 3-123) and cerebellum (fig. 3-126), this extension

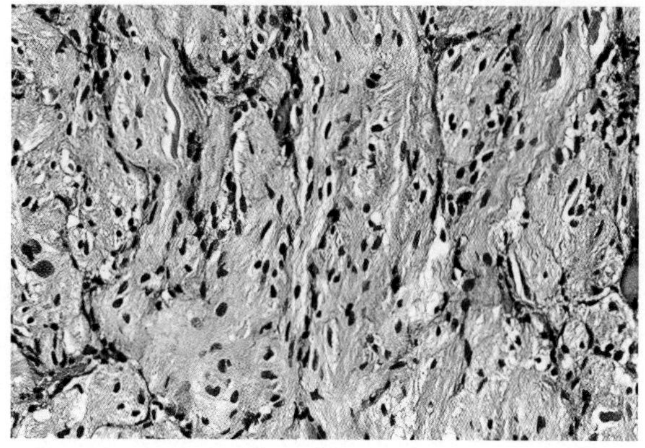

Figure 3-127

PILOCYTIC ASTROCYTOMA

Pilocytic astrocytomas that reach the subarachnoid space often become septated and micronodular.

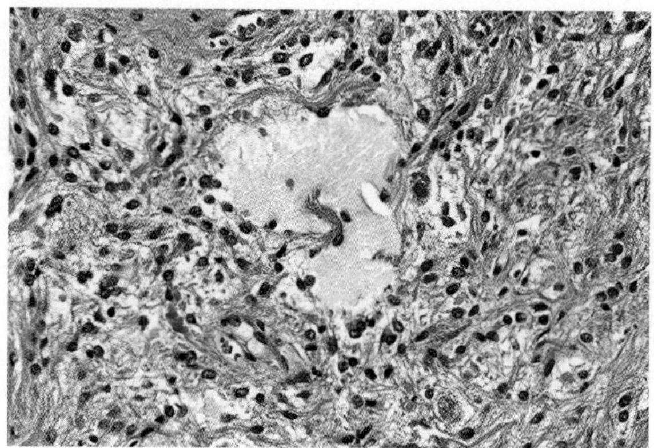

Figure 3-128

PILOCYTIC ASTROCYTOMA

Alternating solid and spongy microcyst-rich areas create the classic "biphasic" pattern.

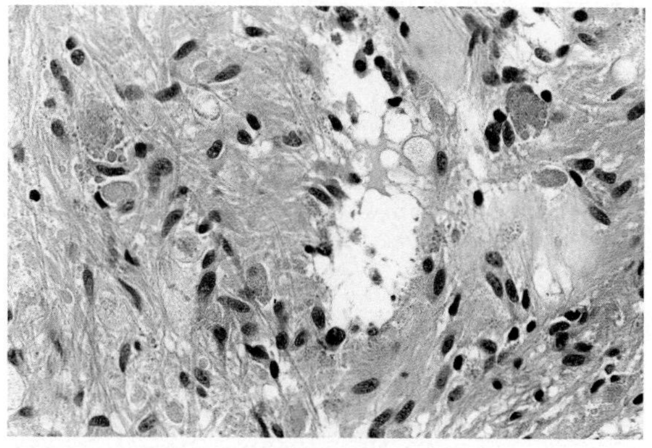

Figure 3-129

PILOCYTIC ASTROCYTOMA

Cell processes filled with protein droplets (eosinophilic granular bodies) are prominent in loose areas.

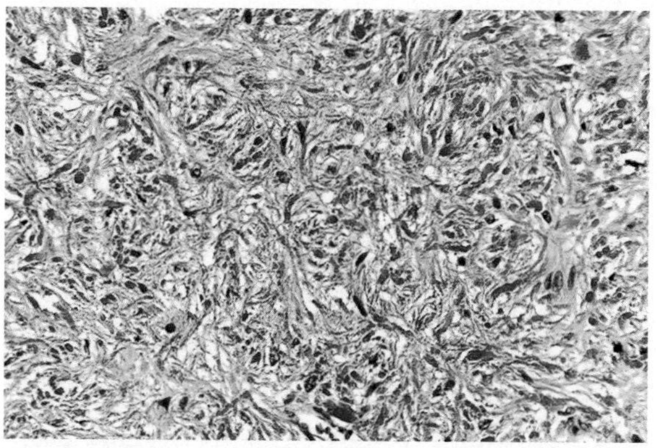

Figure 3-130

PILOCYTIC ASTROCYTOMA

A densely fibrillar mass of piloid tissue with abundant Rosenthal fibers is a variant that closely resembles, and cannot always be distinguished from, piloid gliosis.

must not be misinterpreted as malignant behavior or an indication that a tumor is poised to disseminate in CSF pathways. As the tumor enters the leptomeninges, its cells can become bundled into small lobules within a sometimes considerable fibroblastic reaction (fig. 3-127).

Once termed "juvenile," the classic form of pilocytic astrocytoma is formed of compact tissue, consisting of elongated and highly fibrillated cells, that alternates with loosely knit spongy tissue in which microcysts are prominent (figs. 3-125, 3-126, 3-128, 3-129). The compact tissue contains brightly eosinophilic Rosenthal fibers, whereas either tissue type may contain eosinophilic granular bodies (fig. 3-129).

In addition to this prototypic, "biphasic" architectural pattern, considerable morphologic variation must be recognized if this important tumor is not to be misclassified. The ratio of loose-knit to compact tissue varies greatly: some tumors are largely microcystic and spongy while others are predominantly compact and piloid (figs. 3-130, 3-131).

Occasional pediatric cerebellar astrocytomas are composed of cytologically monomorphous bland cells consistent with pilocytic astrocytoma, but are diffusely infiltrative (fig. 3-132) 207). A myxoid background and an

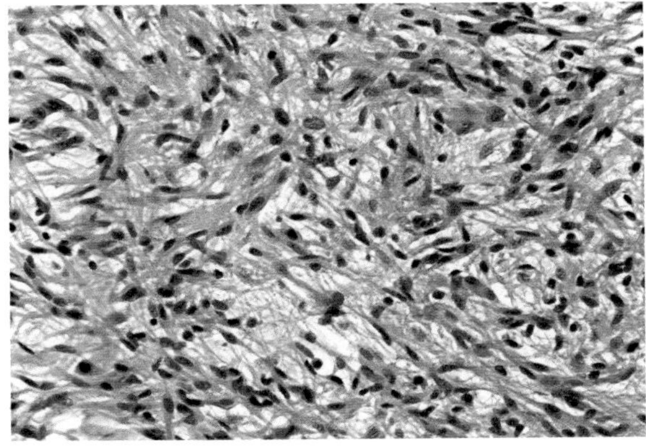

Figure 3-131

PILOCYTIC ASTROCYTOMA

Compact piloid tissue without Rosenthal fibers or microcysts is a pilocytic astrocytoma variant.

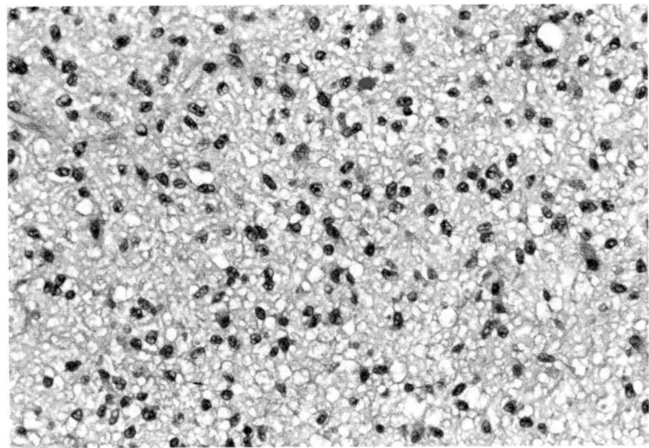

Figure 3-132

PILOCYTIC ASTROCYTOMA

Diffuse pilocytic astrocytomas of the cerebellum closely resemble grade II infiltrating astrocytomas.

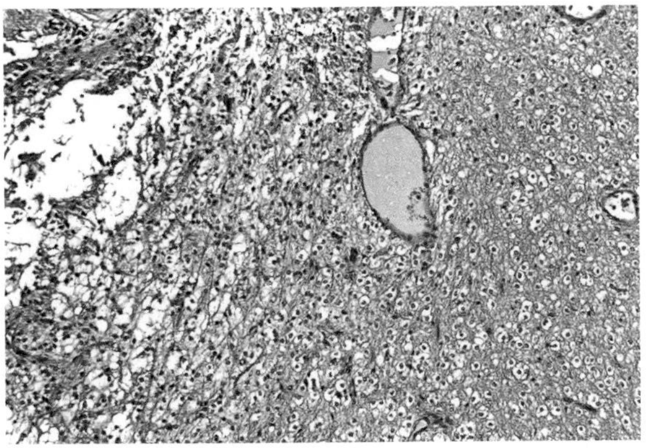

Figure 3-133

PILOCYTIC ASTROCYTOMA

Oligodendroglioma-like areas are common in cerebellar pilocytic astrocytomas. In this case, spongy tissue more typical of pilocytic astrocytoma is present at the left.

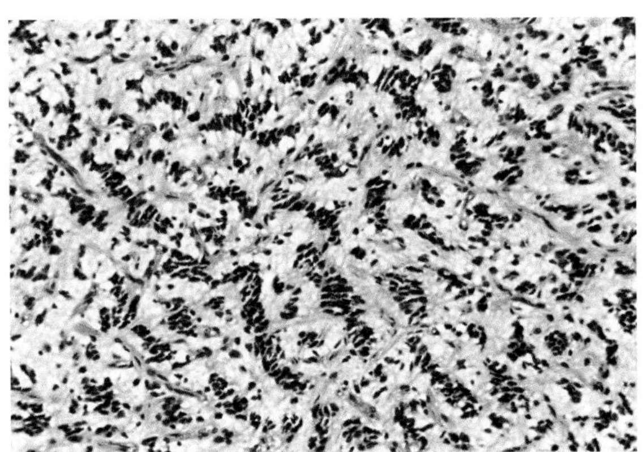

Figure 3-134

PILOCYTIC ASTROCYTOMA

Rows or ribbons of cells are occasionally present.

oligodendroglioma-like appearance are common in such lesions (fig. 3-133). Some of these tumors have a cyst/mural nodule architecture, but others do not. If the lesion has pilocytic astrocytoma features elsewhere, e.g., microcysts, biphasic architecture, Rosenthal fibers, or extension into the subarachnoid space, an overall pilocytic nature is assumed. In the absence of these features, a descriptive diagnosis of well-differentiated astrocytoma is appropriate. The ill-defined concept of "diffuse pilocytic astrocytoma" is discussed under Grading and Histologic Prognostic Factors.

Another pattern of uncertain prognostic significance is clustering and ribboning resembling that of polar spongioblastoma (fig. 3-134). CNS tumors with ribboning or palisading are given in Appendix J.

A monomorphous glioma with piloid cells, myxoid background, and an often perivascular orientation of its cells has been referred to as *infantile pilocytic astrocytoma* (212), *pilomyxoid astrocytoma* (figs. 3-135, 3-136) (213,214), or *pilomyxoid variant of pilocytic astrocytoma* (215). Whether the tumor exists independently of pilocytic astrocytoma can be debated, but not whether it often poses a diagnostic problem, presenting as a monomorphous, somewhat piloid,

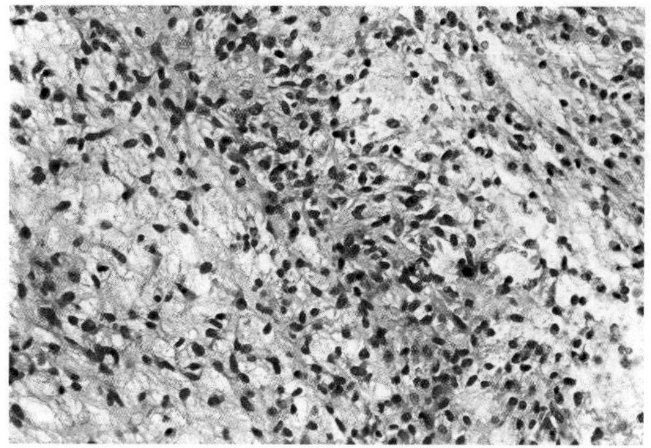

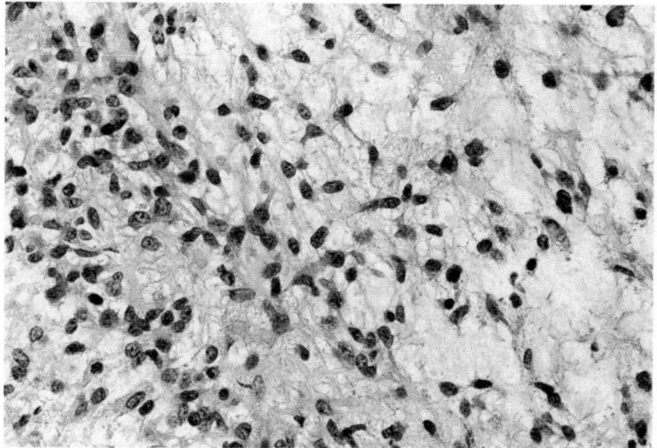

Figure 3-135

PILOMYXOID ASTROCYTOMA/INFANTILE PILOCYTIC ASTROCYTOMA

As seen at low (left) and high (right) magnification, the pilomyxoid lesion is a monomorphous mass of small piloid cells in a mucopolysaccharide matrix. Orientation of the tumor cells to vessels creates a form of perivascular pseudorosette.

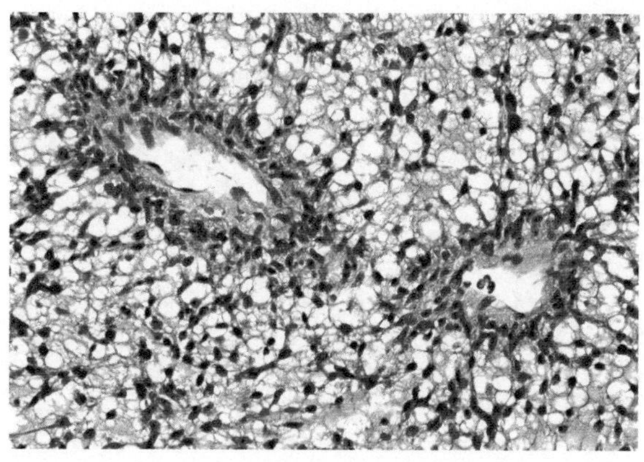

Figure 3-136

PILOMYXOID ASTROCYTOMA/INFANTILE PILOCYTIC ASTROCYTOMA

Monomorphous cells with a perivascular orientation lend an ependymoma-like appearance to the pilomyxoid lesion.

but not classically pilocytic, lesion. The tumor usually occurs at the base of the brain (but also at other sites favored by pilocytic astrocytomas), usually presents in infants and young children, and often becomes large, if not massive, over a period of several months. It remains to be seen if its highly homogeneous bright signal intensity on T2-weighted MRI helps differentiate it from classic pilocytic neoplasms (216). The perivascular orientation of its remarkably monomorphous cells often creates structures somewhat resembling the perivascular pseudorosettes of ependymoma. A cytologic and ultrastructural resemblance to an ependymal relative, the tanycyte, has been suggested (217). Lacking microcysts, Rosenthal fibers, and eosinophilic granular bodies, it is not a typical pilocytic astrocytoma. Necrosis and parenchymal infiltration are more common (215). The MIB-1 rate may be higher than in typical pilocytic astrocytoma. Specimens taken at the time of recurrence show that some of these distinctive tumors appear to evolve or "mature" into a classic pilocytic astrocytoma (fig. 3-137) (215). The pilomyxoid lesion is assigned WHO grade II in light of its overall more aggressive behavior, and CSF dissemination in some cases.

In most instances, the nuclei of classic pilocytic astrocytoma are uniform, with delicate or open chromatin, small nucleoli, and little variation in overall size and shape. Nuclear roundness is typical in microcystic areas rich in "protoplasmic astrocytes," whereas elongation is common in compact piloid regions. Mitoses are often lacking, but may be found in small numbers in a third of pilocytic tumors, particularly in infants.

Considered degenerative in nature, conspicuous hyperchromasia and pleomorphism alone are of no prognostic significance. Multinucleate cells in which nuclei are peripherally arranged like "pennies on a plate" are common (fig. 3-138), particularly in longstanding cerebellar or cerebral hemispheric tumors of young adults. Some pilocytic astrocytomas, especially in patients in the third decade or older, contain markedly

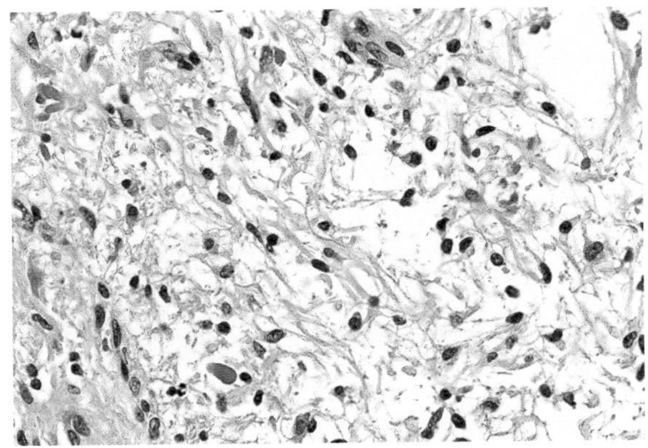

Figure 3-137

PILOMYXOID ASTROCYTOMA/INFANTILE PILOCYTIC ASTROCYTOMA

As illustrated in a recurrence 3 years later of the tumor in figure 3-135, pilomyxoid lesions can evolve into classic pilocytic astrocytomas, here with microcysts and Rosenthal fibers.

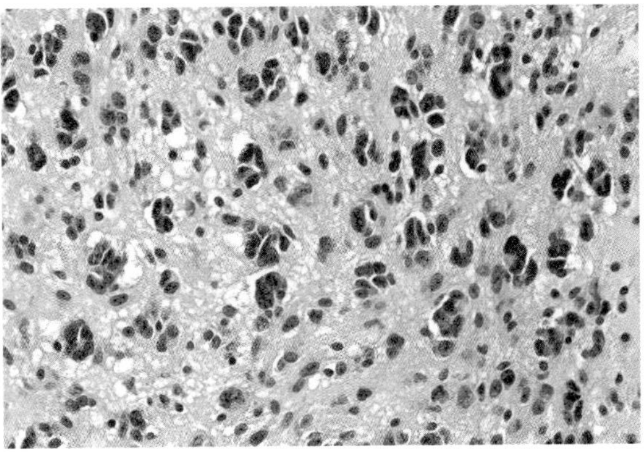

Figure 3-138

PILOCYTIC ASTROCYTOMA

Degenerative atypia and multinucleation are common in longstanding pilocytic lesions.

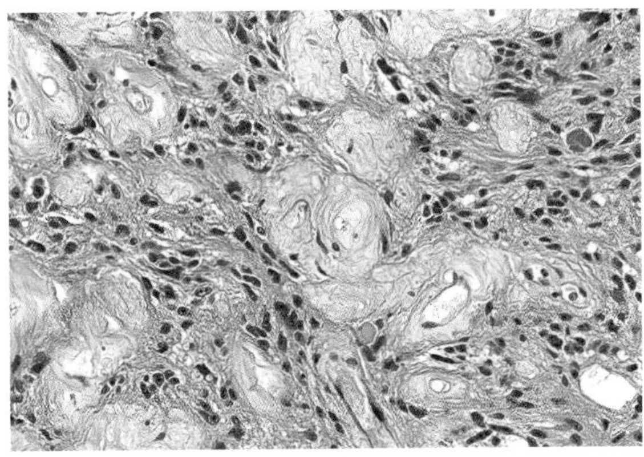

Figure 3-139

PILOCYTIC ASTROCYTOMA

Pilocytic astrocytoma with closely packed sclerotic vessels resembles cavernous angioma.

hyalinized juxtaposed vessels that simulate cavernous angioma (fig. 3-139). Spontaneous hemorrhage, however, is rare.

Histologic features facilitating the diagnosis of pilocytic astrocytoma at any site include the previously mentioned Rosenthal fibers and eosinophilic granular bodies. The Rosenthal fibers are sausage- or corkscrew-shaped bodies that often have one blunt and one tapered end (see figs. 3-130, 3-131, 3-137). Brightly eosinophilic and hyaline, they are blue after Luxol fast blue staining and red with the Masson trichrome method. Rosenthal fibers vary greatly in frequency, being absent in some tumors and abundant in others. They are not diagnostic of neoplasia. Indeed, they are abundant in piloid gliosis surrounding craniopharyngiomas (see fig. 15-9) and about other chronic lesions such as hemangioblastoma (see fig. 7-5), spinal ependymoma (see fig. 3-259), spinal vascular malformations, pineal cyst (see figs. 6-21, 6-22), and even pilocytic astrocytoma itself. In the setting of a pilocytic astrocytoma, it may be difficult, particularly in the cerebellum, to determine whether the Rosenthal fibers are in neoplastic or reactive astrocytes. Rosenthal fibers must be distinguished from astrocytic processes so packed with glial filaments that they appear brightly eosinophilic and hyaline. Such coarse processes are common in pilocytic astrocytomas, but lack the diagnostic significance of Rosenthal fibers.

The eosinophilic granular body (EGB) is composed of delicate to coarse aggregates of hyaline droplets (see fig. 3-129) (218). Although often not obvious in tissue sections, EGBs are located within glial processes. The bodies are also common in pleomorphic xanthoastrocytoma and ganglion cell tumors where, as in pilocytic astrocytomas, their presence suggests a neoplasm that is well circumscribed, slowly growing, and well differentiated. While EGBs can occur in anaplastic tumors, even glioblastoma, a diagnosis of a high-grade tumor should be made with caution in their presence. Appendix M lists features, including EGBs, that suggest the possibility of

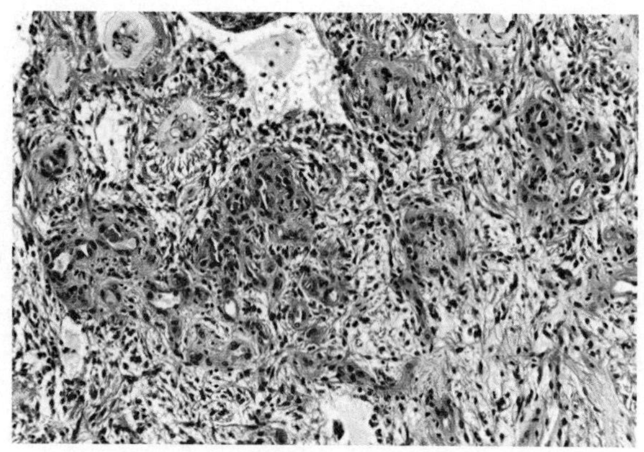

Figure 3-140

PILOCYTIC ASTROCYTOMA

Microvascular proliferation in pilocytic astrocytoma is often loose and telangiectasia-like, but can be compact and glomeruloid as seen here.

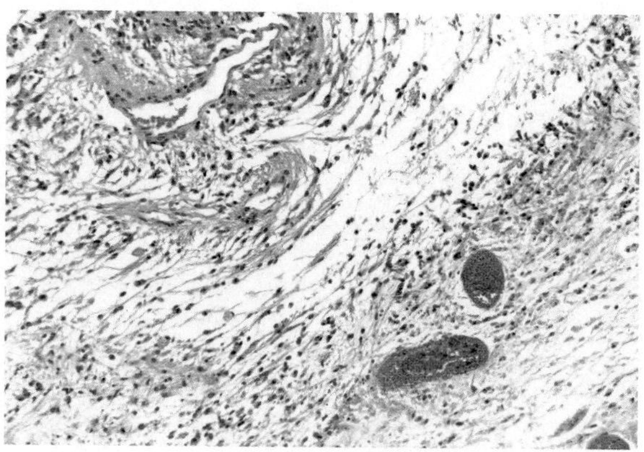

Figure 3-141

PILOCYTIC ASTROCYTOMA

Necrosis, sometimes with a suggestion of pseudopalisading, is present in some paucicellular, but otherwise typical, pilocytic astrocytomas. The infarct-like regions differ from the coagulative tumor necrosis seen in high-grade gliomas.

a well-differentiated or non-neoplastic lesion. Appendix K lists tumors, such as pilocytic astrocytoma, that easily can be overgraded.

Vascular changes are frequent in pilocytic astrocytomas. Glomeruloid vascular proliferation is common in varying degrees, either in the parenchyma (fig. 3-140) or in the cyst wall as a long garland of proliferating vessels. Multilayered endothelial proliferation within larger vessels is rare (205). Although glomeruloid vascular hyperplasia may prompt consideration of glioblastoma or other anaplastic gliomas, it is not in itself prognostically significant. Often, the masses of proliferating vessels are loose, dilated, and more telangiectatic than compactly "glomeruloid."

Necrosis is present in a small percentage of pilocytic astrocytomas. In contrast to glioblastoma, necrotic areas are usually more sharply circumscribed and of low cellularity (fig. 3-141). Although a vague concentration of cells about zones of necrosis may bring glioblastoma to mind, there is usually not the well-developed perinecrotic palisading seen in the latter tumor. The necrosis of pilocytic astrocytoma is also not usually the granular coagulative state seen in glioblastoma, but rather a homogeneous, often hyalinized alteration.

Although pilocytic astrocytomas are macroscopically discrete, they often infiltrate surrounding brain for a distance of millimeters to even centimeters in the case of the cerebellum and visual pathways (208). As a result, what appears to the surgeon as circumscription may not be true at the microscopic level. The peripheral incorporation of preexisting tissue is readily apparent as the tumor enlarges by a process of "creeping substitution." A limited area of the brain is thus gradually taken over so that the axons, well seen after immunostaining for neurofilament protein, and even scattered neuron cell bodies, become trapped within the substance of the mass.

Although encounters with overrun neurons raise the possibility of a ganglion cell tumor, such cells are few and normally formed in most instances. Some typical pilocytic astrocytomas do contain neurons that appear to be part of the neoplasm, not just trapped elements (fig. 3-142). It can be debated whether such a lesion is best treated as ganglioglioma with a prominent pilocytic astrocytoma component, or a pilocytic astrocytoma with a conspicuous neuronal component. It is a form of ganglioglioma in either case.

Another form of infiltration typical of pilocytic astrocytomas is extension into, and dilation of, perivascular spaces. This is of no prognostic significance. Extension into the subarachnoid space is discussed above.

Grading and Histologic Prognostic Factors. Most pilocytic astrocytomas are, and remain, WHO grade I tumors. There are, in fact, no definitions for higher grades, although there are certainly rare lesions with anaplasia and aggressive biologic behavior for which the terms

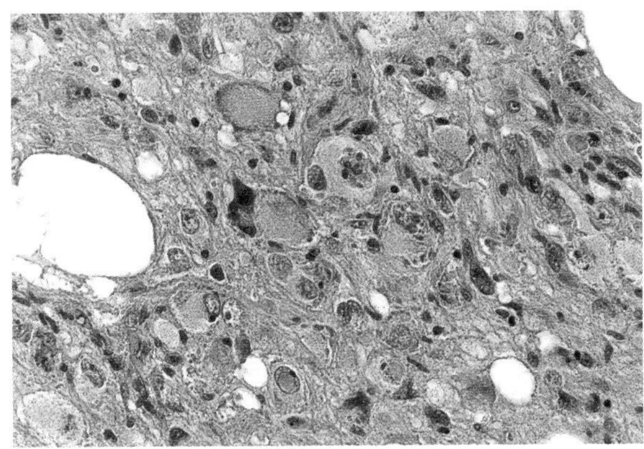

Figure 3-142

PILOCYTIC ASTROCYTOMA

A ganglion cell component is present in occasional cases.

malignant, anaplastic, and atypical are used. There is no consensus, however, on definitions or prognostic significance of such tumors. Most of the limited data concerning outcome in regard to specific histologic factors comes from cerebellar lesions where total excision is often possible, making it difficult to correlate outcome with morphologic features.

Always of concern, although unnecessarily in most cases, are lesions that are otherwise typical but have vascular proliferation and sometimes even necrosis. Intralesional vascular proliferation, of the glomeruloid type, by itself appears to have no prognostic significance. In one study of 107 pilocytic astrocytomas, vascular proliferation, in the absence of increased mitotic activity, was found in 6 patients (219). All were alive at prolonged postoperative intervals. Necrosis is less common, but in the absence of increased mitotic activity or MIB-1 rate, can also be accepted in a grade I pilocytic astrocytoma.

More concern is justified when there is an increase in either mitotic rate or MIB-1 staining index. In the same study of 107 cerebellar pilocytic astrocytomas (219), 4 were felt to be histologically malignant on the basis of an elevated mitotic rate (at least 1 mitosis per 250X microscopic field). Vascular proliferation was often present. In spite of the designation "malignant," not all behave as such, however.

Predicting outcome by the MIB-1 index may be possible since, in one study, significantly inferior outcomes (especially progression-free survival) were noted for patients whose tumors were above 2 percent (220). In another study, the mean MIB-1 index of tumors that later recurred was 6.6 percent as opposed to 3.3 percent in those that did not (221). There is, however, a wide range of MIB-1 results, both intratumoral and intertumoral, making it difficult to prognosticate outcome from MIB-1 rate in a given case, and one study failed to find a predictive value for the MIB-1 index (222).

Unequivocal anaplastic transformation of pilocytic tumors has been reported (219,223), often after a course of many years. Such tumors typically are compact, somewhat fascicular lesions with abundant mitotic activity (fig. 3-143). Although such transformation may occur "spontaneously" (219,224), prior radiotherapy has been administered in most cases (219,223,225,226). As a result, it is difficult to exculpate radiation as an oncogenic influence, particularly when the "recurrent" tumor more closely resembles anaplastic astrocytoma or glioblastoma than pilocytic astrocytoma. Late, postradiation high-grade astrocytomas are increasingly recognized.

Pediatric cerebellar astrocytoma that is diffusely infiltrative is a vexing problem. One study investigated "diffuse cerebellar astrocytomas" in children and found benign biologic behavior irrespective of whether the lesion had classic pilocytic features elsewhere in the specimen (227). Another study found prognostic significance with pilocytic versus fibrillary (diffuse) astrocytomas (228). One investigation found diffuse lesions to be aggressive, but this was largely a study of adults with diffuse grades II, III, and even IV astrocytomas (207).

Another histologic variant with prognostic ramifications is the pilomyxoid lesion. While this may well be an infantile pilocytic astrocytoma (212), it appears more aggressive in terms of growth rate and tendency to seed the neuraxis (213–215). These tumors may stabilize, however; as the years pass some acquire features of classic pilocytic astrocytoma, e.g., microcysts and Rosenthal fibers (figs. 3-135, 3-137).

Frozen Section. Artifacts produced by freezing of pilocytic tumors often degrade architectural and cytologic features. Fortunately, microcysts usually persist, although these must be distinguished from the angular clefts produced by ice crystals. The slightly basophilic or eosinophilic

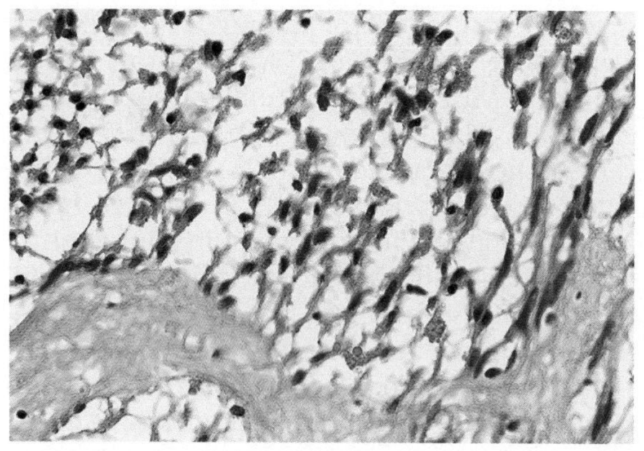

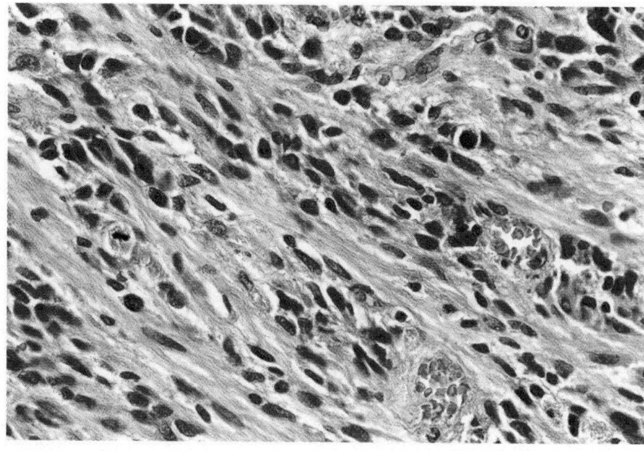

Figure 3-143

PILOCYTIC ASTROCYTOMA WITH ANAPLASTIC TRANSFORMATION

Only rarely do pilocytic astrocytomas undergo anaplastic transformation.
Left: Some areas have the typical microcystic architecture, eosinophilic granular bodies, and "benign" cytologic features.
Right: Elsewhere the tumor is cellular, fascicular, mitosis rich, and unequivocally anaplastic.

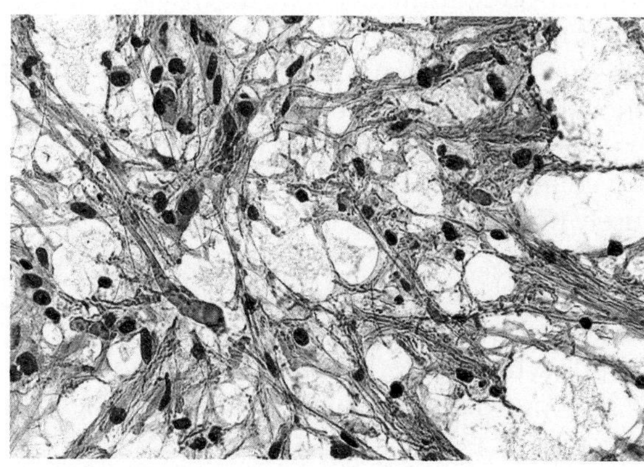

Figure 3-144

PILOCYTIC ASTROCYTOMA: FROZEN SECTION

Piloid cells and Rosenthal fibers are classic features.

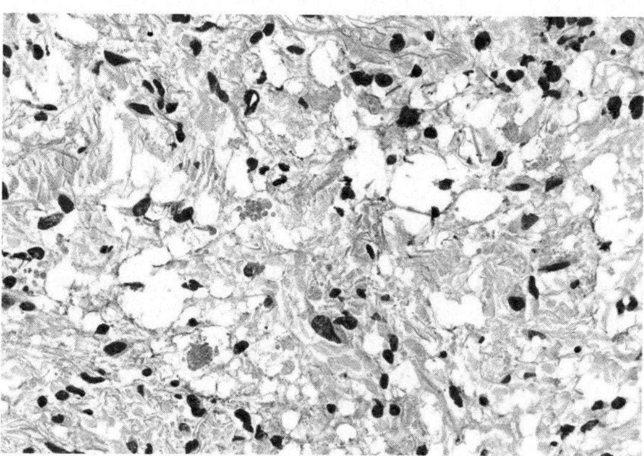

Figure 3-145

PILOCYTIC ASTROCYTOMA: FROZEN SECTION

Loose spongy tissue with eosinophilic granular bodies facilitates identification. The glial processes are often better seen in frozen than paraffin-embedded sections.

proteinaceous content of microcysts, if present, is a helpful diagnostic feature, but is not a specific finding since it can be seen in some well-differentiated infiltrating gliomas of both diffuse astrocytic and oligodendroglial type. Rosenthal fibers (fig. 3-144) and EGBs (fig. 3-145) are readily seen in frozen preparations, and may even be more obvious than in permanent sections. Hyaline cell processes packed with intermediate filaments aid in the identification of pilocytic tumors. Nuclear atypia in the face of virtually no mitotic activity should at least suggest the diagnosis of pilocytic astrocytoma.

Immunohistochemical Findings. Pilocytic astrocytomas are immunoreactive for GFAP (fig. 3-146). The highly fibrillated piloid cells of compact tumors stain more strongly than the fibril-poor cells of microcystic lesions. The pilomyxoid lesion is diffusely and strongly reactive. In regard to the Rosenthal fiber, immunoreactivity for GFAP is restricted to the surface, the body itself is reactive for a lens protein, alphaB-crystallin (218,229–231). EGBs are reactive for alpha-1-antichymotrypsin, and alpha-1-antitrypsin (232).

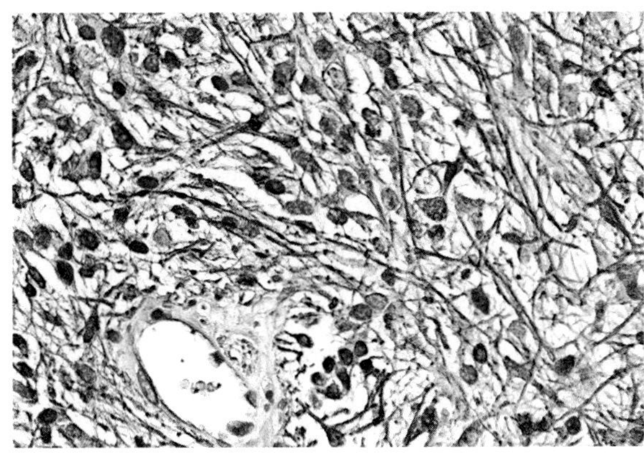

Figure 3-146

PILOCYTIC ASTROCYTOMA

Pilocytic astrocytomas are strongly reactive for GFAP in densely fibrillar areas.

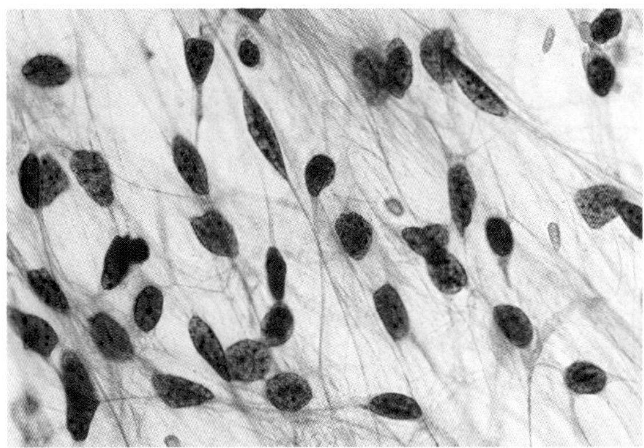

Figure 3-147

PILOCYTIC ASTROCYTOMA: SMEAR PREPARATION

The delicate nuclei and fine bipolar processes of the "hair cells" are seen well.

MIB-1 rates are usually low. In one investigation, the mean, median, and range were 1.1, 0.9, and 0-3.9 percent, respectively (222). Another series restricted to pediatric lesions recorded a mean index of 1.6, with a standard deviation of 1.6 and range of 0.0 to 9.5 percent (220). A study using the proliferative marker bromodeoxyuridine found a higher index in younger patients, suggesting that the growth rate of pilocytic tumors decreases over time (233). MIB-1 rates in pediatric cases do not, however, appear related to age (220).

Staining for apolipoprotein D has been suggested as a possible means to differentiate pilocytic (positive staining) from diffuse astrocytomas (234).

Ultrastructural Findings. Not surprisingly, the elongate piloid cells are filled with intermediate filaments, while the cells of loose textured, microcystic tissue are fibril-poor. Rosenthal fibers are nonfilamentous electron-dense masses that are surrounded by intermediate, or glial, filaments, the basis of peripheral GFAP reactivity. EGBs represent intracellular lipid droplets, myelin figures, and granular debris (235).

One report describes the pilomyxoid lesion as having microvilli, cytoplasmic blebs, rare cilia, vesicles, and coated pits (217), features taken as evidence of tanycytic/ependymal differentiation.

Cytologic Findings. The benign cytologic features and piloid quality of pilocytic astrocytoma are best seen in squash or smear preparations in which uniformity and blandness of nuclei are evident in both loose textured and compact areas (236). Cellular elongation is especially evident in the latter as fine, hair-like processes that may extend across several high-power microscopic fields (fig. 3-147). EGBs and Rosenthal fibers may also be seen; the latter appear as condensations of homogeneous, lumpy material within the processes of neoplastic cells.

Molecular Findings. An exploitable molecular signature of pilocytic astrocytoma would be extremely useful, especially in the distinction from diffuse astrocytoma, grade II. Unfortunately, it is yet to be identified. Loss of *NF1* alleles have been found in pilocytic astrocytomas associated with neurofibromatosis type 1, but only rarely in sporadic lesions (237). Another study found no abnormalities in *NF1* gene expression in sporadic intracranial pilocytic astrocytomas (238). One array comparative genomic hybridization study suggested that, at 0.97 Mb resolution, most pilocytic astrocytomas do not have copy number changes, particularly in younger patients (238a). The significance of changes, most commonly gain of chromosome 5 and/or 6, is unclear.

Differential Diagnosis. The issues are generally straightforward in the optic nerve, since there is little else there that resembles the lesion. Occasionally, the proximal nerve tissue near the chiasm must be evaluated to determine whether there are tumor cells at the resection margin (see fig. 3-124). Identification of cytologically bland tumor cells can be difficult, and a definitive answer may not be possible.

In the optic chiasm and hypothalamus, one must consider the possibility that the tissue is piloid gliosis responding to a craniopharyngioma. The latter tumor usually has a suggestive radiologic profile, a content of cholesterol-rich "motor oil"-like fluid, and unmistakable of cords of adamantinomatous tissue (see fig. 15-9). Nonetheless, the gliotic response can be so remarkable that multiple sections are required before diagnostic tumor tissue is obtained.

Pilocytic astrocytomas of the cerebral hemispheres need to be distinguished from diffuse astrocytoma, ganglion cell tumors, pleomorphic xanthoastrocytoma, and dysembryoplastic neuroepithelial tumor.

The distinction from a grade II diffuse astrocytoma in the cerebral hemispheres is of paramount importance since the prognosis and treatment of patients with these two lesions are so different. The importance of radiologic images cannot be overemphasized. Discreteness, contrast enhancement, and frequent cystic architecture all suggest pilocytic astrocytoma, or other similarly configured lesions. The distinction to the surgeon between pilocytic astrocytoma and grade II diffuse astrocytoma is usually straightforward given the relative demarcation of pilocytic tumors, the tendency to cyst formation, and the localized rather than diffusely infiltrative character.

Histologic differentiation can be very difficult in small specimens and is not always possible. In small specimens, as those from the brain stem, distinctive architectural features may not be present and the pathologist is faced with little but a highly fibrillar, and not particularly piloid, neoplasm. Immunohistochemical staining for neurofilament protein determines the numbers of axons within the mass, which is generally higher in diffuse astrocytoma, but there is considerable overlap between the number of incorporated axons in a diffuse astrocytoma and in the infiltrating edge of a pilocytic astrocytoma. An algorithm combining MIB-1 indices and p53 staining has been suggested to distinguish between these two classes of neoplasm, with the assumption that both MIB-1 rates and percentage of p53-stained nuclei are higher in diffuse astrocytoma (239,240). Another series, however, recorded a high p53 staining index in pilocytic astrocytomas (5 of 7 cases) (241), appearing to compromise the use of p53 staining index as a discriminator between pilocytic and diffuse astrocytomas. The incidence of p53 staining in pilocytic astrocytomas is, in our experience, low.

The resemblance of some pilocytic tumors to high-grade diffuse astrocytoma can be a significant problem in the presence of mitotic activity, vascular proliferation, and occasionally necrosis. This situation puts the pathologist in the unusual position of trying to decide whether the lesion is a grade I pilocytic tumor or a grade III, or even grade IV, astrocytoma. Rosenthal fibers, EGBs, microcystic change, filling of the local subarachnoid space, solid rather than infiltrating architecture, glomeruloid microvascular tufts (rather than hyperplasia of endothelial cells within larger vessels), and hyalinized vasculature become important points that favor pilocytic astrocytoma.

Although usually a histologically distinctive lesion unlikely to be confused with pilocytic astrocytoma, pleomorphic xanthoastrocytoma (PXA) shares several features with it, including young patient age, association with seizures, and frequent cyst/mural nodule configuration. A helpful distinguishing feature of PXA is the superficial, often largely leptomeningeal, position of the mass. Microscopically, this lesion exceeds the pilocytic tumor in its degree of cellularity and pleomorphism. Xanthic astrocytes are not a typical feature of pilocytic tumors, but are common in PXA. A diffusely infiltrative component in the underlying brain is a more common feature of PXA than of pilocytic astrocytoma. The presence of pericellular reticulin in the leptomeningeal element is an additional distinguishing feature of PXA.

Also in the differential diagnosis of lesions in the cerebral hemispheres, particularly if one accepts the histologically diverse lesions referred to as "nonspecific forms" of this entity, is dysembryoplastic neuroepithelial tumor. This difficult issue is discussed with the latter tumor.

A rare neoplasm of the posterior fossa combines features of dysembryoplastic neuroepithelial tumor and pilocytic astrocytoma. This is the rosette-forming glioneuronal tumor of the fourth ventricle, discussed in chapter 4.

In the cerebellum, classic pilocytic astrocytoma is a distinctive entity, with its macrocysts and microcysts. The presence in some examples

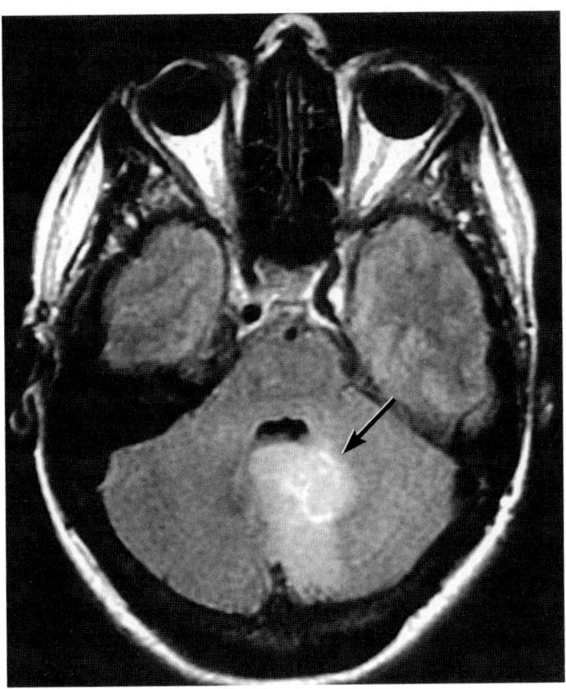

Figure 3-148

DIFFUSE ASTROCYTOMA OF THE CEREBELLUM

Most cerebellar astrocytomas, particularly in children, are pilocytic, but occasional lesions are cerebellar equivalents of cerebral grade II diffuse astrocytoma. As seen in a fluid-attenuated inversion recovery (FLAIR) image, the noncystic tumor has a diffuseness not usually seen in pilocytic lesions. In addition, it did not contrast enhance. The patient, an adult, complained of headaches and unsteadiness.

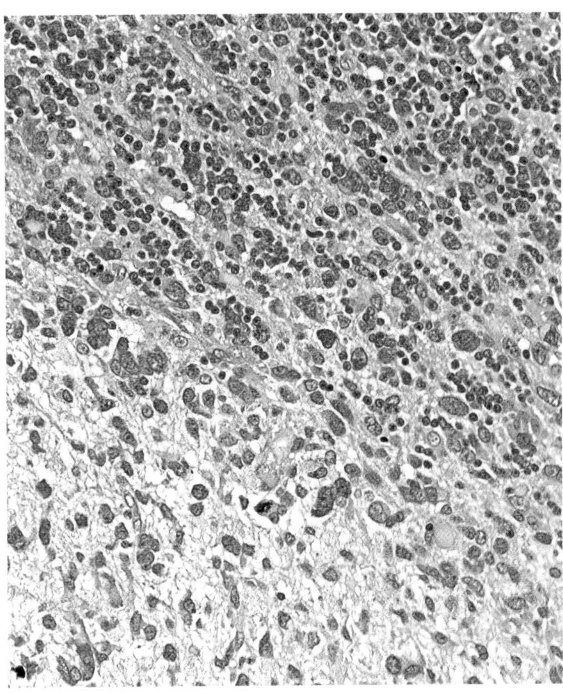

Figure 3-149

DIFFUSE ASTROCYTOMA OF THE CEREBELLUM

Diffuse astrocytomas infiltrate cerebellar tissue, as here in the molecular, Purkinje cell, and internal granular cell layers. Mitoses are more likely in these sometimes grade III tumors than in pilocytic astrocytomas. The degree of nuclear pleomorphism exceeds that of most pilocytic astrocytomas that, when infiltrative, are usually composed of uniform, evenly spaced cells.

of the diffuse pattern described above may cause diagnostic difficulty. An infiltrating WHO grade II, or even grade III, astrocytoma or oligodendroglioma must then be excluded (figs. 3-148, 3-149). Often, the only pilocytic feature is extension into the subarachnoid space. In the absence of features of pilocytic astrocytoma, it may not be possible to distinguish the predominantly "diffuse form" of pilocytic astrocytoma from an ordinary infiltrating, or diffuse, astrocytoma or oligodendroglioma. Clinical and radiographic features then become decisive. Molecular genetics may also come to play a role. Uniformly spaced, cytologically bland nuclei are more consistent with pilocytic astrocytoma.

Truly diffuse, WHO grade II or III astrocytomas are unusual in the cerebellum, and are more common in adults. In one series of cerebellar astrocytomas, the median age for patients with pilocytic astrocytoma was 12 years, whereas the comparable figure for individuals with diffuse astrocytoma was 52 years (207). There is more cytologic atypia and mitotic activity in these lesions than in diffuse pilocytic astrocytomas (fig. 3-149). In any age group, diffuse astrocytomas are more likely to involve the cerebellar peduncles, especially the middle, and then can overlap with brain stem astrocytoma of the diffuse type.

This same issue of pilocytic versus diffuse astrocytoma grades II, III, and even IV occurs for lesions in the brain stem. Diffuse lesions predominate in this location, and few are biopsied in light of their pathognomonic clinical/radiologic features. Surgical specimens thus consist largely of "atypical" lesions such as pilocytic astrocytoma, rare ganglion cell tumors, or diffuse astrocytoma that is unusually exophytic. Clinically and radiologically, a pilocytic astrocytoma should be considered if a tumor is: 1) mesencephalic or medullary rather than pontine; 2) exophytic in growth pattern, extending into the cerebellopontine angle or upward from the dorsal midbrain; 3) discrete rather than infiltrative; 4) cystic rather than solid; or 5) solidly contrast enhancing. Glioblastomas enhance, but as a ring.

Immunohistochemistry for neurofilament protein can be used to demonstrate the number of intratumoral axons and assess the degree of infiltration. Widespread overgrowth of normal tissue is more consistent with diffuse astrocytoma.

In spite of these differences, the distinction here, and at other sites, between pilocytic and diffuse astrocytoma can be difficult, if not impossible. Treatment planning may rest largely upon clinical and radiologic parameters. A period of observation may be appropriate in some cases before chemotherapy or radiation therapy is instituted.

In the spinal cord, the differential diagnosis focuses initially on the exclusion of reactive gliosis and then differentiating pilocytic astrocytoma from other tumors, especially diffusely infiltrating astrocytoma and ganglioglioma. Distinguishing between pilocytic astrocytoma and piloid gliosis in the spinal cord can be especially difficult given the minute specimens often obtained from this slender and function-rich structure. Longstanding intramedullary neoplasms such as ependymoma (fig. 3-150, see fig. 3-259) and hemangioblastoma can induce dense gliosis that is rich in Rosenthal fibers. Assurance from the surgeon that the tissue came from the depths of a solid mass, not its periphery or an accompanying cyst wall, is helpful.

Distinguishing a pilocytic astrocytoma from a diffuse astrocytoma of the spinal cord is similarly difficult and has important therapeutic ramifications. Surgery is usually terminated upon receipt of a diagnosis of diffuse astrocytoma, or even the generic diagnosis "astrocytoma," since this designates a lesion for which surgical cure is usually considered impossible. In contrast, the operation often continues with the goal of complete resection if a diagnosis of pilocytic astrocytoma or ependymoma is rendered. Cautious analysis is required in order to avoid misinterpretation. Radiologic images usually distinguish grade II diffuse astrocytoma and well-circumscribed intramedullary tumors such as pilocytic astrocytoma, ependymoma, and hemangioblastoma.

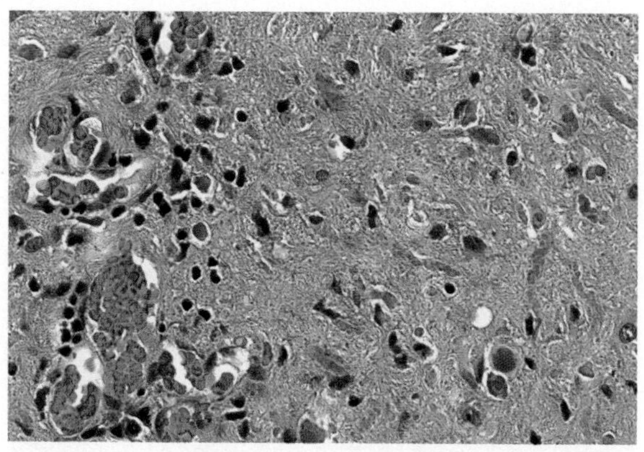

Figure 3-150

PILOID GLIOSIS RESEMBLING PILOCYTIC ASTROCYTOMA

Dense, hemosiderin-rich piloid tissue with many Rosenthal fibers, in this case around a spinal ependymoma, is more consistent with piloid gliosis than pilocytic astrocytoma. Microcystic change is common in pilocytic astrocytomas but is usually absent in piloid gliosis.

Histologically, pilocytic astrocytomas in the spinal cord are similar to those in other locations, although the compact bipolar pattern often predominates, and it may be difficult to find a microcystic element. Since the tumors are typically of long standing, hyalinized vessels are common. Piloid gliosis responding to a differential diagnostic entity, ependymoma, is illustrated in figures 3-150, 3-259.

Differentiating between pilocytic astrocytoma and ganglion cell tumors in any location, whether intracranial or intraspinal, can be complicated (fig. 3-151). They share common clinical and radiographic features, the astrocytic component of ganglion cell tumors is often patently pilocytic, pilocytic astrocytomas can trap normal neurons, chronic reactive piloid gliosis surrounds some ganglion cell tumors, both neoplasms may extend into the subarachnoid space, and some ganglion cell tumors are extensively microcystic. Many neurons in ganglion cell tumors are not phenotypically obviously neuronal, but rather astrocytic in appearance. Nonetheless, under the best of circumstances, ganglion cell tumors are distinctive for their more compact architecture, clustering of

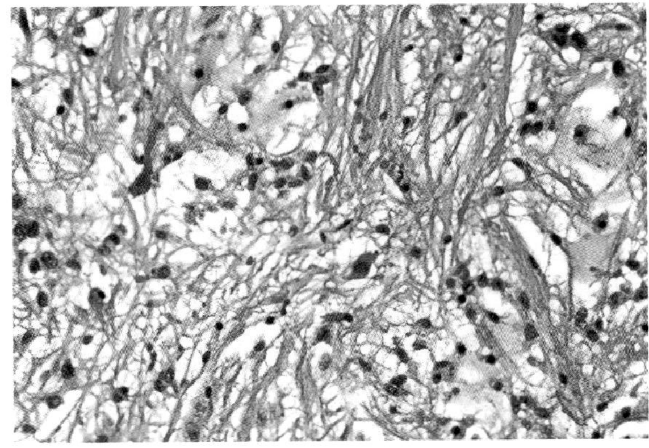

Figure 3-151

GANGLIOGLIOMA RESEMBLING PILOCYTIC ASTROCYTOMA

The neuronal nature of ganglion cell tumors may not be apparent in H&E-stained sections when neurons are small and glia are pilocytic.

abnormal neurons, more numerous EGBs, frequent content of a reticulin-rich stroma, and perivascular chronic inflammation. Immunohistochemistry for neuronal markers often becomes decisive.

Treatment and Prognosis. The natural history of pilocytic astrocytoma at any site is one of slow growth, and many incompletely resected tumors enlarge little, if at all, over periods of 10 or 20 years, or even longer (220). Extent of resection is the principal prognostic variable, and most patients do well without postoperative chemotherapy or radiotherapy (242). Decades can pass with little, if any, change in the tumor's histologic appearance (243). In some cases, rapid "recurrence" is actually due to the reformation of tumoral cysts that can often be drained with impressive clinical improvement.

As with other generally well-circumscribed CNS tumors, total resection is a strong positive prognostic factor (220,244–246). Late "malignant change" can reflect the development of radiation-induced, high-grade diffuse astrocytoma (usually glioblastoma). Only rarely do well-differentiated, entirely typical pilocytic astrocytomas exhibit diffuse spinal leptomeningeal spread (201,244,247–249), but even this does not have the same unequivocally adverse prognostic significance as it does for many other CNS tumors since metastatic deposits may, like the primary neoplasms from which they derive, remain stable over relatively long periods.

The prognosis is especially favorable for patients with tumors of the optic nerve, where complete excision and cure are often possible (203,250). Stability is especially likely in the setting of neurofibromatosis 1, where observation is often the only form of treatment. In this same genetic background, or sometimes even in sporadic lesions, regression is occasionally noted in pilocytic astrocytomas of the optic pathway or other sites (251–253). Recurrences with an ultimately fatal outcome are more likely in chiasmatic/hypothalamic tumors, in which resection is precluded by neuroanatomic realities (220,254,255). Some of these lesions grow very slowly, in many cases due largely to cyst expansion, whereas others continue to enlarge and reach massive proportions.

While their behavior is variable, pilomyxoid astrocytomas, or infantile pilocytic astrocytomas, appear more likely to recur locally and to undergo subarachnoid spread (213–215). Some lesions, however, stabilize, if not convert to classic pilocytic astrocytoma. Hypothalamic/chiasm region tumors reported as pilocytic in young children appear to behave more aggressively (254,255), perhaps in accord with the contribution of the pilomyxoid lesion that is overrepresented at this site and age group.

Due to the relatively discrete nature of many pilocytic astrocytomas, stereotactic excision has been advocated, particularly for deep-seated lesions of the basal ganglia and thalamus (256).

In one series of supratentorial tumors, the 10-year survival rate was 100 percent for 16 patients who had undergone gross total or radical subtotal tumor removal, and 74 percent for 35 patients undergoing subtotal resection or biopsy alone (205). The prognosis is thus favorable, even more so with resection of the mural nodule (245).

The prognosis of patients with cerebellar lesions is also excellent since total excision is feasible in many cases, and even residual tumor remains dormant or grows only slowly (207,220, 257). Histologically, the lesion may change little over a period of decades, and may even regress, becoming less cellular, more sclerotic, and more "degenerative" in appearance. This behavior contrasts with that of the less common, more

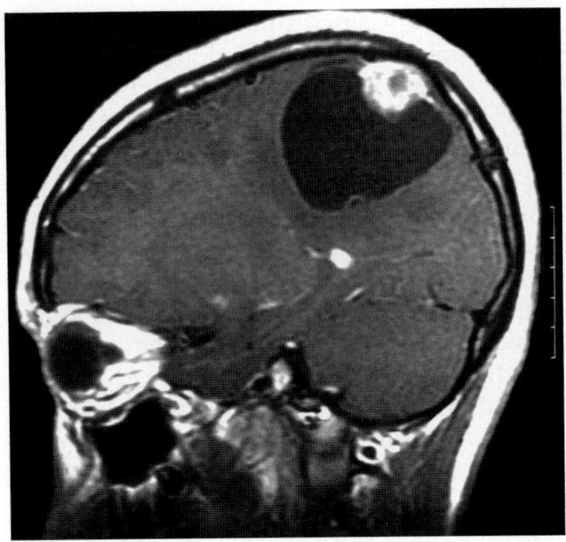

Figure 3-152

PLEOMORPHIC XANTHOASTROCYTOMA

Pleomorphic xanthoastrocytoma (PXA) is typically superficial, solid, and contrast enhancing. Some are cystic, as in this 14-year-old girl with the recent onset of seizures.

infiltrative and aggressive diffuse astrocytomas at this site (207). In one study, the 5-year, 10-year, and 20-year survival rates were 85 percent, 81 percent, and 79 percent for patients with pilocytic lesions and 7 percent, 7 percent, and 7 percent for those with diffuse astrocytomas grades II, III, and IV (207). The vast majority of children with "diffuse cerebellar astrocytomas" who do well actually have pilocytic tumors with a predominantly diffuse pattern of growth (227). One study concluded that subclassification of pediatric low-grade astrocytomas into pilocytic and fibrillary (diffuse) astrocytomas was not prognostically significant (228).

Multiple reoperations may be required to control pilocytic astrocytomas of the brain stem, but the outcome is vastly superior to that of the classic, i.e., diffuse, astrocytoma at this site (210,211).

Only limited data are available regarding the clinical outcome and prognosis of patients with spinal cord pilocytic astrocytomas (258–260). Many are discrete and amenable to gross total removal. Complete resection is not always possible, however, particularly when the tumor extends over many spinal levels.

Little is known about the course of patients presenting initially with histologically malignant pilocytic astrocytoma. In our limited experience, local recurrence and leptomeningeal dissemination may occur.

Pleomorphic Xanthoastrocytoma

Definition. *Pleomorphic xanthoastrocytoma* (PXA) is a superficially situated, often partly leptomeningeal, astrocytic neoplasm with marked cellular pleomorphism and frequent xanthomatous change. PXAs are usually WHO grade II, but there are aggressive and even anaplastic variants for which grading criteria have not been formulated. In the WHO system, a PXA with five or more mitoses is referred to as "pleomorphic astrocytoma with anaplastic features."

Clinical Features. The patients, usually adolescents or young adults, often present with seizures (261–263).

Radiologic Findings. The lesion is usually a superficial, meningocerebral, contrast-enhancing mass that is, in about half of cases, a nodule in the wall of a cyst (fig. 3-152) (263,264). (See Appendix D for a list of cystic CNS tumors.) The temporal lobe is affected in nearly 50 percent of cases. Rare sites include cerebellum (265), spinal cord (266), and retina (267). Mass effects are minor in grade II lesions. Some examples are focally calcified.

Gross Findings. The tumor is a firm, generally superficial, circumscribed nodule or plaque. Although some PXAs appear to lie almost entirely within the leptomeninges, all involve brain parenchyma to some extent. Thus, despite the apparent circumscription of the solid, superficial component, no uniform cleavage plane underlies the entire tumor. An accompanying cyst will contain clear yellow fluid.

Microscopic Findings. The typical histologic features are summarized in Table 3-1 and described below. The growth pattern of PXA varies, making tissue sampling important. Solid portions of the lesion lie in the subarachnoid space (figs. 3-153, 3-154), and extend as tongues of tissue into underlying cortex along perivascular (Virchow-Robin) spaces. The superficial element of PXA can often be identified by its content of large, muscularized arteries such as occur normally in the leptomeninges, but not in CNS parenchyma (figs. 3-153, 3-154). There is also an infiltrating component that both architecturally and cytologically resembles diffuse astrocytoma (discussed below) (see fig. 3-158).

Table 3-1

HISTOLOGIC FEATURES OF PLEOMORPHIC XANTHOASTROCYTOMA

Predominantly superficial location
Infiltrating cortical component
Compact, fascicular, or lobulated architecture[a]
Cellular pleomorphism[a]
Modest or absent mitotic activity[a]
Xanthomatous change[a]
Perivascular and interstitial lymphocytes[a]
Eosinophilic granular bodies[a]
Vascular sclerosis[a]
Abundant reticulin[a]

[a]Features associated with the superficial component.

Predominantly Superficial Location. A superficial component (figs. 3-153, 3-154) is a cardinal feature of PXA. Its compactness or solidity and the lack of infiltrated brain components are shared with other lesions such as pilocytic astrocytoma, subependymal giant cell astrocytoma, ganglion cell tumor of both conventional and desmoplastic infantile variant (see Appendix A for lesions categorized by architectural features). As such, all of these tumors, at least at low magnification, grow by expansion, pressing normal parenchyma aside.

The solid, subarachnoid component typically has a fascicular, nonmicrocystic architecture produced by bundles of somewhat elongated cells (figs. 3-155–3-157). The extent of this fasciculation varies from case to case, but is usually prominent.

Infiltrating Cortical Component. Within the cortex, beneath the solid, nodular element, is a variable zone of infiltrating neoplastic cells that are less pleomorphic than those in the superficial component (fig. 3-158). This component of the tumor closely resembles a grade II infiltrating, or diffuse, astrocytoma, and it can be difficult to distinguish PXA from the far more common infiltrative astrocytoma if there is little in the way of a discrete mass. The compact portion of PXA is rarely calcified, but mineralization of the subjacent cortex is not uncommon.

Cellular Pleomorphism. The degree of cellular polymorphism encountered in PXA is quite variable, and while sometimes marked, is often less marked than the name implies (fig. 3-158).

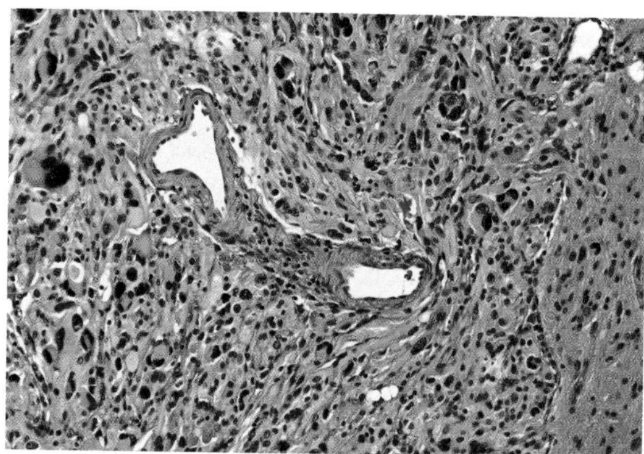

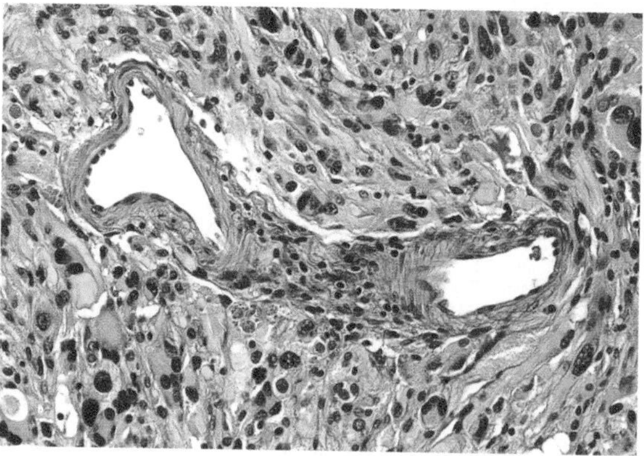

Figure 3-153

PLEOMORPHIC XANTHOASTROCYTOMA

Top: The lesion is usually superficial, if not located principally in the subarachnoid space.

Bottom: A moderate degree of cytologic pleomorphism is typical.

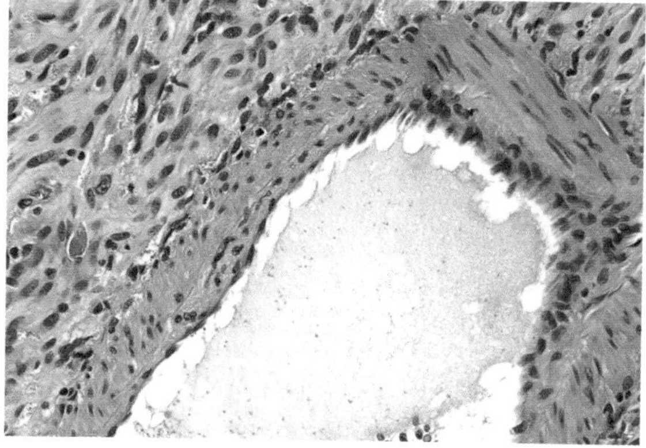

Figure 3-154

PLEOMORPHIC XANTHOASTROCYTOMA

The localization to the subarachnoid space can be inferred from the presence of intratumoral arteries with a degree of muscularization not present in intraparenchymal vessels.

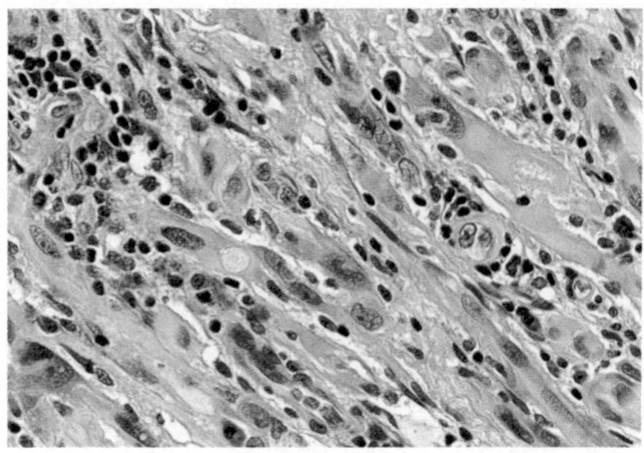

Figure 3-155

PLEOMORPHIC XANTHOASTROCYTOMA

Fascicular architecture and perivascular lymphocytes are common.

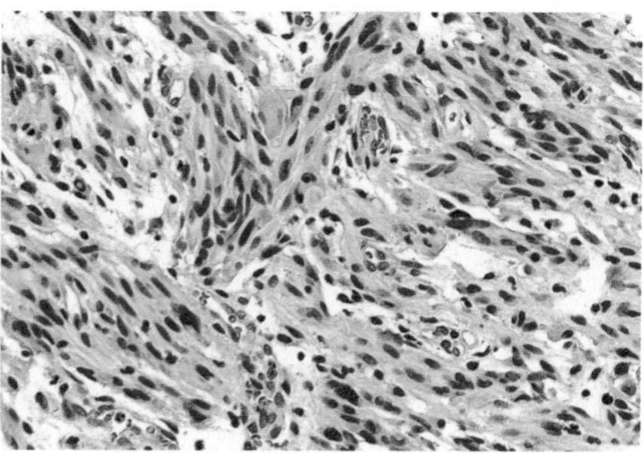

Figure 3-156

PLEOMORPHIC XANTHOASTROCYTOMA

The presence of intersecting bundles of cells is a distinctive feature of some PXAs.

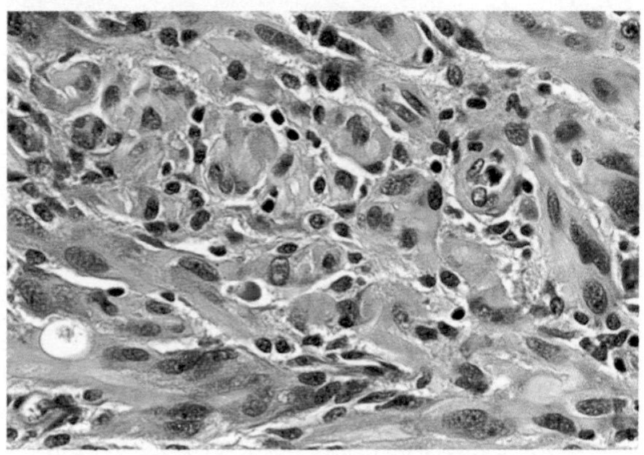

Figure 3-157

PLEOMORPHIC XANTHOASTROCYTOMA

PXAs usually have a moderate to severe degree of nuclear pleo-morphism. Note the compact, non-infiltrative quality of this leptomeningeal component.

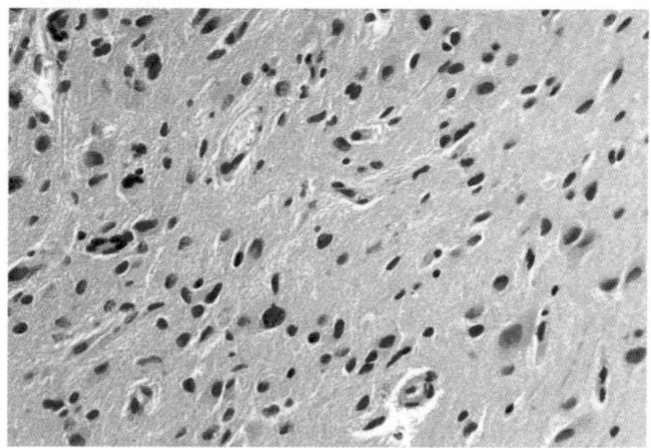

Figure 3-158

PLEOMORPHIC XANTHOASTROCYTOMA

Although the lesion is, overall, well demarcated, there is often an intracortical zone of tumor infiltration that can be misinterpreted as evidence of diffuse astrocytoma.

Indeed, it is only a minor element in some cases. The pleomorphic cells vary in shape from round to irregular or elongate, and possess large, multiple or multilobed, hyperchromatic nuclei. The presence of cytoplasmic nuclear inclusions suggests that the atypia in most PXAs is degenerative in nature.

Modest or Absent Mitotic Activity. In spite of the pleomorphism, mitoses are usually limited in number, if detected at all.

Xanthomatous Change. Xanthic change, expressed as vacuolization of tumor cells (figs. 3-159, 3-160), varies greatly in degree and extent, but since it is often inconspicuous, it is hardly a defining feature of this tumor, its moniker notwithstanding.

Perivascular and Interstitial Lymphocytes. Perivascular lymphocytes, while inconspicuous in many lesions, are part of the constellation of diagnostic features (figs. 3-155, 3-157). Small numbers of interstitial lymphocytes are often present as well.

Eosinophilic Granular Bodies. EGBs vary considerably in number, but are regular findings in the superficial portions of PXA (figs. 3-159, 3-161). They are diagnostically important, since they characterize well-circumscribed, low-grade glial neoplasms as a whole, such as PXA, pilocytic

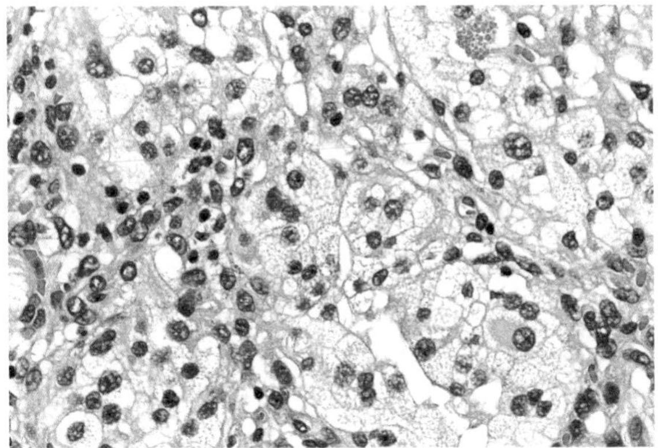

Figure 3-159

PLEOMORPHIC XANTHOASTROCYTOMA

Xanthoma cells are prominent in a minority of lesions. Note the eosinophilic granular body.

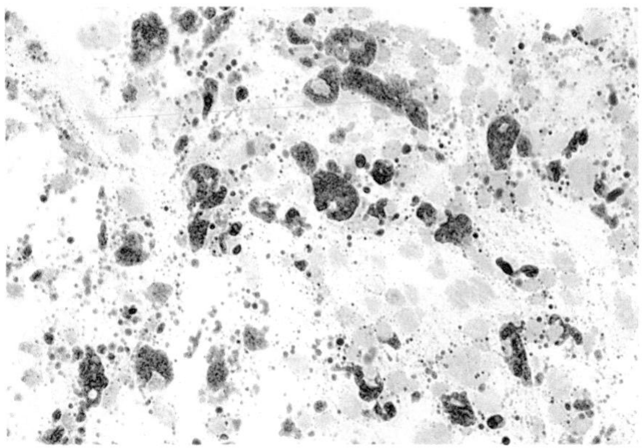

Figure 3-160

PLEOMORPHIC XANTHOASTROCYTOMA

Lipid is well documented after staining with oil red-O.

astrocytoma, and ganglion cell tumor (see Appendix M for a tabulation of features such as EGBs that should suggest a low-grade neoplasm). Variably PAS positive, these aggregates of protein droplets generally indicate chronicity, and a diagnosis of malignancy should be made with caution in their presence. While they can be present in high-grade gliomas and are not a requisite feature for the diagnosis of PXA, we are uncomfortable in making the diagnosis of PXA in their absence.

Vascular Sclerosis. As in other chronic, slowly growing gliomas and neuronal tumors, hyalinized vessels are common in the superficial component.

Abundant Reticulin. Intercellular reticulin staining characterizes several neoplasms that involve the leptomeninges. This is especially prominent in PXA. At least focally, it usually invests cells either individually or in small clusters (fig. 3-162).

Only a small minority of PXAs contain a component of ganglion cells which, by virtue of number, clustering, and cytologic features, appear to be neoplastic rather than simply trapped preexisting elements (fig. 3-163) (268–271). There is no known prognostic significance to these *PXA/gangliogliomas,* but, as is the case for classic PXAs without a ganglion cell component, some examples behave aggressively (269). Tumors with ganglion cells are summarized in Appendix G. A rare PXA is pigmented (fig. 3-164) (272).

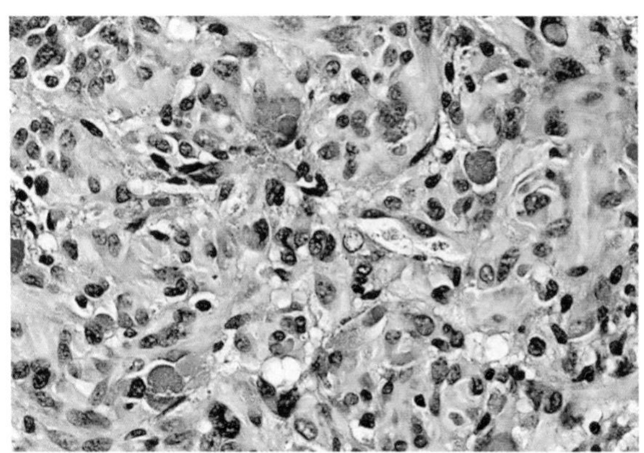

Figure 3-161

PLEOMORPHIC XANTHOASTROCYTOMA

Cells filled with eosinophilic protein droplets (eosinophilic granular bodies) are present in almost all PXAs.

Rosenthal fibers, if any, are present at the periphery of PXAs as a response by the surrounding brain.

Grading. Although the classic PXA is WHO grade II, and curable surgically in many cases, not all PXAs are benign, either histologically or biologically. Occasional tumors clearly have excessive mitotic activity (fig. 3-165), necrosis, and even unequivocal anaplasia (fig. 3-166). Some of these anaplastic lesions arise de novo and others emerge from grade II PXAs (263,273,274). These aggressive variants recur more often locally and undergo more extensive spread in the subarachnoid space. Given the relative infrequency of PXA and the length of follow-up necessary to

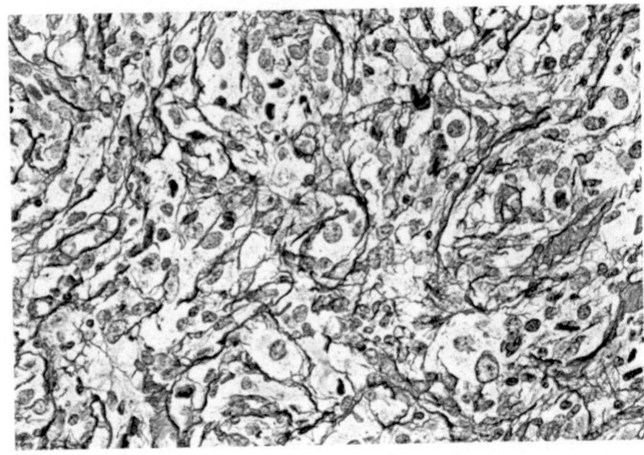

Figure 3-162

PLEOMORPHIC XANTHOASTROCYTOMA

Reticulin is prominent in the superficial component where it invests small groups of, or even individual, cells.

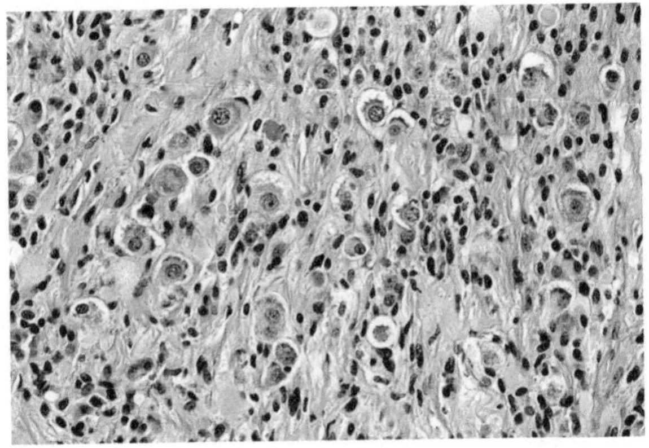

Figure 3-163

PLEOMORPHIC XANTHOASTROCYTOMA

Some PXAs have a ganglion cell component.

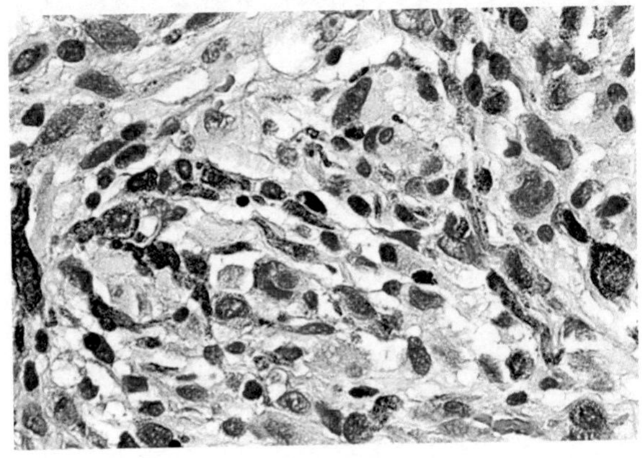

Figure 3-164

PLEOMORPHIC XANTHOASTROCYTOMA

Pigmented cells are unusual PXA components.

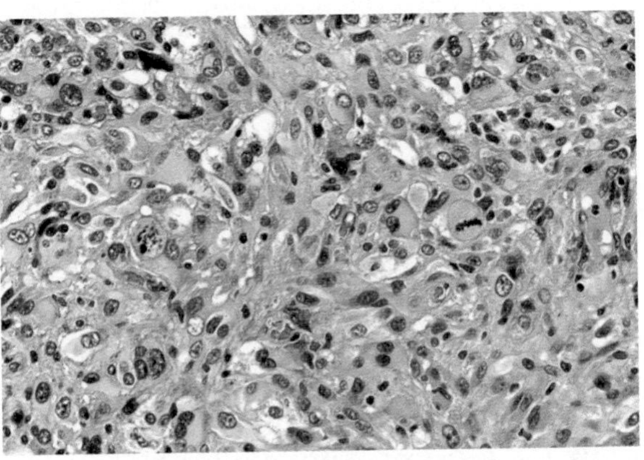

Figure 3-165

PLEOMORPHIC XANTHOASTROCYTOMA

In the absence of overt anaplasia, which is rare, increased mitotic activity is the principal grading parameter in PXAs. This lesion recurred.

assess its biologic behavior, it is difficult to define the prognostic significance of individual histologic features. One study addressing this issue identified both extent of resection and number of mitotic figures as most critical. There was a significantly poorer outcome in patients whose tumors had 5 or more mitoses per 10 high-power fields (263). As an independent predictor, necrosis did not survive multivariate analysis in the latter study, but is reportedly an unfavorable factor (275), and we believe this to be true. At a minimum, its presence should prompt a search for mitoses.

Terminology applicable to proliferative, mitotically active PXAs has not been formulated. There certainly exist high-grade lesions that must be considered biologically grade III, if not grade IV, and deserve postoperative radiotherapy, even if grossly totally resected. It is unclear, however, whether postoperative radiotherapy or chemotherapy is warranted for totally resected lesions with, for example, 5 or more mitoses per 10 high-power fields. For this reason, the WHO was reluctant to stipulate grade III for any lesion with such mitotic activity. The diagnosis of a grade III (malignant) glioma almost certainly requires radiotherapy. Some PXAs with a mitotic index in excess of 5 per 10 high-power fields have been cured by excision alone. The WHO's approach

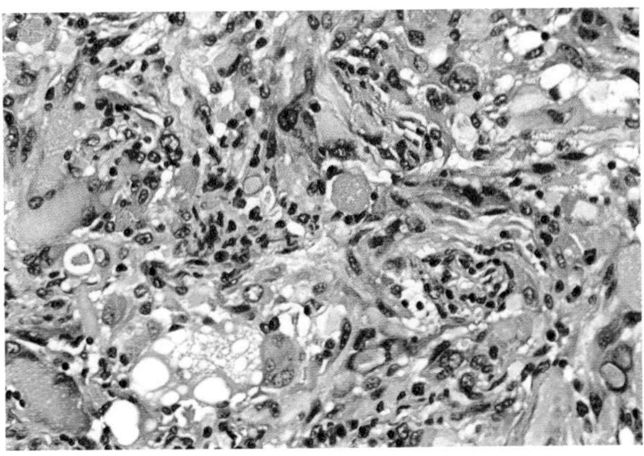

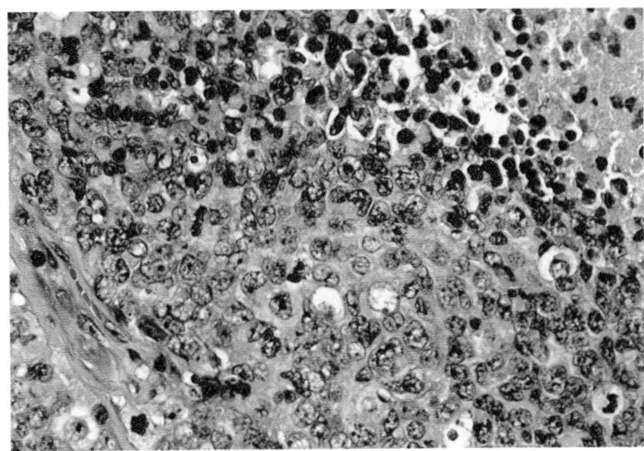

Figure 3-166

PLEOMORPHIC XANTHOASTROCYTOMA WITH ANAPLASTIC TRANSFORMATION

Rare PXAs have both classic and anaplastic features.

Left: The subarachnoid component of this PXA at the time of initial presentation had typical findings, including compact architecture, lymphocytes, xanthomatous change, and eosinophilic granular bodies.

Right: Large areas of the tumor were overtly anaplastic, with high cellularity, many mitoses, and necrosis.

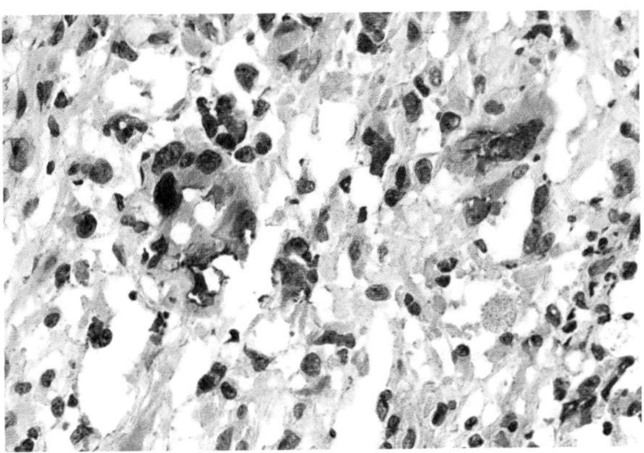

Figure 3-167

PLEOMORPHIC XANTHOASTROCYTOMA: FROZEN SECTION

Cytologic pleomorphism may misrepresent the tumor as glioblastoma at the time of frozen section. Note the eosinophilic granular body.

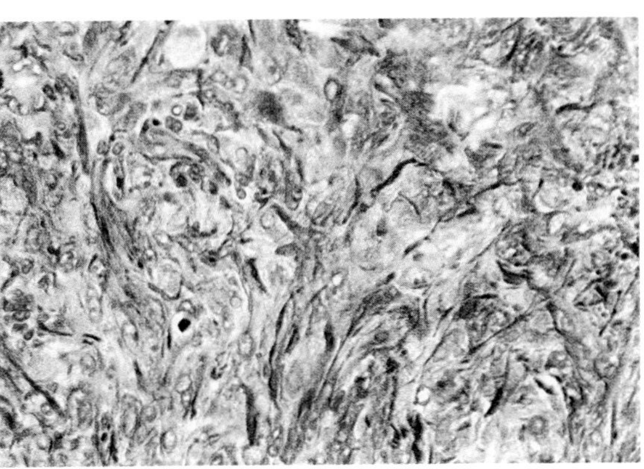

Figure 3-168

PLEOMORPHIC XANTHOASTROCYTOMA

PXAs are GFAP immunoreactiive, although only a minority of cells may be positive.

is to use the term "PXA with anaplastic features" for lesions with 5 or more mitoses per 10 high-power fields or with necrosis (276).

Whether prognostically significant grading by MIB-1 indices is possible remains to be established (263).

Frozen Section. Without an awareness of the entity and its typical radiologic findings, the cellular pleomorphism of PXA can readily be mistaken for that of glioblastoma (fig. 3-167). Lacking, however, are brisk mitotic activity, microvascular proliferation, and necrosis with or without palisading. Other diagnostically helpful features include tissue fragments featuring a sharp tumor-brain interface, xanthic change, perivascular lymphocytic aggregates, and EGBs.

Immunohistochemical Findings. Uniform immunoreactivity for S-100 protein is noted in both the superficial and infiltrating components of PXA (fig. 3-168). One study suggested immunoreactivity for CD34 (277). Staining for GFAP is seen in the superficial component of the tumor, but it may be patchy or weak (278). When present, ganglion cells are appropriately immunoreactive for synaptophysin, chromogranin, neurofilament protein, and microtubule-associated protein 2 (268–270,278). Individual pleomorphic cells, so typical of PXAs, may also be positive for these neuronal markers, but the finding of only isolated cells with no more than a glial/neuronal immunophenotype is not sufficient for a diagnosis of mixed PXA/ganglioglioma. In accord with the rarity of mutations, diffuse immunoreactivity for p53 is uncommon (279,280).

In one study, the mean MIB-1 index of 29 cases was 1.9 percent (263). In 23, the index was equal to or below 2 percent, and only in 6 did it exceed 2 percent (range, 2.9 to 14.0 percent). MIB-1 indices increase during malignant transformation, in one case from 0.1 to 4.9 percent (274), and can be 10 percent or even 20 percent in areas of advanced anaplasia.

Ultrastructural Findings. Some cells contain many intermediate filaments whereas others demonstrate primarily rough endoplasmic reticulum, occasional lysosomes, and lipid droplets (281). Aggregates of complex lysosomes, corresponding to light microscopically evident granular bodies, may be present as well. Collagen is typically scant. Cellular investment by basal lamina is variable but common in superficial tumor components and has been used to suggest an origin of PXA from subpial astrocytes (261). The surfaces of the latter are normally covered by basal lamina as they fan out over the surface of the brain.

Ganglion cells express their neuronal nature in the form of microtubules, dense-core vesicles, and clear vesicles of the synaptic type (281).

Cytologic Findings. Marked nuclear pleomorphism and hyperchromasia are easily seen in cytologic preparations (282,283). EGBs may be harder to find.

Molecular Findings. There is little known of the molecular genetics of PXA, but there appears to be only a low incidence of mutations of *p53* (1 of 47 tumors in one study [279], 3 of 62 in another [280]). Common genetic abnormalities seen in diffuse astrocytomas, e.g., amplification of *EGFR, MDM2,* and *CDK4*, and loss of heterozygosity on chromosomes 9, 17p, and 10 are uncommon, if they occur at all, in PXAs (280).

Differential Diagnosis. Due to obvious therapeutic implications, the principal diagnostic challenge when faced with a PXA is to avoid confusing it with glioblastoma or the rare malignant fibrous histiocytoma. Appendix K lists low-grade neoplasms, such as PXA, that are potentially overgraded, and Appendix O is an algorithm that assures that candidate low-grade, often relatively well-circumscribed lesions are not overlooked. Appendix M details histiologic features suggestive of low-grade, and non-neoplastic, lesions.

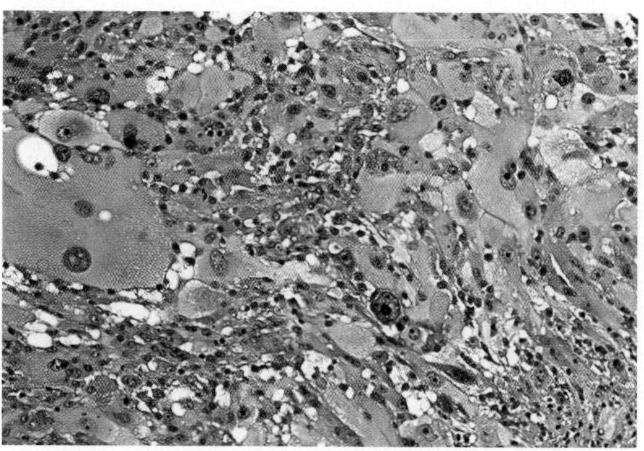

Figure 3-169

GIANT CELL GLIOBLASTOMA RESEMBLING PLEOMORPHIC XANTHOASTROCYTOMA

Flagrant pleomorphism, prominent necrosis, and mitotic activity help distinguish giant cell glioblastoma from PXA.

This task begins with awareness of the clinical and radiographic features which, when classic, are highly suggestive of the diagnosis, and most atypical for the classic glioblastoma. Even with this awareness, distinction from the giant cell variant of glioblastoma (fig. 3-169) can be problematic given the latter's cellular pleomorphism and the often discrete borders, features that it shares with PXA. The classic PXA, however, has EGBs, exhibits a paucity of mitoses, and lacks necrosis. Admittedly, higher-grade PXAs present a more difficult issue. EGBs are foreign to glioblastomas as a whole, and the higher grade PXAs are more fascicular in architecture and likely to have a component of monomorphous, somewhat epithelioid cells, with less cytologic atypia and proliferative activity. Massive cells, in some cases with many nuclei, are more typical of giant cell glioblastoma than PXA.

Immunohistochemically, there may be exploitable differences between PXA and giant cell glioblastoma. Specifically, the latter has a substantial incidence of immunoreactivity for p53, whereas PXA is more often positive focally with neuronal markers such as neurofilament protein, NeuN, and synaptophysin (284). Immunoreactivity of PXA, but not giant cell glioblastoma, for CD34 may be helpful (277).

There are rare cases where the distinction between PXA and giant cell glioblastoma cannot be made, and the diagnosis of a malignant glioma, with an appropriate comment, must suffice. It remains to be seen whether molecular characterization is helpful in this setting. Whereas PXAs, at least in their low-grade form, have a low incidence of mutations in *p53* and p53 protein staining, giant cell glioblastomas often have both. Immunohistochemical positivity in many cells thus favors glioblastoma, although lack of p53 staining does not rule out giant glioblastoma.

Malignant histiocytic tumors, rare CNS lesions, are excluded by the demonstration of diffuse S-100 protein immunoreactivity, as well as variable staining for GFAP, although this may be sparse in PXAs.

The discrete nature of PXA, especially in cyst/mural nodule complex, prompts consideration of a pilocytic astrocytoma, or the radiologically similar ganglion cell tumor discussed in the following paragraph. The degree of cellularity, compactness, and pleomorphism of PXA typically exceeds that of the pilocytic lesion. In addition, the generally spongy pilocytic astrocytoma often exhibits a biphasic pattern wherein compacted, highly fibrillar, elongate cells associated with Rosenthal fibers alternate with a microcystic component composed of multipolar, hypofibrillar, so-called protoplasmic astrocytes associated with granular bodies. When pleomorphic cells are present, they are usually not as prominent as in PXA, and also are not lipidized. Reticulin staining, if any, is much less prominent than in PXA.

The degree of pleomorphism of some ganglion cell tumors clearly overlaps with that of PXA and poses a diagnostic challenge, since both tumors favor younger patients, are typically cystic, and feature abundant reticulin as well as patchy lymphocytic infiltrates. Blurring the distinction is the presence of pleomorphic tumor cells with a neuronal immunophenotype, or rarely, of true neoplastic ganglion cells, in some PXAs and pilocytic astrocytomas. Although PXAs often contain cells superficially resembling bizarre neurons, the latter lack Nissl substance and usually show some degree of GFAP immunoreactivity rather than labeling for neuronal markers such as synaptophysin and neurofilament protein. When ganglion cells are encountered in PXA, they more often represent trapped normal cortical elements than a neoplastic ganglion cell component, as is discussed above. PXAs with ganglion cells are a form of ganglioglioma, but presumably have a somewhat less favorable prognosis than classic gangliogliomas.

Specimens taken entirely from the deeper infiltrating component of PXAs pose a particular problem since the infiltrating tumor cells are, for practical purposes, identical in appearance to cells of diffuse astrocytomas. Reference to neuroimaging studies puts this finding into perspective.

Treatment and Prognosis. The relative infrequency of PXA, the long follow-up required for analysis, and the total resection accomplished in many instances have made it difficult to draw a close parallel between histologic features and biologic behavior. It is particularly difficult to prognosticate in individual cases in which mitoses are more than rare, but fewer than numerous. There is also uncertainty regarding the prognostic significance of deep parenchymal infiltration, or the minimal morphologic criteria for anaplasia or malignant transformation. Nevertheless, overall survival rates of 81 percent and 70 percent have been reported at 5 and 10 years, respectively (263). Patients who succumb do so more often after anaplastic transformation of atypical PXA, or after progression of a PXA that was malignant at first appearance. The frequency of such malignant change is not clear. The initial approach to the treatment of a grade II PXA is generally conservative and includes gross total removal followed by serial radiologic observation. The role of radiotherapy in the treatment of subtotally resected lesions is unknown.

A significant proportion of PXAs recur, some with increased cellularity and mitotic activity (263,273–275). About as often, the lesion undergoes overt malignant transformation to a tumor composed of small, uniform cells exhibiting brisk mitotic activity; conspicuous necrosis, with or without palisading; and loss of intercellular reticulin staining. Interestingly,

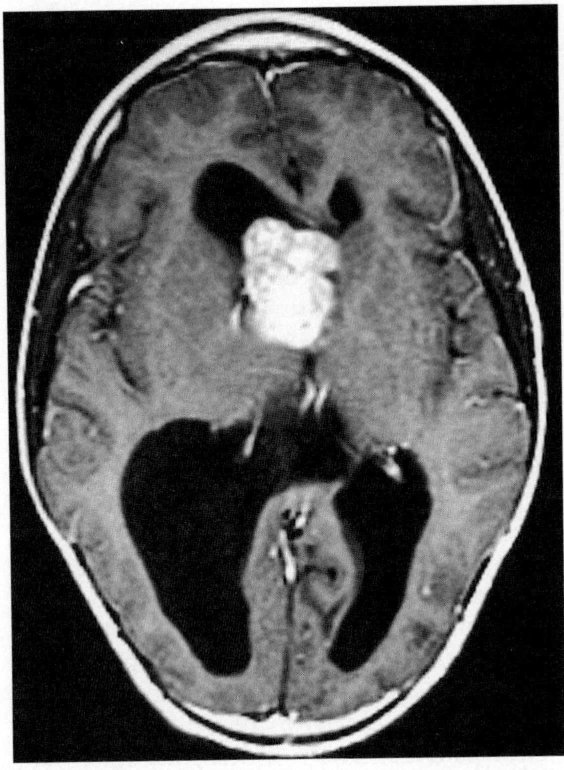

Figure 3-170

SUBEPENDYMAL GIANT CELL ASTROCYTOMA (TUBEROUS SCLEROSIS)

Subependymal giant cell astrocytomas (SEGAs) of tuberous sclerosis are typically bulky, contrast-enhancing masses in the region of the foramen of Monro. As is common with large examples, foraminal obstruction produced hydrocephalus in this case. (Fig. 3-103 from Fascicle 10, 3rd Series.)

microvascular proliferation is not generally a feature. The appearance of well-differentiated typical PXA at remote locations, without recurrence in the original site, has been reported (285).

Subependymal Giant Cell Astrocytoma (Tuberous Sclerosis)

Definition. *Subependymal giant cell astrocytoma* (SEGA) is a WHO grade I, demarcated, mainly intraventricular tumor of large astrocyte- and neuron-like cells that nearly always arises in the setting of tuberous sclerosis. Because of shared astrocytic/neuronal features in some cases, the modifier "astrocytoma" is sometimes omitted and the lesion referred to simply as *subependymal giant cell tumor.*

Clinical Features. Geographically, SEGAs are largely limited to the region of the foramina of Monro, the paramedian channels that conduct CSF from the lateral to the third ventricle (fig. 3-170). The lesions are usually detected in the clinical setting of increased intracranial pressure or during radiographic surveillance in a patient with tuberous sclerosis (286–289). Since the syndrome may be incompletely expressed (forme fruste) and since manifestations may occasionally be limited to the CNS, some patients come to surgery for an intraventricular mass before the underlying genetic abnormality is recognized (286). In a rare patient, there is no clinical or radiologic evidence of tuberous sclerosis (290). Tuberous sclerosis is described in chapter 18.

Although SEGAs appear to evolve from enlargement of hamartomatous subependymal nodules occurring in the walls of the anterior portions of the lateral ventricles (see fig. 18-24), symptomatic tumors are generally restricted to the regions of the foramina of Monro. Infrequently, giant cell tumors in patients with tuberous sclerosis, or without other features of the disorder, are paraventricular, and arise largely within the substance of the brain.

Rarely, malignant glioma and other CNS neoplasms occur in the setting of tuberous sclerosis (291,292). It is not clear whether these are pathogenetically related to the syndrome, or are simply chance occurrences.

Radiologic Findings. Typical intraventricular SEGAs are often bulky, contrast enhancing, and variably calcified (fig. 3-170) (see Appendix C for tumors that are intraventricular and Appendix F for those that are frequently calcified). Smaller, coexistent subependymal hamartomas are even more likely to be densely calcified (see fig. 18-24). The association of a large, symptomatic lesion with smaller, calcified periventricular hamartomas is virtually diagnostic of tuberous sclerosis (293). The demonstration of bright enhancement of cortical tubers with underlying white matter changes on T2-weighted MRI completes the picture (see fig. 18-21).

Gross Findings. The dome-like, broadly based lesions are smooth surfaced, soft to firm, and gray. Densely calcified SEGAs are gritty.

Microscopic Findings. Most SEGAs are demarcated, or only minimally infiltrative of underlying brain parenchyma. As a reflection of their discrete growth pattern, the masses are composed only of tumor cells and a vascular

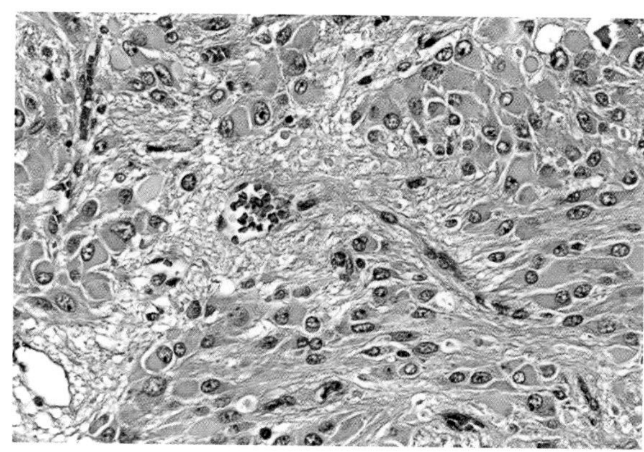
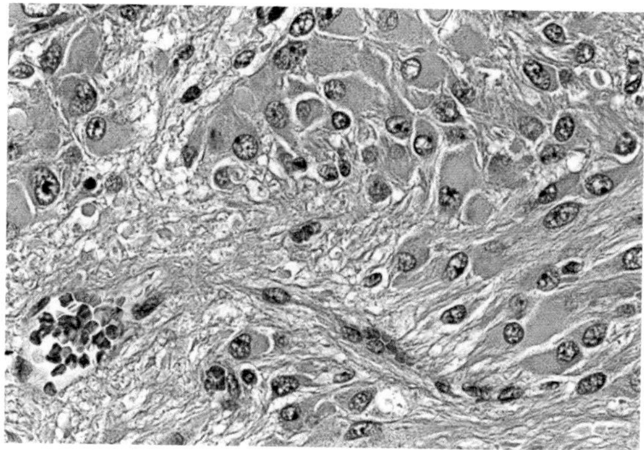

Figure 3-171

SUBEPENDYMAL GIANT CELL ASTROCYTOMA (TUBEROUS SCLEROSIS)

Left: Perivascular fibrillar zones resemble perivascular pseudorosettes of ependymoma.

Right: Higher magnification discloses the fine fibrillarity of the perivascular regions. The cells of SEGA are large, but not giant, and cytologically intermediate between gemistocytes and ganglion cells.

stroma. The incorporated brain parenchyma so characteristic of diffuse astrocytoma is lacking.

Despite their stereotypic clinical associations and radiologic features, SEGAs show considerable microscopic variation. Most have vague perivascular pseudorosettes with large eosinophilic cells disposed in clusters, demarcated by perivascular fibrillarity (figs. 3-171, 3-172). Orderly ependymoma-like pseudorosettes are less frequent and often only focal. Organoid patterns are exceptional.

Although the neoplastic cells are typically spindle to epithelioid in shape, and larger than those of gemistocytic astrocytoma, most are phenotypically very astrocytic. Because of vesicular nuclei and prominent nucleoli, they often resemble neurons. Despite the tumor's name, its cells usually do not reach the truly "giant" proportions seen occasionally in glioblastoma or pleomorphic xanthoastrocytoma.

Despite the distinctively astrocytic quality of the abundant pink cytoplasm of SEGA cells and the accompanying fibrillar background, they differ from reactive or neoplastic gemistocytes. Their eccentric nuclei lie within more copious cytoplasm, and asymmetric processes emanate from the cell surface opposite the nucleus rather than radiating uniformly from the cell circumference. In addition, their vesicular nuclei with nucleoli differ from the densely hyperchromatic nuclei of gemistocytic astrocytes. In addition to this typical cellular pattern, other cells may be spindle-shaped and gathered into broad fascicles (fig. 3-173).

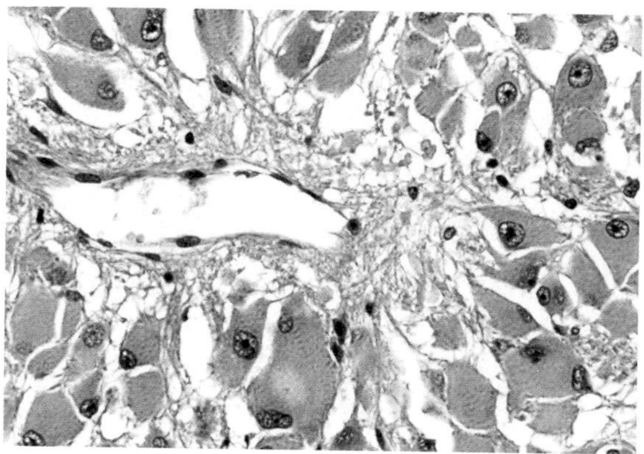

Figure 3-172

SUBEPENDYMAL GIANT CELL ASTROCYTOMA (TUBEROUS SCLEROSIS)

Eccentrically positioned nuclei and glassy cytoplasm are typical of SEGA.

Whether multipolar or bipolar, SEGA cells possess either distinctly eosinophilic or "brick red" cytoplasm. In addition to the somewhat vesicular chromatin and distinct nucleoli, intranuclear cytoplasmic inclusions are common. Mitoses vary in number from case to case, and are present in half of the cases (287). Calcospherites and vascular calcification are frequently present. Scattered mast cells are common. As a rule, vascular proliferation is not a feature. Necrosis is rarely encountered, but should

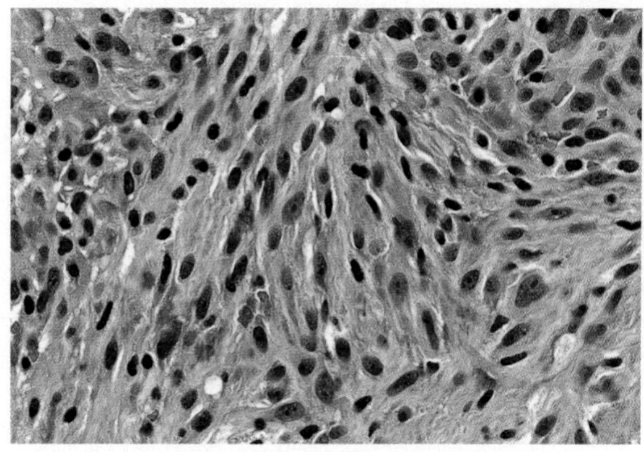

Figure 3-173

SUBEPENDYMAL GIANT CELL ASTROCYTOMA (TUBEROUS SCLEROSIS)

A fascicular architecture dominates in some cases.

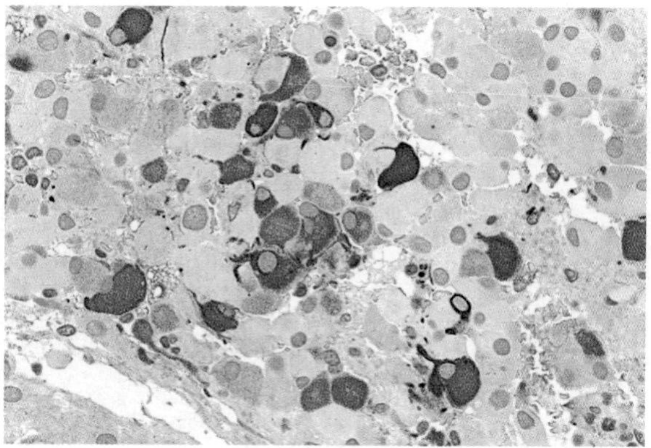

Figure 3-175

SUBEPENDYMAL GIANT CELL ASTROCYTOMA (TUBEROUS SCLEROSIS)

A subset of tumor cells is immunoreactive for neurofilament protein.

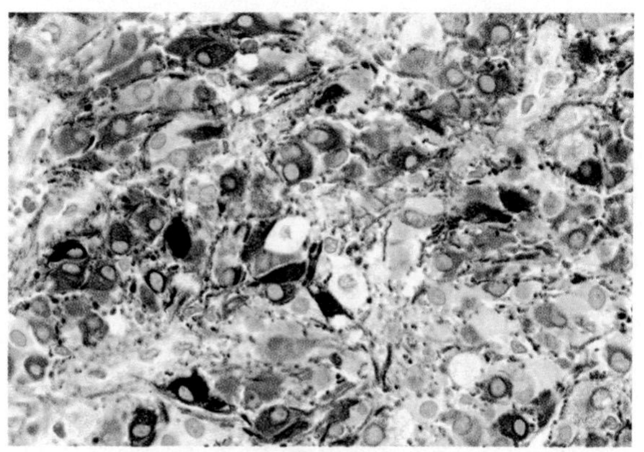

Figure 3-174

SUBEPENDYMAL GIANT CELL ASTROCYTOMA (TUBEROUS SCLEROSIS)

The tumor is variably positive for GFAP.

not prompt reflexive consideration of a high-grade diffuse astrocytoma in an otherwise typical SEGA or in a lesion associated with tuberous sclerosis.

Immunohistochemical Findings. Despite their astrocytic appearance, the cells are variably and often only focally or weakly immunoreactive for GFAP (fig. 3-174) (294,295). S-100 protein reactivity is strong and widespread. Some cells may be reactive for neuronal markers such as neurofilament protein (fig. 3-175), class III beta-tubulin, and even neurotransmitter substances (294,295). As is the case for the large cells of tubers, the giant cells of SEGA are immunoreactive for alpha-beta-crystallin (296). HMB45 immunoreactivity, a frequent feature of the systemic lesions of tuberous sclerosis, is not evident in SEGA or other CNS hamartomas (297,298). In contrast to balloon cells of cortical tubers, the subependymal neoplasm does not stain for tuberin, the gene product of tuberous sclerosis 2 (*TSC2*) (298–300).

The MIB-1 index is low; mean values are about 1 percent (297,300)

Ultrastructural Findings. Intermediate filaments vary greatly in number and disposition within the neoplastic cells of SEGA and their processes (294,301). Some cells are poor in fibrils and have more in the way of small, Nissl substance-like stacks of rough endoplasmic reticulum as well as electron-dense lysosomes. Unlike some of the large cells in tubers, those of SEGA generally lack bona fide features of neuronal differentiation. Nonetheless, secretory granules and processes containing microtubules may be seen (292,294,302).

Cytologic Findings. As they are in histologic sections, processes of the large cells are not omnidirectional as they would be in gemistocytic astrocytoma, but rather emanate from one pole of the prominent glassy cytoplasm (fig. 3-176) (303). Rosette-like structures can create a similarity to ependymoma.

Molecular Findings. Approximately half of patients with tuberous sclerosis have a family history that follows an autosomal dominant

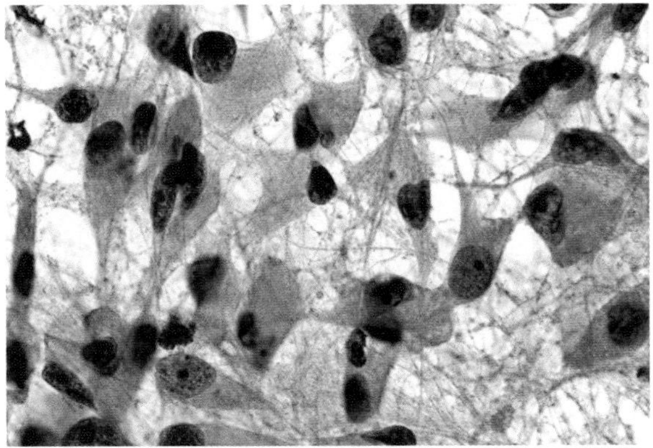

Figure 3-176

SUBEPENDYMAL GIANT CELL ASTROCYTOMA (TUBEROUS SCLEROSIS)

The tumor's typical glassy cytoplasm with processes is evident in smear preparations.

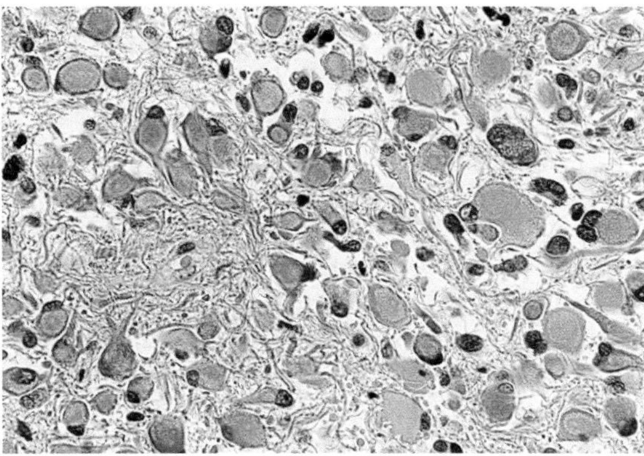

Figure 3-177

GEMISTOCYTIC ASTROCYTOMA RESEMBLING SUBEPENDYMAL GIANT CELL ASTROCYTOMA

While gemistocytic astrocytomas share cytologic features with SEGAs, the former is an infiltrating, intraparenchymal lesion, not exophytic, well circumscribed, and largely intraventricular.

pattern of inheritance, with high penetrance, but with considerable phenotypic variability (288). De novo mutations account for a substantial number of other cases. Tuberous sclerosis has been linked to one of two genes, *TSC1* and *TSC2* on chromosomes 9q and 16p, respectively, with gene products known as hamartin and tuberin. Loss of heterozygosity on chromosome 16p in the region of *TSC2* has been found in a hamartoma and a SEGA (304), but the abnormalities and relative contributions of these two genes in the pathogenesis of SEGA is unclear. One study found that loss of heterozygosity in the region of *TSC1* and *TSC2* occurred in only 1 of 11 SEGAs (305). Chromosomal imbalances, as sought by comparative genomic hybridization, appear to be rare (306).

Differential Diagnosis. Given the variation in morphologic pattern and cytology, the differential diagnosis of SEGA includes a number of lesions. Many can be excluded provisionally by preliminary attention to clinical and radiographic features. Tumors with ganglion cells are listed in Appendix G.

The presence of pink cytoplasm, cell processes, and frequent GFAP positivity are features shared with gemistocytic astrocytoma (fig. 3-177). The latter is an intraparenchymal lesion, not a well-demarcated, exophytic, intraventricular mass. Furthermore, gemistocytic astrocytomas are infiltrative and thus contain residual, overrun parenchymal elements such as myelinated axons, neurons, and reactive astrocytes. Cytologically, neoplastic gemistocytic astrocytes are usually smaller than those of SEGA and are diffusely distributed rather than patterned. Like SEGA cells of the plump sort, they possess a corona of short processes.

Rare SEGAs have brisk mitotic activity and necrosis (287), but despite a resemblance to glioblastoma, such tumors do not appear to behave as a glioblastoma and are best described as SEGA with atypical features.

The vague perivascular orientation of spindle cells and the well-formed pseudorosettes composed of epithelioid cells, as well as the intraventricular location of SEGA, may suggest ependymoma. SEGA cells are considerably larger than those of ependymoma and are overtly astrocytic, and the lesion is overall less cellular. Vesicular nuclei and prominent nucleoli are not features of ependymoma.

Treatment and Prognosis. Serial radiographic studies have established beyond question that subependymal nodules, like SEGAs, are capable of progressive growth (290,307). The postoperative course is uniformly favorable, however, since even large lesions can be excised or debulked. Treatment is thus usually limited to surgical excision, and, although recurrences are seen, death due to the neoplasm is uncommon (287). As noted above, the same favorable outlook accompanies lesions with mitoses and even foci of necrosis (287,300,308).

Desmoplastic Infantile Astrocytoma

In the years both before (309,310) and after (311–313) the description of *desmoplastic infantile ganglioglioma* (DIG), cases were described that were similar to DIG in all aspects save the lack of ganglion cells (313–316). Whether these latter tumors, *desmoplastic cerebral astrocytomas of the infancy,* also known as *desmoplastic infantile astrocytoma* (DIA), are an entity distinct from the DIG or simply an artifact of inadequate tissue sampling can be debated. In any case, DIG and desmoplastic astrocytoma are similar in clinical and behavioral terms and are treated accordingly.

Occasionally, infiltrative, or diffuse, astrocytic neoplasms reach the subarachnoid space, where they can elicit a considerable desmoplastic response. Patients with the tumors are generally older, if not adult, and a prominent intraparenchymal component is usually obvious on neuroimaging. Neither massive cysts, a key feature of DIG/DIA, nor dural attachment is expected in this variant of diffuse astrocytoma. Furthermore, the tissue response is less desmoplastic than that of DIA and usually surrounds cell nests rather than individual cells. Another desmoplastic lesion, gliosarcoma, is discussed later.

Granular Cell Tumor of the Infundibulum

Definition. *Granular cell tumor of the infundibulum* is a tumor of lysosome-rich granular cells that arises in the infundibulum or neurohypophysis. It is WHO grade I.

General Features. Although the term granular cell tumor is applied to a more common lesion of peripheral nerves, including the intracranial segment of the trigeminal nerve, and to a rare variant of diffuse astrocytoma discussed previously, the infundibular or neurohypophysial lesion discussed here is a distinct entity (fig. 3-178). A possibly related infundibular neoplasm, *pituicytoma,* is discussed separately below.

The most common granular cell proliferations are incidental microscopic nodules in the infundibulum or neurohypophysis known as tumorettes or choristomas (figs. 3-179, 3-180) (317–319). The origin or nature of these aggregates, and the rare larger symptomatic versions, has not been firmly established, but is reasonably assigned to regional specialized glia, or pituicytes. The constituent cells merge via transition forms with these infundibular astrocytes at the periphery of the lesion (317). Isolated and clustered granular cells, similar to those of granular cell tumors, are also present in the normal neurohypophysis (320). Scattered cells of the granular cell tumor may contain fine pigment granules resembling those in regional glial cells (321). Ultrastructural studies clearly show normal pituicytes containing varying numbers of electron-dense lysosomes identical to those seen in granular cell tumors (320,322).

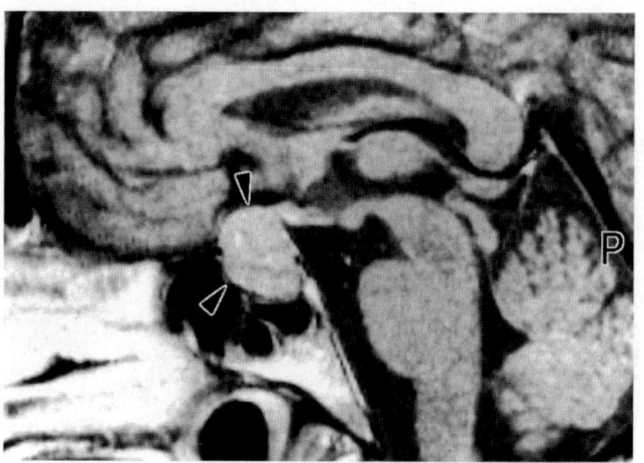

Figure 3-178

GRANULAR CELL TUMOR OF THE INFUNDIBULUM

The infundibular granular cell tumor (arrowheads) is a radiologically nonspecific mass in the sellar region. This example produced visual field changes in a 44-year-old man. (Fig. 3-180 from Fascicle 10, 3rd Series.)

Clinical Features. Some tumors are sufficiently large as to be associated with diminished visual acuity, or, less commonly, endocrine deficiency states (323–327). Such symptomatic tumors occur mainly in adults and show a female predilection (324).

Radiologic Findings. Granular cell tumors are largely suprasellar, discrete, and intensely contrast enhancing (fig. 3-178) (323–328).

Gross Findings. The tumors are soft, discrete, vascular masses that become attached to, but are generally not invasive of, the overlying optic chiasm or hypothalamus.

Microscopic Findings. In cytologic terms, the features of granular cell tumor are identical to those of tumorettes, from which they are presumably derived (fig. 3-181), although large tumors are often septated by a delicate fibrovascular

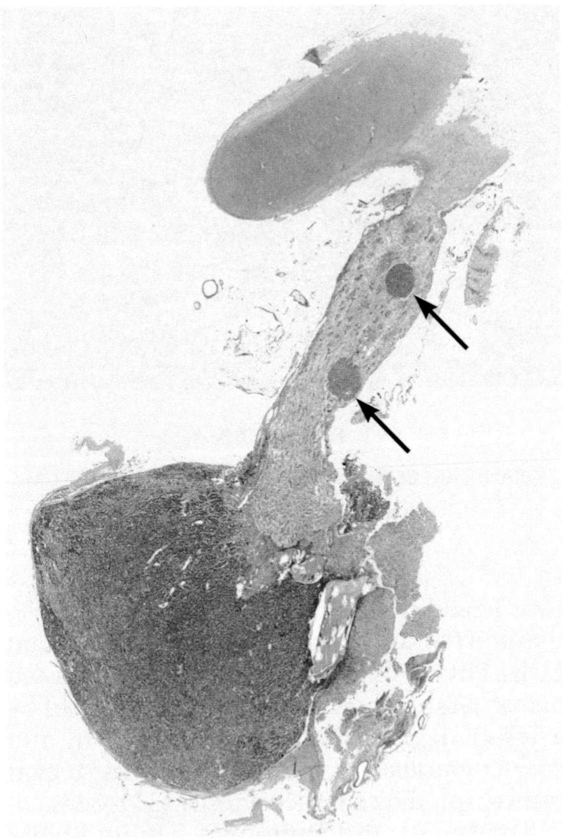

Figure 3-179

INFUNDIBULAR TUMORETTES

Small, round, eosinophilic nodules (arrows) are occasional incidental findings in the infundibulum.

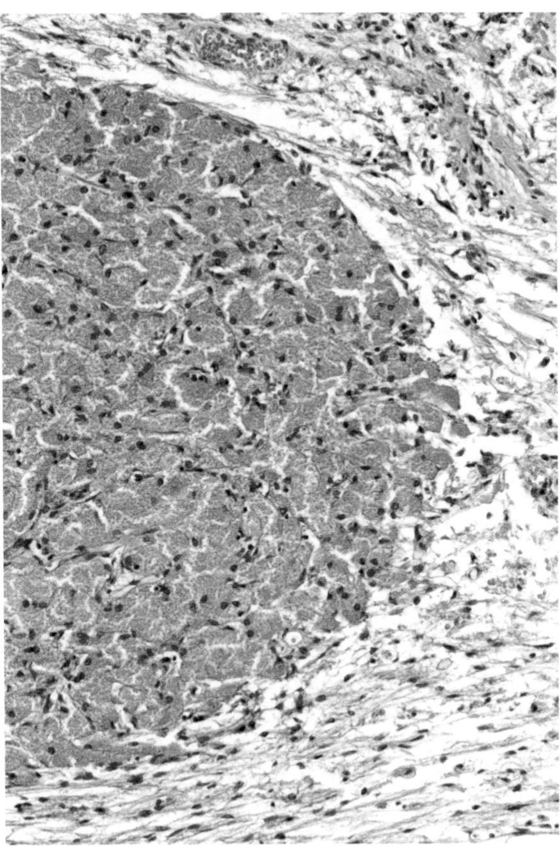

Figure 3-180

INFUNDIBULAR TUMORETTE

A nodule, or tumorette, is a compact, well-circumscribed collection of granulated cells.

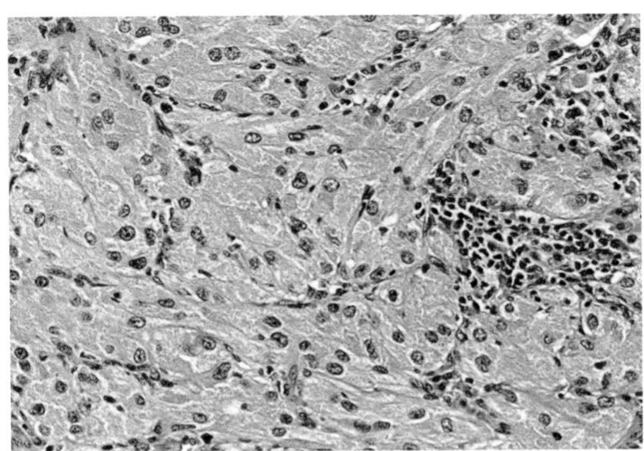

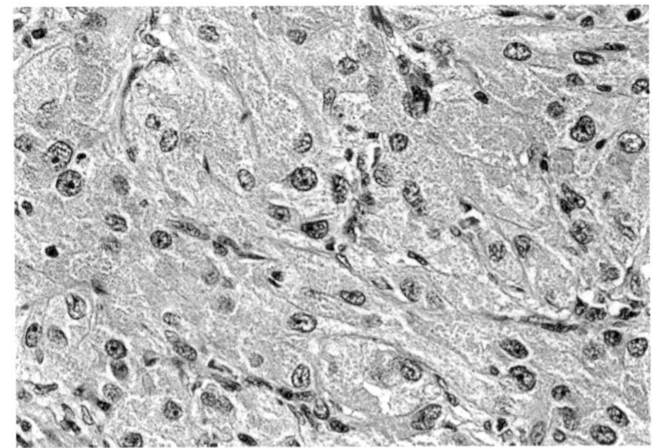

Figure 3-181

GRANULAR CELL TUMOR OF THE INFUNDIBULUM

Left: Infundibular granular cell tumors are compact and lobulated, and often contain perivascular lymphocytes.

Right: The cells are round or slightly elongate, cytologically bland, and granulated, and similar to those in infundibular tumorettes.

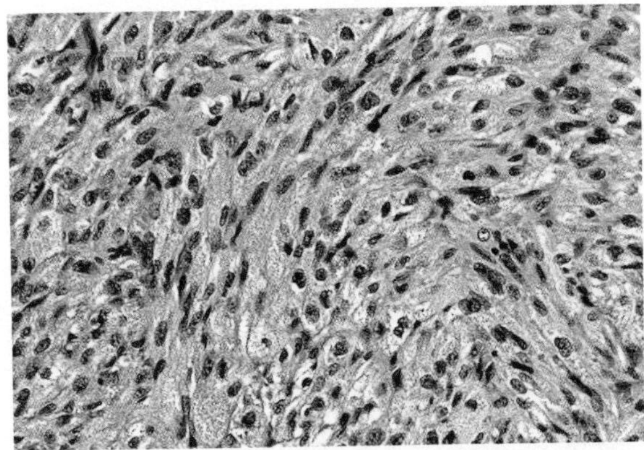

Figure 3-182

GRANULAR CELL TUMOR OF THE INFUNDIBULUM

Some infundibular granular cell tumors are more fascicular, and less granular, than the classic lesion.

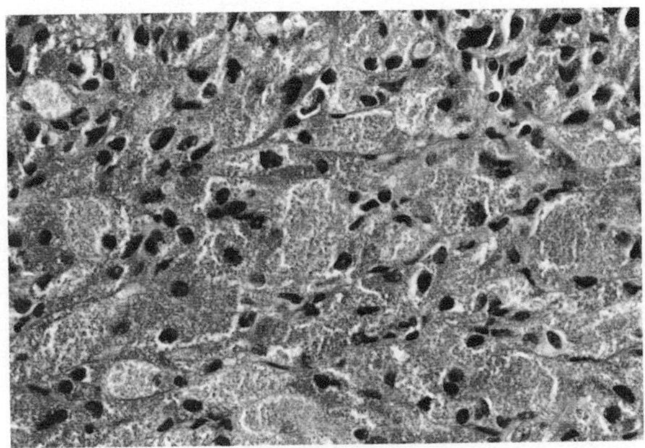

Figure 3-183

GRANULAR CELL TUMOR OF THE INFUNDIBULUM

Cytoplasmic granules are PAS positive.

bands and often exhibit a somewhat spindled or fascicular architecture (fig. 3-182). Somewhat epithelioid, the granular cells possess ample pink cytoplasm of a delicate granular texture that is strongly PAS positive (fig. 3-183). Nuclei are often eccentric, uniformly round with delicate chromatin, and free of mitotic activity. Perivascular lymphoid aggregates are common.

Immunohistochemical Findings. Given the rarity of granular cell tumor, the essentials of its immunoprofile remain to be established. Investigations have variably reported positive or negative reactions for S-100 protein or vimentin (326,329). Although several studies have reported lack of reactivity for GFAP (326,329), its presence was demonstrated by both routine immunohistochemistry and immunoelectron microscopy in one case (330).

Ultrastructural Findings. The granular cells are filled with heterolysosomes but lack intermediate filaments. Basement membranes are scant, if present at all (326,327,330).

Differential Diagnosis. In small specimens, cellular elongation, whether real or artifactual, can resemble that of pilocytic astrocytoma. Even the most spindled elements of a granular cell tumor, however, do not attain the degree of elongation regularly achieved by pilocytic astrocytoma cells. The granular cell tumor also has distinctive granular cytoplasm which contrasts with the markedly fibrillar cytoplasm of compacted portions of the pilocytic tumor and the latter's classic loose-knit, often microcystic architecture. In addition, the astrocytoma is GFAP positive and typically contains Rosenthal fibers and EGBs. Pituicytoma, a related lesion, is discussed below. Granular cell tumor is distinguished by a lesser degree of cellular elongation, and coarser granularity, from another sellar region tumor, spindle cell oncocytoma (331–333).

Treatment and Prognosis. Infundibular granular cell tumor has been reported to grow slowly, if at all, during periods of observation (323,324,326,327). Resection is curative. Postoperative hypopituitarism and persistent diabetes insipidus, present in about half of patients, may require replacement therapy (324).

Pituicytoma

Definition. *Pituicytoma* is a well-differentiated (WHO grade I), intrasellar or suprasellar neoplasm attributed to neoplastic transformation of pituicytes.

Clinical Features. Headaches, visual disturbances, and hypopituitarism are the common expressions of this rare lesion (334,335). One example occurred in conjunction with multiple endocrine neoplasms (336).

Radiologic Findings. Well circumscribed, solid, and contrast enhancing, the tumors are usually a centimeter or two in diameter (fig. 3-184) (334,335,337).

Microscopic Findings. The compact, noninfiltrating mass is composed of variably lobulated sheets of cells with copious cytoplasm and round, uniform nuclei with small nucleoli (figs.

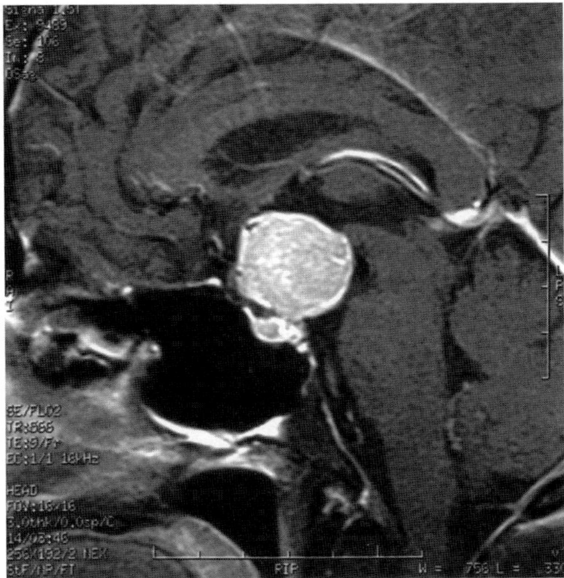

Figure 3-184

PITUICYTOMA

Pituicytomas are solid, suprasellar, and contrast enhancing. This example produced hypopituitarism in a 52-year-old man.

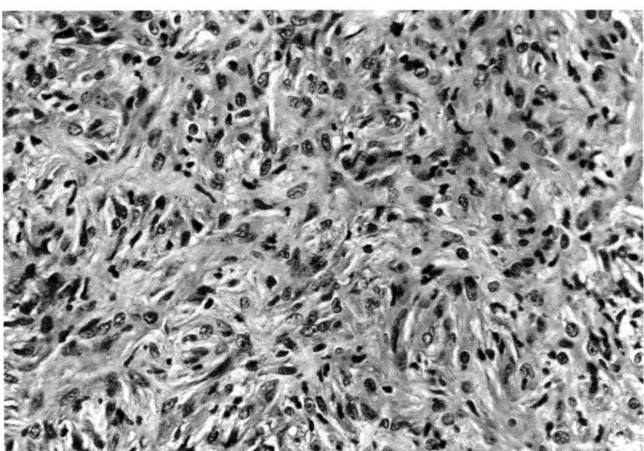

Figure 3-185

PITUICYTOMA

A noninfiltrating architecture, moderate cellularity, and a swirling pattern are typical of this uncommon perisellar lesion.

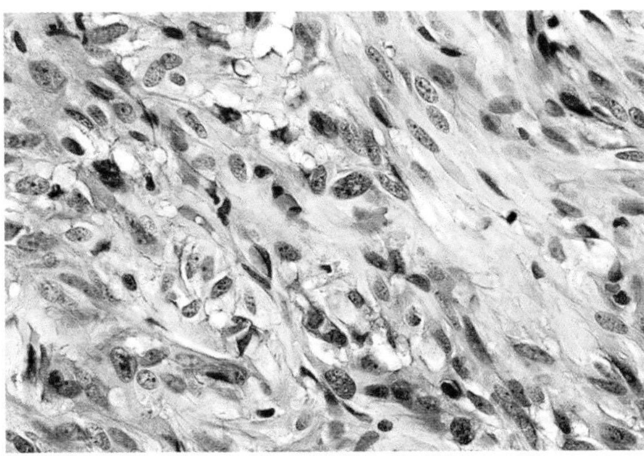

Figure 3-186

PITUICYTOMA

Loosely constructed pituicytomas may be difficult to distinguish from normal infundibulum.

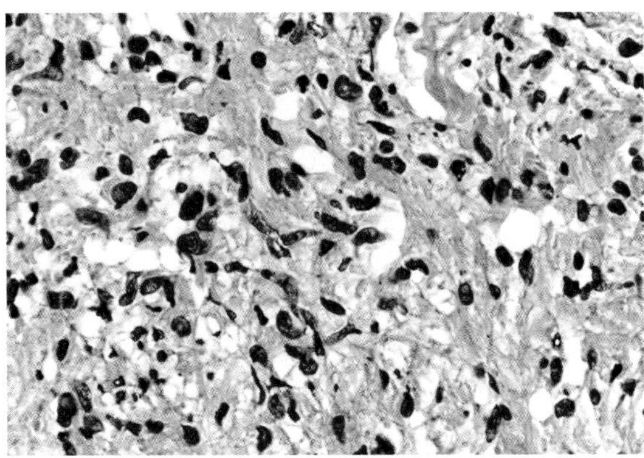

Figure 3-187

PITUICYTOMA: FROZEN SECTION

The lack of distinctive architectural features often precludes a definitive frozen section diagnosis.

3-185, 3-186). A somewhat fascicular or swirling architecture is common. Mitoses are exceptional.

Frozen Section. It is not surprising that a tumor with few distinctive features in permanent sections is largely nondescript in frozen tissues (fig. 3-187). Awareness of the entity is critical.

Immunohistochemical Findings. The tumors are dependably and strongly reactive for S-100 protein and vimentin, but less so for GFAP (fig. 3-188). There may be focal intracytoplasmic reactivity for EMA.

Ultrastructural Findings. The cytoplasm is unstructured other than for intermediate filaments. There are no surface specializations such as microvilli or cilia (334,337). One study, finding "fibrous bodies" and scattered neurosecretory granules, suggested an origin from folliculostellate cells of the adenohypophysis (338). The latter are S-100 protein- or GFAP-positive stellate cells that contribute to the adenohypophyseal follicles. The lesions are to us more

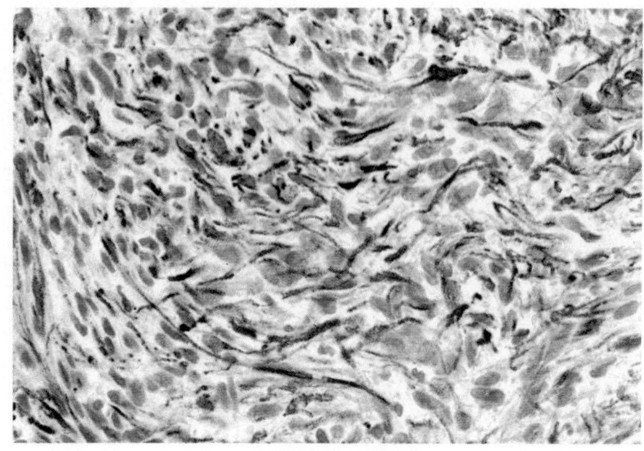

Figure 3-188

PITUICYTOMA

Pituicytomas are immunoreactive for GFAP, although often less intensely than in this case.

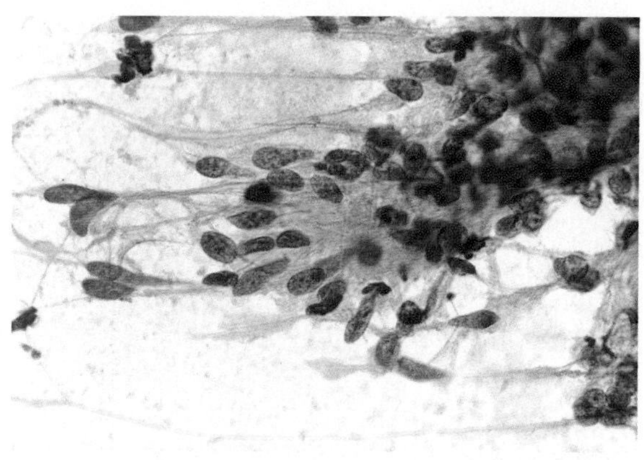

Figure 3-189

PITUICYTOMA: SMEAR PREPARATION

Smearing produces tissue fragments with uniform, somewhat polar cells that have bipolar processes. Nuclei are cytologically benign.

consistent with an origin from neurohypophysial cells, i.e., pituicytes.

Cytologic Findings. Although the cells remain in clumps in smear preparations, the bland "open" nuclei and bipolar processes are seen well (fig. 3-189).

Differential Diagnosis. In some cases, it is difficult to rule out normal infundibulum, since its cellularity can approach that of pituicytoma. Normal tissue is, however, less cellular and contains focal granular axonal swellings known as Herring bodies (fig. 3-190). More compellingly, the normal tissue is diffusely positive for

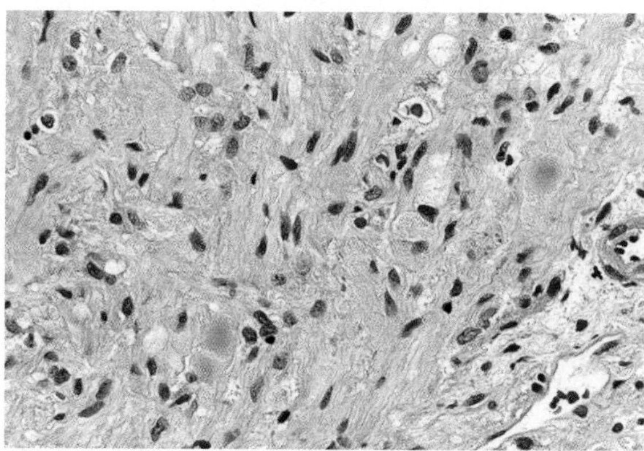

Figure 3-190

NORMAL INFUNDIBULUM

The normal infundibulum is histologically similar to pituicytoma, but is less cellular. The small finely granular structures, Herring bodies are normal infundibular structures not present in pituicytoma.

synaptophysin and neurofilament protein, whereas pituicytomas are negative. In addition, GFAP staining is confined to perivascular regions in the normal infundibulum, but is diffuse, albeit often weak, in pituicytoma.

Among neoplasms, pilocytic astrocytoma is the principal entity in the differential diagnosis. Pituicytomas are more compact, less polar and piloid, and lack microcysts, Rosenthal fibers, and EGBs. A neoplasm possibly related to pituicytoma, granular cell tumor of the infundibulum, is discussed above. Another regional tumor, spindle cell oncocytoma, is similar to pituicytoma in many respects except for the oncocytic phenotype created by an abundance of mitochondria (339–341).

Treatment and Prognosis. The rarity of the lesion precludes a definitive statement about its biologic behavior, but the tumor appears to be curable by total resection. Recurrences may follow subtotal resection (334,335).

OLIGODENDROGLIOMA AND OLIGOASTROCYTOMA

Definition. *Oligodendroglioma* is an infiltrative glioma composed, at least in part, of cells resembling oligodendrocytes.

General Features. Driven in part by the perception that patients with oligodendrogliomas have a significantly better prognosis than those with infiltrative astrocytomas of comparable grade, the prefixes "oligodendro-" and "mixed" have come to dominate surgical pathology reports, and supplant the diagnosis of astrocytoma in some institutions (342). The diagnosis of *mixed glioma* or *oligoastrocytoma* is attractive to pathologists, since a favorable outcome will be attributed to the presence of an oligodendroglioma component whereas overgrowth of the astrocytic component will be used to explain cases that recur rapidly as high-grade astrocytoma. The pathologist is correct in either case. The possibility that the recent increase in incidence of oligodendrogliomas and oligoastrocytomas is, in part, an artifact of overdiagnosis, in an area of pathology that is admittedly extremely subjective, is discussed below in the section Molecular and Cytogenetic Findings.

Clinical Features. Oligodendrogliomas arise throughout the neuraxis, but affect primarily the cerebral hemispheres. While adults are principally affected, occasional cases appear in children (343). Among the latter are those with unusual, and not fully described, spinal lesions composed of small oligodendroglioma-like cells that in some cases disseminate extensively in the spinal and even cranial leptomeninges (344–348). Other similar tumors appear to be leptomeningeal from the onset. Whether any of these are allied with classic cerebral oligodendroglioma is unclear.

Manifestations of oligodendroglioma include any of the usual consequences of an infiltrating, expanding intracranial neoplasm. In light of the lesion's affinity for the cerebral cortex, however, seizures are common at presentation. Large lesions may produce signs and symptoms of increased intracranial pressure.

Radiologic Findings. A preoperative diagnosis of oligodendroglioma is strongly suggested by a largely intracortical lesion on MRI, especially when long segments of the cortical ribbon are affected. One radiologic/pathologic/molecular study suggested that oligodendrogliomas with the 1p/19q codeletion had less well-defined borders than "oligodendrogliomas" without the deletion (349). Oligodendroglioma is especially likely when a band of cortical mineralization undulates in a "gyriform" profile (fig. 3-191) (see Appendix F for a summary of frequently calcified tumors). Calcification is not detected in many small oligodendrogliomas, however. Grade III lesions are large, and while variably enhancing, generally lack the ring profile so typical of glioblastoma.

Table 3-2

HISTOLOGIC FEATURES OF OLIGODENDROGLIOMAS

Affinity for the cerebral cortex
Histologic and cytologic monotony
Round nuclei with small, but prominent, nucleoli
Perinuclear halos producing a "fried-egg" artifact
Microcysts
Delicate capillaries disposed in a "chicken-wire" pattern
Microcalcifications
Palisades or cohesive clusters of cells (in a minority of cases)
Circumscribed nodules of increased cellularity featuring cytologic atypia (in a minority of cases)
Cells with eccentric eosinophilic cytoplasm resembling plump astrocytes (minigemistocytes and gliofibrillary oligodendrocytes)

Gross Findings. Because of their corticocentricity, some oligodendrogliomas are superficial enough to expand the affected gyri (fig. 3-192). As infiltrating gliomas, oligodendrogliomas are macroscopically ill-defined and efface the gray-white junction. Highly cellular grade III tumors are fleshy masses, but generally without the large central area of necrosis of glioblastoma.

Microscopic Findings. The salient histologic features of oligodendrogliomas are summarized in Table 3-2 and discussed below.

Affinity for the Cerebral Cortex. A common architectural feature, seen well at low-magnification, is the lesion's affinity for cortical gray matter (fig. 3-193), which is often notably thickened as it is in MRI scans. Within the cortex, the infiltrating cells are not uniformly distributed but are attracted to neurons (perineuronal satellitosis) and to subpial and perivascular regions (fig. 3-194).

Histologic and Cytologic Monotony. A characteristic, and in many respects, the diagnostic, feature of oligodendroglioma is histologic and cytologic monotony that is seen at low magnification as a uniform blueness to the section, and is confirmed on closer view as sameness of nuclear size and shape (figs. 3-195–3-197).

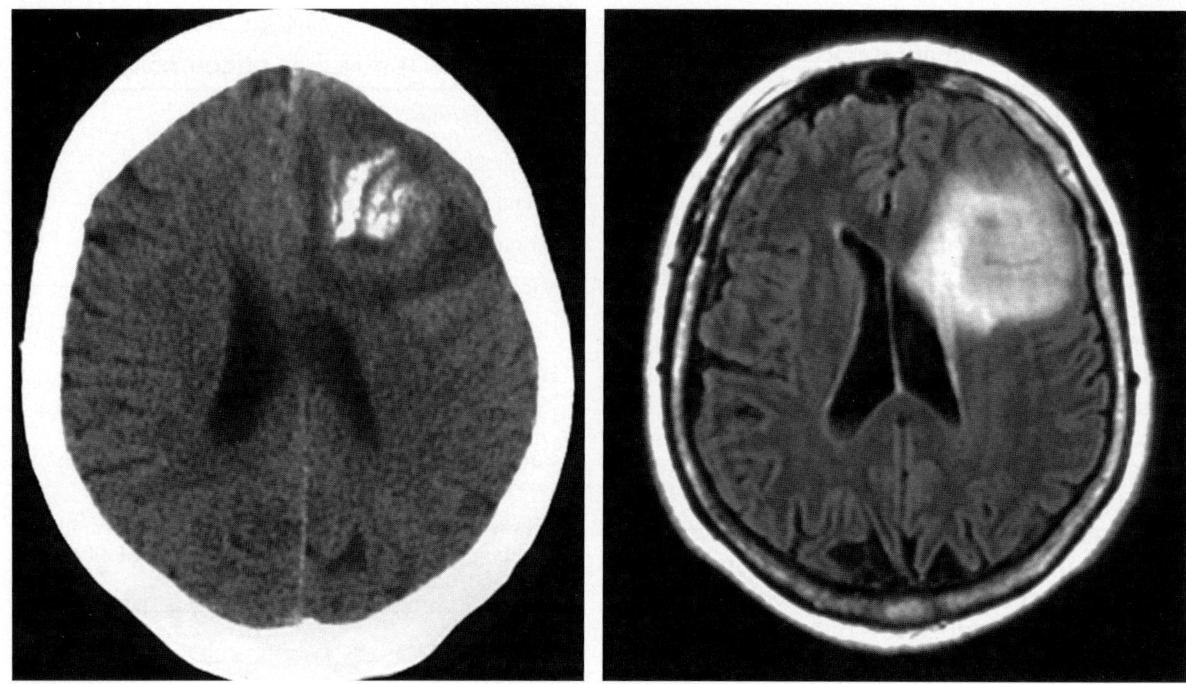

Figure 3-191

OLIGODENDROGLIOMA

Oligodendrogliomas have an affinity for the cerebral cortex where they often induce a curvilinear, "gyriform" band best seen by computerized tomography (CT) (left). Cortical thickening, common in oligodendrogliomas, and underlying edema are better seen in an MRI FLAIR image (right).

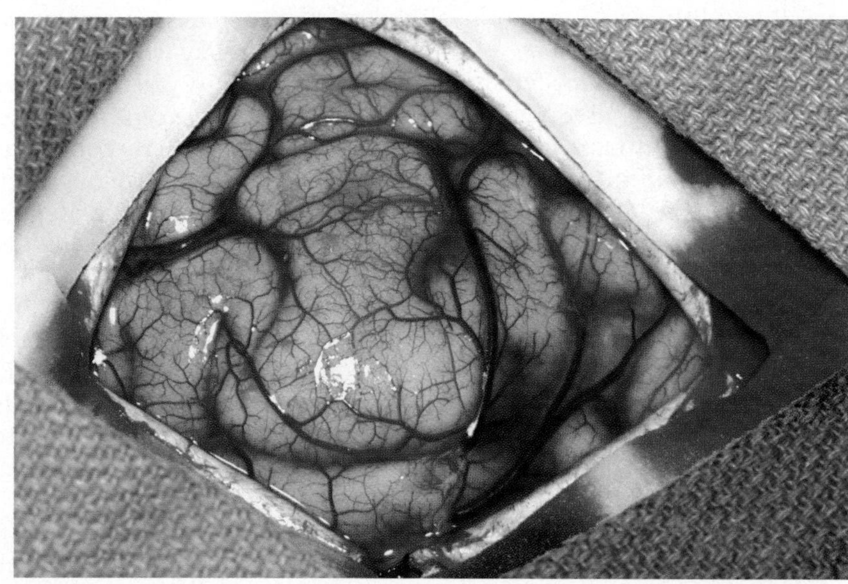

Figure 3-192

OLIGODENDROGLIOMA

Gyral thickening is common in these often superficial neoplasms.

Round Nuclei with Small, but Prominent, Nucleoli. In all grades, the majority of nuclei are round, especially so in grade II lesions (fig. 3-198), and somewhat less so in intermediate to higher-grade lesions (fig. 3-197). In contrast to astrocytomas, the chromatin pattern is more open and bland, and there is often a clearly defined, solitary nucleolus (figs. 3-198, 3-199). Isolated, large, hyperchromatic nuclei may also be present, but not the generalized nuclear pleomorphism present in many high-grade infiltrating astrocytic neoplasms. For high-grade lesions, the distinction from astrocytoma may be dependent, in part, upon ancillary architectural features such as corticocentricity, perineuronal secondary structure formation, relative cellular monomorphism, foci of classic grade II oligodendroglioma, and perhaps the nonspecific finding of microcalcifications. Oligodendrogliomas previously frozen have hyperchromatic, angulated nuclei that mimic those of diffuse astrocytoma (see fig. 3-225).

Perinuclear Halos Producing a "Fried-Egg" Artifact. Oligodendrocytes, whether normal or neoplastic, are susceptible to the autolytic imbibition of water, resulting in the production of a clear perinuclear halo (fig. 3-200). While this well-known artifact is diagnostically valuable, it is not always present, being absent in specimens fixed promptly or used for frozen sections prior to paraffin embedding. Halos are also not

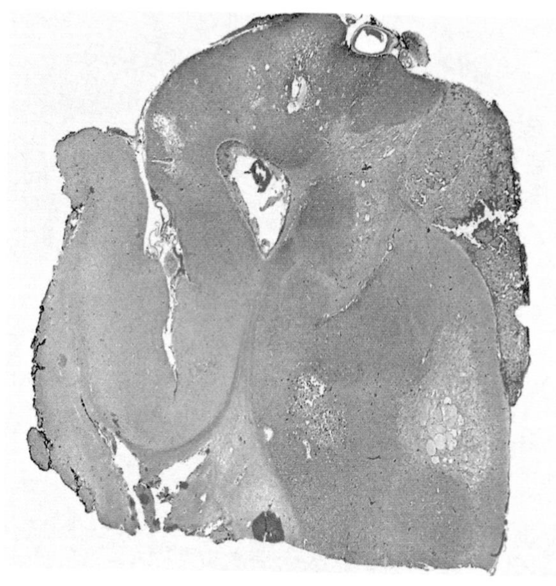

Figure 3-193

OLIGODENDROGLIOMA

Oligodendrogliomas are often corticocentric, as is evident in this whole mount section where the darkly staining lesion thickens the cortical ribbon.

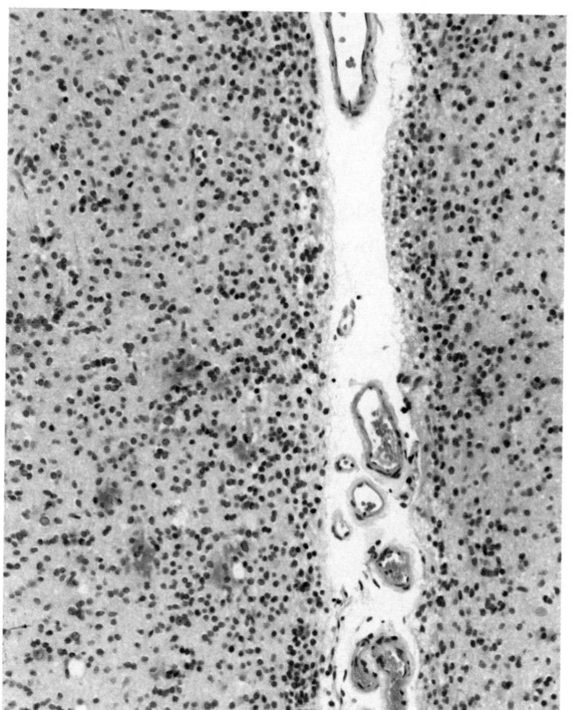

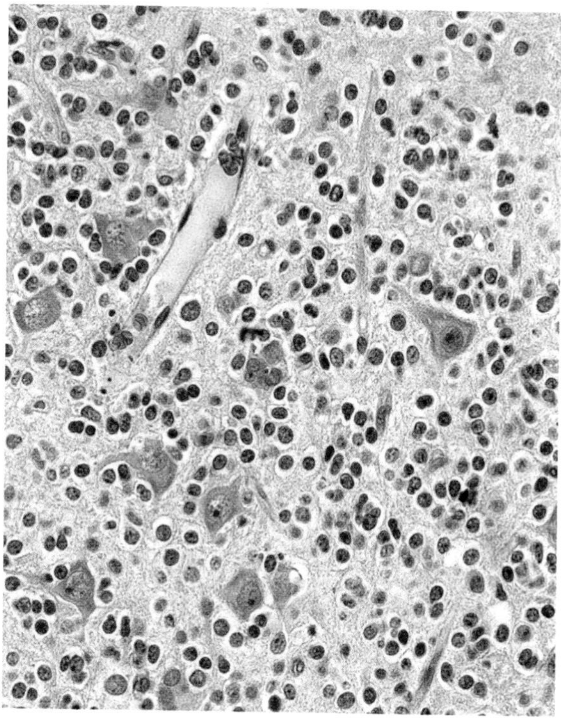

Figure 3-194

OLIGODENDROGLIOMA

Left: Oligodendrogliomas have an affinity for the cerebral cortex, where the cells often concentrate in the subpial region.
Right: Cortical ganglion cells trapped in an oligodendroglioma should not be interpreted as evidence of a ganglion cell tumor.

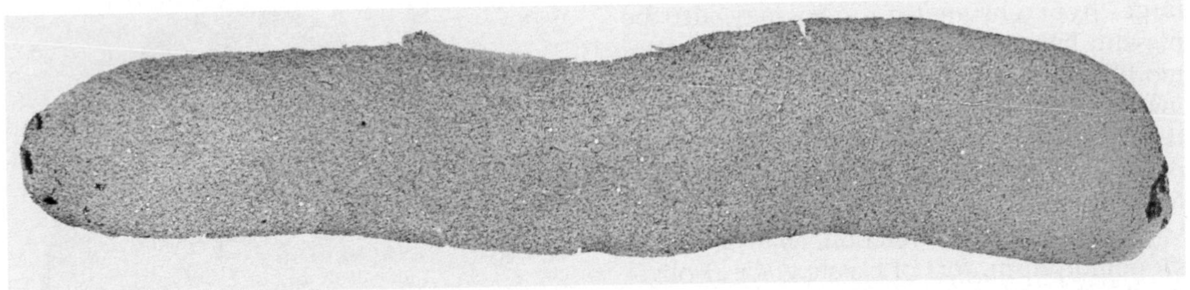

Figure 3-195

OLIGODENDROGLIOMA

The classic monomorphism and uniform cell density of oligodendroglioma is apparent in a needle biopsy specimen at low magnification.

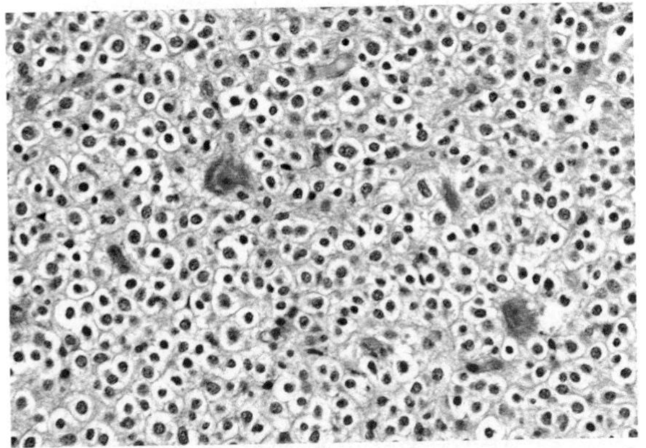

Figure 3-196

OLIGODENDROGLIOMA

The principal features of oligodendroglioma are uniformity in cell spacing, chromasia, and nuclear roundness. Note the trapped cortical neurons common in this glioma that appears drawn to, if not originating in, the cerebral cortex.

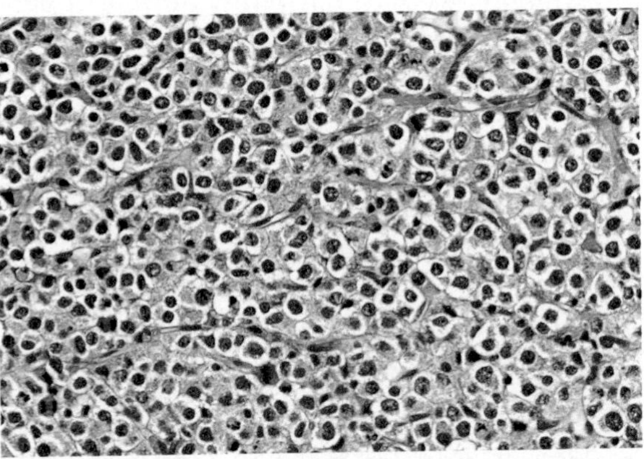

Figure 3-197

OLIGODENDROGLIOMA

The monomorphism of oligodendroglioma extends to mitotically active grade III lesions.

a specific finding, since they are present in some infiltrating astrocytomas, clear cell ependymomas, and neurocytomas.

Microcysts. Particularly common in the cortex are spaces filled with protein-rich fluid known, incorrectly from a semantic standpoint since there is no epithelial lining, as microcysts (fig. 3-201). While often prominent in oligodendrogliomas, microcysts do little to distinguish the lesion from astrocytoma since the latter can be extensively microcystic as well. Microcysts are nonspecific and are found as well in some infiltrating astrocytomas, clear cell ependymomas, and most neurocytic tumors.

Delicate Capillaries Disposed in a "Chicken-Wire" Pattern. Blood vessels in better-differentiated tumors typically consist of short capillary segments arranged geometrically; this arrangement resembles the patterned angulation of "chicken-wire" (fig. 3-202). It is a "soft" finding and of little diagnostic use in the absence cellular monotony and nuclear roundness. In anaplastic lesions, hypertrophy and hyperplasia of vascular cells are common. The glomeruloid vascular tufts so often seen in pilocytic astrocytoma and glioblastoma are usually not prominent.

Microcalcifications. Many oligodendroglial neoplasms exhibit some degree of calcification, at least focally, within the tumor or in surrounding brain (fig. 3-203). In any case, it is mainly intracortical. The deposits take the form

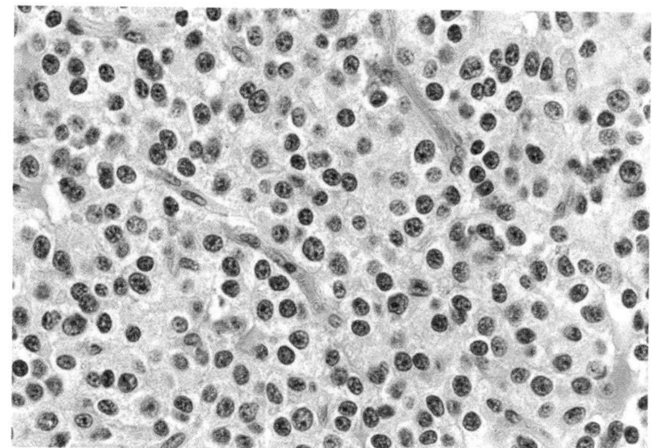

Figure 3-198

OLIGODENDROGLIOMA

Nuclei, especially of grade II lesions, are round and contain small but distinct nucleoli. Delicate capillaries are common in low grade oligodendrogliomas.

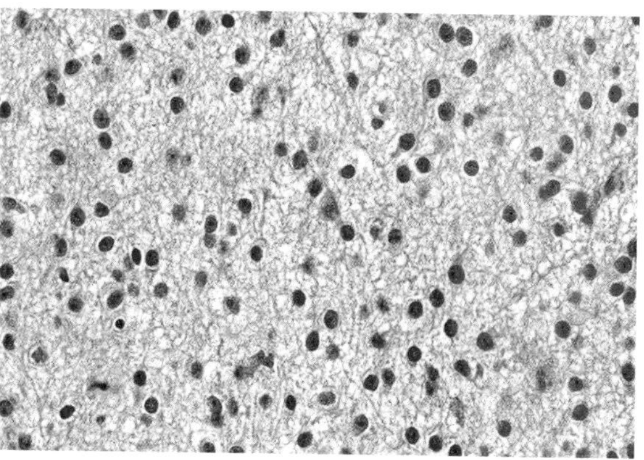

Figure 3-199

OLIGODENDROGLIOMA

Nucleoli are prominent in oligodendrogliomas and useful in the differentiation from astrocytoma, as in this case in which perinuclear halos are absent.

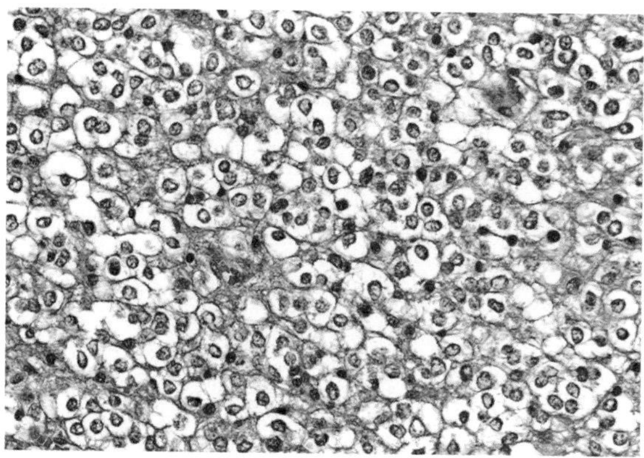

Figure 3-200

OLIGODENDROGLIOMA

Perinuclear halos, round nuclei, and cytologic monotony are classic features of oligodendroglioma. The back-to-back relationship of tumor cells in cellular areas is typical.

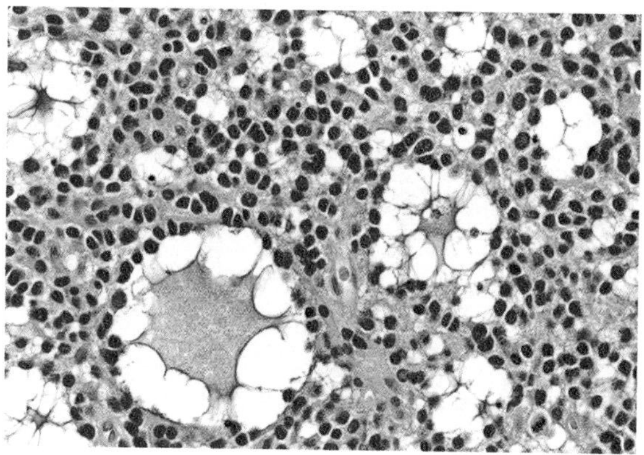

Figure 3-201

OLIGODENDROGLIOMA

Small cavities filled with proteinaceous fluid (microcysts) are common in low and, here, high-grade, oligodendrogliomas.

of laminated calcospherites. Vessel walls are often affected in heavily mineralized cases.

Palisades or Cohesive Clusters of Cells (in a minority of cases). Cellular palisading, or nests of tightly clustered cells, is an infrequent, often focal finding of (usually) grade III lesions (fig. 3-204, see fig. 5-75C). Tumors with ribboning or palisading are summarized in Appendix J.

Circumscribed Nodules of Increased Cellularity Featuring Cytologic Atypia (in a minority of cases). Although most oligodendrogliomas are relatively homogeneous in cell density, in some instances rather circumscribed, hypercellular, clonal nodules are present in a less cellular background (figs. 3-205, 3-206). Cell density, mitotic activity, and MIB-1 labeling indices are all higher within these nodules than they are in surrounding tissues. As a rule, the nodules are, seemingly, half a WHO grade higher than the background neoplasm. While these appear to violate the rule of tumor monotony, they actually confirm it, since the nodules, as miniature oligodendrogliomas, are also uniform in cytologic features and cell density. Similar, although less well-defined, nodules of increased cellularity are also seen in ependymomas (see fig. 3-248).

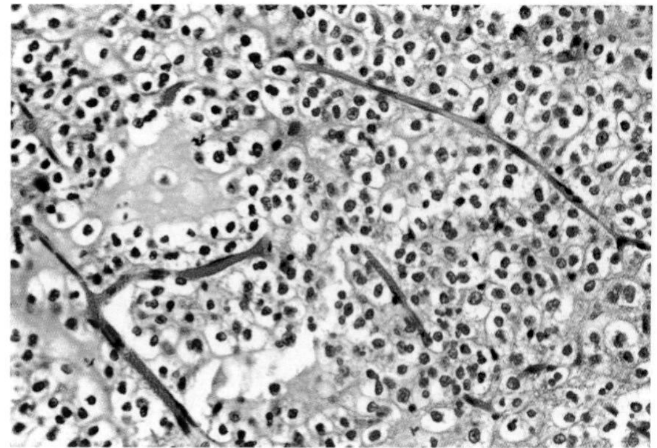

Figure 3-202

OLIGODENDROGLIOMA

Delicate angulated capillary segments create a likeness to "chicken-wire" in many oligodendrogliomas.

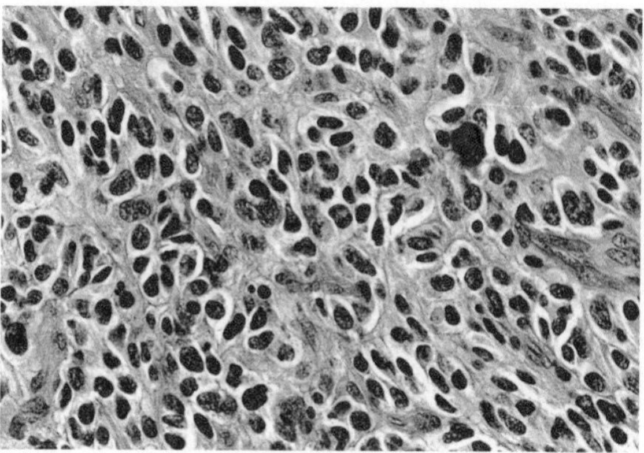

Figure 3-203

OLIGODENDROGLIOMA

Most oligodendrogliomas are calcified, as in this grade III example with considerably more nuclear pleomorphism than is present in a grade II lesion. Molecular testing revealed combined loss of chromosomal arms 1p and 19q.

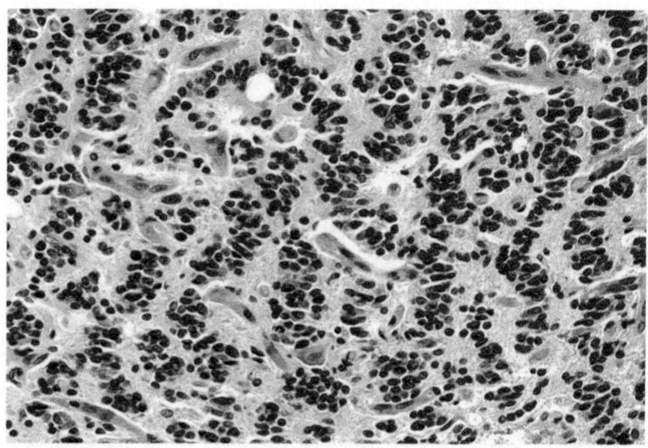

Figure 3-204

OLIGODENDROGLIOMA

Rows, or palisades, of tumor cells enliven some oligodendrogliomas, especially those of grade III.

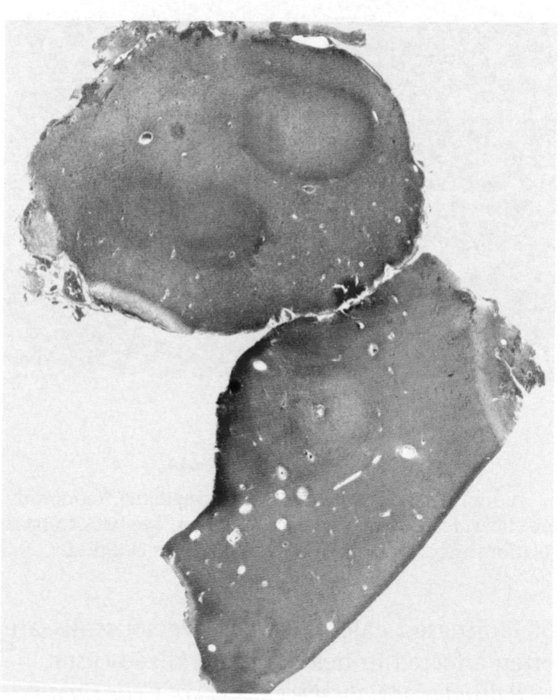

Figure 3-205

OLIGODENDROGLIOMA

Multiple well-circumscribed nodules of increased cellularity are present in some oligodendrogliomas.

Increased cellularity also accompanies malignant transformation in diffuse astrocytic tumors, but these hypercellular zones exhibit greater nuclear pleomorphism, a looser texture, and lesser circumscription.

Cells with Eccentric Eosinophilic Cytoplasm Resembling Plump Astrocytes (Minigemistocytes and Gliofibrillary Oligodendrocytes). Although the classic oligodendroglioma cell has scant cytoplasm, which is inundated with water, otherwise classic tumors may contain varying numbers of cells in which pink cytoplasm is obvious and "astrocytic" in appearance (350). The issue of mixed glioma often arises in this setting. In accord with the increased incidence of these cells in grade III oligodendrogliomas, tumors with such cells are associated, overall, with a less favorable outcome (350). Some cells resemble miniature, process-containing gemistocytes (figs. 3-207, 3-208),

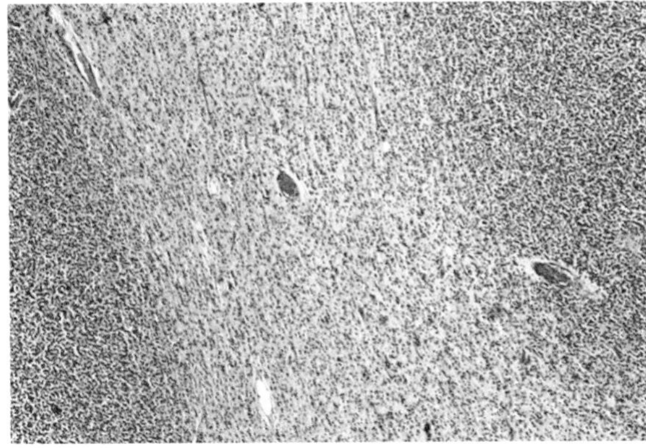

Figure 3-206

OLIGODENDROGLIOMA

Nodules in oligodendrogliomas are denser, but no less monomorphic, than the surrounding parent neoplasm. Two such hypercellular nodules are present here.

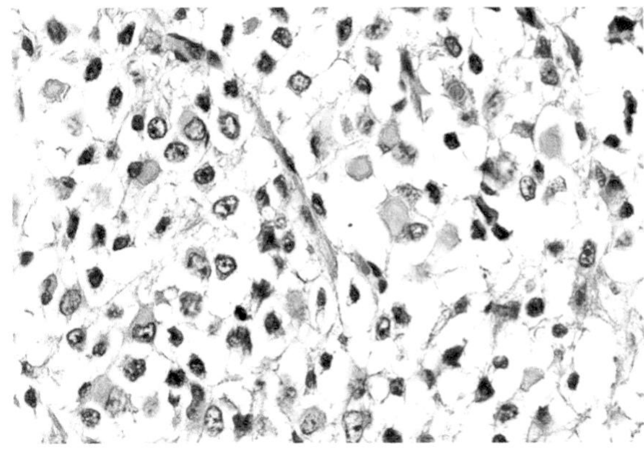

Figure 3-207

OLIGODENDROGLIOMA WITH ASTROCYTE-LIKE CELLS

Some otherwise classic grade II oligodendrogliomas contain scattered cells with eccentric eosinophilic cytoplasm. Cell processes may be present. The cytoplasmic mass is often faintly fibrillar.

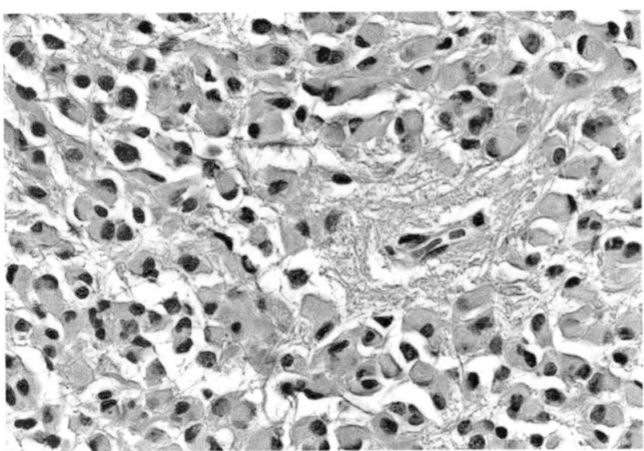

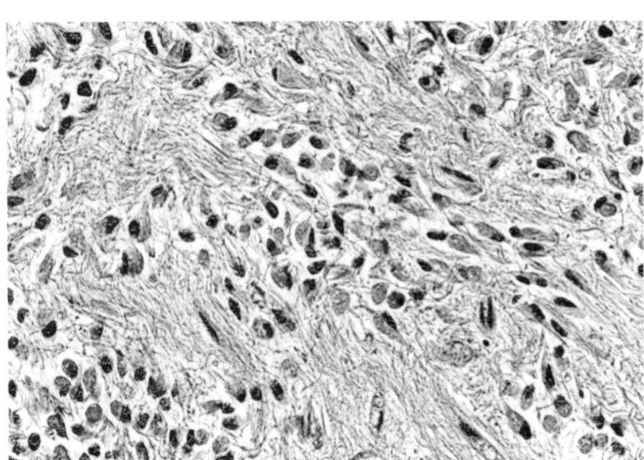

Figure 3-208

OLIGODENDROGLIOMA WITH ASTROCYTE-LIKE CELLS

Left: Small, back-to-back gemistocyte-like cells are present in a lesion that elsewhere was oligodendroglioma, grade III.

Right: As is present elsewhere in the same lesion, an abundant fibrillar background furthers the resemblance to astrocytoma. The term mixed glioma would be used by some for this neoplasm, but the illustrated tissue patterns are common in otherwise classic oligodendrogliomas, usually grade III.

whereas others appear polygonal and have a smooth-surfaced profile, their cytoplasm jutting cap-like in an eccentric manner or disposed about the nucleus (fig. 3-209). The term "minigemistocyte" is applied to the former, whereas "gliofibrillary oligodendrocyte" designates the latter. Both cytologic patterns may be found in the same case, particularly in grade III tumors, and transition forms bridge the two. The terms are sometimes used interchangeably, and both cell types are sometimes referred to as "transitional cells" or "astrocyte-like cells."

The cytoplasm of gliofibrillary oligodendrocytes may be filled with brightly eosinophilic fibrils that form cap-like paranuclear tufts or perinuclear bands. In some cells, the filamentous material seems to accumulate at the expense of a distorted, centrally placed nucleus (fig. 3-210). Collectively, these GFAP-positive cells are most prone to occur in perivascular regions. Some especially fibrillated, epithelioid cells, usually in a grade III lesion, also contain refractile eosinophilic bodies resembling minute Rosenthal fibers (fig. 3-211).

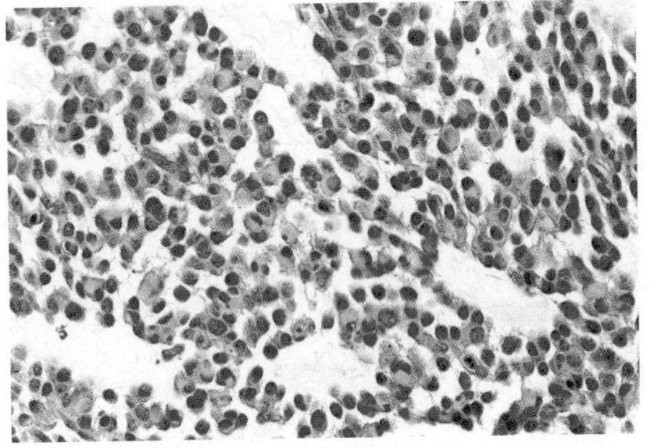

Figure 3-209

OLIGODENDROGLIOMA WITH ASTROCYTE-LIKE CELLS

Small astrocyte-like cells in some oligodendrogliomas, usually of grade III, have well-defined eosinophilic cytoplasm, but without the processes of gemistocytes.

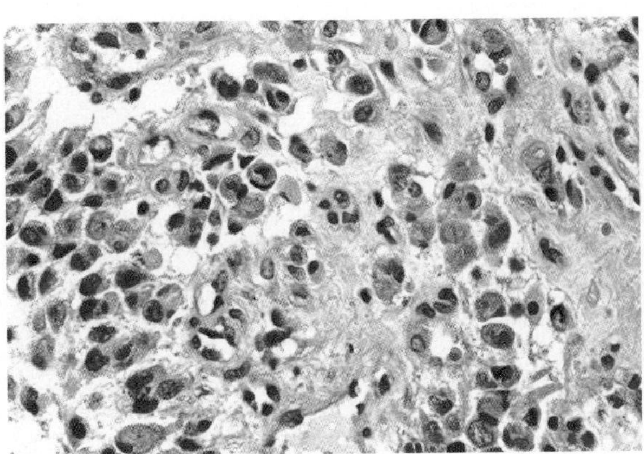

Figure 3-210

OLIGODENDROGLIOMA WITH ASTROCYTE-LIKE CELLS

Bundles of filaments encircle, and appear to strangle, nuclei in some oligodendrogliomas, usually grade III lesions.

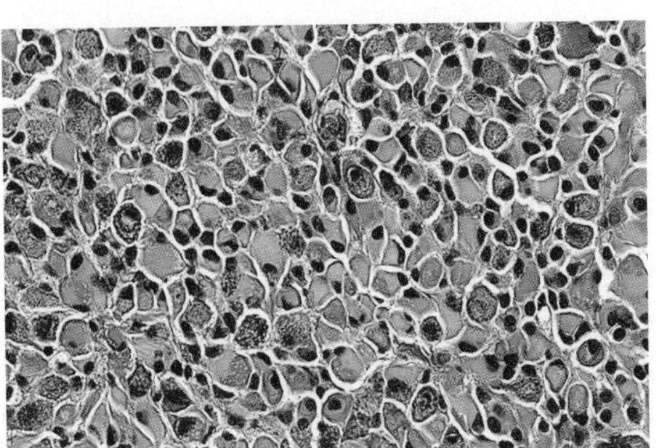

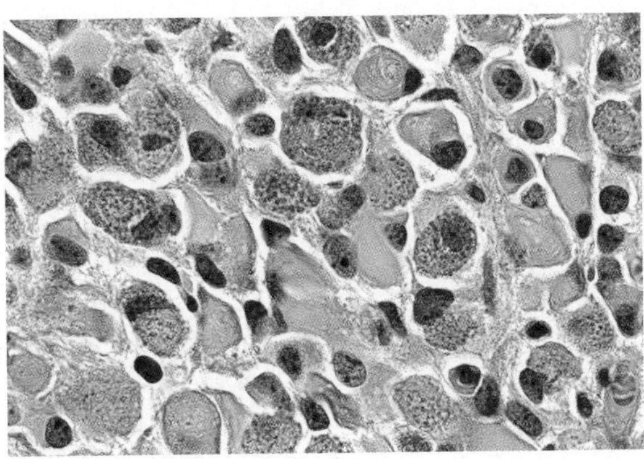

Figure 3-211

OLIGODENDROGLIOMA WITH ASTROCYTE-LIKE CELLS

Left: Cellular lesions with glassy cytoplasm and discrete cell borders give some grade III oligodendrogliomas an epithelioid appearance.

Right: Epithelioid cells in high-grade oligodendrogliomas often have filamentous whorls and small hyaline eosinophilic bodies resembling miniature Rosenthal fibers.

Cellular, higher-grade oligodendrogliomas often contain tumor cells even more astrocytic in terms of their cell profile, cytoplasmic fibrillarity, and degree of process formation, yet retain their round to oval nuclei, as if reluctant to deny their oligodendroglial heritage. In yet other cases, a portion of the tumor is inescapably phenotypically astrocytic. Spindle cell change is present in some, generally grade III, oligodendrogliomas.

The issue of mixed composition is, both practically and conceptually, complex in infiltrating gliomas, i.e., oligodendrogliomas and diffuse astrocytomas, since many are phenotypically mixed or hybrid in cytologic composition. Only uncommonly are there distinct, dichotomous, "biphasic," geographically separated populations of oligodendrocytes and astrocytes for which the term oligoastrocytoma is descriptive and fulfills strict WHO criteria (fig. 3-212) (351).

The terms *mixed glioma, oligoastrocytoma*, and *mixed oligoastrocytoma* are used widely as well for oligodendroglial neoplasms, often loosely defined, with any astrocytic features, or for

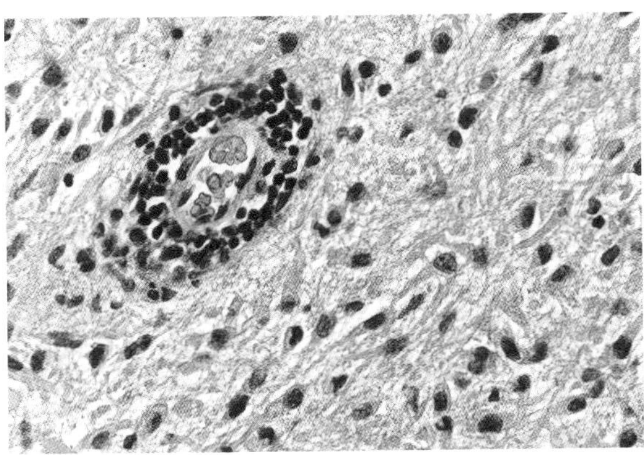
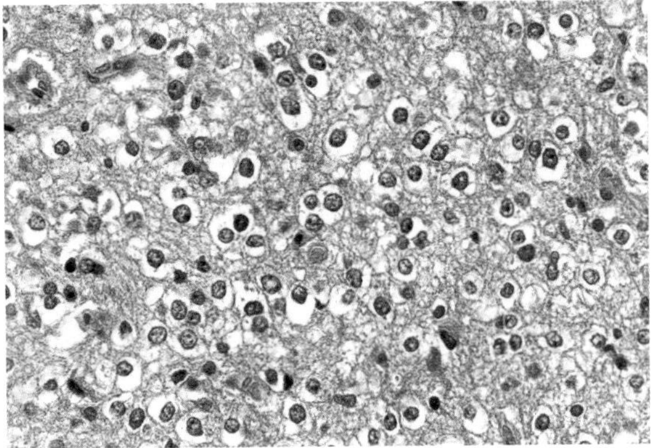

Figure 3-212

OLIGOASTROCYTOMA (MIXED GLIOMA)

Distinct areas of oligodendroglioma (left) and astrocytoma (right) in the same neoplasm fulfill the rigid criteria for oligoastrocytoma.

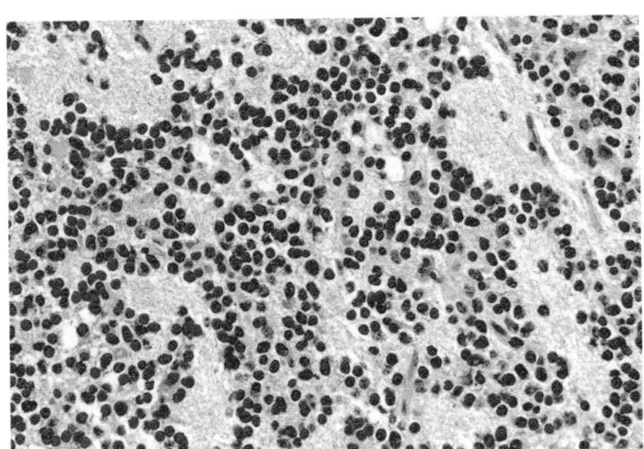
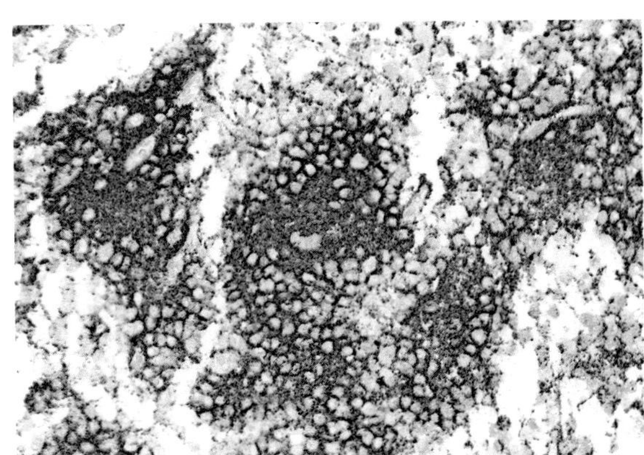

Figure 3-213

OLIGODENDROGLIOMA WITH NEUROCYTIC DIFFERENTIATION

Left: Occasional oligodendrogliomas exhibit neurocytic differentiation, recognized in H&E-stained sections as finely fibrillar, large rosettes or neuropil islands.

Right: Rosettes and islands of neuropil are immunoreactive for synaptophysin.

tumors wherein cells with astrocytic and oligodendroglial features are intermixed. Even more difficult are those in which tumor cells have hybrid features of both astrocytes and oligodendrocytes. It remains to be shown whether such lesions, or even the mixed lesions with distinct areas described above, are mixed at the prognostically significant genetic level. The 1p/19q codeletion has, thus far, been shown to occur in both oligodendroglioma regions and astrocytoma regions, or in neither. The difficult issue of the mixed glioma is discussed in the section on differential diagnosis.

Some oligodendrogliomas, with the chromosome1p/19q loss discussed below, contain foci of neurocytic differentiation in the form of synaptophysin-positive neurocytic rosettes (figs. 3-213, 3-214) or synaptophysin-positive circular areas such as are found in the rosetted glioneuronal tumor discussed in chapter 4 (352).

Reactive astrocytes, a regular accompaniment of all infiltrative gliomas, are readily seen in oligodendrogliomas, where they appear as evenly distributed, GFAP-positive stellate cells with delicate nuclei, moderate amounts of eosinophilic cytoplasm, and long, uniform,

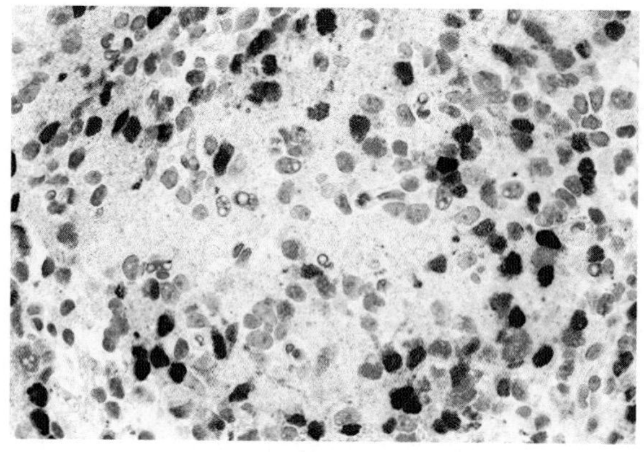

Figure 3-214

OLIGODENDROGLIOMA WITH NEUROCYTIC DIFFERENTIATION

NeuN-positive neurocytes surround the rosette cores in an oligodendroglioma with neurocytic differentiation.

symmetrically radiating processes. These should not prompt a diagnosis of mixed glioma.

Grading. According to the 2007 WHO system, oligodendrogliomas are generally either grade II or III (352a). The concept of oligodendroglioma grade IV or oligodendroglial GBM is discussed below.

The histologic prognostic factors for diffuse astrocytic lesions, as assessed by several studies, include mitotic rate, vascular proliferation, and necrosis, although necrosis does not have as powerful a negative prognostic influence as in the astrocytomas, and in some studies was not an independently significant variable (353–357). The mitotic rate and vascular proliferation remain as principal factors; one study found the former to be more significant (356) but another found vascular proliferation to be the decisive predictor (355). Making comparisons between such studies difficult is the fact that vascular proliferation in the latter study was defined as the state in which all capillaries within at least one low-power field (10X objective) had hyperplastic endothelial cells, and endothelial cell nuclei were in close contact with one another at least focally. This definition overlaps with what others might call hypertrophy, the latter itself having been established as a negative prognostic factor (356). The classic glomeruloid change is, however, what is generally meant by vascular proliferation. In a lesion with other evidence of anaplasia, we accept the hypertrophic form, if pronounced, as a decisive factor for anaplasia, and WHO grade III.

Cytologic atypia is a more subjective grading variable, but is present in certain lesions that have features such as high cellularity, high mitotic index, and vascular proliferation. It is difficult to employ and may be not be overt in difficult lesions at the grade II-III interface.

An obvious grading parameter, mitotic activity, has been related to outcome (354,356,357). Highly significant differences in survival are present when patients are stratified into those with tumors without mitoses and those with mitoses, in any number (356,357). A convenient cutoff point is 6 mitoses per 10 high-power fields: at or above this value the prognosis becomes significantly less favorable (356) and can be used as an indicator of grade III, with or without associated vascular proliferation or hypertrophy.

As with diffuse astrocytomas, age is a powerful predictor of prognosis and in some series the most important factor (356). The presence of radiologic contrast enhancement also has been described as a negative factor (355).

Well-differentiated, grade II lesions thus span a spectrum from paucicellular tumors with only minimal cytologic atypia, to the classic, more cellular examples recognized readily by all. Mitoses, if any, are few, and blood vessels, although somewhat increased in number, remain delicate and, in some cases, chicken-wire in configuration. It is tempting to consider the epileptogenic, very well-differentiated, paucicellular subset of oligodendroglioma, of the type that is difficult to distinguish from dysembryoplastic neuroepithelial tumor, as grade I, but this is not sanctioned by the WHO.

Grade III, malignant or anaplastic oligodendrogliomas are cellular tumors with considerable nuclear hyperchromasia, brisk mitotic activity, and, usually, microvascular proliferation (fig. 3-215) (358). As discussed above, in some cases intraluminal endothelial change is as much hypertrophy as hyperplasia, but we consider this prognostically significant (fig. 3-216). Necrosis may be present, but perinecrotic pseudopalisading is usually absent (fig. 3-217). Not infrequently, oligodendrogliomas fall short of this standard since they have considerable cellularity, cytologic atypia, a focally high MIB-1 rate, but a lower mitotic index than the typical grade

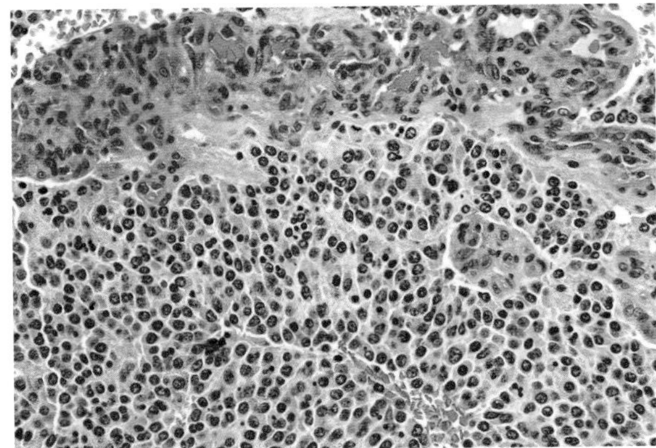

Figure 3-215

ANAPLASTIC (GRADE III) OLIGODENDROGLIOMA

High cellularity, cytologic atypia, and vascular proliferation help define this as a grade III lesion. Sharp cell borders lend the epithelioid appearance that is common in grade III oligodendrogliomas.

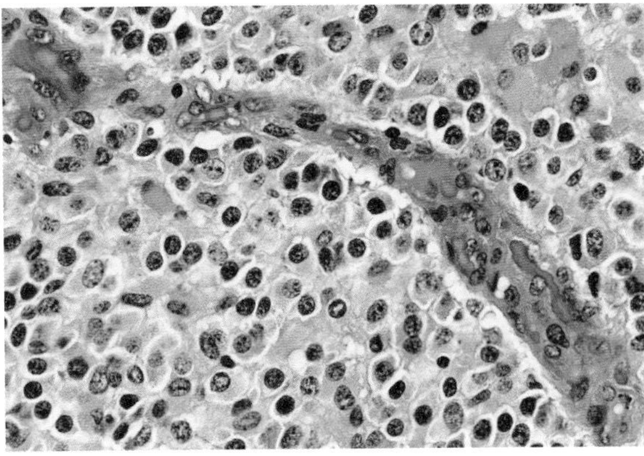

Figure 3-216

ANAPLASTIC (GRADE III) OLIGODENDROGLIOMA

Endothelial cells in some grade III lesions are often prominent because of intravascular hypertrophy and slight hyperplasia.

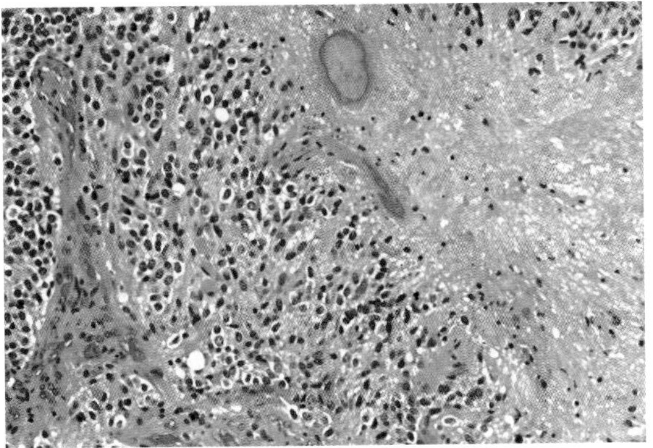

Figure 3-217

ANAPLASTIC (GRADE III) OLIGODENDROGLIOMA

Necrosis is common in grade III lesions. As here, it usually lacks peripheral palisading.

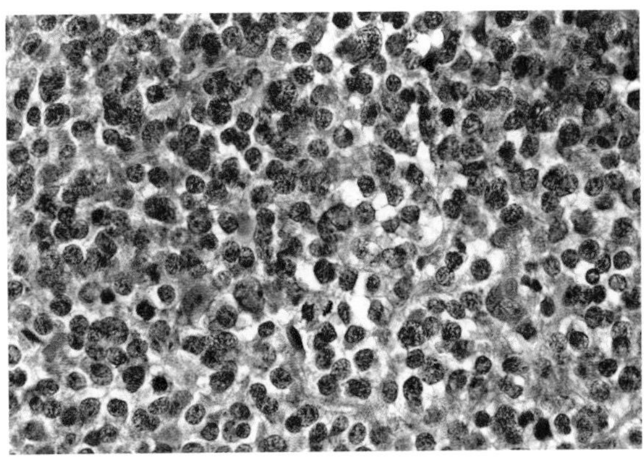

Figure 3-218

ANAPLASTIC (GRADE III) OLIGODENDROGLIOMA

Some grade III lesions are more cellular and mitotically active than grade II lesions, but lack the marked cytologic atypia, astrocyte-like cells, and epithelioid features of other anaplastic oligodendrogliomas.

III lesion and lack microvascular proliferation (fig. 3-218). Others are obviously grade III on the basis of cytologic atypia and mitotic activity alone (figs. 3-219, 3-220). Eosinophilic "astrocyte-like" cytoplasm, either with prominent processes that resemble those of an astrocytoma or well circumscribed giving an epithelioid appearance, is common in anaplastic lesions (figs. 3-219, 3-220). Cells with eosinophilic cytoplasm (minigemistocytes, gliofibrillary oligodendrocytes) are common in grade III oligodendrogliomas, although present in some grade II lesions as well.

Nuclei of grade III lesions vary with the degree of anaplasia. Those with less anaplasia are uniformly round, as expected of an oligodendroglioma, and the diagnosis rests on mitotic activity, MIB-1 rate, and presence of vascular proliferation. Others are histologically malignant but still obviously oligodendroglial because of roundness (more or less) and prominent nucleoli. Truly anaplastic lesions have the nuclear features of high-grade astrocytoma, although these usually retain oligodendroglioma-like round cells, at least in small number. Scattered cells with eosinophilic

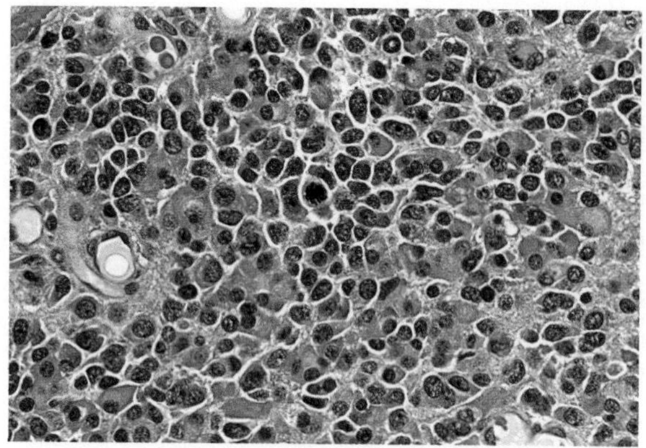

Figure 3-219

ANAPLASTIC (GRADE III) OLIGODENDROGLIOMA

Grade III oligodendrogliomas are usually highly cellular and may retain round nuclei and prominent nucleoli. Discrete cell borders are more typical of oligodendroglioma than astrocytoma.

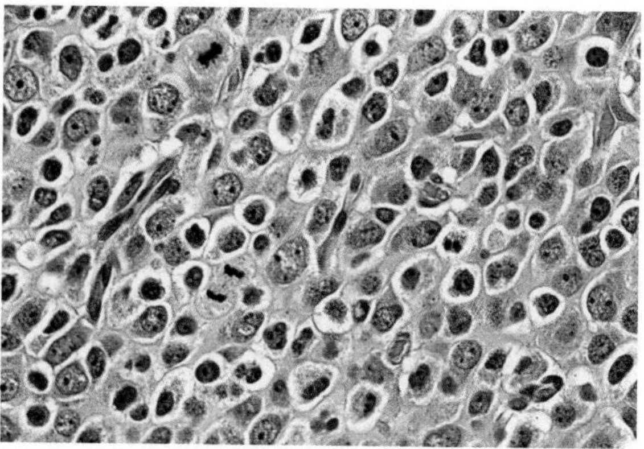

Figure 3-220

ANAPLASTIC (GRADE III) OLIGODENDROGLIOMA

Cytologic atypia, mitotic activity, and prominent epithelioid features characterize some grade III oligodendrogliomas.

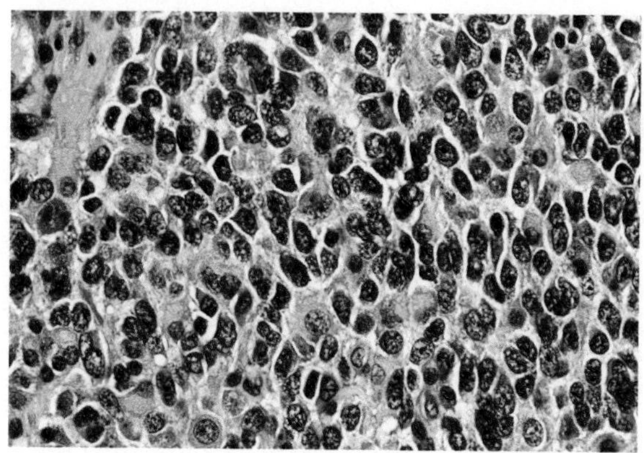

Figure 3-221

ANAPLASTIC (GRADE III) OLIGODENDROGLIOMA

The degree of nuclear pleomorphism in some grade III oligodendrogliomas overlaps with that of high-grade astrocytoma. The artifactual dehiscence of individual cells and, especially, the cells with intracytoplasmic eosinophilic granules (top) are more typical of oligodendroglioma.

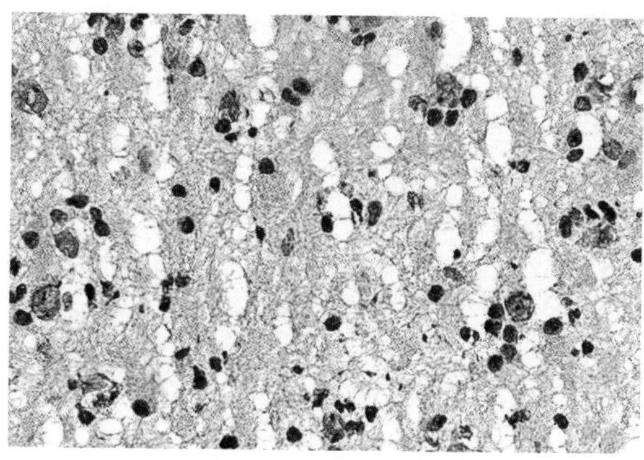

Figure 3-222

OLIGODENDROGLIOMA: FROZEN SECTION

The cytologic detail is better preserved, and ice crystals are largely absent in frozen sections of cortical gray matter, where perineuronal satellitosis is prominent.

cytoplasm, sometimes with small eosinophilic granules, may be present (fig. 3-221).

Despite cytologic malignancy and considerable cellularity, anaplastic lesions retain the overall architectural features of oligodendroglioma while preserving the cytologic uniformity so characteristic of the lesion, albeit with nuclei less perfectly round than those of the grade II lesion. Concomitant areas of better-differentiated WHO grade II tumor are common. The chicken-wire vasculature of the well-differentiated oligodendroglioma tends also to be preserved, but the vessels are more often thickened, and contribute to prominent tumor lobularity in some cases.

The grading of grade II oligodendrogliomas with "clonal nodules" can be as difficult as it is in ependymomas with analogous foci of increased cellularity and cytologic atypia. Generally, the presence of clonal nodules is not sufficient to elevate the grade to III, since the cells are not anaplastic. There are obviously some cases in which the degree of atypia and

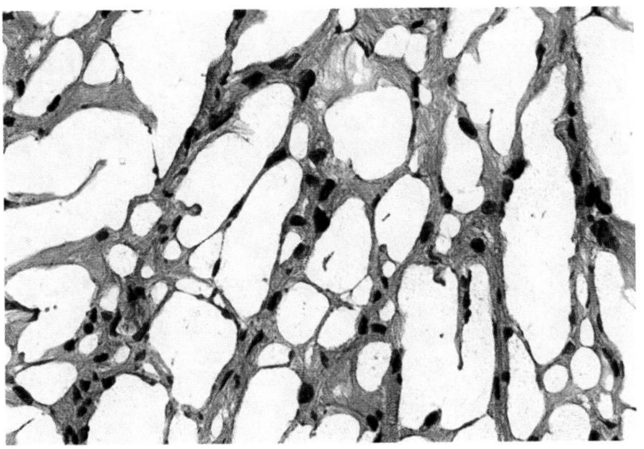

Figure 3-223

GRADE II OLIGODENDROGLIOMA: FROZEN SECTION

Ice crystals complicate the interpretation of oligodendroglioma in edematous white matter.

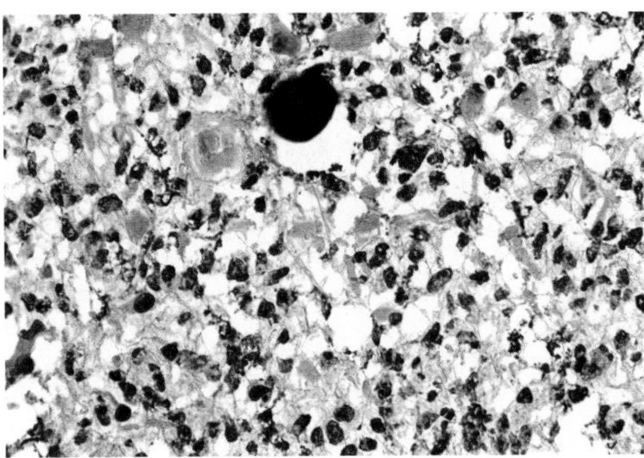

Figure 3-224

GRADE III OLIGODENDROGLIOMA: FROZEN SECTION

Dense cellularity, mitotic activity, and calcification suggest that the lesion is a grade III oligodendroglioma.

presence of vascular proliferation demand a grade III designation.

The relationship of oligodendroglioma to glioblastoma multiforme (astrocytic by definition) remains problematic since there are cases that overlap both lesions. In such lesions, the haloed cells are not, overall, compellingly oligodendroglial, although there are certainly cases that are difficult to assign to either the glioblastoma or anaplastic oligodendroglioma category (see fig. 3-235). We do not use the term glioblastoma when encountering clear-cut oligodendroglial differentiation. At present, the WHO classification does not include a grade IV oligodendroglioma designation. Glioblastomas with oligodendroglial features and high-grade mixed gliomas are discussed in the section on differential diagnosis.

Frozen Section. Rapid processing with freezing robs oligodendroglial tumors of a characteristic and distinctive feature, perinuclear halos, while adding two confusing artifacts in the form of nuclear irregularity and hyperchromasia. Smear preparations are therefore extremely useful. The net effect in frozen sections is to mimic the appearance of astrocytoma, thus, the suspicion of oligodendroglioma must be based largely on architectural features such as prominent cortical involvement with perineuronal satellitosis (fig. 3-222). Frozen sections of edematous white matter are difficult to interpret given the mass of ice crystals that compress cells into bands (fig. 3-223). Nonetheless, the general nuclear roundness, although altered by freezing, are largely retained. Nucleoli are often still obvious. All of the above limitations can be circumvented by smear preparations (see figs. 3-229, 3-230). The diagnosis of a well-differentiated infiltrating glioma suffices. High-grade lesions are recognized, or at least suspected, on the basis of round nuclei, calcification, cells filled with filaments, and mitoses (fig. 3-224).

Oligodendrogliomas in frozen tissues processed for permanent sections have nuclear angulation and condensation that resemble those of grade III diffuse astrocytoma (fig. 3-225).

Immunohistochemical Findings. Although much sought after, no currently available, or at least widely used, antibody reliably and specifically identifies oligodendrocytes in paraffin-embedded material. The antibody Olig 2 stains the nuclei of oligodendroglioma cells, but, unfortunately from a differential diagnosis perspective, also those of infiltrating astrocytoma cells (359).

Minigemistocytes and gliofibrillary oligodendrocytes are usually strongly immunoreactive for GFAP (figs. 3-226, 3-227) (350,360), although the hyaline or densely fibrillar cytoplasmic inclusion body is negative in some cases. Staining is often confined to these globular inclusions and does not involve the entire cytoplasm inclusive of processes. In more anaplastic oligodendrogliomas, both the cytoplasm and short processes are GFAP immunoreactive over wide regions. Such findings make the

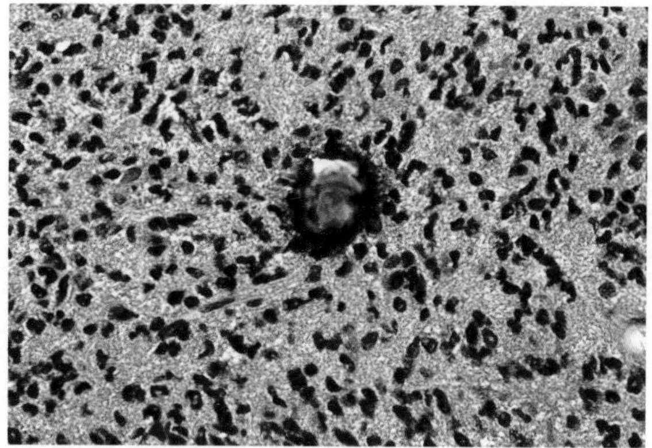

Figure 3-225

OLIGODENDROGLIOMA: FROZEN SECTION CONTROL

Tissues that were previously frozen and then embedded for permanent sections have condensed, irregular nuclei that closely resemble those of diffuse astrocytoma.

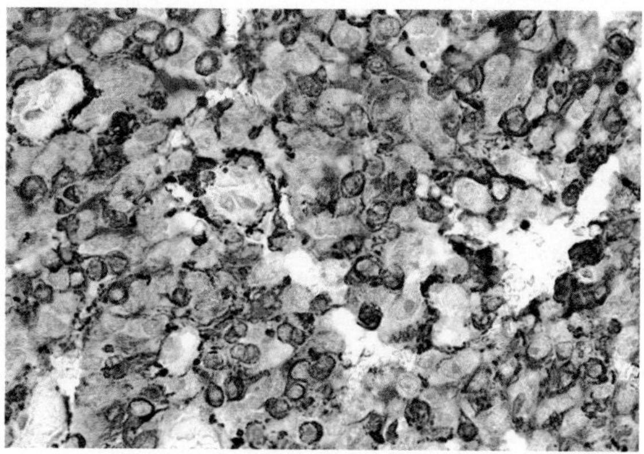

Figure 3-226

OLIGODENDROGLIOMA

The cytoplasm of astrocyte-like cells is usually, but not invariably, positive for GFAP.

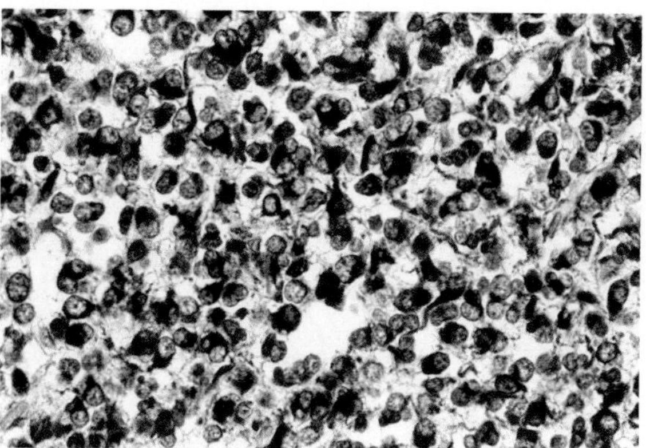

Figure 3-227

OLIGODENDROGLIOMA

Some oligodendrogliomas consist largely of small cells with GFAP-positive cytoplasm. The percentage of positive cells and the intensity of staining can be, paradoxically, higher in oligodendroglioma than astrocytoma. Note the uniform high cellularity, monomorphism, round nuclei, and clearly defined nucleoli typical of oligodendroglioma.

distinction between malignant oligodendroglioma and glioblastoma difficult, if not impossible, particularly in small specimens.

Immunoreactivity for neuronal markers occurs in some cases, including tumors with and without histologic evidence of neuronal/neurocytic differentiation (see figs. 3-213, 3-214). In the former, neuropil-like areas are synaptophysin immunoreactive, and nuclei surrounding the rosettes are positive for NeuN (352). In oligodendrogliomas without neuronal differentiation at the histologic level, tumor cells are immunoreactive for neuronal markers such as NeuN in some cases (360a). One study found that in 3 of 32 cases, tumor cells were reactive for synaptophysin and in 10 of 32 for nonphosphorylated neurofilament protein (361).

Immunoreactivity for p53 is unusual in classic grade II oligodendrogliomas, especially those with the 1p/19q codeletion. It is present in some grade III tumors, but is still unusual.

MIB-1 indices vary from region to region, according to the cellularity and degree of atypia. Three studies of histologically diagnosed oligodendrogliomas concluded that the survival rate was inferior if there was an MIB-1 index greater than 5 percent (357,362,363); another study suggested a cutoff point at greater than 3 percent (364).

Ultrastructural Findings. Neoplastic cells have round nuclei, relatively small quantities of nondescript cytoplasm, and short processes containing microtubules. A classic feature, albeit one observed only in a minority of cases, is pericellular spiral lamination of tumor cell processes reminiscent of those with which oligodendrocytes envelope adjacent axons (365). Scattered small lysosomes should not be mistaken for neurosecretory granules, particularly since the cellular processes of oligodendrocytes contain microtubules, a feature also common to neuritic processes. Although intermediate filament bundles are absent or only few in number

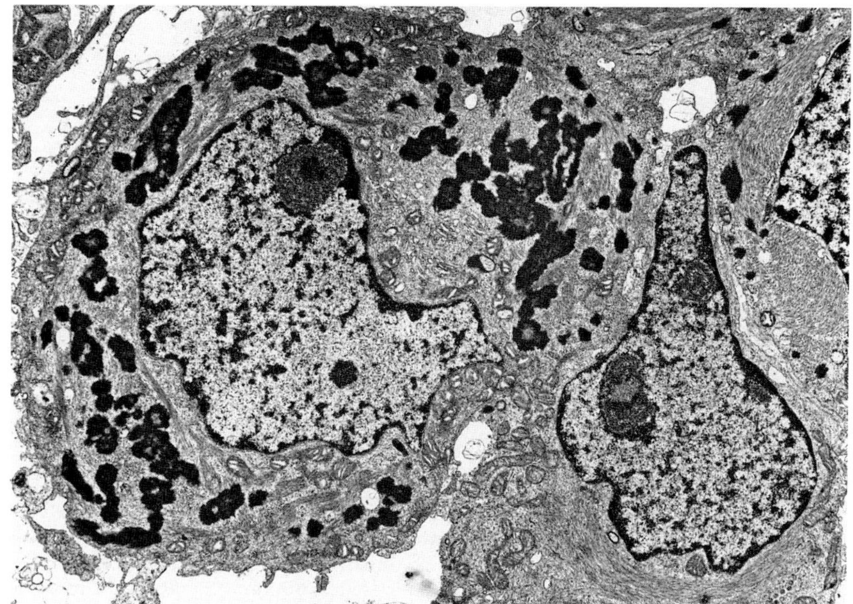

Figure 3-228

OLIGODENDROGLIOMA

Ultrastructurally, the small intracytoplasmic bodies such as those illustrated in figure 3-211 have the marked electron density of a Rosenthal fiber. There is an abundance of intermediate filaments. (Fig. 3-122 from Fascicle 10, 3rd Series.)

in classic oligodendroglioma, many filaments are present within eosinophilic GFAP-positive cells of minigemistocytes and gliofibrillary oligodendrocytes (fig. 3-228) (366, 367). Electron-dense, lumpy intracytoplasmic structures that resemble Rosenthal fibers are common in these densely fibrillated cells (368). Rudimentary neuronal features, including scattered dense core granules and synapse-like junctions, have been identified (369).

Cytologic Findings. Well-differentiated oligodendrogliomas have round nuclei with clearly defined nucleoli (fig. 3-229). There is little, if any, cytoplasm and therefore a paucity of cytoplasmic processes, although there may be more background fibrillarity in smears of oligodendrogliomas than might be expected based on histologic features. Oligodendrogliomas and diffuse astrocytomas are thus not always as different in smears as one might expect. The presence of minigemistocytes and/or gliofibrillary oligodendrocytes, cells easily seen in frozen sections, should prompt consideration of oligodendroglioma, particularly of grade III. Chromatin is coarser in such higher-grade tumors (fig. 3-230). The discohesive, well-defined, and relatively small cells of grade III oligodendroglioma can mimic those of metastatic carcinoma and, especially, malignant lymphoma.

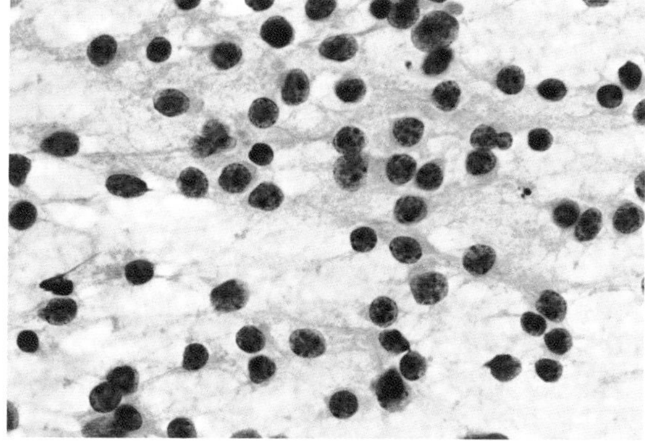

Figure 3-229

GRADE II OLIGODENDROGLIOMA: SMEAR PREPARATION

Round nuclei, small but well-defined nucleoli, and only a slight fibrillar background are typical of grade II oligodendroglioma.

Molecular and Cytogenetic Findings. Grade II and grade III oligodendrogliomas, especially when diagnosed by classic and strict histologic criteria, in most cases have a loss on chromosomes 1p and 19q, generally the entire arm of both (370–383). It appears that the retained chromosomal arms, 1q and 19p, are joined at the centromere to form the derivative chromosome (1;19)(q10;p10) (fig. 3-231) (383a,383b). The duo genetic loss correlates closely with the classic histopathologic features, predicts a response to treatment and survival superior to that associated with astrocytomas. Relevant

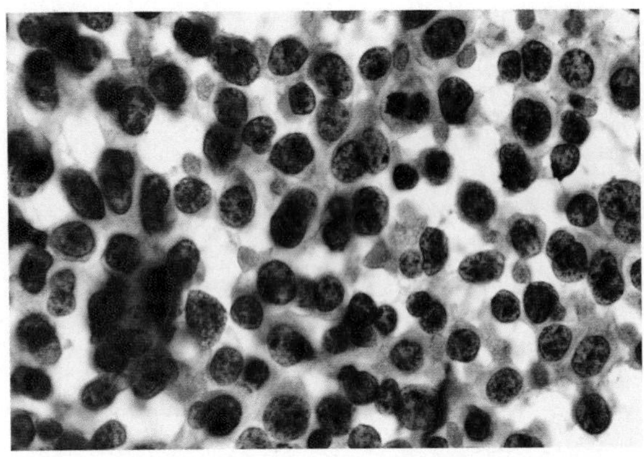

Figure 3-230

GRADE III OLIGODENDROGLIOMA: SMEAR PREPARATION

Nuclei in grade III oligodendrogliomas are coarser than they are in grade II counterparts.

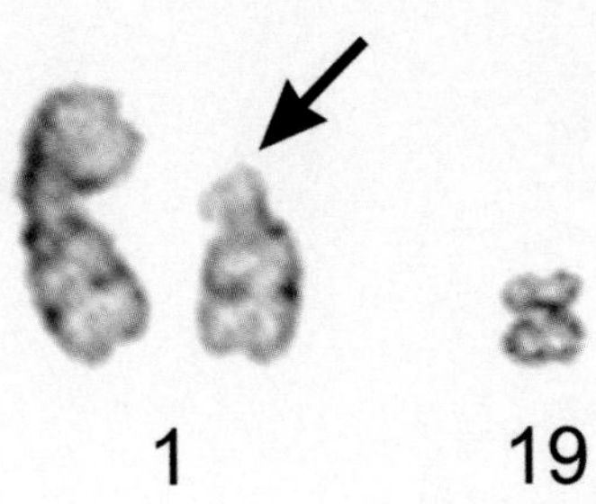

Figure 3-231

OLIGODENDROGLIOMA: CYTOGENETIC FEATURES

The chromosome 1 and 19 complement comprises one chromosome 1, one chromosome 19, and a derivative chromosome composed of 19p and 1q (arrow). The other derivative chromosome, composed of 1p and 19q, is lost. Codeletion of chromosomal arms 1p and 19q is thus the classic finding in FISH or LOH studies. (Courtesy of Dr. Constance A. Griffin, Baltimore, MD.)

tumor suppressor genes on chromosomes 1p and 19q have not been identified, and 1p/19q intact oligodendrogliomas could harbor undetected interstitial deletions or mutations in such genes. A tumor's 1p/19q status is usually determined in paraffin-embedded sections by FISH, a technique that can be applied also to cytologic preparations (383c). The determination can be done also by PCR-based techniques on paraffin-embedded tissue scraped from unstained slides, without the need for control DNA.

The diagnostic criteria for oligodendroglioma have become progressively looser over the past decade and some tumors now diagnosed as oligodendroglioma may be astrocytomas, especially when they have molecular changes such as gain on chromosome 7, mutations of *p53*, and loss of heterozygosity on chromosomes 17p and 10 (342). In one study of 44 cases felt by referring pathologist to be "pure oligodendrogliomas," only 22 were so confirmed by neuropathology review, whereas the other half had more astrocytic features (384). Nineteen of the 22 classic oligodendrogliomas (86 percent) had the 1p/19q codeletion. Of the 22 cases with astrocytic features, only 6 (27 percent) had combined 1p/19q loss. Other studies have also found a much higher correlation with the codeletion and a histologic diagnosis reached with strict criteria (385,385a). Certainly, there is a much greater consensus on the diagnosis in those with the codeletion (383,386,386a).

The molecular profile of pediatric oligodendrogliomas is not clear. Preliminary evidence suggests that combined 1p/19q loss is considerably less common that in adult tumors, particularly in the first decade (387,388). Diagnostically, this is disappointing since it would be useful in distinguishing well-differentiated oligodendroglioma from dysembryoplastic neuroepithelial tumor.

The changes that occur with progression are not as clear as they are in diffuse astrocytomas, but loss of 9p, and perhaps *p16*, appears commonly in grade III oligodendrogliomas (370,371,390).

The issue of distinguishing high-grade oligodendrogliomas from glioblastomas is not resolved, either histologically or genetically. One study of lesions classified as high-grade oligodendrogliomas found some with the 1p/19q codeletion and others with astrocytic features, e.g., gain on chromosome 7 and loss on 10 (391, 392). The latter could reasonably be interpreted as glioblastomas. The histologic, but not molecular, similarity of small cell glioblastomas, which may be calcified, to high-grade oligodendroglioma is discussed earlier in this chapter (see figs. 3-51–3-53).

The significance of chromosome 1p loss without 19q loss is unclear, but one small study found a high incidence of this in a very small series of patients with high-grade gliomas who had prolonged survival (393). Yet another series of oligodendrogliomas with grade IV features found a high incidence of an abnormality generally associated with diffuse astrocytomas: amplification of the platelet-derived growth factor-alpha gene (379). Whether these lesions are simply more malignant examples of the classic grade III oligodendroglioma is not clear, however.

In accord with the rarity of mutations in *p53* in genetically certified (1p/19q codeletion) grade II oligodendrogliomas, diffuse immunostaining for *p53* is exceptional (386). Infiltrating gliomas tend to have either a combined 1p and 19q loss or *p53* mutations, but not both (386,394,395).

Molecular/histopathologic correlations in oligoastrocytomas or mixed gliomas are as unclear as their histologic criteria. Reflecting varying criteria, the incidence of mixed gliomas varies greatly between institutions. In one study, 155 gliomas classified as "mixed gliomas" were reviewed by five experienced neuropathologists. The percentage of lesions felt to be "mixed" by the five ranged from as few as 9 percent to as many as 80 percent (396). Thus far, mixed lesions appear to be genetically either oligodendroglioma or astrocytoma, but not both (397,398). In one study, approximately half of oligoastrocytomas had *p53* mutations and the other half, loss of heterozygosity of 1p/19q (395). It could be argued that these observations challenge the credibility of mixed glioma as a clinicopathologic entity, but the full story of the oligoastrocytoma has not yet been told. In the mixed category, a combined 1p/19q loss is a favorable prognostic factor (396,399,400).

Differential Diagnosis. It is safest to begin the diagnosis of an oligodendroglioma-like lesion by excluding reactive and low-grade noninfiltrating lesions using Appendices E, K, L, M, and O as aids. In essence, this approach asks if the lesion is neoplastic, and, if so, is it really oligodendroglial?

Distinguishing grade II oligodendrogliomas from normal white and gray matter may be a problem, but is accomplished by attention to the distribution of cells and cytologic features. In the cortex, normal oligodendrocytes cluster about neurons, particularly in deep cortical layers where multiple cells may conspicuously surround single neurons (see fig. 2-14). Their numbers also appear to increase in excessively thick sections. The nuclei in neoplastic satellitoses are larger and their chromatin is coarser than in their normal counterparts. Generally, the cells engaged in neoplastic satellitosis are also more numerous and found at higher levels in the cortex. As is illustrated in figure 2-16, small normal neurons in the superficial cerebral cortex often have artifactual perinuclear halos.

One "lesion" characterized by an apparent increase in the number of oligodendrocytes is most often seen in lobectomy specimens removed for seizure control, but is occasionally encountered in other specimens as well (see fig. 2-30). Such a finding raises the issue of a very well-differentiated oligodendroglioma, particularly since the cells cluster about vessels, sometimes several rows deep. There is no cytologic atypia, however. The pathogenesis of this prognostically insignificant finding is unclear. Similar oligodendroglial hypercellularity, perhaps a result of tissue collapse, occurs in the environs of longstanding vascular malformations (401,402).

Important reactive processes that mimic oligodendroglioma are macrophage-rich lesions that include demyelinating disease, cerebral infarct, and primary CNS lymphoma treated with steroids. The brain's response in demyelinating disease includes not only astrocytic hypertrophy, which may simulate astrocytoma, but also an outpouring of macrophages that may resemble oligodendrocytes (fig. 3-232). Lipid-engorged, these phagocytes simulate the fried-egg cells of oligodendroglioma, although the granular cytoplasm of the macrophages contrasts with the water-clear vacuity of oligodendrocytes. Macrophages are particularly evident after PAS staining, but immunohistochemistry for macrophage markers is the primary means to resolve any doubts regarding the nature of the pale cells.

Idiopathic demyelinating disease, with its characteristic neuroradiologic features, is discussed in chapter 19, as is progressive multifocal leukoencephalopathy, an infectious disorder complicated by demyelination. This same chapter also includes a discussion of the other major macrophage-rich process, cerebral infarct. The macrophage-rich state of primary CNS lymphomas treated with corticosteroids is discussed in chapter 12.

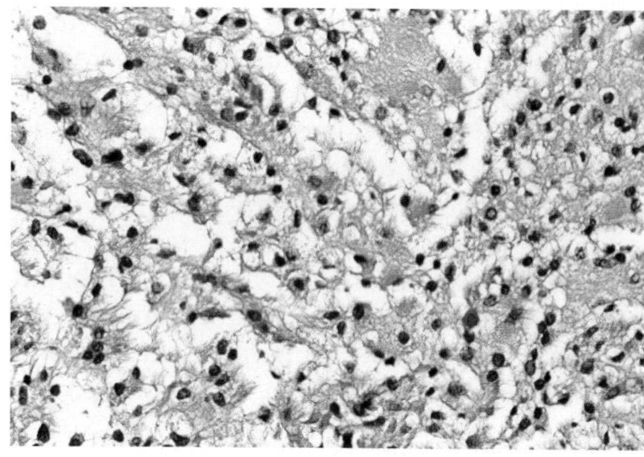

Figure 3-232

DEMYELINATING DISEASE RESEMBLING OLIGODENDROGLIOMA

Demyelinating disease should be the principal initial consideration in the differential diagnosis of grade II oligodendroglioma. The cytoplasm of the macrophages is granular, not clear as would be the case in oligodendrocytes.

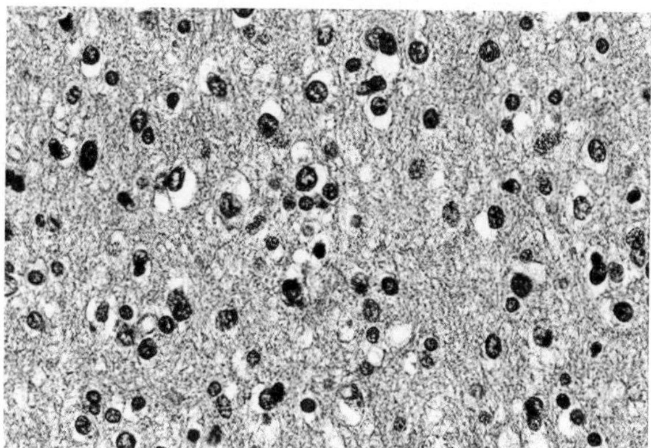

Figure 3-233

DIFFUSE ASTROCYTOMA RESEMBLING OLIGODENDROGLIOMA

Perinuclear halos alone do not an oligodendroglioma make. The nuclear hyperchromatism and polymorphism are characteristic of astrocytoma. There was no loss of either chromosomal arm 1p or 19q. Some pathologists might be tempted to consider this lesion with hybrid features a mixed glioma.

If the lesion under consideration is clearly neoplastic, a number of entities come into the differential diagnosis. Of infiltrating gliomas, particularly those of the cerebral hemispheres, grade II diffuse astrocytoma is the principal contender (fig. 3-233). Even when well differentiated, the latter tumor has more nuclear pleomorphism than is seen in low-grade oligodendroglioma, although this is a subjective distinction much in the eye of the beholder. The nuclei of diffuse astrocytoma cells are often cigar shaped. Furthermore, the cytoplasm of astrocytes varies greatly in amount and appearance. Whereas isolated cells of astrocytoma in intact parenchyma may have little or no discernible cytoplasm and are often GFAP negative, eosinophilic cytoplasm is often seen in areas of higher cellularity. The processes in diffuse astrocytomas produce a fibrillar background that is absent in oligodendrogliomas. The same is true of coarse, asymmetric processes.

Diffuse astrocytomas are also less likely to calcify than oligodendrogliomas and generally lack the chicken-wire capillary pattern. Dense, intracortical calcification with a gyriform profile is most unusual in astrocytomas. Perinuclear halos, while characteristic of oligodendroglioma, are prominent in some astrocytomas. In any case, one must not be beguiled by halos and a touch of nuclear roundness, and demand instead monotonously round nuclei and molecular testing if doubts remain. The level of mitotic activity is helpful since mitoses are uncommon in grade II oligodendrogliomas, particularly in paucicellular lesions. An infiltrating glioma of low cellularity with mitoses is almost always astrocytic.

Some well-differentiated gliomas have features of both astrocytoma (nuclear pleomorphism and hyperchromasia) and oligodendroglioma (roundish nuclei). These tumors are not mixed in the sense of containing two neoplastic cell types, but are "hybrids," whose individual cells have features of both. They can be designated simply "infiltrating glioma, subtype indeterminate," placed in either the oligodendroglial or astrocytic category according to the prejudices of the observer, or assigned to the nebulous category of mixed oligoastrocytoma. Some pathologists try hard to place infiltrating gliomas into either oligodendroglioma or astrocytoma category. Others, facing ambiguous lesions, freely utilize the term *mixed glioma* or *oligoastrocytoma*. Molecular characterization is appropriate for such lesions, while recognizing that mixed glioma as currently defined is a purely histological diagnosis—some such tumors have the 1p/19q codeletion and others do not.

Even when oligodendrogliomas become anaplastic, they retain a monotony of nuclear size

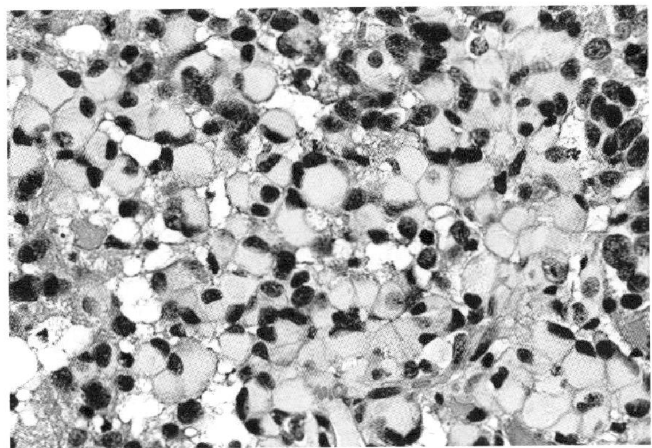

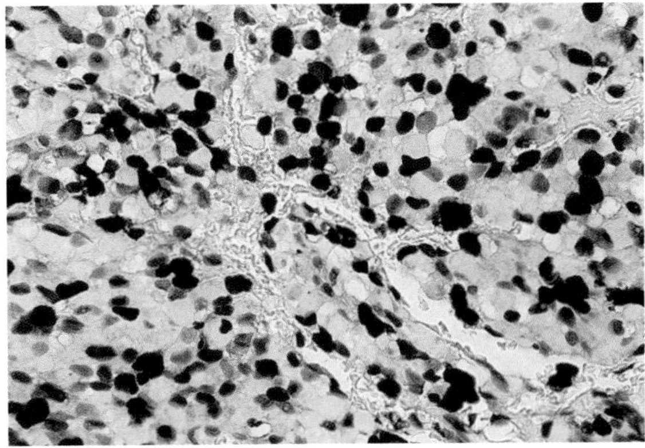

Figure 3-234

GEMISTOCYTIC ASTROCYTOMA RESEMBLING OLIGODENDROGLIOMA

Left: Lesions composed of densely populated small gemistocytes can be difficult to distinguish from oligodendrogliomas with minigemistocytes.

Right: Gemistocytic astrocytomas are often extensively immunopositive for p53, a most unusual reactivity in oligodendrogliomas.

and shape which, when coupled with a scant ring of cytoplasm, distinguishes them from high-grade astrocytomas. Furthermore, since neoplastic oligodendrocytes do not usually project long processes, there is little or no fibrillar background. Any intercellular pink matrix is neuropil that, since it is composed of neuronal rather than glial processes, is synaptophysin rather than GFAP positive. This neuropil formed by neuritic processes is more finely fibrillar than the aggregate of glial process. This is also evident in cellular lesions in which oligodendrocytes are uniformly spaced, or "back-to-back." In diffusely infiltrative astrocytomas disordered fibrillar processes lie between tumor cells.

Some infiltrating gliomas overlap the astrocytoma-oligodendroglioma interface when they are composed of closely packed gemistocytes with rounded nuclei (fig. 3-234, left). These usually have prominent immunoreactivity for p53 (fig. 3-234, right) and in our experience usually do not lose chromosomes 1p and 19q.

The distinction of high-grade oligodendroglioma from glioblastoma (fig. 3-235) poses a problem that is of major prognostic importance. We assign a tumor to the anaplastic oligodendroglioma category if the cells have nuclear uniformity, roundness, and perinuclear halos, or to the glioblastoma category when nuclei are more astrocytic in appearance and the cytoplasm does not appear skirt-like, as in typical oligodendroglioma, or disposed as previously described in minigemistocytic or gliofibrillary cells. In problem cases, it is perfectly acceptable to diagnose only a malignant glioma and explain the diagnostic conundrum in a comment section. The existence of the anaplastic mixed glioma as a prognostically significant histologic entity distinct from glioblastoma was supported in one large study that found that patients with such lesions with necrosis had survivals intermediate between anaplastic astrocytoma and glioblastoma (402a). While the lesions could be considered grade IV mixed gliomas, the 2007 WHO consensus group considered them "glioblastomas with oligodendroglioma features."

The cellular monotony and frequent calcification of small cell glioblastoma and small cell astrocytoma grade III (403) make these a common differential issue (fig. 3-236, left). The astrocytomas have cells with more oval to elongate nuclei (fig. 3-236, right), higher mitotic rates, more conspicuous microvascular proliferation, as well as necrosis with pseudopalisading in the case of glioblastoma. High-grade "oligodendrogliomas" with ring enhancement often have glioblastoma-like molecular changes, and behave like the glioblastomas that they arguably are. There is no combined loss of chromosomal arms 1p and 19q, but rather molecular features of glioblastoma such as loss of chromosome 10 and/or amplification of *EGFR*.

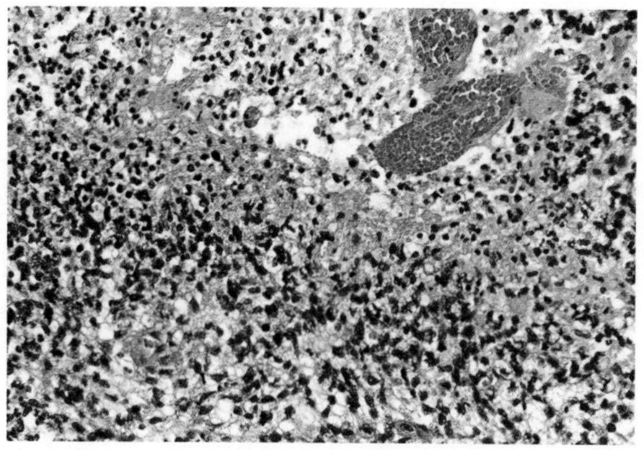
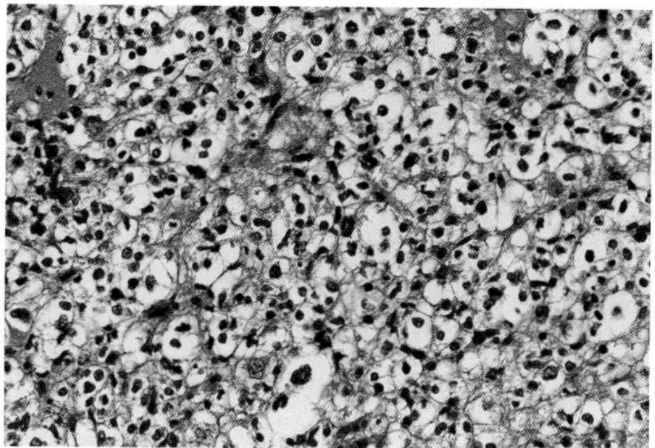

Figure 3-235

GLIOBLASTOMA WITH OLIGODENDROGLIOMA FEATURES

Left: Malignant gliomas with pseudopalisading and oligodendroglioma-like features are a diagnostic challenge.

Right: While perinuclear halos are prominent, the nuclei have the astrocytic qualities of marked hyperchromatism, pleomorphism, and angulation. The lesion was a ring-enhancing mass in a 63-year-old who died 14 months after diagnosis. Histologically, these could be viewed as highly anaplastic mixed gliomas or, as is formulated by the WHO classification, glioblastomas with oligodendroglioma features.

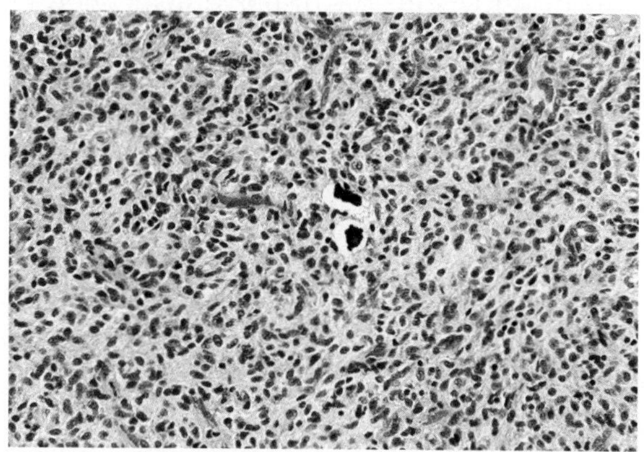
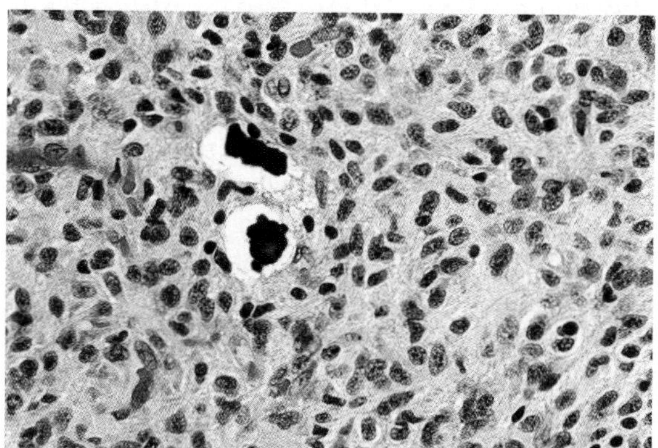

Figure 3-236

SMALL CELL GLIOBLASTOMA RESEMBLING ANAPLASTIC OLIGODENDROGLIOMA

Left: Calcified small cell glioblastomas can be misconstrued as anaplastic oligodendroglioma.

Right: The nuclei are elongated and nucleoli are inconspicuous in glioblastoma, in contrast to those of oligodendroglioma.

Pilocytic astrocytomas often contain patches or sheets of cells that resemble oligodendrocytes, particularly in cerebellar examples (fig. 3-237). Pilocytic astrocytomas, and other well-circumscribed lesions with which oligodendrogliomas can be confused are almost always contrast enhancing. The diagnosis of oligodendroglioma in the face of such a lesion is suspect, as is outlined in Appendix N.

In clear cell ependymoma, radiographic features (contrast enhancement and sharp circumscription), immunohistochemical reactions (dot-like intracytoplasmic EMA staining), and ultrastructural findings (microvilli, junctional complexes, and cilia) are readily evident, despite the tumor's light microscopically oligodendroglial appearance. Histologically, the lesion has a sharp interface with prominent perivascular pseudorosettes of the surrounding brain (GFAP positive), and nuclear grooves (fig. 3-238). Ultrastructural features of ependymal differentiation often withstand paraffin embedding.

The dysembryoplastic neuroepithelial tumor (DNT) often closely mimics well-differentiated

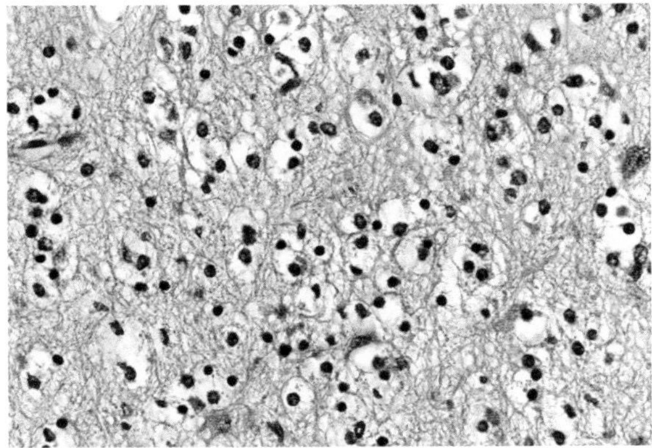

Figure 3-237

PILOCYTIC ASTROCYTOMA RESEMBLING OLIGODENDROGLIOMA

Pilocytic astrocytomas of the cerebellum can contain neoplastic cells indistinguishable from those of oligodendroglioma.

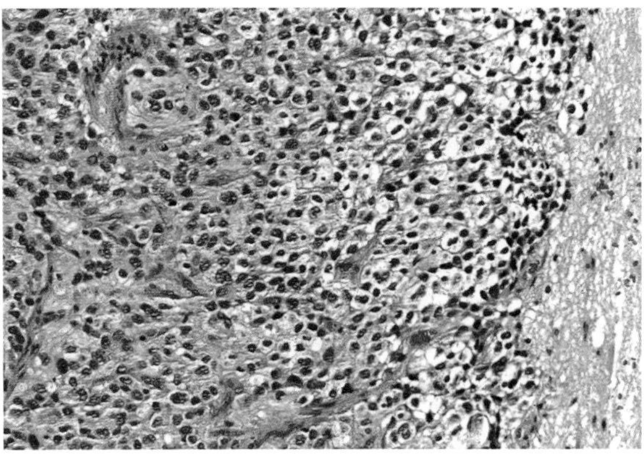

Figure 3-238

CLEAR CELL EPENDYMOMA RESEMBLING OLIGODENDROGLIOMA

Clear cell ependymomas can closely resemble oligodendroglioma. The compact, noninfiltrating architecture with a sharp border is one distinguishing feature.

oligodendroglioma since oligodendroglia-like cells are the DNT's principal component (fig. 3-239) (see Appendix E for oligodendroglioma-like lesions). The identification of DNT is contingent upon receiving a specimen of sufficient size to permit recognition of its telltale nodular architecture and, in some cases, the presence of a loose textured, "specific glioneuronal element." Coupled with a history of partial complex seizures and compatible neuroimaging, even a limited biopsy should bring this entity to mind. Unlike the neuronal satellitosis of oligodendroglioma, neurons in DNT often "float" in a mucoid matrix. Perineuronal satellitosis is uncommon and focal at best. A faintly basophilic mucoid background is seen in some oligodendrogliomas, but is generally more prominent in DNT, as are patterned nodules of oligodendrocyte-like cells that are strongly highlighted with Alcian blue staining. The finding of astrocytic nodules, either piloid or fibrillary, also supports the diagnosis of DNT. Lastly, an advertised, but inconstant, feature of DNT is the presence of cortical dysplasia at the periphery of the lesion. DNTs do not have the combined 1p/19q loss of oligodendrogliomas, but neither do most tumors classified as pediatric oligodendrogliomas.

Cells resembling those of oligodendroglioma are seen in a variety of brain tumors, including nonastrocytic glial neoplasms and metastatic tumors. For example, neurocytic neoplasms so

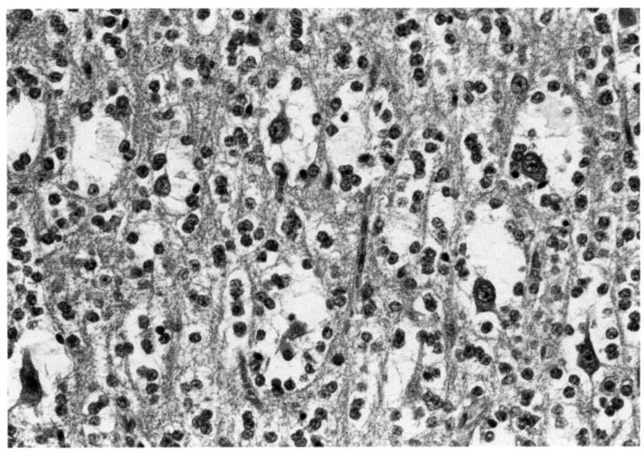

Figure 3-239

DYSEMBRYOPLASTIC NEUROEPITHELIAL TUMOR (DNT) RESEMBLING OLIGODENDROGLIOMA

The oligodendrocyte-like cells of DNT have cytologic and infiltrative properties identical to those of oligodendroglioma. The "floating neurons" aid in a distinction that can be difficult, if not impossible, in some cases.

closely resemble oligodendroglioma that early cases of central neurocytoma were published as "intraventricular oligodendroglioma." Both central and extraventricular neurocytomas have not only striking nuclear uniformity and roundness, but perinuclear halos and calcification as well (fig. 3-240). The recognition of a neurocytic neoplasm is greatly aided by awareness of clinical and radiologic data, by immunohistochemistry that demonstrates synaptophysin reactivity,

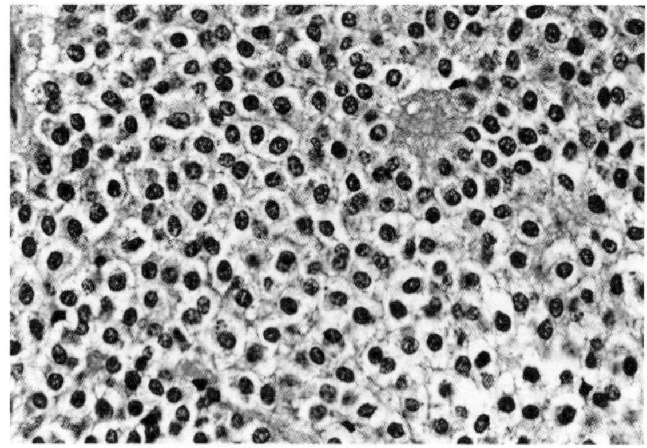

Figure 3-240

CENTRAL NEUROCYTOMA RESEMBLING OLIGODENDROGLIOMA

Cytologic uniformity and round nuclei simulate features of oligodendroglioma.

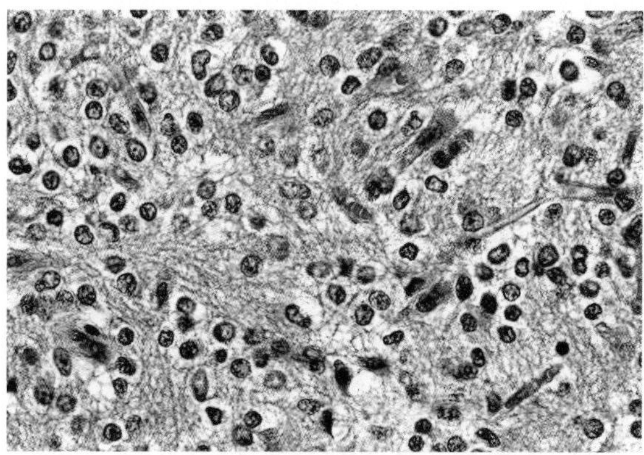

Figure 3-241

EXTRAVENTRICULAR NEUROCYTOMA RESEMBLING OLIGODENDROGLIOMA

Uniformly round nuclei surrounded by clear halos make neurocytomas, extraventricular in this case, similar to oligodendrogliomas.

and by electron microscopy that reveals microtubule-containing processes, clear vesicles, secretory granules, and even synapses. More treacherous are extraventricular neurocytomas that have a looseness or low cell density that creates the impression of an infiltrating lesion (fig. 3-241). The issue is compounded by the presence of focal neurocytic differentiation in some genetically certified oligodendrogliomas, as is discussed above, and by immunoreactivity for neurofilament protein and synaptophysin in some otherwise typical oligodendrogliomas.

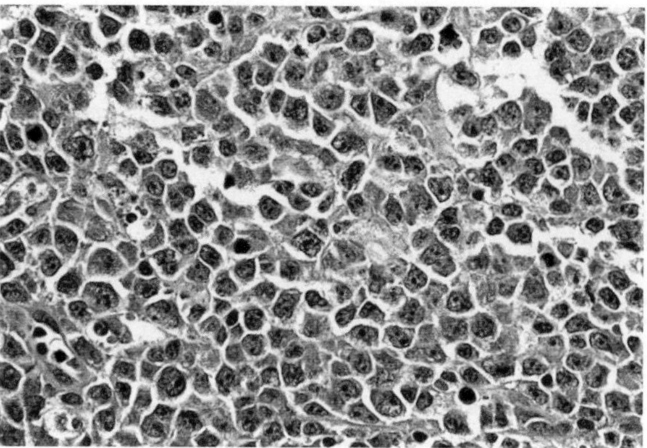

Figure 3-242

PRIMARY CENTRAL NERVOUS SYSTEM (CNS) LYMPHOMA RESEMBLING OLIGODENDROGLIOMA

Round or polygonal cells with discrete cell borders and prominent nucleoli make some primary CNS lymphomas mimics of anaplastic oligodendroglioma.

Densely cellular oligodendrogliomas, particularly those associated with desmoplasia, may simulate metastatic carcinoma or a primary CNS lymphoma (fig. 3-242). Distinguishing features are the infiltrative nature of peripheral portions of oligodendrogliomas, the presence of a better-differentiated oligodendroglioma component, and uniform immunoreactivity for S-100 protein and often, at least focally, for GFAP. Oligodendrogliomas, of course, do not express lymphoid markers.

Treatment and Prognosis. Given their infiltrative nature, oligodendrogliomas are generally considered unresectable, although in the context of modern neuroimaging with its aid in early diagnosis and refined neurosurgical techniques, this may be changing. The fate of patients with small lesions detected by MRI early in the course is still unclear, but the prognosis may be considerably more favorable than previously thought.

The long-term outlook for patients with oligodendrogliomas with the codeletion is not yet clear, but for patients with grade III lesions it appears to be considerably better than that associated with grade III astrocytoma, given the former tumor's responsiveness to radiotherapy and chemotherapy (373,404,405). Some lesions respond dramatically to treatment. In one series of 80 patients with either grade II or grade III lesions, the estimated median postoperative survival period for patients with codeleted

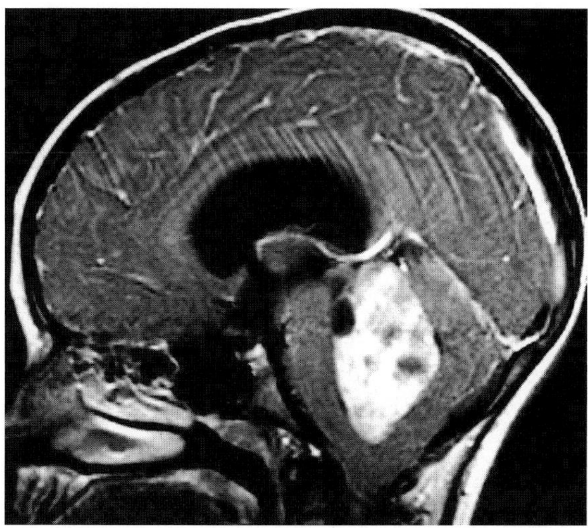

Figure 3-243

EPENDYMOMA

Ependymomas, such as this lesion in the fourth ventricle, are discrete and contrast enhancing. A broad base on the pons is typical. (Fig. 3-130 from Fascicle 10, 3rd Series.)

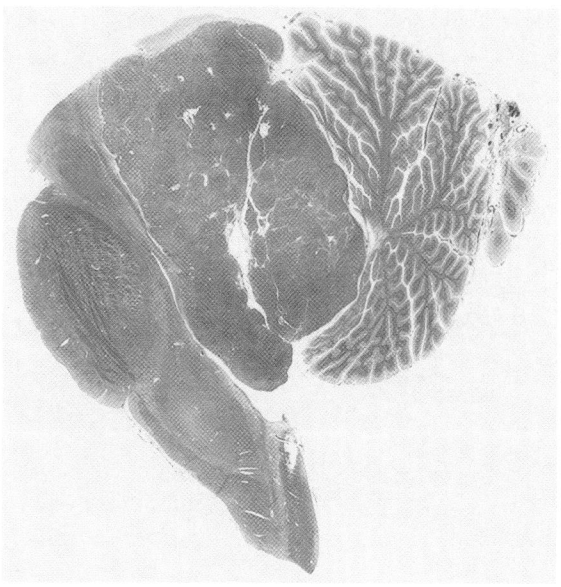

Figure 3-244

EPENDYMOMA

Most ependymomas of the fourth ventricle are broadly and noninvasively based on the pons.

tumors was 14.9 years (390). In contrast, the median survival period for those whose "oligodendrogliomas" lacked the combined 1p/19q loss was an astrocytoma-like 4.7 years.

The relationship of MIB-1 indices to outcome is discussed in Immunohistochemical Findings.

EPENDYMAL NEOPLASMS

Ependymoma

Definition. *Ependymoma* is a cellular-, glial-, or epithelial-appearing tumor that manifests ependymal differentiation, often in the form of perivascular pseudorosettes and, less frequently, true rosettes. CNS neoplasms in which ependymal differentiation is most obvious at the immunohistochemical or ultrastructural level are discussed separately. These lesions include astroblastoma, chordoid glioma of the third ventricle, angiocentric glioma, and papillary tumor of the pineal region.

Clinical Features. Ependymomas arise throughout the neuraxis in intimate association with the ependyma or its remnants. Exceptions include rare ectopic examples involving the presacral or postsacral soft tissue, as well as intracranial tumors as far from the ventricular system as the cerebral cortex and even subarachnoid space. Ependymomas in the fourth ventricle typically occur in children, whereas supratentorial lesions are more common in older individuals (406,407). In one series of adults, supratentorial, intraparenchymal ependymomas were more likely to be anaplastic (WHO grade III) than those at other sites (406).

Supratentorial ependymomas produce symptoms of increased intracranial pressure or symptoms referable to local mass effects. Elevated intracranial pressure due to CSF flow obstruction is the common clinical expression of fourth ventricular ependymomas.

Spinal ependymomas usually affect adults and reside in either the spinal cord proper or filum terminale. In the latter site, they are common in children (408,408a). These tumors induce pain or patterns of motor and sensory deficit reflecting the segmental level of the lesion. Myxopapillary lesions in the filum terminale produce the cauda equina syndrome due to compression of nerve roots, with resultant pain, lower extremity weakness, and sphincter dysfunction.

Rare myxopapillary ependymomas are primary in the presacral or postsacral regions (409,410), and occasionally occur in the sacrum as well.

Radiologic Findings. On CT and MRI scans, the typical ependymoma in the posterior fossa is

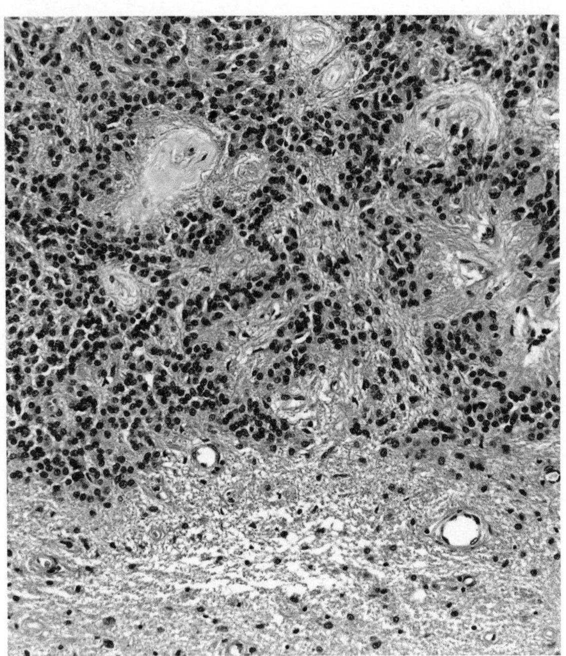

Figure 3-245

EPENDYMOMA

The discreteness of ependymomas extends to the microscopic level, as in this lesion on the floor of the fourth ventricle.

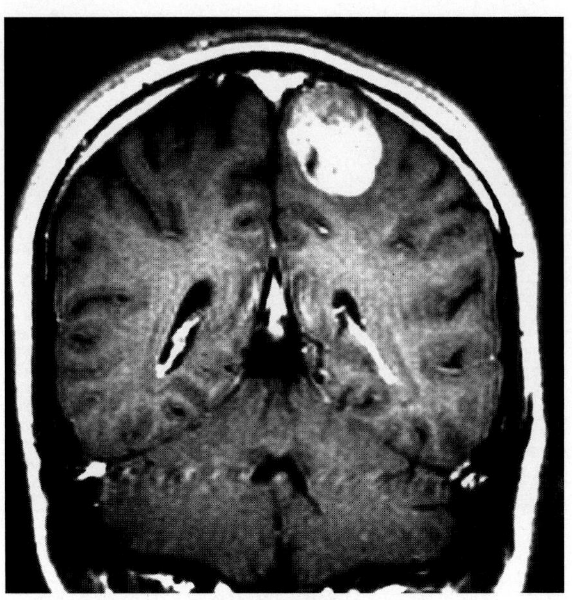

Figure 3-246

EPENDYMOMA

Supratentorial ependymomas are often not intraventricular, and may even be located near the surface of the brain.

a contrast-enhancing mass that has a broad-based origin in the floor of the fourth ventricle, and in some cases, extends along the ventricle's lateral recess (figs. 3-243–3-245) (411). Ependymomas can reach the cervical canal by way of the cisterna magna. Those reaching the cerebellopontine angle, either at presentation or recurrence, can incorporate vessels and cranial nerves, greatly complicating attempts at complete excision.

Supratentorial ependymomas are frequently associated with a cyst, and are often not intraventricular (fig. 3-246) (412,413); these may even occur near the surface of the brain (414). (See Appendix D for a roster of cystic CNS tumors and Appendix C for a list of intraventricular tumors.) Rare lipidized ependymomas are bright on precontrast T1-weighted images (415–417).

Intraspinal ependymomas are sausage-shaped, discrete, contrast-enhancing masses that may be internally cystic or, more commonly, induce a fluid-filled cavity (syrinx) in the adjacent cord. A "cap" of dark signal in T2-weighted and FLAIR images reflects the hemosiderin deposition common with intramedullary ependymomas. (Appendix B considers intraspinal lesions by anatomic site.)

Gross Findings. Supratentorial ependymomas intrude upon the ventricular system but also enlarge centrifugally into the surrounding brain where they present to the surgeon as a fleshy, gray, relatively circumscribed mass. Intraparenchymal lesions are often associated with a cyst. Ependymomas in the posterior fossa usually fill the fourth ventricle from a base on the pons (fig. 3-244), but can be intracerebellar.

Intraspinal ependymomas are gray and soft. While they are usually discrete overall, the surgeon may report a focal loss of the cord-tumor cleavage plane. The more caudally situated myxopapillary variant arises from the filum terminale or conus medullaris, often as a delicately encapsulated "bag" of soft tan tissue (fig. 3-247). Intraoperatively, it may be difficult to distinguish the filum, the delicate extension of the spinal cord, from adjacent nerve roots to which ependymomas and various other regional tumors can become secondarily attached. Myxopapillary tumors occasionally break out of their delicate capsule and seed the subarachnoid space prior to surgery.

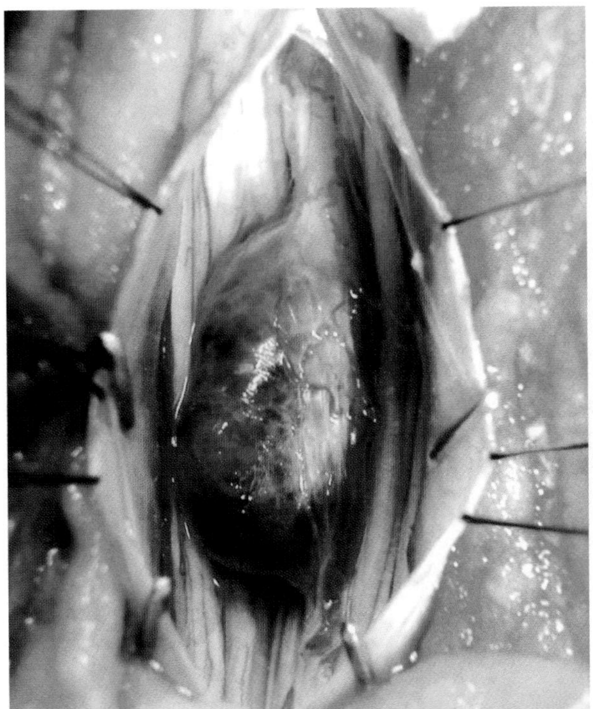

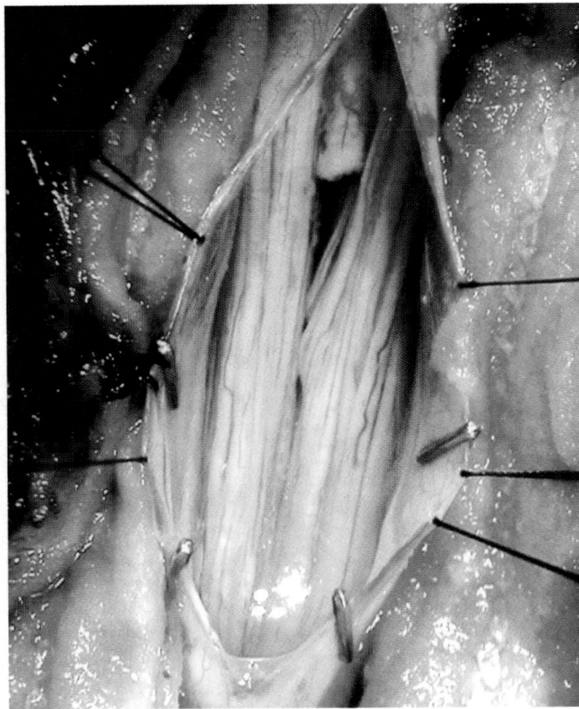

Figure 3-247

MYXOPAPILLARY EPENDYMOMA

Myxopapillary ependymomas are fusiform or globiform masses that almost always arise from the filum terminale (left), whose proximal severed end is seen after resection near the top of the figure (right). The 50-year-old man had complained of back pain and numbness and tingling in the left leg. (Courtesy of Dr. Ziya Gokaslan, Baltimore, MD.)

Microscopic Findings. Ependymomas are usually well-circumscribed masses (fig. 3-245). Spinal intramedullary tumors are sometimes exceptions since these may be intermixed with surrounding parenchyma.

Ependymomas are of three principal types: 1) cellular (with papillary and clear cell variants), 2) tanycytic, and 3) myxopapillary. Uncommon giant cell variants occur with both myxopapillary or cellular lesions, but usually with the former.

The classic *cellular ependymoma* of either brain or spinal cord proper often has extremes of cellularity, beginning with areas that are paucicellular and ending in those that are densely cellular and decidedly blue. Often, both ends of the spectrum are present in the same lesion (fig. 3-248). Throughout, but most obviously in more cellular regions, nuclei respect perivascular zones, leaving an anuclear region composed entirely of tumor cell processes that converge upon vessel walls. Known as perivascular pseudorosettes (figs. 3-249, 3-250), these formations are present, almost by definition, in all cellular ependymomas. In less cellular regions, where the difference in cellularity between tumor tissue and surrounding brain parenchyma is not as great, pseudorosettes are inconspicuous if not absent (figs. 3-251, 3-252). In extreme cases, the fibrillar areas, although they include vessels, are so large that they cannot be considered strictly perivascular. Small gemistocytes, often present in such lesions, add to the confusion by suggesting that the lesion is astrocytic (fig. 3-253). Such tumors are ependymomas, not astrocytomas or mixed gliomas (ependymoastrocytomas).

Noteworthy in cellular lesions, especially those in the posterior fossa, are single, or more often, multiple well-circumscribed areas in which cell density, cytologic atypia, and mitotic activity are all increased over that of the background tumor (figs. 3-248, 3-254). Glomeruloid microvascular proliferation may be present at the periphery of, or occur within, some foci (fig. 3-255). This

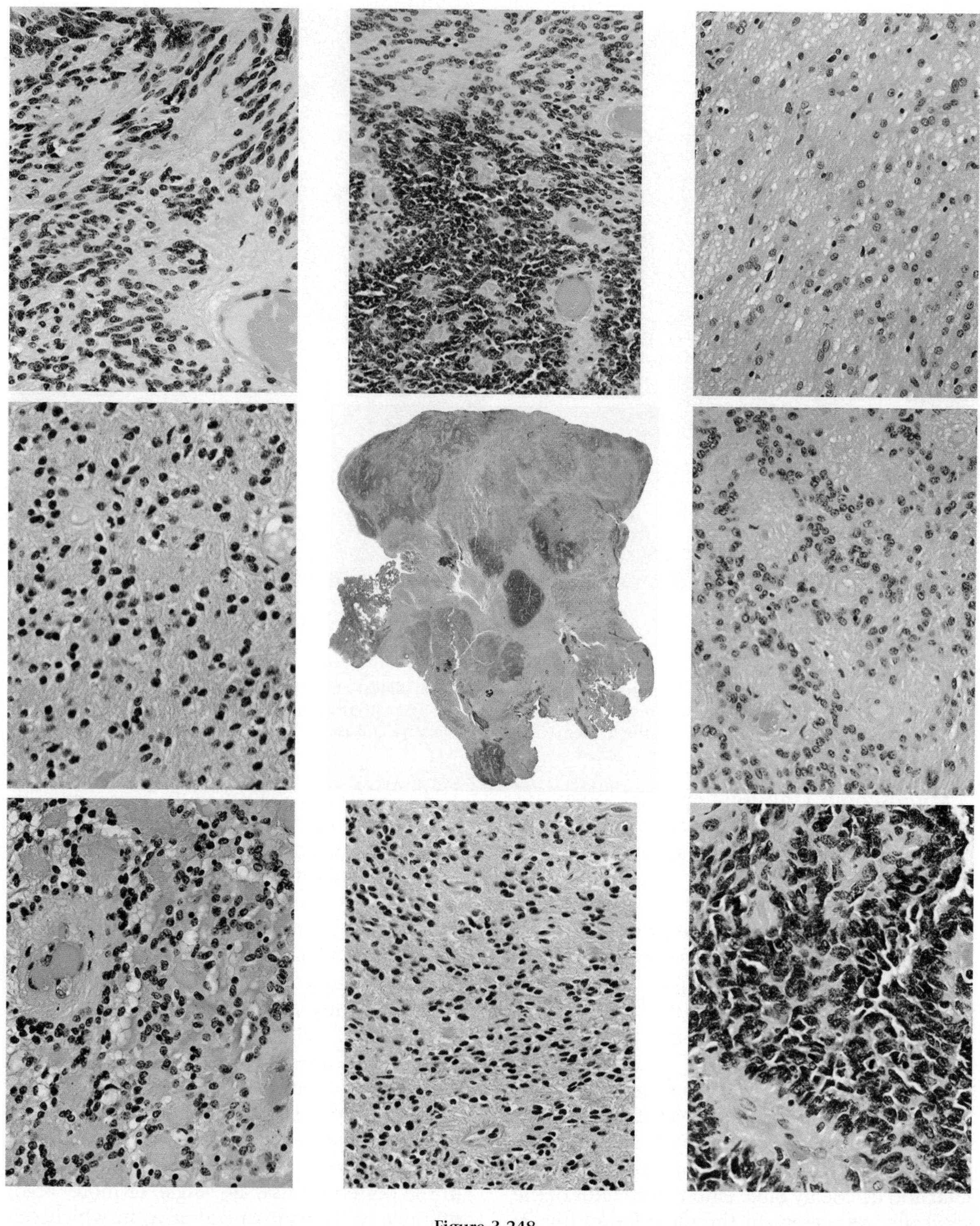

Figure 3-248

EPENDYMOMA

Cellular ependymomas are often composed heterogeneously of cellular and paucicellular areas. Perivascular clearings, perivascular pseudorosettes, are prominent in some areas, but not in others. Well-circumscribed areas of increased cellularity are common, as are structureless fibrillar zones that resemble diffuse astrocytoma. All panels are from the same neoplasm illustrated in a macrosection at the center of the figure.

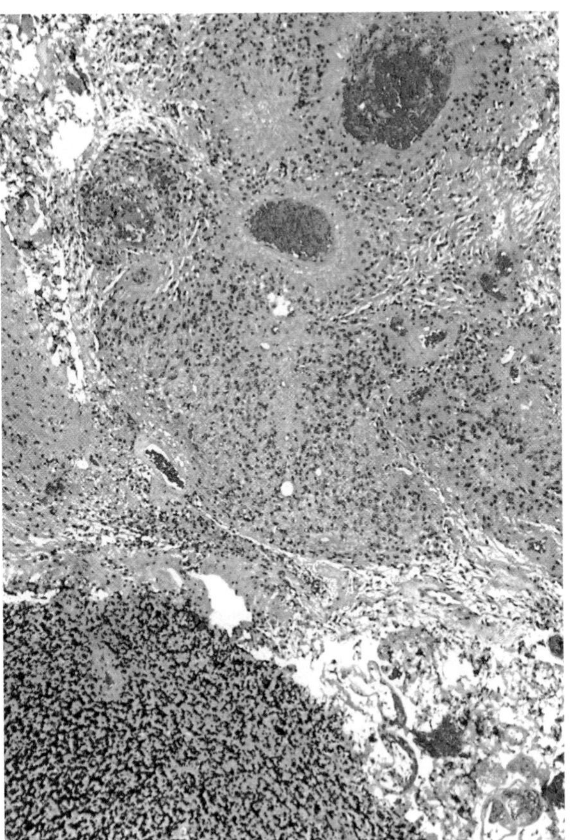

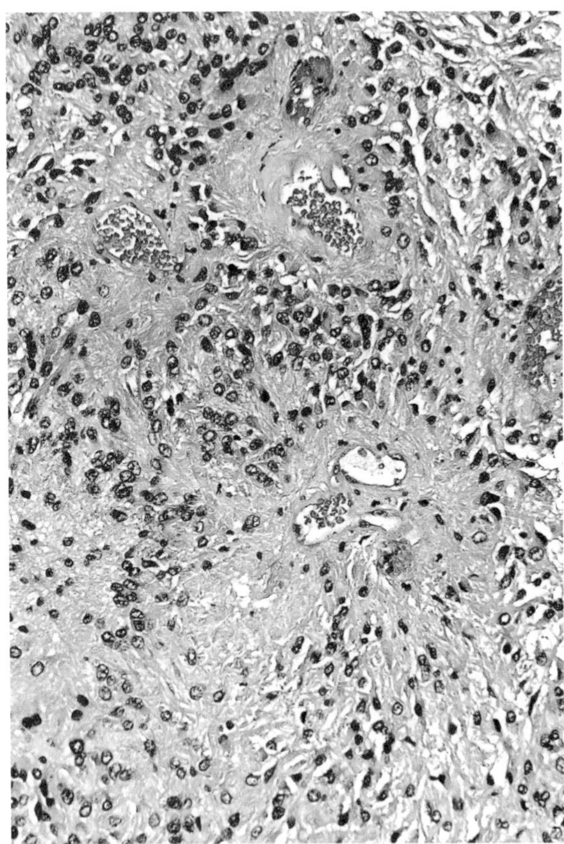

Figure 3-249

EPENDYMOMA

Left: Perivascular rosettes, ependymoma's most distinguishing feature, are robust enough to survive the tissue fragmentation inherent in ultrasonic aspiration. Note the fragment of normal internal granule cell layer in this specimen from the posterior fossa.

Right: The presence of perivascular pseudorosettes is the principal feature that identifies the lesion as ependymoma.

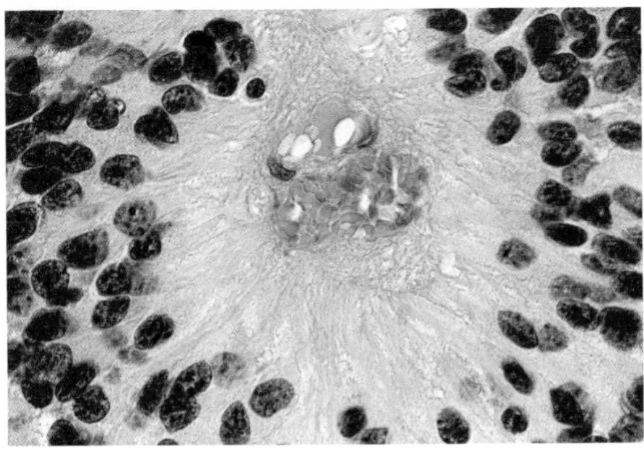

Figure 3-250

EPENDYMOMA

Tumor cell processes converging on the vessel create the perivascular fibrillar zone of ependymoma pseudorosettes.

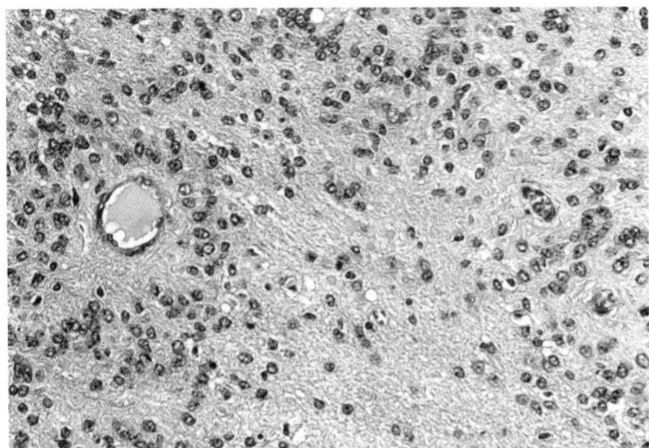

Figure 3-251

EPENDYMOMA

Perivascular pseudorosettes are often inconspicuous in paucicellular regions.

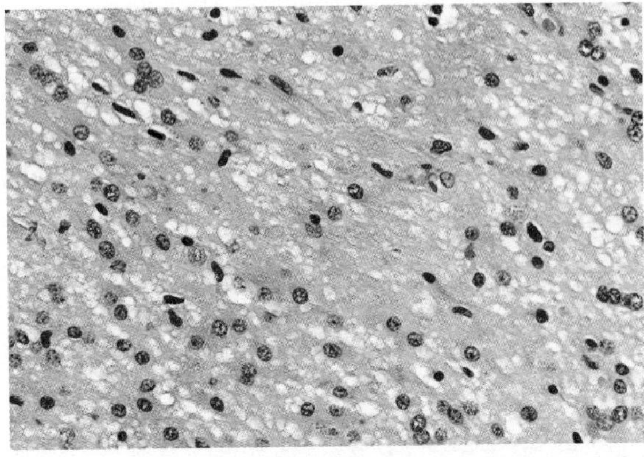

Figures 3-252

EPENDYMOMA

Perivascular pseudorosettes can be absent in astrocytoma-like areas.

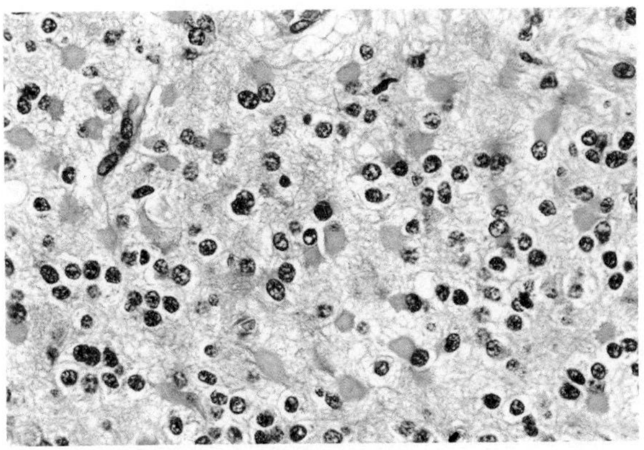

Figure 3-253

EPENDYMOMA

Small gemistocyte-like cells in a fibrillar background create the illusion of astrocytoma.

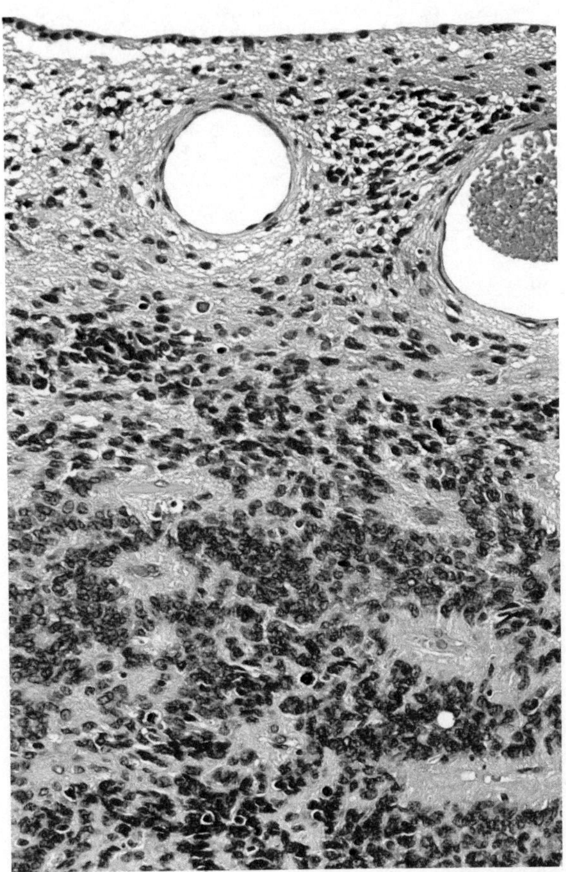

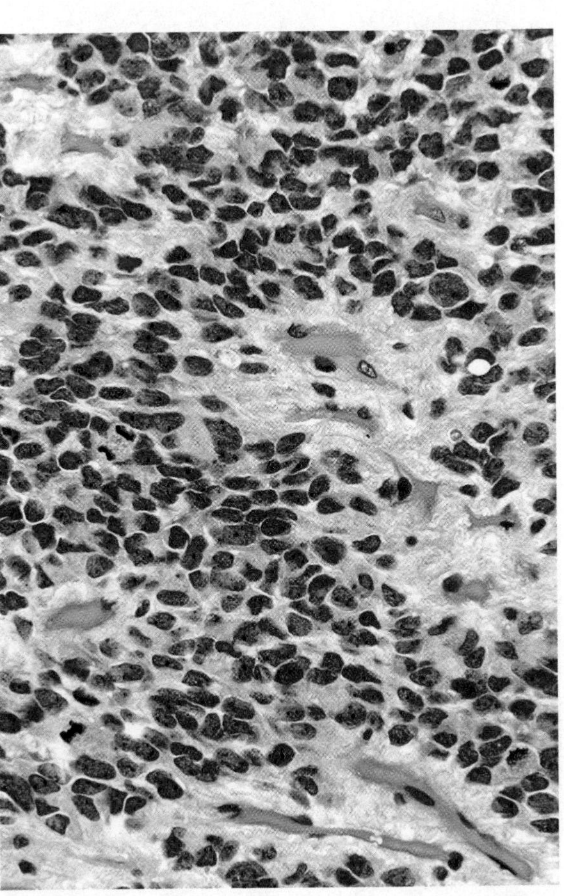

Figure 3-254

EPENDYMOMA

Left: Hypercellular nodules are common in intracranial ependymomas. The surface layer of normal ependyma is often present in ependymomas in the fourth ventricle.

Right: Both mitotic rate and degree of cytologic atypia are higher within the nodules.

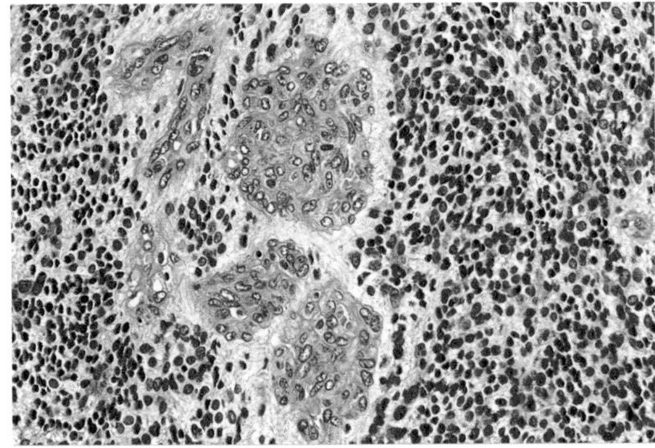

Figure 3-255

ANAPLASTIC (GRADE III) EPENDYMOMA

Grading of ependymomas remains controversial. To us, densely cellular, mitotically active ependymomas, especially with overt vascular proliferation are grade III.

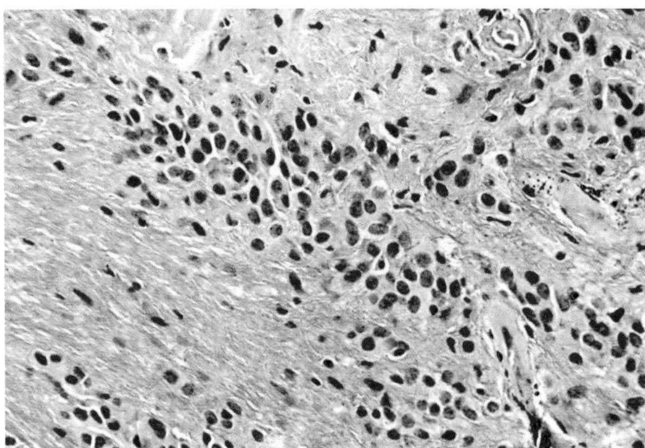

Figure 3-256

EPENDYMOMA OF THE SPINAL CORD

Cellular ependymomas of the spinal cord share the classic features of intracranial lesions, but perivascular pseudorosettes may be less obvious.

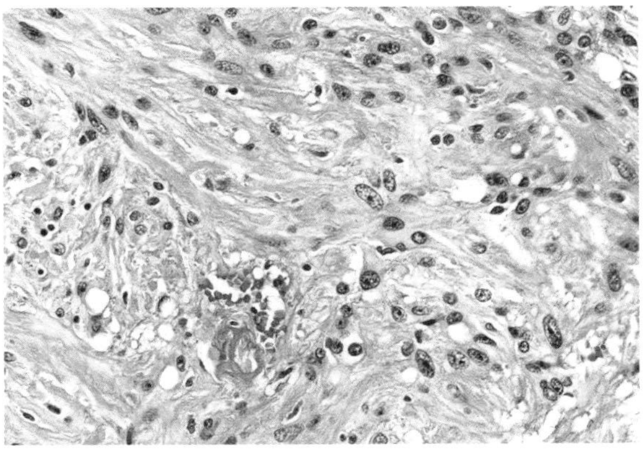

Figure 3-257

EPENDYMOMA OF THE SPINAL CORD

Some tumors are composed of unstructured fibrillar tissue that resembles schwannoma or astrocytoma.

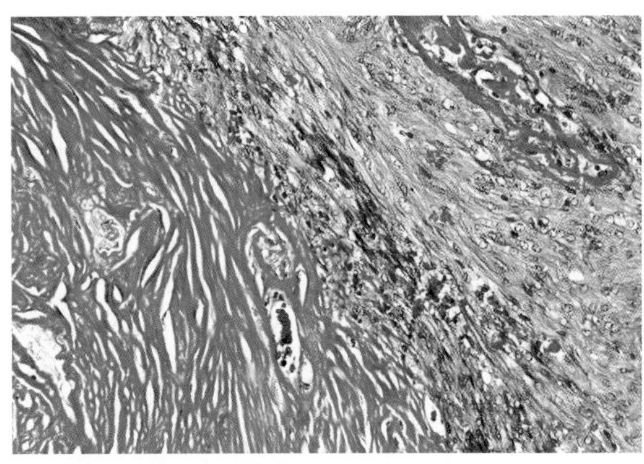

Figure 3-258

EPENDYMOMA OF THE SPINAL CORD

Densely collagenized nodules appear in some spinal ependymomas (Masson trichrome stain).

vascular hyperplasia also may be present throughout ependymomas that are diffusely highly cellular and cytologically atypical.

Spinal ependymomas of the cord proper (myxopapillary lesions of the filum terminale are discussed separately below) are of the cellular or tanycytic (described below) type. Cellular lesions share features with intracranial counterparts, but can be somewhat less well circumscribed, with tongues of tissue that create an ill-defined margin (fig. 3-256). Perivascular pseudorosettes may be inconspicuous (fig. 3-257) and schwannoma or meningioma is often entertained as a diagnosis. Epithelial surfaces, including true ependymal rosettes, are uncommon. Spinal intramedullary ependymomas may also contain nodules of dense, collagenous tissue (fig. 3-258) (418). Intramedullary ependymomas often induce a dense piloid gliosis that can be confused with pilocytic astrocytoma (fig. 3-259). Anaplastic (grade III) examples are rare. Hemosiderin deposits are common.

Epithelial features are variably, and uncommonly, expressed as "true ependymal rosettes" (fig. 3-260), canals, or as simple or elaborate expanses of neoplastic ependymal epithelium.

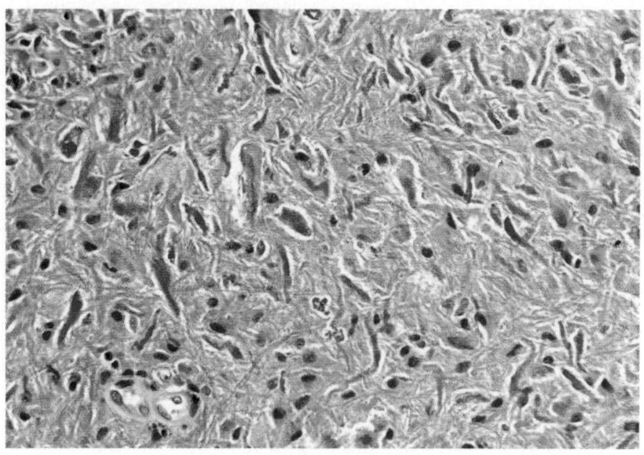

Figure 3-259

PILOID GLIOSIS AROUND SPINAL EPENDYMOMA

Like other chronic, expansile, intramedullary lesions, ependymomas induce piloid gliosis that is rich in Rosenthal fibers. In contrast to pilocytic astrocytoma, microcystic change is lacking. Hemosiderin deposition is common.

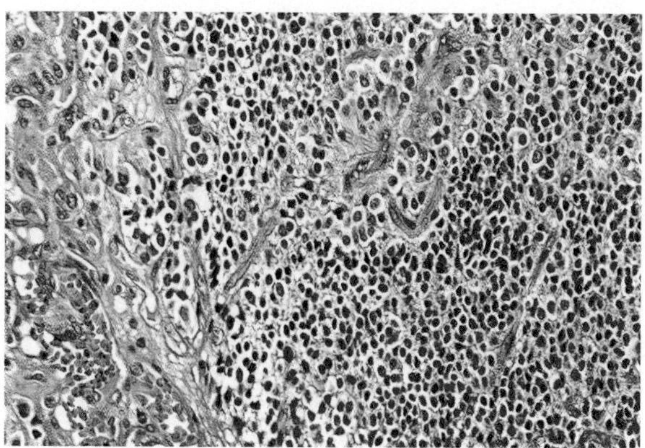

Figure 3-261

CLEAR CELL GRADE III EPENDYMOMA

Round cells with perinuclear halos give this ependymoma variant its resemblance to oligodendroglioma. As is often the case, the lesion was grade III on the basis of increased mitotic activity and the presence of vascular proliferation.

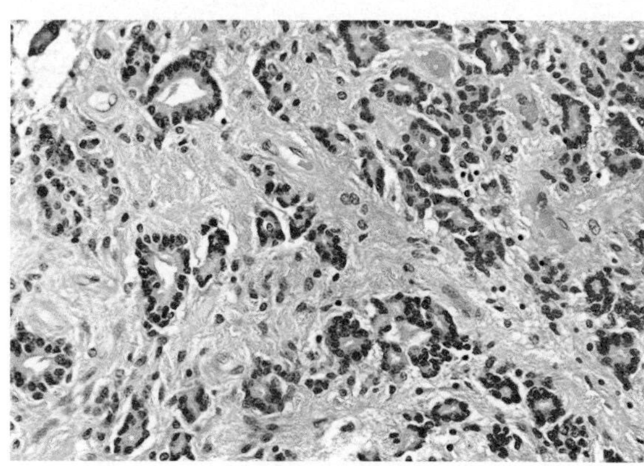

Figure 3-260

EPENDYMOMA

True ependymal rosettes are present in a minority of ependymomas.

As a rule, rosettes and canals of varying size and configuration are present in ependymomas with considerable background fibrillarity, but not usually in tumors that are densely cellular. In addition, normal ependyma often overlies the ventricular surface of any ependymoma, particularly those in the fourth ventricle (fig. 3-254, left). True rosettes range from easily overlooked microscopic clusters of cells with small lumens to large tubules or canals that are inescapably epithelial. At the extreme, some "lumens" are only minute, darkly eosinophilic, weakly PAS-positive, intracytoplasmic inclusions (419,420). Others are so small that immunohistochemistry for EMA or CD99 is necessary for detection, as is discussed below in the section on immunohistochemistry (see fig. 3-281).

Papillary ependymomas or tumors exhibiting extensive surface epithelium are rare and do not usually assume the frond-like, overtly papillary quality of choroid plexus papilloma. They also lack the smooth contoured, light microscopically apparent basement membranes found in choroid plexus tumors, and the collagenous stroma of the latter.

In otherwise typical ependymomas, focal perinuclear halos, in concert with pronounced nuclear uniformity, create a distinctly oligodendroglial appearance. The term *clear cell ependymoma* is applied when these qualities are ubiquitous and fully expressed (figs. 3-261, 3-262) (421–423). Based on mitotic activity and the presence of microvascular proliferation, these tumors are often grade III. The nuclei are typically larger and rounder than those of the other ependymomas. Unlike those of oligodendrogliomas, nuclei of clear cell ependymomas are often clefted or contain pseudoinclusions.

The *tanycytic ependymoma* is an unusual variant that occurs in either brain or spinal cord, more often the latter, and expresses the cellular elongation typical of a specialized form of ependymal cell (419,424,425). This tumor thus differs from classic ependymoma because of its

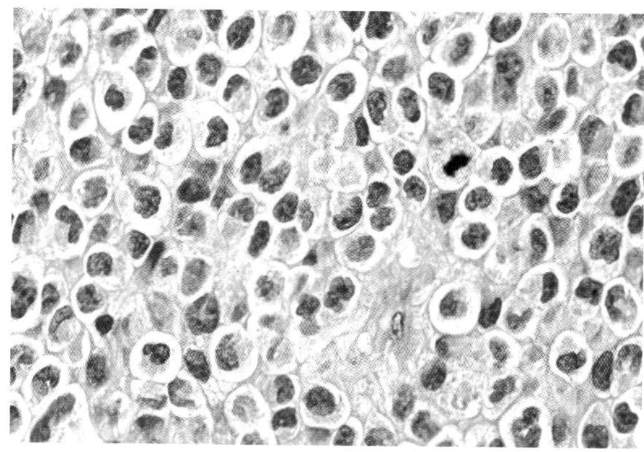

Figure 3-262

CLEAR CELL GRADE III EPENDYMOMA

Nuclear roundness and perinuclear halos create a close resemblance to oligodendroglioma. The nuclei of the clear cell ependymoma are usually clefted or grooved, however.

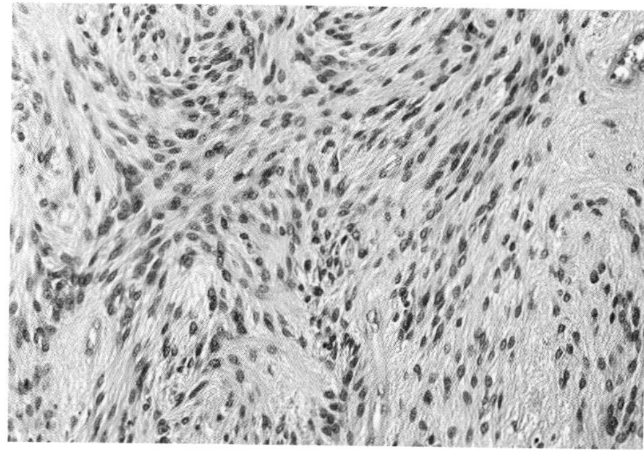

Figure 3-263

TANYCYTIC EPENDYMOMA

Tanycytic ependymomas are fascicular lesions, with only a hint of perivascular pseudorosettes.

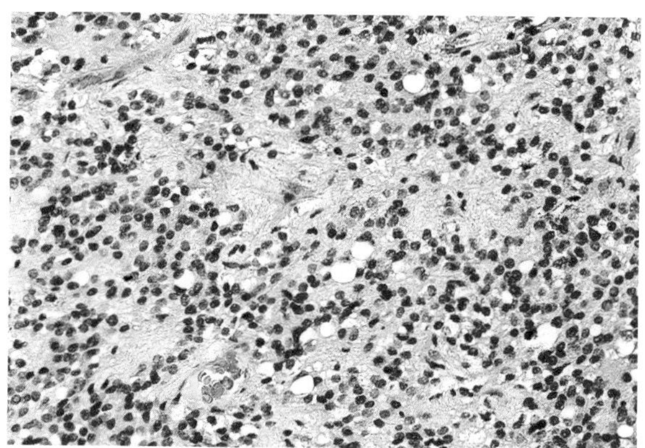

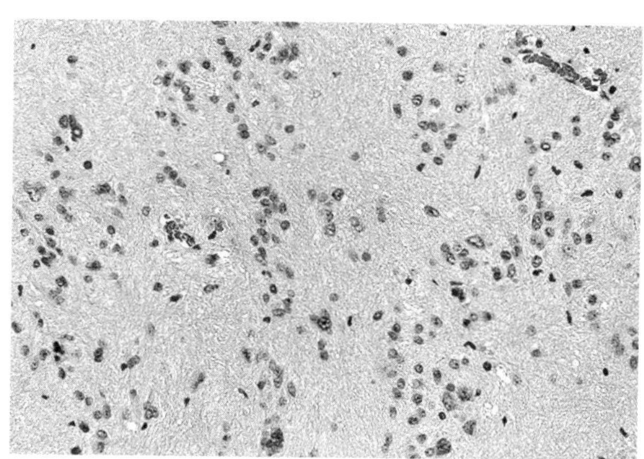

Figure 3-264

COMBINED EPENDYMOMA/SUBEPENDYMOMA

Both ependymoma (left) and subependymoma (right) tissue coexist in some tumors.

bipolar cells with markedly long, highly fibrillated processes. The result is a lesion that somewhat resembles pilocytic astrocytoma or schwannoma (fig. 3-263). Perivascular pseudorosettes, usually ill-defined, are best seen at low magnification. Tanycytic lesions are well differentiated and mitoses are rare.

Particularly in the fourth ventricle of adults, some ependymomas have focal clustering of nuclei in a fibrillar background, i.e., the pattern of subependymoma, in addition to the classic ependymoma tissue (fig. 3-264). These *subependymoma/ependymoma hybrids* are discussed in the next section.

Ependymomas share with a close relative, the choroid plexus papilloma, a rarely expressed capacity for the formation of true melanin (426). In most instances, however, the pigment is a dark, coarsely granular, lipochrome-like substance (lipofuscin) that, although argyrophilic, stains strongly with PAS (427). Heavily *lipidized ependymomas* comprise yet another distinctive variant (fig. 3-265) (415–417). Only occasional ependymomas contain bone or cartilage. Uncommonly, ependymomas contain cells with eosinophilic lysosome-like cytoplasmic granules (fig. 3-266). Ependymomas are variably calcified. An *epithelioid variant* has a prominent clusters of

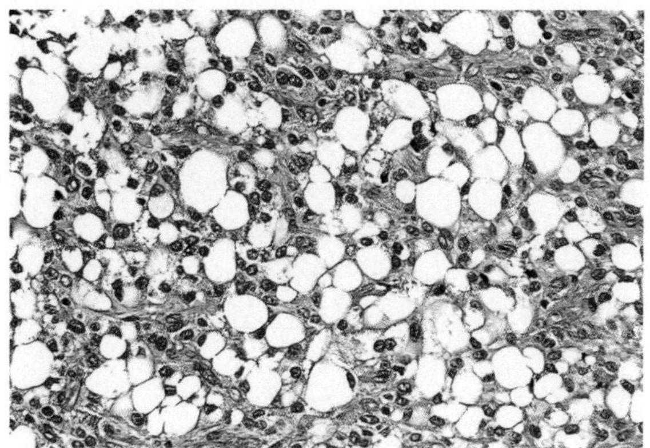

Figure 3-265

EPENDYMOMA

A rare ependymoma is lipidized.

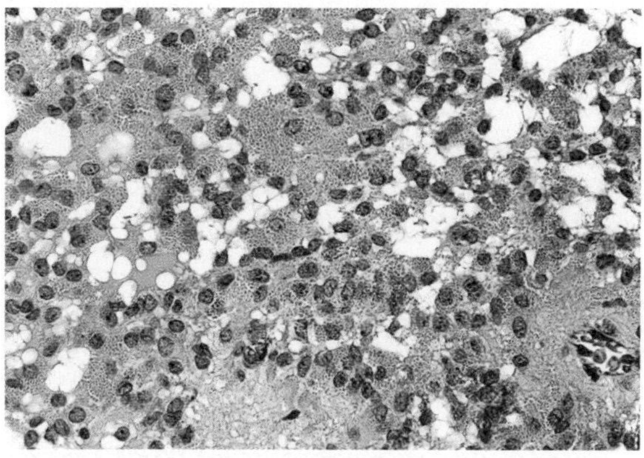

Figure 3-266

EPENDYMOMA

Intracytoplasmic eosinophilic granules are present in some ependymomas.

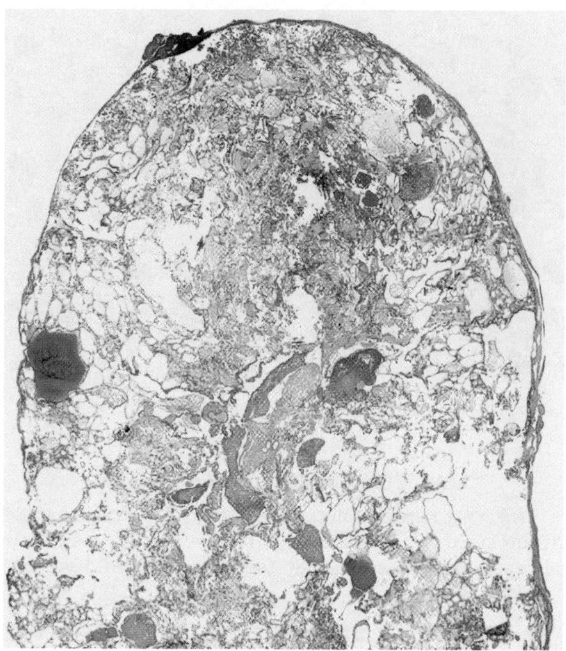

Figure 3-267

MYXOPAPILLARY EPENDYMOMA

As illustrated in a whole mount histologic section, myxopapillary ependymomas are delicately encapsulated, and often variable in tissue density.

cells (428). A provocative feature of rare ependymomas is the synaptophysin-immunoreactive "pale island" similar to that seen in nodular medulloblastomas (429).

Myxopapillary ependymomas are distinctive at low magnification for their delicate capsule and histologic diversity, the latter noteworthy for the pseudopapillary architecture, perivascular and intercellular mucin deposition, and tendency to cellular elongation (figs. 3-267, 3-268) (430,431). Adherence of the cells to the vessels produces a somewhat papillary appearance, but true papillae are infrequent (fig. 3-269). In the conventional lesion, neoplastic cells are either columnar (epithelial) or elongate (glial) and show only minor variation in nuclear shape, size, and degree of chromasia. In some cases, the glial quality produced by elongated cells in a fibrillary background predominates. Such tumors may resemble schwannoma. Mucin, readily highlighted by the PAS or Alcian blue stains (fig. 3-270), is characteristically present near or in the walls of blood vessels, but can accumulate within intercellular microcysts or as extensive extravascular pools that separate the neoplastic cells in a pattern resembling chordoma. It is only focally present in some lesions, and may not be apparent in compact areas of tumors that are elsewhere classically myxopapillary and papillary (fig. 3-271). Some fascicular tumors of the lower spinal cord are free of myxoid substance, at least focally (fig. 3-272), and neither myxopapillary nor papillary (fig. 3-273). An additional feature of myxopapillary lesions, typically those that are less papillary and more solid and, is the presence of round structures (fig. 3-274), which despite their prickly appearance on reticulin stains (fig. 3-275), have been referred to as "balloons."

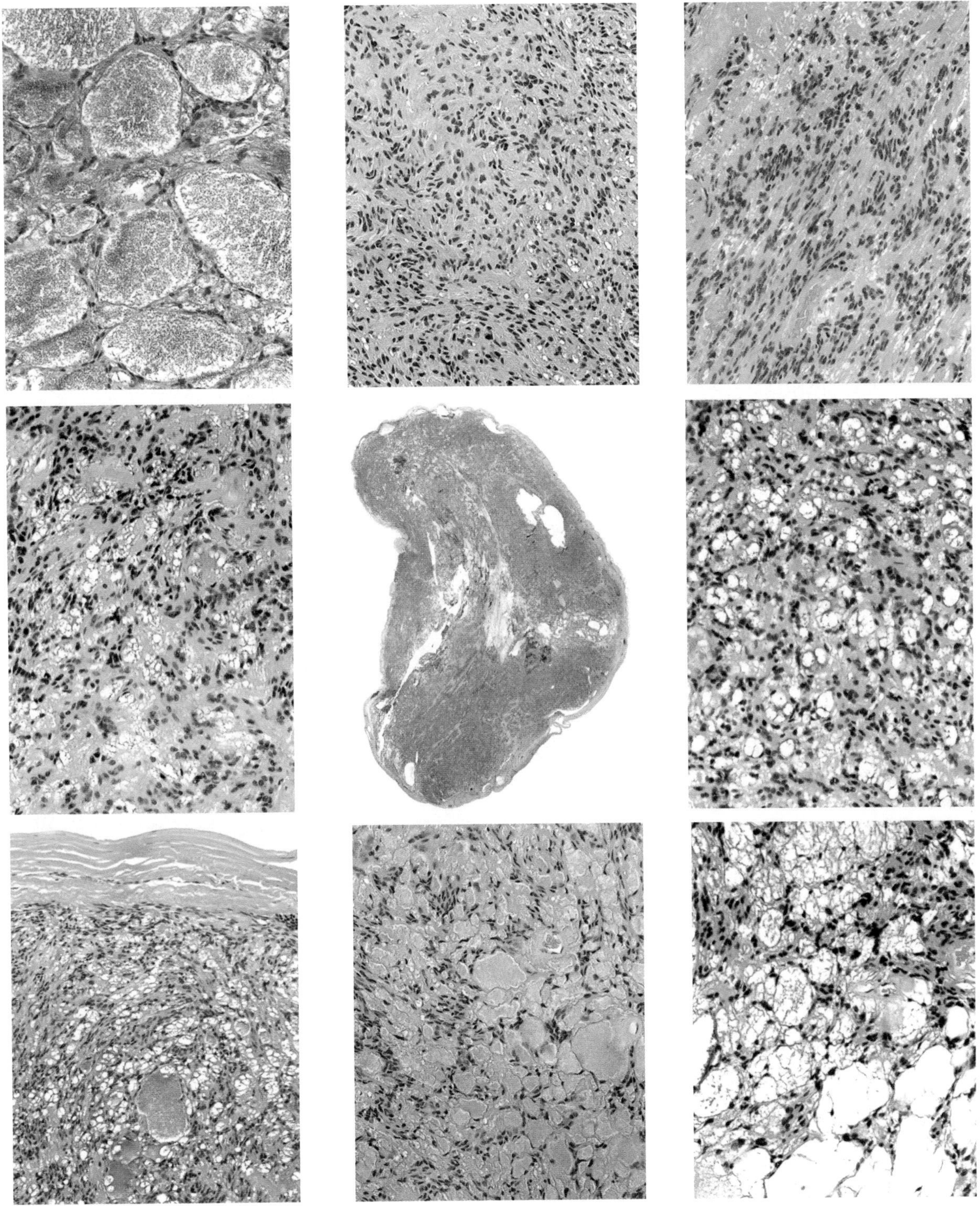

Figure 3-268

MYXOPAPILLARY EPENDYMOMA

The eight panels record the scope of patterns present within the neoplasm illustrated in a whole mount section at the center of the figure. The panels at top left and bottom right show the classic microcystic areas rich in myxoid material. Others demonstrate mucinous but less microcystic regions, and even areas with few myxoid or papillary features.

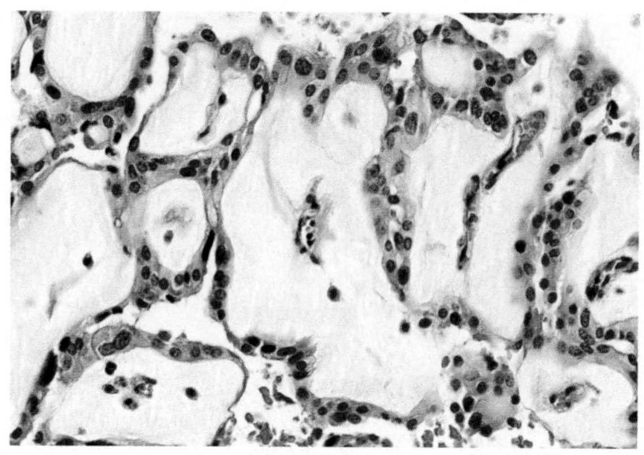

Figure 3-269

MYXOPAPILLARY EPENDYMOMA

Epithelial cells and myxoid material create the classic myxopapillary appearance. As is characteristic, mucin is deposited as a collar around the vessels.

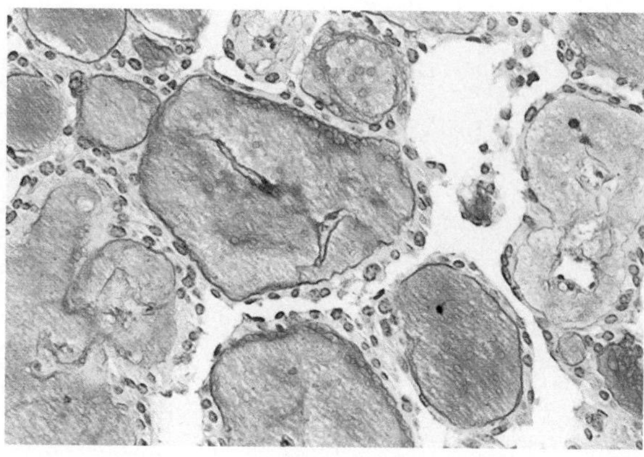

Figure 3-270

MYXOPAPILLARY EPENDYMOMA

The perivascular myxoid material is Alcian blue positive.

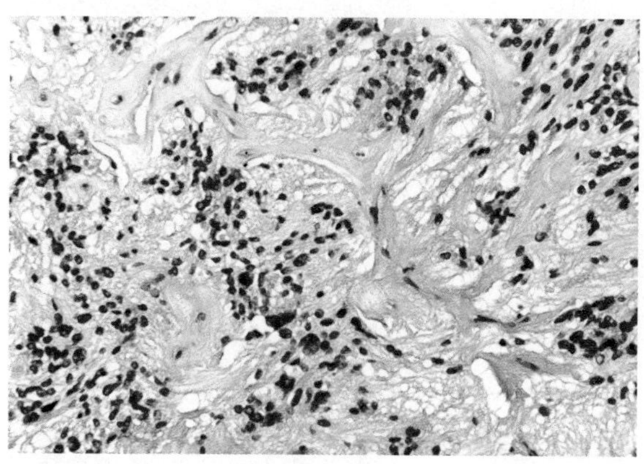

Figure 3-271

EPENDYMOMA

Some ependymomas of the filum terminale have only vague, if any, perivascular pseudorosettes or suggestion of myxoid substance.

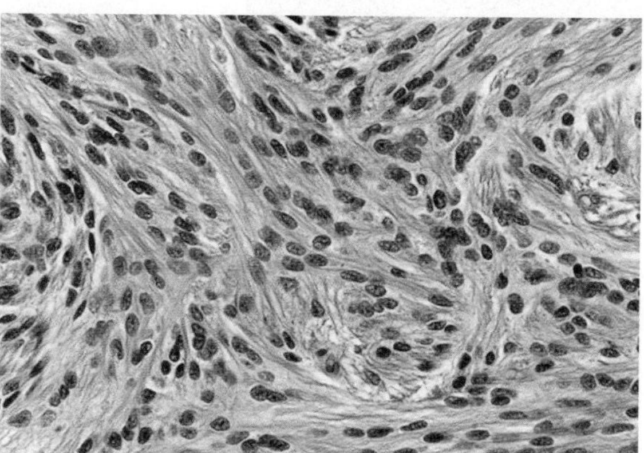

Figure 3-272

EPENDYMOMA

Some caudal ependymomas are fascicular, with few features of ependymoma and none of the myxopapillary variant.

A variant of ependymoma that occurs in the filum terminale (432), rostral spinal cord (433), is a *giant cell ependymoma* known for varying, sometimes severe, degrees of nuclear pleomorphism. This variant is most common in the region of the filum where it occurs as an obviously degenerative phenomenon in an otherwise typical myxopapillary lesion (434). The tumor has a rather compact epithelioid quality that raises the possibility of a high-grade neoplasm such as giant cell glioblastoma, but has few mitoses (fig. 3-276).

The same term, giant cell ependymoma, has been applied as well to intracranial anaplastic ependymomas with an unusual degree of pleomorphism (435,436). Because of their anaplasia, these are to be distinguished from the just-described low-grade spinal lesions with only degenerative atypia.

A papillary neoplasm that, in some respects, combines the histologic features of ependymoma and choroid plexus papilloma has been described as *papillary tumor of the pineal region* (see figs. 6-14, 6-15) (437). The nosologic status of the lesion

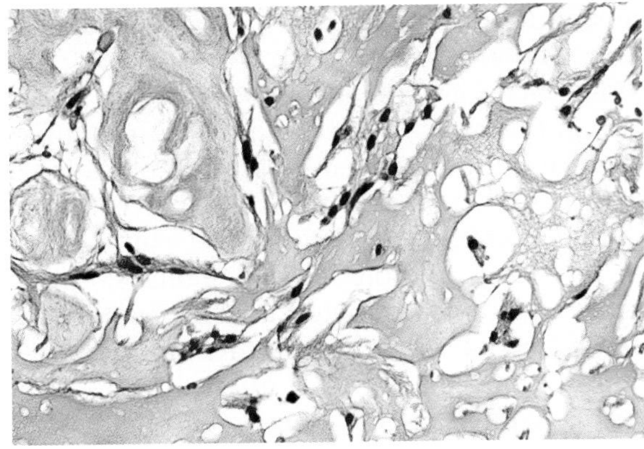

Figure 3-273

MYXOPAPILLARY EPENDYMOMA

Dense collagenous tissue crowds out tumor cells in some longstanding lesions.

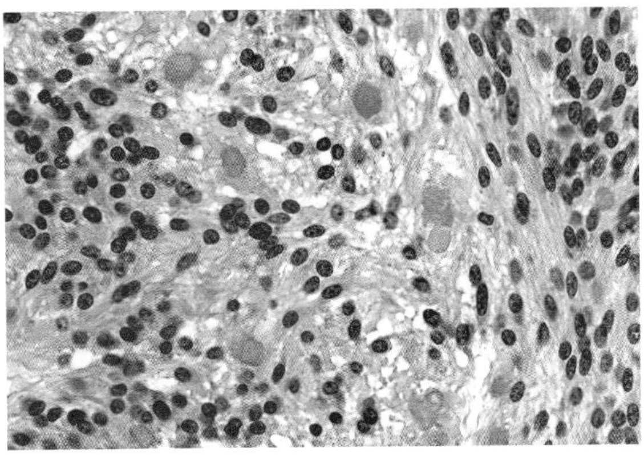

Figure 3-274

"BALLOONS" IN MYXOPAPILLARY EPENDYMOMA

Small, round, hypereosinophilic and faintly fibrillar structures known as "balloons" are present in some myxopapillary ependymomas.

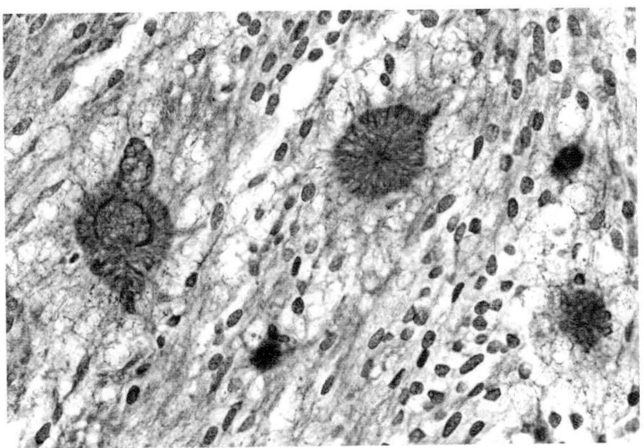

Figure 3-275

BALLOONS IN MYXOPAPILLARY EPENDYMOMA

Balloons are reticulin positive.

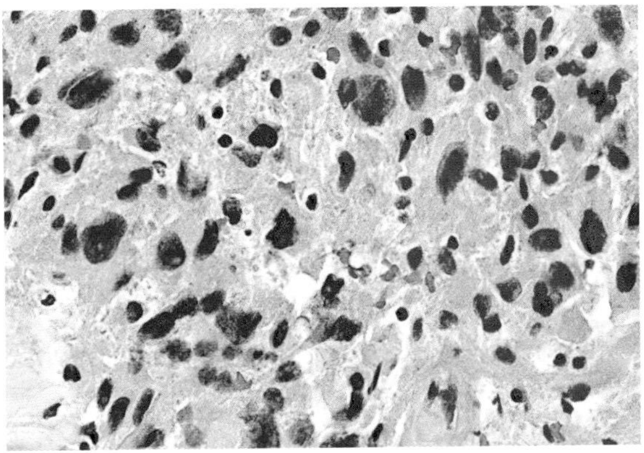

Figure 3-276

GIANT CELL EPENDYMOMA

Considerable pleomorphism and cytomegaly, but no features of anaplasia, define the giant cell variant.

is unclear; the authors describing the lesion speculated about an origin from specialized ependymal cells of the subcommissural organ.

Grading. Since, for practical purposes, myxopapillary lesions, considered WHO grade I, do not undergo anaplastic change, attempts at grading ependymomas have focused upon lesions of the brain and spinal cord proper. Well-differentiated ependymomas in these sites are WHO grade II. Grading is confounded by a number of factors, including: 1) the need to consider supratentorial, infratentorial, and spinal lesions separately; 2) differences in reported outcomes between patients with myxopapillary ependymomas and ependymomas of the spinal cord proper; 3) variable definition of grades in published series; 4) focality of anaplasia in many lesions; 5) prognostic influence of extent of tumor resection; 6) improved surgical techniques that increase the chance of gross total resection; and 7) relatively long follow-up periods necessary to assess the biologic behavior of ependymomas.

These factors aside, there is convincing evidence that highly cellular lesions with brisk mitotic activity as well as microvascular proliferation are more likely to recur (406,438–441), but the issue continues to be debated (442). We consider such lesions as anaplastic, and assign a grade III designation in accordance with the recommendations of the WHO (443). The well differentiated lesion is grade II. While the case for grade III appears to be stronger in the presence of vascular proliferation, high cellularity and more than focal brisk mitotic activity is often sufficient for the diagnosis of anaplastic ependymoma (grade III). Most are intracranial. Spinal ependymomas are usually not anaplastic.

It is generally agreed that cytologic atypia in the absence of these other features is of no prognostic significance. The same is true of necrosis, which is common even in very well-differentiated lesions.

The focality of anaplasia in cellular ependymoma confounds the utility of a simple grading system. As is illustrated in figure 3-248, ependymomas frequently contain small regions of hypercellularity in which perivascular pseudorosettes are less noticeable, and cytologic atypia, nuclear to cytoplasmic ratios, and mitotic rate are all increased. Arcades of proliferating vessels with plump, often multilayered, vascular elements may be seen within or just outside these areas. Ependymomas can readily be assigned to the anaplastic or malignant category if such cellular foci predominate and vascular proliferation is present. The significance of a small focus of anaplasia is unclear, although it may contribute to a somewhat poorer prognosis (444). We accept occasional cellular and atypical foci within the spectrum of well-differentiated ependymoma but can suggest no clear threshold in extent or number beyond which the lesion becomes anaplastic, or grade III.

Only a rare ependymoma progresses to a state of malignancy replete with microvascular proliferation and pseudopalisading necrosis. Since such ependymal tumors are usually circumscribed rather than diffusely invasive, and since the prognosis of patients with these tumors is less predictable than that of those with astrocytic tumors with vascular proliferation and pseudopalisading necrosis, we consider them to be highly anaplastic ependymomas, not glioblastomas. We reserve the latter designation for grade IV astrocytic neoplasms of the infiltrating, diffuse type.

Prognostically meaningful assessment of intracranial examples by MIB-1 indices is supported by several studies, although varying thresholds predictive of recurrence have been reported and the sensitivity of MIB-1 staining may vary from institution to institution. Regional heterogeneity also affects staining indices. In one study, there were significant differences in cumulative survival rates of patients with ependymomas with MIB-1 indices of less than 25 percent versus with those with indices of 25 percent or more (445). A similar division, at 20.5 percent, was determined in another investigation (446). In yet another study, levels above 20 percent were significantly associated with a poorer survival (441). In another report, including myxopapillary lesions, this threshold was considerably lower, at 9 percent, for both overall and progression-free survival (438).

Myxopapillary lesions are considered grade I, although they occasionally recur and even seed the subarachnoid space. The role of MIB-1 indices in prognosis is not clear, but in one small study, the MIB-1 rate was not a reliable predictor of recurrence (434). Occasional myxopapillary ependymomas are mitotically active, have vascular proliferation, and exhibit a high MIB-1 index (447). The significance of these findings for a patient whose lesion has been totally excised is unclear.

Frozen Section. Ependymomas with an epithelial surface are readily recognized in frozen section specimens, but tumors that are less cellular, more fibrillar, and devoid of epithelial features may resemble astrocytoma. Resolution of this diagnostic dilemma is important, since the pathologist's intraoperative diagnosis of an ependymoma will start the surgeon on an exacting course of complete resection, particularly for a lesion in the spinal cord, while balancing this goal against the risk of neurologic deficit. The diagnosis of astrocytoma in the latter setting, especially of the diffuse type, on the other hand, will often bring the operation to an immediate halt since any treatment is radiotherapeutic, not surgical.

Erroneous diagnoses can generally be avoided by viewing the lesion at low magnification, the view in which perivascular pseudorosettes are most apparent (fig. 3-277). Higher magnification is useful for documenting the generally uniform nuclei and lack of overrun, preexisting parenchyma, e.g., neurons and glia. Cytologic features should not be entirely relied upon for diagnosis since neoplastic ependymal cells often look remarkably astrocytic in frozen sections (fig. 3-

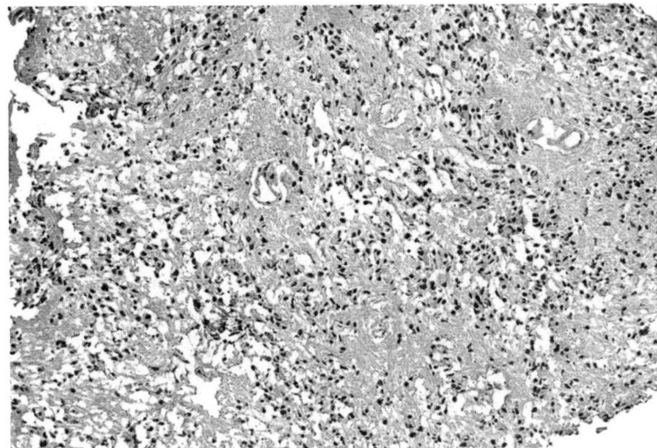

Figure 3-277

EPENDYMOMA: FROZEN SECTION

Perivascular pseudorosettes are recognized best at low magnification, but even then can be indistinct.

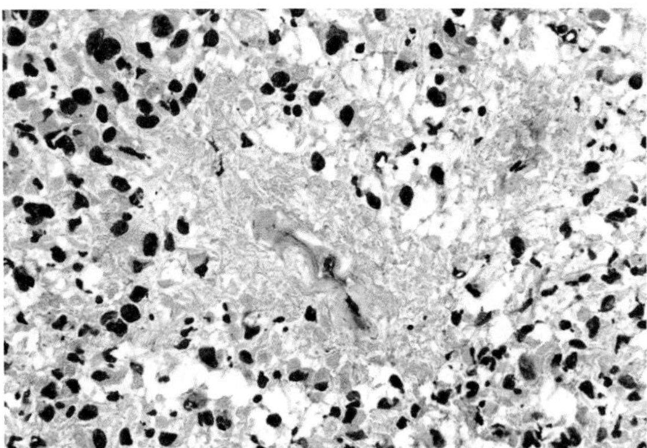

Figure 3-278

EPENDYMOMA: FROZEN SECTION

Pleomorphic nuclei and prominent cytoplasm often create the appearance of astrocytoma in frozen section specimens of ependymoma. Perivascular pseudorosettes are distinguishing features.

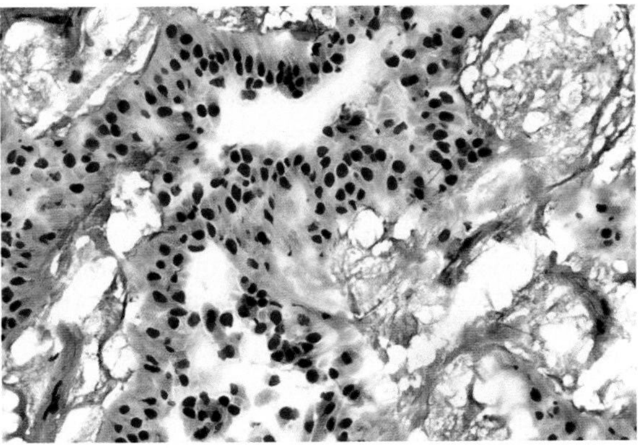

Figure 3-279

MYXOPAPILLARY EPENDYMOMA: FROZEN SECTION

Diagnostic myxopapillary features are often retained in frozen tissues.

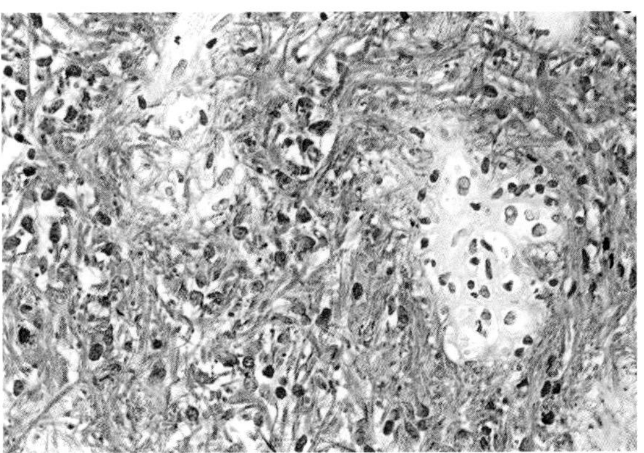

Figure 3-280

EPENDYMOMA

Perivascular pseudorosettes are dependably immunoreactive for GFAP.

278). The epithelial-myxoid dichotomy of the myxopapillary lesion is usually well preserved (fig. 3-279), but some lesions are remarkably nonspecific in their histologic appearance.

Immunohistochemical Findings. All differentiated ependymomas, including those of the myxopapillary type, are characterized by immunoreactivity for GFAP and S-100 protein (fig. 3-280). The former is particularly prominent in fibrillar areas, including perivascular pseudorosettes, in which at least some cell processes are almost always positive. True rosettes, on the other hand, are largely nonreactive. Immunoreactivity for epithelial markers such as EMA and cytokeratins is present on the epithelial surface of some ependymomas (fig. 3-281), or as short segments between adjacent cells. The diffuse cytoplasmic positivity for cytokeratins present in choroid plexus papillomas is lacking. Cytokeratin staining for AE1/AE3 parallels that of GFAP, but staining for other cytokeratins such as CK7, CAM5.2, CK903, and CK20 is focal at best (448). Diffuse staining for these latter markers is not consistent with a diagnosis of ependymoma. Diffuse membrane staining for CD99 has also been reported (449).

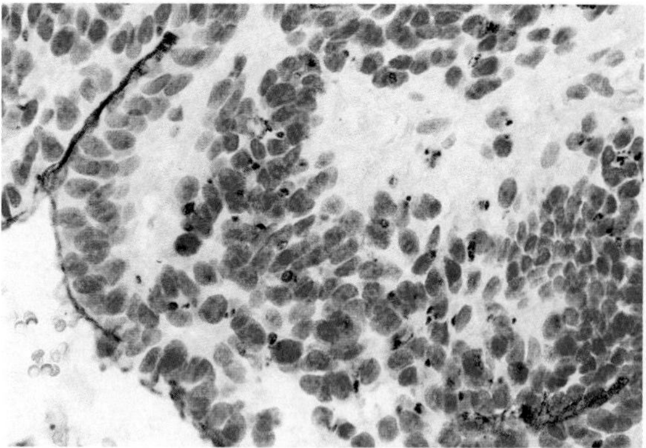

Figure 3-281

EPENDYMOMA

Epithelial surfaces and intracytoplasmic microlumens are sometimes immunoreactive for epithelial membrane antigen (EMA).

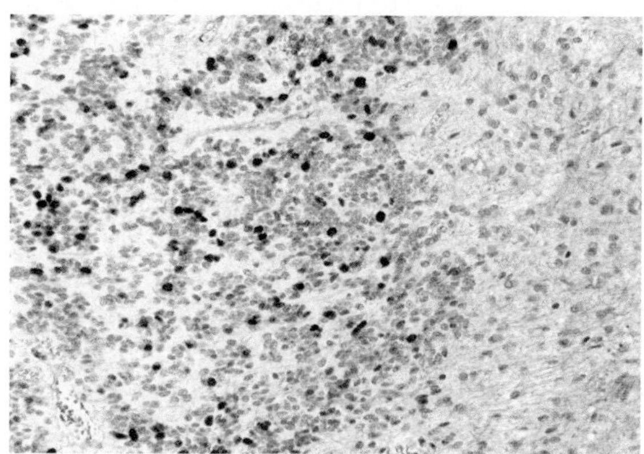

Figure 3-282

EPENDYMOMA

Hypercellular islands have a high MIB-1 index.

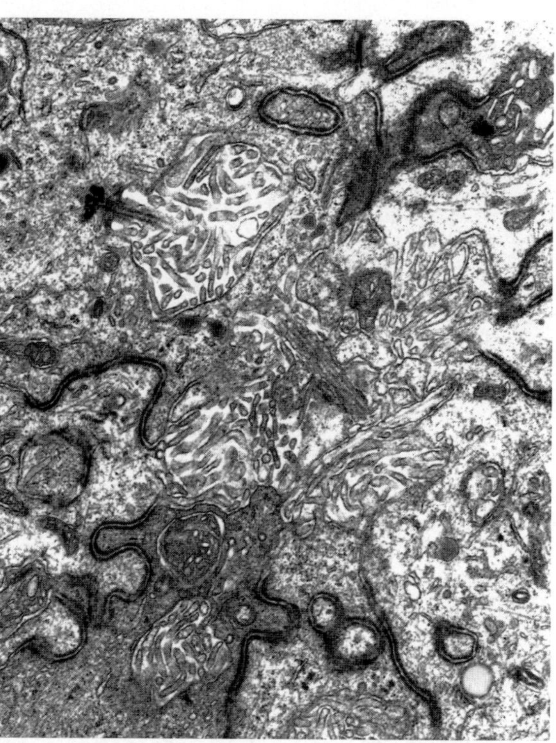

Figure 3-283

EPENDYMOMA

The classic ultrastructural features of ependymoma include extensive, tandem, "zipper-like" intermediate junctions; microvilli-filled lumens; and in this case, a single cilium. The microlumens can be visualized by light microscopy after immunostaining for EMA or CD99.

A distinctive form of staining occurs on the surface of intracellular microlumens with antibodies to EMA (fig. 3-281) and CD99 (420,449–451). At first glance, these structures may seem too small to be real, but they are common, and diagnostically very helpful.

MIB-1 indices vary greatly, as is discussed in the sections on Grading and Prognostic Factors. Rates are considerably higher in the cellular, mitotically active nodules of the cellular variant (fig. 3-282).

Ultrastructural Findings. Apposed cell membranes of ependymoma cells are joined by multiple "zipper-like" junctions without inserting tonofilaments (fig. 3-283). Well-formed terminal bars are evident at apical or luminal portions of cells where ependymal rosettes or canals are formed. The luminal surfaces bristle with microvilli and scattered cilia. Minute lumens having these apical specializations may also contain granular-tubular material (450). Complex desmosomal junctions, microvilli, and cilia leave no doubt as to the ependymal, rather than oligodendroglial, nature of the oligodendroglial-like cells in clear cell ependymomas (452).

Myxopapillary lesions are noted for cellular elongation, paucity of cilia, extensive basal lamina, microtubular aggregates, and flocculent intercellular and perivascular mucin (429,430). Microtubular aggregates within rough endoplasmic reticulin have been reported (453).

True to form, tanycytic lesions have classic ependymoma features: prominent intercellular junctions, microvilli, and minute lumens filled with microvilli and occasional cilia (425).

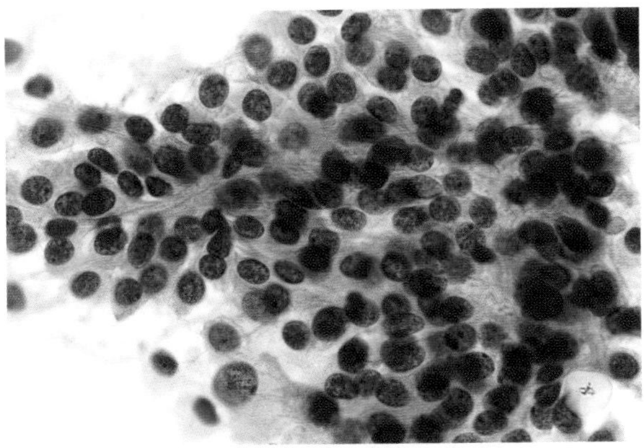

Figure 3-284

EPENDYMOMA: SMEAR PREPARATION

Cells of ependymoma are more cohesive than those of oligodendroglioma or infiltrating astrocytoma. The round, uniform, slightly ovoid nuclei with small nucleoli are typical.

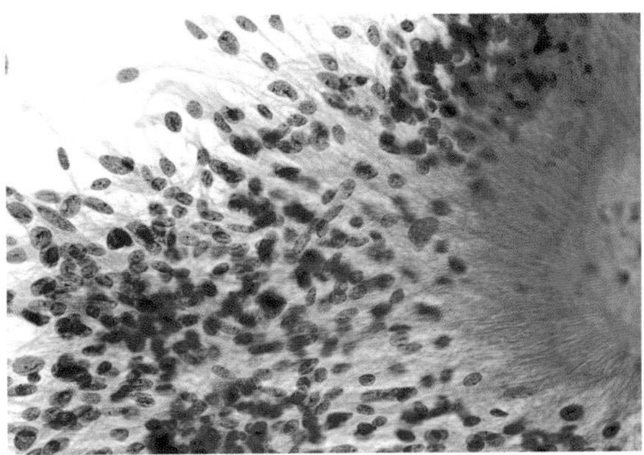

Figure 3-285

EPENDYMOMA: SMEAR PREPARATION

Cells are oriented to blood vessels in this myxopapillary lesion.

Cytologic Findings. Smear or touch preparations of cellular ependymoma demonstrate monomorphous small cells with generally uniform, dark, somewhat elongate nuclei and a small amount of cytoplasm (fig. 3-284) (454). There is often cohesion and a tendency to perivascular arrangement of cells and their processes. The latter is a prominent feature in pseudorosette-rich tumors as well as in myxopapillary ependymomas, which also frequently have a myxoid background (fig. 3-285) (455,456). Hyaline globules surrounded by tumor cells are common. Minute intracellular microlumens are occasionally visible in nonmyxopapillary lesions (457). The fascicular architecture of the tanycytic variant somewhat resembles that of pilocytic astrocytoma and schwannoma (458). Extracellular myxoid substance, often perivascular, is common in myxopapillary lesions.

Molecular and Cytogenetic Findings. Gene expression and molecular profiles of ependymomas vary by the tumors' anatomic sites, and similarities to radial glia point to the latter as a likely precursor, stem, cell (458a). Noteworthy in all studies has been the frequent loss of the chromosomal arm 22q in both intracranial and spinal cord tumors, but especially the latter (459–461). This is not surprising given the occurrence of spinal ependymomas in patients with neurofibromatosis 2 (462). In one study, *NF2* mutations in sporadic ependymomas were restricted to spinal lesions (459). In the same study, loss of heterozygosity in nonspinal ependymomas was not associated with mutation of the retained *NF2* allele. One study of supratentorial clear cell ependymomas found no deletions of *NF2*, but rather deletions of *DAL-1* (18p11.3) in the anaplastic subset of clear cell ependymomas (423). Thus, despite a high incidence of chromosome 22 loss, abnormalities of the *NF2* gene are not invariable. Other candidate genes are being sought (463).

Multiple abnormalities in other chromosomal arms have also been reported, although are not precisely the same in each study. One series of intracranial tumors found frequent gain of 1q and losses on 6q, 9, and 13, the latter two being more frequent in grade III lesions (460). In the same study, gain on chromosome 7 was found exclusively in spinal tumors. Yet another study also found frequent losses on 6q, 4q, 10, and 2q.

Chromosomal imbalances related to tumor progression are not clear. Although losses of chromosomal arms 9p and 13q have been suggested, one study found no relationship between these and histologic grade or biologic behavior (464).

A high incidence of loss on 13q14-q31 has been reported in myxopapillary ependymomas in adults (461), as have concurrent gains on chromosomes 9 and 18 (462).

Differential Diagnosis. While ependymomas seem unlikely to be confused with any other lesion, they are frequent diagnostic problems, particularly in the spinal cord.

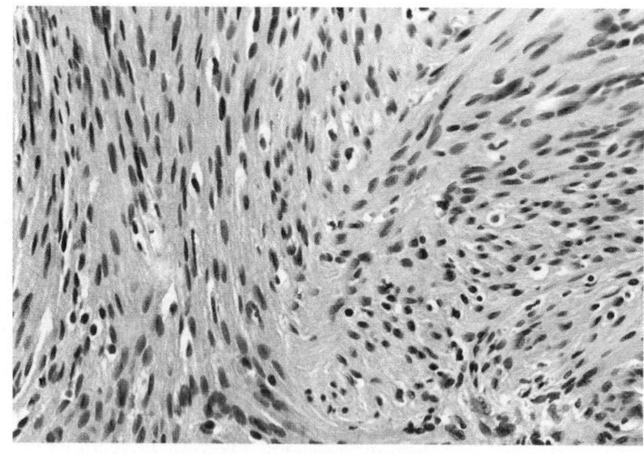

Figure 3-286

SCHWANNOMA RESEMBLING EPENDYMOMA

Some schwannomas with fascicular architecture and vague perivascular fibrillary areas resemble ependymomas.

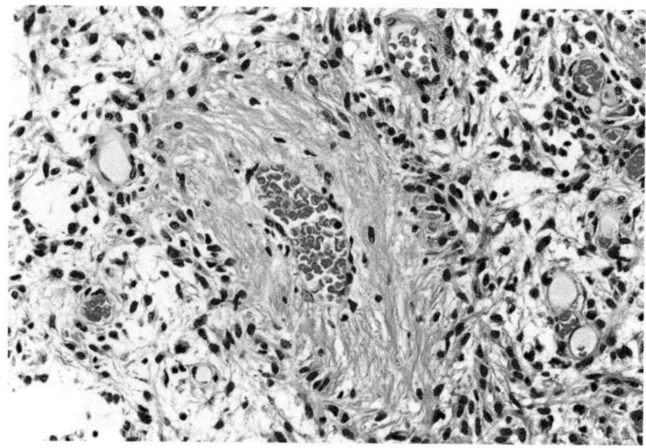

Figure 3-287

PILOCYTIC ASTROCYTOMA RESEMBLING EPENDYMOMA

Perivascular fibrillar zones are common in pilocytic astrocytoma. The loose, microcystic component is typical of pilocytic astrocytoma, not ependymoma.

Cellular ependymomas of the cord or the myxopapillary variant of the filum terminale are usually readily recognized, but there is histologic overlap with schwannoma in some cases (fig. 3-286). In ambiguous cases, schwannomas are identified in part by their diffuse, strong reactivity for S-100 protein, as is discussed in chapter 13. Ependymomas also are immunoreactive, but generally without diffuse avid staining of both nuclei and cytoplasm. Ependymomas, especially the perivascular pseudorosettes, are uniformly GFAP positive. While schwannomas are rarely positive for GFAP to the extent of ependymomas, there are exceptions. Diffuse pericellular reticulin staining or collagen type IV immunoreactivity are typical of schwannoma. The abundant basement membrane material of schwannoma, and the junctional complexes, microvilli, and cilia of ependymomas, are helpful ultrastructural features that survive routine tissue processing and paraffin-embedding. Intraspinal lesions listed by anatomic site are provided in Appendix B.

Tanycytic ependymomas, particularly when intraspinal, are most often confused with schwannoma. The differential between these two entities is discussed above. Myxoid schwannomas that resemble myxopapillary ependymomas are discussed in chapter 13. Meningioma also becomes a suspect intraspinally in the face of a solid, rather epithelioid, nonglial-appearing ependymoma. Meningiomas are EMA positive and GFAP negative.

Ependymomas at any site, but particularly in the fourth ventricle and spinal cord, may need to be distinguished from pilocytic astrocytoma, since the latter may contain perivascular formations that resemble perivascular pseudorosettes (fig. 3-287). These are especially prominent in the so-called pilomyxoid, or infantile juvenile pilocytic, astrocytoma (fig. 3-288). Both classic pilocytic astrocytoma and the pilomyxoid lesion, but particularly the latter, have a loose, spongy, microcystic architecture foreign to ependymoma. Intramedullary piloid gliosis can simulate pilocytic ependymoma (see fig. 3-259).

Neoplastic cells radiating from vessels in paragangliomas of the filum terminale can resemble those of ependymal pseudorosettes (fig. 3-289). Paragangliomas, however, are nested and reticulin rich, more epithelioid, and chromogranin and synaptophysin immunoreactive. Sustentacular cells may be reactive for GFAP, but the chief cells are largely negative. Sustentacular cells are variably S-100 protein positive.

Many cellular ependymomas, particularly those in the posterior fossa, can be confused with diffuse astrocytoma if attention is focused solely upon paucicellular fibrillar areas (see figs. 3-251, 3-252) that sometimes contain small cells closely resembling gemistocytic astrocytes (see fig. 3-253). Ependymomas are sharply defined exophytic masses that do not enlarge the parenchyma of the pons by infiltration as would be the case with a pontine diffuse astrocytoma. Even

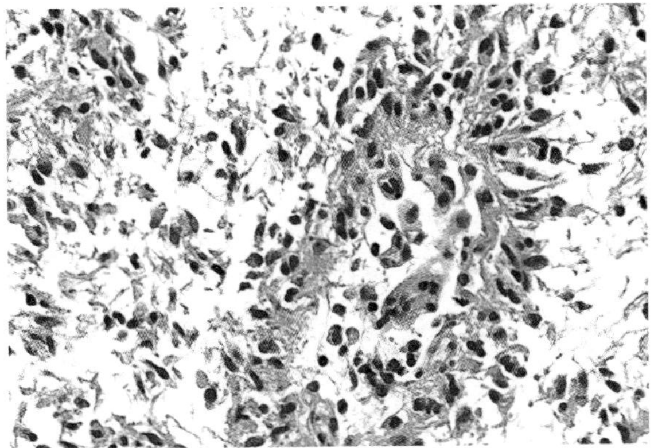

Figure 3-288

PILOMYXOID GLIOMA/INFANTILE PILOCYTIC ASTROCYTOMA RESEMBLING EPENDYMOMA

Pilomyxoid glioma, or infantile pilocytic astrocytoma, has perivascular formations similar to the perivascular pseudorosettes of ependymoma. The loose mucin-rich component with dispersed individual cells is typical of the pilomyxoid lesion, but is not expected in ependymoma.

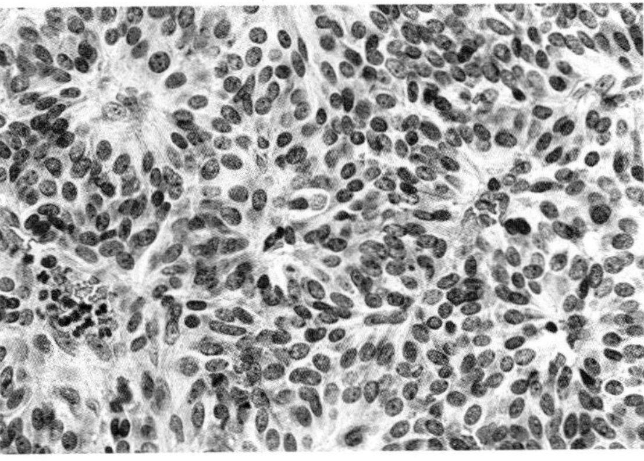

Figure 3-289

PARAGANGLIOMA OF THE FILUM TERMINALE RESEMBLING EPENDYMOMA

Compact, noninfiltrating architecture and perivascular orientation create a similarity to ependymoma.

the most fibrillar of ependymomas exhibits some cellular foci and at least a hint of perivascular pseudorosettes, and, unlike infiltrating astrocytomas, few if any intratumoral axons.

Neurocytic tumors, either central or extraventricular, share the high cellularity and, often, perivascular fibrillar zones of the perivascular pseudorosettes of cellular ependymoma (fig. 3-290, see fig. 4-46). Foci of oligodendroglial-appearing cells, common in neurocytomas, contribute to the difficulty in distinguishing this neuronal tumor from ependymoma of clear cell type. The fibrillar areas in neurocytic tumors are, however, generally larger, more finely fibrillar, more stellate, and less symmetric relative to the vessels than are the pseudorosettes of ependymoma. Since they are composed of tumor cell neurites, they are synaptophysin, rather than GFAP, positive.

Cellular ependymomas of the clear cell type closely resemble oligodendroglioma. In contrast to oligodendroglioma, however, ependymomas are discrete and largely noninfiltrating, and show at least focal GFAP-positive perivascular pseudorosettes. In addition, the ultrastructural features are clearly ependymal. Thus, despite the histologic similarity, such lesions are not oligodendrogliomas, nor do they warrant a diagnosis of mixed glioma (oligoependymoma). The finding of nuclear groves, clefts, and pseudo-

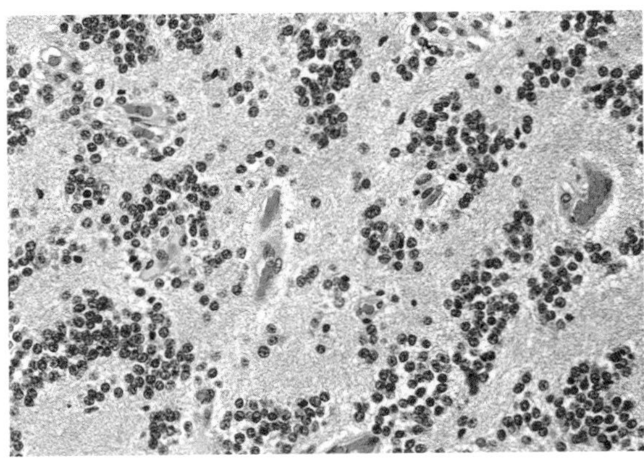

Figure 3-290

CENTRAL NEUROCYTOMA RESEMBLING EPENDYMOMA

Neurocytic tumors often have prominent perivascular fibrillar areas that closely resemble the perivascular pseudorosettes of ependymoma. The fibrillar areas of neurocytoma are positive for synaptophysin, not GFAP.

inclusions should suggest clear cell ependymoma, not oligodendroglioma.

The definitional and practical issues in distinguishing cellular ependymoma from astroblastoma are discussed with the latter tumor.

Markedly cellular ependymomas can resemble such aggressive neoplasms as small cell embryonal tumor, especially when their nuclei are hyperchromatic and mitotically active. The overlap is especially apparent in medulloblastomas with perivascular fibrillar zones (fig. 3-

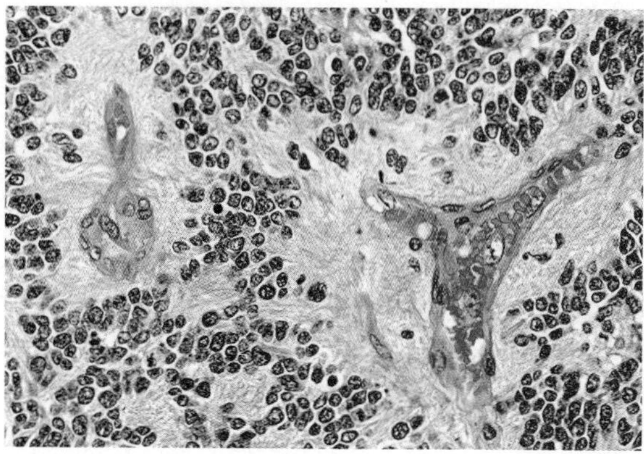

Figure 3-291

MEDULLOBLASTOMA RESEMBLING EPENDYMOMA

Ependymoma-like perivascular fibrillar zones are common in medulloblastoma.

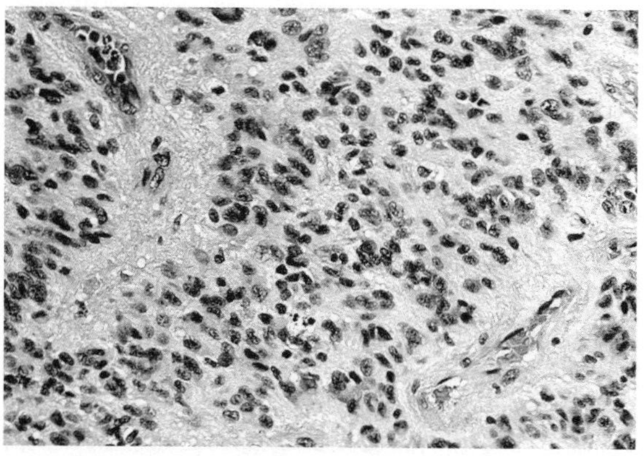

Figure 3-292

SMALL CELL GLIOBLASTOMA MULTIFORME RESEMBLING EPENDYMOMA

Cytologic monomorphism and perivascular fibrillar zones in some glioblastomas create an ependymoma quality.

291, see fig. 5-9). GFAP or synaptophysin immunopositivity becomes decisive in such instances. The conceptual and practical distinction of anaplastic ependymoma from ependymoblastoma is discussed in chapter 5.

A common diagnostic issue supratentorially is the distinction of small cell glioblastoma from ependymoma, since the former often has perivascular fibrillar zones (fig. 3-292). Features identifying glioblastoma are its diffuse tissue infiltration, high mitotic rate, and necrosis with pseudopalisading.

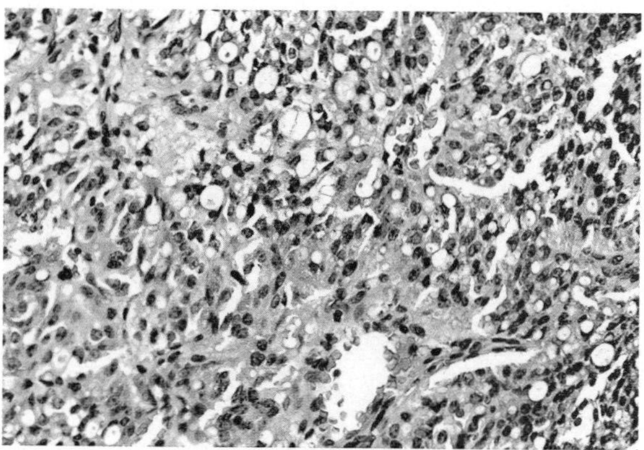

Figure 3-293

PAPILLARY TUMOR OF THE PINEAL REGION RESEMBLING EPENDYMOMA

This uncommon lesion has the overall "feel" of ependymoma but lacks the fibrillar perivascular pseudorosettes. Intracytoplasmic vacuoles or lumens may be prominent.

In the pineal region, the papillary neoplasm of the pineal region, a lesion of uncertain histogenesis, must be considered (fig. 3-293, see figs. 6-14, 6-15). This entity is described in chapter 6.

Treatment and Prognosis. In the past, most intracranial ependymomas recurred in the tumor bed and were ultimately lethal. Current neurosurgical techniques make macroscopically complete resections possible in many cases, and considerably longer survival periods are usual. How often simple excision of intracranial ependymomas is curative remains to be established, but may be more frequent than previously assumed. A small percentage of ependymomas, certainly less than 5 percent, undergo cerebrospinal fluid dissemination (fig. 3-294). Deposits are rarely an isolated finding and usually occur in the context of local recurrence. CSF dissemination is more likely with anaplastic and myxopapillary types (466). In accord with their usual grade of III, patients with clear cell lesions do less well than those with other ependymomas, with early recurrence and even extra-CNS metastases in some cases (422). In adults, one study found an inferior outcome for patients with supratentorial, intraparenchymal lesions (406).

The outlook for patients with spinal ependymomas of the cord proper is very favorable since these often small tumors lend themselves to gross total removal (467). This is also true for myxopapillary ependymomas, of which many

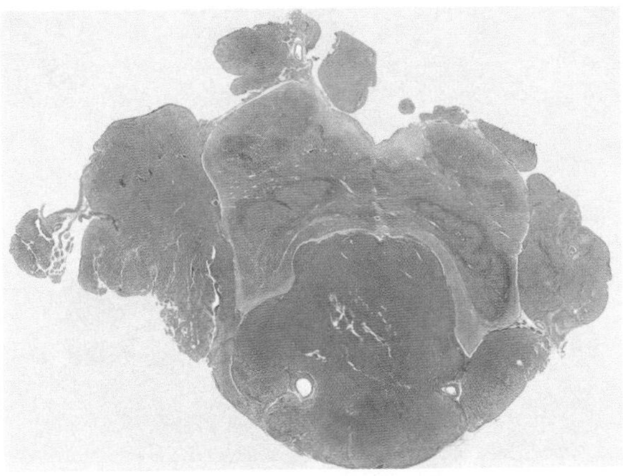

Figure 3-294

CEREBROSPINAL FLUID DISSEMINATION OF CELLULAR EPENDYMOMA

Uncommonly, ependymomas can seed the subarachnoid space, especially after local recurrence.

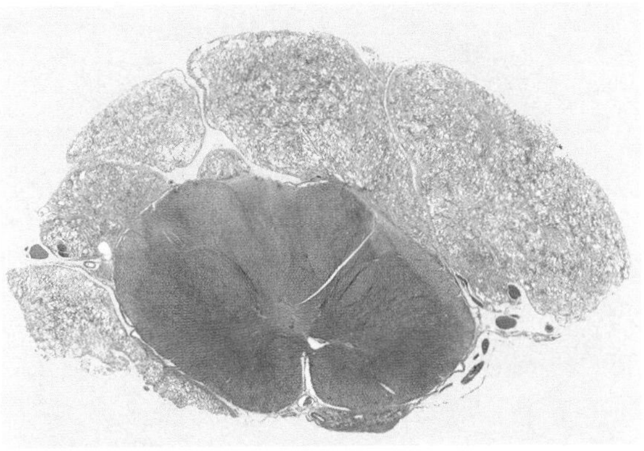

Figure 3-295

CEREBROSPINAL FLUID DISSEMINATION OF MYXOPAPILLARY EPENDYMOMA

While curable in almost all instances, occasional myxopapillary lesions disseminate within the subarachnoid space.

can be totally resected by severing the filum terminale from its origin and dissecting the lesion free from any adherent nerve roots. Local and even widespread dissemination in the spinal subarachnoid space may follow incomplete removal of myxopapillary tumors (fig. 3-295) or, rarely, can occur prior to surgery (408a). Distant metastases, particularly to the lungs, are occasionally observed in those rare examples arising in presacral or postsacral soft tissues (468–470).

Subependymoma

Definition. *Subependymoma* is a highly differentiated, slowly growing, often nodular glioma composed of clusters of ependymal and astrocyte-like cells that produce a densely fibrillar background. Subependymomas are WHO grade I.

Clinical Features. Supratentorial subependymomas are most often encountered incidentally at autopsy as small, smooth-surfaced, glistening lesions projecting into a lateral ventricle from the surface of a caudate nucleus (see fig. 3-298). Infratentorially, they are firm, white, lobulated masses based on the ventricular surface of the pons or medulla.

Only rarely is a subependymoma in this local large enough to produce hydrocephalus, or to undergo intratumoral or intraventricular hemorrhage that precipitates a neurosurgical emergency (figs. 3-296, 3-297) (471). Other symptomatic lesions occur throughout the ventricular system and even in the parenchyma near the cortical surface. Intraventricular tumors are summarized in Appendix C.

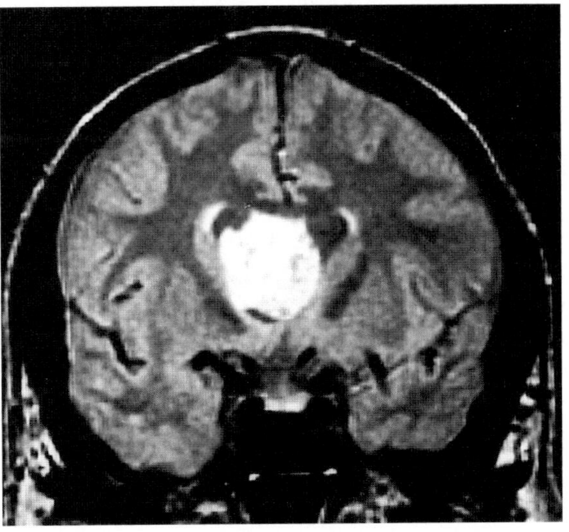

Figure 3-296

SUBEPENDYMOMA

The uncommon, symptom-causing, supratentorial subependymoma, such as this example seen on a proton density MRI, is a discrete intraventricular mass near the foramen of Monro. (Fig. 3-156 from Fascicle 10, 3rd Series.)

Subependymomas in the posterior fossa may compress the brain stem and elicit cranial nerve signs or dysfunction of subjacent respiratory centers (fig. 3-299) (472). CSF obstruction with hydrocephalus is yet another consequence.

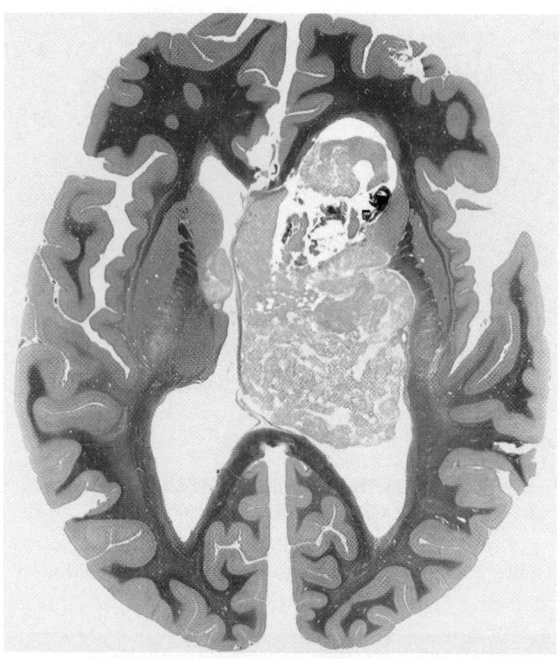

Figure 3-297

SUBEPENDYMOMA

Subependymomas can become sizable masses. Intratumoral hemorrhage was lethal in a 46-year-old woman.

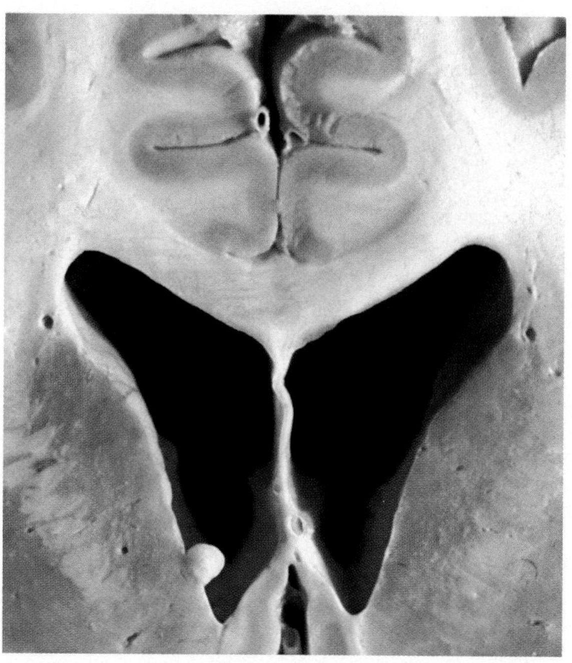

Figure 3-298

SUBEPENDYMOMA

Most subependymomas near the foramen of Monro are found incidentally postmortem as small, smooth-surfaced excrescences on the caudate nucleus.

Subependymomas in the spinal cord cause the generic symptoms of slowly expanding intramedullary masses (473,474).

Radiologic Findings. Symptomatic supratentorial subependymomas are usually large spherical tumors located in and around the region of the foramen of Monro (fig. 3-296), but may appear throughout the ventricular system (475). A few are intraparenchymal, sometimes even near the surface of the brain (476). Unlike other regional neoplasms near the foramen of Monro, such as central neurocytoma and subependymal giant cell astrocytoma, subependymomas exhibit little if any contrast enhancement (477,478).

Subependymomas in the posterior fossa are more likely to enhance and to be calcified than those near the foramen of Monro. Like the related ependymoma, subependymomas in the fourth ventricle may protrude through a foramen of Luschka and reach the cerebellopontine angle. Subependymomas of the fourth ventricle and other frequently calcified tumors are listed in Appendix F.

Gross Findings. Small subependymomas in the lateral ventricles typically appear as smooth-surfaced, dome-shaped, sessile or polypoid excrescences on the caudate nucleus near the foramen of Monro (fig. 3-298). Those that arise from the floor of the fourth ventricle are multinodular and can be gritty (figs. 3-299, 3-300). Spinal subependymomas are discrete and rubbery.

Microscopic Findings. Subependymomas of the lateral ventricles are distinctively nodular, highly fibrillar, and prone to microcystic change (fig. 3-301). The tumor-brain interface is typically sharp. Clustering of nuclei is usually not as prominent as it is in lesions in the fourth ventricle. Nuclei are typically small and may be surrounded by halos that produce an oligodendroglioma- or neurocytoma-like appearance. Nuclear pleomorphism is not uncommon, and occasional mitoses are present. Necrosis is rare and results from degenerative fibrosis and thrombosis of vessels rather than from rapid tumor growth. Hemosiderin deposits are often present in large lesions. Foci of plump, gemistocytic, or elongated fibrillar cells may also be evident.

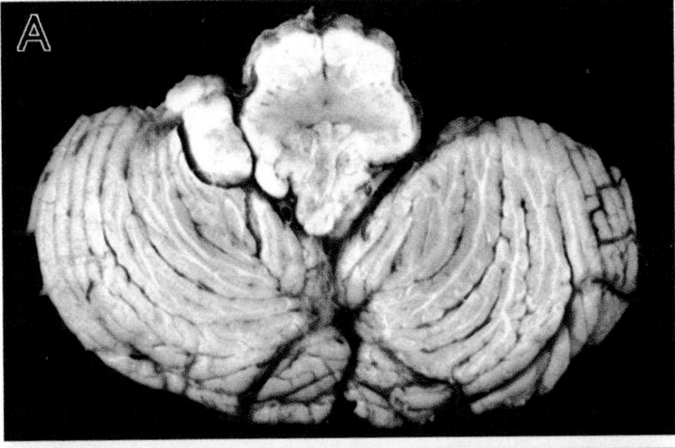

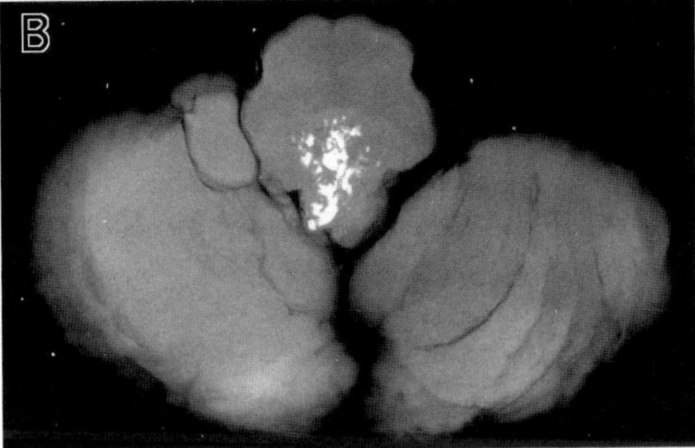

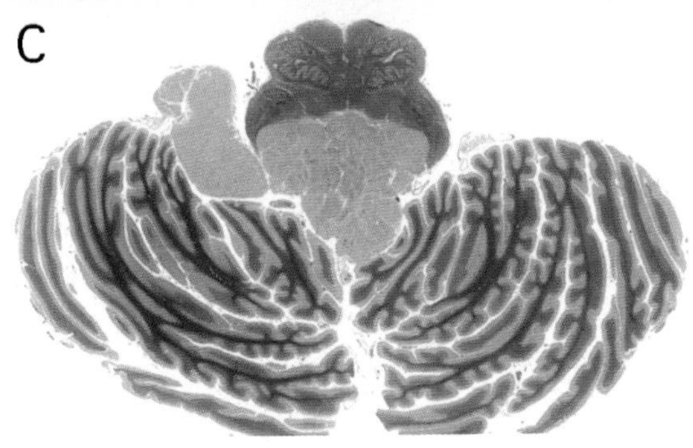

Figure 3-299

SUBEPENDYMOMA

A: Longstanding subependymomas based on the medulla oblongata can fill the ventricle, including its lateral recesses.

B: A radiograph demonstrates the calcium common in these lesions.

C: The histologic section is stained with H&E/Luxol fast blue.

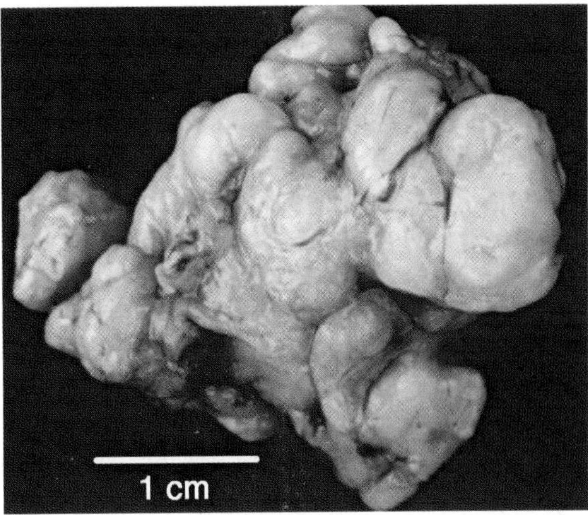

Figure 3-300

SUBEPENDYMOMA

As is well seen in a surgical specimen, fourth ventricular subependymomas are distinctly lobular.

Subependymomas of the fourth ventricle, as well as those of the spinal cord and brain parenchyma distant from the ventricular system, are decidedly lobular, usually calcified, and noted for distinctive clusters of nuclei that sit in a finely fibrillar background. The fibrillar tissue, save for widely scattered nuclei of tumor cells or capillaries, is almost entirely anuclear (figs. 3-302, 3-303). Subependymomas of the fourth ventricle are generally less microcystic than their supratentorial counterparts near the foramen of Monro. Nuclei are also often more uniformly oval and "ependymal" in appearance. Mitotic figures are rare. Pleomorphism is minimal. Scattered Rosenthal fibers may be present.

The relationship of subependymoma to classic ependymoma is highlighted by occasional mixed lesions. Usually, there is a focus of ependymoma in an otherwise typical subependymoma of the fourth ventricular type. Gradations and ill-defined transitions in the degree of cellularity and the prominence of perivascular pseudorosettes are common in such a setting, and there are often areas with features intermediate between those of ependymoma and subependymoma.

Anaplastic transformation of subependymoma remains to be reported, but angiosarcomatous (479) and rhabdomyosarcomatous (480) degeneration have been reported. The tumors in the latter cases become forms of gliosarcoma.

Frozen Section. The distinctive architectural features of subependymoma, i.e., nodularity, clustering of nuclei, and variable microcyst formations, are readily evident in frozen tissues.

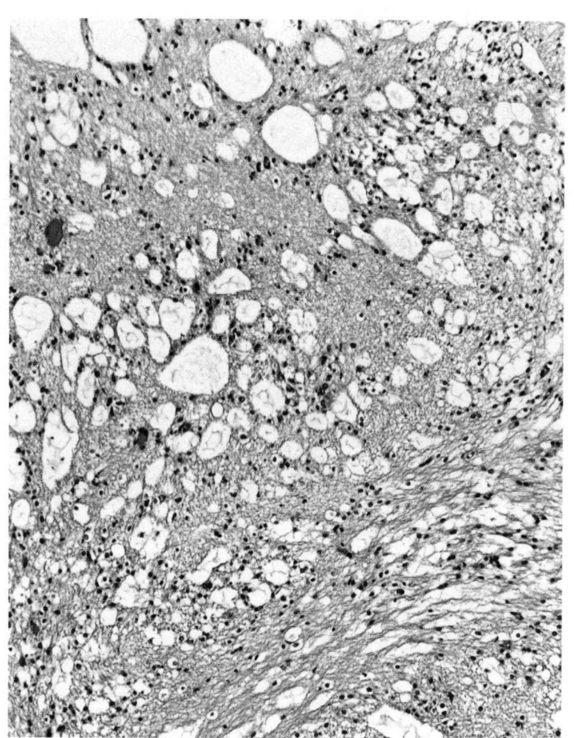

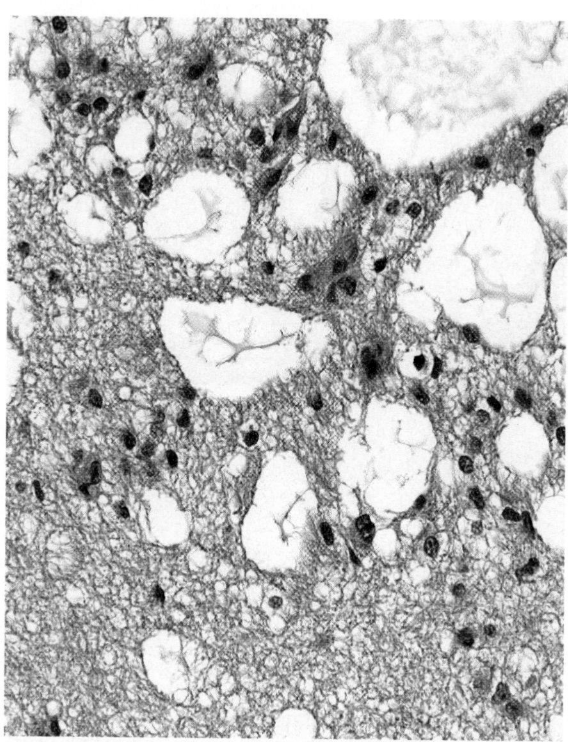

Figure 3-301

SUBEPENDYMOMA

Left: Subependymomas near the foramen of Monro have a distinctive microcystic architecture.

Right: Nuclear pleomorphism and a small amount of perinuclear cytoplasm give the cells a somewhat astrocytic appearance.

Immunohistochemical Findings. Subependymomas are strongly immunoreactive for S-100 protein and GFAP. MIB-1 indices are low, in one series ranging from 0.0 to 1.4 percent (481). In comparison, MIB-1 indices of myxopapillary ependymoma range from 0.0 to 5.5 percent and low-grade cellular ependymoma from 0.0 to 5.4 percent.

Ultrastructural Findings. Subependymomas are composed largely of clustered cells encircled by long processes that are filled with intermediate filaments. Particularly in tumors of the posterior fossa, there is clear evidence of ependymal differentiation in the form of elongated zipper-like junctions, clustered microvilli, and occasional cilia (482).

Cytologic Findings. Microcystic features, when abundant, are seen well in cytologic preparations (483). Highly fibrillar lesions, as of the posterior fossa and spinal cord, often resist disaggregation and remain as tissue fragments similar to those produced on a smear of a schwannoma.

Differential Diagnosis. The characteristic microcystic change, clustered nuclei, and dense fibrillarity produced by glial processes readily distinguish subependymoma in the region of the foramen of Monro from other gliomas. The tumor lacks the cellularity, spindle cells, epithelioid cells, prominent eosinophilic cytoplasm, and vesicular nuclei with prominent nucleoli that characterize subependymal giant cell astrocytoma of tuberous sclerosis.

Subependymomas in the posterior fossa or spinal cord are highly characteristic and generally need only to be differentiated from ependymoma. Unlike ependymoma, fourth ventricular subependymomas generally occur in adults and have clustering of nuclei in an overall hypocellular lesion.

Treatment and Prognosis. Some subependymomas arising in the region of the foramen of Monro may be excised in toto. One study noted no recurrence following gross total resection (472). Although radiotherapy seems reasonable

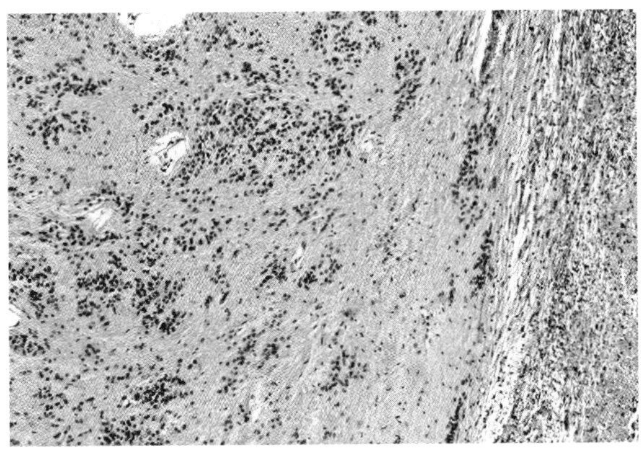

Figure 3-302

SUBEPENDYMOMA

In accord with the macroscopic features, the fourth ventricular lesion is a nodular mass. Note the sharp interface with the medulla (H&E/Luxol fast blue stain).

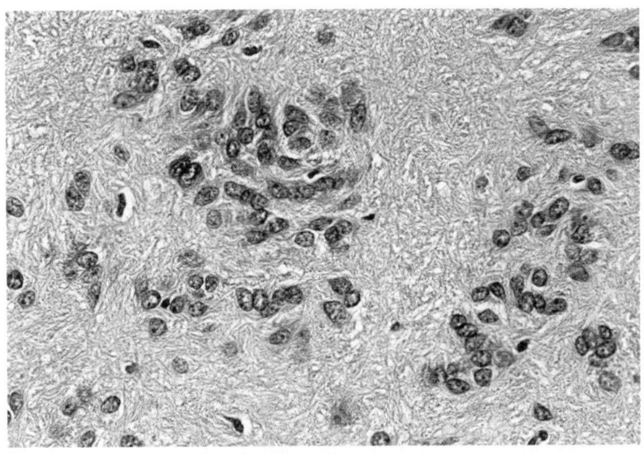

Figure 3-303

SUBEPENDYMOMA

Clustered, cytologically bland nuclei in a finely fibrillar background define this form of subependymoma, which is usually distinct from the more microcystic type seen near the foramen of Monro.

for symptomatic, incompletely resected tumors, or for recurrent lesions, the need for such therapy for asymptomatic, partially resected subependymomas remains to be established (472).

The presence of both subependymoma and ependymoma patterns in the same lesion creates a nosologic and therapeutic dilemma because of the perceived differences in biologic behavior between the two lesions. Subependymomas are believed to be benign (WHO grade I) tumors with little proliferative potential and negligible risk of neuraxis seeding, whereas grade II ependymomas are more likely to recur and can, albeit infrequently, disseminate. Radiotherapy is almost standard for intracranial ependymomas. Since it is unclear whether a focus of ependymoma within an otherwise typical fourth ventricular subependymoma in an adult is a negative prognostic indicator, careful observation, rather than radiotherapy, may be in order. The diagnosis of *mixed ependymoma-subependymoma* is appropriate, with a comment on the uncertainty about prognosis. Treatment is based largely on clinical variables, such as the age of the patient and completeness of excision.

MIXED GLIOMAS

Many ependymomas contain small gemistocytes with glassy cytoplasm and long processes that create the appearance of an astrocytic component. As seen in all but the smallest of specimens, however, these cells or tissues are only part of a larger lesion that is at least focally a classic ependymoma. The term "mixed glioma" is both unnecessary and confusing.

The issues of mixed composition are, both practically and conceptually, more nuanced in infiltrating gliomas, i.e., diffuse astrocytomas and oligodendrogliomas, since many are phenotypically mixed or hybrid in cytologic composition. Only uncommonly does this occur as two distinct, dichotomous, geographically separated populations of oligodendrocytes and astrocytes, for which the term *oligoastrocytoma* is descriptive. Increasingly, the terms mixed glioma, oligoastrocytoma, and mixed oligoastrocytoma are used for neoplasms with any degree of astrocytic, oligodendroglial, or hybrid oligodendroglial-astrocytic features. It remains to be shown, however, whether such lesions, or even the mixed lesions with distinct areas described above, are heterogeneous at the prognostically significant genetic level. This issue is discussed in more detail in the section Oligodendroglioma and Oligoastrocytoma.

Some choroid plexus papillomas consist, in part, of tapered fibrillated cells with intense GFAP positivity, a feature termed glial (ependymal) differentiation (see fig. 3-331). Since such foci are of no known prognostic significance, nor do they affect the classification of lesions, they remain choroid plexus tumors, not mixed gliomas.

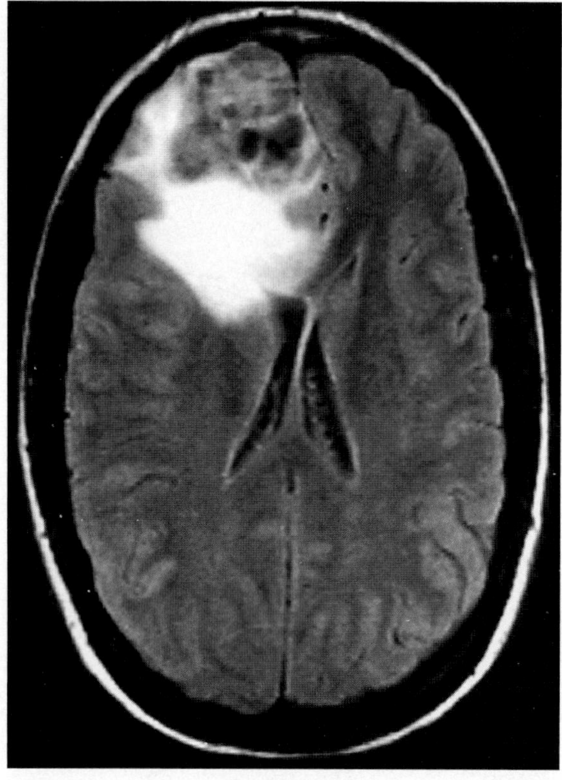

Figure 3-304

ASTROBLASTOMA

Astroblastomas are well-circumscribed, often superficial, discrete masses that usually occur in young adults, in this case a 24-year-old woman.

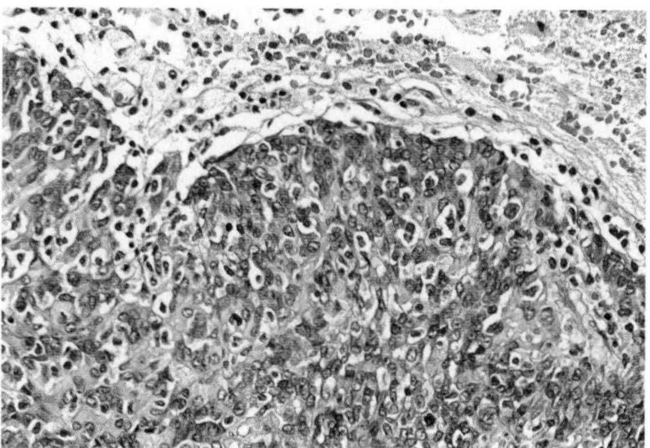

Figure 3-305

ASTROBLASTOMA

In accord with the radiologic findings, the lesion is sharply defined from the surrounding brain. As is typical, the cells have a somewhat epithelioid appearance.

OTHER GLIOMAS

Astroblastoma

Definition. *Astroblastoma* is a rare, typically well-circumscribed glioma characterized by the prominent perivascular orientation of the neoplastic cells, which have short, broad processes. The 2007 WHO consensus group declined to grade astroblastomas for want of sufficient clinicopathologic data.

General Features. Early published accounts did not firmly establish astroblastoma as a clinicopathologic entity, and the term was subsequently applied to a disparate group of neoplasms that included diffuse astrocytoma with focal expression of an astroblastic pattern. We accept as astroblastoma the compact, rather well-circumscribed lesion defined by Bonnin and Rubinstein (484), and described subsequently by others (485,486). Whether such tumors are an entity or merely a variant of ependymoma remains to be seen. If ependymal, the tumor occurs in an unusual locus and exhibits atypical histologic features. Given its ultrastructural phenotype, including both ependymal and astrocytic features, it has been suggested that astroblastoma arises from a close relative of the ependymal cell, the tanycyte (487).

Clinical Features. Although patients of all ages are affected, most astroblastomas occur in children or young adults.

Radiologic Findings. The lesions are usually supratentorial, extraventricular, superficial, and cystic. The contrast-enhancing solid component has a "bubbly" appearance (fig. 3-304) (485,488,489).

Microscopic Findings. Salient characteristics include demarcation, structural homogeneity with little or no infiltration of surrounding normal brain (fig. 3-305), a characteristic type of perivascular pseudorosette, and propensity for vascular hyalinization.

In representative areas, the tumor is architecturally compact, with cells that radiate from the vessels in a manner resembling the pseudorosettes of ependymoma (fig. 3-306). In contrast, however, the constituent processes are broader, shorter, and less tapered, being thus more astrocytic than ependymal. In longitudinal view, the perivascular formations appear ribbon-like. A looser, papillary, or even more astrocytic pattern may also be seen in which

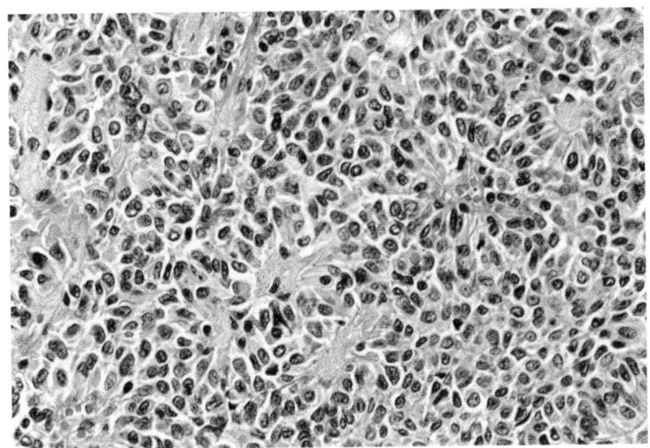

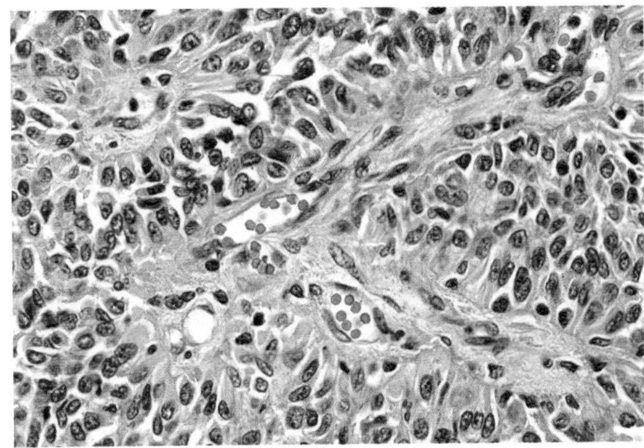

Figure 3-306

ASTROBLASTOMA

Left: Astroblastomas are cellular neoplasms with a perivascular orientation of neoplastic cells.

Right: Although there is overlap in histologic appearance between astroblastoma and ependymoma, the cell processes are somewhat shorter and stouter in astroblastoma.

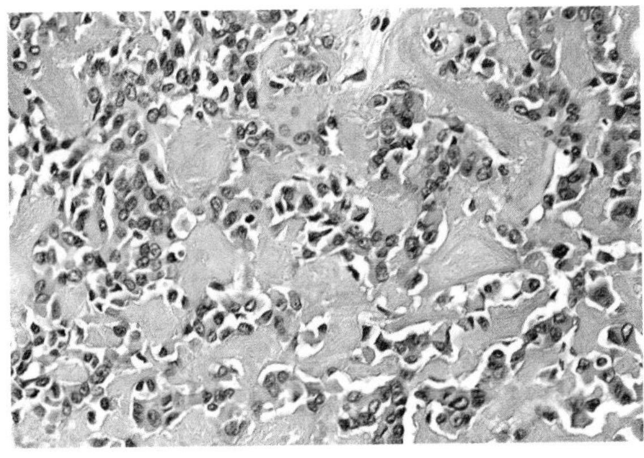

Figure 3-307

ASTROBLASTOMA

Vascular hyalinization is often prominent, with some areas largely replaced by connective tissue.

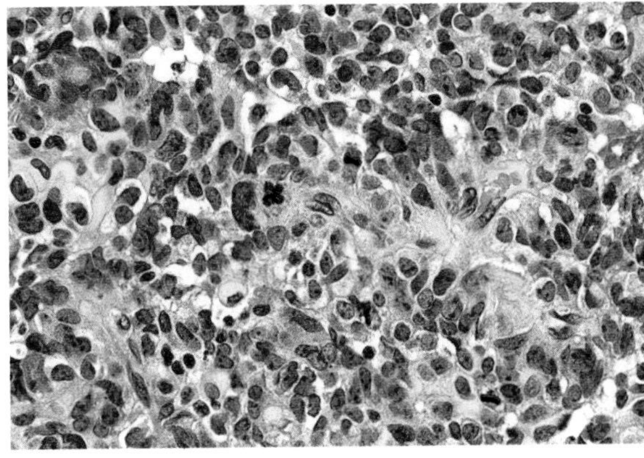

Figure 3-308

ASTROBLASTOMA

Multiple mitoses, some atypical, characterize anaplastic astroblastoma.

dehiscence of cells accentuates the individual radiating processes within the pseudorosettes.

Beginning as delicate sclerosis of vessels, vascular hyalinization may evolve through increasing vascular thickening into large hyalinized areas with little or no intervening tumor tissue (fig. 3-307). In intermediate areas, nests of degenerating neoplastic cells are isolated in a collagenous matrix. The nuclei of astroblastoma cells are larger than those of ependymoma cells, and more uniform than those of diffuse astrocytoma cells wherein variation in cellularity and chromatin texture is the rule.

Like most other gliomas, astroblastomas vary in their degree of differentiation. One study defined histologic features of malignancy as increased cellularity, marked cytologic atypia, and greater than 5 mitoses per 10 high-power fields (fig. 3-308) (485). Necrosis, in some cases with pseudopalisading and vascular proliferation, was present in most of the examples (fig. 3-309). Evolution to high-grade neoplasms with the features of glioblastoma has been noted (484). On balance, however, most astroblastomas, like ependymomas, are placed into two grades, low and high.

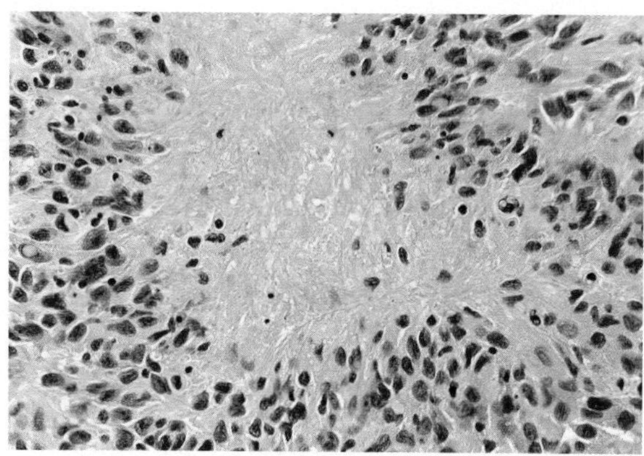

Figure 3-309

ASTROBLASTOMA

Necrosis with pseudopalisading may be present in some anaplastic astroblastomas.

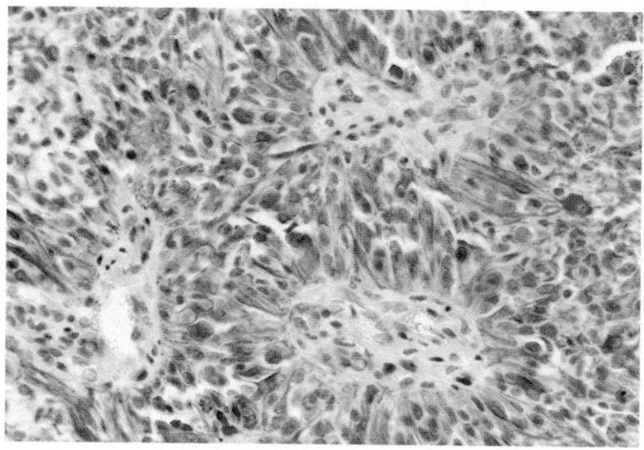

Figure 3-310

ASTROBLASTOMA

Processes within the perivascular rosettes are immunoreactive for GFAP.

Immunohistochemical Findings. The lesion is diffusely positive for vimentin and S-100 protein (485). Cells forming rosettes are often strikingly reactive for GFAP (fig. 3-310), but cells that are strongly positive may be interspersed with ones that are immunonegative. Membrane reactivity for EMA is a focal finding at best.

Ultrastructural Findings. Given the light microscopic similarity between astroblastoma and ependymoma, it is not surprising that there are also reported ultrastructural similarities: surface microvilli and cytoplasmic interdigitations (487). In contrast, however, astroblastomas exhibit fewer cilia and lateral intercellular connections of the intermediate junction type (490). Cytoplasmic intermediate filaments vary in number and distribution. Like ependymoma, cells of astroblastoma rest upon a basal lamina where they abut the vasculature (491).

Molecular and Cytogenetic Findings. One study noted gains on chromosomal arm 20q and chromosome 19, and in lesser incidence, losses of 9q, 10, and X (485). Another found monosomy of chromosomes 10, 21, and 22 (490). Whether these alterations define a pattern characteristic of astroblastoma is unclear. These changes are not typical of cerebral ependymoma or of diffuse astrocytoma.

Differential Diagnosis. Without the presence of tumor demarcation, astroblastoma is difficult to distinguish from other gliomas that have some element of stout, perivascular, glial processes, a feature loosely termed "astroblastic change." By utilizing a strict definition of astroblastoma, the diagnosis is simpler, since astroblastoma is well demarcated, distinctively patterned, and prone to vascular hyalinization.

Given the presence of radiating perivascular processes and the absence of brain invasion, the compact regions of astroblastoma closely resemble ependymoma. The differential diagnosis is complicated by the fact that, like astroblastoma, occasional ependymomas occur away from the ventricle, sometimes even at the surface of the brain. As noted above, however, the perivascular cytoplasmic processes of astroblastoma are robust and broad-based, rather than narrow and tapering. As a result, perivascular pseudorosettes lack the fine fibrillarity that characterizes those of ependymoma, and assume more a papillary, epithelial quality. Although marked vascular sclerosis is not unique to astroblastoma, it is sufficiently characteristic to bring this entity to mind. Ependymomas usually lack such sclerosis.

Astroblastoma may also resemble papillary meningioma (see fig. 9-35) since both are demarcated, often superficially situated, cytologically uniform, rich in perivascular pseudorosettes, and potentially reactive for EMA. Both also occur primarily in young patients. Most papillary meningiomas, however, are clearly dura-based and show areas of clear-cut meningioma. Insofar as any negative result can be conclusive, the meningioma's lack of reactivity for GFAP distinguishes its papillary variant from the immunopositive astroblastoma.

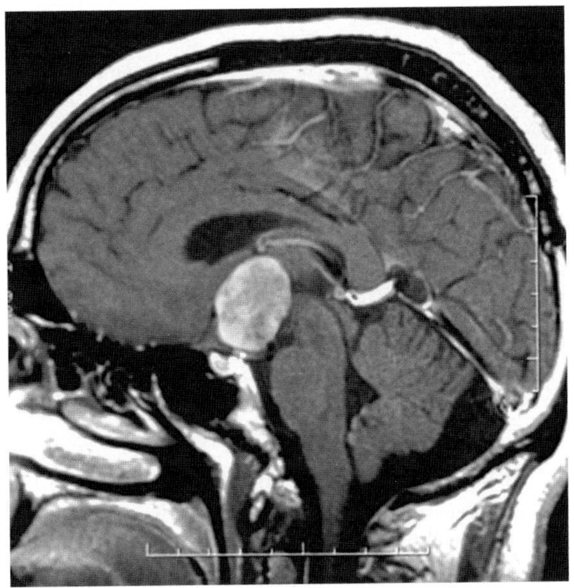

Figure 3-311

CHORDOID GLIOMA OF THE THIRD VENTRICLE

Chordoid gliomas are well-circumscribed, ovoid to circular, homogeneously contrast-enhancing masses within the third ventricle. This example, in a 36-year-old man with memory loss and blurred vision, was firmly adherent to the ventricular wall at the time of surgery.

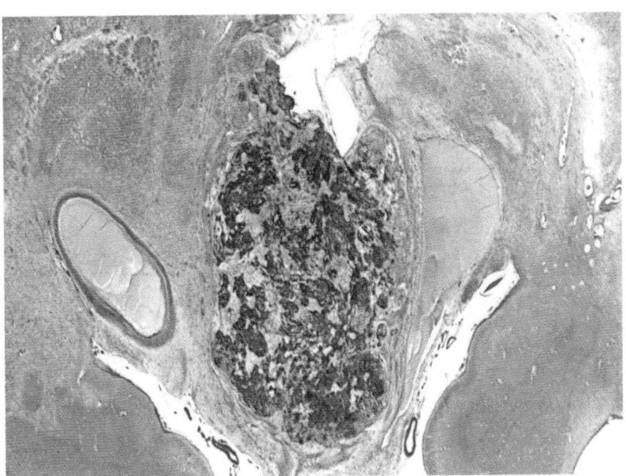

Figure 3-312

CHORDOID GLIOMA OF THE THIRD VENTRICLE

The tumor's location in, and attachment to, the walls of, the third ventricle are anatomic obstacles to resection.

Treatment and Prognosis. Given the rarity and the definitional problems that surround the lesion, the prognosis and appropriate treatment are, at this point, unclear. The demarcation of the lesion, both in well-differentiated and malignant forms, permits gross total resection in most cases. This has been followed, admittedly with short intervals, without recurrences in patients with low-grade astroblastoma (484–486). Most patients with high-grade tumors have been treated with radiotherapy, but only short survival periods of 1 to 2 years were noted (484–486). Recurrences may be discrete and multifocal at the primary site. High-grade examples may seed CSF pathways (486)

Chordoid Glioma of the Third Ventricle

Definition. *Chordoid glioma* is a well-circumscribed glioma of the third ventricle that histologically resembles chordoma and chordoid meningioma. This lesion is WHO grade II.

Clinical Features. The mass obstructs the flow of CSF to produce headaches and mental changes; consequences of hypothalamic compression and dysfunction include amenorrhea and endocrine deficiency states like hypothyroidism. Women are more often affected (492–494). To date, the lesion has been reported only within the third ventricle, but, as a "new" entity, its geographic restriction remains to be confirmed.

Radiologic Findings. The discrete, ovoid mass lies within the suprasellar region and third ventricle (fig. 3-311). On CT and MRI scans, it is homogeneously and avidly contrast enhancing (495). Intraventricular tumors are listed in Appendix C. Study of one small lesion suggests that the tumor could arise from the lamina terminalis (493a).

Gross Findings. The discrete, firm mass lies within the third ventricle and is densely adherent to its walls (fig. 3-312).

Microscopic Findings. The tumor is a well-defined, noninfiltrating mass associated with piloid gliosis in the surrounding brain (fig. 3-313). It is composed of epithelioid cells arranged in short columns or cords that sit in a basophilic mucopolysaccharide background (fig. 3-314). The cells have prominent cytoplasm and bland, round nuclei that are essentially free of mitotic activity. The amount of myxoid and/or collagenous stroma varies but is often conspicuous, as are lymphoplasmacytic infiltrates in which Russell bodies are common (fig. 3-315). Lobular areas lacking the chordoid cell arrangement may be present (fig. 3-316).

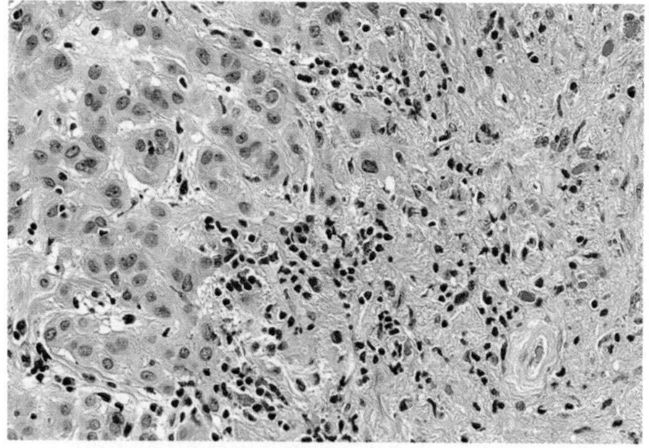

Figure 3-313

CHORDOID GLIOMA OF THE THIRD VENTRICLE

While well circumscribed, chordoid gliomas are broadly adherent to the surrounding brain and do not have a surgically exploitable cleavage plane. The piloid gliosis with Rosenthal fibers is a common response to a mass in this region.

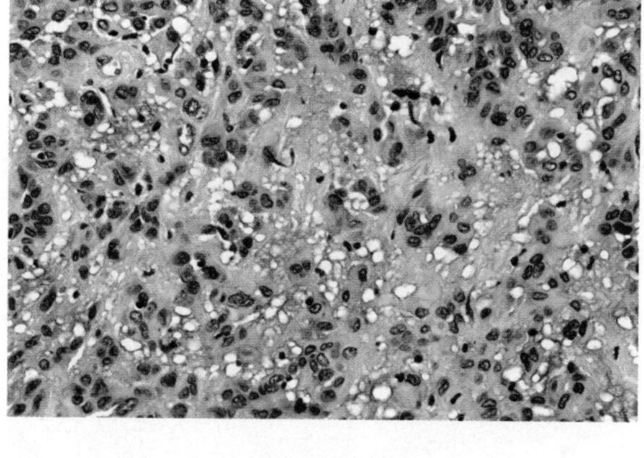

Figure 3-314

CHORDOID GLIOMA OF THE THIRD VENTRICLE

Liver-like cords of tissue separated by blue-staining mucinous material define the lesion.

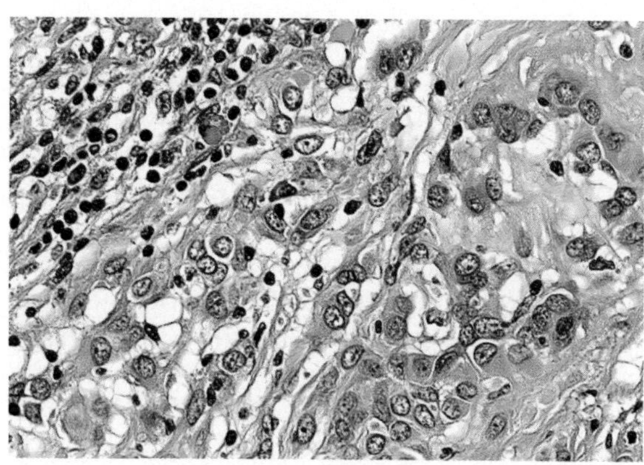

Figure 3-315

CHORDOID GLIOMA OF THE THIRD VENTRICLE

A lymphoplasmacytic infiltrate with Russell bodies is typical.

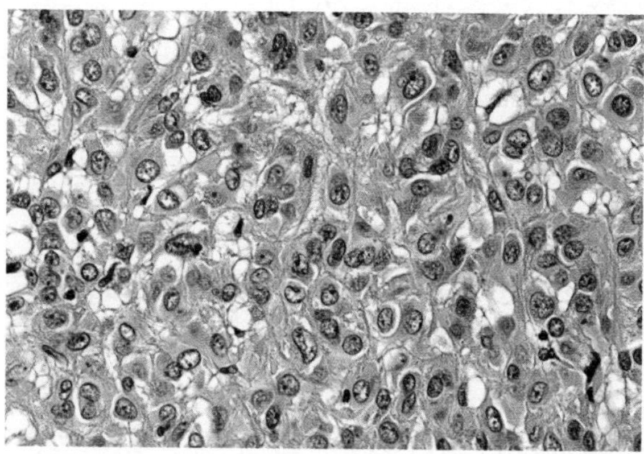

Figure 3-316

CHORDOID GLIOMA OF THE THIRD VENTRICLE

More nested, less chordoid examples may be confused with other entities.

Chondroid metaplasia was reported in a pediatric case (496). A pseudopapillary architecture and cells radiating from blood vessels, reminiscent of ependymoma, have also been observed (497).

Immunohistochemical Findings. The epithelioid cells are variably, but often strongly, reactive for GFAP (fig. 3-317) and vimentin. Many are positive for CD34, with strong labeling of cell membranes. Focal staining for EMA and cytokeratins is present in some cases (492, 494,498,499). The MIB-1 index is low: 1 percent or less (492,494).

Ultrastructural Findings. In accord with their immunoreactivity for GFAP and vimentin, the cells are filled with intermediate filaments. The polarity of the cells, with microvilli and occasionally cilia at an apical end, and many intercellular connections via hemidesmosomes at the other, plus cilia in some cases, lend an ependymoma quality (494,498). These features, and the finding of secretory granules, suggested to some observers that the tumors could arise from the subcommissural organ, a normal posterior third ventricular structure with ependymal features (498).

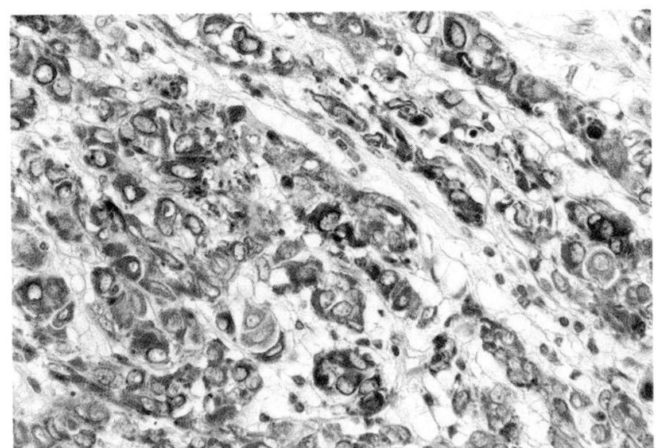

Figure 3-317

CHORDOID GLIOMA OF THE THIRD VENTRICLE

The lesion is GFAP positive.

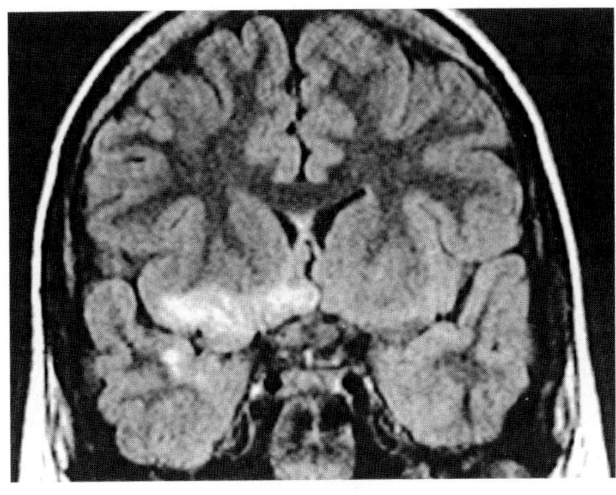

Figure 3-318

ANGIOCENTRIC GLIOMA

The superficial lesion thickens the gyri on the inferior surface of the right temporal lobe (on the left), and extends into the adjacent temporal lobe.

Molecular and Cytogenetic Findings. Little is known of the genetics of chordoid glioma. One recent study found no chromosomal imbalances on comparative genomic hybridization, and no abnormalities in the tumor suppressor genes and oncogenes commonly involved in diffuse astrocytomas, i.e., *p53, CDK2A, EGFR, CDK4,* and *MDM2* (499).

Differential Diagnosis. Given its location and distinctive features, little else enters into the differential in a classic case. The chordoid architecture obviously raises the issue of a clival chordoma, but the lesion's lack of dural/bony involvement and the presence of GFAP positivity as well as a lymphoplasmacytic infiltrate are distinguishing features. Chordoid meningioma is often a consideration, but one that is inconsistent with the lesion's GFAP positivity and its ultrastructural lack of desmosomes. Unlike chordoid meningioma, the glioma has no areas of transition to typical meningioma. The tumor cells somewhat resemble those of a germinoma, but are smaller and lack the large nuclei and prominent nucleoli of the germ cell lesion. Given the location, pituitary adenoma enters the differential diagnosis, but the latter is not as epithelioid, inflammatory, or mucoid as chordoid glioma. Loose-knit areas without the chordoid features can resemble another regional neoplasm, pituicytoma. The latter lacks both a myxoid background and prominent inflammatory component. A third ventricular, epithelial, somewhat papillary lesion, papillary tumor of the pineal region (see figs. 6-14, 6-15) (500), shares some histologic features, but lacks the chordoid architecture, inflammatory infiltrate, and strong GFAP staining of chordoid glioma.

Treatment and Prognosis. Chordoid gliomas are treated principally by surgery, either biopsy or total excision. Given the location and its adherence to the walls of the third ventricle (492, 494), excision may be difficult, with a potential for significant postoperative neurologic deficits if extirpation is pursued aggressively. At present, there is little long-term experience with the lesion, but the tumor appears to grow very slowly (492–494).

Angiocentric Glioma

Definition. *Angiocentric glioma* is an infiltrating glioma with an angiocentric pattern of growth and striking cytologic monomorphism.

Clinical Features. Children and occasionally young adults present with seizures, in some cases of many years duration (500a,500b).

Radiologic Findings. The T2-bright lesions are largely cortical, but may extend into the subjacent white matter (fig. 3-318), reaching as far as the lateral ventricle along a narrow neck (500a,500b). Thus far, cases have not been reported to contrast enhance.

Microscopic Findings. Tumor cells diffuse independently throughout the tissue (fig. 3-319), but coalesce in perivascular formations that resemble pseudorosettes of ependymoma

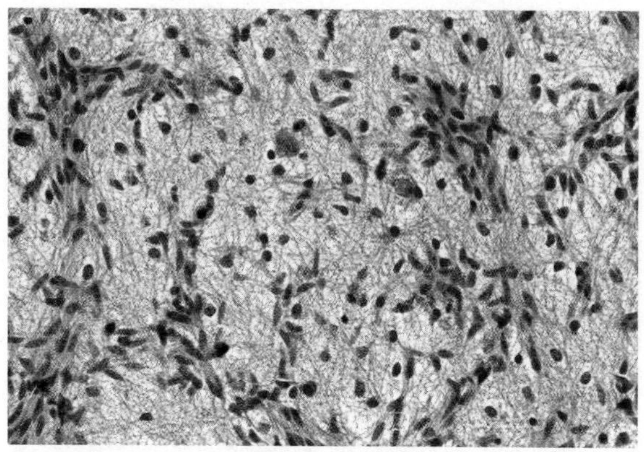

Figure 3-319

ANGIOCENTRIC GLIOMA

The lesions are in some regions diffusely infiltrating, with varying degrees of angiocentricity.

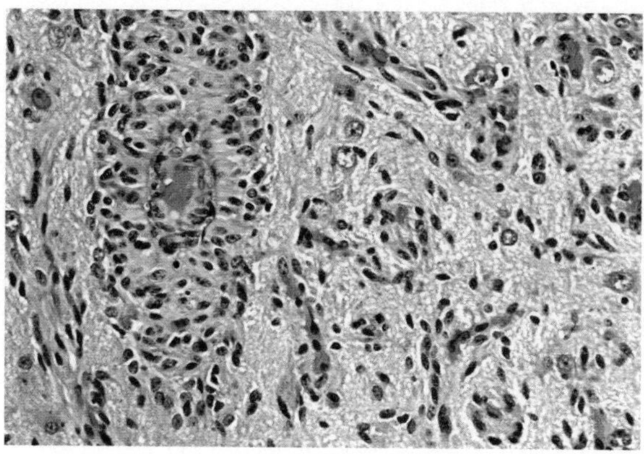

Figure 3-320

ANGIOCENTRIC GLIOMA

Perivascular pseudorosettes around medium-sized vessels simulate those of ependymoma. Tumor cells are attracted to capillary-sized vessels as well.

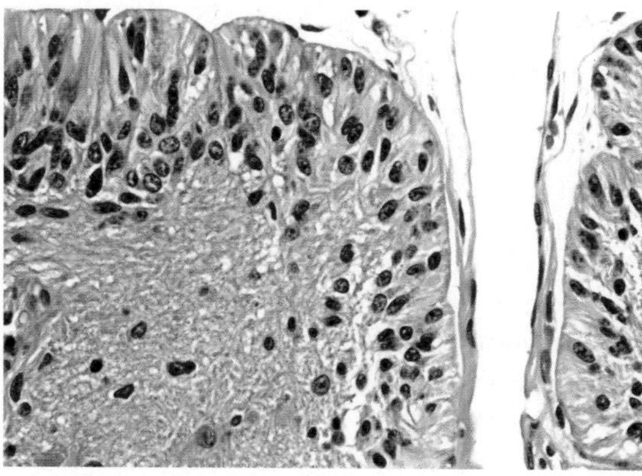

Figure 3-321

ANGIOCENTRIC GLIOMA

The presence of radial arrays of tumor cells in the subpial region is a distinctive feature.

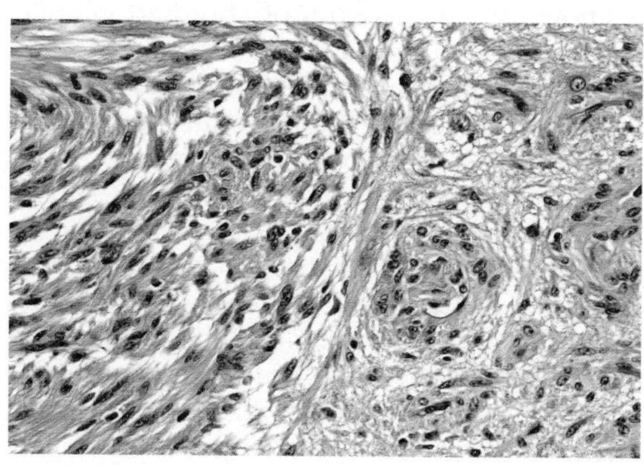

Figure 3-322

ANGIOCENTRIC GLIOMA

Compact masses of tumor cells resemble schwannoma.

and astroblastoma (fig. 3-320). Circular and longitudinal formations are present as well. Radially aligned cells in the subpial region produce prominent palisades (fig. 3-321). An additional "secondary structure" common in infiltrating gliomas, perineuronal satellitosis, is generally absent.

Tumor cells are small, usually bipolar, remarkably uniform in size and shape, and cytologically bland. Dark eosinophilic intracytoplasmic structures, corresponding to EMA-immunoreactive "microlumens" (discussed below), are present in a minority of cases.

Compact masses of elongated tumor cells resembling schwannoma may be present (fig. 3-322). Cyst-like spaces and a mucopolysaccharide background are not uncommon.

Mitoses are sparse if found at all in the basic lesion. One case, upon recurrence, however, had as many as 11 mitoses per 50 high-power fields (500a).

Intratumoral ganglion cells were interpreted as a component of the neoplasm in one study (500b), but merely as trapped elements in another (500a).

Immunohistochemical Findings. The lesions are GFAP positive although the staining may be weak. In some cases, antibodies to EMA decorate short segments of opposed tumor cell

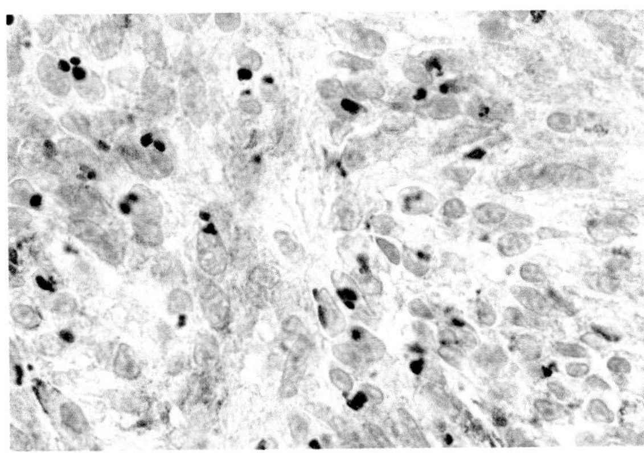

Figure 3-323

ANGIOCENTRIC GLIOMA

Intracytoplasmic "dots" of immunoreactivity for EMA corrrespond to microlumens, as are commonly found in ependymomas.

membranes, or deposit as minute intracytoplasmic "dots" similar to those in ependymomas (fig. 3-323). MIB-1 indices have been low, less than 1 percent, although they were higher in one recurrent lesion (500a).

Ultrastructural Findings. Microlumens, corresponding to the dot-like profiles seen after immunohistochemistry for EMA, suggested ependymal differentiation in one case (500a). Microvilli and runs of intermediate junctions joining adjacent cells were other ependymoma-like features.

Cytologic Findings. Smear preparations demonstrate the typical bipolarity, monomorphism, and angiotrophism of the tumor cells.

Differential Diagnosis. Cellular areas of the tumor with perivascular pseudorosettes resemble ependymoma and astroblastoma. The angiocentric tumor is decidedly infiltrative, not compact, and well-circumscribed, as are these latter two entities.

Looser myxoid areas with angiocentricity can resemble pilocytic astrocytoma, especially pilomyxoid astrocytoma. Distinguishing features are the infiltrating quality of the angiocentric bipolar lesion, as well as its well-formed perivascular pseudorosettes and subpial formations. Created by slender cells separated by myxoid material, perivascular formations in pilomyxoid astrocytomas are less well defined than those of the angiocentric glioma.

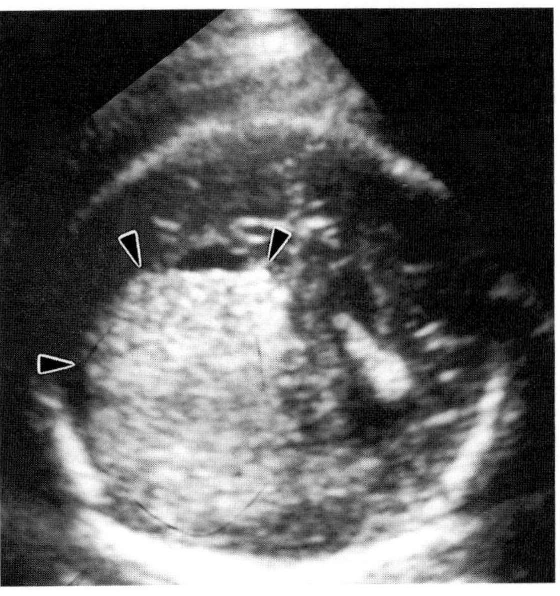

Figure 3-324

CHOROID PLEXUS PAPILLOMA IN A FETUS

Seen here ultrasonographically in utero, the echogenic, and therefore, white, neoplasm (arrowheads) relates to the ventricle on the left. The bright club-shaped region on the right is normal choroid plexus. (Fig. 3-161 from Fascicle 10, 3rd Series.)

Infiltrating gliomas, particularly oligodendroglioma, mimic the lesion when perivascular secondary structures are present, but usually don't have the ependymoma-like pseudorosettes. Perinuclear satellitosis is common in these lesions, but rare in angiocentric glioma. Nuclei of oligodendrogliomas are round; those of the angiocentric glioma are elongated.

Treatment and Prognosis. Gross total resection leads to surgical cure in some cases (500a,500b), and subtotally resected lesions may remain stable for years. One case progressed and was ultimately fatal, however (500a).

CHOROID PLEXUS PAPILLOMA, ATYPICAL PAPILLOMA, AND CARCINOMA

Definition. *Choroid plexus papilloma, atypical papilloma,* and *carcinoma* are papillary neoplasms derived from choroid plexus epithelium. Papillomas are WHO grade I, atypical papillomas are grade II, and carcinomas are grade III (501).

Clinical Features. While occasional examples occur in utero (fig. 3-324), at birth, or so shortly thereafter that a congenital origin is justly assumed (502–505), most plexus tumors

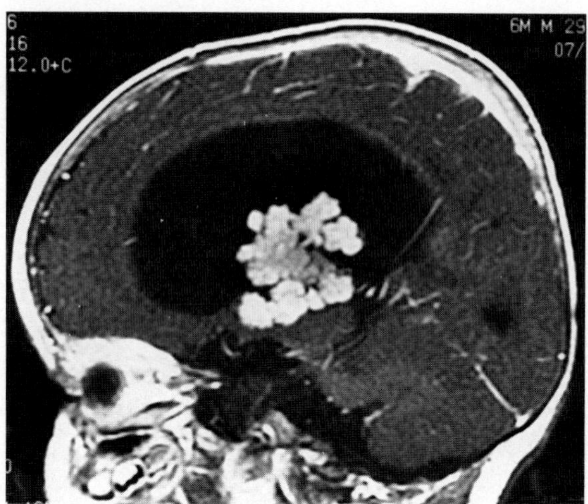

Figure 3-325

CHOROID PLEXUS PAPILLOMA

Some well-differentiated choroid plexus tumors are overtly papillary, as in this example in the lateral and third ventricles of a 7-month-old with hydrocephalus.

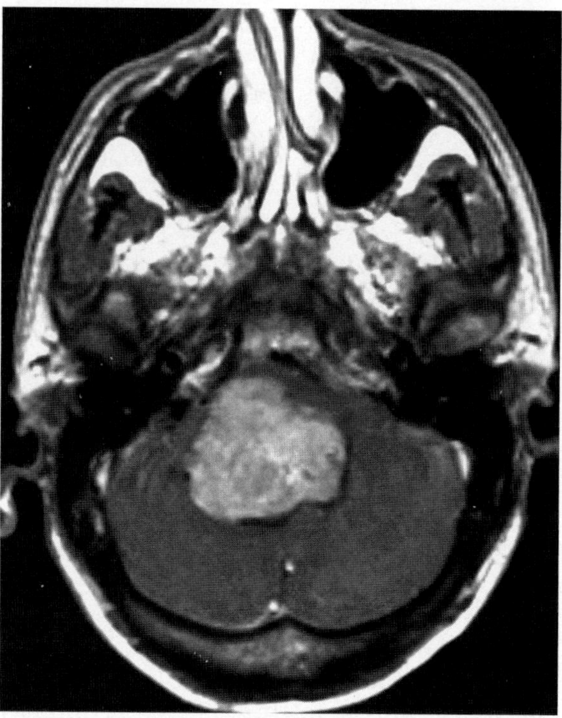

Figure 3-326

CHOROID PLEXUS PAPILLOMA

Choroid plexus tumors in the fourth ventricle are almost always papillomas and occur principally in adults, in this case a 34-year-old man.

are detected in children. The lateral (fig. 3-325) and, less often, third ventricles are favored in this age group (506). The uncommon papillomas in adults are almost always in the fourth ventricle (fig. 3-326). Papillomas from a lateral recess in the choroid plexus may protrude into the cerebellopontine angle through the foramen of Luschka (fig. 3-327) (507,508).

Most papillomas and carcinomas are sporadic, and occur only rarely in cancer-predisposing syndromes such as Li-Fraumeni (509–511) and Aicardi (512). Ectopic neoplasms independent of the choroid plexus have been reported in the suprasellar region, cerebellopontine angle, and even the spinal epidural space (513).

By secreting CSF and obstructing its flow, papillomas often produce hydrocephalus, with all its classic signs and symptoms (506). Macrocephaly appears in the very young. Intraventricular tumors are enumerated in Appendix C.

Radiologic Findings. Both papillomas and carcinomas are discrete, intensely contrast enhancing, and, with the exception of those at the cerebellopontine angle and the exceptions noted above, intraventricular (figs. 3-325, 3-326). Cystic change and parenchymal edema may be prominent (514). Lobular or nearly papillary diagnostic architecture is present in some cases (fig. 3-325) (507,515,516). Tumors of the fourth ventricle may be heavily calcified. Carcinomas are often cystic, invasive, and may undergo dissemination or systemic metastasis (514).

Gross Findings. In situ, papillomas are well-circumscribed, lobular, and in some cases overtly papillary lesions that expand along a broad compressive front (fig. 3-327). Both papillomas and carcinomas are highly vascular, and, because of difficulties with hemostasis, more than one operation may be necessary for total resection.

Microscopic Findings. Choroid plexus tumors span a morphologic spectrum from remarkably well-differentiated papillomas to anaplastic tumors with only a hint of epithelial differentiation. They are thus easy to grade at the extremes, but less so in the mid-range. There are, in addition, frequent intratumoral gradations in cytologic atypia and mitotic activity. Transition of the neoplasm to normal choroid plexus is common as well. Despite this heterogeneity, clinical evolution from a papilloma to carcinoma is unusual (517–519a).

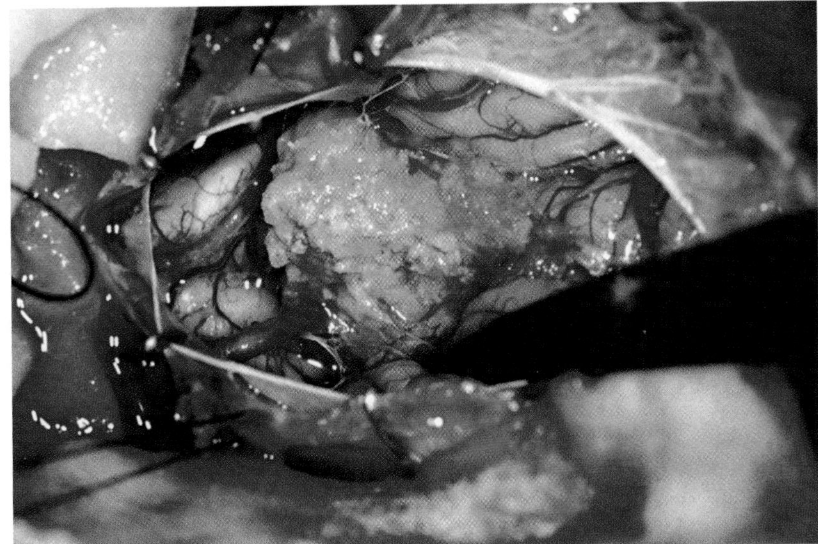

Figure 3-327

CHOROID PLEXUS PAPILLOMA

Arising from the plexus in the lateral recess of the fourth ventricle, papillomas can appear at the cerebellopontine angle. Note the fine macroscopic papillations of this well-differentiated lesion. (Courtesy of Dr. W. E. Krauss, Rochester, MN.)

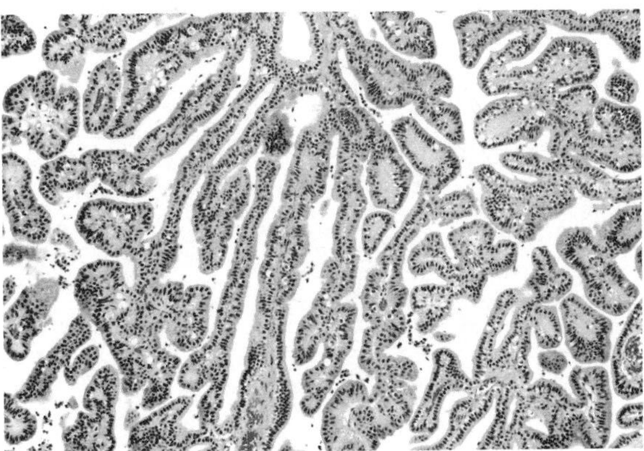

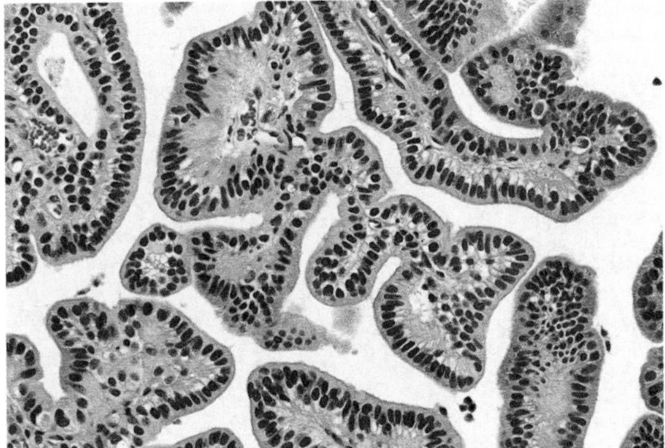

Figure 3-328

CHOROID PLEXUS PAPILLOMA

Left: While cytologically bland, the redundancy of the epithelium helps distinguish the lesion from normal choroid plexus. Right: Critical features of papillomas are cell crowding, increased nuclear to cytoplasmic ratios, and flat rather than "cobblestoned" surfaces.

Most choroid plexus tumors lie at the better-differentiated end of the spectrum and have an orderly layer of columnar epithelium on a basement membrane that overlies delicate fibrovascular stalks. To a varied degree, the cells are crowded, more columnar, more cytologically atypical, and more varied in nuclear size and chromasia than those of the parent tissue (figs. 3-328, 3-329). The epithelium has a linear face in contrast to the "hobnail" or "cobblestone" profile of the normal tissue. Necrosis, although usually a feature of atypical or malignant plexus lesions, is inexplicably present in an occasional otherwise well-differentiated papilloma (520).

While almost all well-differentiated choroid plexus neoplasms are overtly papillary, occasional exceptions form acini or tubules (fig. 3-330) (521,522). Papillomas commonly exhibit focal glial differentiation in the form of elongated GFAP-positive cells with tapering processes (fig. 3-331) or as GFAP positivity in sclerotic lesions (523). Other interesting features are pigmentation (fig. 3-332) (524–526), oncocytic change (fig. 3-333) (523,527,528), calcification, xanthomatous change, bone (529–531), cartilage (532), and extensive sclerosis (fig. 3-334) (523).

Within a group that is intermediate between well-differentiated papilloma and obvious

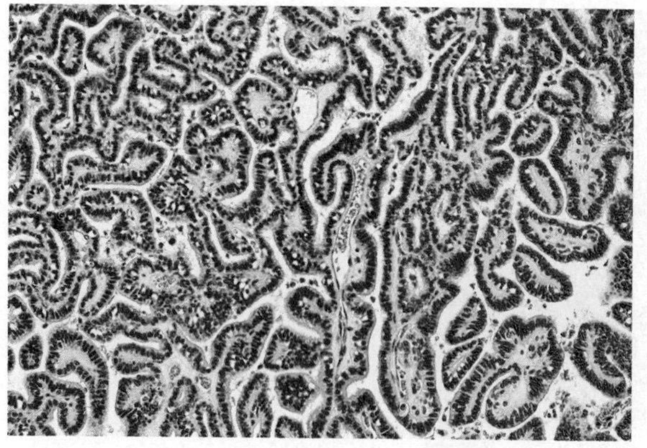

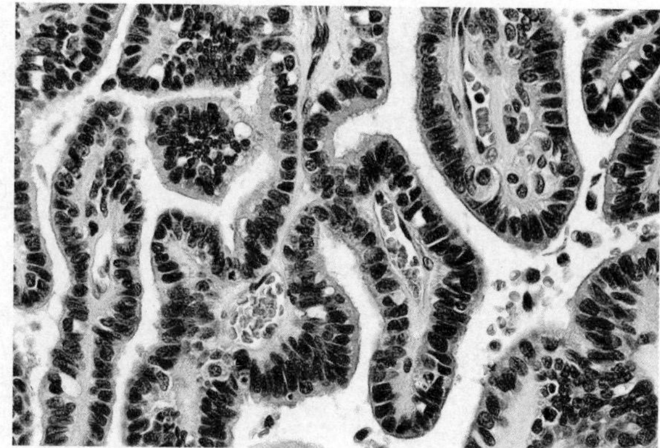

Figure 3-329

CHOROID PLEXUS PAPILLOMA

Left: Even at low magnification, this epithelium has the redundancy and crowding that establishes the diagnosis of a neoplasm.

Right: A high nuclear to cytoplasmic ratio and the degree of cytologic atypia help distinguish such lesions from normal choroid plexus.

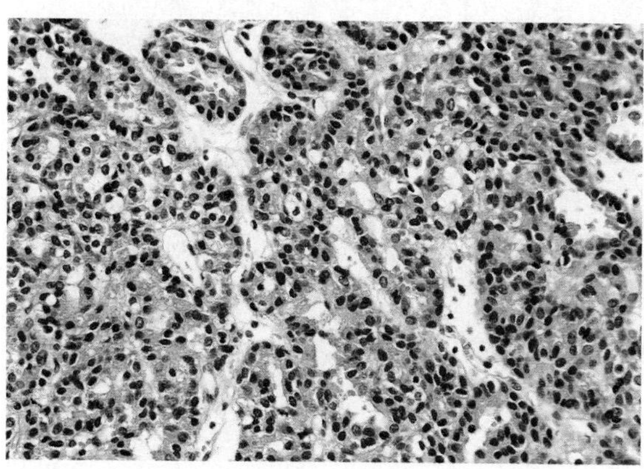

Figure 3-330

ACINAR CHOROID PLEXUS TUMOR

The epithelium in rare choroid plexus tumors turns upon itself to create glands or acini, rather than papillae.

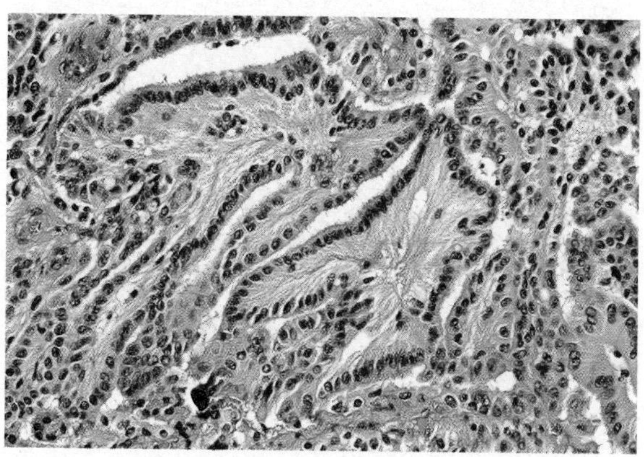

Figure 3-331

CHOROID PLEXUS PAPILLOMA WITH GLIAL DIFFERENTIATION

Extension of cells into long processes is known as glial differentiation.

carcinoma are difficult to grade lesions with significant cytologic atypia, increased nuclear to cytoplasmic ratios, and variable numbers of mitotic figures (figs. 3-335, 3-336). At what point these changes become pronounced enough to establish the diagnosis of carcinoma, especially if they are confined to increased mitotic rate and cytologic atypia, is not clear. Although concern is appropriate regarding the biologic behavior of such intermediate lesions, they are usually designated atypical papillomas, not carcinomas. One study suggested that WHO grade II "atypical papillomas" be distinguished from grade I papillomas on the basis of mitotic activity—2 or more per 10 high-power fields (519a). An upper limit to the number of mitoses in atypical papillomas is not well defined, but there certainly are cases where dividing cells are so numerous that it seems difficult to accept a lesion as only "atypical."

At the low end of the carcinoma spectrum, just above the atypical papilloma, are epithelial lesions that are overtly epithelial, but with cytologic atypia and considerable mitotic activity. Foci of necrosis are not uncommon (fig. 3-337).

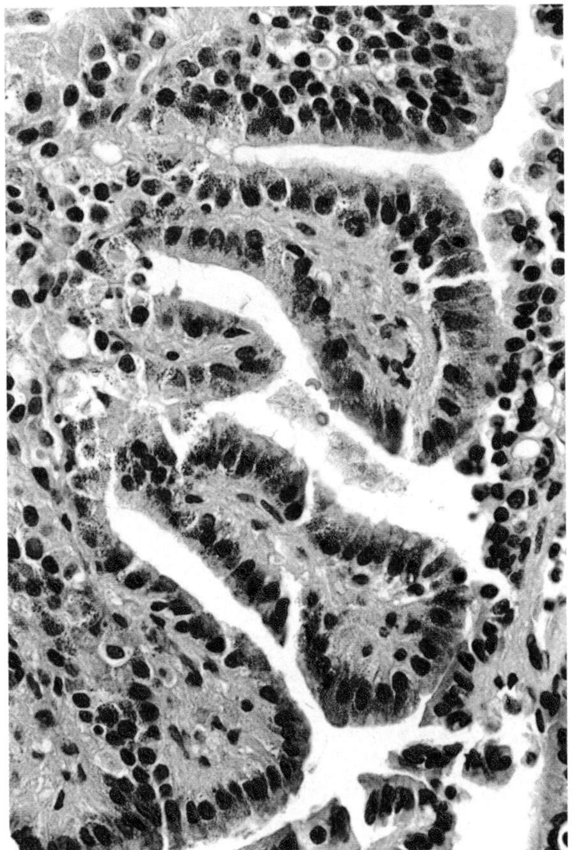

Figure 3-332

PIGMENTED CHOROID PLEXUS PAPILLOMA

A focal "dusting" of fine pigmented granules is not uncommon in papillomas. Only a rare lesion is diffusely melanotic.

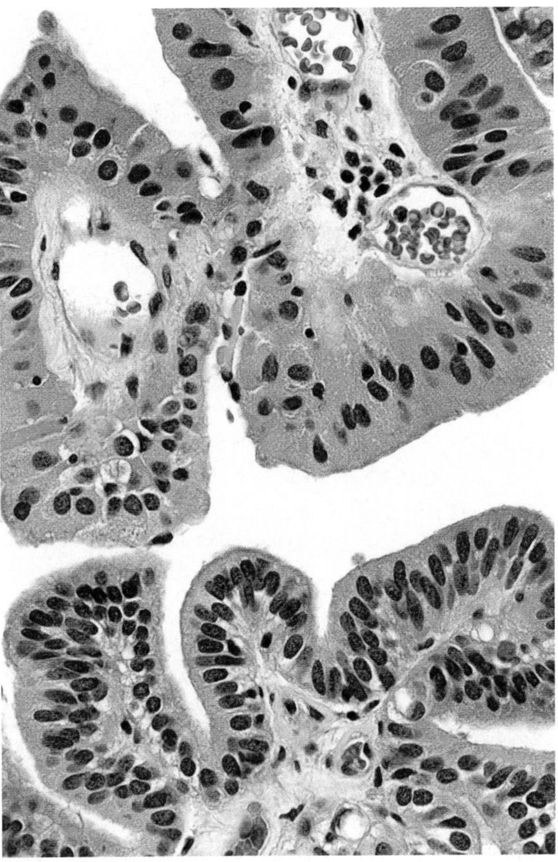

Figure 3-333

CHOROID PLEXUS PAPILLOMA WITH ONCOCYTIC CHANGE

Oncocytic change is focal and prognostically insignificant.

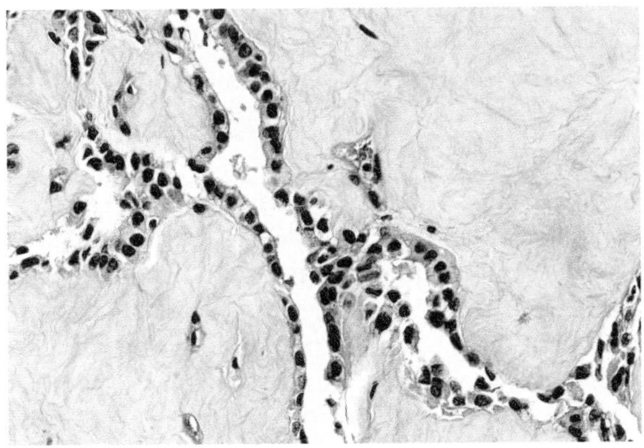

Figure 3-334

CHOROID PLEXUS PAPILLOMA WITH SCLEROSIS

The fibrovascular stroma may be replaced by dense collagenous tissue in longstanding lesions.

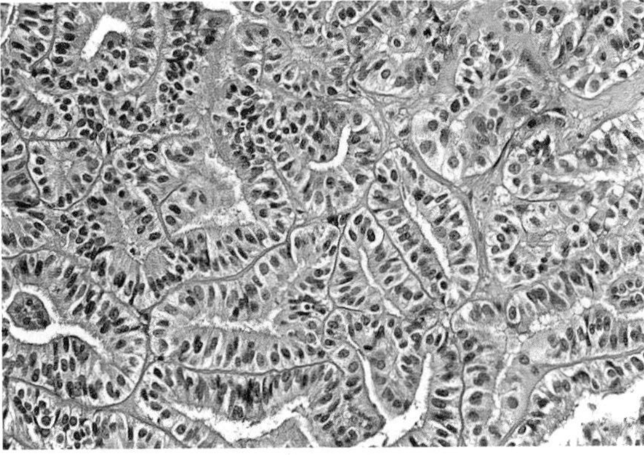

Figure 3-335

ATYPICAL CHOROID PLEXUS PAPILLOMA

Choroid plexus neoplasms with a highly differentiated epithelium, but with scattered mitoses, are difficult to place into either the papilloma or carcinoma category.

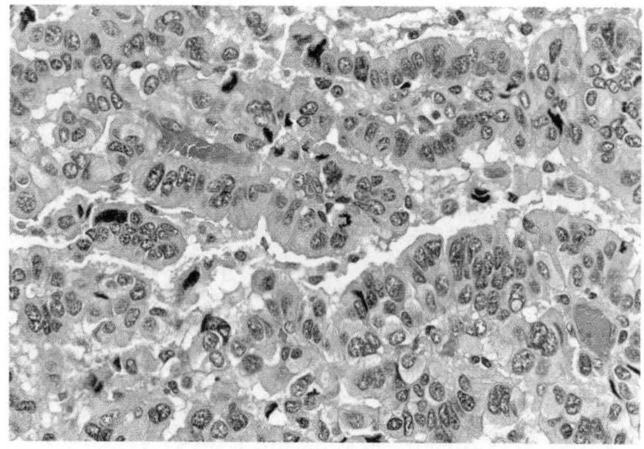

Figure 3-336

CHOROID PLEXUS CARCINOMA

Although the tissue is clearly papillary, its mitotic activity and degree of cytologic atypia exceed those of atypical papilloma.

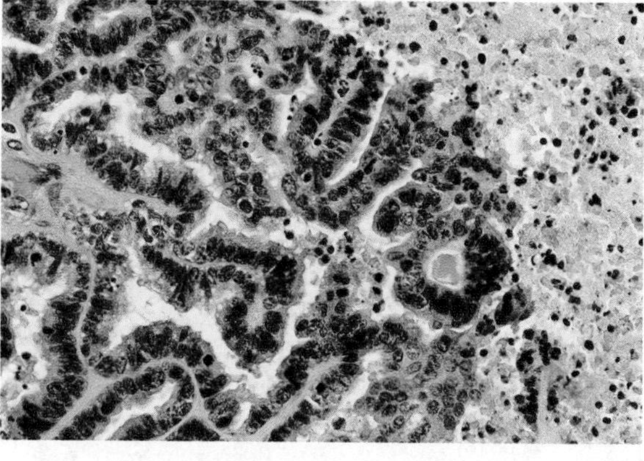

Figure 3-337

CHOROID PLEXUS CARCINOMA

An atypical, mitotically active epithelium in association with focal necrosis is common in carcinoma.

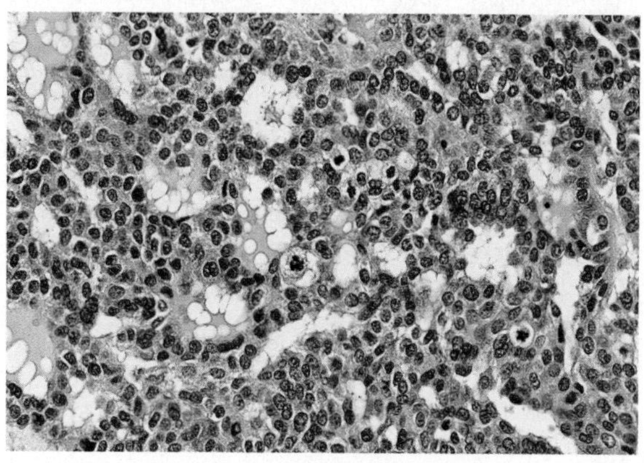

Figure 3-338

CHOROID PLEXUS CARCINOMA

The diagnosis of carcinoma is elementary in mitotically active lesions with both papillary and solid tissue.

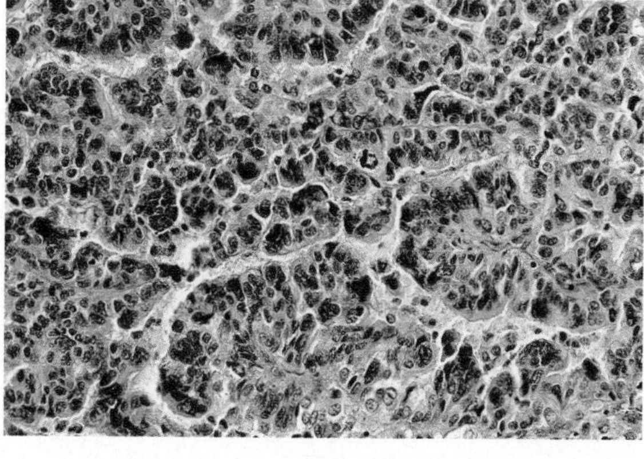

Figure 3-339

CHOROID PLEXUS CARCINOMA

Coarse lobules of epithelial cells are common in anaplastic choroid plexus carcinomas.

Next in the degree of anaplasia are tumors whose cells are no longer constrained to a simple columnar epithelium but create solid masses in variable proportions to a papillary component (fig. 3-338). Then come highly cellular neoplasms showing architectural disarray, complex glands, cribriform arrangements, and only poorly formed papillae (fig. 3-339). Nuclear malignancy is obvious, mitoses are abundant, and necrosis is common. Hyaline protein droplets may be seen in such pleomorphic tumors (fig. 3-340). Finally, there are anaplastic lesions composed of sheets of cells, sometimes with perinuclear halos that lend a somewhat oligodendroglial appearance (fig. 3-341). Other anaplastic tumors with a "jumbled" architecture, including epithelioid cells, resemble atypical teratoid/rhabdoid tumor. Anaplastic areas can also resemble small embryonal tumors.

Both papillomas and carcinomas can invade the brain, and while more likely in the latter (520), does not seem to adversely affect prognosis (533), and certainly is not, in itself, the basis of a "carcinoma" diagnosis. In some cases, the presence of Rosenthal fiber–rich piloid gliosis attests to the chronicity of the lesion.

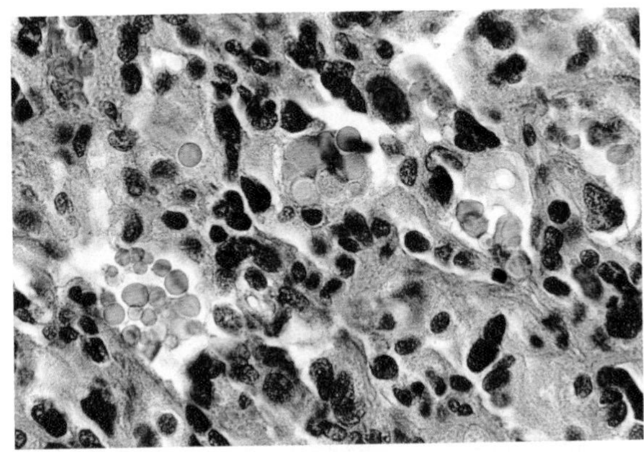

Figure 3-340

CHOROID PLEXUS CARCINOMA

Protein droplets may be present in pleomorphic choroid plexus carcinomas.

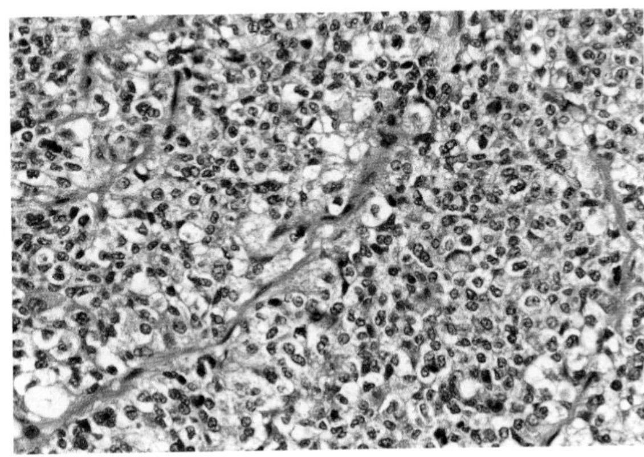

Figure 3-341

CHOROID PLEXUS CARCINOMA

Undifferentiated carcinomas can resemble malignant gliomas, in this case oligodendroglioma.

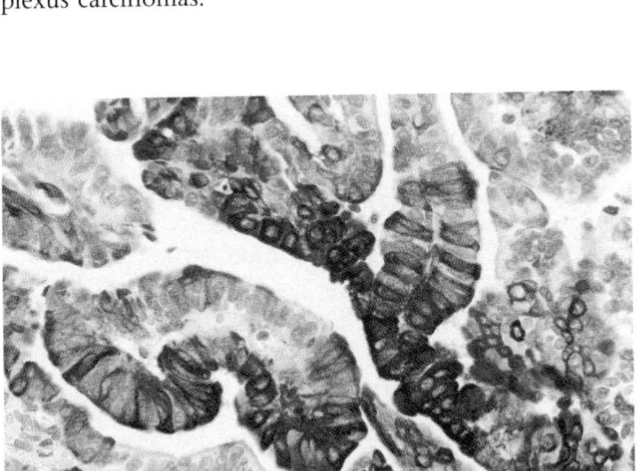

Figure 3-342

CHOROID PLEXUS PAPILLOMA

To a variable degree, most papillomas are immunoreactive for cytokeratins.

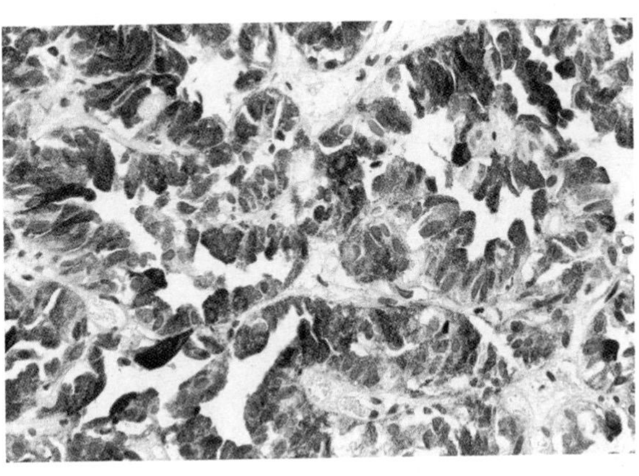

Figure 3-343

CHOROID PLEXUS PAPILLOMA

Although they do not arise from neuronal tissue, papillomas are often immunopositive for synaptophysin.

Immunohistochemical Findings. The immunophenotype of choroid plexus papillomas expresses the hybrid, epithelial-neuroepithelial nature of choroid plexus epithelium. Accordingly, most neoplastic cells are immunoreactive not only for vimentin and cytokeratins (fig. 3-342) (in greater measure for CAM5.2 than AE1/AE3 and for CK7 than CK20), S-100 protein, synaptophysin (fig. 3-343), and, in some cases, EMA (fig. 3-344) and even GFAP (fig. 3-345) (534–538). Foci of ependymal-appearing cells with tapered processes are GFAP positive. Densely sclerotic lesions may be reactive as well (523). In our experience, carcinomas lack GFAP staining and occasionally are nonreactive for S-100 protein. Choroid plexus tumors as a whole are negative for carcinoembryonic antigen. Immunostaining for p53 is considerably more prominent in carcinomas than papillomas (539,540).

Several studies have indicated the diagnostic utility of immunoreactivity of choroid plexus for prealbumin (transthyretin) (536,541). Both the normal plexus and its derived neoplasms are reactive. Since occasional metastatic carcinomas are positive, staining for transthyretin

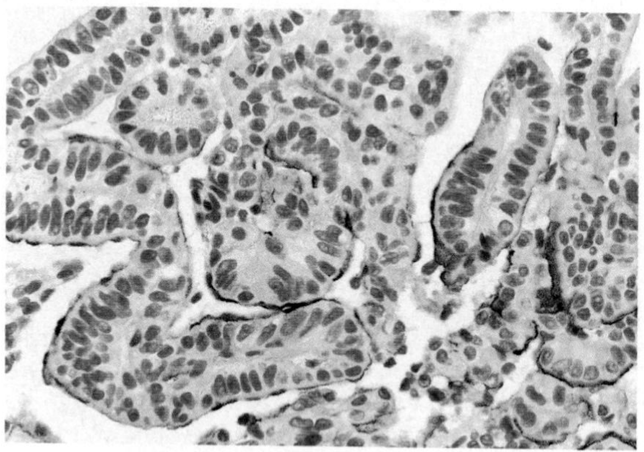

Figure 3-344

CHOROID PLEXUS PAPILLOMA

Immunoreactivity for EMA is not as widespread as it is for cytokeratins.

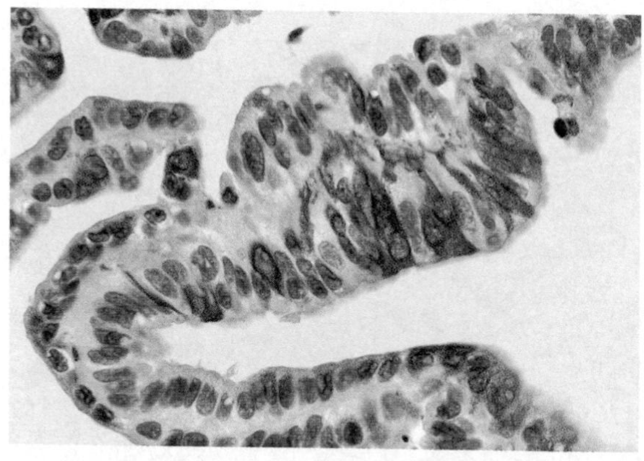

Figure 3-345

CHOROID PLEXUS PAPILLOMA

The epithelium may be focally immunoreactive for GFAP.

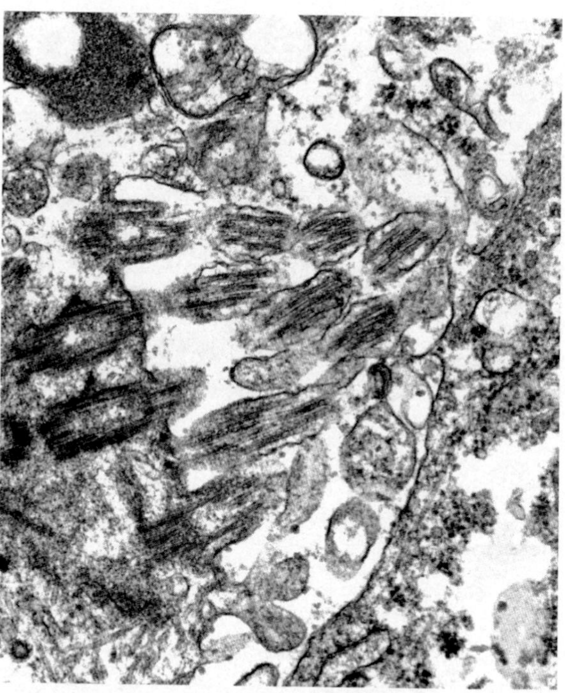

Figure 3-346

CHOROID PLEXUS CARCINOMA

The finding of epithelial features, in this case small lumens filled with cilia, helps establish the epithelial nature of a histologically undifferentiated intraventricular tumor.

cannot be considered entirely as choroid plexus specific; nevertheless, it may be useful (534,542).

MIB-1 indices vary considerably from lesion to lesion and from region to region within a given neoplasm. In one study, the index was less than 10 percent in 73 percent of papillomas, but never above 25 percent (520). No carcinoma had an index less than 10 percent, and 69 percent of malignant tumors had indices greater than 25 percent. Arbitrary cutoff points in terms of mitotic index or MIB-1 rate have not been defined for grading.

Ultrastructural Findings. The papilloma closely resembles normal choroid plexus, since the cells rest upon basal laminae, interdigitate lateral cell membranes, grasp adjacent cells by lateral desmosomes and apical junction complexes, produce occasional bundled intermediate filaments, and project both apical microvilli and scattered cilia (535,543). Mitochondria fill the cytoplasm of oncocytes (523,527, 528). These same features are present, to a limited extent, in choroid plexus carcinomas (fig. 3-346) (518,544).

Cytologic Findings. Papillae and cohesive cell clusters are seen well at low magnification (fig. 3-347). Individual dispersed cells have the typical profile of a columnar epithelial cell, with a tapered end and a blunt pole that contains the nucleus (545). The cells of choroid plexus carcinomas are less organized, less cohesive, and more cytologically atypical (fig. 3-348).

Molecular and Cytogenetic Findings. Cytogenetic changes in papillomas, as detected by FISH or comparative genomic hybridization, are many and include gains of chromosomes 5, 7, 9, 12, 15, and 18, and loss of chromosomes 21 and 22, among others (46). Carcinomas, in

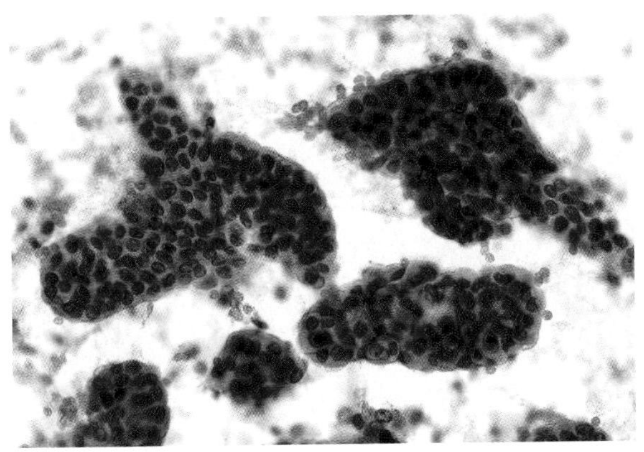

Figure 3-347

CHOROID PLEXUS PAPILLOMA: SMEAR PREPARATION

Cohesive papillary fragments are typical products of the smearing process.

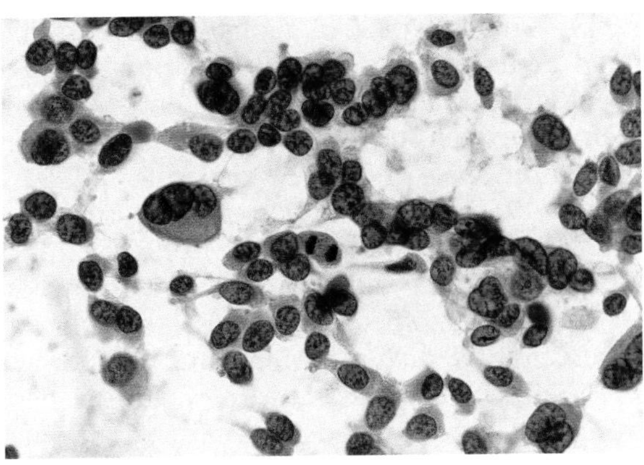

Figure 3-348

CHOROID PLEXUS CARCINOMA: SMEAR PREPARATION

Cells from a carcinoma are notably more atypical and mitotically active than those of a papilloma. Some cells have the tapered shape typical of columnar epithelial cells.

contrast, are more likely to have gain of chromosomes 1, 4, and 14, and loss of chromosomes 5, 15, and 18. How these differing cytogenetic changes relate to tumor progression is not clear, but it has been suggested that carcinomas may not begin as papillomas, although this sequence has been seen histologically, and various degrees of differentiation are common in choroid plexus neoplasms.

Given the similarity of some choroid plexus carcinomas to atypical teratoid/rhabdoid tumors (AT/RT) (547), it is disappointing to the diagnostician that abnormalities of the same gene, *INI1*, have been claimed, although not yet substantiated, in both AT/RT and choroid plexus tumors, particularly the carcinoma (548,549). This issue is discussed below in the section on differential diagnosis, and in chapter 5. Loss of chromosome 22 is common in both choroid plexus neoplasms and AT/RTs (546,547).

SV40 genomic sequences have been sought in light of the oncogenic property of polyoma viruses and the appearance of papillomas in transgenic mice transfected with SV40 (550). They have been found in some tumors (551), but their pathogenetic role, if any, remains unclear.

Differential Diagnosis. In nearly all instances, architectural and cytologic features readily distinguish papilloma from normal choroid plexus (fig. 3-349, see fig. 2-38). The epithelium of the neoplasm is more complicated and more cellular, often by a factor of two or more. As a consequence, the crowded cells are more columnar than the regularly spaced, domed or cobblestone cells of the normal plexus. Nuclear to cytoplasmic ratio and the degree of nuclear atypia are increased, although only slightly in some cases.

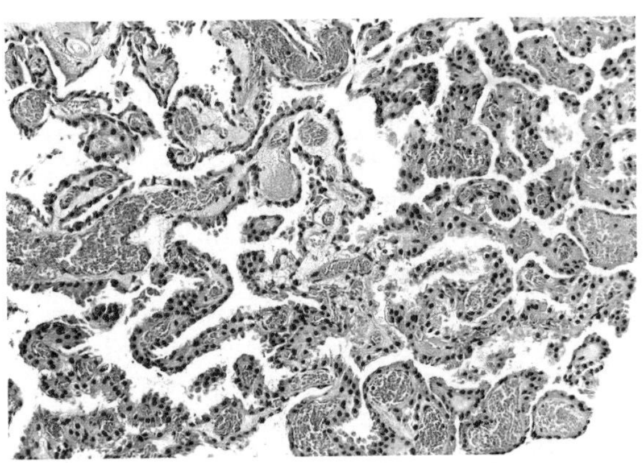

Figure 3-349

NORMAL CHOROID PLEXUS

In contrast to that of even the best-differentiated papilloma, normal choroid plexus epithelium has a "hobnailed" or "cobblestone" profile.

Papillary ependymoma infrequently enters the differential of plexus papilloma since only a rare ependymoma is truly papillary, i.e., with cells supported by fibrovascular stalks. Even then, the ependymoma is characterized not only by epithelial-appearing elements but by nonepithelial, more obviously glial cells whose processes aggregate to form the fibrillar background so typical of gliomas. Highly fibrillated areas are not a feature of papillomas. Ependymomas also lack the prominent basement membrane seen so well on PAS preparations or by immunohistochemistry for laminin (552). Ependymomas, in distinction to papillomas, are widely positive for GFAP. Aside from AE1/AE3, which shows the same pattern of staining as GFAP, ependymomas show far less staining for other keratins, such as CAM5.2 and CK7 (553).

Papillomas in adults must be distinguished from well-differentiated metastatic carcinoma within brain parenchyma. An intraparenchymal papillary lesion is almost certainly a metastasis. Antibodies BerEP4 and HEA25 are useful in distinguishing choroid plexus tumors from metastatic papillary carcinomas; only a rare plexus tumor is positive for either antibody, as is discussed in chapter 20, and illustrated in figures 20-20 and 20-21 (554). Many metastatic neoplasms of pulmonary origin are immunopositive for thyroid transcription factor-1 (TTF-1) (555), whereas choroid plexus tumors are not. Synaptophysin is diffusely positive in many choroid plexus tumors, but may be present in metastases from even a non-neuroendocrine pulmonary primary.

An entirely different diagnostic problem is presented in adults by intraventricular metastatic carcinoma. As noted above, if the lesion is not intraventricular and is clearly malignant, the odds greatly favor metastatic carcinoma. Metastasis to normal choroid plexus is rare but well documented (556). Immunoreactivity for cytokeratins does not distinguish metastatic carcinoma and choroid plexus carcinoma, but when combined with positivity for S-100 protein and synaptophysin, strongly favors a plexus primary, as does the rare occurrence of GFAP staining in better-differentiated examples. Positivity for prealbumin (transthyretin), a regular feature of choroid plexus tumors, has been noted in some metastatic carcinomas (534,542). The utility of BerEP4 and TTF-1 in identifying metastatic carcinomas is discussed above. In some instances, only the diagnosis of high-grade papillary carcinoma can be rendered, with resolution of the problem left to a search for a systemic primary. Infiltration of small glands into the stroma is more prominent in metastases.

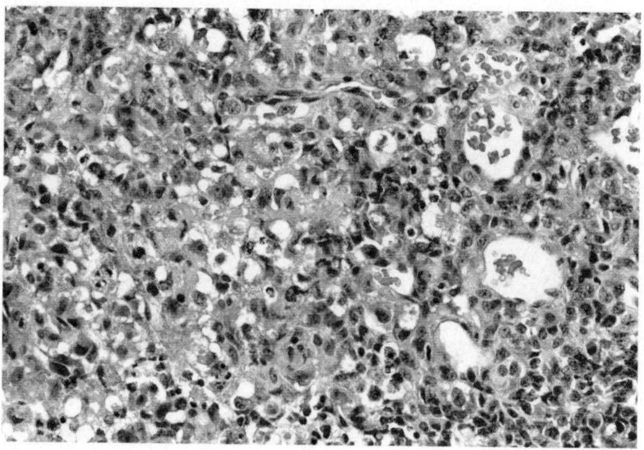

Figure 3-350

ATYPICAL TERATOID/RHABDOID TUMOR (AT/RT) RESEMBLING CHOROID PLEXUS CARCINOMA

The jumbled, vaguely epithelioid tissue of AT/RT can be confused with anaplastic choroid plexus carcinoma.

Solid, nonpapillary variants of choroid plexus carcinoma can resemble AT/RT (fig. 3-350). Plexus carcinoma affects mainly young children but is not as likely as AT/RT to present in infancy. In addition, the rhabdoid lesion often occurs in the posterior fossa, whereas almost all plexus carcinomas are supratentorial and intraventricular. The presence of EMA-positive rhabdoid cells and a polyphenotypic immunohistochemical profile is the defining quality of AT/RT. FISH can be used to seek abnormalities in *INI1* gene in AT/RT, although there are now reports of the same genetic disturbance in choroid plexus carcinomas, as is discussed above in Molecular and Cytogenetic Findings. In spite of this uncertainty, antibodies to INI1 protein are useful in this differential: the nuclei of choroid plexus carcinomas are positive; those of AT/RT are not (557).

Papillary endolymphatic sac tumor (fig. 3-351) needs to be considered in the face of a well-differentiated papillary tumor at the cerebellopontine angle (558). This rare neoplasm arises in adults, often as a component of von Hippel-Lindau syndrome, and is clearly extra-axial since it erodes the petrous ridge from within. Its epithelium is less papillary and simpler than that of a plexus neoplasm.

The distinction between choroid plexus carcinoma and small cell embryonal tumor is not always easy, particularly if the carcinoma is largely undifferentiated, but even then, choroid plexus

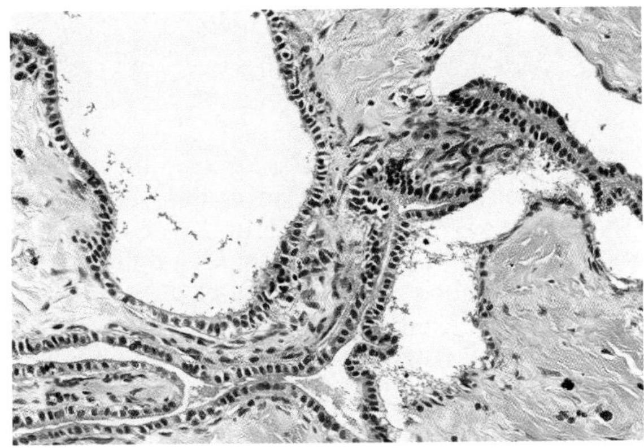

Figure 3-351

PAPILLARY ENDOLYMPHATIC SAC TUMOR RESEMBLING CHOROID PLEXUS CARCINOMA

The differential diagnosis of an epithelial lesion at the cerebellopontine angle should include papillary endolymphatic sac tumor. The epithelium of the latter is simpler, less papillary, and more cuboidal than that of a choroid plexus papilloma.

carcinomas are usually more pleomorphic. Plexus carcinomas are, in addition, almost always supratentorial and intraventricular, as well as keratin immunoreactive.

Anaplastic ependymomas are generally more monomorphic, and epithelial features, if any, include ependymal rosettes rather than papillae. Unlike ependymomas, choroid plexus carcinomas are more strongly cytokeratin reactive and typically GFAP negative.

An uncommon neoplasm that combines features suggestive of both ependymoma and choroid plexus papilloma is papillary tumor of the pineal region (see figs. 3-293, 6-14, 6-15) (559). This is discussed further in chapter 6.

Germ cell tumors in the differential diagnosis are primarily those with an epithelial phenotype and include embryonal carcinoma, endodermal sinus tumor, and immature teratoma. These lesions usually have a distinctive histologic pattern, are placental alkaline phosphatase, and c-kit (CD117) immunoreactive, and, in the case of the endodermal sinus tumor, show reactivity for alpha-fetoprotein.

Treatment and Prognosis. Well-differentiated papillomas are generally amenable to resection and are cured without radiotherapy or chemotherapy (560,561). Despite their location and the passing stream of CSF, the incidence of craniospinal dissemination of papillomas is low, but has occurred (562,563), even in the absence of surgery (564). Whereas most high-grade carcinomas are frankly infiltrative, unresectable, and therefore likely to recur (519), better-differentiated carcinomas are often cured by simple resection. Indeed, total excision is the principal prognostic factor when all carcinomas are considered, although the relationship between total resection and outcome when the degree of anaplasia is considered is not clear (560,565–567). Whether radiotherapy or chemotherapy is necessary for totally resected low-grade carcinomas is unclear (566). Carcinomas may be more invasive and may disseminate through the CSF, but only rarely spread hematogenously to systemic sites (568)

One study suggests that atypical papillomas behave like papillomas rather than carcinomas (561). A larger study also suggested an excellent outlook for patients with atypical papillomas, defined as noncarcinomas with 2 or more mitoses per 10 high-power fields, due to the complete resection that was possible in most instances (519a).

REFERENCES

Diffusely Infiltrating Astrocytic Tumors, Astrocytoma, Anaplastic Astrocytoma, Glioblastoma

1. Albright AL, Guthkelch AN, Packer RJ, Price RA, Rourke LB. Prognostic factors in pediatric brainstem gliomas. J Neurosurg 1986;65:751–5.
2. Albright AL, Price RA, Guthkelch AN. Brain stem gliomas of children. A clinicopathological study. Cancer 1983;52:2313–9.
3. Burger PC. Pathology of brain stem astrocytomas. Pediatr Neurosurg 1996;24:35–40.
4. Fisher PG, Breiter SN, Carson BS, et al. A clinicopathologic reappraisal of brain stem tumor classification. Identification of pilocystic astrocytoma and fibrillary astrocytoma as distinct entities. Cancer 2000;89:1569–76.
5. Amin MR, Kamitani H, Watanabe T, et al. A topographic analysis of the proliferating tumor cells in an autopsied brain with infiltrative thalamic glioma. Brain Tumor Pathol 2002;19:5–10.
6. Burger PC, Cohen KJ, Rosenblum MK, Tihan T. Pathology of diencephalic astrocytomas. Pediatr Neurosurg 2000;32:214–9.
7. Krouwer HG, Prados MD. Infiltrative astrocytomas of the thalamus. J Neurosurg 1995;82:548–57.
8. Di Rocco C, Iannelli A. Bilateral thalamic tumors in children. Childs Nerv Syst 2002;18:440–4.
9. Epstein FJ, Farmer JP, Freed D. Adult intramedullary astrocytomas of the spinal cord. J Neurosurg 1992;77:355–9.
10. Kim MS, Chung CK, Choe G, Kim IH, Kim HJ. Intramedullary spinal cord astrocytoma in adults: postoperative outcome. J Neurooncol 2001;52:85–94.
11. Santi M, Mena H, Wong K, Koeller K, Olsen C, Rushing EJ. Spinal cord malignant astrocytomas. Clinicopathologic features in 36 cases. Cancer 2003;98:554–61.
12. Strik HM, Effenberger O, Schafer O, Risch U, Wickboldt J, Meyermann R. A case of spinal glioblastoma multiforme: immunohistochemical study and review of the literature. J Neurooncol 2000;50:239–43.
13. Bernhardtsen T, Laursen H, Bojsen-Moller M, Gjerris F. Sub-classification of low-grade cerebellar astrocytoma: is it clinically meaningful? Childs Nerv Syst 2003;19:729–35.
14. Gupta V, Goyal A, Sinha S, et al. Glioblastoma of the cerebellum. A report of 3 cases. J Neurosurg Sci 2003;47:157–64; discussion 164–155.
15. Hayostek CJ, Shaw EG, Scheithauer B, et al. Astrocytomas of the cerebellum. A comparative clinicopathologic study of pilocytic and diffuse astrocytomas. Cancer 1993;72:856–69.
16. Albers GW, Hoyt WF, Forno LS, Shratter LA. Treatment response in malignant optic glioma of adulthood. Neurology 1988;38:1071–4.
17. Hoyt WF, Meshel LG, Lessell S, Schatz NJ, Suckling RD. Malignant optic glioma of adulthood. Brain 1973;96:121–32.
18. Millar WS, Tartaglino LM, Sergott RC, Friedman DP, Flanders AE. MR of malignant optic glioma of adulthood. AJNR Am J Neuroradiol 1995;16:1673–6.
19. Krouwer HG, Davis RL, Silver P, Prados M. Gemistocytic astrocytomas: a reappraisal. J Neurosurg 1991;74:399–406.
20. Watanabe K, Tachibana O, Yonekawa Y, Kleihues P, Ohgaki H. Role of gemistocytes in astrocytoma progression. Lab Invest 1997;76: 277–84.
21. Scherer H. Cerebral astrocytomas and their derivatives. Am J Cancer 1940;40:159–98.
22. Burger PC, Kleihues P. Cytologic composition of the untreated glioblastoma with implications for evaluation of needle biopsies. Cancer 1989;63:2014–23.
23. von Deiming A, Burger PC, Ohgaki H, Nakazato Y, Kleihues P. Diffuse astrocytoma. In: Louis DN, Ohgaki H, Wiestler OD, Cavenee W, eds. WHO classification of tumours of the central nervous system. Lyon: IARC Press; 2007. (In press)
24. Kleihues P, Burger PC. Anaplastic astrocytoma. In: Louis DN, Ohgaki H, Wiestler OD, Cavenee W, eds. WHO classification of tumours of the central nervous system. Lyon: IARC Press; 2007. (In press)
25. Kleihues P, Burger P, Cavenee W, et al. Glioblastoma. In: Louis DN, Ohgaki H, Wiestler OD, Cavenee W, eds. WHO classification of tumours of the central nervous system. Lyon: IARC Press; 2007. (In press)
26. Daumas-Duport C, Scheithauer B, O'Fallon J, Kelly P. Grading of astrocytomas. A simple and reproducible method. Cancer 1988;62:2152–65.
27. Kim TS, Halliday AL, Hedley-Whyte ET, Convery K. Correlates of survival and the Daumas-Duport grading system for astrocytomas. J Neurosurg 1991;74:27–37.
28. Giannini C, Scheithauer BW, Burger PC, et al. Cellular proliferation in pilocytic and diffuse astrocytomas. J Neuropathol Exp Neurol 1999; 58:46–53.
29. Hamilton SR, Liu B, Parsons RE, et al. The molecular basis of Turcot's syndrome. N Engl J Med 1995;332:839–47.

30. Brat DJ, James CD, Jedlicka AE, et al. Molecular genetic alterations in radiation-induced astrocytomas. Am J Pathol 1999;154:1431–8.
31. Rappaport ZH, Loven D, Ben-Aharon U. Radiation-induced cerebellar glioblastoma multiforme subsequent to treatment of an astrocytoma of the cervical spinal cord. Neurosurgery 1991;29:606–8.
32. Salvati M, Frati A, Russo N, et al. Radiation-induced gliomas: report of 10 cases and review of the literature. Surg Neurol 2003;60:60–7.
33. Simmons NE, Laws ER Jr. Glioma occurrence after sellar irradiation: case report and review. Neurosurgery 1998;42:172–8.
34. Chang S, Prados MD. Identical twins with Ollier's disease and intracranial gliomas: case report. Neurosurgery 1994;34:903–6.
35. Hofman S, Heeg M, Klein JP, Krikke AP. Simultaneous occurrence of a supra- and an infratentorial glioma in a patient with Ollier's disease: more evidence for non-mesodermal tumor predisposition in multiple enchondromatosis. Skeletal Radiol 1998;27:688–91.
36. Mellon CD, Carter JE, Owen DB. Ollier's disease and Maffucci's syndrome: distinct entities or a continuum. Case report: enchondromatosis complicated by an intracranial glioma. J Neurol 1988;235:376–8.
37. Rawlings CE 3rd, Bullard DE, Burger PC, Friedman AH. A case of Ollier's disease associated with two intracranial gliomas. Neurosurgery 1987;21:400–3.
38. van Nielen KM, de Jong BM. A case of Ollier's disease associated with two intracerebral low-grade gliomas. Clin Neurol Neurosurg 1999; 101:106–10.
39. Tachibana I, Smith JS, Sato K, Hosek SM, Kimmel DW, Jenkins RB. Investigation of germline PTEN, p53, p16(INK4A)/p14(ARF), and CDK4 alterations in familial glioma. Am J Med Genet 2000;92:136–41.
40. McLendon RE, Robinson JS Jr, Chambers DB, Grufferman S, Burger PC. The glioblastoma multiforme in Georgia, 1977-1981. Cancer 1985;56:894–7.
41. Scherer H. Structural developments in gliomas. Am J Cancer 1938;34:333–51.
42. Scherer H. The forms of growth in gliomas and their practical significance. Brain 1940;63:1–35.
43. Coons SW, Pearl DK. Mitosis identification in diffuse gliomas: implications for tumor grading. Cancer 1998;82:1550–5.
44. Kepes JJ, Rubinstein LJ, Chiang H. The role of astrocytes in the formation of cartilage in gliomas. An immunohistochemical study of four cases. Am J Pathol 1984;117:471–83.
45. Cosgrove M, Fitzgibbons PL, Sherrod A, Chandrasoma PT, Martin SE. Intermediate filament expression in astrocytic neoplasms. Am J Surg Pathol 1989;13:141–5.
46. Hsu DW, Louis DN, Efird JT, Hedley-Whyte ET. Use of MIB-1 (Ki-67) immunoreactivity in differentiating grade II and grade III gliomas. J Neuropathol Exp Neurol 1997;56:857–65.
47. McKeever PE, Strawderman MS, Yamini B, Mikhail AA, Blaivas M. MIB-1 proliferation index predicts survival among patients with grade II astrocytoma. J Neuropathol Exp Neurol 1998;57:931–6.
48. Raghavan R, Steart PV, Weller RO. Cell proliferation patterns in the diagnosis of astrocytomas, anaplastic astrocytomas and glioblastoma multiforme: a Ki-67 study. Neuropathol Appl Neurobiol 1990;16:123–33.
49. Kros JM, Stefanko SZ, de Jong AA, van Vroonhoven CC, van der Heul RO, van der Kwast TH. Ultrastructural and immunohistochemical segregation of gemistocytic subsets. Hum Pathol 1991;22:33–40.
50. Scheithauer BW, Bruner JM. Central nervous system tumors. Clin Lab Med 1987;7:157–79.
51. Scheithauer BW, Bruner JM. The ultrastructural spectrum of astrocytic neoplasms. Ultrastruct Pathol 1987;11:535–81.
52. Kosel S, Scheithauer BW, Graeber MB. Genotype-phenotype correlation in gemistocytic astrocytomas. Neurosurgery 2001;48:187–93; discussion 193–184.
53. Peraud A, Kreth FW, Wiestler OD, Kleihues P, Reulen HJ. Prognostic impact of TP53 mutations and P53 protein overexpression in supratentorial WHO grade II astrocytomas and oligoastrocytomas. Clin Cancer Res 2002;8: 1117–24.
54. Reis RM, Hara A, Kleihues P, Ohgaki H. Genetic evidence of the neoplastic nature of gemistocytes in astrocytomas. Acta Neuropathol (Berl) 2001;102:422–5.
55. Hirose Y, Aldape KD, Chang S, Lamborn K, Berger MS, Feuerstein BG. Grade II astrocytomas are subgrouped by chromosome aberrations. Cancer Genet Cytogenet 2003;142:1–7.
56. Wessels PH, Twijnstra A, Kessels AG, et al. Gain of chromosome 7, as detected by in situ hybridization, strongly correlates with shorter survival in astrocytoma grade 2. Genes Chromosomes Cancer 2002;33:279–84.
57. Smith JS, Alderete B, Minn Y, et al. Localization of common deletion regions on 1p and 19q in human gliomas and their association with histological subtype. Oncogene 1999;18: 4144–52.

58. Wessels PH, Hopman AH, Ummelen MI, Krijne-Kubat B, Ramaekers FC, Twijnstra A. Differentiation between reactive gliosis and diffuse astrocytoma by in situ hybridization. Neurology 2001;56:1224–7.
59. Kurtkaya-Yapicier O, Scheithauer BW, Hebrink D, James CD. p53 in nonneoplastic central nervous system lesions: an immunohistochemical and genetic sequencing study. Neurosurgery 2002;51:1246–54; discussion 1254–5.
60. Yaziji H, Massarani-Wafai R, Gujrati M, Kuhns JG, Martin AW, Parker JC Jr. Role of p53 immunohistochemistry in differentiating reactive gliosis from malignant astrocytic lesions. Am J Surg Pathol 1996;20:1086–90.
61. Burger PC. What is an oligodendroglioma? Brain Pathol 2002;12:257–9.
62. Burkhard C, Di Patre PL, Schuler D, et al. A population-based study of the incidence and survival rates in patients with pilocytic astrocytoma. J Neurosurg 2003;98:1170–4.
63. Vertosick FT Jr, Selker RG, Arena VC. Survival of patients with well-differentiated astrocytomas diagnosed in the era of computed tomography. Neurosurgery 1991;28:496–501.
64. McCormack BM, Miller DC, Budzilovich GN, Voorhees GJ, Ransohoff J. Treatment and survival of low-grade astrocytoma in adults—1977-1988. Neurosurgery 1992;31:636–42.
65. Heesters M, Molenaar W, Go GK. Radiotherapy in supratentorial gliomas. A study of 821 cases. Strahlenther Onkol 2003;179:606–14.
66. Ohgaki H, Dessen P, Jourde B, et al. Genetic pathways to glioblastoma: a population-based study. Cancer Res 2004;64:6892–9.
67. Ohgaki H, Kleihues P. Population-based studies on incidence, survival rates, and genetic alterations in astrocytic and oligodendroglial gliomas. J Neuropathol Exp Neurol 2005;64: 479–89.
68. Berger MS, Deliganis AV, Dobbins J, Keles GE. The effect of extent of resection on recurrence in patients with low grade cerebral hemisphere gliomas. Cancer 1994;74:1784–91.
69. Pollack IF, Claassen D, al-Shboul Q, Janosky JE, Deutsch M. Low-grade gliomas of the cerebral hemispheres in children: an analysis of 71 cases. J Neurosurg 1995;82:536–47.
70. Karim AB, Afra D, Cornu P, et al. Randomized trial on the efficacy of radiotherapy for cerebral low-grade glioma in the adult: European Organization for Research and Treatment of Cancer Study 22845 with the Medical Research Council study BRO4: an interim analysis. Int J Radiat Oncol Biol Phys 2002;52:316–24.
71. Minehan KJ, Shaw EG, Scheithauer BW, Davis DL, Onofrio BM. Spinal cord astrocytoma: pathological and treatment considerations. J Neurosurg 1995;83:590–95.
72. Rossitch E Jr, Zeidman SM, Burger PC, et al. Clinical and pathological analysis of spinal cord astrocytomas in children. Neurosurgery 1990; 27:193–6.
73. Stander M, Peraud A, Leroch B, Kreth FW. Prognostic impact of TP53 mutation status for adult patients with supratentorial World Health Organization Grade II astrocytoma or oligoastrocytoma: a long-term analysis. Cancer 2004; 101:1028–35.
74. Perry A, Aldape KD, George DH, Burger PC. Small cell astrocytoma: an aggressive variant that is clinicopathologically and genetically distinct from anaplastic oligodendroglioma. Cancer 2004;101:2318–26.
75. Aldape K, Ballman K, Furth A, et al. Immunohistochemical detection of EGFRvIII in malignant astrocytomas and evaluation of prognostic significance. J Neuropathol Exp Neurol 2004;63:700–7.
76. Smith JS, Tachibana I, Passe SM, et al. PTEN mutation, EGFR amplification, and outcome in patients with anaplastic astrocytoma and glioblastoma multiforme. J Natl Cancer Inst 2001;93:1246–56.
77. Chamberlain MC. Combined-modality treatment of leptomeningeal gliomatosis. Neurosurgery 2003;52:324–30.
78. Chamberlain MC, Silver P, Levin VA. Poorly differentiated gliomas of the cerebellum. A study of 18 patients. Cancer 1990;65:337–40.
79. Dohrmann GJ, Dunsmore RH. Gliobastoma multiforme of the cerebellum. Surg Neurol 1975;3:219–23.
80. Levine SA, McKeever PE, Greenberg HS. Primary cerebellar glioblastoma multiforme. J Neurooncol 1987;5:231–6.
81. Ciappetta P, Salvati M, Capoccia G, Artico M, Raco A, Fortuna A. Spinal glioblastomas: report of seven cases and review of the literature. Neurosurgery 1991;28:302–6.
82. Barnard RO, Geddes JF. The incidence of multifocal cerebral gliomas. A histologic study of large hemisphere sections. Cancer 1987;60: 1519–31.
83. Burger PC, Heinz ER, Shibata T, Kleihues P. Topographic anatomy and CT correlations in the untreated glioblastoma multiforme. J Neurosurg 1988;68:698–704.
84. Daumas-Duport C, Scheithauer BW, Kelly PJ. A histologic and cytologic method for the spatial definition of gliomas. Mayo Clin Proc 1987;62:435–49.

85. Kelly PJ, Daumas-Duport C, Scheithauer BW, Kall BA, Kispert DB. Stereotactic histologic correlations of computed tomography- and magnetic resonance imaging-defined abnormalities in patients with glial neoplasms. Mayo Clin Proc 1987;62:450–9.
86. Iwama T, Yamada H, Sakai N, et al. Correlation between magnetic resonance imaging and histopathology of intracranial glioma. Neurol Res 1991;13:48–54.
87. Johnson PC, Hunt SJ, Drayer BP. Human cerebral gliomas: correlation of postmortem MR imaging and neuropathologic findings. Radiology 1989;170:211–7.
88. Burger PC, Pearl DK, Aldape K, et al. Small cell architecture—a histological equivalent of EGFR amplification in glioblastoma multiforme? J Neuropathol Exp Neurol 2001;60:1099–104.
89. Margetts JC, Kalyan-Raman UP. Giant-celled glioblastoma of brain. A clinico-pathological and radiological study of ten cases (including immunohistochemistry and ultrastructure). Cancer 1989;63:524–31.
90. Rosenblum MK, Erlandson RA, Budzilovich GN. The lipid-rich epithelioid glioblastoma. Am J Surg Pathol 1991;15:925–34.
91. Wyatt-Ashmead J, Kleinschmidt-DeMasters BK, Hill DA, et al. Rhabdoid glioblastoma. Clin Neuropathol 2001;20:248–55.
92. Mork SJ, Rubinstein LJ, Kepes JJ. Patterns of epithelial metaplasia in malignant gliomas. I. Papillary formations mimicking medulloepithelioma. J Neuropathol Exp Neurol Mar 1988; 47:93–100.
93. Mork SJ, Rubinstein LJ, Kepes JJ, Perentes E, Uphoff DF. Patterns of epithelial metaplasia in malignant gliomas. II. Squamous differentiation of epithelial-like formations in gliosarcomas and glioblastomas. J Neuropathol Exp Neurol 1988;47:101–18.
94. Mueller W, Lass U, Herms J, Kuchelmeister K, Bergmann M, von Deimling A. Clonal analysis in glioblastoma with epithelial differentiation. Brain Pathol 2001;11:39–43.
95. Brat DJ, Castellano-Sanchez AA, Hunter SB, et al. Pseudopalisades in glioblastoma are hypoxic, express extracellular matrix proteases, and are formed by an actively migrating cell population. Cancer Res 2004;64:920–7.
96. Brat DJ, Van Meir EG. Vaso-occlusive and prothrombotic mechanisms associated with tumor hypoxia, necrosis, and accelerated growth in glioblastoma. Lab Invest 2004;84: 397–405.
97. Haddad SF, Moore SA, Schelper RL, Goeken JA. Vascular smooth muscle hyperplasia underlies the formation of glomeruloid vascular structures of glioblastoma multiforme. J Neuropathol Exp Neurol 1992;51:488–92.
98. Wesseling P, Schlingemann RO, Rietveld FJ, Link M, Burger PC, Ruiter DJ. Early and extensive contribution of pericytes/vascular smooth muscle cells to microvascular proliferation in glioblastoma multiforme: an immuno-light and immuno-electron microscopic study. J Neuropathol Exp Neurol 1995;54:304–10.
99. Roggendorf W, Strupp S, Paulus W. Distribution and characterization of microglia/macrophages in human brain tumors. Acta Neuropathol (Berl) 1996;92:288–93.
100. Cummings TJ, Hulette CM, Bigner SH, Riggins GJ, McLendon RE. Ham56-immunoreactive macrophages in untreated infiltrating gliomas. Arch Pathol Lab Med 2001;125:637–41.
101. Rossi ML, Hughes JT, Esiri MM, Coakham HB, Brownell DB. Immunohistological study of mononuclear cell infiltrate in malignant gliomas. Acta Neuropathol (Berl) 1987;74:269–77.
102. Burger PC, Mahley MS Jr, Dudka L, Vogel FS. The morphologic effects of radiation administered therapeutically for intracranial gliomas: a postmortem study of 25 cases. Cancer 1979;44:1256–72.
103. Forsyth PA, Kelly PJ, Cascino TL, et al. Radiation necrosis or glioma recurrence: is computer-assisted stereotactic biopsy useful? J Neurosurg 1995;82:436–44.
104. Oh D, Prayson RA. Evaluation of epithelial and keratin markers in glioblastoma multiforme: an immunohistochemical study. Arch Pathol Lab Med 1999;123:917–20.
105. Kepes JJ, Fulling KH, Garcia JH. The clinical significance of "adenoid" formations of neoplastic astrocytes, imitating metastatic carcinoma, in gliosarcomas. A review of five cases. Clin Neuropathol 1982;1:139–50.
106. Martinez-Diaz H, Kleinschmidt-DeMasters BK, Powell SZ, Yachnis AT. Giant cell glioblastoma and pleomorphic xanthoastrocytoma show different immunohistochemical profiles for neuronal antigens and p53 but share reactivity for class III beta-tubulin. Arch Pathol Lab Med 2003;127:1187–91.
107. Meyer-Puttlitz B, Hayashi Y, Waha A, et al. Molecular genetic analysis of giant cell glioblastomas. Am J Pathol 1997;151:853–7.
108. Inda MM, Fan X, Munoz J, et al. Chromosomal abnormalities in human glioblastomas: gain in chromosome 7p correlating with loss in chromosome 10q. Mol Carcinog 2003;36:6–14.
109. Mohapatra G, Bollen AW, Kim DH, et al. Genetic analysis of glioblastoma multiforme provides evidence for subgroups within the grade. Genes Chromosomes Cancer 1998;21:195–206.

110. Fujisawa H, Kurrer M, Reis RM, Yonekawa Y, Kleihues P, Ohgaki H. Acquisition of the glioblastoma phenotype during astrocytoma progression is associated with loss of heterozygosity on 10q25-qter. Am J Pathol 1999;155:387–94.
111. Brat DJ, Castellano-Sanchez A, Kaur B, Van Meir EG. Genetic and biologic progression in astrocytomas and their relation to angiogenic dysregulation. Adv Anat Pathol 2002;9:24–36.
112. Kleihues P, Ohgaki H. Primary and secondary glioblastomas: from concept to clinical diagnosis. Neuro-oncol 1999;1:44–51.
113. Tohma Y, Gratas C, Biernat W, et al. PTEN (MMAC1) mutations are frequent in primary glioblastomas (de novo) but not in secondary glioblastomas. J Neuropathol Exp Neurol 1998;57:684–9.
114. Watanabe K, Tachibana O, Sata K, Yonekawa Y, Kleihues P, Ohgaki H. Overexpression of the EGF receptor and p53 mutations are mutually exclusive in the evolution of primary and secondary glioblastomas. Brain Pathol 1996;6:217–23; discussion 223–4.
115. Fujisawa H, Reis RM, Nakamura M, et al. Loss of heterozygosity on chromosome 10 is more extensive in primary (de novo) than in secondary glioblastomas. Lab Invest 2000;80:65–72.
116. Perry A, Aldape KD, George DH, Burger PC. Small cell astrocytoma: an aggressive variant that is clinicopathologically and genetically distinct from anaplastic oligodendroglioma. Cancer 2004;101:2318–26.
117. Peraud A, Watanabe K, Plate KH, Yonekawa Y, Kleihues P, Ohgaki H. p53 mutations versus EGF receptor expression in giant cell glioblastomas. J Neuropathol Exp Neurol 1997;56:1236–41.
118. Peraud A, Watanabe K, Schwechheimer K, Yonekawa Y, Kleihues P, Ohgaki H. Genetic profile of the giant cell glioblastoma. Lab Invest 1999;79:123–9.
119. Cheng Y, Ng HK, Zhang SF, et al. Genetic alterations in pediatric high-grade astrocytomas. Hum Pathol 1999;30:1284–90.
120. Sung T, Miller DC, Hayes RL, Alonso M, Yee H, Newcomb EW. Preferential inactivation of the p53 tumor suppressor pathway and lack of EGFR amplification distinguish de novo high grade pediatric astrocytomas from de novo adult astrocytomas. Brain Pathol 2000;10:249–59.
120a. Homma T, Fuushima T, Vaccarella S, et al. Correlation among pathology, genotype, and patient outcomes in glioblastoma. J Neuropathol Exp Neurol 2006;65:846–54.
121. Szybka M, Bartkowiak J, Zakrzewski K, Polis L, Liberski P, Kordek R. Microsatellite instability and expression of DNA mismatch repair genes in malignant astrocytic tumors from adult and pediatric patients. Clin Neuropathol 2003;22:180–6.
122. Kriho VK, Yang HY, Moskal JR, Skalli O. Keratin expression in astrocytomas: an immunofluorescent and biochemical reassessment. Virchows Arch 1997;431:139–47.
123. Johnson P, Drazkowski J, Rossman K, Coons S. Cause of death in glioblastoma multiforme. J Neuropathol Exp Neurol 1993;52:289.
124. Silbergeld DL, Rostomily RC, Alvord EC Jr. The cause of death in patients with glioblastoma is multifactorial: clinical factors and autopsy findings in 117 cases of supratentorial glioblastoma in adults. J Neurooncol 1991;10:179–85.
125. Vertosick FT Jr, Selker RG. Brain stem and spinal metastases of supratentorial glioblastoma multiforme: a clinical series. Neurosurgery 1990;27:516–21; discussion 521–2.
126. Armanios M, Grossman S, Yang S, White B, Burger P, Orens J. Transmission of glioblastoma multiforme following bilateral lung transplantation from an affected donor. Case report and review of the literature. Neuro-Oncology. 2004;6:259–63.
127. al-Rikabi AC, al-Sohaibani MO, Jamjoom A, al-Rayess MM. Metastatic deposits of a high-grade malignant glioma in cervical lymph nodes diagnosed by fine needle aspiration (FNA) cytology—case report and literature review. Cytopathology 1997;8:421–7.
128. Ates LE, Bayindir C, Bilgic B, Karasu A. Glioblastoma with lymph node metastases. Neuropathology 2003;23:146–9.
129. Beauchesne P, Soler C, Mosnier JF. Diffuse vertebral body metastasis from a glioblastoma multiforme: a technetium-99m Sestamibi single-photon emission computerized tomography study. J Neurosurg 2000;93:887–90.
130. Dolman CL. Lymph node metastasis as first manifestation of glioblastoma. Case report. J Neurosurg 1974;41:607–9.
131. Gamis AS, Egelhoff J, Roloson G, et al. Diffuse bony metastases at presentation in a child with glioblastoma multiforme. A case report. Cancer 1990;66:180–4.
132. Myers T, Egelhoff J, Myers M. Glioblastoma multiforme presenting as osteoblastic metastatic disease: case report and review of the literature. AJNR Am J Neuroradiol 1990;11:802–3.
133. Newton HB, Rosenblum MK, Walker RW. Extraneural metastases of infratentorial glioblastoma multiforme to the peritoneal cavity. Cancer 1992;69:2149–53.
134. Park CC, Hartmann C, Folkerth R, et al. Systemic metastasis in glioblastoma may represent the emergence of neoplastic subclones. J Neuropathol Exp Neurol 2000;59:1044–50.

135. Schweitzer T, Vince GH, Herbold C, Roosen K, Tonn JC. Extraneural metastases of primary brain tumors. J Neurooncol 2001;53:107–14.
136. Winger MJ, Macdonald DR, Cairncross JG. Supratentorial anaplastic gliomas in adults. The prognostic importance of extent of resection and prior low-grade glioma. J Neurosurg 1989; 71:487–93.
137. Kelly PJ, Hunt C. The limited value of cytoreductive surgery in elderly patients with malignant gliomas. Neurosurgery 1994;34:62–7.
138. Albert FK, Forsting M, Sartor K, Adams HP, Kunze S. Early postoperative magnetic resonance imaging after resection of malignant glioma: objective evaluation of residual tumor and its influence on regrowth and prognosis. Neurosurgery 1994;34:45–60; discussion 60–1.
139. Keles GE, Lamborn KR, Chang SM, Prados MD, Berger MS. Volume of residual disease as a predictor of outcome in adult patients with recurrent supratentorial glioblastomas multiforme who are undergoing chemotherapy. J Neurosurg 2004;100:41–6.
140. Klein R, Molenkamp G, Sorensen N, Roggendorf W. Favorable outcome of giant cell glioblastoma in a child. Report of an 11-year survival period. Childs Nerv Syst 1998;14:288–91.
141. Sabel M, Reifenberger J, Weber RG, Reifenberger G, Schmitt HP. Long-term survival of a patient with giant cell glioblastoma. Case report. J Neurosurg 2001;94:605–11.
142. Yoshimura J, Onda K, Tanaka R, Takahashi H. Clinicopathological study of diffuse type brainstem gliomas: analysis of 40 autopsy cases. Neurol Med Chir (Tokyo) 2003;43:375–82.
143. Simmons ML, Lamborn KR, Takahashi M, et al. Analysis of complex relationships between age, p53, epidermal growth factor receptor, and survival in glioblastoma patients. Cancer Res 2001;61:1122–8.
143a. Mellinghoff IK, Wang MY, Vivanco I, et al. Molecular determinants of the response of glioblastomas to EGRF kinase inhibitors. N Engl J Med 2005:353:201–24.
143b. Hegi ME, Diserens A, Gorlia T, et al. MGMT gene silencing and benefit from temozolomide in glioblastoma. N Engl J Med 2005;352997–1003.
144. Newcomb EW, Cohen H, Lee SR, et al. Survival of patients with glioblastoma multiforme is not influenced by altered expression of p16, p53, EGFR, MDM2 or Bcl-2 genes. Brain Pathol 1998;8:655–67.
145. Waha A, Baumann A, Wolf HK, et al. Lack of prognostic relevance of alterations in the epidermal growth factor receptor-transforming growth factor-alpha pathway in human astrocytic gliomas. J Neurosurg 1996;85:634–41.
146. Schmidt MC, Antweiler S, Urban N, et al. Impact of genotype and morphology on the prognosis of glioblastoma. J Neuropathol Exp Neurol 2002;61:321–8.
146a. Hegi ME, Diserens A, Gorila T, et al. MGMT gene silencing and benefit from temozolomide in glioblastoma. N Engl J Med 2005;352:997–1003.

Gliosarcoma

147. Meis JM, Ho KL, Nelson JS. Gliosarcoma: a histologic and immunohistochemical reaffirmation. Mod Pathol 1990;3:19–24.
148. Perry JR, Ang LC, Bilbao JM, Muller PJ. Clinicopathologic features of primary and postirradiation cerebral gliosarcoma. Cancer 1995;75:2910–8.
149. Paulus W, Jellinger K. Mixed glioblastoma and malignant mesenchymoma, a variety of gliosarcoma. Histopathology 1993;22:277–9.
150. Ozolek JA, Finkelstein SD, Couce ME. Gliosarcoma with epithelial differentiation: immunohistochemical and molecular characterization. A case report and review of the literature. Mod Pathol 2004;17:739–45.
151. Hayashi K, Ohara N, Jeon HJ, et al. Gliosarcoma with features of chondroblastic osteosarcoma. Cancer 1993;72:850–5.
152. Shintaku M, Miyaji K, Adachi Y. Gliosarcoma with angiosarcomatous features: a case report. Brain Tumor Pathol 1998;15:101–5.
153. Haddad SF, Moore SA, Schelper RL, Goeken JA. Smooth muscle can comprise the sarcomatous component of gliosarcomas. J Neuropathol Exp Neurol 1992;51:493–8.
154. Barnard RO, Bradford R, Scott T, Thomas DG. Gliomyosarcoma. Report of a case of rhabdomyosarcoma arising in a malignant glioma. Acta Neuropathol (Berl) 1986;69:23–7.
155. Stapleton SR, Harkness W, Wilkins PR, Uttley D. Gliomyosarcoma: an immunohistochemical analysis. J Neurol Neurosurg Psychiatry 1992;55:728–30.
156. Kepes JJ, Fulling KH, Garcia JH. The clinical significance of "adenoid" formations of neoplastic astrocytes, imitating metastatic carcinoma, in gliosarcomas. A review of five cases. Clin Neuropathol 1982;1:139–50.
157. Mork SJ, Rubinstein LJ, Kepes JJ, Perentes E, Uphoff DF. Patterns of epithelial metaplasia in malignant gliomas. II. Squamous differentiation of epithelial-like formations in gliosarcomas and glioblastomas. J Neuropathol Exp Neurol 1988;47:101–18.
158. Sreenan JJ, Prayson RA. Gliosarcoma. A study of 13 tumors, including p53 and CD34 immunohistochemistry. Arch Pathol Lab Med 1997;121:129–33.
159. Ho KL. Histogenesis of sarcomatous component of the gliosarcoma: an ultrastructural study. Acta Neuropathol (Berl) 1990;81:178–88.

160. Parwani AV, Berman D, Burger PC, Ali SZ. Gliosarcoma: cytopathologic characteristics on fine-needle aspiration (FNA) and intraoperative touch imprint. Diagn Cytopathol 2004;30:77–81.
161. Biernat W, Aguzzi A, Sure U, Grant JW, Kleihues P, Hegi ME. Identical mutations of the p53 tumor suppressor gene in the gliomatous and the sarcomatous components of gliosarcomas suggest a common origin from glial cells. J Neuropathol Exp Neurol 1995;54:651–6.
162. Boerman RH, Anderl K, Herath J, et al. The glial and mesenchymal elements of gliosarcomas share similar genetic alterations. J Neuropathol Exp Neurol 1996;55:973–81.
163. Reis RM, Konu-Lebleblicioglu D, Lopes JM, Kleihues P, Ohgaki H. Genetic profile of gliosarcomas. Am J Pathol 2000;156:425–32.
164. Actor B, Cobbers JM, Buschges R, et al. Comprehensive analysis of genomic alterations in gliosarcoma and its two tissue components. Genes Chromosomes Cancer 2002;34:416–27.
165. Jones H, Steart PV, Weller RO. Spindle-cell glioblastoma or gliosarcoma? Neuropathol Appl Neurobiol 1991;17:177–87.
166. Galanis E, Buckner JC, Dinapoli RP, et al. Clinical outcome of gliosarcoma compared with glioblastoma multiforme: North Central Cancer Treatment Group results. J Neurosurg 1998;89:425–30.
167. Meis JM, Martz KL, Nelson JS. Mixed glioblastoma multiforme and sarcoma. A clinicopathologic study of 26 radiation therapy oncology group cases. Cancer 1991;67:2342–9.
168. Murphy MN, Korkis JA, Robson FC, Sima AA. Gliosarcoma with cranial penetration and extension to the maxillary sinus. J Otolaryngol 1985;14:313–6.

Granular Cell Astrocytic Neoplasms

169. Brat DJ, Scheithauer BW, Medina-Flores R, Rosenblum MK, Burger PC. Infiltrative astrocytomas with granular cell features (granular cell astrocytomas): a study of histopathologic features, grading, and outcome. Am J Surg Pathol 2002;26:750–7.
170. Geddes JF, Thom M, Robinson SF, Revesz T. Granular cell change in astrocytic tumors. Am J Surg Pathol 1996;20:55–63.
171. Kornfeld M. Granular cell glioblastoma: a malignant granular cell neoplasm of astrocytic origin. J Neuropathol Exp Neurol 1986;45:447–62.
172. Chorny JA, Evans LC, Kleinschmidt-DeMasters BK. Cerebral granular cell astrocytomas: a Mib-1, bcl-2, and telomerase study. Clin Neuropathol 2000;19:170–9.
173. Giangaspero F, Cenacchi G. Oncocytic and granular cell neoplasms of the central nervous system and pituitary gland. Semin Diagn Pathol 1999;16:91–7.
174. Castellano-Sanchez AA, Ohgaki H, Yokoo H, et al. Granular cell astrocytomas show a high frequency of allelic loss but are not a genetically defined subset. Brain Pathol 2003;13:185–94.

Gliomatosis Cerebri

175. Herrlinger U, Felsberg J, Kuker W, et al. Gliomatosis cerebri: molecular pathology and clinical course. Ann Neurol 2002;52:390–9.
176. Tancredi A, Mangiola A, Guiducci A, Peciarolo A, Ottaviano P. Oligodendrocytic gliomatosis cerebri. Acta Neurochir (Wien) 2000;142:469–72.
177. Kim DG, Yang HJ, Park IA, et al. Gliomatosis cerebri: clinical features, treatment, and prognosis. Acta Neurochir (Wien) 1998;140:755–62.
178. Vates GE, Chang S, Lamborn KR, Prados M, Berger MS. Gliomatosis cerebri: a review of 22 cases. Neurosurgery 2003;53:261–71.
179. Peretti-Viton P, Brunel H, Chinot O, et al. Histological and MR correlations in gliomatosis cerebri. J Neurooncol 2002;59:249–59.
180. Keene DL, Jimenez C, Hsu E. MRI diagnosis of gliomatosis cerebri. Pediatr Neurol 1999;20:148–51.
181. Shin YM, Chang KH, Han MH, et al. Gliomatosis cerebri: comparison of MR and CT features. AJR Am J Roentgenol 1993;161:859–62.
182. Kros JM, Zheng P, Dinjens WN, Alers JC. Genetic aberrations in gliomatosis cerebri support monoclonal tumorigenesis. J Neuropathol Exp Neurol 2002;61:806–14.

Meningeal Gliomatosis

183. Bailey O. Relation of glioma of the leptomeninges to neuroglia nests. Report of a case of astrocytoma of the leptomeninges. Arch Pathol 1936;21:584–600.
184. Cooper IS, Kernohan JW. Heterotopic glial nests in the subarachnoid space; histopathologic characteristics, mode of origin and relation to meningeal gliomas. J Neuropathol Exp Neurol 1951;10:16–29.
185. Kakita A, Wakabayashi K, Takahashi H, Ohama E, Ikuta F, Tokiguchi S. Primary leptomeningeal glioma: ultrastructural and laminin immunohistochemical studies. Acta Neuropathol (Berl) 1992;83:538–42.
186. Sceats DJ Jr, Quisling R, Rhoton AL Jr, Ballinger WE, Ryan P. Primary leptomeningeal glioma mimicking an acoustic neuroma: case report with review of the literature. Neurosurgery 1986;19:649–54.
187. Armao DM, Stone J, Castillo M, Mitchell KM, Bouldin TW, Suzuki K. Diffuse leptomeningeal oligodendrogliomatosis: radiologic/pathologic correlation. AJNR Am J Neuroradiol 2000;21:1122–6.
188. Chen R, Macdonald DR, Ramsay DA. Primary diffuse leptomeningeal oligodendroglioma. Case report. J Neurosurg 1995;83:724–8.

189. Gilmer-Hill HS, Ellis WG, Imbesi SG, Boggan JE. Spinal oligodendroglioma with gliomatosis in a child. Case report. J Neurosurg 2000;92 (Suppl 1):109–13.
190. Perilongo G, Gardiman M, Bisaglia L, et al. Spinal low-grade neoplasms with extensive leptomeningeal dissemination in children. Childs Nerv Syst 2002;18:505–12.
191. Ushida T, Sonobe H, Mizobuchi H, Toda M, Tani T, Yamamoto H. Oligodendroglioma of the "widespread" type in the spinal cord. Childs Nerv Syst 1998;14:751–5.
192. Baborie A, Dunn EM, Bridges LR, Bamford JM. Primary diffuse leptomeningeal gliomatosis predominantly affecting the spinal cord: case report and review of the literature. J Neurol Neurosurg Psychiatry 2001;70:256–8.
193. Heye N, Iglesias JR, Tonsen K, Graef G, Maier-Hauff K. Primary leptomeningeal gliomatosis with predominant involvement of the spinal cord. Acta Neurochir (Wien) 1990;102:145–8.
194. Kalyan-Raman UP, Cancilla PA, Case MJ. Solitary, primary malignant astrocytoma of the spinal leptomeninges. J Neuropathol Exp Neurol 1983;42:517–21.
195. Ramsay DA, Goshko V, Nag S. Primary spinal leptomeningeal astrocytoma. Acta Neuropathol (Berl) 1990;80:338–41.
196. Corsten LA, Raja AI, Wagner FC Jr. Primary diffuse leptomeningeal gliomatosis. Br J Neurosurg 2001;15:62–6.
197. Kitahara M, Katakura R, Wada T, Namiki T, Suzuki J. Diffuse form of primary leptomeningeal gliomatosis. Case report. J Neurosurg 1985;63:283–7.
198. Yung WA, Horten BC, Shapiro WR. Meningeal gliomatosis: a review of 12 cases. Ann Neurol 1980;8:605–8.

Protoplasmic Astrocytoma

199. Prayson RA, Estes ML. Protoplasmic astrocytoma. A clinicopathologic study of 16 tumors. Am J Clin Pathol 1995;103:705–9.

Pilocytic Astrocytoma

200. Gajjar A, Bhargava R, Jenkins JJ, et al. Low-grade astrocytoma with neuraxis dissemination at diagnosis. J Neurosurg 1995;83:67–71.
201. Hukin J, Siffert J, Cohen H, Velasquez L, Zagzag D, Allen J. Leptomeningeal dissemination at diagnosis of pediatric low-grade neuroepithelial tumors. Neuro-oncol 2003;5:188–96.
202. Pollack IF, Hurtt M, Pang D, Albright AL. Dissemination of low grade intracranial astrocytomas in children. Cancer 1994;73:2869–78.
203. Rush JA, Younge BR, Campbell RJ, MacCarty CS. Optic glioma. Long-term follow-up of 85 histopathologically verified cases. Ophthalmology 1982;89:1213–9.
204. Clark GB, Henry JM, McKeever PE. Cerebral pilocytic astrocytoma. Cancer 1985;56:1128–33.
205. Forsyth PA, Shaw EG, Scheithauer BW, O'Fallon JR, Layton DD Jr, Katzmann JA. Supratentorial pilocytic astrocytomas. A clinicopathologic, prognostic, and flow cytometric study of 51 patients. Cancer 1993;72:1335–42.
206. Cushing H. Experiences with the cerebellar astrocytoma. Surg Gynecol Obstet 1931;52:129–204.
207. Hayostek CJ, Shaw EG, Scheithauer B, et al. Astrocytomas of the cerebellum. A comparative clinicopathologic study of pilocytic and diffuse astrocytomas. Cancer 1993;72:856–69.
208. Coakley KJ, Huston J 3rd, Scheithauer BW, Forbes G, Kelly PJ. Pilocytic astrocytomas: well-demarcated magnetic resonance appearance despite frequent infiltration histologically. Mayo Clin Proc 1995;70:747–51.
209. Burger PC. Pathology of brain stem astrocytomas. Pediatr Neurosurg 1996;24:35–40.
210. Fisher PG, Breiter SN, Carson BS, et al. A clinicopathologic reappraisal of brain stem tumor classification. Identification of pilocystic astrocytoma and fibrillary astrocytoma as distinct entities. Cancer 2000;89:1569–76.
211. Khatib ZA, Heideman RL, Kovnar EH, et al. Predominance of pilocytic histology in dorsally exophytic brain stem tumors. Pediatr Neurosurg 1994;20:2–10.
212. Cottingham S, Boesel C, Yates A. Pilocytic astrocytoma in infants: a distinctive histological pattern. J Neuropathol Exp Neurol 1996;55: 654.
213. Tihan T, Fisher PG, Kepner JL, et al. Pediatric astrocytomas with monomorphous pilomyxoid features and a less favorable outcome. J Neuropathol Exp Neurol 1999;58:1061–8.
214. Komotar RJ, Burger PC, Carson BS, et al. Pilocytic and pilomyxoid hypothalamic/chiasmatic astrocytomas. Neurosurgery 2004;54:72–9; discussion 79–80.
215. Fernandez C, Figarella-Branger D, Girard N, et al. Pilocytic astrocytomas in children: prognostic factors—a retrospective study of 80 cases. Neurosurgery 2003;53:544–53; discussion 554–5.
216. Arslanoglu A, Cirak B, Horska A, et al. MR imaging characteristics of pilomyxoid astrocytomas. AJNR Am J Neuroradiol 2003;24: 1906–8.
217. Fuller CE, Frankel B, Smith M, et al. Suprasellar monomorphous pilomyxoid neoplasm: an ultrastructural analysis. Clin Neuropathol 2001;20:256–62.
218. Murayama S, Bouldin TW, Suzuki K. Immunocytochemical and ultrastructural studies of eosinophilic granular bodies in astrocytic tumors. Acta Neuropathol (Berl) 1992;83:408–14.

219. Tomlinson FH, Scheithauer BW, Hayostek CJ, et al. The significance of atypia and histologic malignancy in pilocytic astrocytoma of the cerebellum: a clinicopathologic and flow cytometric study. J Child Neurol 1994;9:301–10.
220. Bowers DC, Gargan L, Kapur P, et al. Study of the MIB-1 labeling index as a predictor of tumor progression in pilocytic astrocytomas in children and adolescents. J Clin Oncol 2003;21:2968–73.
221. Dirven CM, Koudstaal J, Mooij JJ, Molenaar WM. The proliferative potential of the pilocytic astrocytoma: the relation between MIB-1 labeling and clinical and neuro-radiological follow-up. J Neurooncol 1998;37:9–16.
222. Giannini C, Scheithauer BW, Burger PC, et al. Cellular proliferation in pilocytic and diffuse astrocytomas. J Neuropathol Exp Neurol 1999;58:46–53.
223. Schwartz AM, Ghatak NR. Malignant transformation of benign cerebellar astrocytoma. Cancer 1990;65:333–6.
224. Bernell WR, Kepes JJ, Seitz EP. Late malignant recurrence of childhood cerebellar astrocytoma. Report of two cases. J Neurosurg 1972;37: 470–4.
225. Dirks PB, Jay V, Becker LE, et al. Development of anaplastic changes in low-grade astrocytomas of childhood. Neurosurgery 1994;34:68–78.
226. Ushio Y, Arita N, Yoshimine T, Ikeda T, Mogami H. Malignant recurrence of childhood cerebellar astrocytoma: case report. Neurosurgery 1987;21:251–5.
227. Palma L, Russo A, Celli P. Prognosis of the so-called "diffuse" cerebellar astrocytoma. Neurosurgery 1984;15:315–7.
228. Bernhardtsen T, Laursen H, Bojsen-Moller M, Gjerris F. Sub-classification of low-grade cerebellar astrocytoma: is it clinically meaningful? Childs Nerv Syst 2003;19:729–35.
229. Katsetos CD, Krishna L, Friedberg E, Reidy J, Karkavelas G, Savory J. Lobar pilocytic astrocytomas of the cerebral hemispheres: II. Pathobiology—morphogenesis of the eosinophilic granular bodies. Clin Neuropathol 1994;13:306–14.
230. Goldman JE, Corbin E. Rosenthal fibers contain ubiquitinated alpha B-crystallin. Am J Pathol 1991;139:933–8.
231. Lach B, Sikorska M, Rippstein P, Gregor A, Staines W, Davie TR. Immunoelectron microscopy of Rosenthal fibers. Acta Neuropathol (Berl) 1991;81:503–9.
232. Friedberg E, Katsetos C, Reidv J, et al. Immunolocalization of protease inhibitors alpha-1-antitrypsin and alpha-1-antichymotrypsin in eosinophilic granular bodies of cerebral juvenile pilocytic astrocytomas. J Neuropathol Exp Neurol 1991;50:293.
233. Ito S, Hoshino T, Shibuya M, Prados MD, Edwards MS, Davis RL. Proliferative characteristics of juvenile pilocytic astrocytomas determined by bromodeoxyuridine labeling. Neurosurgery 1992;31:413–8; discussion 419.
234. Hunter S, Young A, Olson J, et al. Differential expression between pilocytic and anaplastic astrocytomas: identification of apolipoprotein D as a marker for low-grade, non-infiltrating primary CNS neoplasms. J Neuropathol Exp Neurol 2002;61:275–81.
235. Scheithauer BW, Bruner JM. The ultrastructural spectrum of astrocytic neoplasms. Ultrastruct Pathol 1987;11:535–81.
236. Teo JG, Ng HK. Cytodiagnosis of pilocytic astrocytoma in smear preparations. Acta Cytol 1998;42:673–8.
237. Kluwe L, Hagel C, Tatagiba M, et al. Loss of NF1 alleles distinguish sporadic from NF1-associated pilocytic astrocytomas. J Neuropathol Exp Neurol 2001;60:917–20.
238. Wimmer K, Eckart M, Meyer-Puttlitz B, Fonatsch C, Pietsch T. Mutational and expression analysis of the NF1 gene argues against a role as tumor suppressor in sporadic pilocytic astrocytomas. J Neuropathol Exp Neurol 2002; 61:896–902.
238a. Jones DT, Ichimura K, Liu L, Pearson DM, Plant K, Collins VP. Genomic analysis of pilocytic astrocytomas at 0.97 Mb resolution shows an increasing tendency toward chromosomal copy number change with age. J Neuropathol Exp Neurol 2006;65:1049–58.
239. Cummings TJ, Provenzale JM, Hunter SB, et al. Gliomas of the optic nerve: histological, immunohistochemical (MIB-1 and p53), and MRI analysis. Acta Neuropathol (Berl) 2000;99:563–70.
240. Tihan T, Davis R, Elowitz E, DiCostanzo D, Moll U. Practical value of Ki-67 and p53 labeling indexes in stereotactic biopsies of diffuse and pilocytic astrocytomas. Arch Pathol Lab Med 2000;124:108–13.
241. Lang FF, Miller DC, Pisharody S, Koslow M, Newcomb EW. High frequency of p53 protein accumulation without p53 gene mutation in human juvenile pilocytic, low grade and anaplastic astrocytomas. Oncogene 1994;9:949–54.
242. Burkhard C, Di Patre PL, Schuler D, et al. A population-based study of the incidence and survival rates in patients with pilocytic astrocytoma. J Neurosurg 2003;98:1170–4.
243. Pagni CA, Giordana MT, Canavero S. Benign recurrence of a pilocytic cerebellar astrocytoma 36 years after radical removal: case report. Neurosurgery 1991;28:606–9.

244. Gajjar A, Sanford RA, Heideman R, et al. Low-grade astrocytoma: a decade of experience at St. Jude Children's Research Hospital. J Clin Oncol 1997;15:2792–9.
245. Palma L, Guidetti B. Cystic pilocytic astrocytomas of the cerebral hemispheres. Surgical experience with 51 cases and long-term results. J Neurosurg 1985;62:811–5.
246. Pollack IF, Claassen D, al-Shboul Q, Janosky JE, Deutsch M. Low-grade gliomas of the cerebral hemispheres in children: an analysis of 71 cases. J Neurosurg 1995;82:536–47.
247. Hukin J, Siffert J, Velasquez L, Zagzag D, Allen J. Leptomeningeal dissemination in children with progressive low-grade neuroepithelial tumors. Neuro-oncol 2002;4:253–60.
248. Kocks W, Kalff R, Reinhardt V, Grote W, Hilke J. Spinal metastasis of pilocytic astrocytoma of the chiasma opticum. Childs Nerv Syst 1989;5:118–20.
249. Obana WG, Cogen PH, Davis RL, Edwards MS. Metastatic juvenile pilocytic astrocytoma. Case report. J Neurosurg 1991;75:972–5.
250. Alvord EC Jr, Lofton S. Gliomas of the optic nerve or chiasm. Outcome by patients' age, tumor site, and treatment. J Neurosurg 1988;68: 85–98.
251. Gallucci M, Catalucci A, Scheithauer BW, Forbes GS. Spontaneous involution of pilocytic astrocytoma in a patient without neurofibromatosis type 1: case report. Radiology 2000; 214:223–6.
252. Parsa CF, Hoyt CS, Lesser RL, et al. Spontaneous regression of optic gliomas: thirteen cases documented by serial neuroimaging. Arch Ophthalmol 2001;119:516–29.
253. Perilongo G, Moras P, Carollo C, et al. Spontaneous partial regression of low-grade glioma in children with neurofibromatosis-1: a real possibility. J Child Neurol 1999;14:352–6.
254. Fouladi M, Wallace D, Langston J, et al. Survival and functional outcome of children with hypothalamic/chiasmatic tumors. Cancer 2003;97:1084–92.
255. Janss AJ, Grundy R, Cnaan A, et al. Optic pathway and hypothalamic/chiasmatic gliomas in children younger than age 5 years with a 6-year follow-up. Cancer 1995;75:1051–9.
256. McGirr SJ, Kelly PJ, Scheithauer BW. Stereotactic resection of juvenile pilocytic astrocytomas of the thalamus and basal ganglia. Neurosurgery 1987;20:447–52.
257. Gjerris F, Klinken L. Long-term prognosis in children with benign cerebellar astrocytoma. J Neurosurg 1978;49:179–84.
258. Minehan K, Scheithauer B, Shaw EG, Onofrio B. Astrocytic tumors of the spinal cord. J Neuropathol Exp Neurol 1993;52:289.
259. Rauhut F, Reinhardt V, Budach V, Wiedemayer H, Nau HE. Intramedullary pilocytic astrocytomas—a clinical and morphological study after combined surgical and photon or neutron therapy. Neurosurg Rev 1989;12:309–13.
260. Rossitch E Jr, Zeidman SM, Burger PC, et al. Clinical and pathological analysis of spinal cord astrocytomas in children. Neurosurgery 1990;27:193–6.

Pleomorphic Xanthoastrocytoma

261. Kepes JJ, Rubinstein LJ, Eng LF. Pleomorphic xanthoastrocytoma: a distinctive meningocerebral glioma of young subjects with relatively favorable prognosis. A study of 12 cases. Cancer 1979;44:1839–52.
262. Kepes JJ. Pleomorphic xanthoastrocytoma: the birth of a diagnosis and a concept. Brain Pathol 1993;3:269–74.
263. Giannini C, Scheithauer BW, Burger PC, et al. Pleomorphic xanthoastrocytoma: what do we really know about it? Cancer 1999;85:2033–45.
264. Lipper MH, Eberhard DA, Phillips CD, Vezina LG, Cail WS. Pleomorphic xanthoastrocytoma, a distinctive astroglial tumor: neuroradiologic and pathologic features. AJNR Am J Neuroradiol 1993;14:1397–404.
265. Wasdahl DA, Scheithauer BW, Andrews BT, Jeffrey RA Jr. Cerebellar pleomorphic xanthoastrocytoma: case report. Neurosurgery 1994;35: 947–50; discussion 950–1.
266. Herpers MJ, Freling G, Beuls EA. Pleomorphic xanthoastrocytoma in the spinal cord. Case report. J Neurosurg 1994;80:564–9.
267. Zarate JO, Sampaolesi R. Pleomorphic xanthoastrocytoma of the retina. Am J Surg Pathol 1999;23:79–81.
268. Furuta A, Takahashi H, Ikuta F, Onda K, Takeda N, Tanaka R. Temporal lobe tumor demonstrating ganglioglioma and pleomorphic xanthoastrocytoma components. Case report. J Neurosurg 1992;77:143–7.
269. Perry A, Giannini C, Scheithauer BW, et al. Composite pleomorphic xanthoastrocytoma and ganglioglioma: report of four cases and review of the literature. Am J Surg Pathol 1997;21:763–71.
270. Powell SZ, Yachnis AT, Rorke LB, Rojiani AM, Eskin TA. Divergent differentiation in pleomorphic xanthoastrocytoma. Evidence for a neuronal element and possible relationship to ganglion cell tumors. Am J Surg Pathol 1996; 20:80–5.
271. Yeh DJ, Hessler RB, Stevens EA, Lee MR. Composite pleomorphic xanthoastrocytoma-ganglioglioma presenting as a suprasellar mass: case report. Neurosurgery 2003;52:1465–8; discussion 1468–9.

272. Kanzawa T, Takahashi H, Hayano M, Mori S, Shimbo Y, Kitazawa T. Melanotic cerebral astrocytoma: case report and literature review. Acta Neuropathol (Berl) 1997;93:200–4.
273. Fouladi M, Jenkins J, Burger P, et al. Pleomorphic xanthoastrocytoma: favorable outcome after complete surgical resection. Neuro-oncol 2001;3:184–92.
274. Prayson RA, Morris HH 3rd. Anaplastic pleomorphic xanthoastrocytoma. Arch Pathol Lab Med 1998;122:1082–6.
275. Pahapill PA, Ramsay DA, Del Maestro RF. Pleomorphic xanthoastrocytoma: case report and analysis of the literature concerning the efficacy of resection and the significance of necrosis. Neurosurgery 1996;38:822–8; discussion 828–9.
276. Kepes J, Louis D, Giannini C, Paulus W. Pleomorphic xanthoastrocytoma. In: Kleihues P, Cavenee W, eds. Pathology and genetics of tumours of the nervous system. Lyon: IARC Press; 2000:52–4.
277. Reifenberger G, Kaulich K, Wiestler OD, Blumcke I. Expression of the CD34 antigen in pleomorphic xanthoastrocytomas. Acta Neuropathol (Berl) 2003;105:358–64.
278. Giannini C, Scheithauer BW, Lopes MB, Hirose T, Kros JM, VandenBerg SR. Immunophenotype of pleomorphic xanthoastrocytoma. Am J Surg Pathol 2002;26:479–85.
279. Giannini C, Hebrink D, Scheithauer BW, Dei Tos AP, James CD. Analysis of p53 mutation and expression in pleomorphic xanthoastrocytoma. Neurogenetics 2001;3:159–62.
280. Kaulich K, Blaschke B, Numann A, et al. Genetic alterations commonly found in diffusely infiltrating cerebral gliomas are rare or absent in pleomorphic xanthoastrocytomas. J Neuropathol Exp Neurol 2002;61:1092–9.
281. Hirose T, Giannini C, Scheithauer BW. Ultrastructural features of pleomorphic xanthoastrocytoma: a comparative study with glioblastoma multiforme. Ultrastruct Pathol 2001;25:469–78.
282. Bleggi-Torres LF, Gasparetto EL, Faoro LN, et al. Pleomorphic xanthoastrocytoma: report of a case diagnosed by intraoperative cytopathological examination. Diagn Cytopathol 2001; 24:120–2.
283. Kobayashi S, Hirakawa E, Haba R. Squash cytology of pleomorphic xanthoastrocytoma mimicking glioblastoma. A case report. Acta Cytol 1999;43:652–8.
284. Martinez-Diaz H, Kleinschmidt-DeMasters BK, Powell SZ, Yachnis AT. Giant cell glioblastoma and pleomorphic xanthoastrocytoma show different immunohistochemical profiles for neuronal antigens and p53 but share reactivity for class III beta-tubulin. Arch Pathol Lab Med 2003;127:1187–91.
285. Haga S, Morioka T, Nishio S, Fukui M. Multicentric pleomorphic xanthoastrocytomas: case report. Neurosurgery 1996;38:1242–4; discussion 1244–5.

Subependymal Giant Cell Astrocytoma (Tuberous Sclerosis)

286. Boesel CP, Paulson GW, Kosnik EJ, Earle KM. Brain hamartomas and tumors associated with tuberous sclerosis. Neurosurgery 1979;4:410–7.
287. Shepherd CW, Scheithauer BW, Gomez MR, Altermatt HJ, Katzmann JA. Subependymal giant cell astrocytoma: a clinical, pathological, and flow cytometric study. Neurosurgery 1991;28:864–8.
288. Short MP, Richardson EP Jr, Haines JL, Kwiatkowski DJ. Clinical, neuropathological and genetic aspects of the tuberous sclerosis complex. Brain Pathol 1995;5:173–9.
289. Torres OA, Roach ES, Delgado MR, et al. Early diagnosis of subependymal giant cell astrocytoma in patients with tuberous sclerosis. J Child Neurol 1998;13:173–7.
290. Sinson G, Sutton LN, Yachnis AT, Duhaime AC, Schut L. Subependymal giant cell astrocytomas in children. Pediatr Neurosurg 1994;20:233–9.
291. Padmalatha C, Harruff RC, Ganick D, Hafez GB. Glioblastoma multiforme with tuberous sclerosis. Report of a case. Arch Pathol Lab Med 1980;104:649–50.
292. Scheithauer B, Reagan T. Neuropathology. In: Gomez M, Sampson J, Whittemore V, eds. Tuberous sclerosis complex. Developmental perspectives in psychiatry. Oxford: Oxford Press; 1999:101–44.
293. Smirniotopoulos JG, Murphy FM. The phakomatoses. AJNR Am J Neuroradiol 1992;13:725–46.
294. Hirose T, Scheithauer BW, Lopes MB, et al. Tuber and subependymal giant cell astrocytoma associated with tuberous sclerosis: an immunohistochemical, ultrastructural, and immunoelectron and microscopic study. Acta Neuropathol (Berl) 1995;90:387–99.
295. Lopes MB, Altermatt HJ, Scheithauer BW, Shepherd CW, VandenBerg SR. Immunohistochemical characterization of subependymal giant cell astrocytomas. Acta Neuropathol (Berl) 1996; 91:368–75.
296. Iwaki T, Wisniewski T, Iwaki A, et al. Accumulation of alpha B-crystallin in central nervous system glia and neurons in pathologic conditions. Am J Pathol 1992;140:345–56.
297. Gyure KA, Prayson RA. Subependymal giant cell astrocytoma: a clinicopathologic study with HMB45 and MIB-1 immunohistochemical analysis. Mod Pathol 1997;10:313–7.

298. Kimura N, Watanabe M, Date F, et al. HMB-45 and tuberin in hamartomas associated with tuberous sclerosis. Mod Pathol 1997;10:952–9.
299. Arai Y, Ackerley CA, Becker LE. Loss of the TSC2 product tuberin in subependymal giant-cell tumors. Acta Neuropathol (Berl) 1999;98:233–9.
300. Kim SK, Wang KC, Cho BK, et al. Biological behavior and tumorigenesis of subependymal giant cell astrocytomas. J Neurooncol 2001;52:217–25.
301. Trombley IK, Mirra SS. Ultrastructure of tuberous sclerosis: cortical tuber and subependymal tumor. Ann Neurol 1981;9:174–81.
302. Nakamura Y, Becker LE. Subependymal giant-cell tumor: astrocytic or neuronal? Acta Neuropathol (Berl) 1983;60:271–7.
303. Altermatt HJ, Scheithauer BW. Cytomorphology of subependymal giant cell astrocytoma. Acta Cytol 1992;36:171–5.
304. Green AJ, Smith M, Yates JR. Loss of heterozygosity on chromosome 16p13.3 in hamartomas from tuberous sclerosis patients. Nat Genet 1994;6:193–6.
305. Henske EP, Scheithauer BW, Short MP, et al. Allelic loss is frequent in tuberous sclerosis kidney lesions but rare in brain lesions. Am J Hum Genet 1996;59:400–6.
306. Rickert CH, Paulus W. No chromosomal imbalances detected by comparative genomic hybridisation in subependymal giant cell astrocytomas. Acta Neuropathol (Berl) 2002;104:206–8.
307. Morimoto K, Mogami H. Sequential CT study of subependymal giant-cell astrocytoma associated with tuberous sclerosis. Case report. J Neurosurg 1986;65:874–7.
308. Chow CW, Klug GL, Lewis EA. Subependymal giant-cell astrocytoma in children. An unusual discrepancy between histological and clinical features. J Neurosurg 1988;68:880–3.

Desmoplastic Infantile Astrocytoma

309. de Chadarevian JP, Pattisapu JV, Faerber EN. Desmoplastic cerebral astrocytoma of infancy. Light microscopy, immunocytochemistry, and ultrastructure. Cancer 1990;66:173–9.
310. Taratuto AL, Monges J, Lylyk P, Leiguarda R. Superficial cerebral astrocytoma attached to dura. Report of six cases in infants. Cancer 1984;54:2505–12.
311. Aydin F, Ghatak NR, Salvant J, Muizelaar P. Desmoplastic cerebral astrocytoma of infancy. A case report with immunohistochemical, ultrastructural and proliferation studies. Acta Neuropathol (Berl) 1993;86:666–70.
312. Louis DN, von Deimling A, Dickersin GR, Dooling EC, Seizinger BR. Desmoplastic cerebral astrocytomas of infancy: a histopathologic, immunohistochemical, ultrastructural, and molecular genetic study. Hum Pathol 1992;23:1402–9.
313. Paulus W, Schlote W, Perentes E, Jacobi G, Warmuth-Metz M, Roggendorf W. Desmoplastic supratentorial neuroepithelial tumours of infancy. Histopathology 1992;21:43–9.
314. Mallucci C, Lellouch-Tubiana A, Salazar C, et al. The management of desmoplastic neuroepithelial tumours in childhood. Childs Nerv Syst 2000;16:8–4.
315. Taratuto A, VandenBerg S, Rorke L. Desmoplastic infantile astrocytoma and ganglioglioma. In: Kleihues P, Cavenee W, eds. Pathology and genetics of tumours of the nervous system. Lyon: IARC Press; 2000.
316. VandenBerg SR. Desmoplastic infantile ganglioglioma and desmoplastic cerebral astrocytoma of infancy. Brain Pathol 1993;3:275–81.

Granular Cell Tumor of the Infundibulum

317. Liss L, Kahn E. Pituicytoma. A tumor of the sella turcica. A clinicopathological study. J Neurosurg 1957;15:481–8.
318. Luse S, Kernohan J. Granular-cell tumors of the stalk and posterior lobe of the pituitary gland. Cancer 1955;8:616–22.
319. Shanklin W. The origin, histology, and senescence of tumorettes in the human neurohypopysis. Acta Anat 1953;18:1–20.
320. Scheithauer B, Horvath E, Kovacs K. Ultrastructure of the neurohypophysis. Micros Res Tech 1992;20:177–86.
321. Jenevein EP. A Neurohypophyseal tumor originating from pituicytes. Am J Clin Pathol 1964; 41:522–6.
322. Takei Y, Seyama S, Pearl GS, Tindall GT. Ultrastructural study of the human neurohypophysis. II. Cellular elements of neural parenchyma, the pituicytes. Cell Tissue Res 1980;205:273–87.
323. Becker DH, Wilson CB. Symptomatic parasellar granular cell tumors. Neurosurgery 1981;8:173–80.
324. Cohen-Gadol AA, Pichelmann MA, Link MJ, et al. Granular cell tumor of the sellar and suprasellar region: clinicopathologic study of 11 cases and literature review. Mayo Clin Proc 2003;78:567–73.
325. Lafitte C, Aesch B, Henry-Lebras F, Fetissof F, Jan M. Granular cell tumor of the pituitary stalk. Case report. J Neurosurg 1994;80:1103–7.
326. Liwnicz BH, Liwnicz RG, Huff JS, McBride BH, Tew JM Jr. Giant granular cell tumor of the suprasellar area: immunocytochemical and electron microscopic studies. Neurosurgery 1984;15:246–51.

327. Schaller B, Kirsch E, Tolnay M, Mindermann T. Symptomatic granular cell tumor of the pituitary gland: case report and review of the literature. Neurosurgery 1998;42:166–70; discussion 170–1.
328. Buhl R, Hugo H, Hempelmann R, Barth H, Mehdorn H. Granular-cell tumour: a rare suprasellar mass. Neuroradiology 2001;43:309–12.
329. Nishioka H, Ii K, Llena JF, Hirano A. Immunohistochemical study of granular cell tumors of the neurohypophysis. Virchows Arch B Cell Pathol Incl Mol Pathol 1991;60:413–7.
330. Vinores SA. Demonstration of glial fibrillary acidic (GFA) protein by electron immunocytochemistry in the granular cells of a choristoma of the neurohypophysis. Histochemistry 1991;96:265–9.
331. Giangaspero F, Cenacchi G. Oncocytic and granular cell neoplasms of the central nervous system and pituitary gland. Semin Diagn Pathol 1999;16:91–7.
332. Roncaroli F, Scheithauer BW, Cenacchi G, et al. 'Spindle cell oncocytoma' of the adenohypophysis: a tumor of folliculostellate cells? Am J Surg Pathol 2002;26:1048–55.
333. Kloub O, Perry A, Tu PH, Lipper M, Lopes MB. Spindle cell oncocytoma of the adenohypophysis: report of two recurrent cases. Am J Surg Pathol 2005;29:247–53.

Pituicytoma

334. Brat DJ, Scheithauer BW, Staugaitis SM, Holtzman RN, Morgello S, Burger PC. Pituicytoma: a distinctive low-grade glioma of the neurohypophysis. Am J Surg Pathol 2000;24:362–8.
335. Figarella-Branger D, Dufour H, Fernandez C, Bouvier-Labit C, Grisoli F, Pellissier JF. Pituicytomas, a mis-diagnosed benign tumor of the neurohypophysis: report of three cases. Acta Neuropathol (Berl) 2002;104:313–9.
336. Schultz AB, Brat DJ, Oyesiku NM, Hunter SB. Intrasellar pituicytoma in a patient with other endocrine neoplasms. Arch Pathol Lab Med 2001;125:527–30.
337. Hurley TR, D'Angelo CM, Clasen RA, Wilkinson SB, Passavoy RD. Magnetic resonance imaging and pathological analysis of a pituicytoma: case report. Neurosurgery 1994;35:314–7.
338. Cenacchi G, Giovenali P, Castrioto C, Giangaspero F. Pituicytoma: ultrastructural evidence of a possible origin from folliculo-stellate cells of the adenohypophysis. Ultrastruct Pathol 2001;25:309–12.
339. Giangaspero F, Cenacchi G. Oncocytic and granular cell neoplasms of the central nervous system and pituitary gland. Semin Diagn Pathol 1999;16:91–7.
340. Roncaroli F, Scheithauer BW, Cenacchi G, et al. 'Spindle cell oncocytoma' of the adenohypophysis: a tumor of folliculostellate cells? Am J Surg Pathol 2002;26:1048–55.
341. Kloub O, Perry A, Tu PH, Lipper M, Lopes MB. Spindle cell oncocytoma of the adenohypophysis: report of two recurrent cases. Am J Surg Pathol 2005;29:247–53.

Oligodendroglioma and Oligoastrocytoma

342. Burger PC. What is an oligodendroglioma? Brain Pathol 2002;12(2):257–9.
343. Razack N, Baumgartner J, Bruner J. Pediatric oligodendrogliomas. Pediatr Neurosurg 1998;28:121–9.
344. Armao DM, Stone J, Castillo M, Mitchell KM, Bouldin TW, Suzuki K. Diffuse leptomeningeal oligodendrogliomatosis: radiologic/pathologic correlation. AJNR Am J Neuroradiol 2000;21: 1122–6.
345. Chen R, Macdonald DR, Ramsay DA. Primary diffuse leptomeningeal oligodendroglioma. Case report. J Neurosurg 1995;83:724–8.
346. Gilmer-Hill HS, Ellis WG, Imbesi SG, Boggan JE. Spinal oligodendroglioma with gliomatosis in a child. Case report. J Neurosurg 2000;92 (Suppl 1):109–13.
347. Perilongo G, Gardiman M, Bisaglia L, et al. Spinal low-grade neoplasms with extensive leptomeningeal dissemination in children. Childs Nerv Syst 2002;18:505–12.
348. Ushida T, Sonobe H, Mizobuchi H, Toda M, Tani T, Yamamoto H. Oligodendroglioma of the "widespread" type in the spinal cord. Childs Nerv Syst 1998;14:751–5.
349. Megyesi JF, Kachur E, Lee DH, et al. Imaging correlates of molecular signatures in oligodendrogliomas. Clin Cancer Res 2004;10:4303–6.
350. Kros JM, Van Eden CG, Stefanko SZ, Waayer-Van Batenburg M, van der Kwast TH. Prognostic implications of glial fibrillary acidic protein containing cell types in oligodendrogliomas. Cancer 1990;66:1204–12.
351. Reifenberger G, Kros J, Burger P, Louis D, Collins V. Oligoastrocytoma. In: Cavenee W, ed. Pathology and genetics of tumours of the nervous system. Lyon: IARC Press; 2000:65–7.
352. Perry A, Scheithauer BW, Macaulay RJ, Raffel C, Roth KA, Kros JM. Oligodendrogliomas with neurocytic differentiation. A report of 4 cases with diagnostic and histogenetic implications. J Neuropathol Exp Neurol 2002;61:947–55.
352a. Reifenberger G, Kros JM, Louis DN, Collins VP. Oligodendroglioma. In: Louis DN, Ohgaki H, Wiestler OD, Cavenee W, eds. WHO classification of tumours of the central nervous system. Lyon: IARC Press; 2007. (In press)

353. Mork SJ, Halvorsen TB, Lindegaard KF, Eide GE. Oligodendroglioma. Histologic evaluation and prognosis. J Neuropathol Exp Neurol 1986;45: 65–78.
354. Burger PC, Rawlings CE, Cox EB, McLendon RE, Schold SC Jr, Bullard DE. Clinicopathologic correlations in the oligodendroglioma. Cancer 1987;59:1345–52.
355. Daumas-Duport C, Tucker ML, Kolles H, et al. Oligodendrogliomas. Part II: A new grading system based on morphological and imaging criteria. J Neurooncol 1997;34:61–78.
356. Giannini C, Scheithauer BW, Weaver AL, et al. Oligodendrogliomas: reproducibility and prognostic value of histologic diagnosis and grading. J Neuropathol Exp Neurol 2001;60:248–62.
357. Reis-Filho JS, Faoro LN, Carrilho C, Bleggi-Torres LF, Schmitt FC. Evaluation of cell proliferation, epidermal growth factor receptor, and bcl-2 immunoexpression as prognostic factors for patients with World Health Organization grade 2 oligodendroglioma. Cancer 2000;88: 862–9.
358. Reifenberger G, Kros J, Burger P, Louis D, Collins V. Anaplastic oligodendroglioma. In: Cavenee W, ed. Pathology and genetics of tumours of the nervous system. Lyon: IARC Press; 2000:62–4.
359. Ligon KL, Alberta JA, Kho AT, et al. The oligodendroglial lineage marker OLIG2 is universally expressed in diffuse gliomas. J Neuropathol Exp Neurol 2004;63:499–509.
360. Herpers MJ, Budka H. Glial fibrillary acidic protein (GFAP) in oligodendroglial tumors: gliofibrillary oligodendroglioma and transitional oligoastrocytoma as subtypes of oligodendroglioma. Acta Neuropathol (Berl) 1984; 64:265–72.
360a. Preusser M, Laggner U, Haberler C, Heinzl H, Budka H, Hainfellner JA. Comparative analysis of NeuN immunoreactivity in primary brain tumors: conclusions for rational use in diagnostic histopatholgy. Histopathology 2006;48: 438–44.
361. Wharton SB, Chan KK, Hamilton FA, Anderson JR. Expression of neuronal markers in oligodendrogliomas: an immunohistochemical study. Neuropathol Appl Neurobiol 1998;24: 302–8.
362. Coons SW, Johnson PC, Pearl DK. The prognostic significance of Ki-67 labeling indices for oligodendrogliomas. Neurosurgery 1997;41: 878–84; discussion 884–5.
363. Dehghani F, Schachenmayr W, Laun A, Korf HW. Prognostic implication of histopathological, immunohistochemical and clinical features of oligodendrogliomas: a study of 89 cases. Acta Neuropathol (Berl) 1998;95:493–504.
364. Heegaard S, Sommer HM, Broholm H, Broendstrup O. Proliferating cell nuclear antigen and Ki-67 immunohistochemistry of oligodendrogliomas with special reference to prognosis. Cancer 1995;76:1809–13.
365. Min KW, Scheithauer BW. Oligodendroglioma: the ultrastructural spectrum. Ultrastruct Pathol 1994;18:47–60.
366. Kros JM, de Jong AA, van der Kwast TH. Ultrastructural characterization of transitional cells in oligodendrogliomas. J Neuropathol Exp Neurol 1992;51:186–93.
367. Kros JM, Stefanko SZ, de Jong AA, van Vroonhoven CC, van der Heul RO, van der Kwast TH. Ultrastructural and immunohistochemical segregation of gemistocytic subsets. Hum Pathol 1991;22:33–40.
368. Radner H, Kleinert R, Vennigerholz F, Denk H. Peculiar changes in Rosenthal fibres in an atypical astrocytoma. Neuropathol Appl Neurobiol 1990;16:171–7.
369. Ng HK, Ko H, Tse C. Immunohistochemical and ultrastructural studies of oligodendrogliomas revealed features of neuronal differentiation. Int J Surg Pathol 1994;2:47–56.
370. Bigner SH, Matthews MR, Rasheed BK, et al. Molecular genetic aspects of oligodendrogliomas including analysis by comparative genomic hybridization. Am J Pathol 1999; 155:375–86.
371. Bigner SH, Rasheed BK, Wiltshire R, McLendon RE. Morphologic and molecular genetic aspects of oligodendroglial neoplasms. Neuro-oncol 1999;1:52–60.
372. Burger PC, Minn AY, Smith JS, et al. Losses of chromosomal arms 1p and 19q in the diagnosis of oligodendroglioma. A study of paraffin-embedded sections. Mod Pathol 2001;14:842–53.
373. Cairncross JG, Ueki K, Zlatescu MC, et al. Specific genetic predictors of chemotherapeutic response and survival in patients with anaplastic oligodendrogliomas. J Natl Cancer Inst 1998;90:1473–9.
374. Ino Y, Betensky RA, Zlatescu MC, et al. Molecular subtypes of anaplastic oligodendroglioma: implications for patient management at diagnosis. Clin Cancer Res 2001;7:839–45.
375. Reifenberger J, Reifenberger G, Liu L, James CD, Wechsler W, Collins VP. Molecular genetic analysis of oligodendroglial tumors shows preferential allelic deletions on 19q and 1p. Am J Pathol 1994;145:1175–90.
376. Reifenberger G, Louis DN. Oligodendroglioma: toward molecular definitions in diagnostic neuro-oncology. J Neuropathol Exp Neurol 2003;62:111–26.
377. Ritland SR, Ganju V, Jenkins RB. Region-specific loss of heterozygosity on chromosome 19 is related to the morphologic type of human glioma. Genes Chromosomes Cancer 1995;12:277–82.
378. Smith JS, Alderete B, Minn Y, et al. Localization of common deletion regions on 1p and 19q in human gliomas and their association with histological subtype. Oncogene 1999;18: 4144–52.

379. Smith JS, Wang XY, Qian J, et al. Amplification of the platelet-derived growth factor receptor-A (PDGFRA) gene occurs in oligodendrogliomas with grade IV anaplastic features. J Neuropathol Exp Neurol 2000;59:495–503.
380. von Deimling A, Louis DN, von Ammon K, Petersen I, Wiestler OD, Seizinger BR. Evidence for a tumor suppressor gene on chromosome 19q associated with human astrocytomas, oligodendrogliomas, and mixed gliomas. Cancer Res 1992;52:4277–9.
381. von Deimling A, Fimmers R, Schmidt MC, et al. Comprehensive allelotype and genetic anaysis of 466 human nervous system tumors. J Neuropathol Exp Neurol 2000;59:544–58.
382. Barbashina V, Salazar P, Holland EC, Rosenblum MK, Ladanyi M. Allelic losses at 1p36 and 19q13 in gliomas: correlation with histologic classification, definition of a 150-kb minimal deleted region on 1p36, and evaluation of CAMTA1 as a candidate tumor suppressor gene. Clin Cancer Res 2005;11:1119–28.
383. McDonald J, See SJ, Tremont I, et al. The prognostic impact of histology and 1p/19q status in anaplastic oligodendroglial tumors. Cancer 2005;104:1468–77.
383a. Griffin CA, Burger P, Morsberger L, et al. Identification of der(1;19)(q10;p10) in five oligodendrogliomas suggests mechanism of concurrent 1p and 19q loss. J Neuropathol Ex Neurol 65:988–94.
383b. Jenkins RB, Blair H, Ballman KV, et al. A t(1;19)(q10;p10) mediates the combined deletions of 1p and 19q and predicts a better prognosis of patients with oligodendroglioma. Cancer Res 2006;66:9852–61.
283c. Scheie D, Andresen PA, Cvancarova M, et al. Fluorescence in situ hybridization (FISH) on touch preparations: a reliable method for detecting loss of heterozygosity at 1p and 19q in oligodendroglial tumors. Am J Surg Pathol 2006;30:828–37.
384. Sasaki H, Zlatescu MC, Betensky RA, et al. Histopathological-molecular genetic correlations in referral pathologist-diagnosed low-grade "oligodendroglioma." J Neuropathol Exp Neurol 2002;61:58–63.
385. Jeuken JW, Sprenger SH, Boerman RH, et al. Subtyping of oligo-astrocytic tumours by comparative genomic hybridization. J Pathol 2001;194:81–7.
385a. McDonald JM, See SJ, Tremont IW, et al. The prognostic impact of histology and 1p/19q status in anaplastic oligodendroglial tumors. Cancer 2005;104:1468–77.
386. Ueki K, Nishikawa R, Nakazato Y, et al. Correlation of histology and molecular genetic analysis of 1p, 19q, 10q, TP53, EGFR, CDK4, and CDKN2A in 91 astrocytic and oligodendroglial tumors. Clin Cancer Res 2002;8:196–201.
386a. McDonald JM, Siew JS, Tremont IW, et al. The prognostic impact o fhistology and 1p/19q status in anaplastic oligodendroglial tumors. Cancer 2005;104:1468–77.
387. Raghavan R, Balani J, Perry A, et al. Pediatric oligodendrogliomas: a study of molecular alterations on 1p and 19q using fluorescence in situ hybridization. J Neuropathol Exp Neurol 2003;62:530–7.
388. Kreiger PA, Okada Y, Simon S, Rorke LB, Louis DN, Golden JA. Losses of chromosomes 1p and 19q are rare in pediatric oligodendrogliomas. Acta Neuropathol (Berl) 2005;109:387–92.
389. Godfraind C, Rousseau E, Ruchoux MM, Scaravilli F, Vikkula M. Tumour necrosis and microvascular proliferation are associated with 9p deletion and CDKN2A alterations in 1p/19q-deleted oligodendrogliomas. Neuropathol Appl Neurobiol 2003;29:462–71.
390. Fallon K, Palmer C, Roth K, et al. Prognostic value of 1p, 19q, 9p, 10q, and EGFR-FISH analyses in recurrent oligodendrogliomas. J Neuropathol Exp Neurol 2004;63):314–22.
391. Jeuken JW, Sprenger SH, Wesseling P, et al. Identification of subgroups of high-grade oligodendroglial tumors by comparative genomic hybridization. J Neuropathol Exp Neurol 1999;58:606–12.
392. Jeuken JW, Nelen MR, Vermeer H, et al. PTEN mutation analysis in two genetic subtypes of high-grade oligodendroglial tumors. PTEN is only occasionally mutated in one of the two genetic subtypes. Cancer Genet Cytogenet 2000;119:42–7.
393. Ino Y, Zlatescu MC, Sasaki H, et al. Long survival and therapeutic responses in patients with histologically disparate high-grade gliomas demonstrating chromosome 1p loss. J Neurosurg 2000;92:983–90.
394. Watanabe T, Nakamura M, Kros JM, et al. Phenotype versus genotype correlation in oligodendrogliomas and low-grade diffuse astrocytomas. Acta Neuropathol (Berl) 2002;103:267–75.
395. Ohgaki H, Kleihues P. Population-based studies on incidence, survival rates, and genetic alterations in astrocytic and oligodendroglial gliomas. J Neuropathol Exp Neurol 2005;64: 479–89.
396. Fuller CE, Schmidt RE, Roth KA, et al. Clinical utility of fluorescence in situ hybridization (FISH) in morphologically ambiguous gliomas with hybrid oligodendroglial/astrocytic features. J Neuropathol Exp Neurol 2003;62: 1118–28.
397. Dong ZQ, Pang JC, Tong CY, Zhou LF, Ng HK. Clonality of oligoastrocytomas. Hum Pathol 2002;33:528–35.
398. Maintz D, Fiedler K, Koopmann J, et al. Molecular genetic evidence for subtypes of oligoastrocytomas. J Neuropathol Exp Neurol 1997;56:1098–104.

399. Smith JS, Perry A, Borell TJ, et al. Alterations of chromosome arms 1p and 19q as predictors of survival in oligodendrogliomas, astrocytomas, and mixed oligoastrocytomas. J Clin Oncol 2000;18:636–45.
400. Felsberg J, Erkwoh A, Sabel MC, et al. Oligodendroglial tumors: refinement of candidate regions on chromosome arm 1p and correlation of 1p/19q status with survival. Brain Pathol 2004;14:121–30.
401. Nazek M, Mandybur TI, Kashiwagi S. Oligodendroglial proliferative abnormality associated with arteriovenous malformation: report of three cases with review of the literature. Neurosurgery 1988;23:781–5.
402. Lombardi D, Scheithauer BW, Piepgras D, Meyer FB, Forbes GS. "Angioglioma" and the arteriovenous malformation-glioma association. J Neurosurg 1991;75:589–596.
402a. Miller CR, Dunham CP, Scheithauer BW, Perry A. Significance of necrosis in grading of anaplastic oligodendroglial tumors: a clinicopathological and genetic study of 916 high-grade gliomas. J Clin Oncol 2006;24:1502.
403. Perry A, Aldape KD, George DH, Burger PC. Small cell astrocytoma: an aggressive variant that is clinicopathologically and genetically distinct from anaplastic oligodendroglioma. Cancer 2004;101:2318–26.
404. Cairncross JG, Macdonald DR, Ramsay DA. Aggressive oligodendroglioma: a chemosensitive tumor. Neurosurgery 1992;31:78–82.
405. Cairncross G, Macdonald D, Ludwin S, et al. Chemotherapy for anaplastic oligodendroglioma. National Cancer Institute of Canada Clinical Trials Group. J Clin Oncol 1994;12:2013–21.

Ependymoma

406. Guyotat J, Signorelli F, Desme S, et al. Intracranial ependymomas in adult patients: analyses of prognostic factors. J Neurooncol 2002;60:255–68.
407. Schwartz TH, Kim S, Glick RS, et al. Supratentorial ependymomas in adult patients. Neurosurgery 1999;44:721–31.
408. Nagib MG, O'Fallon MT. Myxopapillary ependymoma of the conus medullaris and filum terminale in the pediatric age group. Pediatr Neurosurg 1997;26:2–7.
408a. Fassett DR, Pingree J, Kestle JR. The high incidence of tumor dissemination in myxopapillary ependymoma in pediatric patients. Report of five cases and review of the literature. J Neurosurg 2005;102:59–64.
409. Pulitzer DR, Martin PC, Collins PC, Ralph DR. Subcutaneous sacrococcygeal ("myxopapillary") ependymal rests. Am J Surg Pathol 1988;12:672–7.
410. Vagaiwala MR, Robinson JS, Galicich JH, Gralla RJ, Helson L, Beattie EJ Jr. Metastasizing extradural ependymoma of the sacrococcygeal region: case report and review of literature. Cancer 1979;44:326–33.
411. Comi AM, Backstrom JW, Burger PC, Duffner PK. Clinical and neuroradiologic findings in infants with intracranial ependymomas. Pediatric Oncology Group. Pediatr Neurol 1998;18:23–9.
412. Molina OM, Colina JL, Luzardo GD, et al. Extraventricular cerebral anaplastic ependymomas. Surg Neurol 1999;51:630–5.
413. Spoto GP, Press GA, Hesselink JR, Solomon M. Intracranial ependymoma and subependymoma: MR manifestations. AJR Am J Roentgenol 1990;154:837–45.
414. Sato Y, Ochiai H, Yamakawa Y, Nabeshima K, Asada Y, Hayashi T. Brain surface ependymoma. Neuropathology 2000;20:315–8.
415. Chang WE, Finn LS. MR appearance of lipomatous ependymoma in a 5-year-old boy. AJR Am J Roentgenol 2001;177:1475–8.
416. Hirato J, Nakazato Y, Iijima M, et al. An unusual variant of ependymoma with extensive tumor cell vacuolization. Acta Neuropathol (Berl) 1997;93:310–6.
417. Ruchoux MM, Kepes JJ, Dhellemmes P, et al. Lipomatous differentiation in ependymomas: a report of three cases and comparison with similar changes reported in other central nervous system neoplasms of neuroectodermal origin. Am J Surg Pathol 1998;22(3):338–46.
418. Takahashi H, Goto J, Emura I, Honma T, Hasegawa K, Uchiyama S. Lipidized (foamy) tumor cells in a spinal cord ependymoma with collagenous metaplasia. Acta Neuropathol (Berl) 1998;95:421–5.
419. Kawano N, Yagishita S, Oka H, et al. Spinal tanycytic ependymomas. Acta Neuropathol (Berl) 2001;101:43–8.
420. Kawano N, Yasui Y, Utsuki S, Oka H, Fujii K, Yamashina S. Light microscopic demonstration of the microlumen of ependymoma: a study of the usefulness of antigen retrieval for epithelial membrane antigen (EMA) immunostaining. Brain Tumor Pathol 2004;21:17–21.
421. Min KW, Scheithauer BW. Clear cell ependymoma: a mimic of oligodendroglioma: clinicopathologic and ultrastructural considerations. Am J Surg Pathol 1997;21:820–6.
422. Fouladi M, Helton KJ, Dalton J, et al. Clear cell ependymoma: a clinicopathologic and radiographic analysis of 10 cases. Cancer 2003;98:2237–44.
423. Fouladi M, Helton K, Dalton J, et al. Clear cell ependymoma: a clinicopathologic and radiographic analysis of 10 patients. Cancer 2003;98:2232–44.
424. Friede RL, Pollak A. The cytogenetic basis for classifying ependymomas. J Neuropathol Exp Neurol 1978;37:103–18.

425. Langford LA, Barre GM. Tanycytic ependymoma. Ultrastruct Pathol 1997;21:135–42.
426. Rosenblum MK, Erlandson RA, Aleksic SN, Budzilovich GN. Melanotic ependymoma and subependymoma. Am J Surg Pathol 1990;14: 729–36.
427. Chan AC, Ho LC, Yip WW, Cheung FC. Pigmented ependymoma with lipofuscin and neuromelanin production. Arch Pathol Lab Med 2003;127:872–5.
428. Kleinman GM, Zagzag D, Miller DC. Epithelioid ependymoma: a new variant of ependymoma: report of three cases. Neurosurgery 2003;53:743–8.
429. Gessi M, Marani C, Geddes J, Arcella A, Cenacchi G, Glangaspero F. Ependymoma with neuropil-like islands: a case report with diagnostic and histogenetic implications. Acta Neuropathol 2005;109:231–4.
430. Sonneland PR, Scheithauer BW, Onofrio BM. Myxopapillary ependymoma. A clinicopathologic and immunocytochemical study of 77 cases. Cancer 1985;56:883–93.
431. Specht CS, Smith TW, DeGirolami U, Price JM. Myxopapillary ependymoma of the filum terminale. A light and electron microscopic study. Cancer 1986;58:310–7.
432. Zec N, De Girolami U, Schofield DE, Scott RM, Anthony DC. Giant cell ependymoma of the filum terminale. A report of two cases. Am J Surg Pathol 1996;20:1091–101.
433. Fourney DR, Siadati A, Bruner JM, Gokaslan ZL, Rhines LD. Giant cell ependymoma of the spinal cord. Case report and review of the literature. J Neurosurg 2004;100(Suppl 1):75–9.
434. Prayson RA. Myxopapillary ependymomas: a clinicopathologic study of 14 cases including MIB-1 and p53 immunoreactivity. Mod Pathol 1997;10:304–10.
435. Jeon YK, Jung HW, Park SH. Infratentorial giant cell ependymoma: a rare variant of ependymoma. Pathol Res Pract 2004;200:717–25.
436. Brown DF, Chason DP, Schwartz LF, Coimbra CP, Rushing EJ. Supratentorial giant cell ependymoma: a case report. Mod Pathol 1998;11: 398–403.
437. Jouvet A, Fauchon F, Liberski P, et al. Papillary tumor of the pineal region. Am J Surg Pathol 2003;27:505–12.
438. Ho DM, Hsu CY, Wong TT, Chiang H. A clinicopathologic study of 81 patients with ependymomas and proposal of diagnostic criteria for anaplastic ependymoma. J Neurooncol 2001;54:77–85.
439. Merchant TE, Jenkins JJ, Burger PC, et al. Influence of tumor grade on time to progression after irradiation for localized ependymoma in children. Int J Radiat Oncol Biol Phys 2002;53: 52–7.
440. Paulino AC, Wen BC, Buatti JM, et al. Intracranial ependymomas: an analysis of prognostic factors and patterns of failure. Am J Clin Oncol 2002;25:117–22.
441. Ritter AM, Hess KR, McLendon RE, Langford LA. Ependymomas: MIB-1 proliferation index and survival. J Neurooncol 1998;40:51–7.
442. Bouffet E, Perilongo G, Canete A, Massimino M. Intracranial ependymomas in children: a critical review of prognostic factors and a plea for cooperation. Med Pediatr Oncol 1998; 30:319–31.
443. McLendon RE, Wiestler O, Kros JM, Korshunov A, Ng HK. Anaplastic ependymoma. In: Louis DN, Ohgaki H, Wiestler OD, Cavenee W, eds. WHO classification of tumours of the central nervous system. Lyon: IARC Press; 2007. (In press)
444. Nazar GB, Hoffman HJ, Becker LE, Jenkin D, Humphreys RP, Hendrick EB. Infratentorial ependymomas in childhood: prognostic factors and treatment. J Neurosurg 1990;72:408–17.
445. Bennetto L, Foreman N, Harding B, et al. Ki-67 immunolabelling index is a prognostic indicator in childhood posterior fossa ependymomas. Neuropathol Appl Neurobiol 1998;24:434–40.
446. Wolfsberger S, Fischer I, Hoftberger R, et al. Ki-67 immunolabelling index is an accurate predictor of outcome in patients with intracranial ependymoma. Am J Surg Pathol 2004;28:914–20.
447. Awaya H, Kaneko M, Amatya VJ, Takeshima Y, Oka S, Inai K. Myxopapillary ependymoma with anaplastic features. Pathol Int 2003;53: 700–3.
448. Vege KD, Giannini C, Scheithauer BW. The immunophenotype of ependymomas. Appl Immunohistochem Mol Morphol 2000;8:25–31.
449. Choi YL, Chi JG, Suh YL. CD99 immunoreactivity in ependymoma. Appl Immunohistochem Mol Morphol 2001;9:125–9.
450. Kawano N, Ohba Y, Nagashima K. Eosinophilic inclusions in ependymoma represent microlumina: a light and electron microscopic study. Acta Neuropathol (Berl) 2000;99:214–8.
451. Hasselblatt M, Paulus W. Sensitivity and specificity of epithelial membrane antigen staining patterns in ependymomas. Acta Neuropathol (Berl) 2003;106:385–8.
452. Kawano N, Yada K, Yagishita S. Clear cell ependymoma. A histological variant with diagnostic implications. Virchows Arch A Pathol Anat Histopathol 1989;415:467–72.
453. Ho KL. Microtubular aggregates within rough endoplasmic reticulum in myxopapillary ependymoma of the filum terminale. Arch Pathol Lab Med 1990;114:956–60.
454. Ng HK. Cytologic features of ependymomas in smear preparations. Acta Cytol 1994;38:331–4.
455. Kulesza P, Tihan T, Ali SZ. Myxopapillary ependymoma: cytomorphologic characteristics and differential diagnosis. Diagn Cytopathol 2002;26:247–50.
456. Kumar PV. Nuclear grooves in ependymoma. Cytologic study of 21 cases. Acta Cytol 1997; 41:1726–31.

457. Otani M, Fujita K, Yokoyama A, et al. Imprint cytologic features of intracytoplasmic lumina in ependymoma. A report of two cases. Acta Cytol 2001;45:430–4.
458. Dvoracek MA, Kirby PA. Intraoperative diagnosis of tanycytic ependymoma: pitfalls and differential diagnosis. Diagn Cytopathol 2001;24:289–92.
458a. Taylor MD, Poppleton H, Fuller C. Radial glial cells are candidate stem cells of ependymoma. Cancer Cell 2005;8:323–35.
459. Ebert C, von Haken M, Meyer-Puttlitz B, et al. Molecular genetic analysis of ependymal tumors. NF2 mutations and chromosome 22q loss occur preferentially in intramedullary spinal ependymomas. Am J Pathol 1999;155:627–32.
460. Hirose Y, Aldape K, Bollen A, et al. Chromosomal abnormalities subdivide ependymal tumors into clinically relevant groups. Am J Pathol 2001;158:1137–43.
461. Scheil S, Bruderlein S, Eicker M, et al. Low frequency of chromosomal imbalances in anaplastic ependymomas as detected by comparative genomic hybridization. Brain Pathol 2001;11: 133–43.
462. Lee M, Rezai AR, Freed D, Epstein FJ. Intramedullary spinal cord tumors in neurofibromatosis. Neurosurgery 1996;38:32–7.
463. Kraus JA, de Millas W, Sorensen N, et al. Indications for a tumor suppressor gene at 22q11 involved in the pathogenesis of ependymal tumors and distinct from hSNF5/INI1. Acta Neuropathol (Berl) 2001;102:69–74.
464. Rajaram V, Leuthardt EC, Singh PK, et al. 9p21 and 13q14 dosages in ependymomas. A clinicopathologic study of 101 cases. Mod Pathol 2004;17:9–14.
465. Mahler-Araujo MB, Sanoudou D, Tingby O, et al. Structural genomic abnormalities of chromosomes 9 and 18 in myxopapillary ependymomas. J Neuropathol Exp Neurol 2003;62: 927–35.
466. Rezai AR, Woo HH, Lee M, Cohen H, Zagzag D, Epstein FJ. Disseminated ependymomas of the central nervous system. J Neurosurg 1996; 85:618–24.
467. Hanbali F, Fourney DR, Marmor E, et al. Spinal cord ependymoma: radical surgical resection and outcome. Neurosurgery 2002;51:1162–72; discussion 1172–4.
468. Kramer GW, Rutten E, Sloof J. Subcutaneous sacrococcygeal ependymoma with inguinal lymph node metastasis. Case report. J Neurosurg 1988;68:474–7.
469. Miralbell R, Louis DN, O'Keeffe D, Rosenberg AE, Suit HD. Metastatic ependymoma of the sacrum. Cancer 1990;65:2353–5.
470. Wolff M, Santiago H, Duby MM. Delayed distant metastasis from a subcutaneous sacrococcygeal ependymoma. Case report, with tissue culture, ultrastructural observations, and review of the literature. Cancer 1972;30:1046–67.

Subependymoma

471. Changaris DG, Powers JM, Perot PL Jr, Hungerford GD, Neal GB. Subependymoma presenting as subarachnoid hemorrhage: case report. J Neurosurg 1981;55:643–5.
472. Scheithauer BW. Symptomatic subependymoma. Report of 21 cases with review of the literature. J Neurosurg 1978;49:689–96.
473. Dario A, Fachinetti P, Cerati M, Dorizzi A. Subependymoma of the spinal cord: case report and review of the literature. J Clin Neurosci 2001;8:48–50.
474. Jallo GI, Zagzag D, Epstein F. Intramedullary subependymoma of the spinal cord. Neurosurgery 1996;38:251–7.
475. Nishio S, Morioka T, Mihara F, Fukui M. Subependymoma of the lateral ventricles. Neurosurg Rev 2000;23:98–103.
476. Kondziolka D, Bilbao JM. Mixed ependymoma-astrocytoma (subependymoma?) of the cerebral cortex. Acta Neuropathol (Berl) 1988;76:633–7.
477. Nishio S, Morioka T, Suzuki S, Fukui M. Tumours around the foramen of Monro: clinical and neuroimaging features and their differential diagnosis. J Clin Neurosci 2002;9:137–41.
478. Chiechi MV, Smirniotopoulos JG, Jones RV. Intracranial subependymomas: CT and MR imaging features in 24 cases. AJR Am J Roentgenol 1995;165:1245–50.
479. Louis DN, Hedley-Whyte ET, Martuza RL. Sarcomatous proliferation of the vasculature in a subependymoma: a follow-up study of sarcomatous dedifferentiation. Acta Neuropathol (Berl) 1990;80:573–4.
480. Tomlinson FH, Scheithauer BW, Kelly PJ, Forbes GS. Subependymoma with rhabdomyosarcomatous differentiation: report of a case and literature review. Neurosurgery 1991;28: 761–8.
481. Prayson RA, Suh JH. Subependymomas: clinicopathologic study of 14 tumors, including comparative MIB-1 immunohistochemical analysis with other ependymal neoplasms. Arch Pathol Lab Med 1999;123:306–9.
482. Azzarelli B, Rekate HL, Roessmann U. Subependymoma: a case report with ultrastructural study. Acta Neuropathol (Berl) 1977;40:279–82.
483. Inayama Y, Nishio Y, Ishii M, et al. Crush and imprint cytology of subependymoma: a case report. Acta Cytol 2001;45:636–40.

Astroblastoma

484. Bonnin JM, Rubinstein LJ. Astroblastomas: a pathological study of 23 tumors, with a postoperative follow-up in 13 patients. Neurosurgery 1989;25:6–13.
485. Brat DJ, Hirose Y, Cohen KJ, Feuerstein BG, Burger PC. Astroblastoma: clinicopathologic features and chromosomal abnormalities defined by comparative genomic hybridization. Brain Pathol 2000;10:342–52.

486. Thiessen B, Finlay J, Kulkarni R, Rosenblum MK. Astroblastoma: does histology predict biologic behavior? J Neurooncol 1998;40:59–65.
487. Rubinstein LJ, Herman MM. The astroblastoma and its possible cytogenic relationship to the tanycyte. An electron microscopic, immunohistochemical, tissue- and organ-culture study. Acta Neuropathol (Berl) 1989;78:472–83.
488. Baka JJ, Patel SC, Roebuck JR, Hearshen DO. Predominantly extraaxial astroblastoma: imaging and proton MR spectroscopy features. AJNR Am J Neuroradiol 1993;14:946–50.
489. Port JD, Brat DJ, Burger PC, Pomper MG. Astroblastoma: radiologic-pathologic correlation and distinction from ependymoma. AJNR Am J Neuroradiol 2002;23:243–7.
490. Jay V, Edwards V, Squire J, Rutka J. Astroblastoma: report of a case with ultrastructural, cell kinetic, and cytogenetic analysis. Pediatr Pathol 1993;13:323–32.
491. Cabello A, Madero S, Castresana A, Diaz-Lobato R. Astroblastoma: electron microscopy and immunohistochemical findings: case report. Surg Neurol 1991;35:116–21.

Chordoid Glioma of the Third Ventricle

492. Brat DJ, Scheithauer BW, Staugaitis SM, Cortez SC, Brecher K, Burger PC. Third ventricular chordoid glioma: a distinct clinicopathologic entity. J Neuropathol Exp Neurol 1998;57:283–90.
493. Galloway M, Afshar F, Geddes JF. Chordoid glioma: an uncommon tumour of the third ventricle. Br J Neurosurg 2001;15:147–50.
493a. Leeds NE, Lang, FF, Ribalta T, Sawaya R, Fuller GN. Origin of chordoid glioma of the third ventricle. Arch Pathol Lab Med 2006;130:625–36.
494. Pasquier B, Peoc'h M, Morrison AL, et al. Chordoid glioma of the third ventricle: a report of two new cases, with further evidence supporting an ependymal differentiation, and review of the literature. Am J Surg Pathol 2002;26:1330–42.
495. Pomper MG, Passe TJ, Burger PC, Scheithauer BW, Brat DJ. Chordoid glioma: a neoplasm unique to the hypothalamus and anterior third ventricle. AJNR Am J Neuroradiol 2001;22:464–9.
496. Castellano-Sanchez AA, Schemankewitz E, Mazewski C, Brat DJ. Pediatric chordoid glioma with chondroid metaplasia. Pediatr Dev Pathol 2001;4:564–7.
497. Raizer JJ, Shetty T, Gutin PH, et al. Chordoid glioma: report of a case with unusual histologic features, ultrastructural study and review of the literature. J Neurooncol 2003;63:39–47.
498. Cenacchi G, Roncaroli F, Cerasoli S, Ficarra G, Merli GA, Giangaspero F. Chordoid glioma of the third ventricle: an ultrastructural study of three cases with a histogenetic hypothesis. Am J Surg Pathol 2001;25:401–5.
499. Reifenberger G, Weber T, Weber RG, et al. Chordoid glioma of the third ventricle: immunohistochemical and molecular genetic characterization of a novel tumor entity. Brain Pathol 1999;9:617–26.
500. Jouvet A, Fauchon F, Liberski P, et al. Papillary tumor of the pineal region. Am J Surg Pathol 2003;27:505–12.

Angiocentric Glioma

500a. Wang M, Tihan T, Rojiani AM, et al. Monomorphous angiocentric glioma: a distinctive epileptogenic neoplasm with features of infiltrating astrocytoma and ependymoma. J Neuropath Exp Neurol 2005;64:875–81.
500b. Lellouch-Tubiana A, Boddaert N, Bourgeois M, et al. Angiocentric neuroepithelial tumor (ANET): a new epilepsy-related clinicopathological entity with distinctive MRI. Brain Pathol 2005;15:281–6.

Choroid Plexus Papilloma, Atypical Papilloma, and Carcinoma

501. Paulus W, Brandner S. Choroid plexus tumours. In: Louis DN, Ohgaki H, Wiestler OD, Cavenee W, eds. WHO classification of tumours of the central nervous system. Lyon: IARC Press; 2007. (In press.)
502. Anderson DR, Falcone S, Bruce JH, Mejidas AA, Post MJ. Radiologic-pathologic correlation. Congenital choroid plexus papillomas. AJNR Am J Neuroradiol 1995;16:2072–6.
503. Body G, Darnis E, Pourcelot D, Santini JJ, Gold F, Soutoul JH. Choroid plexus tumors: antenatal diagnosis and follow-up. J Clin Ultrasound 1990;18:575–8.
504. Buetow PC, Smirniotopoulos JG, Done S. Congenital brain tumors: a review of 45 cases. AJR Am J Roentgenol 1990;155:587–93.
505. Tomita T, Naidich TP. Successful resection of choroid plexus papillomas diagnosed at birth: report of two cases. Neurosurgery 1987;20:774–9.
506. McEvoy AW, Harding BN, Phipps KP, et al. Management of choroid plexus tumours in children: 20 years experience at a single neurosurgical centre. Pediatr Neurosurg 2000;32: 192–9.
507. Ken JG, Sobel DF, Copeland B, Davis J 3rd, Kortman KE. Choroid plexus papillomas of the foramen of Luschka: MR appearance. AJNR Am J Neuroradiol 1991;12:1201–3.
508. Talacchi A, De Micheli E, Lombardo C, Turazzi S, Bricolo A. Choroid plexus papilloma of the cerebellopontine angle: a twelve patient series. Surg Neurol 1999;51:621–9.
509. Garber JE, Burke EM, Lavally BL, et al. Choroid plexus tumors in the breast cancer-sarcoma syndrome. Cancer 1990;66:2658–60.

510. Yuasa H, Tokito S, Tokunaga M. Primary carcinoma of the choroid plexus in Li-Fraumeni syndrome: case report. Neurosurgery 1993;32: 131–4.
511. Dickens DS, Dothage JA, Heideman RL, Ballard ET, Jubinsky PT. Successful treatment of an unresectable choroid plexus carcinoma in a patient with Li-Fraumeni syndrome. J Pediatr Hematol Oncol 2005;27:46–9.
512. Trifiletti RR, Incorpora G, Polizzi A, Cocuzza MD, Bolan EA, Parano E. Aicardi syndrome with multiple tumors: a case report with literature review. Brain Dev 1995;17:283–5.
513. Kurtkaya-Yapicier O, Scheithauer BW, Van Peteghem KP, Sawicki JE. Unusual case of extradural choroid plexus papilloma of the sacral canal. Case report. J Neurosurg 2002;97 (Suppl 1):102–5.
514. Taylor MB, Jackson RW, Hughes DG, Wright NB. Magnetic resonance imaging in the diagnosis and management of choroid plexus carcinoma in children. Pediatr Radiol 2001;31: 624–30.
515. Koeller KK, Sandberg GD. From the archives of the AFIP. Cerebral intraventricular neoplasms: radiologic-pathologic correlation. Radiographics 2002;22:1473–505.
516. Coates TL, Hinshaw DB Jr, Peckman N, et al. Pediatric choroid plexus neoplasms: MR, CT, and pathologic correlation. Radiology 1989; 173:81–8.
517. Chow E, Jenkins JJ, Burger PC, et al. Malignant evolution of choroid plexus papilloma. Pediatr Neurosurg 1999;31:127–30.
518. Gullotta F, de Melo A. das Karzinom des Plexus chorioideus. Klinische, lichtmikroskopische und elektronenoptische Untersuchungen. Neurochiurgica 1979;22:1–9.
519. Paulus W, Janisch W. Clinicopathologic correlations in epithelial choroid plexus neoplasms: a study of 52 cases. Acta Neuropathol (Berl) 1990;80:635–41.
519a. Jeibmann A, Hasselblatt M, Gerss J, et al. Prognostic implications of atypical histologic features in choroid plexus papilloma. J Neuropathol Exp Neurol 2006;65:1069–73.
520. Tomlinson FA, Scheithauer BN, Hammack JE, Burger PC, Meyer F, Wollan PC. Choroid plexus neoplasia: a correlative clinical and pathologic study. J Neuropathol Exp Neurol 1995;54:423.
521. Ajir F, Chanbusarakum K, Bolles JC. Acinar choroid plexus adenoma of the fourth ventricle. Surg Neurol 1982;17:290–2.
522. Andreini L, Doglioni C, Giangaspero F. Tubular adenoma of choroid plexus: a case report. Clin Neuropathol 1991;10:137–40.
523. Bonnin JM, Colon LE, Morawetz RB. Focal glial differentiation and oncocytic transformation in choroid plexus papilloma. Acta Neuropathol (Berl) 1987;72:277–80.
524. Lana-Peixoto MA, Lagos J, Silbert SW. Primary pigmented carcinoma of the choroid plexus. A light and electron microscopic study. J Neurosurg 1977;47:442–50.
525. Reimund EL, Sitton JE, Harkin JC. Pigmented choroid plexus papilloma. Arch Pathol Lab Med 1990;114:902–5.
526. Watanabe K, Ando Y, Iwanaga H, et al. Choroid plexus papilloma containing melanin pigment. Clin Neuropathol 1995;14:159–61.
527. Kepes JJ. Oncocytic transformation of choroid plexus epithelium. Acta Neuropathol (Berl) 1983;62:145–8.
528. Stefanko SZ, Vuzevski VD. Oncocytic variant of choroid plexus papilloma. Acta Neuropathol (Berl) 1985;66:160–2.
529. Doran SE, Blaivas M, Dauser RC. Bone formation within a choroid plexus papilloma. Pediatr Neurosurg. 1995;23:216–8.
530. Cardozo J, Cepeda F, Quintero M, Mora E. Choroid plexus papilloma containing bone. Acta Neuropathol (Berl) 1985;68:83–5.
531. Duckett S, Osterholm J, Schaefer D, Gonzales C, Schwartzman RJ. Ossified mucin-secreting choroid plexus adenoma: case report. Neurosurgery 1991;29:130–2.
532. Salazar J, Vaquero J, Aranda IF, Menendez J, Jimenez MD, Bravo G. Choroid plexus papilloma with chondroma: case report. Neurosurgery 1986;18:781–3.
533. Levy ML, Goldfarb A, Hyder DJ, et al. Choroid plexus tumors in children: significance of stromal invasion. Neurosurgery 2001;48:303–9.
534. Ang LC, Taylor AR, Bergin D, Kaufmann JC. An immunohistochemical study of papillary tumors in the central nervous system. Cancer 1990;65:2712–9.
535. Gaudio RM, Tacconi L, Rossi ML. Pathology of choroid plexus papillomas: a review. Clin Neurol Neurosurg 1998;100:165–86.
536. Gyure KA, Morrison AL. Cytokeratin 7 and 20 expression in choroid plexus tumors: utility in differentiating these neoplasms from metastatic carcinomas. Mod Pathol 2000;13:638–43.
537. Kepes JJ, Collins J. Choroid plexus epithelium (normal and neoplastic) expresses synaptophysin. A potentially useful aid in differentiating carcinoma of the choroid plexus from metastatic papillary carcinomas. J Neuropathol Exp Neurol 1999;58:398–401.
538. Miettinen M, Clark R, Virtanen I. Intermediate filament proteins in choroid plexus and ependyma and their tumors. Am J Pathol 1986;123:231–40.
539. Jay V, Ho M, Chan F, Malkin D. P53 expression in choroid plexus neoplasms: an immunohistochemical study. Arch Pathol Lab Med 1996;120:1061–5.

540. Carlotti CG, Jr., Salhia B, Weitzman S, et al. Evaluation of proliferative index and cell cycle protein expression in choroid plexus tumors in children. Acta Neuropathol (Berl) 2002; 103:1–10.
541. Herbert J, Cavallaro T, Dwork AJ. A marker for primary choroid plexus neoplasms. Am J Pathol 1990;136:1317–25.
542. Albrecht S, Rouah E, Becker LE, Bruner J. Transthyretin immunoreactivity in choroid plexus neoplasms and brain metastases. Mod Pathol 1991;4:610–4.
543. Matsushima T. Choroid plexus papillomas and human choroid plexus. A light and electron microscopic study. J Neurosurg 1983;59:1054–62.
544. McComb RD, Burger PC. Choroid plexus carcinoma. Report of a case with immunohistochemical and ultrastructural observations. Cancer 1983;51:470–5.
545. Pai R, Kini H, Rao V, Naik R. Choroid plexus papilloma diagnosed by crush cytology. Diagn Cytopathol 2001;25:165–7.
546. Rickert CH, Wiestler OD, Paulus W. Chromosomal imbalances in choroid plexus tumors. Am J Pathol 2002;160:1105–13.
547. Wyatt-Ashmead J, Kleinschmidt-DeMasters B, Mierau GW, et al. Choroid plexus carcinomas and rhabdoid tumors: phenotypic and genotypic overlap. Pediatr Dev Pathol 2001;4:545–9.
548. Gessi M, Giangaspero F, Pietsch T. Atypical teratoid/rhabdoid tumors and choroid plexus tumors: when genetics "surprise" pathology. Brain Pathol 2003;13:409–14.
549. Weber M, Stockhammer F, Schmitz U, von Deimling A. Mutational analysis of INI1 in sporadic human brain tumors. Acta Neuropathol (Berl) 2001;101:479–82.
550. Palmiter RD, Chen HY, Messing A, Brinster RL. SV40 enhancer and large-T antigen are instrumental in development of choroid plexus tumours in transgenic mice. Nature 1985;316: 457–60.
551. Huang H, Reis R, Yonekawa Y, Lopes JM, Kleihues P, Ohgaki H. Identification in human brain tumors of DNA sequences specific for SV40 large T antigen. Brain Pathol 1999;9:33–42.
552. Furness PN, Lowe J, Tarrant GS. Subepithelial basement membrane deposition and intermediate filament expression in choroid plexus neoplasms and ependymomas. Histopathology 1990;16:251–5.
553. Vege KD, Giannini C, Scheithauer BW. The immunophenotype of ependymomas. Appl Immunohistochem Mol Morphol 2000;8:25–31.
554. Gottschalk J, Jautzke G, Paulus W, Goebel S, Cervos-Navarro J. The use of immunomorphology to differentiate choroid plexus tumors from metastatic carcinomas. Cancer 1993;72: 1343–9.
555. Lau SK, Luthringer DJ, Eisen RN. Thyroid transcription factor-1: a review. Appl Immunohistochem Mol Morphol 2002;10:97–102.
556. Raila FA, Bottoms WT Jr, Fratkin JD. Solitary choroid plexus metastasis from a renal cell carcinoma. South Med J 1998;91:1159–62.
557. Judkins AR, Burger PC, Hamilton RL, et al. INI1 protein expression distinguishes atypical teratoid/rhabdoid tumor from choroid plexus carcinoma. J Neuropathol Exp Neurol 2005;64:391–7.
558. Luff DA, Simmons M, Malik T, Ramsden RT, Reid H. Endolymphatic sac tumours. J Laryngol Otol 2002;116:398–401.
559. Jouvet A, Fauchon F, Liberski P, et al. Papillary tumor of the pineal region. Am J Surg Pathol 2003;27:505–12.
560. Wolff JE, Sajedi M, Brant R, Coppes MJ, Egeler RM. Choroid plexus tumours. Br J Cancer 2002;87:1086–91.
561. McGirr SJ, Ebersold MJ, Scheithauer BW, Quast LM, Shaw EG. Choroid plexus papillomas: long-term follow-up results in a surgically treated series. J Neurosurg 1988;69:843–9.
562. Irsutti M, Thorn-Kany M, Arrue P, et al. Suprasellar seeding of a benign choroid plexus papilloma of the fourth ventricle with local recurrence. Neuroradiology 2000;42:657–61.
563. McEvoy AW, Galloway M, Revesz T, Kitchen ND. Metastatic choroid plexus papilloma: a case report. J Neurooncol 2002;56:241–6.
564. Leblanc R, Bekhor S, Melanson D, Carpenter S. Diffuse craniospinal seeding from a benign fourth ventricle choroid plexus papilloma. Case report. J Neurosurg 1998;88:757–60.
565. Berger C, Thiesse P, Lellouch-Tubiana A, Kalifa C, Pierre-Kahn A, Bouffet E. Choroid plexus carcinomas in childhood: clinical features and prognostic factors. Neurosurgery 1998;42:470–5.
566. Fitzpatrick LK, Aronson LJ, Cohen KJ. Is there a requirement for adjuvant therapy for choroid plexus carcinoma that has been completely resected? J Neurooncol 2002;57:123–6.
567. Pencalet P, Sainte-Rose C, Lellouch-Tubiana A, et al. Papillomas and carcinomas of the choroid plexus in children. J Neurosurg 1998;88: 521–8.
568. Valladares JB, Perry RH, Kalbag RM. Malignant choroid plexus papilloma with extraneural metastasis. Case report. J Neurosurg 1980;52: 251–5.

4 NEURONAL AND GLIONEURONAL TUMORS

GANGLIOCYTOMA AND GANGLIOGLIOMA

Definition. *Gangliocytoma* and *ganglioglioma* are neoplasms consisting entirely or in part of large and mature, but dysmorphic, neurons.

General Features. The designation ganglioglioma is applied to tumors with a neoplastic glial component, while the term gangliocytoma is reserved for less common lesions composed exclusively of neurons. Because of transitional forms, the two entities are not always distinct, and the umbrella term *ganglion cell tumor* is usefully encompassing when the neoplastic nature of the glial stroma is uncertain. Gangliocytomas and well-differentiated ganglioglioma are grade I in the World Health Organization (WHO) classification system, whereas gangliogliomas with an anaplastic glial component are grade III (1). The 2007 WHO "Blue Book" authors felt that criteria for a grade II ganglioglioma had yet to be established.

Clinical Features. Ganglion cell tumors arise throughout the neuraxis, including the cerebral hemisphere (fig. 4-1), optic nerve (2), brain stem (fig. 4-2) (3), cerebellum (4), pineal region (5), and spinal cord (6,7), but most are supratentorial, where the temporal lobe is the favored locus. The lesions affect patients of all ages but present principally in those in the first two decades. Seizures or the consequences of increased intracranial pressure are common manifestations (8,9).

Radiologic Findings. Ganglion cell tumors can be either solid or cystic, and in the case of the latter, much of the mass effect is due to the accumulated cyst volume rather than the size of the solid component (fig. 4-1) (9–11). The cyst-mural nodule configuration is shared with

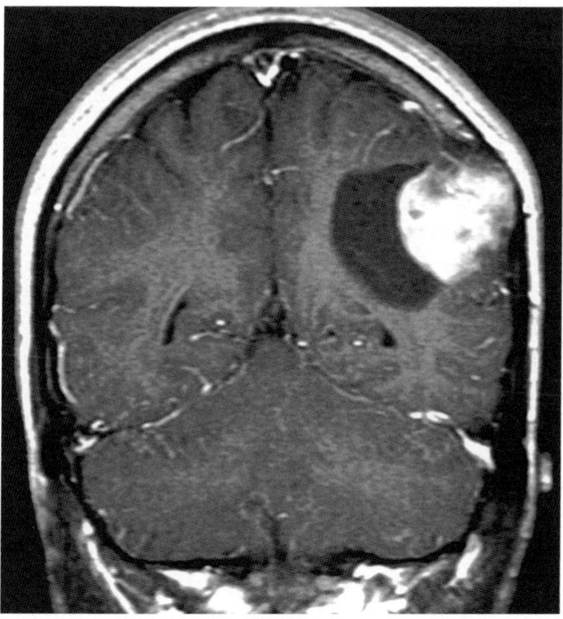

Figure 4-1

GANGLIOGLIOMA

Ganglion cell tumor is usually a discrete mass that is sometimes a nodule in the wall of a cyst, as here in a 19-year-old woman with a history of seizures. The prominent skull erosion is graphic evidence of the lesion's chronicity.

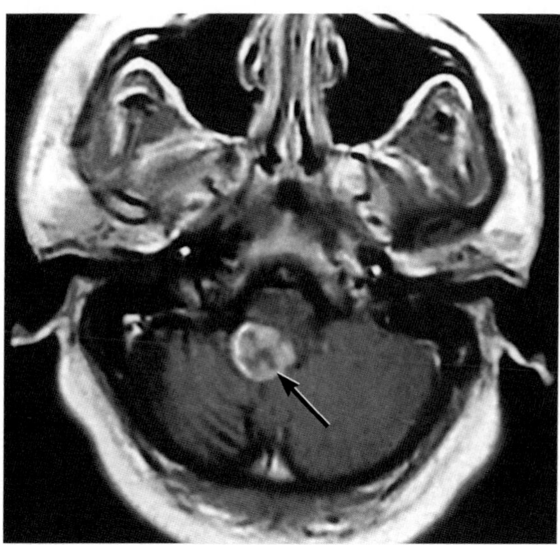

Figure 4-2

GANGLIOGLIOMA

Ganglion cell tumors occur throughout the central nervous system (CNS), including the brain stem, as with this medullary lesion in a 29-year-old man. While sometimes cystic, many are solid, and most are contrast enhancing.

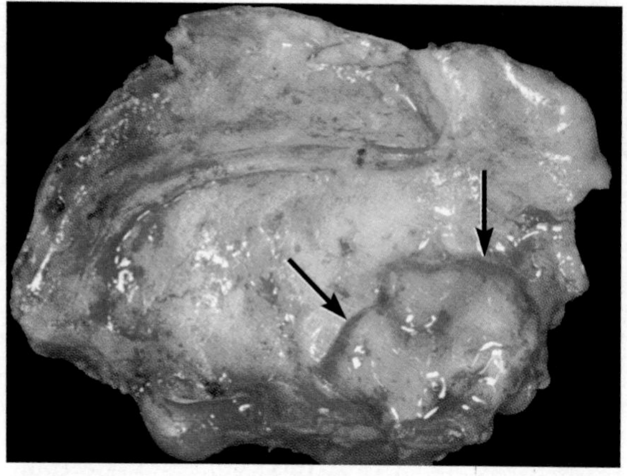

Figure 4-3

GANGLIOGLIOMA

A mural nodule (arrows) with a sector of the cyst wall may be received in a large surgical specimen.

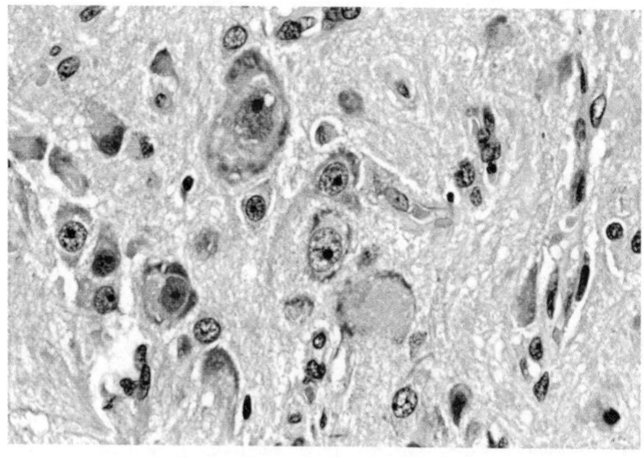

Figure 4-4

GANGLIOCYTOMA

Well-differentiated, paucicellular gangliocytomas can resemble normal gray matter, although the illustrated ganglion cells are too irregular in size, shape, and distribution to be normal cortical components.

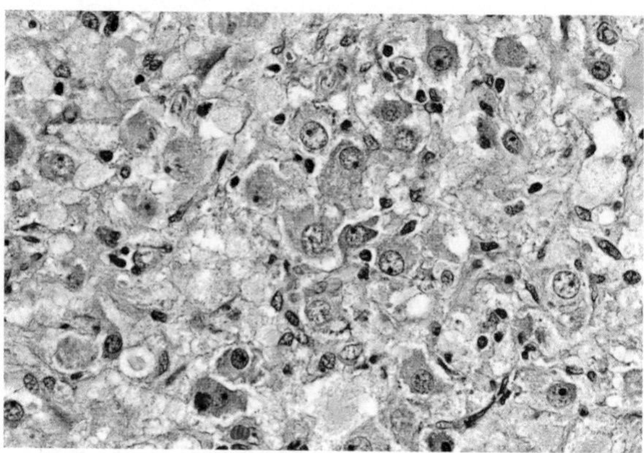

Figure 4-5

GANGLIOCYTOMA

Some gangliocytomas are identified more by high cellularity than by cytologic features.

certain other low-grade neoplasms such as pilocytic astrocytoma, pleomorphic xanthoastrocytoma, and extraventricular neurocytoma, and always is presumptive evidence of a grade I or II neoplasm with limited tissue infiltration. (See Appendix D for frequently cystic lesions, Appendix K for central nervous system [CNS] tumors that can be easily overgraded, and Appendix N for "suspect diagnoses" of surgical neuropathologic specimens.) Even in the absence of a cyst, a solid, well-circumscribed, contrast-enhancing lesion with little mass effect is most consistent with a grade I or II lesion with limited tissue infiltration. Superficial, longstanding ganglion cell tumors can erode the inner table of the skull (fig. 4-1). Calcification is common. Like pleomorphic xanthoastrocytoma and pilocytic astrocytoma, the lesion can give the false impression of malignancy if it spreads within the local subarachnoid space (12).

Gross Findings. Ganglion cell tumors are either solid or cystic, and may be submitted largely intact with a portion of the cyst wall (fig. 4-3). Even noncystic examples are well circumscribed. Ganglion cell tumors may be gritty if extensively calcified.

Microscopic Findings. The most elementary and benign lesion, gangliocytoma, is an essentially hamartomatous process consisting of abnormal neurons in a fibrillar background (figs. 4-4, 4-5). Overall, it resembles gray matter, but is architecturally disorganized and devoid of normal perineuronal oligodendrocytes and stellate reactive astrocytes. The density of ganglion cells is usually low (fig. 4-4) but can be high (fig. 4-5). Perivascular lymphocytes are common; sometimes there are just a few (fig. 4-6), but in other cases many (fig. 4-7). The ganglion cells are abnormally clustered without the orderly distribution and polarity of their normal counterparts. Apical dendrites of normal ganglion cells are well formed and "point" to the pial surface, whereas they are poorly formed and directionally uncoordinated in neoplastic equivalents. Scattered binucleated neurons are often

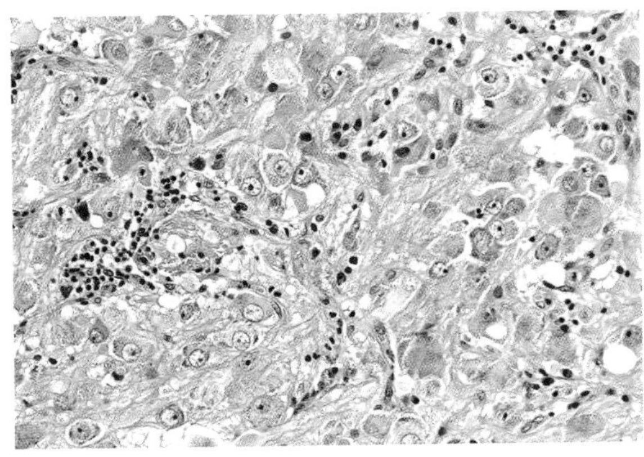

Figure 4-6

GANGLIOCYTOMA

Perivascular lymphocytes, found in most ganglion cell tumors, may be sparse.

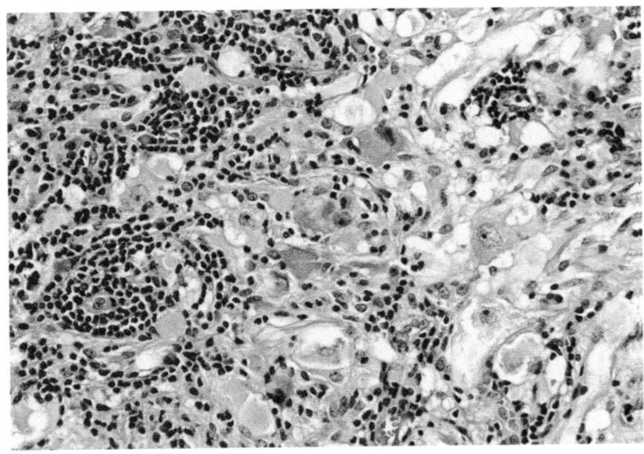

Figure 4-7

GANGLIOCYTOMA

Perivascular lymphocytes are sometimes numerous.

present. Atypia, in the form of large, bizarre, or hyperchromatic nuclei, some with intranuclear cytoplasmic inclusions, is frequent as well. Neoplastic ganglion cells may contain basophilic Nissl substance, but this is usually less well formed than in normal neurons and is often abnormally displaced toward the periphery of the cell. Small neurons and intermediated-sized ganglioid cells are admixed in some instances, but mitotically active neuroblasts are lacking, by definition. Calcospherites are common. Neurofibrillary change and granulovacuolar degeneration are occasionally present in both gangliocytoma and ganglioglioma, particularly in older patients (fig. 4-8) (13).

Histologically heterogeneous, and more difficult to describe than gangliocytoma, ganglioglioma falls into several somewhat overlapping categories. A common variant has a lobular architecture and a reticulin-rich stroma that traps or encircles nests of neoplastic cells (figs. 4-9, 4-10). Some of the latter are obviously neuronal given their vesicular, occasionally multiple, nuclei; prominent nucleoli; and varying amounts of Nissl substance. Others with glassy cytoplasm devoid of Nissl substance may be phenotypically astrocytic.

A second pattern, occurring in isolation or focally in one of the other variants of ganglioglioma, is a microcyst-rich lesion with ganglion cells and often piloid glia (fig. 4-11). The neurons are often smaller and less ganglionic than those of the classic lesion, and identified in part

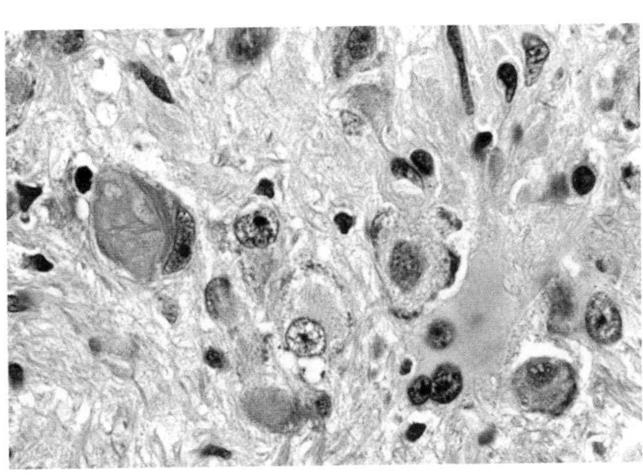

Figure 4-8

GANGLIOCYTOMA WITH NEUROFIBRILLARY CHANGE

Ganglion cells are prone to neurofibrillary change, as seen in the basophilic, faintly fibrillar intracytoplasmic formation in the large neuron towards the left.

by their immunohistochemical features, as is discussed below. A few, at least, are usually neural enough in hematoxylin and eosin (H&E)-stained sections to catch attention. Histologic characteristics include clustering, discrete cell borders, and vesicular nuclei with central nucleoli. In some areas, it is difficult to determine whether the cells around the spaces are neuronal or glial on the basis of their H&E-stained features alone. Other lesions mix ganglion cells and fibril-rich astrocytes into a solid mass with a fibrillar background (fig. 4-12).

Desmoplasia is common in classic gangliogliomas, particularly those that involve the

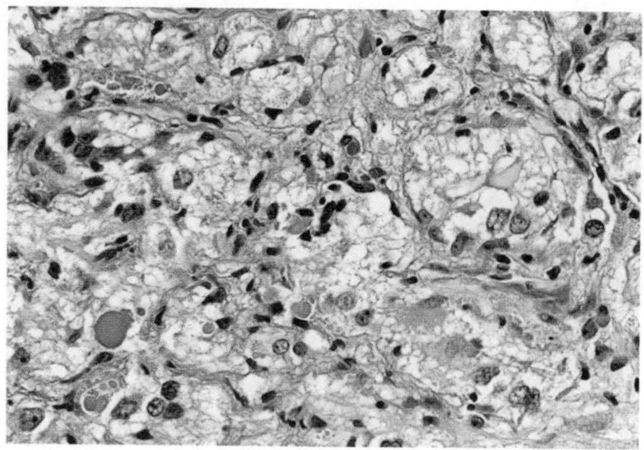

Figure 4-9

GANGLIOGLIOMA

Gangliogliomas assume many forms. The acinus-like grouping present here is a common variant. The hyaline droplets, in small aggregates, form eosinophilic granular bodies (EGBs). The latter usually populate well-circumscribed grade I and II lesions such as ganglion cell tumors, pilocytic astrocytoma, and pleomorphic xanthoastrocytoma.

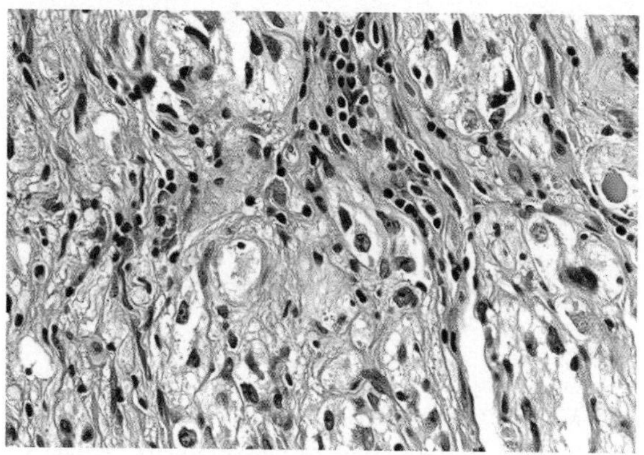

Figure 4-10

GANGLIOGLIOMA

Fibrous stroma, lymphocytes, and EGBs are common features of gangliogliomas. The neurons are smaller and not as obviously ganglionic as in figure 4-5.

subarachnoid space. This connective tissue component varies in extent, being only focal in some tumors and pervasive in others (figs. 4-13, 4-14, left). It is a prominent and requisite feature of the desmoplastic infantile variant that is discussed separately. In either case, ganglion cells may lie entangled in, and sometimes obscured by, trichrome-positive, reticulin-rich matrix (fig. 4-14, right).

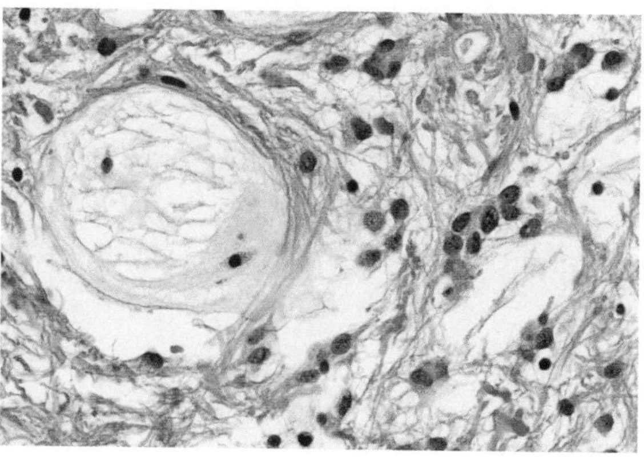

Figure 4-11

GANGLIOGLIOMA

Some extensively microcystic lesions with a pilocytic glial component can be difficult to classify as ganglioglioma (versus pilocytic astrocytoma). In this case, classic ganglioglioma was present elsewhere in the specimen. Immunohistochemistry may be needed in some cases to confirm the presence of ganglion cells or small gangloid cells.

The glial element in ganglioglioma shows not only considerable cytologic diversity, but also wide variation in spatial distribution. In some lesions, particularly those with a lobular architecture, there are GFAP-positive pleomorphic cells with the glassy cytoplasm and short processes of astrocytes of the fibrillary type. These intermingle with ganglion cells and often assume transitional appearances on H&E-stained sections. Piloid astrocytes predominate in other cases. Oligodendroglioma-like cells are common. Even in immunohistochemical preparations, cells initially thought to fit into either the astrocytic or neuronal category may be immunophenotypically noncommittal, while cells with astrocytic features are sometimes synaptophysin positive. It is possible that some large cells stain for both glial and neuronal markers.

Diagnostically important features of ganglion cell tumors, both "-cytomas" and "-gliomas," are focal collections of hyaline protein droplets, or eosinophilic granular bodies (EGBs) (figs. 4-9, 4-10, 4-15). These are common in ganglion cell tumors, pleomorphic xanthoastrocytoma, and pilocytic astrocytoma, and, to a lesser extent, in extraventricular neurocytomas. As with certain other features, such as cyst formation and discrete borders, the presence of EGBs suggests the possibility of a low-grade lesion (see Appendices K, M, and N). In any setting, they

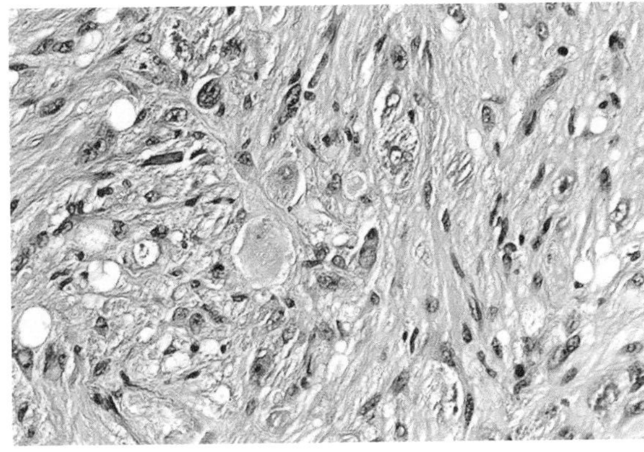

Figure 4-12

GANGLIOGLIOMA

Ganglion cells and astrocytes can be intimately intermixed.

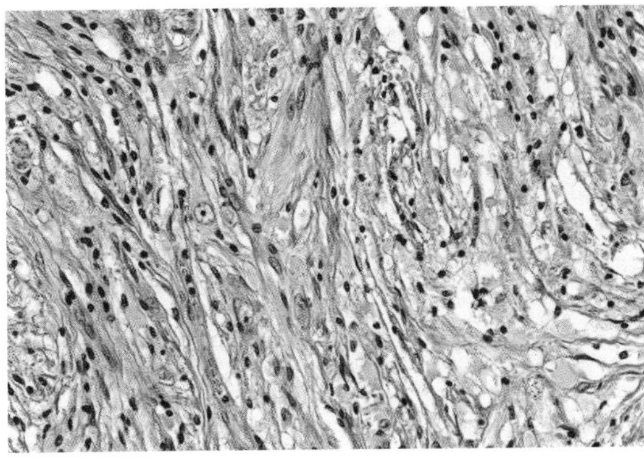

Figure 4-13

GANGLIOGLIOMA

Small ganglion cells are hidden in a dense fibrous stroma.

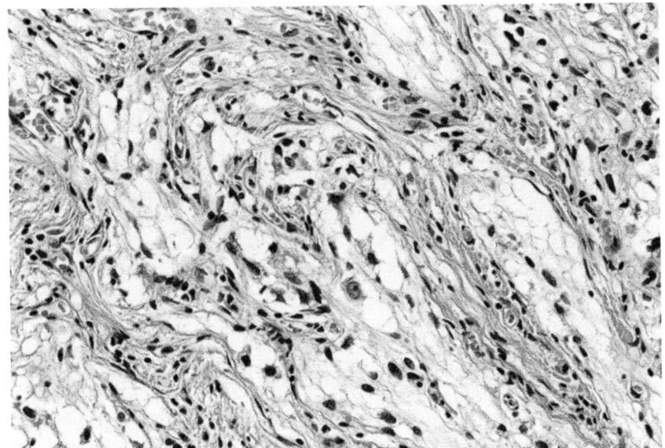

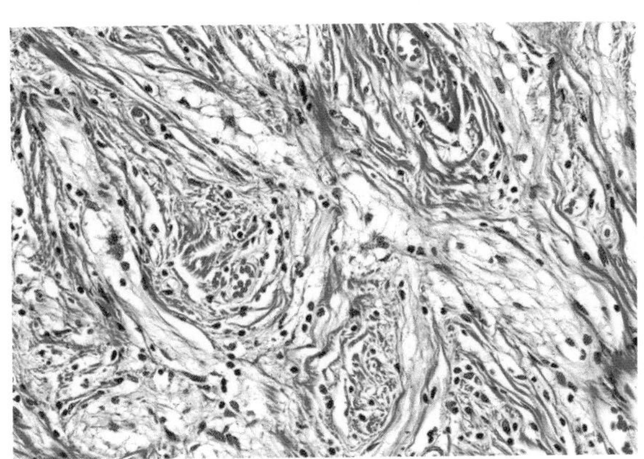

Figure 4-14

GANGLIOGLIOMA

Left: Ganglion cells in desmoplastic tumors are most likely to be found in loose, reticulated areas.
Right: The degree of desmoplasia is more apparent after staining with Masson trichrome.

are presumptive evidence of slow growth, and usually occur in a macroscopically discrete lesion.

Rare mitoses in a bona fide ganglion cell tumor are of no clinical significance, but substantial numbers should suggest that there is anaplastic change in the glial component (rare) or that the lesion is really an infiltrating glioma with trapped neurons (common). Malignant neuronal and glioneuronal tumors, including gangliogliomas with a sarcomatous component, are discussed at the end of this chapter (14).

As in pilocytic astrocytomas, vascular proliferation is not uncommon in gangliogliomas. Most often it is a capillary proliferation in the wall of a cyst (fig. 4-16, top), but large and alarming, although actually innocent, glomeruloid masses also may appear in the parenchyma (fig. 4-16, bottom). Exceptional ganglion cell tumors are melanotic (15,16). Cytologic and architectural abnormalities resembling cortical dysplasia may be present in the cortex adjacent to ganglion cell tumors, thus placing the lesion within an apparent field of abnormality (8), as evidenced by scattered cells within this penumbra that are immunoreactive for CD34 (17,18).

Frozen Section. Individual ganglion cells are structurally labile in frozen tissues and may stain faintly (fig. 4-17). The process of freezing can leave these large cells appearing astrocytic, and even malignant if it exaggerates the degree of nuclear

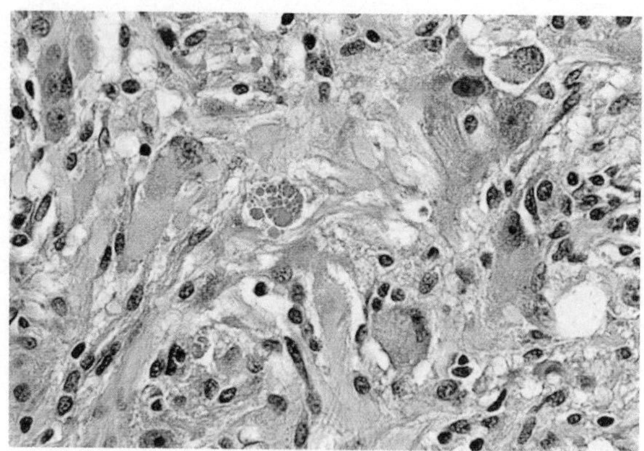

Figure 4-15

GANGLIOGLIOMA

The presence of EGBs should alert the observer to the possibility of a ganglion cell tumor.

pleomorphism and hyperchromasia. A correct interpretation is facilitated by consideration of the clinicoradiologic setting and the low-power architecture of the tumor. A cyst-mural nodule configuration or a solid, discrete, contrast-enhancing lesion with little mass effect, for example, makes the ganglion cell tumor a strong contender, and minimizes the likelihood of an anaplastic tumor. Microscopically, nodular architecture, perivascular cuffs of lymphocytes, and microcalcifications further suggest the diagnosis. The presence of EGBs is especially helpful.

Immunohistochemical Findings. The immunohistochemical profile of most ganglion cell tumors reflects their dimorphous nature. While it has its limitations, immunostaining for synaptophysin is the principal current marker for neurons (11,19). While positivity in some cells is pancytoplasmic, it is a fine granular surface reactivity that is held to be most characteristic (fig. 4-18) (19). Unfortunately, most ganglion cells sit in synaptophysin-positive neuropil and thus it can be difficult to be certain whether circumferential staining belongs to the membrane of the cell in question or only to surrounding neuropil. In addition, strong surface staining for synaptophysin is normal in certain large neurons, as in the spinal cord and brain stem (20–22). Finally, many neurons that are clearly neoplastic have staining that is diffusely cytoplasmic, rather than exclusively surface.

Given these limitations, antibodies to chromogranin are useful adjuncts since they stain

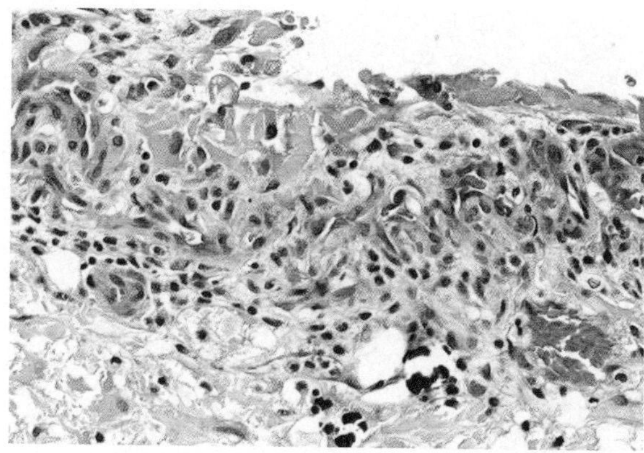

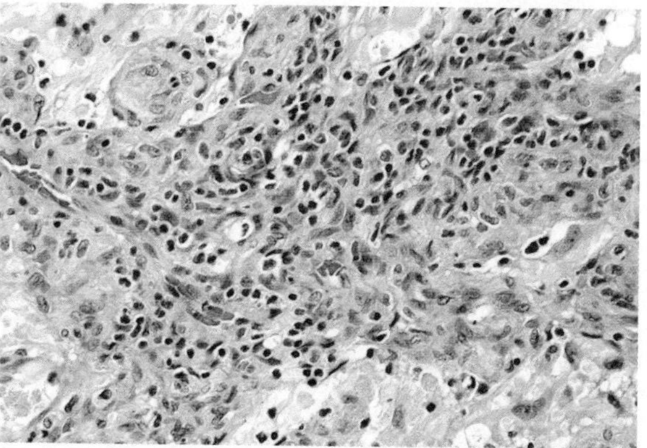

Figure 4-16

VASCULAR PROLIFERATION IN GANGLIOGLIOMA

Top: As in other cystic CNS tumors, the wall of a well-differentiated ganglion cell tumor may contain a uniform layer of proliferating vessels.

Bottom: Intraparenchymal vascular proliferation is not prima fascie evidence of malignancy in gangliogliomas.

the cytoplasm, making the identification of neurons less subjective. Fewer cells are stained, however, and normal cortical ganglion cells are often reactive. The latter often show fine punctate staining limited to the perikarya, whereas the cytoplasm of tumoral ganglion cells usually is more strongly and more diffusely stained. The finely granular product presumably reflects the presence of dense core granules in many ganglion cell tumors (fig. 4-19, see fig. 4-22), 80 percent in one series (23). In general, the degree of immunoreactivity for neuronal markers is proportional to nucleolar size. Ganglion cells are generally positive, whereas cells with small nucleoli may not be. Amphophilia, a reflection of early Nissl accumulation, also parallels reactivity for neuronal markers.

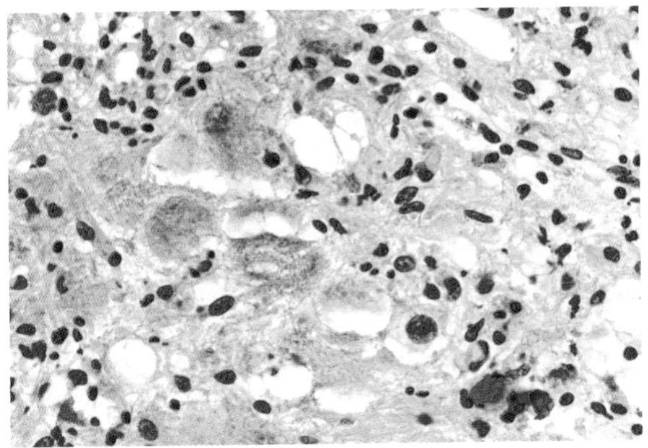

Figure 4-17

GANGLIOGLIOMA: FROZEN SECTION

The cytologic details of ganglion cells often suffer in the freezing process. While clearly nonspecific, the calcification and perivascular lymphocytes should raise the possibility of a ganglion cell tumor, particularly, as in this case, when the lesion is a nodule in the wall of a cyst.

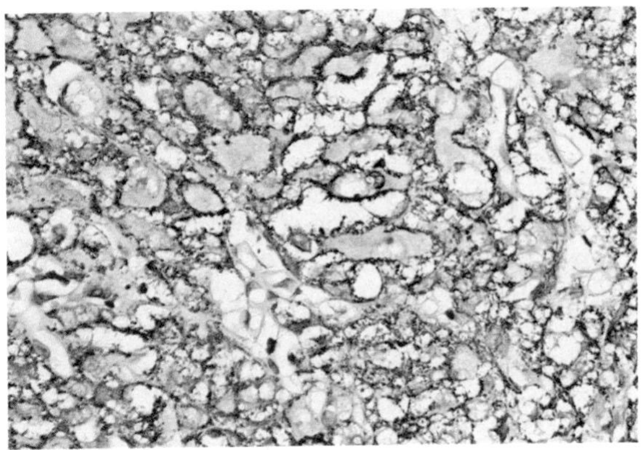

Figure 4-18

GANGLIOGLIOMA

While synaptophysin immunoreactivity may be restricted to the surface of ganglion cells, as it is here, it is often diffusely cytoplasmic.

The antibody, NeuN, advantageously stains the nucleus, and the cytoplasm, of large ganglion cells (24,24a). It is unpredictably reactive in neoplastic tissue, however, and is therefore helpful if positive, but of little significance if not. With the exception of Purkinje cells, normal ganglion cells, as well as cerebellar internal granule cells, are more likely to be stained than are neoplastic neuronal elements.

In some cells, cytoplasmic whorls may be labeled by immunohistochemistry for neurofilament protein, but in most cases cell processes, rather than the perikarya, of ganglion cells are positive. These processes may be difficult to distinguish from normal axons trapped within the lesion.

Ganglion cells may stain for neuropeptides such as vasoactive intestinal peptide (VIP), met-enkephalin, leu-enkephalin, dopamine, tyrosine hydroxylase, serotonin, somatostatin, substance P, and beta-hydrolyase (11,15,23,25,26). More than one substance may be expressed in a single tumor. A neuronal protein, alpha-synuclein, is seen in about half of ganglion cell tumors (27). Tau protein-positive neurofibrillary tangles as well as "neuropil threads" may also be present (13). Two studies suggest that immunoreactivity of ganglion cells for CD34 is common, especially in gangliogliomas (17,18). Stellate astrocytes in the cortex adjacent to the mass are positive for CD34 (17).

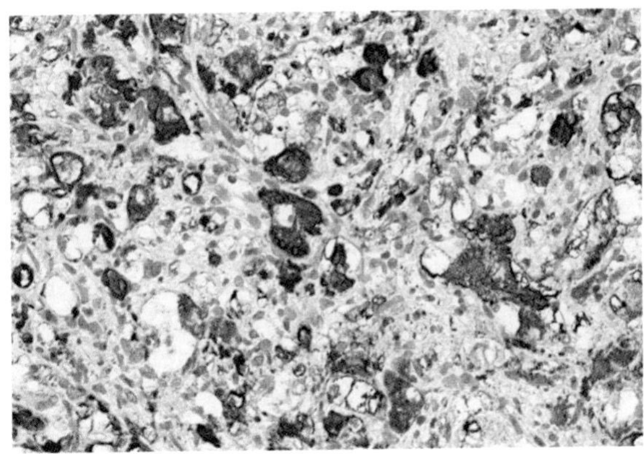

Figure 4-19

GANGLIOGLIOMA

A minority of ganglion cells have fine granular cytoplasmic immunoreactivity for chromogranin.

Astrocytes in ganglion cell tumors are reactive for both S-100 protein and glial fibrillary acidic protein (GFAP) (fig. 4-20), but it may not be possible to decide whether such cells are reactive or neoplastic. In some cases, particularly those where tumor cells have features intermediate between ganglion cells and astrocytes, it appears that the same tumor cells stain for both synaptophysin (fig. 4-21, top) and GFAP (fig. 4-21, bottom). GFAP-positive cells are easy to identify in tumors with a lobular architecture, and are more numerous at the periphery of lobules. The pleomorphism of these cells and their lack

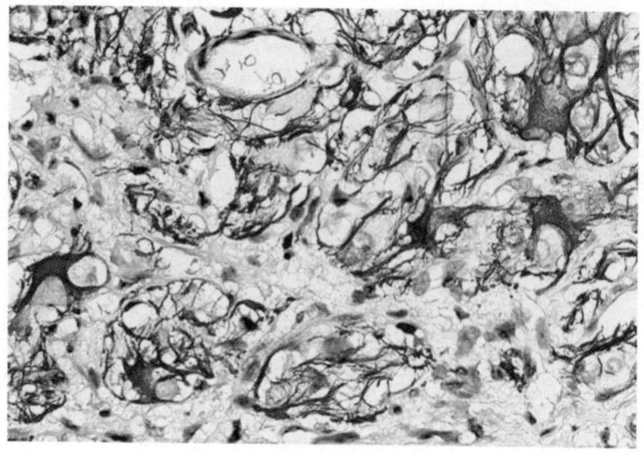

Figure 4-20

GANGLIOGLIOMA

Glial fibrillary acidic protein (GFAP)-positive glia clasp immunonegative neurons in the acinar variant.

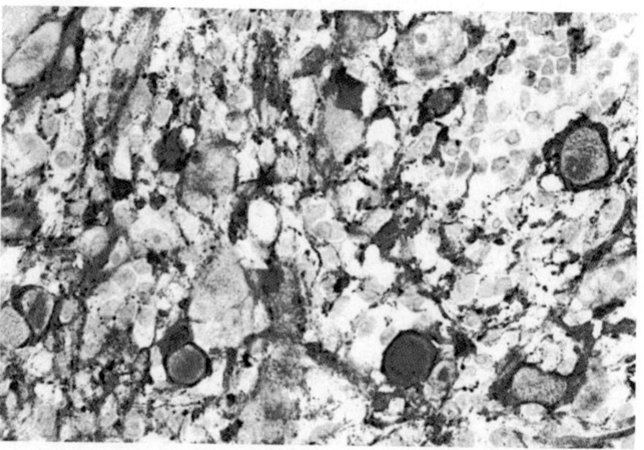

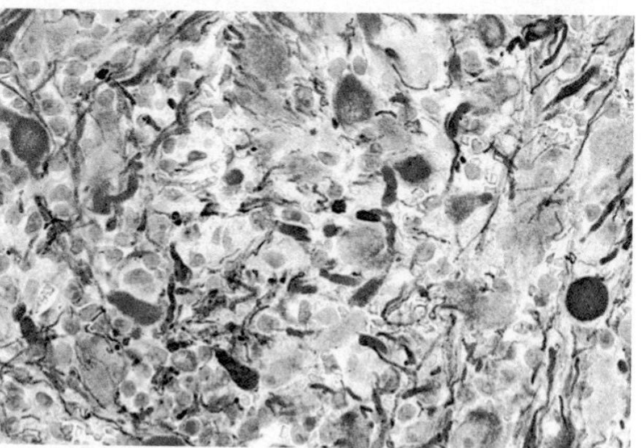

Figure 4-21

GANGLIOGLIOMA

Gangliogliomas with plump cells with features of both neurons and glia have mixed immunoreactivity for synaptophysin (top) and GFAP (bottom). While difficult to prove, it often appears that some cells must be positive for both.

of long, radially symmetric processes are indicators that they are participants in the tumor, not merely reactive bystanders.

Type IV collagen and laminin can be demonstrated in stroma-rich tumors. The latter may be associated with a longer duration of symptoms (28).

The MIB-1 index of the glial component is generally low, with a mean value of 2.7 percent in one series (23), but may be higher in lesions that later recur (23). Staining of the neuronal component for MIB-1 is usually negative, or extremely low.

Ultrastructural Findings. Dense core granules, a distinctive feature of neoplastic ganglion cells (fig. 4-22), are lacking or inconspicuous in most normal CNS neurons. Neurosecretory vesicles and even synapses may be present in cell terminations. Also present are long axonal processes containing microtubules. The astrocytic element has abundant intermediate filaments and, where abutting stroma, basal lamina.

Cytologic Findings. Ganglion cells are better preserved in cytologic preparations than in frozen sections, and may be larger and more rotund than normal ganglion cells in smears (fig. 4-23). The glial component is usually predominant in gangliogliomas. EGBs are present within glial processes in some cases.

Molecular and Cytogenetic Findings. No consistent fractional allelic losses were noted in one series of low-grade gangliogliomas (29).

Differential Diagnosis. CNS tumors with ganglion cells are summarized in Appendix G. Ganglion cell tumors are more often overdiagnosed than underdiagnosed due to the tendency to interpret normal cortical neurons trapped in an infiltrating glioma as neoplastic. Given the affinity of oligodendrogliomas for gray matter, it is precisely this tumor that is most often so misinterpreted (fig. 4-24), although diffuse astrocytomas can be so misconstrued as well. Review of the radiologic findings minimizes the likelihood of this mistake since oligodendrogliomas and infiltrating astrocytomas are "diffuse" areas of brightness on T2-weighted images without contrast enhancement. Ganglion cell tumors, especially gangliogliomas, are discrete, compact, and often contrast enhancing.

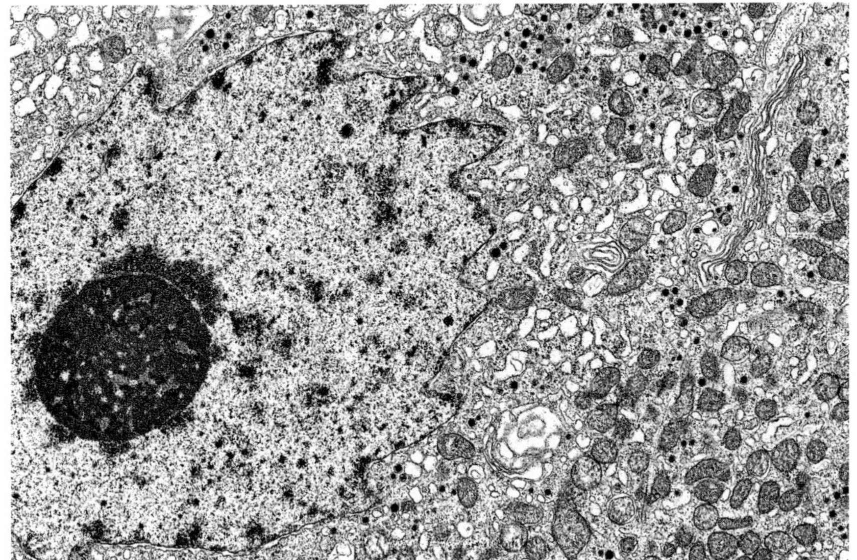

Figure 4-22

GANGLIOGLIOMA

The presence of dense core neurosecretory granules is a classic ultrastructural feature of ganglion cells tumors. (Fig. 4-16 from Fascicle 10, 3rd Series.)

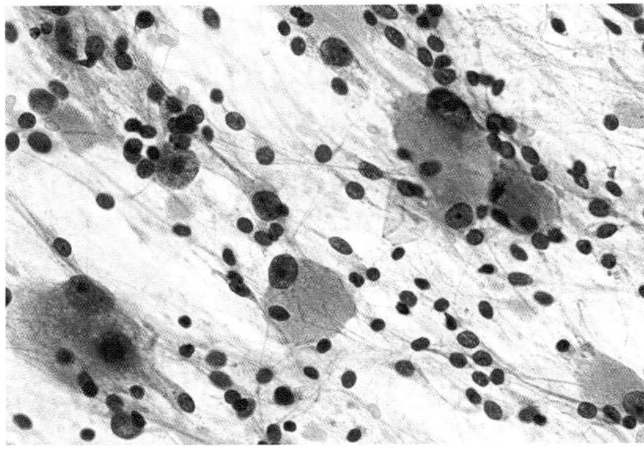

Figure 4-23

GANGLIOCYTOMA: SMEAR PREPARATION

A smear preparation from the specimen illustrated in figure 4-17 preserves the fine cytologic features of the ganglion cells that is lost during freezing. One of the abnormal globose neurons is binucleate.

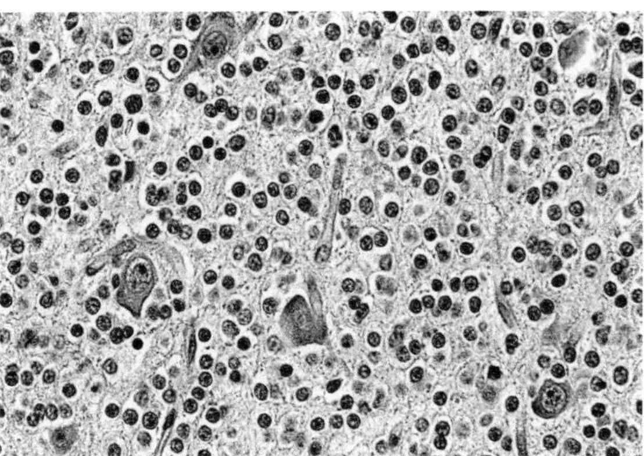

Figure 4-24

OLIGODENDROGLIOMA RESEMBLING GANGLIOGLIOMA

Isolated and highlighted by an infiltrating glioma, trapped cortical ganglion cells can be misinterpreted as neoplastic.

Microscopically, normal neurons trapped in an infiltrating glioma are uniform, fully differentiated and, in favorably cut sections, polar in orientation, with apical dendrites directed towards the pial surface. Tumor cells orbiting ganglion cells (perineuronal satellitosis) are also distinctive of infiltrating gliomas, but not of ganglion cell tumors.

Cortical dysplasia and its cousin the tuber are similar in some respects to gangliocytomas, but are dysplastic abnormalities intrinsic to cerebral cortex and subcortical white matter, not an independent compact mass. Perivascular lymphocytes are common in ganglion cell tumors, but not the other two entities.

An unusual lesion of infiltrating meningeal cells that can simulate gangliocytoma is meningioangiomatosis. Thickened vessels within the affected area trap and isolate islands of cortex whose crowded ganglion cells resemble those of a ganglion cell tumor. Calcification abets the counterfeit (fig. 4-25). Meningioangiomatosis is a plaque-like intracortical process, not a spherical mass.

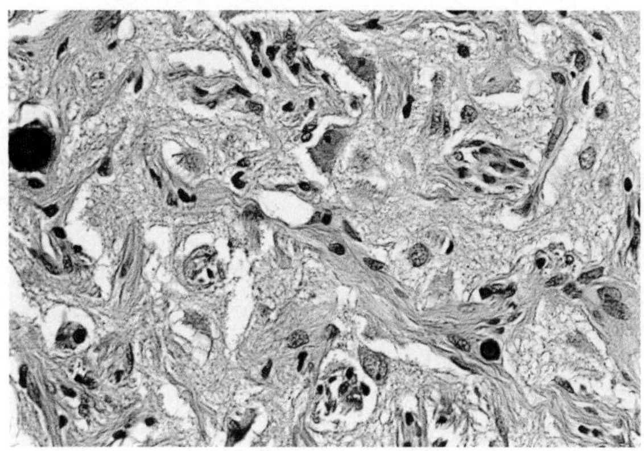

Figure 4-25

MENINGIOANGIOMATOSIS RESEMBLING GANGLIOGLIOMA

Desmoplasia and distorted cortical neurons create the appearance of a ganglion cell tumor.

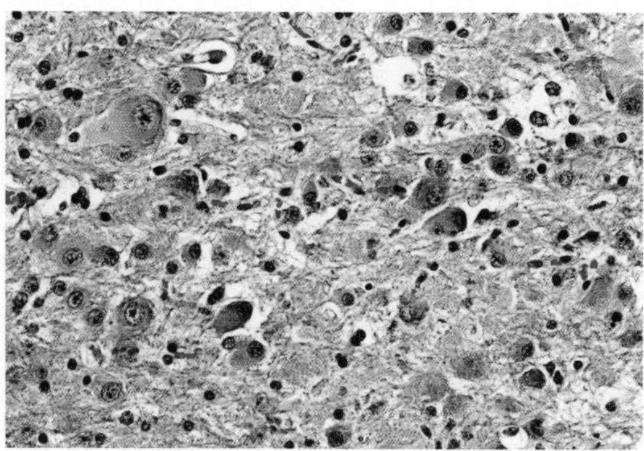

Figure 4-26

LHERMITTE-DUCLOS DISEASE RESEMBLING GANGLIOGLIOMA

Interpreted out of context, the ganglion cells in Lhermitte-Duclos disease can be misinterpreted as those of a conventional ganglion cell tumor.

Although Lhermitte-Duclos disease is a patterned neuronal abnormality confined largely to the sinuous contours of the cerebellar cortex, its distinctive laminar architecture may not be evident in small fragmented specimens, and it may be misinterpreted as a conventional gangliocytoma (fig. 4-26). Ganglion cell tumors of the hypothalamus need to be distinguished from hypothalamic hamartoma and pituitary adenoma with ganglion cell metaplasia.

The architecturally more complicated ganglioglioma should be suspected when confronted by any cellular, compact, polymorphous mass, particularly when cystic and arising in a young patient. The finding of EGBs should always prompt careful scrutiny for neoplastic ganglion cells.

Pilocytic astrocytomas and pleomorphic xanthoastrocytomas (PXA) share radiologic features with ganglion cell tumors, especially ganglioglioma, and may contain large neuron-like astrocytes or even neoplastic ganglion cells in some otherwise conventional cases (see figs. 3-142, 3-163). PXAs have an infiltrating cortical component that, when it overruns normal ganglion cells in the underlying cortex, can further complicate the issue. EGBs are common in both PXA and ganglion cell tumors. The differentiation of ganglion cell tumors, pilocytic astrocytoma, and PXA is discussed further in chapter 3. Immunohistochemistry often becomes decisive. The similarity of subependymal giant cell astrocytoma to ganglion cell tumors is also discussed in chapter 3.

Some neurocytic tumors, usually extraventricular, contain ganglion cells in addition to numerically predominant oligodendroglioma-like neurocytes. Central neurocytomas and their extraventricular counterparts are discussed later in this chapter.

Treatment and Prognosis. Ganglion cell tumors grow very slowly, as is evident from the long seizure history in some patients (23). Not surprisingly, simple, even partial resection often relieves or stabilizes symptoms (30). Recurrence is infrequent, despite histologic evidence of an infiltrative growth pattern at the tumor margin in some cases, but the incidence of recidivism is difficult to determine given the slow-growing, indolent nature of the lesion. In general, cure is not dependent on excision of the entire cyst wall. Grade II gangliogliomas are almost never fatal, although there are rare biologically malignant examples that have undergone anaplastic transformation. Such lesions are rare and generally, but not entirely, limited to the glial component.

DESMOPLASTIC INFANTILE GANGLIOGLIOMA

Definition. *Desmoplastic infantile ganglioglioma* (DIG) is a massive, desmoplastic, supratentorial, superficially situated glioneuronal neoplasm of infancy. DIGs are WHO grade I.

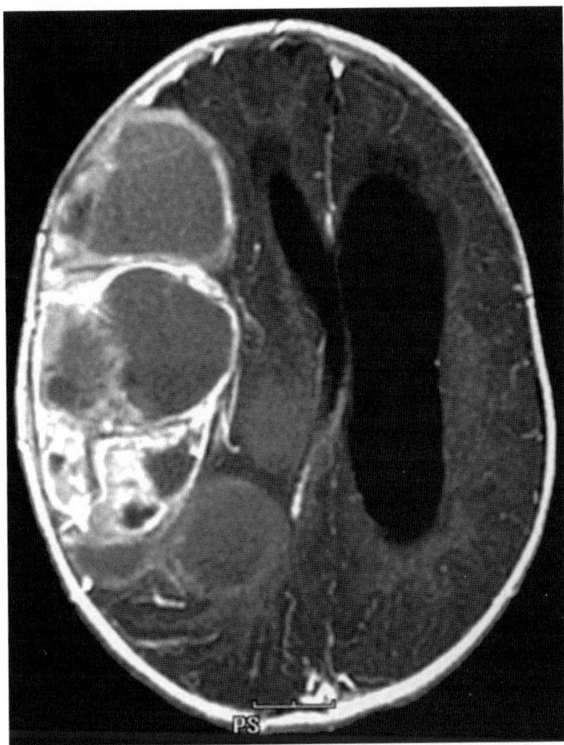

Figure 4-27

DESMOPLASTIC INFANTILE GANGLIOGLIOMA

The lesions are superficial, massive, and cystic, as in this 14-month-old boy with macrocephaly.

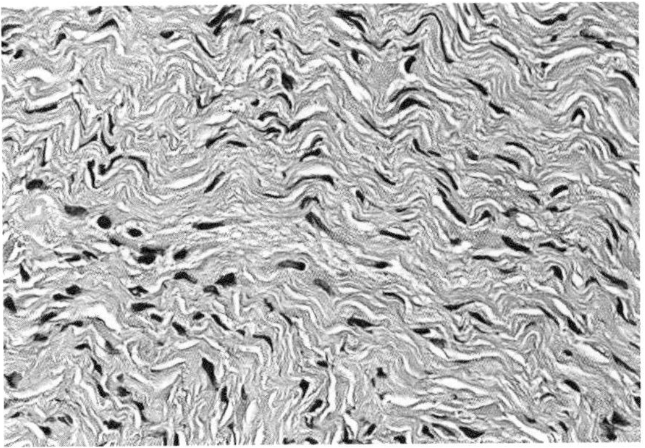

Figure 4-28

DESMOPLASTIC INFANTILE GANGLIOGLIOMA

Densely desmoplastic tissue is highly suggestive of desmoplastic infantile ganglioglioma (DIG).

Clinical Features. Virtually all DIGs are present in patients under the age of 1 year. Usually macrocephaly, and, in some cases, neurologic deficits are present (31–34).

Radiologic Findings. The tumors are supratentorial, superficially situated, and often attached to the dura (fig. 4-27). A large cyst, or cysts, typically underlies the solid, contrast-enhancing component with dark, T2-weighted desmoplastic areas (35).

Gross Findings. The tumors are characteristically massive and, because of the desmoplasia, are sometimes "woody" in texture. Much of the tumor abuts the underlying brain in a compressive manner, but the interface may be obscured focally. Cysts, often sizable and multiple, are common. Large, easily torn vessels may traverse the lesion.

Microscopic Findings. The superficially situated solid component fills and expands the subarachnoid space, from which the lesion reaches into the cortex along the perivascular (Virchow-Robin) spaces (31,32). The tumor's most distinctive component is a collagenous matrix that varies in cellularity, and merges imperceptively with less fibrotic glioneuronal tissue (fig. 4-28). The latter, in turn, often blends with tissue that is considerably more cellular and still less desmoplastic. More cellular portions may be architecturally nondescript or have a slightly swirling or even storiform pattern. Because of this cellularity, especially the high degree discussed below, DIG is easily overgraded. Appendix K lists CNS tumors that are easily overgraded.

The ganglion cells of DIG are often difficult to identify since they may be sparse and small (fig. 4-29), although they sometimes are clustered and easier to detect (fig. 4-30). Some lesions with all the other features of DIG have an overt ganglion cell component, replete with EGBs, identical to those of classic ganglion cell tumors (fig. 4-31). Nissl substance is uncommon in the typical lesion but may be present in DIGs with large ganglion cells. In general, ganglion cells are more often dispersed or regionally concentrated, but not tightly clustered. Because of their nuclear and cytoplasmic features, the cells may not be obviously neurons, and immunohistochemistry is required for identification. By this method, positive ganglion cells may be found in the subarachnoid space well outside the normal cortex, indicating clearly that these are not merely trapped cortical elements.

The glial component varies considerably in its cytologic features. In desmoplastic areas,

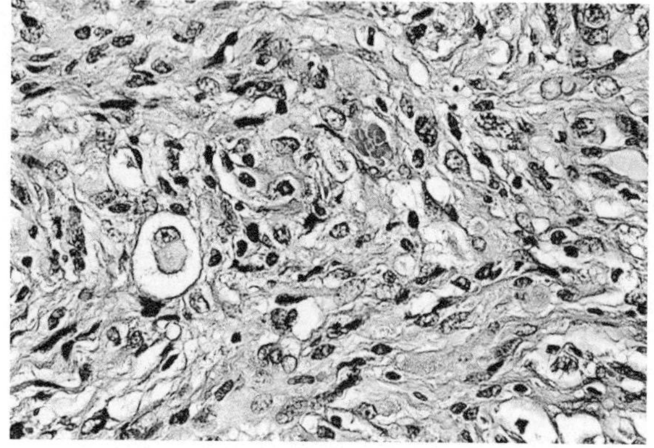

Figure 4-29

DESMOPLASTIC INFANTILE GANGLIOGLIOMA

Ganglion cells are usually smaller than they are in classic gangliocytomas and gangliogliomas.

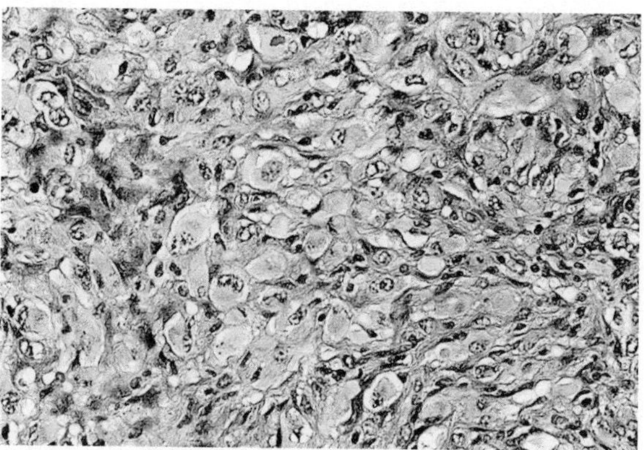

Figure 4-30

DESMOPLASTIC INFANTILE GANGLIOGLIOMA

Neurons, even when small, are readily recognized when concentrated in small groups.

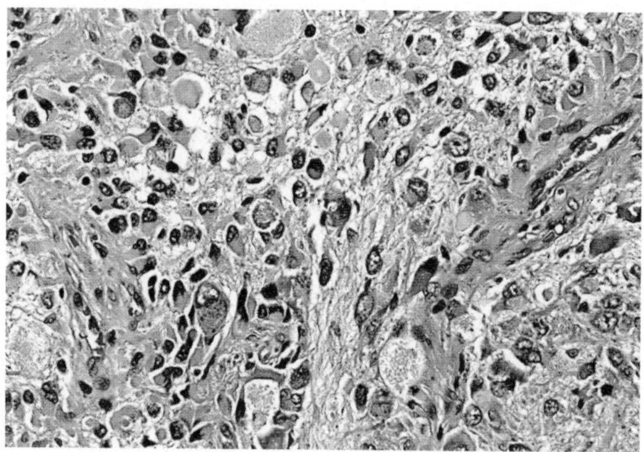

Figure 4-31

DESMOPLASTIC INFANTILE GANGLIOGLIOMA

DIGs with large ganglion cells and EGBs overlap histologically with classic ganglion cell tumors.

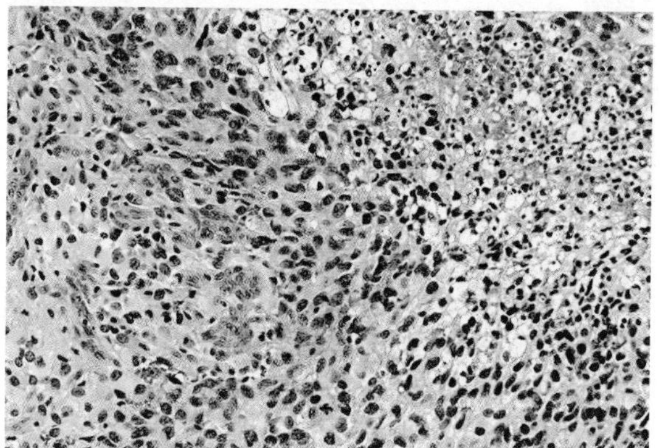

Figure 4-32

DESMOPLASTIC INFANTILE GANGLIOGLIOMA

Surprisingly, given their significance in diffuse astrocytomas, high cellularity, mitotic activity, and even necrosis and vascular proliferation in the DIG are often not prognostically adverse. Treated only with surgery, the lesion, shown radiologically in figure 4-27, has not recurred during a 5-year postoperative interval.

elongated cells are recognized as glial only by their immunoreactivity for GFAP. In less sclerotic regions, there may be cells that, while small, are still obviously gemistocytic. Highly cellular regions contain cells resembling those of small cell glioblastoma.

An alarming feature of some DIGs is a mitotically active small cell element that is accompanied, in some cases, by microvascular proliferation and even necrosis with pseudopalisading (fig. 4-32). The clinical significance of this tissue is discussed in the section, Treatment and Prognosis.

Occasional DIGs have a component of classic ganglioglioma (36). This observation, and the rare DIGs that occur in older patients, suggests that DIG and classic ganglioglioma are sometimes more closely related than is often assumed.

Immunohistochemical Findings. Despite the mesenchymal appearance of the desmoplastic tissue, many of its spindle cells are glial, as evidenced by immunoreactivity for GFAP (fig. 4-33). The small ganglion cells are synaptophysin reactive (fig. 4-34, top) and, often, chromogranin (fig. 4-34, bottom) and NeuN positive. Some studies have suggested a

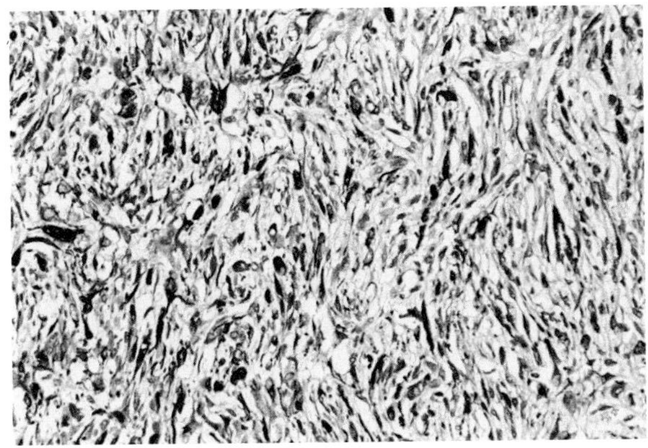

Figure 4-33

DESMOPLASTIC INFANTILE GANGLIOGLIOMA

Spindle cells within the desmoplastic tissue are intensely immunoreactive for GFAP.

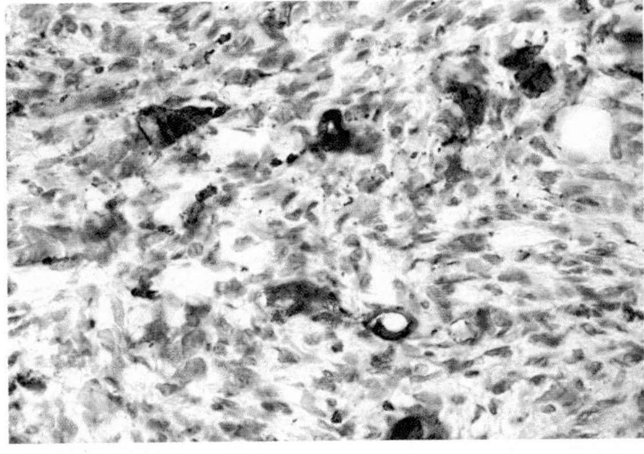

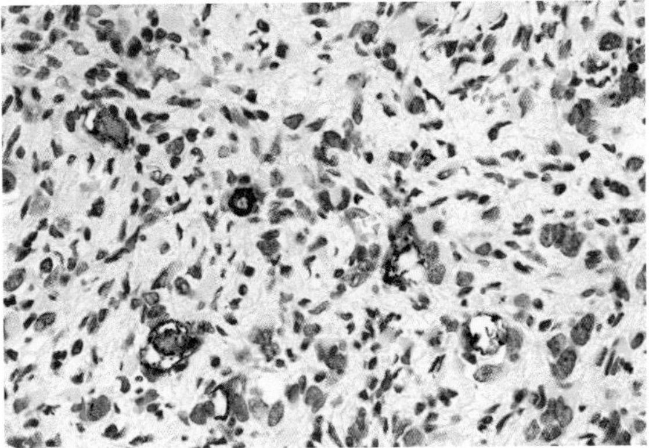

Figure 4-34

DESMOPLASTIC INFANTILE GANGLIOGLIOMA

Small neurons can be identified by their immunoreactivity for synaptophysin (top) and chromogranin (bottom).

Schwann cell component in the form of S-100–positive and GFAP-negative spindle cells (37).

Ultrastructural Findings. The astrocytic element of DIG is appropriately rich in intermediate filaments and often exhibits some degree of pericellular basal lamina formation at the tumor-stroma interface (31). Neoplastic neurons extend microtubule-containing processes, with both vesicles and neurosecretory granules concentrated in their terminations. Cells surrounded by extensive, well-formed basement membrane material have been interpreted by several authors as Schwann cells (36,37). Intercellular collagen is conspicuous.

Differential Diagnosis. The size, superficial location, cystic architecture, and firm desmoplastic component that contains GFAP-positive astrocytes are part of such a distinctive constellation that little else remains in the differential if ganglion cells are identified. When they are not, and this is not uncommon, the tumor falls by default into the category of desmoplastic infantile astrocytoma. If such a tumor is otherwise typical of DIG, then it is generally treated as DIG, i.e., observation alone. Those that are not otherwise classic enter a less specific group that overlaps with ordinary, and generally more aggressive, infiltrative astrocytomas that have reached the subarachnoid space.

PXA shares the superficial location of DIG, but occurs later in life. Even in its most reticulin-rich area, PXAs are not as desmoplastic as DIG, and are more likely to have EGBs and perivascular lymphocytes. Except for anaplastic variants, PXAs lack the small cell component of some DIGs.

In areas where mitotic figures are frequent and cellularity is high, the lesion can resemble meningeal sarcoma, high-grade astrocytoma, and small cell embryonal tumor.

Treatment and Prognosis. Although an extensive dural base and trapped vessels preclude total resection of these often massive lesions in most cases, long survival periods are the rule. Even regression has been observed (38). Death directly due to the tumor is uncommon. The significance of high cellularity and mitotically active tissue in some DIGs is unclear. Although it is difficult to imagine that tumors containing such tissue do not behave aggressively, this seems often to be the case and these features are not necessarily harbingers of recurrence (31–34).

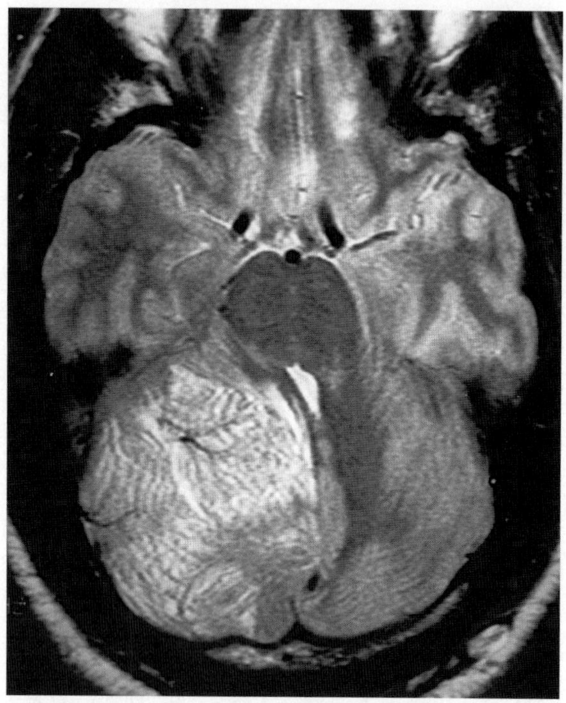

Figure 4-35

DYSPLASTIC GANGLIOCYTOMA (LHERMITTE-DUCLOS DISEASE)

A cerebellar mass with a "striped" profile in T2-weighted images is characteristic.

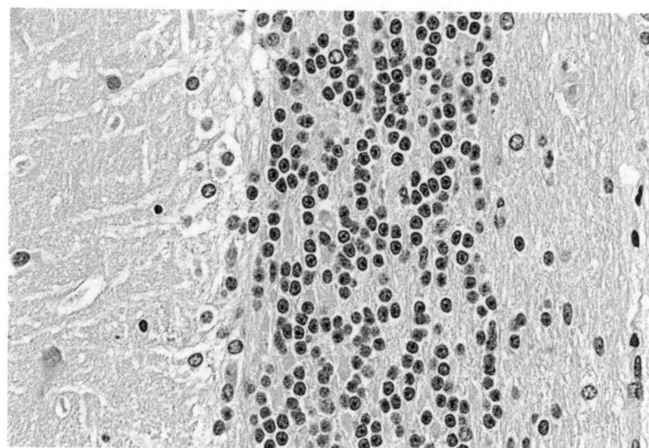

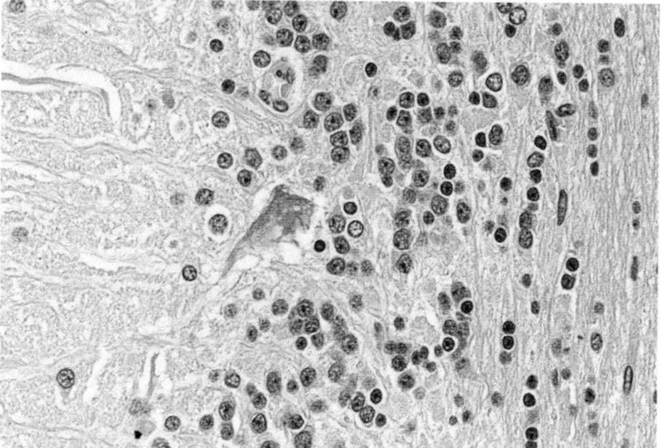

Figure 4-36

DYSPLASTIC GANGLIOCYTOMA (LHERMITTE-DUCLOS DISEASE)

When compared to normal cerebellum (top), the process at the edge of the lesion is limited to a slight enlargement of internal granule cells (bottom).

Occasional cases with DIG features behave aggressively, however, and negate the assumption that this tissue is always benign (39).

DYSPLASTIC CEREBELLAR GANGLIOCYTOMA (LHERMITTE-DUCLOS DISEASE)

Definition. *Dysplastic cerebellar gangliocytoma* is a non-neoplastic lesion composed of cerebellar folia that is massively expanded largely by hypertrophic neurons of the internal granular layer. The non-neoplastic lesion is WHO grade I.

Clinical Features. Most patients with this rare, zonal form of cerebellar enlargement are young adults with signs and symptoms of increased intracranial pressure (40,41). Clinical evidence of cerebellar dysfunction is less common. There is a close association with Cowden's syndrome, a complex discussed in chapter 18 (40,42,43).

Radiologic Findings. Magnetic resonance imaging (MRI) typically shows a unilateral zone of cerebellar enlargement in which thickened folia appear as stripes of increased signal intensity in T2-weighted images (fig. 4-35) (44). Chronic lesions may be calcified.

Gross Findings. In aggregate, the enlarged folia exert a considerable mass effect.

Microscopic Findings. The principal abnormality is massive replacement and expansion of the internal granular layer by large neurons with vesicular nuclei and prominent nucleoli (figs. 4-36–4-38) (45). There is an increase in both the percentage of affected internal granule cells and their size as one moves from the edge to the center of the lesion. There is also a vertical stratification of abnormality such that the deeper internal granule cells, i.e., those adjacent to the white matter, are less likely to be transformed than are those positioned more superficially near the Purkinje cell layer (fig. 4-

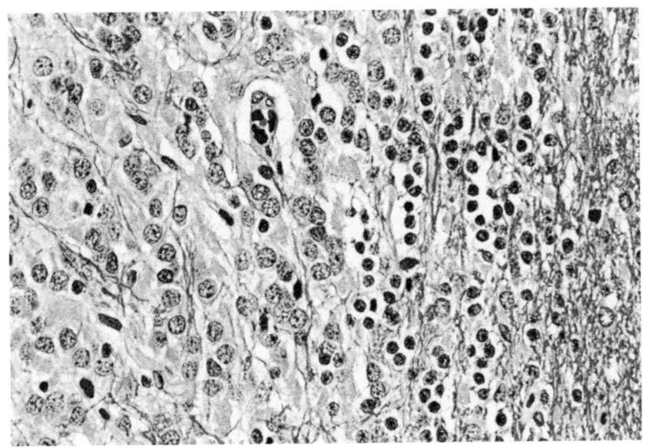

Figure 4-37

DYSPLASTIC GANGLIOCYTOMA (LHERMITTE-DUCLOS DISEASE)

The conversion of internal granule layer (IGL) cells to ganglion cells shows inter- and intraregional variation. As is typical, the IGL cells adjacent to the white matter at the right of the illustration are unaffected, whereas those at the left are enlarged. Nucleoli, which are inconspicuous in normal IGL cells, are obvious in the enlarged counterparts (hematoxylin and eosin [H&E]/Luxol fast blue stain).

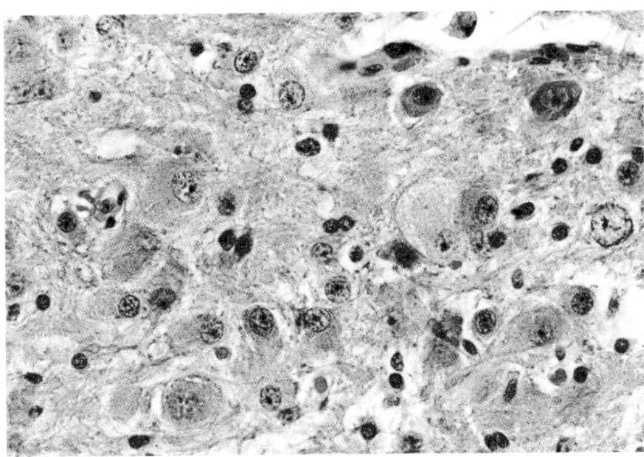

Figure 4-38

DYSPLASTIC GANGLIOCYTOMA (LHERMITTE-DUCLOS DISEASE)

The normal small neurons of the IGL are entirely replaced by ganglion cells in markedly abnormal (advanced) areas.

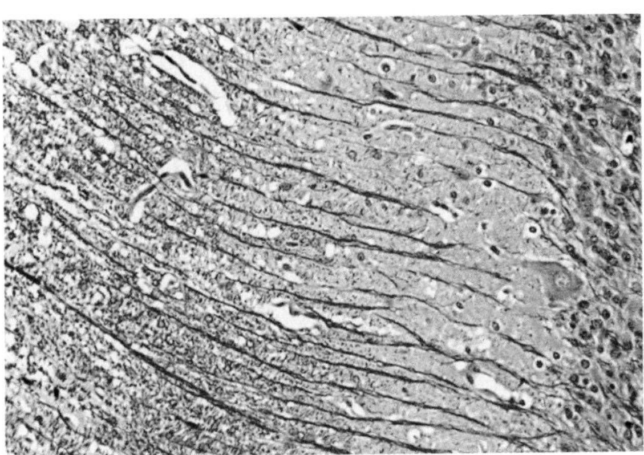

Figure 4-39

DYSPLASTIC GANGLIOCYTOMA (LHERMITTE-DUCLOS DISEASE)

Myelinated axons extend through the molecular layer, a lamina in which they are not normally present (H&E/Luxol fast blue stain).

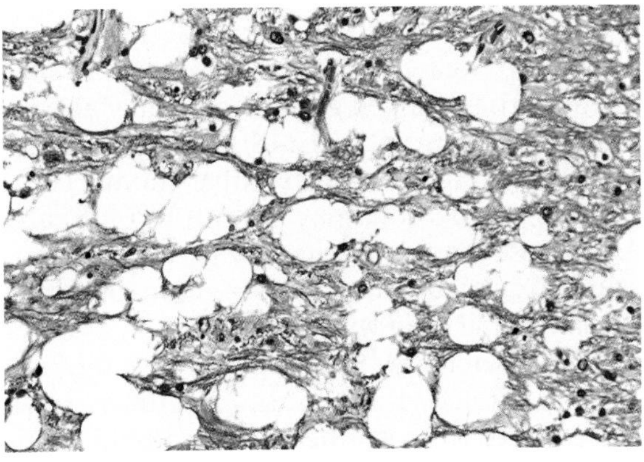

Figure 4-40

DYSPLASTIC GANGLIOCYTOMA (LHERMITTE-DUCLOS DISEASE)

Although it may appear artifactual, vacuolation is a real finding that may be responsible for the prominent stripes in T2-weighted images.

37). Particularly in areas of advanced change, cells are considerably larger than the normal small granular cells, but are generally not as large as Purkinje cells. They are haphazardly oriented and apolar, without the apical dendritic specialization of Purkinje cells. In some cases, however, the cells project myelinated processes that pass through the molecular layer to reach the pial surface (fig. 4-39). Bizarre pleomorphic cells are present in some cases.

The molecular layer and white matter are often distinctly vacuolated (fig. 4-40). This tissue liquid may well explain the bright stripes seen in T2-weighted MRI.

Immunohistochemical Findings. As expected, the abnormal neurons have surface immunoreactivity for synaptophysin (fig. 4-41). Neuronal processes coursing between their cell

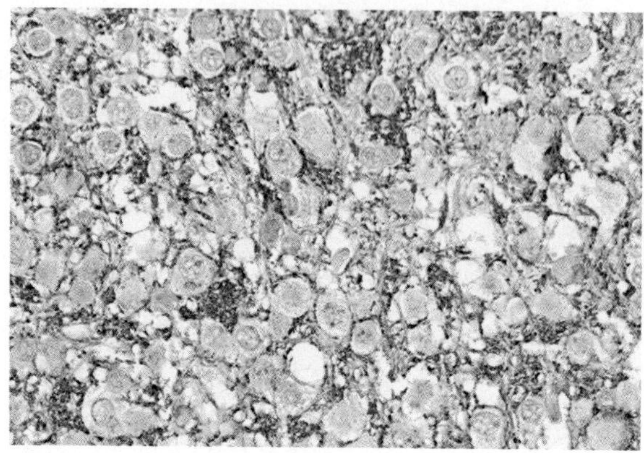

Figure 4-41

DYSPLASTIC GANGLIOCYTOMA (LHERMITTE-DUCLOS DISEASE)

The neurons and neuropil background are immunoreactive for synaptophysin.

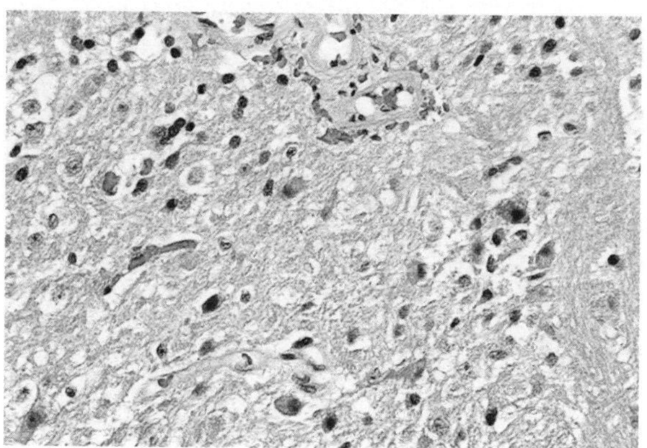

Figure 4-42

DYSPLASTIC GANGLIOCYTOMA (LHERMITTE-DUCLOS DISEASE)

The sharp interface between a dense population of small ganglion cells and white matter is typical of dysplastic gangliocytoma.

bodies are best seen after staining for neurofilament protein. Under conditions of optimal fixation, the enlarged granule cells may also be reactive for this antigen (46). Unlike Purkinje cells, but in common with native granule cells, the enlarged neurons may be positive for NeuN. As a consequence of loss of PTEN inhibition, phosphorylated AKT can be demonstrated in the abnormal neurons (40,43).

Ultrastructural Findings. In the few examples studied, the abnormal neurons have cytoplasmic neurofilaments, microtubule-containing processes, and synapses (47).

Molecular and Cytogenetic Findings. Based on the observation that Lhermitte-Duclos disease (LDD) occurs in the context of Cowden's syndrome, the suspected pathogenetic role of an abnormal *PTEN* gene was confirmed in a mouse model wherein a remarkably close facsimile of LDD was produced when *PTEN* was deleted in the mouse cerebellum (48).

Differential Diagnosis. Given the complicated curvilinear architecture of the cerebellar cortex, the laminar distribution of the lesions may not be obvious in small surgical specimens. As a result, the process can be misinterpreted as gangliocytoma or even astrocytoma (fig. 4-42). Practically and conceptually, dysplastic gangliocytoma of the cerebellum overlaps somewhat with conventional gangliocytoma, but is obviously a dysplastic or malformative process whereas ganglion cell tumors are cellular, tumorous masses. Dysplastic gangliocytoma also has an interface with the white matter that is more linear and abrupt than even the most well-circumscribed ganglion cell tumor.

The histologically more complex ganglioglioma shows greater cellular pleomorphism, a variably collagenous stroma, focal chronic inflammation, frequent cyst formation, extension into the leptomeninges, cells appearing morphologically intermediate between ganglion cells and astrocytes, and a clearly defined glial component. The last is conspicuously absent in dysplastic gangliocytoma. Vacuolation is an additional clue for the latter.

Unlike astrocytoma, the cells of dysplastic gangliocytoma are uniform, larger, rounder, and more neuronal in configuration. Immunohistochemistry resolves the issue. Neuroimaging studies are reassuring, if not diagnostic, in the above contexts. Other intracranial and intraspinal neoplasms with ganglion cells are given in Appendix G.

Treatment and Prognosis. Dysplastic cerebellar gangliocytoma enlarges only slowly and does so as a consequence of hypertrophy and the acquisition of myelin rather than by cell proliferation. Surgical excision relieves increased intracranial pressure and may be curative, although recurrence has been noted (49). Therapeutic intervention is then limited to re-excision.

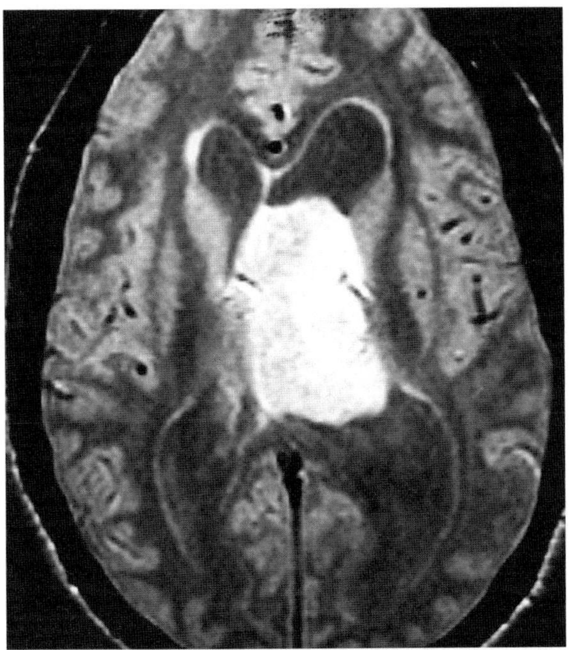

Figure 4-43

CENTRAL NEUROCYTOMA

By the time of surgical intervention, most central neurocytomas are large lesions that straddle the midline and incorporate the septum pellucidum. (Fig. 4-29 from Fascicle 10, 3rd Series.)

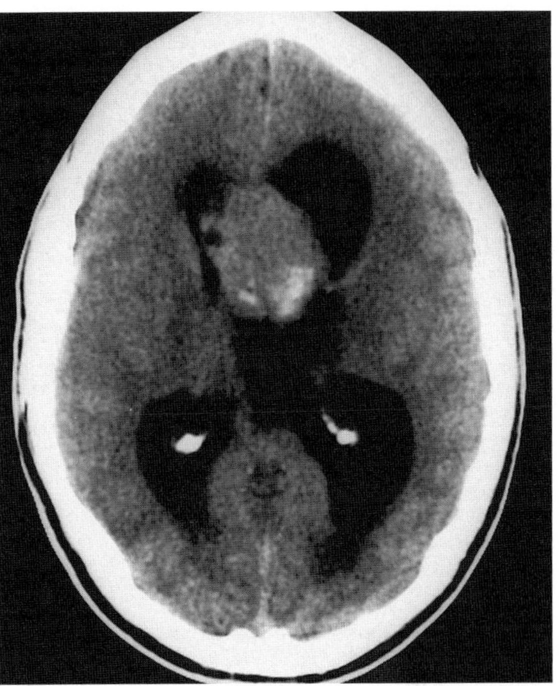

Figure 4-44

CENTRAL NEUROCYTOMA

As seen here by computerized tomography (CT), central neurocytomas are frequently calcified. As is common with this entity, obstruction to the flow of cerebrospinal fluid through the foramen of Monro produced marked hydrocephalus. (Fig. 4-30 from Fascicle 10, 3rd Series.)

NEUROCYTIC NEOPLASMS

Neurocytes are small, but not necessarily immature, neurons that are cytologically uniform, monotonously round, and prone to form oligodendroglial-like perinuclear halos. Neoplasms of these cells comprise a heterogeneous group whose members are variably classified on the basis of location and histologic features. The first-described entity, central neurocytoma, has been joined since by multiple other neurocytic neoplasms, and "new" neurocytic and glioneurocytic entities are sure to follow. Even infiltrating astrocytomas and oligodendrogliomas occasionally express neuronal differentiation.

NEUROCYTIC NEOPLASMS OF THE CEREBRAL HEMISPHERES

Central Neurocytoma

Definition. *Central neurocytoma* is a discrete intraventricular tumor composed of cytologically uniform small neurons. It corresponds to WHO grade II.

Clinical Features. Given its location near the foramen of Monro, neurocytomas obstruct the flow of cerebrospinal fluid and produce signs and symptoms of acutely or chronically increased intracranial pressure. Intratumoral hemorrhage precipitates neurologic symptoms in rare cases (50,51). Central neurocytomas usually occur in young to middle-aged adults, although both children and the elderly are affected (52,53).

Radiologic Findings. The typically large, globular, often calcified mass straddles the midline near the septum pellucidum (fig. 4-43). Heterogeneous contrast enhancement is the rule (54,55). The calcification is best seen by computerized tomography (CT) (fig. 4-44).

Gross Findings. The well-circumscribed, gray, and sometimes gritty mass adheres to the adjacent ventricular surfaces and incorporates the septum pellucidum.

Microscopic Findings. Neurocytomas are small cell neoplasms that, in their solid areas,

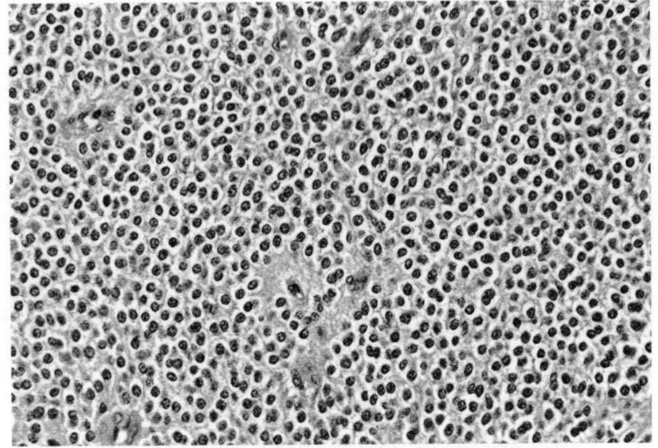

Figure 4-45

CENTRAL NEUROCYTOMA

Central neurocytomas are remarkably monomorphous in cell size, shape, and distribution.

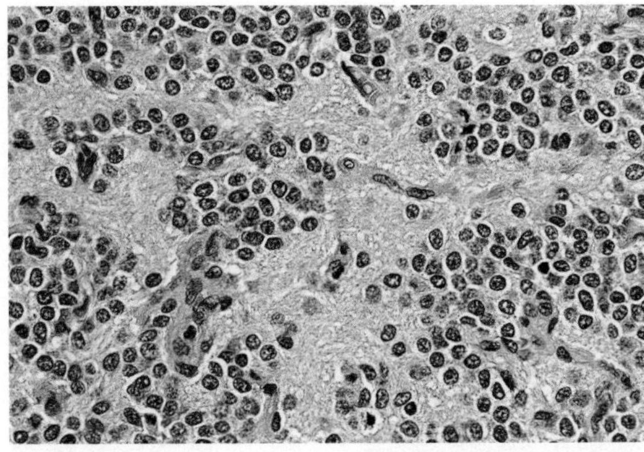

Figure 4-46

CENTRAL NEUROCYTOMA

The tumor is often interrupted focally by finely fibrillar, often perivascular, rosette-like areas composed of tumor cell processes. An ependymoma-like appearance is created.

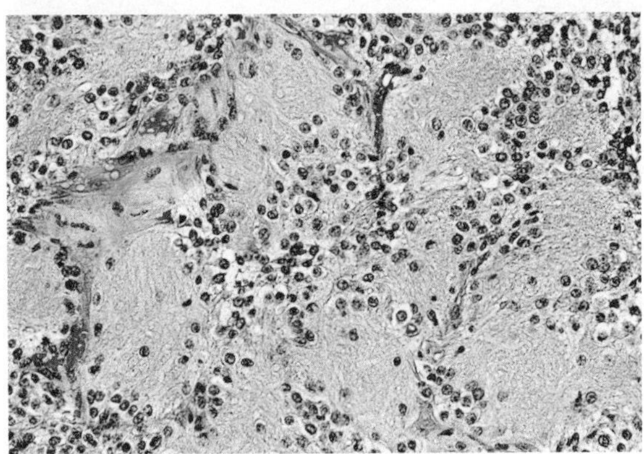

Figure 4-47

CENTRAL NEUROCYTOMA

Large, circular, rosette-like structures lend a decided neuronal/neurocytic quality to some central neurocytomas.

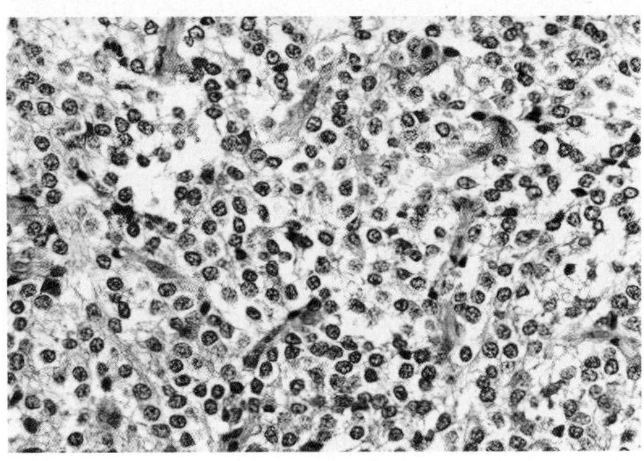

Figure 4-48

CENTRAL NEUROCYTOMA

Halos and round uniform nuclei create an oligodendroglioma-like appearance in many central neurocytomas.

are remarkably uniform in cell density and cytologic features (fig. 4-45). They would have an overall cellular density similar to blue cell tumors were it not for neuropil-like areas or perivascular pseudorosettes, which are similar to, but more finely fibrillar than, those of ependymoma (fig. 4-46). Rosette-like structures resembling pineocytomatous rosettes or large Homer-Wright rosettes are sometimes present as well (fig. 4-47). A neurocytic quality is apparent, uncommonly, when cells stream in a finely fibrillar, process-rich background in a pattern similar to that of some nodular medulloblastomas.

As is characteristic of neurocytic neoplasms in general, nuclei are remarkably uniform in size and shape, have finely speckled chromatin and small nucleoli, and are often surrounded by an artifactual halo (fig. 4-48). Calcospherites add to the remarkable similarity to oligodendroglioma in many cases. The lesions lack both the coarse chromatin of neuroblastomas or other small cell embryonal tumors, and the vesicular chromatin and prominent nucleoli of ganglion cells. Multinucleation and pleomorphism are generally absent. Mitoses are rare or absent in classic lesions, but are present in *atypical neurocytomas*. These are

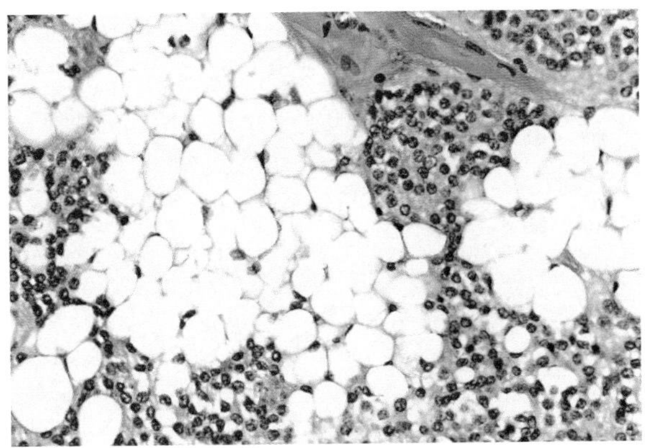

Figure 4-49

CENTRAL LIPONEUROCYTOMA

A rare central neurocytoma is lipidized.

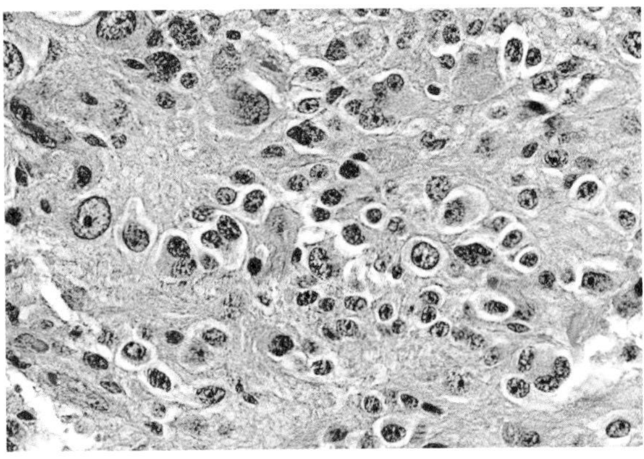

Figure 4-50

GANGLIONEUROCYTOMA

Rare neuronal neoplasms at the foramen of Monro have a component of ganglion cells.

defined by an increased MIB-1 labeling index, as is described below. As is the case with the rare extraventricular neurocytoma, occasional "central" examples are lipidized (fig. 4-49) (55). Neurocytomas laden with lipofuscin-like neuromelanin comprise yet another rare variant (56).

Ganglion cell differentiation is uncommon, although some otherwise typical neurocytic neoplasms at the foramen of Monro contain enough large neurons to warrant the diagnosis *ganglioneurocytoma* (fig. 4-50) (57). Gangliogliomas also appear in this locale (58). In rare cases, ganglion cells appear in central neurocytomas in the interval between two operations (58).

Grading. While central neurocytomas are usually well differentiated, they comprise a spectrum that ends in rare tumors with high mitotic activity, microvascular proliferation, and even necrosis (59–62). Generally neurocytomas that deviate from the classic cytologically bland lesion do so only on the basis of an increased MIB-1 labeling index (see fig. 4-54) (60,62). While scattered mitoses are usually present in such tumors, the tumors are not malignant in cytologic or histologic terms, although vascular proliferation may be present. The typical lesion is WHO grade II.

Grading, therefore, has focused on MIB-1 indices, with a threshold of 2 percent predicting a greater likelihood of recurrence (60,62). A slightly higher cutoff, 3 percent, was suggested in another other study (59). How these values can be applied from institution to institution is not clear given the limited information about interinstitutional variation in MIB-1 indices.

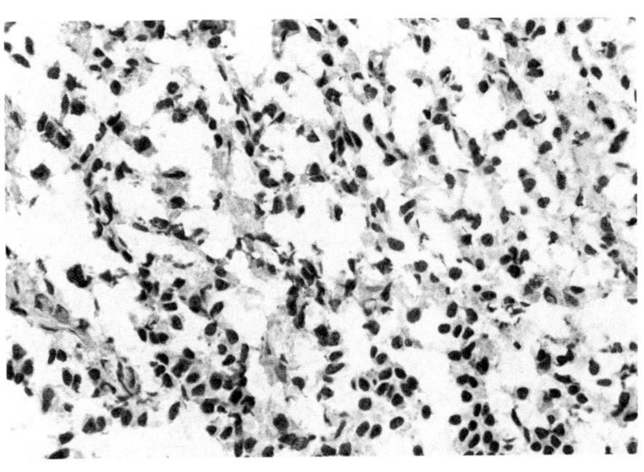

Figure 4-51

CENTRAL NEUROCYTOMA: FROZEN SECTION

Nuclei of neurocytomas loose their roundness and uniformity in frozen sections.

Frozen Section. Because of artifactual pleomorphism and hyperchromatism, neurocytomas lose much of their neurocytic character in frozen sections; they then resemble a generic small cell neoplasm (fig. 4-51). High cellularity and neuropil-like zones can mimic necrosis and bring even malignant tumors such as glioblastoma into the differential diagnosis. Awareness of central neurocytoma and its stereotypic appearance is therefore essential to intraoperative identification. The remarkably round, bland nuclei seen in companion smear preparations help in the diagnosis.

Figure 4-52

CENTRAL NEUROCYTOMA

Finely fibrillar areas are immunoreactive for synaptophysin.

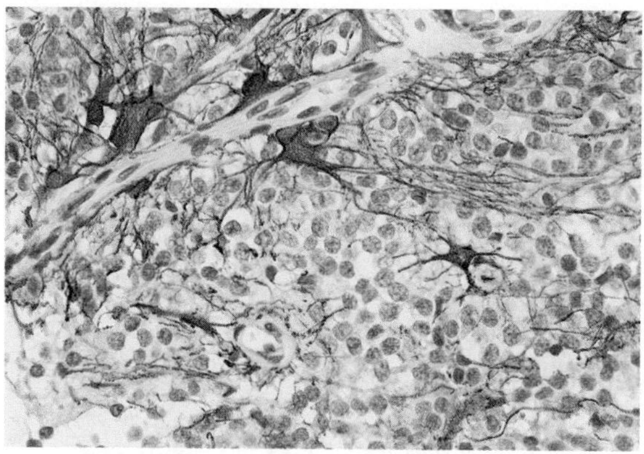

Figure 4-53

CENTRAL NEUROCYTOMA

Although the lesions may have a potential for glial differentiation, GFAP reactivity is usually confined to reactive astrocytes.

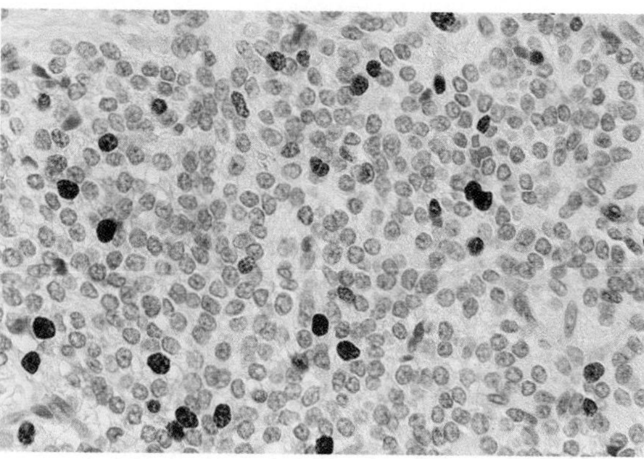

Figure 4-54

ATYPICAL CENTRAL NEUROCYTOMA

Atypical neurocytomas are defined as those with a MIB-1 index greater than 2 percent. In this case it was approximately 7 percent.

Immunohistochemical Findings. Neuropil-like zones are immunoreactive for synaptophysin (fig. 4-52) (63–66), and often for other neuronal epitopes such as class III beta-tubulin and neurofilament protein. Neurocytomas are variably positive for NeuN but are immunonegative for chromogranin. Although reactivity for GFAP is usually confined to scattered reactive, largely perivascular astrocytes (fig. 4-53), tumor cells themselves are focally reactive in some cases (65–67). One investigation utilizing double labeling suggested that some cells express both glial (GFAP) and neuronal markers in vitro, and that the astrocytic immunophenotype increases with passage number (68). Reactivity for retinal S antigen was present in less than half of cases in another study (66).

A low MIB-1 index confirms the benign histologic and cytologic nature of neurocytomas. It is below 2 percent in most instances (59,60, 63), but there is a continuum of indices, terminating in the range of 10 percent or more (fig. 4-54) (59,60,62,69). Mitoses are usually evident in tumors with higher indices. The increased rates that define atypical neurocytomas, and the prognostic significance of elevated indices, are discussed above.

Ultrastructural Findings. The neuropil created by neurocytoma cells is a dense feltwork of cytoplasmic processes that contain microtubules and clear vesicles (fig. 4-55) (50,63,65,67). A few neurosecretory granules are often present within these processes or the perikaryon. Less commonly, there is a high order of differentiation as evidenced by synapses and synapse-like membrane thickenings. When visualized ultrastructurally, fat-containing cells are obviously lipidized tumor cells, not adipocytes (55).

Cytologic Findings. On smears, the tumors appear as individual cells with scant cytoplasm. The remarkably uniform benign nuclei with fine chromatin are those expected of a neurocytic lesion (fig. 4-56).

Molecular and Cytogenetic Findings. As studied by comparative genomic hybridization, gains, especially on chromosomes 2p, 10q, and

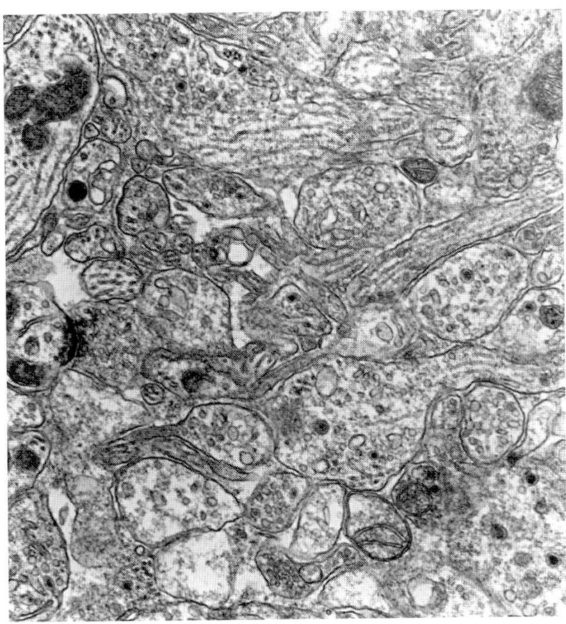

Figure 4-55

CENTRAL NEUROCYTOMA

Fibrillar neuropil-like areas consist of neurotic processes containing microtubules and scattered neurosecretory granules.

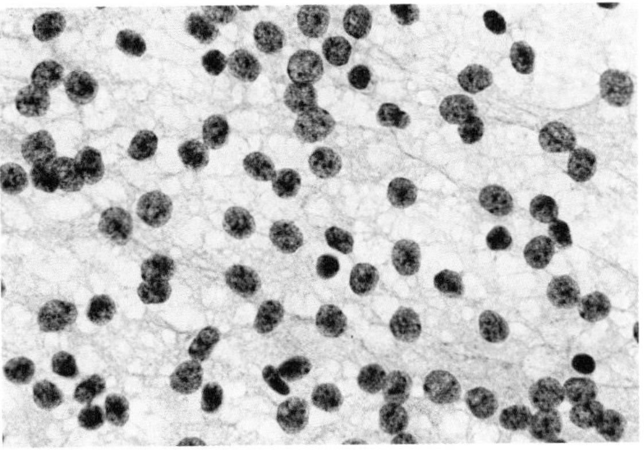

Figure 4-56

CENTRAL NEUROCYTOMA: SMEAR PREPARATION

The marked uniformity in nuclear size, shape, and chromatin configuration is seen best in smear preparations.

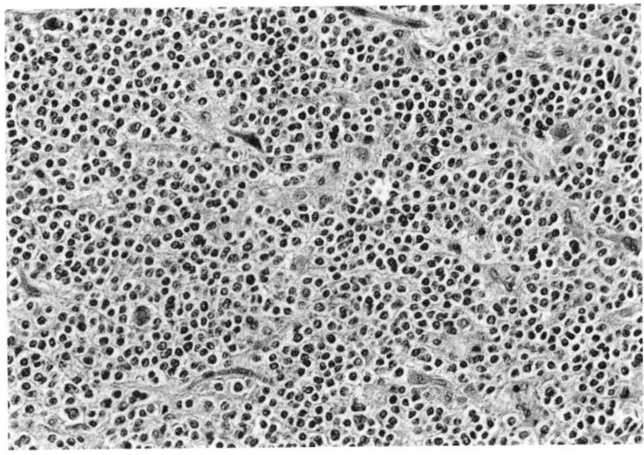

Figure 4-57

OLIGODENDROGLIOMA RESEMBLING CENTRAL NEUROCYTOMA

Highly cellular, compact oligodendrogliomas resemble neurocytic neoplasms. The trapped neurons are more consistent with an infiltrating oligodendroglioma or astrocytoma than central neurocytoma.

18q, have been observed (70). Central neurocytomas do not have the 1p/19q codeletion of oligodendroglioma (71).

Differential Diagnosis. In light of the cellular monotony, perinuclear halos, and frequent calcification of neurocytomas, oligodendroglioma is often the principal consideration (fig. 4-57). Most tumors previously reported as intraventricular oligodendrogliomas are in fact neurocytomas. While the two lesions are similar, diffuse tissue infiltration is lacking in neurocytomas at both radiologic and histologic levels. Thus, unlike oligodendrogliomas, few normal ganglion cells or normal neurofilament-positive axons are present. A phenotypic overlap between oligodendroglioma and neurocytic neoplasia is seen in some oligodendrogliomas with synaptophysin-positive rosettes (see fig. 3-213) (72). While this conceptually blurs the distinction between these two lesions, it does not, from a practical standpoint, detract from the reality of either entity. Oligodendrogliomas are, overall, negative for synaptophysin. The neurocytic lesion does not have the glioma's 1p/19q codeletion. See Appendix E for the differential diagnosis of oligodendroglioma-like lesions, and Appendix M for histologic features, such as an oligodendroglioma-like appearance, that can lead to overdiagnosis of a neoplasm or overgrading of a low-grade tumor. Appendix D gives cystic tumors and Appendix F lists frequently calcified lesions.

Ependymoma enters into the differential diagnosis from both a clinical and radiographic perspective given its paraventricular or largely intraventricular location and tendency to calcify. Most supratentorial ependymomas, however, are not intraventricular, and those that are

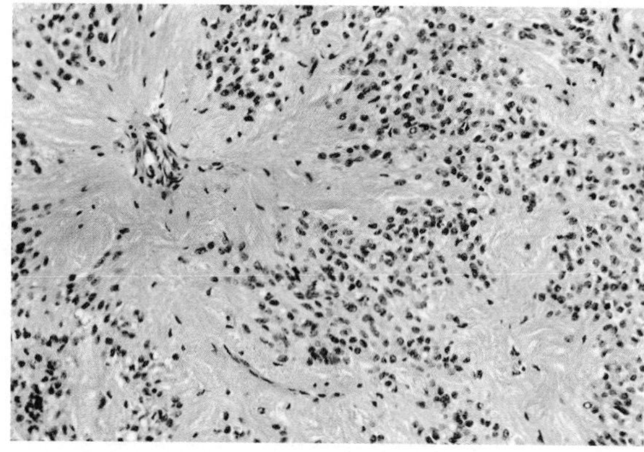

Figure 4-58

EPENDYMOMA RESEMBLING CENTRAL NEUROCYTOMA

Perivascular ependymal pseudorosettes resemble the "neuropil rosettes" that are common in central neurocytomas. Immunohistochemistry for GFAP and synaptophysin may be required to distinguish these two lesions.

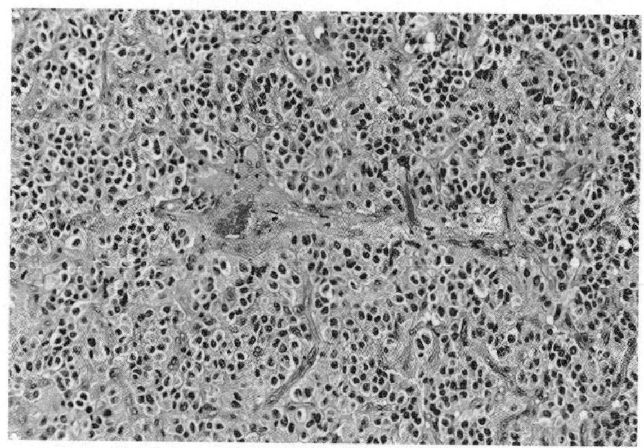

Figure 4-59

CLEAR CELL EPENDYMOMA RESEMBLING CENTRAL NEUROCYTOMA

As compact tumors with round, haloed nuclei, clear cell ependymomas closely resemble central neurocytoma.

intrude unilaterally, whereas neurocytomas straddle the midline and involve the septum pellucidum. Both lesions have finely fibrillar regions, albeit ones that are more strictly perivascular in ependymomas (fig. 4-58). An ependymoma variant that is especially similar to neurocytoma is the clear cell type (fig. 4-59), although its cells are larger, cytologically more atypical, mitotically more active, and often associated with vascular proliferation. Its nuclei are typically clefted, unlike those of central neurocytoma. Ependymal perivascular pseudorosettes are reactive for GFAP, not synaptophysin.

Cerebral neuroblastoma, largely a pediatric tumor, is usually intraparenchymal rather than intraventricular, and highly cellular, cytologically atypical, and mitotically active. The immunohistochemical and electron microscopic features of these two tumors are sufficiently similar to be of little differential diagnostic use.

Dysembryoplastic neuroepithelial tumors of the septum pellucidum, with their oligodendroglioma-like cells, are discussed later in this chapter.

Treatment and Prognosis. Given the size and location of central neurocytomas, total resection is not always possible, leaving open the possibility of regrowth (52,73). The latter comes about only slowly, however, and radiation therapy seems effective in retarding the event, although the long-term outlook for such patients is not yet clear given the rarity of the entity and the lengthy follow-up required to evaluate its behavior (52,74). Not surprisingly, gross total resection is associated with greater overall survival (52). Radiotherapy improves local control after subtotal resection, but not overall survival (52,53). Cerebrospinal fluid spread is rare (73,75,76).

The significance of increased mitotic activity remains to be established, but as is discussed above in the section on grading, MIB-1 labeling indices exceeding 2 percent appear to increase the chances of recurrence. It is not clear how often central neurocytomas progress with time, but an increased MIB-1 rate in a recurrent lesion has been reported at least once (77).

Other Neurocytic Neoplasms of the Cerebral Hemispheres (Extraventricular or Intraparenchymal Neurocytic Neoplasms)

Definition. Other neurocytic neoplasms of the cerebral hemispheres are found at sites other than the lateral ventricles near the foramen of Monro.

General Features. Distinguishing these neoplasms from central neurocytoma located in the lateral ventricle near the foramen of Monro is somewhat arbitrary since, except for location, some are histologically identical. Many, however, are histologically more complex and include varying proportions of ganglionic and glial cells. Sites include the supratentorial compartment (78–82) and fourth ventricle (83,84).

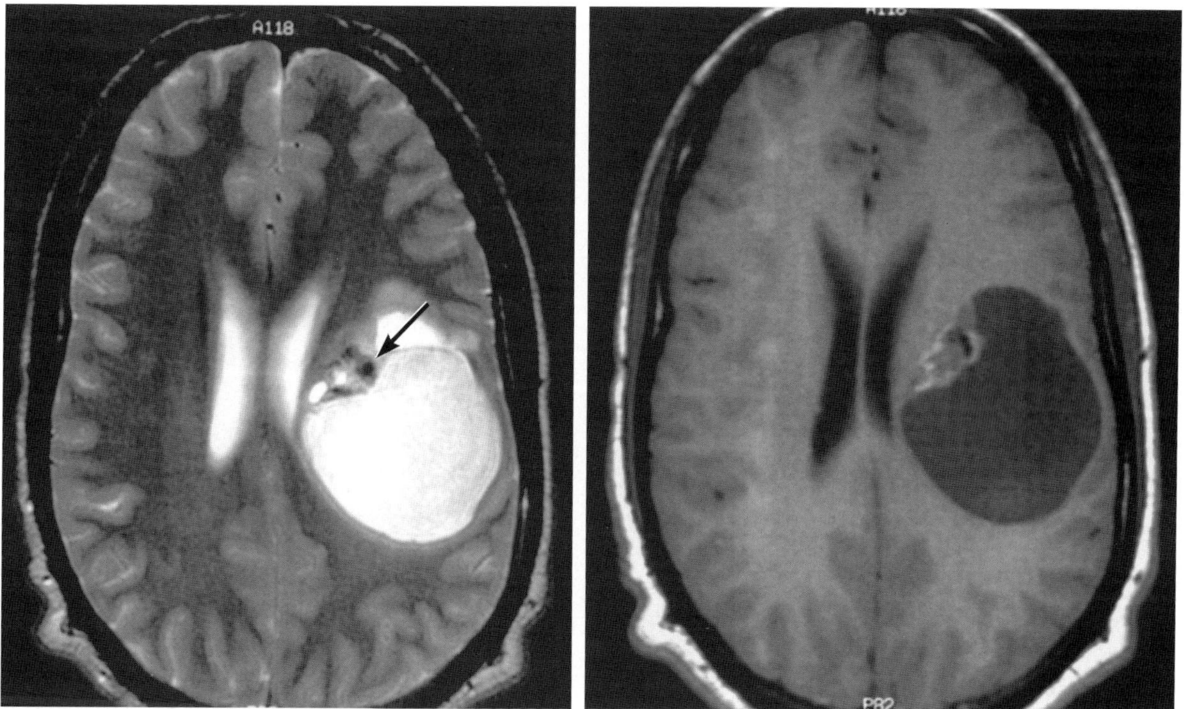

Figure 4-60

EXTRAVENTRICULAR NEUROCYTOMA

A cyst-mural nodule combination is common in discrete grade I and II tumors such as this extraventricular neurocytoma in a 26-year-old. A dark signal consistent with dense calcification is present at the arrow in the T2-weighted image (left). Focal contrast enhancement is seen on the right.

Cerebellar examples, which commonly exhibit adipocyte-like changes and glial differentiation, are discussed separately.

Clinical Features. Patients of all ages are affected. Seizures or expressions of mass effect are the main manifestations.

Radiologic Findings. Occurring throughout the CNS, the tumors are well circumscribed, variably contrast enhancing, and often cystic (fig. 4-60) (78,79,81,83). Some are densely calcified.

Gross Findings. Extraventricular neurocytic tumors are, like their central counterpart, well circumscribed.

Microscopic Findings. Tumor cells in extraventricular neurocytomas are often disposed in a sheet-like growth pattern similar to that of central neurocytoma (figs. 4-61, 4-62), but often contain large neuropil-like areas (figs. 4-63, 4-64). Sometimes the cells are organized into ribbons, large Homer-Wright rosettes, or clusters in the same fine fibrillar background (78,80). Other tumors, or regions thereof, that are less cellular and looser in texture resemble oligodendroglioma or even diffuse astrocytoma. Nuclear uniformity (fig. 4-61) is a cardinal feature,

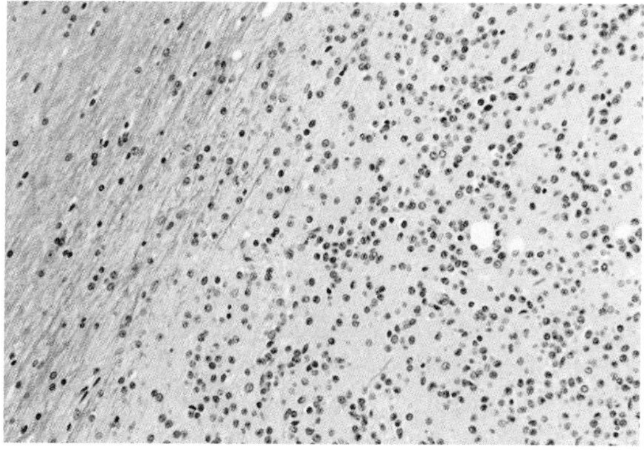

Figure 4-61

EXTRAVENTRICULAR NEUROCYTOMA

A sharp interface with the surrounding parenchyma is typical of neurocytoma and many other grade I and grade II neoplasms.

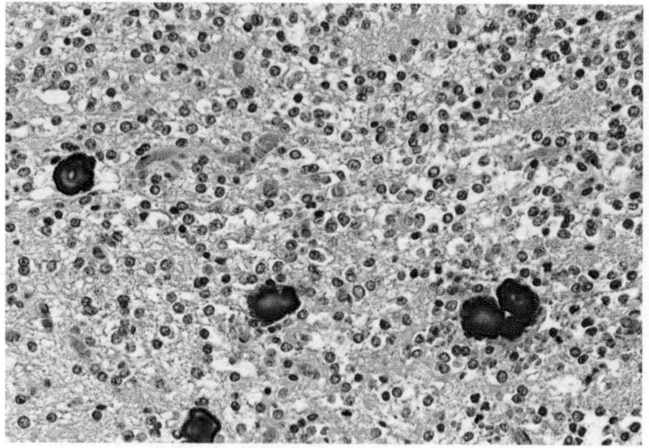

Figure 4-62

EXTRAVENTRICULAR NEUROCYTOMA

High cellularity, perinuclear halos, and calcification combine to create a close likeness to oligodendroglioma.

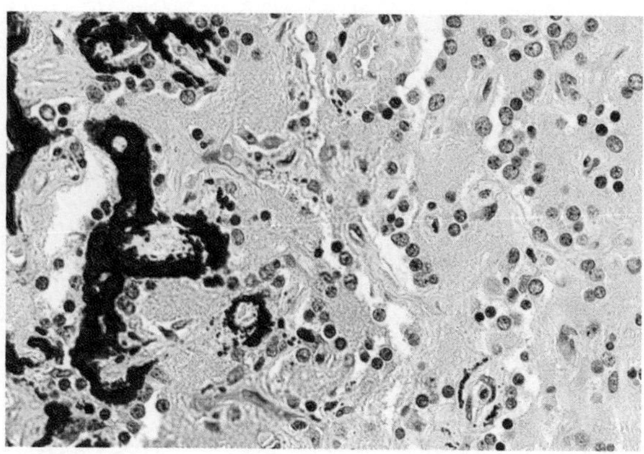

Figure 4-64

EXTRAVENTRICULAR NEUROCYTOMA

Dense calcification and a vague rosetted pattern are common in extraventricular neurocytomas.

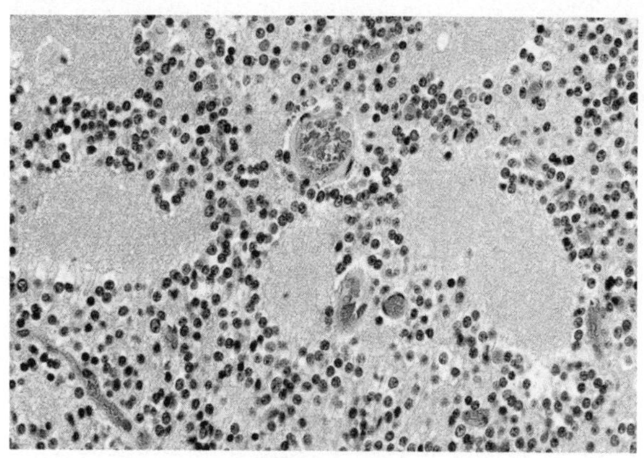

Figure 4-63

EXTRAVENTRICULAR NEUROCYTOMA

Large neurocytic rosettes help identify some extraventricular neurocytomas.

although the synaptophysin positivity of the neoplastic neuropil helps to identify the lesion. In any example, neurocytes, often with perinuclear halos, maintain the nuclear roundness and monotony for which the cell type, and the tumors, are well known. As at the macroscopic level, a cardinal feature is a degree of circumscription that contrasts with the infiltrative quality of oligodendroglioma (fig. 4-61).

Transition to neurons that are smaller than ganglion cells, but larger than neuroblasts or neurocytes, i.e., "ganglioid cells," is not uncommon, and even differentiation to full-blown ganglion cells occurs in some cases, in contrast with central neurocytomas where such maturation is uncommon (78). EGBs may be present in accord with the presence of a glial component. Although GFAP positivity is common, the lesions do not contain astrocytes with pleomorphic nuclei or cytoplasm with coarse processes.

Some extraventricular neurocytomas are calcified, at times heavily (figs. 4-62, 4-64). Hyalinized vessels, as found in many other chronic grade I or II discrete neoplasms, may be present, and should raise suspicion that the tumor is not oligodendroglioma, in spite of the uniformly round nuclei with perinuclear halos. A rare extraventricular neurocytic neoplasm is lipidized.

Grading. As in the case of central neurocytoma, most extraventricular examples are well differentiated, although there is a range of histologic features and MIB-1 labeling indices; rare lesions have overtly aggressive features such as high mitotic rate, necrosis, and vascular proliferation (78). As in central neurocytoma, subdivision into well-differentiated and atypical types has been suggested, with the latter defined as those with necrosis, vascular proliferation, or 3 or more mitoses per 10 high-power fields (78). MIB-1 labeling indices of 5 percent or more are associated with such histologic changes (78).

Immunohistochemical Findings. Diffuse positivity for synaptophysin is the rule (fig. 4-65) (78,80). Some neurocytic tumors are immunoreactive for NeuN (fig. 4-66). About half the tumors have cells that stain for GFAP (78).

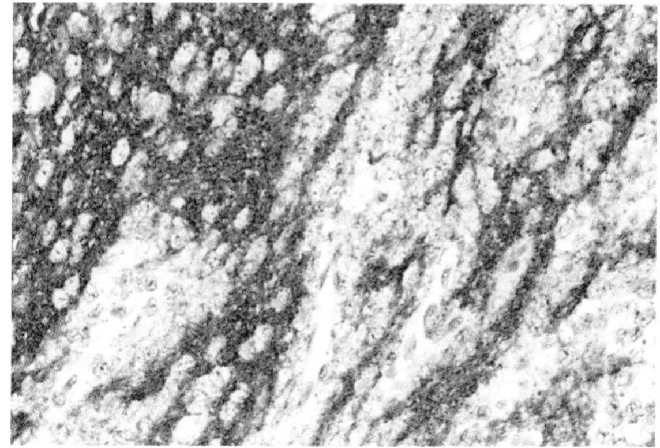

Figure 4-65

EXTRAVENTRICULAR NEUROCYTOMA

Neuropil-like areas are synaptophysin positive.

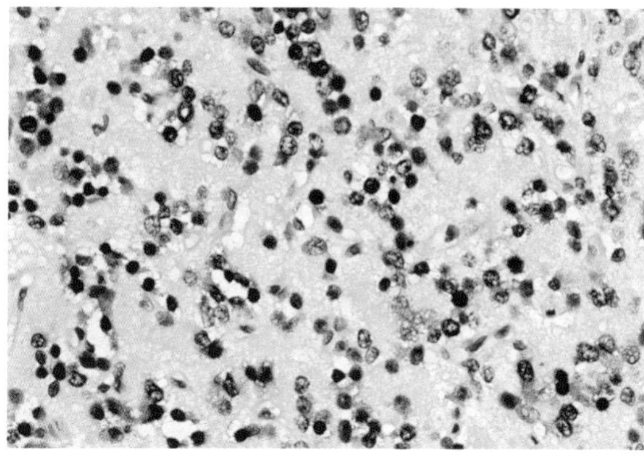

Figure 4-66

EXTRAVENTRICULAR NEUROCYTOMA

The lesions are variably immunoreactive for NeuN, but as in this case, can be diffusely positive.

Molecular Findings. In spite of the histologic similarity to oligodendroglioma, the cytogenetic signature of the latter, losses on chromosomes 1p and 19q, were not found in 10 of 12 extraventricular neurocytomas in one study (85). The nature of the two lesions with the codeletion is not clear.

Differential Diagnosis. See Appendix E for the differential diagnosis of oligodendroglioma-like lesions and Appendix M for histologic features that can lead to overdiagnosis or overgrading of a neoplasm. As is emphasized in Appendix E, a number of neoplastic and non-neoplastic entities enter the differential diagnosis of oligodendroglioma-like lesions.

Oligodendroglioma itself is usually the principal entity in the differential diagnosis. Raising the possibility of extraventricular neurocytoma, however, is the presence of ganglion cells, ganglioid cells, rosettes, and a neuropil-like background denser than that of normal brain. A sharp interface with the surrounding brain and vascular hyalinization are also typical. Radiologically, a well-circumscribed, variably contrast-enhancing, often cystic mass calls the diagnosis of oligodendroglioma into question. Complicating the issue is the occasional expression of neuronal markers, such as synaptophysin, and the presence of neurocytic rosettes in genetically certified oligodendrogliomas (those with codeletion of chromosomes 1p and 19q) (see fig. 3-213) (86).

Treatment and Prognosis. As is the case of central neurocytoma, gross total resection is possible in about half of well-differentiated lesions. Data regarding truly long-term follow-up has yet to be generated, but the overall outlook appears to be favorable (78,80). In one study of 35 cases, none of 14 totally resected lesions recurred, whereas 10 of 19 (45 percent) biopsied or subtotally resected tumors recurred (78). Of the latter 10, 5 were considered atypical by the criteria indicated in the grading section. Total excision was less often possible for atypical tumors than for histologically more benign lesions. The result of radiotherapy in the treatment of recurrence is unclear but has been employed in many instances. The need for and benefits of radiotherapy after initial incomplete resection of an atypical extraventricular neurocytoma has yet to be established.

CEREBELLAR NEUROCYTIC NEOPLASMS

General Features. Neurocytic differentiation in a cerebellar tumor occurs most often in *nodular/desmoplastic medulloblastomas*. Most of these are of the rare, extensively nodular subset.

The discussion here centers upon the rare, non-medulloblastoma, cerebellar neurocytic neoplasms that are distinctive for their cytologic blandness, lipidized components, and frequent glial differentiation. Some have further been described as containing myoid elements, as is discussed below. These combinations have generated appellations including *lipomatous medulloblastoma* (87,88), *neurocytoma* (89), *liponeurocytoma* (90–92), *medullocytoma* (93), *neurolipocytoma*

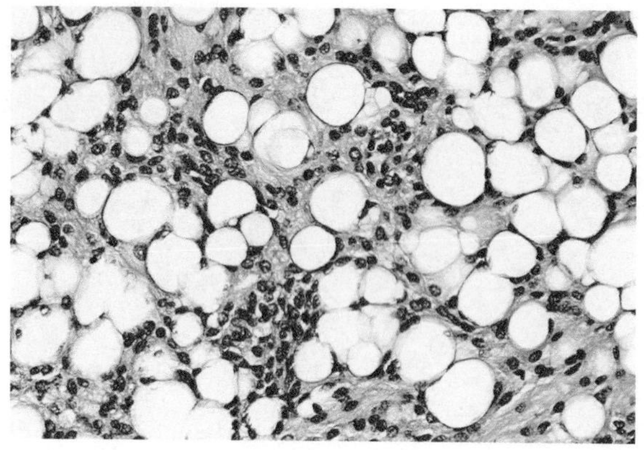

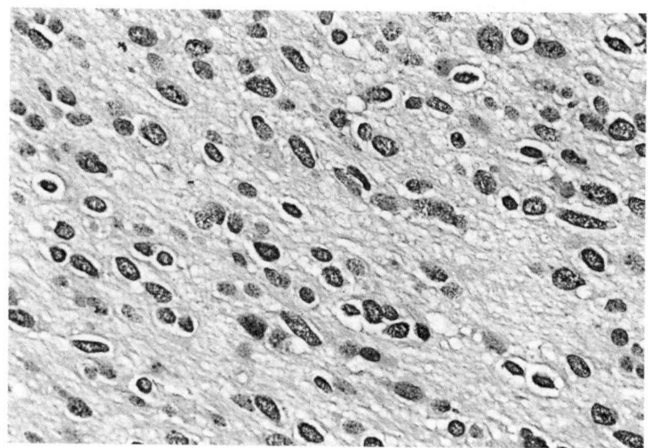

Figure 4-67

CEREBELLAR LIPONEUROCYTOMA

These rare lesions contain lipidized cells (left) and, in some cases, infiltrating glia (right).

(94), *lipomatous glioneurocytoma of the posterior fossa with divergent differentiation* (95), *lipidized mature neuroectodermal tumor of the cerebellum with myoid differentiation* (96), and *neurocytoma/rhabdomyoma (myoneurocytoma)* (97). Whether these are related variants or distinct entities is unclear.

Clinical Features. Almost all of the above WHO grade II lesions present in adults with signs and symptoms of cerebellar dysfunction or effects of hydrocephalus. Only a rare cerebellar nonmedulloblastoma neurocytic tumor appears in childhood (97).

Radiologic Findings. The lesions are well circumscribed and, when heavily lipidized, generate a bright signal on precontrast T1-weighted images (90,91,93,95).

Microscopic Findings. Histologically, all contain densely cellular sheets of small, uniform, synaptophysin-positive cells with round nuclei exhibiting "salt and pepper" chromatin and scant cytoplasm. Lipidization is common (fig. 4-67, left), mitoses are not. Some lesions have a second looser, more infiltrative component of GFAP-positive glia, also histologically bland (fig. 4-67, right). When present, myocytes are round "blasts" or striated "straps" (96).

Immunohistochemical Findings. The lesions are reactive for synaptophysin and GFAP. Either area can be lipidized. Muscle markers highlight myoid components. MIB-1 indices have been variable, but are generally less than 6 percent (98).

Ultrastructural Findings. Ultrastructurally, the lipidized cells, either synaptophysin or GFAP positive, contain nonmembrane bound lipid (91) and are thus lipidized tumor cells, not adipocytes. Neurocytes and their processes contain microtubules and dense core vesicles (88,89,91,95).

Molecular and Cytogenetic Findings. Genetic and expression profiles of cerebellar liponeurocytomas are evidence that the lesions are distinct from medulloblastoma. The lesions appear to lack any of the mutations in genes found variably in medulloblastomas, such as *PTCH*, *β-catenin*, and *APC* (98). Isochromosome 17q is also absent. Gene expression profiles suggested a closer relationship to central neurocytoma than medulloblastoma in the cited study.

Differential Diagnosis. Neurocytic tumors are distinguished from medulloblastoma by their cytologic blandness, lack of nodularity, paucity of mitoses, low MIB-1 index, and the presence of a lipidized component or an infiltrative glial element in some cases.

Treatment and Prognosis. Although recurrences of a "medullocytoma" have been recorded (93,98), the lesions are as a rule biologically indolent. There are, of course, exceptions (99). Given their rarity, appropriate treatment is not clear. Some patients have been irradiated, but others, with generally long survival periods, have not.

NEUROCYTIC NEOPLASMS OF THE SPINAL CORD

Neurocytomas are rare in the spinal cord (100–102).

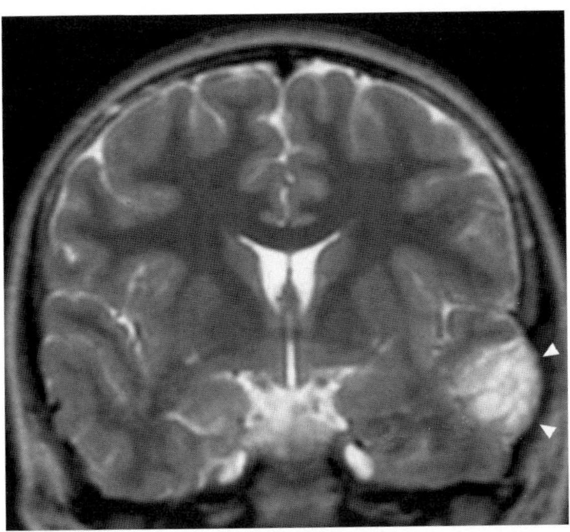

Figure 4-68

DYSEMBRYOPLASTIC NEUROEPITHELIAL TUMOR

Magnetic resonance imaging (MRI) visualizes the mucopolysaccharide-rich, and hence T2-bright, lesion with its distinctive intracortical nodules (arrowheads). As is typical of convexity dysembryoplastic neuroepithelial tumors (DNTs), this lesion has eroded the inner table of the skull. As is also characteristic, there is little other expression of mass effect, and no radiologic evidence of perilesional edema.

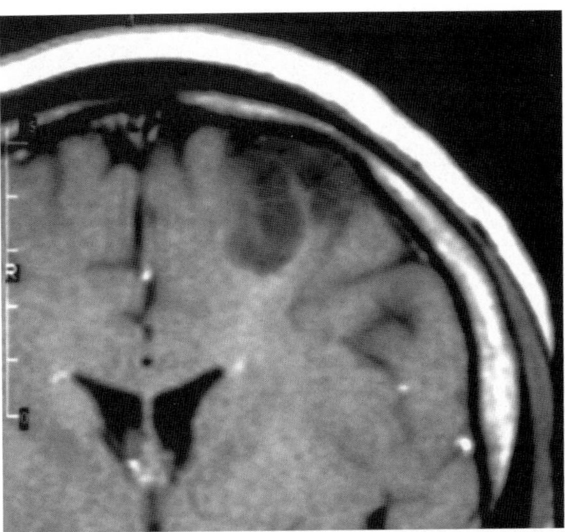

Figure 4-69

DYSEMBRYOPLASTIC NEUROEPITHELIAL TUMOR

Precontrast T1-weighted imaging underscores the intracortical position of the "bubbly" lesion. The longstanding mass has eroded the inner table of the skull.

DYSEMBRYOPLASTIC NEUROEPITHELIAL TUMOR

Definition. *Dysembryoplastic neuroepithelial tumor* (DNT) is a complex, typically multinodular, intracortical cerebral mass.

General Features. While the DNT is considered glioneuronal, defining its lineage has been difficult. Some studies have noted dense core granules in its small oligodendroglia-like cells (103), or report the presence of neural antigens such as NeuN (104). Others, however, have noted immunopositivity for myelin oligodendrocyte glycoprotein (105) or the expression of myelin genes as documented by in situ hybridization (106).

The possibility that the lesion occurs in a "nonspecific form" with histologic features bearing little or no resemblance to the lesion as originally described complicates the definitional issue. This is discussed below in the section on microscopic features. DNT is WHO grade I.

Clinical Features. While most DNTs are supratentorial and intracortical, lesions interpreted as DNT occasionally arise in the caudate nucleus and septum pellucidum (107,108), thalamus and brain stem (109), cerebellum and brain stem (110), and cerebellum (111). One case report of a cerebellar DNT (112) describes a tumor that others interpret as a distinct entity, the rosette-forming glioneuronal tumor of the fourth ventricle.

DNTs usually become symptomatic during the first two decades of life, although years may pass before surgery is undertaken (113–115). Seizures of the partial complex type are the baseline symptom; generalized seizures may occur later in the course.

Radiologic Findings. Favorable planes capture the lesion's intracortical localization and its multiple constituent nodules, best seen at the margins of the lesion, in either precontrast T1-weighted images or in T2-weighted or fluid-attenuated inversion recovery (FLAIR) images (figs. 4-68, 4-69) (116,117). There is usually little if any mass effect, although erosion of the inner table of the skull is common with convexity lesions. A small area, or areas, of intense contrast enhancement, either solid or ring-like, is present in some cases (fig. 4-70). Perilesional cerebral edema is typically absent.

DNTs, or close facsimiles, occasionally appear in the region of the septum pellucidum where they obstruct cerebrospinal fluid flow (fig. 4-71).

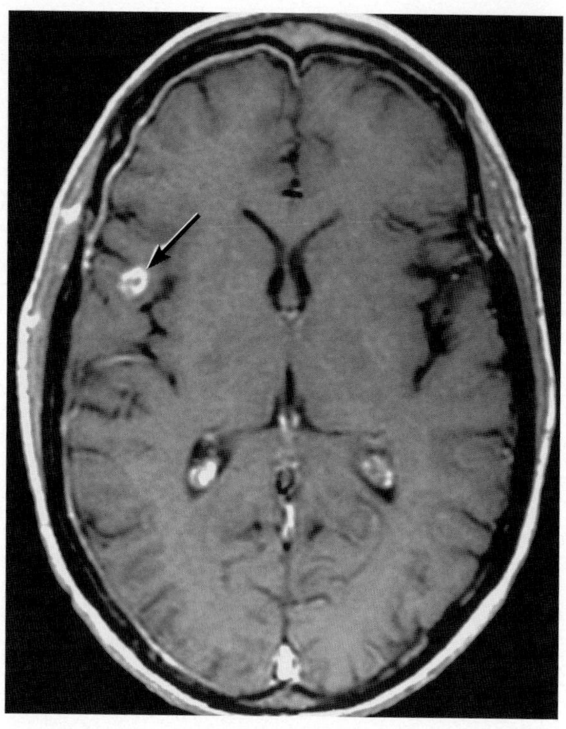

Figure 4-70

DYSEMBRYOPLASTIC NEUROEPITHELIAL TUMOR

A small, often ring-like area of contrast enhancement is present in some tumors, as in this 51-year-old man with a long history of seizures.

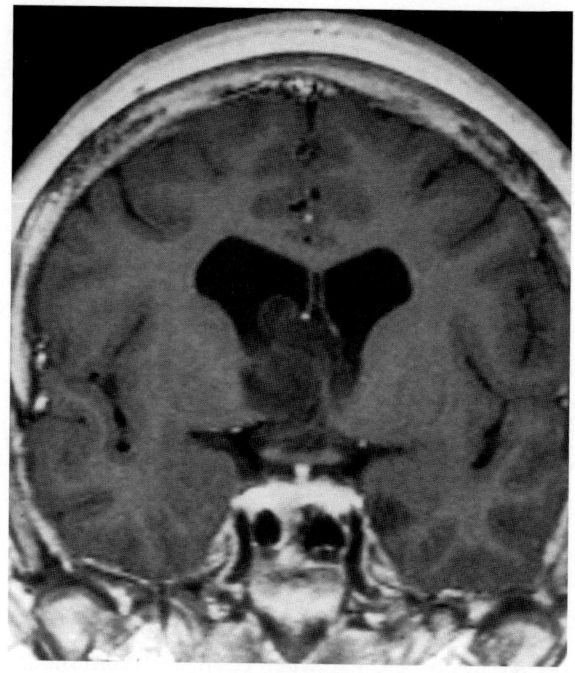

Figure 4-71

DYSEMBRYOPLASTIC NEUROEPITHELIAL TUMOR

DNTs occasionally occur in the region of the septum pellucidum.

Gross Findings. DNTs vary from only millimeters in greatest dimension to those that reach nearly lobar proportions. On balance, most measure several centimeters. Some degree of gyral expansion is generally seen, and the cortical surface may feature blister-like nodules in some fully developed examples or nodules on the cut surface (fig. 4-72). Many DNTs are ill-defined, and only vaguely nodular. Affected segments of cortex are rich in Alcian blue–positive mucopolysaccharide.

Microscopic Findings. The lesion is largely intracortical in most cases, and lacks, at least in concept, the diffuse infiltration of the white matter seen in diffuse astrocytoma and oligodendroglioma. This localization is not always easy to establish, however, especially in small fragmented specimens, and some lesions that appear otherwise to be candidates for DNT infiltrate white matter. In some DNTs, the small oligodendroglial-like cells congregate in the subpial region and even break into the leptomeninges.

The most consistent findings in DNT include multiple intracortical nodules and a distinctive loose-textured internodular element. The nodules range from subtle foci of mild hypercellularity with a faintly basophilic background to larger profiles that in some cases expand the entire cortical ribbon (figs. 4-73–4-75). Regardless of size, the nodules are rich in acid mucopolysaccharide and thus stand out after staining with Alcian blue (fig. 4-73).

The internodular component, also often Alcian blue positive, is known as the specific glioneuronal element, and includes microcystic change and axons extending towards the pial surface ensheathed with oligodendrocyte-like cells (fig. 4-76). Since the linear complexes are directed vertically to the cortical surface, they are often not seen well in small or randomly oriented specimens. A characteristic feature in mucopolysaccharide-rich areas of the specific glioneuronal element, or within nodules, is the presence of large cortical neurons suspended in small pools of mucopolysaccharide ("floating neurons") (figs. 4-77, 4-78).

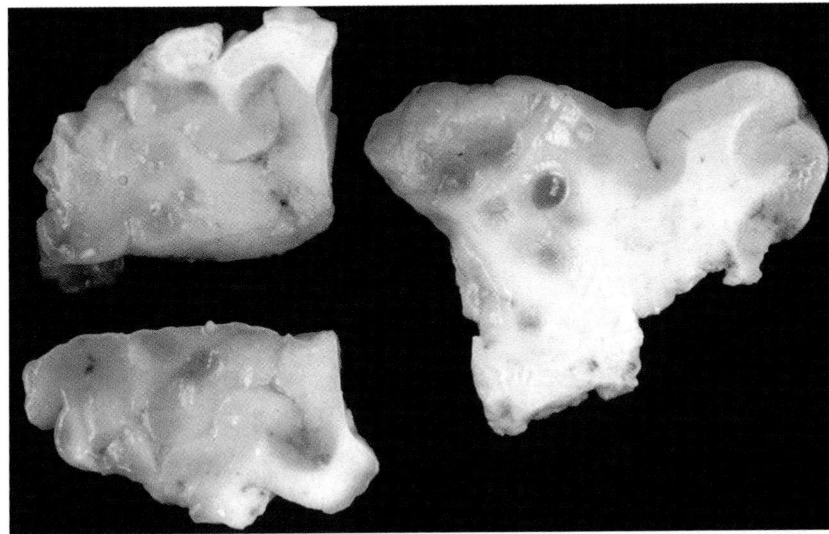

Figure 4-72

DYSEMBRYOPLASTIC NEUROEPITHELIAL TUMOR

Glistening mucopolysaccharide-rich nodules are classic gross features.

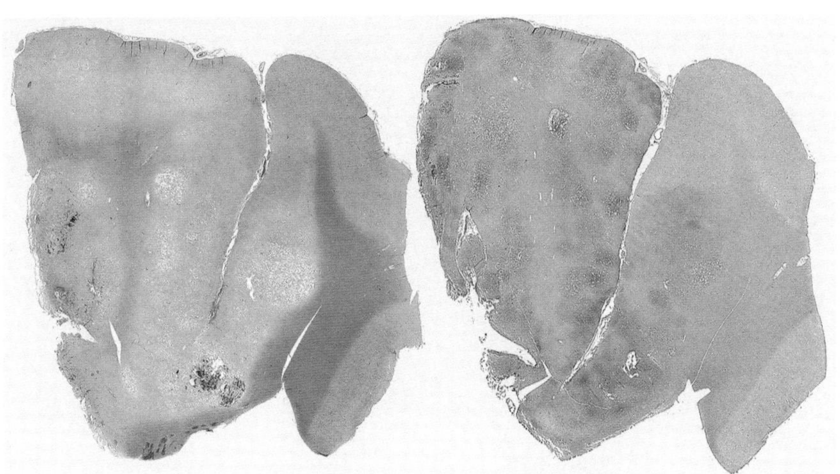

Figure 4-73

DYSEMBRYOPLASTIC NEUROEPITHELIAL TUMOR

Intracortical nodules can be seen in sections stained with H&E (left), but are especially apparent after staining with Alcian blue (right).

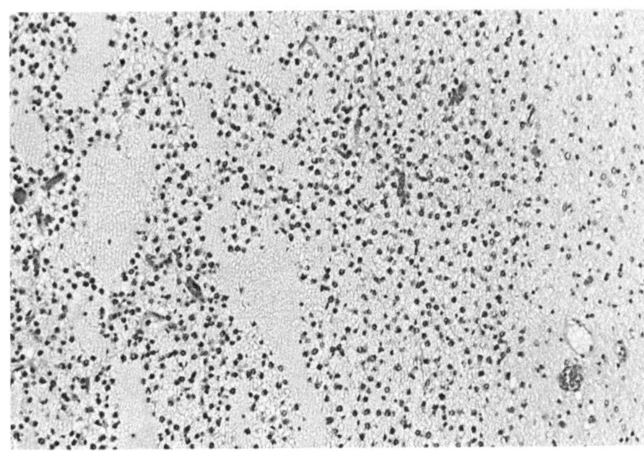

Figure 4-74

DYSEMBRYOPLASTIC NEUROEPITHELIAL TUMOR

Intracortical nodules can be solid.

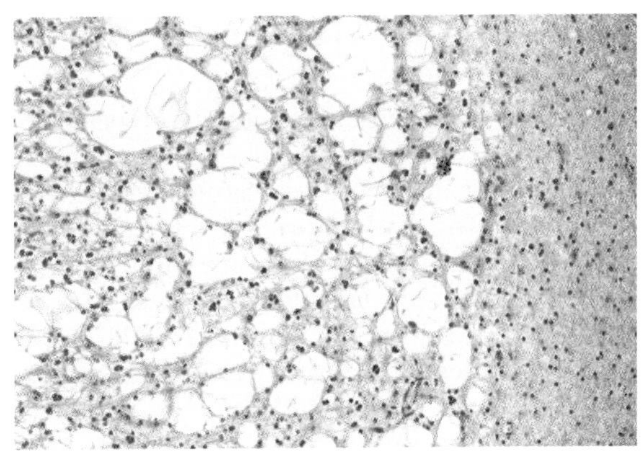

Figure 4-75

DYSEMBRYOPLASTIC NEUROEPITHELIAL TUMOR

Intracortical nodules can be extensively microcystic.

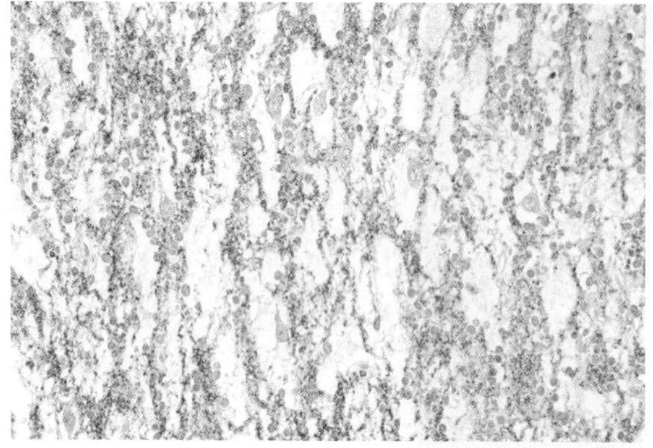

Figure 4-76

DYSEMBRYOPLASTIC NEUROEPITHELIAL TUMOR

Vertical axons, here stained immunohistochemically for synaptophysin, ensheathed in oligodendrocyte-like cells, form the specific glioneuronal element.

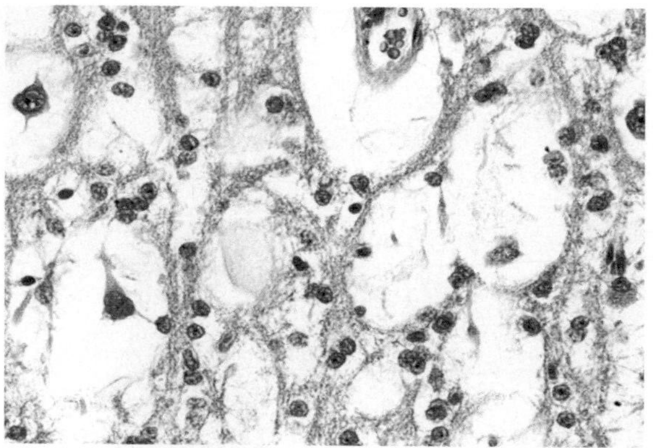

Figure 4-77

DYSEMBRYOPLASTIC NEUROEPITHELIAL TUMOR

Multiple small pools of mucin, many containing ganglion cells, are common in DNTs, especially in intranodular areas.

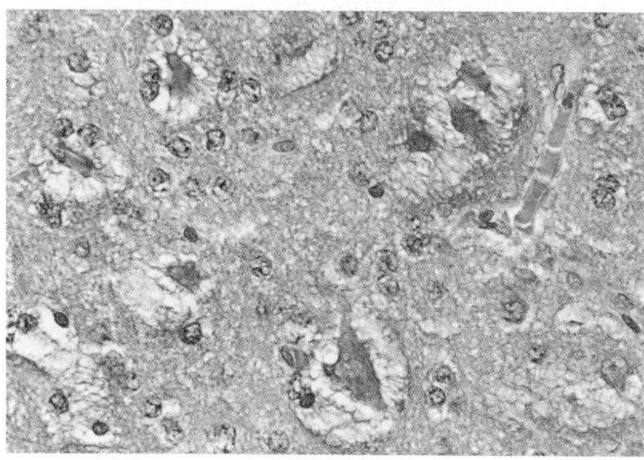

Figure 4-78

DYSEMBRYOPLASTIC NEUROEPITHELIAL TUMOR

Ganglion cells "float" in pools of mucopolysaccharide (Alcian blue stain).

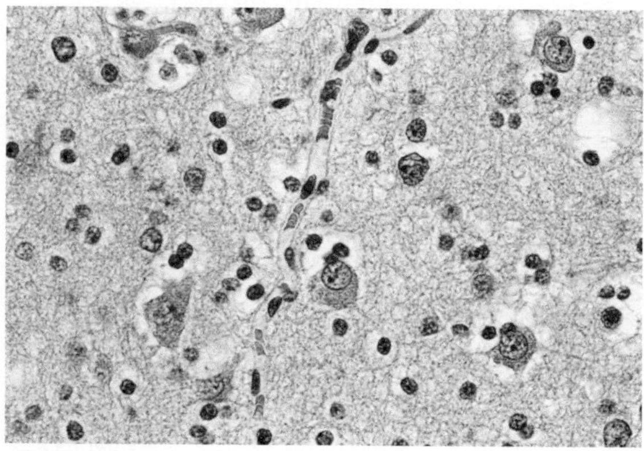

Figure 4-79

DYSEMBRYOPLASTIC NEUROEPITHELIAL TUMOR

Diffusely hypercellular non-nodular areas closely resemble oligodendroglioma, although, as here, perinuclear satellitosis is uncommon in DNTs. A distinction between the two lesions may not be possible in all cases.

The extranodular, diffuse cortical hypercellularity, due to the oligodendroglia-like cells (OLCs), is a diagnostically confusing feature, particularly in regard to the distinction of DNT from oligodendroglioma (fig. 4-79). Unlike oligodendrogliomas, perineuronal satellitosis by OLCs in DNT is uncommon. Furthermore, OLCs are very well differentiated, being more uniform in size and shape than all but the best-differentiated oligodendroglioma cells. The exclusive presence of such diffusely hypercellular tissue has been accepted by some as sufficient evidence of a DNT, even when nodules or the specific glioneuronal element is not present, and even if the lesion affects white matter (118). Using this very liberal approach to the definition, the histologic distinction of DNT from oligodendroglioma becomes problematic, if not impossible. It is difficult enough as it is.

The term *simple DNT* refers to a lesion, or specimen, with only the glioneuronal element. The *complex DNT* has the spectrum of features that includes nodules and, in some cases, cortical dysplasia.

Nodular architecture and cell composition vary greatly. Some lesions are solid and composed

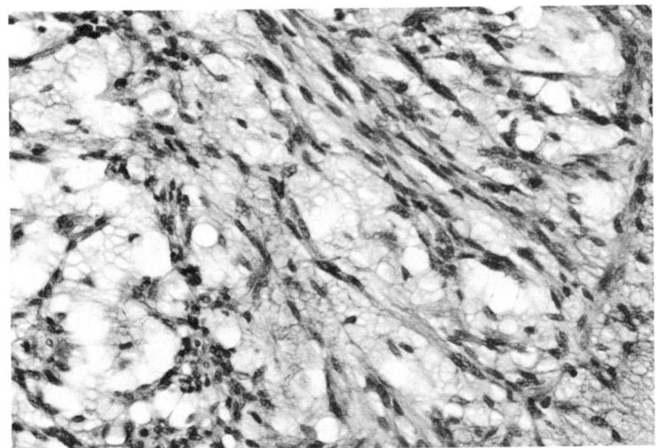

Figure 4-80

DYSEMBRYOPLASTIC NEUROEPITHELIAL TUMOR

Piloid tissue reminiscent of pilocytic astrocytoma is present in some cases.

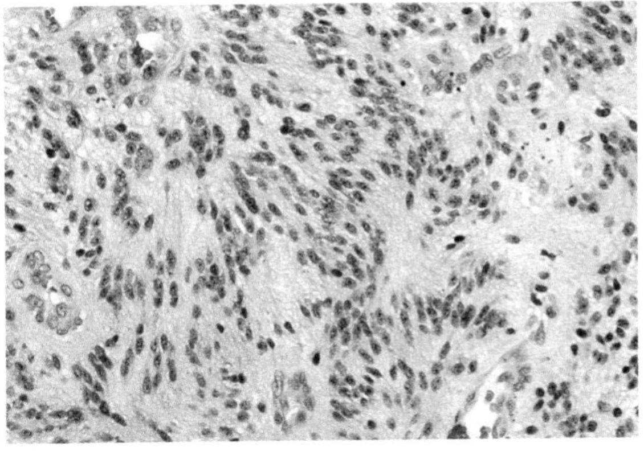

Figure 4-81

DYSEMBRYOPLASTIC NEUROEPITHELIAL TUMOR

Compact tissue with a tendency to ribboning is seen in some DNTs.

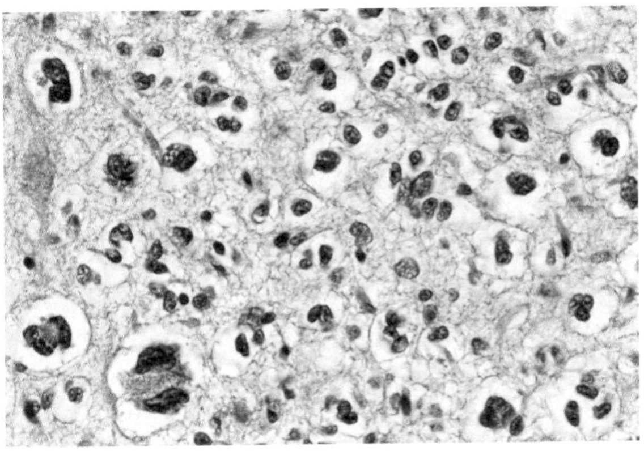

Figure 4-82

DYSEMBRYOPLASTIC NEUROEPITHELIAL TUMOR

Focal groups or lobules of cells with pleomorphic nuclei may be present.

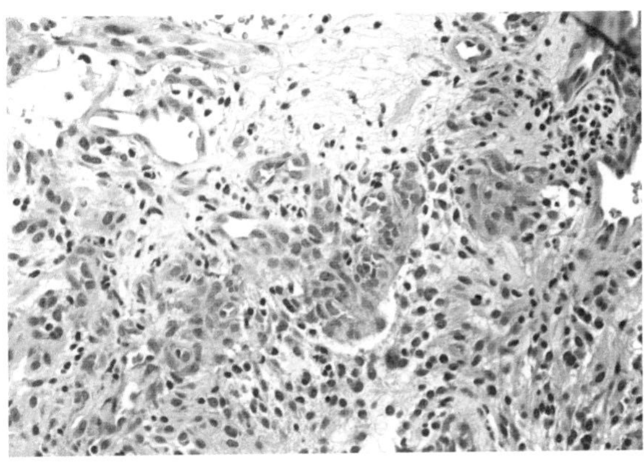

Figure 4-83

DYSEMBRYOPLASTIC NEUROEPITHELIAL TUMOR

Vascular proliferation should not be interpreted as evidence of a high-grade lesion.

of astrocytic cells of fibrillary or even piloid type (figs. 4-80, 4-81), replete with eosinophilic granular bodies. Small or large aggregates of pleomorphic cells are not uncommon (fig. 4-82). The extent to which overtly ganglionic or pilocytic areas are present is discussed below.

DNTs are often, but not invariably, associated with cortical dysplasia, which takes the form of disturbed lamination, neuronal cytologic abnormalities, and general architectural disarray. Within the nodules, ganglion cells are often only trapped. Truly dysplastic cells are more readily recognized at the edge of the lesion, where cellular polymorphism, lack of polarity, and occasional binucleation cannot be simply dismissed as features of normal neurons mechanically affected by nodules or OLC infiltrates.

Paradoxically, given the DNT's indolent nature, garlands of microvascular proliferation are present in some cases, often about an amorphous central area of tissue degeneration (fig. 4-83). The cytologic features of the tumor cells in these regions are no different than elsewhere in the mass. Thus, the vascular reaction is of no clinical significance.

Calcification, if present, is only focal. A melanotic DNT has been reported (119).

Although DNTs are readily, at least in concept, diagnosed in large specimens by the panoply of classic features (complex DNT), partial representation presents a diagnostic challenge that cannot always be met. The issue has become even more problematic with attempts to expand the entity to include a nonspecific form (120), a lesion that is more a clinicopathologic construct than a histologically definable process. This "entity" includes lesions with large amounts of the piloid tissue described above, and even what some would consider pilocytic astrocytoma, in the absence of the above specific features of DNT. The presence of adjacent cortical dysplasia and stability on radiologic imaging are used to help identify this lesion. By this approach, the diagnosis becomes partially dependent on clinical and radiologic features, including: a typically young patient with longstanding partial complex epilepsy and a radiologically discrete intracortical lesion without mass effect, except as related to cystic components, or edema. If these disparate, and in some cases macroscopically bulky components are accepted as part of the DNT spectrum, the spectrum becomes wide indeed.

Frozen Section. Small, markedly uniform OLCs in a mucoid background and floating neurons may be present, but the changes are unlikely to be specific enough for a definitive diagnosis.

Immunohistochemical Findings. In one study, 14 DNTs were assessed immunohistochemically with a spectrum of neuronal and glial markers to delineate the range of differentiation, and in particular, the features, particularly the OLCs (103). Neurofilament protein, class III beta-tubulin, and synaptophysin preparations stained a few OLCs in two, six, and one lesion, respectively. In addition, many OLCs within a single cortical nodule were reactive for class III beta-tubulin. The vast majority of OLCs were strongly S-100 protein positive. GFAP labeled some OLCs in two cases, and one nodule consisted almost entirely of immunoreactive astrocytes, confirming to those authors' satisfaction that the DNT is a mixed glioneuronal lesion. Another study found rare, less than 1 percent, NeuN-positive OLCs in 7 of 16 DNTs, and positive neurons in 13 of 16 (104).

Ultrastructural Findings. In one study, most OLCs resembled oligodendrocytes because of microtubules, prominent Golgi apparatus, and short cell processes (103). Pericellular lamination of cell processes, a characteristic of oligodendroglia, was noted in only one tumor. In two cases, OLCs with astrocytic features contained small numbers of intermediate filaments. In four cases, a few OLCs resembled immature neurons with scant dense core granules or synapses. These findings were interpreted to confirm the glioneuronal nature of DNT, a lesion composed of heterogeneous cells, many resembling oligodendrocytes and a few showing early astrocytic and neuronal differentiation (103).

Cytologic Findings. OLCs are cytologically bland in the extreme (121). Floating neurons can be identified in some cases.

Molecular and Cytogenetic Findings. Despite the similarity to oligodendroglioma, genetic studies of DNT have not found allelic loss on chromosomal arms 1p and 19q (122–125).

Differential Diagnosis. The principal entity in the differential diagnosis of DNT is oligodendroglioma. Since this distinction is based in large part upon recognition of architectural changes, particularly patterned nodules, differentiation may not be possible in small specimens. This is particularly true when OLCs are distributed diffusely rather than collected in nodules. DNT should be suspected, however, even in the presence of a single, round, intracortical, mucin-rich nodule. In DNT, OLCs are no less or more differentiated in nodular than in diffuse areas, whereas intranodular cells in oligodendrogliomas are often of higher grade than the surrounding neoplasm. Like the DNT, the latter are bluer than the surrounding tumor, but because of increased cellularity rather than abundant mucopolysaccharide. In further contrast, nodules of DNT have less rather than more microcystic change. Oligodendrogliomas feature perineuronal satellitosis instead of floating neurons. Patterned arrays of axons with enshrouding OLCs are not a feature in oligodendrogliomas. Cells of the latter lesion satellite about neuronal cell bodies, not axons or dendrites.

Ideally, the lack of the 1p/19q codeletion in DNT (125) would be helpful in separating it from oligodendroglioma. While this may be possible in adults, early reports suggest that these two chromosomal arms are usually retained in pediatric oligodendrogliomas.

In spite of the above differences between DNT and oligodendroglioma, the distinction may be impossible on histologic grounds since the differences are often more apparent on paper than in practice, particularly in the small fragmented specimens that are common with this entity. Immunohistochemistry is also of little use, since, to date, neither DNT nor oligodendroglioma has an exploitable immunohistochemical profile. A high MIB-1 index within a nodule would support the diagnosis of oligodendroglioma, but DNTs can have MIB-1-positive cells.

A descriptive diagnosis of a well-differentiated lesion must thus suffice in some cases, and is often more helpful that it may appear, guiding the clinician to follow rather than treat the patient. Such lesions can always be rebiopsied if they recur, an event favoring a diagnosis of oligodendroglioma. The initial radiologic findings emerge as especially helpful since the classic superficial lesion is distinctively "bubbly" as a reflection of its nodularity. See Appendix E for a tabulation of oligodendroglioma-like lesions.

In contrast to ganglion cell tumor, which usually has a large astrocytic component, DNTs have OLCs and lack other key features of a ganglion cell tumor, e.g., reticulin-rich collagenous stroma, EGBs, perivascular chronic inflammatory cells, and conspicuous neuronal pleomorphism. Rare lesions with distinct components of both DNT and conventional ganglion cell tumor have been described (126,127).

Pilocytic astrocytoma enters into the differential when this glioma contains oligodendroglioma-like areas, or when a DNT contains piloid tissue. An oligodendroglioma appearance in pilocytic astrocytoma is most often seen in cerebellar examples. Features distinguishing pilocytic astrocytoma include its relative lack of infiltration, frequent cystic architecture, contrast enhancement (in most cases), piloid or microcystic pattern (at least focally), and extensive immunoreactivity for GFAP.

In the cerebellum and fourth ventricle, a DNT-like lesion needs to be distinguished both practically and conceptually from the rosette-forming glioneuronal tumor of the fourth ventricle. This uncommon hybrid lesion contains distinctive Homer-Wright–like, synaptophysin-positive rosettes as well as components of pilocytic astrocytoma, or at least a close facsimile of the latter.

The differential diagnosis of nonspecific DNT is either understandably broad if one accepts the entity, or not an issue if one does not. Lesions that might be part of this group include those that are oligodendroglioma-like but also contain an unusual assortment of piloid tissue, ganglion cells, and EGBs. In some instances, the appearance is that of a conventional infiltrative glioma, sometimes with anaplastic features. While the wisdom of the nonspecific DNT concept can be debated, the approach underscores the importance of assuming that chronically epileptogenic, discrete, intracortical lesions with little mass effect are biologically indolent until proven otherwise.

Treatment and Prognosis. DNTs are extremely slow growing and are largely noninfiltrative. Even subtotal resection is reportedly curative, although occasional subtotally resected lesions have been reported to recur, fill in the resection cavity, and require subsequent surgery, or surgeries (128). Given the difficulty in defining the lesion in some cases, claims that a DNT has undergone malignant degeneration have not always been convincing, but it may have happened in at least one instance (129).

OTHER NEURONAL AND GLIONEURONAL TUMORS

Papillary Glioneuronal Tumor

Papillary glioneuronal tumor is a rare mixed glioneuronal lesion that presents in children and adults as a radiologically discrete and, thus far, exclusively supratentorial lesion (130–132). MRI shows a hemispheric, often temporal, contrast-enhancing mass that may be cystic. Like other neurocytic neoplasms, it contains oligodendroglial-like cells, but its distinctive features are hyalinized vessels ensheathed by a rim of cells that are glial (on the basis of immunohistochemistry) and distinguishable from neurocytes that fill intervening zones (fig. 4-84). While neurocytes may be round, uniform, and classically neurocytic, and the glial element dark and cuboidal, the two cell types are not always notably different cytologically (fig. 4-85). Ganglioid cells or frank ganglion cells may be present. Tissue dehiscence creates a pseudopapillary pattern. Reactive, often piloid, gliosis may surround the lesion.

The immediately perivascular cells are GFAP positive (fig. 4-86, left), whereas neuronal elements are reactive for synaptophysin (fig. 4-86, right). Electron microscopy reveals an abundance of

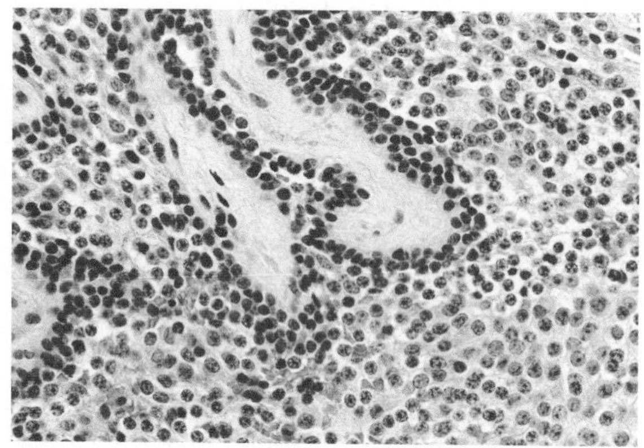

Figure 4-84

PAPILLARY GLIONEURONAL TUMOR

The distinction between the neurocytes, with their "open" nuclei with "salt-and-pepper" chromatin and the darker, immediately perivascular glial cells is obvious in some cases.

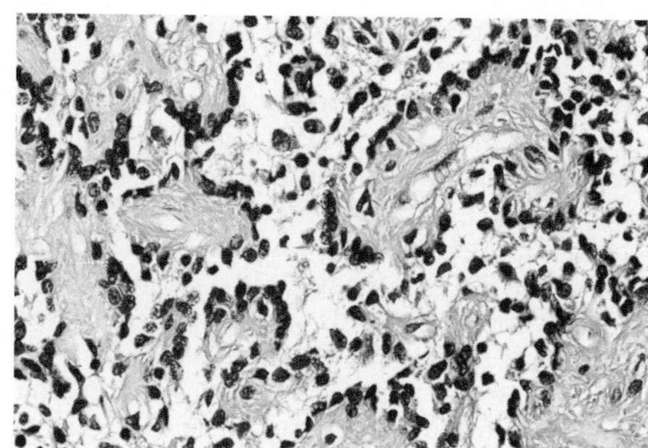

Figure 4-85

PAPILLARY GLIONEURONAL TUMOR

The cytologic distinction between the perivascular glial layer and the neurocytic component is not always apparent as it is in figure 4-84.

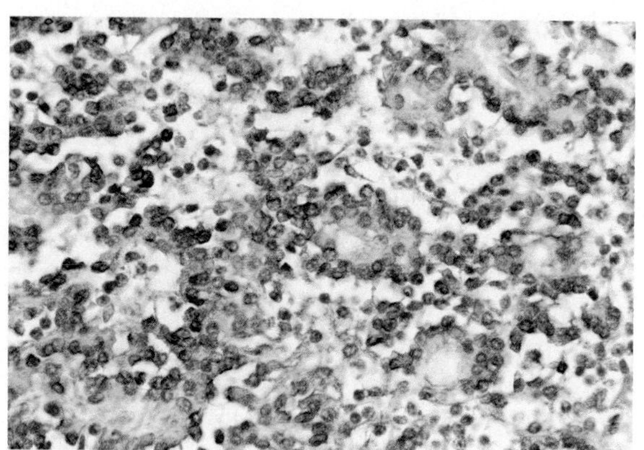

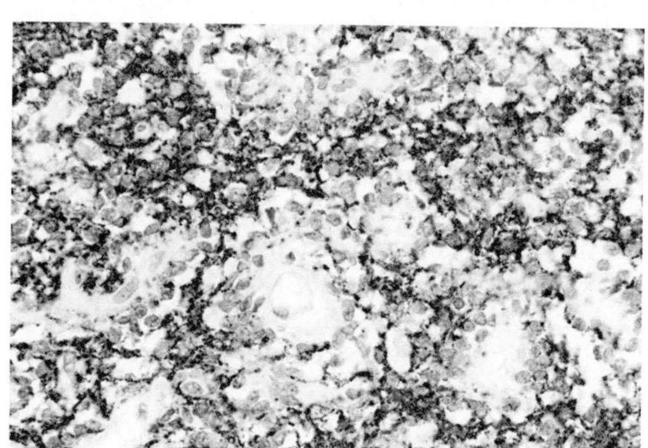

Figure 4-86

PAPILLARY GLIONEURONAL TUMOR

Left: Only cells in direct contact with the vessels are GFAP positive.
Right: Removed from the vessels by one cell layer, the neurocytic component is synaptophysin positive.

cytoplasmic intermediate filaments in the GFAP-positive regions, and microtubule-containing processes, dense core granules, and occasional synapses in the synaptophysin-positive component (130–132).

To date, these lesions have behaved as expected from their neuroradiologic and histologic features. None has recurred following excision alone.

Glioneuronal Tumor with Neuropil-Like ("Rosetted") Islands

On rare occasion, otherwise typical diffuse astrocytomas contain circular patches of synaptophysin-positive "neuropil" surrounded by small ganglion and ganglioid cells (figs. 4-87, 4-88) (133). Both brain (133,134) and spinal cord (135) have been affected by this astrocytoma variant. Molecular findings in one case included gain on chromosome 7q and loss on 9p, both changes common in astrocytic neoplasms of the diffuse or infiltrating type (134).

Based on limited experience, the rosettes do not appear to ameliorate the behavior of the parent lesions. The astrocytomas are aggressive, and are either grade III at onset or later with progression (133).

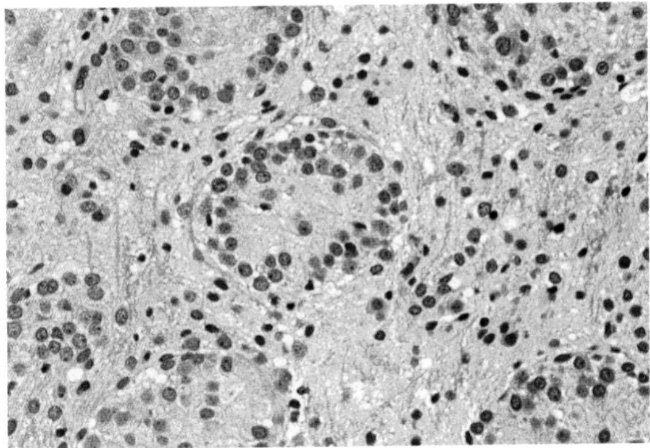

Figure 4-87

GLIONEURONAL TUMOR WITH NEUROPIL-LIKE ("ROSETTED") ISLANDS

Rare, otherwise typical infiltrating gliomas, usually astrocytic, contain circular islands of finely fibrillar "neuropil" surrounded by small ganglion cells or neurocytes.

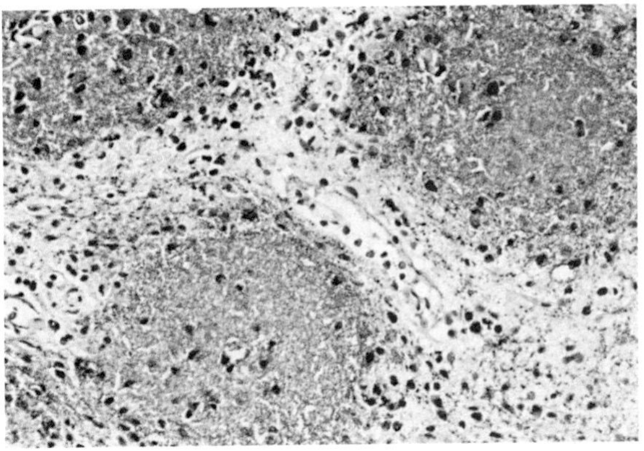

Figure 4-88

GLIONEURONAL TUMOR WITH NEUROPIL-LIKE ("ROSETTED") ISLANDS

Neuropil islands are synaptophysin positive.

Rosette-Forming Glioneuronal Tumor of the Fourth Ventricle

Presenting in children and adults, this is one of many "new" entries to the glioneuronal group. MRI shows a midline lesion affecting mainly the aqueduct, fourth ventricle, and cerebellar vermis (fig. 4-89). Symptoms are the consequence of obstruction to flow of cerebrospinal fluid with resultant hydrocephalus.

The tumor has both neurocytic and glial features (136). The former consists of a circular arrangement of neurocytes that surrounds either a finely fibrillar, synaptophysin-positive, avascular core ("neurocytic rosettes") (figs. 4-90, 4-91) or a small blood vessel. The rosettes resemble somewhat the oligodendroglial rosettes seen in some dysembryoplastic neuroepithelial tumors. The fibrillar component of both types of rosette closely resembles tangled neuronal process (neuropil). Perivascular formations and neurocytic rosettes are immunoreactive for synaptophysin and microtubule-associated protein (MAP-2). Ganglion cells may be present.

The glial component of this distinctive tumor is usually identical to pilocytic astrocytoma, with its elongated cells, microcysts, scattered Rosenthal fibers, and periodic acid–Schiff (PAS)-positive granular bodies. Not surprisingly, staining for GFAP and S-100 protein is regularly seen. Vessels are often glomeruloid in appearance.

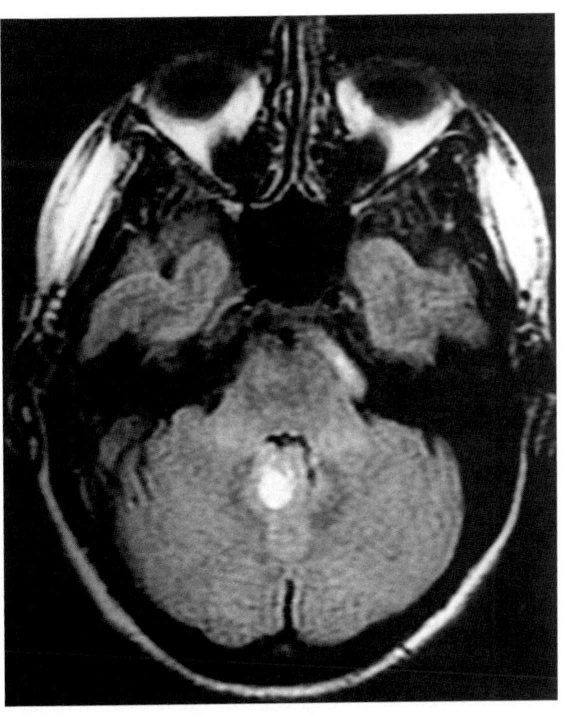

Figure 4-89

ROSSETTE-FORMING GLIONEURONAL TUMOR OF THE FOURTH VENTRICLE

As seen here in a fluid-attenuated inversion recovery (FLAIR) MRI of a 31-year-old woman, the lesion projects into the fourth ventricle from its origin in the cerebellar vermis.

In some areas, this pilocytic element mimics oligodendroglioma. Atypia is minimal, mitoses are absent, and the MIB-1 labeling index is low.

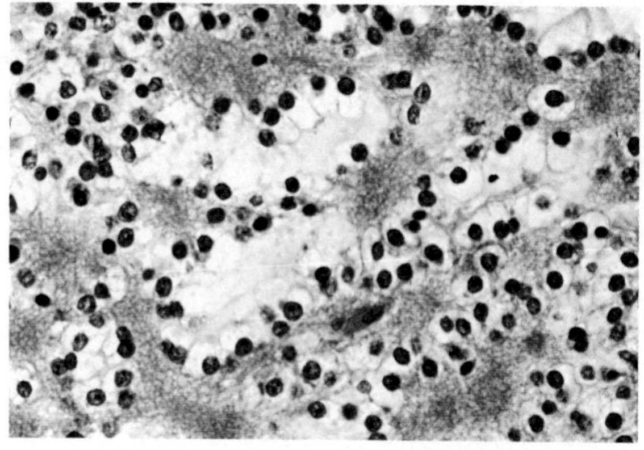

Figure 4-90

ROSSETTE-FORMING GLIONEURONAL TUMOR OF THE FOURTH VENTRICLE

Rosettes, with fine fibrillar cores formed of oligodendrocyte-like cells, are distinctive components of the lesion.

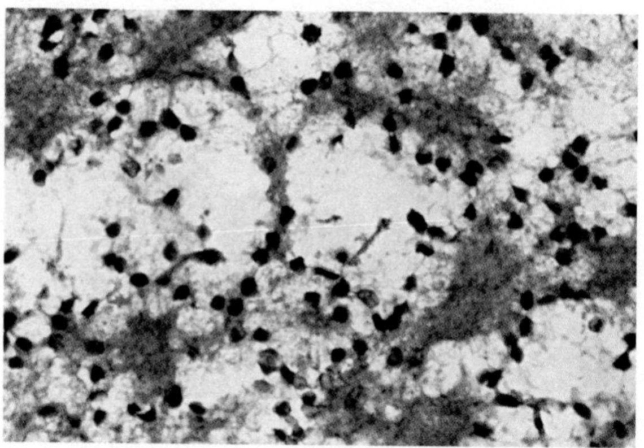

Figure 4-91

ROSSETTE-FORMING GLIONEURONAL TUMOR OF THE FOURTH VENTRICLE

Rosettes are immunoreactive for synaptophysin.

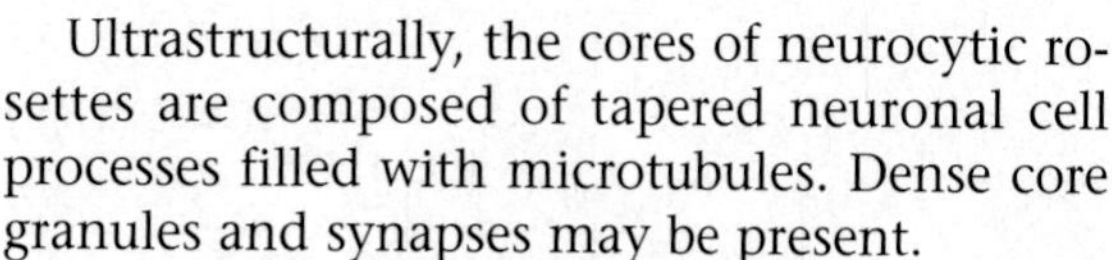

Ultrastructurally, the cores of neurocytic rosettes are composed of tapered neuronal cell processes filled with microtubules. Dense core granules and synapses may be present.

Based on data of a single series of but 11 patients and a report of one case (136), the lesion is biologically indolent, but interval enlargement is possible.

MALIGNANT NEURONAL AND GLIONEURONAL TUMORS

Malignant neuronal and glioneuronal tumors are rare, heterogeneous, difficult to classify, and poorly understood in regard to their biologic behavior. In addition, it is not always clear whether intermediate-grade lesions should be considered neuronal tumors with anaplastic degeneration or differentiating small cell embryonal tumors (primitive neuroectodermal tumor [PNET]).

Otherwise classic gangliogliomas only rarely contain or develop a malignant GFAP-positive component (fig. 4-92) (137–144). The diagnosis of *anaplastic ganglioglioma* should be made cautiously, and only when a clearly defined preexisting ganglion cell tumor is present. Many so-called gangliogliomas, including malignant variants, are simply infiltrating gliomas that trap normal ganglion cells.

Other malignant tumors in this group have a malignant neuronal component (139,145), and some have both malignant neural and glial elements (138). Even myosarcomatous elements may appear (141).

Malignant CNS neuronal neoplasms that do not have a glial component span a spectrum of differentiation akin to the neuroblastoma-differentiating neuroblastoma-ganglioneuroma triad of the peripheral nervous system. It is not clear at what point there are enough small tumor cells for the CNS neoplasm to be considered a malignant neuroblastic tumor, or sufficient neuronal maturation to suggest that therapy can be less aggressive. A similar spectrum with its own ambiguities in grading occurs in pineal parenchymal tumors, where pineocytes, rather than ganglion cells, are the terminal stage of differentiation.

Neurocytic neoplasms are, in part by definition, concentrated at the better-differentiated end of a spectrum. They are generally either well-differentiated or atypical, with the latter defined by increased MIB-1 rate, and perhaps by more mitoses, rather than by cytologic anaplasia. Occasional lesions appear to contain both classic neurocytoma and what are in essence neuroblasts, or cells that are intermediate between neuroblasts and neurocytes.

Other malignant neuronal tumors include rare *malignant dysembryoplastic neuroepithelial tumors* (146) and *giant cell glioblastomas* with scattered, and prognostically insignificant, synaptophysin-positive ganglion cells.

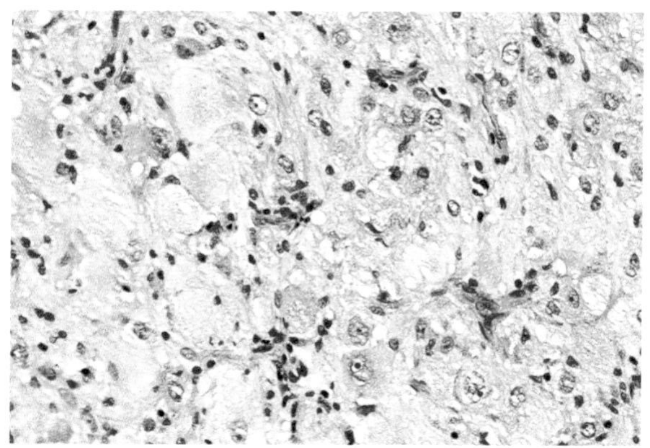

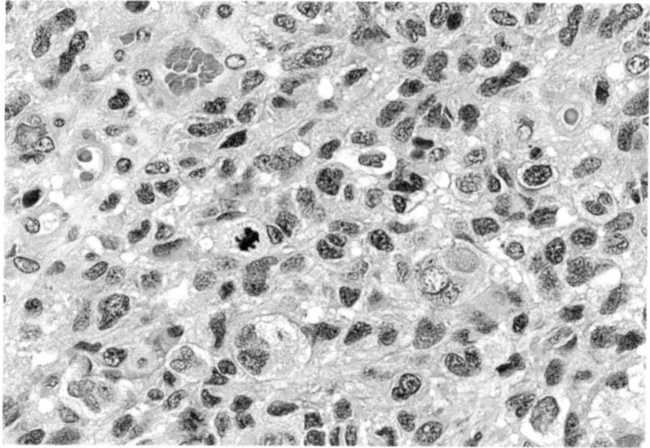

Figure 4-92

ANAPLASTIC GANGLIOGLIOMA

Only rarely do gangliogliomas undergo anaplastic transformation.

Left: The presumed precursor lesion in this case was a ganglion cell tumor with compacted, cytologically abnormal neurons.

Right: Anaplastic transformation added a population of small, mitotically active cells with the cytologic features of anaplastic glia.

REFERENCES

Gangliocytoma and Ganglioglioma

1. Becker AJ, Wiestler OD, Figarella-Branger D, Blumcke I. Ganglioglioma and gangliocytoma. In: Louis DN, Ohgaki H, Wiestler OD, Cavenee W, eds. WHO classification of tumours of the central nervous system. Lyon: IARC Press; 2007. (In press)
2. Liu GT, Galetta SL, Rorke LB, et al. Gangliogliomas involving the optic chiasm. Neurology 1996;46:1669–73.
3. Karamitopoulou E, Perentes E, Probst A, Wegmann W. Ganglioglioma of the brain stem: neurological dysfunction of 16-year duration. Clin Neuropathol 1995;14:162–8.
4. Handa H, Yamagami T, Furuta M. An adult patient with cerebellar ganglioglioma. J Neurooncol 1994;18:183–9.
5. Cho BK, Wang KC, Nam DH, et al. Pineal tumors: experience with 48 cases over 10 years. Childs Nerv Syst 1998;14:53–8.
6. Hamburger C, Buttner A, Weis S. Ganglioglioma of the spinal cord: report of two rare cases and review of the literature. Neurosurgery 1997;41: 1410–6.
7. Miller DJ, McCutcheon IE. Hemangioblastomas and other uncommon intramedullary tumors. J Neurooncol 2000;47:253–70.
8. Prayson RA, Khajavi K, Comair YG. Cortical architectural abnormalities and MIB1 immunoreactivity in gangliogliomas: a study of 60 patients with intracranial tumors. J Neuropathol Exp Neurol 1995;54:513–20.
9. Zentner J, Wolf HK, Ostertun B, et al. Gangliogliomas: clinical, radiological, and histopathological findings in 51 patients. J Neurol Neurosurg Psychiatry 1994;57:1497–502.
10. Castillo M, Davis PC, Takei Y, Hoffman JC Jr. Intracranial ganglioglioma: MR, CT, and clinical findings in 18 patients. AJNR Am J Neuroradiol 1990;11:109–14.
11. Diepholder HM, Schwechheimer K, Mohadjer M, Knoth R, Volk B. A clinicopathologic and immunomorphologic study of 13 cases of ganglioglioma. Cancer 1991;68:2192–201.
12. Tien RD, Tuori SL, Pulkingham N, Burger PC. Ganglioglioma with leptomeningeal and subarachnoid spread: results of CT, MR, and PET imaging. AJR Am J Roentgenol 1992;159:391–3.
13. Brat DJ, Gearing M, Goldthwaite PT, Wainer BH, Burger PC. Tau-associated neuropathology in ganglion cell tumours increases with patient age but appears unrelated to ApoE genotype. Neuropathol Appl Neurobiol 2001;27:197–205.
14. Suzuki H, Otsuki T, Iwasaki Y, et al. Anaplastic ganglioglioma with sarcomatous component: an immunohistochemical study and molecular analysis of p53 tumor suppressor gene. Neuropathology 2002;22:40–7.
15. Hunt SJ, Johnson PC. Melanotic ganglioglioma of the pineal region. Acta Neuropathol (Berl) 1989;79:222–5.

16. Soffer D, Lach B, Constantini S. Melanotic cerebral ganglioglioma: evidence for melanogenesis in neoplastic astrocytes. Acta Neuropathol (Berl) 1992;83:315–23.
17. Blumcke I, Giencke K, Wardelmann E, et al. The CD34 epitope is expressed in neoplastic and malformative lesions associated with chronic, focal epilepsies. Acta Neuropathol (Berl) 1999;97:481–90.
18. Blumcke I, Wiestler OD. Gangliogliomas: an intriguing tumor entity associated with focal epilepsies. J Neuropathol Exp Neurol 2002;61:575–84.
19. Miller DC, Koslow M, Budzilovich GN, Burstein DE. Synaptophysin: a sensitive and specific marker for ganglion cells in central nervous system neoplasms. Hum Pathol 1990;21:271–6.
20. Lindboe CF, Hansen HB. Epiperikaryal synaptophysin reactivity in the normal human central nervous system. Clin Neuropathol 1998;17:237–40.
21. Quinn B. Synaptophysin staining in normal brain: importance for diagnosis of ganglioglioma. Am J Surg Pathol 1998;22:550–6.
22. Zhang PJ, Rosenblum MK. Synaptophysin expression in the human spinal cord. Diagnostic implications of an immunohistochemical study. Am J Surg Pathol 1996;20:273–6.
23. Hirose T, Schneithauer BW, Lopes MB, Gerber HA, Altermatt HJ, VandenBerg SR. Ganglioglioma: an ultrastructural and immunohistochemical study. Cancer 1997;79:989–1003.
24. Wolf H, Buslei R, Schmidt-Kastner R, et al. NeuN: a useful neuronal marker for diagnostic histopathology. J Histochem Cytochem 1996;44:1167–71.
24a. Preusser M, Laggner U, Haberier C, Heinzl H, Budka H, Hainfellner JA. Comparateive analysis of NeuN immunoreactivity ikn primary brain tumors: conclusions for rational use in diagnostic histopathology. Histopathology 2006;48:438–44.
25. Giangaspero F, Burger PC, Budwit DA, Usellini L, Mancini AM. Regulatory peptides in neuronal neoplasms of the central nervous system. Clin Neuropathol 1985;4:111–5.
26. Takahashi H, Wakabayashi K, Kawai K, et al. Neuroendocrine markers in central nervous system neuronal tumors (gangliocytoma and ganglioglioma). Acta Neuropathol (Berl) 1989;77:237–43.
27. Raghavan R, White CL 3rd, Rogers B, Coimbra C, Rushing EJ. Alpha-synuclein expression in central nervous system tumors showing neuronal or mixed neuronal/glial differentiation. J Neuropathol Exp Neurol 2000;59:490–4.
28. Jaffey PB, Mundt AJ, Baunoch DA, et al. The clinical significance of extracellular matrix in gangliogliomas. J Neuropathol Exp Neurol 1996;55:1246–52.
29. von Deimling A, Fimmers R, Schmidt MC, et al. Comprehensive allelotype and genetic anaysis of 466 human nervous system tumors. J Neuropathol Exp Neurol 2000;59:544–8.
30. Khajavi K, Comair YG, Prayson RA, et al. Childhood ganglioglioma and medically intractable epilepsy. A clinicopathological study of 15 patients and a review of the literature. Pediatr Neurosurg 1995;22:181–8.

Desmoplastic Infantile Ganglioglioma

31. VandenBerg SR, May EE, Rubinstein LJ, et al. Desmoplastic supratentorial neuroepithelial tumors of infancy with divergent differentiation potential ("desmoplastic infantile gangliogliomas"). Report on 11 cases of a distinctive embryonal tumor with favorable prognosis. J Neurosurg 1987;66:58–71.
32. VandenBerg SR. Desmoplastic infantile ganglioglioma and desmoplastic cerebral astrocytoma of infancy. Brain Pathol 1993;3:275–81.
33. Mallucci C, Lellouch-Tubiana A, Salazar C, et al. The management of desmoplastic neuroepithelial tumours in childhood. Childs Nerv Syst 2000;16:8–14.
34. Duffner PK, Burger PC, Cohen ME, et al. Desmoplastic infantile gangliogliomas: an approach to therapy. Neurosurgery 1994;34:583–9.
35. Martin DS, Levy B, Awwad EE, Pittman T. Desmoplastic infantile ganglioglioma: CT and MR features. AJNR Am J Neuroradiol 1991;12:1195–7.
36. Komori T, Scheithauer BW, Parisi JE, Watterson J, Priest JR. Mixed conventional and desmoplastic infantile ganglioglioma: an autopsied case with 6-year follow-up. Mod Pathol 2001;14: 720–6.
37. Ng TH, Fung CF, Ma LT. The pathological spectrum of desmoplastic infantile gangliogliomas. Histopathology 1990;16:235–41.
38. Takeshima H, Kawahara Y, Hirano H, Obara S, Niiro M, Kuratsu J. Postoperative regression of desmoplastic infantile gangliogliomas: report of two cases. Neurosurgery 2003;53:979–83; discussion 983–4.
39. De Munnynck K, Van Gool S, Van Calenbergh F, et al. Desmoplastic infantile ganglioglioma: a potentially malignant tumor? Am J Surg Pathol 2002;26:1515–22.

Dysplastic Cerebellar Gangliocytoma (Lhermitte-Duclos Disease)

40. Abel TW, Baker SJ, Fraser MM, et al. Lhermitte-Duclos disease: a report of 31 cases with immunohistochemical analysis of the PTEN/AKT/mTOR pathway. J Neuropathol Exp Neurol 2005;64:341–9.
41. Milbouw G, Born JD, Martin D, et al. Clinical and radiological aspects of dysplastic gangliocytoma (Lhermitte-Duclos disease): a report of two cases with review of the literature. Neurosurgery 1988;22(Pt 1):124–8.

42. Robinson S, Cohen AR. Cowden disease and Lhermitte-Duclos disease: characterization of a new phakomatosis. Neurosurgery 2000;46:371–83.
43. Zhou XP, Marsh DJ, Morrison CD, et al. Germline inactivation of PTEN and dysregulation of the phosphoinositol-3-kinase/Akt pathway cause human Lhermitte-Duclos disease in adults. Am J Hum Genet 2003;73:1191–8.
44. Ashley DG, Zee CS, Chandrasoma PT, Segall HD. Lhermitte-Duclos disease: CT and MR findings. J Comput Assist Tomogr 1990;14:984–7.
45. Ambler M, Pogacar S, Sidman R. Lhermitte-Duclos disease (granule cell hypertrophy of the cerebellum) pathological analysis of the first familial cases. J Neuropathol Exp Neurol 1969; 28:622–47.
46. Yachnis AT, Trojanowski JQ, Memmo M, Schlaepfer WW. Expression of neurofilament proteins in the hypertrophic granule cells of Lhermitte-Duclos disease: an explanation for the mass effect and the myelination of parallel fibers in the disease state. J Neuropathol Exp Neurol 1988;47:206–16.
47. Pritchett PS, King TI. Dysplastic gangliocytoma of the cerebellum—an ultrastructural study. Acta Neuropathol (Berl) 1978;42:1–5.
48. Kwon CH, Zhu X, Zhang J, et al. Pten regulates neuronal soma size: a mouse model of Lhermitte-Duclos disease. Nat Genet 2001;29:404–11.
49. Williams DW 3rd, Elster AD, Ginsberg LE, Stanton C. Recurrent Lhermitte-Duclos disease: report of two cases and association with Cowden's disease. AJNR Am J Neuroradiol 1992;13:287–90.

Central Neurocytoma

50. Hassoun J, Gambarelli D, Grisoli F, et al. Central neurocytoma. An electron-microscopic study of two cases. Acta Neuropathol (Berl) 1982;56:151–6.
51. Smoker WR, Townsend JJ, Reichman MV. Neurocytoma accompanied by intraventricular hemorrhage: case report and literature review. AJNR Am J Neuroradiol 1991;12:765–70.
52. Rades D, Fehlauer F, Lamszus K, et al. Well-differentiated neurocytoma: what is the best available treatment? Neuro-oncol 2005;7:77–83.
53. Schild SE, Scheithauer BW, Haddock MG, et al. Central neurocytomas. Cancer 1997;79:790–5.
54. Chang KH, Han MH, Kim DG, et al. MR appearance of central neurocytoma. Acta Radiol 1993;34:520–6.
55. George DH, Scheithauer BW. Central liponeurocytoma. Am J Surg Pathol 2001;25:1551–5.
56. Ng TH, Wong AY, Boadle R, Compton JS. Pigmented central neurocytoma: case report and literature review. Am J Surg Pathol 1999;23: 1136–40.
57. Nishio S, Takeshita I, Fukui M. Primary cerebral ganglioneurocytoma in an adult. Cancer 1990;66:358–62.
58. Schweitzer JB, Davies KG. Differentiating central neurocytoma. Case report. J Neurosurg 1997;86:543–6.
59. Favereaux A, Vital A, Loiseau H, Dousset V, Caille J, Petry K. Histopathological variants of central neurocytoma: report of 10 cases. Ann Pathol 2000;20:558–63.
60. Soylemezoglu F, Scheithauer BW, Esteve J, Kleihues P. Atypical central neurocytoma. J Neuropathol Exp Neurol 1997;56:551–6.
61. Yasargil MG, von Ammon K, von Deimling A, Valavanis A, Wichmann W, Wiestler OD. Central neurocytoma: histopathological variants and therapeutic approaches. J Neurosurg 1992;76:32–7.
62. Mackenzie IR. Central neurocytoma: histologic atypia, proliferation potential, and clinical outcome. Cancer 1999;85:1606–10.
63. Hassoun J, Soylemezoglu F, Gambarelli D, Figarella-Branger D, von Ammon K, Kleihues P. Central neurocytoma: a synopsis of clinical and histological features. Brain Pathol 1993;3:297–306.
64. Hessler RB, Lopes MB, Frankfurter A, Reidy J, VandenBerg SR. Cytoskeletal immunohistochemistry of central neurocytomas. Am J Surg Pathol 1992;16:1031–8.
65. Ishiuchi S, Tamura M. Central neurocytoma: an immunohistochemical, ultrastructural and cell culture study. Acta Neuropathol (Berl) 1997;94:425–35.
66. Mena H, Morrison AL, Jones RV, Gyure KA. Central neurocytomas express photoreceptor differentiation. Cancer 2001;91:136–43.
67. von Deimling A, Janzer R, Kleihues P, Wiestler OD. Patterns of differentiation in central neurocytoma. An immunohistochemical study of eleven biopsies. Acta Neuropathol (Berl) 1990;79:473–9.
68. Westphal M, Stavrou D, Nausch H, Valdueza JM, Herrmann HD. Human neurocytoma cells in culture show characteristics of astroglial differentiation. J Neurosci Res 1994;38:698–704.
69. Ashkan K, Casey AT, D'Arrigo C, Harkness WF, Thomas DG. Benign central neurocytoma. Cancer 1 2000;89:1111–20.
70. Yin XL, Pang JC, Hui AB, Ng HK. Detection of chromosomal imbalances in central neurocytomas by using comparative genomic hybridization. J Neurosurg 2000;93:77–81.
71. Fujisawa H, Marukawa K, Hasegawa M, et al. Genetic differences between neurocytoma and dysembryoplastic neuroepithelial tumor and oligodendroglial tumors. J Neurosurg 2002;97: 1350–5.

72. Perry A, Scheithauer BW, Macaulay RJ, Raffel C, Roth KA, Kros JM. Oligodendrogliomas with neurocytic differentiation. A report of 4 cases with diagnostic and histogenetic implications. J Neuropathol Exp Neurol 2002;61:947–55.
73. Kulkarni V, Rajshekhar V, Haran RP, Chandi SM. Long-term outcome in patients with central neurocytoma following stereotactic biopsy and radiation therapy. Br J Neurosurg 2002;16:126–32.
74. Schild SE, Scheithauer BW, Haddock MG, et al. Central neurocytomas. Cancer 1997;79:790–5.
75. Eng DY, DeMonte F, Ginsberg L, Fuller GN, Jaeckle K. Craniospinal dissemination of central neurocytoma. Report of two cases. J Neurosurg 1997;86:547–52.
76. Tomura N, Hirano H, Watanabe O, et al. Central neurocytoma with clinically malignant behavior. AJNR Am J Neuroradiol 1997;18:1175–8.
77. Christov C, Adle-Biassette H, Le Guerinel C. Recurrent central neurocytoma with marked increase in MIB-1 labelling index. Br J Neurosurg 1999;13:496–9.

Other Neurocytic Neoplasms of the Cerebral Hemispheres (Extraventricular and Intraparenchymal Neurocytic Neoplasms)

78. Brat DJ, Scheithauer BW, Eberhart CG, Burger PC. Extraventricular neurocytomas: pathologic features and clinical outcome. Am J Surg Pathol 2001;25:1252–60.
79. Brown DM, Karlovits S, Lee LH, Kim K, Rothfus WE, Brown HG. Management of neurocytomas: case report and review of the literature. Am J Clin Oncol 2001;24:272–8.
80. Giangaspero F, Cenacchi G, Losi L, Cerasoli S, Bisceglia M, Burger PC. Extraventricular neoplasms with neurocytoma features. A clinicopathological study of 11 cases. Am J Surg Pathol 1997;21:206–12.
81. Moller-Hartmann W, Krings T, Brunn A, Korinth M, Thron A. Proton magnetic resonance spectroscopy of neurocytoma outside the ventricular region—case report and review of the literature. Neuroradiology 2002;44:230–4.
82. Tortori-Donati P, Fondelli MP, Rossi A, Cama A, Brisigotti M, Pellicano G. Extraventricular neurocytoma with ganglionic differentiation associated with complex partial seizures. AJNR Am J Neuroradiol 1999;20:724–7.
83. Hsu PW, Hsieh TC, Chang CN, Lin TK. Fourth ventricle central neurocytoma: case report. Neurosurgery 2002;50:1365–7.
84. Warmuth-Metz M, Klein R, Sorensen N, Solymosi L. Central neurocytoma of the fourth ventricle. Case report. J Neurosurg 1999;91:506–9.
85. Perry A, Fuller CE, Banerjee R, Brat DJ, Scheithauer BW. Ancillary FISH analysis for 1p and 19q status: preliminary observations in 287 gliomas and oligodendroglioma mimics. Front Biosci 2003;8:a1–9.
86. Perry A, Scheithauer BW, Macaulay RJ, Raffel C, Roth KA, Kros JM. Oligodendrogliomas with neurocytic differentiation. A report of 4 cases with diagnostic and histogenetic implications. J Neuropathol Exp Neurol 2002;61:947–55.

Cerebellar Neurocytic Neoplasms

87. Chimelli L, Hahn MD, Budka H. Lipomatous differentiation in a medulloblastoma. Acta Neuropathol (Berl) 1991;81:471–3.
88. Soylemezoglu F, Soffer D, Onol B, Schwechheimer K, Kleihues P. Lipomatous medulloblastoma in adults. A distinct clinicopathological entity. Am J Surg Pathol 1996;20:413–8.
89. Enam SA, Rosenblum ML, Ho KL. Neurocytoma in the cerebellum. Case report. J Neurosurg 1997;87:100–2.
90. Alkadhi H, Keller M, Brandner S, Yonekawa Y, Kollias SS. Neuroimaging of cerebellar liponeurocytoma. Case report. J Neurosurg 2001; 95:324–31.
91. Jackson TR, Regine WF, Wilson D, Davis DG. Cerebellar liponeurocytoma. Case report and review of the literature. J Neurosurg 2001;95:700–3.
92. Taddei GL, Buccoliero AM, Caldarella A, et al. Cerebellar liponeurocytoma: immunohistochemical and ultrastructural study of a case. Ultrastruct Pathol 2001;25:59–63.
93. Giangaspero F, Cenacchi G, Roncaroli F, et al. Medullocytoma (lipidized medulloblastoma). A cerebellar neoplasm of adults with favorable prognosis. Am J Surg Pathol 1996;20:656–64.
94. Ellison DW, Zygmunt SC, Weller RO. Neurocytoma/lipoma (neurolipocytoma) of the cerebellum. Neuropathol Appl Neurobiol 1993;19:95–8.
95. Alleyne CH Jr, Hunter S, Olson JJ, Barrow DL. Lipomatous glioneurocytoma of the posterior fossa with divergent differentiation: case report. Neurosurgery 1998;42:639–43.
96. Gonzalez-Campora R, Weller RO. Lipidized mature neuroectodermal tumour of the cerebellum with myoid differentiation. Neuropathol Appl Neurobiol 1998;24:397–402.
97. Pal L, Santosh V, Gayathri N, et al. Neurocytoma/rhabdomyoma (myoneurocytoma) of the cerebellum. Acta Neuropathol (Berl) 1998;95:318–23.
98. Horstmann S, Perry A, Reifenberger G, et al. Genetic and expression profiles of cerebellar liponeurocytomas. Brain Pathol 2004;14:281–9.
99. Jenkinson MD, Bosma JJ, Du Plessis D, et al. Cerebellar liponeurocytoma with an unusually aggressive clinical course: case report. Neurosurgery 2003;53:1425–8.

Neurocytic Neoplasms of the Spinal Cord

100. Martin AJ, Sharr MM, Teddy PJ, Gardner BP, Robinson SF. Neurocytoma of the thoracic spinal cord. Acta Neurochir (Wien) 2002;144:823–8.
101. Stapleton SR, David KM, Harkness WF, Harding BN. Central neurocytoma of the cervical spinal cord. J Neurol Neurosurg Psychiatry 1997; 63:119.
102. Stephan CL, Kepes JJ, Arnold P, Green KD, Chamberlin F. Neurocytoma of the cauda equina. Case report. J Neurosurg 1999;90(Suppl 2):247–51.

Dysembryoplastic Neuroepithelial Tumor

103. Hirose T, Scheithauer BW, Lopes MB, VandenBerg SR. Dysembryoplastic neuroepithelial tumor (DNT): an immunohistochemical and ultrastructural study. J Neuropathol Exp Neurol 1994;53:184–95.
104. Wolf HK, Buslei R, Blumcke I, Wiestler OD, Pietsch T. Neural antigens in oligodendrogliomas and dysembryoplastic neuroepithelial tumors. Acta Neuropathol (Berl) 1997;94:436–43.
105. Gyure KA, Sandberg GD, Prayson RA, Morrison AL, Armstrong RC, Wong K. Dysembryoplastic neuroepithelial tumor: an immunohistochemical study with myelin oligodendrocyte glycoprotein. Arch Pathol Lab Med 2000;124:123–6.
106. Wong K, Gyure K, Prayson R, Morrison A, Le T, Armstrong R. Dysembryoplastic neuroepithelial tumor: in situ hybridization of proteolipid protein (plp) messenger ribonucleic acid (mrna). J Neuropathol Exp Neurol 1999;58:542.
107. Cervera-Pierot P, Varlet P, Chodkiewicz JP, Daumas-Duport C. Dysembryoplastic neuroepithelial tumors located in the caudate nucleus area: report of four cases. Neurosurgery 1997;40:1065–70.
108. Baisden BL, Brat DJ, Melhem ER, Rosenblum MK, King AP, Burger PC. Dysembryoplastic neuroepithelial tumor-like neoplasm of the septum pellucidum: a lesion often misdiagnosed as glioma: report of 10 cases. Am J Surg Pathol 2001;25:494–9.
109. Whittle IR, Dow GR, Lammie GA, Wardlaw J. Dsyembryoplastic neuroepithelial tumour with discrete bilateral multifocality: further evidence for a germinal origin. Br J Neurosurg 1999;13: 508–11.
110. Fujimoto K, Ohnishi H, Tsujimoto M, Hoshida T, Nakazato Y. Dysembryoplastic neuroepithelial tumor of the cerebellum and brainstem. Case report. J Neurosurg 2000;93:487–9.
111. Yasha TC, Mohanty A, Radhesh S, Santosh V, Das S, Shankar SK. Infratentorial dysembryoplastic neuroepithelial tumor (DNT) associated with Arnold-Chiari malformation. Clin Neuropathol 1998;17:305–10.
112. Kuchelmeister K, Demirel T, Schlorer E, Bergmann M, Gullotta F. Dysembryoplastic neuroepithelial tumour of the cerebellum. Acta Neuropathol (Berl) 1995;89:385–90.
113. Daumas-Duport C, Scheithauer BW, Chodkiewicz JP, Laws ER Jr, Vedrenne C. Dysembryoplastic neuroepithelial tumor: a surgically curable tumor of young patients with intractable partial seizures. Report of thirty-nine cases. Neurosurgery 1988;23:545–6.
114. Daumas-Duport C. Dysembryoplastic neuroepithelial tumours. Brain Pathol 1993;3:283–95.
115. Taratuto AL, Pomata H, Sevlever G, Gallo G, Monges J. Dysembryoplastic neuroepithelial tumor: morphological, immunocytochemical, and deoxyribonucleic acid analyses in a pediatric series. Neurosurgery 1995;36:474–81.
116. Kuroiwa T, Bergey GK, Rothman MI, et al. Radiologic appearance of the dysembryoplastic neuroepithelial tumor. Radiology 1995;197: 233–8.
117. Ostertun B, Wolf HK, Campos MG, et al. Dysembryoplastic neuroepithelial tumors: MR and CT evaluation. AJNR Am J Neuroradiol 1996;17:419–30.
118. Honavar M, Janota I, Polkey CE. Histological heterogeneity of dysembryoplastic neuroepithelial tumour: identification and differential diagnosis in a series of 74 cases. Histopathology 1999;34:342–56.
119. Elizabeth J, Bhaskara RM, Radhakrishnan VV, Radhakrishnan K, Thomas SV. Melanotic differentiation in dysembryoplastic neuroepithelial tumor. Clin Neuropathol 2000;19:38–40.
120. Daumas-Duport C, Varlet P, Bacha S, Beuvon F, Cervera-Pierot P, Chodkiewicz JP. Dysembryoplastic neuroepithelial tumors: nonspecific histological forms—a study of 40 cases. J Neurooncol 1999;41:267–80.
121. Bleggi-Torres LF, Netto MR, Gasparetto EL, Goncalves ES, Moro M. Dysembryoplastic neuroepithelial tumor: cytological diagnosis by intraoperative smear preparation. Diagn Cytopathol 2002;26:92–4.
122. Fujisawa H, Marukawa K, Hasegawa M, et al. Genetic differences between neurocytoma and dysembryoplastic neuroepithelial tumor and oligodendroglial tumors. J Neurosurg 2002; 97:1350–5.
123. Johnson MD, Vnencak-Jones CL, Toms SA, Moots PM, Weil R. Allelic losses in oligodendroglial and oligodendroglioma-like neoplasms: analysis using microsatellite repeats and polymerase chain reaction. Arch Pathol Lab Med 2003;127:1573–9.

124. Perry A, Fuller CE, Banerjee R, Brat DJ, Scheithauer BW. Ancillary FISH analysis for 1p and 19q status: preliminary observations in 287 gliomas and oligodendroglioma mimics. Front Biosci 2003;8:a1–9.
125. Prayson RA, Castilla EA, Hartke M, Pettay J, Tubbs RR, Barnett GH. Chromosome 1p allelic loss by fluorescence in situ hybridization is not observed in dysembryoplastic neuroepithelial tumors. Am J Clin Pathol 2002;118:512–7.
126. Hirose T, Scheithauer BW. Mixed dysembryoplastic neuroepithelial tumor and ganglioglioma. Acta Neuropathol (Berl) 1998;95:649–54.
127. Prayson RA. Composite ganglioglioma and dysembryoplastic neuroepithelial tumor. Arch Pathol Lab Med 1999;123:247–50.
128. Prayson RA, Morris HH, Estes ML, Comair YG. Dysembryoplastic neuroepithelial tumor: a clinicopathologic and immunohistochemical study of 11 tumors including MIB1 immunoreactivity. Clin Neuropathol 1996;15:47–53.
129. Hammond RR, Duggal N, Woulfe JM, Girvin JP. Malignant transformation of a dysembryoplastic neuroepithelial tumor. Case report. J Neurosurg 2000;92:722–5.

Papillary Glioneuronal Tumor

130. Kim DH, Suh YL. Pseudopapillary neurocytoma of temporal lobe with glial differentiation. Acta Neuropathol (Berl) 1997;94:187–91.
131. Komori T, Scheithauer BW, Anthony DC, et al. Papillary glioneuronal tumor: a new variant of mixed neuronal-glial neoplasm. Am J Surg Pathol 1998;22:1171–83.
132. Bouvier-Labit C, Daniel L, Dufour H, Grisoli F, Figarella–Branger D. Papillary glioneuronal tumour: clinicopathological and biochemical study of one case with 7-year follow up. Acta Neuropathol (Berl) 2000;99:321–6.

Glioneuronal Tumor with Neuropil-Like ("Rosetted") Islands

133. Teo JG, Gultekin SH, Bilsky M, Gutin P, Rosenblum MK. A distinctive glioneuronal tumor of the adult cerebrum with neuropil-like (including "rosetted") islands: report of 4 cases. Am J Surg Pathol 1999;23:502–10.
134. Keyvani K, Rickert CH, von Wild K, Paulus W. Rosetted glioneuronal tumor: a case with proliferating neuronal nodules. Acta Neuropathol (Berl) 2001;101(5):525–8.
135. Harris BT, Horoupian DS. Spinal cord glioneuronal tumor with "rosetted" neuropil islands and meningeal dissemination: a case report. Acta Neuropathol (Berl) 2000;100:575–9.

Rosette-Forming Glioneuronal Tumor of the Fourth Ventricle

136. Komori T, Scheithauer BW, Hirose T. A rosette-forming glioneuronal tumor of the fourth ventricle: infratentorial form of dysembryoplastic neuroepithelial tumor? Am J Surg Pathol 2002; 26:582–91.

Malignant Neuronal and Glioneuronal Tumors

137. Hayashi Y, Iwato M, Hasegawa M, Tachibana O, von Deimling A, Yamashita J. Malignant transformation of a gangliocytoma/ganglioglioma into a glioblastoma multiforme: a molecular genetic analysis. Case report. J Neurosurg 2001;95:138–42.
138. Hirose T, Kannuki S, Nishida K, Matsumoto K, Sano T, Hizawa K. Anaplastic ganglioglioma of the brain stem demonstrating active neurosecretory features of neoplastic neuronal cells. Acta Neuropathol (Berl) 1992;83:365–70.
139. Jay V, Squire J, Becker LE, Humphreys R. Malignant transformation in a ganglioglioma with anaplastic neuronal and astrocytic components. Report of a case with flow cytometric and cytogenetic analysis. Cancer 1994;73: 2862–8.
140. Sasaki A, Hirato J, Nakazato Y, Tamura M, Kadowaki H. Recurrent anaplastic ganglioglioma: pathological characterization of tumor cells. Case report. J Neurosurg 1996;84:1055–9.
141. Suzuki H, Otsuki T, Iwasaki Y, et al. Anaplastic ganglioglioma with sarcomatous component: an immunohistochemical study and molecular analysis of p53 tumor suppressor gene. Neuropathology 2002;22:40–7.
142. Tihan T, Brat D, Goldthwaite P, Burger P. Glioneuronal tumors with malignant histological features. J Neuropathol Exp Neurol. 1999; 58:509.
143. Di Patre PL, Payer M, Brunea M, Delavelle J, De Tribolet N, Pizzolato G. Malignant transformation of a spinal cord ganglioglioma—case report and review of the literature. Clin Neuropathol 2004;23:298–303.
144. Kim NR, Wang KC, Bang JS, et al. Glioblastomatous transformation of ganglioglioma: case report with reference to molecular genetic and flow cytometric analysis. Pathol Int 2003;53: 874–82.
145. McLendon RE, Bentley RC, Parisi JE, et al. Malignant supratentorial glial-neuronal neoplasms: report of two cases and review of the literature. Arch Pathol Lab Med 1997;121:485–92.
146. Hammond RR, Duggal N, Woulfe JM, Girvin JP. Malignant transformation of a dysembryoplastic neuroepithelial tumor. Case report. J Neurosurg 2000;92:722–5.

5 EMBRYONAL TUMORS

INTRODUCTION

The term *embryonal tumor* in this Fascicle is used primarily to denote small blue cell tumors identified by the suffix "-blastoma." Some are undifferentiated, whereas others exhibit cytoarchitectural or immunohistochemical features that place them into one of several more specific categories. Thus, a neoplasm is a medulloepithelioma if medullary-type epithelium is prominent; a neuroblastoma if clearly defined neuroblastic features such as Homer-Wright rosettes are present; and an ependymoblastoma if ependymoblastic rosettes prevail. Although medulloblastomas are recognized in part by location, many have a distinctive nodular pattern of growth rarely seen in supratentorial neoplasms of embryonal type. Given the considerable intertumoral and intratumoral heterogeneity exhibited by embryonal neoplasms arising in an organ in which diverse cell types are derived from common progenitors, it is no surprise that diagnostically challenging mixed and hybrid lesions are common.

The present classification system for embryonal tumors is a useful nosologic framework, since it is based upon histology, immunohistochemistry, and ultrastructure—modalities generally available to diagnostic pathologists. While comparison of the features of these neoplasms to embryonic stages of development formed the original basis for classification of embryonal central nervous system (CNS) neoplasms, and remains an issue germane to their histogenesis, we are not committed to the hypothesis that embryonal neoplasms faithfully recreate stages of embryogenesis.

The term primitive neuroectodermal tumor (PNET) was first applied to a cellular form of embryonal CNS tumor showing differentiation along neuronal, glial, and even mesenchymal lines, as inferred from histologic features in hematoxylin and eosin (H&E)-stained sections alone (1). In retrospect, the lesions may have contained more than one clinicopathologic entity. The term was reintroduced later as the core of histogenetic schemes that asserted that most densely cellular embryonal tumors of the CNS have a common origin from primitive undifferentiated cells, and differ in little but location and type and degree of differentiation (2,3).

Reservations regarding this latter approach appeared quickly, suggesting that embryonal tumors can arise from cells already committed to a pathway of differentiation, not just from undifferentiated cells (4,5). Then, and now, the PNET system was viewed by some as an oversimplification. Even the best-defined member, medulloblastoma, arises along several genetic pathways, and in some cases presumably from a cell type that does not exist elsewhere in the brain, the external granule cell. Molecular genetic features of supratentorial PNET also differ from those of medulloblastoma (6,7) and outcomes appear to be inferior for patients with supratentorial PNET than for those with medulloblastoma (8,9). The current World Health Organization (WHO) classification considers PNET as a restricted group of supratentorial lesions, which includes cerebral neuroblastoma (10).

The general term PNET persists, however, and is often used to include any of the lesions in this chapter with the implicit assumption that the affected patients deserve full neuraxis radiation, patient age permitting. This need for radiation has evolved largely by inference from the behavior of medulloblastoma, a lesion known to disseminate widely in some cases, but has not been clearly established for supratentorial lesions in all their variations (11,12). The distinction between PNET and small cell malignant glioma thus becomes critical in radiation planning, since in glioma, the treated volume is generally confined to the lesion and immediate environs.

The rare peripheral PNET that arises in the CNS is discussed later in this chapter. Most CNS PNETs are unrelated to this entity: they lack the EWS-FLI-1 fusion protein, and do not stain with CD99.

In addition to the classic small cell embryonal tumors, or PNETs, we have included two additional lesions in this chapter. One, atypical teratoid/rhabdoid tumor, is a clearly defined entity; the other, polar spongioblastoma, is not. Atypical teratoid/rhabdoid tumor is included because of the age of affected patients, the malignant behavior, and the small cell component in some cases. The cytologic origin, however, is obscure, and the tumors are not, overall, small blue cell tumors. We have included the disputed entity polar spongioblastoma on even less secure grounds. Pineoblastomas are discussed separately (in chapter 6).

MEDULLOBLASTOMA

Definition. *Medulloblastoma* is a small cell, neuroectodermal tumor of the cerebellum.

General Features. While the histogenesis of medulloblastoma continues to be debated, it is difficult to ignore a possible origin of some tumors from proliferating precursor cells of the external granule cell layer that persist until the end of the first year of life. Neoplasms arise superficially from the external granule cell layer in patched knockout mice (13,14). Gene expression profiles are also consistent with such an origin for desmoplastic/nodular medulloblastomas (15).

Another possible source of origin is from the calbindin-positive cell line that supplies both the stellate and basket cells of the internal granular cell layer, as well as the Purkinje cells (16,17). As opposed to progenitors of the external granular cells that fan out over the cerebellar surface, these cells populate the cerebellum from within by outward or radial migration from the subependymal germinal matrix.

Lastly, minute dysplasias of the cerebellar vermis are yet another possible source of medulloblastoma (18). More than one source is possible.

Clinical Features. Although medulloblastomas generally affect patients in the first two decades of life (19,20), the incidence curve extends late into life (21,22). In children, they are more often median (vermian), whereas in adults, they are more often lateral (hemispheric) (23). Symptoms include effects of cerebellar dysfunction and increased intracranial pressure, and, in some cases, deficits referable to craniospinal dissemination.

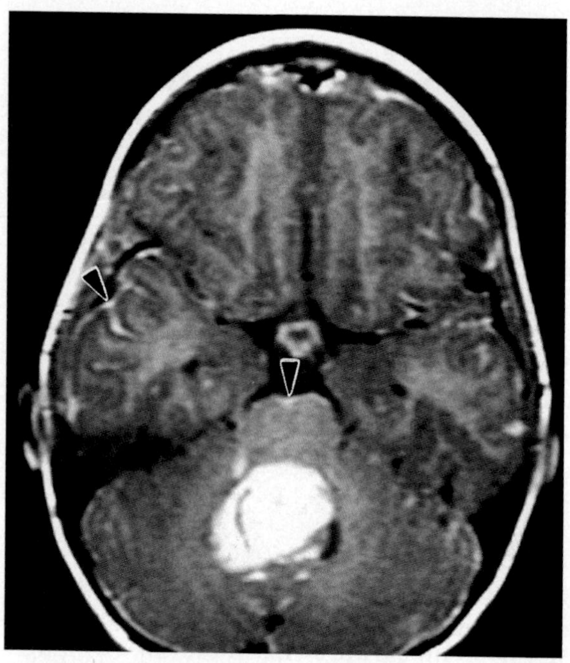

Figure 5-1

MEDULLOBLASTOMA

As in this magnetic resonance image (MRI) of a vermian lesion, medulloblastomas are contrast enhancing, often heterogeneously. Diffuse enhancement in the subarachnoid space (arrowheads) reflects the leptomeningeal dissemination present at diagnosis in this patient. (Fig. 5-14 from Fascicle 10, 3rd Series.)

Most medulloblastomas are sporadic, but rare examples are heritable, occurring in patients with Gorlin's syndrome (mutations in *PTCH*) (24,25), Turcot's syndrome (mutations in *APC*) (25,26), Li-Fraumeni syndrome (mutations in *p53*) (25,27), Coffin-Siris syndrome (28), Rubinstein-Taybi syndrome (29), and in families without a defined genetic abnormality (30). As is discussed in the section Molecular and Cytogenetic Findings, familiarity with the genetic bases of these syndromes has been instrumental in determining molecular changes in sporadic medulloblastomas (19,25).

Radiologic Findings. By magnetic resonance imaging (MRI), medulloblastomas are variably contrast enhancing (fig. 5-1). Leptomeningeal spread can be present at the time of initial presentation. Calcification is uncommon, whereas it is frequent in another posterior fossa tumor in children, ependymoma. In contrast to cerebellar pilocytic astrocytoma, cysts are rare. Large areas of hemorrhage and necrosis are less

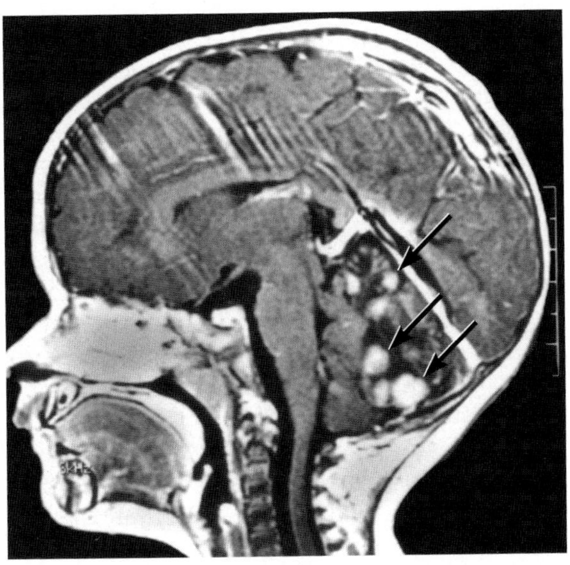

Figure 5-2

MEDULLOBLASTOMA WITH EXTENSIVE NODULARITY

Multiple, discrete, contrast-enhancing nodules (arrows) create a botryoid mass in this rare medulloblastoma variant that usually appears in the first 2 years of life.

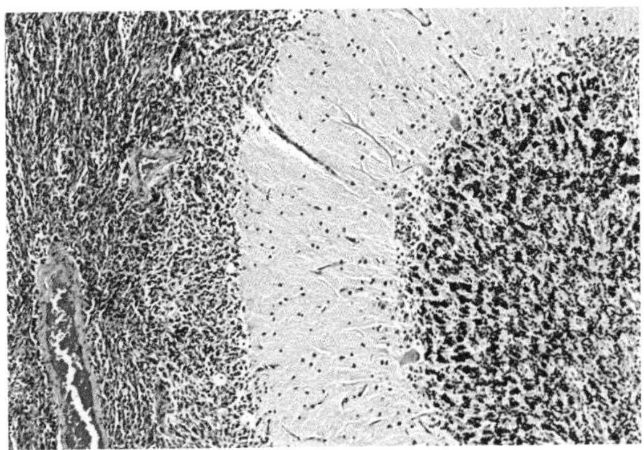

Figure 5-3

MEDULLOBLASTOMA

Medulloblastomas often fill the regional subarachnoid space.

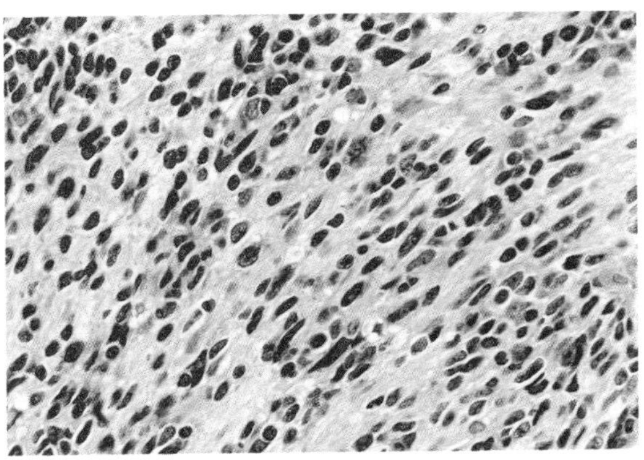

Figure 5-4

MEDULLOBLASTOMA

In zones of infiltration, medulloblastomas may closely resemble infiltrating astrocytomas.

likely to be seen than in atypical teratoid/rhabdoid tumor. A rare variant, "medulloblastoma with extensive nodularity," has a unique multinodular appearance (fig. 5-2) (31).

Gross Findings. The lesions vary in texture; the desmoplastic variant is firm and well circumscribed. Medulloblastomas can exit the cerebellum and, by lateral spread within the subarachnoid space, literally "ice" the pial surface. This is particularly likely in desmoplastic/nodular lesions.

Microscopic Findings. Extension into the subarachnoid space is common, particularly in nodular/desmoplastic medulloblastomas (fig. 5-3). Having reached this compartment, the tumor can reenter the brain along perivascular (Virchow-Robin) spaces or by directly penetrating the parenchyma. Although in large part architecturally solid, the periphery of medulloblastomas, particularly high-grade lesions, may permeate brain parenchyma (fig. 5-4). This infiltration results in a distinctive pattern of palisading within the molecular layer, similar to that seen when stacked Homer-Wright rosettes are cut along their long axes (fig. 5-5). Appendix J summarizes CNS tumors with riboning or palisading.

Only rarely do medulloblastomas undergo extensive or geographic necrosis, and they are among the few tumors other than high-grade gliomas to occasionally exhibit perinecrotic pseudopalisading. Cell death is more often achieved by apoptosis, however, particularly in high-grade lesions. Collagenous stroma is absent in most tumors, but is occasionally conspicuous. Calcification is also uncommon and usually confined to necrotic/apoptotic areas.

While medulloblastomas are densely cellular and "small cell" in appearance, there is considerable intertumoral and intratumoral variation in cytologic and histologic features. Some

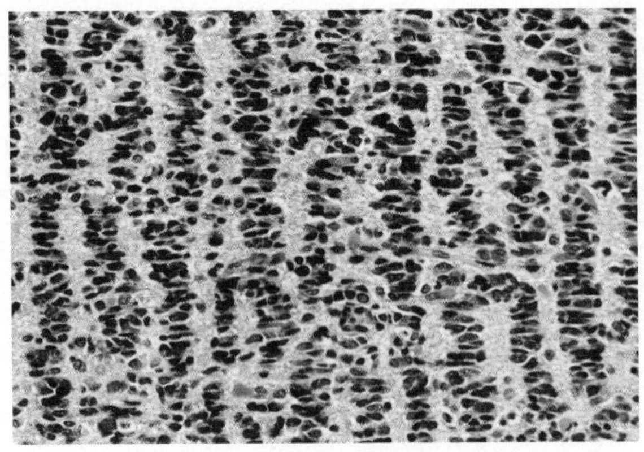

Figure 5-5

MEDULLOBLASTOMA

Columns of neoplastic cells are common, particularly in the molecular layer, where they are aligned perpendicular to the pial surface.

tumors are cytologically bland and mitosis poor, whereas others, either focally or globally, are malignant in the extreme. Such variations in tumor grade are discussed in the section, Anaplastic and Large Cell Medulloblastomas.

Histologic, Cytologic, and Ultrastructural Subtypes. Medulloblastomas exhibit considerable variation in histologic and cytologic features, immunohistochemical expression of neuronal and glial differentiation, and ultrastructural features. The incompletely understood associations between these properties and molecular/cytogenetic features are discussed below. Although medulloblastomas are often divided into nodular/desmoplastic and non-nodular or "classic" types, the tumors are more heterogeneous than suggested by this dichotomous division. The following six categories, admittedly overlapping, are descriptive: 1) classic, i.e., light microscopically undifferentiated, 2) with neuroblastic or neuronal differentiation, 3) with glial and "mixed" or glioneuronal differentiation, 4) with nodular/desmoplastic features, 5) with extensive nodularity, and 6) with anaplastic and/or large cell features.

Rare small blue cell cerebellar neoplasms that differentiate divergently toward muscle melanotic elements, i.e., medullomyoblastoma and melanotic medulloblastoma, are discussed separately.

Classic, or Undifferentiated, Medulloblastoma. The modifier "classic," or "undifferentiated," refers here to the light microscopic appearance of a lesion on routine stains, recognizing that almost all medulloblastomas have a neuronal or neuroblastic phenotype at the immunohistochemical or ultrastructural level. The extent of tissue sampling often determines the proportion of tumors assigned to this default category. Although there are many exceptions, medulloblastomas in children are more often classic, i.e., non-nodular/desmoplastic (23). As with nodular/desmoplastic medulloblastomas, the undifferentiated type varies considerably in regard to the degree of cytologic atypia and mitotic activity (fig. 5-6).

Medulloblastomas with Neuronal/Neuroblastic Differentiation. On the basis of histologic, immunohistochemical, or ultrastructural features, medulloblastomas are basically neuronal/neuroblastic tumors (17). This differentiation is expressed histologically in the form of: 1) Homer-Wright rosettes, 2) mature ganglion cells with or without accompanying ganglioid cells, 3) medium-sized neoplastic cells with neuronal qualities that are more immunohistochemical than histologic, and 4) large finely fibrillar areas (nodules) of synaptophysin-positive neuropil.

Medulloblastomas with neuroblastic (Homer-Wright) rosettes are in every way classic neuroblastic neoplasms (figs. 5-6, right; 5-7). As seen in cross section, the rosettes have a central fibrillar zone composed of delicate cytoplasmic processes emanating from surrounding cells. As might be expected, these cores and the surrounding cells are immunoreactive for synaptophysin and tubulin. Ultrastructurally, the processes represent microtubule- and, to a lesser extent, neurofilament-containing neurites (32). Medulloblastomas with Homer-Wright rosettes are more often of the classic than nodular/desmoplastic type.

Ganglion cells, the ultimate, but uncommon, expression of neuronal differentiation, are eye-catching elements that often lie clustered in small patches of fibrillar neuropil (fig. 5-8). By way of contrast, ganglioid cells are less mature, having smaller only somewhat vesicular nuclei, less conspicuous nucleoli, and no perceptible Nissl substance. Ganglioid cells may be seen, rarely, within the "pale islands" of the desmoplastic/nodular lesions discussed below, as well as in medulloblastomas exhibiting more fully differentiated ganglion cells.

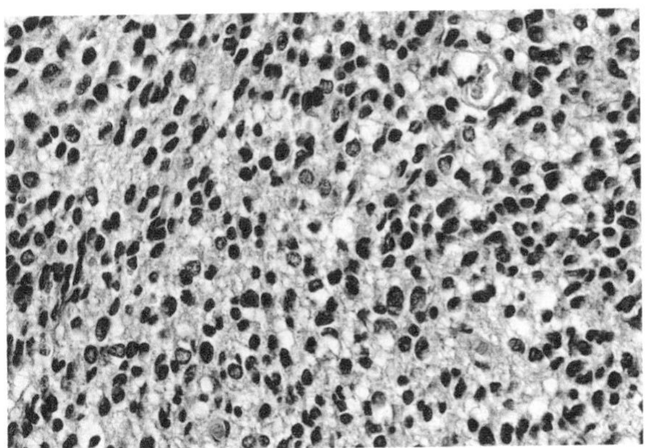

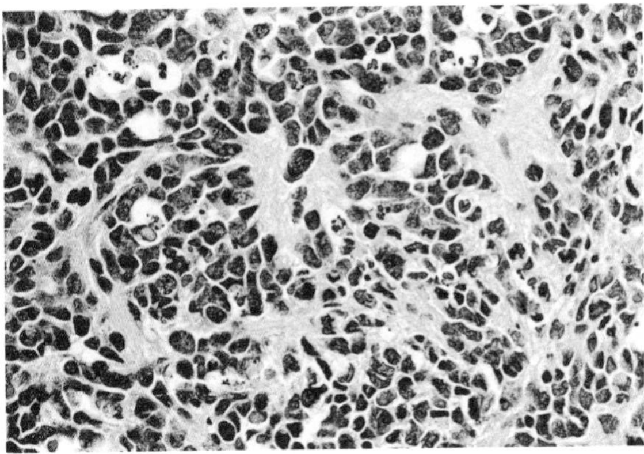

Figure 5-6

MEDULLOBLASTOMA

Medulloblastomas have a wide degree of anaplasia, as illustrated in two areas from the same lesion at the same magnification. Left: While densely cellular, the tumor has only minimal variation in nuclear size and shape in some areas.

Right: Elsewhere, the cells are substantially larger and cytologically more malignant. As is typical, apoptosis and nuclear molding are prominent in the more anaplastic tissue.

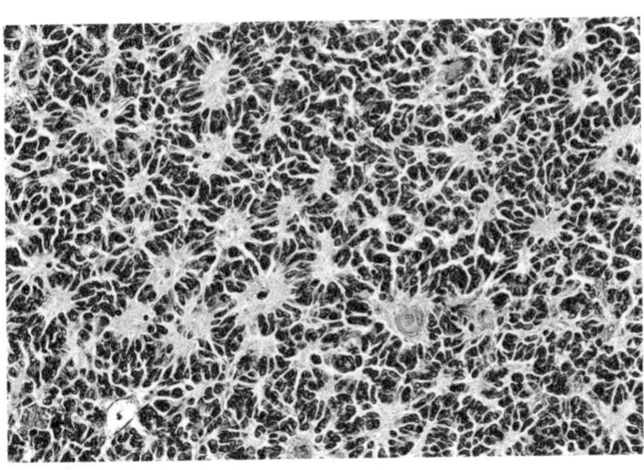

Figure 5-7

MEDULLOBLASTOMA

Homer-Wright rosettes are incontrovertible evidence of neuroblastic differentiation.

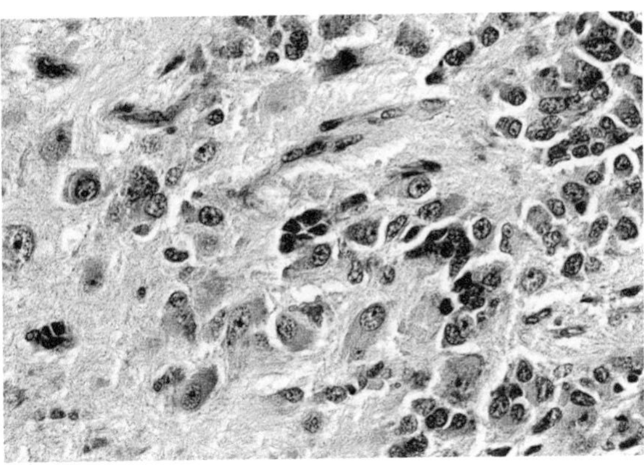

Figure 5-8

MEDULLOBLASTOMA

Differentiation to ganglion cells is uncommon in medulloblastomas. The continuum from small undifferentiated neuroblasts, through intermediate gangliod cells, to large ganglion cells establishes the neoplastic nature of the more mature elements.

In the absence of transitional cells, mature neoplastic neurons must be distinguished from overrun neurons of the Purkinje cell layer or dentate nuclei. Normal ganglion cells are uniform in cytologic appearance, disposed in linear or arcuate arrangements, and often more numerous than their neoplastic counterparts.

A form of neuroblastic differentiation in some medulloblastomas is seen in lobules or small clusters of cells with nuclei that are larger, less dense, and therefore less "primitive" than those of neuroblasts, yet smaller and less phenotypically neuronal than those of ganglioid cells. They too are immunoreactive for synaptophysin.

Particularly in tumors with prominent neuroblastic rosettes, it is common to find large, irregular fibrillar zones composed of tumor cell processes that are, in a sense, a neoplastic counterpart of normal gray matter, or neuropil. Perivascular fibrillar zones are common in such lesions and can closely resemble ependymal perivascular pseudorosettes (fig. 5-9), albeit ones

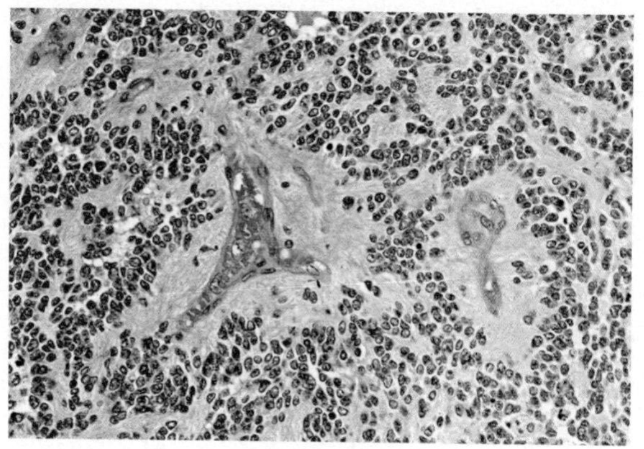

Figure 5-9

MEDULLOBLASTOMA

Perivascular zones of fine fibrillarity can resemble the perivascular pseudorosettes of ependymoma.

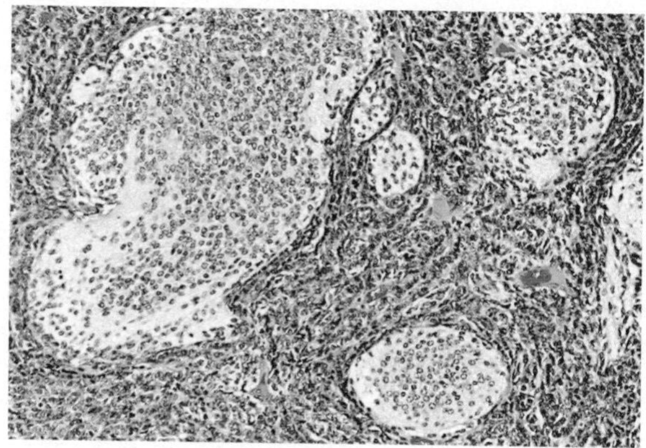

Figure 5-10

NODULAR/DESMOPLASTIC MEDULLOBLASTOMA

Resembling regenerating nodules in a liver are "pale islands" of reduced nuclear density. Neuritic outgrowth of tumor cells is responsible for the fine intranodular fibrillarity that is often most prominent at the edges of the nodules.

positive for synaptophysin rather than glial fibrillary acidic protein (GFAP).

Medulloblastomas with Glial Differentiation or Mixed (Glioneuronal) Differentiation. In most instances, it is difficult to establish the presence of glial differentiation on the basis of H&E-stained sections alone. Although there are occasional cases where it seems obvious because of the presence of cells with copious eosinophilic cytoplasm and coarse processes, medulloblastoma with glial differentiation is generally not a histology-based entity. Immunohistochemistry is thus almost always required, and even then it can be difficult to determine whether immunoreactive cells are neoplastic or reactive.

Due to the lack of generally agreed upon diagnostic criteria, the reported incidence of glial differentiation in medulloblastoma varies considerably. In large part, the issue hinges upon the interpretation of focal immunopositivity for GFAP. For example, pale islands are frequent sites of neoplastic, GFAP-positive astrocytes (see fig. 5-25) (33,34). If such cells are neoplastic, the incidence of astroglial differentiation in medulloblastoma becomes high. Since nodularity is generally considered a manifestation of neuronal differentiation, the concurrence of neoplastic glial cells would place many nodular tumors in the "mixed differentiation" category discussed below. The internodular component in some nodular lesions is extensively positive for GFAP, as is discussed in the section, Immunohistochemical Findings.

Nodular/Desmoplastic Medulloblastoma. The term desmoplastic has been used as a synonym for medulloblastomas with distinctive lucent areas known as "pale islands," those areas with abundant internodular reticulin (figs. 5-10–5-13), and in the literal sense, to those with a collagenous stroma in which dissecting files of neoplastic cells simulate scirrhous carcinoma (fig. 5-14). It is not clear whether truly desmoplastic lesions featuring only dense collagen and files of cells are related to the nodular lesion just described, but most lesions with advanced desmoplasia are at least focally nodular. Medulloblastomas in adults are more often nodular/desmoplastic than non-nodular (23).

The extent to which reticulin deposition is a nonspecific consequence of leptomeningeal involvement or an intrinsic property of the lesion is not clear. Certainly, the most florid desmoplasia is seen in medulloblastomas that invade the leptomeninges.

The pale islands, present in about one third of medulloblastomas, represent foci of neuroblastic or even neurocytic differentiation, and contrast with the cytologically more atypical internodular tissue (32,35,36). The extent of nodularity varies between tumors and even within a given tumor. At one extreme lies the rare "medulloblastoma with extensive nodularity" (described separately below); at the other, the cytologically malignant medulloblastoma

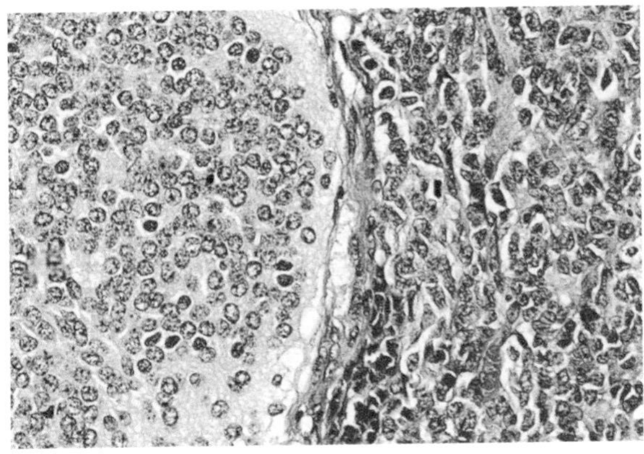

Figure 5-11

NODULAR/DESMOPLASTIC MEDULLOBLASTOMA

Intranodular cells are more uniform and cytologically bland than those of the extranodular compartment.

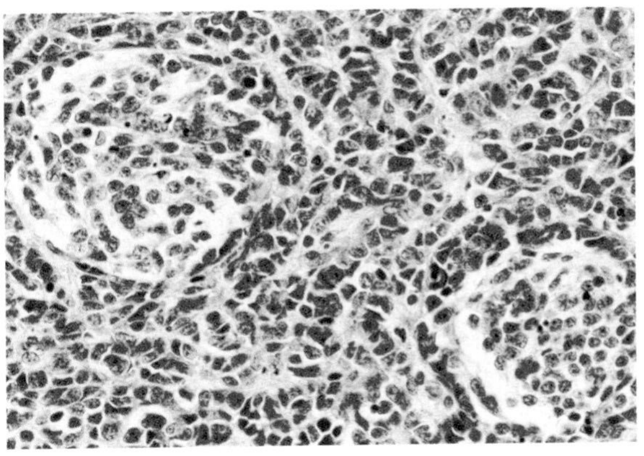

Figure 5-12

NODULAR/DESMOPLASTIC MEDULLOBLASTOMA

Nodules may be small and poorly formed.

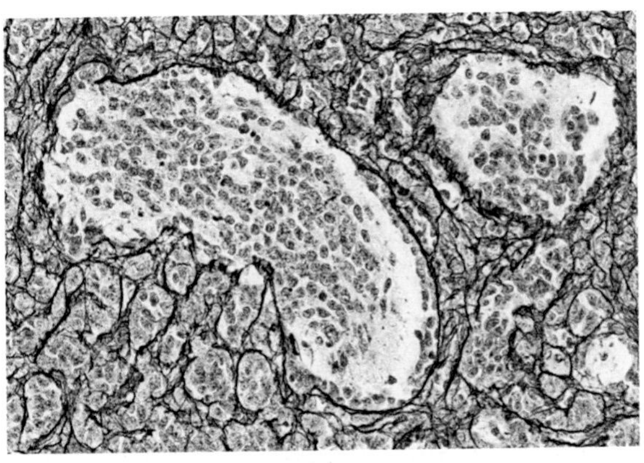

Figure 5-13

NODULAR MEDULLOBLASTOMA

Extranodular tissue is rich in reticulin, which surrounds individual cells, small clusters, or larger nodules.

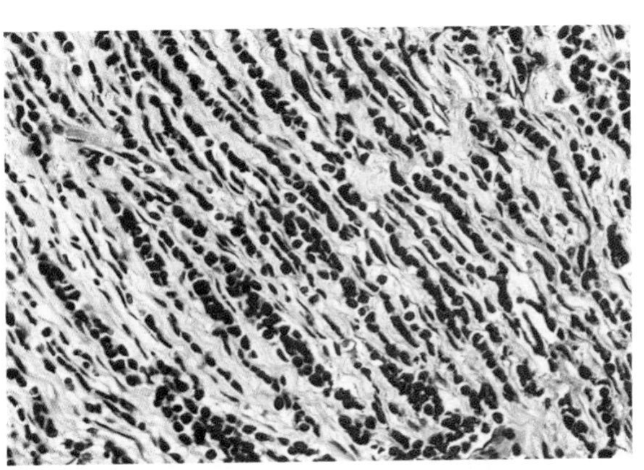

Figure 5-14

MEDULLOBLASTOMA WITH DESMOPLASIA

Often found in conjunction with nodule formation, desmoplastic tissue with columns of tumor cells resembles scirrhous carcinoma.

with nodules that are rudimentary at best (fig. 5-12). Most nodular/desmoplastic tumors fall between these two extremes. Despite the extent to which neuritic processes are formed, intranodular ganglioid cells and frank neurons are only infrequently present.

Cells with perinuclear halos within the islands lend an oligodendroglioma-like appearance, as they do in almost any tumor with neurocytic differentiation. The nodules, although not as lucent, oligodendroglial-appearing, and cytologically benign as in medulloblastomas with extensive nodularity, are also immunoreactive for synaptophysin. Neuronal ultrastructural features include cell processes filled with microtubules and joined by adhesion plaques (32,37).

In contrast to the neurocytic intranodular tissue (pale islands), internodular regions show increased cellularity, a degree of cytologic atypia, and mitotic activity (fig. 5-11). Apoptosis, on the other hand, is less prominent. A delicate rim of reticulin generally surrounds the nodules and pervades the internodular component, where it surrounds cells either individually or in small clusters (fig. 5-13). As is the case for classic, nonnodular medulloblastomas, nodular lesions differ in the degree of cytologic atypia and other

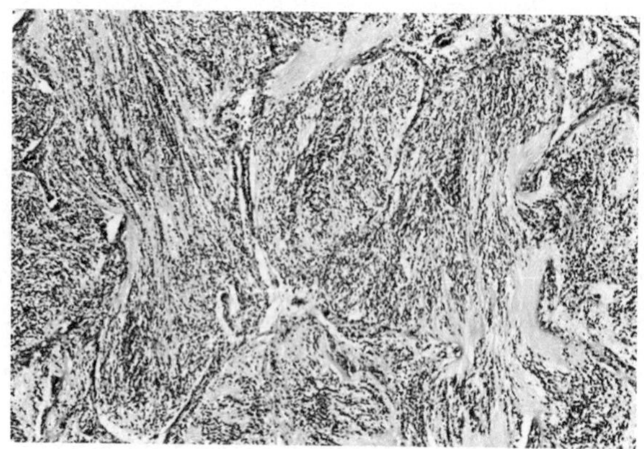

Figure 5-15

MEDULLOBLASTOMA WITH EXTENSIVE NODULARITY

There is little or no extranodular tissue in this rare infantile variant.

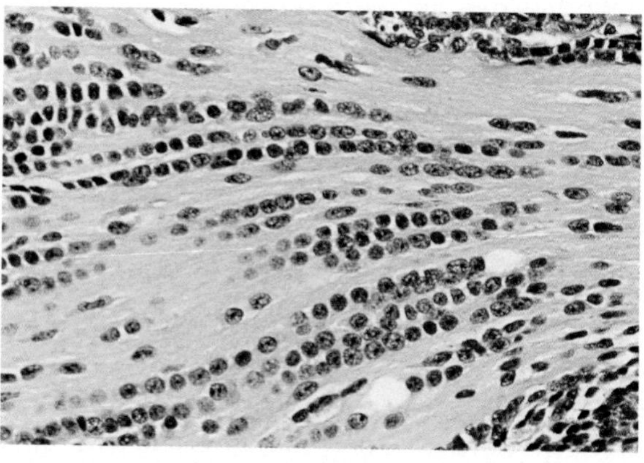

Figure 5-16

MEDULLOBLASTOMA WITH EXTENSIVE NODULARITY

Intranodular tumor cells stream in a neuropil-like background. Such markedly uniform, mitotically inactive cells have the cytologic features of neurocytes.

forms of cytologic malignancy. In some cases, even pale islands contain anaplastic cells.

Occasional medulloblastomas contain less precisely defined pale islands and lack the extensive reticulin network. These lesions may be different from the classic desmoplastic/nodular lesions described above.

Medulloblastoma with Extensive Nodularity. This most recently described medulloblastoma variant consists almost entirely of nodules, with little or no desmoplastic internodular tissue (fig. 5-15) (31). Formerly known as cerebellar neuroblastoma (37,38), it is only the extreme version of the nodular desmoplastic lesions discussed above. This rare lesion (less than 2 percent of all medulloblastomas) is usually seen in the first 2 years of life. Its remarkable radiologic profile is illustrated in figure 5-2.

The extremely nodular pattern, sometimes visible to the naked eye in histologic sections, is highly characteristic because of its large, circular, oval, or parabolic profile of reduced nuclear density. Large areas of neuropil, sometimes focally calcified and containing scattered small ganglion cells, may be present. The streaming of neoplastic neurocytes in a fibrillar background is prominent (fig. 5-16), as it is in some conventional nodular medulloblastomas without the extremes of nodularity present in the extensively nodular variant. The cells are decidedly neurocytic in roundness, blandness, and propensity for the formation of oligodendroglioma-like perinuclear halos.

Ultrastructurally, extensively nodular medulloblastomas show the same neuronal differentiation as the nodules in desmoplastic/nodular medulloblastomas (39).

Large Cell and Anaplastic Medulloblastomas. These represent the extreme in terms of cytologic malignancy in medulloblastoma, and merge through gradual transition from less atypical lesions, either classic or nodular. Approximately 20 to 25 percent of medulloblastomas fall into the combined anaplastic/large cell group.

The designation anaplastic is used with some reservation, and is admittedly difficult to apply to a neoplasm that is already small cell. Nevertheless, it is descriptive and is readily understood to indicate a neoplasm with cytologic features that are malignant over and beyond simple dense cellularity, mitotic activity, and nuclear hyperchromasia (40–45). Thus, nuclear angulation, increased mitotic rate, nuclear molding, and abundant apoptosis are expected in this lesion that fulfills any histopathologist's definition of a malignancy (figs. 5-6, right; 5-17; 5-18). Nuclei are up to threefold larger than those of more ordinary medulloblastomas. The presence of anaplasia, even when advanced, does not preclude differentiation into crude Homer-Wright rosettes (fig. 5-19) or pale islands (fig. 5-20).

Large cell medulloblastomas generally arise in the context of an anaplastic lesion and often appear, at least morphologically, to be the result of incremental tumor progression, since its

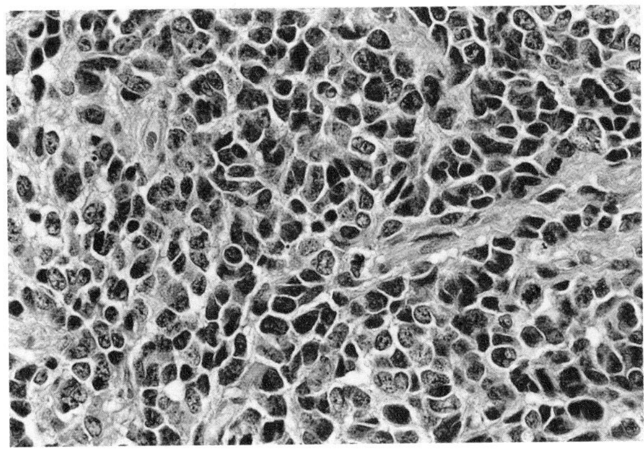

Figure 5-17

ANAPLASTIC MEDULLOBLASTOMA

A moderate degree of anaplasia is expressed as large, irregular, molded nuclei in frequent mitotic division.

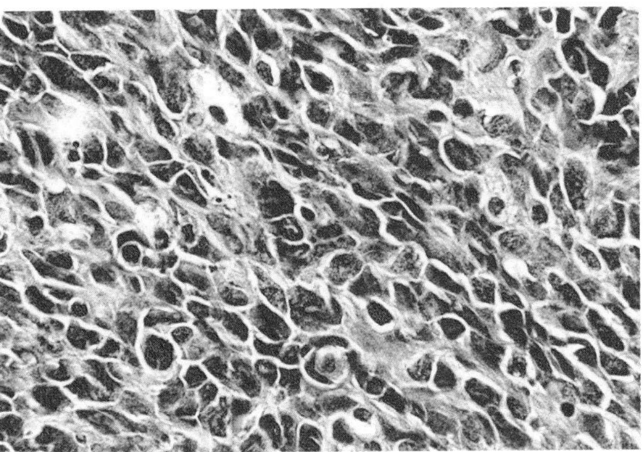

Figure 5-18

ANAPLASTIC MEDULLOBLASTOMA

Severely anaplastic tumor cells are especially large and pleomorphic. Nuclear molding and "cannibalistic" cell wrapping are prominent.

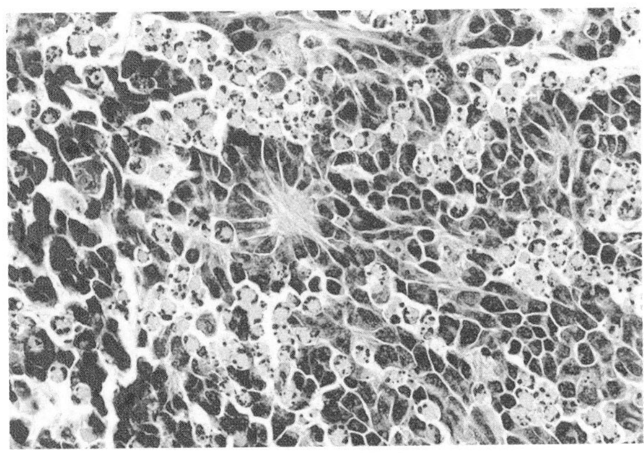

Figure 5-19

ANAPLASTIC MEDULLOBLASTOMA

Advanced anaplasia does not preclude differentiation into Homer-Wright rosettes, albeit ones that are crudely formed. Note the prominent apoptosis and cell wrapping so common in large cell and anaplastic medulloblastomas.

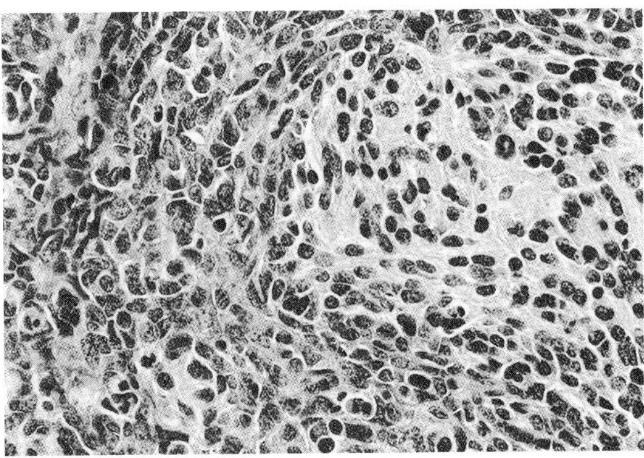

Figure 5-20

ANAPLASTIC MEDULLOBLASTOMA

Nodular/desmoplastic medulloblastomas may be anaplastic, particularly the extranodular component.

cells are often concentrated in homogeneous lobules or nodules within tumor tissue showing the anaplastic features just described (fig. 5-21) (19,40,41,45,46). The cells are not necessarily larger than those of severely anaplastic lesions, and are defined principally by the shape of the nuclei. The round nuclei, with their prominent nucleoli, somewhat resemble those of large cell lymphoma. The borders of these frequently discohesive cells are well defined.

Apoptosis, either as isolated single cells or in large confluent "lakes," is a prominent feature of both large cell and anaplastic lesions (fig. 5-19). Areas of apoptosis may become calcified. Wrapping of one neoplastic cell about another, the latter often apoptotic, is a distinctive and characteristic feature of either anaplastic or large cell medulloblastoma (figs. 5-18, 5-19, 5-21). The cytologic resemblance to the "cannibalism" seen in large cell carcinomas is striking.

The MIB-1 rate is very high in anaplastic areas and lower in better-differentiated areas of the same neoplasm.

Frozen Section. Medulloblastomas are densely cellular tumors that must first be distinguished from normal cerebellum and then

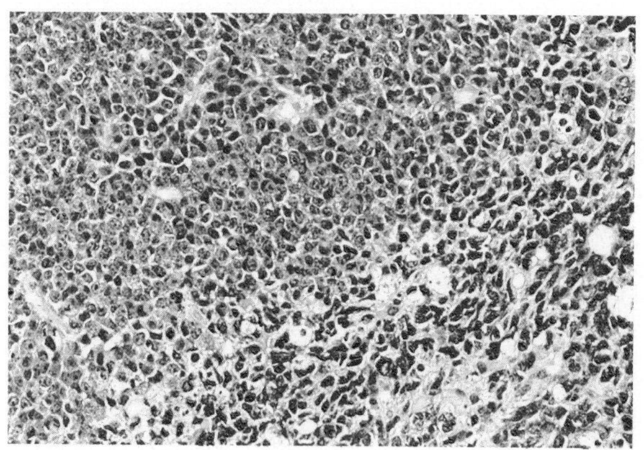

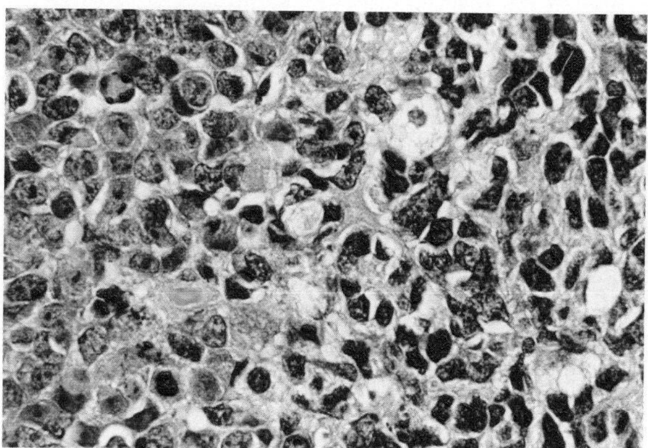

Figure 5-21

LARGE CELL MEDULLOBLASTOMA

Left: The cells of large cell medulloblastoma are both large and round. They are often found as nodules or sheets that emerge from more pleomorphic anaplastic tissue.

Right: In contrast to the irregular, molded, polygonal nuclei in anaplastic medulloblastomas, those of the large cell type are round, or nearly so, and have prominent nucleoli. Cell-cell wrapping is prominent in both anaplastic and large cell tissues.

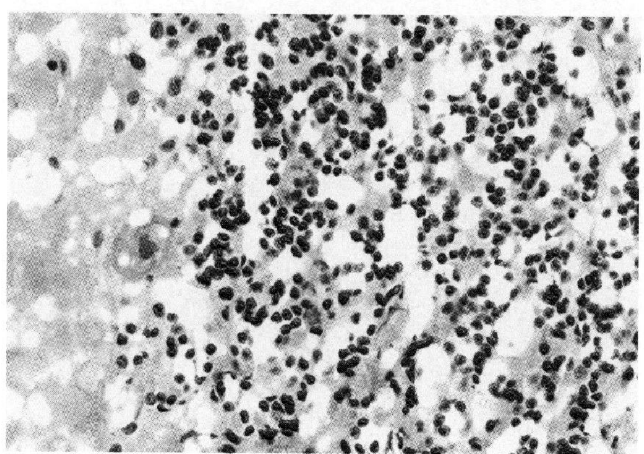

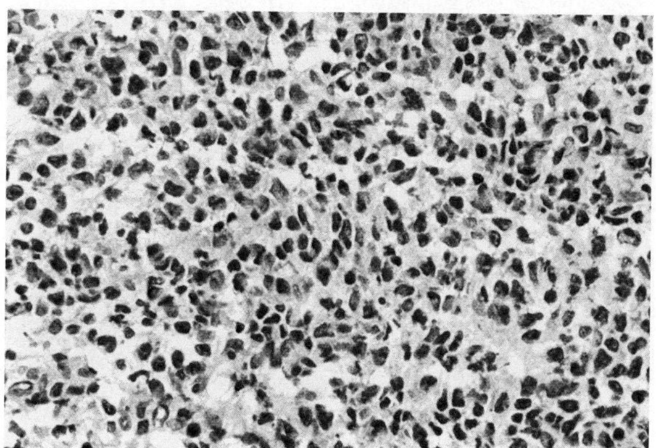

Figure 5-22

NORMAL CEREBELLUM VERSUS MEDULLOBLASTOMA: FROZEN SECTION

Left: Because of its high cellularity and the nuclear distortion induced by the freezing process, the normal internal granular cell can be misinterpreted as medulloblastoma. The almost acellular molecular layer and the presence of even one Purkinje cell are helpful identifying features of normal cerebellum.

Right: In contrast to normal cerebellum, the cells of medulloblastoma are larger and cytologically atypical, and may be mitotically active.

from other neoplasms. Among the latter, ependymoma is the principal entity in the differential diagnosis in children, whereas metastatic carcinoma is the critical issue in adults.

Normal, densely packed, cytoplasm-poor internal granule cells of the normal cerebellum must be considered in the differential diagnosis of any cerebellar tumor (fig. 5-22, left). As described in chapter 2, these small neurons are uniform in size and shape, have a single central nucleolus, and lack proliferative activity. Their identification and distinction from medulloblastoma may be problematic unless both medulloblastoma and normal internal granule cells are available for comparison. Tumor cells, even of nonanaplastic medulloblastomas, are considerably larger and more pleomorphic than normal granule cells (fig. 5-22).

Immunohistochemical Findings. Medulloblastomas are almost always synaptophysin positive, particularly the fibrillar cores of Homer-Wright rosettes, perivascular pseudorosettes,

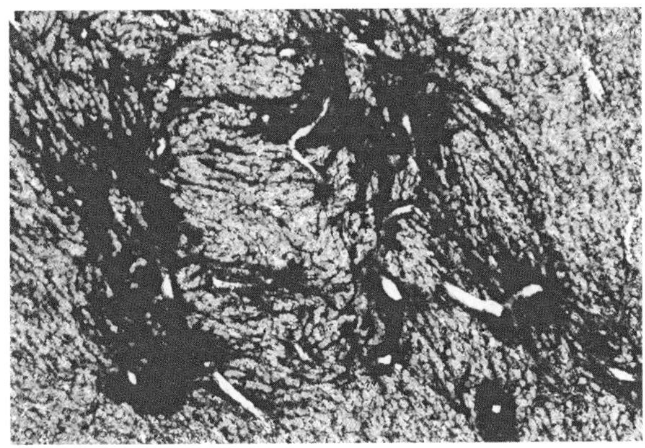

Figure 5-23

MEDULLOBLASTOMA

Stellate areas of finely fibrillar neuropil are immunoreactive for synaptophysin.

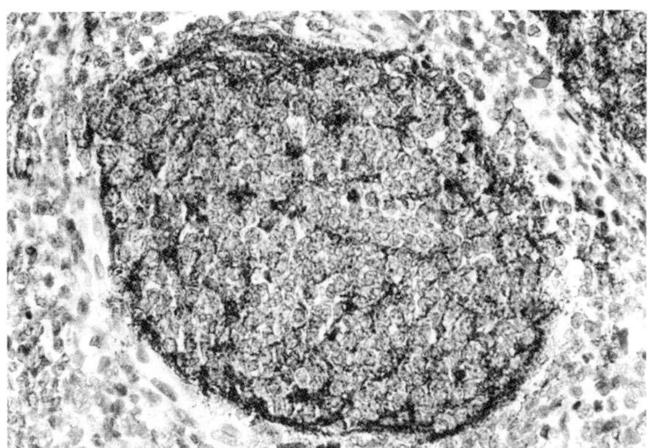

Figure 5-24

MEDULLOBLASTOMA

Intranodular fibrillar areas are strongly immunoreactive for synaptophysin.

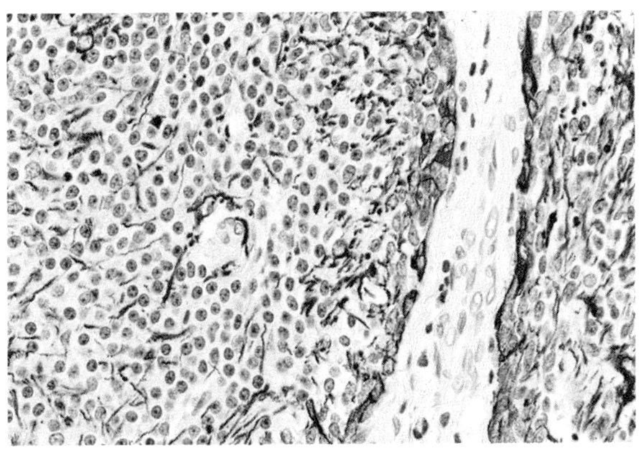

Figure 5-25

MEDULLOBLASTOMA

Nodules contain variable numbers of cells that are long, branched, somewhat radially oriented, and positive for glial fibrillary acidic protein (GFAP).

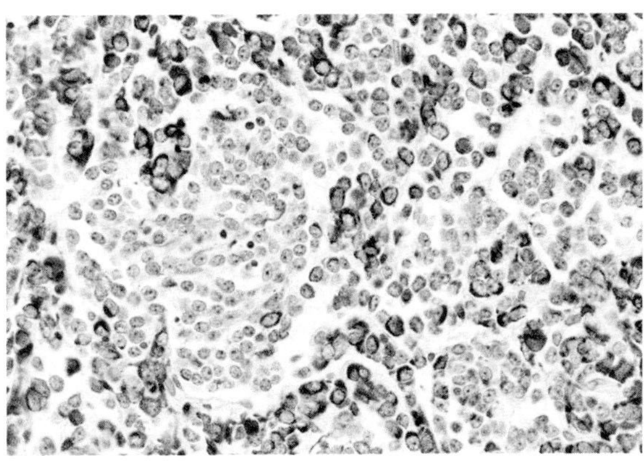

Figure 5-26

MEDULLOBLASTOMA

Extranodular cells are sometimes extensively immunoreactive for GFAP. In contrast to the dendritic intranodular cells illustrated in figure 5-25, extranodular GFAP-positive cells have little cytoplasm and few if any processes.

and pale islands (figs. 5-23, 5-24). The neuroblastic nature of medulloblastoma is often apparent as positivity in even the most desmoplastic areas. The delicately fibrillar, neuropil-like matrix of the islands is also impressively immunoreactive for microtubule-associated protein and class III beta-tubulin (32,35,36). Reactivity for neurofilament protein, although often absent in tissues routinely fixed and processed, is often demonstrated in optimally fixed tissues. Immunoreactivity for the neurotropin receptor p75NTR is present principally in the reticulin-rich intranodular tissue of the nodular/desmoplastic variant (47). Anaplastic and large cell lesions are frequently immunoreactive for synaptophysin and neurofilament protein (41,45). There may be dot-like cytoplasmic reactivity for synaptophysin and chromogranin.

In medulloblastomas with glial differentiation, fibrillar areas are positive for GFAP. The same is true of scattered, branched cells with bland nuclei that are present within pale islands (fig. 5-25) (33,34,36). There is sometimes extensive GFAP positivity in small cells in the extranodular component in some nodular/desmoplastic lesions (fig. 5-26) (35). In contrast to the GFAP-positive intranodular cells, these are small and

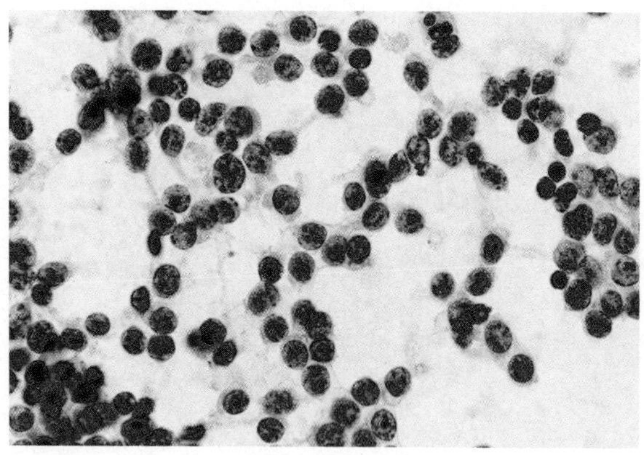

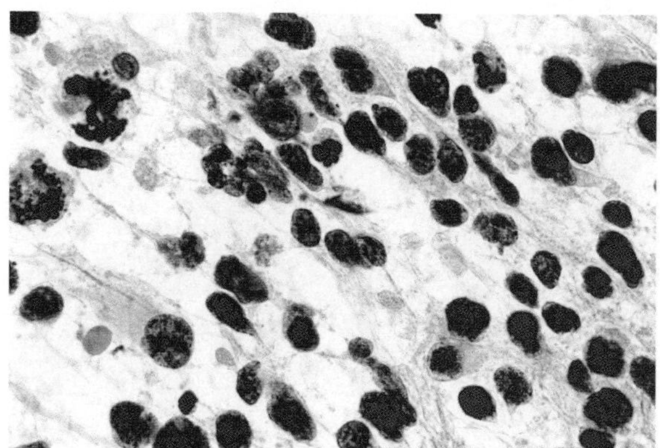

Figure 5-27

MEDULLOBLASTOMA: SMEAR PREPARATIONS

Left: The typical medulloblastoma is composed of small uniform cells that vary little in size, shape, and chromatin density. Right: Pleomorphism, increased cell size, marked cytologic atypia, and high mitotic and apoptotic rates characterize cells from anaplastic medulloblastomas. (See fig. 2-54 for a smear preparation of normal cerebellum.)

cytologically atypical, with immunoreactivity confined to a small amount of perinuclear cytoplasm.

Interpretation of the results of immunostaining for GFAP must be done cautiously since nearly all medulloblastomas contain reactive astrocytes. These are usually stellate in configuration, highly differentiated, and anchored on a vessel. Even when GFAP-reactive cell processes are intimately associated with the nuclei of obviously neoplastic cells, juxtaposition can create the false impression that a process and nucleus belong to the same cell. In many instances, however, even careful up and down focusing fails to resolve the issue and the presence of glial differentiation is neither established nor excluded. Only circumferential perinuclear staining of more than occasional cytologically malignant cells provides evidence for glial differentiation. In general, glial differentiation is focal and does not involve a large enough area to suggest an overtly astrocytic tumor.

Occasional medulloblastomas are immunoreactive for retinal S antigen, a marker of photoreceptor differentiation (48,49).

Cytologic Findings. Relative to most other CNS tumors, the nuclei of medulloblastomas are hyperchromatic, but the degree of pleomorphism and apoptosis varies with tumor grade (fig. 5-27). Nuclei of higher-grade lesions are larger, more pleomorphic, more coarsely constructed, and more often apoptotic. Cell wrapping, or cannibalism, is also more common (50). In only a minority of medulloblastomas with neuroblastic differentiation are Homer-Wright rosettes evident in cytologic preparations (50). The cytologic features of normal cerebellum are illustrated in figure 2-54.

Molecular and Cytogenetic Findings. It is not surprising that multiple cytogenetic abnormalities have been reported in this histologically and cytologically diverse lesion, or that such abnormalities are most numerous in higher-grade lesions, i.e., those with anaplastic and/or large cell features (51). The most common cytogenetic finding, loss of all or part of chromosome 17p, is often associated with duplication and translocation of chromosome 17q (isochromosome 17q) (fig. 5-28) (19,52–59). Gains on chromosomes 1q, 2p, 4p, 7, and 19, as well as losses of chromosomes 10 and 11, are also frequent (52,53,55,57,60,60a). High-level gains of chromosomes 8q and 2p in the region of *N-myc* and *c-myc,* respectively, are found in a small percentage of tumors, particularly those with anaplastic and large cell features (41,51,55,59, 60a,61). The relationship between 17p loss and histologic type, tumor grade, and patient outcome is unclear, although there are suggestions that it is more often present in classic, non-nodular lesions (56), and is perhaps more closely associated with higher-grade tumors (51). A possible association with a poorer outcome is discussed below.

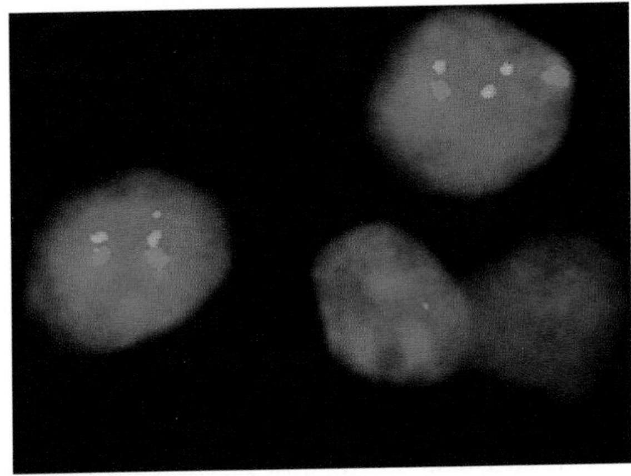

Figure 5-28

MEDULLOBLASTOMA: FLUORESCENCE IN SITU HYBRIDIZATION

Isochromosome 17q, as illustrated by fluorescence in situ hybridization (FISH), is a common genetic finding. Most tumor cells have an additional (red) signal for a marker on chromosome 17q and one fewer signal (green) for a marker on 17p. (Courtesy of Dr. Arie Perry, St. Louis, MO.)

Molecular abnormalities in specific signaling cascades have been sought in accord with the tumor syndromes discussed above (19,25). Loss of heterozygosity (LOH) was found for chromosome 9q22.3-q31, as expected, in medulloblastomas in patients with Gorlin's syndrome and in a subset of sporadic lesions (62). The Gorlin's lesions were nodular/desmoplastic, as were the sporadic lesions with this LOH.

To date, abnormalities have been identified in two major signaling pathways: sonic hedgehog (shh) and wingless (WNT). Most mutations in the shh pathway have been found in the *PTCH1* gene (63–65), with fewer abnormalities in *PTCH2* (66), *SMOH* (smoothened) (67), and *SUFU* (suppressor of fused) (68) genes. Activation of the shh pathway may occur preferentially in nodular/desmoplastic lesions, since mutations in specific pathway members have been noted in these lesions (25,63), Gorlin's syndrome–associated medulloblastomas are nodular (62), and gene expression profiles consistent with activation of this pathway have been identified (15).

In the WNT pathway, mutations have been found in the *β-catenin* (69,70), *axin* (71), and *APC* (72) genes. The presence of mutations in *β-catenin* can be inferred from the translocation of beta-catenin immunoreactivity into the nuclei, since staining is normally confined to the cytoplasm

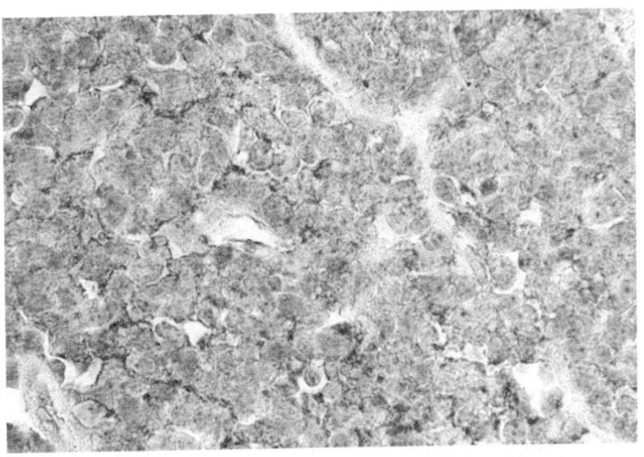

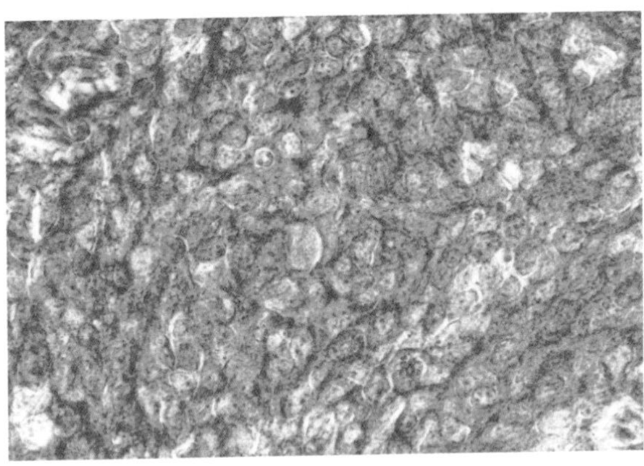

Figure 5-29

MEDULLOBLASTOMA: NUCLEAR TRANSLOCATION OF β-CATENIN

Top: In the absence of an activating mutation in the *$-catenin* gene, the protein is restricted to the cytoplasm.

Bottom: In medulloblastomas with a mutation in the *β-catenin* gene, or certain other members of the WNT signaling pathway, both the cytoplasm and nucleus are immunoreactive for β-catenin protein.

(fig. 5-29) (69). Both amplification and overexpression of the oncogenes *c-myc* and *N-myc* have been documented in medulloblastomas (fig. 5-30) (55,60,73–77); one or both alterations are common in large cell and anaplastic lesions (41,45,51,57,59,61). Overexpression of *c-myc* can also occur in nonanaplastic tumors, with negative, or unfavorable, prognostic significance (see below) (74).

The multiple genetic abnormalities, and their correlation with morphologic features such as anaplasia and large cell features, have suggested that medulloblastomas might progress both genetically and histologically in some cases from better-differentiated tumors with fewer

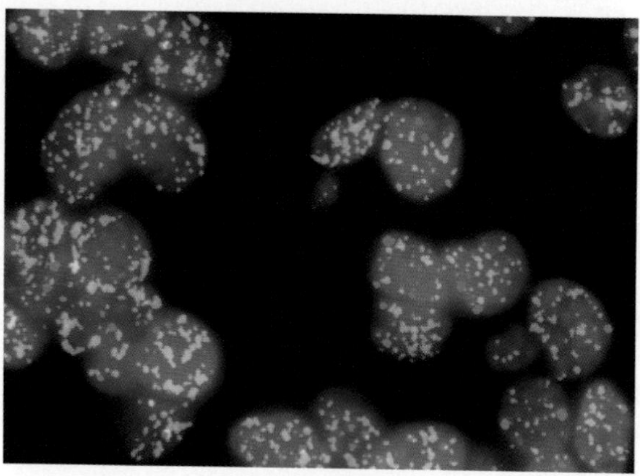

Figure 5-30

MEDULLOBLASTOMA: FLUORESCENCE IN SITU HYBRIDIZATION

Multiple FISH signals for *c-myc* indicate amplification, as is common in large cell/anaplastic medulloblastomas. (Courtesy of Dr. Arie Perry, St. Louis, MO.)

chromosomal abnormalities to anaplastic lesions with multiple alterations (43,51).

Grading and Prognostic Factors. Successful attempts to link histopathology and outcome have focused upon individual cytologic and histologic features, as well as histologic tumor subtype. There is a running debate whether patients with desmoplastic/nodular lesions have, or do not have, a better prognosis; on balance, they seem to have a better prognosis (19,43,78,79). One study of children less than 3 years of age found, overall, a significantly better outcome for those with desmoplastic lesions (80). Lesions with extreme nodularity appear to be less aggressive, although there is as yet little experience with this entity (31,81).

One histologic grading system assigned one point each to the presence of necrosis, desmoplasia, cytoplasmic processes, and mitoses (82). Patients with a low tumor score (a total of 2 points or less) did better. Another, more subjective or "Gestalt" approach, uses both degree of cytologic atypia and a nonquantitated impression of cell size (43,59). Ultimately simplified into a two-tier system, significant differences were noted in the survival experiences between the approximately 20 to 25 percent of patients with anaplastic or large cell lesions (less favorable prognosis) and those with nonanaplastic tumors (43,59). Other studies came to similar conclusions (82a), with one suggesting that the outlook for patients with large cell medulloblastomas may be worse than for those with anaplastic lesions (46).

Grading using mitotic indices also is feasible, as reported in one study quantifying mitoses in areas of maximal activity and finding it to be an independent negative prognostic factor (83). Another study of nondesmoplastic lesions found significant differences in survival when comparing patients whose medulloblastomas had mitotic counts that were above, versus those that were below, the median (46). Proliferation markers are also useful: a negative relationship has been found between the length of survival and the MIB-1 labeling index in areas of maximal staining (84); another study of nondesmoplastic lesions, however, did not find the Ki-67 labeling index to be prognostically significant (46). Immunoreactivity for calbindin is reported to be an unfavorable feature (85).

Grading by the presence or absence of differentiation is reportedly also a means of identifying prognostically unfavorable lesions. In one study, GFAP staining was a negative prognostic factor, becoming more ominous in proportion to the extent of immunoreactivity (86).

At present, prognostication is increasingly focused upon cytogenetic and molecular features, and pathologists can expect increasing demands for molecular studies on fresh frozen tissue. A longstanding issue has been the significance of abnormalities of chromosome 17, and it is still unclear whether these are (59,87, 88) or are not (53,55,89) prognostically unfavorable (19). A more robust correlation with outcome is evident between mRNA levels of the neurotropin receptor *TrkC* in some studies: higher levels are associated with a much more favorable outcome (90,91). Nuclear translocation of β-catenin may also be a favorable prognostic factor (91a).

Other prognostically significant molecular features portending a less favorable outcome are overexpression and amplification of *c-myc* and possibly of *N-myc* (55,59,74–77,90), both more common in anaplastic and large cell lesions. Yet another proposed system takes both *TrkC* and *c-myc* levels into consideration, dividing medulloblastomas into three categories: 1) high *TrkC*, low *c-myc;* 2) high *TrkC*, high *c-myc*; and 3) low *TrkC*, high *c-myc* (90). The prognosis deteriorates as one progresses through these combinations.

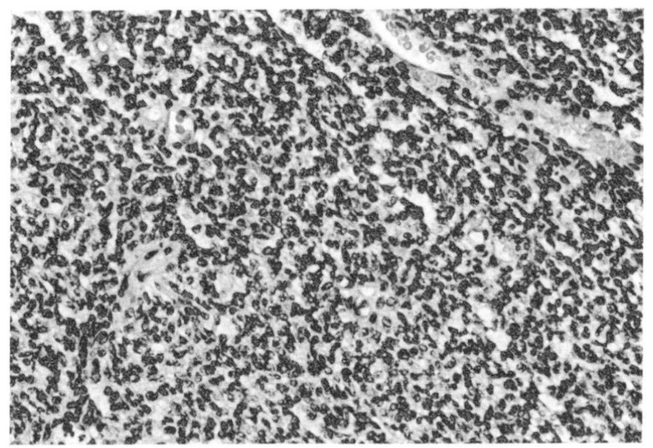
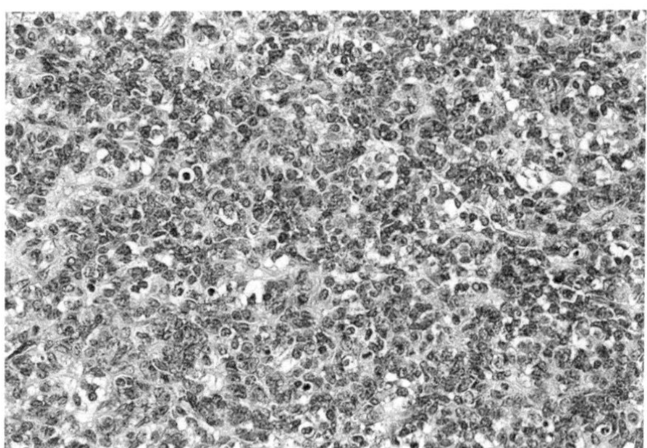

Figure 5-31

ATYPICAL TERATOID/RHABDOID TUMORS SIMULATING MEDULLOBLASTOMA

Left: Small cell areas of atypical teratoid/rhabdoid tumor (AT/RT) can closely resemble medulloblastoma.

Right: The classic appearance of AT/RT resembles anaplastic medulloblastoma, but has a more "jumbled" architecture and lacks the flagrant anaplasia of the high-grade medulloblastoma.

Although mutations in *p53* are uncommon, immunopositivity and high intensity staining (but not percentage of positive cells) were found to be prognostically adverse in at least two studies (92,93). Another mechanism for *p53* inactivation, amplification of *MDM2*, was found also to be prognostically unfavorable in adults with medulloblastomas (94). One study found that abnormalities in the *p53-ARF* pathway were perhaps more common in anaplastic and large cell lesions (95). Another study suggested that amplification of *hTERT* (telomerase), found in slightly less than half of medulloblastomas, was prognostically unfavorable (96). Lastly, a high ErbB2 protein level has been a negative prognostic factor (83,97).

Differential Diagnosis. The diagnosis of medulloblastoma is usually straightforward when faced with a densely cellular, mitotically active, small cell tumor from the cerebellum in a child. A nodular pattern greatly simplifies the task. Nevertheless, alternative diagnoses must be considered, particularly in the very young and in adults.

Before embarking on a tumor A versus tumor B differential diagnosis at the time of frozen section, the distinction from the normal internal granule cell layer needs to be resolved. As illustrated in figures 5-22, left, 2-50, and 2-51, small cells of this layer can be misinterpreted as neoplastic, although they are usually distinguished by their smaller size and lack of cytologic atypia.

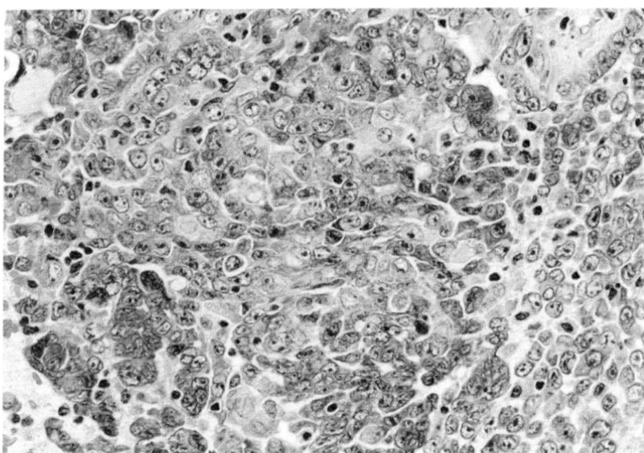

Figure 5-32

ATYPICAL TERATOID/RHABDOID TUMOR RESEMBLING ANAPLASTIC/LARGE CELL MEDULLOBLASTOMA

Rhabdoid cells with large nuclei and macronucleoli resemble cells of large cell medulloblastoma. Scattered cells with a cytoplasmic mass of filaments are present in AT/RT, but not in anaplastic/large cell medulloblastoma. The cytoplasm is almost nonexistent in large cell medulloblastoma, and certainly does not displace the nucleus as it does in AT/RT.

In infants, atypical teratoid/rhabdoid tumor (AT/RT) must be excluded, since small cell areas of the latter can be indistinguishable from conventional medulloblastoma (fig. 5-31, left). Furthermore, the large pale cells of AT/RT somewhat resemble those of large cell or anaplastic medulloblastoma (figs. 5-31, right; 5-32). The issue is usually settled by the complex immunophenotype of AT/RT. Lingering uncertainties can be resolved by molecular studies, such as

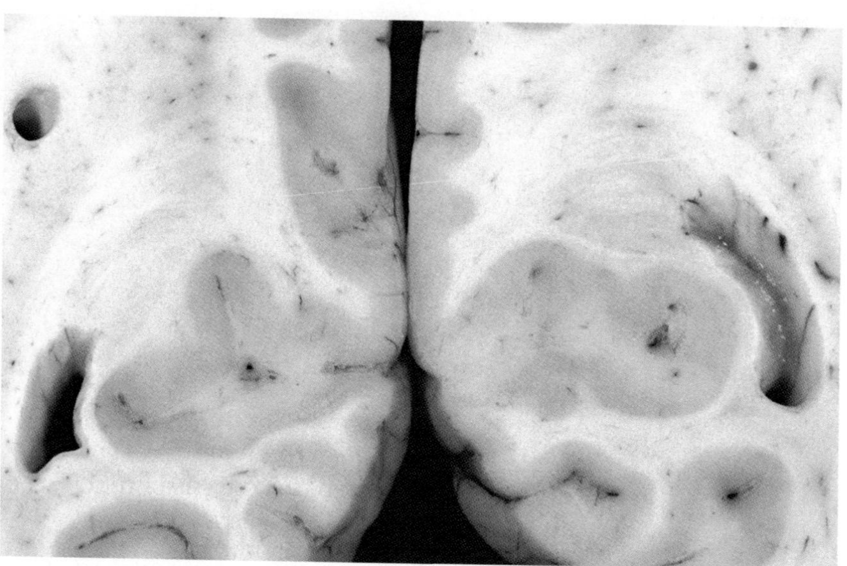

Figure 5-33

MEDULLOBLASTOMA WITH LEPTOMENINGEAL DISSEMINATION

Disseminated neoplasm can fill the subarachnoid space, as here over the visual cortex. (Courtesy of Dr. Caterina Giannini, Rochester, MN.)

fluorescence in situ hybridization (FISH) analysis for deletion of at least one copy of the *INI1* gene on chromosome 22q, which is present in many, but not all, cases of AT/RT (98). An immunohistochemical method for detecting INI1 protein is discussed in the section Atypical Teratoid/Rhabdoid Tumor, and illustrated in figure 5-70.

In principle, ependymomas are distinguished by their usual origin from the floor of the fourth ventricle rather than within the cerebellum, although this distinction is less than absolute since some ependymomas originate from the superior or lateral walls of the fourth ventricle and then often appear as intracerebellar masses. Medulloblastomas are more uniformly hypercellular than ependymomas and lack their widespread fibrillarity and perivascular pseudorosettes, both of which are GFAP positive in ependymomas. As previously noted, some medulloblastomas have perivascular fibrillar zones similar to ependymal perivascular pseudorosettes except for their synaptophysin rather than GFAP positivity (see fig. 5-23). Ependymomas are more likely to be calcified.

A somewhat different diagnostic problem arises when faced with a densely cellular posterior fossa tumor featuring nuclear elongation and a fibrillar background, suggesting a high-grade diffuse astrocytoma (see fig. 5-4). The problem is readily resolved if the cells are negative for GFAP and positive for neuronal markers, but persists if the cells lack staining for both GFAP and synaptophysin. Such tumors can be difficult to assign to either the medulloblastoma or glioma category. Diffuse parenchymal infiltration is more consistent with astrocytoma but is present in some medulloblastomas.

Metastatic small cell carcinoma, discussed with that tumor, is in the differential diagnosis of adults. In adults, one must also consider the rare cerebellar neurocytoma when confronted with a densely cellular, but cytologically bland, mitosis-poor lesion, especially if lipidized.

Treatment and Prognosis. Conventional treatment of patients with medulloblastoma includes craniospinal radiotherapy, which takes a toll on a child's intellectual and physical development. Nonetheless, it has met with considerable success considering the fate of untreated patients. At present, over half of patients survive 5 years or more. Neuraxis radiation greatly lessens the chance of craniospinal spread, a frequent occurrence in untreated patients. Treatment with chemotherapy alone may be feasible in the very young in whom radiation therapy is especially likely to produce adverse effects (80).

Although most recurrences are local, neuraxis dissemination always remains a possibility (figs. 5-33–5-35). Only occasional tumors metastasize to extracranial sites such as lung, lymph nodes, and bone (fig. 5-36) (99). Dissemination to the peritoneum can occur through a shunt (100).

Late recurrences of the original tumor (101, 102) or the eruption of a radiation-induced

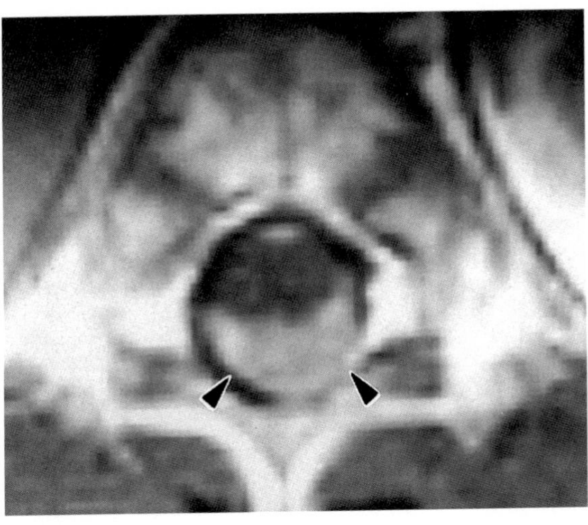

Figure 5-34

MEDULLOBLASTOMA WITH LEPTOMENINGEAL SPREAD

Spinal dissemination, a feared event, is apparent in this MRI, as a white "cap" (arrowheads) that overlies the thoracic cord. (Fig. 5-34 from Fascicle 10, 3rd Series.)

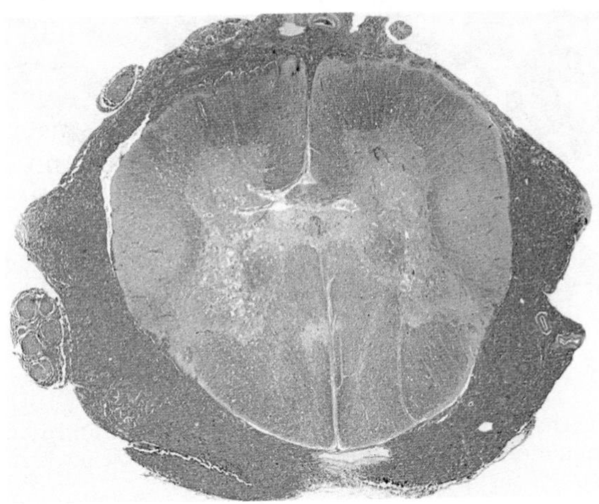

Figure 5-35

MEDULLOBLASTOMA WITH LEPTOMENINGEAL SPREAD

Although recurrence is usually detected first at the primary site, dissemination is craniospinal in some patients. The metastatic neoplasm here fills the subarachnoid space where it encases spinal nerve roots and the anterior spinal artery.

malignant glioma within the radiation field (103, 104) interjects an element of pessimism regarding the ultimate rate of cure. Late meningiomas, also radiation-induced, are another possible complication (105). Even occasional induced schwannomas appear in long-term survivors.

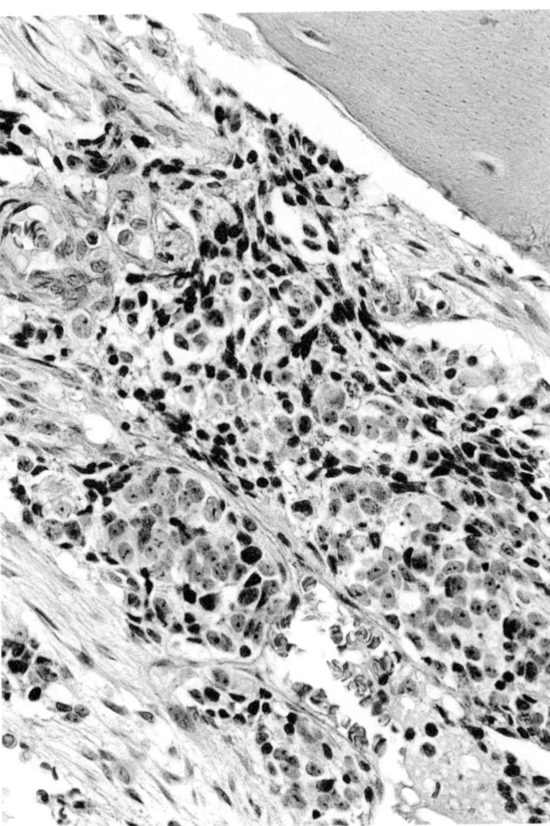

Figure 5-36

MEDULLOBLASTOMA WITH BONY METASTASES

Osseous metastases, while unusual, are more common than with most other primary central nervous system (CNS) neoplasms.

MEDULLOMYOBLASTOMA AND MELANOTIC MEDULLOBLASTOMA

Given their location and high cellularity, these rare tumors are generally classified as medulloblastoma variants, although it is possible that they are pathologic entities in their own right. They have, in any case, distinctive tissue patterns foreign to standard medulloblastomas. Occasional small cell cerebellar neoplasms bridge medullomyoblastoma and melanotic medulloblastoma by containing both myoid and pigmented cells (106,107). Cerebellar tumors with myogenic and teratomatous components may be difficult to classify (108,109).

Medullomyoblastomas occur most frequently in children, usually boys (109–115); adults are only rarely affected (116). Considerable variation is seen in the relative degree in which myoblasts

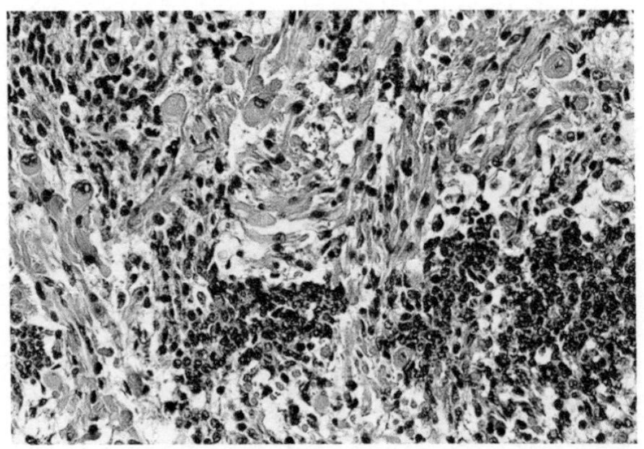

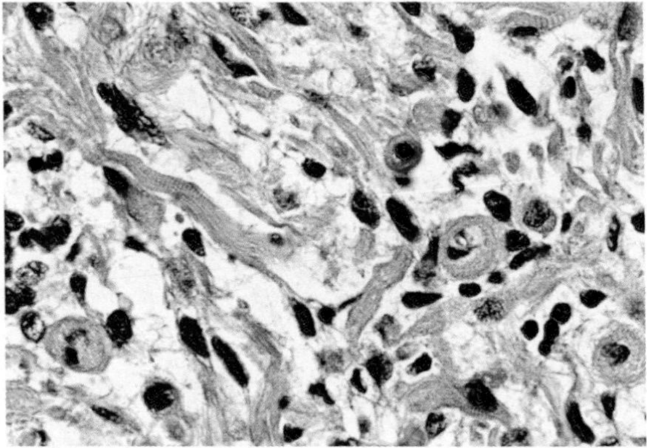

Figure 5-37

MEDULLOMYOBLASTOMA

Left: Cells with prominent cytoplasm are suspiciously myoid even at low magnification.
Right: Cross striations seen at high magnification confirm the impression of a myogenic neoplasm.

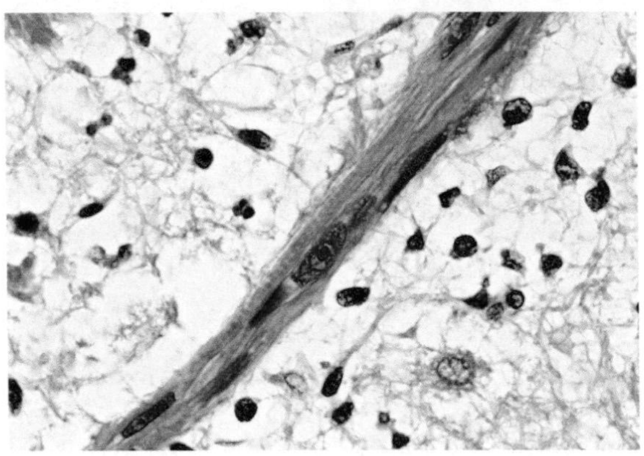

Figure 5-38

MEDULLOMYOBLASTOMA

Medullomyoblastomas can have paucicellular areas traversed by long myocytes.

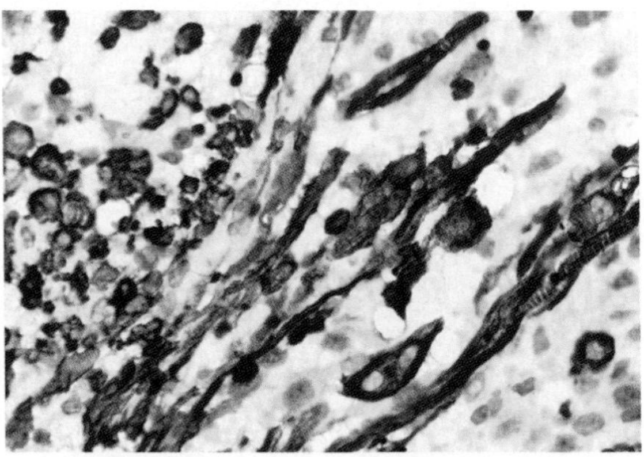

Figure 5-39

MEDULLOMYOBLASTOMA

The presence of a myoid component is confirmed by immunohistochemistry for actin or other muscle markers.

or more mature myocytes are represented. The former cells usually predominate, but cross-striated "strap" forms, or even multinucleate myotubes, can weave and crisscross through cellular and hypocellular regions (figs. 5-37, 5-38). Paucicellular areas are sometimes extensive. Desmin, myoglobin, and muscle-specific actin are diagnostically useful immunomarkers since cross striations may be difficult to find (fig. 5-39).

The undifferentiated-appearing element of these tumors consists of small "blue cells" variably organized in terms of histologic pattern and cell density. In addition to synaptophysin immunoreactivity, both neuroblastic and ganglionic differentiation have been reported (110,113,114,117), as have GFAP-positive neoplastic glia (110,112,118). Anaplastic and large cell features appear in some cases (119)

One study failed to find the molecular changes common in medulloblastomas, i.e., losses of chromosomal arms 17p, 1q, and 9q (110). Another, however, found a common feature of medulloblastomas, isochromosome 17q, in two large cell/anaplastic medullomyoblastomas (120).

The differential diagnosis should include cerebellar neurocytoma with myoid differentiation. These cytologically bland lesions, which usually occur in adults, are discussed in chapter 4.

Melanotic medulloblastoma is characterized by the presence of small clusters or tubules of

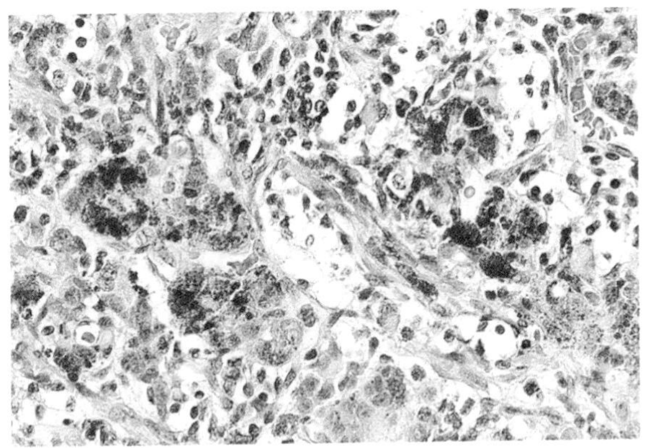

Figure 5-40

MELANOTIC MEDULLOBLASTOMA

Small tubules are the pigmented components of this rare lesion.

pigmented epithelial cells that contain finely granular black pigment identified ultrastructurally as "true," i.e., melanosome-based melanin (fig. 5-40) (121–126). Pigmented/melanotic tumors are listed in Appendix H. As in conventional medulloblastomas, mitotic rates vary. Synaptophysin reactivity is evident in the small, more typical medulloblastoma cells. Aggressive behavior, including craniospinal dissemination, is the rule (121,123,124,126).

MEDULLOEPITHELIOMA

Definition. *Medulloepithelioma* is a neoplasm composed largely or in part of epithelium resembling that of the embryonic neural tube.

Clinical Features. Most medulloepitheliomas present during the first 5 years of life, and many appear before 1 year (127). Even congenital examples have been reported (128). The lesions occur both supratentorially (127,129–133) and infratentorially (127,128,134–136). The former are often deep-seated, near the ventricular system. Unusual loci include optic nerve (137) and the region of the sella turcica (138). The majority are large masses that elevate intracranial pressure or produce neurologic abnormalities in accord with their location (127).

Radiologic Findings. Medulloepitheliomas are generally circumscribed and contrast enhancing, and sometimes cystic and calcified.

Gross Findings. The lesions are generally discrete, solid, gray-tan tumors that, although basically soft, may be modified by cystic change,

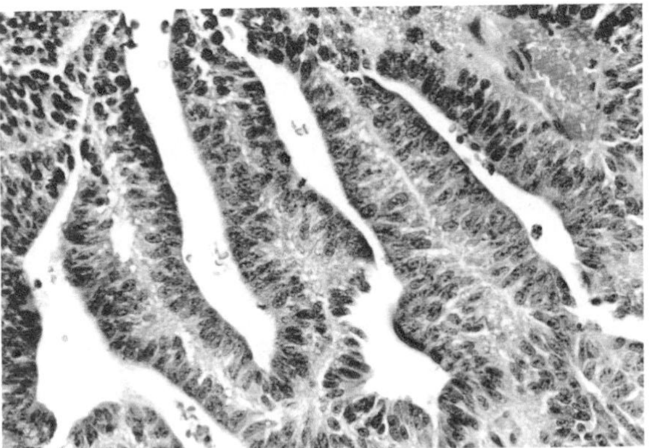

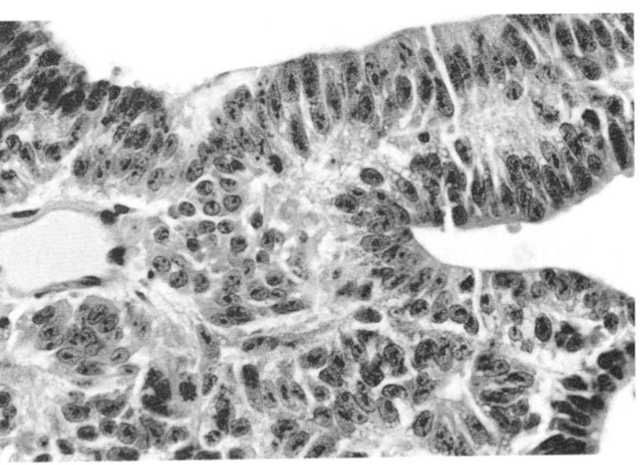

Figure 5-41

MEDULLOEPITHELIOMA

Top: Cavities or canals are lined by tall epithelium that resembles the primitive epithelium of the developing brain.

Bottom: Mitoses occur near the luminal surface, as is typical of both normal and neoplastic medullary epithelium.

hemorrhage, calcification, fibrosis, or, on rare occasion, ossification.

Microscopic Findings. The hallmark of medulloepithelioma is a pseudostratified, primitive-appearing epithelium disposed in glands, tubules, or canals (fig. 5-41). The extent to which this is represented varies considerably: prominent in some cases and focal in others. Markedly cellular regions of undifferentiated-appearing neoplasm without specific architectural or immunohistochemical features are present.

The epithelium is mitotically active and delineated by a well-defined, periodic acid–Schiff (PAS)- or collagen type IV-positive basement membrane on its outer surface, and an ill-defined amorphous pseudomembrane on its internal or luminal side (fig. 5-42). Mitotic figures often lie adjacent to the lumen, as they do in the

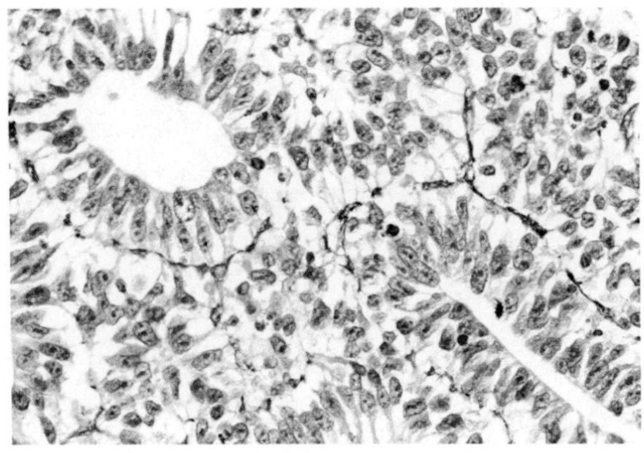

Figure 5-42

MEDULLOEPITHELIOMA

Demonstrated here by immunostaining for collagen type IV, basement membrane material underlies the epithelium.

epithelium of the normal developing neural tube. Small cytoplasmic blebs protrude from luminal cell surfaces in some lesions.

In addition to its signature epithelium, medulloepitheliomas may exhibit glial, neuronal, and even mesenchymal differentiation. Astrocytic components, if present, may exhibit a spectrum of maturation. Ependymal components also vary considerably in appearance, from ependymoblastoma to, less commonly, more mature-appearing typical ependymoma. Ependymoblastoma features are common and may cause definitional and diagnostic confusion with the neoplastic medullary epithelium of medulloepithelioma (130).

Neuronal components vary in extent and maturation. In some instances, they consist largely of neuroblasts whose identification requires immunohistochemical confirmation when Homer-Wright rosettes are not in evidence. Immature or well-formed ganglion cells are seen in other tumors. Occasional mesenchymal components include bone, cartilage, muscle, and adipose tissue (129).

Immunohistochemical Findings. The epithelium of medulloepithelioma is generally immunonegative for S-100 protein, GFAP, and synaptophysin, whereas the neoplasm's differentiated components, e.g., neurons, are appropriately reactive for synaptophysin, tubulin, and microtubule-associated protein. GFAP reactivity is limited to glial elements (133,139,140).

Ultrastructural Findings. Given the rarity of medulloepithelioma, information is limited regarding its ultrastructural features. As expected, the primitive columnar epithelium rests upon a basal lamina and its component cells are joined by intermediate junctions (133,136).

Differential Diagnosis. It may be difficult to distinguish medulloepithelioma from ependymoblastoma. In concept, rosettes of ependymoblastoma possess a clearly defined lumen, but lack an outer and inner membrane, as well as a relationship between mitoses and the luminal surface. Mitoses in ependymoblastoma are abluminal. Apical cytoplasmic blebs are also not seen. CNS tumors with riboning or palisading are itemized in Appendix J, and illustrated in figures 5-74 and 5-75.

The differential diagnosis also includes choroid plexus carcinoma. The distinction is not difficult since the epithelium of the latter is papillary, not linear. The lesion also lacks the divergent differentiation so often seen in medulloepitheliomas. Choroid plexus tumors are immunoreactive for S-100 protein and cytokeratins.

Since medullary epithelium is a common feature of immature teratomas, this germ cell tumor is included in the differential diagnosis. Distinguishing features of immature teratoma include fetal-appearing tissues of other germ layers, the concurrence of embryonal carcinoma or endodermal sinus tumor elements, and immunoreactivity for placental alkaline phosphatase and other markers such as alpha-fetoprotein and carcinoembryonic antigen.

Another rare small cell embryonal tumor with rosettes, neuroblastic tumor with abundant neuropil and true rosettes, is discussed separately.

Treatment and Prognosis. Most patients die within a year of diagnosis (127). Appropriate therapy remains to be determined. Cerebrospinal dissemination is not uncommon, but systemic metastases to bone, lymph nodes, or other viscera are rare.

NEUROBLASTOMA AND GANGLIONEUROBLASTOMA

Definition. *Neuroblastoma* and *ganglioneuroblastoma* are small cell neoplasms with neuroblastic differentiation. Prominent ganglion cell differentiation is present in ganglioneuroblastomas.

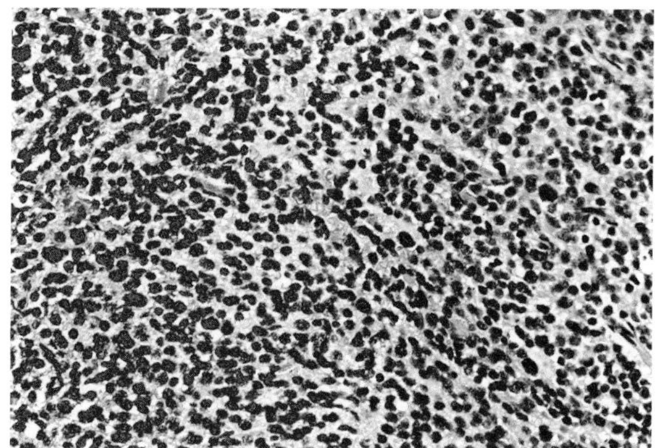

Figure 5-43

NEUROBLASTOMA

Cerebral neuroblastomas are densely cellular and in large areas, often architecturally nonspecific.

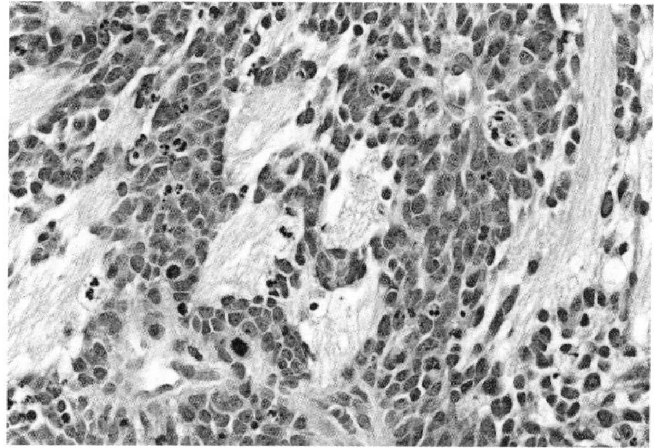

Figure 5-44

NEUROBLASTOMA

Patches of finely fibrillar neoplastic neuropil help identify the neuroblastic nature of the lesion. Like normal gray matter, it is immunoreactive for synaptophysin.

Clinical Features. Most patients present in the first decade of life, usually before the age of 5 years, although occasional lesions occur in young adults (141–143).

Radiologic Findings. Neuroblastomas are large, discrete, contrast enhancing, often cystic, and sometimes calcified (144). They usually lie deep within a cerebral hemisphere.

Gross Findings. The lesions are circumscribed, gray, fleshy, and variably desmoplastic. Cystic change and necrosis are common.

Microscopic Findings. Cerebral neuroblastomas vary in their histologic appearance, but all are small cell neoplasms with architectures that vary from patternless sheets (fig. 5-43) to a pattern with rosettes (fig. 5-44), ribbons, palisades (fig. 5-45), or fascicles. When pronounced, nuclear palisading resembles that in polar spongioblastoma (145–147). Appendix J and figures 5-74 and 5-75 summarize and illustrate CNS tumors with ribboning or palisading.

In many cerebral neuroblastomas, there is an almost defining fine fibrillarity that interrupts otherwise dense sheets of tumor cells. This background, which represents the aggregate of neuronal cell processes (neurites), includes the cores of rosettes (fig. 5-44), anuclear perivascular zones, and larger less structured areas. Large areas of fibrillarity, when circumscribed, can be viewed either as extensions of the perivascular pattern or as markedly exaggerated Homer-Wright rosettes.

Cells within densely cellular regions often exhibit variation in nuclear shape, chromatin content, texture, and mitotic activity. In the absence of neuronal differentiation, nucleoli are inconspicuous. Some tumors show little nuclear pleomorphism and mitotic activity, whereas others are overtly malignant, both histologically and cytologically.

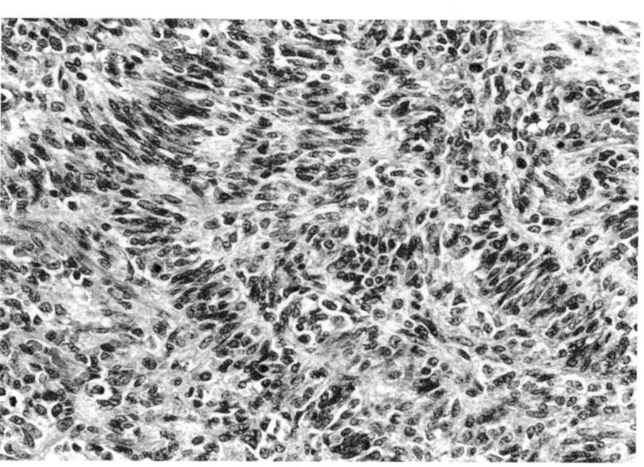

Figure 5-45

NEUROBLASTOMA

The presence of cellular palisades is a distinctive feature in a minority of neuroblastomas.

The connective tissue in neuroblastoma and ganglioneuroblastoma varies from scant to abundant and is the basis of the classification of these tumors into classic, transitional, and desmoplastic forms (143). Collagen and reticulin in the intermediate or transitional variant produces a lobular pattern, but not with the precision and extent seen in medulloblastomas with extensive

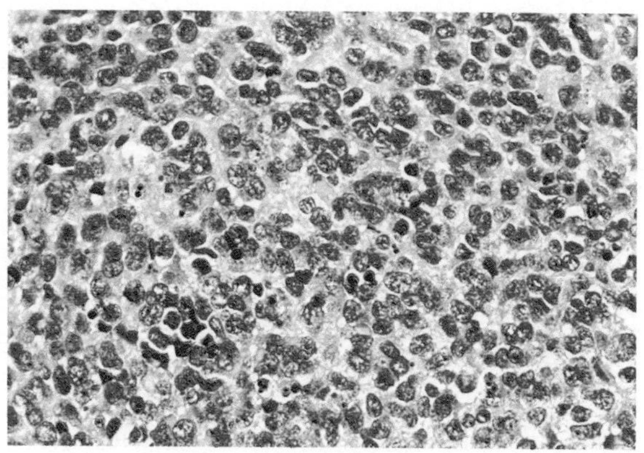

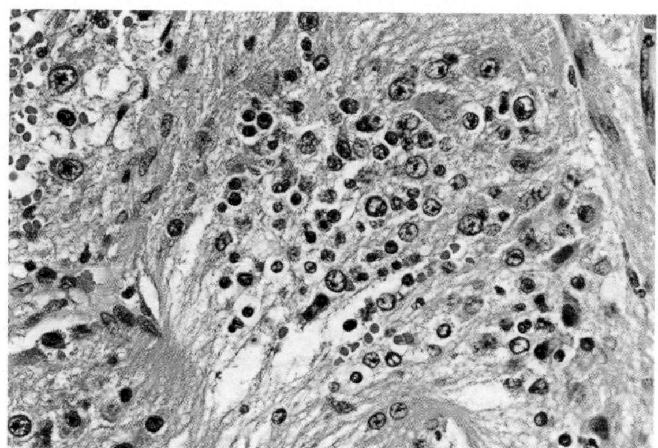

Figure 5-46

GANGLIONEUROBLASTOMA

The spectrum of differentiation in a rare CNS tumor extends from neuroblasts (left) to ganglion cells (right). The smaller ganglioid cells help exclude the possibility that the ganglion cells are merely trapped normal elements.

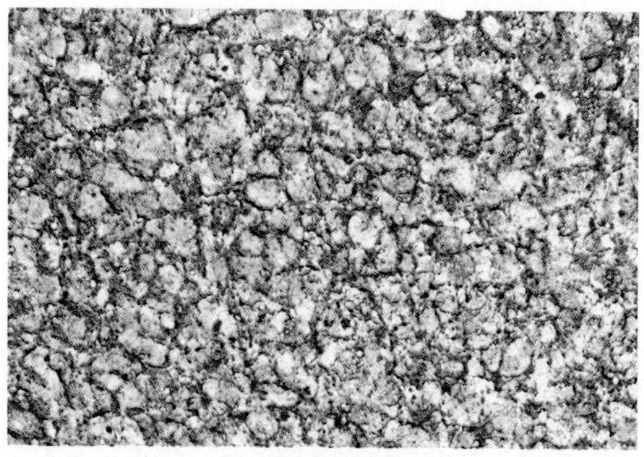

Figure 5-47

NEUROBLASTOMA

The fibrillar areas of neuropil are strongly immunoreactive for synaptophysin.

nodularity. As with medulloblastoma, a connective tissue response is most conspicuous when a tumor reaches the leptomeninges.

Focal or partial differentiation to neurons is not rare but is less common than in neuroblastomas arising in peripheral autonomic ganglia (fig. 5-46) (148,149). Although ganglion cells may be found in some central neuroblastomas, maturation proceeds only as far as the intermediate ganglioid cell stage in most instances. Cells with increased cytoplasm, minimal Nissl substance at best, and emerging nucleoli typically sit in a finely fibrillar, synaptophysin-positive neuropil background. Advanced differentiation to ganglion cells is present in ganglioneuroblastoma.

Lesions intermediate between neuroblastoma and ganglioneuroblastoma may be difficult to classify and grade given the variable percentages of small, medium, and large cells. At what arbitrary point a neoplasm has a high enough proportion of small cells to be considered neuroblastoma, or of ganglion cells to be a ganglion cell tumor, is not clear. The most primitive cells are presumably the principal proliferating element but may make up only a small percentage of the lesion. Little is known about the biologic behavior of such lesions, particularly in regard to optimal postoperative treatment (148,149).

Immunohistochemical Findings. Highly fibrillar areas, including cores of Homer-Wright rosettes and cytoplasm of ganglioid cells and neuroblasts, are immunoreactive for synaptophysin (fig. 5-47). Immunopositivity for neurofilament protein, heavily dependent upon optimal fixation and advanced differentiation, may be lacking or restricted to rare neuronal cell bodies or processes. As in ganglion cell tumors, the nuclei of relatively mature neurons may be immunoreactive for NeuN, but experience is limited with this antibody in neuroblastomas. Generally, it is more positive in normal than in neoplastic cells.

Since normal gray matter is synaptophysin positive, it is essential that this cause for reactivity is excluded. This can be difficult to do, but is the likely explanation if normal ganglion cells are present or if the immunoreactivity is restricted to a band-like area.

Some degree of immunoreactivity for GFAP, noted in virtually all cases of neuroblastoma and ganglioneuroblastoma, is usually attributable to scattered reactive astrocytes whose long processes concentrate about vessels. These cells, which can be very large, are especially prominent at the periphery of the lesion where the processes may be linearly arrayed between streams of infiltrating tumor cells. In other neuroblastomas (or PNETs with neuroblastic differentiation), the presence of neoplastic astrocytes is incontrovertible, since GFAP positivity appears in the small neoplastic cells with an eccentric skirt of pink cytoplasm. Such tumors fall into the category of embryonal tumors with mixed neuronal and glial differentiation, described later in this chapter. In some cases, even within the center of the lesion, the neoplastic versus reactive nature of the glial component may be difficult to determine, particularly when only a feltwork of GFAP-positive processes is seen. It is not clear whether a glial component has prognostic significance.

Ultrastructural Findings. Neoplastic neuroblasts are noted for their high nuclear to cytoplasmic ratio and microtubule-containing bipolar processes. Neurosecretory granules or vesicles are present, although usually are sparse (145,150–152). Synapse formation is generally lacking.

Cytologic Findings. In smear preparations, neuritic processes, in aggregate, form a limited fibrillar background. The round to somewhat elongated nuclei are small and dark.

Molecular and Cytogenetic Features. In contrast to medulloblastoma, loss of chromosome 17p is exceptional in supratentorial neuroblastomas, if it occurs at all (153–155). Loss of chromosome 3p12.3-p14 has been noted (155). Mutations in *p53* were not observed in 12 supratentorial PNETs (154).

Differential Diagnosis. Because malignant gliomas are generally treated with only local radiation, but neuroblastomas, as PNETs, are treated with local and neuraxis radiation, the distinction between these two classes of neoplasm is extremely important. Unfortunately, this distinction can be difficult, if not impossible, and there are few absolutes. Features favoring a glioma are: 1) diffuse infiltration, particularly with "secondary structures" such as perineuronal satellitosis; 2) necrosis with pseudopalisading (although this is by no means an absolute criterion); 3) nuclear pleomorphism; 4) microvascular proliferation; and 5) presence of better-differentiated areas. A small cell embryonal tumor is favored by: 1) occurrence in a very young patient; 2) radiologically well-circumscribed lesion; 3) fibrous stroma; and 4) many small cells with an intensely GFAP-positive, small "skirt" of cytoplasm in the subset of small cell embryonal tumors with glial differentiation.

A distinctive lesion, neuroblastic tumor with abundant neuropil and true rosettes, is discussed in the following section.

Central neurocytoma, a neuronal neoplasm related to neuroblastoma, is mentioned briefly above and is discussed in chapter 4. Like neuroblastoma, neurocytoma is cellular, but does not appear "primitive" since it exhibits remarkable cytologic uniformity consisting of benign-appearing, uniform, round, mitotically inactive nuclei with delicate "salt and pepper" chromatin and inconspicuous micronucleoli. Central neurocytomas typically appear in a lateral ventricle near the septum pellucidum in the region of the foramen of Monro, and generally in young adults.

Treatment and Prognosis. There is little available information regarding prognostic factors, either clinical or histologic, for patients with neuroblastoma. In one study, the only significant factor was patient age, with recurrences more likely in younger patients (141). Full maturation of a neuroblastoma to a ganglion cell tumor, as occasionally occurs with peripheral neuroblastoma, is rare (149). Extracranial metastases have been reported (156,157).

NEUROBLASTIC TUMOR WITH ABUNDANT NEUROPIL AND TRUE ROSETTES

Since the pattern discussed here, *neuroblastic tumor with abundant neuropil and true rosettes*, appears in published discussions of both medulloepithelioma and ependymoblastoma, it is not clear whether the lesion is a variant of other small cell embryonal tumors or is an entity of its own. In its fully expressed form, however, this lesion is certainly distinctive and does not fulfill classic definitions of either of these two entities (158). The tissue pattern is more common than that of classic medulloepithelioma or ependymoblastoma. The tumor

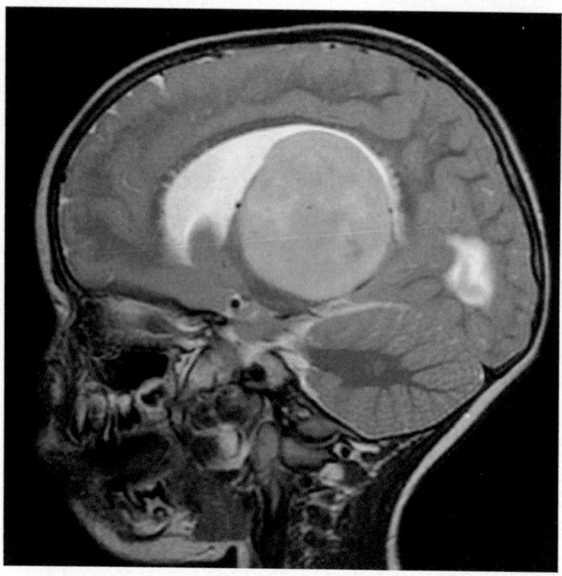

Figure 5-48

NEUROBLASTIC TUMOR WITH ABUNDANT NEUROPIL AND TRUE ROSETTES

As in this 2-year-old, the lesion typically occurs in the young, and in some cases as a well-circumscribed mass.

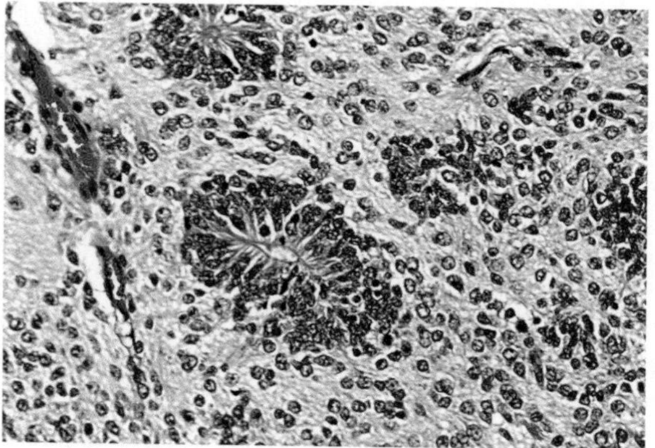

Figure 5-49

NEUROBLASTIC TUMOR WITH ABUNDANT NEUROPIL AND TRUE ROSETTES

Well-formed epithelial rosettes arise directly out of a finely fibrillar, neuropil-like background.

appears in the few first years of life as a large, and seemingly, well circumscribed mass (fig. 5-48).

It contains regions of small cell embryonal tumor; abundant, finely fibrillar, synaptophysin-positive neuropil; and lumen-containing rosettes (fig. 5-49). Distinctively, the rosettes usually have a small circular or slit-like lumen, or only a minute granular core without a discernible lumen, and often arise directly within the neuropil-like areas.

The neoplasm is malignant. Survival periods are generally less than 2 years.

EPENDYMOBLASTOMA

Definition and General Features. The term *ependymoblastoma* denotes a rare, small cell embryonal neoplasm with ependymoblastic rosettes (159,160). Although in concept and practice ependymoblastoma and ependymoma are distinguishable in most instances, sometimes they are not, and we have seen anaplastic ependymomas evolve into tumors with "primitive" cytologic features and slit-like ependymoblastic rosettes. There is also histologic overlap with medulloepithelioma. Very few examples of ependymoblastoma have been described.

Clinical and Radiologic Findings. Although rare ependymoblastomas occur in adults, most are encountered in the first 5 years of life (159, 160), or even congenitally (161). Any level of the neuraxis may be affected, but the cerebrum is the most common site.

Given the rarity of the entity, its radiographic features have not been systematically studied. The tumor should appear similar to other embryonal tumors, i.e., large, discrete, contrast enhancing, and deeply situated.

Gross Features. Like other densely cellular embryonal neoplasms, ependymoblastomas are well circumscribed, fleshy, and often highly vascular.

Microscopic Findings. The distinctive feature is the presence of ependymoblastic rosettes within fields of undifferentiated cells (fig. 5-50). These multilayered rosettes have small, round to slit-like lumens. Mitoses are abluminal. Chromatin is coarse and nucleoli are distinct. Smaller versions of these rosettes, with their prominent internal limiting membrane, can closely resemble the Flexner-Wintersteiner rosettes of retinoblastoma.

Immunohistochemical Findings. Experience with this entity is limited, but the rosettes of ependymoblastoma, unlike those of ordinary ependymoma, are reportedly negative for GFAP (162).

Ultrastructural Findings. Reports of the ultrastructural features of ependymoblastoma are few. They describe rosettes formed by cells with

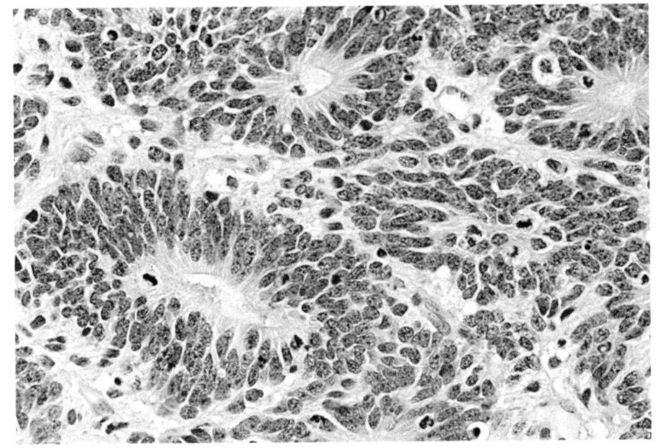

Figure 5-50

EPENDYMOBLASTOMA

The presence of slit-like rosettes in malignant, mitotically active tissue is the lesion's defining feature.

apical microvilli, cilia with basal bodies, and apical junctional complexes (162,163).

Differential Diagnosis. Distinguishing ependymoblastoma from anaplastic ependymoma depends on finding primitive rosettes of ependymoblastic type in association with a monotonous undifferentiated cell background. It is this association of malignant small cell neoplasm with well-formed but mitotically active true rosettes that, in concept, distinguishes ependymoblastoma as an entity. As conventional ependymomas become more anaplastic, they usually give up their rosette-forming capacity and evolve into densely cellular neoplasms with only focal perivascular fibrillar zones (pseudorosettes). Vascular proliferation accompanies anaplasia and becomes an additional feature of differential diagnostic importance, since it is not typical of ependymoblastoma or other embryonal tumors. Nevertheless, occasional high-grade ependymomas contain slit-like true rosettes identical to those in ependymoblastomas.

In most cases, the distinction of the ependymoblastoma from the medulloepithelioma is not difficult and centers primarily upon recognizing the morphology of the epithelial components. The epithelial structures of medulloepithelioma are more complicated and rest on an external basement membrane. Their nuclei more fully occupy the full thickness of the epithelium, and thus lack the luminal nuclei-free zone of ependymoblastic rosettes. In addition, medulloepitheliomas are, in the majority of cases, complex neoplasms exhibiting divergent differentiation toward glia, neurons, and even mesenchyme. Nonetheless, the distinction of medulloepithelioma from ependymoblastoma may be difficult when surgical sampling is limited or in cases in which ependymal or ependymoblastic differentiation is present. The issue is complicated by the presence of ependymoblastic rosettes in occasional medulloepitheliomas and other small cell embryonal tumors of mixed type.

An additional lesion that overlaps with both ependymoblastoma and medulloepithelioma is neuroblastic tumor with abundant neuropil and true rosettes, described in the previous section. The latter lesion is distinctive for its prominent, finely fibrillar neuropil.

Treatment and Prognosis. Ependymoblastoma is an aggressive tumor with a tendency to craniospinal dissemination. Most patients die within 1 year of surgery. Little information exists regarding the efficacy of radiotherapy or chemotherapy (159).

TUMORS OF THE TRILATERAL RETINOBLASTOMA SYNDROME

When a small cell embryonal tumor occurs as an intracranial midline mass in conjunction with the bilateral, hereditary form of retinoblastoma and either the pineal or suprasellar regions are affected, the complex lesion is termed *trilateral retinoblastoma* (164–166). The intracranial blastomas are densely cellular masses that, like their ocular counterparts, are replete with Homer-Wright and Flexner-Wintersteiner rosettes.

MISCELLANEOUS SMALL CELL EMBRYONAL TUMORS OF THE BRAIN AND SPINAL CORD

Largely Undifferentiated Small Cell Embryonal Neoplasms

These lesions vary considerably in histologic pattern and nuclear characteristics. Some are composed of closely packed, cytologically uniform cells, whereas others are frankly malignant, with coarse chromatin, frequent mitoses, and either single cell or geographic necrosis. If these lesions are immunoreactive for synaptophysin, it can be debated whether the lesion is best considered neuroblastoma or a histologically undifferentiated embryonal tumor. It remains to be seen what impact better tissue

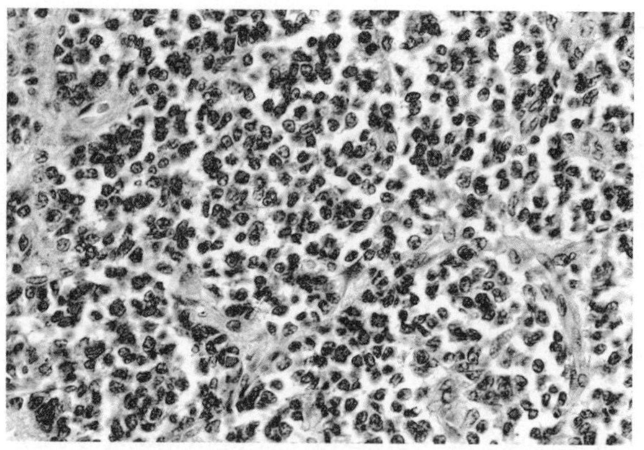

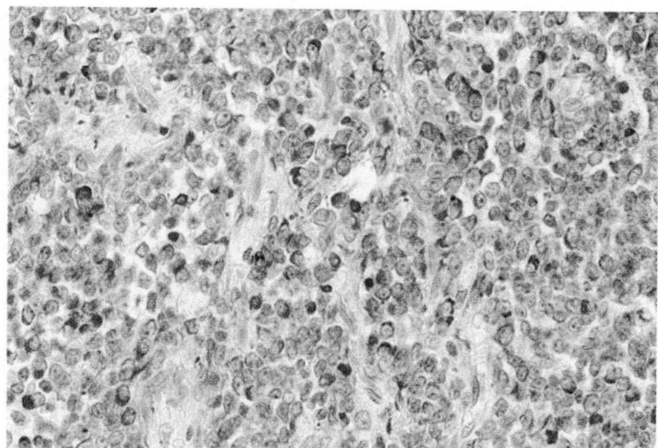

Figure 5-51

SMALL CELL EMBRYONAL NEOPLASM WITH GLIAL DIFFERENTIATION

Left: There may be little in the appearance of the lesion after staining with hematoxylin and eosin (H&E) to suggest glial differentiation.

Right: Immunopositivity for GFAP is necessary to establish glial differentiation.

sampling, ultrastructural characterization, improved fixation, molecular techniques, and newer staining methods will have upon the ratio of "undifferentiated" and "differentiated" embryonal tumors.

Small Cell Embryonal Tumors with Glial Differentiation

Histologically, these lesions vary in appearance but are in large part diffusely cellular, monomorphous, and round cell in composition (fig. 5-51, left). Most are highly cellular and have cytoplasm so scanty that the possibility of glial differentiation may not be evident. Reactivity for GFAP, usually confined to the immediate perinuclear region, may nevertheless be impressive (fig. 5-51, right). Immunopositive fields may alternate with expanses of nonreactivity.

Embryonal tumors with glial differentiation must be distinguished from both embryonal tumors of neuroblastic type and anaplastic gliomas. Immunostaining for synaptophysin differentiates the former although the staining result may be difficult to interpret if the tumor is situated in the cortex since the latter is strongly immunoreactive as well.

Small Cell Embryonal Tumors with Mixed Neuronal and Glial Differentiation

The prototypic embryonal tumor with mixed neuronal/neuroblastic and glial differentiation is medulloepithelioma when it expresses its potential for divergent differentiation. Neuroblastic tumors that contain ependymoblastic rosettes are also mixed. Neuronal medulloblastomas, usually nodular neoplasms with glial differentiation, fit into the mixed neuronal/glial category but are considered separately.

On rare occasion, other mixed neuronal/glial embryonal tumors arise in the supratentorial compartment. Analogous to medulloblastoma, these tumors are densely cellular, and often contain Homer-Wright rosettes, synaptophysin-positive perivascular zones of fibrillarity, and some GFAP-reactive cells (167). In yet other extracerebellar embryonal tumors, both the neuronal and glial components are better differentiated (168).

Small Cell Embryonal Neoplasms with Mesenchymal Differentiation

A desmoplastic response is common in many embryonal tumors, such as neuroblastoma and medulloblastoma, as well as the PNET defined in 1973 (169). Densely cellular CNS neoplasms occurring outside the cerebellum may also exhibit differentiation along myoid lines (170). Some of these tumors are considered neuroepithelial tumors with muscle differentiation by some or rhabdomyosarcomas by others, the object of disagreement hinging upon the neuroectodermal versus the mesenchymal character

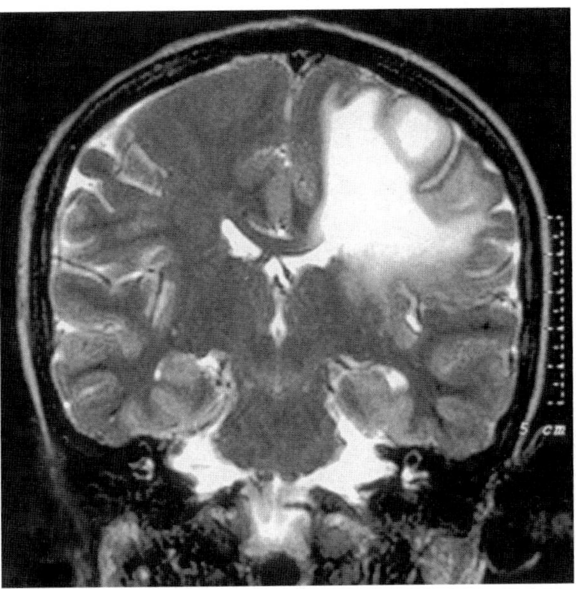

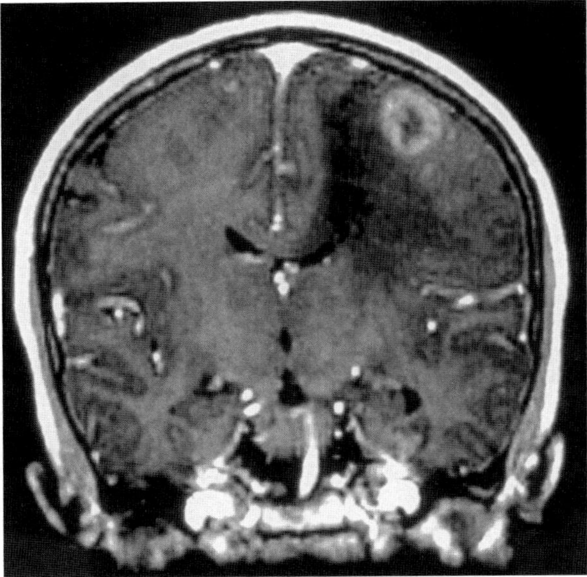

Figure 5-52

MALIGNANT NEOPLASM FOLLOWING TREATMENT OF ACUTE LYMPHOCYTIC LEUKEMIA

Seven years after treatment of acute lymphocytic leukemia (ALL), which included cranial irradiation, this 13-year-old girl presented with a seizure and this edema-producing (left), contrast-enhancing (right), parietal lobe mass.

of the undifferentiated-appearing component. The presence of a synaptophysin- or GFAP-positive component (170) helps establish the lesion as small cell embryonal tumor rather than rhabdomyosarcoma.

Peripheral Primitive Neuroectodermal Tumor/Ewing's Sarcoma

While most intracranial small cell embryonal tumors are PNETs of the central type, occasional lesions fulfill criteria, including the t(11;22)(q24;q12) translocation, for peripheral PNET/Ewing's sarcoma (171).

Malignant Tumor Following Prophylactic Irradiation for Acute Lymphocytic Leukemia

These cerebral lesions (fig. 5-52), which have mutations in *K-ras*, are densely cellular, mitotically active lesions that may exhibit neuroblastic neuronal features, glial differentiation, or both (172). A distinctive ribboning may be present (fig. 5-53). This and other lesions with ribboning are given in Appendix J. Most lesions in this clinical setting lack the features of this PNET, and are infiltrative high-grade gliomas or frank glioblastomas.

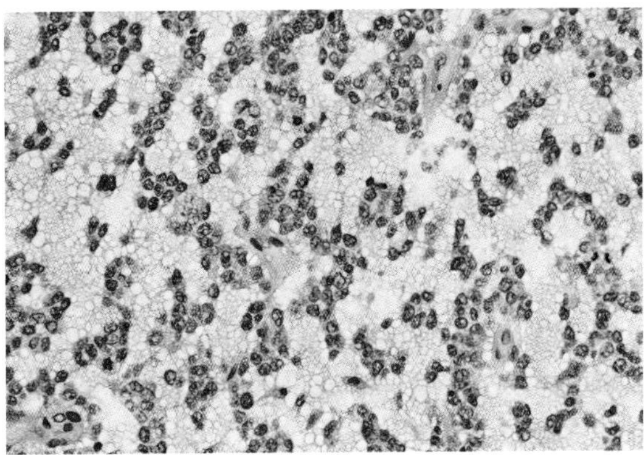

Figure 5-53

MALIGNANT NEOPLASM FOLLOWING TREATMENT OF ACUTE LYMPHOCYTIC LEUKEMIA

While the histopathologic features vary, the tumors often have cords or clusters of epithelioid cells.

OLFACTORY NEUROBLASTOMA (ESTHESIONEUROBLASTOMA)

Olfactory neuroblastoma, also called *esthesioneuroblastoma,* forms a polypoid mass that presents as epistaxis or nasal obstruction, often longstanding, and which usually affects adults (173, 174). Involvement of the cribriform plate is so typical that in its absence a diagnosis of olfactory

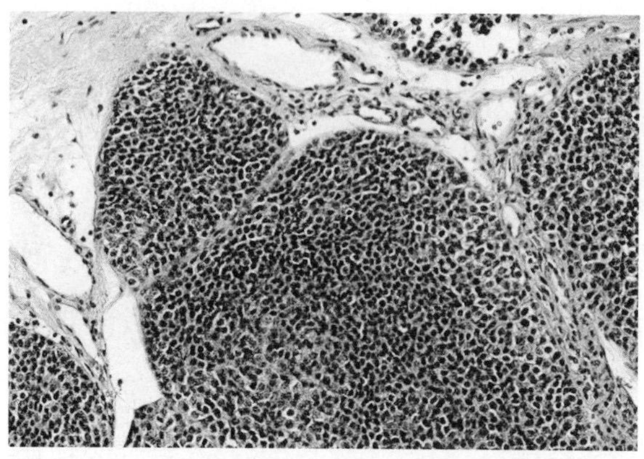

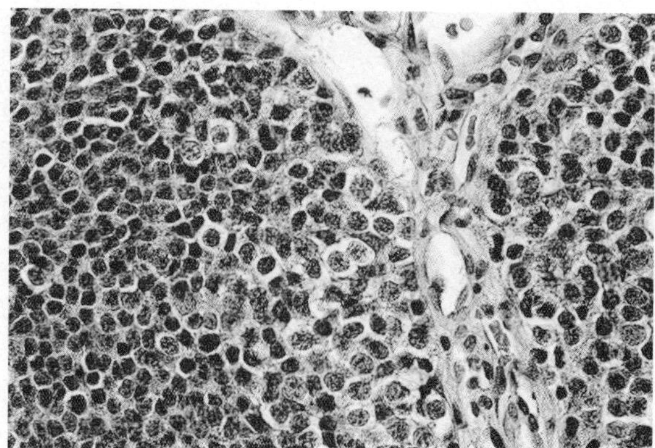

Figure 5-54

OLFACTORY NEUROBLASTOMA

Left: The classic lesion expands in the submucosa as a multilobed mass.
Right: Well-circumscribed lobules are often composed of round, uniform cells with little mitotic activity.

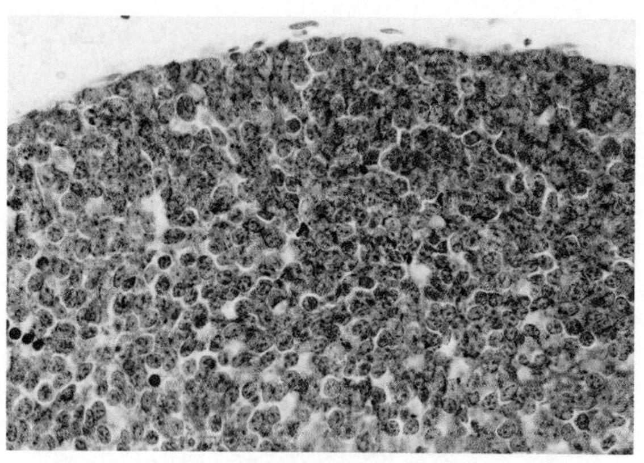

Figure 5-55

OLFACTORY NEUROBLASTOMA

Olfactory neuroblastomas are immunoreactive for synaptophysin.

neuroblastoma should be questioned. Subfrontal intracranial extension may produce a midline intradural mass that elevates the frontal lobes (175). Olfactory neuroblastoma is discussed in detail in the Armed Forces Institute of Pathology (AFIP) Tumor Fascicle *Tumors of the Upper Aerodigestive Tract and Ear* (174).

At low magnification, two principal growth patterns are seen: smooth-contoured cellular nests and lobules within a vascular stroma (fig. 5-54). Less often, there is diffuse sheet-like growth supported by delicate capillaries. At higher magnification, cellular monomorphism is conspicuous. Although a suggestion of rosetting is often evident, well-formed Homer-Wright rosettes are present in only a minority of lesions. Flexner-Wintersteiner–like rosettes, also termed olfactory rosettes, are rare. As in other neuroblastic tumors, olfactory neuroblastomas usually exhibit a delicate fibrillary background composed of neuroblastic processes: in essence, a neoplastic neuropil. Ganglion cell differentiation is infrequent. The monomorphous cells are small with round nuclei and delicate, stippled chromatin. Nucleoli are inconspicuous. Necrosis is uncommon.

Key among the immunoreactivities of olfactory neuroblastoma is the high frequency of positivity of the nested cells for neuronal markers such as synaptophysin (fig. 5-55) and of the sustentacular cells at the edge of the lobules for S-100 protein (fig. 5-56). Scattered S-100 protein–positive cells are also present internally within lobules. Other neuronal markers include neurofilament protein, microtubule-associated protein, and class III beta-tubulin.

The differential diagnosis includes such entities as such as carcinoma of various types (sinonasal undifferentiated [SNUC] and neuroendocrine), sarcomas, and malignant melanoma (174).

Although pituitary adenomas are often included in the differential diagnosis of olfactory neuroblastoma, and vice versa, the two lesions occur in regions that are close, but geographically distinct. Olfactory neuroblastomas, even when large, do not involve, or especially enlarge,

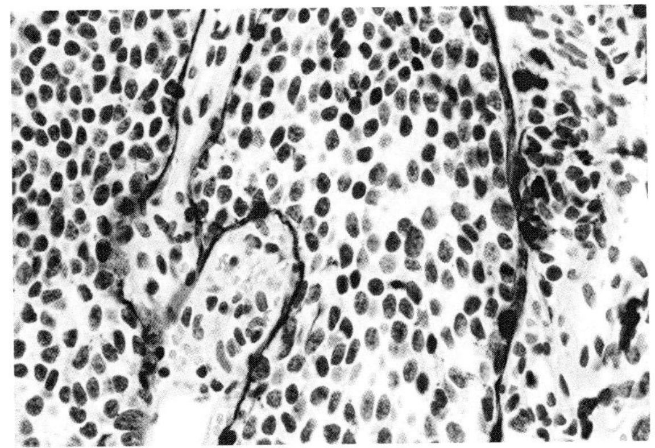

Figure 5-56

OLFACTORY NEUROBLASTOMA

Slender, S-100 protein–positive sustentacular cells circumscribe lobules.

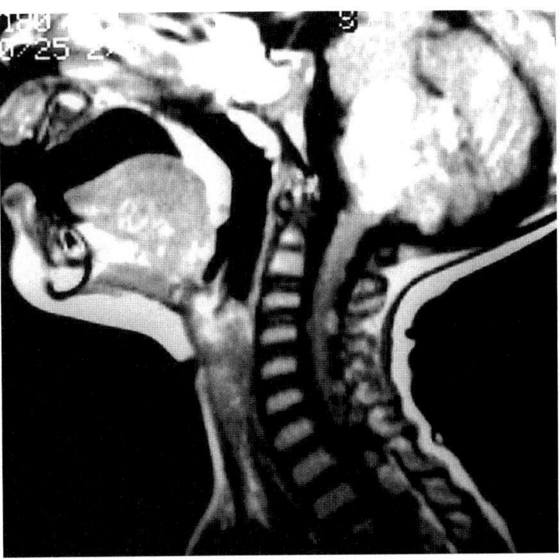

Figure 5-57

ATYPICAL TERATOID/RHABDOID TUMOR

As in this 19-month-old boy who presented with a mass centered on the brain stem, AT/RTs often arise in the posterior fossa. The drop metastases anterior to the medulla and cervical spinal cord were noted at the time of presentation. (Fig. 7-11 from Fascicle 10, 3rd Series.)

the sella turcica. Adenomas have a more neuroendocrine appearance and are more likely to be immunoreactive for chromogranin, but staining for pituitary hormones may be necessary for reassurance.

ATYPICAL TERATOID/RHABDOID TUMOR

Definition. *Atypical teratoid/rhabdoid tumor* (AT/RT) is a highly malignant, immunohistochemically polyphenotypic neoplasm of uncertain origin.

General Features. The CNS lesion described here is related histologically and molecularly to rhabdoid tumor of the kidney and other extrarenal, but non-CNS, counterparts. A germline mutation involving the *INI1* locus can be established in some cases of primary renal or CNS neoplasms, either alone or in combination (176). Small cell tumors of the cerebrum, pineal gland, and cerebellum have been reported in the setting of renal rhabdoid tumors, but it is not clear if these are small cell embryonal tumors or AT/RTs without the classic large, pale cells (177, 178), although largely anecdotal experience is that they are AT/RTs. In a set of monozygous twins, one infant had a cerebellar neoplasm with the appearance of medulloblastoma, the other a systemic and spinal epidural AT/RT (179).

Clinical Features. Almost all patients with AT/RT are below the age of 2 years, and often are less than 1 year (180–183). Boys are more often affected. Only rarely does the tumor occur in adults (184,185).

Radiologic Findings. The characteristically large, rapidly enlarging, sometimes massive tumors occur throughout the CNS (fig. 5-57) (186), but are, in our experience, most frequent in the posterior fossa. In contrast to medulloblastomas, infratentorial AT/RTs are often more extra-axial, such as at the cerebellopontine angle, than deeply seated within the cerebellar parenchyma. Large areas of hemorrhage and necrosis are also more common in AT/RTs than in medulloblastoma. AT/RTs may be multifocal at the time of diagnosis (187).

Gross Findings. The bulky, fleshy to firm lesions are often extensively necrotic.

Microscopic Findings. The cellular sheets of neoplastic cells are interrupted by fibrovascular septa that are often broad and edematous (fig. 5-58). Necrosis, frequently with dystrophic calcification, is common (fig. 5-59).

In spite of the designation, the lesion is usually not composed solely or even largely of cells with overtly rhabdoid features, but rather a jumble of medium to large cells with nuclei of moderate density and clearly visible cytoplasm (fig. 5-60). Some lesions do contain rhabdoid cells with large, round or reniform nuclei;

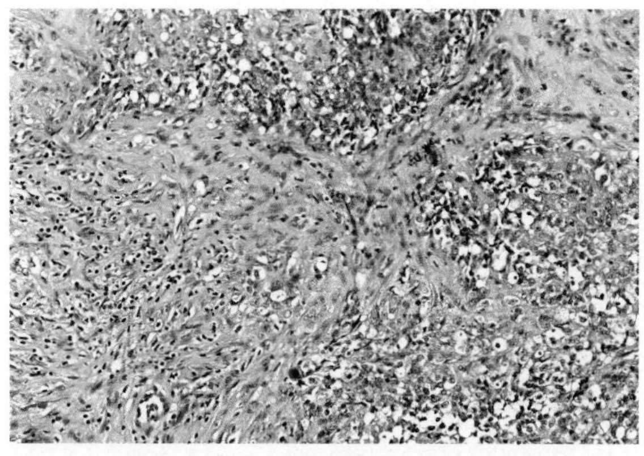

Figure 5-58

ATYPICAL TERATOID/RHABDOID TUMOR

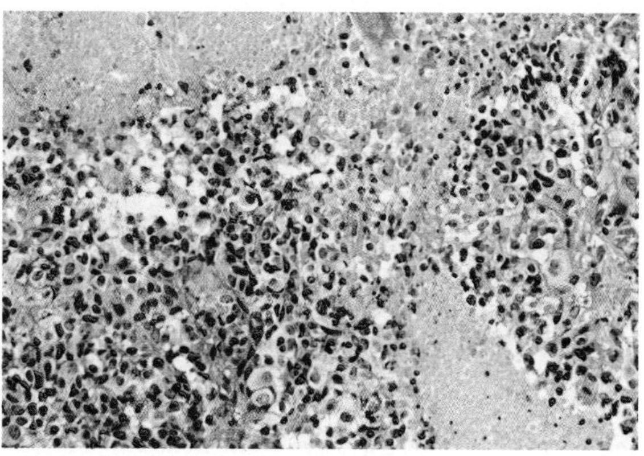

Figure 5-59

ATYPICAL TERATOID/RHABDOID TUMOR

Necrosis is common in AT/RT.

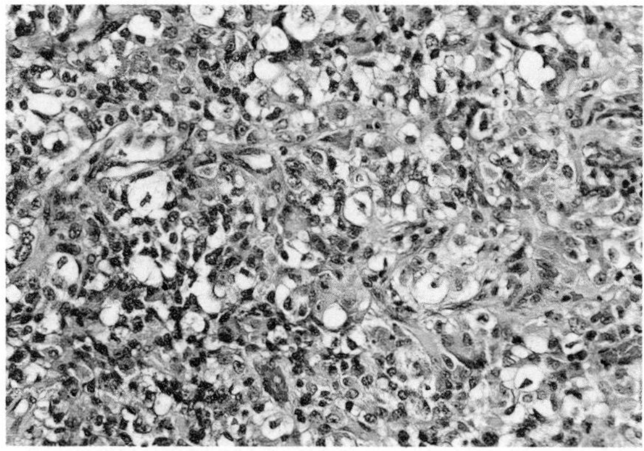

Figure 5-60

ATYPICAL TERATOID/RHABDOID TUMOR

The prototypical lesion has jumbled architecture with a prominent component of large pale cells.

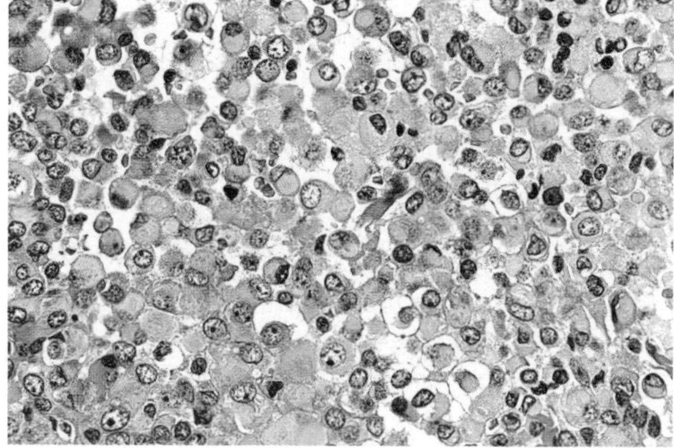

Figure 5-61

ATYPICAL TERATOID/RHABDOID TUMOR

Tissue so overtly and uniformly rhabdoid is unusual in AT/RT.

prominent nucleoli; and cytoplasmic inclusion-like mass of filaments (fig. 5-61), but even these tumors are not usually rhabdoid throughout.

Some AT/RTs are predominantly "small cell" and can then be difficult to distinguish from an embryonal tumor such medulloblastoma. The cells, albeit closely packed, are more pleomorphic and jumbled than the typical medulloblastoma (fig. 5-62). Fascicular sarcoma-like areas are common (fig. 5-63).

Small nests and cords of cells can resemble either chordoma or the trabecular pattern of renal rhabdoid tumor (fig. 5-64). Vacuolated cells are frequent either as sheets or as individual forms; the latter gives a "starry sky" appearance.

In spite of its epithelial immunohistochemical markers, the tumor only uncommonly shows any histologic differentiation, and then only crude glands (fig. 5-65), Flexner-Wintersteiner rosettes, or primitive squamous epithelium.

Immunohistochemical Findings. The undifferentiated histologic appearance belies a complex immunophenotype, with frequent positivity for GFAP (fig. 5-66, left), epithelial membrane antigen (fig. 5-66, right), cytokeratins, smooth muscle actin, and vimentin (180, 181,188,189). Reactivity for desmin is sometimes present, but is uncommon, and strap cells are not present. The loss of nuclear immunoreactivity for the INI1 protein, and its aid in the

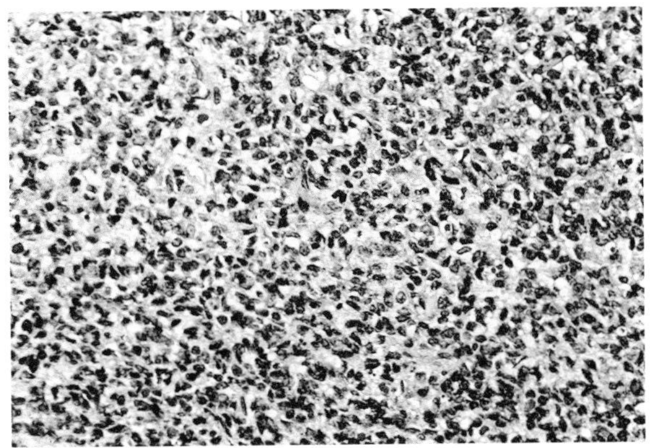

Figure 5-62

ATYPICAL TERATOID/RHABDOID TUMOR

A small cell component can closely mimic medullo-

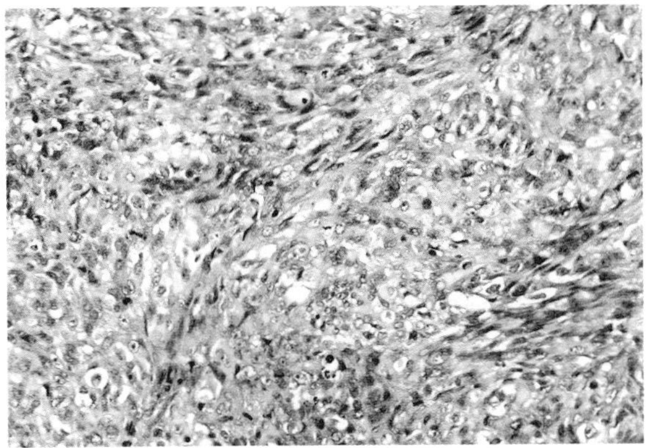

Figure 5-63

ATYPICAL TERATOID/RHABDOID TUMOR

Tightly packed fascicles give a sarcoma-like appearance

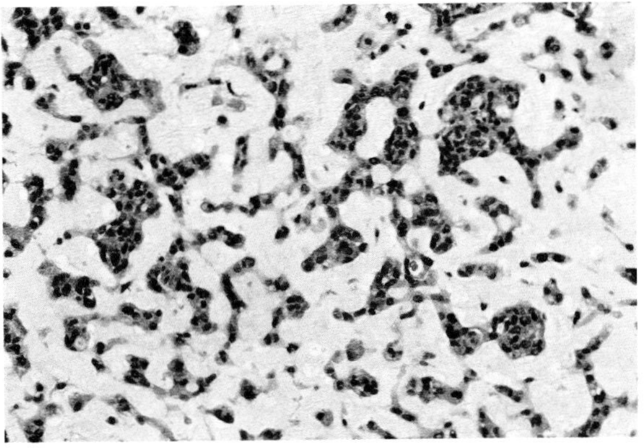

Figure 5-64

ATYPICAL TERATOID/RHABDOID TUMOR

Cords and narrow trabeculae are uncommon features.

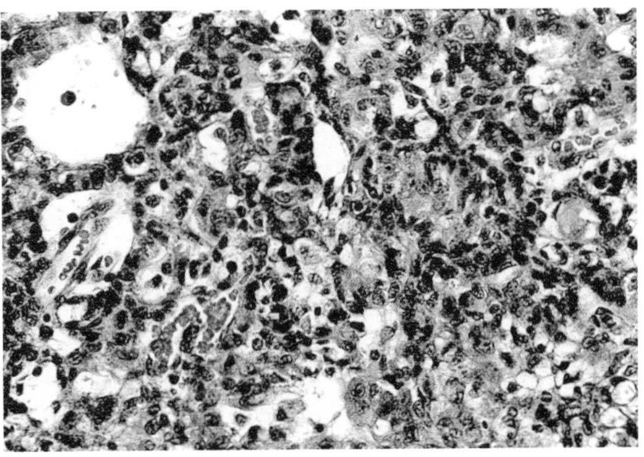

Figure 5-65

ATYPICAL TERATOID/RHABDOID TUMOR

Crude attempts to form an epithelial surface are occasionally present.

differential diagnosis is discussed below in the section, Differential Diagnosis (see fig. 5-71).

Positivity for neuronal markers is less common than for epithelial markers and GFAP in our experience, but others have found frequent positivity for neurofilament protein (188). Synaptophysin positivity is present occasionally.

Most individual cells are immunoreactive for only one antigen, but some are positive for two (190). The MIB-1 index is high, ranging in one study from 10 to 28 percent (191).

Ultrastructural Findings. Features include masses of intermediate filaments that create the cytoplasmic inclusion (fig. 5-67) (175,186,192, 193). Neurosecretory granules are observed in some cases (186).

Cytologic Findings. The cytologic polymorphism and large pale cells are seen well in smear preparations (fig. 5-68).

Molecular and Cytogenic Findings. The tumor is ascribed to mutations in the gene *INI1* on chromosome 22q, the same gene that appears to be responsible for renal rhabdoid tumors (173, 194,195). FISH is useful in the diagnosis by demonstrating loss at this locus in many cases (fig. 5-69) (185,196). In the case of mitotic recombination, however, there is disomy, but still loss of heterozygosity since the two chromosomes harbor the same mutated allele. Mutational analysis is necessary in such cases if a molecular confirmation is desired.

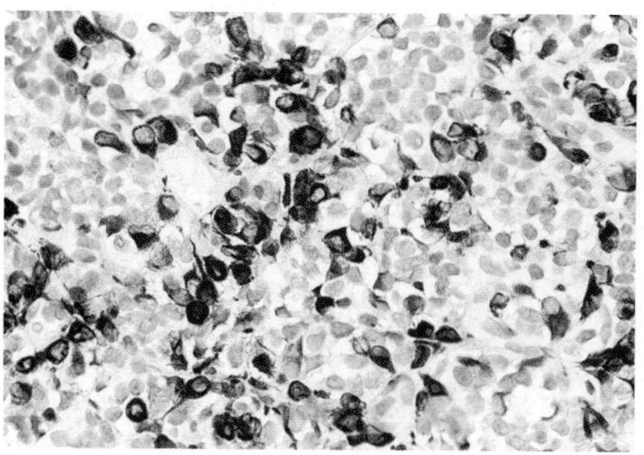

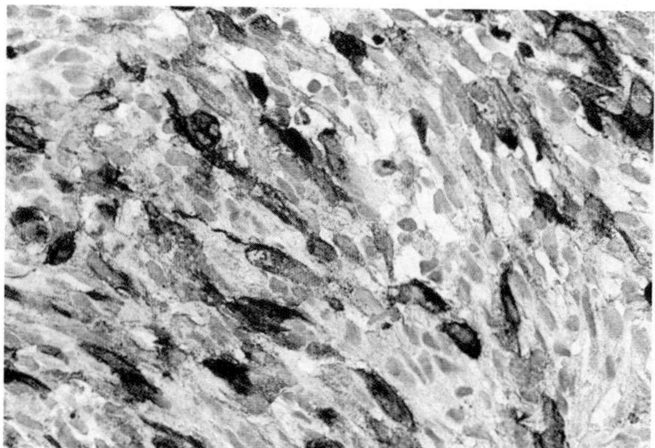

Figure 5-66

ATYPICAL TERATOID/RHABDOID TUMOR

AT/RTs are immunohistochemically polyphenotypic. These figures illustrate reactivity for GFAP (left) and epithelial membrane antigen (right).

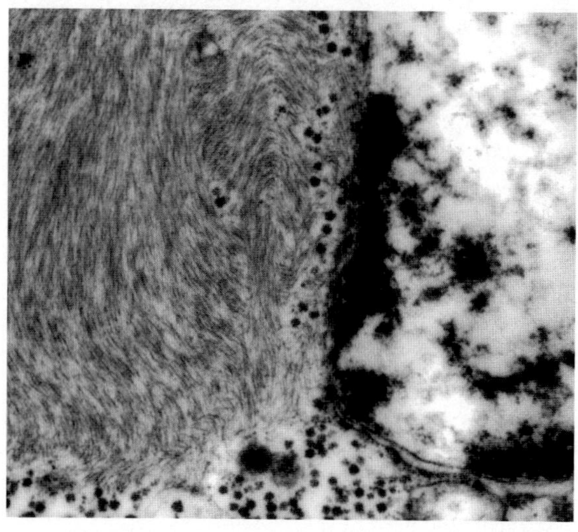

Figure 5-67

ATYPICAL TERATOID/RHABDOID TUMOR

An aggregate of intermediate filaments creates the cytoplasmic inclusion of the rhabdoid cell. (Courtesy of Dr. Kathryn A. Kolquist, Marshfield, WI.)

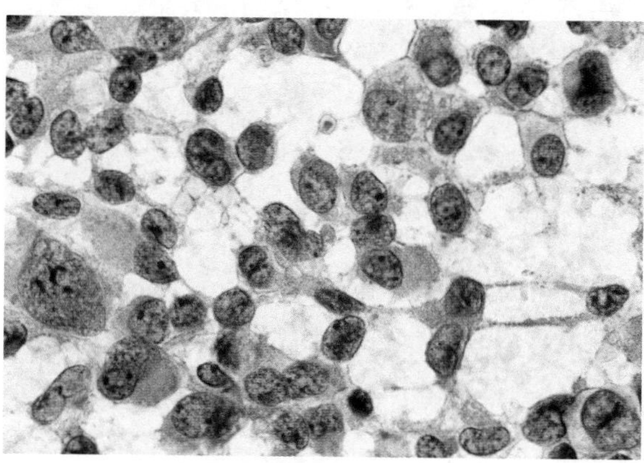

Figure 5-68

ATYPICAL TERATOID/RHABDOID TUMOR: SMEAR PREPARATION

The round bland nuclei and focal rhabdoid cytoplasmic features are seen well in cytologic preparations.

Differential Diagnosis. AT/RT should be carefully considered in the face of a cellular malignant CNS tumor occurring within the first 3 years of life, especially during the first 18 months. In the posterior fossa, the differential diagnosis logically focuses on medulloblastoma of either anaplastic/large cell or conventional type (fig. 5-70) (194). In one study of tumors originally diagnosed as medulloblastoma, the molecular results prompted review of the histologic sections, with reclassification of some "medulloblastomas" as AT/RT (194). Anaplastic and large cell medulloblastomas thus overlap with AT/RT, but generally are of higher grade in terms of mitotic activity and degree of nuclear hyperchromatism. Apoptosis is also extremely frequent and cell-cell wrapping is more prominent than in AT/RT. In addition, most AT/RTs have a distinctive jumbled assortment of large, pale cells in addition to any PNET-like areas. Nevertheless, the diagnosis often awaits the results of immunohistochemistry or molecular studies (196). Although deletion of chromosome

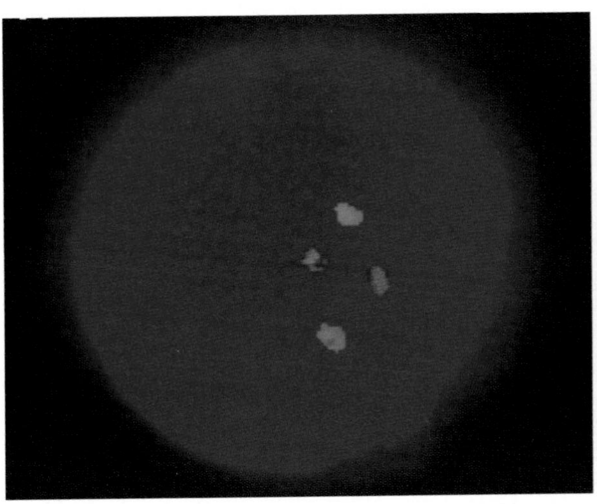
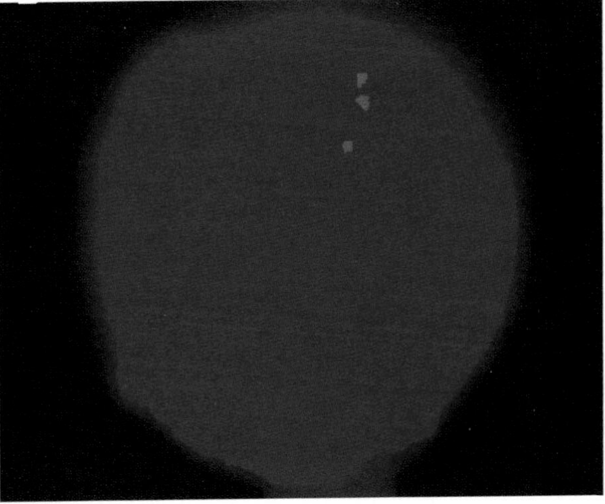

Figure 5-69

ATYPICAL TERATOID/RHABDOID TUMOR: FLUORESCENCE IN SITU HYBRIDIZATION

FISH can be used diagnostically to distinguish AT/RT from other small cell lesions such as medulloblastoma.

Left: A normal cell has the normal two signals for *INI1* at 22q11.2 (red) and two signals for Ewing's sarcoma locus (*EWS*) at 22q12 (green).

Right: An AT/RT cell has two signals for *EWS*, but only one for *INI1*. (Courtesy of Dr. Jaclyn Biegel, Philadelphia, PA.)

22q11.2 is not found in all cases of AT/RT, FISH has been useful in distinguishing this neoplasm from medulloblastoma (196). The distinction between AT/RT and medulloblastoma may be effected by immunohistochemistry for the INI1 protein (197). Staining of tumor cell nuclei is retained in medulloblastomas (fig. 5-71, left), but lost in AT/RT (fig. 5-71, right) (197,198).

The lateral alignment of cells to each other about septa or vessels can lend a vaguely epithelial appearance that simulates choroid plexus carcinoma (fig. 5-72). Immunostaining for the INI1 protein is useful in the distinction (198).

Treatment and Prognosis. AT/RT is highly malignant, with rapid local recurrence and cerebrospinal fluid spread in many cases (180, 181,183). Exceptionally, the neoplasm spreads systemically (199). It remains to be seen whether intensive therapy improves this grim outlook (182,183).

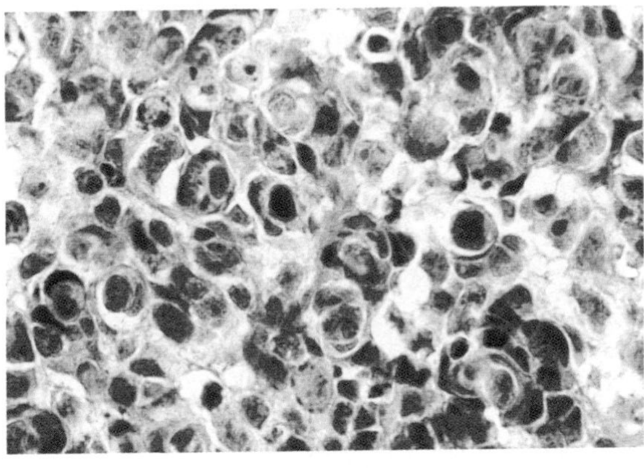

Figure 5-70

LARGE CELL MEDULLOBLASTOMA RESEMBLING ATYPICAL TERATOID/RHABDOID TUMOR

Large cell medulloblastoma can resemble AT/RT, although the nuclei of the medulloblastoma are coarser and more atypical. Cell wrapping is also more prominent.

MELANOTIC NEUROECTODERMAL TUMOR OF INFANCY

Melanotic neuroectodermal tumor of infancy (MNTI), or *melanotic progonoma,* is a rare neoplasm that generally arises in the maxilla (200–202). Involvement of the cranial vault or other extracranial bones is rare (200,201,203). Even rarer are similar lesions that arise in the brain (200,201, 204). Histologically comparable tumors in the pineal region are referenced in chapter 6.

This gray-brown tumor expands surrounding bone, compressing and invading tissue when large. It contains both pigmented epithelium, as well as islands of neuronal cells that vary from

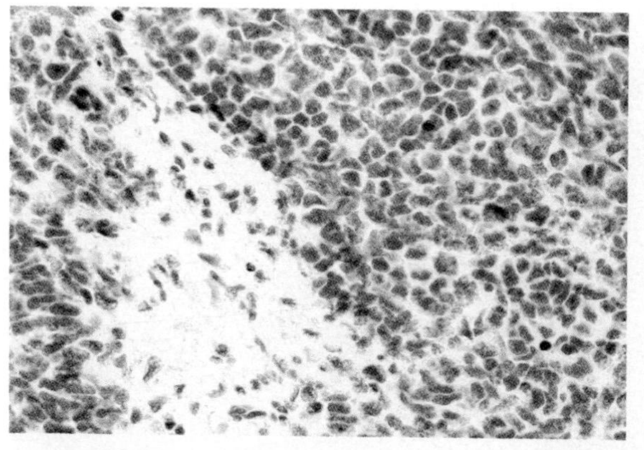

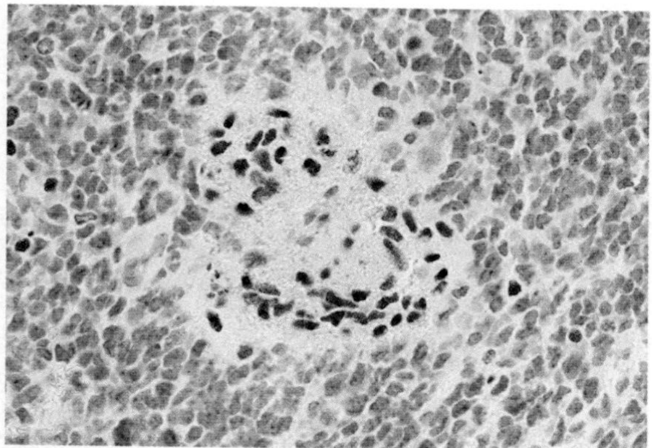

Figure 5-71

IMMUNOHISTOCHEMICAL DISTINCTION BETWEEN MEDULLOBLASTOMA AND ATYPICAL TERATOID/RHABDOID TUMOR

Left: The nuclei of both tumor cells and stroma are immunoreactive with antibodies to INI1 protein in a medulloblastoma.
Right: Only stromal and reactive cells are positive in AT/RT; tumor cells are negative. (Courtesy of Dr. Alexander R. Judkins, Philadelphia, PA.)

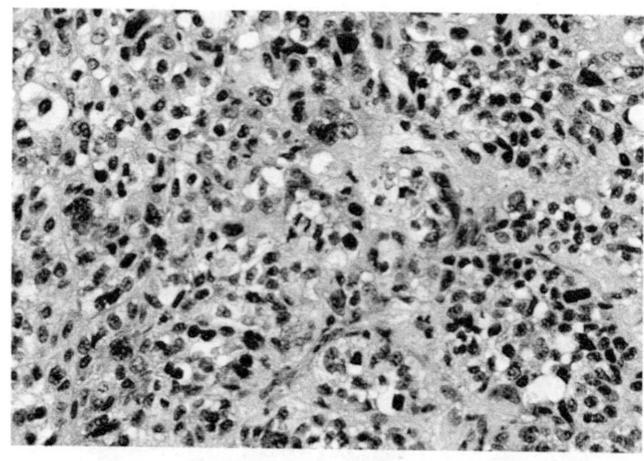

Figure 5-72

CHOROID PLEXUS CARCINOMA RESEMBLING ATYPICAL TERATOID/RHABDOID TUMOR

Solid regions within anaplastic choroid plexus carcinomas can resemble AT/RT. Elsewhere, this intraventricular lesion was distinctly papillary.

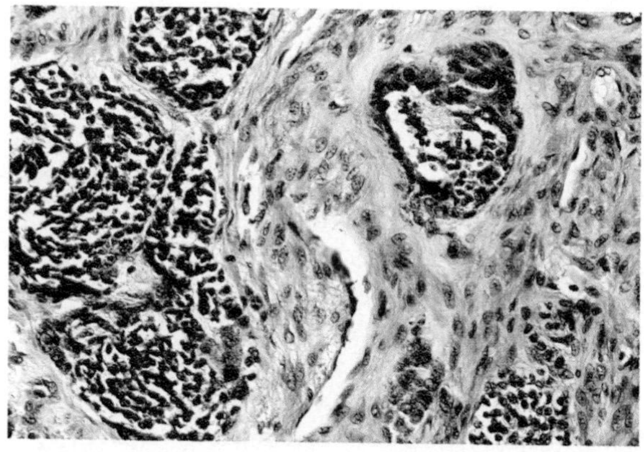

Figure 5-73

MELANOTIC NEUROECTODERMAL TUMOR OF INFANCY

As in this maxillary lesion, the classic lesion is composed of nests of small dark neuronal cells surrounded by delicate pigmented epithelium.

neuroblastic (fig. 5-73) to better-differentiated small neurons in a fibrillar neuropil background. The pigmented cells form clusters, glands or tubules, or long simple surfaces that surround lobules of the neuronal component. The latter can be mature to the extent that small ganglion cells are present (fig. 5-74) (200,201).

Classic MNTI is locally invasive, but often amenable to resection and cure, depending on location (200,201).

POLAR SPONGIOBLASTOMA

Definition. *Polar spongioblastoma* is a neoplasm composed of elongated bipolar cells arrayed in prominent ribbons or palisades.

General Features. Opinions as to whether polar spongioblastoma (fig. 5-75) represents a clinicopathologic entity (205) were aired in deliberations on the 2000 World Health Organization (WHO) classification and grading system. A majority of the pathologists considered the lesion

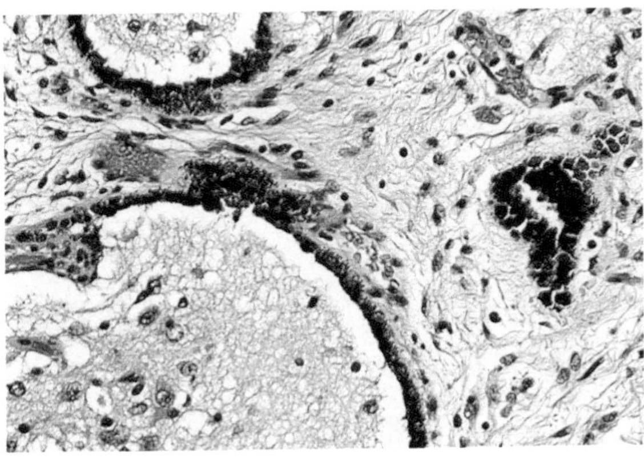

Figure 5-74

MELANOTIC NEUROECTODERMAL TUMOR OF INFANCY

The neural component of some tumors can be well differentiated. In this intracranial lesion from an 11-month-old, the neural nodules contain small ganglion cells in a neuropil matrix.

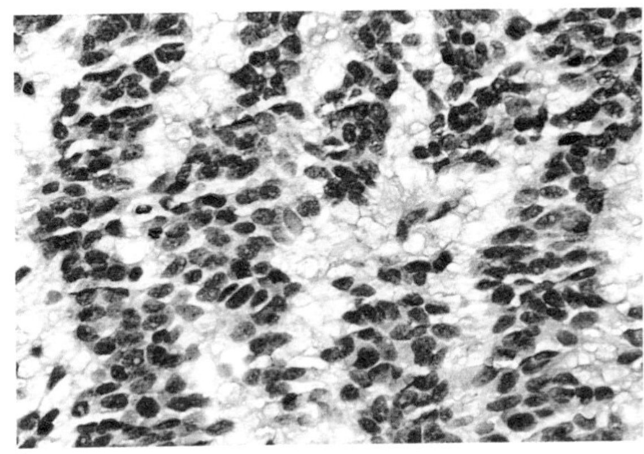

Figure 5-75

POLAR SPONGIOBLASTOMA

Mitotically active cells in prominent ribbons or palisades define the lesion. Solid areas may be present.

as only a tissue pattern, and not an entity, since palisading or ribboning is prominent in a variety of low- and high-grade CNS tumors, as summarized in Appendix J. These include medulloblastoma (fig. 5-76A), neuroblastoma (fig. 5-76B), oligodendroglioma (fig. 5-76C), and pilocytic astrocytoma (fig. 5-76D) (206).

There are, nevertheless, rare lesions, usually in children, in which a rhythmic architecture is exclusive, and there is no histologic justification for assigning the lesion to any of the above tumors with palisading. Even rarer are such lesions that are histologically or biologically malignant, as would appear to justify the prefix "primitive" that is sometimes added to the designation (205–207). Discredited or not, the diagnosis of polar spongioblastoma is very attractive in such a setting.

Clinical Features. In pure form, this rare neoplasm/tissue pattern generally occurs in the first decade of life. The cerebrum (207–209), cerebellum, brain stem, and spinal cord have been affected (206,210).

Gross Findings. Polar spongioblastoma is sufficiently rare that no specific gross characteristics have become evident. Circumscription has been noted in most reported cases. Leptomeningeal seeding may opacify the leptomeninges (205,207).

Microscopic Findings. This highly distinctive tumor is seen as regiments of palisaded cells assembling in long columns over several microscopic fields (fig. 5-74). The columns of nuclei are interrupted by highly fibrillar zones that are linear in longitudinal profile and circular when viewed on end. Thin and tapered, some cell processes extend from nuclei to nearby vessel walls. Accordingly, a delicate blood vessel is apparent in the center of many fibrillar areas. Nuclei are oval and chromatin rich, but not particularly pleomorphic. Viewed in cross section, perinuclear clearing may mimic the appearance of oligodendroglioma.

Mitotic activity has been variable and some cases appear histologically and clinically benign, or at least not malignant (208–210). As such, it does not appear to deserve the descriptor primitive, PNET, or small cell embryonal tumor.

Immunohistochemical Findings. One case was largely synaptophysin positive (207). The principal immunoreactivity in another was for S-100 protein, with staining negative for neurofilament protein (209). The tumor had no mitoses, however, and its classification as polar spongioblastoma could be questioned.

Ultrastructural Findings. Ultrastructural studies are too few to formulate diagnostic criteria. Several lesions have had decidedly neuronal features, with processes filled with microtubules and neurosecretory granules (207–209).

Differential Diagnosis. In its most overt expression, the histologic pattern of polar spongioblastoma is so distinctive that the issue is therefore more nosologic than morphologic. It

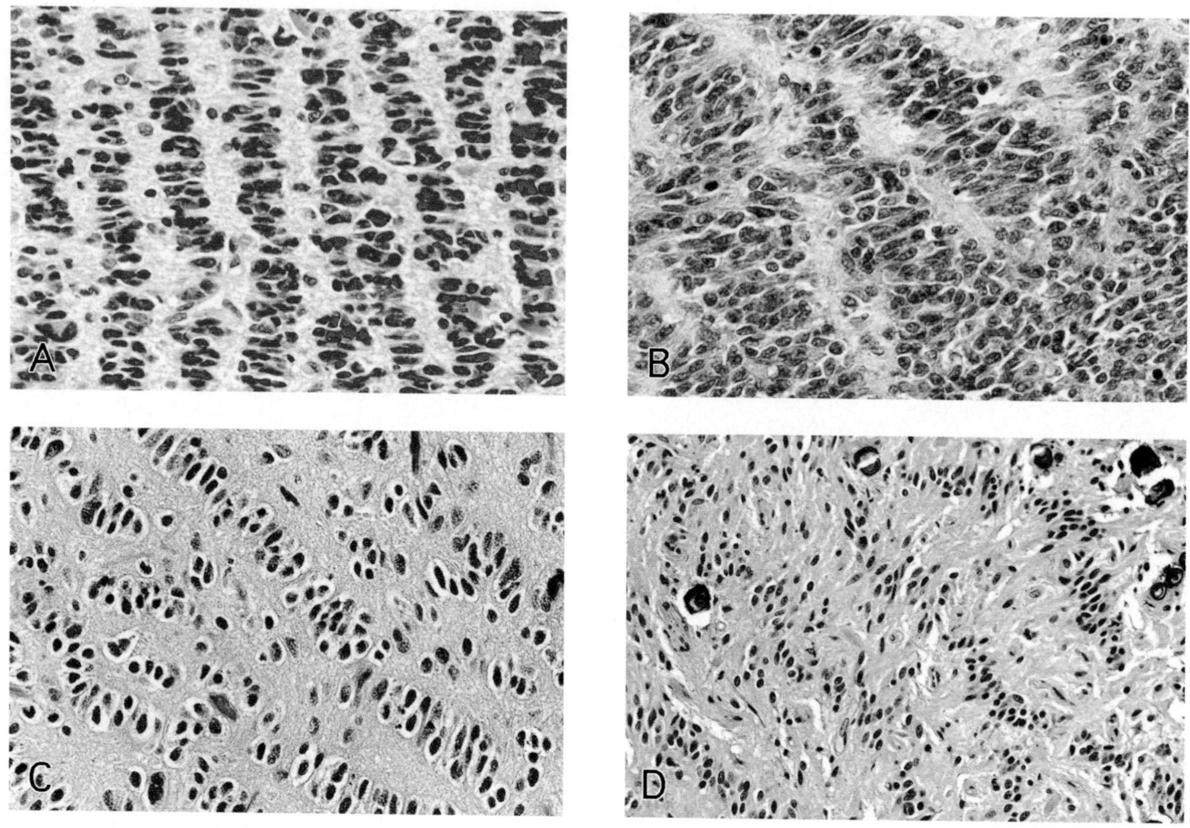

Figure 5-76

OTHER CNS NEOPLASMS WITH RIBBONING

CNS neoplasms with the potential for ribboning or palisading include medulloblastoma (A), neuroblastoma (B), oligodendroglioma (C), and pilocytic astrocytoma (D).

is important to scan the nonpalisaded areas for evidence of the more ordinary tumors listed above in which the spongioblastoma pattern is known to occur.

Treatment and Prognosis. The prognosis is unclear given definitional issues and look-alike lesions of varying biologic aggressiveness. Within the reported group of lesions with rhythmic pattern, some have behaved aggressively (205–207). Others that were histologically benign behaved accordingly, at least in the short term (208–210). In the case of a "pure," mitotically active lesion, it is appropriate to point out both the possibility of local recurrence and distant dissemination as well as to transmit present uncertainties regarding the classification and biologic behavior of this interesting tumor.

REFERENCES

Introduction

1. Hart MN, Earle KM. Primitive neuroectodermal tumors of the brain in children. Cancer 1973;32: 890–7.
2. Becker LE, Hinton D. Primitive neuroectodermal tumors of the central nervous system. Hum Pathol 1983;14:538–50.
3. Rorke LB. The cerebellar medulloblastoma and its relationship to primitive neuroectodermal tumors. J Neuropathol Exp Neurol 1983;42:1–15.
4. Rubinstein LJ. Embryonal central neuroepithelial tumors and their differentiating potential. A cytogenetic view of a complex neuro-oncological problem. J Neurosurg 1985;62:795–805.
5. Rubinstein LJ. A commentary on the proposed revision of the World Health Organization classification of brain tumors for childhood brain tumors. Cancer 1985;56(Suppl 7):1887–8.
6. Burnett ME, White EC, Sih S, von Haken MS, Cogen PH. Chromosome arm 17p deletion analysis reveals molecular genetic heterogeneity in supratentorial and infratentorial primitive neuroectodermal tumors of the central nervous system. Cancer Genet Cytogenet 1997;97:25–31.
7. Nicholson JC, Ross FM, Kohler JA, Ellison DW. Comparative genomic hybridization and histological variation in primitive neuroectodermal tumours. Br J Cancer 1999;80:1322–31.
8. Paulino AC, Melian E. Medulloblastoma and supratentorial primitive neuroectodermal tumors: an institutional experience. Cancer 1999; 86:142–8.
9. Reddy AT, Janss AJ, Phillips PC, Weiss HL, Packer RJ. Outcome for children with supratentorial primitive neuroectodermal tumors treated with surgery, radiation, and chemotherapy. Cancer 2000;88:2189–93.
10. Rorke L, Hart M, McLendon R. Supratentorial primitive neuroectodermal tumour (PNET). In: Kleihues P, Cavenee W, eds. Pathology and genetics of the nervous system. Lyon: IARC Press; 2000:141–4.
11. Marec-Berard P, Jouvet A, Thiesse P, Kalifa C, Doz F, Frappaz D. Supratentorial embryonal tumors in children under 5 years of age: an SFOP study of treatment with postoperative chemotherapy alone. Med Pediatr Oncol 2002;38:83–90.
12. Yang HJ, Nam DH, Wang KC, Kim YM, Chi JG, Cho BK. Supratentorial primitive neuroectodermal tumor in children: clinical features, treatment outcome and prognostic factors. Childs Nerv Syst 1999;15:377–83.

Medulloblastoma

13. Goodrich LV, Milenkovic L, Higgins KM, Scott MP. Altered neural cell fates and medulloblastoma in mouse patched mutants. Science 1997; 277:1109–13.
14. Hahn H, Wojnowski L, Zimmer AM, Hall J, Miller G, Zimmer A. Rhabdomyosarcomas and radiation hypersensitivity in a mouse model of Gorlin syndrome. Nat Med 1998;4:619–22.
15. Pomeroy SL, Tamayo P, Gaasenbeek M, et al. Prediction of central nervous system embryonal tumour outcome based on gene expression. Nature 2002;415:436–42.
16. Katsetos CD, Burger PC. Medulloblastoma. Semin Diagn Pathol 1994;11:85–97.
17. Katsetos CD, Del Valle L, Legido A, de Chadarevian JP, Perentes E, Mork SJ. On the neuronal/neuroblastic nature of medulloblastomas: a tribute to Pio del Rio Hortega and Moises Polak. Acta Neuropathol (Berl) 2003;105:1–13.
18. Yachnis AT, Rorke LB, Trojanowski JQ. Cerebellar dysplasias in humans: development and possible relationship to glial and primitive neuroectodermal tumors of the cerebellar vermis. J Neuropathol Exp Neurol 1994;53:61–71.
19. Ellison D. Classifying the medulloblastoma: insights from morphology and molecular genetics. Neuropathol Appl Neurobiol 2002;28: 257–82.
20. Packer RJ, Cogen P, Vezina G, Rorke LB. Medulloblastoma: clinical and biologic aspects. Neuro-oncol 1999;1:232–50.
21. Chan AW, Tarbell NJ, Black PM, et al. Adult medulloblastoma: prognostic factors and patterns of relapse. Neurosurgery 2000;47:623–31; discussion 631–2.
22. Kunschner LJ, Kuttesch J, Hess K, Yung WK. Survival and recurrence factors in adult medulloblastoma: the M.D. Anderson Cancer Center experience from 1978 to 1998. Neuro-oncol 2001;3:167–73.
23. Sarkar C, Pramanik P, Karak AK, et al. Are childhood and adult medulloblastomas different? A comparative study of clinicopathological features, proliferation index and apoptotic index. J Neurooncol 2002;59:49–61.
24. Evans DG, Farndon PA, Burnell LD, Gattamaneni HR, Birch JM. The incidence of Gorlin syndrome in 173 consecutive cases of medulloblastoma. Br J Cancer 1991;64:959–61.
25. Taylor MD, Mainprize TG, Rutka JT. Molecular insight into medulloblastoma and central nervous system primitive neuroectodermal tumor biology from hereditary syndromes: a review. Neurosurgery 2000;47:888–901.

26. Hamilton SR, Liu B, Parsons RE, et al. The molecular basis of Turcot's syndrome. N Engl J Med 1995;332:839–47.
27. Guran S, Tunca Y, Imirzalioglu N. Hereditary TP53 codon 292 and somatic P16INK4A codon 94 mutations in a Li-Fraumeni syndrome family. Cancer Genet Cytogenet 1999;113:145–51.
28. Rogers L, Pattisapu J, Smith RR, Parker P. Medulloblastoma in association with the Coffin-Siris syndrome. Childs Nerv Syst 1988;4:41–4.
29. Taylor MD, Mainprize TG, Rutka JT, Becker L, Bayani J, Drake JM. Medulloblastoma in a child with Rubenstein-Taybi syndrome: case report and review of the literature. Pediatr Neurosurg 2001;35:235–8.
30. von Koch CS, Gulati M, Aldape K, Berger MS. Familial medulloblastoma: case report of one family and review of the literature. Neurosurgery 2002;51:227–33.
31. Giangaspero F, Perilongo G, Fondelli MP, et al. Medulloblastoma with extensive nodularity: a variant with favorable prognosis. J Neurosurg 1999;91:971–7.
32. Katsetos CD, Liu HM, Zacks SI. Immunohistochemical and ultrastructural observations on Homer Wright (neuroblastic) rosettes and the "pale islands" of human cerebellar medulloblastomas. Hum Pathol 1988;19:1219–27.
33. Gould VE, Jansson DS, Molenaar WM, et al. Primitive neuroectodermal tumors of the central nervous system. Patterns of expression of neuroendocrine markers, and all classes of intermediate filament proteins. Lab Invest 1990;62:498–509.
34. Herpers MJ, Budka H. Primitive neuroectodermal tumors including the medulloblastoma: glial differentiation signaled by immunoreactivity for GFAP is restricted to the pure desmoplastic medulloblastoma ("arachnoidal sarcoma of the cerebellum"). Clin Neuropathol 1985;4:12–8.
35. Eberhart CG, Kaufman WE, Tihan T, Burger PC. Apoptosis, neuronal maturation, and neurotrophin expression within medulloblastoma nodules. J Neuropathol Exp Neurol 2001;60: 462–9.
36. Katsetos CD, Herman MM, Frankfurter A, et al. Cerebellar desmoplastic medulloblastomas. A further immunohistochemical characterization of the reticulin-free pale islands. Arch Pathol Lab Med 1989;113:1019–29.
37. Shin WY, Laufer H, Lee YC, Aftalion B, Hirano A, Zimmerman HM. Fine structure of a cerebellar neuroblastoma. Acta Neuropathol (Berl) 1978;42:11–3.
38. Pearl GS, Takei Y. Cerebellar "neuroblastoma": nosology as it relates to medulloblastoma. Cancer 1981;47:772–9.
39. Nishio S, Inamura T, Morioka T, et al. Cerebellar neuroblastoma in an infant. Clin Neurol Neurosurg 2000;102:52–7.
40. Giangaspero F, Rigobello L, Badiali M, et al. Large-cell medulloblastomas. A distinct variant with highly aggressive behavior. Am J Surg Pathol 1992;16:687–93.
41. Brown HG, Kepner JL, Perlman EJ, et al. "Large cell/anaplastic" medulloblastomas: a Pediatric Oncology Group Study. J Neuropathol Exp Neurol 2000;59:857–65.
42. Eberhart CG, Kepner JL, Goldthwaite PT, et al. Histopathologic grading of medulloblastomas: a Pediatric Oncology Group study. Cancer 2002;94:552–60.
43. Eberhart CG, Burger PC. Anaplasia and grading in medulloblastomas. Brain Pathol 2003;13: 376–85.
44. Eberhart CG, Kratz J, Wang Y, et al. Histopathological and molecular prognostic factors in medulloblastoma: c-myc, N-myc, TrkC and anaplasia. J Neuropathol Exp Neurol 2004;63:441–9.
45. Leonard JR, Cai DX, Rivet DJ, et al. Large cell/ anaplastic medulloblastomas and medullomyoblastomas: clinicopathological and genetic features. J Neurosurg 2001;95:82–8.
46. McManamy CS, Lamont JM, Taylor RE, et al. Morphophenotypic variation predicts clinical behavior in childhood non-desmoplastic medulloblastomas. J Neuropathol Exp Neurol 2003;62:627–32.
47. Buhren J, Christoph AH, Buslei R, Albrecht S, Wiestler OD, Pietsch T. Expression of the neurotrophin receptor p75NTR in medulloblastomas is correlated with distinct histological and clinical features: evidence for a medulloblastoma subtype derived from the external granule cell layer. J Neuropathol Exp Neurol 2000;59:229–40.
48. Czerwionka M, Korf HW, Hoffmann O, Busch H, Schachenmayr W. Differentiation in medulloblastomas: correlation between the immunocytochemical demonstration of photoreceptor markers (S-antigen, rod-opsin) and the survival rate in 66 patients. Acta Neuropathol (Berl) 1989;78:629–36.
49. Kramm CM, Korf HW, Czerwionka M, Schachenmayr W, de Grip WJ. Photoreceptor differentiation in cerebellar medulloblastoma: evidence for a functional photopigment and authentic S-antigen (arrestin). Acta Neuropathol (Berl) 1991;81:296–302.
50. Kumar PV, Hosseinzadeh M, Bedayat GR. Cytologic findings of medulloblastoma in crush smears. Acta Cytol 2001;45:542–6.
51. Eberhart CG, Kratz JE, Schuster A, et al. Comparative genomic hybridization detects an increased number of chromosomal alterations in large cell/anaplastic medulloblastomas. Brain Pathol 2002;12:36–44.

52. Biegel JA, Rorke LB, Janss AJ, Sutton LN, Parmiter AH. Isochromosome 17q demonstrated by interphase fluorescence in situ hybridization in primitive neuroectodermal tumors of the central nervous system. Genes Chromosomes Cancer 1995;14:85–96.
53. Biegel JA, Janss AJ, Raffel C, et al. Prognostic significance of chromosome 17p deletions in childhood primitive neuroectodermal tumors (medulloblastomas) of the central nervous system. Clin Cancer Res 1997;3:473–8.
54. Bigner SH, Mark J, Friedman HS, Biegel JA, Bigner DD. Structural chromosomal abnormalities in human medulloblastoma. Cancer Genet Cytogenet 1988;30:91–101.
55. Michiels EM, Weiss MM, Hoovers JM, et al. Genetic alterations in childhood medulloblastoma analyzed by comparative genomic hybridization. J Pediatr Hematol Oncol 2002;24:205–10.
56. Nicholson J, Wickramasinghe C, Ross F, Crolla J, Ellison D. Imbalances of chromosome 17 in medulloblastomas determined by comparative genomic hybridisation and fluorescence in situ hybridisation. Mol Pathol 2000;53:313–9.
57. Reardon DA, Michalkiewicz E, Boyett JM, et al. Extensive genomic abnormalities in childhood medulloblastoma by comparative genomic hybridization. Cancer Res 1997;57:4042–7.
58. Russo C, Pellarin M, Tingby O, et al. Comparative genomic hybridization in patients with supratentorial and infratentorial primitive neuroectodermal tumors. Cancer 1999;86:331–9.
59. Lamont JH, McManamy CS, Pearson AD, Clifford SC, Ellison DW. Combined histopathologic and molecular cytogenetic stratification of medulloblastoma patients. Clin Cancer Res 2004;10:5482–93.
60. Bayani J, Zielenska M, Marrano P, et al. Molecular cytogenetic analysis of medulloblastomas and supratentorial primitive neuroectodermal tumors by using conventional banding, comparative genomic hybridization, and spectral karyotyping. J Neurosurg 2000;93:437–48.
60a. Rossi MR, Conroy J, McQuaid D, Nowak NJ, Rutka JT, Cowell JK. Array CGH analysis of pediatric medulloblastomas. Genes Chromosomes Cancer 2006;45:290–303.
61. Reardon DA, Jenkins JJ, Sublett JE, Burger PC, Kun LK. Multiple genomic alterations including N-myc amplification in a primary large cell medulloblastoma. Pediatr Neurosurg 2000;32: 187–91.
62. Schofield D, West DC, Anthony DC, Marshal R, Sklar J. Correlation of loss of heterozygosity at chromosome 9q with histological subtype in medulloblastomas. Am J Pathol 1995;146:472–80.
63. Pietsch T, Waha A, Koch A, et al. Medulloblastomas of the desmoplastic variant carry mutations of the human homologue of Drosophila patched. Cancer Res 1997;57:2085–8.
64. Raffel C, Jenkins RB, Frederick L, et al. Sporadic medulloblastomas contain PTCH mutations. Cancer Res 1997;57:842–5.
65. Zurawel RH, Allen C, Chiappa S, et al. Analysis of PTCH/SMO/SHH pathway genes in medulloblastoma. Genes Chromosomes Cancer 2000;27:44–51.
66. Smyth I, Narang MA, Evans T, et al. Isolation and characterization of human patched 2 (PTCH2), a putative tumour suppressor gene in basal cell carcinoma and medulloblastoma on chromosome 1p32. Hum Mol Genet 1999;8:291–7.
67. Lam CW, Xie J, To KF, et al. A frequent activated smoothened mutation in sporadic basal cell carcinomas. Oncogene 1999;18:833–6.
68. Taylor MD, Liu L, Raffel C, et al. Mutations in SUFU predispose to medulloblastoma. Nat Genet 2002;31:306–10.
69. Eberhart CG, Tihan T, Burger PC. Nuclear localization and mutation of beta-catenin in medulloblastomas. J Neuropathol Exp Neurol 2000;59:333–7.
70. Zurawel RH, Chiappa SA, Allen C, Raffel C. Sporadic medulloblastomas contain oncogenic beta-catenin mutations. Cancer Res 1998;58: 896–9.
71. Dahmen RP, Koch A, Denkhaus D, et al. Deletions of AXIN1, a component of the WNT/wingless pathway, in sporadic medulloblastomas. Cancer Res 2001;61:7039–43.
72. Huang H, Mahler-Araujo BM, Sankila A, et al. APC mutations in sporadic medulloblastomas. Am J Pathol 2000;156:433–7.
73. Tomlinson FH, Jenkins RB, Scheithauer BW, et al. Aggressive medulloblastoma with high-level N-myc amplification. Mayo Clin Proc 1994;69: 359–65.
74. Aldosari N, Bigner SH, Burger PC, et al. MYCC and MYCN oncogene amplification in medulloblastoma. A fluorescence in situ hybridization study on paraffin sections from the Children's Oncology Group. Arch Pathol Lab Med 2002;126:540–4.
75. Grotzer MA, Hogarty MD, Janss AJ, et al. MYC messenger RNA expression predicts survival outcome in childhood primitive neuroectodermal tumor/medulloblastoma. Clin Cancer Res 2001;7:2425–33.
76. Herms J, Neidt I, Luscher B, et al. C-MYC expression in medulloblastoma and its prognostic value. Int J Cancer 2000;89:395–402.
77. Scheurlen WG, Schwabe GC, Joos S, Mollenhauer J, Sorensen N, Kuhl J. Molecular analysis of childhood primitive neuroectodermal tumors defines markers associated with poor outcome. J Clin Oncol 1998;16:2478–85.
78. Chatty EM, Earle KM. Medulloblastoma. A report of 201 cases with emphasis on the relationship of histologic variants to survival. Cancer 1971;28:977–83.

79. Sure U, Berghorn WJ, Bertalanffy H, et al. Staging, scoring and grading of medulloblastoma. A postoperative prognosis predicting system based on the cases of a single institute. Acta Neurochir (Wien) 1995;132:59–65.
80. Rutkowski S, Bode U, Deinlein F, et al. Treatment of early childhood medulloblastoma by postoperative chemotherapy alone. N Engl J Med 2005;352:978–86.
81. Schofield DE, Yunis EJ, Geyer JR, Albright AL, Berger MS, Taylor SR. DNA content and other prognostic features in childhood medulloblastoma. Proposal of a scoring system. Cancer 1992;69:1307–14.
82. Kopelson G, Linggood RM, Kleinman GM. Medulloblastoma. The identification of prognostic subgroups and implications for multimodality management. Cancer 1983;51:312–9.
82a. Giangaspero F, Wellek S, Masuoka J, Gessi M, Kleihues P, Ohgaki H. Stratification of medulloblastoma on the basis of histopathological grading. Act Neuropathol 2006;112:5–12.
83. Gilbertson RJ, Jaros E, Perry RH, Kelly PJ, Lunec J, Pearson AD. Mitotic percentage index: a new prognostic factor for childhood medulloblastoma. Eur J Cancer 1997;33:609–15.
84. Grotzer MA, Geoerger B, Janss AJ, Zhao H, Rorke LB, Phillips PC. Prognostic significance of Ki-67 (MIB-1) proliferation index in childhood primitive neuroectodermal tumors of the central nervous system. Med Pediatr Oncol 2001;36:268–73.
85. Pelc K, Vincent S, Ruchoux MM, et al. Calbindin-d(28k): a marker of recurrence for medulloblastomas. Cancer 2002;95:410–9.
86. Janss AJ, Yachnis AT, Silber JH, et al. Glial differentiation predicts poor clinical outcome in primitive neuroectodermal brain tumors. Ann Neurol 1996;39:481–9.
87. Batra SK, McLendon RE, Koo JS, et al. Prognostic implications of chromosome 17p deletions in human medulloblastomas. J Neurooncol 1995;24:39–45.
88. Pan E, Pellarin M, Holmes E, et al. Isochromosome 17q is a negative prognostic factor in poor-risk childhood medulloblastoma patients. Clin Cancer Res 2005;11:4733–40.
89. Emadian SM, McDonald JD, Gerken SC, Fults D. Correlation of chromosome 17p loss with clinical outcome in medulloblastoma. Clin Cancer Res 1996;2:1559–64.
90. Grotzer MA, Janss AJ, Fung K, et al. TrkC expression predicts good clinical outcome in primitive neuroectodermal brain tumors. J Clin Oncol 2000;18:1027–35.
91. Segal RA, Goumnerova LC, Kwon YK, Stiles CD, Pomeroy SL. Expression of the neurotrophin receptor TrkC is linked to a favorable outcome in medulloblastoma. Proc Natl Acad Sci U S A 1994;91:12867–71.
91a. Ellison DW, Onilude OE, Lindsey JC. beta-Catenin status predicts a favorable outcome in childhood medulloblastoma: the United Kingdom Children's Cancer Study Group Brain Tumour Committee. J Clin Oncol 2005; 23:7951–7.
92. Burns AS, Jaros E, Cole M, Perry R, Pearson AJ, Lunec J. The molecular pathology of p53 in primitive neuroectodermal tumours of the central nervous system. Br J Cancer 2002;86:1117–23.
93. Woodburn RT, Azzarelli B, Montebello JF, Goss IE. Intense p53 staining is a valuable prognostic indicator for poor prognosis in medulloblastoma/central nervous system primitive neuroectodermal tumors. J Neurooncol 2001;52:57–62.
94. Giordana MT, Duo D, Gasverde S, et al. MDM2 overexpression is associated with short survival in adults with medulloblastoma. Neuro-oncol 2002;4:115–22.
95. Frank AJ, Hernan R, Hollander A, et al. The TP53-ARF tumor suppressor pathway is frequently disrupted in large/cell anaplastic medulloblastoma. Brain Res Mol Brain Res 2004;121:137–40.
96. Fan X, Wang Y, Kratz J, et al. hTERT gene amplification and increased mRNA expression in central nervous system embryonal tumors. Am J Pathol 2003;162:1763–9.
97. Gajjar A, Hernan R, Kocak M, et al. Clinical, histopathologic, and molecular markers of prognosis: toward a new disease risk stratification system for medulloblastoma. J Clin Oncol 2004;22:984–93.
98. Bruch LA, Hill DA, Cai DX, Levy BK, Dehner LP, Perry A. A role for fluorescence in situ hybridization detection of chromosome 22q dosage in distinguishing atypical teratoid/rhabdoid tumors from medulloblastoma/central primitive neuroectodermal tumors. Hum Pathol 2001;32:156–62.
99. Eberhart CG, Cohen KJ, Tihan T, Goldthwaite PT, Burger PC. Medulloblastomas with systemic metastases: evaluation of tumor histopathology and clinical behavior in 23 patients. J Pediatr Hematol Oncol 2003;25:198–203.
100. Magtibay PM, Friedman JA, Rao RD, Buckner JC, Cliby WA. Unusual presentation of adult metastatic peritoneal medulloblastoma associated with a ventriculoperitoneal shunt: a case study and review of the literature. Neuro-oncol 2003;5:217–20.
101. Amagasaki K, Yamazaki H, Koizumi H, Hashizume K, Sasaguchi N. Recurrence of medulloblastoma 19 years after the initial diagnosis. Childs Nerv Syst 1999;15:482–5.
102. Nishio S, Morioka T, Takeshita I, Fukui M. Medulloblastoma: survival and late recurrence after the Collins' risk period. Neurosurg Rev 1997;20:245–9.

103. Nakamizo A, Nishio S, Inamura T, et al. Evolution of malignant cerebellar astrocytoma at the site of a treated medulloblastoma: report of two cases. Acta Neurochir (Wien) 2001;143:697–700.
104. Pearl GS, Mirra SS, Miles ML. Glioblastoma multiforme occurring 13 years after treatment of a medulloblastoma. Neurosurgery 1980;6: 546–51.
105. Duffner PK, Krischer JP, Horowitz ME, et al. Second malignancies in young children with primary brain tumors following treatment with prolonged postoperative chemotherapy and delayed irradiation: a Pediatric Oncology Group study. Ann Neurol 1998;44:313–6.

Medullomyoblastoma and Melanotic Medulloblastoma

106. Duinkerke SJ, Slooff JL, Gabreels FJ, Renier WO, Thijssen HO, Biesta JH. Melanotic rhabdomyomedulloblastoma or teratoid tumour of the cerebellar vermis. Clin Neurol Neurosurg 1981;83: 29–33.
107. Kalimo H, Paljarvi L, Ekfors T, Pelliniemi LJ. Pigmented primitive neuroectodermal tumor with multipotential differentiation in cerebellum (pigmented medullomyoblastoma). A case with light- and electron-microscopic, and immunohistochemical analysis. Pediatr Neurosci 1987;13:188–95.
108. Chowdhury C, Roy S, Mahapatra AK, Bhatia R. Medullomyoblastoma. A teratoma. Cancer 1985;55:1495–500.
109. Mahapatra AK, Sinha AK, Sharma MC. Medullomyoblastoma. A rare cerebellar tumour in children. Childs Nerv Syst 1998;14:312–6.
110. Bergmann M, Pietsch T, Herms J, Janus J, Spaar HJ, Terwey B. Medullomyoblastoma: a histological, immunohistochemical, ultrastructural and molecular genetic study. Acta Neuropathol (Berl) 1998;95:205–12.
111. Cheema ZF, Cannon TC, Leech R, Brennan J, Adesina A, Brumback RA. Medullomyoblastoma: case report. J Child Neurol 2001;16:598–9.
112. Dickson DW, Hart MN, Menezes A, Cancilla PA. Medulloblastoma with glial and rhabdomyoblastic differentiation. A myoglobin and glial fibrillary acidic protein immunohistochemical and ultrastructural study. J Neuropathol Exp Neurol 1983;42:639–47.
113. Lata M, Mahapatra AK, Sarkar C, Roy S. Medullomyoblastoma. A case report. Indian J Cancer 1989;26:240–6.
114. Smith TW, Davidson RI. Medullomyoblastoma. A histologic, immunohistochemical, and ultrastructural study. Cancer 1984;54:323–32.
115. Stahlberger R, Friede RL. Fine structure of myomedulloblastoma. Acta Neuropathol (Berl) 1977;37:43–8.
116. Rao C, Friedlander ME, Klein E, Anzil AP, Sher JH. Medullomyoblastoma in an adult. Cancer 1990;65:157–63.
117. Schiffer D, Giordana MT, Pezzotta S, Pezzulo T, Vigliani MC. Medullomyoblastoma: report of two cases. Childs Nerv Syst 1992;8:268–72.
118. Holl T, Kleihues P, Yasargil MG, Wiestler OD. Cerebellar medullomyoblastoma with advanced neuronal differentiation and hamartomatous component. Acta Neuropathol (Berl) 1991;82:408–13.
119. Fuller C, Meyer R, Schmidt R, et al. Large cell/anaplastic medullomyoblastoma: report of six cases with clinicopathologic and radiographic correlation. J Neuropathol Exp Neurol 2003;62: 581.
120. Leonard JR, Cai DX, Rivet DJ, et al. Large cell/anaplastic medulloblastomas and medullomyoblastomas: clinicopathological and genetic features. J Neurosurg 2001;95:82–8.
121. Best PV. A medulloblastoma-like tumour with melanin formation. J Pathol 1973;110:109–11.
122. Boesel CP, Suhan JP, Sayers MP. Melanotic medulloblastoma. Report of a case with ultrastructural findings. J Neuropathol Exp Neurol 1978;37:531–43.
123. Dolman CL. Melanotic medulloblastoma. A case report with immunohistochemical and ultrastructural examination. Acta Neuropathol (Berl) 1988;76:528–31.
124. Fowler M, Simpson DA. A malignant melanin-forming tumour of the cerebellum. J Pathol Bacteriol 1962;84:307–11.
125. Jimenez CL, Carpenter BF, Robb IA. Melanotic cerebellar tumor. Ultrastruct Pathol 1987;11: 751–9.
126. Sung JH, Mastri AR, Segal EL. Melanotic medulloblastoma of the cerebellum. J Neuropathol Exp Neurol 1973;32:437–45.

Medulloepithelioma

127. Molloy PT, Yachnis AT, Rorke LB, et al. Central nervous system medulloepithelioma: a series of eight cases including two arising in the pons. J Neurosurg 1996;84:430–6.
128. Treip C. A congenital medulloepithelioma of the midbrain. J Pathol Bacteriol 1957;74:357–63.
129. Auer RN, Becker LE. Cerebral medulloepithelioma with bone, cartilage, and striated muscle. Light microscopic and immunohistochemical study. J Neuropathol Exp Neurol 1983;42:256–67.
130. Deck JH. Cerebral medulloepithelioma with maturation into ependymal cells and ganglion cells. J Neuropathol Exp Neurol 1969;28(3):442–54.
131. Karch SB, Urich H. Medulloepithelioma: definition of an entity. J Neuropathol Exp Neurol 1972;31:27–53.

132. Scheithauer BW, Rubinstein LJ. Cerebral medulloepithelioma. Report of a case with multiple divergent neuroepithelial differentiation. Childs Brain 1979;5:62–71.
133. Troost D, Jansen GH, Dingemans KP. Cerebral medulloepithelioma—electron microscopy and immunohistochemistry. Acta Neuropathol (Berl) 1990;80:103–7.
134. Best PV. Posterior fossa medulloepithelioma. Report of a case. J Neurol Sci 1974;22:511–18.
135. Khoddami M, Becker LE. Immunohistochemistry of medulloepithelioma and neural tube. Pediatr Pathol Lab Med 1997;17:913–25.
136. Pollak A, Friede RL. Fine structure of medulloepithelioma. J Neuropathol Exp Neurol 1977; 36:712–25.
137. Chidambaram B, Santosh V, Balasubramaniam V. Medulloepithelioma of the optic nerve with intradural extension—report of two cases and a review of the literature. Childs Nerv Syst 2000;16(6):329–33.
138. Pang LM, Roebuck DJ, Ng HK, Chan YL. Sellar and suprasellar medulloepithelioma. Pediatr Radiol 2001;31:594–6.
139. Caccamo DV, Herman MM, Rubinstein LJ. An immunohistochemical study of the primitive and maturing elements of human cerebral medulloepitheliomas. Acta Neuropathol (Berl) 1989;79:248–54.
140. Cruz-Sanchez FF, Rossi ML, Hughes JT, Moss TH. Differentiation in embryonal neuroepithelial tumors of the central nervous system. Cancer 1991;67:965–76.

Neuroblastoma and Ganglioneuroblastoma

141. Bennett JP Jr, Rubinstein LJ. The biological behavior of primary cerebral neuroblastoma: a reappraisal of the clinical course in a series of 70 cases. Ann Neurol 1984;16:21–7.
142. Berger MS, Edwards MS, Wara WM, Levin VA, Wilson CB. Primary cerebral neuroblastoma. Long-term follow-up review and therapeutic guidelines. J Neurosurg 1983;59:418–23.
143. Horten BC, Rubinstein LJ. Primary cerebral neuroblastoma. A clinicopathological study of 35 cases. Brain 1976;99:735–56.
144. Klisch J, Husstedt H, Hennings S, von Velthoven V, Pagenstecher A, Schumacher M. Supratentorial primitive neuroectodermal tumours: diffusion-weighted MRI. Neuroradiology 2000;42:393–8.
145. Dehner LP, Abenoza P, Sibley RK. Primary cerebral neuroectodermal tumors: neuroblastoma, differentiated neuroblastoma, and composite neuroectodermal tumor. Ultrastruct Pathol 1988;12:479–94.
146. Langford LA, Camel MH. Palisading pattern in cerebral neuroblastoma mimicking the primitive polar spongioblastoma. An ultrastructural study. Acta Neuropathol (Berl) 1987;73:153–9.
147. Ojeda VJ, Spagnolo DV, Vaughan RJ. Palisades in primary cerebral neuroblastoma simulating so-called polar spongioblastoma. A light and electron microscopical study of an adult case. Am J Surg Pathol 1987;11:316–22.
148. Ahdevaara P, Kalimo H, Torma T, Haltia M. Differentiating intracerebral neuroblastoma: report of a case and review of the literature. Cancer 1977;40:784–8.
149. Torres LF, Grant N, Harding BN, Scaravilli F. Intracerebral neuroblastoma. Report of a case with neuronal maturation and long survival. Acta Neuropathol (Berl) 1985;68:110–4.
150. Azzarelli B, Richards DE, Anton AH, Roessman U. Central neuroblastoma. Electron microscopic observations and catecholamine determinations. J Neuropathol Exp Neurol 1977;36:384–97.
151. Rhodes RH, Davis RL, Kassel SH, Clague BH. Primary cerebral neuroblastoma: a light and electron microscopic study. Acta Neuropathol (Berl) 1978;41(2):119–24.
152. Yagishita S, Itoh Y, Chiba Y, Yuda K. Cerebral neuroblastoma. Virchows Arch A Pathol Anat Histol 1978;381:1–11.
153. Burnett ME, White EC, Sih S, von Haken MS, Cogen PH. Chromosome arm 17p deletion analysis reveals molecular genetic heterogeneity in supratentorial and infratentorial primitive neuroectodermal tumors of the central nervous system. Cancer Genet Cytogenet 1997;97:25–31.
154. Kraus JA, Felsberg J, Tonn JC, Reifenberger G, Pietsch T. Molecular genetic analysis of the TP53, PTEN, CDKN2A, EGFR, CDK4 and MDM2 tumour-associated genes in supratentorial primitive neuroectodermal tumours and glioblastomas of childhood. Neuropathol Appl Neurobiol 2002;28:325–33.
155. Nicholson JC, Ross FM, Kohler JA, Ellison DW. Comparative genomic hybridization and histological variation in primitive neuroectodermal tumours. Br J Cancer 1999;80:1322–31.
156. Henriquez AS, Robertson DM, Marshall WJ. Primary neuroblastoma of the central nervous system with spontaneous extracranial metastases. Case report. J Neurosurg 1973;38:226–31.
157. Sakaki S, Mori Y, Motozaki T, Nakagawa K, Matsuoka K. A cerebral neuroblastoma with extracranial metastases. Surg Neurol 1981;16:53–60.

Neuroblastic Tumor with Abundant Neuropil and True Rosettes

158. Eberhart CG, Brat DJ, Cohen KJ, Burger PC. Pediatric neuroblastic brain tumors containing abundant neuropil and true rosettes. Pediatr Dev Pathol 2000;3:346–52.

Ependymoblastoma

159. Mork SJ, Rubinstein LJ. Ependymoblastoma. A reappraisal of a rare embryonal tumor. Cancer 1985;55:1536–42.
160. Rubinstein LJ. The definition of the ependymoblastoma. Arch Pathol 1970;90:35–45.
161. Lorentzen M, Hagerstrand I. Congenital ependymoblastoma. Acta Neuropathol (Berl) 1980; 49:71–4.
162. Cruz-Sanchez FF, Haustein J, Rossi ML, Cervos-Navarro J, Hughes JT. Ependymoblastoma: a histological, immunohistological and ultrastructural study of five cases. Histopathology 1988;12:17–27.
163. Langford LA. The ultrastructure of the ependymoblastoma. Acta Neuropathol (Berl) 1986;71: 136–41.

Tumors of the Trilateral Retinoblastoma Syndrome

164. Bader JL, Meadows AT, Zimmerman LE, et al. Bilateral retinoblastoma with ectopic intracranial retinoblastoma: trilateral retinoblastoma. Cancer Genet Cytogenet 1982;5:203–13.
165. Paulino AC. Trilateral retinoblastoma: is the location of the intracranial tumor important? Cancer 1999;86:135–41.
166. Provenzale JM, Weber AL, Klintworth GK, McLendon RE. Radiologic-pathologic correlation. Bilateral retinoblastoma with coexistent pinealoblastoma (trilateral retinoblastoma). AJNR Am J Neuroradiol 1995;16:157–65.

Miscellaneous Small Cell Embryonal Tumors of the Brain and Spinal Cord

167. Tang TT, Harb JM, Mork SJ, Sty JR. Composite cerebral neuroblastoma and astrocytoma. A mixed central neuroepithelial tumor. Cancer 1985;56:1404–12.
168. Gambarelli D, Hassoun J, Choux M, Toga M. Complex cerebral tumor with evidence of neuronal, glial and Schwann cell differentiation: a histologic, immunocytochemical and ultrastructural study. Cancer 1982;49:1420–8.
169. Hart MN, Earle KM. Primitive neuroectodermal tumors of the brain in children. Cancer 1973;32:890–7.
170. Abenoza P, Wick MR. Primitive cerebral neuroectodermal tumor with rhabdomyoblastic differentiation. Ultrastruct Pathol 1986;10: 347–54.
171. Dedeurwaerdere F, Giannini C, Sciot R, et al. Primary peripheral PNET/Ewing's sarcoma of the dura: a clinicopathologic entity distinct from central PNET. Mod Pathol 2002;15:673–8.
172. Brustle O, Ohgaki H, Schmitt HP, Walter GF, Ostertag H, Kleihues P. Primitive neuroectodermal tumors after prophylactic central nervous system irradiation in children. Association with an activated K-ras gene. Cancer 1992;69:2385–92.

Olfactory Neuroblastoma (Esthesioneuroblastoma)

173. Mills SE, Frierson HF Jr. Olfactory neuroblastoma. A clinicopathologic study of 21 cases. Am J Surg Pathol 1985;9:317–27.
174. Mills S, Gaffey M, Frierson H. Tumors of the upper aerodigestive tract and ear. Atlas of Tumor Pathology, 3rd Series, Fascicle 26. Washington, DC: Armed Forces Institute of Pathology; 2000.
175. Meneses MS, Thurel C, Mikol J, et al. Esthesioneuroblastoma with intracranial extension. Neurosurgery 1990;27:813–20.

Atypical Teratoid/Rhabdoid Tumor

176. Biegel JA, Zhou JY, Rorke LB, Stenstrom C, Wainwright LM, Fogelgren B. Germ-line and acquired mutations of INI1 in atypical teratoid and rhabdoid tumors. Cancer Res 1999;59:74–9.
177. Bonnin JM, Rubinstein LJ, Palmer NF, Beckwith JB. The association of embryonal tumors originating in the kidney and in the brain. A report of seven cases. Cancer 1984;54:2137–46.
178. Howat AJ, Gonzales MF, Waters KD, Campbell PE. Primitive neuroectodermal tumour of the central nervous system associated with malignant rhabdoid tumour of the kidney: report of a case. Histopathology 1986;10:643–50.
179. Fernandez C, Bouvier C, Sevenet N, et al. Congenital disseminated malignant rhabdoid tumor and cerebellar tumor mimicking medulloblastoma in monozygotic twins: pathologic and molecular diagnosis. Am J Surg Pathol 2002;26:266–70.
180. Rorke LB, Packer RJ, Biegel JA. Central nervous system atypical teratoid/rhabdoid tumors of infancy and childhood: definition of an entity. J Neurosurg 1996;85:56–65.
181. Burger PC, Yu IT, Tihan T, et al. Atypical teratoid/rhabdoid tumor of the central nervous system: a highly malignant tumor of infancy and childhood frequently mistaken for medulloblastoma: a Pediatric Oncology Group study. Am J Surg Pathol 1998;22:1083–92.
182. Hilden JM, Watterson J, Longee DC, et al. Central nervous system atypical teratoid tumor/rhabdoid tumor: response to intensive therapy and review of the literature. J Neurooncol 1998;40:265–75.
183. Hilden J, Meerbaum S, Burger P, et al. Central nervous system atypical teratoid/rhabdoid tumor: results of therapy in children enrolled in a registry. J Clin Oncol 2004;22:2877–84.
184. Lutterbach J, Liegibel J, Koch D, Madlinger A, Frommhold H, Pagenstecher A. Atypical teratoid/rhabdoid tumors in adult patients: case report and review of the literature. J Neurooncol 2001;52:49–56.

185. Raisanen J, Biegel JA, Hatanpaa KJ, Judkins A, White CL, Perry A. Chromosome 22q deletions in atypical teratoid/rhabdoid tumors in adults. Brain Pathol 2005;15:23–8.
186. Hanna SL, Langston JW, Parham DM, Douglass EC. Primary malignant rhabdoid tumor of the brain: clinical, imaging, and pathologic findings. AJNR Am J Neuroradiol 1993;14:107–15.
187. Bambakidis NC, Robinson S, Cohen M, Cohen AR. Atypical teratoid/rhabdoid tumors of the central nervous system: clinical, radiographic and pathologic features. Pediatr Neurosurg 2002;37:64–70.
188. Behring B, Bruck W, Goebel HH, et al. Immunohistochemistry of primary central nervous system malignant rhabdoid tumors: report of five cases and review of the literature. Acta Neuropathol (Berl) 1996;91:578–86.
189. Ho DM, Hsu CY, Wong TT, Ting LT, Chiang H. Atypical teratoid/rhabdoid tumor of the central nervous system: a comparative study with primitive neuroectodermal tumor/medulloblastoma. Acta Neuropathol (Berl) 2000;99: 482–8.
190. Bouffard JP, Sandberg GD, Golden JA, Rorke LB. Double immunolabeling of central nervous system atypical teratoid/rhabdoid tumors. Mod Pathol 2004;17:679–83.
191. Berrak SG, Ozek MM, Canpolat C, et al. Association between DNA content and tumor suppressor gene expression and aggressiveness of atypical teratoid/rhabdoid tumors. Childs Nerv Syst 2002;18:485–91.
192. Biggs PJ, Garen PD, Powers JM, Garvin AJ. Malignant rhabdoid tumor of the central nervous system. Hum Pathol 1987;18:332–7.
193. Chou SM, Anderson JS. Primary CNS malignant rhabdoid tumor (MRT): report of two cases and review of literature. Clin Neuropathol 1991; 10:1–10.
194. Biegel JA, Fogelgren B, Zhou JY, et al. Mutations of the INI1 rhabdoid tumor suppressor gene in medulloblastomas and primitive neuroectodermal tumors of the central nervous system. Clin Cancer Res 2000;6:2759–63.
195. Biegel JA, Tan L, Zhang F, Wainwright L, Russo P, Rorke LB. Alterations of the hSNF5/INI1 gene in central nervous system atypical teratoid/rhabdoid tumors and renal and extrarenal rhabdoid tumors. Clin Cancer Res 2002;8:3461–7.
196. Bruch LA, Hill DA, Cai DX, Levy BK, Dehner LP, Perry A. A role for fluorescence in situ hybridization detection of chromosome 22q dosage in distinguishing atypical teratoid/rhabdoid tumors from medulloblastoma/central primitive neuroectodermal tumors. Hum Pathol 2001;32:156–62.
197. Judkins AR, Mauger JH, HT A, Rorke L, Biegel J. Immunohistochemical analysis of hSNF5/INI1 in pediatric CNS neoplasms. Am J Surg Pathol 2004;28:644–50.
198. Judkins AR, Burger PC, Hamilton RL, et al. INI1 protein expression distinguishes atypical teratoid/rhabdoid tumor from choroid plexus carcinoma. J Neuropathol Exp Neurol 2005;64: 391–7.
199. Guler E, Varan A, Soylemezoglu F, et al. Extraneural metastasis in a child with atypical teratoid rhabdoid tumor of the central nervous system. J Neurooncol 2001;54:53–6.

Melanotic Neuroectodermal Tumor of Infancy

200. Kapadia SB, Frisman DM, Hitchcock CL, Ellis GL, Popek EJ. Melanotic neuroectodermal tumor of infancy. Clinicopathological, immunohistochemical, and flow cytometric study. Am J Surg Pathol 1993;17:566–73.
201. Parizek J, Nemecek S, Cernoch Z, Heger L, Nozicka Z, Spacek J. Melanotic neuroectodermal neurocranial tumor of infancy of extra-intra- and subdural right temporal location: CT examination, surgical treatment, literature review. Neuropediatrics 1986;17:115–23.
202. Nozicka Z, Spacek J. Melanotic neuroectodermal tumor of infancy with highly differentiated neural component. Light and electron microscopic study. Acta Neuropathol (Berl) 1978;44:229–33.
203. Yu JS, Moore MR, Kupsky WJ, Scott RM. Intracranial melanotic neuroectodermal tumor of infancy: two case reports. Surg Neurol 1992;37:123–9.
204. Stowens D, Lin TH. Melanotic progonoma of the brain. Hum Pathol 1974;5:105–13.

Polar Spongioblastoma

205. Rubinstein LJ. Presidential address. Cytogenesis and differentiation of primitive central neuroepithelial tumors. J Neuropathol Exp Neurol 1972;31:7–26.
206. Schiffer D, Cravioto H, Giordana MT, Migheli A, Pezzulo T, Vigliani MC. Is polar spongioblastoma a tumor entity? J Neurosurg 1993;78: 587–91.
207. Ng HK, Tang NL, Poon WS. Polar spongioblastoma with cerebrospinal fluid metastases. Surg Neurol 1994;41:137–42.
208. de Chadarevian JP, Guyda HJ, Hollenberg RD. Hypothalamic polar spongioblastoma associated with the diencephalic syndrome. Ultrastructural demonstration of a neuro-endocrine organization. Virchows Arch A Pathol Anat Histopathol 1984;402:465–74.
209. Jansen GH, Troost D, Dingemans KP. Polar spongioblastoma: an immunohistochemical and electron microscopical study. Acta Neuropathol (Berl) 1990;81:228–32.
210. Schochet SS Jr, Violett TW, Nelson J, Pelofsky S, Barnes PA. Polar spongioblastoma of the cervical spinal cord: case report. Clin Neuropathol 1984;3:225–7.

6 TUMORS OF PINEAL REGION

Although both neurosensory and glial components of the pineal gland are potential sources of neoplasms, almost all tumors arise from pineocytes, the specially modified neurons distantly related to retinal photoreceptors. Miscellaneous tumors include papillary tumor of the pineal region and gliomas. The spectrum of germ cell neoplasms, many occurring in the pineal gland, is considered separately in chapter 8. Pineal cysts are included in this chapter, rather than with other cysts in chapter 16, since they may be confused with pineocytoma and gliomas.

PINEAL PARENCHYMAL TUMORS

Pineal parenchymal tumors (PPTs) comprise a spectrum of increasing malignancy and biologic aggressiveness. Within this continuum, it has been difficult to establish correlations between pathologic features and outcome in individual patients, since the tumors are rare, specimens are often small and potentially unrepresentative, and treatments have varied from institution to institution and from decade to decade (1). Grading criteria for this tumor spectrum are unsettled, particularly for PPTs of intermediate grade.

The grading system used here is that of the World Health Organization (WHO). Based largely on features seen in routine hematoxylin and eosin (H&E)-stained sections, it recognizes three grades: pineocytoma (grade I), PPT of intermediate differentiation (grades II and III), and pineoblastoma (grade IV) (2–5). Not surprisingly, the spectrum of differentiation has its parallels at the ultrastructural level (6).

This grading system is based to a large extent on personal and anecdotal experience, and the assumption that increases in cellularity and mitotic activity are prognostically unfavorable. The role of MIB-1 indices is largely unexplored.

Support for the above grading system comes from a detailed study that found prognostic factors to include: number of mitoses, presence of necrosis, and positive staining for neurofilament protein (recognized in H&E-stained sections as neuropil-like fibrillarity) (2). These elements were used in a four-tier grading system and correlated with clinical variables, including outcome (1,2). Grade 1 lesions were equivalent to rosette-forming (pineocytomatous) and/or pleomorphic pineocytomas that stain for neuronal markers (synaptophysin, neurofilament protein, chromogranin) and do not show mitotic activity or necrosis. Grade 2 tumors exhibited immunoreactivity for neuronal markers, fewer than 6 mitoses per 10 high-power fields, and no necrosis. Grade 3 tumors had either no or only weak immunostaining and fewer than 6 mitoses per 10 high-power fields, or positive immunostaining and 6 or more mitoses. Grade 4 tumors, or pineoblastomas, were mitotically active small cell tumors with little if any immunostaining for neurofilament protein. How this translates into the three-tiered WHO system is not clear, particularly which of the grade 3 tumors are pineoblastomas in the WHO approach. Certainly, the malignant of the grade 3 lesions, i.e., those with 6 or more mitoses per 10 high-power fields, would be considered pineoblastoma by most observers.

The 2007 WHO "Blue Book" recognized that individual tumors in the important and diagnostically difficult group of intermediate lesions correspond to either WHO grade II or III (5). The authors stated, however, that definite grading criteria had not been established.

Pineocytoma

Clinical Features. *Pineocytomas* occur primarily in mid to late adulthood with signs referable to compression of the quadrigeminal plate. These include paresis of upward gaze (Perinaud sign) and obstruction of cerebrospinal fluid (CSF) flow with resultant hydrocephalus and often changes in mental status (7–9). Intratumoral hemorrhage is rare ("pineal apoplexy"). No sex predilection is apparent.

Radiologic Findings. The tumors are hypodense on T1- and hyperintense on T2-weighted

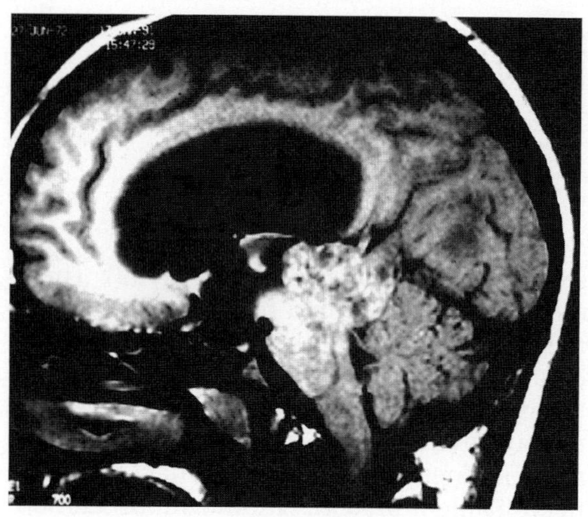

Figure 6-1

PINEOCYTOMA

Pineocytomas are discrete, contrast-enhancing masses that often indent the posterior third ventricle. (Fig. 6-1 from Fascicle 10, 3rd Series.)

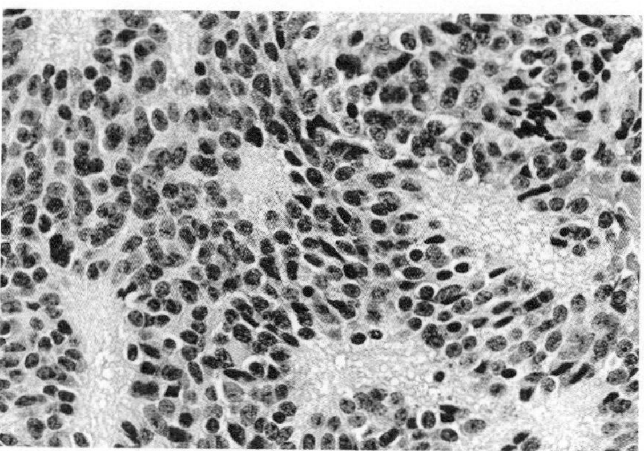

Figure 6-2

PINEOCYTOMA

Pineocytomas are distinctive for lobularity, pineocytomatous rosettes, and pale nuclei with prominent nucleoli.

magnetic resonance images (MRIs). Discrete, often round, and usually contrast enhancing, the masses remain localized to the pineal region or posterior third ventricle (fig. 6-1). Most measure less than 3 cm. Calcification and cystic change are reported, but in our experience, these changes, particularly macrocystic change, are features of pineal cysts, not pineocytomas. In contrast to germ cell tumors, pineocytomas compress the adjacent quadrigeminal plate more by expansion than invasion (10).

Gross Findings. The soft, homogeneous, and finely granular tumors displace rather than infiltrate adjacent brain parenchyma and leptomeninges. Calcification and scant hemorrhage may be seen, but necrosis is lacking.

Microscopic Findings. The classic and diagnostically inescapable WHO grade I pineocytoma is a moderately cellular neoplasm with multiple pineocytomatous rosettes. The latter consist of large, localized feltworks of delicate processes circumscribed by tumor cell nuclei (fig. 6-2). Such rosettes are larger and more irregular than neuroblastic (Homer-Wright) rosettes, and lack the lumens of Flexner-Wintersteiner, true ependymal, and ependymoblastic rosettes. Unlike ependymal pseudorosettes, perivascular processes in pineocytoma do not wheel about a vessel, but create a zone of fine processes resembling neuropil of normal gray matter. Processes, including those about vessels, typically have terminal expansions, a feature best seen after staining with the Bielschowsky method. Clusters, confluent sheets, or ribbons of neoplastic cells may also be seen. Small specimens may include only solid, nonrosette-forming tissue. The nuclei of pineocytoma cells are of moderate size, i.e., somewhat larger than those of pineoblastoma, and generally have a lower chromatin density as well as inconspicuous nucleoli. The mitotic index of pineocytomas is low, in most cases zero (1,2).

A minority of pineocytomas have an alarming degree of cellular and nuclear pleomorphism, but a low MIB-1 index and rare if any mitoses (fig. 6-3) (11). Well-formed ganglion cells or slightly less well-developed ganglioid cells are sometimes present in such lesions. On occasion, ganglion cells are present in otherwise typical pineocytomas, especially those with prominent rosettes (fig. 6-4). Although pineocytomas may be calcified, the mineralization is usually scalloped with laminated "brain sand" (remnants of the normal gland).

The presence of glial differentiation in PPTs is difficult to establish, since the astrocytic stroma of the normal pineal gland is intensely positive for glial fibrillary acidic protein (GFAP) and remains so when overrun by a PPT. In addition, gliomas arising primarily in the pineal region can distort and overrun its parenchyma, producing a mass resembling pineocytoma with

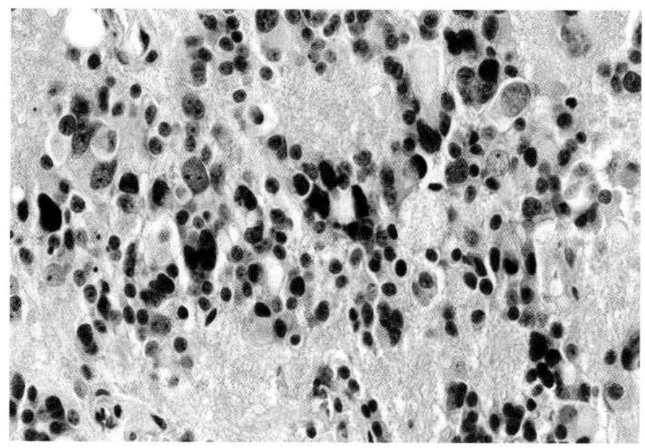

Figure 6-3

PINEOCYTOMA

Considerable nuclear pleomorphism obscures the pineocytomatous rosettes and, therefore, the well-differentiated nature of some pineocytomas.

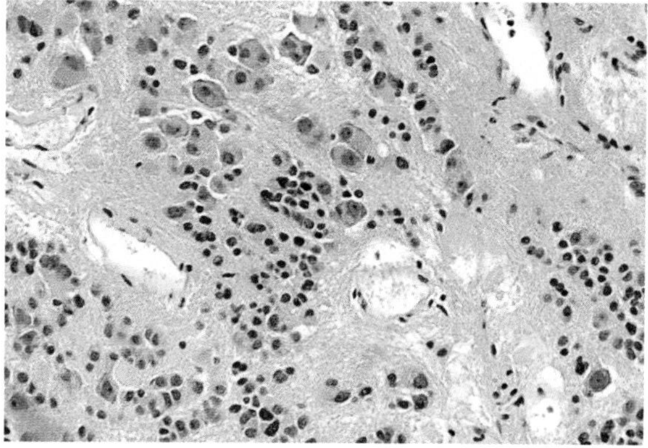

Figure 6-4

PINEOCYTOMA

Ganglion cell differentiation is uncommon in pineocytomas.

a glial component. The diagnosis of glial differentiation in pineal parenchymal tumors thus requires the finding of substantial glial tissue showing nuclear pleomorphism, glassy cytoplasm, and immunoreactivity for GFAP (12). Few lesions satisfy these criteria.

Immunohistochemical Findings. In accord with the neuronal nature of pineocytoma, perinuclear cytoplasm, processes, and pineocytomatous rosettes are immunoreactive for synaptophysin (fig. 6-5), chromogranin (see fig. 6-8, bottom), and in some cases, neurofilament protein (1,2,13). The last reagent highlights club-shaped terminations of cell processes, particularly around vessels. Not surprisingly, there may be reactivity for class III beta-tubulin, tau protein, and PGP9.5. Immunoreactivity for rhodopsin and retinal S antigen have also been noted, both in pineocytoma (14–17) and in PPTs of higher grade (15,18,19). Some cells may be S-100 protein positive. Reactivity for GFAP, a much more specific glial marker, is prominent in interstitial reactive astrocytes (20). As previously noted, the MIB-1 labeling index is low, often less than 1 percent (16).

Ultrastructural Findings. The well-differentiated, organelle-rich cells of pineocytoma feature neuronal characteristics, including processes containing microtubules, synapses, synaptic ribbons, and occasional dense core as well as clear vesicles. The latter two are more numerous in the termination of processes than in the cytoplasm (6,12,13,16,21–23). Paired, twisted filaments may be diagnostic of pineocytoma (24). Junctional complexes are inconspicuous. Photoreceptor differentiation takes the form of cytoplasmic annulate lamellae, synaptic ribbons, vesicle crowned rodlets, and cilia with a 9+0 microtubule configuration (6,12,13). The finding of intermediate filament-rich processes suggests the presence of astrocytes, but true astrocytic differentiation in these tumors is hard to confirm at the ultrastructural level.

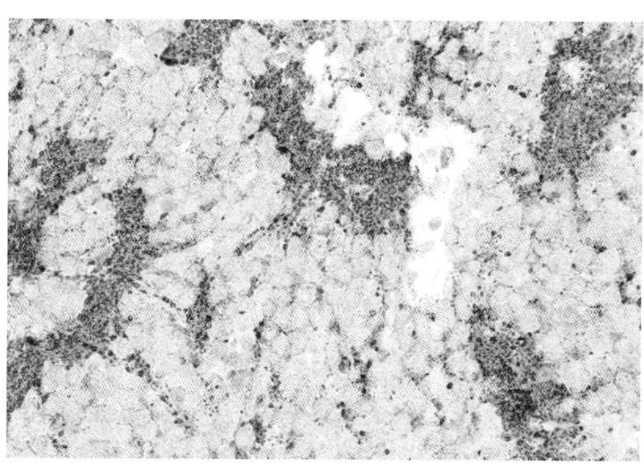

Figure 6-5

PINEOCYTOMA

The centers of pineocytomatous rosettes are immunoreactive for synaptophysin.

Molecular and Cytogenetic Findings. Although one cytogenetic study found no chromosomal abnormalities in three pineocytomas

(25), multiple abnormalities were identified in two other investigations (26,27).

Differential Diagnosis. In the absence of large, synaptophysin-positive pineocytomatous rosettes, the diagnosis begins with establishing the neoplastic nature of the tissue. Normal pineal parenchyma, as seen in the wall of the non-neoplastic pineal cyst, is the principal consideration (fig. 6-6). This distinction is not always simple, since the normal gland varies somewhat in histologic pattern and may resemble a neoplasm, as is discussed in chapter 2 (see fig. 2-45) and in the section on pineal cysts later in this chapter (see figs. 6-21, 6-22). Two typical features of the normal gland are lobules and the tendency for cell processes at the lobule periphery to form perivascular expansions. These are also clearly seen in the peripheral, glandular layer of pineal cyst, which is smoothly demarcated from the inner or glial layer.

Papillary tumor of the pineal region is discussed separately later in this chapter (28).

Treatment and Prognosis. Interpreting the literature relative to prognosis in PPTs poses a problem given the variable grading criteria used in published series. If only those tumors with pineocytomatous rosettes and lack of mitotic activity are accepted as pineocytomas, the outcome is very favorable (2,7). Occasional lesions recur locally, but the potential for CSF seeding is low (15). In a recently reported series of PPTs, no pineocytomas metastasized and the 5-year survival rate was 86 percent (8). The prognosis for patients with pineocytomas with divergent differentiation does not differ substantially from that of patients with conventional pineocytoma, but data are limited (15,29).

Pineal Parenchymal Tumor of Intermediate Differentiation

In the continuum from pineocytoma to pineoblastoma, there is an increase in cell density and the cytologic expression of anaplasia and mitotic index, and a decrease in nuclear size, amount of cytoplasm, and degree of immunoreactivity for neurofilament protein (2,7,15). Where the intermediate lesions fit in this spectrum is clearly arbitrary in concept and difficult to apply in individual cases, particularly when evaluated in the small specimens that are common from this region.

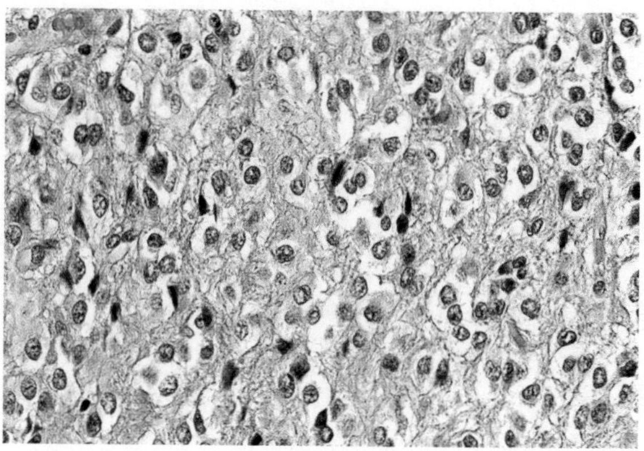

Figure 6-6

NORMAL PINEAL GLAND

Especially when relatively alobular, normal pineal parenchyma can be readily mistaken for a neoplasm.

If pineocytomas are defined as amitotic tumors with pineocytomatous rosettes and low MIB-1 labeling indices, then the nonpineoblastic remainder of PPTs falls into the intermediate category. Reflecting uncertainties of grading, the relative frequencies of pineocytoma and *pineal parenchymal tumor of intermediate differentiation* (PPTID) vary in the literature. One study of 30 PPTs showed pineocytomas to predominate (9 versus 4) (8), whereas in another series, the opposite was true (11 pineocytomas versus 39 PPTIDs) (2). In the latter study, there was a considerable range of mitotic counts, from 0 to 16 per 10 high-power fields, emphasizing the difficulty of defining intermediate lesions in general, and PPTs in particular. Tumors with high mitotic counts are best considered pineoblastoma.

As defined in the WHO classification, "PPTIDs are either diffuse (neurocytoma-like) or lobulated tumours characterized by moderately high cellularity, mild to moderate nuclear atypia, and low to moderate mitotic activity" (fig. 6-7) (5). Nuclei are somewhat smaller and denser than those in pineocytomas. One factor complicating the grading of PPTs is intratumoral variation in grade; some tumors show transition between pineocytoma and PPTID, or between PPTID and pineoblastoma (see fig. 6-11).

Immunohistochemically, PPTIDs, like pineocytoma and most pineoblastomas, are positive for both synaptophysin (fig. 6-8, top) and

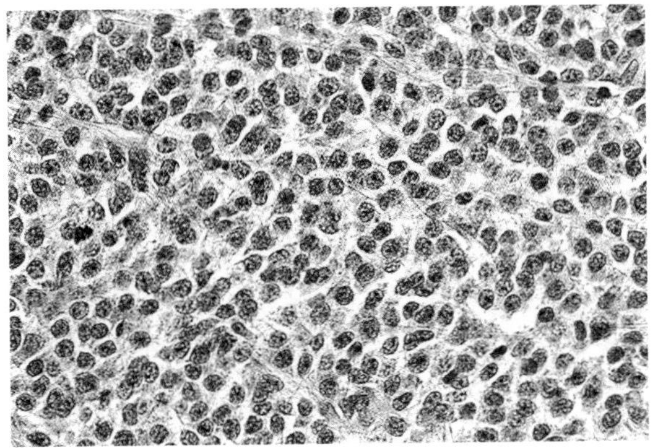

Figure 6-7

PINEAL PARENCHYMAL TUMOR OF INTERMEDIATE DIFFERENTIATION

Such tumors are composed of cells that are cytologically less malignant than those of pineoblastoma, but more closely packed and mitotically active than those of pineocytoma.

chromogranin (fig. 6-8, bottom). Not surprisingly, photoreceptor differentiation has been documented by the finding of interphotoreceptor retinoid-binding protein and its mRNA (19). In one study, the mean MIB-1 labeling index was 3.5 percent (16).

Most patients with PPTID are treated with radiotherapy and, in some cases, with chemotherapy as well. The prognosis is difficult to estimate given definitional problems with the entity, but as a group, patients with PPTID do less well than those with pineocytoma and better than those with pineoblastoma (1,2,30). Fauchon et al. (1) found a 61 percent 10-year survival rate for patients with lesions defined as grade 2 and 5-year survival rates of 74 and 39 percent for those with grade 2 and 3 tumors. Occasional PPTIDs spread distantly in the CSF spaces (7).

Pineoblastoma

Definition. *Pineoblastoma* is a WHO grade IV, medulloblastoma-like, small blue cell tumor arising in the pineal gland.

Clinical Features. Pineoblastomas typically occur in children, but occasional lesions do present in adults (30–32). Rarely, they occur in association with bilateral retinoblastoma, a complex termed *trilateral retinoblastoma* (see chapter 5).

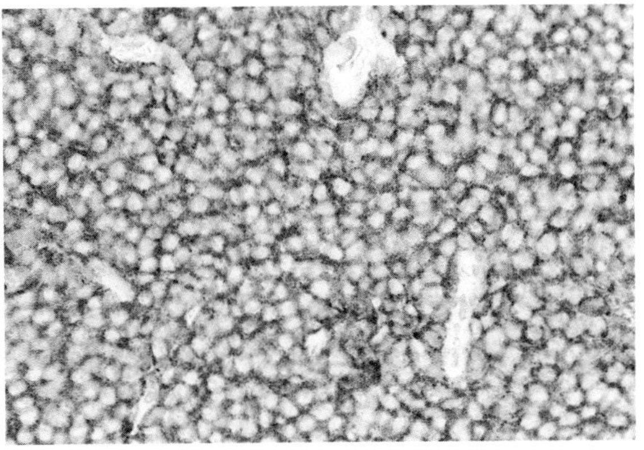

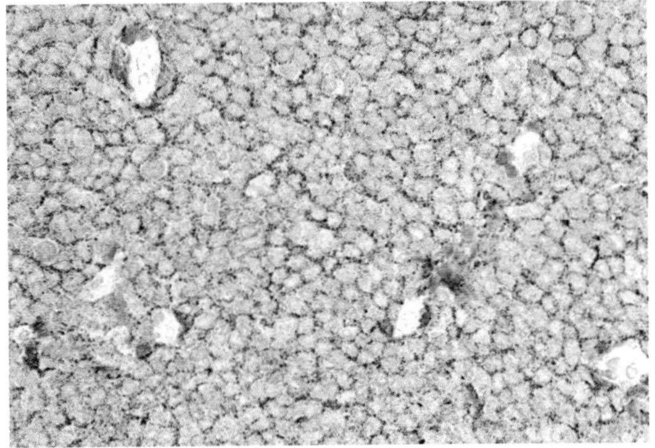

Figure 6-8

PINEAL PARENCHYMAL TUMOR OF INTERMEDIATE DIFFERENTIATION

As with pineocytomas and most pineoblastomas, this lesion is immunoreactive for synaptophysin (top) and chromogranin (bottom).

Radiologic Findings. As seen with computerized tomography (CT) and MRI scans, these are contrast-enhancing lesions that, unlike pineocytoma, are likely to invade surrounding brain parenchyma. Local or distant CSF spread is present at the time of presentation in some cases.

Gross Findings. Pineoblastomas are gray-tan and soft, and often infiltrate surrounding parenchyma and leptomeninges. Having reached the subarachnoid space and regional cisterns or gained access to the ventricular system, they can expand widely to form an opaque, sheet-like infiltrate. Only an occasional pineoblastoma is deceptively well circumscribed.

Microscopic Findings. Pineoblastomas are small blue cell neoplasms (fig. 6-9), but as with medulloblastomas, not all are equally "malignant" in the degree of cytologic atypia and

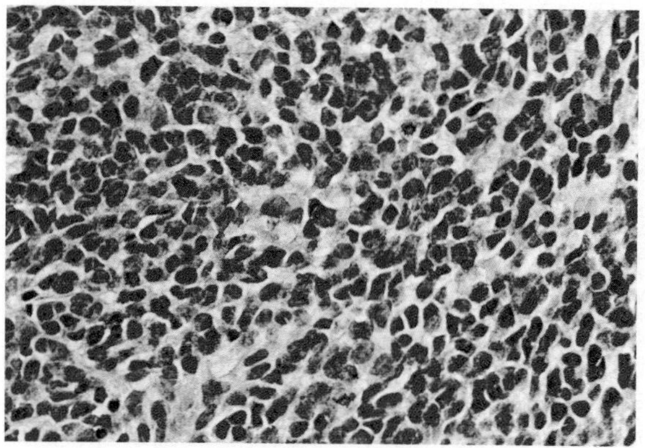

Figure 6-9

PINEOBLASTOMA

Pineoblastomas usually consist of patternless sheets of small undifferentiated cells.

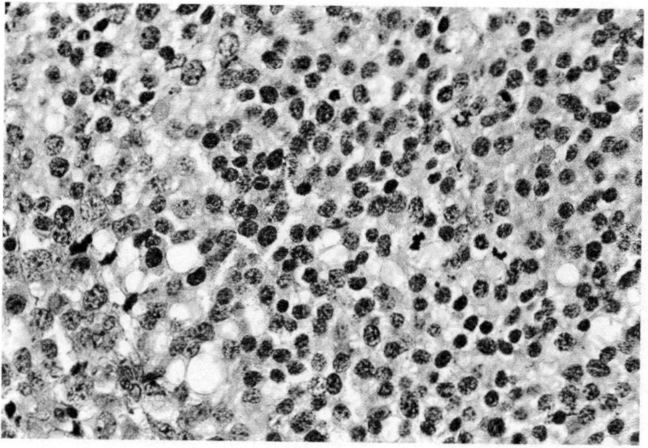

Figure 6-11

PINEAL PARENCHYMAL TUMOR OF INTERMEDIATE DIFFERENTIATION AND PINEOBLASTOMA

Areas of both lesions coexist.

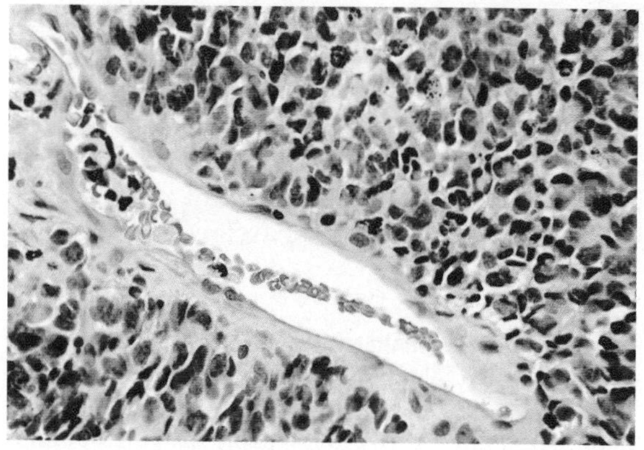

Figure 6-10

PINEOBLASTOMA

As in some medulloblastomas, large cell/anaplastic change occurs in some pineoblastomas.

mitotic activity. Some, while densely blue, have relatively uniform nuclei with restrained mitotic activity, whereas others are cytologically malignant and mitotically energetic. Large cell/anaplastic features, as occur in medulloblastomas, can be present as well (fig. 6-10). Necrosis is common. Vascular proliferation may be present.

Pineoblastomas sometimes exhibit a better-differentiated component of PPTID (fig. 6-11). Only rarely does pineoblastoma coexist with a pineocytoma element. If the latter is suspected, one must exclude the possibility that the well-differentiated component is actually normal pineal gland. Occasional pineoblastomas do, however, appear to arise by malignant transformation of a pineocytoma (32).

An eye-catching feature in some pineoblastomas is retinoblastomatous differentiation, a not surprising event given the photoreceptor function of the pineal gland in lower animals, and the occurrence of photoreceptor differentiation in the developing human pineal gland (33). The most common expression of tumoral retinal differentiation is the presence of Flexner-Wintersteiner rosettes, in which neoplastic cells radially surround a small lumen (fig. 6-12). Encountered more often in textbooks than in practice are more graphic reminders of the gland's photoreceptor origins, known as "fleurettes." These are wavy cytoplasmic extensions that project into small lumens surrounded by cells joined by terminal bars (12,34,35). Neuroblastic (Homer-Wright) rosettes are also occasionally present (fig. 6-13).

Melanin-containing cells, usually in the form of small tubules, are noted in some tumors, both with and without associated retinal lesions. This finding also reflects the ocular nature of the primitive pineal gland and a transient phase in human embryogenesis during which pigmented cells are present (33).

Immunohistochemical Findings. Most pineoblastomas are immunoreactive for synaptophysin, to varying degrees, and less so for chromogranin and neurofilament protein (2). Reactivity for retinal S antigen is present in some cases (15,18). The mean MIB-1 index was 6.5 percent in one study (16).

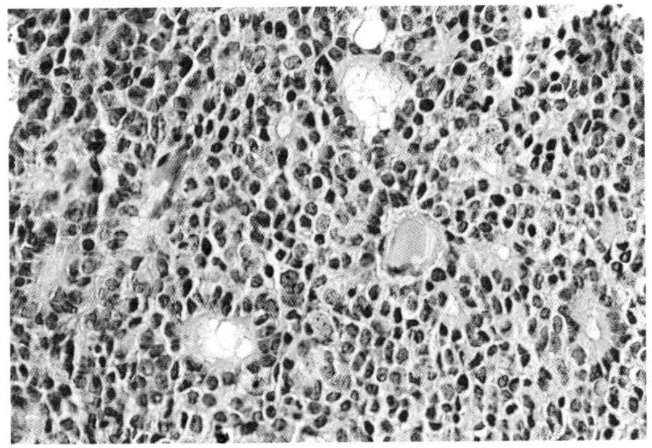

Figure 6-12

PINEOBLASTOMA

Neuroblastic (Homer-Wright) rosettes are present in a minority of pineoblastomas.

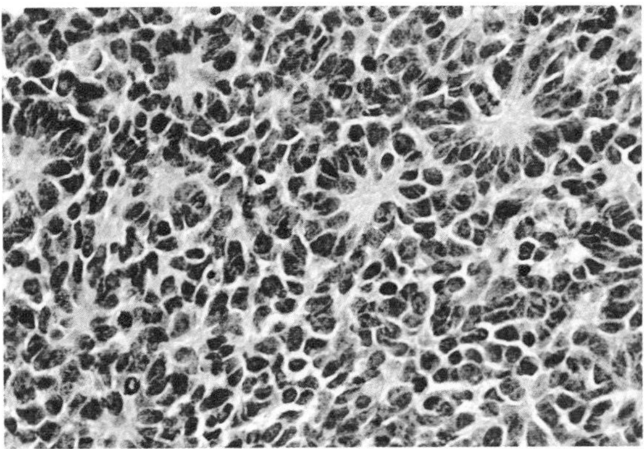

Figure 6-13

PINEOBLASTOMA

As a reminder of the photoreceptor origins of the pineal gland, Flexner-Wintersteiner rosettes are present in some pineoblastomas.

Ultrastructural Findings. Although little information is available regarding the ultrastructural features of pineoblastoma, the cells are largely undifferentiated. The cytoplasm is scant and contains polyribosomes, few profiles of rough endoplasmic reticulum, and scattered mitochondria, in addition to lysosomes. Nonetheless, some have abortive processes, scant microtubules, and rare granules or vesicles (6). Synapses are generally lacking. Club-shaped cell processes with a 9+0 microtubular architecture have been described (13,36).

Molecular and Cytogenetic Findings. Genetic susceptibility to pineoblastoma occurs in association with familial (bilateral) retinoblastoma, resulting in trilateral retinoblastoma (18, 37). A rare example was reported in association with familial adenomatous polyposis (38). Multiple chromosomal gains and losses have been found, but imbalances have been principally gains of 1q and losses of 22 (25).

Differential Diagnosis. The features distinguishing pineoblastoma from PPTID are discussed with the latter. In essence, the distinction hinges upon cellularity, cytologic atypia, mitotic activity, and presence of necrosis. According to Fauchon and Jouvet (1,2), more than 6 mitoses per 10 high-power fields (in a lesion without neurofilament staining) places the lesion in the pineoblastoma category (1,2). Patients with their grade 3 tumor had a poor (39 percent) 5-year survival rate; this tumor would, by some, be considered pineoblastoma.

The distinction from medulloblastoma depends in part upon location, since it is not always possible on the basis of histopathologic findings alone. Flexner-Wintersteiner rosettes are much more likely to occur in pineoblastoma.

The rare pineal anlage tumor combines primitive small cell tumor, Flexner-Wintersteiner rosettes, cartilage, pigmented epithelium, and skeletal muscle (39–41).

Treatment and Prognosis. As a rapidly proliferating, small cell neuronal lesion, the response of pineoblastoma to postoperative therapy is in some respects similar to that of medulloblastoma (7,8,42). On balance, the outcome for patients with pineoblastoma is more favorable than for those with other supratentorial small cell tumors (PNETs). This is, of course, with the exception of infants, who do poorly, since therapy is necessarily restricted to chemotherapy (42–44). As with medulloblastoma, preoperative stage is an important factor, and inferior outcomes are seen in patients with disseminated disease (7,8,30). One study in which children were treated with radiotherapy and chemotherapy recorded 3-year survival and progression-free rates of 73 and 61 percent, respectively (42). A 3-year progression-free survival rate of 61 percent was noted in another study (44).

PAPILLARY TUMOR OF THE PINEAL REGION

Papillary tumor presents as an obstructing mass in the pineal region and often the posterior third ventricle (45,45a). To date, only adults

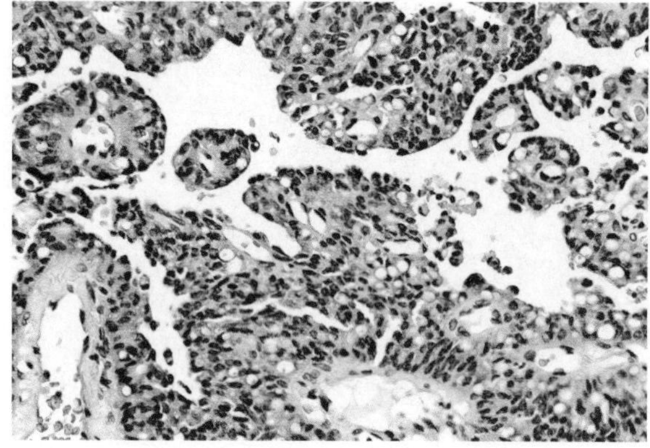

Figure 6-14

PAPILLARY TUMOR OF THE PINEAL REGION

Tissue dehiscence produces a pseudopapillary appearance.

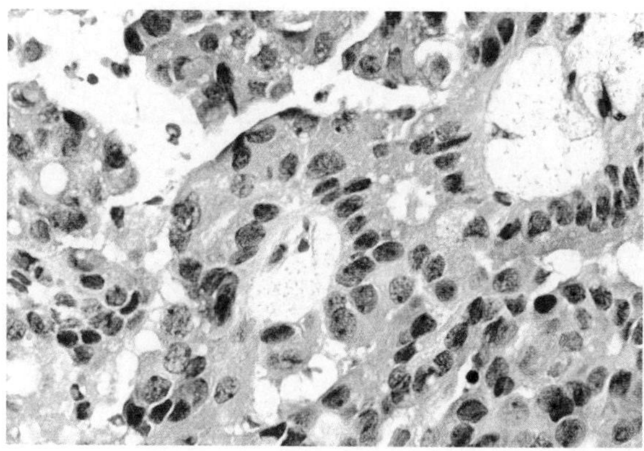

Figure 6-15

PAPILLARY TUMOR OF THE PINEAL REGION

Epithelial surfaces are present focally.

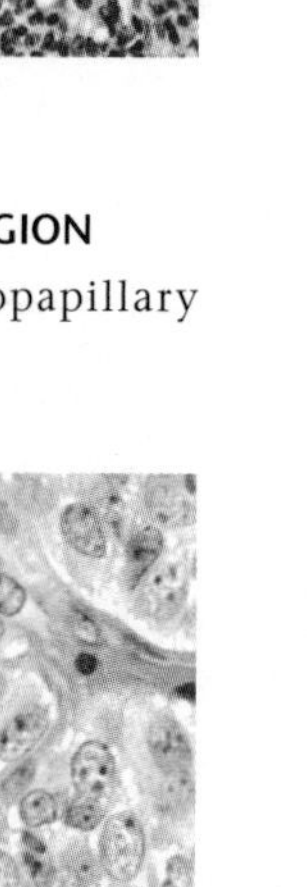

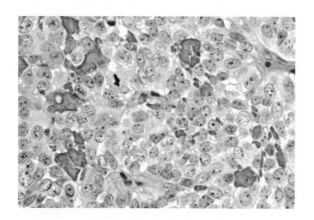

Figure 6-16

PAPILLARY TUMOR OF THE PINEAL REGION

Distinct cell borders lend an epithelial quality in densely cellular areas. Mitoses are common.

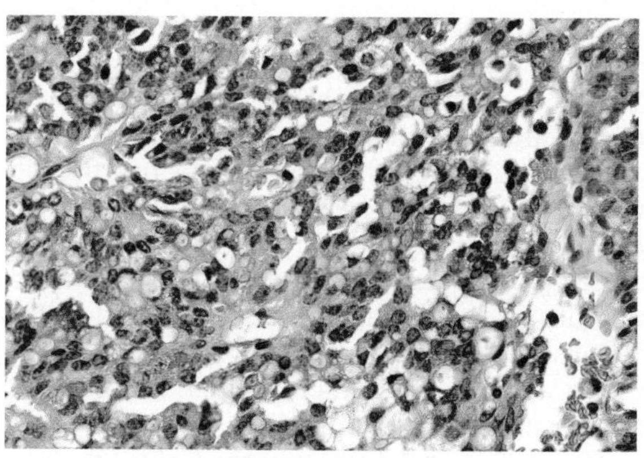

Figure 6-17

PAPILLARY TUMOR OF THE PINEAL REGION

Small intracytoplasmic vacuoles are a frequent feature of this rare pineal region neoplasm.

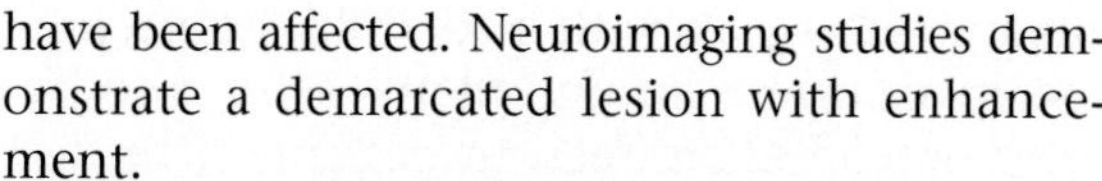

have been affected. Neuroimaging studies demonstrate a demarcated lesion with enhancement.

Histologically, the mass is a diffusely cellular lesion with pseudopapillae (fig. 6-14) and short segments of epithelium (fig. 6-15). Discrete cell borders reinforce the epithelial nature of the lesions, albeit sometimes only focally (fig. 6-16). Cytoplasmic vacuoles sometimes create a signet ring appearance (fig. 6-17). Mitoses range from 0 to 10 per 10 high-power fields. Vessels may be hyalinized but show no vascular proliferation. Limited necrosis is a common feature.

Immunohistochemically, the lesion is predictably positive for cytokeratins (fig. 6-18), and NSE, generally but weakly for EMA, and variably for S-100 protein, synaptophysin, and chromogranin. Reactivity for GFAP, neurofilament protein, and retinal S antigen is not observed (45).

Ultrastructurally, the tumor cells are organelle-rich and have some ependymoma-like features, including "zipper-like" junctions and microlumens filled with microvilli; cilia are rare. Neuronal features include clear and coated vesicles, rare dense core granules, and microtubules.

The biologic behavior of papillary tumor of the pineal region is unclear, but recurrence and

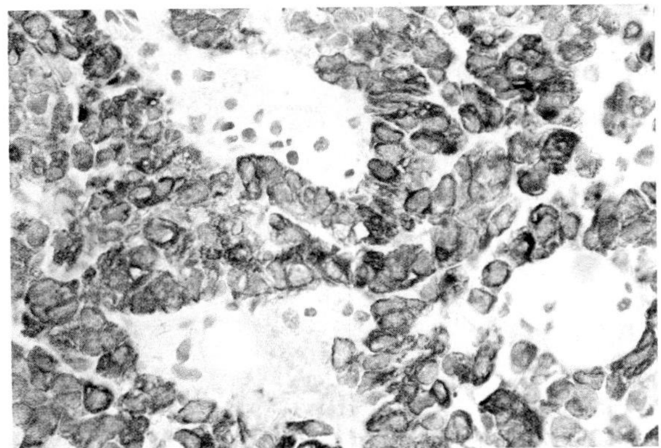

Figure 6-18

PAPILLARY TUMOR OF THE PINEAL REGION

The lesions are immunoreactive for cytokeratins, here CAM5.2.

even leptomeningeal dissemination have been reported (45). Local recurrence is common in spite of radiotherapy, and in one series the 5-year overall and progression-free survival rates were 73 and 27 percent, respectively (45a). Although the tumor shares ependymal and neuroendocrine features, its derivation remains unsettled. An origin in the subcommissural organ is one possibility (45).

PINEAL CYST

Definition. *Pineal cyst* is an intrapineal, nonneoplastic glial cyst.

Clinical Features. Pineal cysts most often are incidental radiographic (fig. 6-19) (46,47) or postmortem findings (fig. 6-20). Only a minority are large enough to compress the aqueduct and produce signs of increased intracranial pressure, or to impinge upon the quadrigeminal plate and disturb upward gaze (Parinaud sign) (47–52). Symptomatic cysts most often occur in young adults, whereas the far more common asymptomatic examples are found in all age groups. Women are more often affected in either case (47,53).

Radiologic Findings. The lesions are midline, well circumscribed, and 1 to 3 cm in greatest dimension. The residual pineal gland is attenuated over much of the cyst surface, but its bulk, often partly calcified after adolescence, is displaced posteroinferiorly. The proteinaceous fluid content typically generates a high (white) signal intensity on proton density or T2-weighted MRIs (fig. 6-19) (47,49,54,55). The dark signal of hemosiderin in T2-weighted images is evidence of remote hemorrhage in many cases (55,56). Blood layers within the cyst may

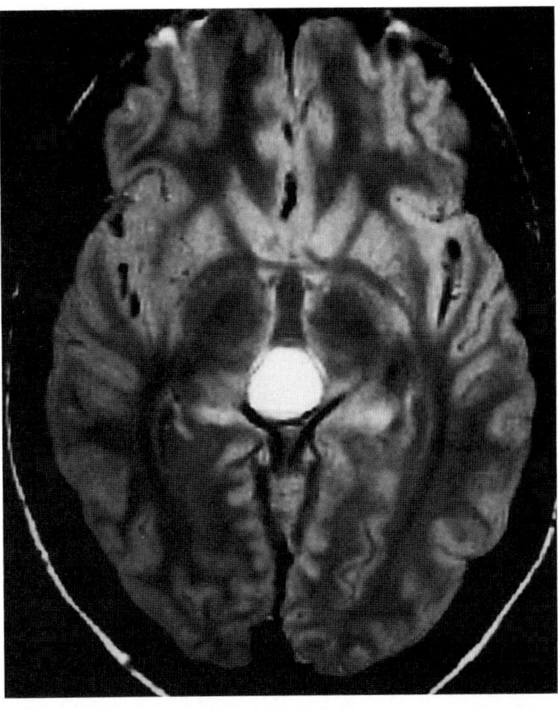

Figure 6-19

PINEAL CYST

Pineal cysts are smoothly contoured masses that often indent the posterior third ventricle. The encysted fluid appears white in a proton density image.

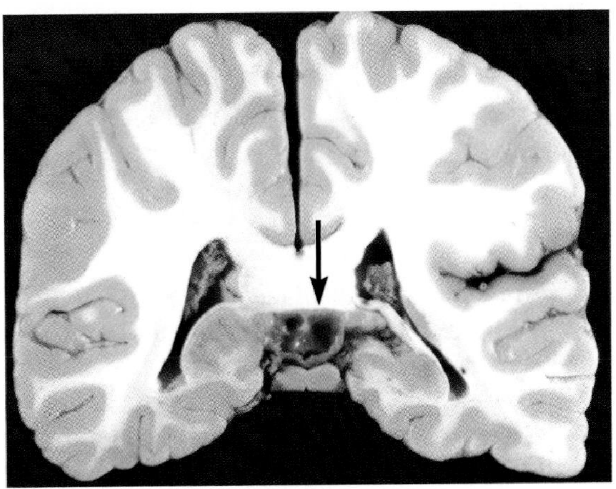

Figure 6-20

PINEAL CYST

Usually discovered incidentally, pineal cysts (arrow) are discrete, midline masses with a smooth glistening lining.

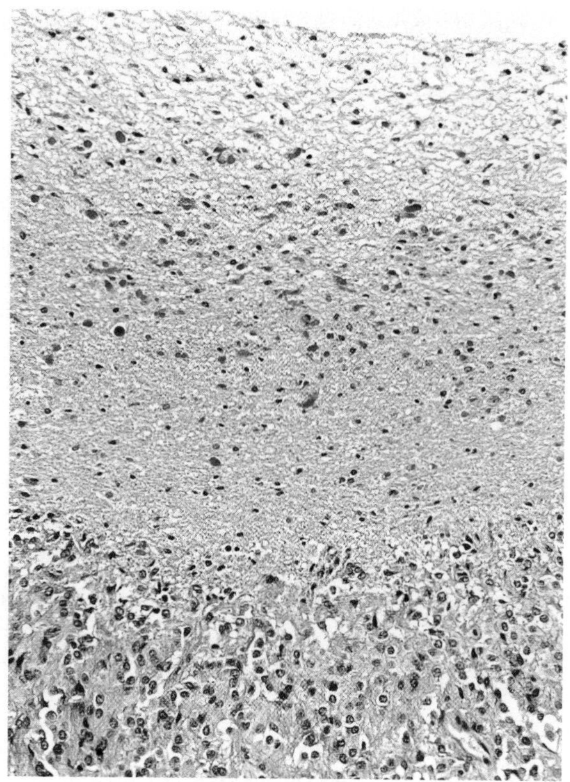

Figure 6-21

PINEAL CYST

A sharp interface between a paucicellular glial layer and often attenuated normal pineal gland is characteristic of pineal cyst. Rosenthal fibers and, in some cases, hemosiderin are common in the glial component.

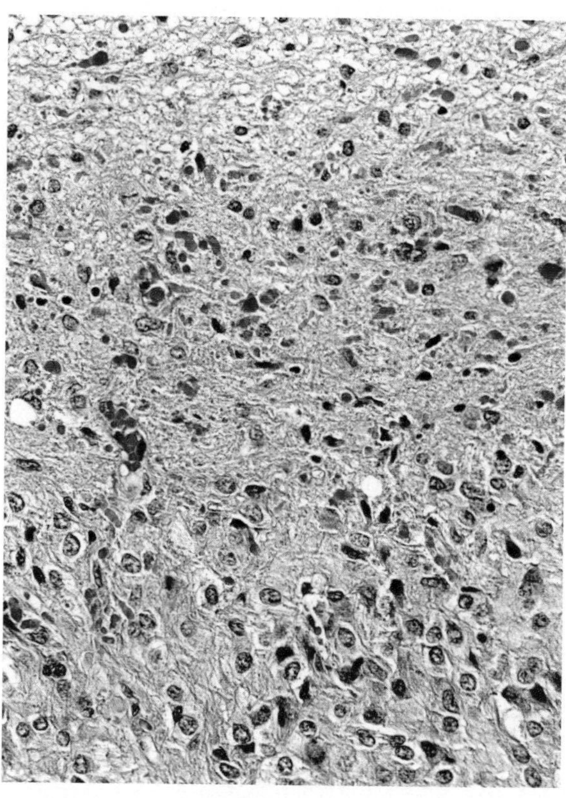

Figure 6-22

PINEAL CYST

Higher magnification confirms the presence of Rosenthal fibers, which are often abundant in the glial layer. Note the deceptive similarity between normal pineal parenchyma and pineocytoma.

produce a fluid-fluid level. Cystic lesions are summarized in Appendix D.

Gross Findings. The smooth-walled cysts both attenuate pineal parenchyma over much of their surface and displace it eccentrically (fig. 6-20). The smooth inner aspect of the cyst is often cinnamon in color.

Microscopic Findings. Pineal cysts are lined by a uniform layer of hypocellular, densely fibrillar, glial tissue in which Rosenthal fibers often abound (figs. 6-21, 6-22). Siderophages are also common. An embryologic diverticulum of the third ventricle deeply indents the normal pineal gland, and may be the origin of some cysts in this region (57), but pineal cysts typically lack an ependymal lining.

External to the glial layer is a demarcated zone of vaguely lobular, sometimes calcified, pineal parenchyma. Attenuated and architecturally altered, it can readily be misdiagnosed as pineocytoma (see figs. 6-2, 6-21, 6-22). Prominent GFAP-positive astrocytes are characteristic (see figs. 2-47–2-49). A delicate and often incomplete covering of leptomeningeal connective tissue comprises the third and outer layer of the cyst.

Differential Diagnosis. Pineal cysts must be differentiated from both gliomas and PPTs. In one series of 24 cases, 8 were interpreted as pineocytoma and 3 as pilocytic astrocytoma (47). Only two specimens were diagnosed initially as pineal cyst. Such errors can be dramatically reduced by recognition of the entity, and with special attention to the gross features of the specimen. Allowing the specimen to float intact in formalin permits it to regain its telltale cystic shape. Once fixed, it can be cut to best demonstrate the diagnostic three-layered architecture, in particular, the sharp interface between pineal parenchyma and the glial component of the cyst wall.

Cut en face instead of perpendicular, the glial layer may appear to be the bulk of the specimen. As such, the highly fibrillated glial lining may resemble pilocytic astrocytoma. This is especially true when Rosenthal fibers and granular bodies are abundant. In contrast to glioma, however, the astrocytic layer of pineal cysts is composed of highly fibrillated, bipolar astrocytes and lacks the loose, microcystic element so often present in pilocytic astrocytomas.

Pineocytoma also enters into the differential diagnosis, since the pineal parenchyma is a notorious mimic of a neoplasm. The often uniform lobularity of the normal pineal gland, absence of pineocytomatous rosettes, and abundance of corpora arenacea ("brain sand") are all features of normal pineal tissue (see figs. 6-2, 6-3, 2-45). GFAP-positive astrocytes with long processes are typical of normal pineal gland (see fig. 2-47), and not of pineocytoma. Lastly, the finding of a sharp interface between a layer of lobular, synaptophysin-positive pineal tissue and a gliotic lamina containing scattered Rosenthal fibers and granular bodies is more characteristic of pineal cyst. The presence of a macrocyst, particularly with a fluid-fluid level, is highly suggestive of pineal cyst. Pineocytomas are rarely macrocystic.

Pineal cysts and other non-neoplastic lesions that can be interpreted as a neoplasm are given in Appendix L. "Suspect diagnoses" such as a pineocytoma or glioma in the face of a cystic pineal mass are given in Appendix N.

Treatment and Prognosis. The long-term outcome for patients undergoing resection is excellent. Recurrences have not been reported after complete resection. No data are available regarding the capacity of subtotally resected cysts to recur. The alternative approach of cyst drainage, rather than excision, remains to be evaluated (58,59). Asymptomatic, incidentally encountered cysts tend to remain stable. Relatively few enlarge and some may even regress (46).

OTHER NEOPLASMS AND NON-NEOPLASTIC MASSES IN THE PINEAL REGION

The occasional gliomas arising in the pineal gland are usually astrocytomas, either diffuse or pilocytic type (60–63). The latter are common in the paraventricular regions. Other miscellaneous neoplasms and non-neoplastic masses in the pineal region include: arachnoid cyst (64), cavernous angioma (65,66), craniopharyngioma (67,68), epidermoid cyst (64,69, 70), ganglioglioma (71,72), melanotic ganglioglioma (73), ganglioneuroblastoma (74), melanotic progonoma (75,76), granulocytic sarcoma (77), hemangiopericytoma (78), melanocytic neoplasms (79–82), meningioma (64,83–85), and metastatic carcinoma (86–88).

REFERENCES

Pineocytoma, Pineal Parenchymal Tumor of Intermediate Differentiation, and Pineoblastoma

1. Fauchon F, Jouvet A, Paquis P, et al. Parenchymal pineal tumors: a clinicopathological study of 76 cases. Int J Radiat Oncol Biol Phys 2000;46:959–68.
2. Jouvet A, Saint-Pierre G, Fauchon F, et al. Pineal parenchymal tumors: a correlation of histological features with prognosis in 66 cases. Brain Pathol 2000;10:49–60.
3. Mena H, Nakazato Y, Jouvet A, Scheithauer B. Pineocytoma. In: Kleihues P, Cavenee W, eds. Pathology and genetics of tumours of the nervous system. Lyon: IARC Press; 2000:118–20.
4. Mena H, Nakazato Y, Jouvet A, Scheithauer B. Pineoblastoma. In: Kleihues P, Cavenee W, eds. Pathology and genetics of tumours of the nervous system. Lyon: IARC Press; 2000:116–8.
5. Nakazato Y, Jouvet A, Scheithauer B. Pineal parenchymal tumour of intermediate differentiation. In: Louis DN, Ohgaki H, Wiestler OD, Cavenee W, eds. WHO classification of tumours of the central nervous system. Lyon: IARC Press; 2007. (In press)
6. Min KW, Scheithauer BW, Bauserman SC. Pineal parenchymal tumors: an ultrastructural study with prognostic implications. Ultrastruct Pathol 1994;18:69–85.
7. Schild SE, Scheithauer BW, Schomberg PJ, et al. Pineal parenchymal tumors. Clinical, pathologic, and therapeutic aspects. Cancer 1993;72: 870–80.

8. Schild SE, Scheithauer BW, Haddock MG, et al. Histologically confirmed pineal tumors and other germ cell tumors of the brain. Cancer 1996;78:2564–71.
9. Vaquero J, Ramiro J, Martinez R, Bravo G. Neurosurgical experience with tumours of the pineal region at Clinica Puerta de Hierro. Acta Neurochir (Wien) 1992;116:23–32.
10. Satoh H, Uozumi T, Kiya K, et al. MRI of pineal region tumours: relationship between tumours and adjacent structures. Neuroradiology 1995;37:624–30.
11. Kuchelmeister K, von Borcke IM, Klein H, Bergmann M, Gullotta F. Pleomorphic pineocytoma with extensive neuronal differentiation: report of two cases. Acta Neuropathol (Berl) 1994;88:448–53.
12. Herrick MK, Rubinstein LJ. The cytological differentiating potential of pineal parenchymal neoplasms (true pinealomas). A clinicopathological study of 28 tumours. Brain 1979;102:289–320.
13. Jouvet A, Fevre-Montange M, Besancon R, et al. Structural and ultrastructural characteristics of human pineal gland, and pineal parenchymal tumors. Acta Neuropathol (Berl) 1994;88:334–48.
14. Korf HW, Klein DC, Zigler JS, Gery I, Schachenmayr W. S-antigen-like immunoreactivity in a human pineocytoma. Acta Neuropathol (Berl) 1986;69:165–7.
15. Mena H, Rushing EJ, Ribas JL, Delahunt B, McCarthy WF. Tumors of pineal parenchymal cells: a correlation of histological features, including nucleolar organizer regions, with survival in 35 cases. Hum Pathol 1995;26:20–30.
16. Numoto RT. Pineal parenchymal tumors: cell differentiation and prognosis. J Cancer Res Clin Oncol 1994;120:683–90.
17. Perentes E, Rubinstein LJ, Herman MM, Donoso LA. S-antigen immunoreactivity in human pineal glands and pineal parenchymal tumors. A monoclonal antibody study. Acta Neuropathol (Berl) 1986;71:224–7.
18. Donoso LA, Rorke LB, Shields JA, Augsburger JJ, Brownstein S, Lahoud S. S-antigen immunoreactivity in trilateral retinoblastoma. Am J Ophthalmol 1987;103:57–62.
19. Lopes MB, Gonzalez-Fernandez F, Scheithauer BW, VandenBerg SR. Differential expression of retinal proteins in a pineal parenchymal tumor. J Neuropathol Exp Neurol 1993;52:516–24.
20. Yamane Y, Mena H, Nakazato Y. Immunohistochemical characterization of pineal parenchymal tumors using novel monoclonal antibodies to the pineal body. Neuropathology 2002;22:66–76.
21. Hassoun J, Gambarelli D, Peragut JC, Toga M. Specific ultrastructural markers of human pinealomas. A study of four cases. Acta Neuropathol (Berl) 1983;62:31–40.
22. Nielsen SL, Wilson CB. Ultrastructure of a "pineocytoma." J Neuropathol Exp Neurol 1975;34:148–58.
23. Scheithauer BW. Pathobiology of the pineal gland with emphasis on parenchymal tumors. Brain Tumor Pathol 1999;16:1–9.
24. Hassoun J, Devictor B, Gambarelli D, Peragut JC, Toga M. Paired twisted filaments: a new ultrastructural marker of human pinealomas? Acta Neuropathol (Berl) 1984;65:163–5.
25. Rickert CH, Simon R, Bergmann M, Dockhorn-Dworniczak B, Paulus W. Comparative genomic hybridization in pineal parenchymal tumors. Genes Chromosomes Cancer 2001;30:99–104.
26. Bello MJ, Rey JA, de Campos JM, Kusak ME. Chromosomal abnormalities in a pineocytoma. Cancer Genet Cytogenet 1993;71:185–6.
27. Rainho CA, Rogatto SR, de Moraes LC, Barbieri-Neto J. Cytogenetic study of a pineocytoma. Cancer Genet Cytogenet 1992;64:127–32.
28. Jouvet A, Fauchon F, Liberski P, et al. Papillary tumor of the pineal region. Am J Surg Pathol 2003;27:505–12.
29. Rubinstein LJ. Cytogenesis and differentiation of pineal neoplasms. Hum Pathol 1981;12:441–8.
30. Lutterbach J, Fauchon F, Schild SE, et al. Malignant pineal parenchymal tumors in adult patients: patterns of care and prognostic factors. Neurosurgery 2002;51:44–55; discussion 55–6.
31. Chang SM, Lillis-Hearne PK, Larson DA, Wara WM, Bollen AW, Prados MD. Pineoblastoma in adults. Neurosurgery 1995;37:383–90; discussion 390–1.
32. Saito R, Shirane R, Oku T, et al. Surgical treatment of a mixed pineocytoma/pineoblastoma in a 72-year-old patient. Acta Neurochir (Wien) 2002;144:389–93.
33. Min KW, Seo IS, Song J. Postnatal evolution of the human pineal gland. An immunohistochemical study. Lab Invest 1987;57:724–8.
34. Sobel RA, Trice JE, Nielsen SL, Ellis WG. Pineoblastoma with ganglionic and glial differentiation: report of two cases. Acta Neuropathol (Berl) 1981;55:243–6.
35. Stefanko SZ, Manschot WA. Pinealoblastoma with retinoblastomatous differentiation. Brain 1979;102:321–32.
36. Kline KT, Damjanov I, Katz SM, Schmidek H. Pineoblastoma: an electron microscopic study. Cancer 1979;44:1692–9.
37. Elias WJ, Lopes MB, Golden WL, Jane JA Sr, Gonzalez-Fernandez F. Trilateral retinoblastoma variant indicative of the relevance of the retinoblastoma tumor-suppressor pathway to medulloblastomas in humans. J Neurosurg 2001;95:871–8.

38. Ikeda J, Sawamura Y, van Meir EG. Pineoblastoma presenting in familial adenomatous polyposis (FAP): random association, FAP variant or Turcot syndrome? Br J Neurosurg 1998;12:576–8.
39. McGrogan G, Rivel J, Vital C, Guerin J. A pineal tumour with features of "pineal anlage tumour." Acta Neurochir (Wien) 1992;117:73–7.
40. Raisanen J, Vogel H, Horoupian DS. Primitive pineal tumor with retinoblastomatous and retinal/ciliary epithelial differentiation: an immunohistochemical study. J Neurooncol 1990;9:165–70.
41. Schmidbauer M, Budka H, Pilz P. Neuroepithelial and ectomesenchymal differentiation in a primitive pineal tumor ("pineal anlage tumor"). Clin Neuropathol 1989;8:7–10.
42. Cohen BH, Zeltzer PM, Boyett JM, et al. Prognostic factors and treatment results for supratentorial primitive neuroectodermal tumors in children using radiation and chemotherapy: a Childrens Cancer Group randomized trial. J Clin Oncol 1995;13:1687–96.
43. Duffner PK, Cohen ME, Sanford RA, et al. Lack of efficacy of postoperative chemotherapy and delayed radiation in very young children with pineoblastoma. Pediatric Oncology Group. Med Pediatr Oncol 1995;25:38–44.
44. Jakacki RI, Zeltzer PM, Boyett JM, et al. Survival and prognostic factors following radiation and/or chemotherapy for primitive neuroectodermal tumors of the pineal region in infants and children: a report of the Childrens Cancer Group. J Clin Oncol 1995;13:1377–83.

Papillary Tumor of the Pineal Region

45. Jouvet A, Fauchon F, Liberski P, et al. Papillary tumor of the pineal region. Am J Surg Pathol 2003;27:505–12.
45a. Févre-Montange M, Hasselblatt M, Figarella-Branger D, et al. Prognosis and histopathologic features in papillary tumors of the pineal region: a retrospective multicenter study of 31 cases. J Neuropathol Exp Neurol 2006;65:1004–11.

Pineal Cyst

46. Barboriak DP, Lee L, Provenzale JM. Serial MR imaging of pineal cysts: implications for natural history and follow-up. AJR Am J Roentgenol 2001;176:737–43.
47. Fain JS, Tomlinson FH, Scheithauer BW, et al. Symptomatic glial cysts of the pineal gland. J Neurosurg 1994;80:454–60.
48. Fetell MR, Bruce JN, Burke AM, et al. Non-neoplastic pineal cysts. Neurology 1991;41:1034–40.
49. Fleege MA, Miller GM, Fletcher GP, Fain JS, Scheithauer BW. Benign glial cysts of the pineal gland: unusual imaging characteristics with histologic correlation. AJNR Am J Neuroradiol 1994;15:161–6.
50. Mena H, Armonda RA, Ribas JL, Ondra SL, Rushing EJ. Nonneoplastic pineal cysts: a clinicopathologic study of twenty-one cases. Ann Diagn Pathol 1997;1:11–8.
51. Michielsen G, Benoit Y, Baert E, Meire F, Caemaert J. Symptomatic pineal cysts: clinical manifestations and management. Acta Neurochir (Wien) 2002;144:233–42.
52. Wisoff JH, Epstein F. Surgical management of symptomatic pineal cysts. J Neurosurg 1992;77:896–900.
53. Sawamura Y, Ikeda J, Ozawa M, Minoshima Y, Saito H, Abe H. Magnetic resonance images reveal a high incidence of asymptomatic pineal cysts in young women. Neurosurgery 1995;37:11–6.
54. Mamourian AC, Towfighi J. Pineal cysts: MR imaging. AJNR Am J Neuroradiol 1986;7:1081–6.
55. Mukherjee KK, Banerji D, Sharma R. Pineal cyst presenting with intracystic and subarachnoid haemorrhage: report of a case and review of the literature. Br J Neurosurg 1999;13:189–92.
56. Osborn RE, Deen HG, Kerber CW, Glass RF. A case of hemorrhagic pineal cyst: MR/CT correlation. Neuroradiology 1989;31:187–9.
57. Cooper E. The human pineal gland and pineal cysts. J Anat 1932;67:28–46.
58. Musolino A, Cambria S, Rizzo G, Cambria M. Symptomatic cysts of the pineal gland: stereotactic diagnosis and treatment of two cases and review of the literature. Neurosurgery 1993;32:315–20; discussion 320–1.
59. Stern JD, Ross DA. Stereotactic management of benign pineal region cysts: report of two cases. Neurosurgery 1993;32:310–4.

Other Neoplasms and Non-Neoplastic Masses in the Pineal Region

60. DeGirolami U, Armbrustmacher VW. Juvenile pilocytic astrocytoma of the pineal region: report of a case. Cancer 1982;50:1185–8.
61. Barnett DW, Olson JJ, Thomas WG, Hunter SB. Low-grade astrocytomas arising from the pineal gland. Surg Neurol 1995;43:70–6.
62. Snipes GJ, Horoupian DS, Shuer LM, Silverberg GD. Pleomorphic granular cell astrocytoma of the pineal gland. Cancer 1992;70:2159–65.
63. Vaquero J, Ramiro J, Martinez R. Glioblastoma multiforme of the pineal region. J Neurosurg Sci 1990;34:149–50.
64. Chandy MJ, Damaraju SC. Benign tumours of the pineal region: a prospective study from 1983 to 1997. Br J Neurosurg 1998;12:228–33.
65. Donati P, Maiuri F, Gangemi M, Gallicchio B, Sigona L. Cavernous angioma of the pineal region. J Neurosurg Sci 1992;36:155–60.

66. Fukui M, Matsuoka S, Hasuo K, Numaguchi Y, Kitamura K. Cavernous hemangioma in the pineal region. Surg Neurol 1983;20:209–15.
67. Solarski A, Panke ES, Panke TW. Craniopharyngioma in the pineal gland. Arch Pathol Lab Med 1978;102:490–1.
68. Usanov EI, Hatomkin DM, Nikulina TA, Gorban NA. Craniopharyngioma of the pineal region. Childs Nerv Syst 1999;15:4–7.
69. Konovalov AN, Spallone A, Pitzkhelauri DI. Pineal epidermoid cysts: diagnosis and management. J Neurosurg 1999;91:370–4.
70. MacKay CI, Baeesa SS, Ventureyra EC. Epidermoid cysts of the pineal region. Childs Nerv Syst 1999;15:170–8.
71. Chang YL, Lin SZ, Chiang YH, Liu MY, Lee WH. Pineal ganglioglioma with premature thelarche. Report of a case and review of the literature. Childs Nerv Syst 1996;12:103–6.
72. Faillot T, Sichez JP, Capelle L, Kujas M, Fohanno D. Ganglioglioma of the pineal region: case report and review of the literature. Surg Neurol 1998;49:104–8.
73. Hunt SJ, Johnson PC. Melanotic ganglioglioma of the pineal region. Acta Neuropathol (Berl) 1989;79:222–5.
74. Tanaka M, Shibui S, Nomura K, Nakanishi Y. Pineal ganglioneuroblastoma in an adult. J Neurooncol 1999;44:169–73.
75. Gorhan C, Soto-Ares G, Ruchoux MM, Blond S, Pruvo JP. Melanotic neuroectodermal tumour of the pineal region. Neuroradiology 2001;43:944–7.
76. Rickert CH, Probst-Cousin S, Blasius S, Gullotta F. Melanotic progonoma of the brain: a case report and review. Childs Nerv Syst 1998;14:389–93.
77. Voessing R, Berthold F, Richard KE, Thun F, Schroeder R, Krueger GR. Primary myeloblastoma of the pineal region. Clin Neuropathol 1992;11:11–5.
78. Sell JJ, Hart BL, Rael JR. Hemangiopericytoma: a rare pineal mass. Neuroradiology 1996;38: 782–4.
79. Czirjak S, Vitanovic D, Slowik F, Magyar A. Primary meningeal melanocytoma of the pineal region. Case report. J Neurosurg 2000;92:461–5.
80. Rubino GJ, King WA, Quinn B, Marroquin CE, Verity MA. Primary pineal melanoma: case report. Neurosurgery 1993;33:511–5.
81. Suzuki T, Yasumoto Y, Kumami K, et al. Primary pineal melanocytic tumor. Case report. J Neurosurg 2001;94:523–7.
82. Yamane K, Shima T, Okada Y, et al. Primary pineal melanoma with long-term survival: case report. Surg Neurol 1994;42:433–7.
83. Haque M, Ohata K, Tsuyuguchi N, Sakamoto S, Hara M. A case of pineal region meningioma without dural attachment, presented with bilateral hearing impairment. Acta Neurochir (Wien) 2002;144:209–11.
84. Konovalov AN, Spallone A, Pitzkhelauri DI. Meningioma of the pineal region: a surgical series of 10 cases. J Neurosurg 1996;85:586–90.
85. Madawi AA, Crockard HA, Stevens JM. Pineal region meningioma without dural attachment. Br J Neurosurg 1996;10:305–7.
86. Halpert B, Erickson EE, Fields WS. Intracranial involvement from carcinoma of the lung. Arch Pathol 1960;69:93–103.
87. Lauro S, Trasatti L, Capalbo C, Mingazzini PL, Vecchione A, Bosman C. Unique pineal gland metastasis of clear cell renal carcinoma: case report and review of the literature. Anticancer Res 2002;22:3077–9.
88. Weber P, Shepard KV, Vijayakumar S. Metastases to pineal gland. Cancer 1989;63:164–5.

7 HEMANGIOBLASTOMA

GENERAL PATHOLOGIC FEATURES

Definition. *Hemangioblastoma* is a capillary-rich neoplasm that contains variably lipidized stromal or interstitial cells.

Clinical Features. Hemangioblastomas arise either in the setting of von Hippel-Lindau (VHL) disease (see figs. 18-15–18-18) or as solitary sporadic lesions without extracerebellar stigmata or a family history (1,2). The latter are considerably more common. VHL disease, discussed in chapter 18, includes intracranial or intraspinal hemangioblastoma (see figs. 18-15, 18-19), often multiple (3–5); retinal hemangioblastoma (see fig. 18-16); cystic lesions of the liver, kidney, pancreas, and epididymis; and benign and malignant renal cell tumors. Papillary endolymphatic sac tumor of the ear is another, albeit rare, concomitant of VHL (see figs. 3-351, 16-17) (4,6).

Sporadic hemangioblastomas are tumors of adulthood, generally occurring between ages 30 and 65 years, and are somewhat more common in men. Diagnosis at a younger age is more likely in the setting of VHL disease (2). The cerebellum is the most frequently affected site, followed distantly in incidence by the medulla (7) and spinal cord (8–10). Involvement of cranial or spinal nerve roots, particularly when multiple, is especially likely in patients with VHL disease (see fig. 18-19), but can occur sporadically as well (11–13). Also associated with VHL disease are rare hemangioblastomas in the supratentorial meninges (14). Polycythemia, a result of the tumoral production of erythropoietin, occurs in approximately 10 percent of patients (15–17).

Radiologic Findings. Hemangioblastomas in the cerebellum and spinal cord are often associated with a cyst, whose size relative to the contrast-enhancing nodule varies greatly from case to case (fig. 7-1) (18,19). Hemangioblastoma and other cystic lesions are listed in Appendix D. The nodule is dominant in some cases but minute, if identified at all, in others. The vascularity of hemangioblastoma is evident angiographically, as well as by magnetic resonance imaging (MRI), in which flow voids are common. When small, spinal lesions may be more obviously localized to the posterior aspect of the cord and, as is typical of hemangioblastoma in general, are related to the pial surface (see fig. 18-18) (18).

Gross Findings. Cerebellar hemangioblastomas are highly vascular masses that typically abut the leptomeninges. Spinal hemangioblastomas are intramedullary, discrete, and, as at other sites, usually in contact with the leptomeninges posteriorly (see fig. 18-18). Prominent leptomeningeal blood vessels closely simulate those of a vascular malformation (fig. 7-2). A large syrinx dwarfs the tumor in some cases.

On cut section, the dark red, spongy tumor exudes blood upon compression. The degree of yellowness, if any, reflects the lipid content of the lesion's stromal cells (fig. 7-3).

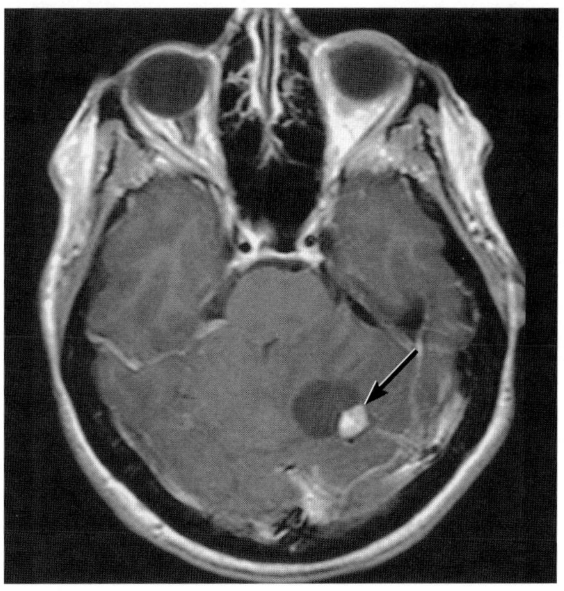

Figure 7-1

HEMANGIOBLASTOMA

The prototypic hemangioblastoma is a cystic cerebellar lesion with a contrast-enhancing mural nodule (arrow).

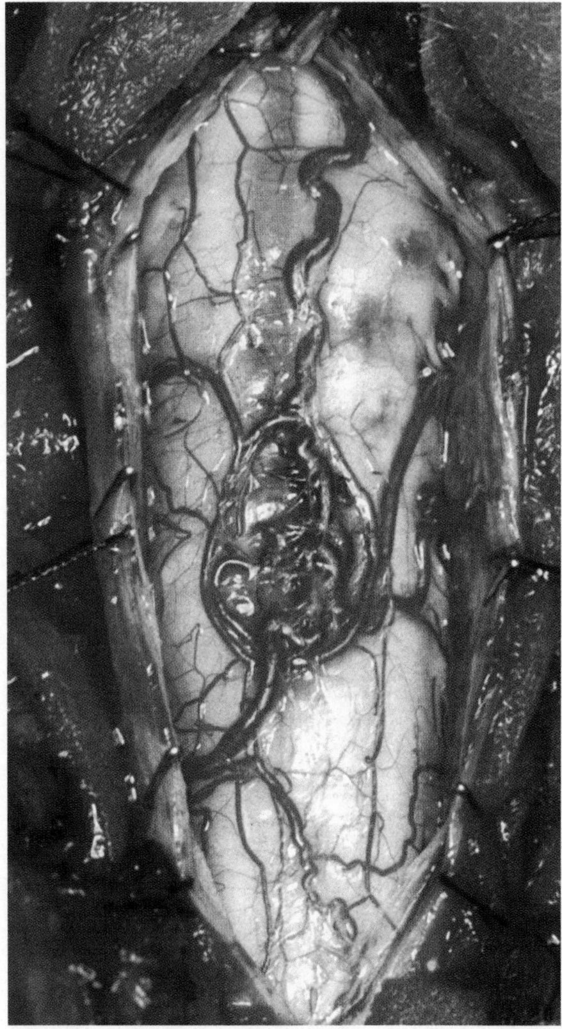

Figure 7-2

HEMANGIOBLASTOMA

The rich superficial vascularity of intramedullary hemangioblastoma simulates that of an arteriovenous malformation. (Courtesy of Dr. Ziya Gokaslan, Baltimore, MD.)

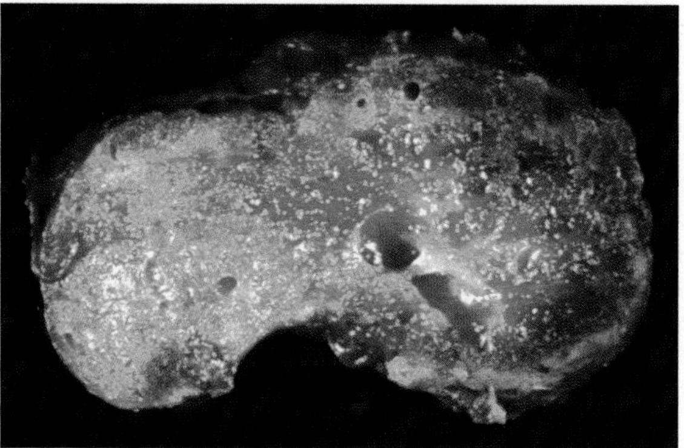

Figure 7-3

HEMANGIOBLASTOMA

The highly vascular, lipidized lesions are understandably both red and yellow.

Microscopic Findings. At low magnification, hemangioblastomas often vary considerably in their cell density, with highly cellular areas alternating with paucicellular regions composed largely of dilated vessels and cyst-like spaces (fig. 7-4). The cyst wall and the brain in direct contact with the lesion consist of a dense layer nonneoplastic piloid gliosis in which Rosenthal fibers are usually abundant (fig. 7-5).

While tumor vascularity is almost always prominent, in some cases predominant, stromal cells are the essence of the lesion. The variation in their cytologic appearance and number relative to vascular elements underlies the lesion's heterogeneity (figs. 7-6, 7-7). The adjective *reticular* is used for hemangioblastomas that are composed predominantly of small nests and individual cells, and are therefore especially rich in reticulin. Larger lobules of cells predominate in the *cellular variant.* Stromal cells vary in size depending partly upon their lipid content. The lipid, extracted during tissue processing, leaves clear cytoplasmic vacuoles that may be fine or coarse. When nonvacuolated and smooth bordered, the cells can be sufficiently epithelioid to suggest metastatic carcinoma (fig. 7-7A–C). Cytoplasmic hyaline droplets are present in a minority of cases (fig. 7-6F).

The nuclei of stromal cells vary considerably in size and chromatin density. Most are medium-sized, round, or reniform, but large hyperchromatic nuclei often lie scattered throughout the lesion or congregate at its periphery (fig. 7-6A,B). Nuclear uniformity is more the rule in the reticular variant. Mitoses are uncommon and of no prognostic significance.

The vasculature of hemangioblastoma consists of large feeding and draining vessels, as well as innumerable intervening capillaries. The proportion and distribution of the vessels underlie the lesions's reticular or cellular architecture. Reticulin highlights the vasculature in the highly vascular reticular variant (fig. 7-8), but is sparse and limited to the capillaries surrounding expanded lobules in the cellular variant. Best seen

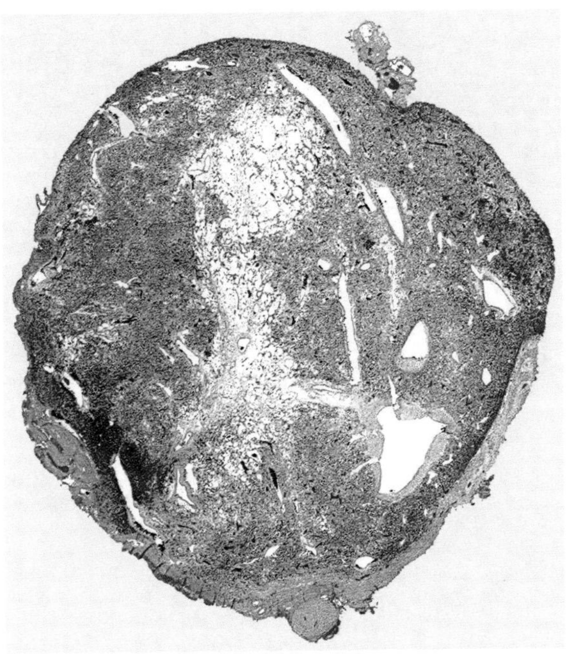

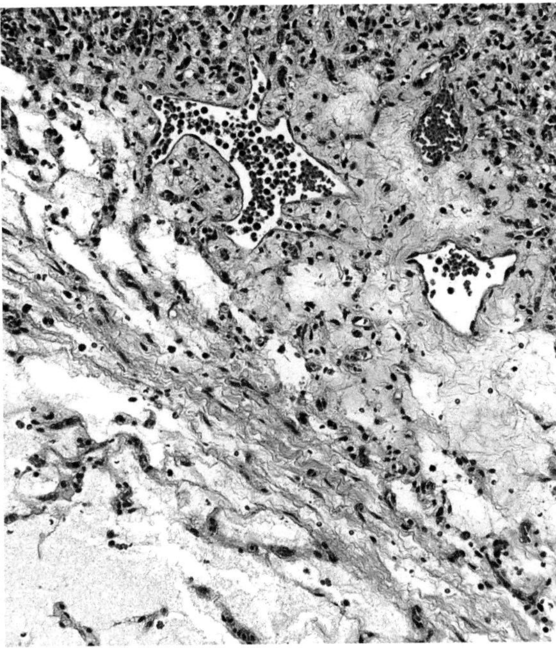

Figure 7-4

HEMANGIOBLASTOMA

As seen in a whole mount histologic section (left) and photomicrograph (right), the highly vascular lesion is often heterogeneous in its degree of cellularity.

in plastic-embedded semi-thin sections or by electron microscopy, smaller vessels consist of both endothelial cells and pericytes (see fig. 7-14).

Mast cells, a common feature of hemangioblastoma (20), are uniformly distributed throughout the tumor and readily demonstrated with the toluidine blue or Giemsa stain (fig. 7-9). Aside from occasional meningiomas and the giant cell astrocytoma of tuberous sclerosis, this is one of the few brain tumors to contain a significant number of mast cells. Occasional foci of extramedullary erythropoiesis are presumably the result of the tumoral production of erythropoietin that underlies the erythrocytosis seen in some patients (21).

A special variant of hemangioblastoma, the so-called *angioglioma*, consists of both hemangioblastoma and what appears to be a glial element such as astrocytoma, ependymoma, or oligodendroglioma (22). At present, no consensus has been reached regarding the nature of such tumors. Since the term "angioglioma" has also been applied to highly vascular gliomas (23), it is devoid of specific meaning.

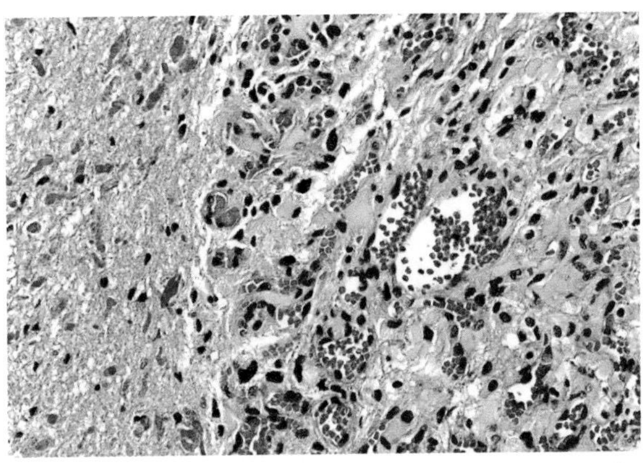

Figure 7-5

HEMANGIOBLASTOMA

A zone of piloid gliosis with Rosenthal fibers commonly surrounds these slow-growing lesions. Note the sharp interface between the lesion and the surrounding cerebellum.

No differences have been noted in the histologic features of sporadic and VHL syndrome-associated lesions (24).

Frozen Section. Fibrillarity and nuclear pleomorphism are accentuated in frozen sections

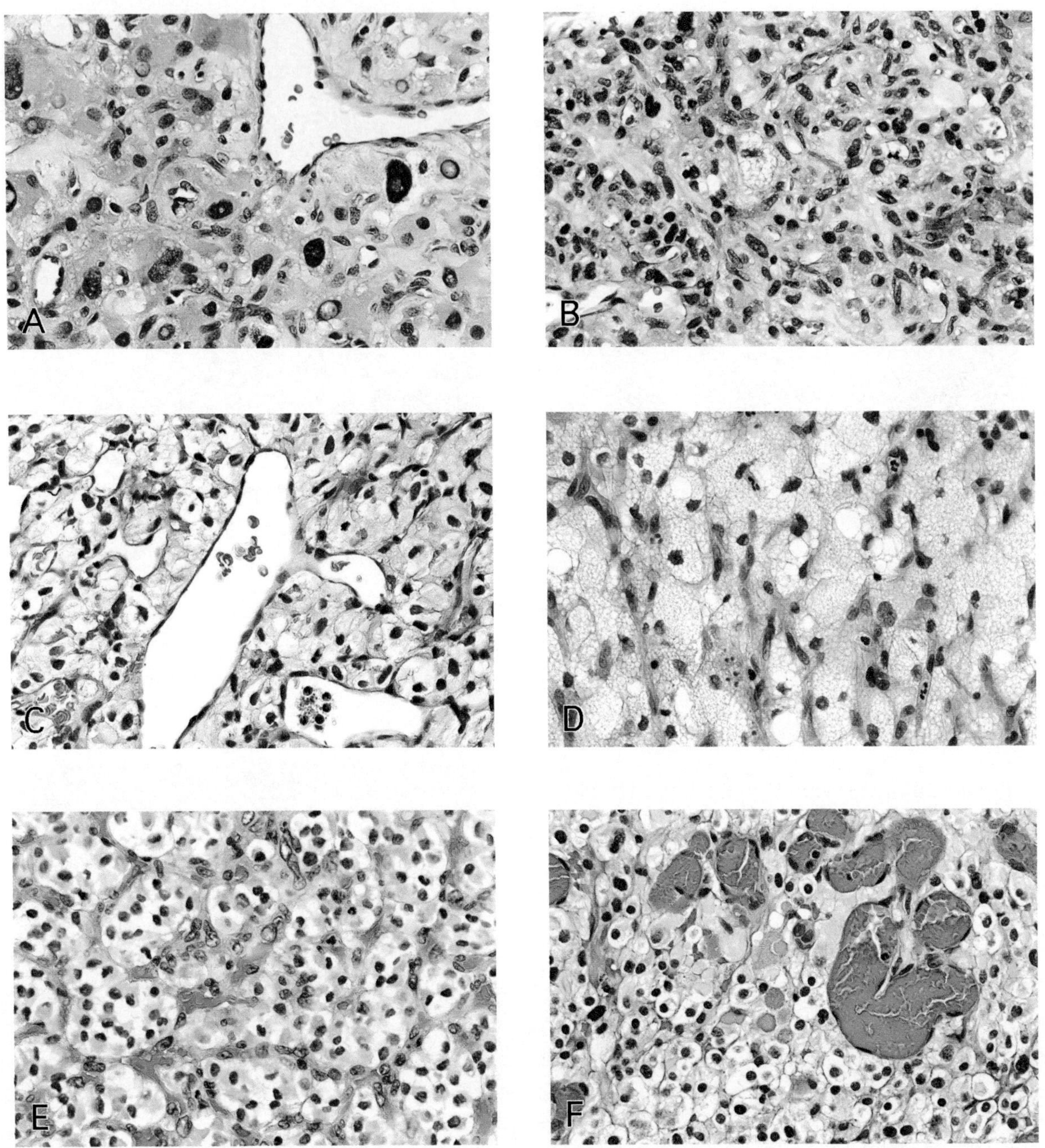

Figure 7-6

HEMANGIOBLASTOMA

A: Rich vascularity, dark pleomorphic nuclei, and scattered vacuolated cells are common elements in hemangioblastoma.

B: The presence of vacuolated cells in a compact, highly vascular, cerebellar or spinal mass is very suggestive of hemangioblastoma.

C: Hemangioblastomas can closely resemble metastatic clear cell carcinoma when stromal cells are uniform in size and shape.

D: Cytoplasmic lipid vacuoles are very small in some cases.

E: Cytoplasmic clearing gives some hemangioblastomas an oligodendroglioma-like appearance.

F: Distinct cell borders lend an epithelial appearance. Hyaline globules may be present in such lesions, as here.

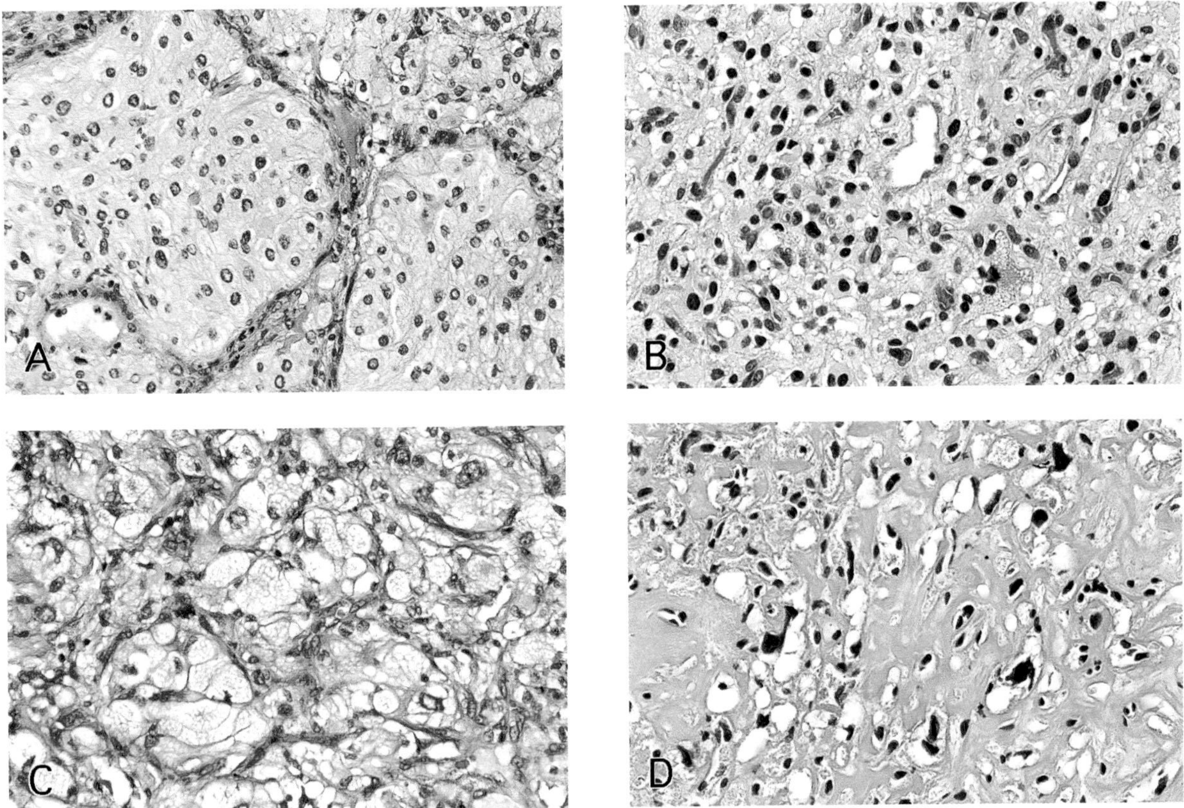

Figure 7-7

HEMANGIOBLASTOMA

A: Large lobules internally devoid of capillaries characterize the cellular variant.

B: Diffuse sheets with inconspicuous capillaries create a glioma-like appearance in some cases.

C: With a nested, almost glandular architecture some hemangioblastomas simulate renal cell carcinoma. The cytoplasmic vacuolation is more consistent with hemangioblastoma.

D: Hemangioblastomas can become densely sclerotic.

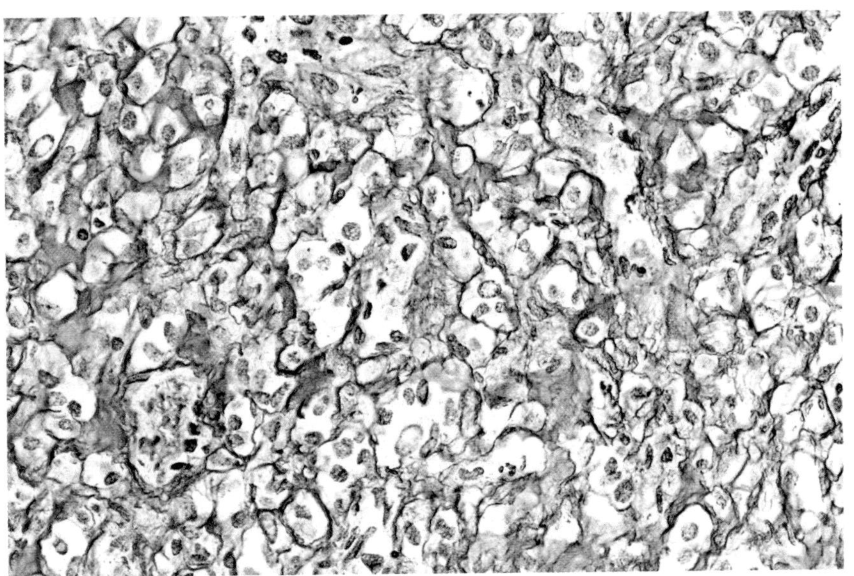

Figure 7-8

HEMANGIOBLASTOMA

Reticulin is abundant in the tightly nested reticular variant.

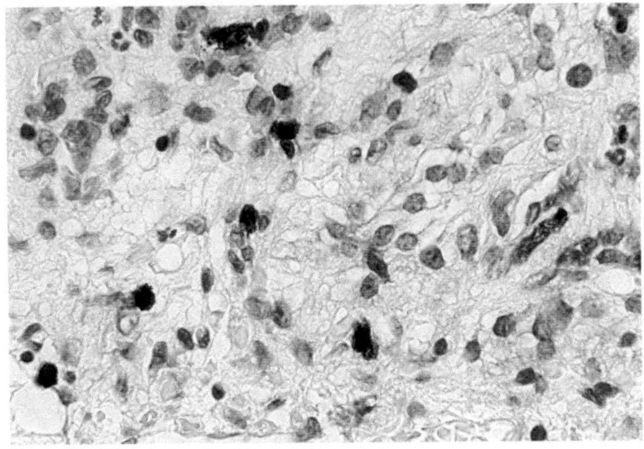

Figure 7-9

HEMANGIOBLASTOMA

Some hemangioblastomas are rich in mast cells (Giemsa stain).

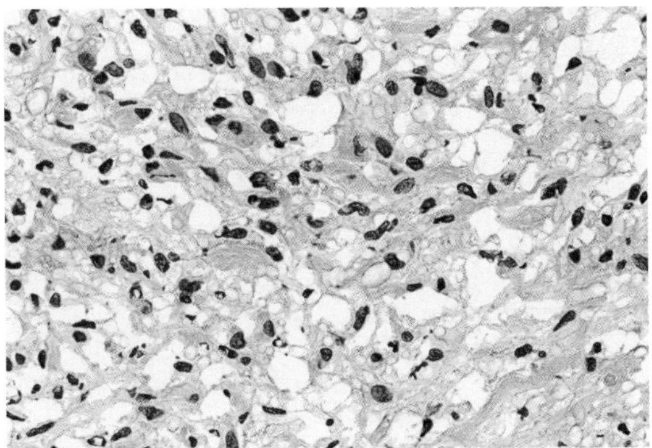

Figure 7-10

HEMANGIOBLASTOMA: FROZEN SECTION

Often devoid of specific architectural features in frozen sections, hemangioblastoma can be readily misinterpreted as astrocytoma.

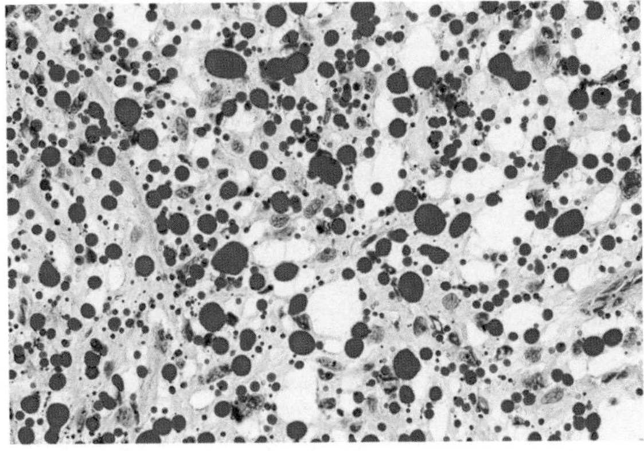

Figure 7-11

HEMANGIOBLASTOMA: FROZEN SECTION STAINED FOR FAT

The content of fat in the stromal cells is confirmed by staining with oil red-O.

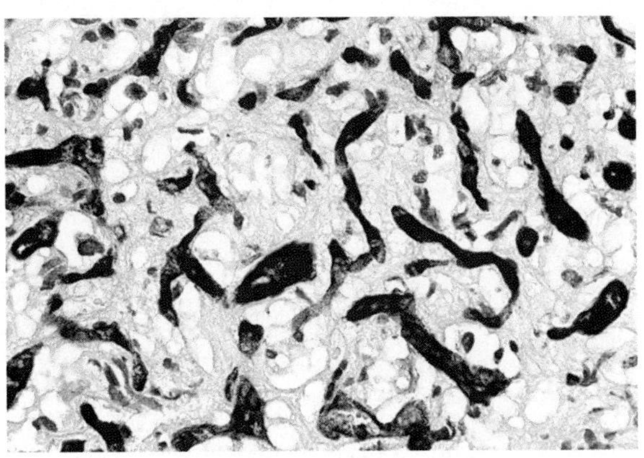

Figure 7-12

HEMANGIOBLASTOMA

Immunostaining for CD31 underscores the abundance of non-neoplastic capillaries. Stromal cells are negative for vascular markers.

to a degree that creates a remarkable likeness to malignant glioma, albeit one with few if any mitoses (fig. 7-10). As a result, a firm diagnosis is often made with trepidation, even when clinical, radiologic, and intraoperative findings leave little room for diagnostic alternatives. Rich vascularity, mast cells, scattered pleomorphic nuclei, lack of mitotic activity, and presence of lipid (fig. 7-11) all support the diagnosis. Although lipid is unusual in gliomas, it can be present in metastatic renal cell carcinoma, a component of VHL syndrome and a principal in the differential diagnosis of hemangioblastoma.

Immunohistochemical Findings. Despite the impressive vascularity of hemangioblastoma, its stromal cells are immunonegative for vascular markers such as CD31 (fig. 7-12) and factor VIII–related antigen (von Willebrand factor) (25,26). Immunoreactivity of stromal cells for inhibin, if confirmed, could become an important marker of the lesion, particularly in its distinction from metastatic carcinoma (27), but one study suggests that inhibin can be positive in renal cell carcinoma as well (28). Although stromal cells generally do not contain glial fibrillary acidic protein (GFAP), they are sometimes

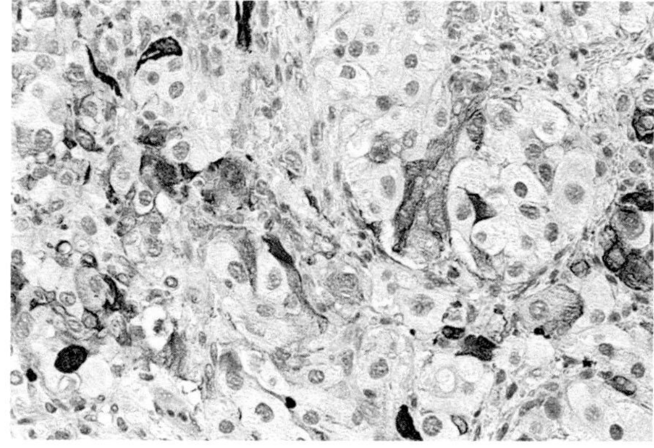

Figure 7-13

HEMANGIOBLASTOMA

Stromal cells are immunoreactive for glial fibrillary acidic protein (GFAP) in some hemangioblastomas.

unequivocally, although not uniformly, positive for this marker, particularly in lobular, cellular lesions (fig. 7-13) (29–31). The nature of these tumors is unclear, but those with otherwise classic features can be placed in the hemangioblastoma group.

Immunoreactivity for neuron-specific enolase (NSE) has been noted, but more specific neuronal markers such as neurofilament protein and synaptophysin are usually negative (25,32,33). Stromal cells are variably positive for S-100 protein and immunonegative for epithelial markers such as keratin and epithelial membrane antigen (EMA) (25,30,34). There are, however, occasional hemangioblastomas that are focally immunoreactive EMA. In one study, immunoreactivity of some vacuolated stromal cells for factor XIIIa, alpha-1-antitrypsin, and alpha-1-antichymotrypsin suggested that the stromal cells could be fibrohistiocytic (35). Immunoreactivity of both mast cells and stromal cells for erythropoietin has been observed (17,34). Many stromal and endothelial cells are immunopositive for vascular endothelial growth factor (VEGF) protein (36).

Hemangioblastomas have a low MIB-1 labeling index, usually less than 3 percent, as is described in the section, Differential Diagnosis.

Ultrastructural Findings. Stromal cells lack specific organelles as well as intercellular attachments (fig. 7-14) (20,37). Glycogen particles and lipid droplets vary greatly in number.

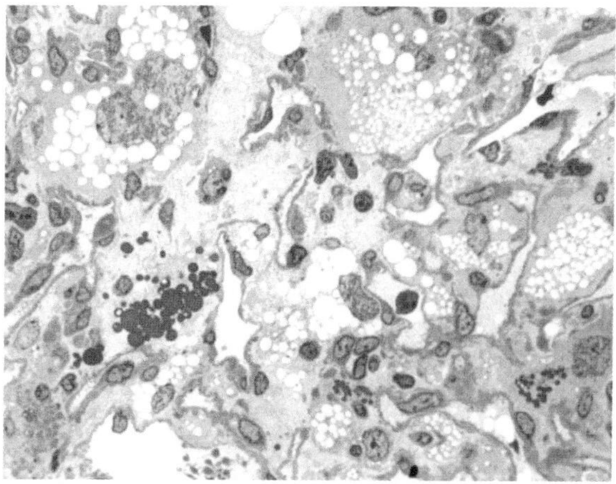

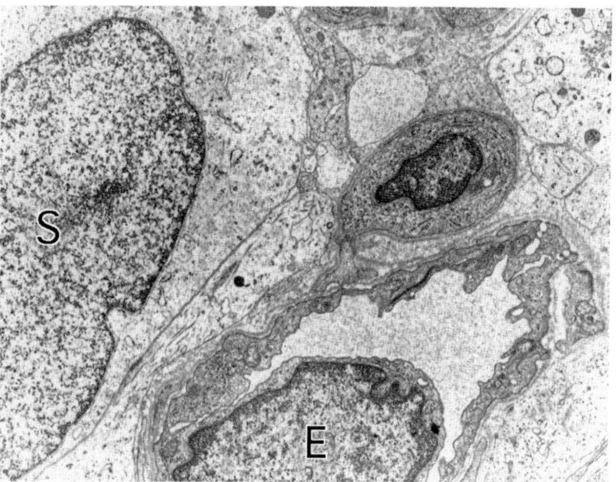

Figure 7-14

HEMANGIOBLASTOMA

The content of lipid and the extravascular location of the stromal cells are readily seen in both a plastic-embedded semithin section (top) and an electron micrograph (bottom). (S = stromal cell, E = endothelial cell).

Cytologic Findings. Like schwannomas, hemangioblastomas often resist even forceful attempts to disperse individual cells (38), but the oft-vacuolated stromal cells may be evident at the edges of the tissue fragments (fig. 7-15).

Molecular Findings. Given the part it plays in expression of the VHL syndrome, it is not surprising that mutations in the *VHL* gene are the rule in syndrome-associated hemangioblastomas (39,40). Since mutations are not found in all sporadic tumors, the genetic mechanisms underlying these lesions may differ (39–43). A small percentage of patients with ostensibly sporadic hemangioblastomas have germline mutations in the *VHL* gene (40). Comparative

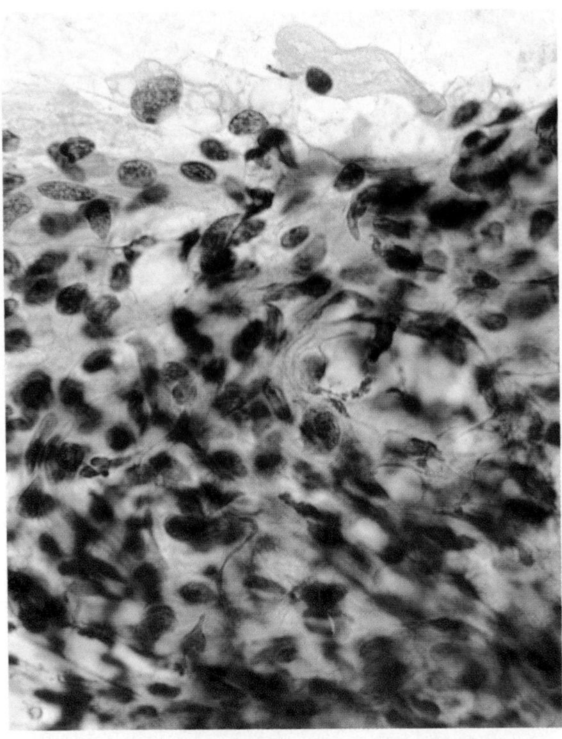

Figure 7-15

HEMANGIOBLASTOMA: SMEAR PREPARATION

Although the tissue resists dispersal of individual cells, there may be enough separation at a fragment's edge (top) to visualize the lesion's characteristic lipid-laden stromal cells.

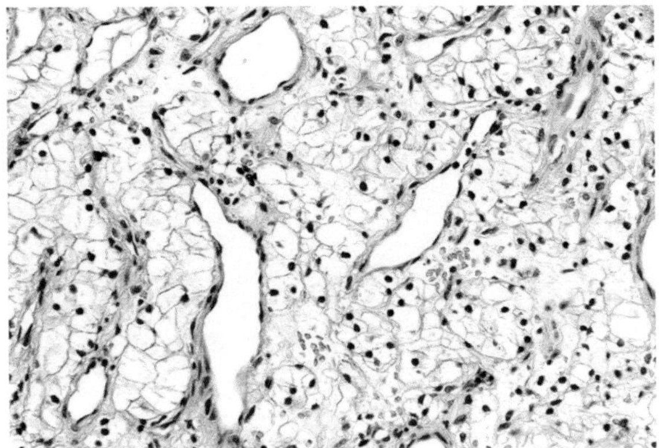

Figure 7-16

RENAL CELL CARCINOMA SIMULATING HEMANGIOBLASTOMA

Lobular renal clear cell carcinomas resemble hemangioblastomas of the cellular variant. The lacey appearance, due to somewhat sinuous cell borders, gives the carcinoma a distinctive appearance. Nuclei are also generally more monomorphic than those of hemangioblastoma.

genomic hybridization has demonstrated chromosome imbalances, specifically, losses on chromosomes 3, 6, 9, and 18, and gain on 19 (44–46). Microdissection and in situ hybridization studies have confirmed that it is the stromal cell and not the vessel that is the neoplastic component (42,47). Without the inhibitory effect of normal VHL protein, stromal cells overexpress VEGF, with resultant angiogenesis and the abundant vasculature for which the tumor is so well known (48).

DIFFERENTIAL DIAGNOSIS

The recognition of hemangioblastoma depends in large part upon simple awareness of the lesion and its typical radiographic and macroscopic features. The diagnosis is readily and confidently made in many instances, but with considerable difficulty in others. Hemangioblastomas and other low-grade lesions that can be overgraded are given in Appendix K.

A nonrepresentative biopsy of the gliotic cyst wall or adjacent brain (see fig. 7-5) should not be confused with cerebellar or spinal pilocytic astrocytoma, another often cystic tumor with a contrast-enhancing mural nodule. Unlike hemangioblastoma, cerebellar astrocytomas affect primarily the young.

Hemangioblastomas dwarfed by the cyst may escape both radiologic and surgical detection. Such lesions may be confused with the so-called simple cerebellar cyst (49).

At any site, but particularly in the spinal cord, a nibbling biopsy of surrounding parenchyma may obtain only gliotic tissue that can closely resemble diffuse astrocytoma of grade II or III if histologically nondescript or if gemistocytes are present. A discrete contrast-enhancing mass by MRI makes a diagnosis of low-grade infiltrative glioma highly unlikely. Nonetheless, there may be a similarity between hemangioblastoma and diffuse astrocytoma in frozen sections (see fig. 7-10).

Particularly when a hemangioblastoma lacks an accompanying cyst, the alternate diagnosis of metastatic clear cell carcinoma becomes a serious issue (fig. 7-16). This is especially relevant in light of the association of hemangioblastoma with renal cell carcinoma (RCC) in patients with VHL disease. The issue is confounded further by the rare metastasis of RCC to a

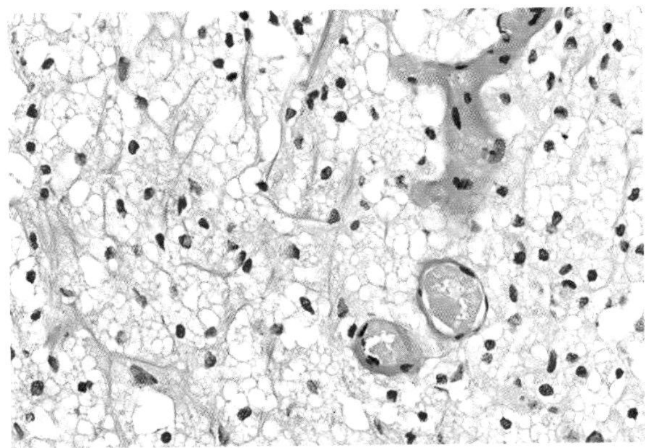

Figure 7-17

CEREBELLAR INFARCT SIMULATING HEMANGIOBLASTOMA

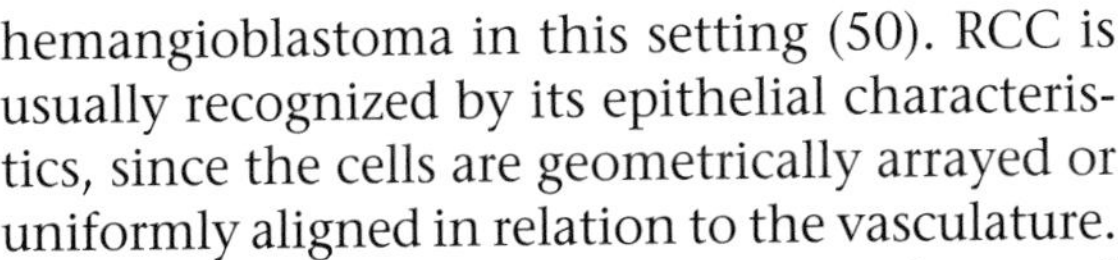

Sheets of closely apposed macrophages can resemble lobules of hemangioblastoma, or even metastatic clear cell carcinoma.

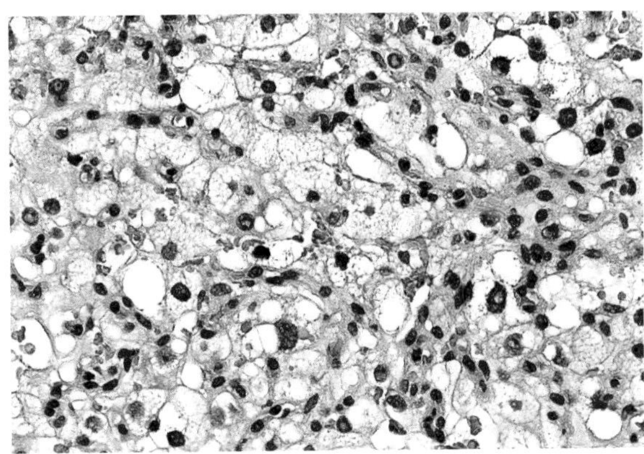

Figure 7-18

MICROCYSTIC MENINGIOMA SIMULATING HEMANGIOBLASTOMA

Xanthomatous areas of microcystic meningioma can closely resemble hemangioblastoma.

hemangioblastoma in this setting (50). RCC is usually recognized by its epithelial characteristics, since the cells are geometrically arrayed or uniformly aligned in relation to the vasculature.

The diffuse cytoplasmic clearing of stromal cells in RCC is due to their content of both glycogen and lipid. Although hemangioblastomas are lipid rich, they also contain considerable glycogen, thus precluding their distinction from RCC on the basis of simple periodic acid–Schiff (PAS) and diastase preparations alone. Reticulin is usually more abundant in hemangioblastomas, in which it surrounds individual cells or small clusters to an extent not usually seen in RCC.

The level of mitotic activity is not as much a diagnostic aid in separating hemangioblastoma from RCC as one might think since mitoses may be infrequent in RCC and disconcertingly numerous in occasional hemangioblastomas. MIB-1 staining can be useful, however. In one study, the labeling index of 27 hemangioblastomas was 2.8 ± 2.4 percent (mean and standard deviation) and 28.8 ± 25.6 percent in 5 metastatic renal cell carcinomas (51). Only one metastasis had a low labeling index (4 percent), within the range of hemangioblastomas. Another study found a mean MIB-1 index of 1.1 percent (range, 0.4 to 2.1 percent) in sporadic hemangioblastomas, and 0.4 percent (range, 0.03 to 0.9 percent) in VHL-associated tumors (36). Ultimately, immunoreactivity for tumor markers often becomes decisive, since RCC is reactive for EMA and CD10, whereas hemangioblastomas generally are not (24,34). Some RCCs are immunoreactive for the antibody RCC, but the utility and reliability of this marker is not clear. The specificity of inhibin staining, often patchy, is also unclear, but has been suggested as a marker of hemangioblastoma (27).

Macrophages bloated with cellular debris, as in a cerebellar infarct, can simulate lipid-laden stromal cells (fig. 7-17). Infarcts lack the reticulin network, and are extensively immunoreactive for macrophage markers.

Supratentorially, hemangioblastomas must be distinguished from microcystic meningioma (fig. 7-18), although the latter is overwhelmingly more common in this locale. This distinctive meningioma subtype, despite frequent xanthic change, has more hyalinized vessels, little if any reticulin, and at least focal immunoreactivity for EMA.

Rare, but diagnostically challenging in its distinction from hemangioblastoma, is capillary hemangioma (fig. 7-19). The latter lesion has gradations of vessels from larger, even cavernous, channels through intermediate smaller forms to minute capillaries. Unlike hemangioblastoma, there are no vacuolated interstitial cells. Capillary angiomas occur throughout the neuraxis (see chapter 10).

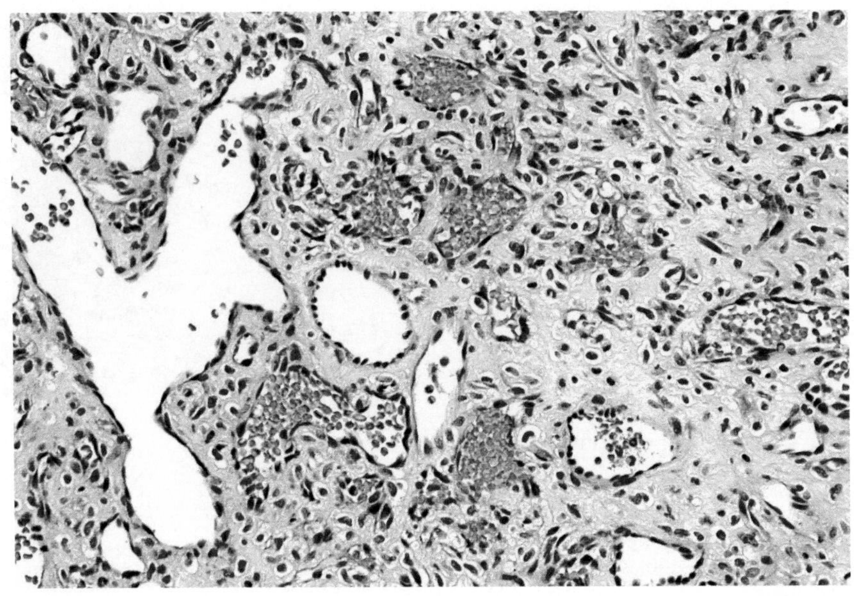

Figure 7-19

CAPILLARY ANGIOMA SIMULATING HEMANGIOBLASTOMA

The highly vascular angioma shares many features with hemangioblastoma, but not the presence of interstitial cells.

TREATMENT AND PROGNOSIS

Surgical excision is the treatment of choice; complete resection is possible in most cases. The role of radiotherapy is limited, but it may be employed for nonresectable or recurrent tumors of the medulla or high cervical spinal cord (52).

Recurrences may be seen in patients with or without VHL disease (2), and new lesions continue to appear in VHL patients (1,53). Dissemination in the cerebrospinal space or the presence of multiple primaries is rare (see fig. 18-19) (54–56). Patients with VHL disease succumb either to RCC or to multiple hemangioblastomas (2). No difference has been noted in the rate of growth of sporadic hemangioblastoma and those occurring in VHL disease (24).

REFERENCES

1. Conway JE, Chou D, Clatterbuck RE, Brem H, Long DM, Rigamonti D. Hemangioblastomas of the central nervous system in von Hippel-Lindau syndrome and sporadic disease. Neurosurgery 2001;48:55–62; discussion 62–3.
2. Niemela M, Lemeta S, Summanen P, et al. Long-term prognosis of haemangioblastoma of the CNS: impact of von Hippel-Lindau disease. Acta Neurochir (Wien) 1999;141:1147–56.
3. Baker KB, Moran CJ, Wippold FJ 2nd, et al. MR imaging of spinal hemangioblastoma. AJR Am J Roentgenol 2000;174:377–82.
4. Richard S, David P, Marsot-Dupuch K, Giraud S, Beroud C, Resche F. Central nervous system hemangioblastomas, endolymphatic sac tumors, and von Hippel-Lindau disease. Neurosurg Rev 2000;23:1–22; discussion 23–4.
5. Wanebo JE, Lonser RR, Glenn GM, Oldfield EH. The natural history of hemangioblastomas of the central nervous system in patients with von Hippel-Lindau disease. J Neurosurg 2003;98:82–94.
6. Horiguchi H, Sano T, Toi H, Kageji T, Hirokawa M, Nagahiro S. Endolymphatic sac tumor associated with a von Hippel-Lindau disease patient: an immunohistochemical study. Mod Pathol 2001;14:727–32.
7. Wang C, Zhang J, Liu A, Sun B. Surgical management of medullary hemangioblastoma. Report of 47 cases. Surg Neurol 2001;56:218–26; discussion 226–7.
8. Miller DJ, McCutcheon IE. Hemangioblastomas and other uncommon intramedullary tumors. J Neurooncol 2000;47:253–70.
9. Roonprapunt C, Silvera VM, Setton A, Freed D, Epstein FJ, Jallo GI. Surgical management of isolated hemangioblastomas of the spinal cord. Neurosurgery 2001;49:321–7; discussion 327–8.

10. Van Velthoven V, Reinacher PC, Klisch J, Neumann HP, Glasker S. Treatment of intramedullary hemangioblastomas, with special attention to von Hippel-Lindau disease. Neurosurgery 2003;53:1306–13; discussion 1313–4.
11. Giannini C, Scheithauer BW, Hellbusch LC, et al. Peripheral nerve hemangioblastoma. Mod Pathol 1998;11:999–1004.
12. Raghavan R, Krumerman J, Rushing EJ, et al. Recurrent (nonfamilial) hemangioblastomas involving spinal nerve roots: case report. Neurosurgery 2000;47:1443–8.
13. Escott EJ, Kleinschmidt-DeMasters BK, Brega K, Lillehei KO. Proximal nerve root spinal hemangioblastomas: presentation of three cases, MR appearance, and literature review. Surg Neurol 2004;61:262–73; discussion 273.
14. Iplikcioglu AC, Yaradanakul V, Trakya U. Supratentorial haemangioblastoma: appearances on MR imaging. Br J Neurosurg 1997;11:576–8.
15. Horton JC, Harsh GRt, Fisher JW, Hoyt WF. Von Hippel-Lindau disease and erythrocytosis: radioimmunoassay of erythropoietin in cyst fluid from a brainstem hemangioblastoma. Neurology 1991;41:753–4.
16. So CC, Ho LC. Polycythemia secondary to cerebellar hemangioblastoma. Am J Hematol 2002;71:346–7.
17. Tachibana O, Yamashima T, Yamashita J. Immunohistochemical study of erythropoietin in cerebellar hemangioblastomas associated with secondary polycythemia. Neurosurgery 1991; 28:24–6.
18. Chu BC, Terae S, Hida K, Furukawa M, Abe S, Miyasaka K. MR findings in spinal hemangioblastoma: correlation with symptoms and with angiographic and surgical findings. AJNR Am J Neuroradiol 2001;22:206–17.
19. Ho VB, Smirniotopoulos JG, Murphy FM, Rushing EJ. Radiologic-pathologic correlation: hemangioblastoma. AJNR Am J Neuroradiol 1992;13:1343–52.
20. Ho KL. Ultrastructure of cerebellar capillary hemangioblastoma. II. Mast cells and angiogenesis. Acta Neuropathol (Berl) 1984;64:308–18.
21. Zec N, Cera P, Towfighi J. Extramedullary hematopoiesis in cerebellar hemangioblastoma. Neurosurgery 1991;29:34–7.
22. Bonnin JM, Pena CE, Rubinstein LJ. Mixed capillary hemangioblastoma and glioma. A redefinition of the "angioglioma." J Neuropathol Exp Neurol 1983;42:504–16.
23. Lombardi D, Scheithauer BW, Piepgras D, Meyer FB, Forbes GS. "Angioglioma" and the arteriovenous malformation-glioma association. J Neurosurg 1991;75:589–96.
24. Neumann H, Eggert H, Weigel K, Friedburg H, Wiestler O, Schollmeyer P. Hemangioblastomas of the central nervous system. A 10-year study with special reference to von Hippel-Lindau syndrome. J Neurosurg. 1989;70:24–30.
25. Frank TS, Trojanowski JQ, Roberts SA, Brooks JJ. A detailed immunohistochemical analysis of cerebellar hemangioblastoma: an undifferentiated mesenchymal tumor. Mod Pathol 1989; 2:638–51.
26. McComb RD, Jones TR, Pizzo SV, Bigner DD. Localization of factor VIII/von Willebrand factor and glial fibrillary acidic protein in the hemangioblastoma: implications for stromal cell histogenesis. Acta Neuropathol (Berl) 1982;56: 207–13.
27. Hoang MP, Amirkhan RH. Inhibin alpha distinguishes hemangioblastoma from clear cell renal cell carcinoma. Am J Surg Pathol 2003;27: 1152–6.
28. Jung SM, Kuo TT. Immunoreactivity of CD10 and inhibin alpha in differentiating hemangioblastoma of the brain from renal cell carcinoma. Mod Pathol 2004;17(Suppl 1):317A.
29. Deck JH, Rubinstein LJ. Glial fibrillary acidic protein in stromal cells of some capillary hemangioblastomas: significance and possible implications of an immunoperoxidase study. Acta Neuropathol (Berl) 1981;54:173–81.
30. Mills SE, Ross GW, Perentes E, Nakagawa Y, Scheithauer BW. Cerebellar hemangioblastoma: immunohistochemical distinction from metastatic renal cell carcinoma. Surg Pathol 1990;3: 121–32.
31. Kepes JJ, Rengachary SS, Lee SH. Astrocytes in hemangioblastomas of the central nervous system and their relationship to stromal cells. Acta Neuropathol (Berl) 1979;47:99–104.
32. Becker I, Paulus W, Roggendorf W. Histogenesis of stromal cells in cerebellar hemangioblastomas. An immunohistochemical study. Am J Pathol 1989;134:271–5.
33. Feldenzer JA, McKeever PE. Selective localization of gamma-enolase in stromal cells of cerebellar hemangioblastomas. Acta Neuropathol (Berl) 1987;72:281–5.
34. Hufnagel TJ, Kim JH, True LD, Manuelidis EE. Immunohistochemistry of capillary hemangioblastoma. Immunoperoxidase-labeled antibody staining resolves the differential diagnosis with metastatic renal cell carcinoma, but does not explain the histogenesis of the capillary hemangioblastoma. Am J Surg Pathol 1989;13:207–16.
35. Nemes Z. Fibrohistiocytic differentiation in capillary hemangioblastoma. Hum Pathol 1992;23: 805–10.

36. Miyagami M, Katayama Y. Long-term prognosis of hemangioblastomas of the central nervous system: clinical and immunohistochemical study in relation to recurrence. Brain Tumor Pathol 2004;21:75–82.
37. Chaudhry AP, Montes M, Cohn GA. Ultrastructure of cerebellar hemangioblastoma. Cancer 1978;42(4):1834–50.
38. Ortega L, Jimenez-Heffernan JA, Perna C. Squash cytology of cerebellar haemangioblastoma. Cytopathology 2002;13:184–5.
39. Gijtenbeek JM, Jacobs B, Sprenger SH, et al. Analysis of von Hippel-Lindau mutations with comparative genomic hybridization in sporadic and hereditary hemangioblastomas: possible genetic heterogeneity. J Neurosurg 2002;97:977–82.
40. Olschwang S, Richard S, Boisson C, et al. Germline mutation profile of the VHL gene in von Hippel-Lindau disease and in sporadic hemangioblastoma. Hum Mutat 1998;12:424–30.
41. Kanno H, Shuin T, Kondo K, et al. Somatic mutations of the von Hippel-Lindau tumor suppressor gene and loss of heterozygosity on chromosome 3p in human glial tumors. Cancer Res 1997;57:1035–8.
42. Lee JY, Dong SM, Park WS, et al. Loss of heterozygosity and somatic mutations of the VHL tumor suppressor gene in sporadic cerebellar hemangioblastomas. Cancer Res 1998;58:504–8.
43. Oberstrass J, Reifenberger G, Reifenberger J, Wechsler W, Collins VP. Mutation of the von Hippel-Lindau tumour suppressor gene in capillary haemangioblastomas of the central nervous system. J Pathol 1996;179:151–6.
44. Lemeta S, Aalto Y, Niemela M, et al. Recurrent DNA sequence copy losses on chromosomal arm 6q in capillary hemangioblastoma. Cancer Genet Cytogenet 2002;133:174–8.
45. Sprenger SH, Gijtenbeek JM, Wesseling P, et al. Characteristic chromosomal aberrations in sporadic cerebellar hemangioblastomas revealed by comparative genomic hybridization. J Neurooncol 2001;52:241–7.
46. Lemeta S, Pylkkanen L, Sainio M, et al. Loss of heterozygosity at 6q is frequent and concurrent with 3p loss in sporadic and familial capillary hemangioblastomas. J Neuropathol Exp Neurol 2004;63:1072–9.
47. Stratmann R, Krieg M, Haas R, Plate KH. Putative control of angiogenesis in hemangioblastomas by the von Hippel-Lindau tumor suppressor gene. J Neuropathol Exp Neurol 1997;56: 1242–52.
48. Wizigmann-Voos S, Breier G, Risau W, Plate KH. Up-regulation of vascular endothelial growth factor and its receptors in von Hippel-Lindau disease-associated and sporadic hemangioblastomas. Cancer Res 1995;55:1358–64.
49. Silverberg G. Simple cysts of the cerebellum. J Neurosurg 1971;35:320–7.
50. Hamazaki S, Nakashima H, Matsumoto K, Taguchi K, Okada S. Metastasis of renal cell carcinoma to central nervous system hemangioblastoma in two patients with von Hippel-Lindau disease. Pathol Int 2001;51:948–53.
51. Brown DF, Gazdar AF, White CL 3rd, Yashima K, Shay JW, Rushing EJ. Human telomerase RNA expression and MIB-1 (Ki-67) proliferation index distinguish hemangioblastomas from metastatic renal cell carcinomas. J Neuropathol Exp Neurol 1997;56:1349–55.
52. Smalley SR, Schomberg PJ, Earle JD, Laws ER Jr, Scheithauer BW, O'Fallon JR. Radiotherapeutic considerations in the treatment of hemangioblastomas of the central nervous system. Int J Radiat Oncol Biol Phys 1990;18:1165–71.
53. de la Monte SM, Horowitz SA. Hemangioblastomas: clinical and histopathological factors correlated with recurrence. Neurosurgery 1989; 25:695–8.
54. Mohan J, Brownell B, Oppenheimer DR. Malignant spread of haemangioblastoma: report on two cases. J Neurol Neurosurg Psychiatry 1976; 39:515–25.
55. Reyns N, Assaker R, Louis E, Lejeune JP. Leptomeningeal hemangioblastomatosis in a case of von Hippel-Lindau disease: case report. Neurosurgery 2003;52:1212–5; discussion 1215–6.
56. Weil RJ, Vortmeyer AO, Zhuang Z, et al. Clinical and molecular analysis of disseminated hemangioblastomatosis of the central nervous system in patients without von Hippel-Lindau disease. Report of four cases. J Neurosurg 2002;96:775–87.

8 GERM CELL TUMORS

GENERAL PATHOLOGIC FEATURES

Clinical Features. Intracranial germ cell tumors are typically midline lesions that principally affect the pineal or suprasellar regions (sometimes synchronously) (fig. 8-1). Occasional examples arise more laterally in the basal ganglia or thalamus (1–3). Only rarely is the spinal cord a primary site (4–6).

Germ cell tumors in the pineal gland and basal ganglia usually affect males, but no sex predilection is observed with suprasellar lesions. Precocious puberty is common with pineal region tumors, which can also result in paresis of upward gaze (Perinaud's syndrome) by compressing the quadrigeminal plate. Suprasellar tumors, usually germinomas, often cause diabetes insipidus, an endocrine deficiency state that is an unusual feature of other neoplastic lesions occurring in this region; it is common with Langerhans cell histiocytosis.

Germ cell tumors are associated with the production of alpha-fetoprotein, beta-human chorionic gonadotrophin (β-HCG), and carcinoembryonic antigen (CEA). Levels of these substances in cerebrospinal fluid (CSF) or blood are useful for diagnosis and monitoring response to treatment. High levels of alpha-fetoprotein and β-HCG imply the presence of yolk sac tumor and choriocarcinoma, respectively. Low levels of alpha-fetoprotein may be seen in teratomas. Modest elevation of β-HCG is not de facto evidence of choriocarcinoma, however, since syncytiotrophoblastic cells producing this substance are found in some germinomas and other germ cell tumors.

Teratomas, which may occur in utero or in the neonatal period, arise throughout the craniospinal axis. Some are so massive that they replace the brain or extend into adjacent cervical or cephalic soft tissues (7,8).

The vast majority of germ cell tumors occur in otherwise normal individuals. Klinefelter's syndrome, with its chronically elevated gonadotropin levels, is a rare predisposing factor (9,10). There are also isolated accounts of germ cell tumors in association with Down's syndrome (11) and neurofibromatosis 1 (12). The lesions have also occurred in siblings (13). Rarely, central nervous system (CNS) germ cell tumors are associated with second lesions in the gonads (11).

Radiologic Features. Most intracranial germ cell tumors are well circumscribed and contrast enhancing (fig. 8-1). Suprasellar lesions can be based in the infundibulum (14). Teratomas are understandably heterogeneous in their signal characteristics, with intralesional cysts being especially common (15,16). Germ cell tumors are prone to subependymal spread along the walls of the lateral ventricles (see fig. 8-27).

Gross Findings. Depending upon type, germ cell tumors vary greatly in appearance. CNS germinomas are often firm if not tough; embryonal carcinoma and yolk sac tumors may

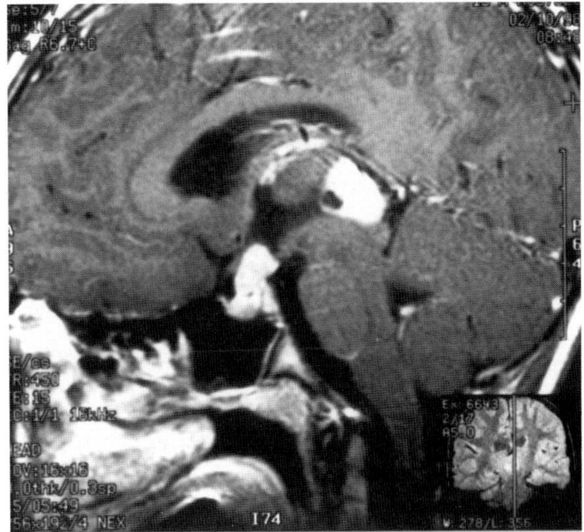

Figure 8-1

GERMINOMA

Intracranial germinomas usually occur in the suprasellar or pineal regions, sometimes simultaneously, as in this 15-year-old male. As is common in patients with suprasellar germinomas, the patient presented with diabetes insipidus.

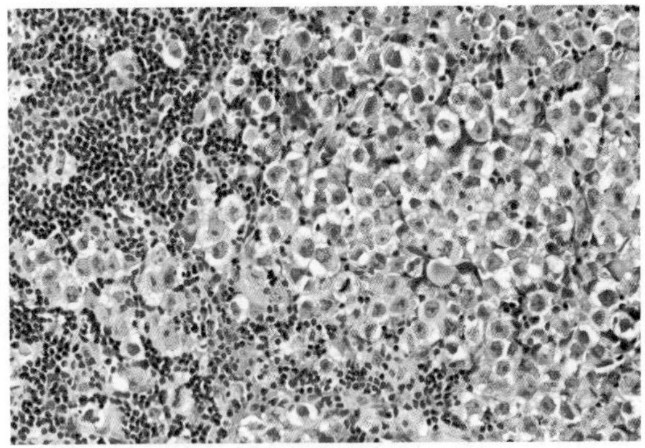

Figure 8-2

GERMINOMA

The diagnosis is elementary in the presence of large, pale, cytologically malignant tumor cells and lymphoid aggregates.

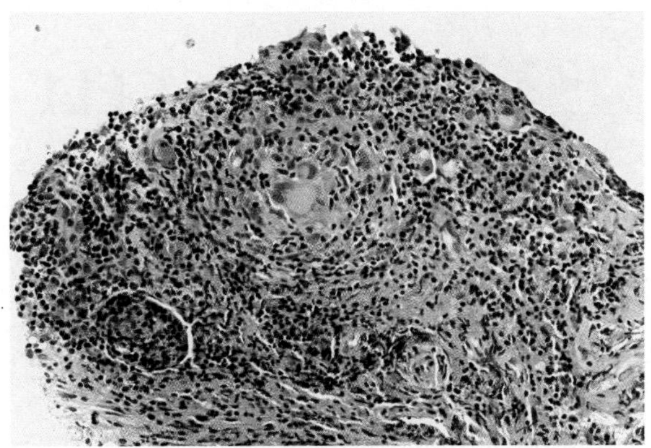

Figure 8-4

GERMINOMA

A dense lymphocytic infiltrate, sometimes with focal granulomas, can dominate the specimen. A neoplastic cell is present just to the right of the granuloma.

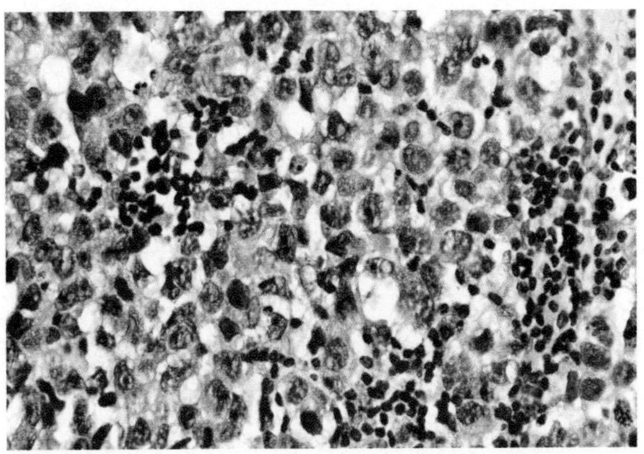

Figure 8-3

GERMINOMA

Because of their susceptibility to handling and processing artifacts, germinomas often do not appear as "classic" as in the previous figure.

have necrosis; choriocarcinomas are typically hemorrhagic; and teratomas are often multicystic. Understandably, mixed germ cell tumors are heterogenous in appearance.

Microscopic and Immunohistochemical Findings. There are five histologic subtypes of intracranial germ cell neoplasms: germinoma, embryonal carcinoma, yolk sac tumor, teratoma (immature, mature, and with malignant transformation), and choriocarcinoma. Combinations of these elements comprise "mixed germ cell tumors." Particularly common among the latter are immature teratoma or yolk sac carcinoma with an element of germinoma. The histopathology of germ cell tumors is fully described in two Third Series Fascicles, *Tumors of the Ovary, Maldeveloped Gonads, Fallopian Tube, and Broad Ligament* (17) and *Tumors of the Testis, Adnexa, Spermatic Cord, and Scrotum* (18).

Germinoma

As best seen in a sizable specimen, unfortunately rare, *germinomas* are composed of sheets and lobules of large cells with often indistinct cell membranes and somewhat vacuolated, glycogen-rich cytoplasm. Their round, vesicular nuclei contain prominent, often irregular or bar-shaped nucleoli. Mitotic activity is brisk (figs. 8-2, 8-3). A chronic inflammatory element is common, either about lobules or admixed with tumor cells. While scanty in some cases, it predominates in others and largely obscures the neoplastic infiltrate. Occasionally, the presence of noncaseating granulomas suggests an infectious rather than a neoplastic process (fig. 8-4) (19). Syncytiotrophoblastic giant cells are present in a minority of cases.

Germinoma cells are noted for their surface immunoreactivity for c-kit (CD117) (figs. 8-5, 8-6) and placental alkaline phosphatase (PLAP) (20,21). Reactivity is not always apparent, particularly in small specimens previously subjected to frozen section, or in tumors with a prominent inflammatory component. Immunostaining for

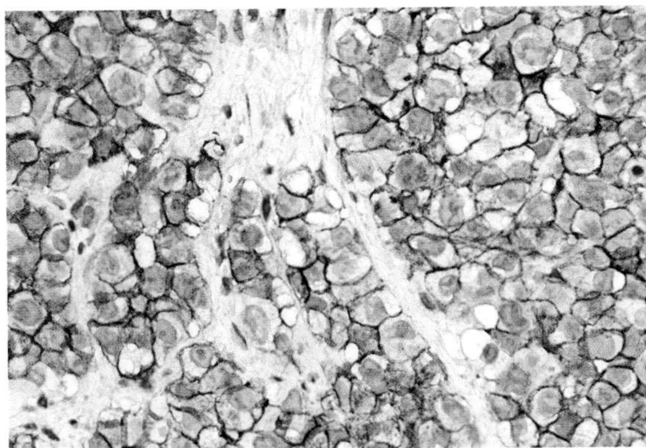

Figure 8-5

GERMINOMA

Although the cytologic appearance of the tumor cells in this case leaves little doubt about their nature, immunoreactivity for c-kit supports the diagnosis.

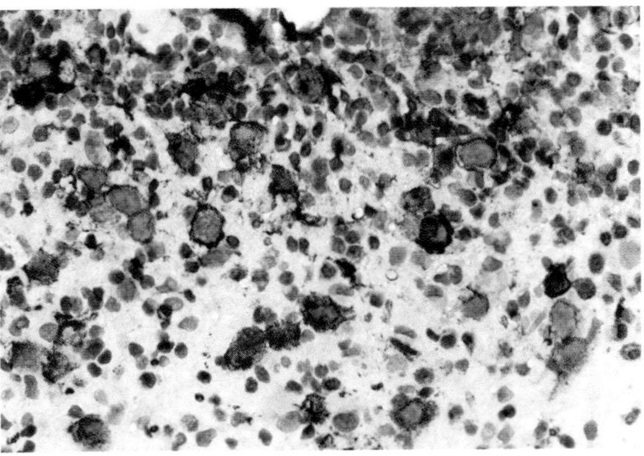

Figure 8-6

GERMINOMA

Immunohistochemistry for c-kit confirms the diagnosis by marking large, round malignant cells in a prominent lymphoid infiltrate.

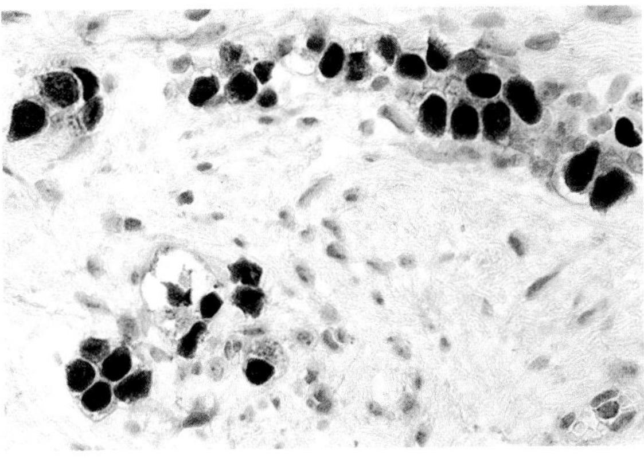

Figure 8-7

GERMINOMA

Antibodies to OCT4 identify germinoma cells, as here in a desmoplastic variant. (Courtesy of Dr. Marc K. Rosenblum, New York, NY.)

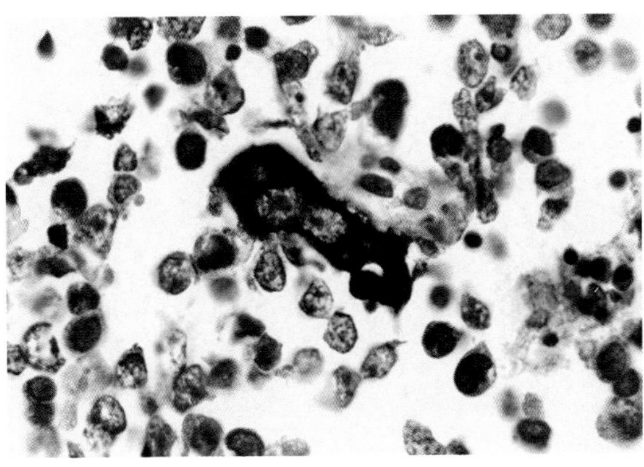

Figure 8-8

GERMINOMA

Isolated syncytiotrophoblastic cells immunopositive for beta-human chorionic gonadotropin (β-HCG) are common in germinomas.

c-kit is increasingly used for diagnosis, since such staining is both crisper and more reliable than that of PLAP. Another antibody, OCT4, appears sensitive for seminoma and germinoma (fig. 8-7), as well as for embryonal carcinoma as is described below (22). Antibodies to the transcription factor NANOG stain the nuclei of germinoma cells (22a). When present, syncytiotrophoblasts are strongly reactive for β-HCG (fig. 8-8) and cytokeratins (20,21). As noted above, elevated levels of the former may be seen in germinoma and are thus not indicative of choriocarcinoma.

The classic spectrum of histologic features described above is often lacking due to the small size of the specimen obtained in deep and close quarters, mechanical crush artifacts, and the effects of freezing. In some instances, only isolated large cells are seen lying in a chronically inflamed, T lymphocyte-rich fibrous stroma (fig. 8-9). Given the clinical picture and radiologic findings, such biopsies are strongly suggestive of germinoma but may fall below the threshold of diagnostic criteria. Periodic acid–Schiff (PAS) stain may be used to highlight the individual, glycogen-rich tumor cells. Alternatively,

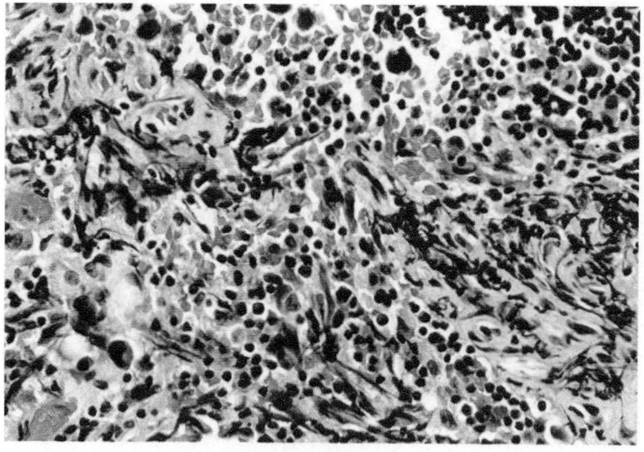

Figure 8-9

GERMINOMA

The tumor cells are particularly susceptible to crush artifact, particularly in small specimens from the pineal or suprasellar region.

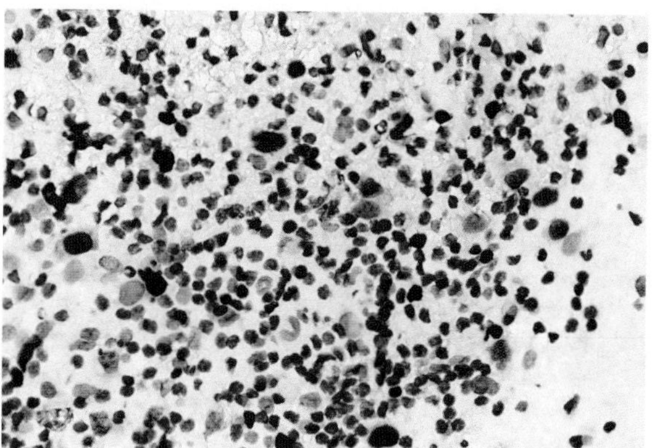

Figure 8-10

GERMINOMA

The very high MIB-1 index of germinomas is useful in the identification of neoplastic cells.

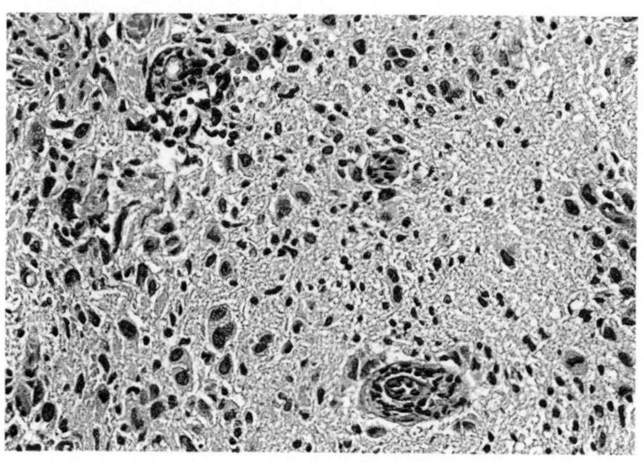

Figure 8-11

GERMINOMA

While typically encountered as clusters or sheets of tumor cells with an accompanying inflammatory infiltrate, germinomas can diffusely infiltrate in the fashion of a glioma.

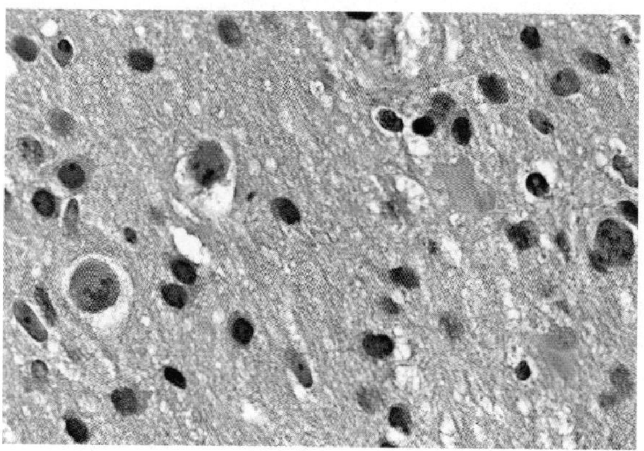

Figure 8-12

GERMINOMA

Infiltrating germinoma cells can resemble normal ganglion cells, but are cytologically atypical and would have a high MIB-1 rate.

immunostaining for c-kit (CD117), placental alkaline phosphatase, OCT4, or NANOG also may demonstrate their presence. Reactivity for the proliferation marker MIB-1 can be helpful since reactivity usually survives the artifacts mentioned above. While not specific for neoplastic cells, a high labeling index in cytologically suspicious large cells strongly suggests the diagnosis (fig. 8-10). Even these procedures may be fruitless, however, leaving no options other than rebiopsy, simple observation, or treatment in the absence of a firm histologic diagnosis. The detection in of germ cell markers in the CSF serves to fortify a presumptive histologic diagnosis.

Although germinomas are often visualized as solid, lymphocyte-rich masses, they may also diffusely infiltrate CNS parenchyma, particularly at the base of the brain (figs. 8-11, 8-12). Immunohistochemical analysis of germ cell markers and MIB-1 help identify dispersed cells in such cases (fig. 8-13).

In frozen sections, the large neoplastic cells of germinoma may fall victim to artifacts, thus disappearing into the background (fig. 8-14). Cytologic touch or smear preparations may help

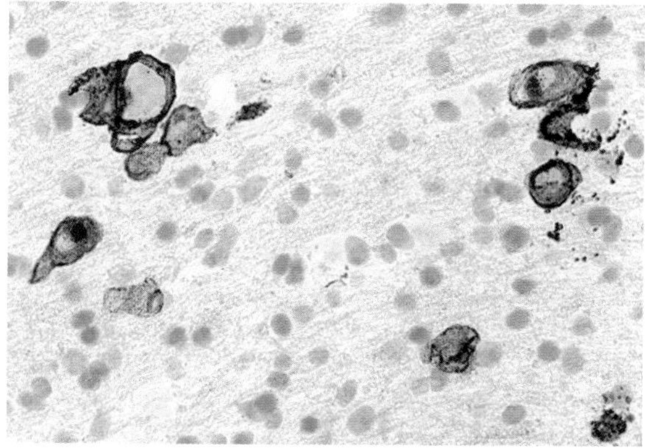

Figure 8-13

GERMINOMA

Immunohistochemistry, in this case for c-kit, is useful in identifying isolated intraparenchymal tumor cells.

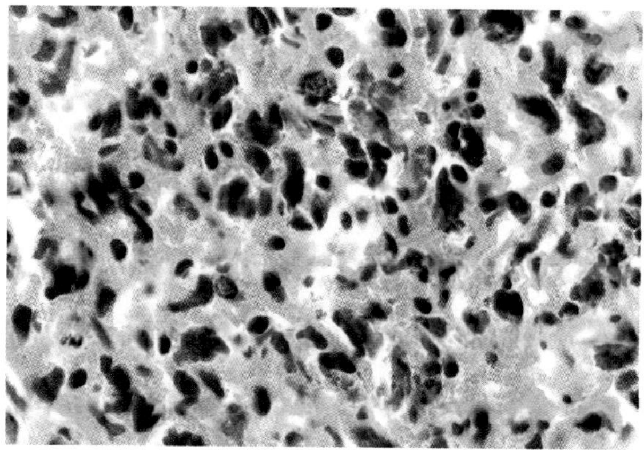

Figure 8-14

GERMINOMA: FROZEN SECTION

Freezing artifacts can induce cytologic polymorphism resembling that of a high-grade glioma.

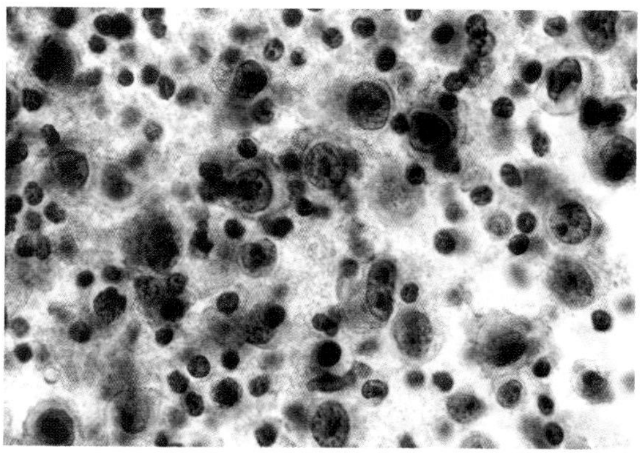

Figure 8-15

GERMINOMA: SMEAR PREPARATION

The dichotomous population of small reactive lymphocytes and large neoplastic cells is seen well in cytologic preparations.

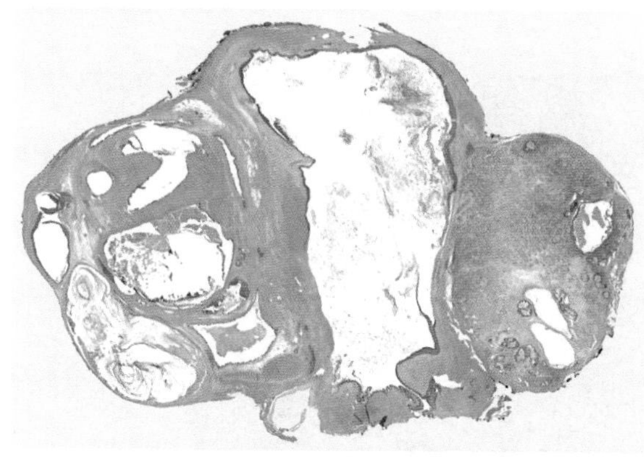

Figure 8-16

MATURE TERATOMA

As seen in a whole mount histologic specimen, mature teratomas are usually multicystic.

resolve the issue in such cases (fig. 8-15). Although of no prognostic significance, evidence of epithelial differentiation or "early carcinomatous transformation" of germinoma may be seen at the ultrastructural level (23).

Teratoma

Teratomas occur as both mature and immature types. *Mature teratomas* are usually cystic (fig. 8-16) and carry differentiation to its terminal state, although in a disorganized manner that creates a caricature of the body (figs. 8-17, 8-18). Intracranial *fetus-in-fetu* is the ultimate expression (24,25).

Immature teratomas are more common than their mature counterpart. They consist of fetal-appearing tissue of the type seen in products of conception. In this melange, developing neuroectodermal structures such as embryonic medullary neuroepithelium (fig. 8-19), retina, and choroid plexus are regularly seen. Large areas of disorganized CNS parenchyma are also common (fig. 8-20). Mesenchymal elements include cellular nondescript or myxoid stroma, islands of cartilage occasionally undergoing ossification (fig. 8-19), and intersecting bands of smooth or striated muscle. Ectoderm and endoderm are represented by skin and cutaneous adnexa, and

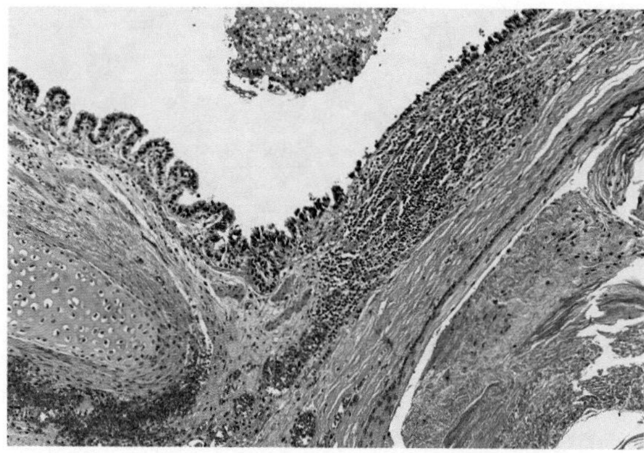

Figure 8-17

MATURE TERATOMA

Impressive evidence of the differentiating potential of teratomas is this juxtaposition of skin and a miniature bronchus.

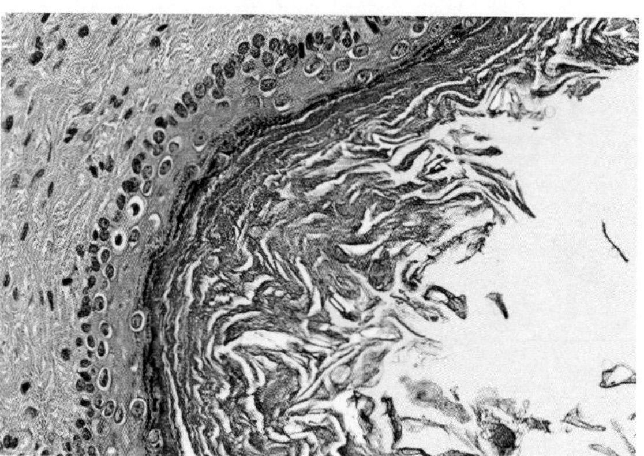

Figure 8-18

MATURE TERATOMA

Squamous epithelium in this lesion is seen to mature from a basal layer, to cells with keratohyaline granules, to anuclear squames.

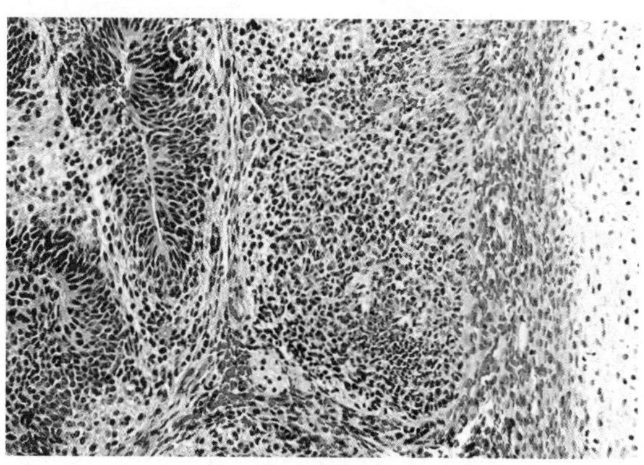

Figure 8-19

IMMATURE TERATOMA

Closed segments of medullary epithelium and cartilage are common in immature teratomas.

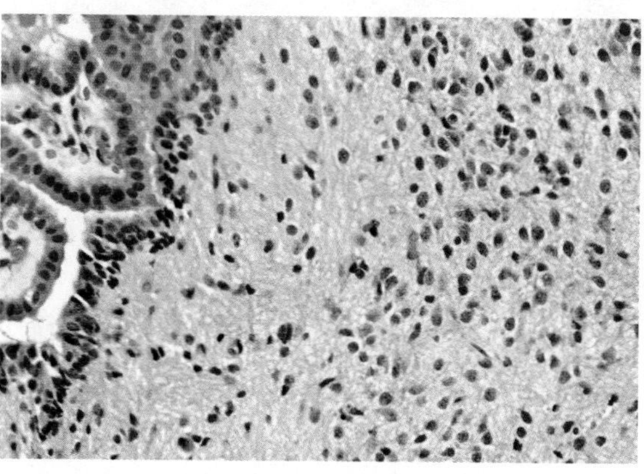

Figure 8-20

IMMATURE TERATOMA

Central nervous system tissue is common in immature teratomas.

by respiratory, pancreatic, or intestinal tissue. The last may be accompanied by a remarkably well-differentiated muscularis. The designation immature teratoma is applied to occasional teratomas containing both immature and mature elements. It has been suggested that spontaneous maturation is a part of the natural history of immature teratoma in the CNS (26).

Concise, descriptive terms are preferable in cases in which teratomas are associated with patently malignant germ cell components. Thus, the term *embryonal carcinoma with teratoma* is more informative than the once used designation "teratocarcinoma." Similarly, teratomas with frankly malignant tissues of conventional somatic type, such as carcinoma or sarcoma, are better characterized as *teratoma with adenocarcinoma* or *teratoma with rhabdomyosarcoma* than simply "malignant teratoma." These are collectively referred to as teratomas with malignant transformation or teratomas with somatic malignant components.

Given their spectrum of differentiation, teratomas express reactivity for epithelial (keratin,

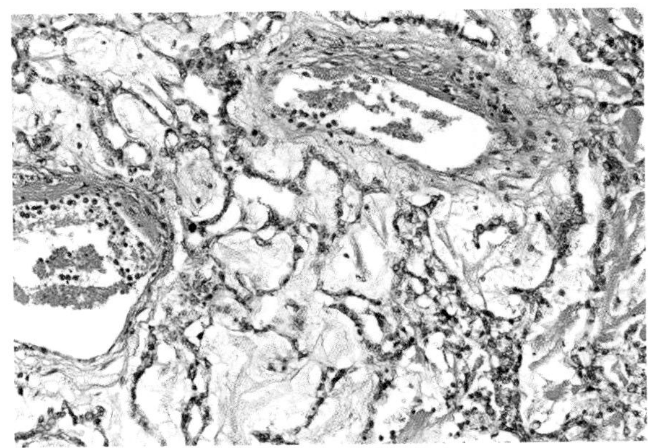

Figure 8-21

YOLK SAC TUMOR

A lacey network of delicate epithelium comprises the vitelline pattern of yolk sac tumor.

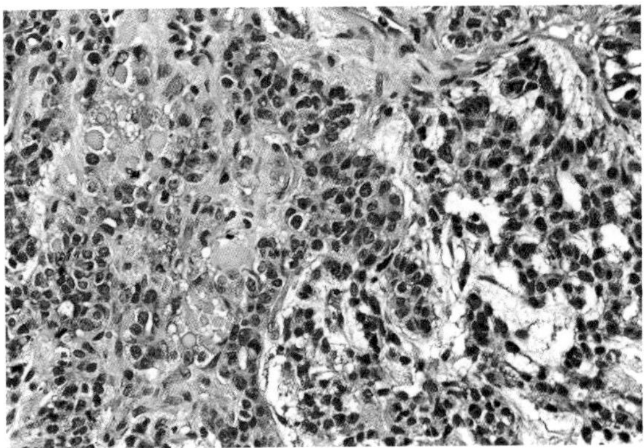

Figure 8-22

YOLK SAC TUMOR

Strands of epithelium and hyaline droplets are common features of yolk sac tumor.

epithelial membrane antigen [EMA]) and mesenchymal markers. The epithelial component is often positive for CEA (20,21). Neuroepithelial markers (S-100 protein, glial fibrillary acidic protein [GFAP], synaptophysin, neurofilament protein) are frequently demonstrated. Immature epithelium can be reactive for alpha-fetoprotein (20).

Yolk Sac Tumor

Also termed *endodermal sinus tumor, yolk sac tumor* follows germinoma and teratoma in incidence, and rarely occurs in pure form (20). An epithelial tumor, it consists variably of compact sheets, ribbons, cords, or papillae. A loose-knit "vitelline" pattern is also common (fig. 8-21). Diagnostic Schiller-Duval bodies, when present, are identified as tufts of neoplastic epithelium-covered vessels projecting into clear spaces lined by similar cells. Also diagnostic of yolk sac tumor are PAS-positive/diastase-resistant cytoplasmic and extracellular eosinophilic droplets (fig. 8-22). These, as well as the tumor cells themselves, are immunoreactive for alpha-fetoprotein (fig. 8-23), both in the epithelium and in eosinophilic droplets (20,21), and are variably positive for keratins (CK7 positive, CK20 negative). They are EMA negative.

Yolk sac tumor is commonly admixed with other malignant germ cell elements, especially germinoma (fig. 8-24) and immature teratoma.

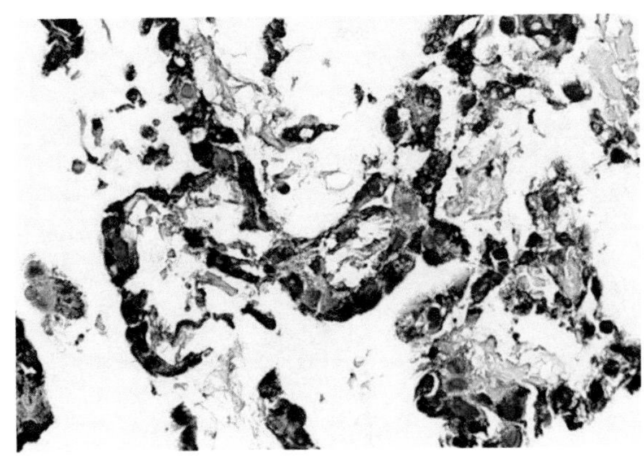

Figure 8-23

YOLK SAC TUMOR

Yolk sac tumors are immunoreactive for alpha-fetoprotein.

Embryonal Carcinoma

Although relatively common as a component in mixed germ cell tumors, *embryonal carcinoma* rarely occurs in pure form (fig. 8-25) (27). In typical cases, its large epithelial cells form cohesive sheets and glands. In exceptional cases, the presence of plate-like miniature embryos, or embryoid bodies, is a prominent feature. In the absence of epithelial arrangements, the lesion may closely resemble germinoma since both have large vacuolated cells.

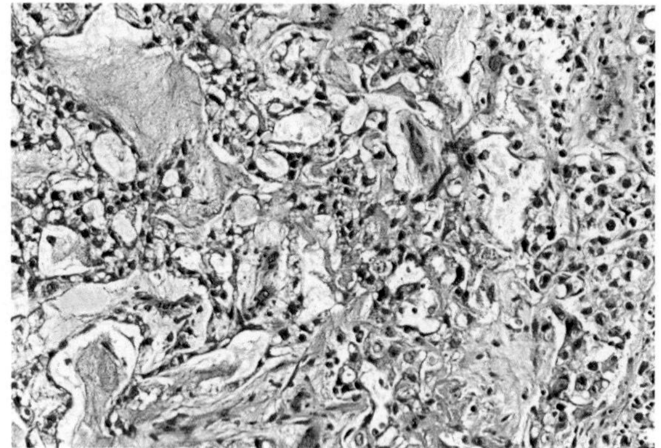

Figure 8-24

MIXED YOLK SAC TUMOR AND GERMINOMA

Yolk sac tumors are often admixed with other germ cell tumor elements, especially germinoma.

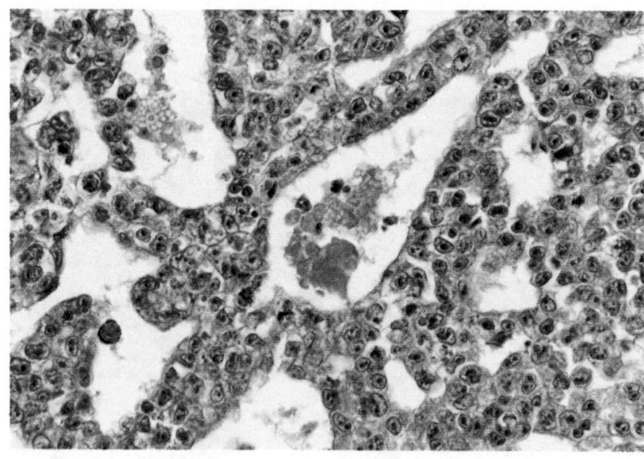

Figure 8-25

EMBRYONAL CARCINOMA

Embryonal carcinoma presents as sheets, thick cords, or gland-like arrangements of primitive-appearing, cytologically malignant cells.

Embryonal carcinomas are reactive for CD30 and have the same cytokeratin and EMA profile as yolk sac tumors. There is little information as yet in regard to CNS examples, but OCT4 marks testicular embryonal carcinoma and seminomas (fig. 8-26) (22).

Choriocarcinoma

Choriocarcinoma, the rarest of intracranial germ cell tumors (28–30), is composed of two cell types, syncytiotrophoblasts and cytotrophoblasts. A bilaminar arrangement of these constituents is the essential diagnostic feature of

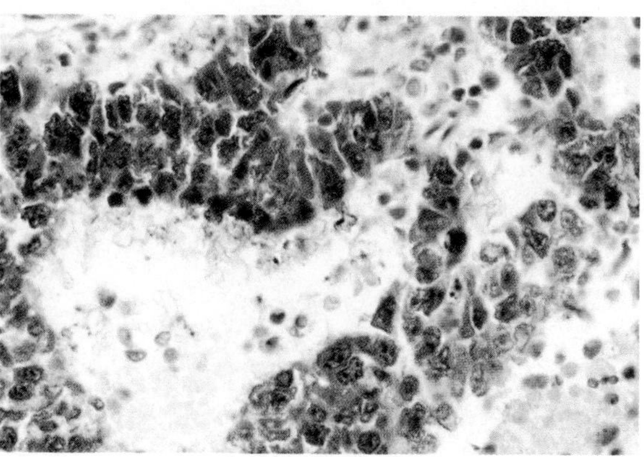

Figure 8-26

EMBRYONAL CARCINOMA

The nuclei of embryonal carcinoma cells stain immunohistochemically for OCT4. (Courtesy of Dr. Marc K. Rosenblum, New York, NY.)

choriocarcinoma. Single or clustered syncytiotrophoblastic cells may be seen in other germ cell tumors, particularly germinoma and immature teratoma.

Choriocarcinomas are immunoreactive for β-HCG, human placental lactogen (HPL), and cytokeratins.

TREATMENT AND PROGNOSIS

In light of the rarity of germ cell tumors, evolving refinements in surgical techniques and progress in chemotherapy, it is difficult to provide meaningful information regarding histologic subtypes and their prognosis. No doubt the single most important factor relating to prognosis is histologic tumor type. Tumor stage seems to also play a role. With respect to the latter, tumors may be spread via ventriculoperitoneal shunts (10), and can metastasize distantly in the absence of surgical manipulation or shunting.

It is generally agreed that, given their chemosensitivity and radiosensitivity, germinomas are often curable. In one series, patients with pineal germinomas had an 80 percent 5-year survival rate (31). Another series of intracranial germinomas recorded 10- and 15-year survival rates of 93 and 81 percent, respectively (32). Although the presence in germinomas of syncytiotrophoblasts does not equate with a diagnosis of choriocarcinoma, it has been associated with an increased incidence of local

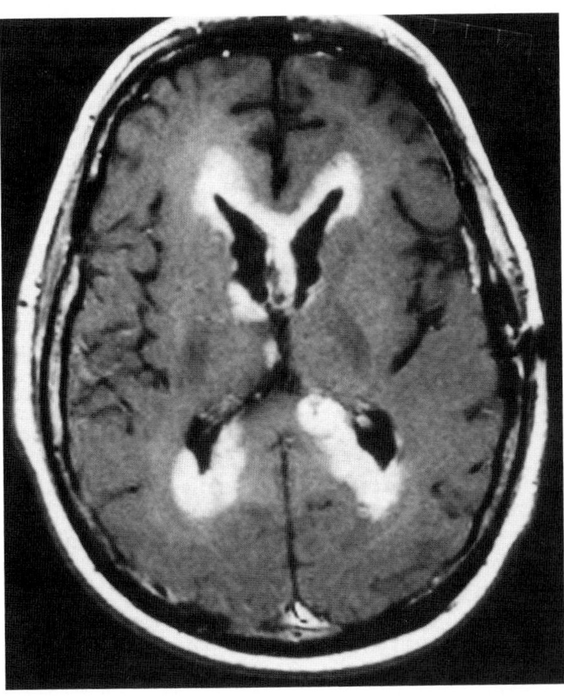

Figure 8-27

MIXED MALIGNANT GERM CELL TUMOR WITH SUBEPENDYMAL SPREAD

Especially at recurrence, but sometimes at initial presentation, germ cell tumors spread extensively in the walls of the lateral ventricles.

recurrence (33,34) and of extra-CNS metastasis (35). The lesions can recur as a higher-grade germ cell tumor that was not seen in the original specimen (36). Recurrent and sometimes primary germinomas seed the subependymal region of the lateral ventricles (fig. 8-27).

A favorable outlook is associated with mature teratomas. Due to their benign nature, firm texture, and demarcation, they lend themselves to complete resection (37). A 5-year survival rate for patients with pineal examples was 86 percent, and 67 percent for those with immature teratomas (31). Although one would assume that immature teratomas, when well circumscribed, would be amenable to complete resection and cure, this is not always the case (20,37). Such lesions can recur as immature teratoma or as a more malignant tissue type not seen in the original specimen. Maturation of immature to mature teratoma can occur (26).

Nongerminomatous malignant germ cell tumors with components of embryonal carcinoma, yolk sac tumor, or choriocarcinoma are highly malignant and capable of metastatic spread, more often cerebrospinal than systemic (20,31). One study reported a 5-year survival rate of 17 percent for patients with pure nongerminomatous malignant germ cell tumors (31), another a 3-year survival rate of 27 percent for patients with tumors at all intracranial sites combined (32). The outlook may not always be bleak for alpha-fetoprotein–producing tumors (38). Treatment of some histologically confirmed malignant lesions is successful, at least initially, since only teratoma is found in the recurrent lesion (39).

REFERENCES

1. Higano S, Takahashi S, Ishii K, Matsumoto K, Ikeda H, Sakamoto K. Germinoma originating in the basal ganglia and thalamus: MR and CT evaluation. AJNR Am J Neuroradiol 1994;15:1435–41.
2. Kim DI, Yoon PH, Ryu YH, Jeon P, Hwang GJ. MRI of germinomas arising from the basal ganglia and thalamus. Neuroradiology 1998;40:507–11.
3. Sugimoto K, Nakahara I, Nishikawa M. Bilateral metachronous germinoma of the basal ganglia occurring long after total removal of a mature pineal teratoma: case report. Neurosurgery 2002;50:613–6; discussion 616–7.
4. Miller DJ, McCutcheon IE. Hemangioblastomas and other uncommon intramedullary tumors. J Neurooncol 2000;47:253–70.
5. Miyauchi A, Matsumoto K, Kohmura E, Doi T, Hashimoto K, Kawano K. Primary intramedullary spinal cord germinoma. Case report. J Neurosurg 1996;84:1060–1.
6. Slagel DD, Goeken JA, Platz CA, Moore SA. Primary germinoma of the spinal cord: a case report with 28-year follow-up and review of the literature. Acta Neuropathol (Berl) 1995;90:657–9.
7. Chien YH, Tsao PN, Lee WT, Peng SF, Yau KI. Congenital intracranial teratoma. Pediatr Neurol 2000;22:72–4.
8. Lanzino G, Kaptain GJ, Jane JA, Lin KY. Successful excision of a large immature teratoma involving the cranial base: report of a case with long-term follow-up. Neurosurgery 1998;42:389–93.

9. Arens R, Marcus D, Engelberg S, Findler G, Goodman RM, Passwell JH. Cerebral germinomas and Klinefelter syndrome. A review. Cancer 1988;61:1228–31.
10. Jennings MT, Gelman R, Hochberg F. Intracranial germ-cell tumors: natural history and pathogenesis. J Neurosurg 1985;63:155–67.
11. Hashimoto T, Sasagawa I, Ishigooka M, et al. Down's syndrome associated with intracranial germinoma and testicular embryonal carcinoma. Urol Int 1995;55:120–2.
12. Wong TT, Ho DM, Chang TK, Yang DD, Lee LS. Familial neurofibromatosis 1 with germinoma involving the basal ganglion and thalamus. Childs Nerv Syst 1995;11:456–8.
13. Wakai S, Segawa H, Kitahara S, et al. Teratoma in the pineal region in two brothers. Case reports. J Neurosurg 1980;53:239–43.
14. Oishi M, Morii K, Okazaki H, Tamura T, Tanaka R. Neurohypophyseal germinoma traced from its earliest stage via magnetic resonance imaging: case report. Surg Neurol 2001;56:236–41.
15. Korogi Y, Takahashi M, Ushio Y. MRI of pineal region tumors. J Neurooncol 2001;54:251–61.
16. Liang L, Korogi Y, Sugahara T, et al. MRI of intracranial germ-cell tumours. Neuroradiology 2002; 44:382–8.
17. Scully R, Young R, Clement P. Tumors of the ovary, maldeveloped gonads, fallopioan tube, and broad ligament. Atlas of Tumor Pathology, 3rd Series, Fascicle 23. Washington, DC: Armed Forces Institute of Pathology; 1996.
18. Ulbright T, Amim M, Young R. Tumors of the testis, adnexa, spermatic cord, and sacrum. Atlas of Tumor Pathology, 3rd Series, Fascicle 25. Washington, DC: Armed Forces Institute of Pathology; 1997.
19. Kraichoke S, Cosgrove M, Chandrasoma PT. Granulomatous inflammation in pineal germinoma. A cause of diagnostic failure at stereotaxic brain biopsy. Am J Surg Pathol 1988;12:655–60.
20. Bjornsson J, Scheithauer BW, Okazaki H, Leech RW. Intracranial germ cell tumors: pathobiological and immunohistochemical aspects of 70 cases. J Neuropathol Exp Neurol 1985;44:32–46.
21. Ho DM, Liu HC. Primary intracranial germ cell tumor. Pathologic study of 51 patients. Cancer 1992;70:1577–84.
22. Jones TD, Ulbright TM, Eble JN, Baldridge LA, Cheng L. OCT4 staining in testicular tumors: a sensitive and specific marker for seminoma and embryonal carcinoma. Am J Surg Pathol 2004;28:935–40.
22a. Santagata S, Hornick JL, Ligon KL. Comparative analysis of germ cell transcription factors in CNS germinoma reveals diagnostic utility of NANOG. Am J Surg Pathol 2006;30:1613–18.
23. Min KW, Scheithauer BW, Bauserman SC. Pineal parenchymal tumors: an ultrastructural study with prognostic implications. Ultrastruct Pathol 1994;18:69–85.
24. Afshar F, King TT, Berry CL. Intraventricular fetus-in-fetu. Case report. J Neurosurg 1982;56:845–9.
25. Yang ST, Leow SW. Intracranial fetus-in-fetu: CT diagnosis. AJNR Am J Neuroradiol 1992;13:1326–9.
26. Shaffrey ME, Lanzino G, Lopes MB, Hessler RB, Kassell NF, VandenBerg SR. Maturation of intracranial immature teratomas. Report of two cases. J Neurosurg 1996;85:672–6.
27. Packer RJ, Sutton LN, Rorke LB, et al. Intracranial embryonal cell carcinoma. Cancer 1984;54: 520–4.
28. Bjornsson J, Scheithauer BW, Leech RW. Primary intracranial choriocarcinoma: a case report. Clin Neuropathol 1986;5:242–5.
29. Kawakami Y, Yamada O, Tabuchi K, Ohmoto T, Nishimoto A. Primary intracranial choriocarcinoma. J Neurosurg 1980;53:369–74.
30. Sievers EL, Berger M, Geyer JR. Long-term survival of a patient with primary sellar choriocarcinoma with pulmonary metastases: a case report. Med Pediatr Oncol 1996;26:293–5.
31. Schild SE, Scheithauer BW, Haddock MG, et al. Histologically confirmed pineal tumors and other germ cell tumors of the brain. Cancer 1996;78:2564–71.
32. Matsutani M, Sano K, Takakura K, et al. Primary intracranial germ cell tumors: a clinical analysis of 153 histologically verified cases. J Neurosurg 1997;86:446–55.
33. Uematsu Y, Tsuura Y, Miyamoto K, Itakura T, Hayashi S, Komai N. The recurrence of primary intracranial germinomas. Special reference to germinoma with STGC (syncytiotrophoblastic giant cell). J Neurooncol 1992;13:247–56.
34. Utsuki S, Kawano N, Oka H, Tanaka T, Suwa T, Fujii K. Cerebral germinoma with syncytiotrophoblastic giant cells: feasibility of predicting prognosis using the serum hCG level. Acta Neurochir (Wien) 1999;141:975–7; discussion 977–8.
35. Akai T, Iizuka H, Kadoya S, Nojima T. Extraneural metastasis of intracranial germinoma with syncytiotrophoblastic giant cells—case report. Neurol Med Chir (Tokyo) 1998;38:574–7.
36. Ono N, Isobe I, Uki J, Kurihara H, Shimizu T, Kohno K. Recurrence of primary intracranial germinomas after complete response with radiotherapy: recurrence patterns and therapy. Neurosurgery 1994;35:615–20; discussion 620–1.
37. Sawamura Y, Kato T, Ikeda J, Murata J, Tada M, Shirato H. Teratomas of the central nervous system: treatment considerations based on 34 cases. J Neurosurg 1998;89:728–37.
38. Itoyama Y, Kochi M, Kuratsu J, et al. Treatment of intracranial nongerminomatous malignant germ cell tumors producing alpha-fetoprotein. Neurosurgery 1995;36:459–64; discussion 464–6.
39. Friedman JA, Lynch JJ, Buckner JC, Scheithauer BW, Raffel C. Management of malignant pineal germ cell tumors with residual mature teratoma. Neurosurgery 2001;48:518–22; discussion 522–3.

9 MENINGIOMA

Meningioma is a neoplasm derived from meningothelial (arachnoidal) cells. Although occasional meningiomas occur after cranial radiation (1–5) or arise in the background of neurofibromatosis 2 (NF2) (see figs. 18-7–18-10), most arise from causes unknown. The tumors typically occur in adults, but children are sometimes affected (6–8). NF2 should be suspected in the latter age group (8,9). Occasional examples are associated with other estrogen-dependent tumors, including cancers of the breast and endometrium (10,11). The association between meningioma and meningioangiomatosis (12,13) is discussed later in the chapter as well as in chapter 17.

MENINGIOMAS AT SPECIFIC ANATOMIC SITES

Meningiomas of the Intracranial Meninges

In this most common location, symptoms are expressions of an expanding intracranial mass, i.e., focal neurologic deficits, increased intracranial pressure, and seizures. As a consequence of the extra-axial position of these meningiomas, seizures are less common than they are in patients with intraparenchymal tumors such as gliomas and metastases. Intracranial grade I meningiomas are more common in women (ratio of 3 to 2), whereas grade II and grade III lesions predominate in men. Intracranial meningiomas are common incidental findings (14).

Radiologically, intracranial meningiomas are usually globular, highly vascular, contrast-enhancing, and dura-based tumors (figs. 9-1, 9-2). A wedge-shaped tongue of neoplastic tissue or a cluster of small vessels lying within the angle between the tumor and the attached dura produces a radiographically useful, although nonspecific, focus of contrast enhancement known as the "dural tail" (fig. 9-1; see fig. 9-10). Occasional tumors, particularly those of the sphenoid ridge, diffusely carpet the dura as a lesion termed *meningioma en plaque*. Some meningiomas, particularly the microcystic variant, are associated with a large intratumoral or peritumoral cyst (15,16) (see Appendix D for cystic central nervous system [CNS] tumors). Microcystic lesions are often attached to the pial surface, and sometimes partially if not entirely embedded within the brain. As a consequence, peritumoral edema is common.

Unlike most other CNS tumors, which are hyperintense (bright or white) to cortex in T2-weighted images, meningiomas are usually isointense (gray) or even hypointense (dark) when densely fibrotic and/or heavily calcified (fig. 9-2) (17).

Grade II (atypical) and grade III (anaplastic) meningiomas that attach to the pia typically incite underlying cerebral edema (18,19),

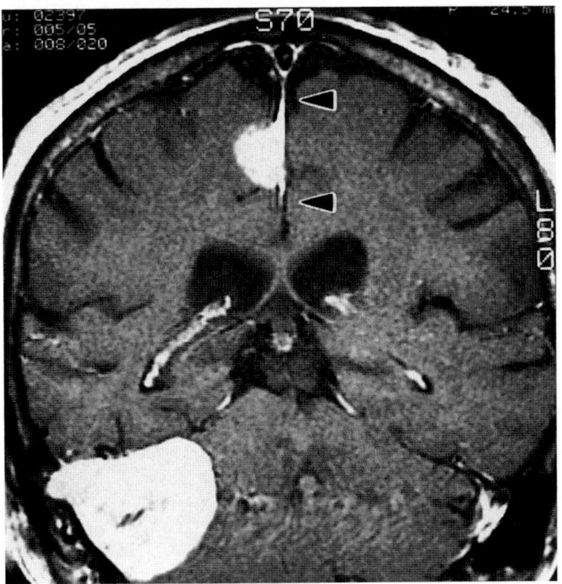

Figure 9-1

MENINGIOMA

Meningiomas are generally dura based and contrast enhancing. Triangular areas of enhancement (arrowheads) that extend along the dura from the edge of the falcine lesion form the diagnostically helpful, albeit nonspecific, "dural tail sign." (Fig. 9-1 from Fascicle 10, 3rd Series.)

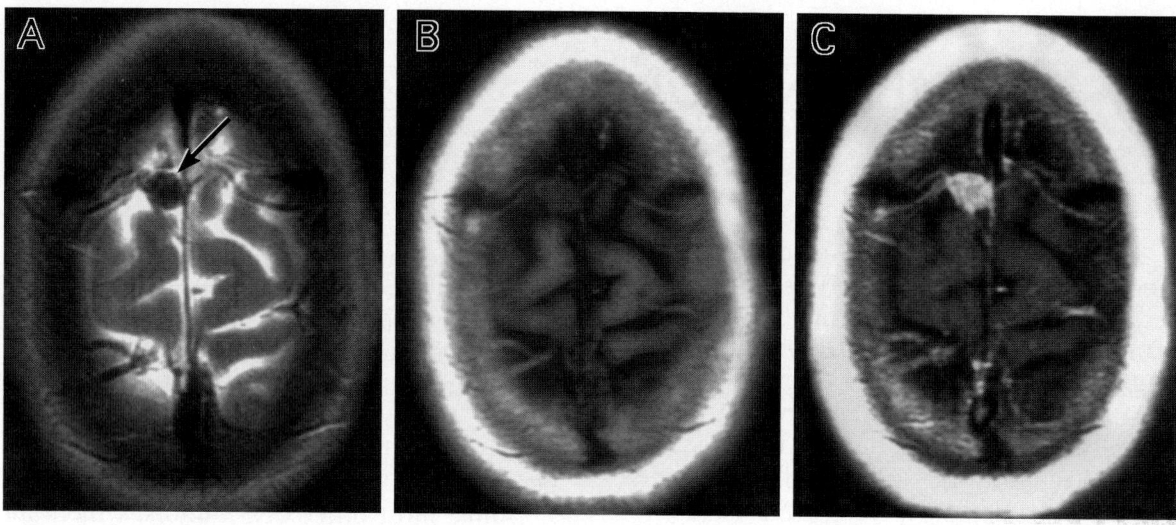

Figure 9-2

MENINGIOMA

The classic magnetic resonance imaging (MRI) features of meningiomas are illustrated in these three panels of an incidental lesion. Because of cellularity, fibrous tissue, calcification, or a combination thereof, meningiomas are usually dark or isointense to cortex in T2-weighted images (A, arrow). The lesions may be thus inconspicuous in precontrast T1-weighted views because of their isointensity to cortex (B), although they are almost always contrast enhancing (C).

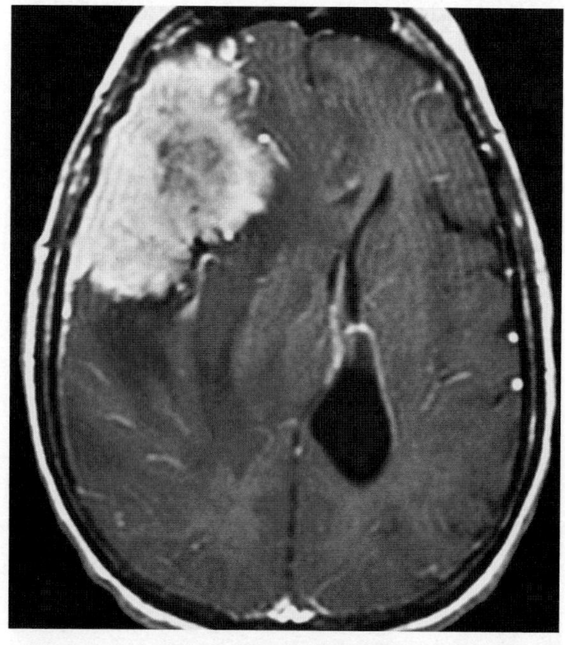

Figure 9-3

GRADE II MENINGIOMA

The irregular surfaces of many grade II and grade III meningiomas contrast with the smooth profile of the typical grade I lesion. Although not well seen in this contrast-enhanced TI-weighted image, considerable cerebral edema is common in the two higher-grade lesions, or any meningioma that attaches to the pial surface.

although the latter can occur with ordinary grade I lesions. As a correlate, edema appears to be more common in meningiomas with a higher MIB-1 index (20). Invasive lesions may have a telltale irregular surface (fig. 9-3).

Although arachnoidal in origin, meningiomas are intimately related to, and may freely infiltrate, the dura and the skull (figs. 9-4–9-6). The result is often cranial hyperostosis consisting of bony spicules that radiate from the outer, and to a lesser extent, the inner table (fig. 9-6). After penetrating the calvarium, such tumors elevate or penetrate the galea and produce a scalp mass that may be the neoplasm's first clinical expression. Primary calvarial intraosseous meningiomas are described below.

Since the choroid plexus originates as an invagination of vessels and meninges along the choroidal fissure, it is not surprising that meningothelial cells may be found within the plexus stroma (see fig. 2-39), or that they occasionally give rise to intraventricular meningiomas. The third (21), lateral (22,23), and fourth (24,25) ventricles may be affected. (See Appendix C for intraventricular tumors.) Occasional meningiomas appear in the pineal region (26).

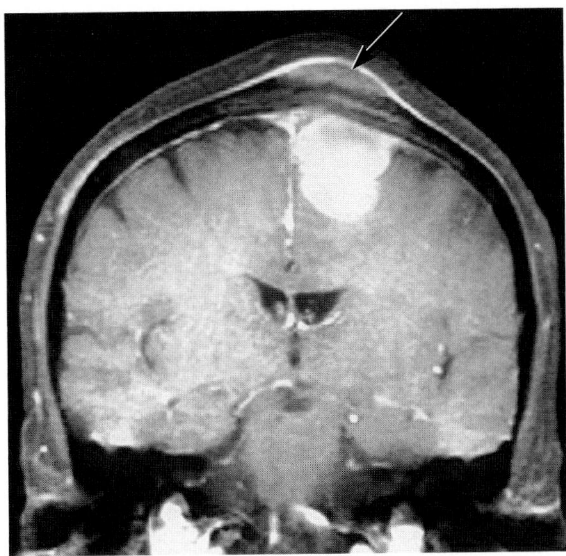

Figure 9-4

MENINGIOMA

After penetrating the skull, some meningiomas enlarge into a scalp mass (arrow). Such behavior, in itself, does not warrant the designation grade II or III.

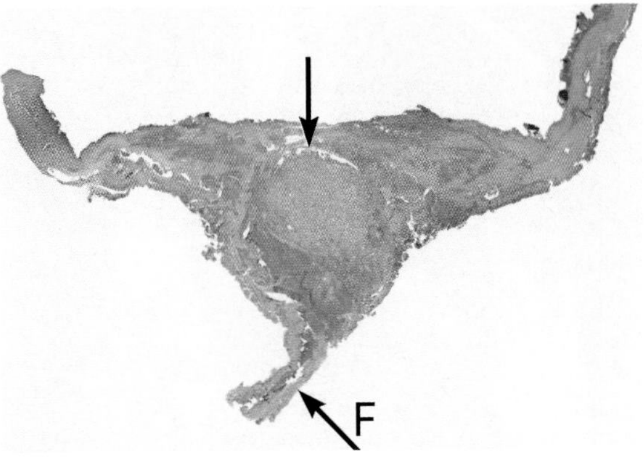

Figure 9-5

MENINGIOMA INVADING SUPERIOR SAGITTAL SINUS

Parasagittal meningiomas of any grade are prone to invade, and even occlude, the superior sagittal sinus (arrow). When totally obstructed, as in this case, the sinus may be surgically excised. This was a chordoid, grade II, variant (F = falx).

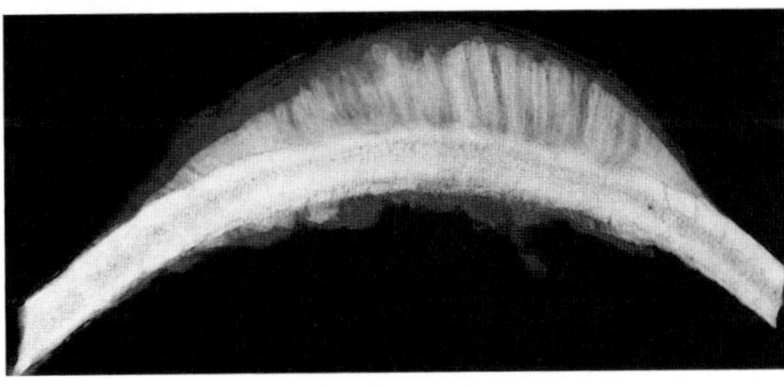

Figure 9-6

MENINGIOMA

Meningiomas that penetrate the skull often induce bony spicules that radiate from the outer and, to a lesser extent, from the inner table. (Fig. 9-3, bottom from Fascicle 10, 3rd Series.)

Meningiomas of the Optic Nerve Sheath

Visual loss, strabismus, and ptosis are expressions of the uncommon meningiomas of the optic nerve sheath (27–31). Whereas some of these tumors are primarily intradural, others lie largely free within orbital soft tissues.

Intraspinal Meningiomas

Intradural and extramedullary intraspinal meningiomas expand at the expense of the adjacent spinal cord and produce the expected segmental neurologic deficits (fig. 9-7) (32). Cervicothoracic segments are most often affected, and except for those of the clear cell type, meningiomas are rare in the nether, lumbosacral, region. Most spinal meningiomas are situated ventrally or laterally near the nerve root exit zone, an area where meningothelial cells are normally concentrated. Dense calcification is common. (See Appendix F for a list of calcified tumors.)

Spinal meningiomas are only rarely intramedullary. In contrast to intracranial counterparts, intraspinal meningiomas rarely involve surrounding osseous structures. The relative predominance in women is higher than with intracranial lesions, at almost 10 to 1.

Epidural and Intraosseous Meningiomas

The principal component of these uncommon meningiomas lies outside the dura, either intracranially or intraspinally. Given the apposition of cranial dura to the inner table of the skull, the intracranial examples are often osseoinvasive and may arise from within the bone without an intradural component (fig. 9-8). The spinal counterparts, in contrast, rarely involve bone.

Ectopic Meningiomas

Ectopic meningiomas arise at a variety of sites: cranial bones (including diploe) (33,34), ear and temporal bone (35), skin (36), lungs (37–

Figure 9-7

MENINGIOMA

Intraspinal meningiomas are discrete, intradural, and extramedullary. This example produced myelopathy and back pain in a 51-year-old policeman. (Courtesy of Dr. Ziya Gokaslan, Baltimore, MD.)

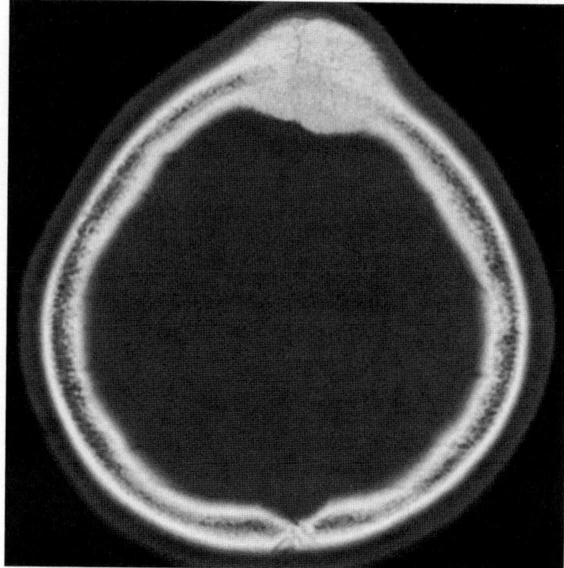

Figure 9-8

MENINGIOMA

Some meningiomas arise primarily within the skull, as did this osteoblastic lesion seen by computerized tomography (CT). There was no intradural component.

40), sinonasal tract (41), mediastinum (42), and peripheral nerves (43). Although some result from direct extension along soft tissue planes or peripheral nerves, such a mechanism obviously fails to explain meningiomas affecting the lungs or mediastinum. In addition to macroscopic meningiomas in the lungs, there are dispersed minute clusters of cells with histologic, immunohistochemical, and ultrastructural features of meningothelial cells (44). Molecular studies on microdissected specimens suggest that the isolated pulmonary lesions may not be neoplasms, whereas multifocal lesions may be neoplasms, or at least their immediate precursors (45).

GROSS FINDINGS

In situ, most grade I meningiomas push the leptomeninges before them with a margin that serves as a cleavage plane between the tumor and the adjacent brain. Microcystic variants are an exception, and not infrequently, are broadly attached to the cortical surface. These and, especially, grades II and III meningiomas of any subtype that are attached to the pia, may not always be removed cleanly, particularly if there is cortical invasion. Surgeons may thus anticipate the diagnosis of a grade II or III lesion in the face of a postexcision raw cortical surface. Recurrent tumors are less well defined and more likely to adhere to the brain and incorporate nerves and vessels.

Most meningiomas are soft, discrete, smooth-surfaced masses broadly attached to the dura (fig. 9-9). Fibrous lesions are firmer, more discrete, even more smooth surfaced, and tougher. Lesions of the microcystic subtype are more likely to be macrocystic, and attached to the brain. Calcified or ossified lesions may be gritty. Occasional lipidized meningiomas are bright yellow, whereas myxomatous tumors are gray and semigelatinous.

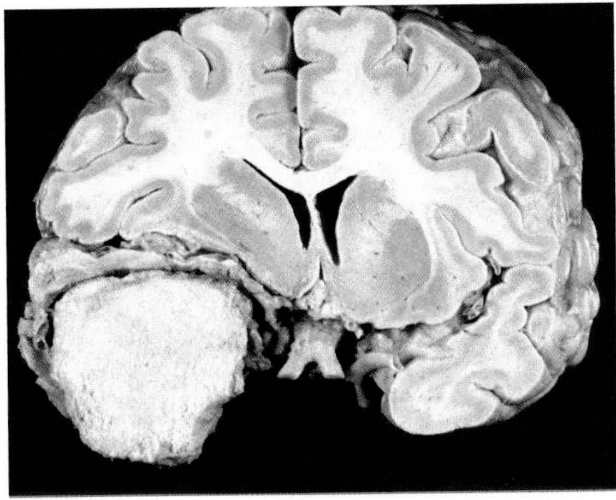

Figure 9-9

MENINGIOMA

The classic meningioma is a soft, spherical mass that displaces rather than infiltrates.

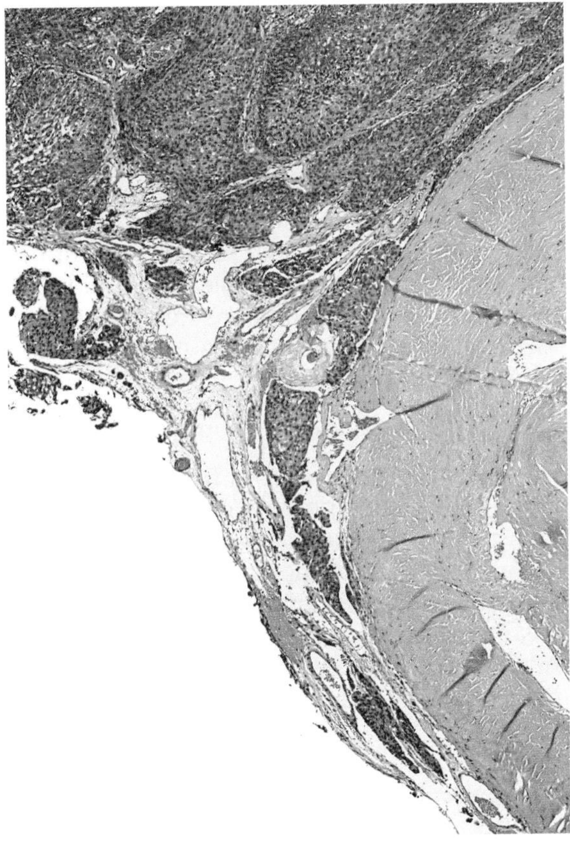

Figure 9-10

MENINGIOMA

A trail of tumor (dural tail) often extends for a short distance along the dura.

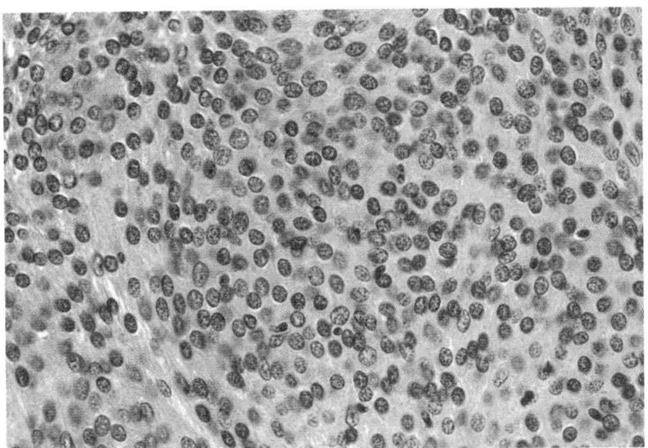

Figure 9-11

MENINGOTHELIAL MENINGIOMA

Indistinct cell borders create the syncytial appearance.

HISTOLOGIC SUBTYPES

Meningiomas are histologically so diverse that no verbal description does justice to their heterogeneity. Most demonstrate patterns corresponding to at least one of the categories described below, although only a few occur in pure form. Some variants have unfavorable prognostic implications and are therefore defined as grade II or III, as is discussed below. The following classification is that of the World Health Organization (WHO) (46).

Whatever the type, the tumor is known for a wedge of tumor tissue, corresponding to the radiologic dural tail, that extends along the dura (fig. 9-10), although only a collection of thin-walled vessels is present in some cases.

Meningothelial Meningioma

Meningothelial meningioma consists of varying sized lobules of neoplastic cells with indistinct cell borders that can create a syncytial appearance (fig. 9-11) although only a rare meningioma is exclusively syncytial. Within these sheets, the neoplastic cells demonstrate classic cytologic features of meningothelium, i.e., round to oval nuclei, delicate chromatin, small solitary nucleoli, and frequent nuclear-cytoplasmic invaginations and/or homogenization. The invaginations, also termed pseudoinclusions, appear as round, circumscribed, and intranuclear areas surrounded by marginated chromatin. They can be found in almost any meningioma variant. The

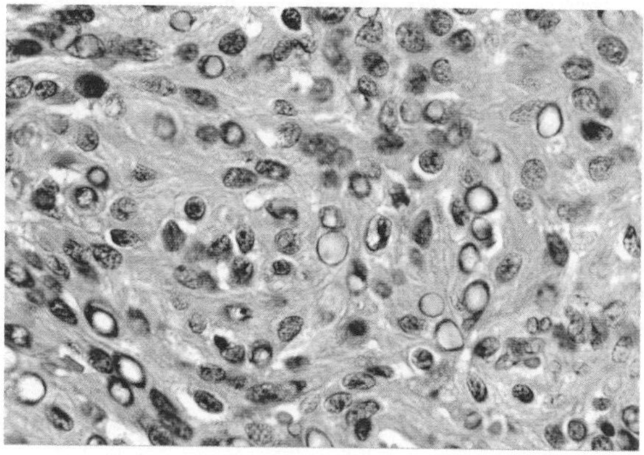

Figure 9-12

MENINGOTHELIAL MENINGIOMA

Nuclear invaginations, or pseudoinclusions, are common in meningiomas.

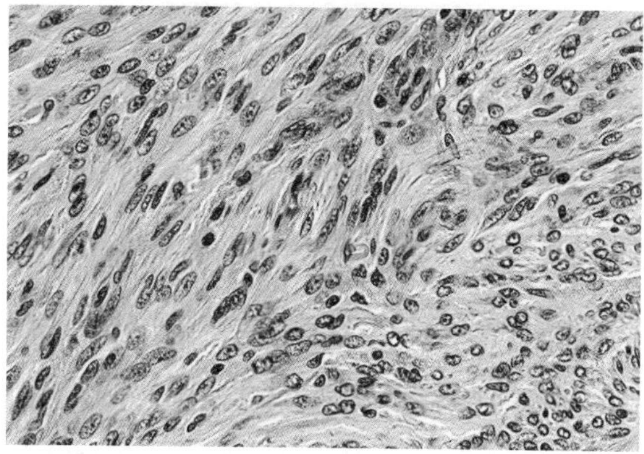

Figure 9-13

FIBROUS MENINGIOMA

Elongated cells, fascicular architecture, and a paucity of whorls and psammoma bodies are typical of the fibrous subtype.

homogenization, at least as common as pseudoinclusions, is a gray homogenization or partial clearing of the nuclei due to glycogen deposition (fig. 9-12). Fibrous tissue is typically scant in meningothelial meningiomas.

Fibrous (Fibroblastic) Meningioma

Fibrous (fibroblastic) meningioma is more likely to occur in pure form than the meningotheliomatous variant, although there are no consensus criteria for distinguishing fibrous from transitional lesions with many elongated cells (see below). Fibrous meningioma is generally less cellular and consists of elongated cells in a collagen-rich matrix (fig. 9-13). Somewhat arbitrarily, the designation "fibrous" may also be applied when cellular elongation is pronounced despite a relative lack of collagen. Some fibrous lesions exhibit a fascicular or storiform architecture (fig. 9-14). Nuclei of fibrous tumor cells are somewhat hyperchromatic and much more elongated than those of the meningothelial-type cells. Unlike transitional tumors, whorls, psammoma bodies, and intranuclear pseudoinclusions are infrequent, but calcification of fibrous stroma or the vasculature may be prominent.

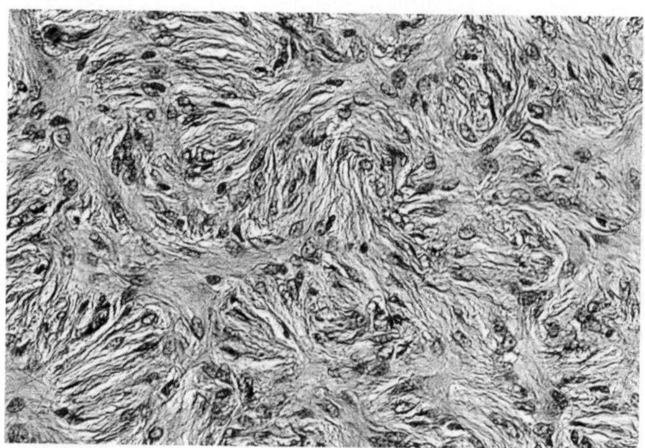

Figure 9-14

FIBROUS MENINGIOMA

Meningiomas with a storiform pattern are included within the fibrous subtype.

Transitional Meningioma

Transitional meningioma is the quintessential meningioma, with prominent lobules, whorls, psammoma bodies, and collagenized vessels (fig. 9-15). The centers of the small lobules are often syncytial whereas elongated fibroblast-like cells stream from the periphery. Such lesions are therefore considered "transitional" between meningothelial and fibrous types.

Psammomatous Meningioma

Psammoma bodies are so prominent in some meningiomas as to warrant the designation *psammomatous* (fig. 9-16). The term is not limited to tumors of a specific cytologic type, although most are transitional. Such tumors favor the spinal meninges. Heavily mineralized, paucicellular examples may be biologically indolent.

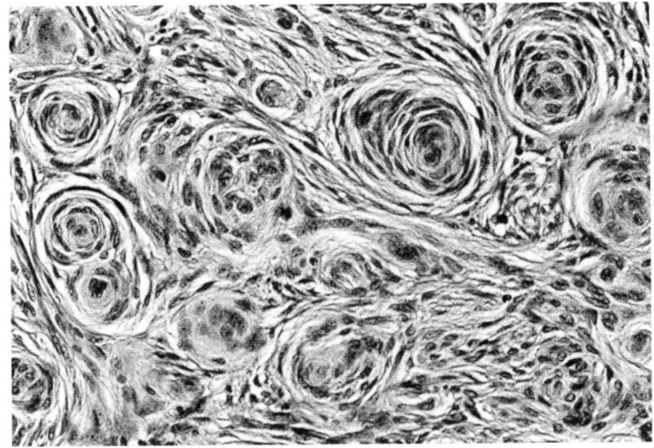

Figure 9-15

TRANSITIONAL MENINGIOMA

Whorls and psammoma bodies are classic to this common variant.

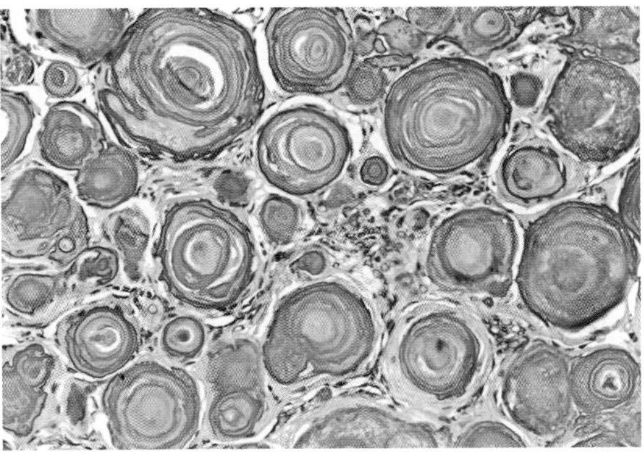

Figure 9-16

PSAMMOMATOUS MENINGIOMA

Some meningiomas are composed largely of whorls and psammoma bodies, with only the latter remaining in some "burnt out" examples.

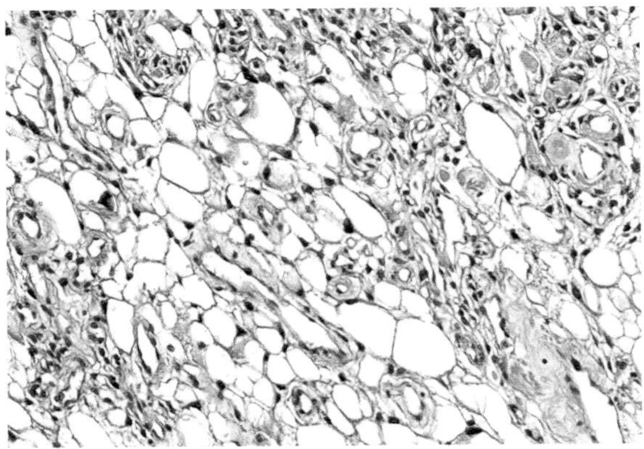

Figure 9-17

MICROCYSTIC MENINGIOMA

A lacey, vacuolated, glioma-like background helps define the microcystic variant.

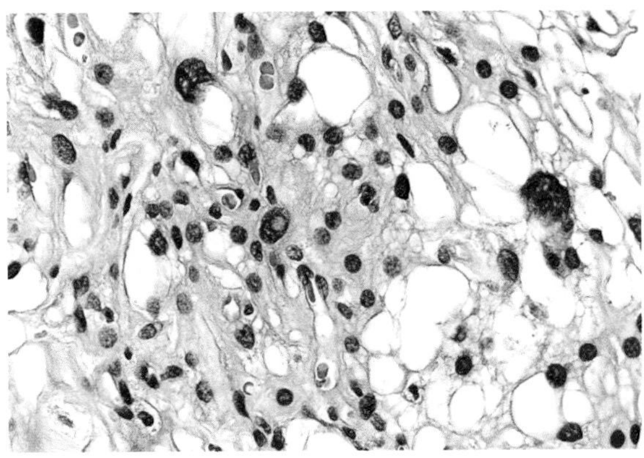

Figure 9-18

MICROCYSTIC MENINGIOMA

Vacuolated cells and focal nuclear pleomorphism are classic features of microcytic meningiomas. Whorls and psammoma bodies are usually sparse, if present at all.

Microcystic Meningioma

Histologically variable, *microcystic meningiomas* generally are hypocellular, with a conspicuous loose-textured quality, considerable vascular hyalinization, and scattered cells with large hyperchromatic and pleomorphic nuclei (figs. 9-17, 9-18) (47–49). The cobweb microcystic quality of the tumor is the result of elongated cell processes enclosing intercellular fluid-filled spaces and neoplastic cells with xanthomatous change. This type of meningioma can closely resemble hemangioblastoma because of vascularity, foamy cells, and scattered large pleomorphic nuclei (fig. 9-18; see fig. 7-6A). Closely apposed, hyalinized vessels are present in some cases (fig. 9-19). It is in this type that carcinoma-like cells, albeit sans mitoses and anaplasia, are most likely to be found (fig. 9-20). Considerable nuclear pleomorphism is present in some cases (fig. 9-21). No prognostic significance is assigned to the microcystic variant, although they are more likely than other grade I lesions to attach broadly to the surface of the brain. They are also over-represented among meningiomas with large cysts.

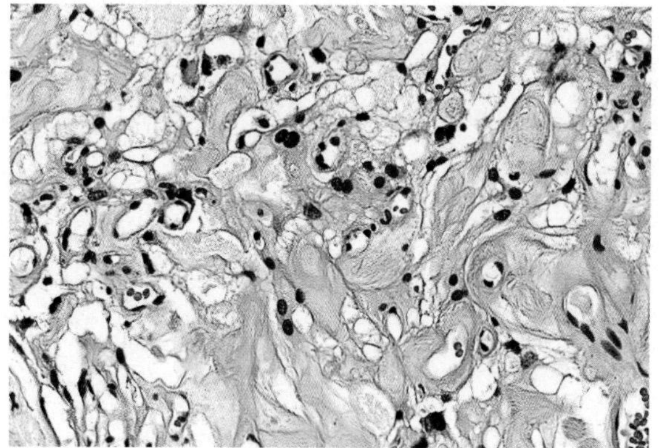

Figure 9-19

MICROCYSTIC MENINGIOMA

Hyalinized vessels are often prominent in microcystic meningiomas.

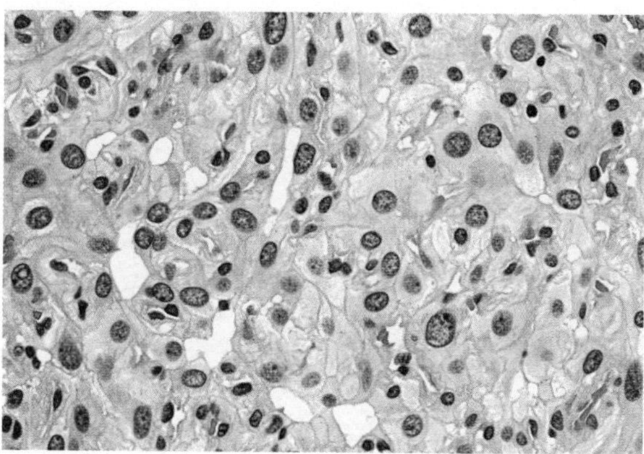

Figure 9-20

MICROCYSTIC MENINGIOMA

Nuclear pleomorphism and distinct cell borders lend an epithelial appearance to some meningiomas, usually of the microcystic type.

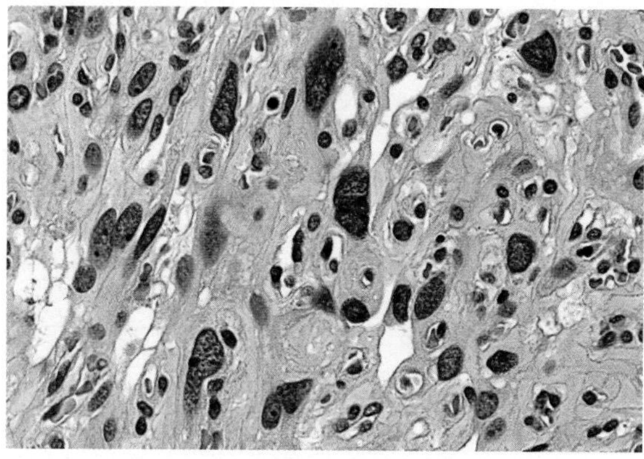

Figure 9-21

MICROCYSTIC MENINGIOMA

While considerable nuclear pleomorphism can occur in other meningioma subtypes, it is most common in association with the microcystic variant.

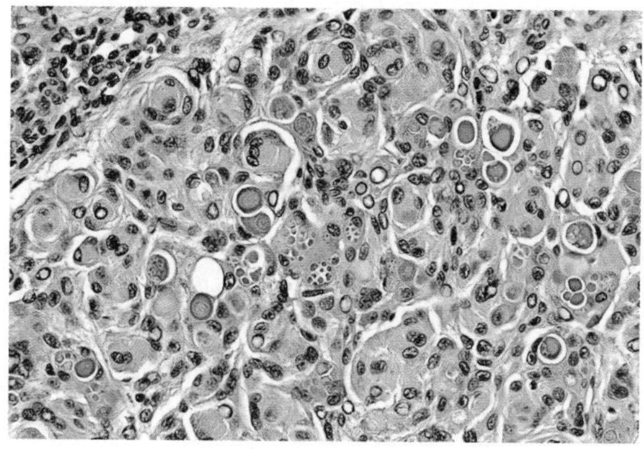

Figure 9-22

SECRETORY MENINGIOMA

The presence of single or multiple, hyaline, eosinophilic, cytoplasmic inclusions is the signature of the secretory variant.

Secretory Meningioma

Discrete hyaline inclusions, known as pseudopsammoma bodies, occur in a minority of otherwise ordinary meningiomas, generally of the meningothelial/transitional type. Meningiomas with such bodies are designated as *secretory* (50–56). One study suggested a particularly high incidence in women, and a frequent origin over the frontal lobes or sphenoid ridge (52). The eosinophilic, hyaline, often multiple, intracytoplasmic structures (fig. 9-22) must be distinguished from the larger, laminated, basophilic, extracellular psammoma bodies that generally originate in the center of a whorl. Pseudopsammoma bodies are intracellular, sometimes multiple, red rather than blue, and independent of whorls. They are periodic acid–Schiff (PAS) positive and diastase resistant (fig. 9-23). No prognostic significance is attached to secretory meningioma, although these lesions are usually well differentiated and, in principle, may be associated with a better prognosis. Elevated serum levels of carcinoembryonic antigen (CEA) have been reported in several cases in which pseudopsammoma bodies were abundant (55). Peritumoral edema is prominent around some secretory meningiomas (56).

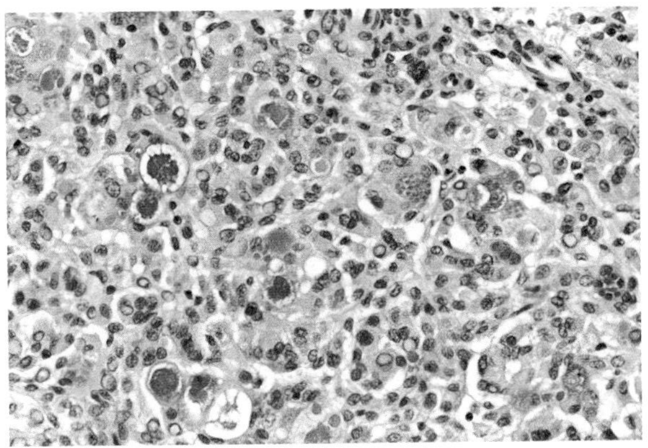

Figure 9-23

SECRETORY MENINGIOMA

Pseudopsammoma bodies are periodic acid–Schiff (PAS) positive.

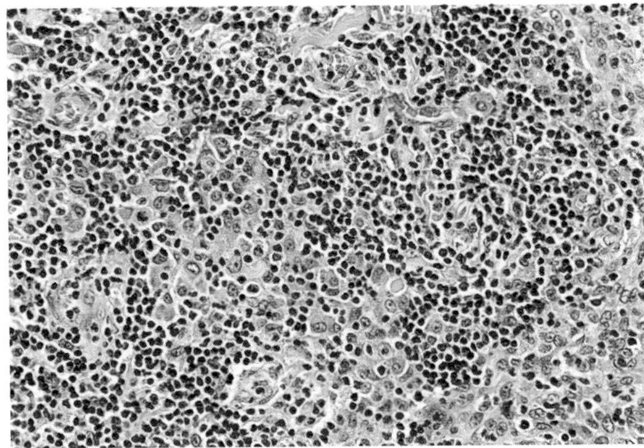

Figure 9-24

LYMPHOPLASMACYTE-RICH MENINGIOMA

The tumor cells often have a somewhat epithelioid appearance, mimicking the histiocytes in inflammatory pseudotumors such as Rosai-Dorfman disease.

Lymphoplasmacyte-Rich Meningioma

These rare meningiomas elicit a pronounced chronic inflammatory response consisting of lymphocytes and plasma cells in varying proportion (fig. 9-24) (57,58). They can dominate the lesion and obscure the neoplastic component. Russell bodies are frequent; germinal centers may be present.

Angiomatous Meningioma

Meningiomas of almost any type can be highly vascularized, and the nonspecific term *angiomatous meningioma* is sometimes applied (fig. 9-25). This description should be distinguished from the obsolete term "angioblastic" that designated a disparate group, but principally lesions that are now considered meningeal hemangiopericytoma. Like the microcystic type, angiomatous lesions can resemble hemangiopericytoma. Angiomatous meningiomas are usually well-differentiated (grade I) lesions (59).

Chordoid Meningioma

Chordoid meningioma is an infrequent variant that consists of nests and cords of epithelioid cells separated by basophilic extracellular material (fig. 9-26) (60,61). Chordoid meningiomas are, by definition, grade II. While occasional meningiomas are entirely and overtly chordoid at presentation, chordoid features may evolve in a nonchordoid grade I or grade II lesion (fig. 9-27). Classic features of meningioma, such as whorls and psammoma bodies, are scant, if present at all, in examples with advanced chordoid features. The prognostic significance of only focal chordoid change is not clear. The 2007 WHO "Blue Book" suggests that the diagnosis of this subtype should be made when the pattern is "predominant" (61a).

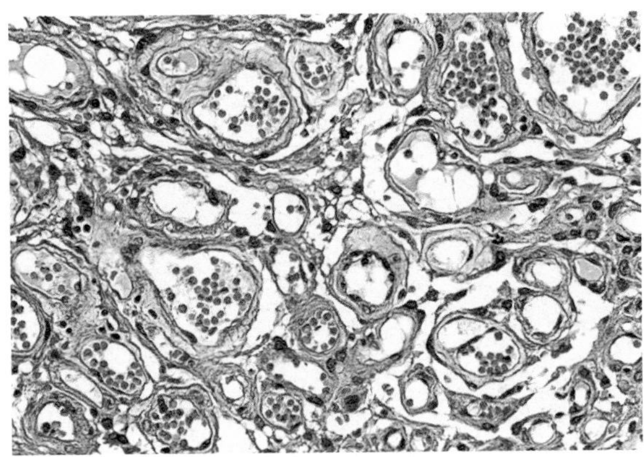

Figure 9-25

ANGIOMATOUS MENINGIOMA

Meningiomas with a high ratio of vessels to tumor cells are sometimes designated angiomatous.

Clear Cell Meningioma

The rare *clear cell meningioma* is composed of cells with clear cytoplasm (fig. 9-28) (24,62–64). Whorls, if present at all, are focal and poorly formed (fig. 9-29). Psammoma bodies are not expected. Thus, clear cell meningiomas are generally clear cell throughout, unlike chordoid and

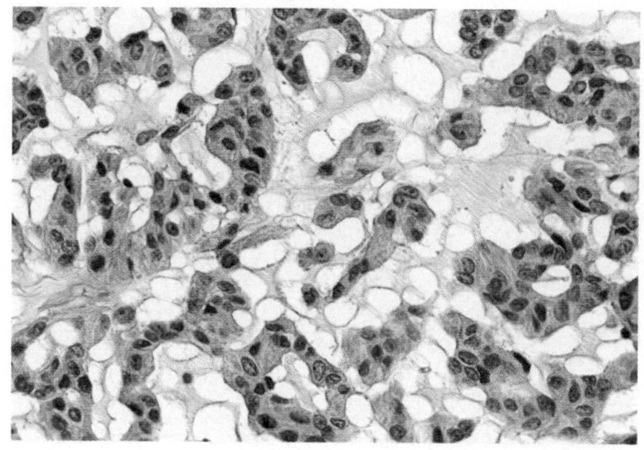

Figure 9-26

CHORDOID MENINGIOMA

Rows and clusters of epithelioid cells in a myxoid background produce the chordoid pattern of this grade II meningioma variant.

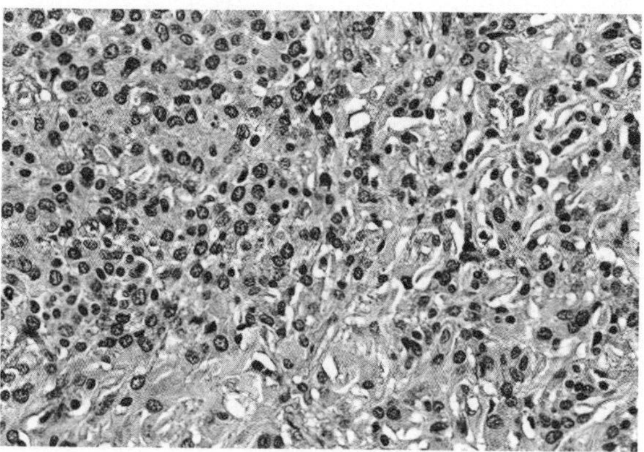

Figure 9-27

CHORDOID MENINGIOMA

Chordoid features often appear focally in an "ordinary," nonchordoid meningioma, in this case a tumor that was grade II on the basis of mitotic activity. By definition, chordoid meningiomas are grade II.

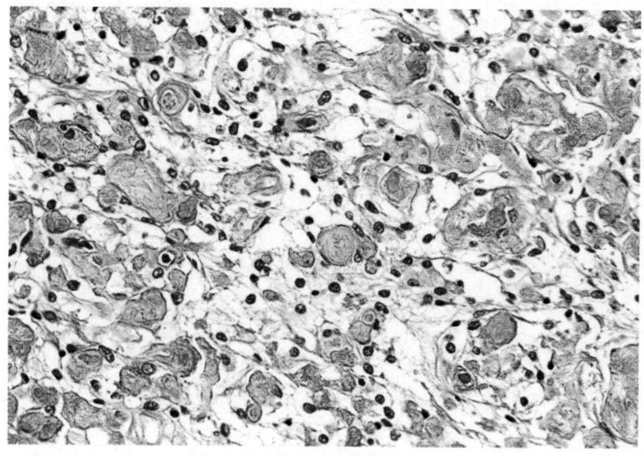

Figure 9-28

CLEAR CELL MENINGIOMA

Clear cell meningiomas have a monomorphous population of round cells with clear or vacuolated cytoplasm, but usually few specific architectural features to suggest meningioma. Small concentric knots of collagen are common. Clear cell meningiomas are grade II.

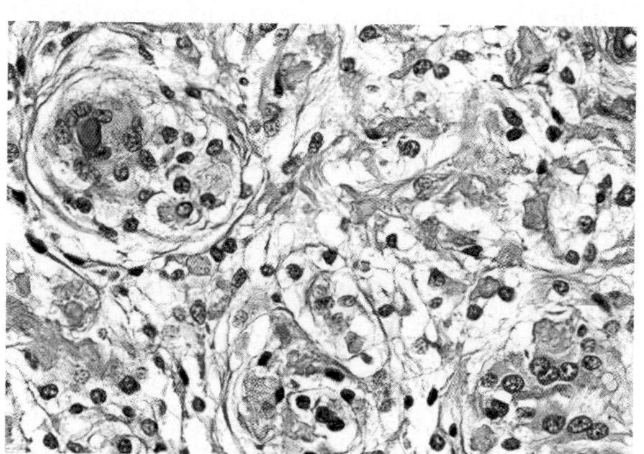

Figure 9-29

CLEAR CELL MENINGIOMA

The presence of whorls and psammoma bodies is a welcome, but uncommon, finding in clear cell meningiomas, which have otherwise sometimes ambiguous features.

rhabdoid variants in which these tissue patterns are often accompanied by areas of lower-grade "conventional" meningioma. Bands of connective tissue course through the mass, and some lesions are densely and almost exclusively sclerotic (fig. 9-30). The bland cytologic features and low mitotic activity belie the propensity of this variant to recur. It is, therefore, assigned grade II. The glycogen-rich cytoplasmic contents are PAS positive (fig. 9-31).

Clear cell meningiomas are often subtentorial, e.g., cerebellopontine angle or lumbosacral meninges, and may not be dura based.

Rhabdoid Meningioma

Rhabdoid cells, with a rather discrete hyaline or faintly fibrillar cytoplasmic mass, are present either focally or globally in a small percentage of meningiomas (fig. 9-32) (65–68). Almost all have an elevated mitotic rate or other prognostically unfavorable features. Such *rhabdoid meningiomas* are grade III. The significance of only

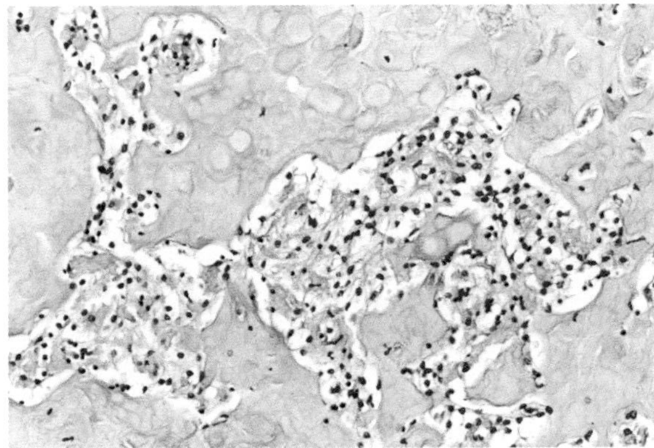

Figure 9-30

CLEAR CELL MENINGIOMA

Extensive fibrosis is common in this unusual variant.

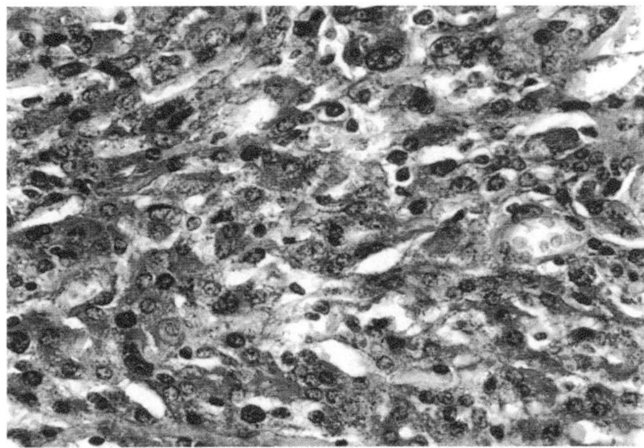

Figure 9-31

CLEAR CELL MENINGIOMA

Much of the clear cytoplasm is PAS positive.

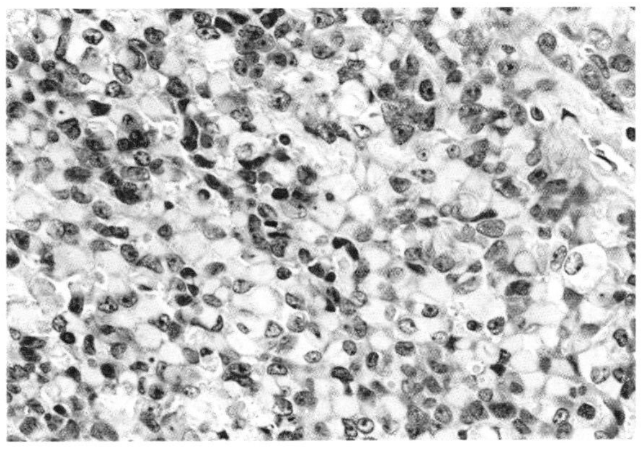

Figure 9-32

RHABDOID MENINGIOMA

A cytoplasmic mass displacing a nucleus with a prominent nucleolus characterizes rhabdoid meningioma (grade III). As is often the case, mitotic figures are easy to find.

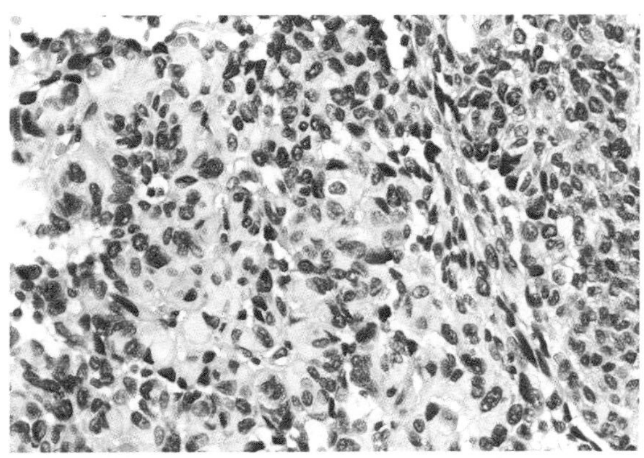

Figure 9-33

MENINGIOMA WITH PAPILLARY AND RHABDOID FEATURES

Papillary and rhabdoid features are combined in some anaplastic meningiomas. In this case, a perivascular orientation of tumor cells classic to the papillary type, as seen here, is combined with rhabdoid features elsewhere in the tumor (see figure 9-34).

focal rhabdoid change in an otherwise grade I or II lesion is unclear. The 2007 WHO "Blue Book" suggests that the diagnosis of this subtype should be made when the pattern is "predominant" (61a). Some tumors combine the features of the rhabdoid and papillary types (figs. 9-33, 9-34).

Papillary Meningioma

Papillary meningiomas are rare, sometimes pediatric, lesions that have monotonous neoplastic cells that are meningothelial in appearance, but with nuclei generally rounder and more uniform than those of the typical meningothelial lesion (69–71). Scattered mitoses are often present. The distinctive, albeit sometimes subtle, feature is a perivascular orientation of tumor cells that mimics the perivascular pseudorosettes in ependymoma (fig. 9-35). In contrast to those of ependymoma, however, the cells that approximate the vessel are, in most instances, separated by delicate reticulin fibers in a perivascular stellate pattern (fig. 9-36). The term papillary meningioma should be reserved for this variant and not applied to a meningioma with no other distinctive feature than the formation of papillae produced by artifactual tissue

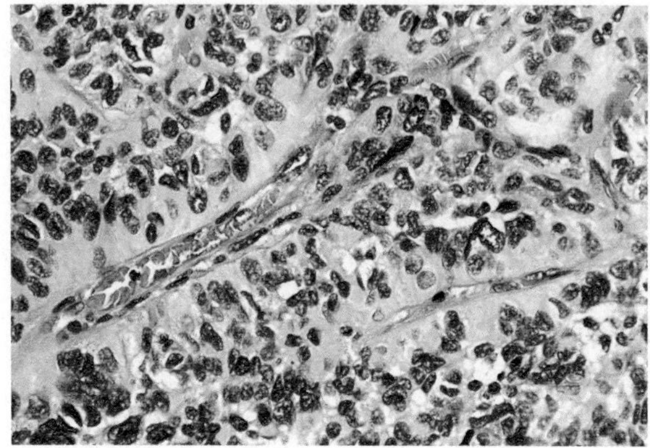

Figure 9-34

MENINGIOMA WITH PAPILLARY AND RHABDOID FEATURES

The perivascular orientation of tumor cells illustrated in figure 9-33 is associated with rhabdoid features in the same tumor.

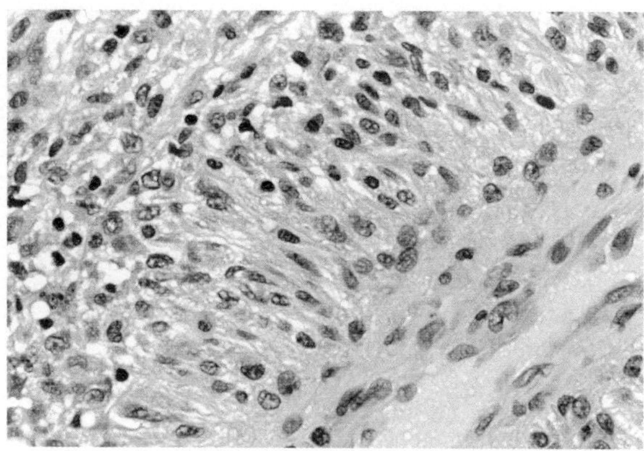

Figure 9-35

PAPILLARY MENINGIOMA

A form of perivascular pseudorosette defines this rare variant. Although often cytologically bland, this rare lesion is aggressive and assigned grade III.

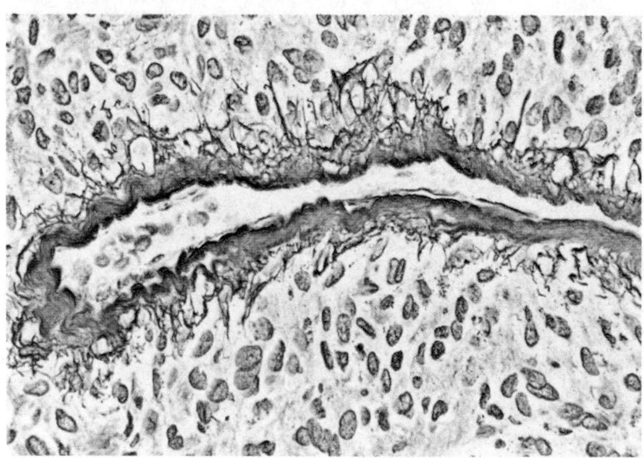

Figure 9-36

PAPILLARY MENINGIOMA

Spikes of reticulin radiating from the vessels are common in the papillary variant.

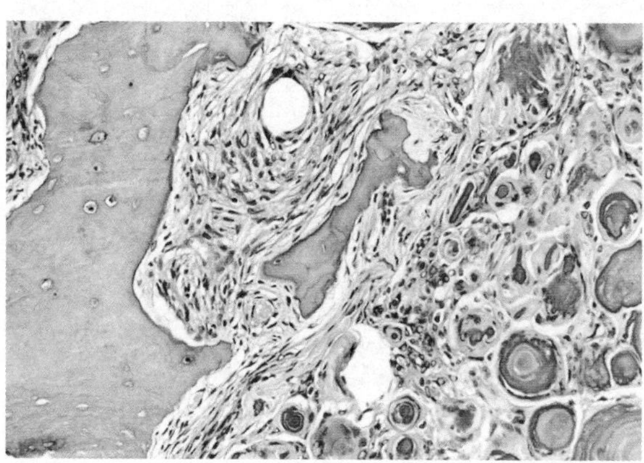

Figure 9-37

METAPLASTIC MENINGIOMA WITH BONE FORMATION

Metaplastic bone is especially common in extensively psammomatous lesions.

dehiscence during processing, or to highly atypical tumors in which necrosis results in the formation of viable perivascular cuffs of cells. Papillary meningiomas are grade III.

Metaplastic Variants

Metaplastic variants are seen occasionally as meningotheliomatous and transitional lesions that produce such mesenchymal elements as bone (fig. 9-37) and fat (fig. 9-38) (72). Bone is most likely in highly calcified lesions with abundant psammoma bodies and mineralization of fibrous stroma. Metaplastic bone must be distinguished from the reactive bone overrun by meningiomas invading the skull. Myxoid metaplasia is another rare metaplastic feature (fig. 9-39) (73,74). A rare meningioma has oncocytic features (75).

Rare grade I meningiomas have a proliferation of reticulin-invested pericyte-like cells that can be quite prominent (76,77). The tumors seem to be associated with a high incidence of peritumoral edema.

A rare form of "collision" tumor affecting the nervous system consists of a meningioma that either abuts or is intermingled with a glioma

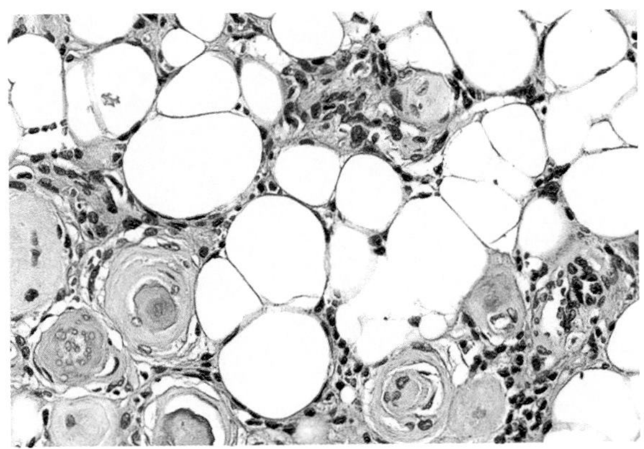

Figure 9-38

METAPLASTIC (LIPIDIZED) MENINGIOMA

Lipid-laden cells closely resembling adipocytes are common in meningiomas. Ultrastructural and immunohistochemical studies suggest that these are lipidized tumor cells, not adipocytes.

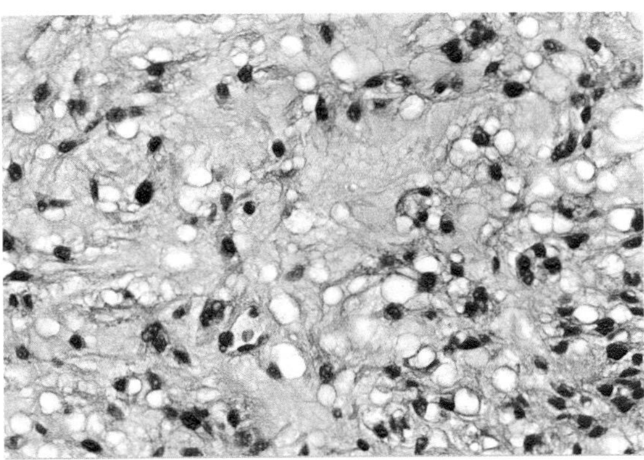

Figure 9-39

METAPLASTIC (MYXOID) MENINGIOMA

Individual cells or small clusters in a myxoid background create one type of metaplastic meningioma.

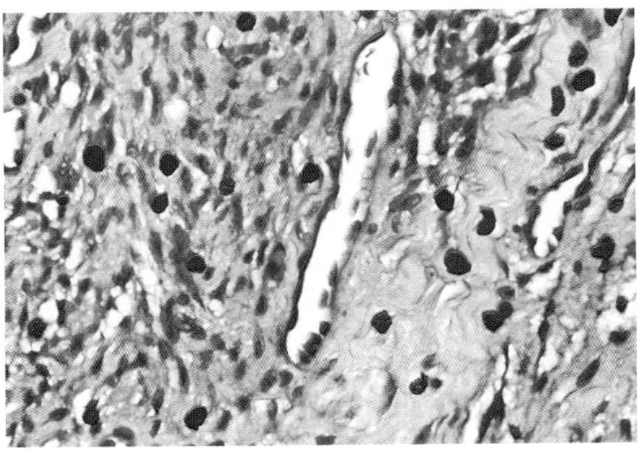

Figure 9-40

MENINGIOMA WITH MAST CELLS

Mast cells are abundant in some meningiomas.

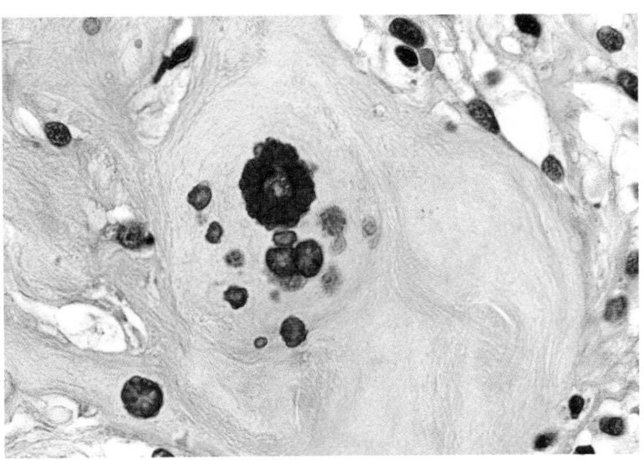

Figure 9-41

MENINGIOMA WITH "PETALOID" TYROSINE-RICH CRYSTALS

Tyrosine-rich crystals are deposited inexplicably in some meningiomas, as here in a clear cell variant.

(78,79). While coincidence seems responsible in most cases, there is of course speculation that the two lesions are related by more than chance. A meningioma-schwannoma confrontation occurs occasionally at the cerebellopontine angle in patients with NF1 (see figs. 18-8–18-10).

Mast cells are abundant in some meningiomas (fig. 9-40) (80). Petal-like ("petaloid"), tyrosine-rich crystals are a rare finding (fig. 9-41) (81), as are melanocytes that colonize meningiomas involving extracranial soft tissue (82).

Rare meningiomas are host to metastatic carcinoma (fig. 9-42) (83–86) or leukemia (87). These should be contrasted with rare meningiomas with glandular or pseudoglandular differentiation (fig. 9-43) (88).

HISTOLOGIC GRADING

The grading of meningiomas has evolved from the simple "brain invasion = malignancy" approach to a multiplex paradigm that assigns grades on the basis of tumor subtype and histologic features (46). While not a criterion in the WHO meningioma grading system, the MIB-1 index is also used by some in prognostication.

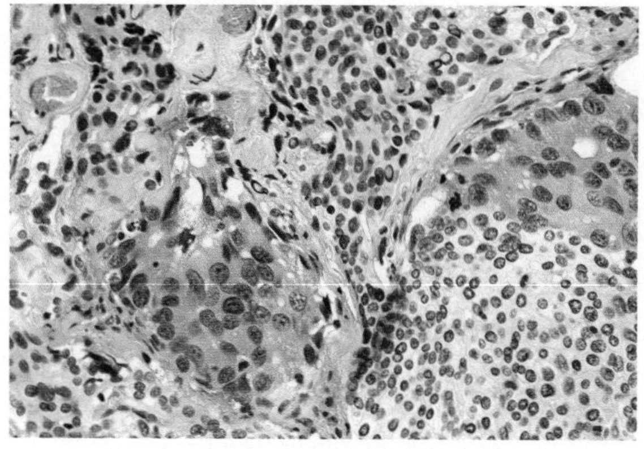

Figure 9-42

CARCINOMA METASTATIC TO MENINGIOMA

Nests of large, mitotically active cells are readily distinguished from the small, bland cells of the host meningioma. The primary of this adenocarcinoma was the lung.

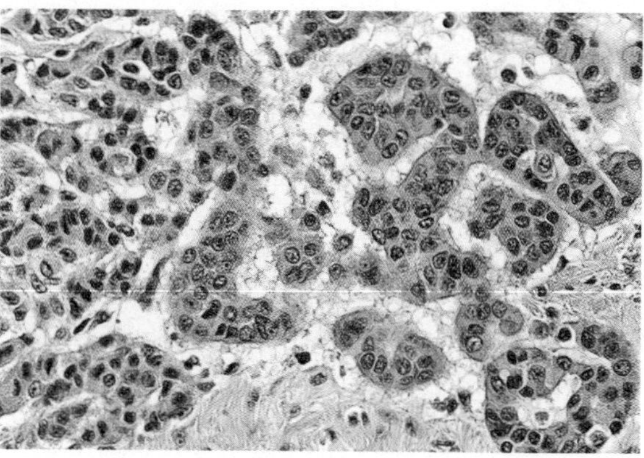

Figure 9-43

MENINGIOMA WITH PSEUDOGLANDULAR PATTERN

Gland-like structures are occasionally present in meningiomas. A gradual transition between the epithelioid structures and classic meningioma distinguishes these formations from a metastatic carcinoma, as is illustrated in figure 9-42.

Table 9-1

GRADING OF MENINGIOMAS

Grade I ("typical") meningioma: meningioma without features of grade II or III lesions
Grade II ("atypical") meningioma
With brain invasion, and/or
Four or more, but fewer than 20 mitoses per 10 high-power fields, and/or
Three or more of the following:
increased cellularity
small cell change
prominent nucleoli
loss of lobular architecture ("sheeting")
necrosis not explainable by preoperative tumor embolization, and/or
Chordoid or clear cell subtype
Grade III ("anaplastic") meningioma
Overt anaplasia, and/or
Twenty or more mitoses per 10 high-power fields, and/or
Rhabdoid or papillary subtype

Summarized in Table 9-1, the WHO grading system of meningiomas is the culmination of clinicopathological studies from many institutions that have identified histologic grading parameters and histologic subtypes with more aggressive behavior (89–94). The latter include chordoid, clear cell, papillary, and rhabdoid variants. Oncocytic lesions may be more aggressive too, but only a few cases have been reported (75).

Brain Invasion

The typical meningioma has a smooth surface, delicate fibrous capsule, and respectful, noninvasive relationship to the pia. Occasional lesions are broadly adherent to, but not invasive of, the brain. Small avulsed fragments of cerebral cortex may remain attached to the surface of any meningioma, especially ones that are large. This is not brain invasion.

Invasion of the brain is present when cohesive tongues of neoplastic tissue in continuity with the main tumor break through the pia to involve underlying cortex. Invasive meningiomas often have a serrated outer surface with multiple peg-like extensions in which small foci of necrosis are common (figs. 9-44, 9-45; see fig. 9-53). Such extensions may be only somewhat irregular in configuration but can be frankly ragged. In some cases, a broad front of tumor-brain attachment is associated with downward extension along the vessels in the Virchow-Robin spaces. This likely increases the odds of local recurrence, but does not technically qualify as parenchymal invasion. Simple superficial extension of a plug of tumor along a perivascular space thus does not qualify as invasion, although it may admittedly be difficult in an individual case to determine whether brain involvement is limited to irregular perivascular extension or whether the process has breached the pia and

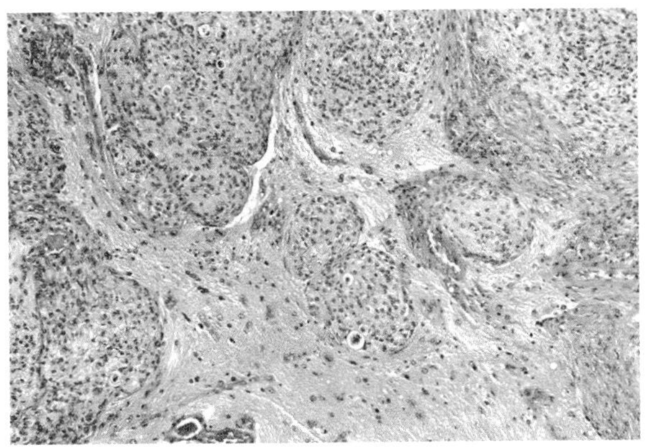

Figure 9-44

MENINGIOMA: BRAIN INVASION

Tongues of cohesive cells extending into the cortex are considered evidence of invasion. Single cell infiltration is rare.

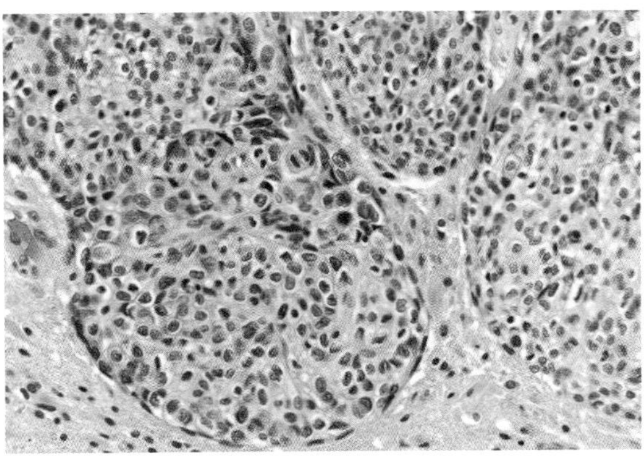

Figure 9-45

MENINGIOMA: BRAIN INVASION

Invasion of the brain establishes grade II. As is common, the lesion is otherwise unremarkable, and in this case has prominent whorls.

is truly neuroinvasive. Many pathologists do not attempt this distinction and equate irregular or complex perivascular extension with invasion, assuming that this is an adverse prognostic factor. Brain invasion is the prognostic equivalent of atypia based on criteria such as mitotic rate, nucleolar size, etc. Although the equivalent of a grade II designation, not all brain invasive meningiomas are histologically atypical (grade II). Brain invasion can be mentioned as a finding, in addition to the grade based on the conventional histologic grading criteria noted above.

The relationship of the tumor to the brain is facilitated by glial fibrillary acidic protein (GFAP) staining since the latter detects narrow tongues of brain within an invading tumor front that are easily overlooked in hematoxylin and eosin (H&E)-stained sections (fig. 9-46). Although there is usually little response, invasive neoplasms occasionally incite exuberant reactive gliosis.

Infiltration of dura, bone (fig. 9-47), and extracranial soft tissue (fig. 9-48) may have little clinical importance in resectable lesions affecting the cranial vault, but often has a major, and prognostically unfavorable, effect in tumors at the skull base. Although it is not always predictive of recurrence, it is a worrisome finding in certain locations. Parasagittal lesions often infiltrate and ultimately may occlude the superior sinus. During this gradual process collaterals can develop so that the affected sinus may be sacrificed surgically without resultant cortical infarction (see fig. 9-5).

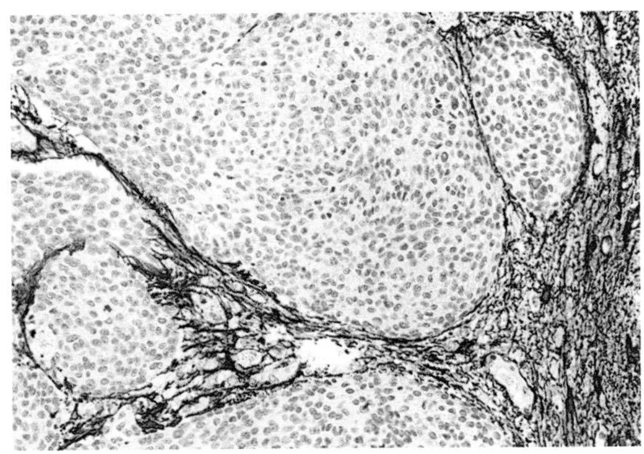

Figure 9-46

MENINGIOMA: BRAIN INVASION

Glial fibrillary acidic protein (GFAP) staining helps establish the presence of brain invasion by identifying brain tissue between tongues of the neoplasm.

Increased Mitotic Rate

The number of mitoses in meningiomas varies greatly from case to case. They are difficult to find in the classic well-differentiated lesion, and more than a few are unusual (fig. 9-49). Four or more (but less than 20) per 10 high-power fields indicates grade II, and 20 or more satisfies a requirement for grade III (anaplastic meningioma).

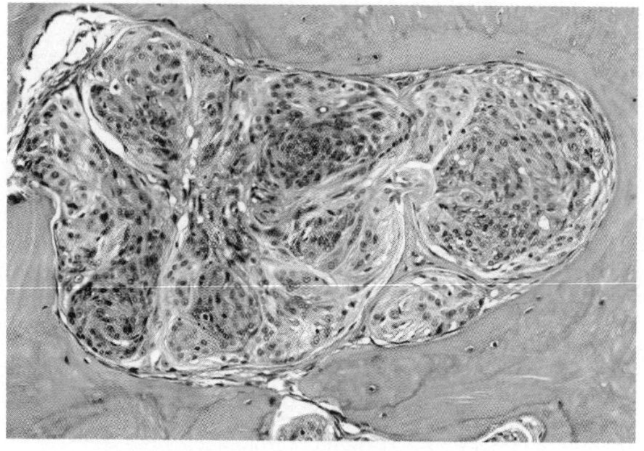

Figure 9-47

MENINGIOMA: SKULL INVASION

Invasion of the overlying skull can occur in meningiomas of all grades.

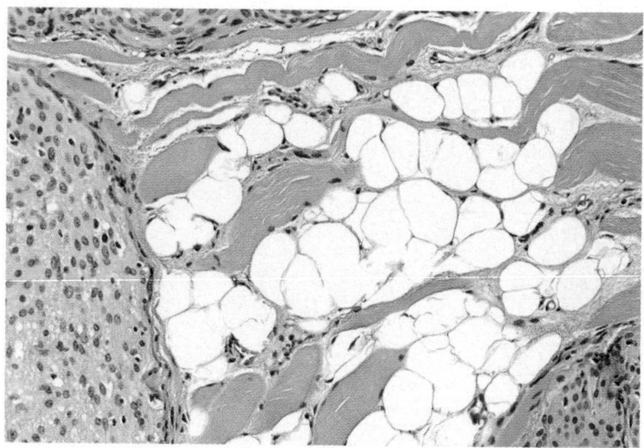

Figure 9-48

MENINGIOMA: SOFT TISSUE INVASION

After penetrating the skull, meningiomas are free to invade cranial soft tissues.

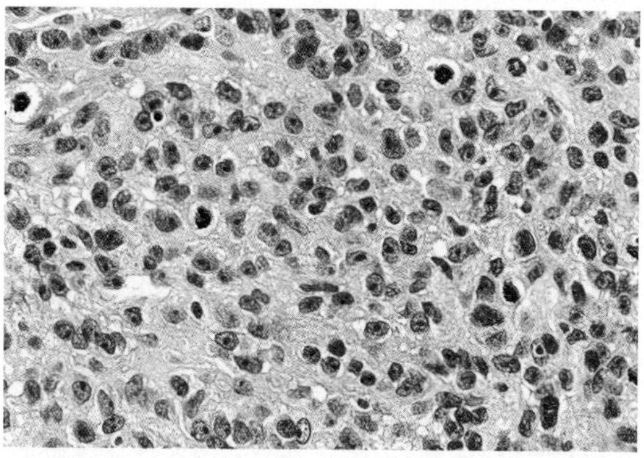

Figure 9-49

MENINGIOMA: INCREASED MITOTIC ACTIVITY

Mitoses of this number far exceed those of a grade I lesion.

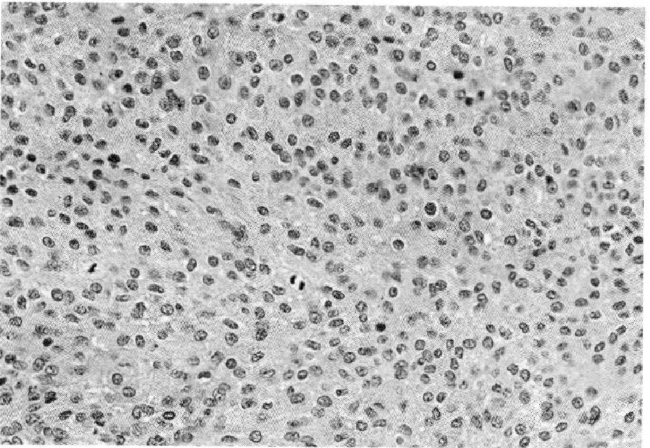

Figure 9-50

MENINGIOMA: LOSS OF LOBULARITY

The presence of diffuse, unstructured sheets of cells is one index of a potentially more aggressive lesion.

Loss of Lobularity

Loss of lobularity, or "sheeting," refers to the presence of sheet-like areas without the lobularity, frequent collagenous septa, and perivascular thickening of the typical meningioma (fig. 9-50).

Prominent Nucleoli

Nucleoli in typical grade I meningiomas are small. They become prominent in grade II and grade III lesions (fig. 9-51).

Increased Cellularity

Although it can be quantified by determining the number of nuclei along the diameter of a microscopic field, increased cellularity is generally assayed subjectively as being greater than that of the typical grade I meningioma.

Necrosis

Necrosis usually occurs as small foci, sometimes surrounded by pseudopalisading (fig. 9-52). Minute foci of necrosis are especially frequent in the peg-like extensions from the surface of the tumor at the tumor-brain interface (fig. 9-53). Large geographic areas of necrosis are possible, but these should suggest a result of preoperative embolization, which, in contrast to the usual tumor necrosis, affects large geographic areas and embolic material may be present (95) (fig. 9-54). In other cases, necrotic areas are small, punched

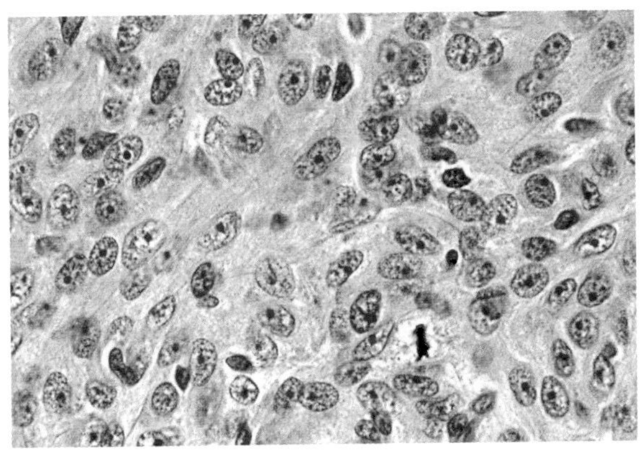

Figure 9-51

MENINGIOMA: PROMINENT NUCLEOLI

Nucleoli are prominent in most grade II, and essentially all grade III, meningiomas.

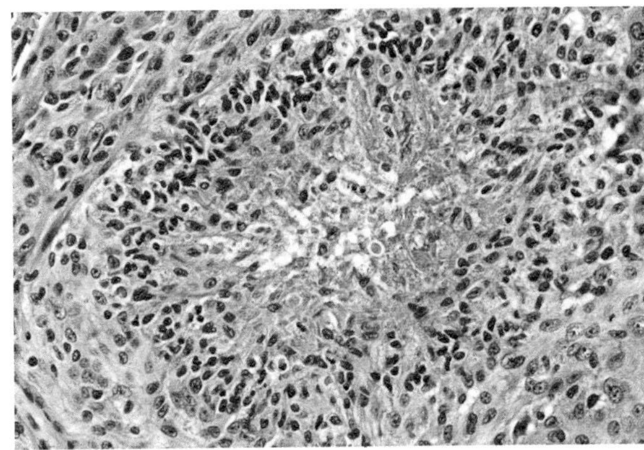

Figure 9-52

MENINGIOMA: NECROSIS

Small foci of necrosis, sometimes with peripheral pseudopalisading, are common in grade II and grade III meningiomas.

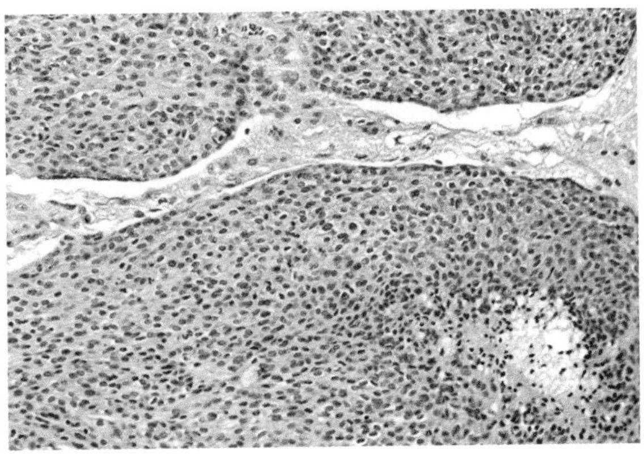

Figure 9-53

MENINGIOMA: NECROSIS

Necrosis is common in tongues of tumor that project from the surface of the neoplasm into the underlying brain.

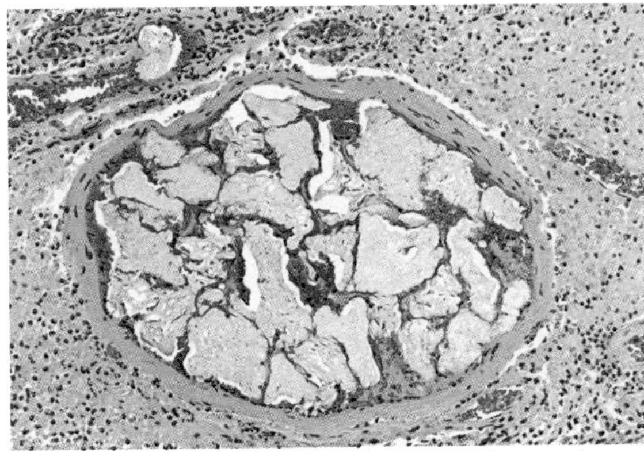

Figure 9-54

NECROSIS SECONDARY TO EMBOLIZATION

Extensive necrosis was present in this lesion after embolization. Intravascular particles of the embolic agent polyvinyl alcohol form a distinctive mosaic.

out foci with an acute, monophasic appearance. Mitoses and an elevated MIB-1 index are often present in perinecrotic regions.

Small Cell Change

Small cell change refers to focal, often multiple, regions in which both the size of the overall cell and that of the nucleus are reduced (fig. 9-55).

Increased MIB-1 Index

Although the MIB-1 index is not part of the WHO grading system, there is a prognostically significant, and inverse, relationship between MIB-1 indices and outcome (96–98). There is, of course, overlap in values between cases that ultimately do and do not recur, and this detracts somewhat from the value of the technique (99). While there is no sanctioned threshold that, when exceeded, mandates grade II or III, a level of approximately 4 percent or more in lesions that have been totally resected was associated with a reduced recurrence free survival in one study (fig. 9-56) (98). A similar level, greater than 3 percent, was found to be prognostically unfavorable in another study (100). As mentioned above, MIB-1 indices can be higher around foci of necrosis (101), although this issue has been debated (49,95).

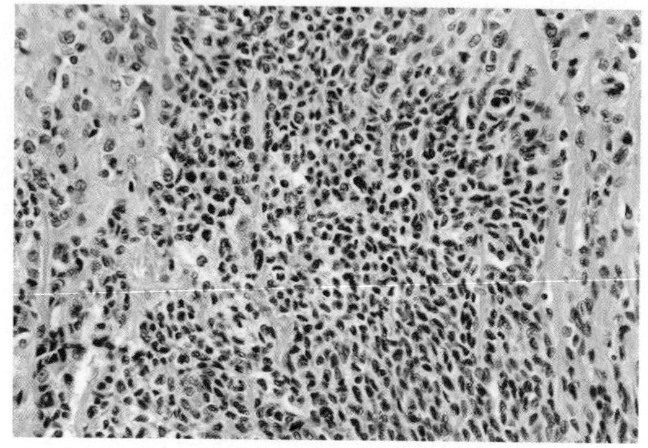

Figure 9-55

MENINGIOMA: SMALL CELL CHANGE

Aggregates of small, condensed cells are known as small cell change. Despite their appearance, the cells are no more proliferative or mitotically active than are surrounding cells.

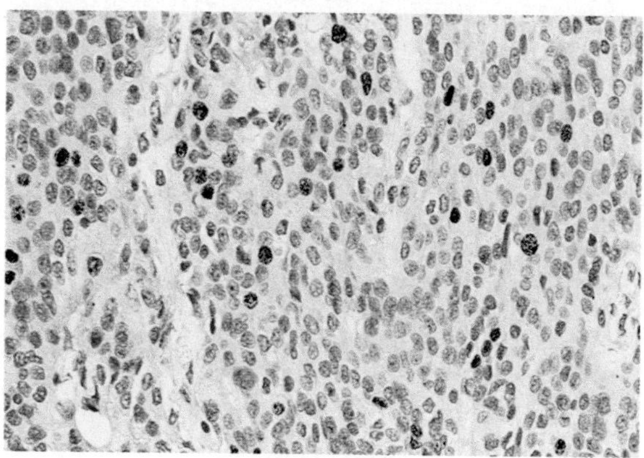

Figure 9-56

MENINGIOMA: INCREASED MIB-1 INDEX

This degree of activity, approximately 7 percent, exceeds the usual rate of a grade I lesion.

As with proliferation indices in general, sampling is often an issue. Should one attempt an overall MIB-1 estimate by counting random fields, or count only areas of higher indices? When a random sampling method was employed, the mean MIB-1 indices were 1.15, 3.33, and 9.45 for meningioma grades I, II, and III, respectively (96). Higher levels, 2.2, 5.5, and 16.3, were found when the highest labeling areas were selected. The former correlated better with outcome.

Anaplasia

Most meningiomas are, and remain, well differentiated, or at the most become grade II. Only rarely do they have or acquire features that fully justify the term anaplasia (fig. 9-57). Grade III "anaplastic" meningiomas are so assigned on the basis of increased mitoses, not on cytologic changes that would not, in themselves, justify the term. Truly anaplastic meningiomas are often rather sarcomatoid in architecture, and formed of cells with prominent nucleoli. Necrosis is almost always present and mitoses are abundant. Some meningiomas become atypical or anaplastic over time, as sequential recurrences show a progressive loss of the meningothelial phenotype. In rare instances, various forms of sarcoma, such as rhabdomyosarcoma, appear (102). Some lesions have carcinoma- or melanoma-like features.

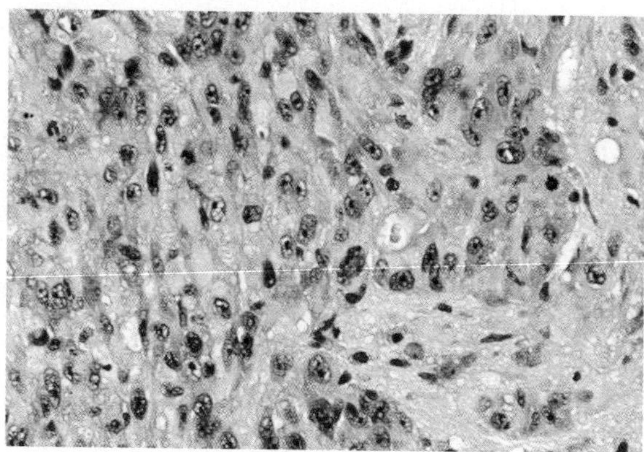

Figure 9-57

MENINGIOMA: ANAPLASIA

Cytologic atypia severe enough to warrant the designation anaplasia is present in only a small percentage of meningiomas.

Other Features

Other factors that are associated with a less favorable clinical outcome, but not utilized directly in grading, include extent of resection and lack of immunostaining for progesterone receptors (103,104). The latter has been found to predict an increased likelihood of recurrence.

Grading in the context of prior embolization is complicated by the fact that embolization is associated with augmented nucleolar size, mitotic activity, and MIB-1 rate. Such lesions, in concept, could be overgraded, although one detailed analysis concluded that the lesions could be meaningfully graded using the same criteria as nonembolized lesions (95).

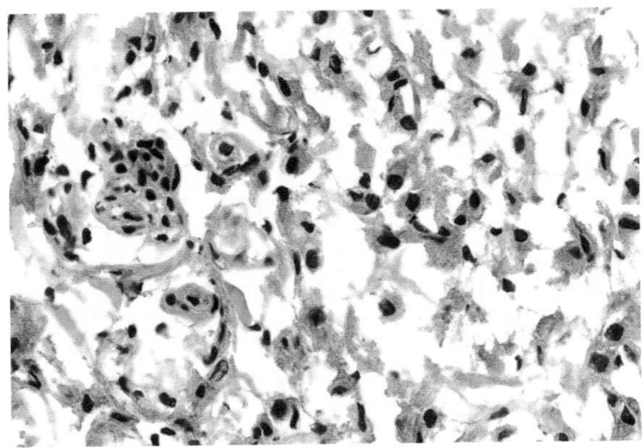

Figure 9-58

MENINGIOMA: FROZEN SECTION

The presence of whorls and psammoma bodies, no matter how small or focal, is a welcome finding in meningiomas that are otherwise architecturally nondescript.

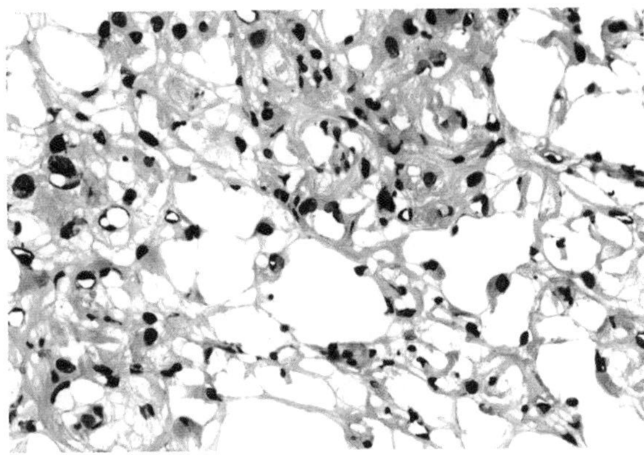

Figure 9-59

MICROCYSTIC MENINGIOMA: FROZEN SECTION

A meningothelial origin may not be apparent in these spongy, glioma-like lesions with few, if any, whorls or psammoma bodies.

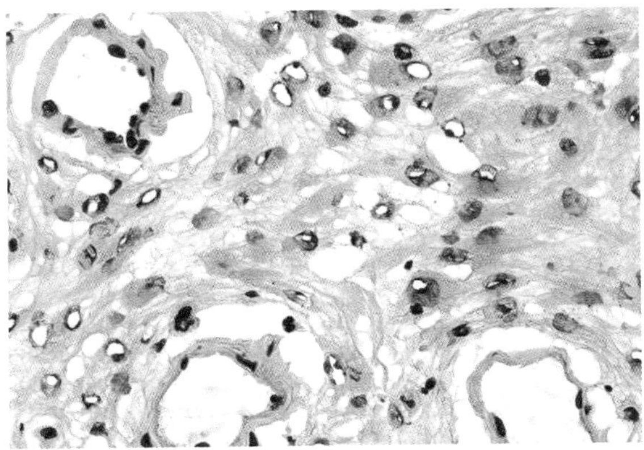

Figure 9-60

MENINGIOMA RESEMBLING ASTROCYTOMA: FROZEN SECTION

Gemistocyte-like cells can create the appearance of astrocytoma.

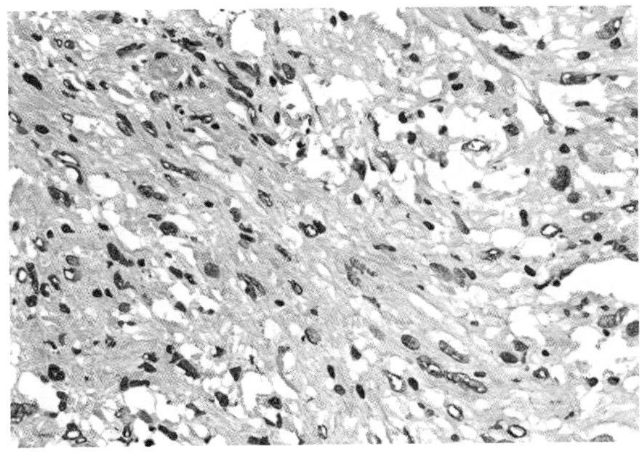

Figure 9-61

FIBROUS MENINGIOMA: FROZEN SECTION

The fascicular fibrous subtype without whorls or psammoma bodies is difficult to distinguish from schwannoma.

FROZEN SECTION

Lobules, whorls, and psammoma bodies, while classic and distinctive features, are not always conspicuous and may even be absent in frozen sections of a lesion with all the clinical and radiologic features of meningioma (fig. 9-58). In their absence, alternative diagnoses must be considered, including metastatic carcinoma, glioma, and schwannoma, among others. Carcinoma is entertained when nuclear pleomorphism and hyperchromatism are exaggerated by freezing, whereas glioma seems a likely diagnosis for paucicellular, especially microcystic, lesions with a glioma-like looseness and disarray (fig. 9-59), or when gemistocyte-like cells are present (fig. 9-60). Fibrous lesions are similar to schwannoma (fig. 9-61). Artifacts of freezing are especially prominent in microcystic meningiomas in which the reticular texture, nuclear pleomorphism and hyperchromasia, and paucity of meningothelial features can be confusing. Reference to smear preparations usually relieves concern about almost any lesion other than meningioma. Hemangiopericytomas have less cytoplasm, less prominent nucleoli, and usually lack intranuclear pseudoinclusions.

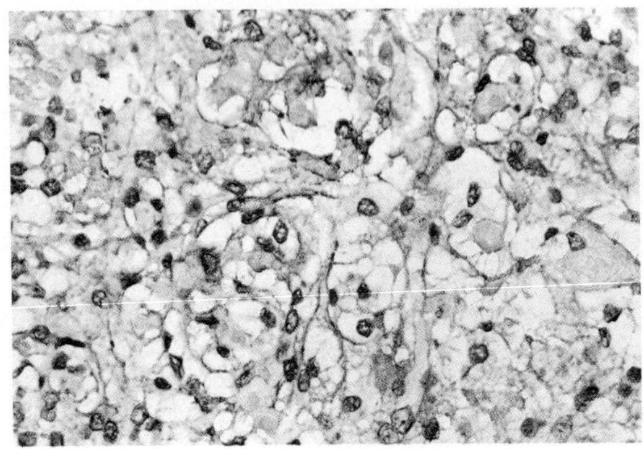

Figure 9-62

MENINGIOMA

As in this clear cell lesion, surface staining with epithelial membrane antigen (EMA) is typical of meningiomas of any subtype.

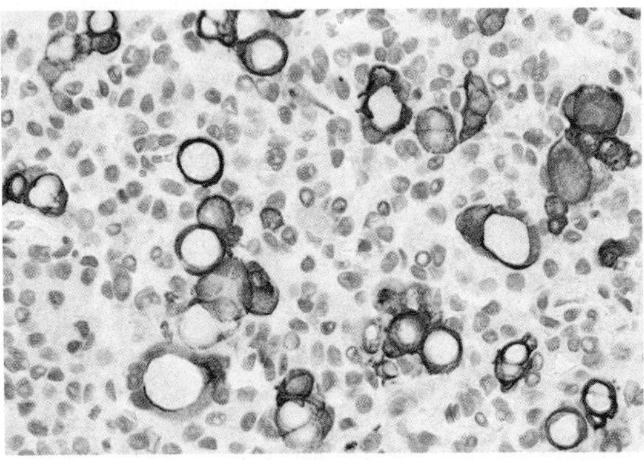

Figure 9-63

MENINGIOMA

Pseudopsammoma bodies are immunoreactive for cytokeratins (here AE1/AE3).

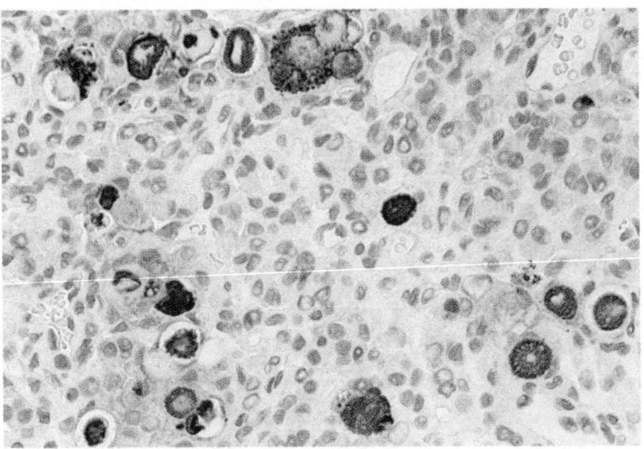

Figure 9-64

MENINGIOMA

The pseudopsammoma body is positive for carcinoembryonic antigen (CEA).

IMMUNOHISTOCHEMICAL FINDINGS

The most diagnostically useful marker of meningioma is a membranous pattern of immunoreactivity for epithelial membrane antigen (EMA) (fig. 9-62) (105–108). Positivity is often more marked in meningothelial and transitional lesions, and often subdued in fibrous, clear cell, and papillary lesions. Focal cytoplasmic staining with some membrane accentuation may be seen also in some Schwann cell tumors, an important differential diagnostic entity (108). EMA staining is often focal at best in chordoid meningiomas. In accord with the arachnoid origin, the lesions are immunoreactive for the gap junction proteins connexin 26 and 43 (109).

Diffuse immunoreactivity for vimentin, a nondiagnostic feature in view of the differential, is typical of all forms of meningioma. Somewhat over 50 percent of meningiomas are reactive for S-100 protein, but the reaction is typically patchy and less intense than it is in schwannomas.

A cytokeratin reaction is typically present in secretory meningiomas, with staining limited to cells surrounding pseudopsammoma bodies (fig. 9-63) (107). The body itself is positive for CEA (fig. 9-64) (50,108). Interestingly, these same cells are usually negative for vimentin. Many other meningiomas are reactive for cytokeratins, depending on the specific keratin polypeptide and subtype of meningioma. Most meningiomas are, for example, immunoreactive for CK18, whereas none are positive or CK20 (110). In the same study, anaplastic meningiomas were all positive for CK18, and often for other cytokeratins as well, but were negative for CK20 (110). Another study found common positivity for AE1/AE3, CAM5.2, and pankeratins in "malignant meningiomas" (111).

Although meningiomas are generally GFAP negative, there are of course exceptions, particularly in filament-rich rhabdoid lesions, where scattered cells may be positive (67,112, 113). Immunoreactivity for progesterone receptors is common in well-differentiated grade I lesions, but is less prominent as the grade

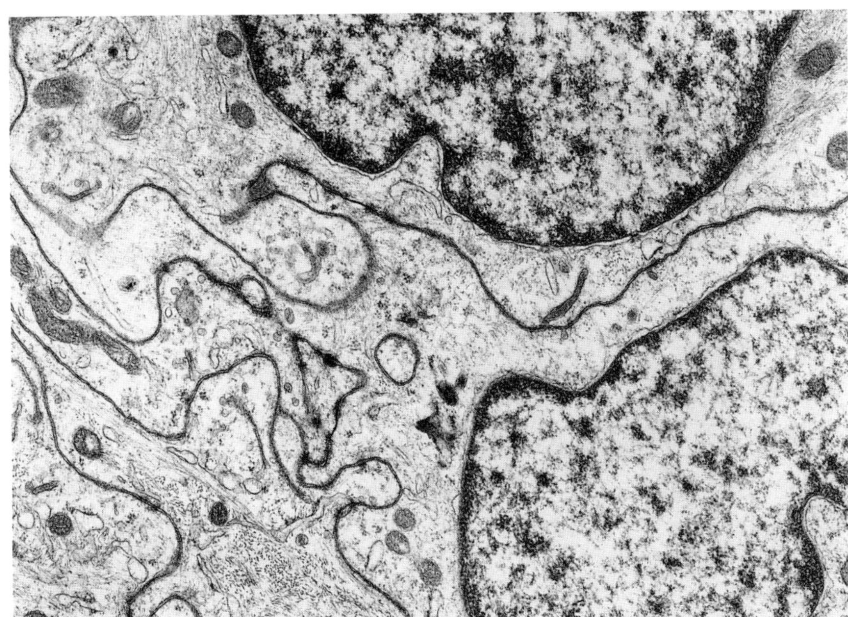

Figure 9-65

MENINGIOMA

Syncytial meningioma cells have interdigitating cell borders, well-formed desmosomes, and abundant intermediate filaments. (Fig. 9-28 from Fascicle 10, 3rd Series.)

increases (103). The results of staining with MIB-1 are discussed above.

ULTRASTRUCTURAL FINDINGS

Classic meningiomas, particularly meningotheliomatous or transitional types, are noted for interdigitating processes, cytoplasmic intermediate filaments in varying number and distribution, and well-formed desmosomes (figs. 9-65, 9-66). Desmosomes and interdigitating cytoplasmic processes are the most telling features in cases in which electron microscopy is required for diagnosis.

The cells of overtly fibrous meningiomas exhibit few intercellular junctions, since they are separated from one another by varying quantities of collagen. In addition, cytoplasmic filaments are often sparse, and a vague condensation of intercellular matrix partially invests the cells. With the addition of abundant collagen, fibrous meningiomas assume a distinctly mesenchymal appearance.

Microcystic tumors are characterized by elongated, often fibril-poor cell processes joined terminally by well-formed desmosomes (fig. 9-67). Flocculent electron-lucent material fills the intercellular spaces.

The pseudopsammoma bodies of secretory meningiomas are masses of minute vesicles and debris within an intracellular space whose luminal surface exhibits numerous short microvilli (fig. 9-68) (51,54). Multiple acini may be present within the same cell. Desmosomes connect adjacent cells. Bundles of tonofilaments surrounding the pseudopsammoma bodies explain the cytokeratin immunoreactivity.

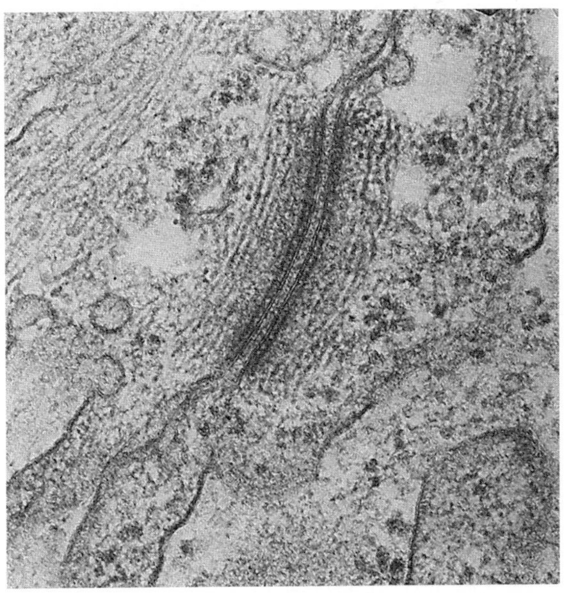

Figure 9-66

MENINGIOMA

Desmosomes are a classic feature of meningiomas and other epithelial lesions.

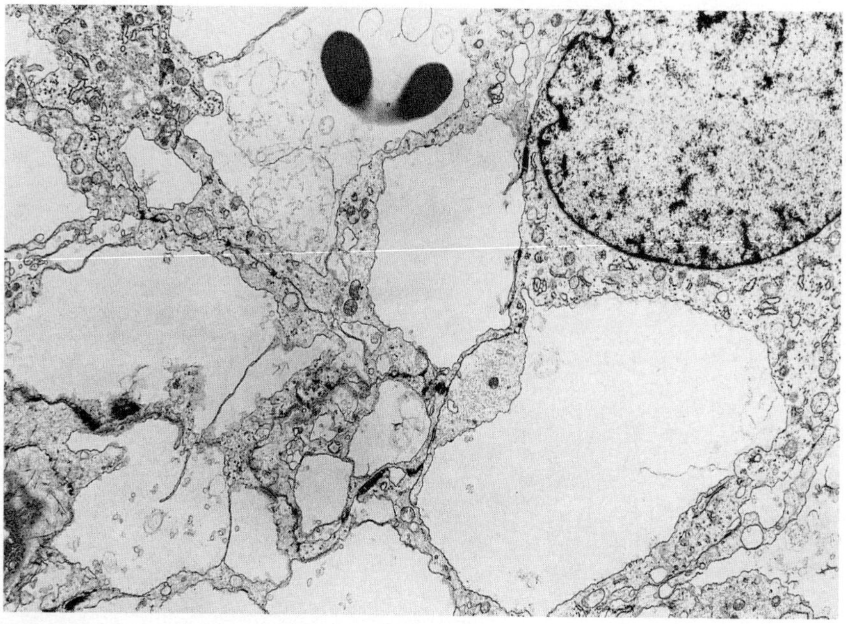

Figure 9-67

MICROCYSTIC MENINGIOMA

Desmosome-connected cytoplasmic extensions surround microcystic spaces. (Fig. 9-29 from Fascicle 10, 3rd Series.)

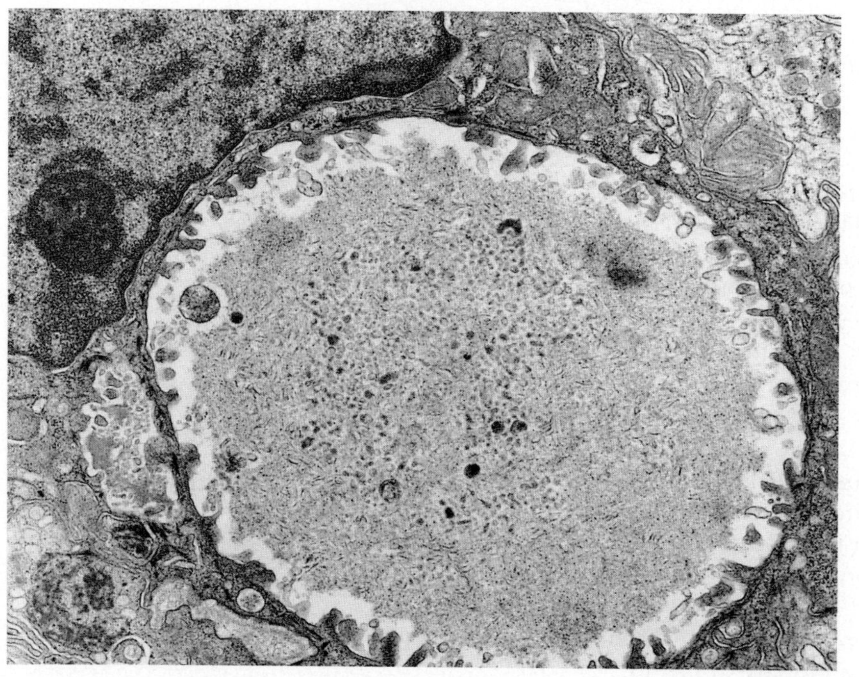

Figure 9-68

SECRETORY MENINGIOMA

The pseudopsammoma body is a microvilli-lined space filled with membranous and amorphous debris. (Fig. 9-30 from Fascicle 10, 3rd Series.)

CYTOLOGIC FINDINGS

Despite remarkable intertumoral variation in histologic features, meningiomas look remarkably similar at the cytologic level. With the exception of overtly fibrous meningiomas, neoplastic meningothelial cells have slightly oval nuclei, delicate chromatin, and solitary discrete micronucleoli (fig. 9-69). Intranuclear pseudoinclusions are readily seen. Especially characteristic is the copious, bipolar, delicate, often somewhat folded or "twisted" cytoplasm. The resemblance of these cells to squames should not be surprising given the immunohistochemical and ultrastructural epithelial phenotype of meningothelial cells. Higher-grade lesions have coarser nuclei and more prominent nucleoli (fig. 9-70). Psammoma bodies make the diagnosis

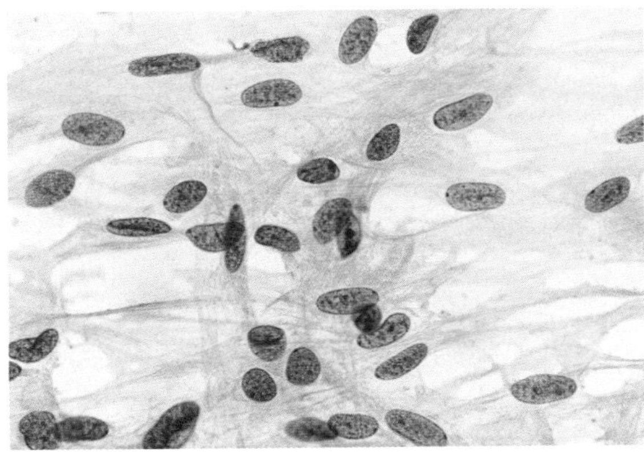

Figure 9-69

GRADE I MENINGIOMA: SMEAR PREPARATION

Almost regardless of subtype, grade I meningiomas shed cells with conspicuous cytoplasm, round delicate nuclei with small precisely defined nucleoli, and (in women) a Barr body.

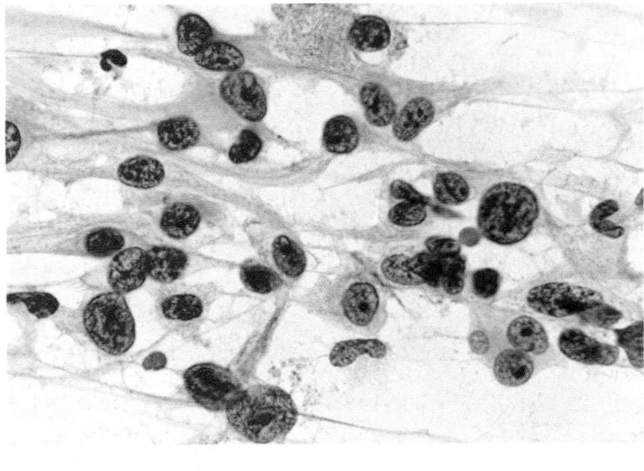

Figure 9-70

GRADE II MENINGIOMA: SMEAR PREPARATION

In comparison to a grade I lesion, nuclei of grade II meningiomas are usually coarser and have larger nucleoli.

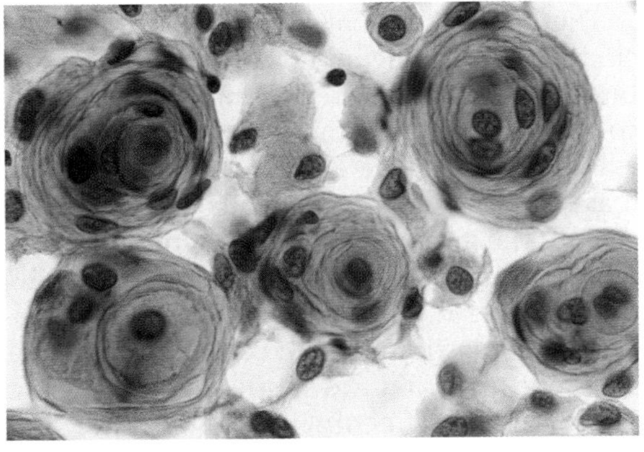

Figure 9-71

MENINGIOMA: SMEAR PREPARATION

The diagnosis is elementary in meningiomas with this degree of whorl formation.

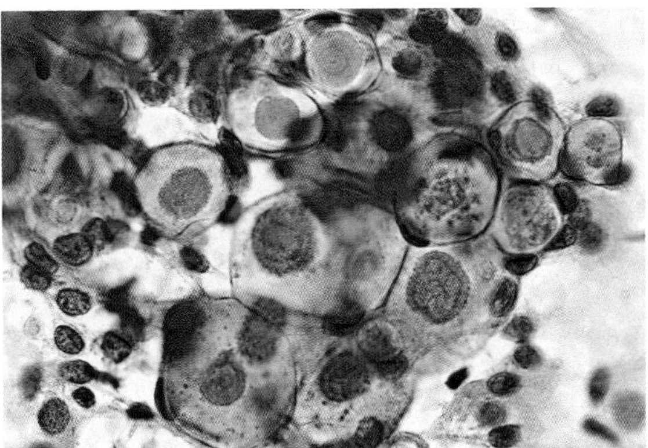

Figure 9-72

SECRETORY MENINGIOMA: SMEAR PREPARATION

Brightly eosinophilic, discrete, and intracytoplasmic pseudopsammoma bodies are seen clearly in smear preparations.

elementary (fig. 9-71). Spherical, eosinophilic, and intracytoplasmic, pseudopsammoma bodies enliven smears from secretory variants (fig. 9-72).

MOLECULAR AND CYTOGENETIC FEATURES

A critical and early genetic event responsible for NF2-associated tumors (104,114,115) and an as yet undetermined, but substantial, percentage of sporadic lesions is inactivation of the *NF2* gene on chromosome 22q12 (fig. 9-73) (104,116,117). Loss of expression of the gene product, merlin, is the result (104,118). The cytogenetic equivalent is loss, usually monosomy, of chromosome 22 (119,120). The incidence of *NF2* mutations is substantially higher in fibroblastic and transitional variants than in meningothelial types (116,117).

Beginning with most grade I lesions with a normal karyotype or monosomy 22, there is a spectrum of molecular and cytogenetic changes. While both losses and gains occur, most attention has been directed at the former. How these relate to tumor progression is not clear, but abnormalities on/of 1p, 14q (fig. 9-74), 18q, 22q19,

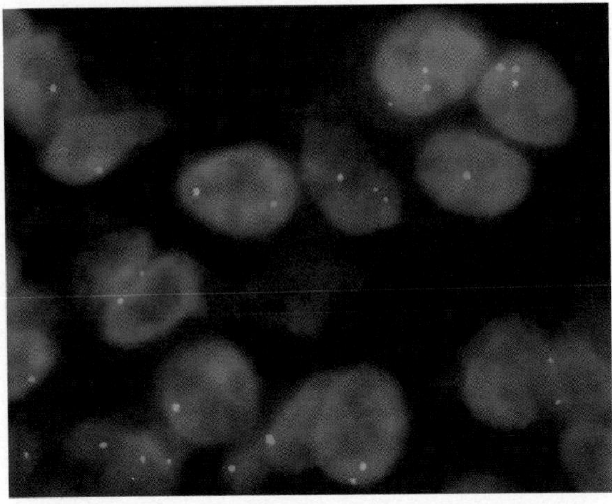

Figure 9-73

MENINGIOMA

Loss of one copy of the *NF2* gene is apparent in this fluorescence in situ hybridization study. Most cells have the normal two signals for *DAL-1* (18p11) (green), but only one for *NF2* (22q12) (red). (Courtesy of Dr. Arie Perry, St. Louis, MO.)

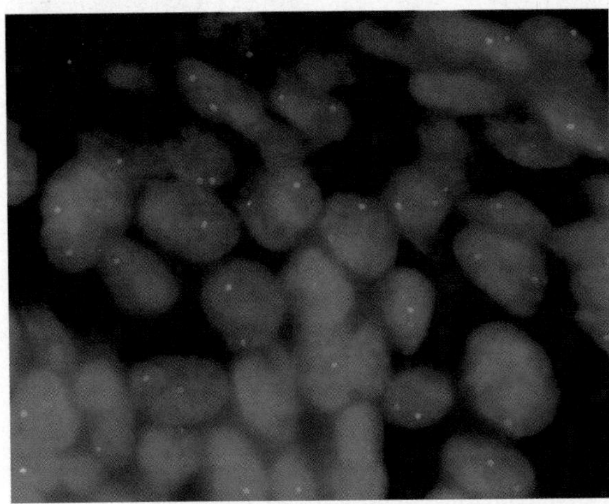

Figure 9-74

GRADE II MENINGIOMA

Loss of chromosomal arms 1p and 14q is common in atypical and anaplastic meningiomas. Probes here for 1p32 (green) and 14q32 (red) show only one signal each, instead of the normal two. (Courtesy of Dr. Arie Perry, St. Louis, MO.)

among others, may be related to progression (118–127). Candidate genes such *p16* (9p) are being evaluated (124,127,128). Abnormalities on 9p21, 6q, 10, and 17q may correlate with anaplasia (116,124,127–129).

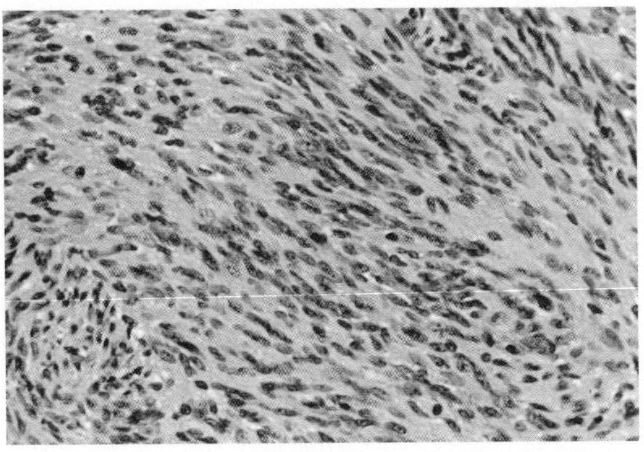

Figure 9-75

SCHWANNOMA RESEMBLING MENINGIOMA

Cellular schwannomas with a fascicular architecture can closely resemble fibrous meningioma.

DIFFERENTIAL DIAGNOSIS

The differential diagnosis is broad, and includes both non-neoplastic and neoplastic entities. Inflammatory pseudotumors such as sarcoidosis, Rosai-Dorfman disease, and plasma cell granuloma often present as a meningioma-like mass. These entities are discussed in chapter 19.

The cellular elongation of schwannoma mimics that of meningioma, particularly the fibrous type (fig. 9-75), but the long club-shaped nuclei of neoplastic Schwann cells usually can be distinguished from the shorter, more fusiform nuclei of fibrous meningioma. Smear preparations are ideally suited to resolve this common differential issue. Tissue from schwannoma largely resists dispersal as individual cells, remaining as tissue fragments even when forcefully squeezed or smeared (see figs. 13-18, 13-19). Histologically, meningiomas lack the juxtaposition of a compact fascicular pattern with spongy areas (Antoni A and B patterns). Cellular palisading, as in Verocay bodies, is often lacking in acoustic tumors, but when present is highly characteristic. Although meningiomas may be reactive for S-100 protein, they generally lack the diffuse positivity and nuclear positivity of schwannomas. Schwannomas, on the other hand, lack membranous staining for EMA, and reactivity, if any, is diffusely cytoplasmic.

The differential diagnosis of fibrous meningioma focuses on solitary fibrous tumor (fig. 9-76) and hemangiopericytoma. Both are

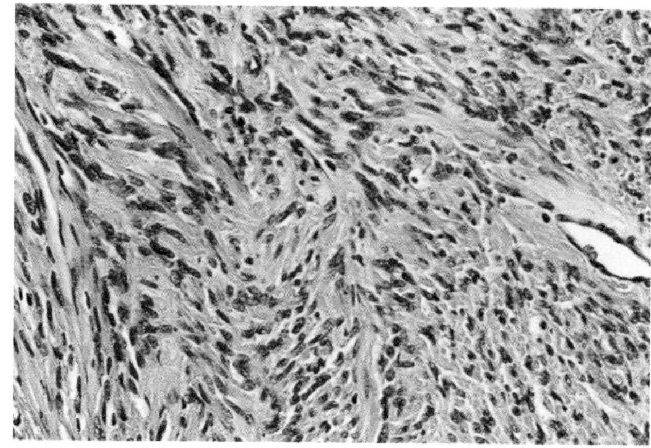

Figure 9-76

SOLITARY FIBROUS TUMOR RESEMBLING MENINGIOMA

Bands of eosinophilic collagen help distinguish solitary fibrous tumor from fibrous meningioma.

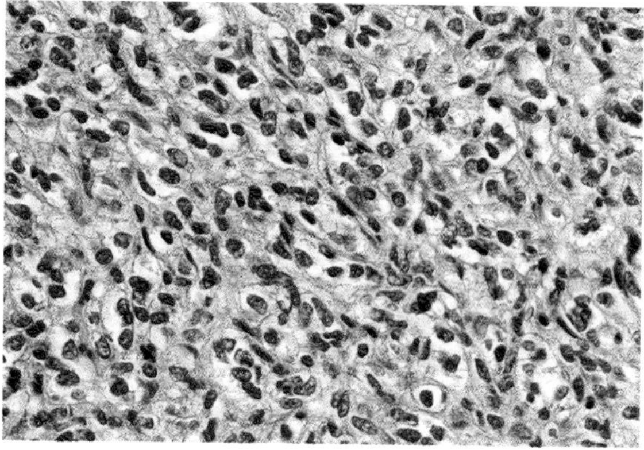

Figure 9-77

HEMANGIOPERICYTOMA RESEMBLING MENINGIOMA

Fascicular tissue of medium cellularity allows some hemangiopericytomas to resemble fibrous meningioma.

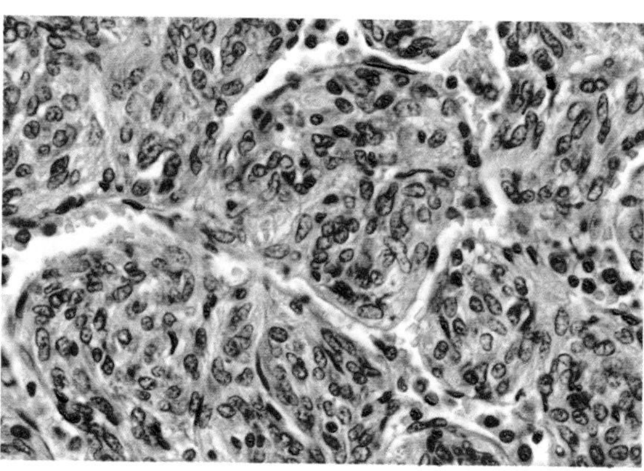

Figure 9-78

HEMANGIOPERICYTOMA RESEMBLING MENINGIOMA

Whorl-like small lobules create a likeness to meningioma.

discussed in chapter 10. Solitary fibrous tumor is recognized by its brightly eosinophilic bands of collagen, lack of whorls and psammoma bodies, and almost defining immunoreactivity for CD34, and often bcl2. It is negative for EMA.

Hemangiopericytoma (HPC) has a high cellularity, "staghorn" vascular channels, swirling pattern of oval or somewhat elongated cells, variable but often brisk mitotic activity, abundance of intercellular reticulin, and lack of reactivity for EMA. Occasional meningiomas share the cellular and staghorn pattern but are identified by their tight whorls and extensive nuclear clearing. Occasional meningeal HPCs lack classic architectural or cytologic features and can then resemble cellular or fibrous meningiomas (fig. 9-77). Other HPCs have whorls (fig. 9-78), albeit not the tight concentric versions of meningioma. The distinction is aided by histochemistry and immunohistochemistry since HPCs lack membranous staining for EMA, although positivity may be scant or absent in atypical cellular meningiomas (106,108). Abundant reticulin is common in HPCs, but is typically scant in meningiomas. Ultrastructurally, HPCs have pericellular basal lamina-like material and lack both cellular interdigitation and desmosomal attachments, although there may be scattered rudimentary junctions. Occasional cellular primary meningeal neoplasms lack specific architectural or cytologic features of either meningioma or HPC. These are usually locally aggressive and prone to recurrence.

Hemangioblastomas can arise, albeit with great rarity, in the supratentorial space and then simulate microcystic meningioma (compare figs. 7-6, 7-18, 9-18). The similarity results from a mutual content of vacuolated cells, conspicuous vascularity, and variable nuclear pleomorphism. In contrast to hemangioblastoma, most microcystic meningiomas lack a uniformly high capillary density and exhibit a membranous pattern of EMA staining. Wanting in meningiomas is the hemangioblastoma's abundant reticulin network.

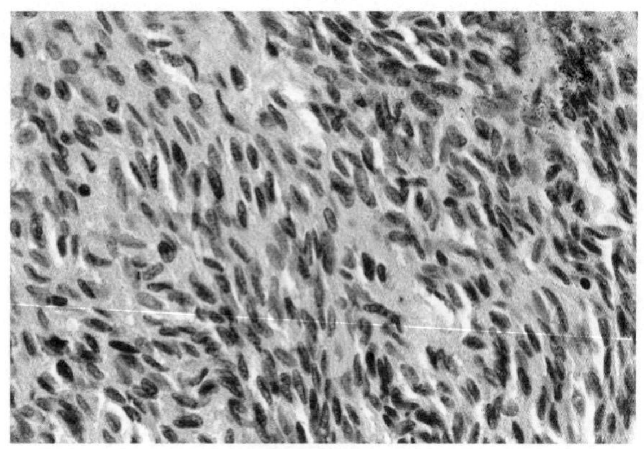

Figure 9-79

MELANOCYTOMA RESEMBLING MENINGIOMA

Cytologically bland amelanotic melanocytomas can closely simulate fibrous meningiomas. A high index of suspicion helps exclude melanocytic neoplasms of both low and high grade.

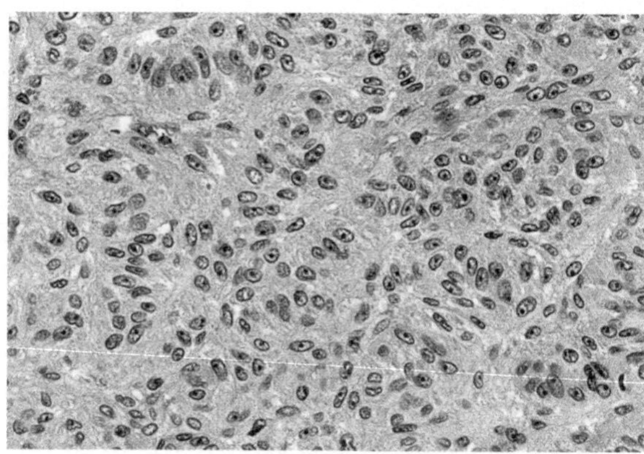

Figure 9-80

MELANOCYTOMA RESEMBLING MENINGIOMA

Cytologically neoplastic melanocytes can closely resemble meningothelial cells, although a nested architecture is more typical of melanocytoma.

Pilocytic astrocytomas resemble microcystic meningiomas, but immunohistochemistry for EMA and GFAP resolves the issue. Astroblastoma, a generally superficial lesion, has a perivascular orientation of cells that resembles that of papillary meningioma. The two are distinguished by the astroblastoma's intraparenchymal position and its GFAP positivity. The meningioma is extra-axial and EMA positive.

As is discussed in chapter 14, cytologically bland melanocytomas can closely resemble meningioma (figs. 9-79, 9-80). Immunohistochemistry may be required.

The globular configuration typical of metastatic carcinoma to the dura makes this lesion a common differential diagnostic problem. Cytologic preparations are particularly helpful, since most meningiomas exhibit classic features that quickly dispel consideration of a malignancy. Carcinoma remains a consideration only in meningiomas with advanced anaplasia. Nuclear pleomorphism and hyperchromasia alone are of limited diagnostic significance since some benign meningiomas contain bizarre atypical cells. Such tumors often lack or have only sparse mitotic activity. EMA staining will not distinguish the two, although it is usually more intense in carcinomas. Cytokeratins, while infrequently expressed in grade I meningiomas, can be quite prominent in grade III variants, although staining for CK20 appears to absent (110). Epithelial markers such as BerEP4 can be used to identify the carcinoma (111).

A diagnostic contender in the differential diagnosis of anaplastic meningioma is primary meningeal sarcoma, although most malignant meningiomas retain, at least focally, some features indicative of their meningothelial origin. Meningeal fibrosarcomas often have a distinctive "herringbone" architecture not found in meningiomas. Pericellular reticulin is a prominent feature of such lesions and is not typical of malignant meningioma. Where no meningothelial component is evident, or was known to exist, a diagnosis of sarcoma becomes tenable, but a meningothelial origin cannot always be excluded readily. Ultrastructural findings may help show an absence of meningothelial features. Immunohistochemistry may be useful since, in concept, anaplastic meningiomas are reactive for EMA. In practice, however, positivity may be sparse or absent in all but the better-differentiated examples.

TREATMENT AND PROGNOSIS

Since most meningiomas lend themselves to surgical removal, surgery is the primary therapeutic modality, and a principal prognostic factor. Resection is especially likely for lesions overlying the cerebral convexities and for those in the intraspinal compartment. Large tumors at the base of the skull, en plaque lesions, and meningiomas encompassing major vessels are obvious exceptions. Vascular invasion is a common finding; the superior sagittal sinus is most often affected (see fig. 9-5). The rare occurrence of distant metastases, usually to the lungs, is not known to correlate with vascular invasion at the primary site.

Prognostication in specific cases is imprecise and depends upon histologic parameters, tumor location, and extent of resection (92,93, 130,131). Using the grading criteria on which the current WHO system is based, and combining two large series (94,93) for a total of 643 Mayo Clinic primary meningiomas stratified by WHO criteria, the 10-year recurrence free and overall survival rates are approximately: grade I (70 and 80 percent), grade II (60 and 34 percent), and grade III (~0 and 0 percent) (Lohse CM, personal communication, 2004). Death may result from mass effects, parenchymal destruction, or bacterial meningitis acquired through defects in the skull and dura.

The tumors recur locally and sometimes disseminate intracranially. The mechanisms of such recurrence are unclear but may be due to an unsuspected field of meningioma beyond the initial mass (132). Molecular studies have been useful in establishing the clonal nature of some multiple lesions (133–135), and have found that recurrent meningiomas are clonal in regard to the initial lesion, suggesting true recurrence and not a second primary (135).

Molecular features of grade II and III meningiomas are discussed above. One study suggested that chromosome 14 status in combination with age and histologic grade provided a better prognostic stratification (136).

Well-differentiated meningiomas metastasize to distant sites such as lung (137) or bone (138). Meningiomas of the papillary (70) and rhabdoid (67) types are likely to do the same.

REFERENCES

1. Sadetzki S, Flint-Richter P, Ben-Tal T, Nass D. Radiation-induced meningioma: a descriptive study of 253 cases. J Neurosurg 2002;97:1078–82.
2. Salvati M, Caroli E, Brogna C, Orlando ER, Delfini R. High-dose radiation-induced meningiomas. Report of five cases and critical review of the literature. Tumori 2003;89:443–7.
3. Starshak RJ. Radiation-induced meningioma in children: report of two cases and review of the literature. Pediatr Radiol 1996;26:537–41.
4. Joachim T, Ram Z, Rappaport ZH, et al. Comparative analysis of the NF2, TP53, PTEN, KRAS, NRAS and HRAS genes in sporadic and radiation-induced human meningiomas. Int J Cancer 2001;94:218–21.
5. Al-Mefty O, Topsakal C, Pravdenkova S, Sawyer JR, Harrison MJ. Radiation-induced meningiomas: clinical, pathological, cytokinetic, and cytogenetic characteristics. J Neurosurg 2004;100: 1002–13.
6. Deen HG Jr, Scheithauer BW, Ebersold MJ. Clinical and pathological study of meningiomas of the first two decades of life. J Neurosurg 1982;56:317–22.
7. Ferrante L, Acqui M, Artico M, Mastronardi L, Fortuna A. Paediatric intracranial meningiomas. Br J Neurosurg 1989;3:189–96.
8. Perry A, Dehner LP. Meningeal tumors of childhood and infancy. An update and literature review. Brain Pathol 2003;13:386–408.
9. Evans DG, Birch JM, Ramsden RT. Paediatric presentation of type 2 neurofibromatosis. Arch Dis Child 1999;81:496–9.
10. Custer BS, Koepsell TD, Mueller BA. The association between breast carcinoma and meningioma in women. Cancer 2002;94:1626–35.
11. Jacobs DH, McFarlane MJ, Holmes FF. Female patients with meningioma of the sphenoid ridge and additional primary neoplasms of the breast and genital tract. Cancer 1987;60:3080–2.
12. Giangaspero F, Guiducci A, Lenz FA, Mastronardi L, Burger PC. Meningioma with meningioangiomatosis: a condition mimicking invasive meningiomas in children and young adults: report of two cases and review of the literature. Am J Surg Pathol 1999;23:872–5.
13. Sinkre P, Perry A, Cai D, et al. Deletion of the NF2 region in both meningioma and juxtaposed meningioangiomatosis: case report supporting a neoplastic relationship. Pediatr Dev Pathol 2001;4:568–72.

14. Nakamura M, Roser F, Michel J, Jacobs C, Samii M. The natural history of incidental meningiomas. Neurosurgery 2003;53:62–70; discussion 70–1.
15. Bowen JH, Burger PC, Odom GL, Dubois PJ, Blue JM. Meningiomas associated with large cysts with neoplastic cells in the cysts walls. Report of two cases. J Neurosurg 1981;55:473–8.
16. Odake G. Cystic meningioma: report of three patients. Neurosurgery 1992;30:935–40.
17. Elster AD, Challa VR, Gilbert TH, Richardson DN, Contento JC. Meningiomas: MR and histopathologic features. Radiology 1989;170(Pt 1):857–62.
18. Mahmood A, Caccamo DV, Tomecek FJ, Malik GM. Atypical and malignant meningiomas: a clinicopathological review. Neurosurgery 1993; 33:955–63.
19. Tamiya T, Ono Y, Matsumoto K, Ohmoto T. Peritumoral brain edema in intracranial meningiomas: effects of radiological and histological factors. Neurosurgery 2001;49:1046–51; discussion 1051–2.
20. Ide M, Jimbo M, Yamamoto M, Umebara Y, Hagiwara S, Kubo O. MIB-1 staining index and peritumoral brain edema of meningiomas. Cancer 1996;78:133–43.
21. Renfro M, Delashaw JB, Peters K, Rhoton E. Anterior third ventricle meningioma in an adolescent: a case report. Neurosurgery 1992;31: 746–50; discussion 750.
22. Fornari M, Savoiardo M, Morello G, Solero CL. Meningiomas of the lateral ventricles. Neuroradiological and surgical considerations in 18 cases. J Neurosurg 1981;54:64–74.
23. Nakamura M, Roser F, Bundschuh O, Vorkapic P, Samii M. Intraventricular meningiomas: a review of 16 cases with reference to the literature. Surg Neurol 2003;59:491–503; discussion 503–4.
24. Carlotti CG Jr, Neder L, Colli BO, et al. Clear cell meningioma of the fourth ventricle. Am J Surg Pathol 2003;27:131–5.
25. Perry RD, Parker GD, Hallinan JM. CT and MR imaging of fourth ventricular meningiomas. J Comput Assist Tomogr 1990;14:276–80.
26. Konovalov AN, Spallone A, Pitzkhelauri DI. Meningioma of the pineal region: a surgical series of 10 cases. J Neurosurg 1996;85:586–90.
27. Clark WC, Theofilos CS, Fleming JC. Primary optic nerve sheath meningiomas. Report of nine cases. J Neurosurg 1989;70:37–40.
28. Karp LA, Zimmerman LE, Borit A, Spencer W. Primary intraorbital meningiomas. Arch Ophthalmol 1974;91:24–8.
29. Liu JK, Forman S, Hershewe GL, Moorthy CR, Benzil DL. Optic nerve sheath meningiomas: visual improvement after stereotactic radiotherapy. Neurosurgery 2002;50:950–5; discussion 955–7.
30. Marquardt MD, Zimmerman LE. Histopathology of meningiomas and gliomas of the optic nerve. Hum Pathol 1982;13:226–35.
31. Saeed P, Rootman J, Nugent RA, White VA, Mackenzie IR, Koornneef L. Optic nerve sheath meningiomas. Ophthalmology 2003;110:2019–30.
32. Salvati M, Artico M, Lunardi P, Gagliardi FM. Intramedullary meningioma: case report and review of the literature. Surg Neurol 1992;37:42–5.
33. Lee WH, Tu YC, Liu MY. Primary intraosseous malignant meningioma of the skull: case report. Neurosurgery 1988;23:505–8.
34. Oka K, Hirakawa K, Yoshida S, Tomonaga M. Primary calvarial meningiomas. Surg Neurol 1989;32:304–10.
35. Thompson LD, Bouffard JP, Sandberg GD, Mena H. Primary ear and temporal bone meningiomas: a clinicopathologic study of 36 cases with a review of the literature. Mod Pathol 2003;16: 236–45.
36. Lopez DA, Silvers DN, Helwig EB. Cutaneous meningiomas—a clinicopathologic study. Cancer 1974;34:728–44.
37. Chumas JC, Lorelle CA. Pulmonary meningioma. A light- and electron-microscopic study. Am J Surg Pathol 1982;6:795–801.
38. Kemnitz P, Spormann H, Heinrich P. Meningioma of lung: first report with light and electron microscopic findings. Ultrastruct Pathol 1982;3:359–65.
39. Kodama K, Doi O, Higashiyama M, Horai T, Tateishi R, Nakagawa H. Primary and metastatic pulmonary meningioma. Cancer 1991;67:1412–7.
40. Robinson PG. Pulmonary meningioma. Report of a case with electron microscopic and immunohistochemical findings. Am J Clin Pathol 1992;97:814–7.
41. Thompson LD, Gyure KA. Extracranial sinonasal tract meningiomas: a clinicopathologic study of 30 cases with a review of the literature. Am J Surg Pathol 2000;24:640–50.
42. Wilson AJ, Ratliff JL, Lagios MD, Aguilar MJ. Mediastinal meningioma. Am J Surg Pathol 1979;3:557–62.
43. Coons SW, Johnson PC. Brachial plexus meningioma, report of a case with immunohistochemical and ultrastructural examination. Acta Neuropathol (Berl) 1989;77:445–8.
44. Gaffey MJ, Mills SE, Askin FB. Minute pulmonary meningothelial-like nodules. A clinicopathologic study of so-called minute pulmonary chemodectoma. Am J Surg Pathol 1988;12:167–75.

45. Ionescu D, Sasatomi E, Aldeeb D, et al. Pulmonary meningothelial-like nodules. A genotypic comparison with meningiomas. Am J Surg Pathol 2004;28:207–14.
46. Louis D, Scheithauer B, Budka H, von Deimling A, Kepes J. Meningiomas. In: Kleihues P, Cavenee W, eds. Pathology and genetics of tumours of the central nervous system. Lyon: IARC Press; 2000:176–84.
47. Kleinman GM, Liszczak T, Tarlov E, Richardson EP Jr. Microcystic variant of meningioma: a light-microscopic and ultrastructural study. Am J Surg Pathol 1980;4:383–9.
48. Michaud J, Gagne F. Microcystic meningioma. Clinicopathologic report of eight cases. Arch Pathol Lab Med 1983;107:75–80.
49. Ng HK, Poon WS, Goh K, Chan MS. Histopathology of post-embolized meningiomas. Am J Surg Pathol 1996;20:1224–30.
50. Alguacil-Garcia A, Pettigrew NM, Sima AA. Secretory meningioma. A distinct subtype of meningioma. Am J Surg Pathol 1986;10:102–11.
51. Budka H. Hyaline inclusions (pseudopsammoma bodies) in meningiomas: immunocytochemical demonstration of epithelial-like secretion of secretory component and immunoglobulins A and M. Acta Neuropathol (Berl) 1982; 56:294–8.
52. Buhl R, Hugo HH, Mihajlovic Z, Mehdorn HM. Secretory meningiomas: clinical and immunohistochemical observations. Neurosurgery 2001;48:297–301; discussion 301–2.
53. Font RL, Croxatto JO. Intracellular inclusions in meningothelial meningioma. A histochemical and ultrastructural study. J Neuropathol Exp Neurol 1980;39:575–83.
54. Kepes JJ. The fine structure of hyaline inclusions (pseudopsammoma bodies) in meningiomas. J Neuropathol Exp Neurol 1975;34:282–94.
55. Louis DN, Hamilton AJ, Sobel RA, Ojemann RG. Pseudopsammomatous meningioma with elevated serum carcinoembryonic antigen: a true secretory meningioma. Case report. J Neurosurg 1991;74:129–32.
56. Probst-Cousin S, Villagran-Lillo R, Lahl R, Bergmann M, Schmid KW, Gullotta F. Secretory meningioma: clinical, histologic, and immunohistochemical findings in 31 cases. Cancer 1997;79:2003–15.
57. Gi H, Nagao S, Yoshizumi H, et al. Meningioma with hypergammaglobulinemia. Case report. J Neurosurg 1990;73:628-9.
58. Horten BC, Urich H, Stefoski D. Meningiomas with conspicuous plasma cell-lymphocytic components: a report of five cases. Cancer 1979;43:258–64.
59. Hasselblatt M, Nolte K, Paulus W. Angiomatous meningioma. A clinicopathologic study of 38 cases. Am J Surg Pathol 2004;28:390–3.
60. Couce ME, Aker FV, Scheithauer BW. Chordoid meningioma: a clinicopathologic study of 42 cases. Am J Surg Pathol 2000;24:899–905.
61. Kepes JJ, Chen WY, Connors MH, Vogel FS. "Chordoid" meningeal tumors in young individuals with peritumoral lymphoplasmacellular infiltrates causing systemic manifestations of the Castleman syndrome. A report of seven cases. Cancer 1988;62:391–406.
61a. Perry A, Louis DN, Scheithauer BW, Budka H, von Deimling A. Meningiomas. In: Louis DN, Ohgaki H, Wiestler OD, Cavenee W, eds. WHO classification of tumours of the central nervous system Lyon: IARC Press; 2007. (In press)
62. Heth JA, Kirby P, Menezes AH. Intraspinal familial clear cell meningioma in a mother and child. Case report. J Neurosurg 2000;93(Suppl 2):317–21.
63. Jallo GI, Kothbauer KF, Silvera VM, Epstein FJ. Intraspinal clear cell meningioma: diagnosis and management: report of two cases. Neurosurgery 2001;48:218–21; discussion 221–2.
64. Zorludemir S, Scheithauer BW, Hirose T, Van Houten C, Miller G, Meyer FB. Clear cell meningioma. A clinicopathologic study of a potentially aggressive variant of meningioma. Am J Surg Pathol 1995;19:493–505.
65. Bannykh SI, Perry A, Powell HC, Hill A, Hansen LA. Malignant rhabdoid meningioma arising in the setting of preexisting ganglioglioma: a diagnosis supported by fluorescence in situ hybridization. Case report. J Neurosurg 2002;97: 1450–5.
66. Kepes JJ, Moral LA, Wilkinson SB, Abdullah A, Llena JF. Rhabdoid transformation of tumor cells in meningiomas: a histologic indication of increased proliferative activity: report of four cases. Am J Surg Pathol 1998;22:231–8.
67. Perry A, Scheithauer BW, Stafford SL, Abell-Aleff PC, Meyer FB. "Rhabdoid" meningioma: an aggressive variant. Am J Surg Pathol 1998;22: 1482–90.
68. Parwani AV, Mikolaenko I, Eberhart CG, Burger PC, Rosenthal DL, Ali SZ. Rhabdoid meningioma: cytopathologic findings in cerebrospinal fluid. Diagn Cytopathol 2003;29:297–9.
69. Ludwin SK, Rubinstein LJ, Russell DS. Papillary meningioma: a malignant variant of meningioma. Cancer 1975;36:1363–73.
70. Pasquier B, Gasnier F, Pasquier D, Keddari E, Morens A, Couderc P. Papillary meningioma. Clinicopathologic study of seven cases and review of the literature. Cancer 1986;58:299–305.

71. Piatt JH Jr, Campbell GA, Oakes WJ. Papillary meningioma involving the oculomotor nerve in an infant. Case report. J Neurosurg 1986;64:808–12.
72. Roncaroli F, Scheithauer BW, Laeng RH, Cenacchi G, Abell-Aleff P, Moschopulos M. Lipomatous meningioma: a clinicopathologic study of 18 cases with special reference to the issue of metaplasia. Am J Surg Pathol 2001;25: 769–75.
73. Begin LR. Myxoid meningioma. Ultrastruct Pathol 1990;14:367–74.
74. Harrison JD, Rose PE. Myxoid meningioma: histochemistry and electron microscopy. Acta Neuropathol (Berl) 1985;68:80–2.
75. Roncaroli F, Riccioni L, Cerati M, et al. Oncocytic meningioma. Am J Surg Pathol 1997;21:375–82.
76. Mirra SS, Miles ML. Unusual pericytic proliferation in a meningotheliomatous meningioma: an ultrastructural study. Am J Surg Pathol 1982;6:573–80.
77. Robinson JC, Challa VR, Jones DS, Kelly DL Jr. Pericytosis and edema generation: a unique clinicopathological variant of meningioma. Neurosurgery 1996;39:700–6; discussion 706–7.
78. Nagashima C, Nakashio K, Fujino T. Meningioma and astrocytoma adjacent in the brain. J Neurosurg 1963;20:995–9.
79. Sackett JF, Stenwig JT, Songsirikul. Meningeal and glial tumors in combination. Neuroradiology 1974;7:153–60.
80. Popovic EA, Lyons MK, Scheithauer BW, Marsh WR. Mast cell-rich convexity meningioma presenting as chronic subdural hematoma: case report and review of the literature. Surg Neurol 1994;42:8–13.
81. Couce ME, Perry A, Webb P, Kepes JJ, Scheithauer BW. Fibrous meningioma with tyrosine-rich crystals. Ultrastruct Pathol 1999;23:341–5.
82. Nestor SL, Perry A, Kurtkaya O, et al. Melanocytic colonization of a meningothelial meningioma: histopathological and ultrastructural findings with immunohistochemical and genetic correlation: case report. Neurosurgery 2003;53:211–4; discussion 214–5.
83. Cserni G, Bori R, Huszka E, Kiss AC. Metastasis of pulmonary adenocarcinoma in right sylvian secretory meningioma. Br J Neurosurg 2002; 16:66–8.
84. Doron Y, Gruszkiewicz J. Metastasis of invasive carcinoma of the breast to an extradural meningioma of the cranial vault. Cancer 1987;60: 1081–4.
85. Watanabe T, Fujisawa H, Hasegawa M, et al. Metastasis of breast cancer to intracranial meningioma: case report. Am J Clin Oncol 2002;25: 414–7.
86. Zon LI, Johns WD, Stomper PC, et al. Breast carcinoma metastatic to a meningioma. Case report and review of the literature. Arch Intern Med 1989;149:959–62.
87. Sonet A, Hustin J, De Coene B, et al. Unusual growth within a meningioma (leukemic infiltrate). Am J Surg Pathol 2001;25:127–30.
88. Kepes JJ, Goldware S, Leoni R. Meningioma with pseudoglandular pattern. A case report. J Neuropathol Exp Neurol 1983;42:61–8.
89. de la Monte SM, Flickinger J, Linggood RM. Histopathologic features predicting recurrence of meningiomas following subtotal resection. Am J Surg Pathol 1986;10:836–43.
90. Jaaskelainen J, Haltia M, Laasonen E, Wahlstrom T, Valtonen S. The growth rate of intracranial meningiomas and its relation to histology. An analysis of 43 patients. Surg Neurol 1985;24:165–72.
91. Jaaskelainen J, Haltia M, Servo A. Atypical and anaplastic meningiomas: radiology, surgery, radiotherapy, and outcome. Surg Neurol 1986;25: 233–42.
92. Jaaskelainen J. Seemingly complete removal of histologically benign intracranial meningioma: late recurrence rate and factors predicting recurrence in 657 patients. A multivariate analysis. Surg Neurol 1986;26:461–9.
93. Perry A, Scheithauer BW, Stafford SL, Lohse CM, Wollan PC. "Malignancy" in meningiomas: a clinicopathologic study of 116 patients, with grading implications. Cancer 1999;85:2046–56.
94. Perry A, Stafford SL, Scheithauer BW, Suman VJ, Lohse CM. Meningioma grading: an analysis of histologic parameters. Am J Surg Pathol 1997;21:1455–65.
95. Perry A, Chicoine MR, Filiput E, Miller JP, Cross DT. Clinicopathologic assessment and grading of embolized meningiomas: a correlative study of 64 patients. Cancer 2001;92:701–11.
96. Nakasu S, Li DH, Okabe H, Nakajima M, Matsuda M. Significance of MIB-1 staining indices in meningiomas: comparison of two counting methods. Am J Surg Pathol 2001;25:472–8.
97. Ohta M, Iwaki T, Kitamoto T, Takeshita I, Tateishi J, Fukui M. MIB1 staining index and scoring of histologic features in meningioma. Indicators for the prediction of biologic potential and postoperative management. Cancer 1994;74:3176–89.
98. Perry A, Stafford SL, Scheithauer BW, Suman VJ, Lohse CM. The prognostic significance of MIB-1, p53, and DNA flow cytometry in completely resected primary meningiomas. Cancer 1998;82:2262–9.

99. Abramovich CM, Prayson RA. Histopathologic features and MIB-1 labeling indices in recurrent and nonrecurrent meningiomas. Arch Pathol Lab Med 1999;123:793–800.
100. Hsu DW, Efird JT, Hedley-Whyte ET. MIB-1 (Ki-67) index and transforming growth factor-alpha (TGF alpha) immunoreactivity are significant prognostic predictors for meningiomas. Neuropathol Appl Neurobiol 1998;24:441–52.
101. Paulus W, Meixensberger J, Hofmann E, Roggendorf W. Effect of embolisation of meningioma on Ki-67 proliferation index. J Clin Pathol 1993;46:876–7.
102. Ferracini R, Poggi S, Frank G, et al. Meningeal sarcoma with rhabdomyoblastic differentiation: case report. Neurosurgery 1992;30:782–5.
103. Hsu DW, Efird JT, Hedley-Whyte ET. Progesterone and estrogen receptors in meningiomas: prognostic considerations. J Neurosurg 1997;86:113–20.
104. Perry A, Cai DX, Scheithauer BW, et al. Merlin, DAL-1, and progesterone receptor expression in clinicopathologic subsets of meningioma: a correlative immunohistochemical study of 175 cases. J Neuropathol Exp Neurol 2000;59:872–9.
105. Artlich A, Schmidt D. Immunohistochemical profile of meningiomas and their histological subtypes. Hum Pathol 1990;21:843–9.
106. Perry A, Scheithauer BW, Nascimento AG. The immunophenotypic spectrum of meningeal hemangiopericytoma: a comparison with fibrous meningioma and solitary fibrous tumor of meninges. Am J Surg Pathol 1997;21:1354–60.
107. Theaker JM, Gatter KC, Esiri MM, Fleming KA. Epithelial membrane antigen and cytokeratin expression by meningiomas: an immunohistological study. J Clin Pathol 1986;39:435–9.
108. Winek RR, Scheithauer BW, Wick MR. Meningioma, meningeal hemangiopericytoma (angioblastic meningioma), peripheral hemangiopericytoma, and acoustic schwannoma. A comparative immunohistochemical study. Am J Surg Pathol 1989;13:251–61.
109. Arishima H, Sato K, Kubota T. Immunohistochemical and ultrastructural study of gap junction proteins connexin 26 and 43 in human arachnoid villi and meningeal tumors. J Neuropathol Exp Neurol 2002;61:1048–55.
110. Miettinen M, Paetau A. Mapping of the keratin polypeptides in meningiomas of different types: an immunohistochemical analysis of 463 cases. Hum Pathol 2002;33:590–8.
111. Liu Y, Sturgis CD, Bunker M, et al. Expression of cytokeratin by malignant meningiomas: diagnostic pitfall of cytokeratin to separate malignant meningiomas from metastatic carcinoma. Mod Pathol 2004;17:1129–33.
112. Hojo H, Abe M. Rhabdoid papillary meningioma. Am J Surg Pathol 2001;25:964–9.
113. Su M, Ono K, Tanaka R, Takahashi H. An unusual meningioma variant with glial fibrillary acidic protein expression. Acta Neuropathol (Berl) 1997;94:499–503.
114. Lamszus K, Vahldiek F, Mautner VF, et al. Allelic losses in neurofibromatosis 2-associated meningiomas. J Neuropathol Exp Neurol 2000;59:504–12.
115. Lamszus K, Lachenmayer L, Heinemann U, et al. Molecular genetic alterations on chromosomes 11 and 22 in ependymomas. Int J Cancer 2001;91:803–8.
116. Lamzus K. Meningioma pathology, genetics, and biology. J Neuropathol Exp Neurol 2004; 63:275–86.
117. Perry A, Gutmann DH, Reifenberger G. Molecular pathogenesis of meningiomas. J Neurooncol 2004;70:183–202.
118. Perry A, Gutmann DH, Reifenberger G. Molecular pathogenesis of meningiomas. J Neurooncol 2004;70:183–202.
119. Lekanne Deprez RH, Riegman PH, van Drunen E, et al. Cytogenetic, molecular genetic and pathological analyses in 126 meningiomas. J Neuropathol Exp Neurol 1995;54:224–35.
120. Zang KD. Meningioma: a cytogenetic model of a complex benign human tumor, including data on 394 karyotyped cases. Cytogenet Cell Genet 2001;93:207–20.
121. Cai DX, Banerjee R, Scheithauer BW, Lohse CM, Kleinschmidt-Demasters BK, Perry A. Chromosome 1p and 14q FISH analysis in clinicopathologic subsets of meningioma: diagnostic and prognostic implications. J Neuropathol Exp Neurol 2001;60:628–36.
122. Ketter R, Henn W, Niedermayer I, et al. Predictive value of progression-associated chromosomal aberrations for the prognosis of meningiomas: a retrospective study of 198 cases. J Neurosurg 2001;95:601–7.
123. Menon AG, Rutter JL, von Sattel JP, et al. Frequent loss of chromosome 14 in atypical and malignant meningioma: identification of a putative 'tumor progression' locus. Oncogene 1997;14:611–6.
124. Perry A, Banerjee R, Lohse CM, Kleinschmidt-DeMasters BK, Scheithauer BW. A role for chromosome 9p21 deletions in the malignant progression of meningiomas and the prognosis of anaplastic meningiomas. Brain Pathol 2002; 12:183–90.

125. Simon M, von Deimling A, Larson JJ, et al. Allelic losses on chromosomes 14, 10, and 1 in atypical and malignant meningiomas: a genetic model of meningioma progression. Cancer Res 1995;55:4696–701.
126. Vagner-Capodano AM, Grisoli F, Gambarelli D, Sedan R, Pellet W, De Victor B. Correlation between cytogenetic and histopathological findings in 75 human meningiomas. Neurosurgery 1993;32:892–900; discussion 900.
127. Weber RG, Bostrom J, Wolter M, et al. Analysis of genomic alterations in benign, atypical, and anaplastic meningiomas: toward a genetic model of meningioma progression. Proc Natl Acad Sci U S A 1997;94:14719–24.
128. Bostrom J, Meyer-Puttlitz B, Wolter M, et al. Alterations of the tumor suppressor genes CDKN2A (p16(INK4a)), p14(ARF), CDKN2B (p15(INK4b)), and CDKN2C (p18(INK4c)) in atypical and anaplastic meningiomas. Am J Pathol 2001;159:661–9.
129. Buschges R, Ichimura K, Weber RG, Reifenberger G, Collins VP. Allelic gain and amplification on the long arm of chromosome 17 in anaplastic meningiomas. Brain Pathol 2002; 12:145–53.
130. Kallio M, Sankila R, Hakulinen T, Jaaskelainen J. Factors affecting operative and excess long-term mortality in 935 patients with intracranial meningioma. Neurosurgery 1992;31:2–12.
131. Mirimanoff RO, Dosoretz DE, Linggood RM, Ojemann RG, Martuza RL. Meningioma: analysis of recurrence and progression following neurosurgical resection. J Neurosurg 1985;62:18–24.
132. Borovich B, Doron Y, Braun J, et al. Recurrence of intracranial meningiomas: the role played by regional multicentricity. Part 2: Clinical and radiological aspects. J Neurosurg 1986;65:168–71.
133. Stangl AP, Wellenreuther R, Lenartz D, et al. Clonality of multiple meningiomas. J Neurosurg 1997;86:853–8.
134. von Deimling A, Kraus JA, Stangl AP, et al. Evidence for subarachnoid spread in the development of multiple meningiomas. Brain Pathol 1995;5:11–4.
135. von Deimling A, Larson J, Wellenreuther R, et al. Clonal origin of recurrent meningiomas. Brain Pathol 1999;9:645–50.
136. Maillo A, Orfao A, Sayagues JM, et al. New classification scheme for the prognostic stratification of meningioma on the basis of chromosome 14 abnormalities, patient age, and tumor histopathology. J Clin Oncol 2003;21:3285–95.
137. Som PM, Sacher M, Strenger SW, Biller HF, Malis LI. "Benign" metastasizing meningiomas. AJNR Am J Neuroradiol 1987;8:127–30.
138. Lee YY, Wen-Wei Hsu R, Huang TJ, Hsueh S, Wang JY. Metastatic meningioma in the sacrum: a case report. Spine 2002;27:E100–3.

10 TUMORS AND TUMOR-LIKE LESIONS OF MESENCHYMAL TISSUE

VASCULAR MALFORMATIONS

Arteriovenous Malformation

Arteriovenous malformations (AVMs) are usually large, threatening, medusa-like lesions with a potential for imminent rupture (figs. 10-1, 10-2). Most arise over the cerebral convexity in the distribution of the middle cerebral artery, and others lie deep within the brain, sometimes with an effluent that engorges and dilates the deep venous, or galenic, system. In children, such arteriovenous shunting may enlarge the great vein of Galen (aneurysm of the great vein of Galen) and even produce cardiac decompensation. In adults, AVMs most often produce seizures, focal neurologic deficits, progressive signs of increased intracranial pressure, or sudden and catastrophic consequences of acute hemorrhage (fig. 10-1). One long-term follow-up study of unoperated AVMs suggested a 4 percent annual rate of major hemorrhage (1).

Spinal arteriovenous lesions are either an intraparenchymal malformation morphologically identical to those that occur in the brain, or a dural or extradural arteriovenous fistula whose draining veins become distended serpentine channels on the surface of the cord (fig. 10-3). Treatment of the latter lesion is directed at the offending dural fistula rather than the enlarged, but otherwise innocent, leptomeningeal veins (2). Retrograde pressure from such lesions may produce vascular enlargement and thickening that simulate an intramedullary vascular malformation. Resultant gliosis and parenchymal destruction may be responsible for some cases of Foix-Alajouanine syndrome (3–6). Dura-based arteriovenous fistulas also occur intracranially.

Microscopically, AVMs are composed of vessels that vary greatly in type (arteries, veins, and transitional vessels) and in caliber (figs. 10-4, 10-5) (7). Those at the lesion's core, or "nidus," are often more irregular in size and cross-sectional configuration and less sclerotic than those of cavernous angioma. Muscularization is variable, ranging from the delicate lamina of myocytes in large veins to prominent, often irregular layers in abnormal arteries. Variation in thickness of the media is seen well on trichrome preparations as well as with immunostains for

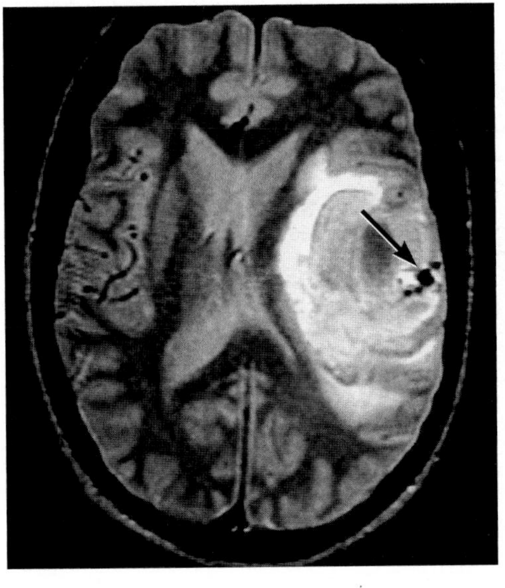

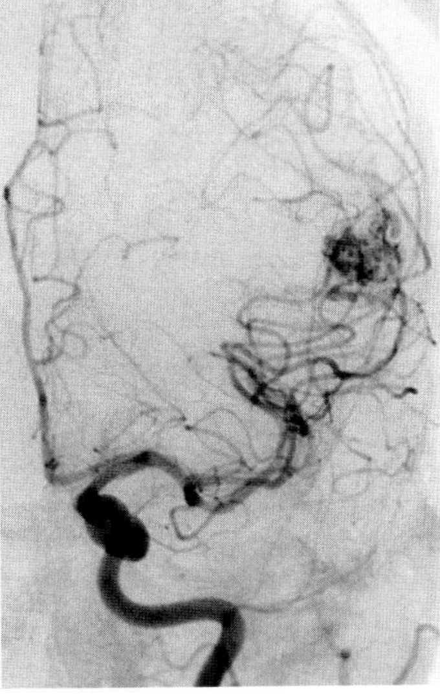

Figure 10-1

ARTERIOVENOUS MALFORMATION

Massive hemorrhage is a risk of arteriovenous malformations (AVMs), as in this 51-year-old man who presented acutely with aphasia and hemiparesis. Dark flow voids superficial to the hematoma (left, arrow) suggested an AVM, a presumption confirmed by angiography (right).

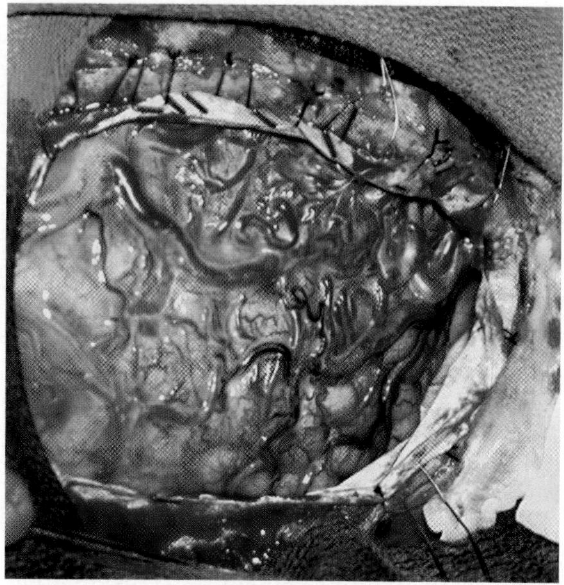

Figure 10-2

ARTERIOVENOUS MALFORMATION

The malformative nature of the lesion is evident at the time of surgery. Individual, large draining veins conduct dark venous blood, red arterial blood, or, simultaneously, laminar streams of both.

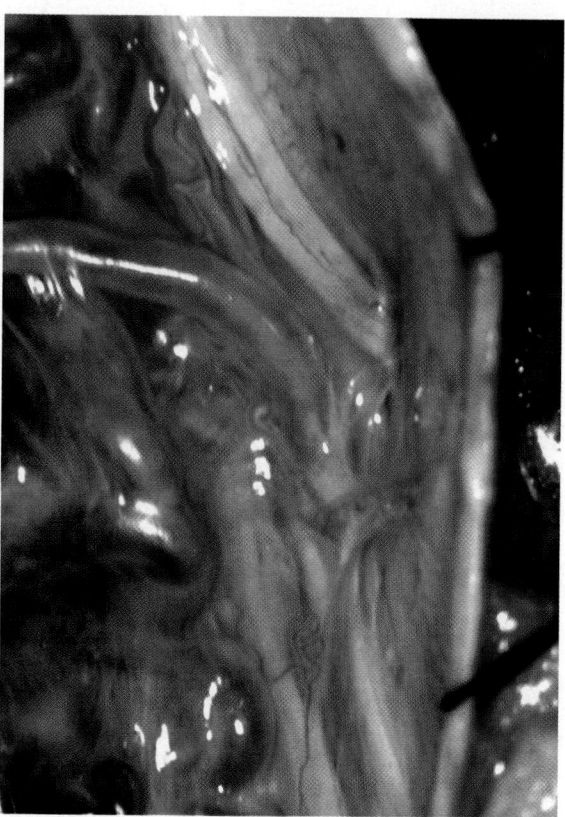

Figure 10-3

ARTERIOVENOUS MALFORMATION

The arteriovenous shunt in some spinal AVMs lies within the dura (right). The intradural component, on the left, is composed only of large draining veins. (Courtesy of Dr. Allan Friedman, Durham, NC.)

actin (8). Also distinctive are large "cushions" of smooth muscle that project into the lumens of abnormal vessels. A well-formed elastic membrane is seen in the arterial component.

Although the most central portion of an AVM may exclude preexisting brain parenchyma, peripheral feeding and draining vessels are often separated by ferruginated and gliotic central nervous system (CNS) parenchyma. Crowding of oligodendroglial cells in some cases is due to either collapse of the white matter from myelin loss or the inclusion of malformative-appearing nodules of white matter within the lesion (9,10).

In an effort to reduce vascularity and facilitate surgical resection, AVMs may be embolized preoperatively and synthetic material may then fill lumens (fig. 10-6). Depending upon the nature of the substance and the postembolization interval, a foreign body response may be elicited (11,12).

Assisted by angiographic and intraoperative data, a diagnosis of vascular malformation is usually obvious in histologic sections. Nonetheless, there are pitfalls. For example, normal vessels of the subarachnoid space may be compacted in

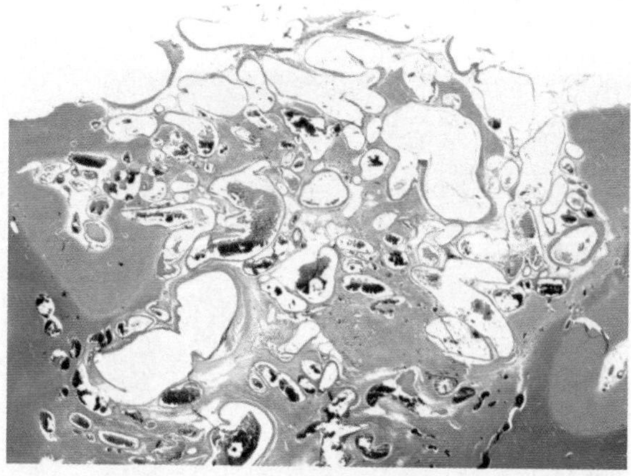

Figure 10-4

ARTERIOVENOUS MALFORMATION

AVMs consist of a jumble of small and large vessels, typically with considerable intervening parenchyma (hematoxylin and eosin [H&E]/Luxol fast blue stain).

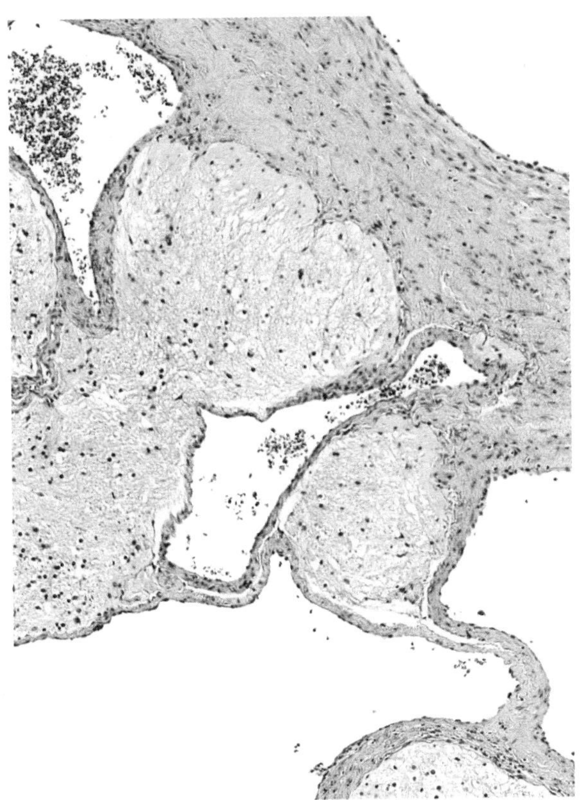

Figure 10-5

ARTERIOVENOUS MALFORMATION

Vessels of an AVM vary in caliber and thickness. Note the intervening parenchyma.

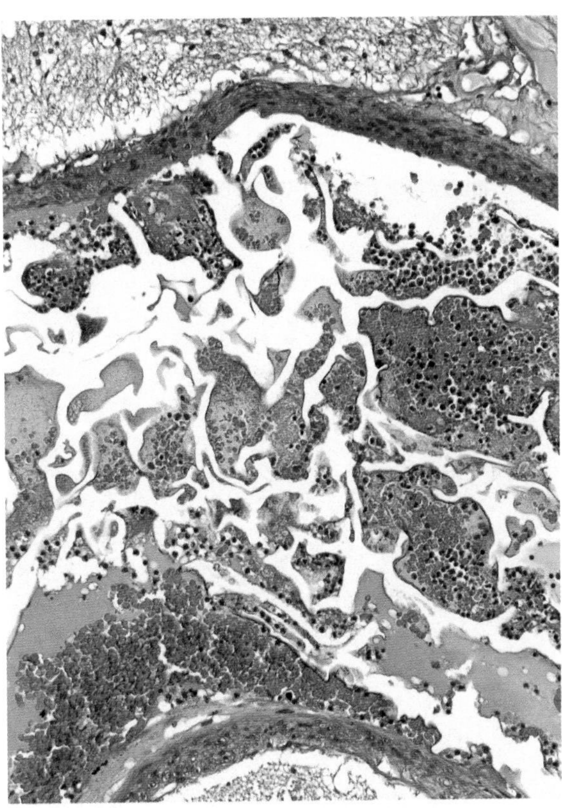

Figure 10-6

ARTERIOVENOUS MALFORMATION

AVMs may be embolized preoperatively to facilitate resection.

such a way by the biopsy procedure as to resemble a vascular malformation (see fig. 2-9). In contrast to those of AVM, however, such compacted subarachnoid vessels are individually normal in structure and degree of muscularization, and thus are easily differentiated from those of both AVM and cavernous angioma.

Although the noncommittal term *vascular malformation* is appropriate in cases in which histology is indeterminate, the presence of arterial shunting during angiography is strong clinical evidence for an AVM. Surgeons rightfully object to the diagnosis of cavernous angioma in the presence of such shunting since cavernous angiomas are usually angiographically occult (13).

Cavernous Angioma

The classic *cavernous angioma*, or *cavernoma*, is a compact mass that is more likely to present with seizures than with focal neurologic symptoms or signs of increased intracranial pressure (14,15). Hemorrhages are relatively common and certainly can be symptomatic, but they are not usually the apoplectic events with devastating mass effects or intraventricular rupture that can occur with AVMs. The majority of cavernous angiomas affect the cerebrum, although the brain stem, cerebellum, and spinal cord can be affected; even the leptomeninges are not exempt. Multiple lesions are frequent. In the United States, there appears be an increased incidence of cavernous angioma among Hispanic-Americans (16). A responsible mutation in *KRIT1* has been identified in this population (17). Some cavernous angiomas are clearly familial (18). A variant is radiation induced (19–21).

In magnetic resonance images (MRIs), the aggregated vascular profiles of cavernous angioma yield a "popcorn"-like profile surrounded by a dark corona of low signal intensity on T2-

weighted images due to hemosiderin deposition (figs. 10-7, 10-8) (13). This combination of a central mass and perilesional halo is almost diagnostic. A neighboring large vein similar to that seen in venous angiomas attends the lesion in some cases. Cavernous angiomas are angiographically occult (13). The macroscopic appearance mirrors the radiographic profile, i.e., a discrete compact mass of vascular profiles (figs. 10-9, 10-10) with a rusty rim of gliotic, hemosiderin-rich parenchyma (fig. 10-8).

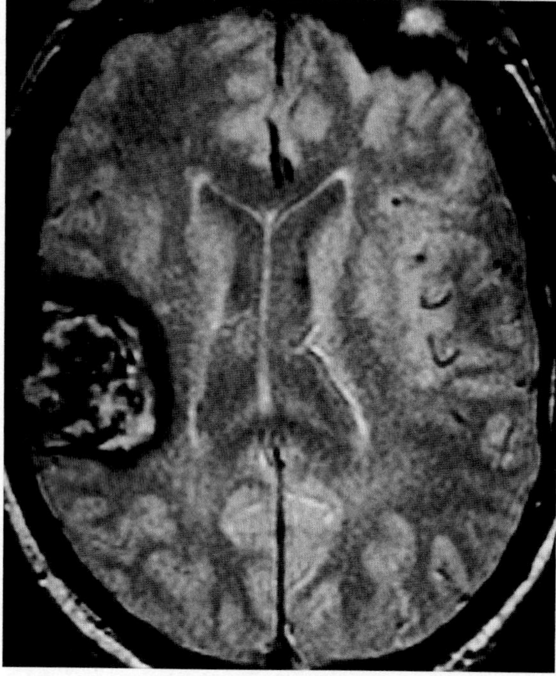

Figure 10-7

CAVERNOUS ANGIOMA

As seen in a proton density magnetic resonance image (MRI), a central mass of vascular profiles is surrounded by a dark rim of hemosiderin-rich tissue. This appearance is typical of cavernous angioma.

Microscopically, cavernous angiomas consist of a honeycomb of compact vessels that vary considerably in caliber and degree of collagenization (fig. 10-11). The vessels are free of both smooth muscle and elastic lamellae (15). The vessels are often closely apposed, leaving little if any interstitial parenchyma in most cases, but others are more loosely constructed and CNS parenchyma is present. Both compact and racemose malformations, composed of the typical hyaline vessels of cavernous angiomas, are common among angiographically occult vascular malformations (13). Hemosiderin discolors the surrounding parenchyma and burdens the cytoplasm of both macrophages and astrocytes (fig. 10-12). Advanced sclerosis and heavy calcification typify the chronic lesion.

Although the distinction between cavernous angioma and capillary telangiectasis may seem trivial to the surgical pathologist, there has been considerable debate concerning the relationship between these two lesions (22). Some suggest that cavernous angiomas evolve from telangiectases. Only a few lesions exhibit features of both (13).

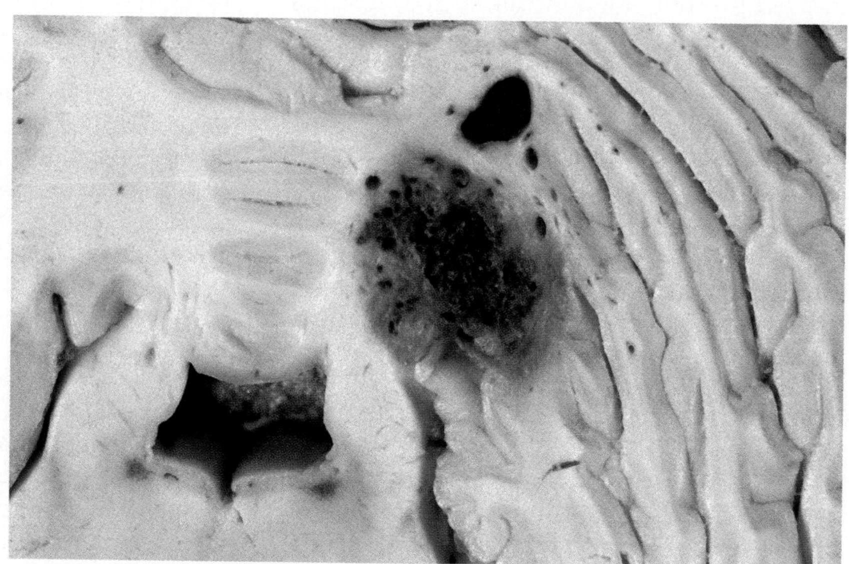

Figure 10-8

CAVERNOUS ANGIOMA

Cavernous angiomas are well circumscribed and surrounded by hemosiderin-rich parenchyma.

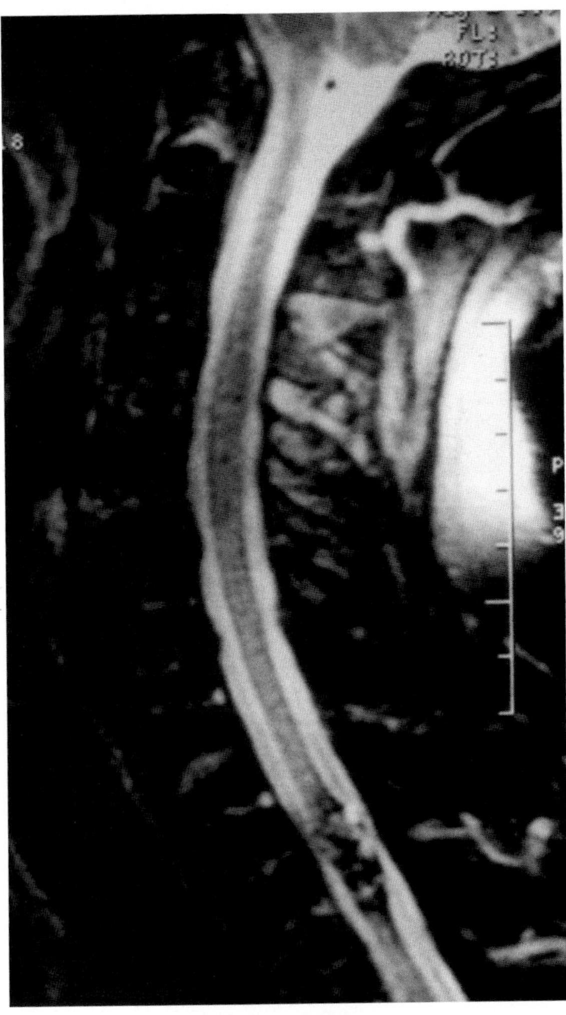

Figure 10-9

CAVERNOUS ANGIOMA

Cavernous angiomas are common in the spinal cord where, as in the brain, the entity is suggested by the dark signal generated by hemosiderin in T2-weighted images.

Venous Malformation

Consisting solely of venous channels without the capillaries of telangiectases, *venous angiomas* occur both in the brain and spinal cord. They are common as incidental radiologic (figs. 10-13, 10-14) or postmortem (fig. 10-15) findings. Only a rare example is symptomatic (23,24).

Angiographically, the lesions appear as veins with a "caput medusa" profile (figs. 10-14) (25, 26). In the spinal cord, serpentine enlargement of the constituent veins must be distinguished from the normally prominent venous channels

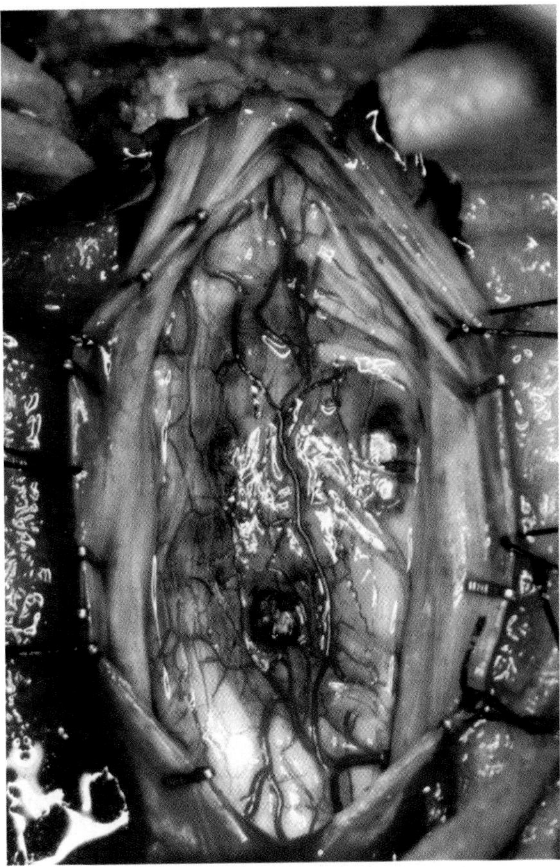

Figure 10-10

CAVERNOUS ANGIOMA

This discrete vascular mass produced bilateral upper extremity weakness and difficulty walking in a 45-year-old man. (Courtesy of Dr. Ziya Gokaslan, Baltimore, MD.)

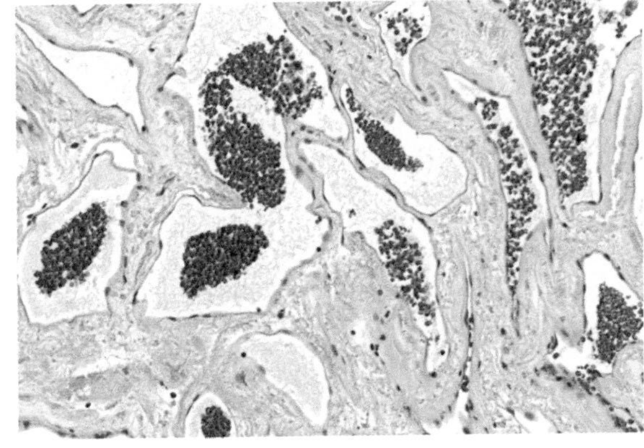

Figure 10-11

CAVERNOUS ANGIOMA

Closely apposed, structureless hyaline vessels are characteristic of cavernous angiomas.

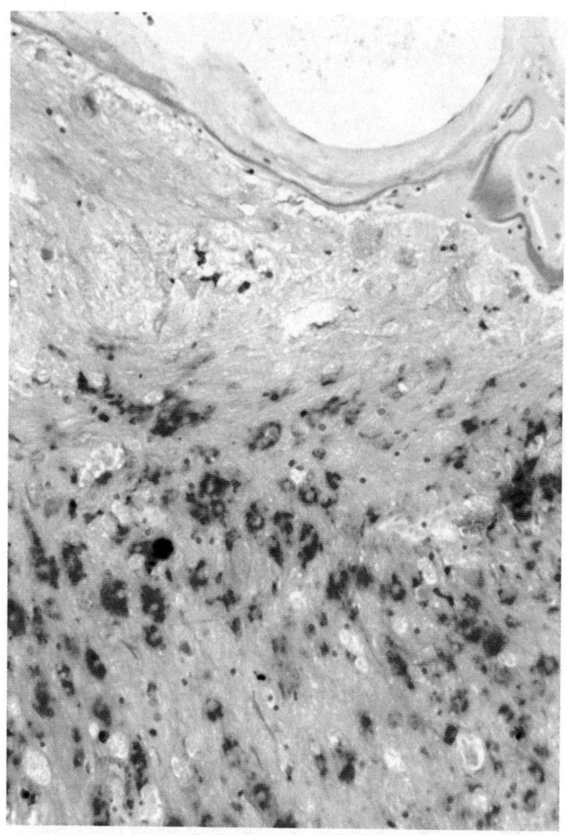

Figure 10-12

CAVERNOUS ANGIOMA

Perilesional parenchymal staining for iron is characteristic of cavernous angiomas.

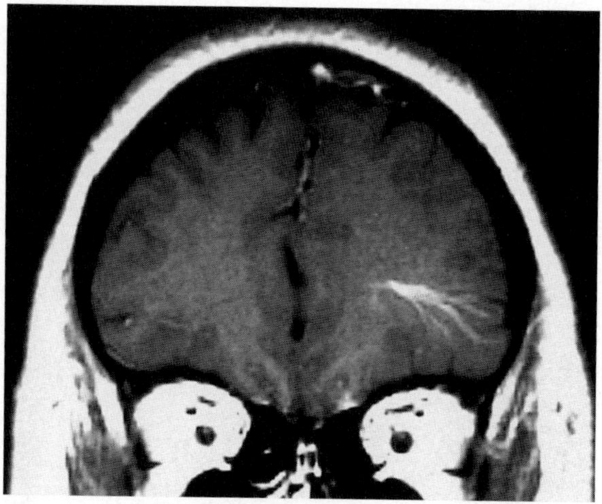

Figure 10-13

VENOUS ANGIOMA

As seen on the right in a gadolinium-enhanced MRI scan, venous malformations are common incidental neuroradiologic findings. The feeding vessels pass from right to left of the illustration where they enter a draining vein. (Fig. 10-8 from Fascicle 10, 3rd Series.)

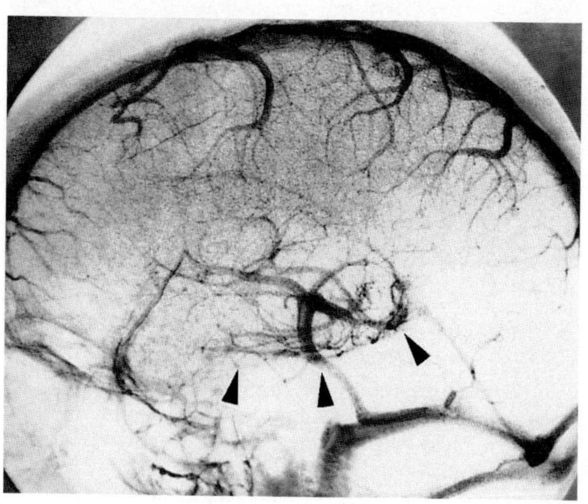

Figure 10-14

VENOUS ANGIOMA

Angiographically, venous malformations are multiple feeding vessels that, by converging on a draining vein, produce a "caput medusae" profile (arrows). (Fig. 10-9 from Fascicle 10, 3rd Series.)

on the dorsal surface in the lumbar region, and the secondarily enlarged draining veins that accompany AVMs and dural arteriovenous fistulae. Histologically, the lesion is formed of ectatic venous channels featuring a media but lacking an elastic lamina. As a rule, there is no associated parenchymal degeneration. The lesion's thin-walled channels lie within brain parenchyma, free of gliosis or hemosiderosis.

Capillary Telangiectasis

Capillary telangiectasis is usually an incidental radiologic (fig. 10-16) or postmortem finding consisting of a small blush or fine stippling, usually in the pons (27,28). The lesion is generally of little clinical significance. Microscopically, the delicate vessels are devoid of smooth muscle, contain only a small amount of collagen, and lie within a small zone of otherwise unremarkable brain parenchyma. The caliber of the constituent capillaries varies considerably due to aneurysmal dilations (fig. 10-17) that are best seen in thick-tissue preparations (29). There is no gliosis or evidence of past hemorrhage.

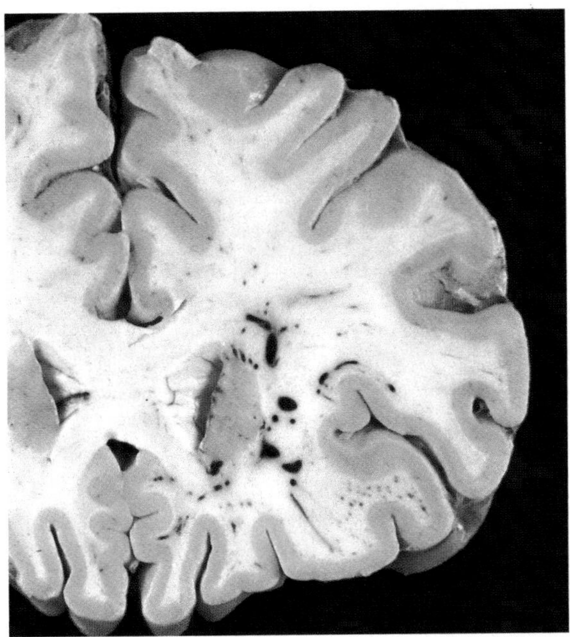

Figure 10-15

VENOUS ANGIOMA

Common as an incidental finding, venous angiomas consist of large veins in otherwise normal brain.

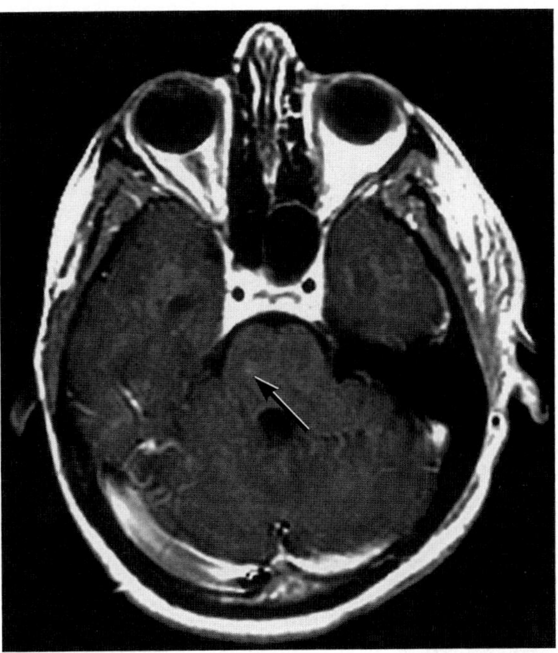

Figure 10-16

CAPILLARY TELANGIECTASIS

Capillary telangiectasis is a common incidental neuroradiologic or postmortem finding (arrow). As here in its most common, pontine location, the asymptomatic lesion (arrow) is no more than a "blush" in a contrast-enhanced MRI.

VASCULAR MALFORMATIONS IN NEUROCUTANEOUS SYNDROMES

Sturge-Weber Syndrome

Sturge-Weber syndrome has both a cutaneous vascular "nevus" and a leptomeningeal angioma of the venous type (30,31). Independent involvement of the skin or leptomeninges may be a "forme fruste" expression.

The cutaneous capillary-venous lesion typically stains the face over a variable area, but the latter always includes a portion of skin innervated by the ophthalmic division of the trigeminal nerve. Not only may mucosal, gingival, and orbital soft tissue be involved, but angiomas of the retina and choroid plexus are seen with some frequency. Concurrent involvement of the skin of the neck or upper body is less common. Associated brain lesions are usually ipsilateral to the facial abnormality, and generally include the parieto-occipital region. Bilaterality is observed in some cases; cerebellar or brain stem involvement is relatively uncommon.

Radiologically, leptomeningeal angiomas produce thickening and contrast enhancement of the subarachnoid space in association with atrophy of the underlying brain (32). Cortical calcification, when present, produces a classic "tram track" pattern (fig. 10-18).

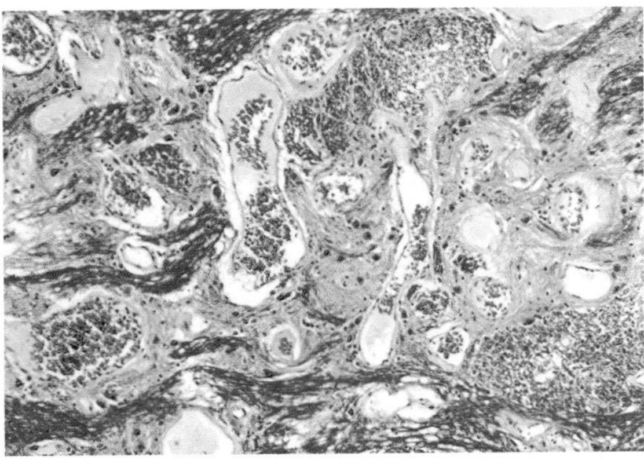

Figure 10-17

CAPILLARY TELANGIECTASIS

Capillary telangiectases are variably sized, thin-walled channels set in otherwise normal parenchyma (H&E/Luxol fast blue stain).

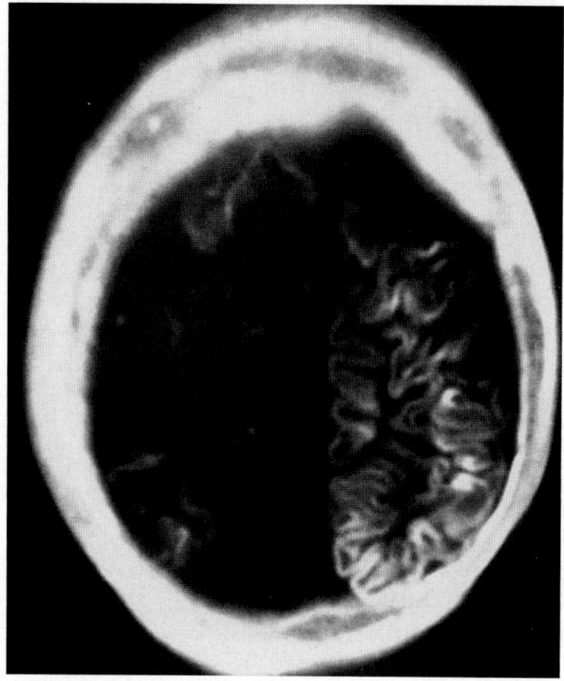

Figure 10-18

STURGE-WEBER SYNDROME

As seen by computerized tomography (CT), the intracortical calcification produces a sinuous or gyriform pattern of radiologic density. As here, the process is often asymmetric. (Fig. 18-21 from Fascicle 10, 3rd Series.)

Figure 10-19

STURGE-WEBER SYNDROME

Seizure control sometimes entails surgical resection of the epileptogenic cortex. The dark red coloration is produced by the diffuse vascular malformation. (Courtesy of Dr. Harold J. Hoffman, Toronto, Canada.)

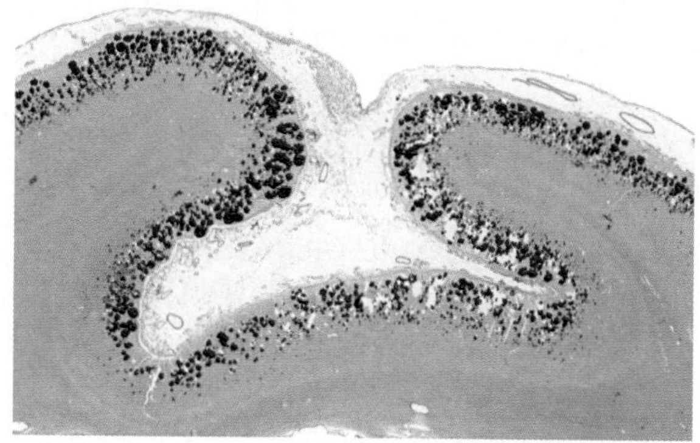

Figure 10-20

STURGE-WEBER SYNDROME

A lamina of concretions in the superficial cortex is responsible for the radiologic appearance of the lesion illustrated in figure 10-18.

Clinically, the intracranial lesions frequently induce seizures whose control may require surgical resection of the calcified epileptogenic cortex (fig. 10-19) (33–36). With time, brain atrophy ensues on the side of the leptomeningeal angioma, with resultant motor deficits in some cases. In many instances, diminished intellect, if not frank dementia, occurs. Cognitive deficits are understandably more pronounced with bilateral cerebral involvement (37).

Macroscopically, the affected region is hypervascularized and red (fig. 10-19). The underlying brain may exhibit particulate calcification but is less affected. The dura is normal.

Microscopically, the abnormal vessels reside within the subarachnoid space and are thin-walled channels largely devoid of smooth muscle (figs. 10-20–10-22). Calcospherites, although initially localized to perivascular regions, later become so abundant within the upper cortex that they escape precise localization (fig. 10-21).

Mesencephalo-Oculo-Facial Angiomatosis (Wyburn-Mason Syndrome)

Mesencephalo-oculo-facial angiomatosis, also known as *Wyburn-Mason syndrome,* is a rare neurocutaneous syndrome that includes: 1) an intracranial AVM, typically of the midbrain; 2) a homolateral facial vascular nevus in the trigeminal distribution; and 3) a retinal vascular malformation. The jaws and soft tissues of the face may be affected as well. The intracranial lesion may occur along the visual pathway as well as in the region of the midbrain (38–40). Despite

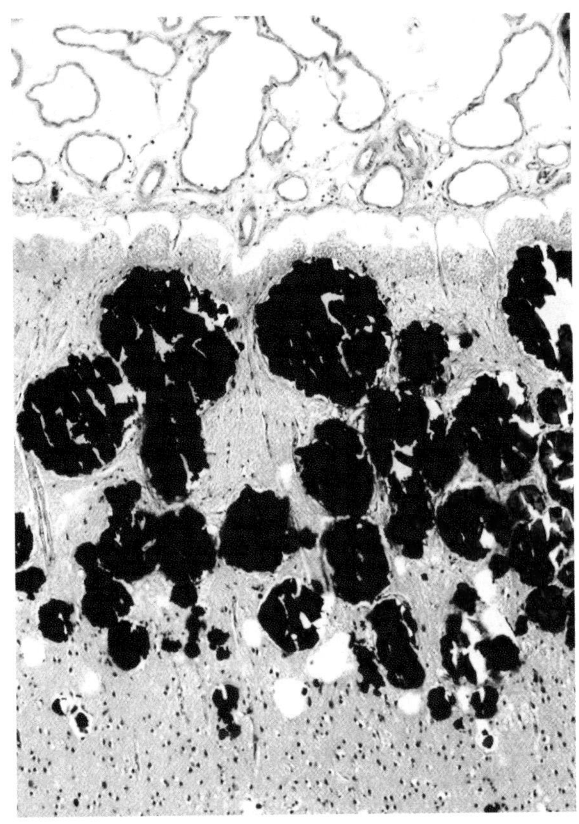

Figure 10-21

STURGE-WEBER SYNDROME

The cortex is filled with calcospherites in advanced cases. There is a venous malformation in the subarachnoid space.

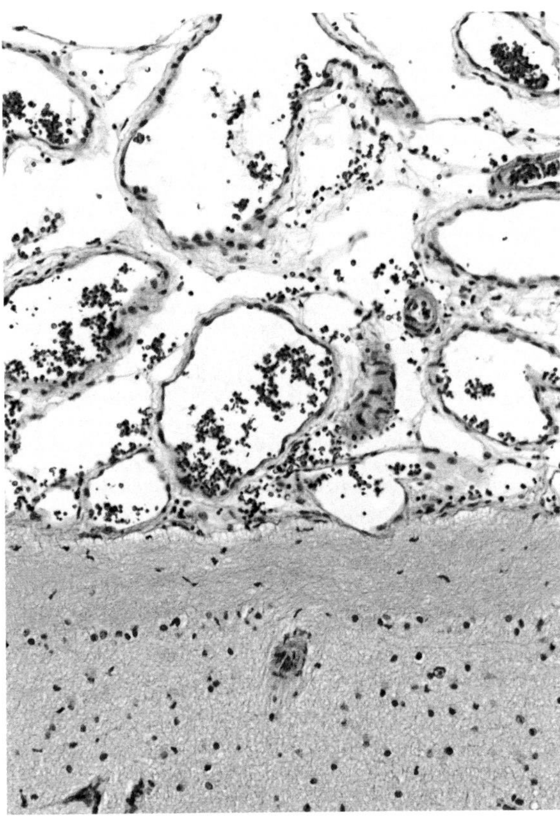

Figure 10-22

STURGE-WEBER SYNDROME

The leptomeningeal vascular malformation fills the subarachnoid space with small veins.

similarities of lesion distribution, the focal mass of vessels, accompanied by arteriovenous shunting, differs from the diffuse, superficially situated, vascular anomaly of Sturge-Weber disease. Little is known about the heredity of mesencephalo-oculo-facial angiomatosis, although an autosomal dominant mechanism has been suggested (40).

Hereditary Hemorrhagic Telangiectasia (Rendu-Osler-Weber Disease)

Hereditary hemorrhagic telangiectasia, also known as *Rendu-Osler-Weber disease,* is inherited as an autosomal dominant disorder with high penetrance. The condition is characterized by multifocal telangiectases, most notably dermal, mucosal, and visceral, associated with recurrent bleeding (41–44). Epistaxis, as a consequence of nasal involvement, is a common symptom. CNS involvement by the disease is seen in approximately 15 percent of cases and takes the form of telangiectases of the cerebrum, brain stem, or spinal cord, as well as other, sometimes multiple, intracranial vascular malformations. Subarachnoid, intracerebral, or intraventricular hemorrhage may result. Cerebral ischemic or inflammatory lesions (abscesses) secondary to embolism may result from the peripheral vascular anomalies (42) that result in pulmonary arteriovenous shunts.

VASCULAR NEOPLASMS AND TUMOR-LIKE LESIONS

Papillary Endothelial Hyperplasia (Masson's Vegetant Hemangioendothelioma)

Papillary endothelial hyperplasia, also known as *Masson's vegetant hemangioendothelioma,* is an uncommon reactive endothelial proliferation that can occur in virtually any vessel of the body, usually a vein. In the CNS, most lesions are

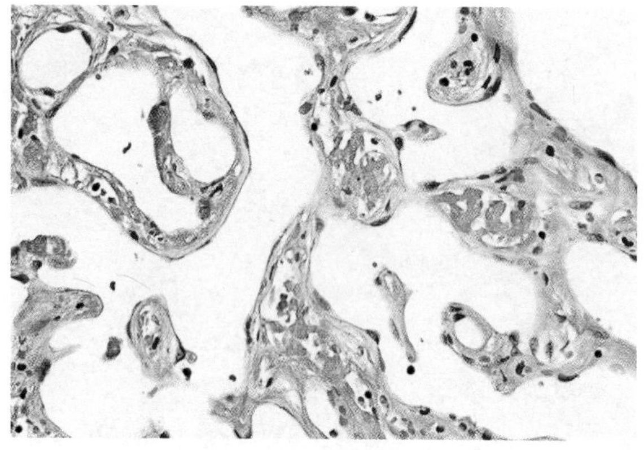

Figure 10-23

PAPILLARY ENDOTHELIAL HYPERPLASIA

Multiple papillary formations with fibrin cores covered by endothelial cells create an unusual intracranial mass lesion.

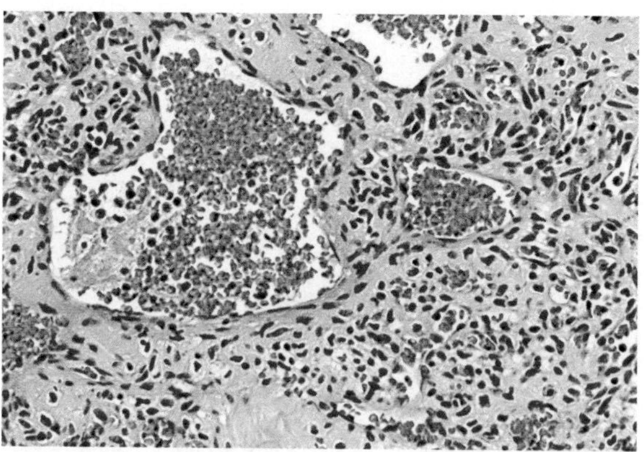

Figure 10-24

CAPILLARY HEMANGIOMA

Compact lobules of capillaries and large draining veins are typical of capillary hemangioma.

incidental findings associated with vascular malformations or hematomas. On occasion, they form symptomatic masses (45,46). The lesion begins as an endothelial ingrowth within a thrombus. With continued growth, the thrombus is subdivided into minute clumps covered by endothelium. Finally, papillae with central cores of collagen fuse to form a complex anastomotic pattern (fig. 10-23). Microscopically, the proliferation lacks the lobular architecture of hemangioma, although it does consist of endothelial cells accompanied by pericytes. Significant cytologic atypia and mitotic activity are lacking. Papillary endothelial hyperplasia is not to be mistaken for an endothelial neoplasm, particularly angiosarcoma.

Capillary Hemangioma

Capillary hemangioma, or *cellular hemangioma*, once termed *hemangioendothelioma*, is a rare tumor in the CNS that can affect brain or spinal cord (fig. 10-24, see fig. 7-19) (47–49), as well as proximal spinal nerve root (50). The occurrence of multiple hemangiomas (*hemangiomatosis*) is rare (51). Histologically, the lesion consists of mature capillaries and in some cases large draining vessels. The proliferation often features a lobular architecture at low magnification. The lobules consist of crowded minute capillaries and a "feeder" or draining vessel, all lined by a thin layer of endothelial cells surrounded by pericytes. An extensive reticulin network extends to even the smallest capillary.

Immunohistochemically, endothelial cells (CD31 and CD34 positive) and pericytes (smooth muscle actin positive) are represented; ultrastructural studies confirm the presence of a dual cell population (47,48). Although histologically benign, subtotal resection leaves open the possibility of recurrence.

Epithelioid Hemangioendothelioma

Conceptually, *epithelioid hemangioendothelioma* is viewed as an endothelial tumor of low- to intermediate-grade malignancy, since its behavior is more favorable than that of angiosarcoma. In the CNS, this rare lesion is parenchymal, occurring more often in the cerebral hemispheres (52–55). Patients vary in age from neonates to elderly. With computerized tomography (CT) and MRI, the lesions are demarcated, contrast enhancing, and surrounded by significant edema. Macroscopically, the bulky, demarcated tumors vary from soft and thrombus-like to firm and gray-red. Focal calcification may be observed.

Histologically, epithelioid hemangioendotheliomas are composed of short strands or solid nests of rounded to elongated endothelial cells. Vascular channels are not a feature of the central portion of the tumor. Instead, there are intracytoplasmic lumens in the form of vacuoles, some of which compress nuclei. Stains for mucin are negative in these structures. Tumoral stroma varies from myxoid to hyaline or chondroid appearing.

Immunohistochemically, the tumors may be more reliably and uniformly positive for CD31 than for CD34 or factor VIII. Ultrastructural features are those of the endothelium. The differential diagnosis includes glioma, metastatic carcinoma, cardiac myxoma, cartilaginous tumors, and chordoid meningioma.

Although the lesions can behave aggressively, there is little information about the prognosis for patients with CNS examples. Spindled (56) and polymorphous (57) hemangioendotheliomas also occur in the CNS.

Hemangiopericytoma

Definition. *Hemangiopericytoma* (HPC) is a malignant neoplasm presumably derived from pericytes.

General Comments. The classic CNS HPC has distinctive pathologic features and aggressive clinical behavior that earned it the status of an entity distinguished from meningioma on histologic, clinical, and genetic grounds. However, the very existence of the HPC has become an issue in the nosology of soft tissue tumors, with the assertion that the lesion is part of the spectrum of solitary fibrous tumors (58). By this approach, reticulin-rich, densely cellular, CD34-negative lesions with staghorn vessels sit at one end of the continuum, whereas biologically less aggressive collagenous tumors of lower cellularity with "wire-like" collagen and diffuse immunoreactivity for CD34 lie at the other. If so, then intermediate lesions become major diagnostic and prognostic issues since CNS lesions with classic HPC features behave, overall, as low-grade sarcomas, whereas solitary fibrous tumors (SFTs) are usually benign. While recognizing the possibility that the two are variations of the same entity, HPCs and SFTs are described separately in this chapter.

Clinical Features. HPCs comprise approximately 2 percent of meningeal tumors. Most occur in adults. Unlike meningiomas, they show no female predilection, and may be slightly more common in men. The tumors are largely intracranial, but spinal examples also occur (58a). Both supratentorial and, to a lesser extent, infratentorial meninges are affected (59–61).

Radiologic Findings. On CT and MRI, these dura-based, extra-axial tumors are lobulated and contrast enhancing (fig. 10-25) (62). Unlike meningiomas, with their broad-based dural attachment and smooth contour, HPCs often have a narrower base and a more prominent lobularity. They lack calcification and, particularly when recurrent, tend to invade and destroy adjacent bone, in contrast to meningiomas, which typically have a hyperostotic effect. Angiographically, HPCs are highly vascular.

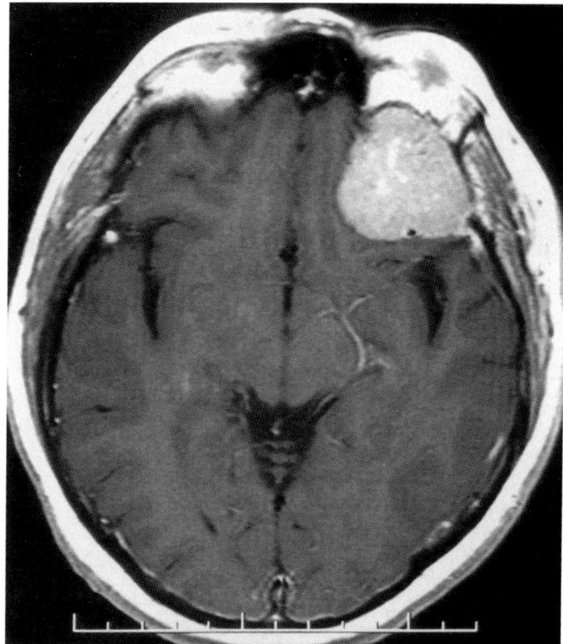

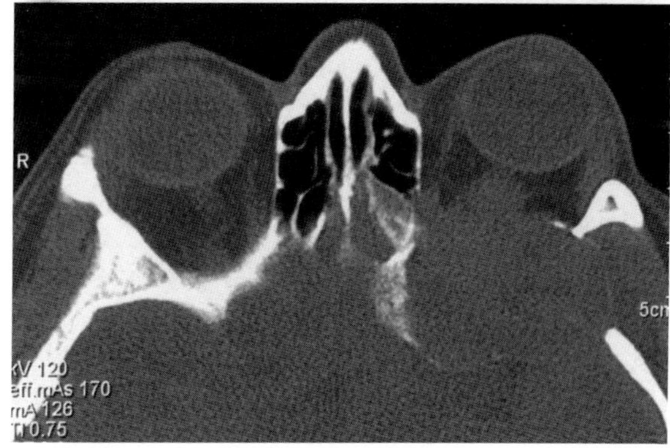

Figure 10-25

HEMANGIOPERICYTOMA

Hemangiopericytomas (HPCs) are well-circumscribed, intensely enhancing, and usually dura-based lesions (top). Unlike meningiomas, HPCs often erode adjacent bone (bottom). Proptosis called attention to this large lesion in a 47-year-old woman.

Gross Findings. Lobulated, gray-pink, and fleshy, the lesions bleed profusely upon surgical

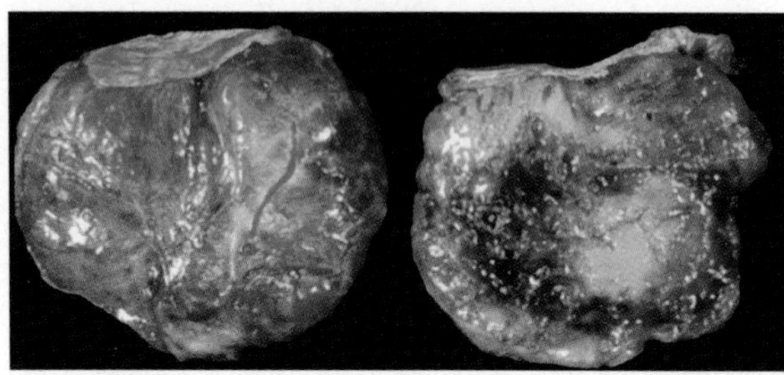

Figure 10-26

HEMANGIOPERICYTOMA

Although often more fleshy and vascular than meningiomas, the discreteness and dural base of HPCs create a close resemblance in many cases. Vessels are prominent both on and within the lesion.

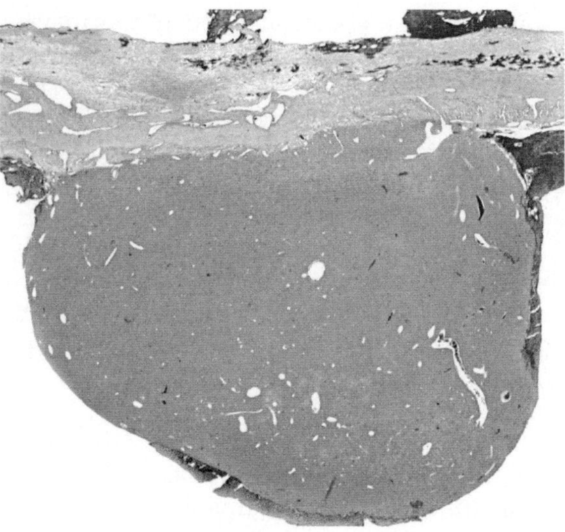

Figure 10-27

HEMANGIOPERICYTOMA

Meningeal HPCs often have a uniform blueness that is highly suggestive of the entity, even at low magnification. Classic "staghorn vessels" are also evident.

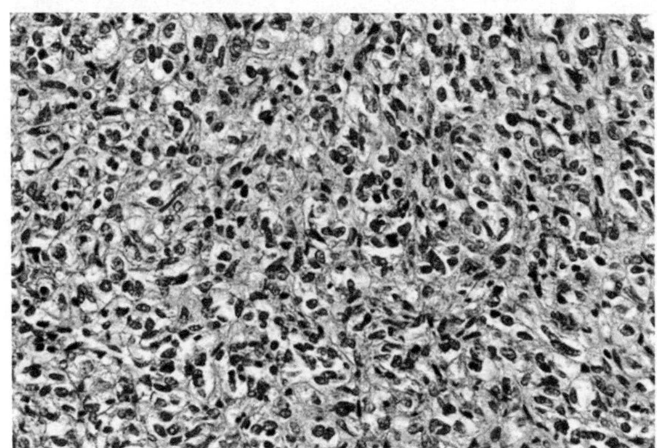

Figure 10-28

HEMANGIOPERICYTOMA

HPCs are cellular lesions without the whorls, lobularity, calcification, and thickened hyaline vessels of meningioma.

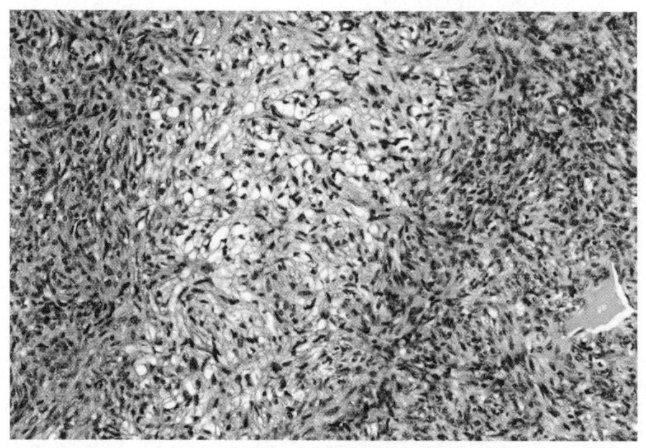

Figure 10-29

HEMANGIOPERICYTOMA

Focal areas of reduced cell density are common.

incision (fig. 10-26). Gaping vessels may be apparent on cut section.

Microscopic Findings. Even a cursory low-magnification examination of the typically uniformly blue-staining sections suggests the diagnosis of HPC (figs. 10-27, 10-28). Patches of hypocellularity often punctuate the tumor (fig. 10-29). Foci of necrosis are uncommon but may be seen in high-grade variants (discussed below). Slit-like vascular spaces are apparent which, when large and branched, assume the classic "staghorn" pattern (fig. 10-30).

HPCs are typically highly cellular lesions with a distinctive swirling or "turbulent" pattern in which the cells appear to writhe in a reticulin-rich stroma (figs. 10-31, 10-32). As a rule, reticulin invests individual or small clusters of cells, but occasional examples are reticulin sparse. While the cells are often elongated, if not focally spindle-shaped, they may also be plump and epithelioid, and thus more similar to those of meningioma (fig. 10-33). Small, poorly formed whorls are occasionally present (fig. 10-34).

Considerable morphologic variation exists in terms of tumor cellularity and degree of nuclear atypia. Most often, the nuclei are oval and plump, and have moderate chromatin density. Pleomorphic nuclei are only an occasional and

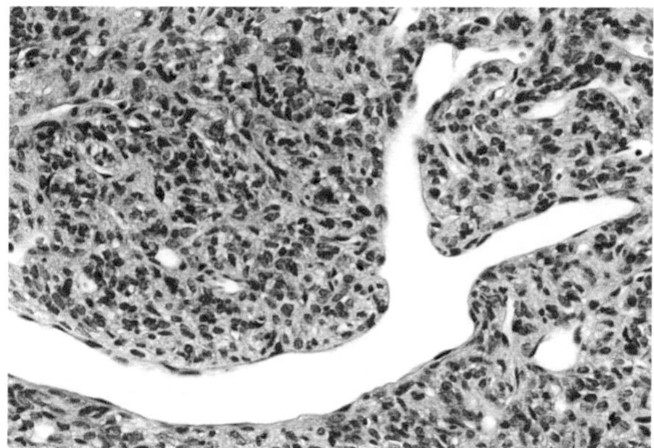

Figure 10-30

HEMANGIOPERICYTOMA

The slit-like vessels comprising the staghorn vasculature characteristic of HPC consist of little more than a thin layer of endothelial cells.

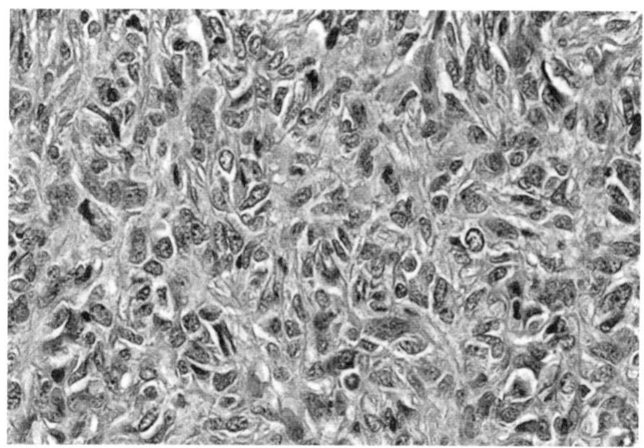

Figure 10-31

HEMANGIOPERICYTOMA

Cytologic disarray creates a sense of cellular turbulence.

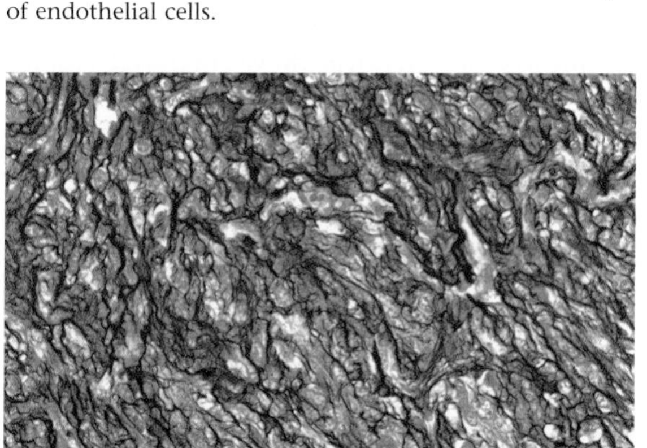

Figure 10-32

HEMANGIOPERICYTOMA

Reticulin typically surrounds either individual cells or small groups of cells.

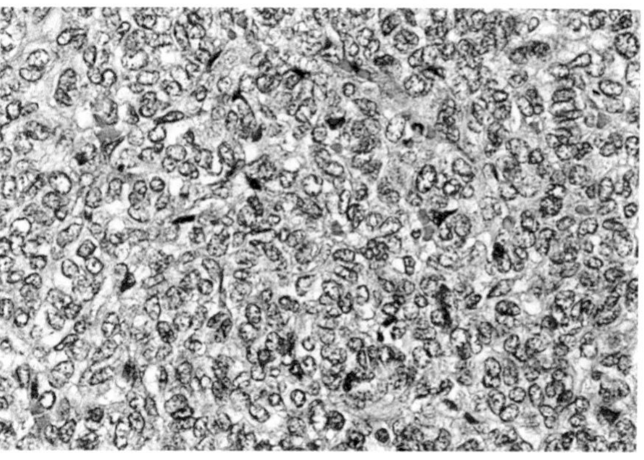

Figure 10-33

HEMANGIOPERICYTOMA

The plump, round cells of some HPCs are similar to those of meningioma.

focal finding. Nucleoli are inconspicuous. Mitotic activity is highly variable, but is found readily in most examples.

Histologic grading of HPCs has succeeded in defining differentiated and anaplastic categories (63). Tumors in the latter are defined as exhibiting either necrosis or more than 5 mitoses per 10 high-power microscopic fields as well as two or more of the following features: hemorrhage, moderate to marked cytologic atypia, and moderate to high cellularity (63).

Frozen Section. In contrast to meningioma, the main entity in the differential diagnosis, HPCs tend to be more densely cellular and lack well-formed whorls, cytoplasmic-nuclear inclusions, and collagenized vessels (fig. 10-35). Patches of hypocellularity and staghorn vasculature are helpful features of HPC.

Immunohistochemical Findings. In one large study of differentiated and anaplastic HPCs (64), the tumors were often (85 percent) strongly immunoreactive for vimentin, showed scattered cells that stained for factor XIIIa (78 percent), had reactivity for leu-7 (70 percent), and had staining, often weak, for CD34 (33 percent) (fig. 10-36). Unlike meningiomas, they were negative for epithelial membrane antigen (EMA) except focally in rare exceptions (64,65). In comparison

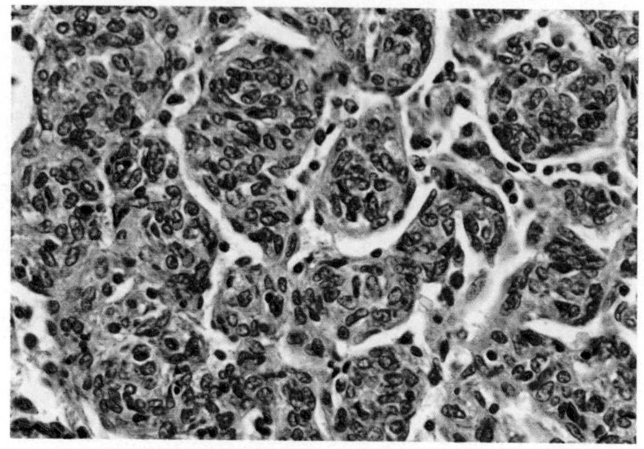

Figure 10-34

HEMANGIOPERICYTOMA

Small whorls can simulate meningioma, but are loosely wound and free of psammoma bodies.

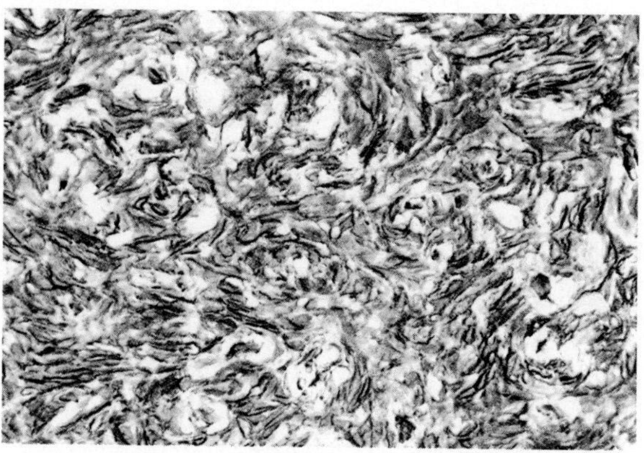

Figure 10-36

HEMANGIOPERICYTOMA

Approximately one third of HPCs are immunoreactive for CD34 in a patchy distribution.

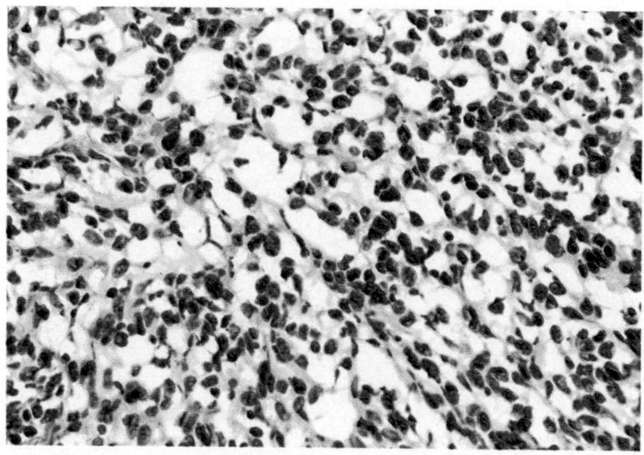

Figure 10-35

HEMANGIOPERICYTOMA: FROZEN SECTION

Coarser chromatin, lack of well-formed whorls, and a general paucity of fibrous tissue help distinguish HPC from meningoma.

to SFT, HPCs are much less likely to have strong diffuse CD34 staining, but there is overlap in some cases (65). Although normal pericytes are immunoreactive for smooth muscle actin at many sites, muscle antigens have not been found in HPCs (66). The basement membrane material responsible for the often positive reticulin staining is immunoreactive with antibodies to collagen type IV (67). The significance of some scant desmin or cytokeratin staining in a small minority of cases is unknown (64).

Ultrastructural Findings. The neoplastic cells vary in differentiation from pericytic to myogenic to fibroblastic. In many cases, they are surrounded by amorphous electron-dense material. Somewhat resembling basal lamina, this material is the ultrastructural equivalent of reticulin in histologic preparations (63,67). Rudimentary intercellular junctions may be seen, but not well-formed desmosomes or complex interdigitations of cell membranes (fig. 10-37). Spherical, whorled masses of intermediate filaments (vimentin) may be present within the cytoplasm.

Cytologic Findings. The cells of HPC disperse readily. Their nuclei are more uniformly ovoid and exhibit somewhat greater chromatin density than those of meningioma. The cytoplasm is less abundant and less wispy (fig. 10-38).

Cytogenetic and Molecular Findings. Cytogenetic abnormalities in HPC include nonconstitutional balanced chromosome anomalies of 7p15 and 12q (68). Distinguishing HPC from meningiomas is the presence of *NF2* mutations in about a third of meningiomas. No such mutations have been reported in HPCs (69).

Differential Diagnosis. The classic HPC, with its high degree of cellularity, nuclear uniformity, staghorn vasculature, hypocellular patches, and pericellular pattern of reticulin is a distinctive neoplasm unlikely to be confused with other lesions. Care must be taken, however, to distinguish it from meningioma if either lobules or a suggestion of whorl formation is present. In most cases, the two lesions are readily identified on architectural and cytologic grounds alone. HPCs lack the tight concentric whorls, psammoma bodies, nuclear-cytoplasmic

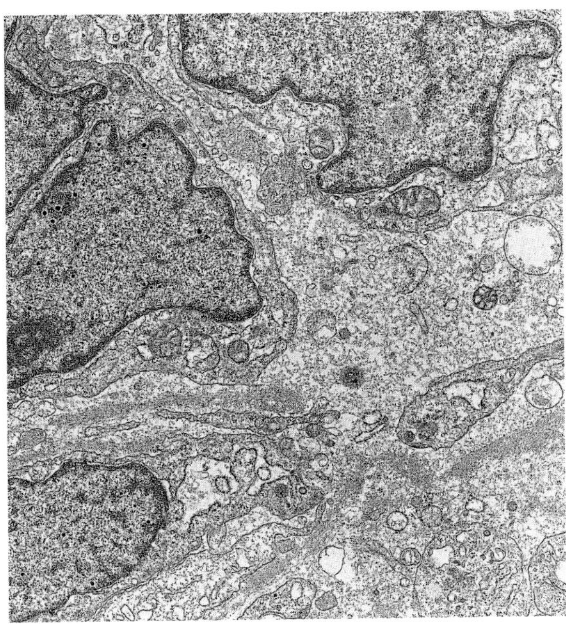

Figure 10-37

HEMANGIOPERICYTOMA

Cells of HPC are separated by amorphous basement membrane-like material. Unlike meningiomas, cell membranes do not interdigitate or form desmosomal connections. (Fig. 10-19 from Fascicle 10, 3rd Series.)

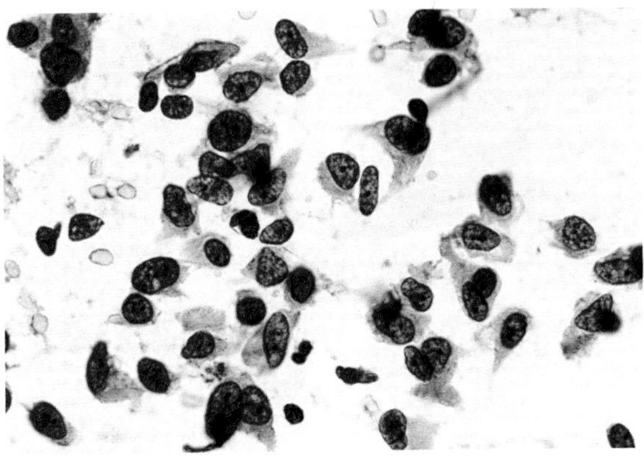

Figure 10-38

HEMANGIOPERICYTOMA: SMEAR PREPARATION

HPCs lack the delicate chromatin and often prominent cytoplasm of meningiomas.

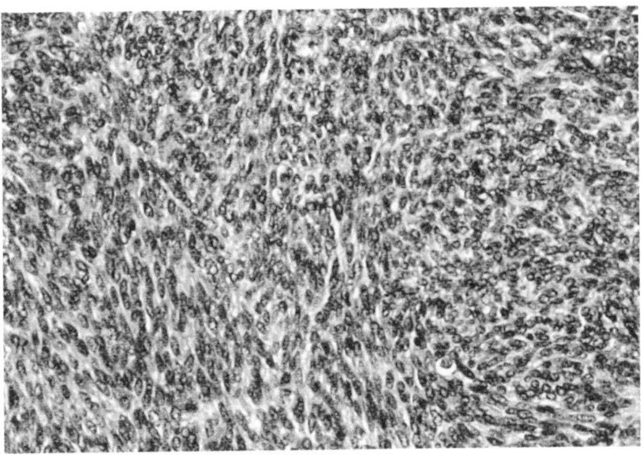

Figure 10-39

SOLITARY FIBROUS TUMOR RESEMBLING HEMANGIOPERICYTOMA

The cell density of some solitary fibrous tumors overlaps with that of HPC.

inclusions, and sclerotic vessels so common in meningiomas. Nonetheless, occasional HPCs contain eddies of cells that simulate whorls. Such clusters generally lack the compact concentric arrangement of flattened cells that characterize meningioma. While meningiomas exhibit considerable variation in nuclear morphology, the nuclei of HPCs are usually more uniform, plump, and hyperchromatic; furthermore, they lack the nuclear pseudoinclusions so typical of meningothelial cells. Slit-like vascular spaces are more common in HPCs. Cellular elongation and collagen production in some HPCs simulate fibrous meningioma, a tumor variably EMA and S-100 protein immunoreactive. This meningioma variant, when well sampled, often exhibits areas of more obvious meningothelial differentiation.

A mesenchymal entity to be distinguished from HPC is solitary fibrous tumor (SFT) (fig. 10-39). This distinction can be difficult, and the suggestion has even been made that HPC and SFT are one and the same, or at least different ends of a spectrum. On one end would be classic HPC with abundant pericellular reticulin and absent CD34 immunoreactivity, on the other would be reticulin-free, histologically classic SFT with diffuse strong staining for CD34. There is ongoing discussion and debate over this issue that is of considerable importance. HPCs are usually considered sarcomas and SFTs are usually cured by simple excision.

Histologically, SFT is distinguished by the universal cellular elongation and "wire-like" bands of brightly eosinophilic collagen. Pericellular reticulin is absent in these areas, with only coarse collagen in its stead. Strong immunoreactivity for CD34 is almost required for the

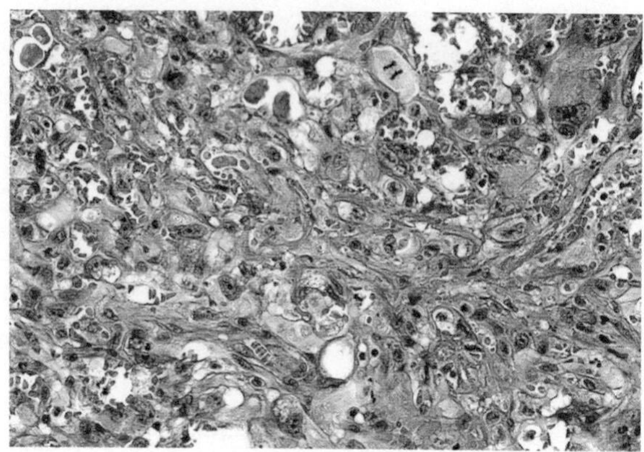

Figure 10-40

ANGIOSARCOMA

The tumoral vasculature consists of plump, cytologically atypical, and mitotically active cells.

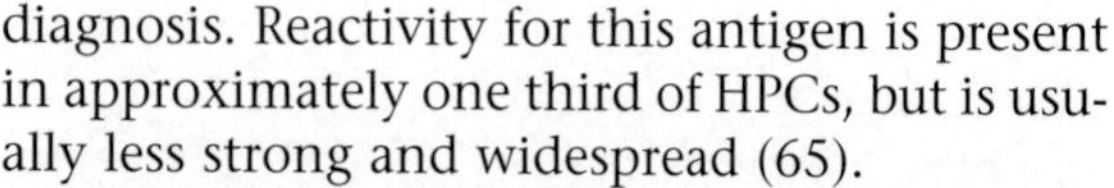

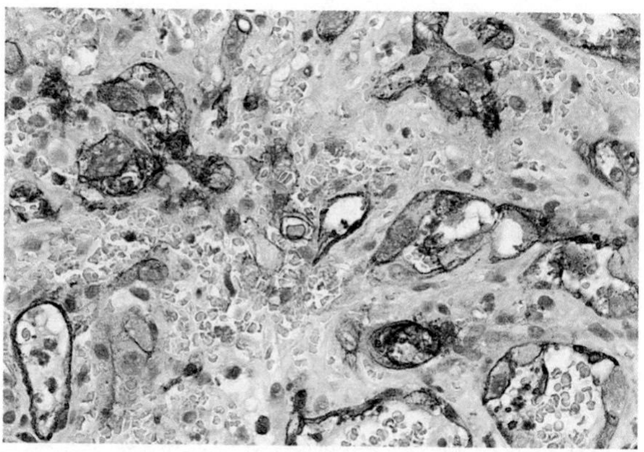

Figure 10-41

ANGIOSARCOMA

Angiosarcomas are immunoreactive for CD31.

diagnosis. Reactivity for this antigen is present in approximately one third of HPCs, but is usually less strong and widespread (65).

Although rare, mesenchymal chondrosarcoma (see fig. 10-70) enters into the differential diagnosis of HPC since its densely cellular, small cell component has a decidedly hemangiopericytoma-like appearance. The presence of cartilaginous islands is the obvious distinguishing feature.

Treatment and Prognosis. The ease with which HPCs often appear to be enucleated surgically belies their high rate of local recurrence and tendency to late leptomeningeal spread and/or distant metastasis despite radiotherapy. The probability of recurrence is approximately 65 percent at 5 years and 90 percent at 12 years, and of metastasis, approximately 80 percent at 12 years (61). Favored metastatic sites include bone, liver, lung, CNS, and the abdominal cavity (59–61,63). One study has indicated that radiotherapy improves progression-free survival and increases overall survival (61). Overall survival of patients with variably treated tumors in that series was approximately 20 percent at 20 years. Another series found a 5-year disease-free survival rate of 89 percent (60).

Differentiated and anaplastic HPCs, as defined above (63), have differing rates of recurrence (63 versus 85 percent). Of patients dead of disease, median survival was 12 years with differentiated lesions and 5 years with anaplastic tumors (63).

Angiosarcoma

Angiosarcomas rarely occur as primary tumors in the CNS (70–72). Some ostensibly solitary or multiple angiosarcomas of the brain are found later to be a component of systemic disease when the tumor is discovered in other organs. It is difficult in such instances to determine which was the primary lesion. Most angiosarcomas occur in the brain, few in the meninges. Patients range in age from neonates to the elderly. As expected, given their frankly malignant nature, angiosarcomas are contrast-enhancing lesions associated with edema and mass effect. Macroscopically, the specimen is often frankly hemorrhagic.

Microscopically, angiosarcomas invariably exhibit some element of vascular architecture. In these areas, lumens are poorly formed anastomosing channels or tubules, not well-defined vessels (fig. 10-40). The neoplastic cells vary from elongated to plump and have variable cytologic atypia and mitotic activity. A significant number of tumors are in part undifferentiated. Reticulin is abundant. The vascular nature of the neoplasm is readily confirmed by immunostaining of the cells for CD31 (fig. 10-41), CD34, factor VIII-related antigen, or *Ulex europaeus* agglutinin (72).

Too few cases of CNS angiosarcoma have been reported to permit generalization regarding their biologic behavior. In one series of eight patients, five died of disease within 4 months and two young patients with 3.0- and 8.5-year follow-ups appeared to have been cured by resection (72); no histologic features were predictive of tumor behavior.

BENIGN NONVASCULAR MESENCHYMAL TUMORS AND TUMOR-LIKE LESIONS

Solitary Fibrous Tumor

Definition. *Solitary fibrous tumor* (SFT) is a usually spindle cell neoplasm composed of fibroblastic cells in which strip-like bands of collagen are present.

General Features. The origin of the lesion is not clear, but some could be related to CD34-positive fibroblasts in normal dura (73). The possible relationship between SFT and HPC is discussed with the latter.

Clinical Features. SFTs almost always occur in adults, and like meningiomas, favor females. The lesions produce mass effects or neurological deficits related to tumor site (74–77). Nearly all tumors are meningeal, but occasional examples are intraventricular or spinal intramedullary (76,78). All levels and compartments of the neuraxis are affected, but the posterior fossa and spinal cord appear to be more often involved than is the case for meningioma (76,77,79).

Radiologic Findings. The tumors are well circumscribed, contrast enhancing, and primarily dura based (fig. 10-42). As such, they resemble meningiomas, but may lack a "dural tail" sign. Densely fibrotic examples are dark on T2-weighted or fluid-attenuated inversion recovery (FLAIR) imaging. Mass effects may be minor.

Gross Findings. The well-circumscribed, occasionally delicately encapsulated lesions are rubbery to firm in texture. On cut section, they are solid and gray-tan, with a somewhat nodular surface. Gross total removal is usually possible.

Microscopic Findings. The lesion is composed of spindle- to oval-shaped cells arranged in short fascicles, vague whorls, or sheets (figs. 10-43, 10-44). Intercellular stroma consists of dense, parallel, wire-like bands of brightly eosinophilic collagen that separates individual or clustered cells. Cellularity varies considerably, from densely compact, fascicular tissue composed exclusively of neoplastic cells to areas in which collagen predominates. Lesions of lower cellularity can resemble meningiomas (fig. 10-45). The neoplastic cells have indistinct cell borders, and nuclei with finely dispersed chromatin and inconspicuous nucleoli. Mitoses are uncommon generally, but may be apparent in small numbers in more cellular areas. Unlike HPC, there is little or no pericellular reticulin, only weakly staining or nonstaining mature collagen. About one third of tumors show regions of increased cellularity (76).

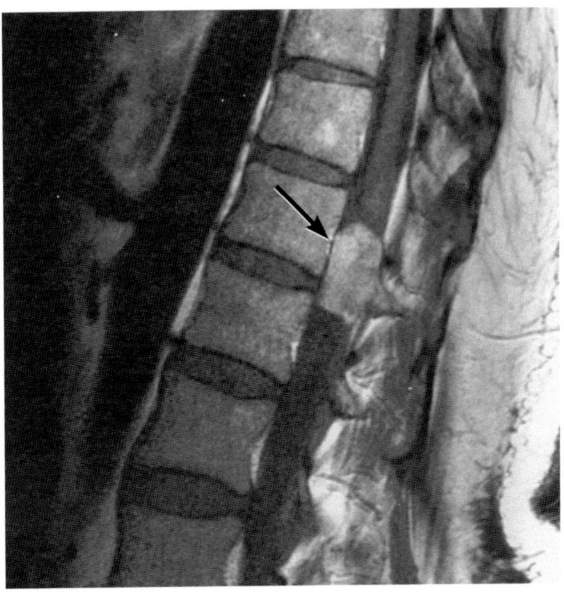

Figure 10-42

SOLITARY FIBROUS TUMOR

Whether intracranial or intraspinal, solitary fibrous tumors (SFTs) are usually extra-axial, well-circumscribed, meningioma-like masses (arrow).

Occasional SFTs are highly cellular and exhibit malignant features that include brisk mitotic activity and/or necrosis (fig. 10-46). Such features may characterize the entire tumor or appear in transition from more typical SFT. Anaplastic features may be widespread or appear focally in transition from more typical SFT. Such tumors occur at all sites in which SFTs arise but, at present, no precise histopathologic criteria of malignancy as related to negative outcome have been established.

Immunohistochemical Findings. Almost by definition, the lesions are extensively immunopositive for CD34, and often for BCL2 (fig. 10-47) (74,76,80), although there are occasional lesions that seem to be exceptions. A significant percentage of lesions are also CD99 reactive. EMA and S-100 protein stains are negative.

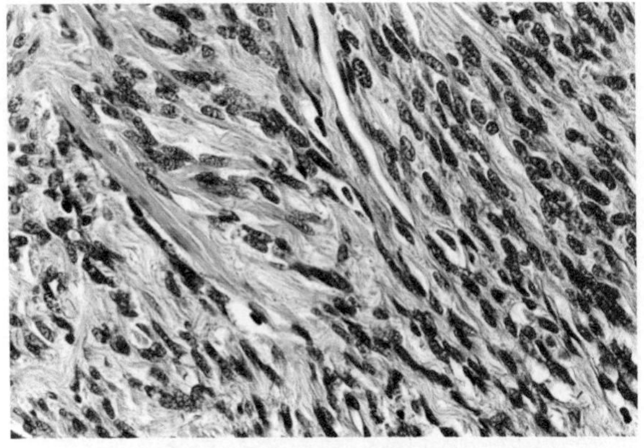

Figure 10-43

SOLITARY FIBROUS TUMOR

Sheets of tumor interrupted by brightly eosinophilic bands of "wire-like" collagen are characteristic.

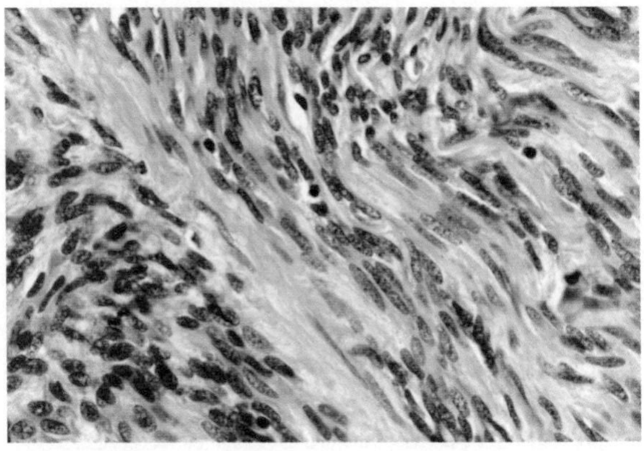

Figure 10-44

SOLITARY FIBROUS TUMOR

The lesion's fusiform nuclei are uniform in size, shape, and chromatin content.

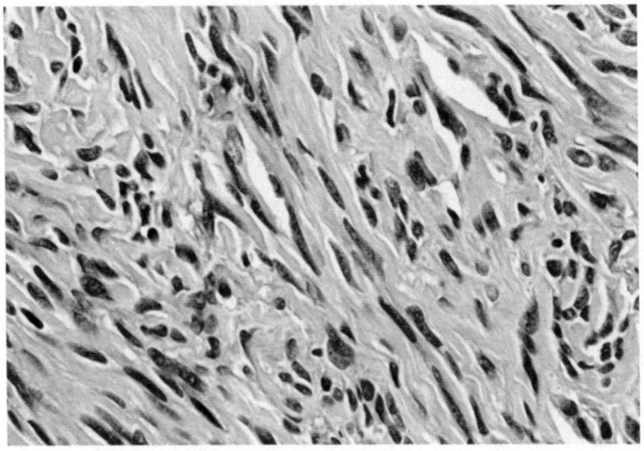

Figure 10-45

SOLITARY FIBROUS TUMOR

Less cellular examples simulate fibrous meningioma, although there is often more intercellular collagen in the SFT.

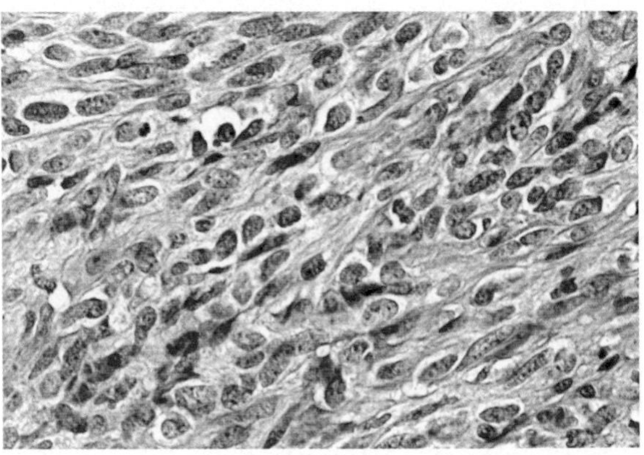

Figure 10-46

SOLITARY FIBROUS TUMOR

Only rarely is there sufficient cellularity and mitotic activity to warrent the designation anaplastic SFT.

Differential Diagnosis. The differential diagnosis between SFT and HPC is discussed above, and between SFT and fibrous meningioma in chapter 9 (74,80).

Treatment and Prognosis. Based on our current understanding, most SFTs are benign and cured by gross total resection (74,76,77). In one large series of 18 cases, 15 underwent gross total removal alone; none recurred (76). Three subtotally resected tumors were also irradiated; of these, two typical SFTs recurred at 3 and 5 years and one "anaplastic" tumor recurred at 1 year. The significance of foci of hypercellularity and scattered mitoses is not clear, but these may not predict recurrence (76). Too few cases of anaplastic SFT have been reported to draw conclusions regarding optimal therapy and prognosis (76).

Lipoma (Lipomatous Hamartoma)

Definition. *Lipoma* is a malformative or occasionally neoplastic mass composed of mature adipose tissue. The former is more appropriately called *lipomatous hamartoma.*

Clinical and Radiologic Findings. As errors of embryogenesis, not neoplasms, intracranial "lipomas" generally occur at midline sites such as the anterior corpus callosum (fig. 10-48), quadrigeminal plate (fig. 10-49), and base of the

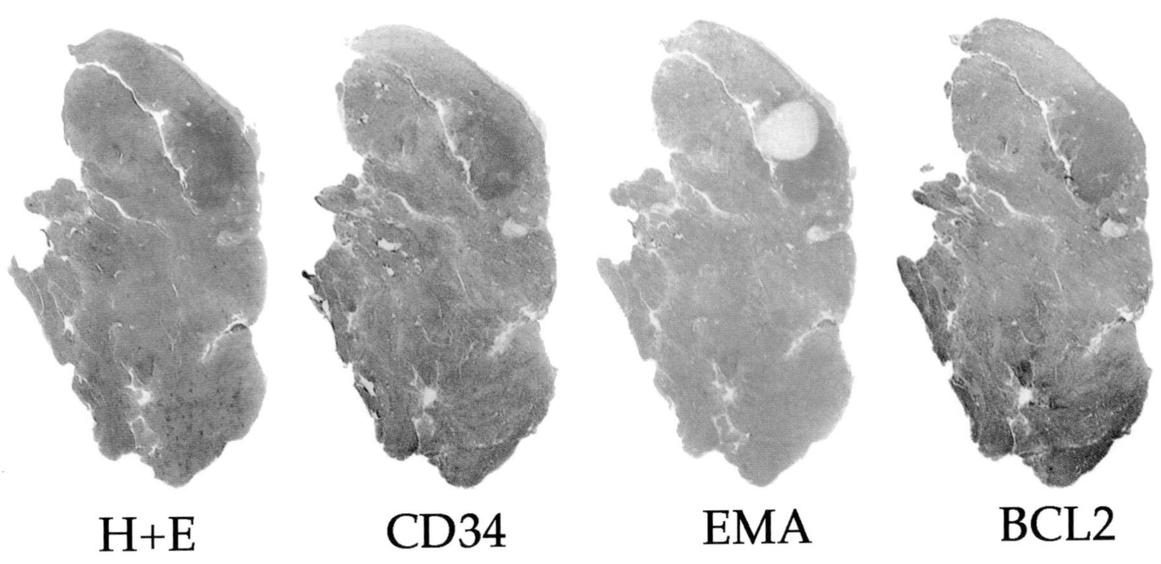

Figure 10-47

SOLITARY FIBROUS TUMOR

Almost by definition, SFTs are strongly and diffusely immunoreactive for CD34 and BCL2. In contrast to meningiomas, they are negative for epithelial membrane antigen (EMA).

brain (81–83). Callosal examples may be massive, and are frequently associated with partial or even complete agenesis of this preeminent commissure. Small lipomas, sometimes with a myoid component, that occur on cranial nerves are discussed in chapter 13 (figs. 12-32–13-35).

In light of their typical location and imaging characteristics, intracranial lipomas usually are diagnosed with confidence on MRI because of their extremely high (white) signal in precontrast T1-weighted MRIs (figs. 10-48, 10-49A) (83).

Known as *lipomeningocele* or *leptomyelolipoma*, intraspinal, malformative, non-neoplastic masses of adipose tissue occur at lumbosacral levels, where they are often associated with neural tube defects (84). Intradural spinal lipomas at more rostral levels are usually extramedullary (85) but, rarely, can be intramedullary (86). These more cephalad masses are usually unassociated with congenital abnormalities and may be neoplasms, or at least can enlarge, albeit slowly.

The extremely rare *liposarcoma* of the CNS is discussed in the section, Other Sarcomas. A diffuse hypertrophy or enlargement of spinal epidural adipose tissue (*epidural lipomatosis*) is a rare complication of steroid therapy or endogenous steroid excess (87).

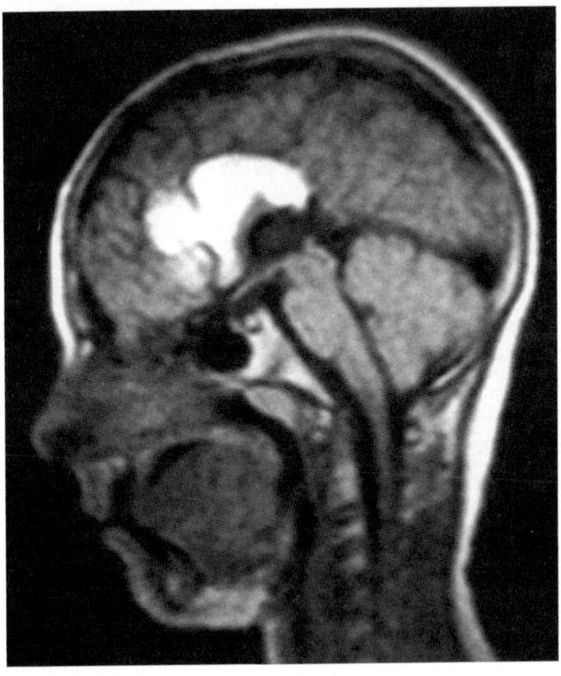

Figure 10-48

LIPOMA

The region of the corpus callosum is a common site for intracranial lipomas, which generate a bright signal in precontrast T1-weighted images. (Fig. 10-25 from Fascicle 10, 3rd Series.)

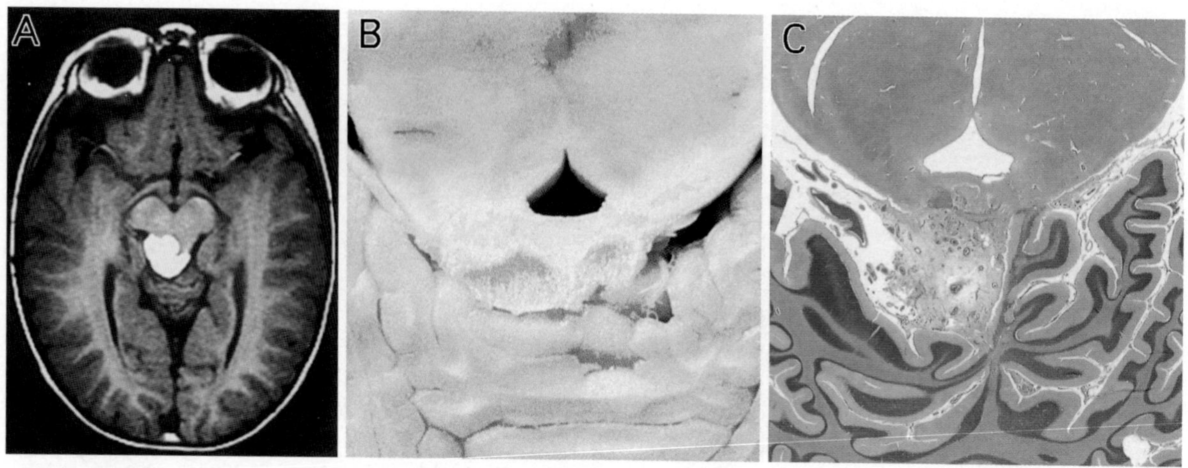

Figure 10-49

LIPOMA

Lipomas are diagnosed with great reliability given their MRI signal characteristics (A). The fatty lesions are appropriately bright yellow (B) and densely adherent to or intimately admixed with surrounding structures (C). The tectal plate is a common site for incidental examples.

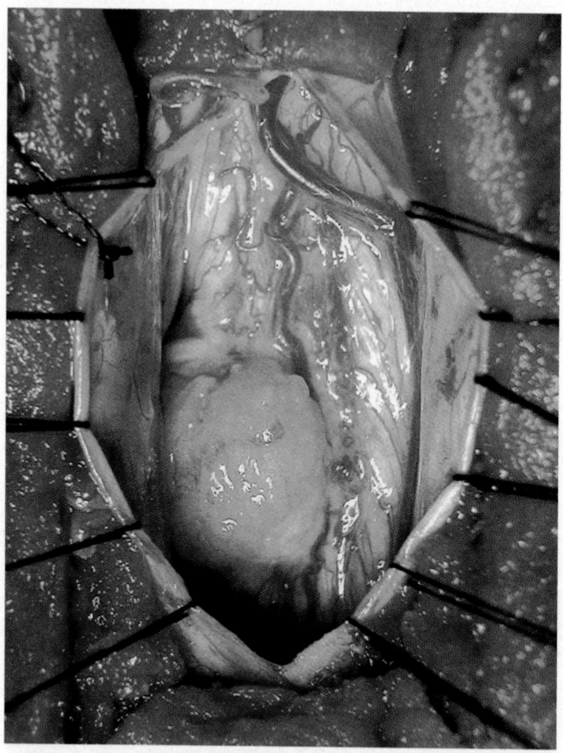

Figure 10-50

LIPOMA

Unlike the more clearly malformative intracranial lipomatous lesions, some of those occurring in the confines of the spinal canal behave more like a tumor. This lesion caused buttock pain in a 52-year-old woman. (Courtesy of Dr. Ziya Gokaslan, Baltimore, MD.)

Gross Findings. The bright yellow lesions entangle local vasculature and cranial nerves, and adhere to adjacent brain (figs. 10-49B, 10-50) or spinal cord.

Microscopic Findings. Lipomas are masses of mature adipose tissue, which although macroscopically demarcated, are unencapsulated and incorporate regional cranial nerves, spinal nerves, and blood vessels (figs. 10-49C, 10-51). Some callosal lipomas contain hyaline cartilage. *Lipomeningoceles* may contain a variety of elements including peripheral nerves, fibrous tissue, and striated muscle. Spinal extradural lipomas may be markedly vascular (*angiolipoma*) (88–90), as are some intraspinal examples (91–93).

Treatment and Prognosis. Surgical intervention is most often directed toward intraspinal lesions, more often as part of a repair of a neural tube defect than as an effort to remove an isolated true lipoma. Surgery for intracranial lesions is infrequent since the majority are incidental findings and total excision is neither necessary nor possible (83).

Chondroma and Osteochondroma

Chondromas are benign, slowly growing tumors that generally arise from the skull base (94) or dura (fig. 10-52) (95–99). In some cases, the tumors are a consequence of the systemic predisposition inherent in patients with Ollier's disease

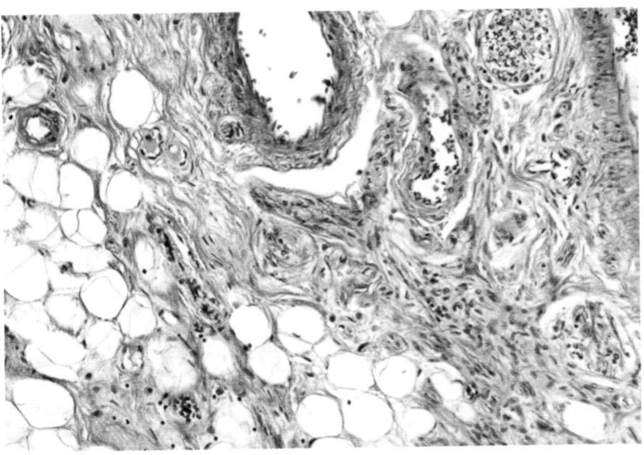

Figure 10-51

LIPOMA

Lipomas are intimately involved with regional structures, as in this tectal lesion also illustrated in figure 10-49. Note the trapped fibers of the trochlear nerve (H&E/Luxol-fast blue stain).

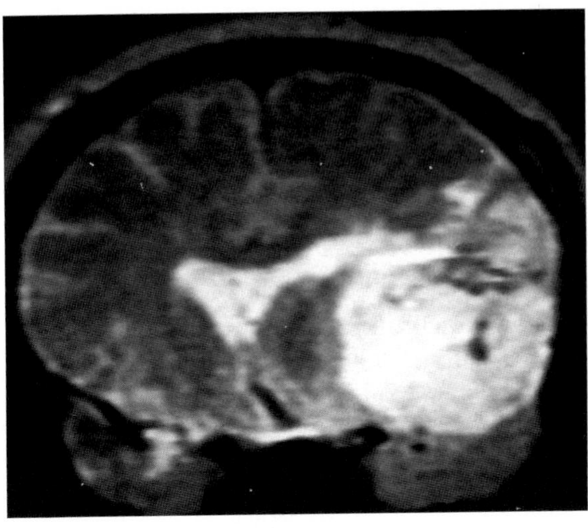

Figure 10-52

CHONDROMA

Intracranial chondromas are discrete masses usually arising from the dura. (Fig. 10-30 from Fascicle 10, 3rd Series.)

(100) or Maffucci's syndrome (101). At any site, chondromas are circumscribed, bosselated, exophytic masses composed of well-differentiated, cytologically benign hyaline cartilage (fig. 10-53). Rare lesions containing both cartilage and bone are designated *osteochondroma* (102).

Total excision may not be possible for skull-based lesions. Although they are slow growing, delayed clinical recurrence is possible. Cure is possible for dura-based tumors.

Fibroma and Fibromatosis

Fibromas are rare, often bulky, discrete masses arising either from the meninges (103,104) or within the brain parenchyma (105,106). Moderate in cellularity, they are composed of elongated, cytologically benign cells in an eosinophilic collagenous matrix. The observation of marked immunoreactivity for S-100 protein in one case (104), and the ultrastructural finding of desmosome-like junctions in another (107), leave open the issue of histogenesis and permit some heterogeneity of the lesions grouped under the designation "fibroma."

Fibromas must be distinguished from *fibromatoses*. The latter are reactive and progress naturally through an early cellular phase to a late collagen-rich lesion. Infiltrative and potentially aggressive, fibromatoses are prone to recur. They most often originate in the dura to grow within the craniospinal or epidural space. Some occur

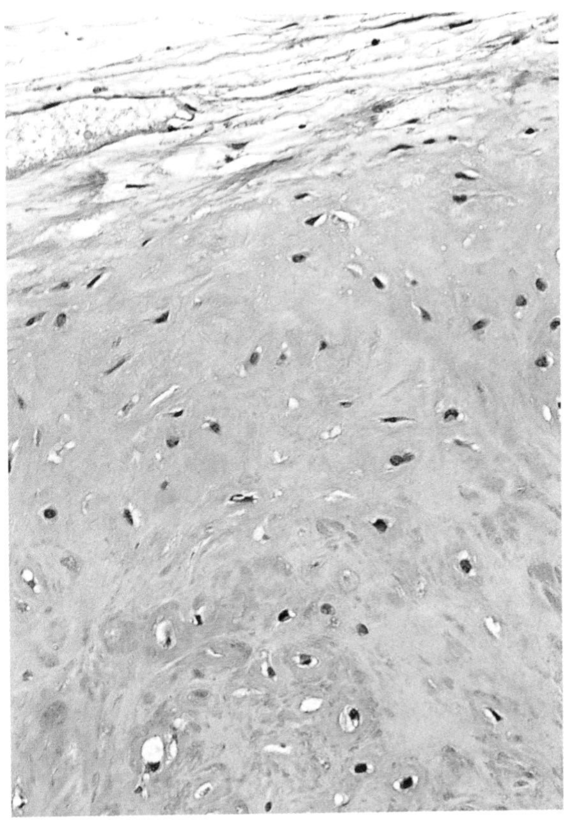

Figure 10-53

CHONDROMA

Chondromas consist of very well-differentiated hyaline cartilage.

spontaneously (108) and others arise at an operative site (109,110). Early, proliferative lesions should not be mistaken for low-grade fibrosarcoma. Unlike fibromatoses, fibrosarcomas consist of more atypical cells disposed in long fascicles or arranged in a herringbone pattern.

Other Benign Mesenchymal Tumors and Tumor-Like Lesions

Other uncommon benign mesenchymal lesions include *fibrous xanthoma* (111), *chondromyxoid fibroma* (112), and *leiomyoma* (113). Benign lesions of adipose tissue and/or muscle, both striated and smooth, on cranial nerves are discussed in chapter 13.

MALIGNANT NONVASCULAR MESENCHYMAL TUMORS AND TUMOR-LIKE LESIONS

Chordoma

Definition. *Chordoma* is a malignant notochordal tumor occurring along the axial skeleton in the region of the embryonic notochord, particularly at its cranial and sacral extremes.

General Features. Chordomas arise from the archipelago of notochordal remnants, from its rostral reaches in the region of the sella turcica to its caudal extreme in the sacrum. The tumors are unevenly distributed along this axis; approximately half arise in the sacrum, one third in the spheno-occipital region or clivus (114, 115), and the remainder in articulating vertebrae (fig. 10-54) (115–117). A notochordal origin seems inescapable in light of histologic, immunohistochemical, and ultrastructural features common to chordomas and notochordal remnants. Among the latter is *ecchordosis physaliphora,* a lesion often found incidentally at autopsy as a small translucent nodule of notochordal tissue projecting from the clivus by a delicate pedicle and secondarily attached to the basilar artery (figs. 10-55, 10-56).

From the ecchordosis noted above, there is a spectrum of larger, intracranial or intraspinal lesions exhibiting at least in part the histologic features of chordoma. Such intradural lesions have variously been termed *giant ecchordosis* or *intradural chordoma* (118–120).

Occasional spinal lesions are both extravertebral and extradural (121). By whatever designation, giant ecchordosis or chordoma, the prognosis of patients with extraosseous lesions appears much more favorable than for those with intraosseous chordomas.

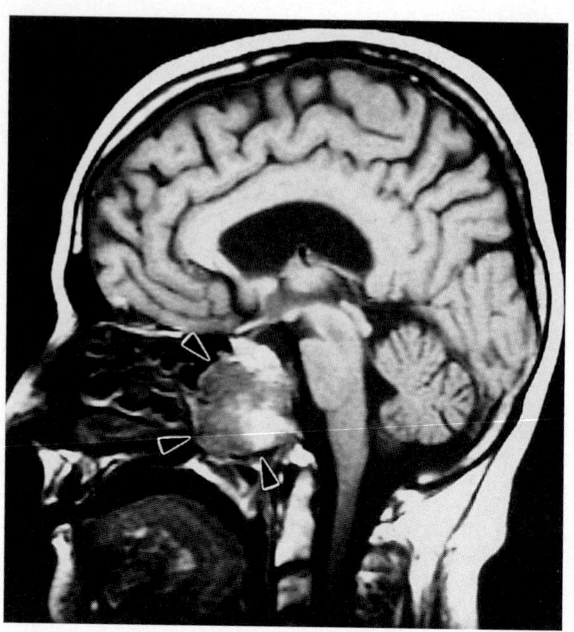

Figure 10-54

CHORDOMA

Cranial chordomas (arrowheads) typically arise from the clivus as bulky masses that destroy bone, trap cranial nerves, and compress the overlying brain. (Fig. 10-32 from Fascicle 10, 3rd Series.)

Entirely intraosseous, nondestructive vertebral lesions with ecchordosis-like features (*giant notochordal hamartoma*) and other chordoma-like tissues are discussed under Differential Diagnosis. Chordomas are described also in the Fourth Series Fascicle, *Tumors of the Bones and Joints* (122).

Clinical Features. Chordomas usually occur in adults, but tumors in children have been reported (123). Intracranial lesions straddle the midline and erode the clivus, and generally produce headaches and often unilateral cranial nerve palsies, particularly diplopia. Sacral chordomas are destructive intrasacral tumors that produce pain, sphincter disturbances, and neurologic symptoms from pressure upon regional nerve roots (fig. 10-57).

Gross Findings. Chordomas are lobulated, mucoid, often hemorrhagic masses that permeate and destroy bone (fig. 10-57). Chondroid

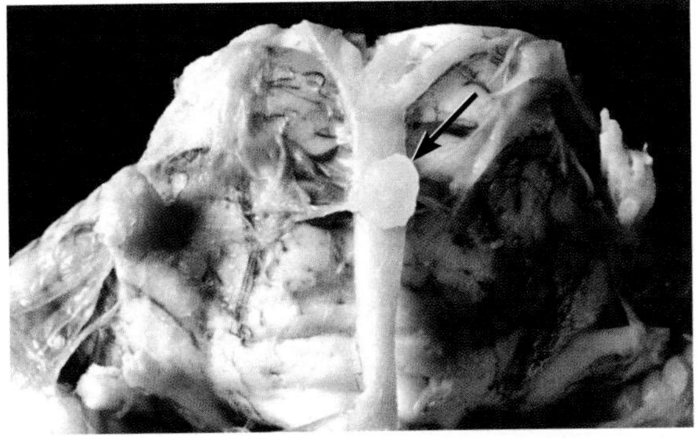

Figure 10-55

ECCHORDOSIS PHYSALIPHORA

Small gelatinous, translucent nodules of notochordal tissue (arrow) are occasionally present over the basilar artery.

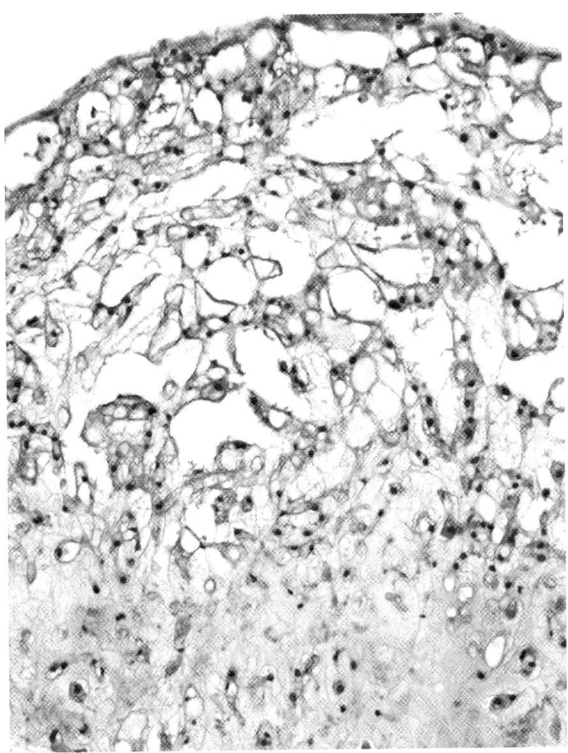

Figure 10-56

ECCHORDOSIS PHYSALIPHORA

Epithelioid cells in a myxoid background closely resemble those of chordoma.

regions or calcifications are encountered primarily in clival examples.

Microscopic Findings. The macroscopic lobularity is reflected at the microscopic level, wherein tumor lobules are separated by stromal septa (figs. 10-58, 10-59). At higher magnification, great histologic variability can be expected. Commonly, the neoplastic cells are epithelial in appearance and strung out in rows or cords within a mucoid matrix (figs. 10-59, 10-60). Their cytologic appearance varies from cells with nonvacuolated eosinophilic cytoplasm, through ones containing a single large vacuole and having a signet ring cell appearance, to multivacuolated cells with bubbly cytoplasm (physaliphorous cells) (fig.10-61). The latter are a classic feature of chordoma and are particularly well seen in smear preparations (see fig. 10-66). The cells of chordomas contain relatively uniform, bland-appearing nuclei with dense chromatin. Mitoses are rare or nonexistent. The tumor matrix varies in appearance. Extracellular mucus is common in chordomas

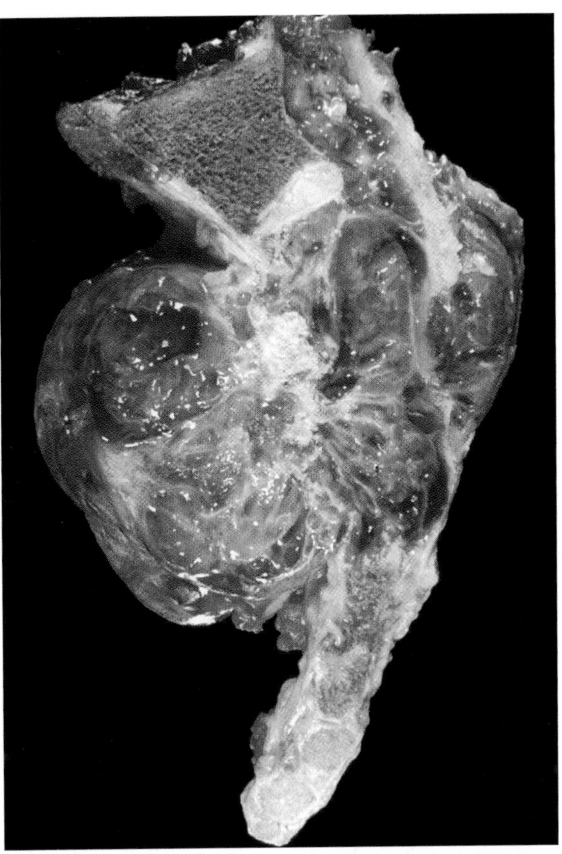

Figure 10-57

CHORDOMA

Sacral chordomas are bulky, lobular, and gelatinous. (Courtesy of Dr. Ziya Gokaslan, Baltimore, MD.)

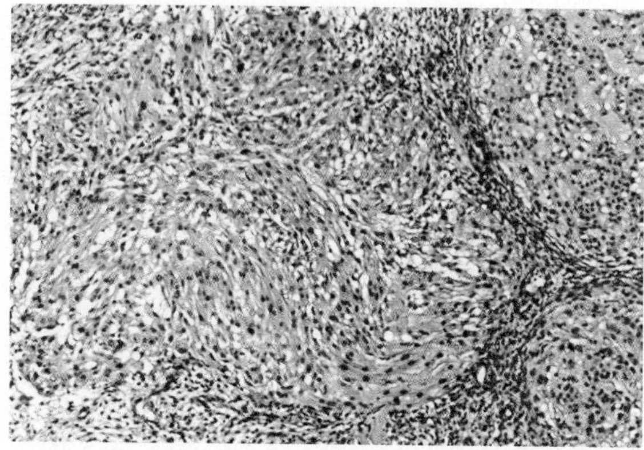

Figure 10-58

CHORDOMA

Chordomas consist of cellular, mucopolysaccharide-rich lobules associated with a fibrous stroma.

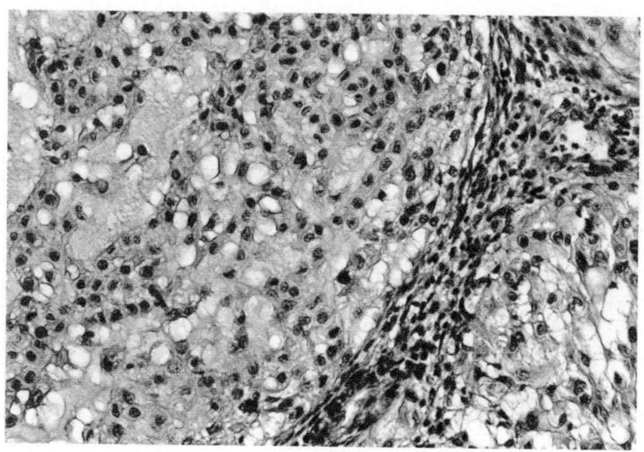

Figure 10-59

CHORDOMA

Chordomas are lobular, mucin-rich, and epithelial.

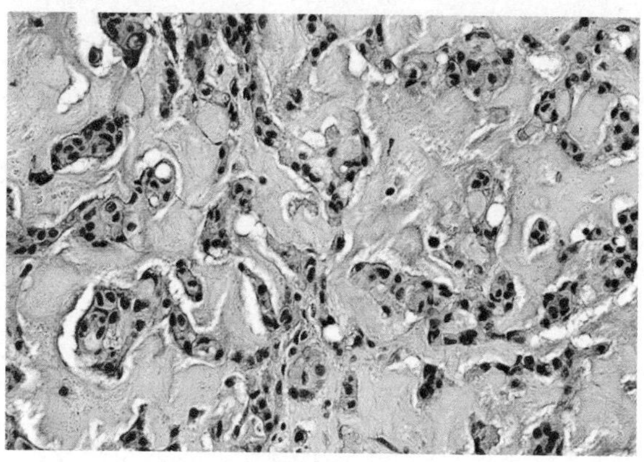

Figure 10-60

CHORDOMA

The epithelial features of chordomas are often apparent as ribbons or cords of cells with discrete cell borders.

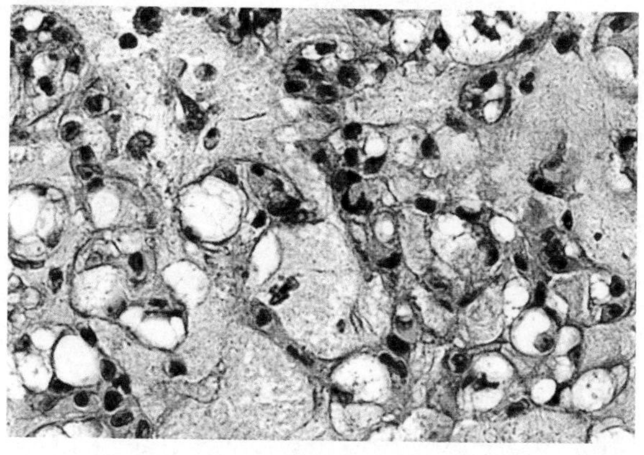

Figure 10-61

CHORDOMA

Tumor cells with mucin-filled vacuoles are termed physaliphorous cells.

of any pattern, and may be a dominant feature (fig. 10-62). It contains hyaluronidase-resistant sulfated mucopolysaccharides.

Only a minority of chordomas, particularly clival lesions, exhibit focal or widespread cartilaginous differentiation and are termed *chondroid chordoma* (fig. 10-63) (124,125). This variant is further discussed in the Differential Diagnosis and Treatment and Prognosis sections.

So-called *dedifferentiated chordomas* are biphasic tumors composed of both typical chordoma and a pleomorphic sarcomatous component. In most instances, the latter resembles malignant fibrous histiocytoma, but fibrosarcoma, osteosarcoma, and high-grade chondrosarcoma elements can appear (fig. 10-64). Sarcomatous components are typically a feature of multiply recurrent chordomas, irradiated or not (126–130).

Immunohistochemical Findings. The epithelial phenotype of classic and chondroid chordomas is expressed as reactivity for keratins and EMA (fig. 10-65). Half of classic chordomas and 85 percent of chondroid chordomas (see below) are positive for S-100 protein (131–135). Dedifferentiated chordomas lack reactivity for epithelial markers.

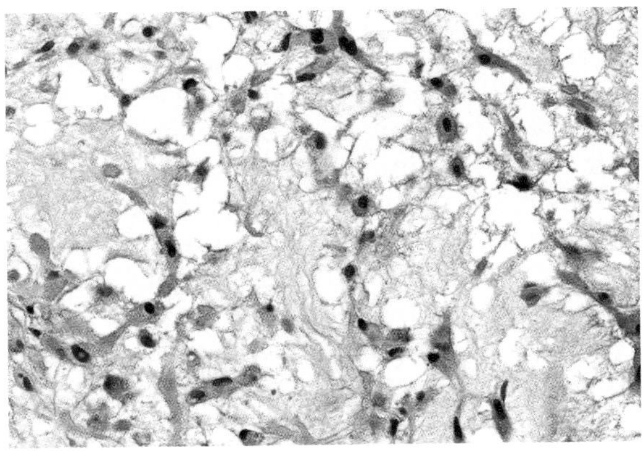

Figure 10-62

CHORDOMA

Extracellular matrix rich in mucopolysaccharides dominates some tumors.

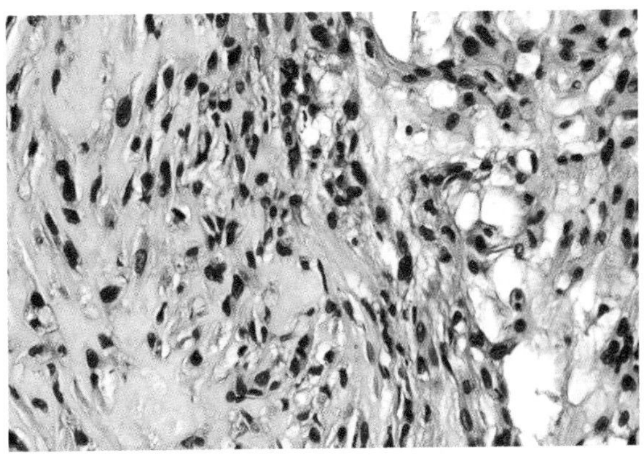

Figure 10-63

CHONDROID CHORDOMA

Cartilaginous differentiation is most often seen in clival chordomas.

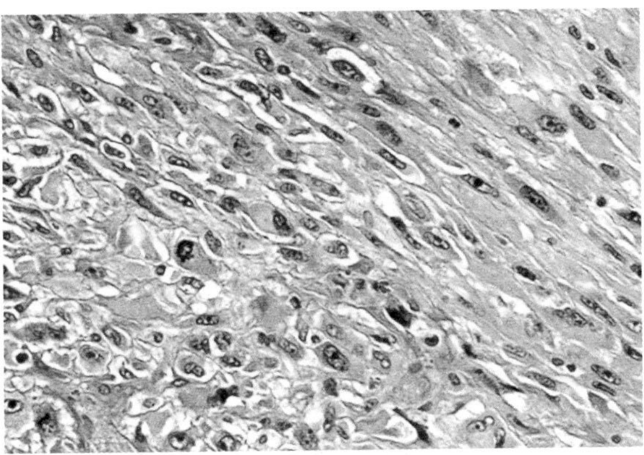

Figure 10-64

CHORDOMA: ANAPLASTIC TRANSFORMATION

Anaplastic degeneration in a chordoma is expressed as spindle cell transformation, high cellularity, and mitotic activity.

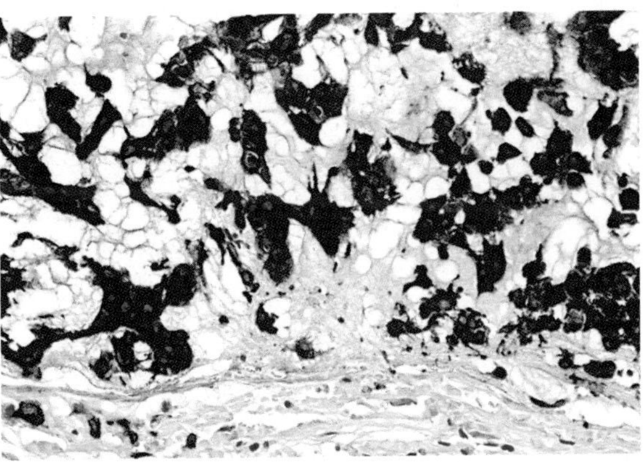

Figure 10-65

CHORDOMA

As indicated by immunoreactivity for EMA, chordomas are essentially epithelial neoplasms.

Cytologic Findings. Epithelioid and bubbly physaliphorous cells sit in a mucinous matrix (fig. 10-66) (136).

Ultrastructural Findings. As in notochordal tissue and in ecchordosis, the epithelial qualities of classic chordoma are obvious at the ultrastructural level, in the form of tonofilament bundles and desmosomes (137,138). Abundant cytoplasmic mucus vacuoles and extracellular mucus are also prominent features.

Differential Diagnosis. Benign notochordal tissue occurs in two forms, and neither should be interpreted as chordoma. The first comprises small rests that commonly sit in the center of intervertebral discs. The constituent cords of cells in a mucopolysaccharide matrix closely resemble those of chordoma. By site and size alone, however, these rests are unlikely to be confused with chordoma. One study suggests that these rests are immunoreactive for CD18, whereas chordomas are not (139).

The second form of notochordal tissue is usually discovered incidentally in radiologic studies in the course of a workup for back pain (fig. 10-67). These lesions, which do not destroy trabecular or cortical bone, have been variously termed giant notochordal hamartoma (140), giant vertebral notochordal rest (141), benign notochordal lesion (142), and benign notochordal tumor (139). Microscopically, these are small, circumscribed, but not encapsulated, masses of vacuolated cells that resemble intertrabecular adipose tissue (fig. 10-68, left). In contrast to notochordal rests and chordomas, a mucopolysaccharide matrix is absent. Hyaline droplets may be present. Immunostains for keratins and EMA are strongly positive (fig. 10-68, right). Useful in the separation from chordoma is that one of the keratins, CK18, is negative in chordomas (139). Although these lesions may well be benign notochordal neoplasms, they do not require excision. One case, however, did evolve into a chordoma (142).

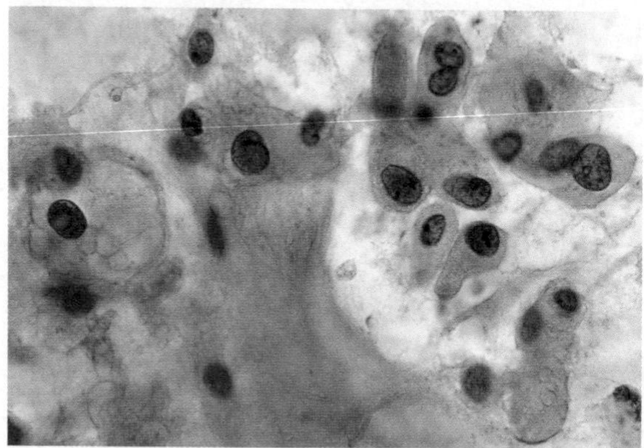

Figure 10-66

CHORDOMA: SMEAR PREPARATION

Vacuolated physaliphorous cells and more epithelioid-appearing cells with distinct cell borders are typical.

Classic chordoma, with its cords of epithelial cells and distinctive physaliphorous cells lying in a mucoid matrix, is a highly distinctive neoplasm posing little diagnostic problem. When its cells sit in small lacunae within chondroid matrix (chondroid chordoma), however, the similarity to low-grade chondrosarcoma becomes an issue. The distinction may be particularly difficult when a classic chordomatous component is not present. Nonetheless, chondrosarcoma cells have fewer epithelial features than chordoma and have smaller, darker nuclei. Their cytoplasm generally lacks the marked vacuolization seen in chordomas. At the immunohistochemical level, chondrosarcomas do not stain for keratins or EMA. Both tumors may be S-100 protein immunoreactive, but this is more often the case for chondrosarcomas.

Taking a simplistic approach, some observers consider chondroid chordomas to be simply chondrosarcomas. One comparative immunohistochemical study, however, found convincing cytokeratin and EMA reactivity in 32

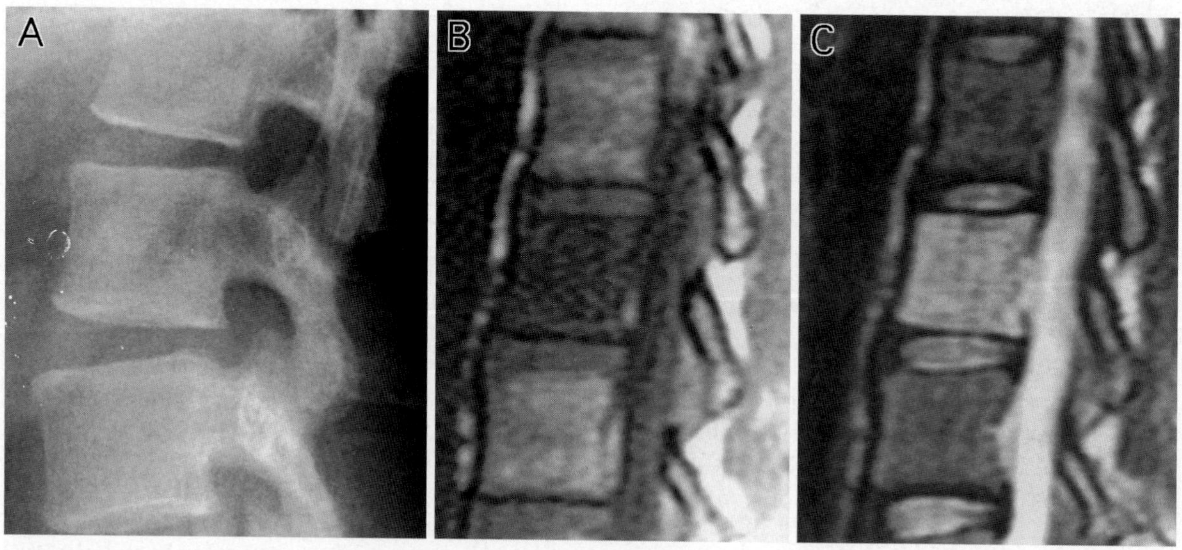

Figure 10-67

GIANT NOTOCHORDAL HAMARTOMA

Conventional radiographs show only minimal sclerosis in the central vertebral body (A). The entirely intraosseous lesion is dark on T1-weighted MRI (B) and has a bright signal on T2-weighted imaging (C).

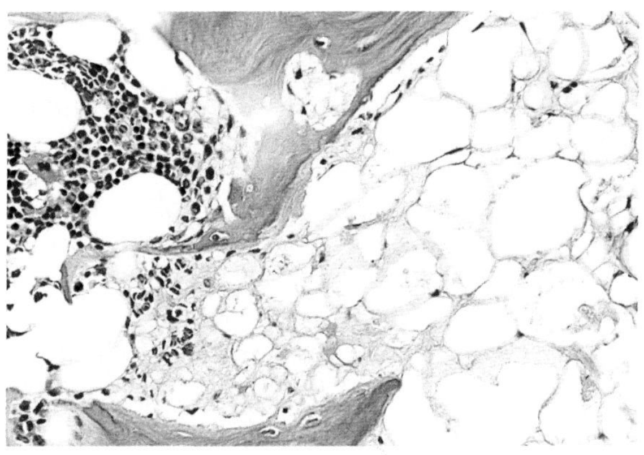
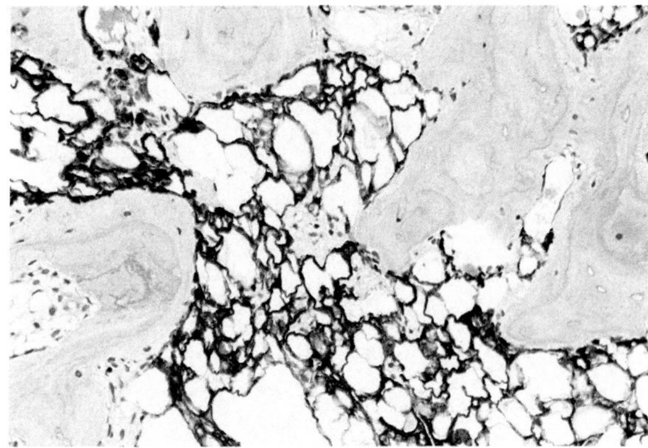

Figure 10-68

GIANT NOTOCHORDAL HAMARTOMA

Left: Such hypocellular lesions more closely resemble ecchordosis than chordoma. This particular example could easily be mistaken for intertrabecular fat.

Right: Immunoreactivity for cytokeratins is in keeping with the lesion's notochordal origin.

percent of cartilaginous tumors of the skull base. By this analysis, these epithelium-marking tumors are chordomas, whereas the remainder are chondrosarcomas (133).

Chordoma must also be distinguished from metastatic mucinous adenocarcinoma. Histochemically, the mucins in these lesions differ. The mucin in chordoma is acid mucin whereas it is neutral, epithelial-type mucin positive for periodic acid–Schiff (PAS) in carcinomas. Mucicarmine may stain both tumors. Routine immunohistochemistry may be misleading given the reactivity for keratins and EMA in chordomas. Stains for carcinoembryonic antigen are of use in that chordomas do not express this antigen. At the ultrastructural level, chordomas also lack microvilli-containing acini.

Treatment and Prognosis. Given the obstacles to total resection, patients with chordomas have a high mortality rate. In a series of 36 patients with skull-based lesions, mean survival was 4.1 years; only 1 patient survived 10 years (124). In one study of sacrococcygeal lesions, the estimated 5- and 10-year overall survival rates were 88 percent and 49 percent (30). In another series of chordomas, of both sacrum and mobile spine, the estimated 5-, 10-, 15-, and 20-year survival rates were 84, 64, 52, and 52 percent, respectively (116).

At any site, wide surgical excision at the time of initial surgery offers the best chance of cure. For subtotal resections, the addition of radiotherapy is beneficial. Proton beam therapy offers targeting advantages over conventional radiation. Advances in surgical techniques and in the use of reconstructive devices now make total resection of some chordomas (sacral and vertebral) more feasible and improve the outlook (143,144). Chordomas are noted for late recurrence and occasional metastasis to lung, lymph nodes, and skin (145,146).

A better prognosis is reportedly associated with chondroid chordoma (124). That this is in part an artifact of patient age is suggested by a study showing that at 5 years, all patients younger than 40 years, including those with classic and chondroid tumors, were alive (133). In contrast, of patients over 40 years, only 22 percent of those with classic chordoma and 38 percent of those with chondroid chordoma were alive. There were no differences in survival between patients with keratin-positive tumors (chondroid chordoma) and keratin-negative lesions (chondrosarcoma).

Attempts to grade chordomas have not been successful. According to one study of irradiated skull-based chordomas, however, predictors of shortened overall survival include female sex, tumor necrosis in the pretreatment biopsy, and tumor volume exceeding 70 mL (147). As noted above, tumors that are entirely extraosseous may be cured by excision.

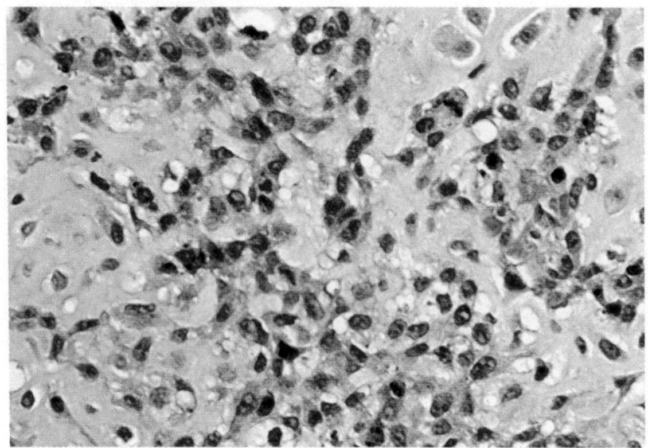

Figure 10-69

CHONDROSARCOMA

Mitotic activity and the degree of cytologic atypia distinguish this chondrosarcoma from chondroma.

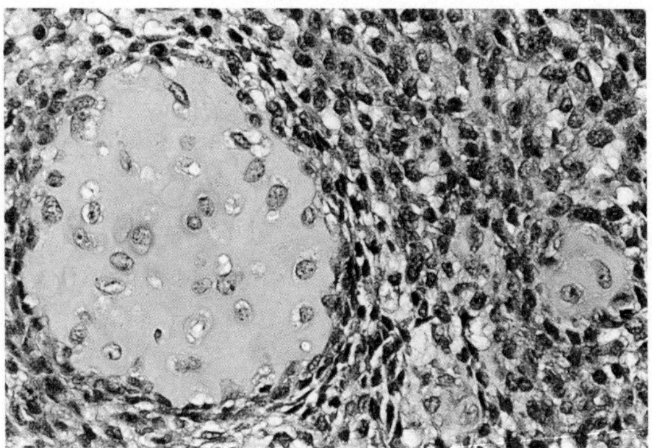

Figure 10-70

MESENCHYMAL CHONDROSARCOMA

Mesenchymal chondrosarcomas combine a small cell, hemangiopericytoma-like background with well-defined islands of hyaline cartilage.

Chondrosarcoma

Intracranial *chondrosarcomas* (fig. 10-69) arise from either the dura (148–153) or the skull base. Given their often bulky nature, a precise origin is difficult to establish at the latter site. Some lesions appear to be radiation induced (154). Most are classic chondrosarcomas, but the myxoid variant is also encountered (153). Mesenchymal chondrosarcoma is described separately below.

In the skull, the tumor should be distinguished from chondroma and chondroid chordoma. Given the well-differentiated characteristics of most intracranial chondrosarcomas, the distinction from a chondroma, a hypocellular lesion showing little atypia, may be difficult. Features favoring sarcoma include osseous permeation, myxoid change, cytologic malignancy, and abundant multinucleation. Unlike chondroid chordoma, chondrosarcomas lack chordomatous elements and are nonreactive for keratins and EMA. In addition, they are always positive for S-100 protein. Given the frequent resectability of chondroma and chondrosarcoma of dura, their distinction may not be critical.

It is difficult to generalize about the course of patients with totally excised dura-based chondrosarcomas (155). Skull-based chondrosarcomas often cannot be totally excised, and are quite likely to recur.

Mesenchymal Chondrosarcoma

Mesenchymal chondrosarcoma is a rare, aggressive lesion that affects the nervous system by direct extension from a nearby osseous primary (cranial or spinal), from an origin in dura, or by arising directly within brain parenchyma (156–161). Most patients present in the first three decades of life.

Mesenchymal chondrosarcomas are bulky, solid neoplasms with an unmistakable histologic pattern characterized by well-circumscribed islands of cytologically bland or atypical hyaline cartilage. The cartilage is suspended in a densely cellular field of small cells showing variable mitotic activity (fig. 10-70). The latter cells may be round, oval, or slightly spindle shaped and thus resemble those of HPC, a key entity in the differential diagnosis. The likeness is furthered by an irregular or staghorn-like vasculature and an abundance of intercellular reticulin. Thorough specimen sampling may be necessary to find the tell-tale cartilaginous islands since they are, with rare exception, only a small part of the tumor.

When seen alone, the small cell component may also mimic Ewing's sarcoma. The small cells of both tumors are glycogen rich, but Ewing's sarcoma lacks spindle cells and, of course, cartilaginous islands. Small cell osteosarcoma features lace-like deposits of osteoid rather than circumscribed cartilaginous islands.

Clinical experience with cranial or spinal mesenchymal chondrosarcomas is limited, but frequent recurrence and distant metastasis is the rule. Radiation and chemotherapy may be useful.

Rhabdomyosarcoma

Definition. *Rhabdomyosarcoma* is a malignant neoplasm composed of rhabdomyoblasts.

General Features. Given the frequency with which primary intracranial rhabdomyosarcomas exhibit an undifferentiated small cell component, the distinction of these tumors from embryonal neuroepithelial tumors with rhabdomyoblastic differentiation is both a conceptual and practical issue. Some observers interpret largely undifferentiated neoplasms with rhabdomyoblastic features as primary neuroepithelial neoplasms; others classify the same tumors as embryonal rhabdomyosarcomas. This issue is of therapeutic significance and becomes relevant in reference to medullomyoblastoma, a rare posterior fossa tumor discussed below.

Rhabdomyosarcomatous components may be seen in malignant germ cell tumors of the pineal and suprasellar regions, specifically in teratomas with malignant transformation (i.e., with somatic-type elements) (162). Skeletal muscle differentiation is rarely seen in gliosarcomas (163) or in other gliomas such as subependymoma (164). CNS lesions exhibiting muscle differentiation are summarized in Appendix I.

Clinical and Radiologic Findings. Rhabdomyosarcomas arise throughout the neuraxis, primarily in children but occasionally in adults. In the posterior fossa, the lesion should be distinguished from the rare medullomyoblastoma, a task that is not always possible (165–168). Rhabdomyosarcomas of the cerebral hemispheres or spinal cord (169–172) have a better claim to be entirely sarcomatous and independent of an underlying small cell embryonal tumor (primitive neuroectodermal tumor [PNET]).

The possibility that CNS involvement represents extension from a rhabdomyosarcoma in a paranasal sinus should be considered in the face of diffuse leptomeningeal involvement.

Gross Findings. Most lesions are fleshy and many are partly hemorrhagic. Some are attached to the dura whereas others are leptomeningeal (173). Diffuse leptomeningeal lesions have also been described (174,175). The relationship, if any, of meningeal rhabdomyosarcoma to the foci of skeletal muscle occasionally found in normal leptomeninges (176,177) is unclear.

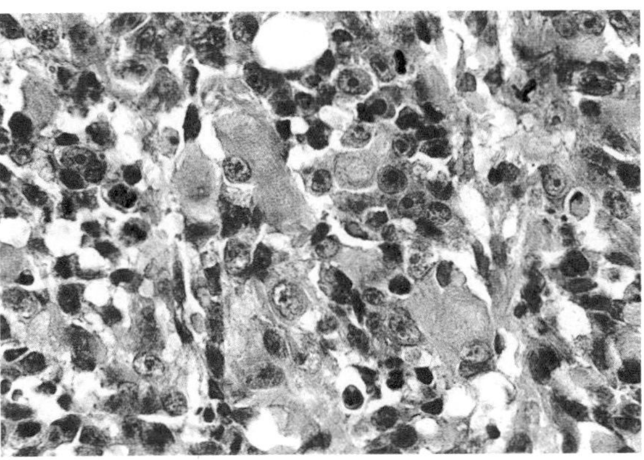

Figure 10-71

RHABDOMYOSARCOMA

Globular rhabdomyoblasts and striated strap cells compete for attention in rhabdomyosarcomas.

Microscopic Findings. Most rhabdomyosarcomas are densely cellular, consisting of small cells with eccentric cytoplasm, strap cells, or both. The filament-rich cytoplasm is often brightly eosinophilic with the hematoxylin and eosin (H&E) stain. Cross striations are most obvious in strap cells (fig. 10-71). Well-differentiated myotubes are less often present.

Immunohistochemical Findings. Myoid cells are positive for myoglobin, myosin, muscle-specific actin (fig. 10-72), and desmin (178). Myogenin reliably labels myoid cell nuclei. Stains for neuroectodermal markers such as GFAP and synaptophysin are negative.

Ultrastructural Findings. The ultrastructural findings mirror the degree of differentiation seen at the light microscopic level. Orderly, elongated arrays of well-formed sarcomeres characterize strap cells, whereas disorganized masses of thick and thin filaments with only occasional Z-bands fill the cytoplasm of globular myoblasts (fig. 10-73) (166,167,172,174,178).

Differential Diagnosis. Supratentorially, the presence of rhabdomyoblasts in a cellular and mitotically active neoplasm raises a nosologic problem. Is the lesion a primary rhabdomyosarcoma or a poorly differentiated embryonal neoplasm (PNET) with rhabdomyoblastic differentiation?

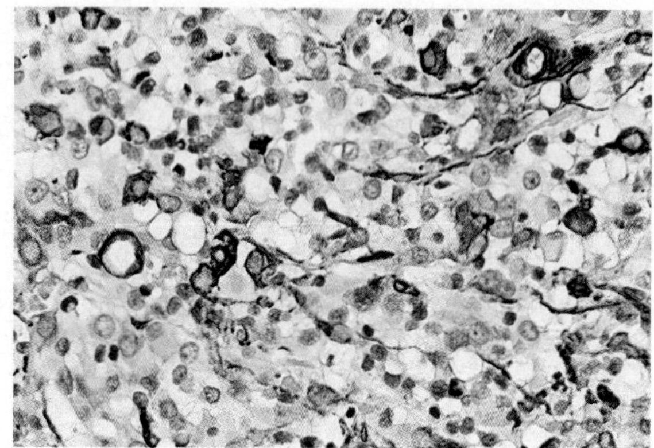

Figure 10-72

RHABDOMYOSARCOMA

Immunohistochemistry for actin can be used to confirm the presence of myoid differentiation.

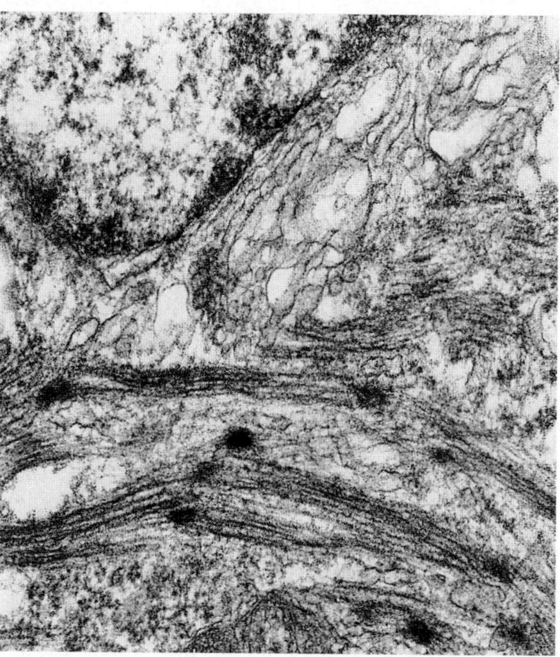

Figure 10-73

RHABDOMYOSARCOMA

Thick and thin filaments and abortive Z-band formation are diagnostic of rhabdomyoblastic differentiation. (Fig. 10-41 from Fascicle 10, 3rd Series.)

The latter is very rare and occurs primarily in the cerebellum (medullomyoblastoma).

Confirmation of an underlying PNET depends upon identifying neuroectodermal features in the undifferentiated cells, a process requiring immunocytologic or ultrastructural study. Immunostains may succeed in demonstrating synaptophysin or GFAP positivity in undifferentiated-appearing cells. The simplest approach is staining for synaptophysin, a marker of many small cell neuroectodermal tumors.

Treatment and Prognosis. The results of treatment have been disappointing: most patients with rhabdomyosarcoma are dead within 2 years despite aggressive radiotherapy and chemotherapy (167–169). There are some, however, who have survived without residual or recurrent disease for longer periods (167,170).

Fibrosarcoma

Definition. *Fibrosarcoma* is a neoplasm attributed to the neoplastic transformation of fibroblasts.

Clinical Features. Within the CNS, fibrosarcomas may appear as the mesenchymal component of gliosarcoma (discussed in chapter 3), as radiation-induced neoplasms (fig. 10-74) (179–182), and rarely, as spontaneously occurring tumors (fig. 10-75) (183–185). Of the last, some arise in childhood (186,187). Fibrosarcomas arising in the sellar region following radiation for pituitary adenoma (fig. 10-74) (179–181) are often intimately associated with residual or recurrent adenoma. The postirradiation latency period for cranial and intracranial sarcomas in one study ranged from 4 to 15 years (179).

Gross Findings. Fibrosarcomas are either dura-based or arise in the brain, presumably from leptomeningeal infoldings. The latter can, by extending diffusely along perivascular Virchow-Robin spaces, produce macroscopic hypertrophy of affected gyri (fig. 10-76).

Microscopic Findings. The histologic features of fibrosarcomas affecting the CNS are the same as those at any other body site: a cellular mitotically active lesion in which a reticulin-rich, fascicular, "herringbone" pattern is common (figs. 10-77, 10-78). A sclerosing epithelioid variant has been reported (188,189). In contrast to the diffuse single cell infiltration of the parenchyma inherent in diffuse astrocytomas and most glioblastomas, fibrosarcomas are compact masses that extend into surrounding brain and along blood vessels (fig. 10-79).

Differential Diagnosis. Given the rarity of CNS fibrosarcomas, exclusion of gliosarcoma is essential. The search for a neoplastic glial

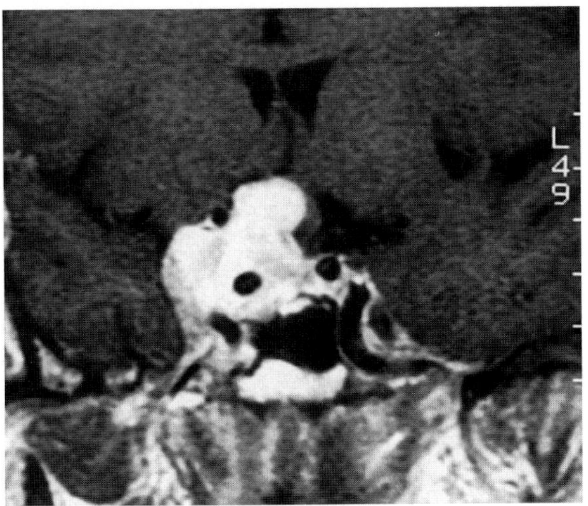

Figure 10-74

RADIATION-INDUCED FIBROSARCOMA

Radiation therapy is often implicated in fibrosarcomas of the sellar region, as in this individual treated 7 years earlier for pituitary adenoma.

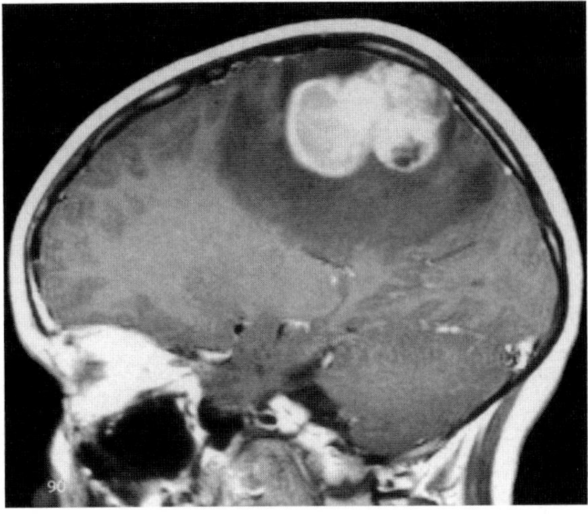

Figure 10-75

FIBROSARCOMA

Fibrosarcomas are discrete, edema-generating masses that are usually dura based. Such was the case with this right parietal mass in a 9-year-old girl.

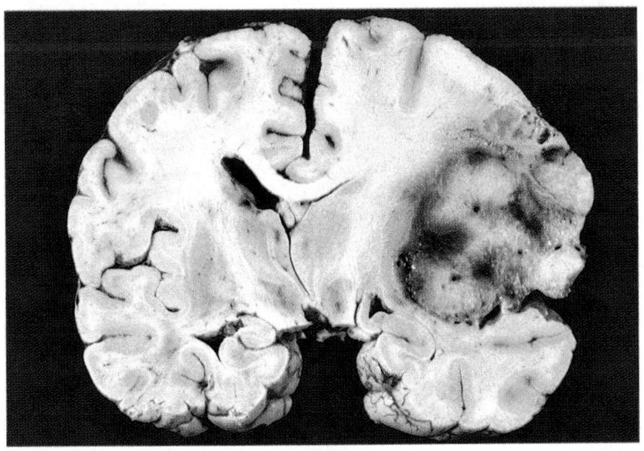

Figure 10-76

FIBROSARCOMA

Fibrosarcomas of the central nervous system (CNS) are often large, fleshy masses with a macroscopic appearance similar to that of fibrosarcomas occurring at any site.

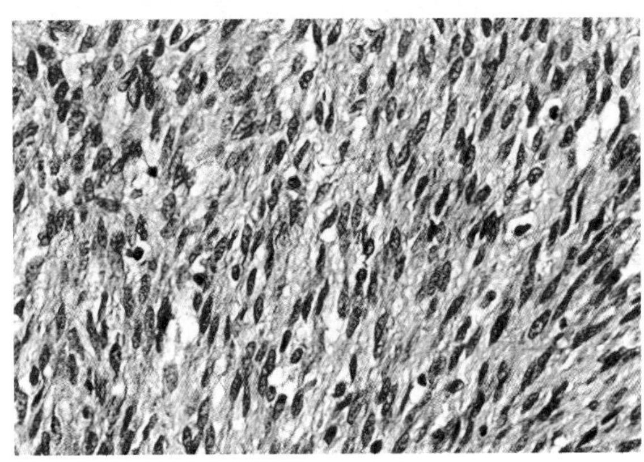

Figure 10-77

FIBROSARCOMA

Compact, intersecting fascicles of densly packed spindle-shaped cells produce the classic "herringbone" pattern of fibrosarcoma.

element has it pitfalls, however, since sarcomas may trap small islands of reactive astrocytes.

As noted in the discussion of malignant meningioma, it is not always possible to determine whether a malignant meningeal spindle cell neoplasm has arisen de novo or has evolved by anaplastic transformation of a meningioma. The presence of a herringbone pattern suggests a primary fibrosarcoma rather than sarcomatous change in a meningioma. Pericellular reticulin, virtually a defining feature of fibrosarcoma (fig. 10-80), is generally absent in malignant meningioma. Membrane staining for EMA is frequent in malignant meningioma but lacking in fibrosarcoma. S-100 protein reactivity is common in fibrous meningioma, and helps exclude fibrosarcoma. Also helpful is generous tissue sampling to search for whorls or psammoma bodies in better-differentiated portions of the tumor. Lastly, a past history of meningioma speaks for itself,

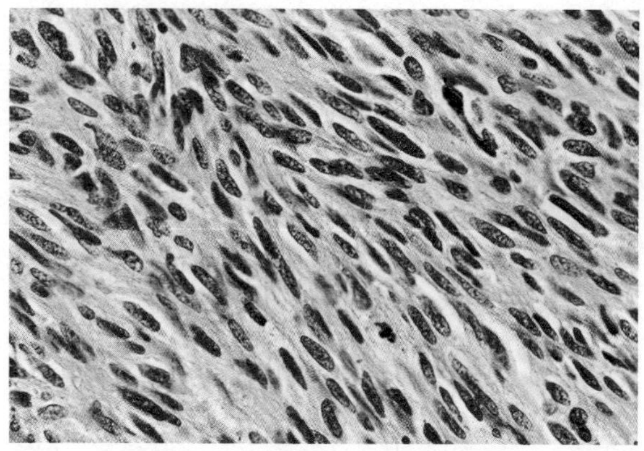

Figure 10-78

FIBROSARCOMA

The cells of fibrosarcoma are elongated, cytologically atypical, and mitotically active.

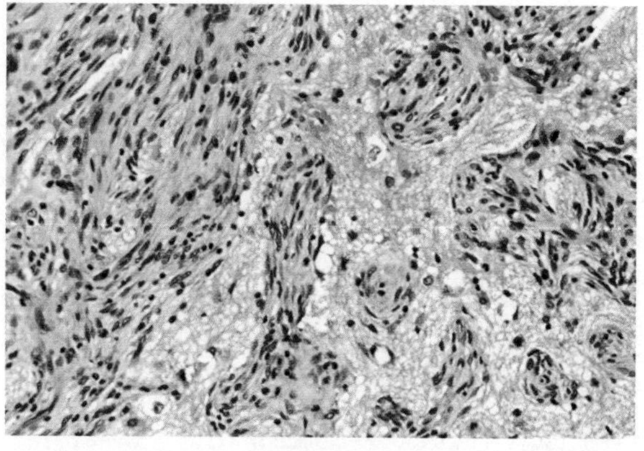

Figure 10-79

FIBROSARCOMA

Although they are not usually diffusely invasive of CNS tissue, fibrosarcomas extend freely along perivascular spaces.

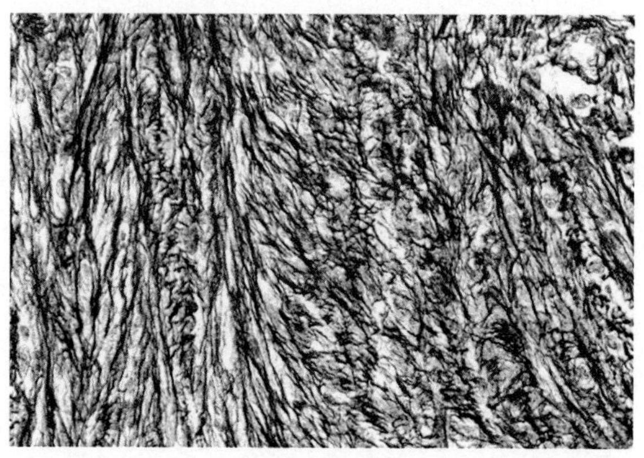

Figure 10-80

FIBROSARCOMA

The presence of pericellular reticulin is a consistent finding in fibrosarcomas.

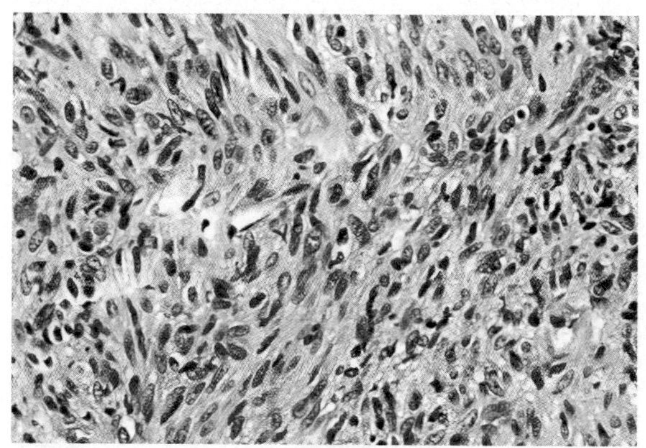

Figure 10-81

LEIOMYOSARCOMA

A fascicular, spindle cell tumor of the CNS occurring in the setting of immunocompromise should always prompt consideration of a smooth muscle tumor, more often leiomyosarcoma than leiomyoma.

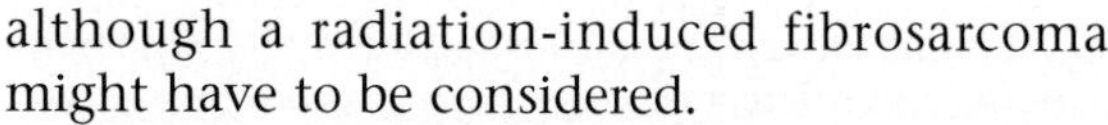

although a radiation-induced fibrosarcoma might have to be considered.

In infants, the differential diagnosis includes desmoplastic infantile ganglioglioma, as is discussed in chapter 4.

Leiomyosarcoma

In the CNS, these rare lesions occur almost exclusively in the setting of immunosuppression, generally in patients with acquired immunodeficiency syndrome (AIDS) (190–193). The tumors are masses of spindle cells that resemble fibrous meningioma or spindled melanocytoma (fig. 10-81). The Masson trichrome stain shows the expected red cytoplasm characteristic of smooth muscle cells (fig. 10-82). Immunoreactivity for smooth muscle actin and desmin is diagnostic (fig. 10-83). In the context of immunosuppression, the lesions are reliably positive for Epstein-Barr virus using in situ hybridization (fig. 10-84) (191–193). While immunosuppression-associated smooth muscle tumors are generally classified as sarcomas, they are often low grade, and some are even histologically and clinically benign.

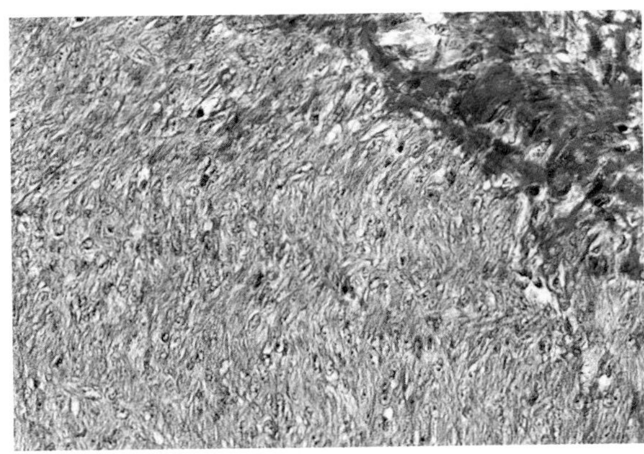

Figure 10-82

LEIOMYOSARCOMA

Cytoplasm that stains red with the Masson trichrome stain is typical of smooth muscle tumors.

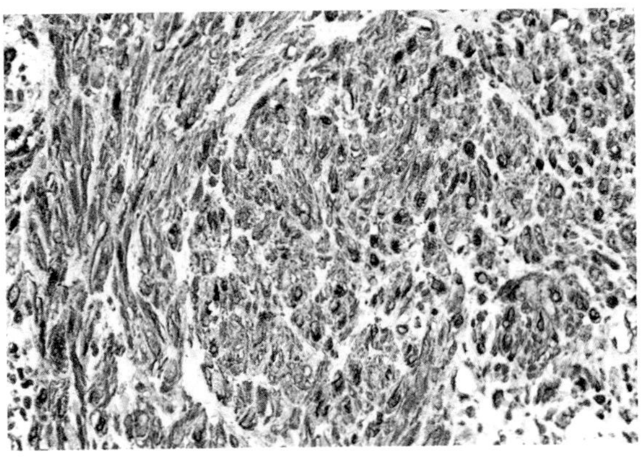

Figure 10-83

LEIOMYOSARCOMA

Immunohistochemistry for muscle markers, in this case smooth muscle actin, confirms the diagnosis.

Malignant Fibrous Histiocytoma

Although rare, spontaneously occurring *malignant fibrous histiocytomas* arise in both the cranial and spinal compartments. Most primary examples originate in the meninges (194–196), but rare examples lie entirely within the brain substance (195,197–199). Some are radiation induced (200,201).

Other Sarcomas

At systemic sites, a significant proportion of soft tissue sarcomas are undifferentiated lesions that defy precise classification (202). Some small cell undifferentiated sarcomas with diffuse leptomeningeal spread (primary meningeal sarcomatosis) present the pathologist with the challenge of distinguishing them from microscopically similar small cell embryonal tumors (PNET). Small cell sarcomas usually affect children, are to some extent reticulin rich, and lack immunoreactivity for all neuroglial markers (203,204).

Among the least common primary intracranial or intraspinal sarcomas are *liposarcoma* (205), *epithelioid sarcoma* (206), *xanthosarcoma*, (207) and *osteosarcoma*. Osteosarcomas can arise in the skull as a consequence of radiation therapy or Paget's disease (208,209), but can appear primarily in the meninges (208) or brain (210–212) without a recognized inciting agent. More commonly, they are part of a gliosarcoma.

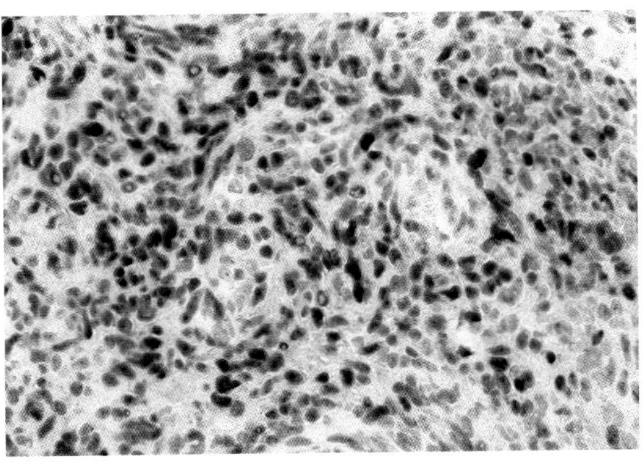

Figure 10-84

LEIOMYOSARCOMA

In situ hybridization for Epstein-Barr virus is usually positive in acquired immunodeficiency syndrome (AIDS)-associated smooth muscle tumors of the CNS.

REFERENCES

Arteriovenous Malformation, Cavernous Angioma, Venous Angioma, and Capillary Telangiectasis

1. Ondra SL, Troupp H, George ED, Schwab K. The natural history of symptomatic arteriovenous malformations of the brain: a 24-year follow-up assessment. J Neurosurg 1990;73:387–91.
2. Rosenblum B, Oldfield EH, Doppman JL, Di Chiro G. Spinal arteriovenous malformations: a comparison of dural arteriovenous fistulas and intradural AVM's in 81 patients. J Neurosurg 1987;67:795–802.
3. Asai J, Hayashi T, Fujimoto T, Suzuki R. Exclusively epidural arteriovenous fistula in the cervical spine with spinal cord symptoms: case report. Neurosurgery 2001;48:1372–6.
4. Criscuolo GR, Oldfield EH, Doppman JL. Reversible acute and subacute myelopathy in patients with dural arteriovenous fistulas. Foix-Alajouanine syndrome reconsidered. J Neurosurg 1989;70:354–9.
5. Kataoka H, Miyamoto S, Nagata I, Ueba T, Hashimoto N. Venous congestion is a major cause of neurological deterioration in spinal arteriovenous malformations. Neurosurgery 2001;48:1224–30.
6. Versari PP, D'Aliberti G, Talamonti G, Branca V, Boccardi E, Collice M. Progressive myelopathy caused by intracranial dural arteriovenous fistula: report of two cases and review of the literature. Neurosurgery 1993;33:914–9.
7. Sato S, Kodama N, Sasaki T, Matsumoto M, Ishikawa T. Perinidal dilated capillary networks in cerebral arteriovenous malformations. Neurosurgery 2004;54:163–70.
8. Mandybur TI, Nazek M. Cerebral arteriovenous malformations. A detailed morphological and immunohistochemical study using actin. Arch Pathol Lab Med 1990;114:970–3.
9. Lombardi D, Scheithauer BW, Piepgras D, Meyer FB, Forbes GS. "Angioglioma" and the arteriovenous malformation-glioma association. J Neurosurg 1991;75:589–96.
10. Nazek M, Mandybur TI, Kashiwagi S. Oligodendroglial proliferative abnormality associated with arteriovenous malformation: report of three cases with review of the literature. Neurosurgery 1988;23:781–5.
11. Germano IM, Davis RL, Wilson CB, Hieshima GB. Histopathological follow-up study of 66 cerebral arteriovenous malformations after therapeutic embolization with polyvinyl alcohol. J Neurosurg 1992;76:607–14.
12. Vinters HV, Lundie MJ, Kaufmann JC. Long-term pathological follow-up of cerebral arteriovenous malformations treated by embolization with bucrylate. N Engl J Med 1986;314:477–83.
13. Tomlinson FH, Houser OW, Scheithauer BW, Sundt TM Jr, Okazaki H, Parisi JE. Angiographically occult vascular malformations: a correlative study of features on magnetic resonance imaging and histological examination. Neurosurgery 1994;34:792–800.
14. Casazza M, Broggi G, Franzini A, et al. Supratentorial cavernous angiomas and epileptic seizures: preoperative course and postoperative outcome. Neurosurgery 1996;39:26–32; discussion 32–4.
15. Simard JM, Garcia-Bengochea F, Ballinger WE, Jr., Mickle JP, Quisling RG. Cavernous angioma: a review of 126 collected and 12 new clinical cases. Neurosurgery 1986;18:162–72.
16. Rigamonti D, Hadley MN, Drayer BP, et al. Cerebral cavernous malformations. Incidence and familial occurrence. N Engl J Med 1988;319: 343–7.
17. Zhang J, Clatterbuck RE, Rigamonti D, Dietz HC. Mutations in KRIT1 in familial cerebral cavernous malformations. Neurosurgery 2000;46:1272–9.
18. Laurans MS, DiLuna ML, Shin D, et al. Mutational analysis of 206 families with cavernous malformations. J Neurosurg 2003;99:38–43.
19. Gaensler EH, Dillon WP, Edwards MS, Larson DA, Rosenau W, Wilson CB. Radiation-induced telangiectasia in the brain simulates cryptic vascular malformations at MR imaging. Radiology 1994;193:629–36.
20. Narayan P, Barrow DL. Intramedullary spinal cavernous malformation following spinal irradiation. Case report and review of the literature. J Neurosurg 2003;98(Suppl 1):68–72.
21. Pozzati E, Giangaspero F, Marliani F, Acciarri N. Occult cerebrovascular malformations after irradiation. Neurosurgery 1996;39:677–82; discussion 682–4.
22. Rigamonti D, Johnson PC, Spetzler RF, Hadley MN, Drayer BP. Cavernous malformations and capillary telangiectasia: a spectrum within a single pathological entity. Neurosurgery 1991; 28:60–4.
23. Lindquist C, Guo WY, Karlsson B, Steiner L. Radiosurgery for venous angiomas. J Neurosurg 1993;78:531–6.
24. Malik GM, Morgan JK, Boulos RS, Ausman JI. Venous angiomas: an underestimated cause of intracranial hemorrhage. Surg Neurol 1988;30:350–8.

25. Rigamonti D, Spetzler RF, Medina M, Rigamonti K, Geckle DS, Pappas C. Cerebral venous malformations. J Neurosurg 1990;73:560–4.
26. Toro VE, Geyer CA, Sherman JL, Parisi JE, Brantley MJ. Cerebral venous angiomas: MR findings. J Comput Assist Tomogr 1988;12:935–40.
27. Farrell DF, Forno LS. Symptomatic capillary telangiectasis of the brainstem without hemorrhage. Report of an unusual case. Neurology 1970;20(4):341–6.
28. Lee RR, Becher MW, Benson ML, Rigamonti D. Brain capillary telangiectasia: MR imaging appearance and clinicohistopathologic findings. Radiology 1997;205:797–805.
29. Blackwood W. Two cases of benign cerebral telangiectasis. J Pathol 1941;52:209–12.

Sturge-Weber Syndrome

30. Comi AM. Pathophysiology of Sturge-Weber syndrome. J Child Neurol 2003;18:509–16.
31. Peterman AF, Hayles AB, Dockerty MB, Love JG. Encephalotrigeminal angiomatosis (Sturge-Weber disease); clinical study of thirty-five cases. J Am Med Assoc 1958;167:2169–76.
32. Lin DD, Barker PB, Kraut MA, Comi A. Early characteristics of Sturge-Weber syndrome shown by perfusion MR imaging and proton MR spectroscopic imaging. AJNR Am J Neuroradiol 2003;24:1912–5.
33. Devlin AM, Cross JH, Harkness W, et al. Clinical outcomes of hemispherectomy for epilepsy in childhood and adolescence. Brain 2003; 126(Pt 3):556–66.
34. Ito M, Sato K, Ohnuki A, Uto A. Sturge-Weber disease: operative indications and surgical results. Brain Dev 1990;12:473–7.
35. Kossoff EH, Buck C, Freeman JM. Outcomes of 32 hemispherectomies for Sturge-Weber syndrome worldwide. Neurology 2002;59:1735–8.
36. Ogunmekan AO, Hwang PA, Hoffman HJ. Sturge-Weber-Dimitri disease: role of hemispherectomy in prognosis. Can J Neurol Sci 1989;16:78–80.
37. Bebin EM, Gomez MR. Prognosis in Sturge-Weber disease: comparison of unihemispheric and bihemispheric involvement. J Child Neurol 1988;3:181–4.

Mesencephalo-Oculo-Facial Angiomatosis (Wyburn-Mason Syndrome) and Hereditary Hemorrhagic Telangiectasia (Rendu-Osler-Weber Disease)

38. Patel U, Gupta SC. Wyburn-Mason syndrome. A case report and review of the literature. Neuroradiology 1990;31:544–6.
39. Theron J, Newton TH, Hoyt WF. Unilateral retinocephalic vascular malformations. Neuroradiology 1974;7:185–96.
40. Wyburn-Mason R. Arteriovenous aneurysm of mid-brain and retina, facial naevi and mental changes. Brain 1943;66:163–203.
41. Aesch B, Lioret E, de Toffol B, Jan M. Multiple cerebral angiomas and Rendu-Osler-Weber disease: case report. Neurosurgery 1991;29:599–602.
42. Sobel D, Norman D. CNS manifestations of hereditary hemorrhagic telangiectasia. AJNR Am J Neuroradiol 1984;5:569–73.
43. Willinsky RA, Lasjaunias P, Terbrugge K, Burrows P. Multiple cerebral arteriovenous malformations (AVMs). Review of our experience from 203 patients with cerebral vascular lesions. Neuroradiology 1990;32:207–10.
44. Oda M, Takahashi JA, Hashimoto N, Koyama T. Rendu-Osler-Weber disease with a giant intracerebral varix secondary to a high-flow pial AVF: case report. Surg Neurol 2004;61:353–6.

Papillary Endothelial Hyperplasia (Masson's Vegetant Hemangioendothelioma)

45. Avellino AM, Grant GA, Harris AB, Wallace SK, Shaw CM. Recurrent intracranial Masson's vegetant intravascular hemangioendothelioma. Case report and review of the literature. J Neurosurg 1999;91:308–12.
46. Stoffman MR, Kim JH. Masson's vegetant hemangioendothelioma: case report and literature review. J Neurooncol 2003;61:17–22.

Capillary Hemangioma

47. Pearl GS, Takei Y, Tindall GT, O'Brien MS, Payne NS, Hoffman JC. Benign hemangioendothelioma involving the central nervous system: "strawberry nevus" of the neuraxis. Neurosurgery 1980;7:249–56.
48. Pearl GS, Takei Y. Hemangioendothelioma of the neuraxis: an ultrastructural study. Neurosurgery 1982;11:486–90.
49. Roncaroli F, Scheithauer BW, Krauss WE. Capillary hemangioma of the spinal cord. Report of four cases. J Neurosurg 2000;93(Suppl 1):148–51.
50. Roncaroli F, Scheithauer BW, Krauss WE. Hemangioma of spinal nerve root. J Neurosurg 1999;91(Suppl 2):175–80.
51. Roncaroli F, Scheithauer BW, Deen HG Jr. Multiple hemangiomas (hemangiomatosis) of the cauda equina and spinal cord. Case report. J Neurosurg 2000;92(Suppl 2):229–32.

Hemangioendothelioma

52. Chan YL, Ng HK, Poon WS, Cheung HS. Epithelioid haemangioendothelioma of the brain: a case report. Neuroradiology 2001;43:848–50.

53. Chow LT, Chow WH, Fong DT. Epithelioid hemangioendothelioma of the brain. Am J Surg Pathol 1992;16:619–25.
54. Nora FE, Scheithauer BW. Primary epithelioid hemangioendothelioma of the brain. Am J Surg Pathol 1996;20:707–14.
55. Taratuto AL, Zurbriggen G, Sevlever G, Saccoliti M. Epithelioid hemangioendothelioma of the central nervous system. Immunohistochemical and ultrastructural observations of a pediatric case. Pediatr Neurosci 1988;14:11–4.
56. Abdullah JM, Mutum SS, Nasuha NA, Biswal BM, Ariff AR. Intramedullary spindle cell hemangioendothelioma of the thoracic spinal cord—case report. Neurol Med Chir (Tokyo) 2002;42:259–63.
57. Roncaroli F, Scheithauer BW, Papazoglou S. Primary polymorphous hemangio-endothelioma of the spinal cord. Case report. J Neurosurg 2001;95(Suppl 1):93–5.

Hemangiopericytoma

58. Gengler C, Guillou L. Solitaryfibrous tumour and haemangiopericytoma: evolution of a concept. Histopathology 2006;48:63–74.
58a. Betchen S, Schwartz A, Black C, Post K. Intradural hemangiopericytoma of the lumbar spine: case report. Neurosurgery 2002;50:654–7.
59. Dufour H, Metellus P, Fuentes S, et al. Meningeal hemangiopericytoma: a retrospective study of 21 patients with special review of postoperative external radiotherapy. Neurosurgery 2001;48: 756–62; discussion 762–3.
60. Ecker RD, Marsh WR, Pollock BE, et al. Hemangiopericytoma in the central nervous system: treatment, pathological features, and long-term follow up in 38 patients. J Neurosurg 2003;98:1182–7.
61. Guthrie BL, Ebersold MJ, Scheithauer BW, Shaw EG. Meningeal hemangiopericytoma: histopathological features, treatment, and long-term follow-up of 44 cases. Neurosurgery 1989;25:514–22.
62. Chiechi MV, Smirniotopoulos JG, Mena H. Intracranial hemangiopericytomas: MR and CT features. AJNR Am J Neuroradiol 1996;17:1365–71.
63. Mena H, Ribas JL, Pezeshkpour GH, Cowan DN, Parisi JE. Hemangiopericytoma of the central nervous system: a review of 94 cases. Hum Pathol 1991;22:84–91.
64. Perry A, Scheithauer BW, Nascimento AG. The immunophenotypic spectrum of meningeal hemangiopericytoma: a comparison with fibrous meningioma and solitary fibrous tumor of meninges. Am J Surg Pathol 1997;21:1354–60.
65. Tihan T, Viglione M, Rosenblum MK, Olivi A, Burger PC. Solitary fibrous tumors in the central nervous system. A clinicopathologic review of 18 cases and comparison to meningeal hemangiopericytomas. Arch Pathol Lab Med 2003;127: 432–9.
66. Porter PL, Bigler SA, McNutt M, Gown AM. The immunophenotype of hemangiopericytomas and glomus tumors, with special reference to muscle protein expression: an immunohistochemical study and review of the literature. Mod Pathol 1991;4(1):46–52.
67. D'Amore ES, Manivel JC, Sung JH. Soft-tissue and meningeal hemangiopericytomas: an immunohistochemical and ultrastructural study. Hum Pathol 1990;21:414–23.
68. Herath SE, Stalboerger PG, Dahl RJ, Parisi JE, Jenkins RB. Cytogenetic studies of four hemangiopericytomas. Cancer Genet Cytogenet 1994; 72:137–40.
69. Joseph JT, Lisle DK, Jacoby LB, et al. NF2 gene analysis distinguishes hemangiopericytoma from meningioma. Am J Pathol 1995;147:1450–5.

Angiosarcoma

70. Charman HP, Lowenstein DH, Cho KG, DeArmond SJ, Wilson CB. Primary cerebral angiosarcoma. Case report. J Neurosurg 1988; 68(5):806–10.
71. Kuratsu J, Seto H, Kochi M, Itoyama Y, Uemura S, Ushio Y. Metastatic angiosarcoma of the brain. Surg Neurol 1991;35(4):305–9.
72. Mena H, Ribas JL, Enzinger FM, Parisi JE. Primary angiosarcoma of the central nervous system. Study of eight cases and review of the literature. J Neurosurg 1991;75(1):73–6.

Solitary Fibrous Tumor

73. Cummings TJ, Burchette JL, McLendon RE. CD34 and dural fibroblasts: the relationship to solitary fibrous tumor and meningioma. Acta Neuropathol (Berl) 2001;102:349–54.
74. Carneiro SS, Scheithauer BW, Nascimento AG, Hirose T, Davis DH. Solitary fibrous tumor of the meninges: a lesion distinct from fibrous meningioma. A clinicopathologic and immunohistochemical study. Am J Clin Pathol 1996;106:217–24.
75. Martin AJ, Fisher C, Igbaseimokumo U, Jarosz JM, Dean AF. Solitary fibrous tumours of the meninges: case series and literature review. J Neurooncol 2001;54:57–69.
76. Tihan T, Viglione M, Rosenblum MK, Olivi A, Burger PC. Solitary fibrous tumors in the central nervous system. A clinicopathologic review of 18 cases and comparison to meningeal hemangiopericytomas. Arch Pathol Lab Med 2003;127:432–9.

77. Caroli E, Salvati M, Orlando ER, Lenzi J, Santoro A, Giangaspero F. Solitary fibrous tumors of the meninges: report of four cases and literature review. Neurosurg Rev 2004;27:246–51.
78. Kocak A, Cayli SR, Sarac K, Aydin NE. Intraventricular solitary fibrous tumor: an unusual tumor with radiological, ultrastructural, and immunohistochemical evaluation: case report. Neurosurgery 2004;54:213–7.
79. Pizzolitto S, Falconieri G, Demaglio G. Solitary fibrous tumor of the spinal cord: a clinicopathologic study of two cases. Ann Diagn Pathol 2004;8:268–75.
80. Perry A, Scheithauer BW, Nascimento AG. The immunophenotypic spectrum of meningeal hemangiopericytoma: a comparison with fibrous meningioma and solitary fibrous tumor of meninges. Am J Surg Pathol 1997;21:1354–60.

Lipoma (Lipomatous Hamartoma)

81. Budka H. Intracranial lipomatous hamartomas (intracranial "lipomas"). A study of 13 cases including combinations with medulloblastoma, colloid and epidermoid cysts, angiomatosis and other malformations. Acta Neuropathol (Berl) 1974;28:205–22.
82. Truwit CL, Barkovich AJ. Pathogenesis of intracranial lipoma: an MR study in 42 patients. AJR Am J Roentgenol 1990;155:855–65.
83. Spallone A, Pitskhelauri DI. Lipomas of the pineal region. Surg Neurol 2004;62:52–9.
84. Harrison MJ, Mitnick RJ, Rosenblum BR, Rothman AS. Leptomyelolipoma: analysis of 20 cases. J Neurosurg 1990;73:360–7.
85. Heary RF, Bhandari Y. Intradural cervical lipoma in a neurologically intact patient: case report. Neurosurgery 1991;29:468–72.
86. Lee M, Rezai AR, Abbott R, Coelho DH, Epstein FJ. Intramedullary spinal cord lipomas. J Neurosurg 1995;82:394–400.
87. Quint DJ, Boulos RS, Sanders WP, Mehta BA, Patel SC, Tiel RL. Epidural lipomatosis. Radiology 1988;169:485–90.
88. Anson JA, Cybulski GR, Reyes M. Spinal extradural angiolipoma: a report of two cases and review of the literature. Surg Neurol 1990;34:173–8.
89. Klisch J, Spreer J, Bloss HG, Baborie A, Hubbe U. Radiological and histological findings in spinal intramedullary angiolipoma. Neuroradiology 1999;41:584–7.
90. Mascalchi M, Arnetoli G, Dal Pozzo G, Canavero S, Pagni CA. Spinal epidural angiolipoma: MR findings. AJNR Am J Neuroradiol 1991;12:744–5.
91. Preul MC, Leblanc R, Tampieri D, Robitaille Y, Pokrupa R. Spinal angiolipomas. Report of three cases. J Neurosurg 1993;78(2):280–6.
92. Preul MC, Leblanc R. MRI in the diagnosis of spinal extradural angiolipoma. Br J Neurosurg 1993;7:328.
93. Wilkins PR, Hoddinott C, Hourihan MD, Davies KG, Sebugwawo S, Weeks RD. Intracranial angiolipoma. J Neurol Neurosurg Psychiatry 1987;50:1057–9.

Chondroma and Osteochondroma

94. Dutton J. Intracranial solitary chondroma. Case report. J Neurosurg 1978;49:460–3.
95. Colpan E, Attar A, Erekul S, Arasil E. Convexity dural chondroma: a case report and review of the literature. J Clin Neurosci 2003;10:106–8.
96. Kurt E, Beute GN, Sluzewski M, van Rooij WJ, Teepen JL. Giant chondroma of the falx. Case report and review of the literature. J Neurosurg 1996;85:1161–4.
97. Mapstone TB, Wongmongkolrit T, Roessman U, Ratcheson RA. Intradural chondroma: a case report and review of the literature. Neurosurgery 1983;12:111–4.
98. Nakazawa T, Inoue T, Suzuki F, Nakasu S, Handa J. Solitary intracranial chondroma of the convexity dura: case report. Surg Neurol 1993;40:495–8.
99. Nakayama M, Nagayama T, Hirano H, Oyoshi T, Kuratsu J. Giant chondroma arising from the dura mater of the convexity. Case report and review of the literature. J Neurosurg 2001;94:331–4.
100. Traflet RF, Babaria AR, Barolat G, Doan HT, Gonzalez C, Mishkin MM. Intracranial chondroma in a patient with Ollier's disease. Case report. J Neurosurg 1989;70:274–6.
101. Chakrabortty S, Tamaki N, Kondoh T, Kojima N, Kamikawa H, Matsumoto S. Maffucci's syndrome associated with intracranial enchondroma and aneurysm: case report. Surg Neurol 1991;36:216–20.
102. Nagai S, Yamamoto N, Wakabayashi K, et al. Osteochondroma arising from the convexity dura mater. Case illustration. J Neurosurg 1998;88:610.

Fibroma and Fibromatosis

103. Palma L, Spagnoli LG, Yusuf MA. Intracerebral fibroma: light and electron microscopic study. Acta Neurochir (Wien) 1985;77:152–6.
104. Reyes-Mugica M, Chou P, Gonzalez-Crussi F, Tomita T. Fibroma of the meninges in a child: immunohistological and ultrastructural study. Case report. J Neurosurg 1992;76:143–7.
105. Koos WT, Jellinger K, Sunder-Plassmann M. Intracerebral fibroma in an 11-month-old infant. Case report. J Neurosurg 1971;35:77–81.

106. Llena JF, Chung HD, Hirano A, Feiring EH, Zimmerman HM. Intracerebellar "fibroma." Case report. J Neurosurg 1975;43:98–101.
107. Hirano A, Llena JF, Chung HD. Fine structure of a cerebellar "fibroma." Acta Neuropathol (Berl) 1975;32:175–86.
108. Friede RL, Pollak A. Neurosurgical desmoid tumors: presentation of four cases with a review of the differential diagnoses. J Neurosurg 1979; 50:725–32.
109. Mitchell A, Scheithauer BW, Ebersold MJ, Forbes GS. Intracranial fibromatosis. Neurosurgery 1991;29:123–6.
110. Quest DO, Salcman M. Fibromatosis presenting as a cranial mass lesion; case report. J Neurosurg 1976;44:237–40.

Other Benign Mesenchymal Tumors and Tumor-Like Lesions

111. Kepes JJ, Kepes M, Slowik F. Fibrous xanthomas and xanthosarcomas of the meninges and the brain. Acta Neuropathol (Berl) 1973;23:187–99.
112. Wolf DA, Chaljub G, Maggio W, Gelman BB. Intracranial chondromyxoid fibroma. Report of a case and review of the literature. Arch Pathol Lab Med 1997;121:626–30.
113. Kroe DJ, Hudgins WR, Simmons JC, Blackwell CF. Primary intrasellar leiomyoma. Case report. J Neurosurg 1968;29:189–91.

Chordoma

114. Forsyth PA, Cascino TL, Shaw EG, et al. Intracranial chordomas: a clinicopathological and prognostic study of 51 cases. J Neurosurg 1993;78:741–7.
115. O'Neill P, Bell BA, Miller JD, Jacobson I, Guthrie W. Fifty years of experience with chordomas in southeast Scotland. Neurosurgery 1985;16:166–70.
116. Bergh P, Kindblom LG, Gunterberg B, Remotti F, Ryd W, Meis-Kindblom JM. Prognostic factors in chordoma of the sacrum and mobile spine: a study of 39 patients. Cancer 2000;88:2122–34.
117. York JE, Kaczaraj A, Abi-Said D, et al. Sacral chordoma: 40-year experience at a major cancer center. Neurosurgery 1999;44:74–80.
118. Gunnarsson T, Leszniewski W, Bak J, Davidsson L. An intradural cervical chordoma mimicking a neurinoma. Case illustration. J Neurosurg 2001;95:144.
119. Warnick RE, Raisanen J, Kaczmar T Jr, Davis RL, Prados MD. Intradural chordoma of the tentorium cerebelli. Case report. J Neurosurg 1991; 74:508–11.
120. Wolfe JT 3rd, Scheithauer BW. "Intradural chordoma" or "giant ecchordosis physaliphora"? Report of two cases. Clin Neuropathol 1987;6: 98–103.
121. Tomlinson FH, Scheithauer BW, Miller GM, Onofrio BM. Extraosseous spinal chordoma. Case report. J Neurosurg 1991;75:980–4.
122. Unni KK, Inwards CY, Bridge, JA, Kindblom, L. Tumors of the bones and joints. Atlas of Tumor Pathology, 4th Series, Fascicle 2. Washington, DC: Armed Forces Institute of Pathology; 2005.
123. Coffin CM, Swanson PE, Wick MR, Dehner LP. Chordoma in childhood and adolescence. A clinicopathologic analysis of 12 cases. Arch Pathol Lab Med 1993;117:927–33.
124. Heffelfinger MJ, Dahlin DC, MacCarty CS, Beabout JW. Chordomas and cartilaginous tumors at the skull base. Cancer 1973;32:410–20.
125. Dahlin D. Chordoma: a study of fifty-nine cases. Cancer 1952;5:1170–8.
126. Hruban RH, May M, Marcove RC, Huvos AG. Lumbo-sacral chordoma with high-grade malignant cartilaginous and spindle cell components. Am J Surg Pathol 1990;14:384–9.
127. Hruban RH, Traganos F, Reuter VE, Huvos AG. Chordomas with malignant spindle cell components. A DNA flow cytometric and immunohistochemical study with histogenetic implications. Am J Pathol 1990;137:435–47.
128. Miettinen M, Karaharju E, Jarvinen H. Chordoma with a massive spindle-cell sarcomatous transformation. A light- and electron-microscopic and immunohistological study. Am J Surg Pathol 1987;11:563–70.
129. Morimitsu Y, Aoki T, Yokoyama K, Hashimoto H. Sarcomatoid chordoma: chordoma with a massive malignant spindle-cell component. Skeletal Radiol 2000;29:721–5.
130. Tomlinson FH, Scheithauer BW, Forsythe PA, Unni KK, Meyer FB. Sarcomatous transformation in cranial chordoma. Neurosurgery 1992;31:13–8.
131. Meis JM, Giraldo AA. Chordoma. An immunohistochemical study of 20 cases. Arch Pathol Lab Med 1988;112:553–6.
132. Miettinen M, Lehto VP, Dahl D, Virtanen I. Differential diagnosis of chordoma, chondroid, and ependymal tumors as aided by anti-intermediate filament antibodies. Am J Pathol 1983;112:160–9.
133. Mitchell A, Scheithauer BW, Unni KK, Forsyth PJ, Wold LE, McGivney DJ. Chordoma and chondroid neoplasms of the spheno-occiput. An immunohistochemical study of 41 cases with prognostic and nosologic implications. Cancer 1993;72:2943–9.

134. Salisbury JR, Isaacson PG. Demonstration of cytokeratins and an epithelial membrane antigen in chordomas and human fetal notochord. Am J Surg Pathol 1985;9:791–7.
135. Salisbury JR, Isaacson PG. Distinguishing chordoma from chondrosarcoma by immunohistochemical techniques. J Pathol 1986;148:251–2.
136. Kfoury H, Haleem A, Burgess A. Fine-needle aspiration biopsy of metastatic chordoma: case report and review of the literature. Diagn Cytopathol 2000;22:104–6.
136. Kay S, Schatzki PF. Ultrastructural observations of a chordoma arising in the clivus. Hum Pathol 1972;3:403–13.
138. Murad TM, Murthy MS. Ultrastructure of a chordoma. Cancer 1970;25:1204–15.
139. Yamaguchi T, Suzuki S, Ishiiwa H, Shimizu K, Ueda Y. Benign notochordal cell tumors: a comparative histological study of benign notochordal cell tumors, classic chordomas, and notochordal vestiges of fetal intervertebral discs. Am J Surg Pathol 2004;28:756–61.
140. Mirra JM, Brien EW. Giant notochordal hamartoma of intraosseous origin: a newly reported benign entity to be distinguished from chordoma. Report of two cases. Skeletal Radiol 2001;30:698–709.
141. Kyriakos M, Totty WG, Lenke LG. Giant vertebral notochordal rest: a lesion distinct from chordoma: discussion of an evolving concept. Am J Surg Pathol 2003;27:396–406.
142. Yamaguchi T, Yamato M, Saotome K. First histologically confirmed case of a classic chordoma arising in a precursor benign notochordal lesion: differential diagnosis of benign and malignant notochordal lesions. Skeletal Radiol 2002;31:413–8.
143. Baratti D, Gronchi A, Pennacchioli E, et al. Chordoma: natural history and results in 28 patients treated at a single institution. Ann Surg Oncol 2003;10:291–6.
144. Bosma JJ, Pigott TJ, Pennie BH, Jaffray DC. En bloc removal of the lower lumbar vertebral body for chordoma. Report of two cases. J Neurosurg 2001;94(Suppl 2):284–91.
145. Chambers PW, Schwinn CP. Chordoma. A clinicopathologic study of metastasis. Am J Clin Pathol 1979;72:765–76.
146. Delank KS, Kriegsmann J, Drees P, Eckardt A, Eysel P. Metastasizing chordoma of the lumbar spine. Eur Spine J 2002;11:167–71.
147. O'Connell JX, Renard LG, Liebsch NJ, Efird JT, Munzenrider JE, Rosenberg AE. Base of skull chordoma. A correlative study of histologic and clinical features of 62 cases. Cancer 1994; 74:2261–7.

Chondrosarcoma

148. Bosma JJ, Kirollos RW, Broome J, Eldridge PR. Primary intradural classic chondrosarcoma: case report and literature review. Neurosurgery 2001;48:420–3.
149. Cybulski GR, Russell EJ, D'Angelo CM, Bailey OT. Falcine chondrosarcoma: case report and literature review. Neurosurgery 1985;16:412–5.
150. Forbes RB, Eljamel MS. Meningeal chondrosarcomas, a review of 31 patients. Br J Neurosurg 1998;12:461–4.
151. Lee YY, Van Tassel P, Raymond AK. Intracranial dural chondrosarcoma. AJNR Am J Neuroradiol 1988;9:1189–93.
152. Oruckaptan HH, Berker M, Soylemezoglu F, Ozcan OE. Parafalcine chondrosarcoma: an unusual localization for a classical variant. Case report and review of the literature. Surg Neurol 2001;55:174–9.
153. Salcman M, Scholtz H, Kristt D, Numaguchi Y. Extraskeletal myxoid chondrosarcoma of the falx. Neurosurgery 1992;31:344–8.
154. Bernstein M, Perrin RG, Platts ME, Simpson WJ. Radiation-induced cerebellar chondrosarcoma. Case report. J Neurosurg 1984;61:174–7.
155. Hassounah M, Al-Mefty O, Akhtar M, Jinkins JR, Fox JL. Primary cranial and intracranial chondrosarcoma. A survey. Acta Neurochir (Wien) 1985;78:123–32.

Mesenchymal Chondrosarcoma

156. Cho BK, Chi JG, Wang KC, Chang KH, Choi KS. Intracranial mesenchymal chondrosarcoma: a case report and literature review. Childs Nerv Syst 1993;9:295–9.
157. Oliveira AM, Scheithauer BW, Salomao DR, Parisi JE, Burger PC, Nascimento AG. Primary sarcomas of the brain and spinal cord: a study of 18 cases. Am J Surg Pathol 2002;26:1056–63.
158. Ranjan A, Chacko G, Joseph T, Chandi SM. Intraspinal mesenchymal chondrosarcoma. Case report. J Neurosurg 1994;80:928–30.
159. Rushing EJ, Armonda RA, Ansari Q, Mena H. Mesenchymal chondrosarcoma: a clinicopathologic and flow cytometric study of 13 cases presenting in the central nervous system. Cancer 1996;77:1884–91.
160. Scheithauer BW, Rubinstein LJ. Meningeal mesenchymal chondrosarcoma: report of 8 cases with review of the literature. Cancer 1978;42: 2744–52.
161. Swanson PE, Lillemoe TJ, Manivel JC, Wick MR. Mesenchymal chondrosarcoma. An immunohistochemical study. Arch Pathol Lab Med 1990;114:943–8.

Rhabdomyosarcoma

162. Bjornsson J, Scheithauer BW, Okazaki H, Leech RW. Intracranial germ cell tumors: pathobiological and immunohistochemical aspects of 70 cases. J Neuropathol Exp Neurol 1985;44:32–46.
163. Stapleton SR, Harkness W, Wilkins PR, Uttley D. Gliomyosarcoma: an immunohistochemical analysis. J Neurol Neurosurg Psychiatry 1992;55:728–30.
164. Tomlinson FH, Scheithauer BW, Kelly PJ, Forbes GS. Subependymoma with rhabdomyosarcomatous differentiation: report of a case and literature review. Neurosurgery 1991;28:761–8.
165. Lopes de Faria J. Rhabdomyosarcoma of cerebellum. Arch Pathol 1957;63:234–8.
166. Hinton DR, Halliday WC. Primary rhabdomyosarcoma of the cerebellum—a light, electron microscopic, and immunohistochemical study. J Neuropathol Exp Neurol 1984;43:439–49.
167. Olson JJ, Menezes AH, Godersky JC, Lobosky JM, Hart M. Primary intracranial rhabdomyosarcoma. Neurosurgery 1985;17:25–34.
168. Taratuto AL, Molina HA, Diez B, Zuccaro G, Monges J. Primary rhabdomyosarcoma of brain and cerebellum. Report of four cases in infants: an immunohistochemical study. Acta Neuropathol (Berl) 1985;66:98–104.
169. Bradford R, Crockard HA, Isaacson PG. Primary rhabdomyosarcoma of the central nervous system: case report. Neurosurgery 1985;17:101–4.
170. Celli P, Cervoni L, Maraglino C. Primary rhabdomyosarcoma of the brain: observations on a case with clinical and radiological evidence of cure. J Neurooncol 1998;36:259–67.
171. Min KW, Gyorkey F, Halpert B. Primary rhabdomyosarcoma of the cerebrum. Cancer 1975;35:1405–11.
172. Yagishita S, Itoh Y, Chiba Y, Fujino H. Primary rhabdomyosarcoma of the cerebrum. An ultrastructural study. Acta Neuropathol (Berl) 1979;45:111–5.
173. Korinthenberg R, Edel G, Palm D, Muller KM, Brandt M, Muller RP. Primary rhabdomyosarcoma of the leptomeninx. Clinical, neuroradiological and pathological aspects. Clin Neurol Neurosurg. 1984;86:301–5.
174. Smith MT, Armbrustmacher VW, Violett TW. Diffuse meningeal rhabdomyosarcoma. Cancer 1981;47:2081–6.
175. Xu F, De Las Casas LE, Dobbs LJ Jr. Primary meningeal rhabdomyosarcoma in a child with hypomelanosis of Ito. Arch Pathol Lab Med 2000;124:762–5.
176. Fix SE, Nelson J, Schochet SS Jr. Focal leptomeningeal rhabdomyomatosis of the posterior fossa. Arch Pathol Lab Med 1989;113:872–3.
177. Hoffman SF, Rorke LB. On finding striated muscle in the brain. J Neurol Neurosurg Psychiatry 1971;34:761–4.
178. Masuzawa T, Shimabukuro H, Kamoshita S, Sato F. The ultrastructure of primary cerebral rhabdomyosarcoma. Acta Neuropathol (Berl). 1982;56:307–10.

Fibrosarcoma

179. Chang SM, Barker FG 2nd, Larson DA, Bollen AW, Prados MD. Sarcomas subsequent to cranial irradiation. Neurosurgery 1995;36:685–90.
180. Coppeto JR, Roberts M. Fibrosarcoma after proton-beam pituitary ablation. Arch Neurol 1979;36:380–1.
181. Pages A, Pages M, Ramos J, Benezech J. Radiation-induced intracranial fibrochondrosarcoma. J Neurol 1986;233:309–10.
182. Schrantz JL, Araoz CA. Radiation induced meningeal fibrosarcoma. Arch Pathol 1972;93:26–31.
183. Gaspar LE, Mackenzie IR, Gilbert JJ, et al. Primary cerebral fibrosarcomas. Clinicopathologic study and review of the literature. Cancer 1993;72:3277–81.
184. McDonald P, Guha A, Provias J. Primary intracranial fibrosarcoma with intratumoral hemorrhage: neuropathological diagnosis with review of the literature. J Neurooncol 1997;35:133–9.
185. Oliveira AM, Scheithauer BW, Salomao DR, Parisi JE, Burger PC, Nascimento AG. Primary sarcomas of the brain and spinal cord: a study of 18 cases. Am J Surg Pathol 2002;26:1056–63.
186. Bisogno G, Roganovic J, Carli M, et al. Primary intracranial fibrosarcoma. Childs Nerv Syst 2002;18:648–51.
187. Cai N, Kahn LB. A report of primary brain fibrosarcoma with literature review. J Neurooncol 2004;68:161–7.
188. Antonescu CR, Rosenblum MK, Pereira P, Nascimento AG, Woodruff JM. Sclerosing epithelioid fibrosarcoma: a study of 16 cases and confirmation of a clinicopathologically distinct tumor. Am J Surg Pathol 2001;25:699–709.
189. Bilsky MH, Schefler AC, Sandberg DI, Dunkel IJ, Rosenblum MK. Sclerosing epithelioid fibrosarcomas involving the neuraxis: report of three cases. Neurosurgery 2000;47:956–60.

Leiomyosarcoma

190. Bejjani GK, Stopak B, Schwartz A, Santi R. Primary dural leiomyosarcoma in a patient infected with human immunodeficiency virus: case report. Neurosurgery 1999;44:199–202.

191. Brown HG, Burger PC, Olivi A, Sills AK, Barditch-Crovo PA, Lee RR. Intracranial leiomyosarcoma in a patient with AIDS. Neuroradiology 1999;41:35–9.
192. Litofsky NS, Pihan G, Corvi F, Smith TW. Intracranial leiomyosarcoma: a neuro-oncological consequence of acquired immunodeficiency syndrome. J Neurooncol 1998;40:179–83.
193. Ritter AM, Amaker BH, Graham RS, Broaddus WC, Ward JD. Central nervous system leiomyosarcoma in patients with acquired immunodeficiency syndrome. Report of two cases. J Neurosurg 2000;92:688–92.

Malignant Fibrous Histiocytoma

194. Kalyanaraman UP, Taraska JJ, Fierer JA, Elwood PW. Malignant fibrous histiocytoma of the meninges. Histological, ultrastructural, and immunocytochemical studies. J Neurosurg 1981;55:957–62.
195. Sima AA, Ross RT, Hoag G, Rozdilsky B, Diocee M. Malignant intracranial fibrous histiocytomas. Histologic, ultrastructural and immunohistochemical studies of two cases. Can J Neurol Sci 1986;13:138–45.
196. Swamy KS, Shankar SK, Asha T, Reddy AK. Malignant fibrous histiocytoma arising from the meninges of the posterior fossa. Surg Neurol 1986;25:18–24.
197. Oliveira AM, Scheithauer BW, Salomao DR, Parisi JE, Burger PC, Nascimento AG. Primary sarcomas of the brain and spinal cord: a study of 18 cases. Am J Surg Pathol 2002;26:1056–63.
198. Roosen N, Cras P, Paquier P, Martin JJ. Primary thalamic malignant fibrous histiocytoma of the dominant hemisphere causing severe neuropsychological symptoms. Clin Neuropathol 1989;8: 16–21.
199. Schrader B, Holland BR, Friedrichsen C. Rare case of a primary malignant fibrous histiocytoma of the brain. Neuroradiology 1989;31: 177–9.
200. Chang SM, Barker FG 2nd, Larson DA, Bollen AW, Prados MD. Sarcomas subsequent to cranial irradiation. Neurosurgery 1995;36:685–90.
201. Gonzalez-Vitale JC, Slavin RE, McQueen JD. Radiation-induced intracranial malignant fibrous histiocytoma. Cancer 1976;37:2960–3.

Other Sarcomas

202. Oliveira AM, Scheithauer BW, Salomao DR, Parisi JE, Burger PC, Nascimento AG. Primary sarcomas of the brain and spinal cord: a study of 18 cases. Am J Surg Pathol 2002;26:1056–3.
203. Budka H, Pilz P, Guseo A. Primary leptomeningeal sarcomatosis. Clinicopathological report of six cases. J Neurol 1975;211:77–93.
204. Thibodeau LL, Ariza A, Piepmeier JM. Primary leptomeningeal sarcomatosis. Case report. J Neurosurg 1988;68:802–5.
205. Kothandaram P. Dural liposarcoma associated with subdural hematoma. Case report. J Neurosurg 1970;33:85–7.
206. Kurtkaya-Yapicier O, Scheithauer BW, Dedrick DJ, Wascher TM. Primary epithelioid sarcoma of the dura: case report. Neurosurgery 2002;50:198–202; discussion 202–3.
207. Kepes JJ, Kepes M, Slowik F. Fibrous xanthomas and xanthosarcomas of the meninges and the brain. Acta Neuropathol (Berl) 1973;23: 187–99.
208. Lam M, Malik G, Chason J. Osteosarcoma of meninges. Clinical, light, and ultrastructural observations of a case. Am J Surg Pathol 1981;5: 203–8.
209. Young HA, Hardy DG, Ashleigh R. Osteogenic sarcoma of the skull complicating Paget's disease: case report. Neurosurgery 1983;12:454–7.
210. Bauman GS, Wara WM, Ciricillo SF, Davis RL, Zoger S, Edwards MS. Primary intracerebral osteosarcoma: a case report. J Neurooncol 1997;32:209–13.
211. Reznik M, Lenelle J. Primary intracerebral osteosarcoma. Cancer 1991;68:793–7.
212. Sipos EP, Tamargo RJ, Epstein JI, North RB. Primary intracerebral small-cell osteosarcoma in an adolescent girl: report of a case. J Neurooncol 1997;32:169–74.

11 PARAGANGLIOMA

GENERAL PATHOLOGIC FEATURES

Definition. *Paraganglioma* is a neuroendocrine neoplasm that is composed largely of paraganglion chief cells.

Clinical Features. Paragangliomas affecting the central nervous system (CNS) almost always arise from the filum terminale, although exceptions occasionally appear on a nerve root (1,2) or in the sellar and suprasellar regions (3,4). Jugulotympanic paragangliomas generally extend from their origin in the temporal bone (5).

Patients, usually adults, with the prototypic lesion of the filum terminale present with radicular pain, sensorimotor deficits, and urinary or fecal incontinence ("cauda equina syndrome").

Radiologic Findings. Paraganglioma is well circumscribed and contrast enhancing radiologically (fig. 11-1) (6,7). Rarely, the lesion is densely calcified (8).

Gross Findings. The delicately encapsulated mass is soft, well-circumscribed, and often supplied by prominent feeder vessels (fig. 11-2). As

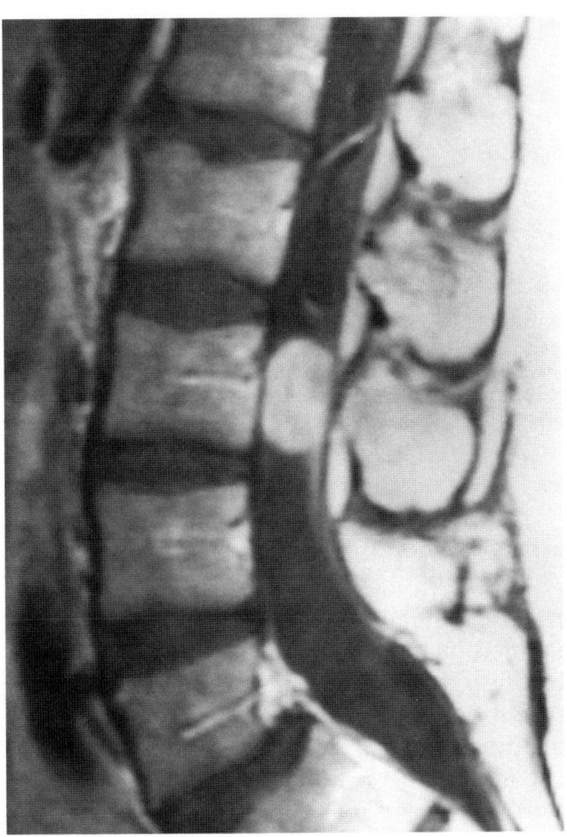

Figure 11-1

PARAGANGLIOMA

As in this L3 tumor in a 45-year-old man, intraspinal paragangliomas are discrete, contrast enhancing, and almost always located in the cauda equina region.

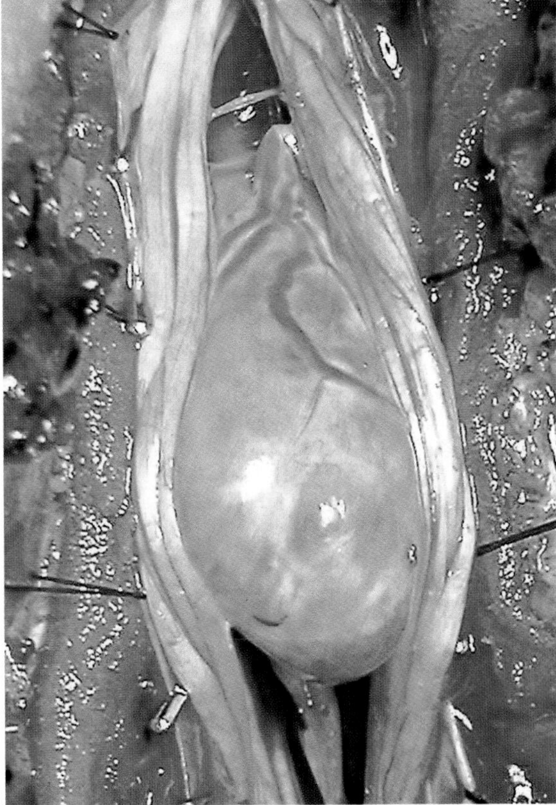

Figure 11-2

PARAGANGLIOMA

Paragangliomas are well-circumscribed masses. This 47-year-old man had lower back and leg pain. (Courtesy of Dr. Ziya Gokaslan, Baltimore, MD.)

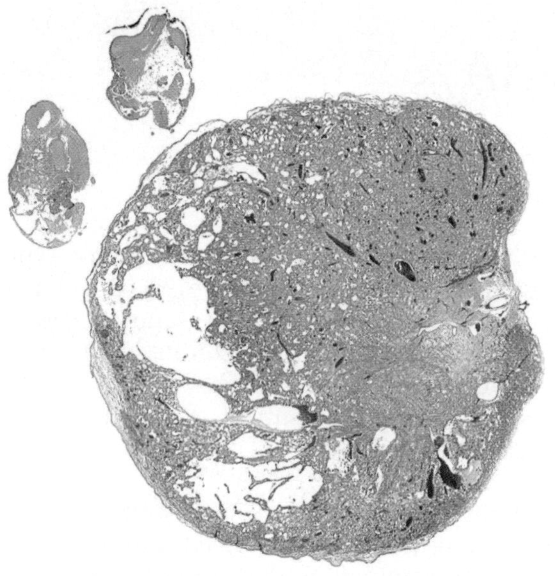

Figure 11-3

PARAGANGLIOMA

A whole mount section illustrates the delicate capsule and variability of cellularity of the typical spinal paraganglioma. The two smaller profiles at the top left are the proximal and distal surgical margins of the filum terminale.

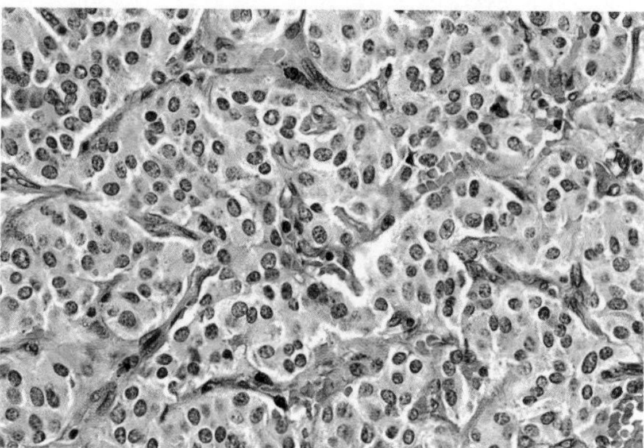

Figure 11-4

PARAGANGLIOMA

The prototypical paraganglioma has a distinctive acinar pattern, finely granular cytoplasm with epithelioid features, and delicate nuclei with stippled "salt and pepper" chromatin.

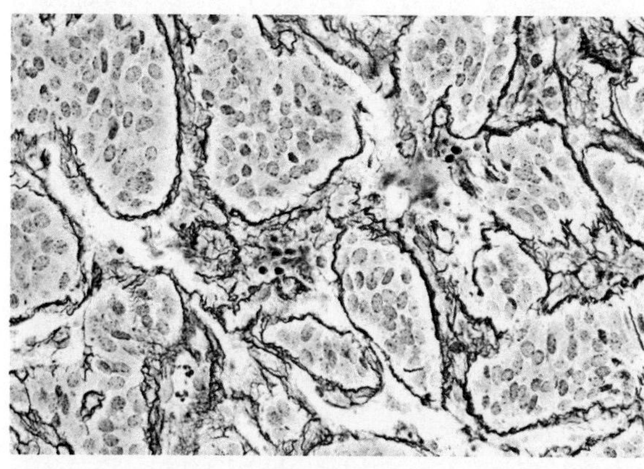

Figure 11-5

PARAGANGLIOMA

Reticulin staining emphasizes the acinar architecture of the lesion.

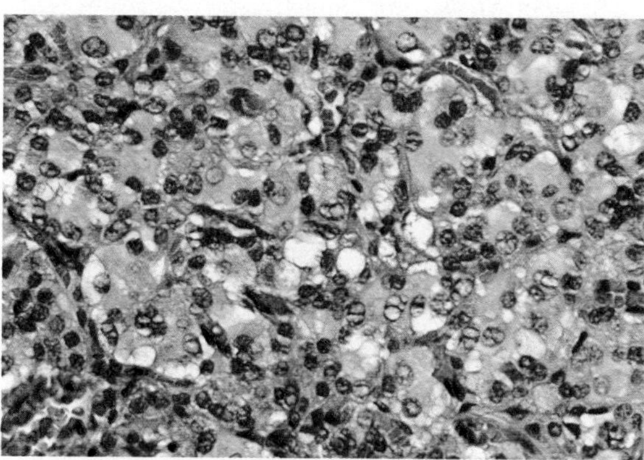

Figure 11-6

PARAGANGLIOMA

The diagnosis of paraganglioma may not come to mind in some compact examples with few acini.

is seen well in whole mount histologic sections, paragangliomas are enclosed by a delicate capsule that can become calcified (fig. 11-3).

Microscopic Findings. The tumor is composed of uniform cells with demarcated borders that are arranged in large lobules or smaller nests ("Zellballen") (fig. 11-4). The nests are defined by a circumferential lamina of reticulin (fig. 11-5). Some paragangliomas are more solidly cellular with only a subtle acinar architecture (figs. 11-6, 11-7); prominent ribboning occurs in others (fig. 11-8). (See Appendix J for this and other CNS tumors with ribboning or palisading.) Occasional pseudopapillae are artifacts of tissue preparation.

The aggregated chief cells are surrounded by capillaries geometrically disposed throughout the lesion. Sustentacular cells, consistent but inconspicuous components, form a flattened layer that

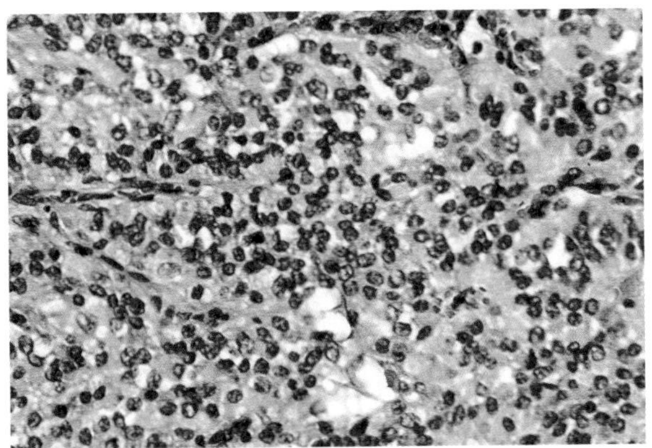

Figure 11-7

PARAGANGLIOMA

In some areas, acini are sparse if present at all.

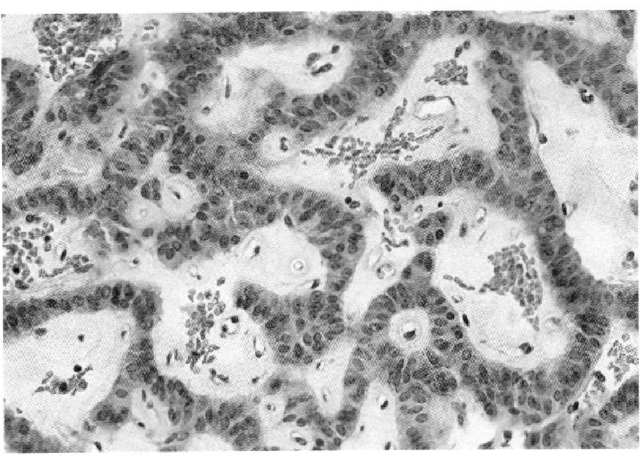

Figure 11-8

PARAGANGLIOMA

A ribboning architecture is present in a minority of paragangliomas.

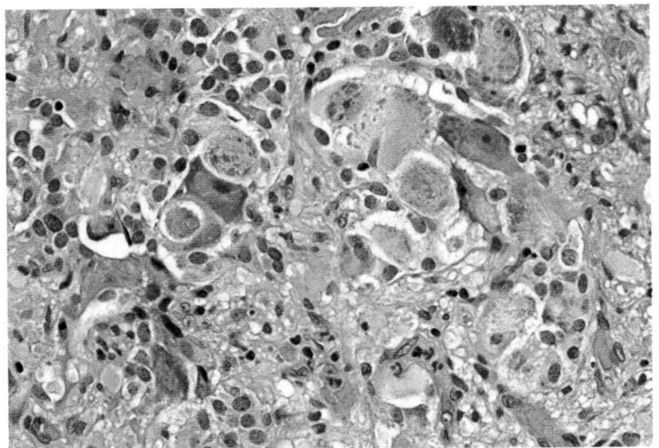

Figure 11-9

PARAGANGLIOMA

Ganglion cells, in some cases, appear to arise through transitional forms from chief cells.

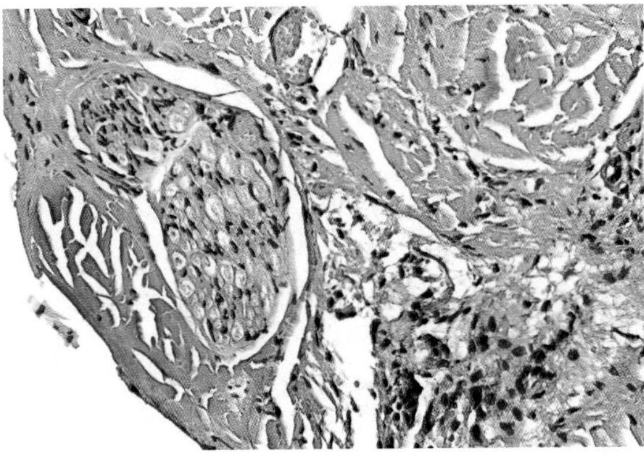

Figure 11-10

SURGICAL MARGIN (FILUM TERMINALE) OF PARAGANGLIOMA

The filum terminale, here free of tumor, consists largely of fibrous connective tissue, nerve roots, and a small amount of central nervous system (CNS) tissue (bottom right). Remnants of the central canal, not seen here, are common.

encompasses the lobules and Zellballen. Such cells are best seen with immunohistochemical stains for S-100 protein, as is discussed below.

Nuclear pleomorphism ("endocrine atypia") is generally mild and, like the finding of occasional mitotic figures, is clinically insignificant. The cytoplasm is eosinophilic and finely granular. Oncocytic change is uncommon and rarely extensive. Ganglion cells, present in nearly half of paragangliomas affecting the CNS, are fully mature elements that lie in clusters within a connective tissue matrix containing aligned Schwann cell processes or are intra-acinar and in transition to chief cells (fig. 11-9).

The specimen may arrive attached to the filum terminale, whose proximal and distal ends may be sampled as surgical margins (fig. 11-10).

Frozen Section. The acinar architecture may not be as obvious in frozen section specimens as it is in permanent sections, but epithelioid cytologic features are retained (fig. 11-11).

Immunohistochemical Findings. The chief cells are immunoreactive for synaptophysin and chromogranin (fig. 11-12), with considerable intercellular variability in the intensity of

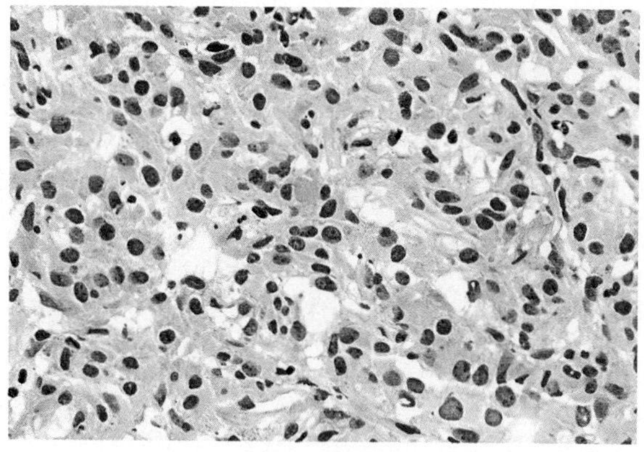

Figure 11-11

PARAGANGLIOMA: FROZEN SECTION

The typical acinar pattern may be partially obscured.

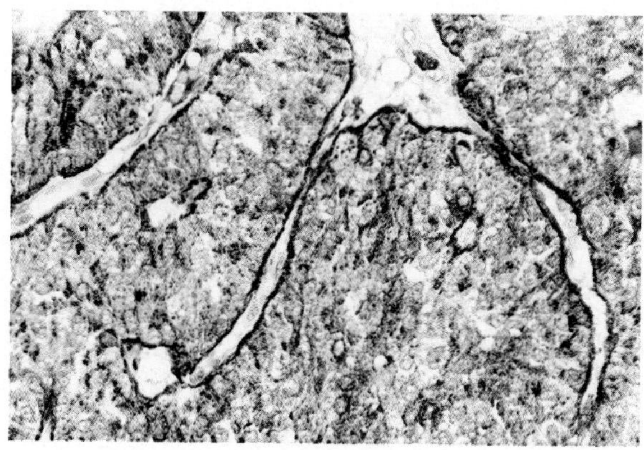

Figure 11-12

PARAGANGLIOMA

The chief cells are immunoreactive for chromogranin.

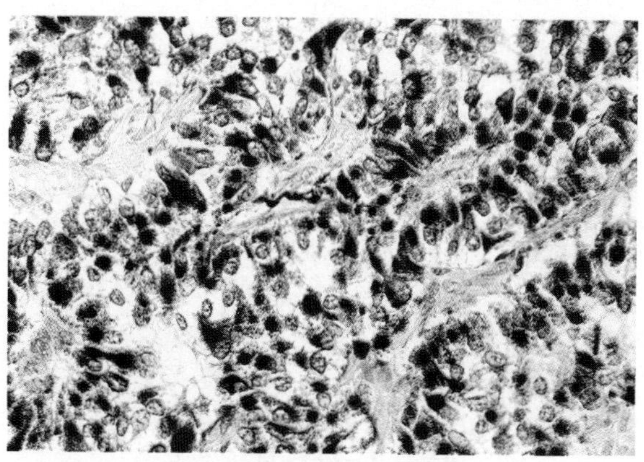

Figure 11-13

PARAGANGLIOMA

Immunoreactivity for cytokeratins, in this example for CAM5.2, should not be interpreted as evidence of an epithelial lesion such as carcinoma.

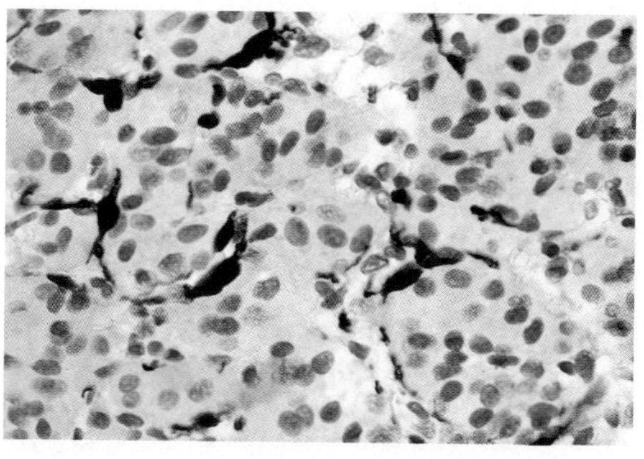

Figure 11-14

PARAGANGLIOMA

Small, bipolar S-100-positive sustentacular cells incompletely invest some acini.

staining for the latter. Neurofilament stains often show paranuclear globular reactivity corresponding in distribution to intermediate filament whorls (see below). Some react for cytokeratins (fig. 11-13). Sustentacular cells are uniformly reactive for S-100 protein (fig. 11-14), and often show glial fibrillary acidic protein (GFAP) positivity as well. Trapped GFAP-positive astrocytes, usually perivascular or subcapsular in location, may also be present in tumors of the filum terminale.

Although primary spinal paragangliomas are unassociated with endocrine symptoms, a variety of hormones and neurotransmitter substances, including somatostatin, serotonin, and metenkephalin, may be demonstrated (2,9).

Ultrastructural Findings. The epithelial-appearing tumor cells are noted for their variable content of dense core (neurosecretory) granules and organelles of hormone manufacture, including well-developed Golgi complexes and rough endoplasmic reticulum (2,9). Paranuclear intermediate filament whorls may also be present, as are intercellular junctions and basal lamina in areas where cells abut stroma. The elongated processes of agranular, electron-dense

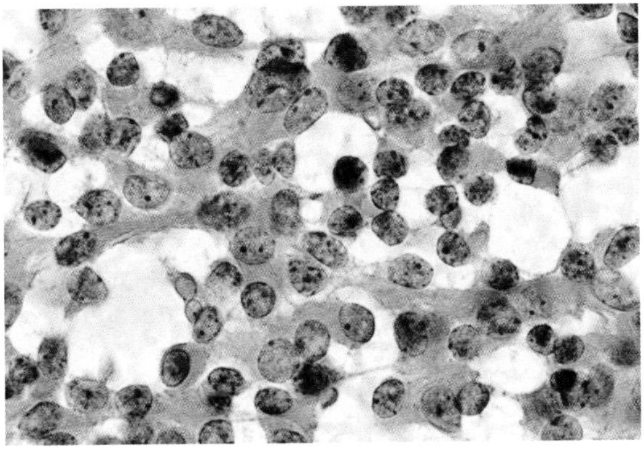

Figure 11-15

PARAGANGLIOMA: SMEAR PREPARATION

Nuclei with delicate salt and pepper chromatin are typical of neuroendocrine tumors such as paraganglioma. The circumferential, sharply defined cytoplasm without processes helps exclude ependymoma.

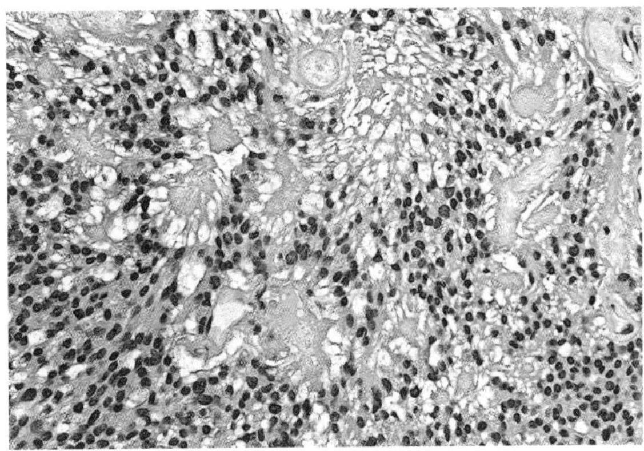

Figure 11-16

MYXOPAPILLARY EPENDYMOMA RESEMBLING PARAGANGLIOMA

Myxopapillary ependymomas with epithelioid cytologic features can simulate paraganglioma. Immunohistochemistry may be necessary to resolve the issue.

sustentacular cells surround Zellballen and occasionally embrace small groups of chief cells.

Cytologic Findings. The nuclei of the chief cells, with their crisp nuclear membrane, delicate stippled chromatin, and small nucleoli, are typical of neuroendocrine neoplasms (fig. 11-15). A rim of cytoplasm without processes is present. Larger hyperchromatic nuclei may also be seen ("endocrine atypia").

DIFFERENTIAL DIAGNOSIS

The differential diagnosis of intraspinal lesions by anatomic site is given in Appendix B. Given the clinical history and intraoperative findings, the principal differential diagnostic consideration is the far more common ependymoma (fig. 11-16). The resemblance of these two lesions is close when the perivascular processes of paragangliomas are fine and glioma-like, or when mucoid vascular change furthers the similarity. In general, however, paragangliomas have cells that are more uniform and epithelial appearing because of their well-defined cell borders and relationship to capillary walls. The cytoplasm is faintly granular and the chromatin speckled in a distinctive "salt and pepper" pattern. The loose texture and sweeping cell processes so characteristic of ependymomas are not a feature. The significance of lesions that appear to have both paraganglionic and ependymal differentiation is unclear (10). Immunoreactivity for chromogranin is a decisive finding in paraganglioma. Conversely, diffuse GFAP reactivity, a feature of ependymoma, essentially excludes paraganglioma.

TREATMENT AND PROGNOSIS

Almost all paragangliomas of the cauda equina region are well circumscribed and resectable. The small percentage that cannot be resected may recur (2,7). Spread within the CNS and distant metastases are rare (1,11,12). Systemic metastases have not been reported.

REFERENCES

1. Moran CA, Rush W, Mena H. Primary spinal paragangliomas: a clinicopathological and immunohistochemical study of 30 cases. Histopathology 1997;31:167–73.
2. Sonneland PR, Scheithauer BW, LeChago J, Crawford BG, Onofrio BM. Paraganglioma of the cauda equina region. Clinicopathologic study of 31 cases with special reference to immunocytology and ultrastructure. Cancer 1986;58:1720–35.
3. Bilbao JM, Horvath E, Kovacs K, Singer W, Hudson AR. Intrasellar paraganglioma associated with hypopituitarism. Arch Pathol Lab Med 1978;102:95–8.
4. Flint EW, Claassen D, Pang D, Hirsch WL. Intrasellar and suprasellar paraganglioma: CT and MR findings. AJNR Am J Neuroradiol 1993; 14:1191–3.
5. Sen C, Hague K, Kacchara R, Jenkins A, Das S, Catalano P. Jugular foramen: microscopic anatomic features and implications for neural preservation with reference to glomus tumors involving the temporal bone. Neurosurgery 2001; 48:838–47; discussion 847–8.
6. Aggarwal S, Deck JH, Kucharczyk W. Neuroendocrine tumor (paraganglioma) of the cauda equina: MR and pathologic findings. AJNR Am J Neuroradiol 1993;14:1003–7.
7. Sundgren P, Annertz M, Englund E, Stromblad LG, Holtas S. Paragangliomas of the spinal canal. Neuroradiology 1999;41:788–94.
8. Sharma A, Gaikwad SB, Goyal M, Mishra NK, Sharma MC. Calcified filum terminale paraganglioma causing superficial siderosis. AJR Am J Roentgenol 1998;170:1650–2.
9. Hirose T, Sano T, Mori K, et al. Paraganglioma of the cauda equina: an ultrastructural and immunohistochemical study of two cases. Ultrastructural Pathol 1988;12:235–43.
10. Caccamo DV, Ho KL, Garcia JH. Cauda equina tumor with ependymal and paraganglionic differentiation. Hum Pathol 1992;23:835–8.
11. Roche PH, Figarella-Branger D, Regis J, Peragut JC. Cauda equina paraganglioma with subsequent intracranial and intraspinal metastases. Acta Neurochir (Wien) 1996;138:475–9.
12. Strommer KN, Brandner S, Sarioglu AC, Sure U, Yonekawa Y. Symptomatic cerebellar metastasis and late local recurrence of a cauda equina paraganglioma. J Neurosurg 1995;83:166–9.

12 TUMORS OF LYMPHOID AND HEMATOPOIETIC TISSUE

PRIMARY TUMORS

Primary Central Nervous System Lymphoma

Definition. *Primary central nervous system lymphoma* (PCNSL) is a malignant lymphoma arising primarily within the parenchyma of the brain or spinal cord.

General Features. PCNSLs are mainly intraparenchymal, high-grade B-cell lesions of large cell type. Only uncommonly are they principally leptomeningeal. Rare low-grade B-cell lesions of mucosa-associated lymphoid tissue (MALT) type, or marginal zone lymphomas, are curiously oriented to the dura (1–3). Only a rare example of Burkitt's lymphoma is encountered in the central nervous system (CNS) (4). T-cell lymphomas of the CNS comprised only 8 of 220 (3.6 percent) cases in one study (5) and 3 of 60 (4.8 percent) in another (6), but the overall incidence may be even lower. True histiocytic lymphomas (histiocytic sarcoma) are even less common (6–8). Anaplastic large cell lymphomas of T- or occasionally null-cell derivation are also rare (9,10). Hodgkin's lymphoma is not unusual as a secondary epidural lesion, but is exquisitely rare as a PCNSL (11,12). Amyloidomas, and their clonal nature, are discussed separately.

Clinical Features. PCNSLs arise sporadically in otherwise healthy individuals, usually after the age of 60 years (13–15), or appear in immunologically compromised, and generally younger, individuals such as those with acquired immunodeficiency syndrome (AIDS) (16). In contrast to secondary lymphomas, or those of the MALT type that target the meninges (17), PCNSLs are almost always intraparenchymal, and are only occasionally associated with detectable systemic disease at the time of presentation (18). Given their deep-seated nature, surgical specimens are usually obtained by stereotactically directed needle biopsy (19,20). A number of investigators (21–24), but not all (25), believe there is an increasing incidence of sporadic tumors.

Prior to, or at the time of, diagnosis of PCNSL, approximately 20 percent of patients have ocular involvement, as evaluated by slit lamp examination or on magnetic resonance imaging (MRI) (13,26–28). At the time of autopsy, approximately 5 percent of patients with PCNSL have a secondary visceral focus of involvement, although such "peripheralized" CNS lymphomas rarely cause symptoms (29).

Radiologic Findings. Sporadic PCNSLs may be unifocal (fig. 12-1) or multifocal (fig. 12-2), with the latter particularly common in immunosuppressed patients such as those with AIDS. Most PCNSLs are supratentorial and often involve the subependymal region (fig. 12-2) (30),

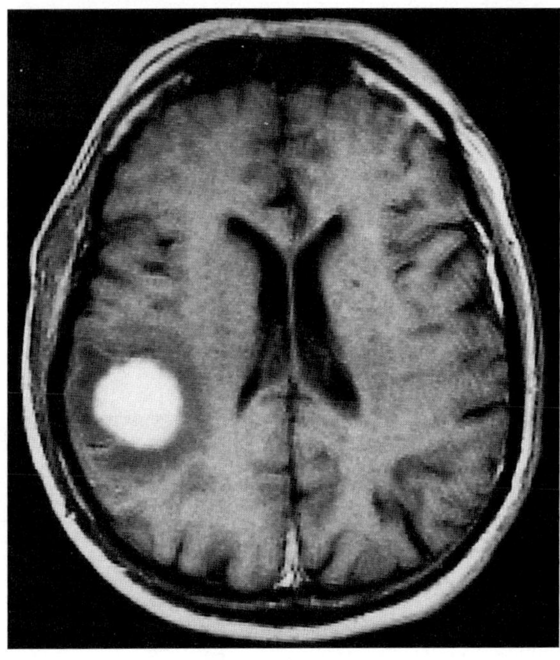

Figure 12-1

PRIMARY CENTRAL NERVOUS SYSTEM (CNS) LYMPHOMA

As sporadic lesions, primary central nervous system lymphomas (PCNSLs) contrast enhance homogeneously without the dark center of glioblastomas and most metastases. (Fig. 12-1 from Fascicle 10, 3rd Series.)

although occasional examples arise in the cerebellum, brain stem, or, rarely, spinal cord (31–33). Primary lymphomas of the intradural peripheral nervous system, e.g., in nerve roots of the cauda equina, are discussed in the section Neurolymphomatosis.

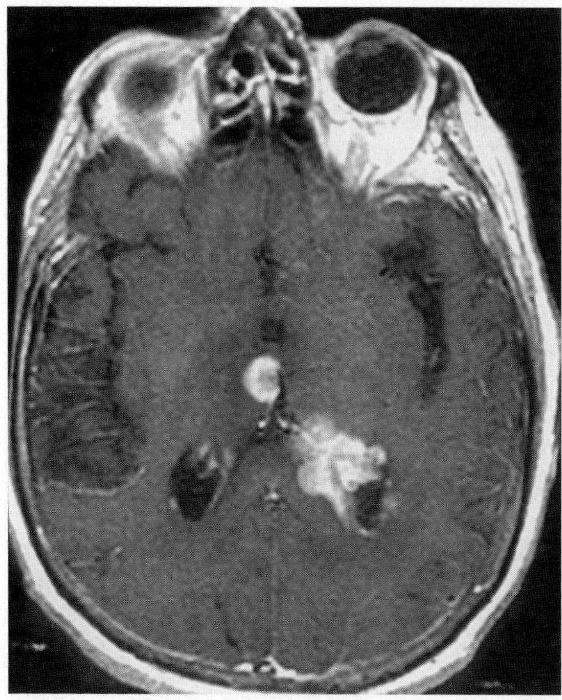

Figure 12-2

PRIMARY CNS LYMPHOMA

Multiple lesions, especially when subependymal, are highly suggestive of PCNSL, particularly in an older patient such as this 67-year-old man.

Sporadic PCNSLs are generally homogeneously contrast enhancing without the ring configuration so characteristic of glioblastoma and many metastatic carcinomas (figs. 12-1, 12-2). Because of their necrotic core, AIDS-associated lesions, in contrast, frequently have a central dark signal circumscribed by an enhancing ring that is composed of highly cellular tumor.

PCNSLs often respond dramatically to corticosteroid therapy and may literally, albeit transiently, disappear from view (34–36).

Gross Findings. The lesions are soft, gray, somewhat ill-defined, and usually deep seated. In some instances, they transit the corpus callosum in a fashion reminiscent of infiltrating glioma (fig. 12-3). AIDS-associated PCNSLs are often multiple and frequently centrally necrotic (fig. 12-4). MALT lymphomas are usually dura-based (see fig. 12-12, top).

Microscopic Findings. Unlike many systemic lymphomas, PCNSLs are diffuse rather than nodular, and highly infiltrative of brain parenchyma. A most distinctive architectural feature is angiocentricity or angioinvasiveness that at low magnification lends a patchy blueness to the slide. The cells laminate within vessel walls and adventitia in a "tree-ring" fashion that is highlighted with reticulin staining (figs. 12-5–12-7).

The typical PCNSL forms dense infiltrates in addition to less cellular areas of infiltration. The cells are larger than macrophages, and at least twice the size of normal lymphocytes. Nuclei are round to oval, occasionally notched, and contain prominent nucleoli (fig. 12-8). Mitotic

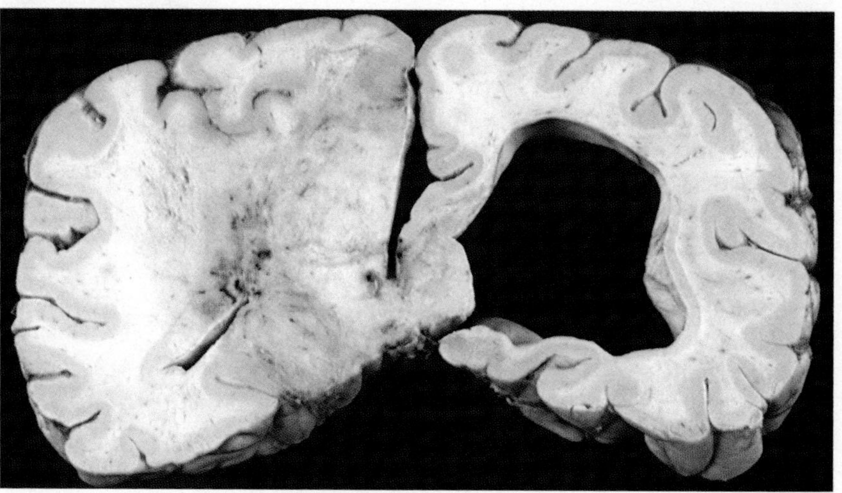

Figure 12-3

PRIMARY CNS LYMPHOMA

PCNSLs are fleshy masses that, like gliomas, can cross the midline through the corpus callosum. Unlike glioblastoma, there is no large core of necrosis in this tumor in a 56-year-old man.

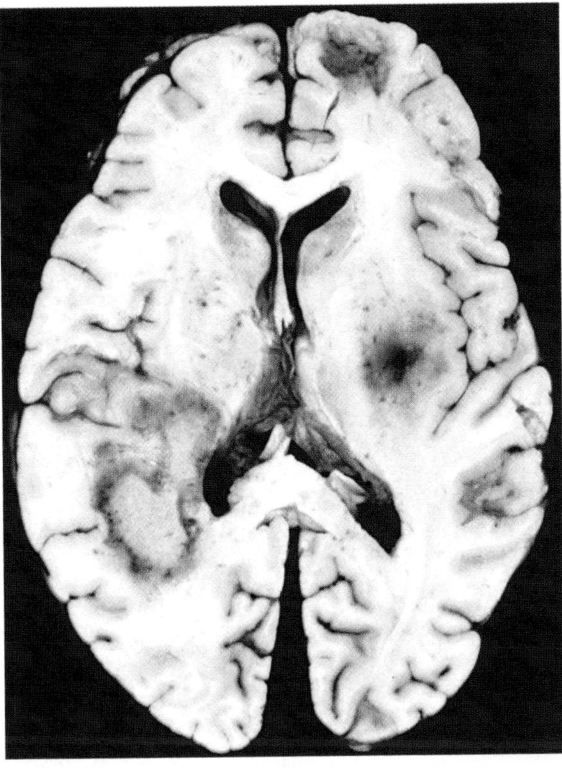

Figure 12-4

PRIMARY CNS LYMPHOMA

Multiple lesions are common in patients with acquired immunodeficiency syndrome (AIDS). Central necrosis (in the large lesion) is a frequent feature of PCNSLs arising in the setting of immunosuppression.

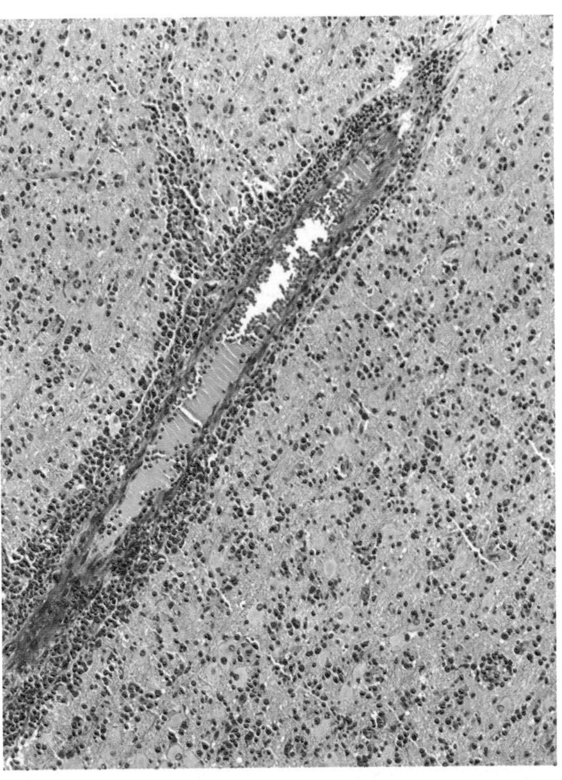

Figure 12-5

PRIMARY CNS LYMPHOMA

PCNSLs are decidedly angiocentric, but also have a diffusely infiltrating component similar to that of oligodendroglioma and diffuse astrocytoma.

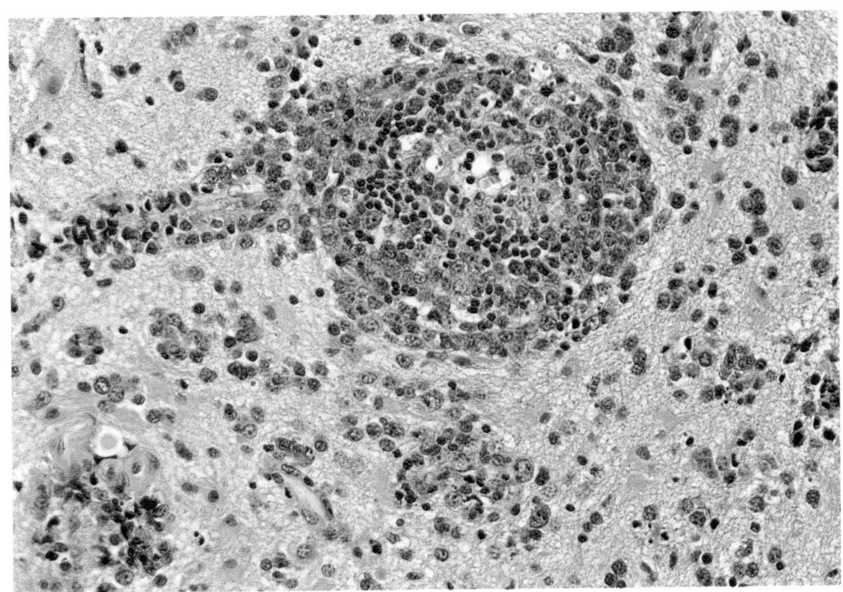

Figure 12-6

PRIMARY CNS LYMPHOMA

Tumor cells are attracted to both small and large vessels. Small, well-differentiated lymphocytes are often admixed with tumor cells.

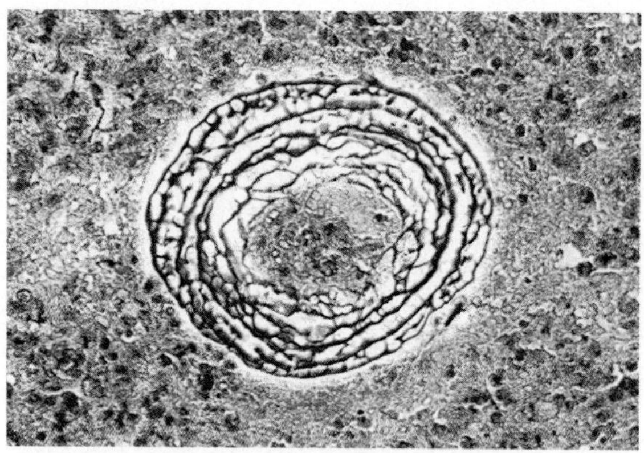

Figure 12-7

PRIMARY CNS LYMPHOMA

The presence of concentric, vascular and perivascular, tumor-containing lamellae of reticulin is a classic, and almost diagnostic, feature of PCNSL.

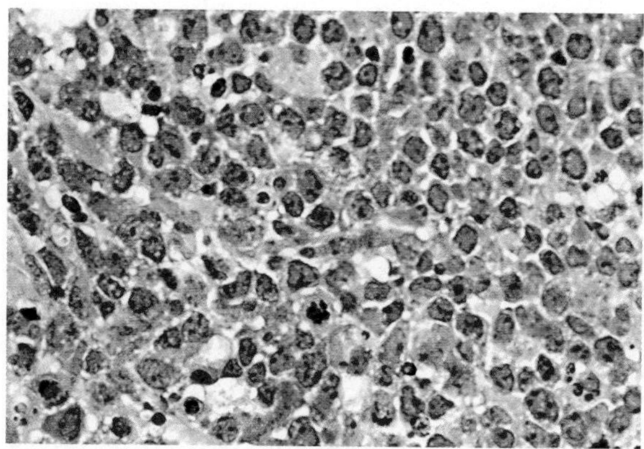

Figure 12-8

PRIMARY CNS LYMPHOMA

In their most cellular regions, PCNSLs form solid, sometimes somewhat epithelioid, sheets of back-to-back cells. Mitotic activity and apoptosis are common.

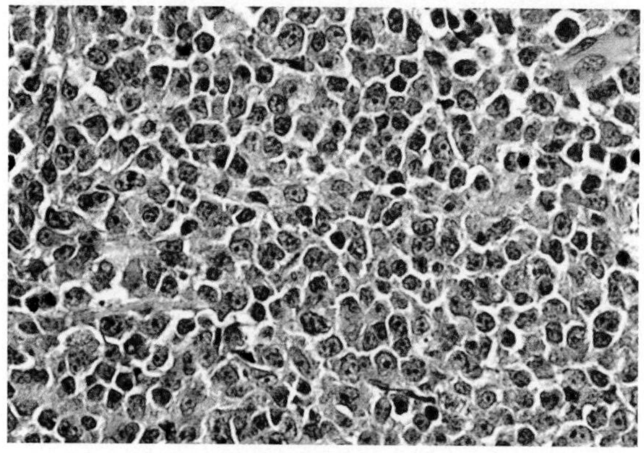

Figure 12-9

PRIMARY CNS LYMPHOMA

Plasmacytoid features are present in some PCNSLs.

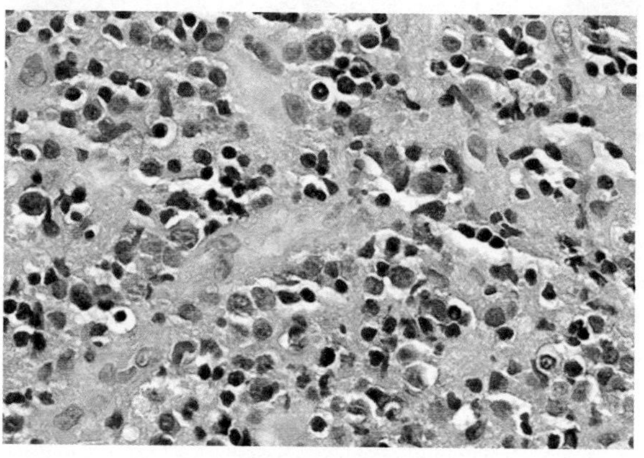

Figure 12-10

PRIMARY CNS LYMPHOMA

Small lymphocytes and apoptotic bodies can dilute the concentration of large neoplastic cells and create the appearance of a glioma.

and apoptotic activity are frequent. The cytoplasm is often poorly delineated in tissue sections, but the nuclear to cytoplasmic ratios are clearly high. A few lesions have either immunoblastic or overtly plasmacytoid cytologic features (fig. 12-9). In some areas, tumor cells are accompanied by reactive astrocytes, histiocytes, and small lymphocytes (fig. 12-10). Tumor cells are separated by collagenized vessels in some lesions (fig. 12-11).

MALT lymphomas, B-cell and usually dura-based, are of lower grade than the classic large cell type (fig. 12-12) (17). T-cell tumors, which are rare, are typically composed of cells smaller than those of large B-cell lymphoma, and, when rich in reactive lymphocytes, can simulate an inflammatory process (fig. 12-13) (37).

Anaplastic large cell lymphomas are blatantly malignant at the cytologic level. The cells are large, round to oval or polygonal and have bizarre pleomorphic nuclei (9,10). The epithelioid cytologic features, when coupled with cohesive growth, may suggest carcinoma.

At their margins, PCNSLs permeate the parenchyma while retaining their distinctive angiocentricity (fig. 12-14), but there may be

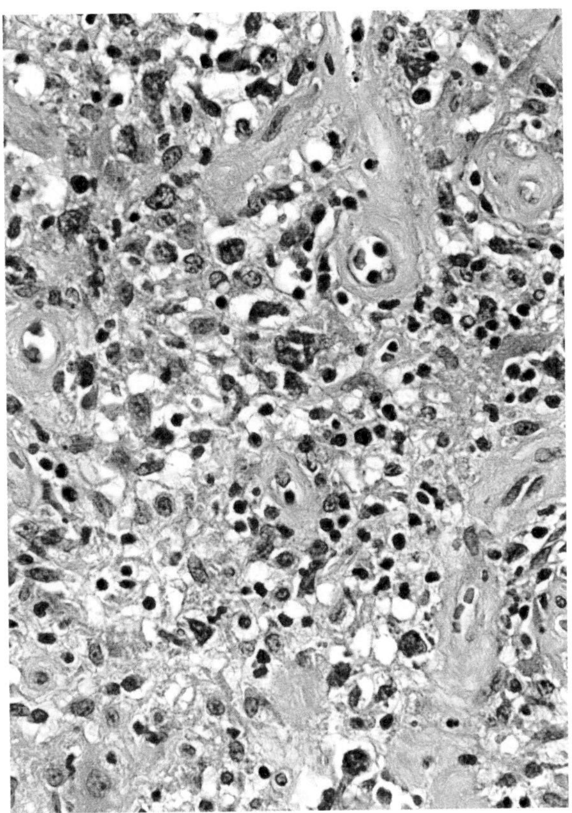

Figure 12-11

PRIMARY CNS LYMPHOMA

Pleomorphism, reactive astrocytes, and sclerotic vessels can combine to obscure the lymphoid nature of the lesion.

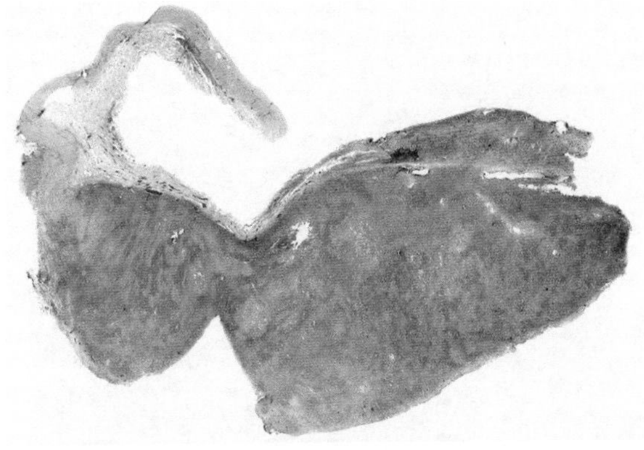

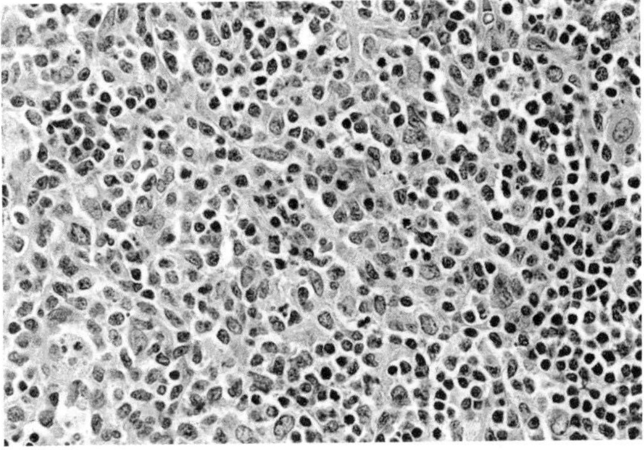

Figure 12-12

PRIMARY CNS LYMPHOMA: MUCOSA ASSOCIATED LYMPHOID TISSUE (MALT) TYPE

Top: The rare MALT PCNSL usually involves the dura. Bottom: The cells of MALT lymphomas, in contrast to those of the classic large cell PCNSL, are smaller, may have more irregular nuclear membranes, and have less cellular atypia.

regions with diffuse, glioma-like infiltration without suggestive architectural features of lymphoma (fig. 12-15). The neoplastic cells in some cases are accompanied, or even overshadowed, by reactive cells, both histiocytic and astrocytic. The B-cell immunophenotype and MIB-1 labeling index are useful in identifying sparse individual cells (see fig. 12-24).

Prebiopsy corticosteroid therapy for the treatment of edema often has a profound, apoptosis-based effect on tumor cells, vastly reducing their number (fig. 12-16). In extreme cases, only a sea of macrophages remains in a picture that superficially resembles demyelinating disease or infarct (26,34). This can hinder, if not preclude, a histologic diagnosis (fig. 12-17). Rebiopsy after a period off steroids may be required.

Despite their cytologic atypia and high mitotic activity, PCNSLs lymphomas lack vascular proliferation and perinecrotic palisading.

Frozen Section. Unwanted effects of freezing include artifactual nuclear angulation and general obscuring of cytologic details. The spurious pleomorphism results in an appearance easily mistaken for that of a malignant glioma. Nevertheless, patchy cellularity due largely to angiocentricity, apoptosis, and large vesicular nuclei with prominent nucleoli usually remain as compelling arguments in favor of PCNSL. Both mitotic and apoptotic rates are substantially higher than in most glioblastomas (fig. 12-18).

Immunohistochemical and In Situ Hybridization Findings. Most PCNSLs are positive for CD20, as well for other B-cell markers such as CD5 (mature B cell), CD79A (pan B cell) and CD22 (mature B cell), but are negative for CD3 (figs. 12-19, 12-20). There is considerable variation,

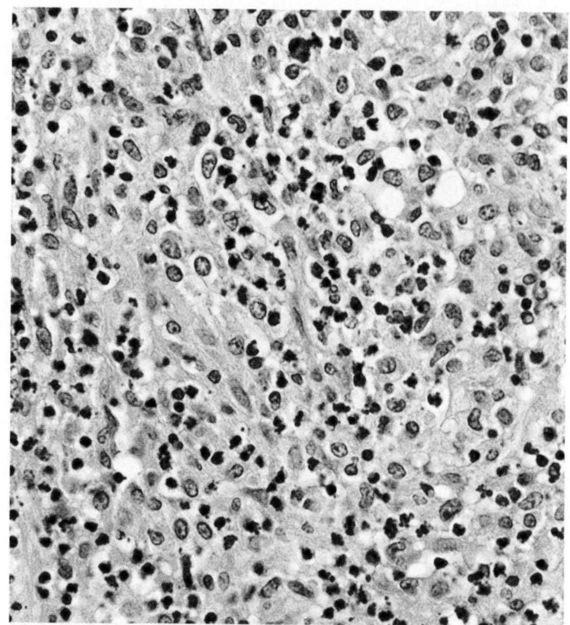

Figure 12-13

PRIMARY CNS LYMPHOMA: T-CELL TYPE

The mixed cell composition and frequent paucity of large tumor cells lend an inflammatory quality to many T-cell lymphomas.

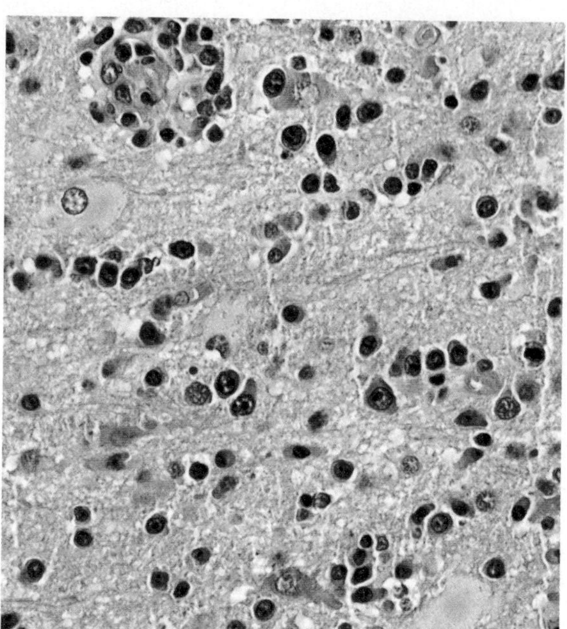

Figure 12-14

PRIMARY CNS LYMPHOMA

Plasmacytoid cytologic features help differentiate some diffusely infiltrating PCNSLs from gliomas.

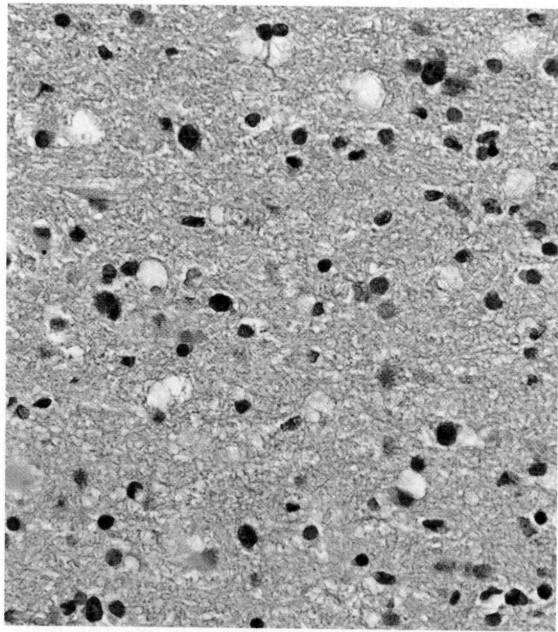

Figure 12-15

PRIMARY CNS LYMPHOMA

Widely scattered atypical cells at the edge of the lesion can be difficult to recognize as lymphoid in hematoxylin and eosin (H&E)-stained sections.

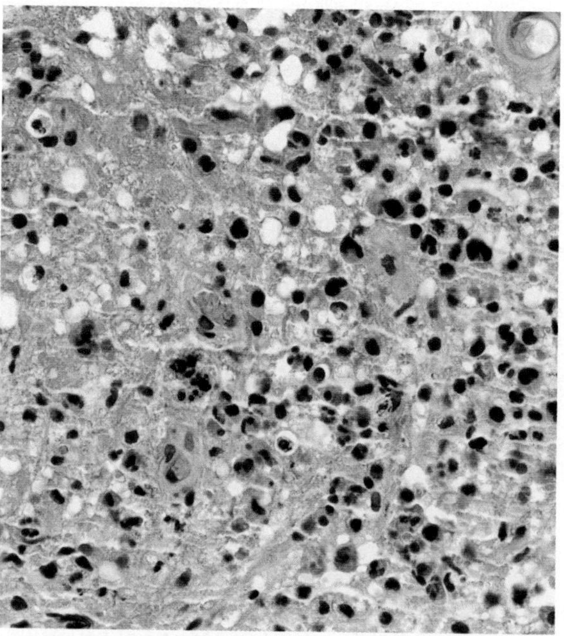

Figure 12-16

PRIMARY CNS LYMPHOMA
TREATED WITH CORTICOSTEROIDS

By inducing apoptosis, preoperative corticosteroids can dramatically alter the histologic appearance of PCNSL.

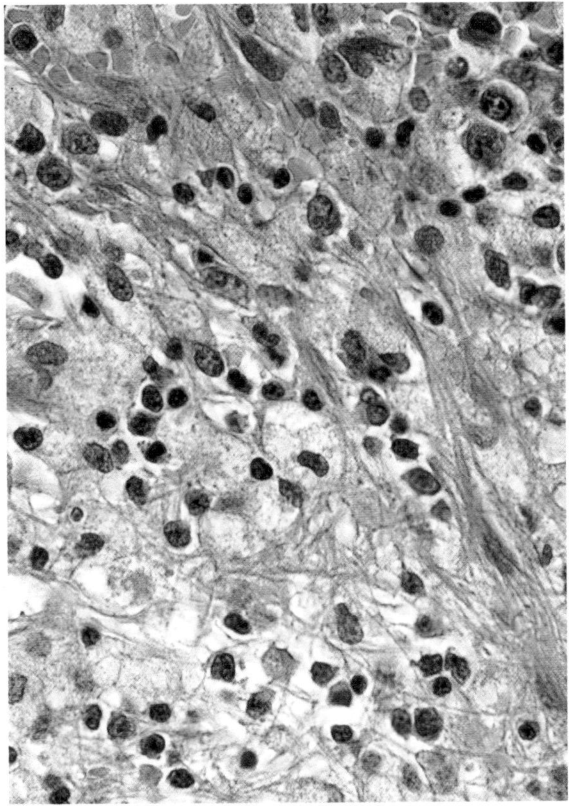

Figure 12-17

PRIMARY CNS LYMPHOMA TREATED WITH CORTICOSTEROIDS

PCNSLs treated with steroids may largely, if not entirely, consist of macrophages. Only a few tumor cells remain in this case.

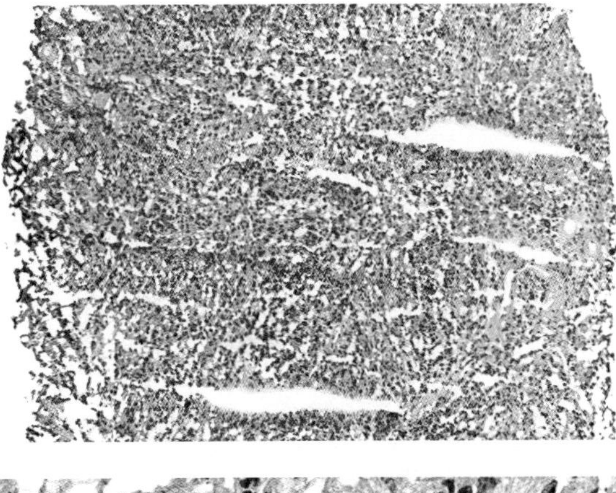

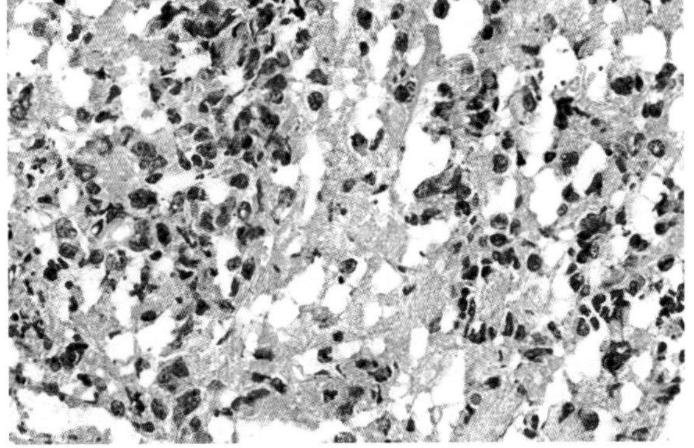

Figure 12-18

PRIMARY CNS LYMPHOMA: FROZEN SECTION

Top: As a consequence of their angiocentricity, PCNSLs often have a distinctive variegated quality at low-power magnification.

Bottom: Even in frozen sections, PCNSLs retain their typical nuclear roundness and prominent nucleoli, which help distinguish them from gliomas. Angiocentricity and prominent apoptosis are additional features that should bring PCNSL to mind.

however, in the number of CD3-positive reactive T cells. These are often limited in degree and are diagnostically nondistracting (fig. 12-21, top), but may visually compete with (fig. 12-21, bottom), or even obscure, neoplastic cells in some cases (T-cell–rich B-cell lymphoma) (fig. 12-22). T-cell lesions are immunoreactive for CD3 (fig. 12-23).

The rare anaplastic large cell lymphoma may lack CD45 staining, and is often epithelial membrane antigen (EMA) immunopositive (9,10). Most have a characteristic 2:5 translocation and produce the fusion product ALK-1 which can be detected immunohistochemically (9).

In AIDS-associated PCNSLs and in post-transplant lymphoproliferative disease, tumor cells are positive for Epstein-Barr virus (EBV) by immunohistochemistry (latent membrane protein) or, with greater sensitivity, by situ hybridization for EBV-encoded mRNA (see fig. 12-27) (38–43).

PCNSLs often have an MIB-1 rate exceeding 50 percent, a level substantially higher than that of almost all glioblastomas (fig. 12-24).

Ultrastructural Findings. PCNSLs are noted for their lack of specific ultrastructural features: the absence of intermediate filaments, specific organelles, and intercellular junctions (44). Nonetheless, the presence of "nuclear pockets" is a common feature of large cell lymphomas. In addition, plasmacytoid differentiation is accompanied by a characteristic pattern of chromatin aggregation and by prominence of both paranuclear Golgi apparatus and rough endoplasmic reticulum.

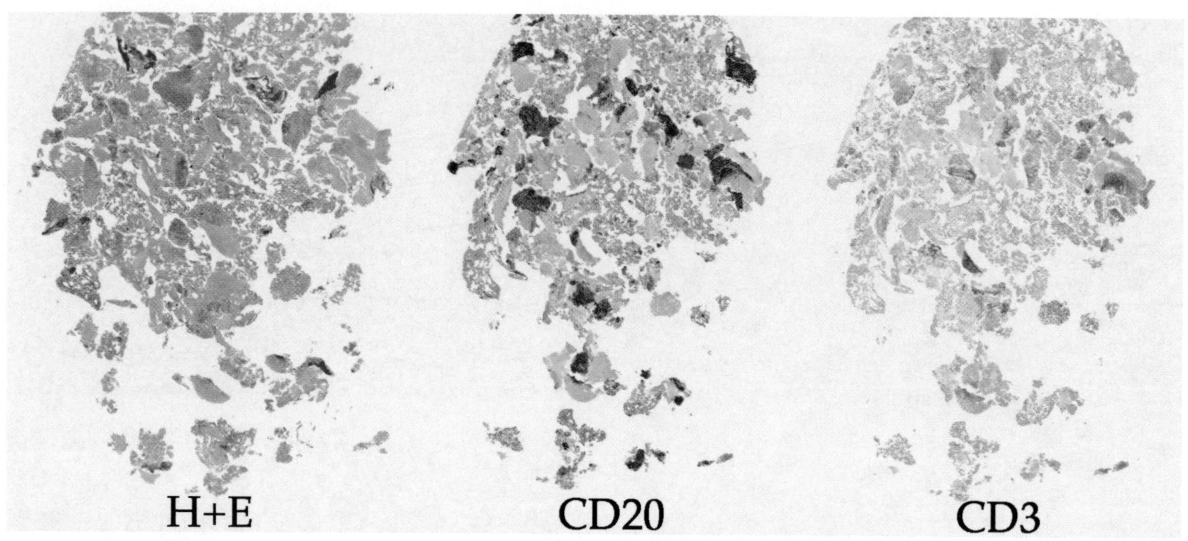

Figure 12-19

PRIMARY CNS LYMPHOMA

As seen in a macrosection, tumor-containing tissue fragments principally stain for CD20, but also variably, and usually to a lesser degree, for CD3 in proportion to the percentage of reactive T cells.

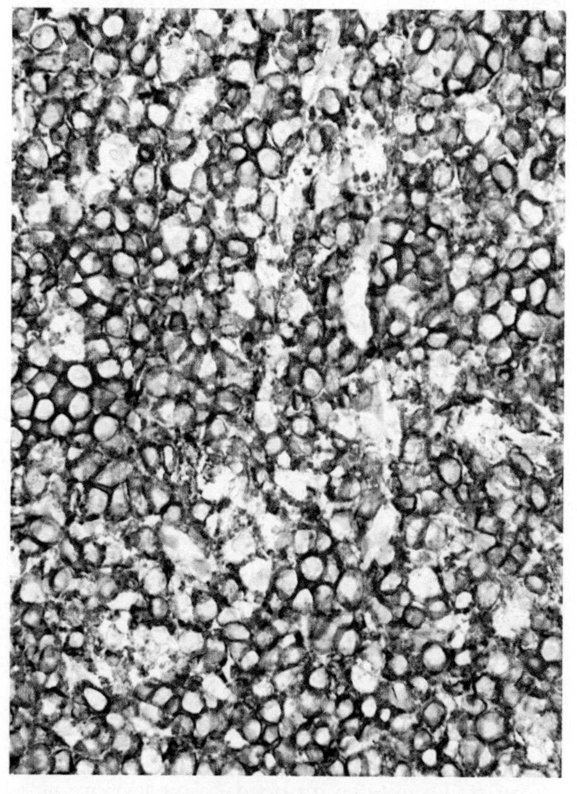

Figure 12-20

PRIMARY CNS LYMPHOMA

The tumor seen in figures 12-19 and 12-20 has in some areas an almost pure population of CD20-positive B cells.

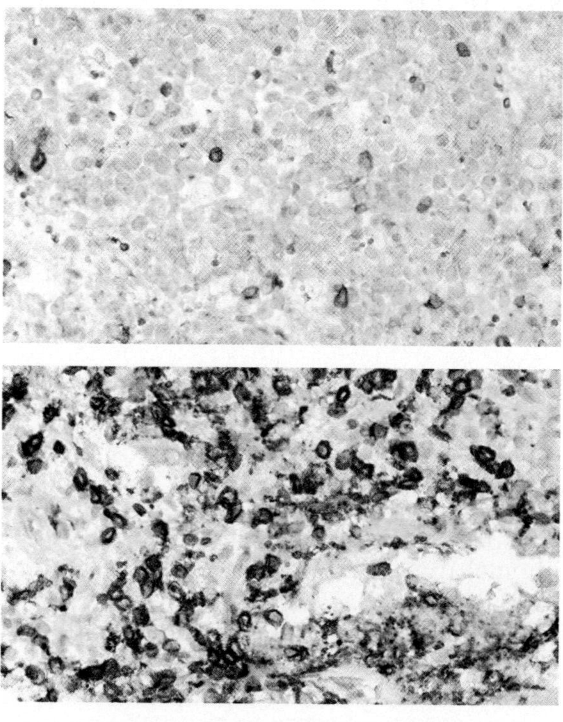

Figure 12-21

PRIMARY CNS LYMPHOMA

The tumor seen in figure 12-19 exhibits the varying ratio of neoplastic B and reactive T cells present in many PCNSLs. CD3-positive T cells (top) are rare in some areas. Elsewhere (bottom), B and T cells are present in approximately equal numbers.

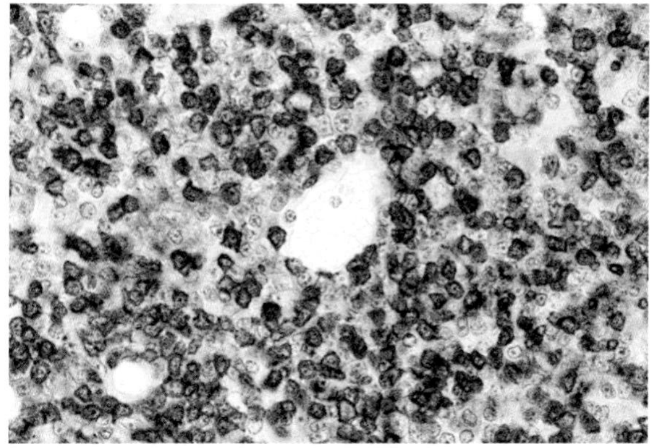

Figure 12-22

PRIMARY CNS LYMPHOMA

Unstained neoplastic B cells may be hidden in the background of T cells (T-cell–rich B-cell lymphoma) (UCHL-1 stain).

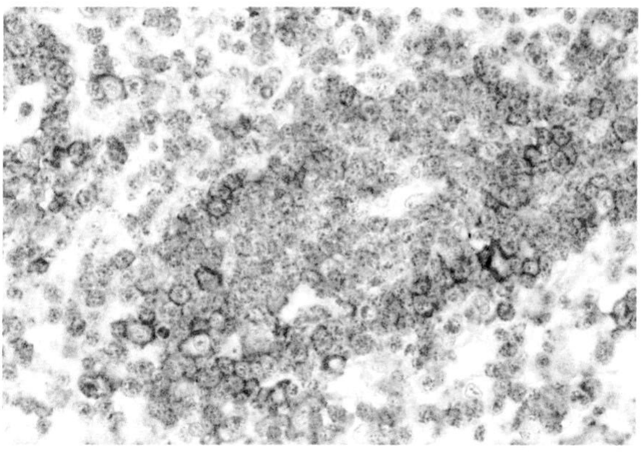

Figure 12-23

PRIMARY CNS LYMPHOMA

Only a rare PCNSL, such as this CD3-positive tumor, is of T-cell type.

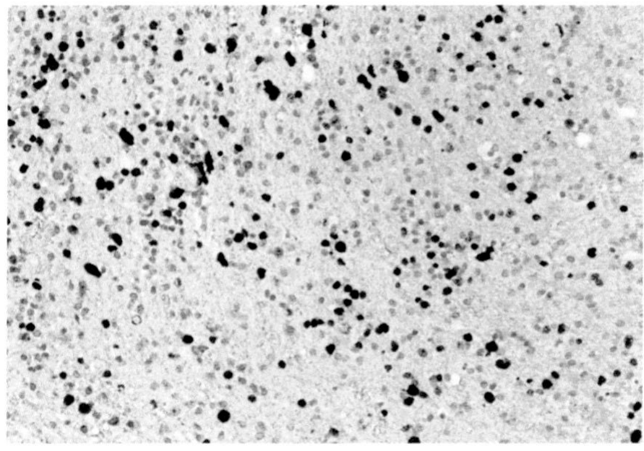

Figure 12-24

PRIMARY CNS LYMPHOMA

A very high MIB-1 rate in a diffusely infiltrating lesion should prompt consideration of PCNSL.

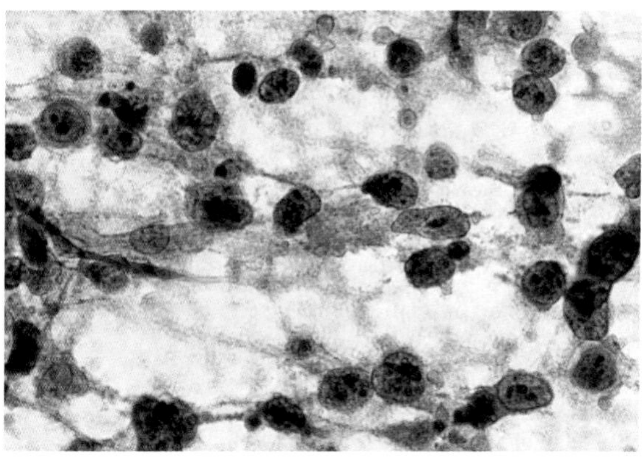

Figure 12-25

PRIMARY CNS LYMPHOMA: SMEAR PREPARATION

Dispersed cells with little cytoplasm and round or indented nuclei with prominent nucleoli are typical of PCNSL.

Cytologic Findings. Since lymphoid cells lack intercellular junctions, they shed individually, unlike metastatic carcinomas in which cells remain clumped and cohesive, or malignant gliomas, whose cells lie entangled in their own processes in tissue fragments. The discrete cell borders, vesicular nuclei, frequent apoptosis, and prominent nucleoli of PCNSL are well seen (fig. 12-25) (20).

Differential Diagnosis. Although classic lesions were diagnosed in the pre-immunohistochemistry era on the basis of features in hematoxylin and eosin (H&E)-stained sections and reticulin staining, immunohistochemistry is now the decisive technique for diagnosis in most cases. It remains for the pathologist to entertain the possibility of PCNSL when presented with one of its several look-alikes.

PCNSLs have diffusely infiltrating processes that incorporate neurons and reactive astrocytes to an extent far greater than do metastatic carcinomas. While some carcinomas extend along vessels, they do not form perivascular, reticulin-rich lamellae.

While PCNSLs share the invasive quality of glioblastoma, lymphomas lack the perinecrotic pseudopalisading and vascular proliferation that

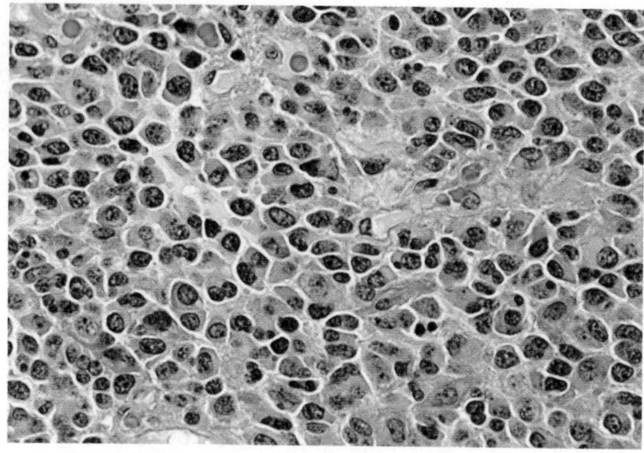

Figure 12-26

ANAPLASTIC OLIGODENDROGLIOMA RESEMBLING PRIMARY CNS LYMPHOMA

High-grade oligodendrogliomas with epithelioid features mimic PCNSL.

almost define this glioma. Abundant eosinophilic cytoplasm with processes is also alien to lymphomas. Glioblastomas are only uncommonly angiocentric, and almost never angioinvasive.

One infiltrating glioma with a cytologic resemblance to PCNSL is the occasional anaplastic oligodendroglioma, which has somewhat epithelioid cells with large, round nuclei and prominent nucleoli (fig. 12-26). These oligodendrogliomas, however, lack the infiltrate of small lymphocytes and histiocytes that attends PCNSLs. The glioma is also not angiocentric, nor does it have the vascular reticulin pattern.

In light of the marked cytologic atypia of most PCNSLs, their distinction from chronic inflammatory lesions usually poses no difficulty. Exceptions include uncommon examples of well-differentiated lymphoma, especially T-cell types, as well as high-grade lymphomas that are dominated by a T-cell inflammatory infiltrate (see fig. 12-10). Idiopathic inflammatory lesions composed of a dense lymphoplasmacytic infiltrate often exhibit minor atypia. Although sometimes difficult to distinguish from lymphoma in H&E-stained sections alone, inflammatory infiltrates vary in cell composition, are polyclonal, and usually contain mature plasma cells. Nonetheless, the diagnosis of lymphoma may not be possible without resorting to ancillary studies. These include procedures to detect gene rearrangements or in situ hybridization for EBV in the context of a lymphoid lesion with atypical cells. This is done with the recognition that a gene rearrangement is, in itself, not sufficient evidence for a diagnosis of PCNSL.

Treatment and Prognosis. Although PCNSLs are often exquisitely sensitive to corticosteroid treatment, and may disappear from radiologic images for weeks, months, or even years, the effect is temporary. The same is true of radiotherapy and chemotherapy. Aggressive treatment with methotrexate is now widely employed in place of irradiation, and sometimes with good results. Nonetheless, the majority of patients with PCNSL die within 2 years (45,46). Occasional patients are long-term survivors, but cures are rare. As a consequence of disease extent, failure is often multicentric in both brain parenchyma and meninges. A diffuse subependymal pattern of spread is often seen.

MALT lymphomas, in contrast, are low-grade lesions for which radiotherapy and chemotherapy appear to produce complete remissions (17).

Post-Transplant Lymphoproliferative Disease

Post-transplant lymphoproliferative disease is a lymphoid proliferation or lymphoma that develops as a consequence of immunosuppression, usually in the setting of solid organ or bone marrow transplantation. It comprises a spectrum from early EBV-driven polyclonal proliferations resembling infectious mononucleosis to EBV-positive or -negative lymphomas of primarily B, or to a lesser extent T-, cell type (fig. 12-27). Involvement of the CNS is infrequent (47–49).

Lymphomatoid Granulomatosis

Lymphomatoid granulomatosis is a systemic lesion that most observers consider an EBV-driven malignant lymphoma of B-cell type. The CNS is symptomatically involved in approximately 25 percent of cases (50). Only on occasion is lymphomatoid granulomatosis limited to the CNS at the time of presentation (51–53). Whether secondary or primary, CNS symptoms are in evidence and the lesions are typically multifocal on neuroradiologic studies. Lymphomatoid granulomatosis has been reported in the setting of AIDS (54).

Macroscopically, the lesions are often hemorrhagic and necrotic. Microscopically, they consist of a mixed inflammatory infiltrate associated with necrosis, typical lymphocytes,

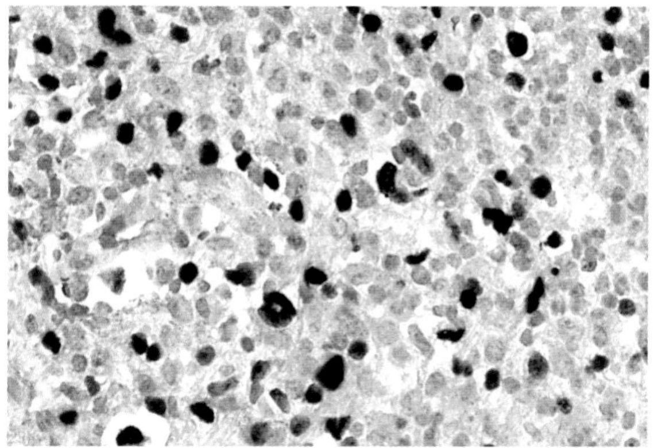

Figure 12-27

POST-TRANSPLANT LYMPHOPROLIFERATIVE DISEASE: IN SITU HYBRIDIZATION FOR EPSTEIN-BARR VIRUS

These lesions are typically positive for the Epstein-Barr virus.

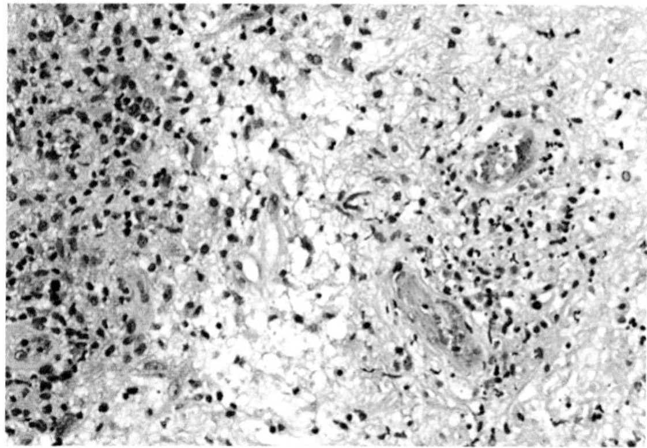

Figure 12-28

LYMPHOMATOID GRANULOMATOSIS

Necrosis and small reactive lymphocytes may dominate the lesion.

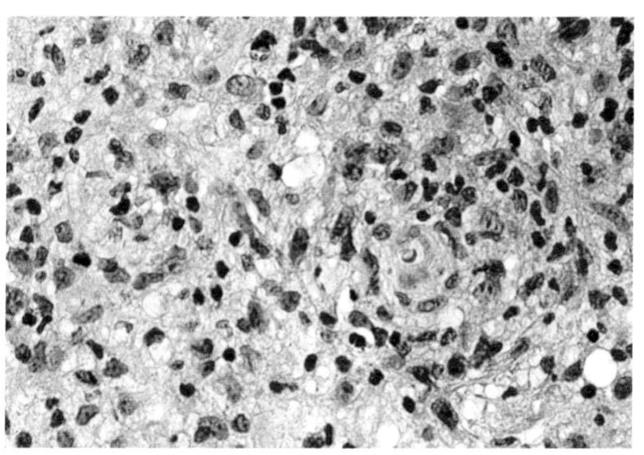

Figure 12-29

LYMPHOMATOID GRANULOMATOSIS

Large perivascular atypical cells may be sparse.

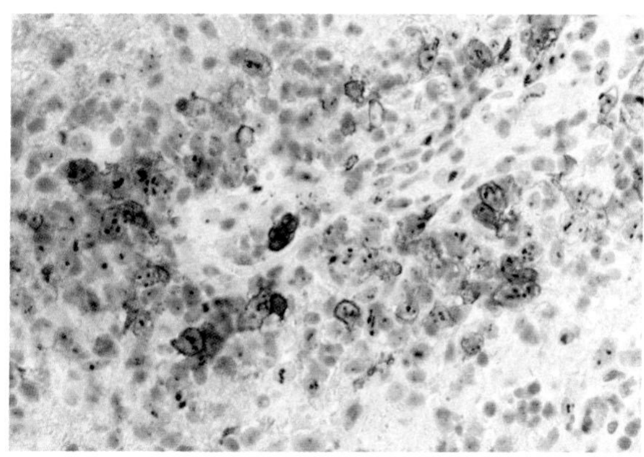

Figure 12-30

LYMPHOMATOID GRANULOMATOSIS

Large atypical cells are immunoreactive for CD20.

plasmacytoid lymphocytes, and macrophages (fig. 12-28). The infiltrate is characteristically both invasive and destructive of vessels. Multiple concentric laminae of perivascular reticulin and atypical cells, a typical feature of malignant lymphoma, may be present. T cells predominate, but scattered atypical large B cells are present perivascularly (fig. 12-29). Perivascular infarction and hemorrhage are common. Histologically more obvious B-cell lymphoma infiltrates, apparently evolving from the polymorphous infiltrate that characterizes lymphomatoid granulomatosis, have also been reported in the setting of AIDS (54).

The differential diagnosis focuses primarily on distinguishing lymphomatoid granulomatosis from reactive lymphoplasmacytic proliferations. Histologically and immunohistochemically, this depends on finding atypical B cells (figs. 12-29, 12-30). With these identified, in situ hybridization for EBV-encoded mRNA becomes diagnostic (fig. 12-31), and most pathologists are unwilling to make the diagnosis in the absence of this positivity. In the face of therapeutic immunosuppression, the disease is considered post-transplant lymphoproliferative syndrome. Immunoglobulin gene rearrangements are generally detected in cases of lymphomatoid granulomatosis in which cytologically atypical large cells, necrosis, and numerous EBV-positive cells are present. In contrast, cases showing a paucity of these features may not have detectable clonality and may not be distinguishable from a reactive lymphoplasmacytic infiltrate.

The appropriate therapy for lymphomatoid granulomatosis remains to be established. Some patients have responded to chemotherapy and radiotherapy; steroid therapy has successfully, if temporarily, arrested the disease in others.

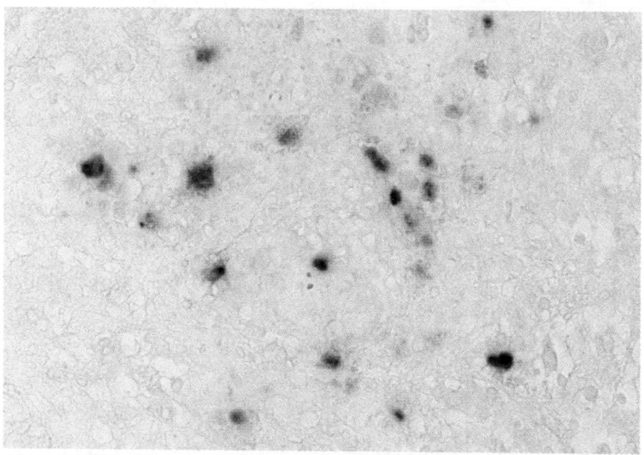

Figure 12-31

LYMPHOMATOID GRANULOMATOSIS

In situ hybridization for Epstein-Barr virus is almost required for the diagnosis.

Neurolymphomatosis

The term *neurolymphomatosis* describes lymphomas that are inexplicably attracted to, and even centered upon, peripheral nerves, either extradural or intradural. Such lesions are, in a sense, primary to the peripheral nervous system, i.e., "PPNSLs." Numerous nerves are often involved. Intradurally, the cauda equina is most often affected (fig. 12-32) (55,56).

Plasmacytoma

Uncommon solitary intracranial *plasmacytomas* can either be dura-based and form a meningioma-like mass (fig. 12-33) (57–61), or arise within the skull (61,62). A few have been intrasellar and have mimicked pituitary adenoma, both grossly and microscopically (63). Only rare plasmacytomas arise within the substance of the brain (64).

Histologically, the lesions vary from ones composed of very well-differentiated plasma cells (fig. 12-34, left) to tumors with obvious cytologic atypia. Immunohistochemically, plasma cell tumors differ from lymphoma in being nonreactive for

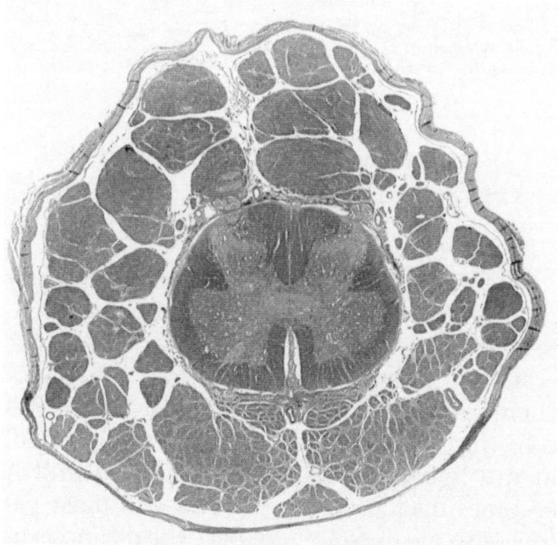

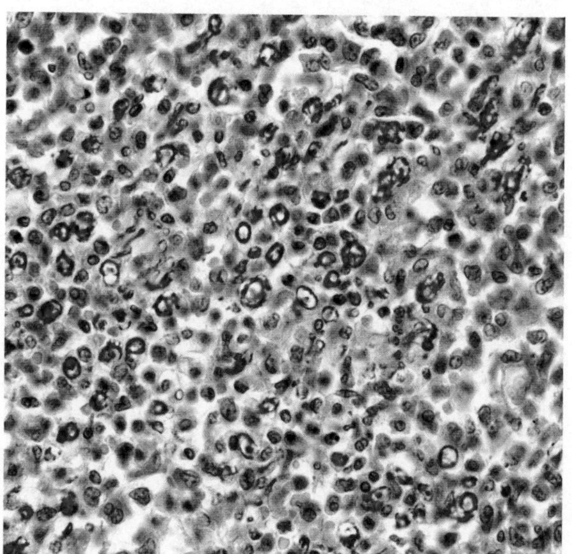

Figure 12-32

NEUROLYMPHOMATOSIS

Left: Lymphomas can target peripheral nerves such as those of the cauda equina.

Right: Scattered myelinated axons are recognized among the tumor cells after staining with H&E/Luxol fast blue.

leukocyte common antigen (LCA), but positive for CD138 (fig. 12-34, right). Usually demonstrable in paraffin-embedded tissue is expression of monotypic immunoglobulin light (kappa, lambda) chains. Normal and neoplastic plasma cells are conspicuously positive for EMA, a feature that should not be misinterpreted as evidence of carcinoma.

Since well-differentiated plasmacytomas may be harbingers of multiple myeloma, the truly isolated nature of a lesion is always open to question. In some cases, however, follow-up of 5 years or more has failed to reveal local recurrence or dissemination (57–59). One series found that lesions at the skull base more often evolve into myeloma than dura-based lesions (61). In the same study, marked cytologic atypia and a high MIB-1 rate predicted rapid recurrence and aggressive behavior.

Amyloidoma

On rare occasion, tumorous masses of amyloid present as extra- or intra-axial lesions in patients in whom there are no systemic manifestations of a plasma cell dyscrasia or of amyloidosis. Sites of involvement have included brain parenchyma (65,66), trigeminal ganglion (67,68), and skull base (69,70). Parenchymal lesions have centered upon periventricular white matter.

Histologically, the amyloid may be deposited about vessels in concentric lamellae, or simply as independent or cohesive spherical masses. A small number of lymphocytes and plasma cells may be present, but nothing to suggest a frank neoplasm. Characterization of the amyloid and the plasma cells indicates clonality and heavy chain gene rearrangements (65). Invariably, the amyloid has been of the ALγ type (65). Despite their neoplastic nature, these indolent lesions are not prone to relentless enlargement, although the natural history of the disease is not clear.

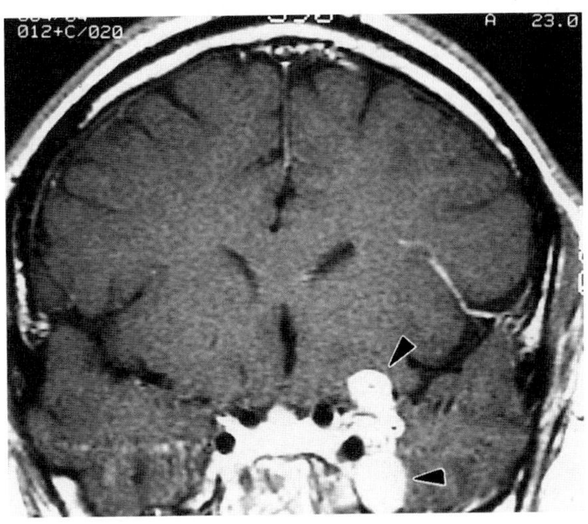

Figure 12-33

PLASMACYTOMA

Intracranial plasmacytomas are usually dura-based, dome-shaped masses (arrowheads) that mimic meningiomas. (Fig. 12-12 from Fascicle 10, 3rd Series.)

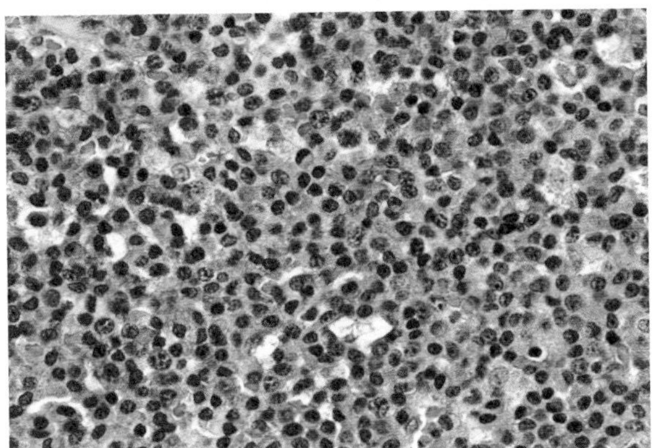

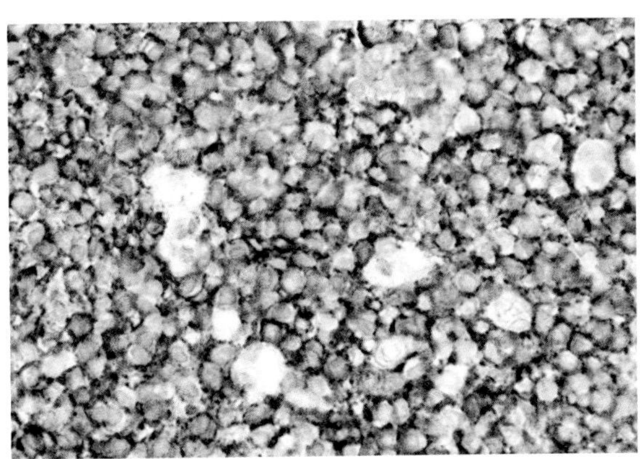

Figure 12-34

PLASMACYTOMA

Left: Some plasmacytomas are densely cellular masses with only subtle plasmacytic qualities.
Right: Immunostaining for CD138 confirms the plasmacytic nature of the cells.

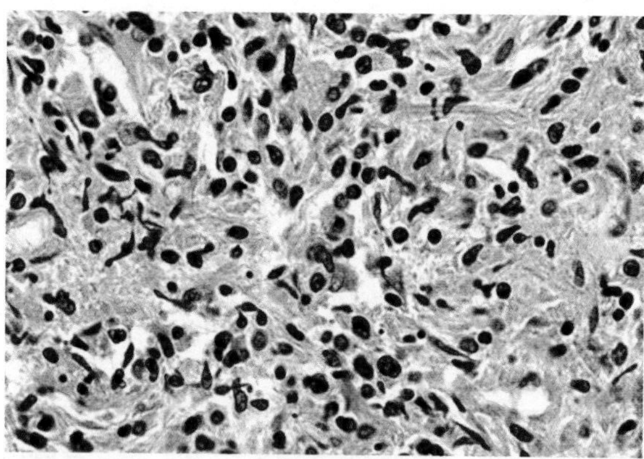
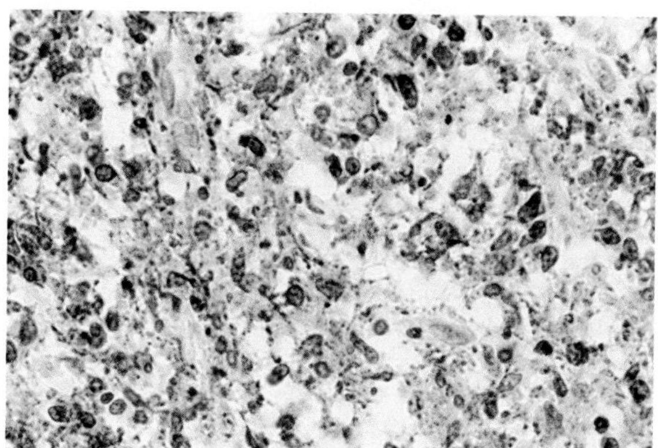

Figure 12-35

MICROGLIOMATOSIS

Left: A most rare and incompletely characterized lesion, microgliomatosis is composed of elongated cells that, at least in part, resemble microglia in their reactive, "rod cell" configuration.

Right: Microgliomatosis is immunoreactive for macrophage markers such as HAM56.

MALT lymphoma has been associated with amyloid deposition (71). Reactive amyloid deposits have occurred in the setting of chronic multiple sclerosis (72)

Microgliomatosis

If microglia are endogenous tissue histiocytes equivalent to Kupffer cells or fixed histiocytes in the spleen, their neoplastic transformation should produce very rare examples of a true histiocytic tumor. We are aware of only a solitary case published in the immunohistochemical era (73). In this yet to be confirmed analysis, neoplastic microglia appear to be highly infiltrative, a property expected of a cell that, even in a non-neoplastic state, moves freely within brain substance in response to injury. The nuclei vary from round to markedly elongated and tend to be slightly angulated or twisted (fig. 12-35, left). Thus, the cells can resemble mature microglia in their activated "rod cell" form, as is seen in response to injury. The cells are immunoreactive for such macrophage markers as HAM56 (fig. 12-35, right) and Ricinus communis agglutinin-1 (73).

SECONDARY TUMORS

Secondary Malignant Lymphomas

Intravascular Lymphomatosis. This largely intravascular malignant lymphoma affects any number of nonlymphoreticular organs, such as skin, kidney, and adrenal gland, but has a particular affinity for the vasculature of the CNS. Historically, the lesion was usually diagnosed postmortem, however, it is now increasingly recognized during life. CNS involvement is often accompanied by systemic involvement, but the CNS is occasionally affected in apparent isolation.

Intravascular lymphomatosis usually presents in adulthood with diffuse or focal neurologic signs as the consequence of ischemia produced by "sludging" or outright occlusion of cerebrospinal vessels (74–76). Resulting infarcts are therefore common (77). An associated rash is a common consequence of cutaneous involvement, and renal and adrenal infiltrates are also frequent. Cerebral symptoms are often nonfocal and include dementia. Patchy, bright infiltrates are seen on T2-weighted MRI, some of which are contrast enhancing; infarcts are also seen (78,79).

Macroscopically, abnormalities are generally limited to infarcts of varying size. Microscopically, the lesion is a striking, but often focal, intravascular proliferation of cytologically malignant lymphocytes, generally of large cell type (fig. 12-36). Vesicular nuclei with prominent nucleoli are therefore typical. The number of neoplastic cells and their degree of atypia vary. Not only may involved vessels be hard to find, but only several, or even just one or two, neoplastic cells may be present in a vessel cross section (fig. 12-37). As in the case of PCNSLs, neoplastic cells may be present beneath

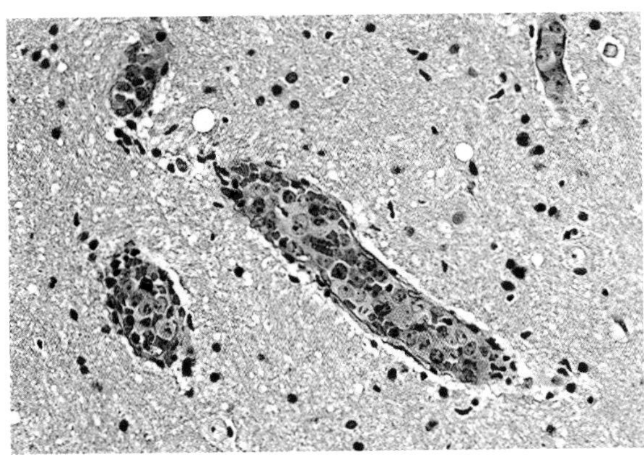

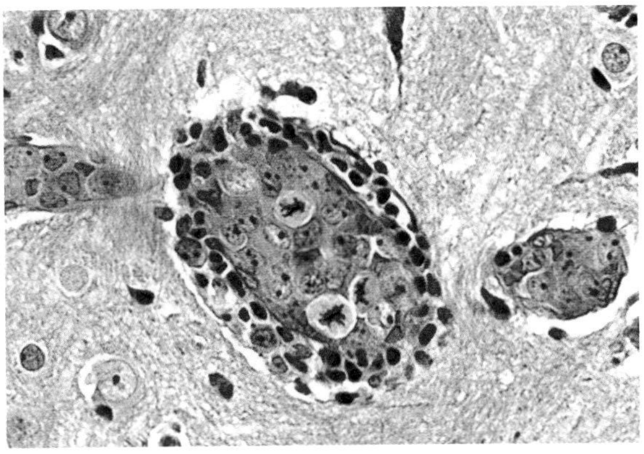

Figure 12-36

INTRAVASCULAR LYMPHOMATOSIS

Left: When distended with large, markedly atypical cells, affected vessels are conspicuous even at low magnification. Right: Cells with large round nuclei, prominent nucleoli, and frequent mitotic figures are typical.

the endothelium or within vessel walls where they can incite a fibrous reaction, but all remain strictly intravascular. Mitoses are often abundant and the MIB-1 rate is very high, often 50 percent or more. Immunohistochemically, a B-cell phenotype is the rule (fig. 12-38), although T-cell lesions have been described (80).

The differential diagnosis is limited. Intravascular lymphomatosis must be distinguished from the far more common PCNSL. The latter is largely a parenchymal process and, although angiotropic, does not spread intravascularly. Inflammatory processes should also be considered, but their cells are not as cytologically atypical, and are perivascular rather than intravascular.

Although steroid therapy has a beneficial short-term effect, producing dramatic clinical improvement, most patients survive less than 2 years, and many less than 1 year from the onset of symptoms (74,75). At autopsy, many systemic sites are affected, particularly skin, kidney, heart, lung, prostate gland, and adrenal gland and other endocrine organs.

Intraparenchymal and Leptomeningeal Lymphomas. Secondary involvement of intradural structures by malignant lymphoma is generally confined to the leptomeninges and to cranial and spinal nerve roots. CNS involvement typically occurs late in the course of diffuse high-grade lymphomas, usually at a time when extranodal sites such as bone marrow have become affected. As in meningeal carcinomatosis, symptoms reflect increased intracranial pressure and involvement of cranial nerves.

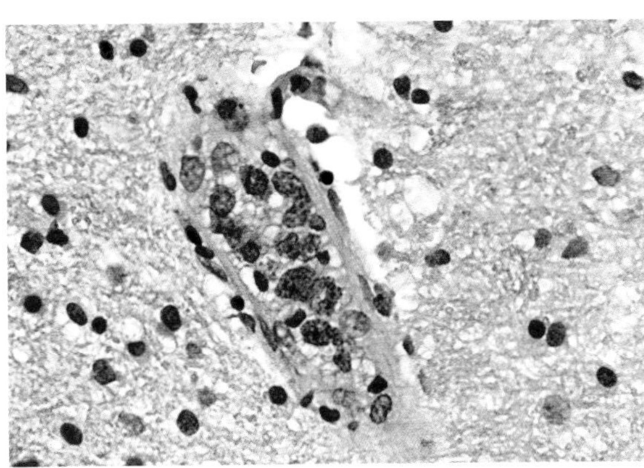

Figure 12-37

INTRAVASCULAR LYMPHOMATOSIS

The intravascular tumor cells are not always as large and atypical as illustrated in figure 12-36.

Macroscopically, little may be seen other than hypertrophy of cranial and spinal nerve roots. Microscopically, the neoplastic cells proliferate within the leptomeninges, extend into the Virchow-Robin spaces, and permeate nerves (81–84).

Spinal Epidural Lymphoma. Occasional lymphomas, either of non-Hodgkin's or less often of Hodgkin's type, involve the spinal epidural space and compress the spinal cord. In some cases, particularly with Hodgkin's lymphoma, the underlying disease process has been recognized previously at other body sites (85).

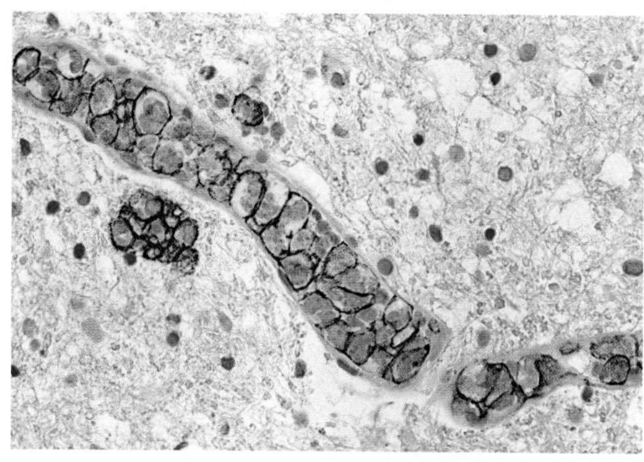

Figure 12-38

INTRAVASCULAR LYMPHOMATOSIS

As evidenced by immunoreactivity for CD20, most intravascular lymphomas are of B-cell type.

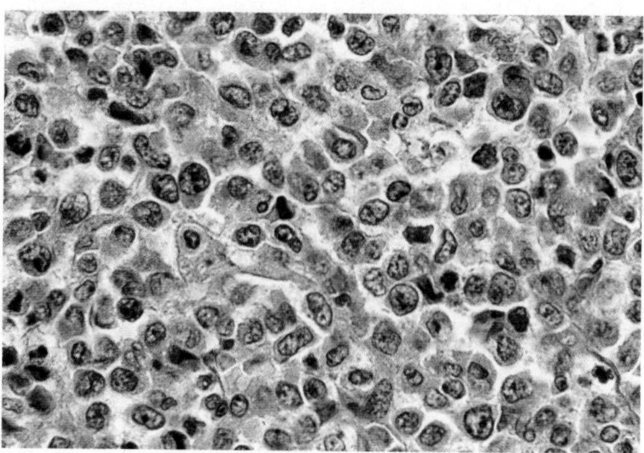

Figure 12-39

GRANULOCYTIC SARCOMA

A mixed cell population with nuclear variations lends a "granulocytic" appearance. The lesion occurred in the meninges of a 36-year-old man with chronic myelogenous leukemia in accelerated phase who presented with a superficial mass that radiologically resembled a subdural hematoma. Instead, an extensive subdural tumor mass was found.

In other patients, the neoplasm makes its first appearance in the intraspinal space (86–91), and an extraspinal focus may never be found. It remains unclear whether such ostensibly isolated lymphomas are truly primary in the epidural space or have spread from an occult visceral or lymphoreticular primary focus (89).

Spinal epidural lymphomas are noted for their lack of osseous involvement, in contrast to the frequent osteolytic or osteoblastic consequences of spinal metastatic carcinoma.

Leukemia

Secondary involvement of the CNS in *leukemia* occurs in several forms: 1) meningeal involvement, 2) intraparenchymal hemorrhagic foci containing intravascular and perivascular aggregates of immature or blast forms, and 3) as a mass in the setting of acute myeloid leukemia known as *extramedullary myeloid cell tumor*, *granulocytic sarcoma*, or *chloroma*.

Meningeal leukemia most often occurs in the setting of acute disease, particularly of lymphoblastic type. Headaches and cranial nerve dysfunction result from obstruction of the flow of cerebrospinal fluid and from infiltration of cranial nerves (92–94). Although the infiltrate is confined to subarachnoid and Virchow-Robin spaces early in the course, with progression it can penetrate the pia to gain access to brain parenchyma (95). As is the case with primary lymphoma, neoplastic cells sometimes infiltrate vessel walls to become entwined in a lamellar network of reticulin.

In cases associated with markedly elevated leukocyte counts, as in blast crisis, cerebral vessels may become occluded by neoplastic cells (leukostasis) (92,96–98). Thereafter, the neoplastic cells may proliferate locally to form sizable masses. Superimposed hemorrhage may abruptly and fatally enlarge such foci. Macroscopically, these hemorrhages contain a distinctive central gray core of neoplastic cells. These mass lesions may occur de novo or in association with acute myelogenous leukemia (99–102). Microscopically, there are myeloblasts, more differentiated cells (promyelocytes), and mature neutrophils (fig. 12-39). Stains for myeloperoxidase, lysozyme, and chloroacetate esterase are critical to the recognition of the tumor.

When a myeloid sarcoma occurs as an isolated lesion without any evidence of leukemia, an unusual event, radiotherapy may result in prolonged survival. Usually, however, the lesion is a harbinger of a leukemia that shortly makes its appearance, or is unappreciated.

REFERENCES

Primary Central Nervous System Lymphoma

1. Goetz P, Lafuente J, Revesz T, Galloway M, Dogan A, Kitchen N. Primary low-grade B-cell lymphoma of mucosa-associated lymphoid tissue of the dura mimicking the presentation of an acute subdural hematoma. Case report and review of the literature. J Neurosurg 2002;96:611–4.
2. Kumar S, Kumar D, Kaldjian EP, Bauserman S, Raffeld M, Jaffe ES. Primary low-grade B-cell lymphoma of the dura: a mucosa associated lymphoid tissue-type lymphoma. Am J Surg Pathol 1997;21:81–7.
3. Lehman NL, Horoupian DS, Warnke RA, Sundram UN, Peterson K, Harsh GR. Dural marginal zone lymphoma with massive amyloid deposition: rare low-grade primary central nervous system B-cell lymphoma. Case report. J Neurosurg 2002;96:368–72.
4. Kobayashi H, Sano T, Ii K, Hizawa K. Primary Burkitt-type lymphoma of the central nervous system. Acta Neuropathol (Berl) 1984;64:12–4.
5. Bataille B, Delwail V, Menet E, et al. Primary intracerebral malignant lymphoma: report of 248 cases. J Neurosurg 2000;92:261–6.
6. Tomlinson FH, Kurtin PJ, Suman VJ, et al. Primary intracerebral malignant lymphoma: a clinicopathological study of 89 patients. J Neurosurg 1995;82:558–66.
7. Cheuk W, Walford N, Lou J, et al. Primary histiocytic lymphoma of the central nervous system: a neoplasm frequently overshadowed by a prominent inflammatory component. Am J Surg Pathol 2001;25:1372–9.
8. Sun W, Nordberg ML, Fowler MR. Histiocytic sarcoma involving the central nervous system: clinical, immunohistochemical, and molecular genetic studies of a case with review of the literature. Am J Surg Pathol 2003;27:258–65.
9. George DH, Scheithauer BW, Aker FV, et al. Primary anaplastic large cell lymphoma of the central nervous system: prognostic effect of ALK-1 expression. Am J Surg Pathol 2003;27:487–93.
10. Havlioglu N, Manepalli A, Galindo L, Sotelo-Avila C, Grosso L. Primary Ki-1 (anaplastic large cell) lymphoma of the brain and spinal cord. Am J Clin Pathol 1995;103:496–9.
11. Ashby MA, Barber PC, Holmes AE, Freer CE, Collins RD. Primary intracranial Hodgkin's disease. A case report and discussion. Am J Surg Pathol 1988;12:294–9.
12. Herrlinger U, Klingel K, Meyermann R, et al. Central nervous system Hodgkin's lymphoma without systemic manifestation: case report and review of the literature. Acta Neuropathol (Berl) 2000;99:709–14.
13. Hochberg FH, Miller DC. Primary central nervous system lymphoma. J Neurosurg 1988;68: 835–53.
14. Miller DC, Hochberg FH, Harris NL, Gruber ML, Louis DN, Cohen H. Pathology with clinical correlations of primary central nervous system non-Hodgkin's lymphoma. The Massachusetts General Hospital experience 1958–1989. Cancer 1994;74:1383–97.
15. O'Neill BP, Illig JJ. Primary central nervous system lymphoma. Mayo Clin Proc 1989;64:1005–20.
16. Baumgartner JE, Rachlin JR, Beckstead JH, et al. Primary central nervous system lymphomas: natural history and response to radiation therapy in 55 patients with acquired immunodeficiency syndrome. J Neurosurg 1990;73(2):206–11.
17. Tu P, Giannini C, Judkins A, et al. Clinicopathologic and genetic profile of intracranial marginal zone lymphoma: a primary low-grade CNS lymphoma that mimics meningioma. J Clin Oncol 2005;23:5718–27.
18. O'Neill BP, Dinapoli RP, Kurtin PJ, Habermann TM. Occult systemic non-Hodgkin's lymphoma (NHL) in patients initially diagnosed as primary central nervous system lymphoma (PCNSL): how much staging is enough? J Neurooncol 1995;25:67–71.
19. O'Neill BP, Kelly PJ, Earle JD, Scheithauer B, Banks PM. Computer-assisted stereotaxic biopsy for the diagnosis of primary central nervous system lymphoma. Neurology 1987;37:1160–4.
20. Sherman ME, Erozan YS, Mann RB, et al. Stereotactic brain biopsy in the diagnosis of malignant lymphoma. Am J Clin Pathol 1991;95:878–83.
21. Eby NL, Grufferman S, Flannelly CM, Schold SC Jr, Vogel FS, Burger PC. Increasing incidence of primary brain lymphoma in the US. Cancer 1988;62:2461–5.
22. Cote TR, Manns A, Hardy CR, Yellin FJ, Hartge P. Epidemiology of brain lymphoma among people with or without acquired immunodeficiency syndrome. AIDS/Cancer Study Group. J Natl Cancer Inst 1996;88:675–9.
23. Nuckols JD, Liu K, Burchette JL, McLendon RE, Traweek ST. Primary central nervous system lymphomas: a 30-year experience at a single institution. Mod Pathol 1999;12:1167–73.

24. Olson JE, Janney CA, Rao RD, et al. The continuing increase in the incidence of primary central nervous system non-Hodgkin lymphoma: a surveillance, epidemiology, and end results analysis. Cancer 2002;95:1504–10.
25. Hao D, DiFrancesco LM, Brasher PM, et al. Is primary CNS lymphoma really becoming more common? A population-based study of incidence, clinicopathological features and outcomes in Alberta from 1975 to 1996. Ann Oncol 1999;10:65–70.
26. Alderson L, Fetell MR, Sisti M, Hochberg F, Cohen M, Louis DN. Sentinel lesions of primary CNS lymphoma. J Neurol Neurosurg Psychiatry 1996;60:102–5.
27. Kuker W, Herrlinger U, Gronewaller E, Rohrbach JM, Weller M. Ocular manifestation of primary nervous system lymphoma: what can be expected from imaging? J Neurol 2002;249:1713–6.
28. Chan CC. Primary intraocular lymphoma: clinical features, diagnosis, and treatment. Clin Lymphoma 2003;4:30–1.
29. Brown MT, McClendon RE, Gockerman JP. Primary central nervous system lymphoma with systemic metastasis: case report and review. J Neurooncol 1995;23:207–21.
30. Roman-Goldstein SM, Goldman DL, Howieson J, Belkin R, Neuwelt EA. MR of primary CNS lymphoma in immunologically normal patients. AJNR Am J Neuroradiol 1992;13:1207–13.
31. Herrlinger U, Weller M, Kuker W. Primary CNS lymphoma in the spinal cord: clinical manifestations may precede MRI detectability. Neuroradiology 2002;44:239–44.
32. Lee DK, Chung CK, Kim HJ, et al. Multifocal primary CNS T cell lymphoma of the spinal cord. Clin Neuropathol 2002;21:149–55.
33. Miller DJ, McCutcheon IE. Hemangioblastomas and other uncommon intramedullary tumors. J Neurooncol 2000;47:253–70.
34. Geppert M, Ostertag CB, Seitz G, Kiessling M. Glucocorticoid therapy obscures the diagnosis of cerebral lymphoma. Acta Neuropathol (Berl) 1990;80:629–34.
35. Gray RS, Abrahams JJ, Hufnagel TJ, Kim JH, Lesser RL, Spencer DD. Ghost-cell tumor of the optic chiasm. Primary CNS lymphoma. J Clin Neuroophthalmol 1989;9:98–104.
36. Vaquero J, Martinez R, Rossi E, Lopez R. Primary cerebral lymphoma: the "ghost tumor." Case report. J Neurosurg 1984;60:174–6.
37. Choi JS, Nam DH, Ko YH, et al. Primary central nervous system lymphoma in Korea: comparison of B- and T-cell lymphomas. Am J Surg Pathol 2003;27:919–28.
38. Auperin I, Mikolt J, Oksenhendler E, et al. Primary central nervous system malignant non-Hodgkin's lymphomas from HIV-infected and non-infected patients: expression of cellular surface proteins and Epstein-Barr viral markers. Neuropathol Appl Neurobiol 1994;20:243–52.
39. Bashir RM, Hochberg FH, Wei MX. Epstein-Barr virus and brain lymphomas. J Neurooncol 1995;24:195–205.
40. Camilleri-Broet S, Davi F, Feuillard J, et al. AIDS-related primary brain lymphomas: histopathologic and immunohistochemical study of 51 cases. The French Study Group for HIV-Associated Tumors. Hum Pathol 1997;28:367–74.
41. Morgello S. Pathogenesis and classification of primary central nervous system lymphoma: an update. Brain Pathol 1995;5:383–93.
42. Rouah E, Rogers BB, Wilson DR, Kirkpatrick JB, Buffone GJ. Demonstration of Epstein-Barr virus in primary central nervous system lymphomas by the polymerase chain reaction and in situ hybridization. Hum Pathol 1990;21:545–50.
43. Roychowdhury S, Peng R, Baiocchi RA, et al. Experimental treatment of Epstein-Barr virus-associated primary central nervous system lymphoma. Cancer Res 2003;63:965–71.
44. Hirano A. A comparison of the fine structure of malignant lymphoma and other neoplasms in the brain. Acta Neuropathol Suppl (Berl) 1975;Suppl 6:141–5.
45. Damek DM. Primary central nervous system lymphoma. Curr Treat Options Neurol 2003;5: 213–22.
46. Plasswilm L, Herrlinger U, Korfel A, et al. Primary central nervous system (CNS) lymphoma in immunocompetent patients. Ann Hematol 2002;81:415–23.

Post-Transplant Lymphoproliferative Disease

47. Nalesnik MA, Jaffe R, Starzl TE, et al. The pathology of posttransplant lymphoproliferative disorders occurring in the setting of cyclosporine A-prednisone immunosuppression. Am J Pathol 1988;133:173–92.
48. Ferry JA, Jacobson JO, Conti D, Delmonico F, Harris NL. Lymphoproliferative disorders and hematologic malignancies following organ transplantation. Mod Pathol 1989;2:583–92.
49. Penn I. The changing pattern of posttransplant malignancies. Transplant Proc 1991;23(Pt 2):1101–3.

Lymphomatoid Granulomatosis

50. Katzenstein AL, Carrington CB, Liebow AA. Lymphomatoid granulomatosis: a clinicopathologic study of 152 cases. Cancer 1979;43:360–73.

51. Kleinschmidt-DeMasters BK, Filley CM, Bitter MA. Central nervous system angiocentric, angiodestructive T-cell lymphoma (lymphomatoid granulomatosis). Surg Neurol 1992;37:130–7.
52. Mizuno T, Takanashi Y, Onodera H, et al. A case of lymphomatoid granulomatosis/angiocentric immunoproliferative lesion with long clinical course and diffuse brain involvement. J Neurol Sci 2003;213:67–76.
53. Schmidt BJ, Meagher-Villemure K, Del Carpio J. Lymphomatoid granulomatosis with isolated involvement of the brain. Ann Neurol 1984;15: 478–81.
54. Anders KH, Latta H, Chang BS, Tomiyasu U, Quddusi AS, Vinters HV. Lymphomatoid granulomatosis and malignant lymphoma of the central nervous system in the acquired immunodeficiency syndrome. Hum Pathol 1989;20:326–34.

Neurolymphomatosis

55. Baehring JM, Damek D, Martin EC, Betensky RA, Hochberg FH. Neurolymphomatosis. Neurooncology 2003;5:104–15.
56. Mauney M, Sciotto CG. Primary malignant lymphoma of the cauda equina. Am J Surg Pathol 1983;7:185–90.

Plasmacytoma

57. Kohli CM, Kawazu T. Solitary intracranial plasmacytoma. Surg Neurol 1982;17:307–12.
58. Krivoy S, Gonzalez JE, Cespedes G, Walzer I. Solitary cerebral falx plasmacytoma. Surg Neurol 1977;8:222–4.
59. Krumholz A, Weiss HD, Jiji VH, Bakal D, Kirsh MB. Solitary intracranial plasmacytoma: two patients with extended follow-up. Ann Neurol 1982;11:529–32.
60. Mancardi GL, Mandybur TI. Solitary intracranial plasmacytoma. Cancer 1983;51:2226–33.
61. Schwartz TH, Rhiew R, Isaacson SR, Orazi A, Bruce JN. Association between intracranial plasmacytoma and multiple myeloma: clinicopathological outcome study. Neurosurgery 2001;49:1039–44; discussion 1039–44.
62. Du Preez JH, Branca EP. Plasmacytoma of the skull: case reports. Neurosurgery 1991;29:902–6.
63. Losa M, Terreni MR, Tresoldi M, et al. Solitary plasmacytoma of the sphenoid sinus involving the pituitary fossa: a case report and review of the literature. Surg Neurol 1992;37:388–93.
64. Wisniewski T, Sisti M, Inhirami G, Knowles DM, Powers JM. Intracerebral solitary plasmacytoma. Neurosurgery 1990;27:826–9.

Amyloidoma

65. Laeng RH, Altermatt HJ, Scheithauer BW, Zimmermann DR. Amyloidomas of the nervous system: a monoclonal B-cell disorder with monotypic amyloid light chain lambda amyloid production. Cancer 1998;82:362–74.
66. Lee J, Krol G, Rosenblum M. Primary amyloidoma of the brain: CT and MR presentation. AJNR Am J Neuroradiol 1995;16:712–4.
67. Matsumoto T, Tani E, Fukami M, Kaba K, Yokota M, Hoshii Y. Amyloidoma in the gasserian ganglion: case report. Surg Neurol 1999;52:600–3.
68. O'Brien TJ, McKelvie PA, Vrodos N. Bilateral trigeminal amyloidoma: an unusual case of trigeminal neuropathy with a review of the literature. Case report. J Neurosurg 1994;81:780–3.
69. Ferreiro JA, Bhuta S, Nieberg RK, Verity MA. Amyloidoma of the skull base. Arch Pathol Lab Med 1990;114:974–6.
70. Unal F, Hepgul K, Bayindir C, Bilge T, Imer M, Turantan I. Skull base amyloidoma. Case report. J Neurosurg 1992;76:303–6.
71. Lehman NL, Horoupian DS, Warnke RA, Sundram UN, Peterson K, Harsh GR. Dural marginal zone lymphoma with massive amyloid deposition: rare low-grade primary central nervous system B-cell lymphoma. Case report. J Neurosurg 2002;96:368–72.
72. Nennesmo I, Bogdanovic N, Petren AL, Fredrikson S. Multiple sclerosis and amyloid deposits in the white matter of the brain. Acta Neuropathol (Berl) 1997;93:205–9.

Microgliomatosis

73. Hulette CM. Microglioma, a histiocytic neoplasm of the central nervous system. Mod Pathol 1996;9:316–9.

Intravascular Lymphomatosis

74. Baehring JM, Longtine J, Hochberg FH. A new approach to the diagnosis and treatment of intravascular lymphoma. J Neurooncol 2003;61: 237–48.
75. Beristain X, Azzarelli B. The neurological masquerade of intravascular lymphomatosis. Arch Neurol 2002;59:439–43.
76. Glass J, Hochberg FH, Miller DC. Intravascular lymphomatosis. A systemic disease with neurologic manifestations. Cancer 1993;71:3156–64.
77. Dubas F, Saint-Andre JP, Pouplard-Barthelaix A, Delestre F, Emile J. Intravascular malignant lymphomatosis (so-called malignant angioendotheliomatosis): a case confined to the lumbosacral spinal cord and nerve roots. Clin Neuropathol 1990;9:115–20.
78. Martin-Duverneuil N, Mokhtari K, Behin A, Lafitte F, Hoang-Xuan K, Chiras J. Intravascular malignant lymphomatosis. Neuroradiology 2002;44:749–54.

79. Williams RL, Meltzer CC, Smirniotopoulos JG, Fukui MB, Inman M. Cerebral MR imaging in intravascular lymphomatosis. AJNR Am J Neuroradiol 1998;19:427–31.
80. Sepp N, Schuler G, Romani N, et al. "Intravascular lymphomatosis" (angioendotheliomatosis): evidence for a T-cell origin in two cases. Hum Pathol 1990;21:1051–8.

Intraparenchymal and Leptomeningeal Lymphomas

81. Griffin JW, Thompson RW, Mitchinson MJ, De Kiewiet JC, Welland FH. Lymphomatous leptomeningitis. Am J Med 1971;51:200–8.
82. Liang R, Chiu E, Loke SL. Secondary central nervous system involvement by non-Hodgkin's lymphoma: the risk factors. Hematol Oncol 1990;8:141–5.
83. MacKintosh FR, Colby TV, Podolsky WJ, et al. Central nervous system involvement in non-Hodgkin's lymphoma: an analysis of 105 cases. Cancer 1982;49:586–95.
84. Young RC, Howser DM, Anderson T, Fisher RI, Jaffe E, DeVita VT Jr. Central nervous system complications of non-Hodgkin's lymphoma. The potential role for prophylactic therapy. Am J Med 1979;66:435–43.

Spinal Epidural Lymphoma

85. Friedman M, Kim TH, Panahon AM. Spinal cord compression in malignant lymphoma. Treatment and results. Cancer 1976;37:1485–91.
86. Alameda F, Pedro C, Besses C, et al. Primary epidural lymphoma. Case report. J Neurosurg 2003;98(Suppl 2):215–7.
87. Epelbaum R, Haim N, Ben-Shahar M, Ben-Arie Y, Feinsod M, Cohen Y. Non-Hodgkin's lymphoma presenting with spinal epidural involvement. Cancer 1986;58:2120–4.
88. Grant J, Kaech D, Jones D. Spinal cord compression as the first presentation of lymphoma—a review of 15 cases. Histopathology 1986;10: 1191–202.
89. Lyons MK, O'Neill BP, Marsh WR, Kurtin PJ. Primary spinal epidural non-Hodgkin's lymphoma: report of eight patients and review of the literature. Neurosurgery 1992;30:675–80.
90. Raco A, Cervoni L, Salvati M, Delfini R. Primary spinal epidural non-Hodgkin's lymphomas in childhood: a review of 6 cases. Acta Neurochir (Wien) 1997;139:526–8.
91. Salvati M, Cervoni L, Artico M, Raco A, Ciappetta P, Delfini R. Primary spinal epidural non-Hodgkin's lymphomas: a clinical study. Surg Neurol 1996;46:339–43; discussion 343–4.

Leukemia

92. Bojsen-Moller M, Nielsen JL. CNS involvement in leukaemia. An autopsy study of 100 consecutive patients. Acta Pathol Microbiol Immunol Scand [A] 1983;91:209–16.
93. Demopoulos A, DeAngelis LM. Neurologic complications of leukemia. Curr Opin Neurol 2002;15:691–9.
94. Price R, Johnson W. The central nervous system in childhood leukemia: I. The arachnoid. Cancer 1973;31:520–33.
95. Azzarelli V, Roessmann U. Pathogenesis of central nervous system infiltration in acute leukemia. Arch Pathol Lab Med 1977;101:203–5.
96. Freireich EJ, Thomas LB, Frei E 3rd, Fritz RD, Forkner CE Jr. A distinctive type of intracerebral hemorrhage associated with "blastic crisis" in patients with leukemia. Cancer 1960;13:146–54.
97. Kawanami T, Kurita K, Yamakawa M, Omoto E, Kato T. Cerebrovascular disease in acute leukemia: a clinicopathological study of 14 patients. Intern Med 2002;41:1130–4.
98. Moore EW, Thomas LB, Shaw RK, Freireich EJ. The central nervous system in acute leukemia: a postmortem study of 117 consecutive cases, with particular reference to hemorrhages, leukemic infiltrations, and the syndrome of meningeal leukemia. Arch Intern Med 1960;105:451–68.
99. van Veen S, Kluin PM, de Keizer RJ, Kluin-Nelemans HC. Granulocytic sarcoma (chloroma). Presentation of an unusual case. Am J Clin Pathol 1991;95:567–71.
100. Ohta K, Kondoh T, Yasuo K, Kohsaka Y, Kohmura E. Primary granulocytic sarcoma in the sphenoidal bone and orbit. Childs Nerv Syst 2003;19:674–9.
101. Parker K, Hardjasudarma M, McClellan RL, Fowler MR, Milner JW. MR features of an intracerebellar chloroma. AJNR Am J Neuroradiol 1996;17:1592–4.
102. Fitoz S, Atasoy C, Yavuz K, Gozdasoglu S, Erden I, Akyar S. Granulocytic sarcoma. Cranial and breast involvement. Clin Imaging 2002;26:166–9.

13 TUMORS OF INTRACRANIAL AND INTRASPINAL PERIPHERAL NERVES

CONVENTIONAL SCHWANNOMA

Definition. *Conventional schwannoma* is a benign tumor composed entirely of well-differentiated Schwann cells.

Clinical Features. Schwannomas occur at all ages, although children are only rarely affected (1), and at all levels of the neuraxis. Outside the setting of neurofibromatosis 2 (NF2), almost all schwannomas are solitary.

Anatomic Considerations. Peripheral schwannomas are discussed in detail in the Armed Forces Institute of Pathology (AFIP) tumor Fascicle, *Tumors of the Peripheral Nervous System* (2). The discussion here is limited to schwannomas affecting the cranial nerves, central nervous system (CNS) parenchyma, and spinal nerve roots.

Intracranial schwannomas most often arise from the vestibular division of the eighth cranial nerve near or within the vestibular ganglion (fig. 13-1). Strictly speaking, therefore, they are not "acoustic," as suggested by a common moniker, but "vestibular," a designation that is increasingly employed. Other cranial nerves, such as the fifth and seventh, are much less commonly affected. Hearing loss and tinnitus are common symptoms of vestibular schwannomas of any size, whereas large examples produce cerebellar dysfunction or other cranial nerve deficits. Bilaterality is a feature in patients with NF2 (see fig. 18-6).

Schwannomas arising in brain parenchyma are rare, but occasionally appear in the cerebrum (fig. 13-2), cerebellum, and brain stem (3–9). Amputation neuroma-like lesions around rare brain stem infarcts suggest that perivascular peripheral nerve ingrowth could be a source of parenchymal schwannomas (10). The finding of aberrant peripheral nerve fibers in the medulla of occasional patients supports this view (11).

Intraspinal extramedullary schwannomas, like their cranial nerve counterpart, favor sensory nerves and therefore usually arise on dorsal roots. Most sporadic tumors are solitary, whereas multiplicity prevails in patients with NF2 (see fig. 18-11). Symptomatic masses in the latter setting are often accompanied by small, subclinical, intraneural Schwann cell proliferations collectively termed *schwannosis* (see fig. 18-12). NF2 is discussed in chapter 18.

Intraspinal schwannomas of nerve roots occur at all levels but most affect the lumbosacral and cauda equina regions (figs. 13-3, 13-4). The lesions produce generic symptoms depending on their location, but pain is common at any site (12). Expanding more freely in the intraspinal and extraspinal compartments than within

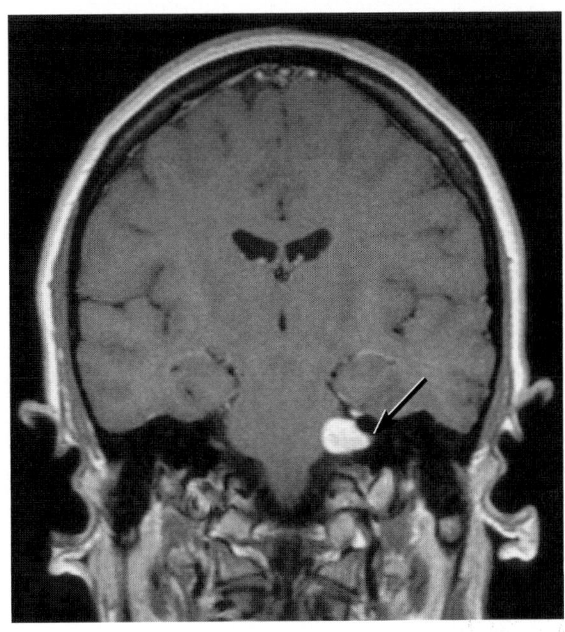

Figure 13-1

VESTIBULAR SCHWANNOMA

A contrast-enhancing mass with a nipple-like profile (arrow) is characteristic of vestibular schwannoma.

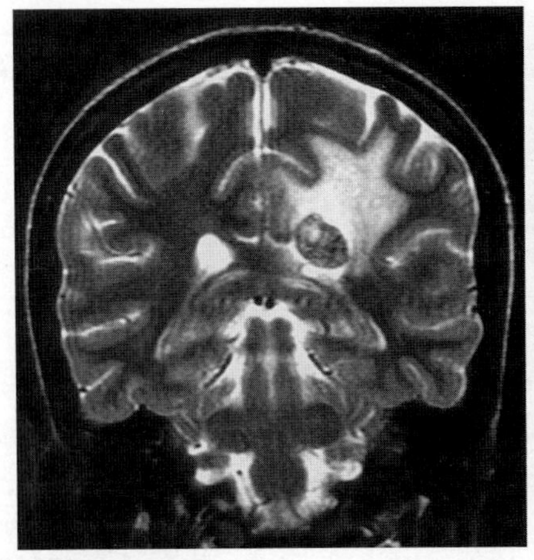

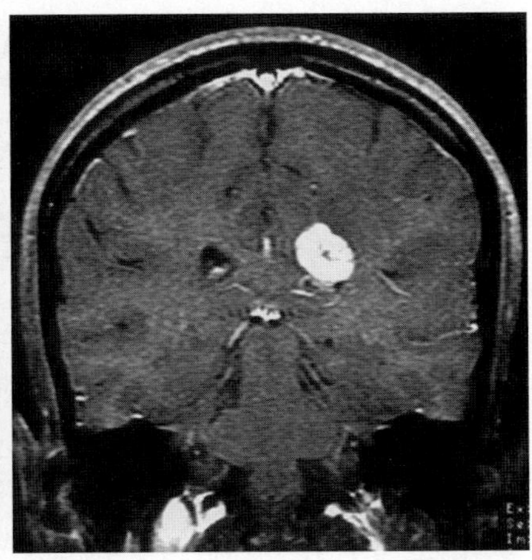

Figure 13-2

INTRAPARENCHYMAL SCHWANNOMA

Only rarely do schwannomas appear within the brain, as here in the frontal lobe of a 10-year-old girl. The lesion was found incidentally in a computerized tomography (CT) scan performed because of head trauma near the lateral ventricle on the right. The contrast-enhancing lesion is surrounded by a bright zone of edema on a T2-weighted magnetic resonance image (MRI).

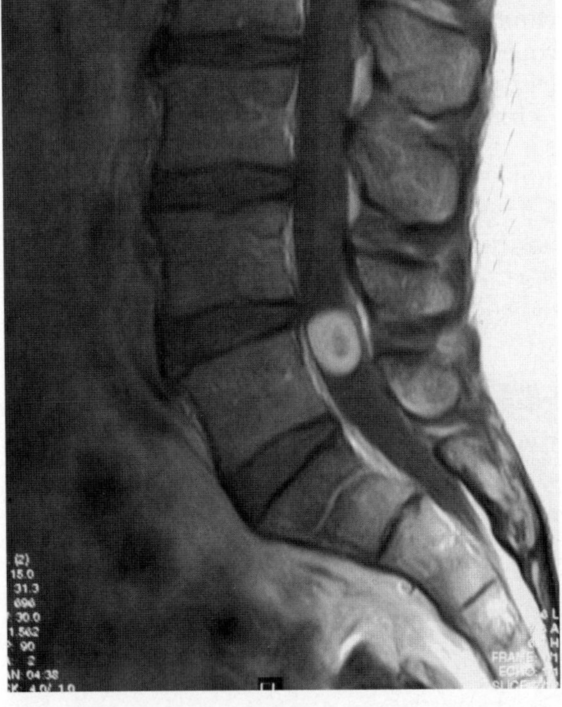

Figure 13-3

SCHWANNOMA OF SPINAL NERVE ROOT

Schwannomas of spinal nerve roots are discrete and contrast enhancing, as illustrated in a 57-year-old man who had pain in the leg when coughing or sneezing. An intraoperative view of the lesion is seen in figure 13-4.

the confines of the dural root sleeve or neural foramen, some lesions assume the classic "dumbbell" configuration. Schwannomas arising from lumbosacral nerve roots can become massive *giant sacral schwannomas* that erode bone and displace the rectum (13).

Spinal intramedullary schwannomas are rare (12,14–21) and presumably arise from myelinated peripheral nerve fibers that accompany blood vessels into the parenchyma, or from Schwann cells near the dorsal root entry zone (19,22). A distinction between extramedullary and intramedullary spinal schwannomas is not always simple since nerve root tumors invaginating the cord may become partially intramedullary, or, conversely, an intramedullary lesion may reach the subarachnoid space (23). The differential diagnosis of spinal lesions by anatomic site is given in Appendix B.

Radiologic Findings. Vestibular schwannomas are discrete contrast-enhancing masses that, finding little resistance to expansion in the proximal portion of the nerve, mushroom into the intracranial compartment at the cerebellopontine angle. A distinctive radiologic profile is thus generated (fig. 13-1). Larger examples are associated with enlargement or flaring of the internal auditory canal. Such meatal enlargement is rare in other extra-axial tumors such as meningioma and provides strong presumptive evidence of the diagnosis. Spinal schwannomas may be dumbbell shaped when there are both intraspinal and paraspinal components. Intratumoral cysts are found in a minority of lesions in an incidence that is generally proportional to tumor size (24). As with schwannomas at any site, intraparenchymal schwannomas are well circumscribed and contrast enhancing (figs. 13-2, 13-3).

Gross Findings. Schwannomas arising from cranial or spinal nerve roots are typically well circumscribed and far more often globular than

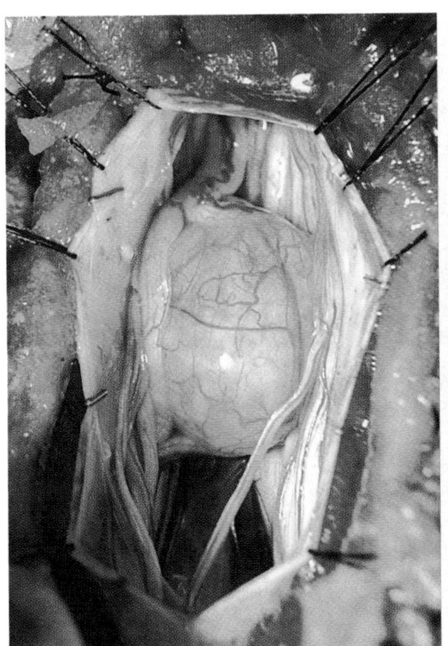

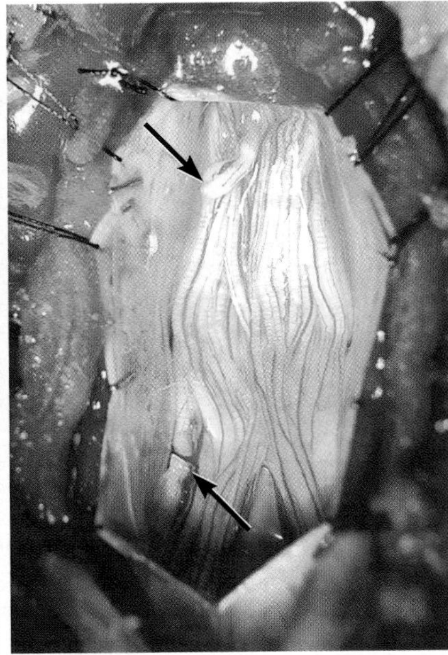

Figure 13-4

SCHWANNOMA OF SPINAL NERVE ROOT

Left: Intradural extramedullary spinal schwannomas is a discrete mass attached to a nerve root.

Right: The arrows indicate the proximal and distal surgical margins of the parent nerve. (Courtesy of Dr. Ziya Gokaslan, Baltimore, MD.)

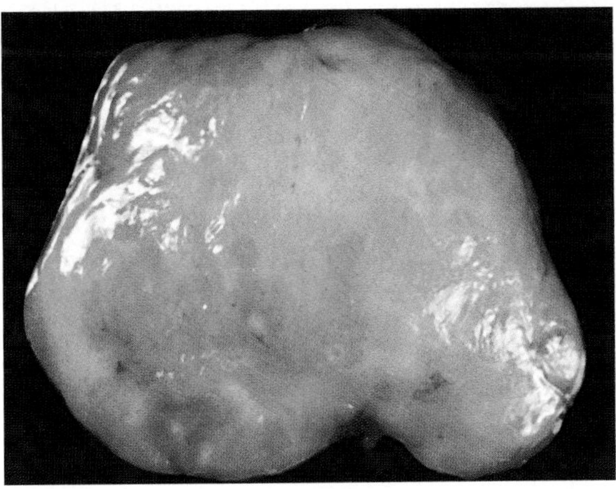

Figure 13-5

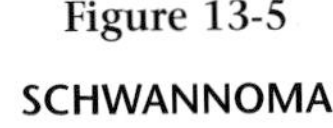

SCHWANNOMA

Lipid in the Antoni B component tints some schwannomas yellow.

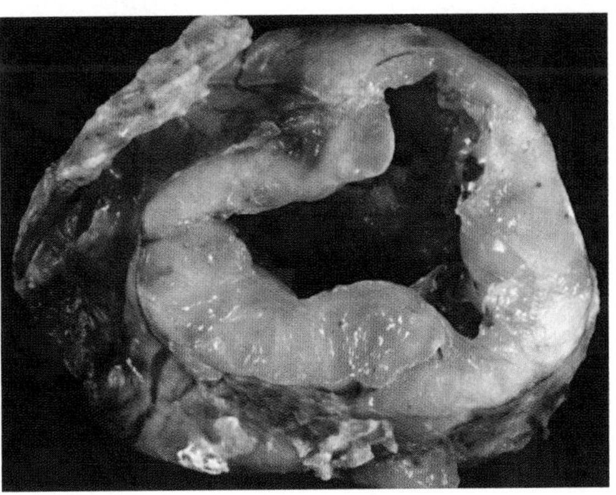

Figure 13-6

SCHWANNOMA

Schwannomas can be cystic, particularly larger examples such as this lesion of the cauda equina.

fusiform in configuration (fig. 13-4). The parent nerve may be detected within the substance of small or early lesions but can disappear as a macroscopic structure when attenuated over the surface of larger tumors. The discrete appearance is the result of a complete collagenous capsule (25). Due to lipid deposition, the cut surface is often yellow (fig. 13-5). Cystic degeneration is common in large tumors but also can be present in small lesions, particularly those in the cauda equina (fig. 13-6).

Microscopic Findings. Classic schwannomas are often biphasic, with areas of compact fascicular Antoni A tissue and loose-textured Antoni B tissue; the latter has a distinctive, somewhat degenerative, appearance (fig. 13-7). Antoni A tissue consists of elongated bipolar cells disposed in fascicles, since cell borders are

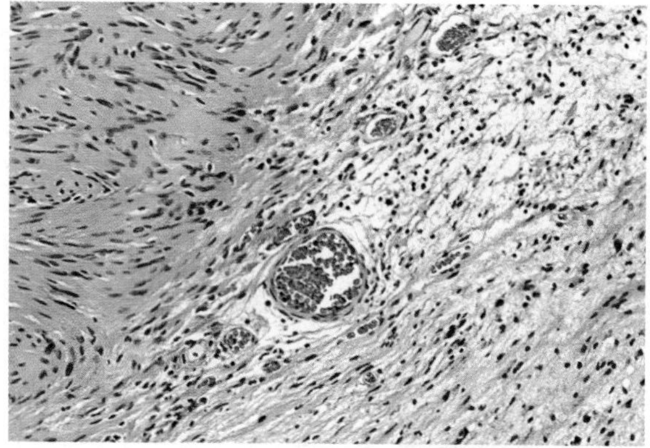

Figure 13-7

SCHWANNOMA

The typical schwannoma has compact Antoni A tissue abutting a loose-textured Antoni B component.

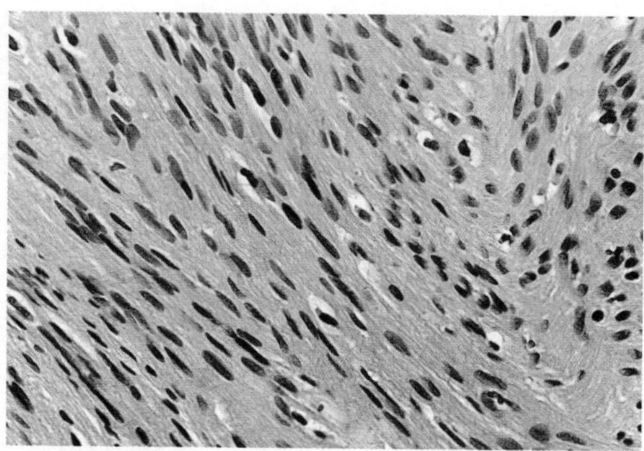

Figure 13-8

SCHWANNOMA

Fascicular Antoni A tissue contains cells with long club-shaped nuclei.

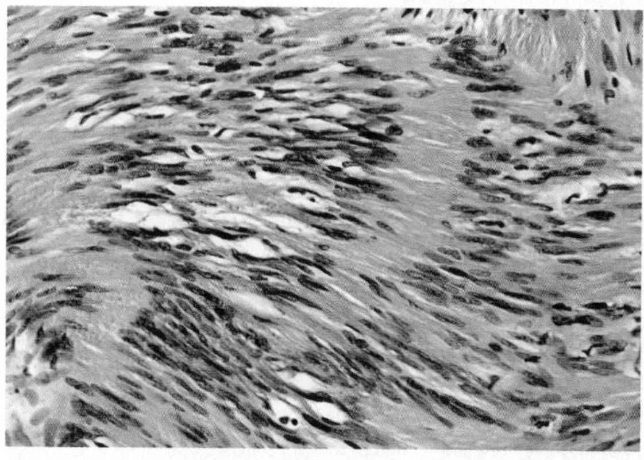

Figure 13-9

SCHWANNOMA

The presence of Verocay bodies in Antoni A tissue is a distinctive feature of schwannoma, although they are often not as prominent as in this intracranial lesion.

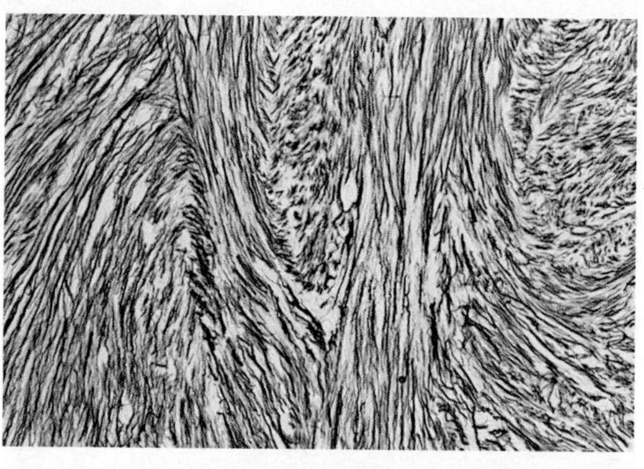

Figure 13-10

SCHWANNOMA

Pericellular reticulin is abundant in schwannoma, but absent in a common differential diagnostic entity, meningioma.

obscured at the light microscopic level. Long club-shaped nuclei are typical (fig. 13-8). Verocay bodies, striking palisades resulting from stacked arrays of nuclei alternating with anucleate zones composed of cell processes, are a common feature of Antoni A tissue (fig. 13-9). Verocay bodies are usually absent or only rudimentary in intracranial schwannomas. Pericellular reticulin corresponding to pericellular basement membranes is best seen in Antoni A tissue (fig. 13-10).

Cells in Antoni B tissue generally have round, condensed nuclei and indistinct cytoplasm. Such cells may superficially resemble lymphocytes. Other degenerative changes, and helpful diagnostic features, are hyalinization of blood vessels, perivascular accumulation of hemosiderin-laden macrophages (fig. 13-11), and cysts. The last are most common in large lesions and are often lined by pseudoepithelial tumor cells. So-called *ancient schwannomas* often have nuclear pleomorphism and hyperchromasia (degenerative nuclear atypia) unassociated with proliferative activity (fig. 13-12). Some schwannomas acquire a mucinous stroma (fig. 13-13). A rarity is intraspinal schwannoma with rhabdomyoblastic differentiation (26).

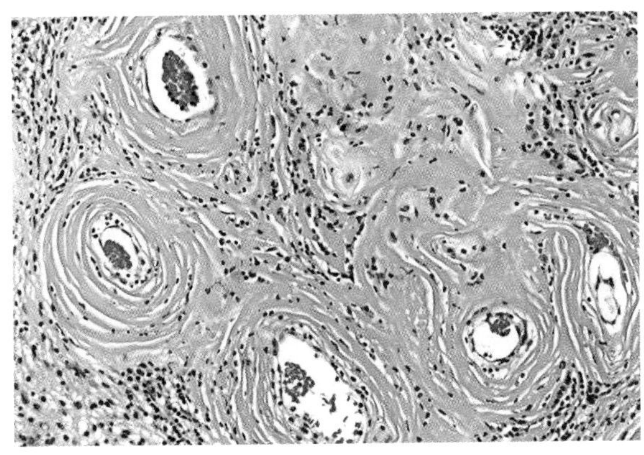

Figure 13-11

SCHWANNOMA

Vascular hyalinization is a frequent expression of chronicity.

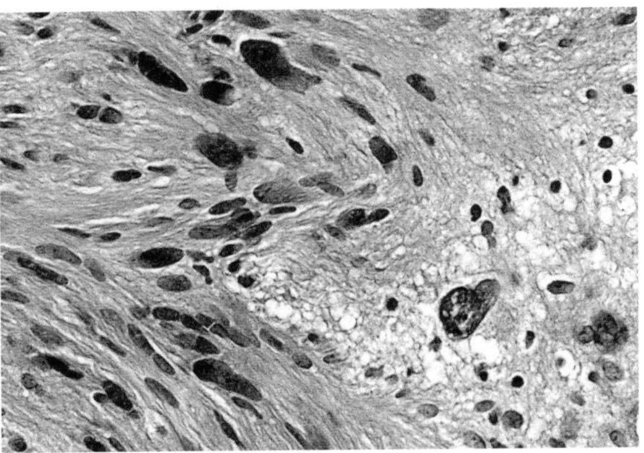

Figure 13-12

SCHWANNOMA

Degenerative nuclear atypia in the form of nuclear pleomorphism and hyperchromatism is a common feature in schwannomas of long duration.

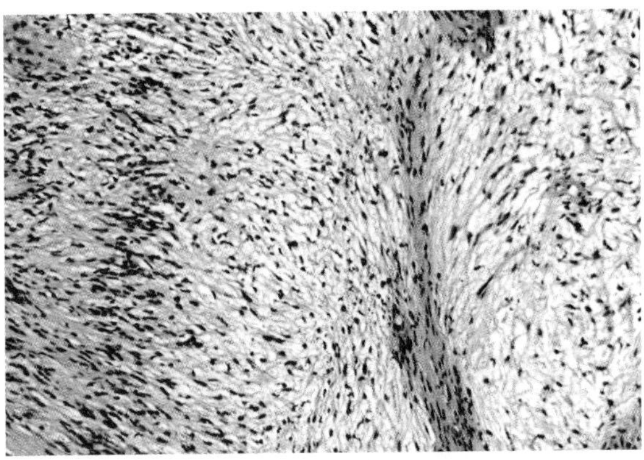

Figure 13-13

SCHWANNOMA

Extensive myxoid change alters some schwannomas. A residual Verocay body is present.

Mitotic figures are rare in conventional schwannomas. A distinct subset of schwannomas is hypercellular, consists mainly of Antoni A tissue, and shows variable mitotic activity. Termed *cellular schwannomas*, these must not be misdiagnosed as malignant peripheral nerve sheath tumors (MPNSTs). Cellular schwannomas are discussed separately.

A collision tumor, composed of schwannoma and meningioma at the cerebellopontine angle in the setting of NF2, is discussed in chapter 18 (see figs. 18-8–8-10).

Intraparenchymal schwannomas are, overall, well circumscribed but tongues of tissue may extend into the brain where chronic gliosis is common (fig. 13-14). Although the tumor is macroscopically discrete, some axons will be seen within the mass after staining for neurofilament protein.

Frozen Section. The small fragments often presented to pathologists consist of vaguely fascicular tissue that would be nondescript were it not for the classic, long, club-shaped nuclei (fig. 13-15). The presence of a thick capsule, hyaline vessels, and both Antoni A and B tissues simplifies the diagnosis.

Immunohistochemical Findings. Schwannomas are strongly and uniformly reactive for S-100 protein (see fig. 18-10, right), vimentin, and often leu-7 (27). Variable staining for glial fibrillary acidic protein (GFAP) is common (28). Delicate intercellular staining of basement membrane for collagen type IV or laminin is the immunohistochemical equivalent of the pattern seen in reticulin preparations (fig. 13-16).

The MIB-1 labeling indices for vestibular schwannomas are typically low, a mean value of 1.86 percent being reported in one series of unilateral sporadic lesions (29). Indices for tumors that were either stable or growing were 1.11 and 3.17 percent, respectively. Yet another study comparing vestibular schwannomas in NF2 patients with those of non-NF2 patients found higher labeling indices in the syndrome-associated tumors (1.7 versus 0.95 percent) (30).

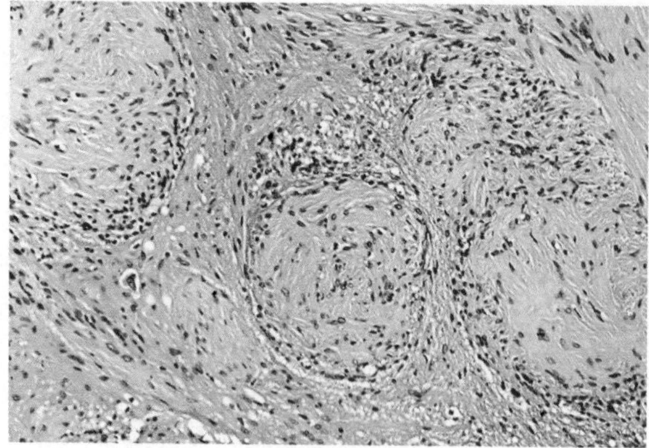

Figure 13-14

INTRAPARENCHYMAL SCHWANNOMA

Nodules of fascicular Antoni A tissue are juxtaposed with gliotic central nervous system (CNS) tissue in intraparenchymal schwannomas.

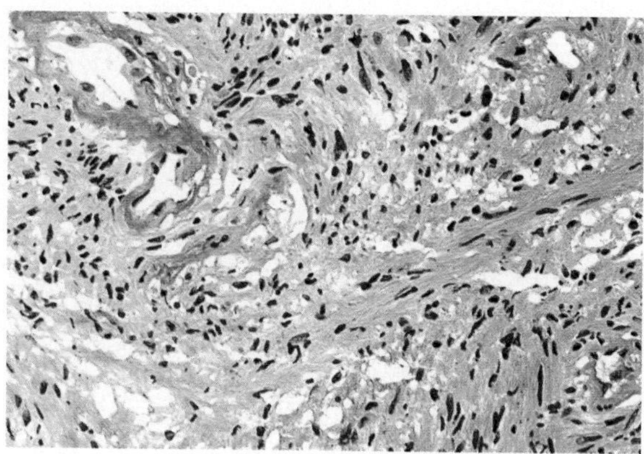

Figure 13-15

SCHWANNOMA: FROZEN SECTION

Although the features are nonspecific, the fascicular architecture and the long club-shaped nuclei are characteristic.

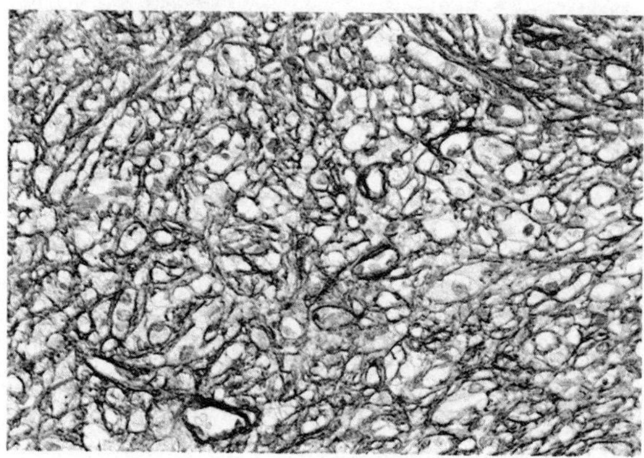

Figure 13-16

SCHWANNOMA

The basement membrane material responsible for the reticulin staining can be readily demonstrated with immunohistochemistry for collagen type IV or laminin.

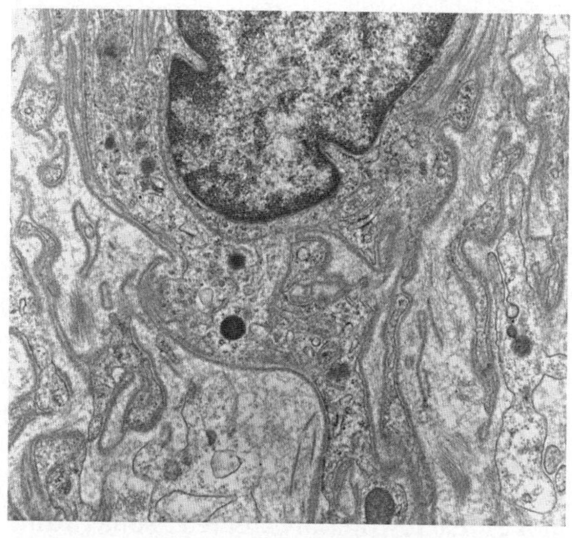

Figure 13-17

SCHWANNOMA

As in this vestibular lesion, ubiquitous basement membrane material covers cell surfaces.

Ultrastructural Findings. The cells of schwannomas vary in configuration: bipolar in Antoni A tissue and stellate in Antoni B. Processes interweave in the former and occasionally surround aggregates of intercellular collagen. Cells are surrounded by often duplicated basal lamina (fig. 13-17). Long-spacing collagen, in the form of Luse bodies, is common.

Cytologic Findings. Even forceful smearing produces only tissue fragments, with little if any separation into individual cells (fig. 13-18). Nevertheless, the characteristic long, frequently club-shaped nuclei of Antoni A tissue are seen well with up and down focusing (fig. 13-19). These are readily distinguished from those of meningioma (fig. 13-20). Antoni B tissue dissociates more readily, since its nuclei are round and its processes somewhat stellate.

Molecular Findings. In keeping with the occurrence of schwannomas in patients with NF2, loss of heterozygosity on chromosome 22 in

Figure 13-18

SCHWANNOMA: SMEAR PREPARATION

Resisting disaggregation, cells of schwannomas remain defiantly as thick tissue fragments of Antoni A tissue.

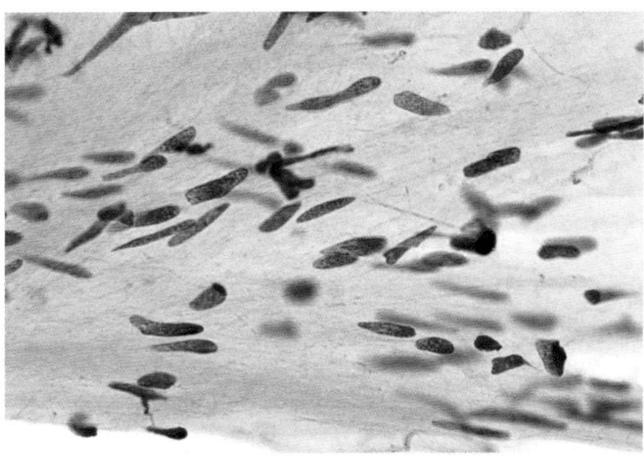

Figure 13-19

SCHWANNOMA: SMEAR PREPARATION

The cells of Antoni A tissue have distinctive long club-shaped nuclei.

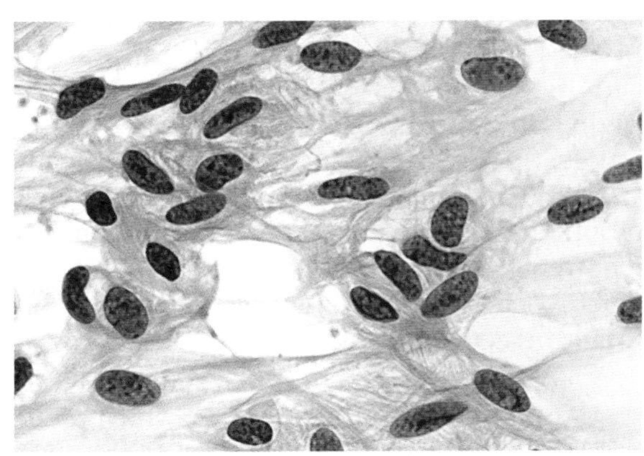

Figure 13-20

MENINGIOMA: SMEAR PREPARATION

In contrast to schwannomas, meningiomas disaggregate readily, leaving small fragments and individual cells with plate-like or wispy cytoplasm. Nuclei are oval and delicate.

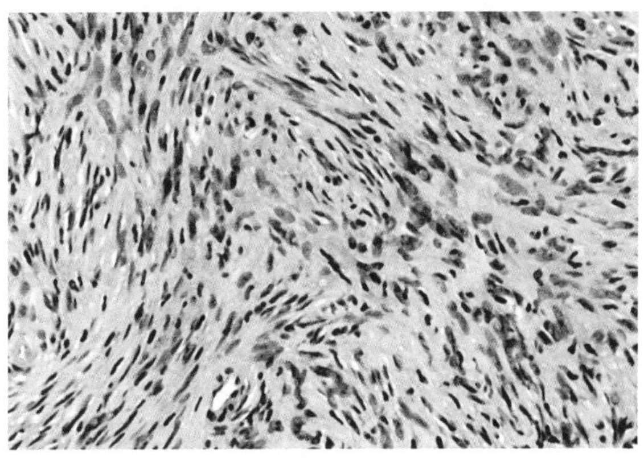

Figure 13-21

MENINGIOMA RESEMBLING SCHWANNOMA

Meningiomas with a fascicular architecture closely resemble schwannomas composed exclusively of Antoni A tissue.

the region of the *NF2* gene is a common feature (31,32).

Differential Diagnosis. Despite different origins, i.e., meninges versus nerve sheath, the shared cellular elongation makes fibrous meningioma and schwannoma a common differential diagnostic duo (fig. 13-21). Smear preparations are very helpful since, unlike schwannoma, the cells of meningioma readily dissociate to reveal their characteristic features (figs. 13-19, 13-20). Immunohistochemistry can be applied if doubts remain, although fibrous meningiomas are variably S-100 protein positive in up to 80 percent of cases. More convincing is a membranous pattern of immunoreactivity for epithelial membrane antigen (EMA) in meningiomas. The occasional EMA reactivity in schwannomas is typically cytoplasmic rather than membranous. Meningioma cells lack pericellular basal lamina, a feature characteristic of schwannoma as shown by reticulin staining and immunoreactivity for collagen IV and laminin.

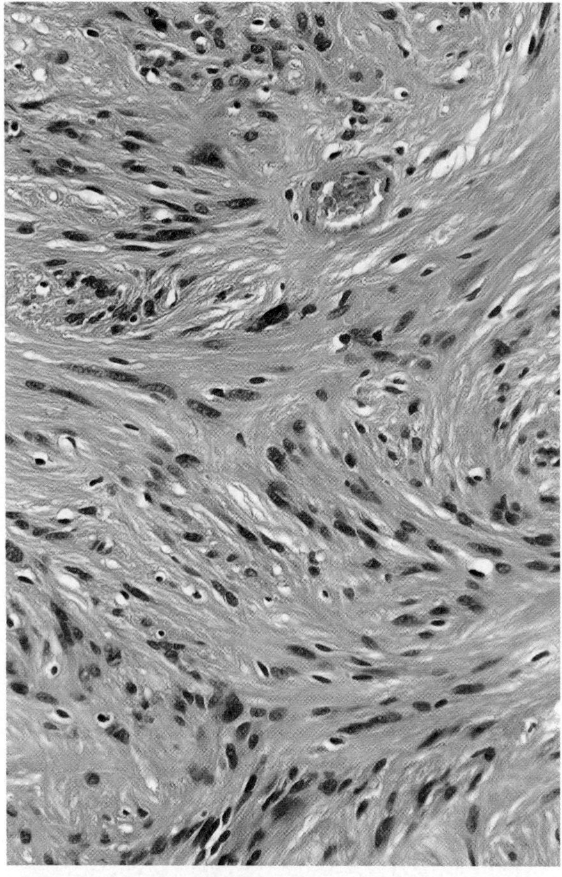

Figure 13-22

INTRASPINAL EPENDYMOMA RESEMBLING SCHWANNOMA

An unstructured fascicular architecture creates a likeness to schwannoma in some ependymomas.

Especially in the lumbar region, ependymomas with a fascicular architecture and poorly developed perivascular pseudorosettes can be mistaken for schwannoma (fig. 13-22). Smear preparations of ependymoma show small oval nuclei in cells with bipolar, often perivascular, processes, whereas schwannomas have only clumps of cells with long nuclei. Unlike schwannomas, ependymomas are reticulin poor and intensely and universally GFAP positive. Schwannomas, while generally patchy, if reactive at all for GFAP, are sometimes surprisingly positive. Expression of collagen IV and laminin is stronger in schwannomas.

In a limited biopsy, schwannomas can resemble pilocytic astrocytomas. The presence of a capsule favors schwannoma. Reactivity for GFAP does not resolve the problem, since schwannomas often have at least some immunoreactivity as well. Intercellular reticulin staining and immunoreactivity for collagen IV and laminin readily settles the issue in favor of schwannoma. The distinction of conventional schwannoma from cellular schwannoma and low-grade MPNST is summarized in Table 13-1.

Treatment and Prognosis. Most craniospinal schwannomas are cured surgically. In a series of vestibular schwannomas, 29 (10 percent) of 298 tumors required reoperation for recurrence (33). Morphologic study of the tumors undergoing recurrence related lobulated contour on magnetic resonance imaging (MRI), high cellularity, and nuclear pleomorphism with increased incidence of recurrence ($p = 0.001$). Whether these cellular tumors corresponded to cellular schwannomas (see below) was not addressed. In any event, despite no differences in mitotic activity, the mean MIB-1 labeling index was higher in the recurring group, 2.28 percent (range, 0.1 to 8.6 percent), versus 0.59 percent (range, 0 to 1.5 percent) in the nonrecurring group. In one series of trigeminal nerve tumors, recurrence-free survival rates were 50 percent and 92 percent for patients with subtotally and totally resected lesions, respectively (34).

Malignant schwannomas arising in transition from previously well-differentiated schwannomas, i.e., true malignant schwannomas, are extremely rare (35,36). The relationship of anaplastic transformation to prior radiation therapy is unclear (37).

CELLULAR SCHWANNOMA

Definition. *Cellular schwannoma* is a benign, well-differentiated schwannoma of high cellularity composed largely of Antoni A tissue and devoid of Verocay bodies. The proliferative activity is highly variable.

General Features. Since its original description (38), several studies have confirmed the benign, but potentially recurring although nonmetastasizing nature of this schwannoma variant (39–41).

Clinical Features. Cellular schwannomas are solitary lesions that represent approximately 5 percent of benign peripheral nerve sheath tumors (39). The maximal incidence is in the

Table 13-1

DIFFERENTIAL DIAGNOSIS OF CONVENTIONAL SCHWANNOMA, CELLULAR SCHWANNOMA, AND MALIGNANT PERIPHERAL NERVE SHEATH TUMOR (MPNST)

Findings	Conventional Schwannoma	Cellular Schwannoma	MPNST
Gross	Usually globoid encapsulated tumor composed of homogeneous light tan tissue; may be cystic or hemorrhagic and show yellow patches; no gross necrosis	Usually globoid encapsulated tumor; firmer than classic schwannoma and homogeneously tan; occasional patches of yellow, but no gross necrosis	Fusiform or globoid, pseudocapsulated (infiltrative of surrounding tissues); firm, cream-tan, usually grossly necrotic tumor
Microscopic	Antoni A and B areas with Verocay bodies; hyalinized thick-walled blood vessels and lipid-laden histiocytes; mitotic figures infrequent; rarely see malignant transformation	Mainly hypercellular Antoni A tissue with cells arranged in fascicles or whorls; may show marked hyperchromasia and nuclear pleomorphism; notable are lymphoid deposits in capsule or perivascular areas, thick-walled blood vessels, and collections of lipid-laden histiocytes; rare foci of necrosis; mitoses not uncommon but usually number no more than 4 per 10 hpf[a]	Markedly hypercellular, fasciculated, spindle cell tumor generally consisting of cells of uniform size and pronounced hyperchromasia; geographic necrosis and mitotic counts in excess of 4 per 10 hpf are common; epithelioid cells predominate in about 5 percent of tumors and 15 percent show heterologous glandular or sarcomatous elements
Immunohistochemical	Diffuse and strong expression of S-100 protein	Diffuse and strong expression of S-100 protein	S-100 protein expression in scattered cells of 50 to 70 percent of cases
Electron Microscopy	Well-differentiated cells with long, often interlacing cytoplasmic processes coated by basement membrane on their free surfaces; intercellular long-spacing collagen common	Similar to classic schwannoma; increased cellularity; more nuclear atypia and occasional residual arrays of long, basement membrane-coated cytoplasmic processes; long-spacing collagen less common	Poorly differentiated cells with more pleomorphic nuclei, thick cytoplasmic processes, and sometimes patchy basement membrane; long-spacing collagen rarely seen
Clinical Behavior	May cause bone erosion and can recur if incompletely excised; rare reported examples with malignant transformation that followed an aggressive clinical course	May cause bone erosion and recur if incompletely excised; thus far, no clinically malignant examples	Has a proclivity to invade and destroy nearby soft tissues, recur locally, and metastasize distantly (usually to lung); about 90 percent are high-grade lesions

[a] hpf = high-power fields.

fourth decade of life; only 5 percent occur in childhood or adolescence. Cellular schwannomas show a significant tendency to involve the paravertebral regions of the mediastinum, retroperitoneum, and pelvis (39). Of paravertebral lesions, approximately one third have a dumbbell configuration. In one large series, nearly 10 percent of tumors were intracranial, some involving cranial nerves five and eight (39). Cellular schwannomas show no significant association with NF1 or NF2.

Gross Findings. Approximately one third of cellular schwannomas arise from a recognizable nerve; they are globular and eccentric to the nerve rather than fusiform in configuration. Only occasional examples are multinodular or plexiform. Rubbery in texture, their cut surface is tan and often solid. Foci of hemorrhage may be seen, but gross necrosis is not. Even in areas of bone erosion, tumor margins are smooth.

Microscopic Findings. Aside from being highly cellular and lacking a significant Antoni

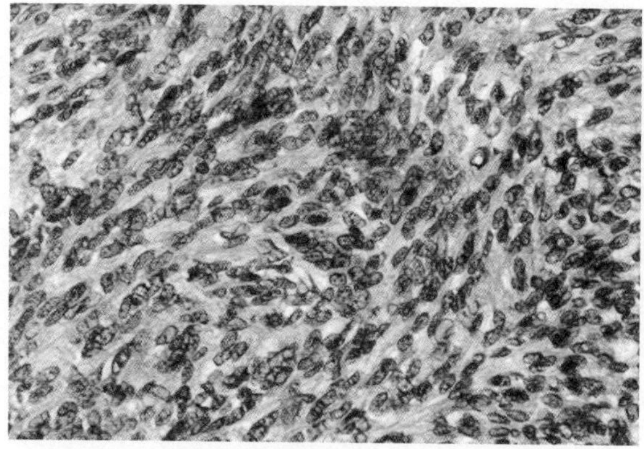

Figure 13-23

CELLULAR SCHWANNOMA

Cellular schwannomas are densely cellular and vaguely fascicular, with little of the reassuring Antoni B architecture.

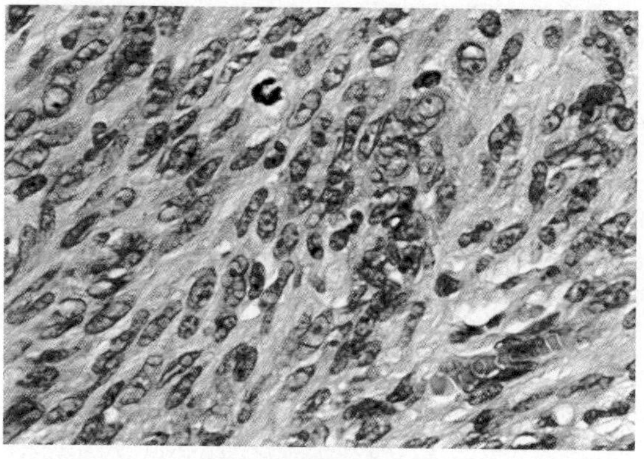

Figure 13-24

CELLULAR SCHWANNOMA

Scattered mitoses in a cellular lesion featuring other classic features of schwannoma are not evidence of malignant peripheral nerve sheath tumor.

B component (fig. 13-23), cellular schwannomas exhibit many features of conventional schwannoma. These include the presence of well-formed capsules and hyalinization of tumor vessels. Cell patterns vary from interlacing fascicles to occasional storiform or whorled arrangements. Cytologic features are typically schwannian, with bipolar tumor cells featuring elongate club-shaped nuclei, eosinophilic fibrillar cytoplasm, and interlacing processes. When present, nuclear pleomorphism is minor. Collections of lipid-laden histiocytes are frequent,

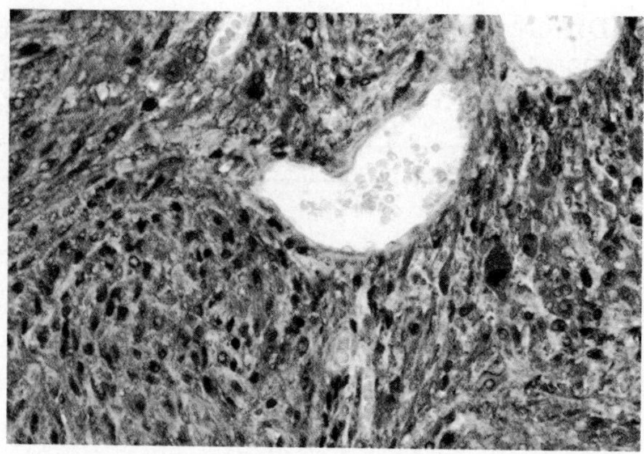

Figure 13-25

CELLULAR SCHWANNOMA

Cellular schwannomas are diffusely positive for S-100 protein.

as is subcapsular chronic inflammation. Microfoci of necrosis are infrequent and unassociated with pseudopalisading. Mitotic figures are present in the majority of tumors and usually range from 1 to 4 per 10 high-power fields (fig. 13-24). Greater mitotic activity is occasionally seen, but this does not indicate malignancy.

Immunohistochemical Findings. As with conventional schwannomas, cellular schwannomas show uniformly strong staining for S-100 protein (fig. 13-25), as well as pericellular staining of basal lamina for collagen IV and laminin. Approximately half show GFAP immunoreactivity, which may be strong. EMA staining is limited to the tumor capsule, specifically its content of residual perineurial cells. Neurofilament protein stains show residual axons, usually within the tumor capsule or in subcapsular portions of the tumor. In one large series, the MIB-1 labeling index was 6 percent in nonrecurring lesions and 8 percent in tumors that recurred (39).

Ultrastructural Findings. Cellular schwannomas are well-differentiated neoplasms showing the full spectrum of Schwann cell features (38–40,42,43). Conspicuous among these are elongate, interlacing tumor cell processes covered by a continuous, occasionally duplicated, basement membrane. Cells are joined by poorly developed intercellular junctions. Intercellular matrix varies in quantity, particularly in collagen content. Long-spacing collagen (Luse bodies) is less common (15 percent) than in conventional schwannomas.

Differential Diagnosis. Cellular schwannoma must be distinguished from low-grade MPNSTs. Key features for this differential appear in Table 13-1.

Neurofibromas are less cellular, have smaller nuclei, and, while S-100 positive, lack the uniform reactivity of schwannomas. Fibrosarcomas are S-100 protein negative. Meningiomas are EMA positive, and less uniformly, if at all, immunoreactive for S-100 protein. They are negative for GFAP, collagen IV, and laminin.

Treatment and Prognosis. In series dealing primarily with peripherally situated cellular schwannomas, the recurrence rate was 5 percent, and fully 90 percent were amenable to gross total removal (43). Data are sparse on the outcome for patients with intracranial or intraspinal lesions, but there appears to be a significantly higher rate of recurrence than attends conventional schwannomas (39). To date, no cellular schwannoma has metastasized or proven lethal.

MELANOTIC SCHWANNOMA

Definition. *Melanotic schwannoma* is an often circumscribed nerve sheath tumor composed of melanin-producing Schwann cells.

Clinical Features. Morphologically, these rare tumors occur in two near equally represented forms: nonpsammomatous and psammomatous. Nonpsammomatous tumors more often involve nerve roots or dorsal root ganglia, while psammomatous tumors more often affect visceral sites, particularly the alimentary tract.

While nonpsammomatous melanotic lesions favor spinal nerve roots (44–49), they also affect cranial nerves (50). Some are intraparenchymal, in either brain stem (51) or spinal cord (52,53). As with conventional nerve root schwannomas, it can be difficult in some cases to determine the precise site of origin when the lesion is both leptomeningeal and partially embedded in the cord.

Half of psammomatous tumors are associated with Carney's complex, an autosomal dominant disorder (54–58). Patients with the complex present at a mean age of 27 years, a full decade earlier than patients with nonpsammomatous tumors and lacking the syndrome. Carney's complex features lentiginous pigmentation (65 percent); myxomas of the heart (65 percent), skin (25 percent), or breast (20 percent); endocrine overactivity (10 percent); and blue nevi (10 percent). Pigmentation involves primarily the face and, in females, the external genitalia. Endocrinopathy includes Cushing's syndrome (25 percent), associated with pigmented nodular adrenocortical disease; sexual precocity (30 percent), resulting from large cell calcifying Sertoli cell tumor of the testis; and acromegaly (8 percent), due to pituitary adenoma. Multifocality of psammomatous melanotic schwannomas approaches 20 percent. Most patients with multiple tumors (83 percent) have Carney's complex.

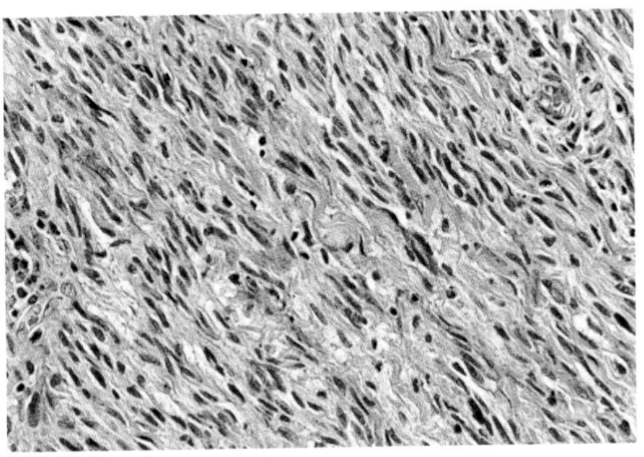

Figure 13-26

MELANOTIC SCHWANNOMA

Melanotic schwannomas are usually composed largely of fascicular Antoni A tissue. The pronounced cellular elongation in this case helps distinguish the lesion from melanocytoma.

Gross Findings. Melanotic schwannomas vary greatly in size. Tar black or brown pigmentation may be uniform or unevenly distributed. The tumors are generally enveloped by a thin layer of connective tissue, which may be interrupted with associated infiltration of surrounding soft tissue. Widespread invasiveness and bone destruction are generally associated with malignant examples.

Psammomatous tumors may show gross subcapsular calcification and even metaplastic bone formation. Lesions affecting spinal nerve roots are often dumbbell shaped (55).

Microscopic Findings. Most nonpsammomatous melanotic schwannomas are highly cellular and composed of spindle to epithelioid cells closely packed and arranged in lobules, fascicles, or cellular whorls (fig. 13-26). Multinucleation is often present. Nuclei are usually round and contain a small, distinct nucleolus. Nuclear-cytoplasmic pseudoinclusions are common. Melanin pigmentation of tumor cells varies considerably. In some instances, it obscures nuclear details. Heavily pigmented melanophages are frequent.

Psammomatous tumors are characterized by the often focal finding of laminated calcospherites. About half of psammomatous tumors have large cytoplasmic vacuoles that resemble mature adipose tissue.

Unlike conventional and cellular schwannomas, vessels in melanotic tumors are thin-walled rather than thickened and hyalinized.

Approximately 10 to 15 percent of melanotic schwannomas are malignant. Features of such tumors include widespread soft tissue invasion, large violaceous nucleoli, increased mitotic activity including abnormal mitoses, and broad zones of necrosis.

Immunohistochemical Findings. Melanotic schwannomas are immunoreactive for vimentin, S-100 protein, and HMB45. Collagen IV or laminin immunoreactive basement membrane more often surrounds clusters of neoplastic cells than individual cells. Reactivity for GFAP has not been reported.

Ultrastructural Findings. Melanotic schwannomas of spinal nerve or sympathetic ganglia are composed of spindle-shaped or plump cells, some with interdigitating processes. Intercellular junctions are rudimentary. Basement membrane, often reduplicated, coats the free surface. Melanosomes of stages 2 to 4 are evident. Extracellular long-spacing collagen, surface micropinocytotic vesicles, cytoplasmic intermediate filaments, and glycogen may also be seen (48,59,60).

Differential Diagnosis. There are a limited number of lesions in the differential diagnosis. Melanotic CNS neoplasms are summarized in Appendix H. Included is metastatic melanoma, which is distinguished by its dendritic-appearing cells, cytologic malignancy, lack of psammoma bodies and adipose-like cells, and rare basement membrane formation.

It may be difficult to differentiate meningeal melanocytomas from nonpsammomatous melanotic schwannoma since the latter can have a spindle cell architecture, and both are immunoreactive for melanocytic markers and S-100 protein. Melanocytomas have less abundant collagen IV immunoreactivity, and lack pericellular basement membranes and long-spacing collagen.

Treatment and Prognosis. Most melanotic schwannomas are benign, but approximately 15 percent of both nonpsammomatous and psammomatous tumors result in death from disease, and nearly all have histologically malignant features (44,45,49,50,61). Since melanotic schwannomas may be multiple, distinguishing a second primary lesion from a metastasis may be difficult (55). Patients with Carney's complex have additional factors contributing to morbidity and mortality, such as cardiac myxomas and endocrinopathy.

NEUROFIBROMA

Definition. *Neurofibroma* is a well-differentiated nerve sheath tumor composed predominantly of Schwann cells and, to a lesser extent, fibroblasts and perineurial-like cells. Residual myelinated or unmyelinated axons are often present.

General Features. Neurofibromas occur in a variety of architectural types, including localized cutaneous, diffuse cutaneous, localized intraneural, plexiform, and massive soft tissue examples. As isolated lesions, plexiform and massive soft tissue tumors are almost pathognomonic of NF1, although there are cases in which the syndrome is not identified, at least at the time of surgery (62). Neurofibromas are discussed in AFIP tumor Fascicle, *Tumors of the Peripheral Nervous System* (63). As lesions affecting the CNS, neurofibromas are much less common than schwannomas.

Clinical Features. In the general population, neurofibromas are far more often sporadic than associated with neurofibromatosis. Unlike schwannomas, neurofibromas rarely arise intradurally, and more often affect pericranial or paraspinous soft tissue. Only an occasional example traverses the dura to become partly intraspinal or intracranial. Multiple nerve involvement, plexiform architecture, and massive soft tissue lesions are associated closely with NF1 (figs. 13-27, 13-28; see fig. 18-1).

Radiologic Findings. The often high mucopolysaccharide content of neurofibromas shows bright on T2-weighted MRIs. A decreased signal in transverse orientation at the center of intraneural neurofibromas (reflecting the less affected, less mucinous parent nerve) produces a "target sign."

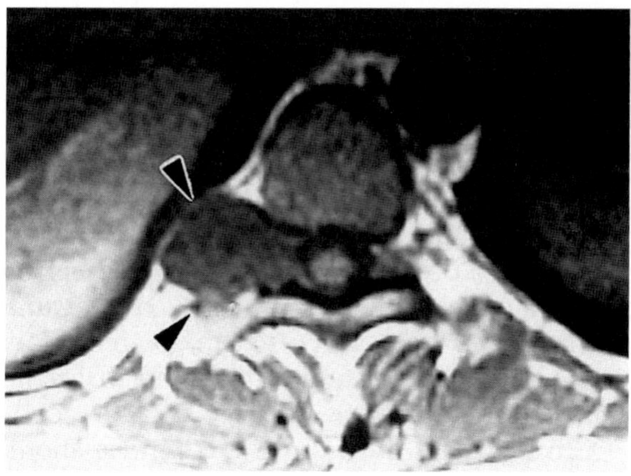

Figure 13-27

NEUROFIBROMA

Neurofibroma of a nerve root (arrowhead) can enter, or arise within, the intraspinal compartment and compress the spinal cord. (Fig. 13-13 from Fascicle 10, 3rd Series.)

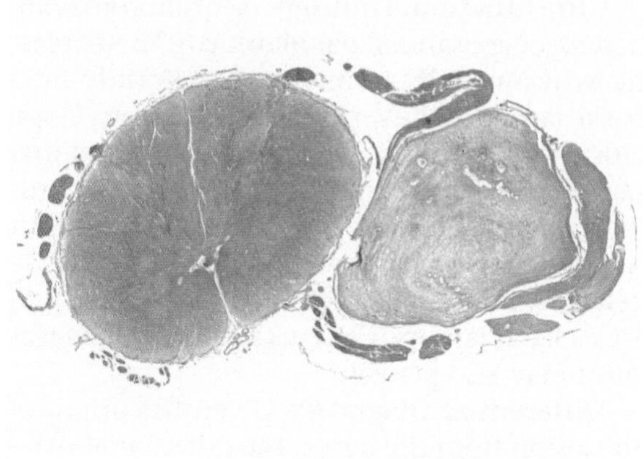

Figure 13-28

NEUROFIBROMA

Spinal nerve root neoplasms in patients with neurofibromatosis 1 are usually neurofibromas.

Gross Findings. Localized intraneural neurofibromas are soft to firm, fusiform expansions of a nerve still delicately enveloped by what remains of its perineurium and epineurium. Depending upon the content of stromal mucin and collagen, cut surfaces vary from translucent to opaque and gray to tan. Degenerative changes such as cysts and hemorrhage are lacking.

By definition, plexiform tumors involve multiple fascicles. In a branching nerve, these typically form a "bag of worms." In a nonbranching nerve such as the sciatic, much of its length is transformed into a thick rope-like mass. Morphologic overlap occurs; for example, massive soft tissue neurofibromas often have a plexiform element.

Microscopic Findings. Neurofibromas consist in large part of Schwann cells arrayed in wavy bundles separated by a loose, mucoid interstitial matrix (fig. 13-29). Axons, with or without their myelin sheaths, pass through the tumor, reflecting the intimate relationship of the neoplasm to the parent nerve. A small number of mast cells lie within the substance of the tumor. Mitoses are scant in ordinary neurofibromas.

There are two histologic variants of neurofibroma: *cellular neurofibroma*, characterized by patchy or widespread hypercellularity and low level mitotic activity, and *atypical neurofibroma*, showing degenerative nuclear atypia and little or no proliferative activity. Only the cellular variant is of importance in that it may be mistaken for low-grade MPNST. This differential is discussed with the latter tumor. Malignant transformation of neurofibromas is discussed below.

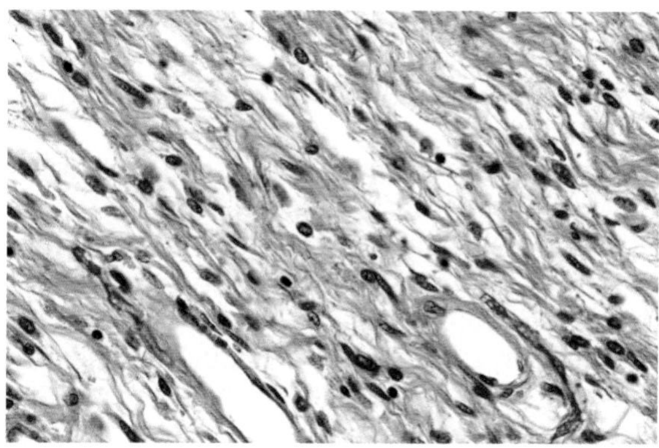

Figure 13-29

NEUROFIBROMA

Wavy neoplastic cells in a myxoid matrix are classic features of neurofibroma. The scattered myelin sheaths in this section are stained with hematoxylin and eosin (H&E)/ Luxol fast blue.

Immunohistochemical Findings. In keeping with their significant Schwann cell content (64), all neurofibromas are partially immunoreactive for S-100 protein (65). The fibroblastic component exhibits a vimentin reaction alone, whereas perineurial-like cells are occasionally positive for EMA (66). Variable staining for CD34 is also seen. Neurofilament protein reactivity is limited to overrun axons.

Ultrastructural Findings. Neurofibromas are heterogeneous in cellular makeup (67). Neoplastic Schwann cells are invested uniformly by a basal lamina. Their processes often envelope clusters of stromal collagen fibers. Occasional perineurial cells exhibit electron-dense cytoplasm and pericellular basal lamina in addition to numerous pinocytotic vesicles. Fibroblastic cells feature rough endoplasmic reticulum and lack a basal lamina. Cells with intermediate features may also be seen.

Differential Diagnosis. Given the origin of the lesion from the nerve, the differential diagnosis of neurofibroma necessarily focuses upon schwannoma. The latter is eccentric to the parent nerve, has a fibrous capsule, is somewhat firm, and may in part be cystic. On cut section, many schwannomas are yellow due to lipid accumulation, and generally lack the gray mucinous quality common in neurofibromas. Occasional solitary neurofibromas are globular rather than fusiform, while some schwannomas have a plexiform quality.

Histologically, most schwannomas have compact cellularity, both Antoni A and Antoni B tissues, vascular hyalinization, and perivascular hemosiderin deposits. Neurofibromas are more often loose textured, their wavy cells lying in a myxoid stroma. Residual axons pass freely and singly through the bulk of neurofibromas, but remain more aggregated and peripherally situated in schwannomas.

Treatment and Prognosis. The goal of surgery in neurofibromas is resection of the mass or at least decompression of the spinal cord when the latter is impinged upon by nerve root examples. Additional lesions may arise during ensuing years in patients with NF1. Malignant transformation occurs in only a small proportion of neurofibromas, including approximately 2 percent of plexiform tumors.

MALIGNANT PERIPHERAL NERVE SHEATH TUMOR

Definition. *Malignant peripheral nerve sheath tumor* (MPNST) is a malignant neoplasm arising from or differentiating towards cells intrinsic to peripheral nerves. Specifically excluded are tumors of epineurial soft tissue and endothelial tumors originating from peripheral nerve vasculature.

General Features. Uncertainty regarding the nature of the cells contributing to these tumors is reflected in the various terms previously applied to MPNST: neurogenic sarcoma, neurofibrosarcoma, and malignant schwannoma. The descriptive designation, malignant peripheral nerve sheath tumor, is preferred since these tumors should not be considered variants of fibrosarcoma or malignant counterparts of schwannoma. MPNSTs are discussed in greater detail in the AFIP tumor Fascicle, *Tumors of the Peripheral Nervous System* (68).

Clinical Features. MPNSTs comprise about 5 percent of malignant soft tissue neoplasms. Most arise between ages 20 and 50 years; approximately 15 percent occur in childhood or adolescence. An association with NF1 is seen in 50 to 60 percent of tumors (69). Widely accepted criteria for the diagnosis include: 1) origin in a peripheral nerve; 2) origin in a benign peripheral nerve tumor; 3) development in the setting of NF1 and expression of typical MPNST histologic features; or 4) histologic, immunohistochemical, and ultrastructural features of MPNST despite lack of association with NF1. Half of MPNSTs are derived from neurofibromas, a feature more often associated with NF1. The other half arise de novo in normal peripheral nerves. Extremely rare MPNSTs arise in transition from schwannoma (70,71).

Paraspinous tumors show a distinct tendency to central extension (fig. 13-30). In one series of 25 paraspinous MPNSTs, 10 encroached upon the spinal column, 4 showed vertebral body involvement, 2 reached a vertebral foramen by intraneural extension, 2 featured epidural involvement with and without spinal cord compression, and 2 extended intradurally (72).

MPNSTs originating in paraspinal nerve segments are clearly more numerous than tumors arising primarily from intradural portions of a spinal nerve. Intracranial MPNSTs most often occur de novo and from the fifth cranial nerve (73–75). Eighth cranial nerve examples are less frequent (76,77). Primary intracerebral MPNSTs are rare (78).

Radiologic Findings. On both CT and MRI, irregular tumor margins and lack of signal homogeneity are typical of MPNST. Nonetheless, many are radiologically indistinguishable from benign nerve sheath tumors.

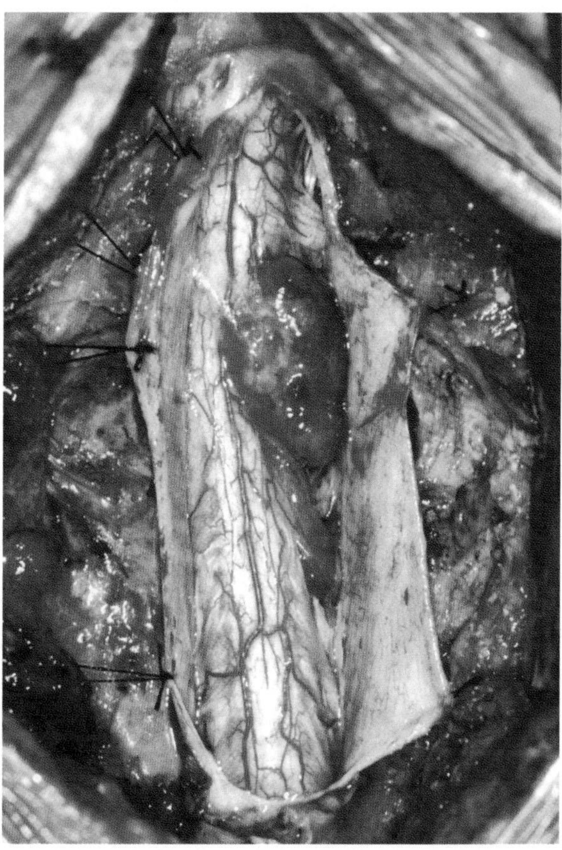

Figure 13-30

MALIGNANT PERIPHERAL NERVE SHEATH TUMOR

Occasional tumors of paraspinal nerves or nerve roots reach the intradural compartment.

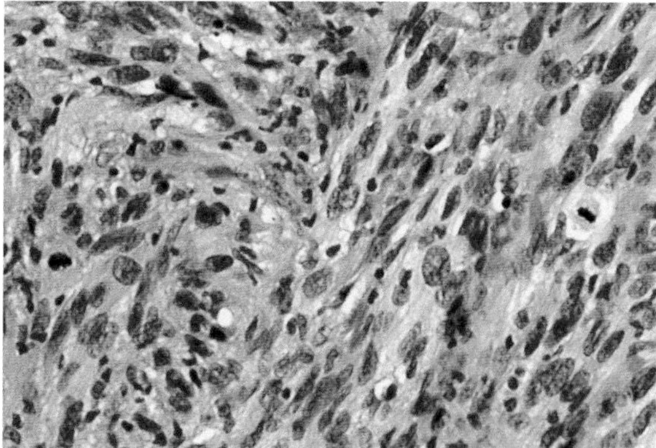

Figure 13-31

MALIGNANT PERIPHERAL NERVE SHEATH TUMOR

The tumors are cellular, often fascicular, and mitotically active.

Gross Findings. MPNSTs vary in appearance with grade. Low-grade tumors resemble neurofibroma: they are tan on cut section. High-grade lesions are fleshy, necrotic, and hemorrhagic. Pseudocapsules of the latter are derived from infiltrated soft tissue. Nerve-based tumors tend to be fusiform, whereas those lacking a nerve association are globular. MPNSTs arising in plexiform neurofibromas may be associated with additional microscopic foci of anaplastic transformation. As in any MPNST or neurofibroma suspected of malignant transformation, tissue sampling is critical. One microsection per centimeter of greatest tumor dimension is suggested. Due to the frequency of intraneural spread, assessment of nerve margins is important.

Microscopic Findings. Fully 85 percent of MPNSTs are high-grade tumors that are markedly cellular and have brisk mitotic activity, cellular pleomorphism, nuclear enlargement and hyperchromasia, as well as necrosis (fig. 13-31). Given the rarity of intracranial and intraspinal MPNSTs, this percentage is necessarily derived from experience with systemic examples.

The majority of tumors are composed of spindle cells, although epithelioid variants are not uncommon. Tissue patterns range from sheet-like to herringbone, storiform, or myxoid. Perivascular sparing of viable cells is common within expanses of necrosis.

About 15 percent of MPNSTs are low grade. Their cytologic features vary considerably, ranging from foci of moderate cellularity in which nuclei are only somewhat hyperchromatic and mitotically active, to regions with unmistakable features of malignancy. Although the benign-malignant "breakpoint" can be difficult to establish, there are criteria (1). The distinction of cellular neurofibroma from low-grade MPNST depends upon a triad of findings: 1) definite cellular crowding, 2) general nuclear enlargement (at least three times the size of neurofibroma nuclei), and 3) hyperchromasia. Spindle cell lesions are most common, but epithelioid and even clear cell tumors do occur (69,79). In any one tumor, cytologic and histologic pattern variation is the rule. Distinctive features found in some lesions include divergent mesenchymal (skeletal muscle, cartilage) or epithelial differentiation (69,76,77,80–84).

Immunohistochemical Findings. All MPNSTs are vimentin reactive, which also attests to tumor immunoviability. Approximately 65 percent show some degree of immunostaining for S-100 protein, leu-7, or myelin basic protein (85). Using a battery of stains increases the likelihood of detecting the expression of at least one marker. S-100 protein immunoreactivity is sometimes especially strong in malignant epithelioid schwannomas. Tumors with divergent differentiation to epithelium (malignant glandular schwannoma) and to cartilage or muscle (malignant Triton tumor) show commensurate staining for epithelial (keratin, EMA, carcinoembryonic antigen, chromogranin) and myoid (desmin, muscle specific actin, sarcomeric actin, myogenin) markers, respectively. Reactivity for GFAP is negative in all but the best-differentiated, low-grade MPNSTs (86).

Ultrastructural Findings. Although the fine structure of MPNSTs reflects their degree of differentiation, the preponderance of high-grade, undifferentiated-appearing tumors makes ultrastructural study of marginal diagnostic value. Tumors of intermediate differentiation often exhibit schwannian features, i.e., arrays of relatively thick, occasionally intersecting, cytoplasmic processes; rudimentary junctions; and focal basal lamina formation. These rudimentary features of nerve sheath differentiation are also seen in MPNSTs with perineurial differentiation in which pinocytotic vesicle formation is an additional feature (87). Rhabdomyoblastic features (sarcomere formation) may be seen in malignant Triton tumors (83).

Differential Diagnosis. The features distinguishing MPNSTs from neurofibroma include mitotic rate, hypercellularity, nuclear size exceeding by threefold that of neurofibroma, and hyperchromasia.

The distinction from cellular schwannoma is addressed in Table 13-1. In brief, the latter lesion has: 1) a thick capsule; 2) hyalinized vessels; 3) uniform and strong S-100 immunoreactivity; and 4) presence of pericellular basement lamina as established by histochemical (reticulin), immunohistochemical (collagen type IV, laminin), or ultrastructural means.

Treatment and Prognosis. The essential treatment of MPNST is surgical. Wide en bloc resection is preferred in tumors of soft tissue. Unfortunately, due to both the complexity of the craniospinal anatomy and to the tumors' capacity for infiltrative growth, it is often not possible to entirely resect MPNSTs. Local recurrence can be reduced by intraoperative or postoperative radiation therapy. Recurrence and metastasis rates are somewhat site dependent; the figure for both in paraspinous tumors is approximately 65 percent (72). To date, chemotherapy has had little to offer.

Although occasional long survival periods have been noted, most patients die of tumor (69,88,89). In two major series with long-term follow-up, death due to disease progression occurred in 63 and 68 percent of patients (69,88). Reported 5- and 10-year survival rates in these series ranged from 34 to 52 percent and 23 to 34 percent, respectively. Most series have shown no significant difference in outcome between patients with and without NF1.

LIPOMA AND RHABDOMYOMA (HAMARTOMA, CHORISTOMA, ECTOMESENCHYMOMA) OF INTRACRANIAL PERIPHERAL NERVES

These uncommon and variably constituted lesions typically arise on the eighth cranial nerve (90–96), but can also affect the second (97), third (98), fifth (99,100), or seventh nerves (95,101). The lipid content in some *lipomas* is useful in distinguishing them preoperatively from the overwhelmingly more common vestibular schwannoma (92,93,96). Fat has a bright signal in precontrast T1-weighted MRIs whereas schwannomas are bright only postcontrast (fig. 13-32).

Microscopically, the lesions contain adipose tissue alone (fig. 13-33) (92–94), adipose tissue plus smooth muscle (95,97), or adipose tissue and striated muscle (fig. 13-34) (90,91,101). Some reports describing lesions with striated muscle do not comment on the presence or absence of adipose tissue (98,100,102). The term *rhabdomyoma* is used for such processes, especially for bulky lesions with mass effect, although they are malformative rather than neoplastic (98,100). Surgery is curative, but its role is less clearly defined for some hamartomatous/lipomatous lesions that are intermingled with normal cranial nerves.

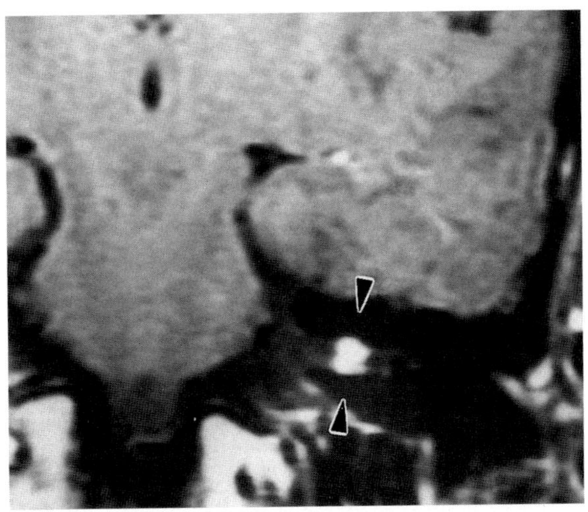

Figure 13-32

LIPOMA OF THE EIGHTH CRANIAL NERVE

Because of their content of lipid, lipomas of the eighth cranial nerve (arrowheads) are bright in precontrast T1-weighted images. Schwannomas require gadolinium administration to produce a bright T1 signal. (Fig. 10-28 from Fascicle 10, 3rd Series.)

Microscopically, mesenchymal elements coexist peacefully with myelinated axons and ganglion cells.

OTHER TUMORS OF INTRACRANIAL AND INTRASPINAL PERIPHERAL NERVES

Other intracranial tumors of peripheral nerves include: 1) *granular cell tumors* similar to those occurring at peripheral sites (103–105) but distinct from the granular cell tumors of the infundibulum and granular cell astrocytomas, 2) *neurothekeoma* (106), 3) *perineurioma* (107), 4) *neuronal hamartoma* (108), and 5) *glioblastoma* (109). Other intraspinal peripheral nerve tumors include *hemangioblastoma* (110–112), *paraganglioma* (113,114), *peripheral neuroectodermal tumor* (115), *melanocytoma* (116), *ependymomas* (117), *pilocytic astrocytoma* (118), *granular cell tumor* (119), and *adrenal adenoma* (120). Involvement of intradural nerve roots in lymphoma is discussed in chapter 12 (see fig. 12-32). Non-neoplastic enlargement of spinal nerve roots can occur with hypertrophic neuropathy (121) and chronic inflammatory demyelinating neuropathy.

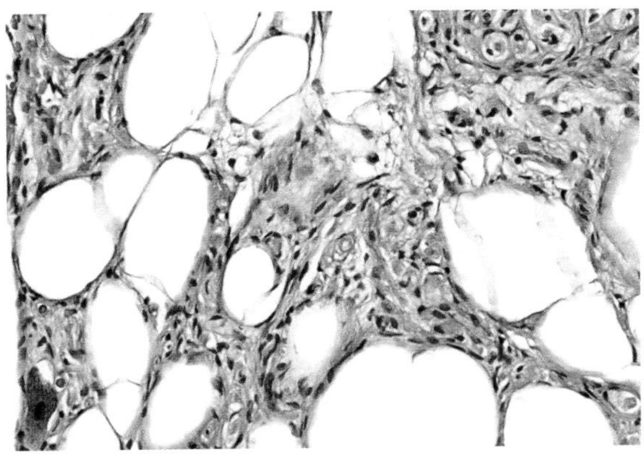

Figure 13-33

LIPOMA OF THE EIGHTH CRANIAL NERVE

Adipocytes mix freely with myelinated nerve fibers and a ganglion cell.

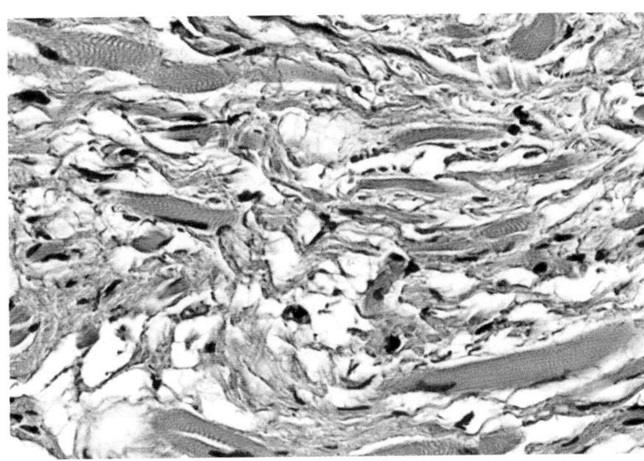

Figure 13-34

RHABDOMYOMA OF THE EIGHTH CRANIAL NERVE

Striated muscle cells are so well differentiated as to be obvious even in a frozen section.

REFERENCES

Conventional Schwannoma

1. Allcutt DA, Hoffman HJ, Isla A, Becker LE, Humphreys RP. Acoustic schwannomas in children. Neurosurgery 1991;29:14–8.
2. Scheithauer B, Woodruff J, Erlandson R. Tumors of the peripheral nervous system. Atlas of Tumor Pathology, 3rd Series, Fascicle 24. Washington, DC: Armed Forces Institute of Pathology. 1999:105–76.
3. Aryanpur J, Long DM. Schwannoma of the medulla oblongata. Case report. J Neurosurg 1988;69:446–9.
4. Auer RN, Budny J, Drake CG, Ball MJ. Frontal lobe perivascular schwannoma. Case report. J Neurosurg 1982;56:154–7.
5. Cruz-Sanchez F, Cervos-Navarro J, Kashihara M, Ferszt R. Intracerebral neurinomas in a case of von Recklinghausen's disease (neurofibromatosis). Clin Neuropathol 1987;6:174–8.
6. Di Biasi C, Trasimeni G, Iannilli M, Polettini E, Gualdi GF. Intracerebral schwannoma: CT and MR findings. AJNR Am J Neuroradiol 1994;15:1956–8.
7. Lin J, Feng H, Li F, Zhao B, Guo Q. Intraparenchymal schwannoma of the medulla oblongata. Case report. J Neurosurg 2003;98:621–4.
8. Tran-Dinh HD, Soo YS, O'Neil P, Chaseling R. Cystic cerebellar schwannoma: case report. Neurosurgery 1991;29:296–300.
9. Zagardo MT, Castellani RJ, Rees JH, Rothman MI, Zoarski GH. Radiologic and pathologic findings of intracerebral schwannoma. AJNR Am J Neuroradiol 1998;19:1290–3.
10. Payan H, Levine S. Focal axonal proliferation in pons (central neurinoma). Association with cystic encephalomalacia. Arch Pathol 1965;79:501–4.
11. Demyer W. Aberrant peripheral nerve fibres in the medulla oblongata of man. J Neurol Neurosurg Psychiatry 1965;28:121–3.
12. Conti P, Pansini G, Mouchaty H, Capuano C, Conti R. Spinal neurinomas: retrospective analysis and long-term outcome of 179 consecutively operated cases and review of the literature. Surg Neurol 2004;61:34–44.
13. Abernathey CD, Onofrio BM, Scheithauer B, Pairolero PC, Shives TC. Surgical management of giant sacral schwannomas. J Neurosurg 1986;65:286–95.
14. Acciarri N, Padovani R, Riccioni L. Intramedullary melanotic schwannoma. Report of a case and review of the literature. Br J Neurosurg 1999;13:322–5.
15. Herregodts P, Vloeberghs M, Schmedding E, Goossens A, Stadnik T, D'Haens J. Solitary dorsal intramedullary schwannoma. Case report. J Neurosurg 1991;74:816–20.
16. Melancia JL, Pimentel JC, Conceicao I, Antunes JL. Intramedullary neuroma of the cervical spinal cord: case report. Neurosurgery 1996;39: 594–8.
17. Marchese MJ, McDonald JV. Intramedullary melanotic schwannoma of the cervical spinal cord: report of a case. Surg Neurol 1990;33:353–5.
18. Miller DJ, McCutcheon IE. Hemangioblastomas and other uncommon intramedullary tumors. J Neurooncol 2000;47:253–70.
19. Riggs HE, Clary WU. A case of intramedullary sheath cell tumor of the spinal cord; consideration of vascular nerves as a source of origin. J Neuropathol Exp Neurol 1957;16:332–6.
20. Ross DA, Edwards MS, Wilson CB. Intramedullary neurilemomas of the spinal cord: report of two cases and review of the literature. Neurosurgery 1986;19:458–64.
21. Solomon RA, Handler MS, Sedelli RV, Stein BM. Intramedullary melanotic schwannoma of the cervicomedullary junction. Neurosurgery 1987;20:36–8.
22. Kamiya M, Hashizume Y. Pathological studies of aberrant peripheral nerve bundles of spinal cords. Acta Neuropathol (Berl) 1989;79:18–22.
23. Gorman PH, Rigamonti D, Joslyn JN. Intramedullary and extramedullary schwannoma of the cervical spinal cord—case report. Surg Neurol 1989;32:459–62.
24. Tali ET, Yuh WT, Nguyen HD, et al. Cystic acoustic schwannomas: MR characteristics. AJNR Am J Neuroradiol 1993;14:1241–7.
25. Hasegawa M, Fujisawa H, Hayashi Y, Tachibana O, Kida S, Yamashita J. Surgical pathology of spinal schwannomas: a light and electron microscopic analysis of tumor capsules. Neurosurgery 2001;49:1388–92; discussion 1392–3.
26. Kurtkaya-Yapicier O, Scheithauer BW, Woodruff JM, Wenger DD, Cooley AM, Dominique D. Schwannoma with rhabdomyoblastic differentiation: a unique variant of malignant triton tumor. Am J Surg Pathol 2003;27:848–53.
27. Winek RR, Scheithauer BW, Wick MR. Meningioma, meningeal hemangiopericytoma (angioblastic meningioma), peripheral hemangiopericytoma, and acoustic schwannoma. A comparative immunohistochemical study. Am J Surg Pathol 1989;13:251–61.

28. Memoli VA, Brown EF, Gould VE. Glial fibrillary acidic protein (GFAP) immunoreactivity in peripheral nerve sheath tumors. Ultrastruct Pathol 1984;7:269–75.
29. Niemczyk K, Vaneecloo FM, Lecomte MH, et al. Correlation between Ki-67 index and some clinical aspects of acoustic neuromas (vestibular schwannomas). Otolaryngol Head Neck Surg 2000;123:779–83.
30. Antinheimo J, Haapasalo H, Seppala M, Sainio M, Carpen O, Jaaskelainen J. Proliferative potential of sporadic and neurofibromatosis 2-associated schwannomas as studied by MIB-1 (Ki-67) and PCNA labeling. J Neuropathol Exp Neurol 1995;54:776–82.
31. Irving RM, Moffat DA, Hardy DG, Barton DE, Xuereb JH, Maher ER. Somatic NF2 gene mutations in familial and non-familial vestibular schwannoma. Hum Mol Genet 1994;3:347–50.
32. Irving RM, Moffat DA, Hardy DG, et al. A molecular, clinical, and immunohistochemical study of vestibular schwannoma. Otolaryngol Head Neck Surg 1997;116:426–30.
33. Hwang SK, Kim DG, Paek SH, et al. Aggressive vestibular schwannomas with postoperative rapid growth: clinicopathological analysis of 15 cases. Neurosurgery 2002;51:1381–90; discussion 1390–1.
34. Gwak HS, Hwang SK, Paek SH, Kim DG, Jung HW. Long-term outcome of trigeminal neurinomas with modified classification focusing on petrous erosion. Surg Neurol 2003;60:39–48; discussion 48.
35. Mikami Y, Hidaka T, Akisada T, Takemoto T, Irei I, Manabe T. Malignant peripheral nerve sheath tumor arising in benign ancient schwannoma: a case report with an immunohistochemical study. Pathol Int 2000;50:156–61.
36. Woodruff JM, Selig AM, Crowley K, Allen PW. Schwannoma (neurilemoma) with malignant transformation. A rare, distinctive peripheral nerve tumor. Am J Surg Pathol 1994;18:882–95.
37. Bari ME, Forster DM, Kemeny AA, Walton L, Hardy D, Anderson JR. Malignancy in a vestibular schwannoma. Report of a case with central neurofibromatosis, treated by both stereotactic radiosurgery and surgical excision, with a review of the literature. Br J Neurosurg 2002;16:284–9.

Cellular Schwannoma

38. Woodruff JM, Godwin TA, Erlandson RA, Susin M, Martini N. Cellular schwannoma: a variety of schwannoma sometimes mistaken for a malignant tumor. Am J Surg Pathol 1981;5:733–44.
39. Casadei GP, Scheithauer BW, Hirose T, Manfrini M, Van Houton C, Wood MB. Cellular schwannoma. A clinicopathologic, DNA flow cytometric, and proliferation marker study of 70 patients. Cancer 1995;75:1109–19.
40. Fletcher CD, Davies SE, McKee PH. Cellular schwannoma: a distinct pseudosarcomatous entity. Histopathology 1987;11:21–35.
41. Seppala MT, Haltia MJ. Spinal malignant nerve-sheath tumor or cellular schwannoma? A striking difference in prognosis. J Neurosurg 1993;79:528–32.
42. Deruaz JP, Janzer RC, Costa J. Cellular schwannomas of the intracranial and intraspinal compartment: morphological and immunological characteristics compared with classical benign schwannomas. J Neuropathol Exp Neurol 1993;52:114–8.
43. White W, Shiu MH, Rosenblum MK, Erlandson RA, Woodruff JM. Cellular schwannoma. A clinicopathologic study of 57 patients and 58 tumors. Cancer 1990;66:1266–75.

Melanotic Schwannoma

44. Cras P, Ceuterick-de Groote C, Van Vyve M, Vercruyssen A, Martin JJ. Malignant pigmented spinal nerve root schwannoma metastasizing in the brain and viscera. Clin Neuropathol 1990;9:290–4.
45. Graham DI, Paterson A, McQueen A, Milne JA, Urich H. Melanotic tumours (blue naevi) of spinal nerve roots. J Pathol 1976;118:83–9.
46. Gregorios JB, Chou SM, Bay J. Melanotic schwannoma of the spinal cord. Neurosurgery 1982;11(Pt 1):57–60.
47. Hisaoka M, Ohta H, Haratake J, Horie A. Melanocytic schwannoma in the spinal canal. Acta Pathol Jpn 1991;41:685–8.
48. Mennemeyer RP, Hallman KO, Hammar SP, Raisis JE, Tytus JS, Bockus D. Melanotic schwannoma. Clinical and ultrastructural studies of three cases with evidence of intracellular melanin synthesis. Am J Surg Pathol 1979;3:3–10.
49. Roytta M, Elfversson J, Kalimo H. Intraspinal pigmented schwannoma with malignant progression. Acta Neurochir (Wien) 1988;95:147–54.
50. Dastur D, Sinh G, Pandya S. Melanotic tumor of the acoustic nerve. J Neurosurg 1967;27:166–70.
51. Solomon RA, Handler MS, Sedelli RV, Stein BM. Intramedullary melanotic schwannoma of the cervicomedullary junction. Neurosurgery 1987;20:36–8.
52. Acciarri N, Padovani R, Riccioni L. Intramedullary melanotic schwannoma. Report of a case and review of the literature. Br J Neurosurg 1999;13:322–5.
53. Marchese MJ, McDonald JV. Intramedullary melanotic schwannoma of the cervical spinal cord: report of a case. Surg Neurol 1990;33:353–5.

54. Carney JA, Gordon H, Carpenter PC, Shenoy BV, Go VL. The complex of myxomas, spotty pigmentation, and endocrine overactivity. Medicine (Baltimore) 1985;64:270–83.
55. Carney JA. Psammomatous melanotic schwannoma. A distinctive, heritable tumor with special associations, including cardiac myxoma and the Cushing syndrome. Am J Surg Pathol 1990;14:206–22.
56. Carney JA, Stratakis CA. Epithelioid blue nevus and psammomatous melanotic schwannoma: the unusual pigmented skin tumors of the Carney complex. Semin Diagn Pathol 1998;15: 216–24.
57. Kirschner LS, Sandrini F, Monbo J, Lin JP, Carney JA, Stratakis CA. Genetic heterogeneity and spectrum of mutations of the PRKAR1A gene in patients with the Carney complex. Hum Mol Genet 2000;9:3037–46.
58. Stergiopoulos SG, Stratakis CA. Human tumors associated with Carney complex and germline PRKAR1A mutations: a protein kinase A disease! FEBS Lett 2003;546:59–64.
59. Jensen OA, Bretlau P. Melanotic schwannoma of the orbit. Immunohistochemical and ultrastructural study of a case and survey of the literature. APMIS 1990;98:713–23.
60. Krausz T, Azzopardi JG, Pearse E. Malignant melanoma of the sympathetic chain: with a consideration of pigmented nerve sheath tumours. Histopathology 1984;8:881–94.
61. Fu YS, Kaye G, Lattes R. Primary malignant melanocytic tumors of the sympathetic ganglia, with an ultrastructural study of one. Cancer 1975;36:2029–41.

Neurofibroma

62. McCarron KF, Goldblum JR. Plexiform neurofibroma with and without associated malignant peripheral nerve sheath tumor: a clinicopathologic and immunohistochemical analysis of 54 cases. Mod Pathol 1998;11:612–7.
63. Scheithauer B, Woodruff J, Erlandson R. Tumors of the peripheral nervous system. Atlas of Tumor Pathology, 3rd Series, Fascicle 24. Washington, DC: Armed Forces Institute of Pathology; 1999:177–218.
64. Johnson MD, Glick AD, Davis BW. Immunohistochemical evaluation of Leu-7, myelin basic-protein, S100-protein, glial-fibrillary acidic-protein, and LN3 immunoreactivity in nerve sheath tumors and sarcomas. Arch Pathol Lab Med 1988;112:155–60.
65. Hirose T, Sano T, Hizawa K. Ultrastructural localization of S-100 protein in neurofibroma. Acta Neuropathol (Berl) 1986;69:103–10.
66. Perentes E, Nakagawa Y, Ross GW, Stanton C, Rubinstein LJ. Expression of epithelial membrane antigen in perineurial cells and their derivatives. An immunohistochemical study with multiple markers. Acta Neuropathol (Berl) 1987;75:160–65.
67. Erlandson RA, Woodruff JM. Role of electron microscopy in the evaluation of soft tissue neoplasms, with emphasis on spindle cell and pleomorphic tumors. Hum Pathol 1998;29:1372–81.

Malignant Peripheral Nerve Sheath Tumor

68. Scheithauer B, Woodruff J, Erlandson R. Tumors of the peripheral nervous system. Atlas of Tumor Pathology, 3rd Series, Fascicle 24. Washington, DC: Armed Forces Institute of Pathology; 1999:303–72.
69. Ducatman BS, Scheithauer BW, Piepgras DG, Reiman HM, Ilstrup DM. Malignant peripheral nerve sheath tumors. A clinicopathologic study of 120 cases. Cancer 1986;57:2006–21.
70. Mikami Y, Hidaka T, Akisada T, Takemoto T, Irei I, Manabe T. Malignant peripheral nerve sheath tumor arising in benign ancient schwannoma: a case report with an immunohistochemical study. Pathol Int 2000;50:156–61.
71. Woodruff JM, Selig AM, Crowley K, Allen PW. Schwannoma (neurilemoma) with malignant transformation. A rare, distinctive peripheral nerve tumor. Am J Surg Pathol 1994;18:882–95.
72. Kourea HP, Bilsky MH, Leung DH, Lewis JJ, Woodruff JM. Subdiaphragmatic and intrathoracic paraspinal malignant peripheral nerve sheath tumors: a clinicopathologic study of 25 patients and 26 tumors. Cancer 1998;82:2191–203.
73. Levy WJ, Ansbacher L, Byer J, Nutkiewicz A, Fratkin J. Primary malignant nerve sheath tumor of the gasserian ganglion: a report of two cases. Neurosurgery 1983;13:572–6.
74. Liwnicz BH. Bilateral trigeminal neurofibrosarcoma. Case report. J Neurosurg 1979;50:253–6.
75. Tegos S, Georgouli G, Gogos C, Polythothorakis J, Sanidas V, Mavrogiorgos C. Primary malignant schwannoma involving simultaneously the right Gasserian ganglion and the distal part of the right mandibular nerve. Case report. J Neurosurg Sci 1997;41:293–7.
76. Best PV. Malignant triton tumour in the cerebellopontine angle. Report of a case. Acta Neuropathol (Berl) 1987;74:92–6.
77. Han DH, Kim DG, Chi JG, Park SH, Jung HW, Kim YG. Malignant triton tumor of the acoustic nerve. Case report. J Neurosurg 1992;76:874–7.
78. Stefanko SZ, Vuzevski VD, Maas AI, van Vroonhoven CC. Intracerebral malignant schwannoma. Acta Neuropathol (Berl) 1986;71:321–5.

79. DiCarlo EF, Woodruff JM, Bansal M, Erlandson RA. The purely epithelioid malignant peripheral nerve sheath tumor. Am J Surg Pathol 1986;10:478–90.
80. Brooks JS, Freeman M, Enterline HT. Malignant "Triton" tumors. Natural history and immunohistochemistry of nine new cases with literature review. Cancer 1985;55:2543–9.
81. Christensen WN, Strong EW, Bains MS, Woodruff JM. Neuroendocrine differentiation in the glandular peripheral nerve sheath tumor. Pathologic distinction from the biphasic synovial sarcoma with glands. Am J Surg Pathol 1988;12: 417–26.
82. Ducatman BS, Scheithauer BW. Postirradiation neurofibrosarcoma. Cancer 1983;51:1028–33.
83. Woodruff JM, Chernik NL, Smith MC, Millett WB, Foote FW Jr. Peripheral nerve tumors with rhabdomyosarcomatous differentiation (malignant "Triton" tumors). Cancer 1973;32:426–39.
84. Woodruff JM, Christensen WN. Glandular peripheral nerve sheath tumors. Cancer 1993;72: 3618–28.
85. Wick MR, Swanson PE, Scheithauer BW, Manivel JC. Malignant peripheral nerve sheath tumor. An immunohistochemical study of 62 cases. Am J Clin Pathol 1987;87:425–33.
86. Giangaspero F, Fratamico FC, Ceccarelli C, Brisigotti M. Malignant peripheral nerve sheath tumors and spindle cell sarcomas: an immunohistochemical analysis of multiple markers. Appl Pathol 1989;7:134–44.
87. Hirose T, Sano T, Hizawa K. Heterogeneity of malignant schwannomas. Ultrastruct Pathol 1988;12:107–16.
88. Hruban RH, Shiu MH, Senie RT, Woodruff JM. Malignant peripheral nerve sheath tumors of the buttock and lower extremity. A study of 43 cases. Cancer 1990;66:1253–65.
89. Seppala MT, Haltia MJ. Spinal malignant nerve-sheath tumor or cellular schwannoma? A striking difference in prognosis. J Neurosurg 1993; 79:528–32.

Lipoma and Rhabdomyoma (Hamartoma, Choristoma, Ectomesenchymoma) of Intracranial Peripheral Nerves

90. Apostolides PJ, Spetzler RF, Johnson PC. Ectomesenchymal hamartoma (benign "ectomesenchymoma") of the VIIIth nerve: case report. Neurosurgery 1995;37:1204–7.
91. Carvalho GA, Matthies C, Osorio E, Samii M. Hamartomas of the internal auditory canal: report of two cases. Neurosurgery 2003;52:944–9.
92. Cohen TI, Powers SK, Williams DW 3rd. MR appearance of intracanalicular eighth nerve lipoma. AJNR Am J Neuroradiol 1992;13:1188–90.
93. Saunders JE, Kwartler JA, Wolf HK, Brackmann DE, McElveen JT Jr. Lipomas of the internal auditory canal. Laryngoscope 1991;101:1031–7.
94. Singh SP, Cottingham SL, Slone W, Boesel CP, Welling DB, Yates AJ. Lipomas of the internal auditory canal. Arch Pathol Lab Med 1996;120:681–3.
95. Smith MM, Thompson JE, Thomas D, et al. Choristomas of the seventh and eighth cranial nerves. AJNR Am J Neuroradiol 1997;18:327–9.
96. Tankere F, Vitte E, Martin-Duverneuil N, Soudant J. Cerebellopontine angle lipomas: report of four cases and review of the literature. Neurosurgery 2002;50:626–31; discussion 631–2.
97. Giannini C, Reynolds C, Leavitt JA, et al. Choristoma of the optic nerve: case report. Neurosurgery 2002;50:1125–8.
98. Lee JI, Nam DH, Kim JS, et al. Intracranial oculomotor nerve rhabdomyoma. J Neurosurg 2000;93:715.
99. Lena G, Dufour T, Gambarelli D, Chabrol B, Mancini J. Choristoma of the intracranial maxillary nerve in a child. Case report. J Neurosurg 1994;81:788–91.
100. Zwick DL, Livingston K, Clapp L, Kosnik E, Yates A. Intracranial trigeminal nerve rhabdomyoma/choristoma in a child: a case report and discussion of possible histogenesis. Hum Pathol 1989;20:390–2.
101. Vandewalle G, Brucher JM, Michotte A. Intracranial facial nerve rhabdomyoma. Case report. J Neurosurg 1995;83:919–22.
102. van Leeuwen JP, Pruszczynski M, Marres HA, Grotenhuis JA, Cremers CW. Unilateral hearing loss due to a rhabdomyoma in a six-year-old child. J Laryngol Otol 1995;109:1186–9.

Other Tumors of Intracranial and Intraspinal Peripheral Nerves

103. Carvalho GA, Lindeke A, Tatagiba M, Ostertag H, Samii M. Cranial granular-cell tumor of the trigeminal nerve. Case report. J Neurosurg 1994;81:795–8.
104. Chimelli L, Symon L, Scaravilli F. Granular cell tumor of the fifth cranial nerve: further evidence for Schwann cell origin. J Neuropathol Exp Neurol 1984;43:634–42.
105. Rao TV, Puri R, Reddy GN. Intracranial trigeminal nerve granular cell myoblastoma. Case report. J Neurosurg 1983;59:706–9.
106. Paulus W, Warmuth-Metz M, Sorensen N. Intracranial neurothekeoma (nerve-sheath myxoma). Case report. J Neurosurg 1993;79: 280–2.

107. Giannini C, Scheithauer BW, Steinberg J, Cosgrove TJ. Intraventricular perineurioma: case report. Neurosurgery 1998;43:1478–81; discussion 1481–2.
108. Patel N, Moss T, Coakham H. Neuronal hamartoma of the trigeminal sensory root associated with trigeminal neuralgia. J Neurosurg 2000;93: 514.
109. Reifenberger G, Bostrom J, Bettag M, Bock WJ, Wechsler W, Kepes JJ. Primary glioblastoma multiforme of the oculomotor nerve. Case report. J Neurosurg 1996;84:1062–6.
110. Ishikawa E, Matsumura A, Matsumaru Y, et al. Intratumoral hemorrhage due to hemangioblastoma arising from a cervical nerve root—a case report. J Clin Neurosci 2002;9:713–6.
111. McEvoy AW, Benjamin E, Powell MP. Haemangioblastoma of a cervical sensory nerve root in Von Hippel-Lindau syndrome. Eur Spine J 2000;9:434–6.
112. Raghavan R, Krumerman J, Rushing EJ, et al. Recurrent (nonfamilial) hemangioblastomas involving spinal nerve roots: case report. Neurosurgery 2000;47:1443–8.
113. Sonneland PR, Scheithauer BW, LeChago J, Crawford BG, Onofrio BM. Paraganglioma of the cauda equina region. Clinicopathologic study of 31 cases with special reference to immunocytology and ultrastructure. Cancer 1986; 58:1720–35.
114. Sundgren P, Annertz M, Englund E, Stromblad LG, Holtas S. Paragangliomas of the spinal canal. Neuroradiology 1999;41(10):788–94.
115. Yavuz AA, Yaris N, Yavuz MN, Sari A, Reis AK, Aydin F. Primary intraspinal primitive neuroectodermal tumor: case report of a tumor arising from the sacral spinal nerve root and review of the literature. Am J Clin Oncol 2002;25:135–9.
116. Goyal A, Sinha S, Singh AK, Tatke M, Kansal A. Lumbar spinal meningeal melanocytoma of the l3 nerve root with paraspinal extension: a case report. Spine 2003;28:E140–2.
117. Moser FG, Tuvia J, LaSalla P, Llana J. Ependymoma of the spinal nerve root: case report. Neurosurgery 1992;31:962–4.
118. Philipson MR, Timothy J, Chakrobarthy A, Towns G. Pilocytic astrocytoma of a spinal nerve root. Case report. J Neurosurg 2002; 97(Suppl 1):110–2.
119. Haku T, Hosoya T, Hayashi M, Adachi M, Konno M, Ogino T. Granular cell tumor of the spinal nerve root: MR findings. Radiat Med 2002;20:137–40.
120. Mitchell A, Scheithauer BW, Sasano H, Hubbard EW, Ebersold MJ. Symptomatic intradural adrenal adenoma of the spinal nerve root: report of two cases. Neurosurgery 1993;32:658–61; discussion 661–2.
121. Kretzer RM, Burger PC, Tamargo RJ. Hypertrophic neuropathy of the cauda equina: case report. Neurosurgery 2004;54:515–9.

14 MELANOCYTIC NEOPLASMS

PRIMARY LEPTOMENINGEAL MELANOCYTIC NEOPLASMS

General Features. Primary melanocytic neoplasms are derived from leptomeningeal melanocytes. Rare intracranial and intraspinal tumors are presumed to arise from normal melanocytes, which are most numerous in the leptomeninges at the base of the brain and upper cervical cord (see figs. 2-6, 2-7). Melanotic schwannoma and its psammomatous variant are discussed in chapter13. Appendix H summarizes melanotic central nervous system (CNS) neoplasms.

Clinical Features. Primary melanocytic tumors occur both as discrete masses (1–25) and diffuse proliferations (26–30). Both are leptomeningeal in derivation, even those that are intraventricular (31) or appear in the pineal region (23,32–34). The discrete forms occur sporadically in patients of all ages.

Occasional meningeal melanocytic neoplasms are associated with a pigmented facial cutaneous and ocular lesion, the nevus of Ota (4,5,22). The complex is known as neurocutaneous melanosis when the cutaneous lesions are multiple and/or extensive and is discussed separately below.

Radiologic Findings. Densely pigmented lesions have a distinctive profile on magnetic resonance imaging (MRI) of high signal intensity (white) on precontrast T1-weighted images and low signal intensity (dark) on T2-weighted images (fig. 14-1) (6,9,10,13,15,31). Many melanocytic neoplasms are not sufficiently pigmented to generate this MRI profile and are approached surgically in anticipation of meningioma.

Gross Findings. Discrete meningeal melanocytic neoplasms arise within both intracranial and intraspinal compartments. Intracranially, the base of the brain, often in the posterior fossa, is favored. The degree of pigmentation ranges from shades of gray to black as tar.

Diffuse meningeal melanocytic neoplasms vary from a faint dusky clouding of the leptomeninges to dense black replacement of the subarachnoid space.

Microscopic Findings. *Discrete leptomeningeal melanocytic neoplasms* occur along a spectrum from well-differentiated lesions termed *melanocytoma* (6,7,16,18,21,24,25,35) to frankly malignant tumors designated *melanoma* (1–3,6,12, 19). A histologically intermediate group bridges the two extremes (6).

Melanocytomas vary histologically from spindle cell tumors resembling spindle A melanoma of the eye (figs. 14-2, 14-3) to neoplasms that are more obviously epithelioid in appearance (fig. 14-4). Purely epithelioid examples are unusual and raise the question of a more aggressive lesion. Spindle cell lesions have been likened to blue nevi since they vary in pigmentation and have nuclei that are elongated and

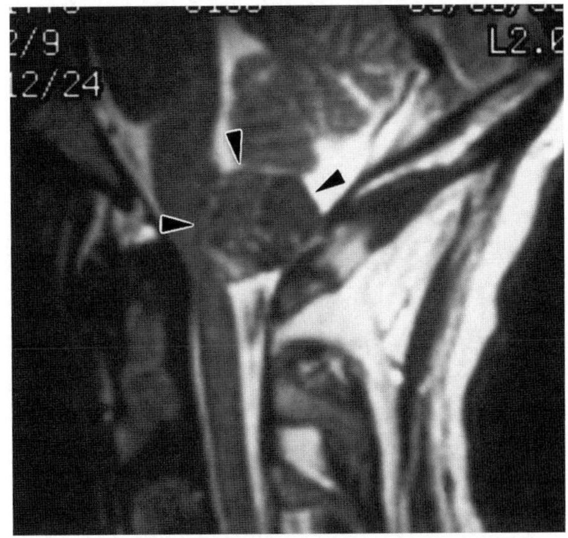

Figure 14-1

PRIMARY LEPTOMENINGEAL MALIGNANT MELANOMA

Highly pigmented lesions (arrowheads), as here in a 52-year-old woman, generate a dark signal intensity in T2-weighted images. They also have bright signal in precontrast T1-weighted images. (Fig. 14-1 from Fascicle 10, 3rd Series.)

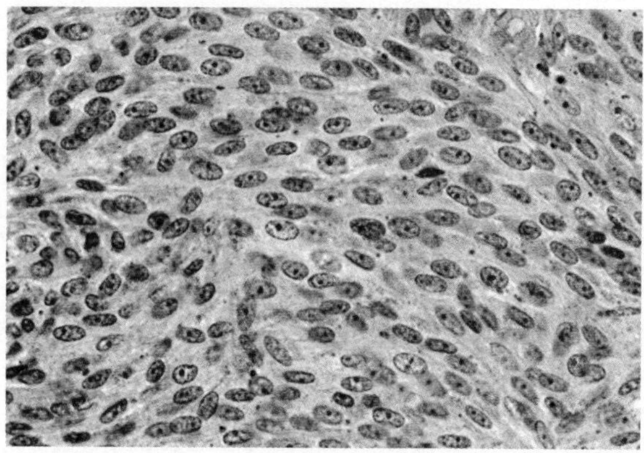

Figure 14-2

MELANOCYTOMA

Melanocytomas often have a fascicular architecture with interwoven fascicles of spindle-shaped cells. The resemblance to meningioma is obvious.

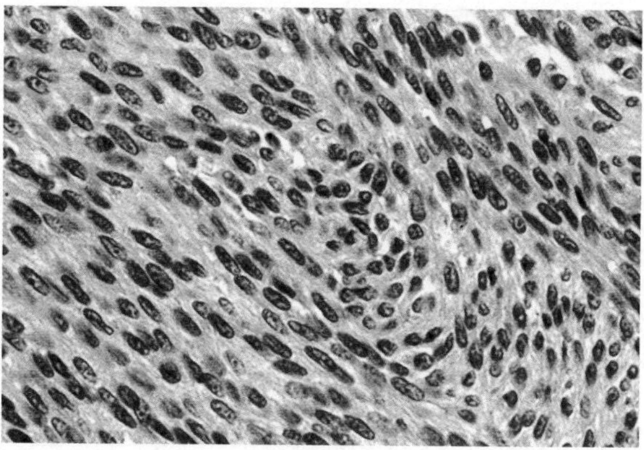

Figure 14-3

MELANOCYTOMA

Cytologically bland, amelanotic examples with a fascicular architecture can be difficult to recognize as melanocytic neoplasms.

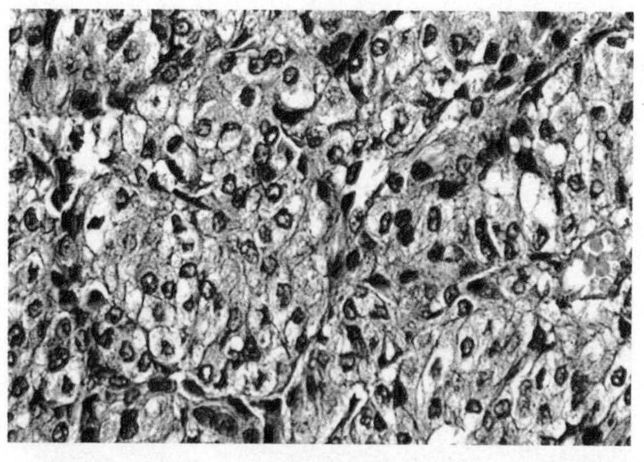

Figure 14-4

MELANOCYTOMA

A lobular, nested architecture is seen in some melanocytomas.

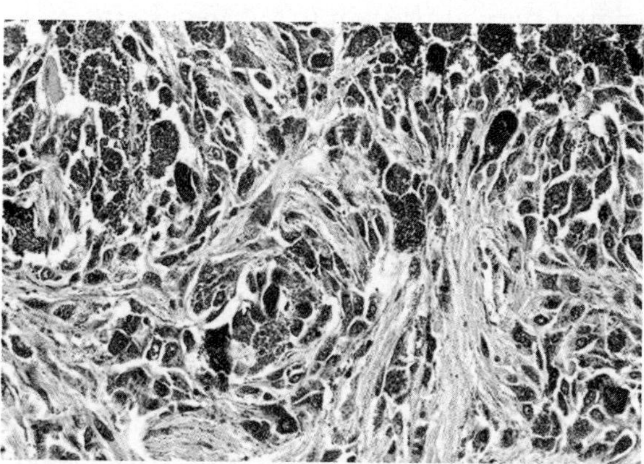

Figure 14-5

MELANOCYTOMA

Dense pigmentation obscures cytologic details in some melanocytomas.

often delicate (16,36). Nests and small lobules are common, particularly in epithelioid melanocytomas. Nuclear features are so obscured by pigment that peroxide bleaching is required for cytologic assessment (fig. 14-5). Oncocytic change has been reported (11).

Nuclei are cytologically bland, with one small nucleolus per cell. Nuclear grooves are common. At most, the lesions feature only rare mitoses (1 per 10 high-power fields) but the majority of lesions have none (6). In one study, the mean MIB-1 index for 17 melanocytomas ranged from 0 to 2.0 percent with a mean of 0.5 percent (6).

Intermediate-grade tumors are uncommon and difficult to define; firm criteria have not been established. These tumors are often cellular, have a sheet-like architecture, are composed of cells with an epithelioid appearance and bland nuclei with delicate nucleoli, and have only rare mitoses (fig. 14-6). Intermediate tumors often have the "feel" of a melanoma but lack clearly defined features of malignancy. The MIB-1 index of three intermediate-grade tumors in one study ranged from 1 to 4 percent, and the mitotic index from 1 to 3 per 10 high-power fields (6). The tumor can invade the brain in

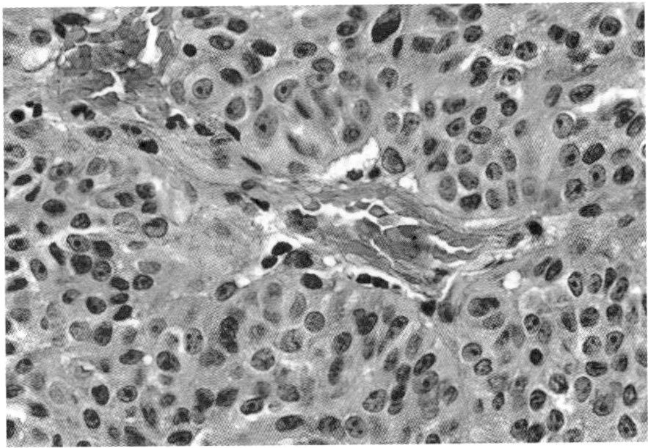

Figure 14-6

MELANOCYTIC NEOPLASM OF INTERMEDIATE DIFFERENTIATION

Melanocytic lesions with prominent nucleoli but only rare mitoses are best included in the intermediate category.

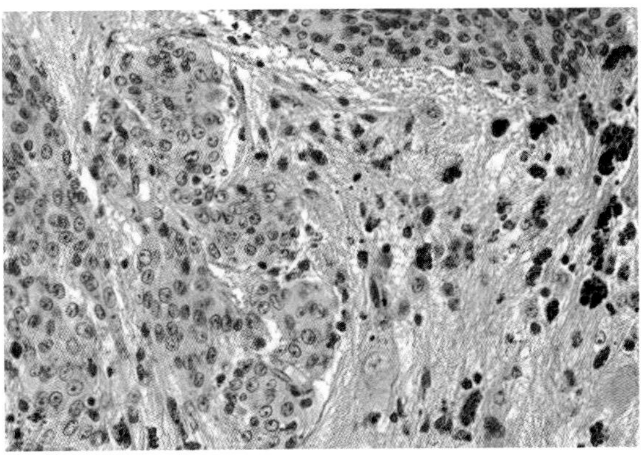

Figure 14-7

MELANOCYTIC NEOPLASM OF INTERMEDIATE DIFFERENTIATION

The prognostic significance of brain invasion is unclear but, similar to meningioma, should be an unfavorable feature. Scattered mitoses elsewhere in the lesion also supported its grading as an intermediate lesion.

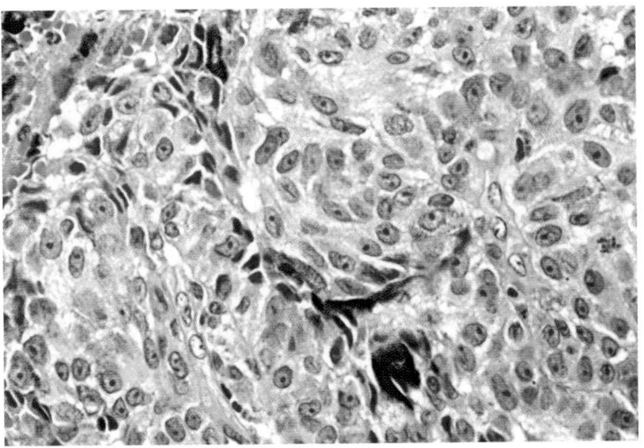

Figure 14-8

MALIGNANT MELANOMA

Nested, mitotically active pigmented cells with prominent nucleoli are common in melanomas at any site.

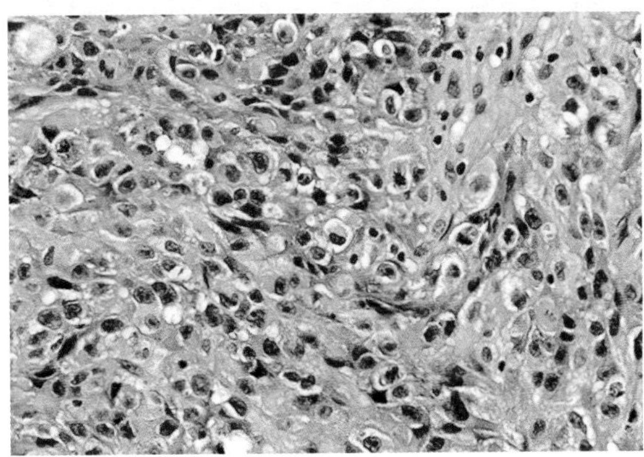

Figure 14-9

MALIGNANT MELANOMA

Meningeal masses of large round cells with prominent nucleoli and scattered mitoses need to be evaluated for a possible melanocytic origin.

the form of tongues of tissue similar to those of brain-invasive meningiomas (fig. 14-7). When melanocytomas invade, they do so along perivascular spaces for the most part, and often elicit a chronic piloid gliosis with Rosenthal fibers. It is unclear whether brain invasion is a criterion for an intermediate grade in a lesion that is otherwise melanocytoma. It was so employed in the above study (6).

Primary malignant melanomas vary in cytologic appearance much like those at other sites (figs. 14-8, 14-9). Their cells vary from spindled to epithelioid with prominent nucleoli. Mitoses range from scant to numerous. Necrosis may be extensive. The degree of pigmentation is variable. Transition to a diffuse growth pattern, satellite nodules in surrounding meninges, and invasion of brain or spinal cord parenchyma are common. Balloon cell features may be seen (1). In the series cited above, the MIB-1 labeling index ranged from 2 to 15 percent (mean, 7.8 percent) and the mitotic index from 2 to 15 per 10 high-power fields (6).

Diffuse leptomeningeal melanocytic neoplasms (melanocytosis/melanomatosis) vary in cytologic appearance from well-differentiated with nevoid cytologic features to unequivocally malignant proliferations for which only the term *malignant melanoma* is appropriate (fig. 14-10) (3,26, 27,28,30). Some diffuse tumors, despite biologic aggressiveness, are morphologically intermediate in their cytologic appearance and, as is the case for some discrete lesions, defy precise histologic classification as either benign or malignant.

Immunohistochemical Findings. Melanocytic tumors are reactive for vimentin and lack epithelial markers. Staining for S-100 protein varies, but is often strong. Markers for melanocytic neoplasms, such as HMB45, Melan-A (fig. 14-11, left), and microphthalmia transcription factor (MITF) (fig. 14-11, right) may be positive (6,7,11).

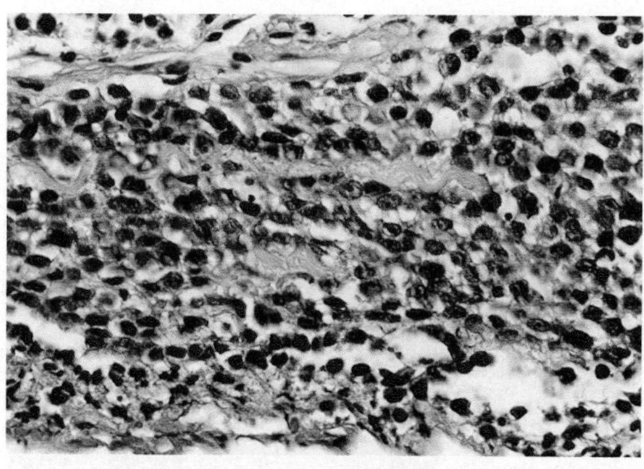

Figure 14-10

MENINGEAL MELANOMATOSIS

Amelanotic, malignant melanocytes can fill the subarachnoid space in the absence of a primary tumor mass, as in this 1-year-old child.

Ultrastructural Findings. As expected, neoplastic melanocytes lack junctions and contain melanosomes in varying stages of pigmentation (2,5,8,14,16,17,18,20,24,25,28). Unlike Schwann cell tumors, melanomas lack diffuse pericellular basal lamina, but clusters of cells in melanocytomas may be circumscribed by delicate reticulin fibers.

Differential Diagnosis. The possibility of a melanocytic neoplasm is self-evident if a lesion is pigmented, but the entity may never be considered if it is not, particularly in the face of a monomorphous, histologically benign, meningioma-like appearance (see fig. 9-80).

Meningiomas, particularly those with a nested architecture, can closely simulate melanocytic neoplasms (fig. 14-12), but are generally less monomorphic, have more parenchymal fibrous connective tissue and hyalinized vessels, and are often calcified. Immunopositivity for melanocyte markers and lack of epithelial membrane antigen staining resolve the issue.

A more difficult problem is distinguishing melanocytic tumors from melanotic schwannomas. The distinction usually becomes an issue in the intraspinal compartment where melanotic schwannomas are most likely to occur. The cells of the latter often have a more

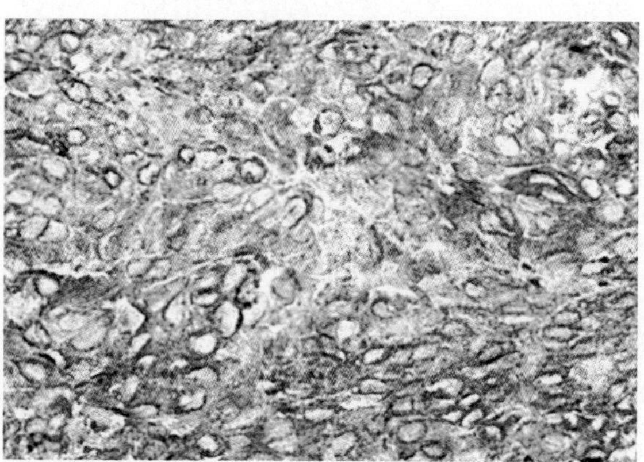

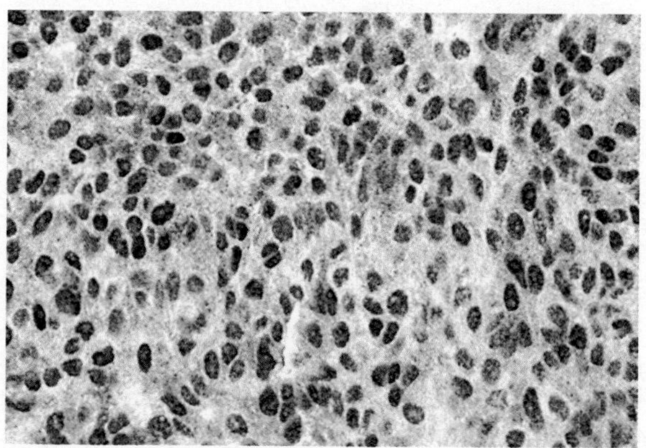

Figure 14-11

MELANOCYTOMA

Meningeal melanocytic neoplasms are immunoreactive for melanocytic markers such as Melan-A (left) and microphthalmia transcription factor (right).

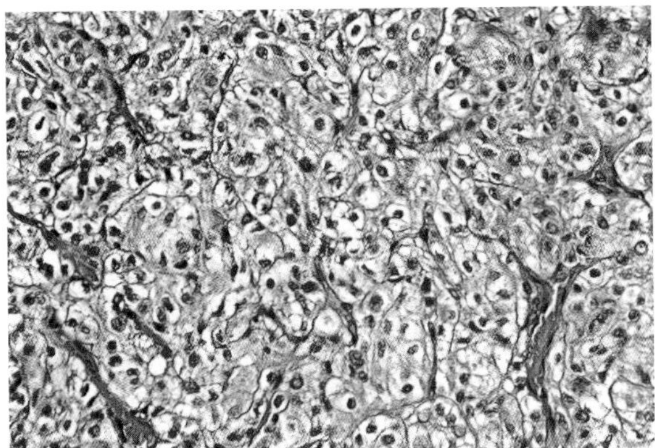

Figure 14-12

MENINGIOMA RESEMBLING MELANOCYTOMA

Meningioma is the principal differential entity for amelanotic melanocytoma. The nested architecture of this meningioma is similar to that of many melanocytomas.

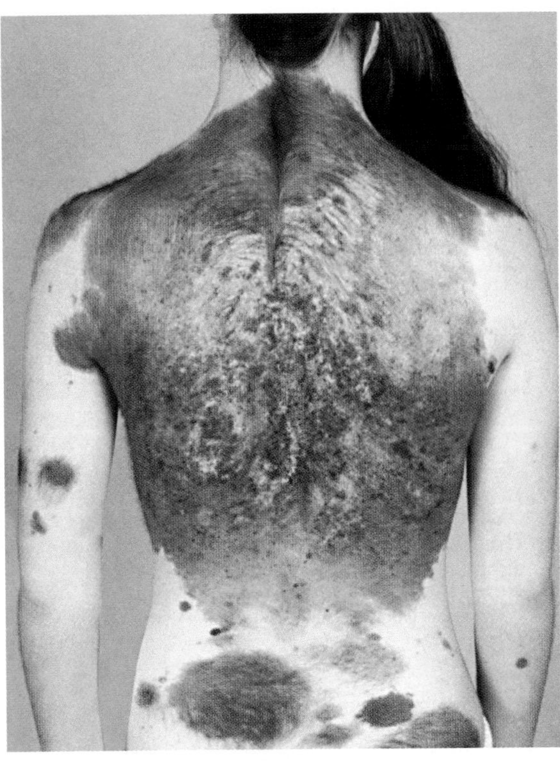

Figure 14-13

NEUROCUTANEOUS MELANOSIS

The large "bathing trunk" nevus in a 13-year-old is one expression of this syndrome. The ultimately lethal meningeal melanoma is shown in figure 14-14. (Courtesy of Dr. E. Hsu, Ottawa, Canada.)

spindled appearance and exhibit a greater ratio of cytoplasm and extracellular material to nuclear volume than is apparent in melanocytic tumors. Extensive pericellular basal laminae, characteristic ultrastructural features of schwannomas, are light microscopically highlighted by histochemical staining for reticulin or immunostains for type IV collagen or laminin. Since pigmented schwannomas exhibit HMB45 staining, it is doubtful that the demonstration of other melanoma-associated antigens will be useful in distinguishing melanocytic neoplasms from morphologically similar pigmented Schwann cell tumors.

It may be difficult, if not impossible, to rule out the possibility that the CNS lesion is a melanoma metastasis from an occult extracranial primary on the basis of histologic features alone.

Treatment and Prognosis. The biologic behavior of these lesions is not clear, although experience with discrete melanocytic tumors suggests that the prognosis is, at least in part, related to the histologic findings. Overtly malignant lesions behave as expected, whereas totally excised melanocytomas can be cured in many instances (6). Subtotally resected melanocytomas remain stable or enlarge, and a number of cases designated melanocytoma, albeit in some cases with mitoses, have recurred (5,8,14,25). Rare melanocytomas can transform to melanoma (1,36). The outlook may be less favorable, however, for infiltrative lesions, even if designated melanocytoma by cytologic criteria (6). Diffuse melanocytic lesions are generally aggressive, regardless of cytologic features.

NEUROCUTANEOUS MELANOSIS

Neurocutaneous melanosis is defined as the association of a diffuse leptomeningeal melanocytic proliferation with pigmented cutaneous lesions, generally large patches of hyperpigmentation or congenital hairy nevi that usually involve the head, neck, back, or buttocks (fig. 14-13) (37–44). An association with the ocular nevus of Ota has also been described (45). Simultaneous, but independent, melanocytic proliferations involving the leptomeninges and skin may be congenital and inherited on an autosomal dominant basis (Touraine's syndrome).

Neurocutaneous melanosis presents clinically in the first 2 years of life. Manifestations of the

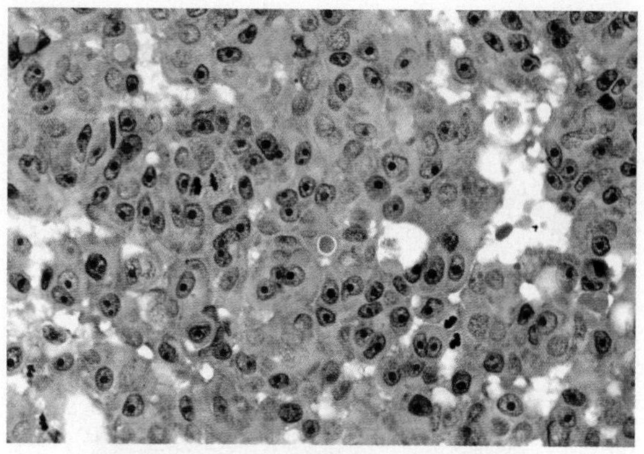

Figure 14-14

NEUROCUTANEOUS MELANOSIS

Large nucleoli, coarse chromatin, and mitoses all attest to the malignancy of this widely disseminated meningeal lesion in the patient illustrated in figure 14-13.

CNS component include hydrocephalus due to obstruction of cerebrospinal fluid flow, seizures, intracranial hemorrhage, and symptomatic infiltration of cranial or spinal nerves. The leptomeninges may be diffusely contrast enhancing.

Histologically, the lesions vary considerably from very well-differentiated melanocytic neoplasms to unequivocal malignant melanomas (fig. 14-14). In many instances, they are of intermediate grade.

The prognosis of patients with neurocutaneous melanosis is poor, even in the absence of melanomatous transformation. The solitary, discrete, cranial or spinal melanocytoma has not been specifically identified as part of the syndrome.

REFERENCES

Melanocytoma, Melanocytic Neoplasm of Intermediate Differentiation, and Malignant Melanoma

1. Adamek D, Kaluza J, Stachura K. Primary balloon cell malignant melanoma of the right temporo-parietal region arising from meningeal naevus. Clin Neuropathol 1995;14:29–32.
2. Aichner F, Schuler G. Primary leptomeningeal melanoma. Diagnosis by ultrastructural cytology of cerebrospinal fluid and cranial computed tomography. Cancer 1982;50:1751–6.
3. Allcutt D, Michowiz S, Weitzman S, et al. Primary leptomeningeal melanoma: an unusually aggressive tumor in childhood. Neurosurgery 1993;32:721–9.
4. Balmaceda CM, Fetell MR, O'Brien JL, Housepian EH. Nevus of Ota and leptomeningeal melanocytic lesions. Neurology 1993;43:381–6.
5. Botticelli AR, Villani M, Angiari P, Peserico L. Meningeal melanocytoma of Meckel's cave associated with ipsilateral Ota's nevus. Cancer 1983;51:2304–10.
6. Brat DJ, Giannini C, Scheithauer BW, Burger PC. Primary melanocytic neoplasms of the central nervous system. Am J Surg Pathol 1999;23:745–54.
7. Chow M, Clarke DB, Maloney WJ, Sangalang V. Meningeal melanocytoma of the planum sphenoidale. Case report and review of the literature. J Neurosurg 2001;94:841–5.
8. Clarke DB, Leblanc R, Bertrand G, Quartey GR, Snipes GJ. Meningeal melanocytoma. Report of a case and a historical comparison. J Neurosurg 1998;88:116–21.
9. Czarnecki EJ, Silbergleit R, Gutierrez JA. MR of spinal meningeal melanocytoma. AJNR Am J Neuroradiol 1997;18:180–2.
10. Faro SH, Koenigsberg RA, Turtz AR, Croul SE. Melanocytoma of the cavernous sinus: CT and MR findings. AJNR Am J Neuroradiol 1996;17:1087–90.
11. Gelman BB, Trier TT, Chaljub G, Borokowski J, Nauta HJ. Oncocytoma in melanocytoma of the spinal cord: case report. Neurosurgery 2000;47:756–9.
12. Haddad FS, Jamali AF, Rebeiz JJ, Fahl M, Haddad GF. Primary malignant melanoma of the gasserian ganglion associated with neurofibromatosis. Surg Neurol 1991;35:310–6.
13. Hamasaki O, Nakahara T, Sakamoto S, Kutsuna M, Sakoda K. Intracranial meningeal melanocytoma. Neurol Med Chir (Tokyo) 2002;42:504–9.
14. Jellinger K, Bock F, Brenner H. Meningeal melanocytoma. Report of a case and review of the literature. Acta Neurochir (Wien) 1988;94:78–87.
15. Kurita H, Segawa H, Shin M, et al. Radiosurgery of meningeal melanocytoma. J Neurooncol 2000;46:57–61.

16. Lach B, Russell N, Benoit B, Atack D. Cellular blue nevus ("melanocytoma") of the spinal meninges: electron microscopic and immunohistochemical features. Neurosurgery 1988;22:773–80.
17. Limas C, Tio FO. Meningeal melanocytoma ("melanotic meningioma"). Its melanocytic origin as revealed by electron microscopy. Cancer 1972;30:1286–94.
18. Litofsky NS, Zee CS, Breeze RE, Chandrasoma PT. Meningeal melanocytoma: diagnostic criteria for a rare lesion. Neurosurgery 1992;31:945–8.
19. Lopez-Castilla J, Diaz-Fernandez F, Soult JA, Munoz M, Barriga R. Primary leptomeningeal melanoma in a child. Pediatr Neurol 2001;24:390–2.
20. Maiuri F, Iaconetta G, Benvenuti D, Lamaida E, De Caro ML. Intracranial meningeal melanocytoma: case report. Surg Neurol 1995;44:556–61.
21. O'Brien TF, Moran M, Miller JH, Hensley SD. Meningeal melanocytoma. An uncommon diagnostic pitfall in surgical neuropathology. Arch Pathol Lab Med 1995;119:542–6.
22. Rahimi-Movaghar V, Karimi M. Meningeal melanocytoma of the brain and oculodermal melanocytosis (nevus of Ota): case report and literature review. Surg Neurol 2003;59:200–10.
23. Suzuki T, Yasumoto Y, Kumami K, et al. Primary pineal melanocytic tumor. Case report. J Neurosurg 2001;94:523–7.
24. Uematsu Y, Yukawa S, Yokote H, Itakura T, Hayashi S, Komai N. Meningeal melanocytoma: magnetic resonance imaging characteristics and pathological features. Case report. J Neurosurg 1992;76:705–9.
25. Winston KR, Sotrel A, Schnitt SJ. Meningeal melanocytoma. Case report and review of the clinical and histological features. J Neurosurg 1987;66:50–7.
26. Crisp DE, Thompson JA. Primary malignant melanomatosis of the meninges. Clinical course and computed tomographic findings in a young child. Arch Neurol 1981;38:528–9.
27. Gaetani P, Martelli A, Sessa F, Zappoli F, Rodriguez R, Baena. Diffuse leptomeningeal melanomatosis of the spinal cord: a case report. Acta Neurochir (Wien) 1993;121:206–11.
28. Nakamura Y, Becker LE. Meningeal tumors of infancy and childhood. Pediatr Pathol 1985;3:341–58.
29. Painter TJ, Chaljub G, Sethi R, Singh H, Gelman B. Intracranial and intraspinal meningeal melanocytosis. AJNR Am J Neuroradiol 2000;21:1349–53.
30. Pirini MG, Mascalchi M, Salvi F, et al. Primary diffuse meningeal melanomatosis: radiologic-pathologic correlation. AJNR Am J Neuroradiol 2003;24:115–8.
31. Arbelaez A, Castillo M, Armao DM. Imaging features of intraventricular melanoma. AJNR Am J Neuroradiol 1999;20:691–3.
32. Czirjak S, Vitanovic D, Slowik F, Magyar A. Primary meningeal melanocytoma of the pineal region. Case report. J Neurosurg 2000;92:461–5.
33. Rubino GJ, King WA, Quinn B, Marroquin CE, Verity MA. Primary pineal melanoma: case report. Neurosurgery 1993;33:511–5.
34. Yamane K, Shima T, Okada Y, et al. Primary pineal melanoma with long-term survival: case report. Surg Neurol 1994;42:433–7.
35. Ahluwalia S, Ashkan K, Casey AT. Meningeal melanocytoma: clinical features and review of the literature. Br J Neurosurg 2003;17:347–51.
36. Ochiai H, Nakano S, Miyahara S, Goya T, Wakisaka S, Kinoshita K. Magnetic resonance imaging of a malignant transformation of an intracranial cellular blue nevus. A case report. Surg Neurol 1992;37:371–3.

Neurocutaneous Melanosis

37. Akinwunmi J, Sgouros S, Moss C, Grundy R, Green S. Neurocutaneous melanosis with leptomeningeal melanoma. Pediatr Neurosurg 2001; 35:277–9.
38. Chu WC, Lee V, Chan YL, et al. Neurocutaneous melanomatosis with a rapidly deteriorating course. AJNR Am J Neuroradiol 2003;24:287–90.
39. Di Rocco F, Sabatino G, Koutzoglou M, Battaglia D, Caldarelli M, Tamburrini G. Neurocutaneous melanosis. Childs Nerv Syst 2004;20:23–8.
40. Faillace WJ, Okawara SH, McDonald JV. Neurocutaneous melanosis with extensive intracerebral and spinal cord involvement. Report of two cases. J Neurosurg 1984;61:782–5.
41. Kaplan AM, Itabashi HH, Hanelin LG, Lu AT. Neurocutaneous melanosis with malignant leptomeningeal melanoma. A case with metastases outside the nervous system. Arch Neurol 1975; 32:669–71.
42. Makin GW, Eden OB, Lashford LS, et al. Leptomeningeal melanoma in childhood. Cancer 1999;86:878–86.
43. Reed WB, Becker SW Sr, Becker SW Jr, Nickel WR. Giant pigmented nevi, melanoma, and leptomeningeal melanocytosis: a clinical and histopathological study. Arch Dermatol 1965;91:100–19.
44. Slaughter JC, Hardman JM, Kempe LG, Earle KM. Neurocutaneous melanosis and leptomeningeal melanomatosis in children. Arch Pathol 1969;88:298–304.
45. Sagar HJ, Ilgren EB, Adams CB. Nevus of Ota associated with meningeal melanosis and intracranial melanoma. Case report. J Neurosurg 1983;58:280–3.

15 CRANIOPHARYNGIOMAS

ADAMANTINOMATOUS CRANIOPHARYNGIOMA

General Features. *Adamantinomatous craniopharyngioma* is a neoplasm that closely resembles adamantinoma of the jaw, but is even more similar to a close relative of the latter, keratinizing and calcifying odontogenic cyst (1). Misplaced odontogenic epithelium or a related developmental aberration is therefore an attractive possibility for tumor source.

Clinical Features. This numerically predominant form of craniopharyngioma usually presents as a suprasellar mass with endocrine, visual, and psychological abnormalities. Most patients are in the first two decades of life, but occasional lesions appear later, even in the elderly. Unusual locations include the sphenoid sinus (2,3), cerebellopontine angle (4), pineal region (5), and posterior fossa (6,7). Occasional tumors produce a chemical meningitis by spilling their irritating, cholesterol-rich contents into the cerebrospinal fluid (8,9).

Radiologic Findings. The tumor is a lobulated, solid and cystic, contrast-enhancing, intrasellar or suprasellar mass. The considerable protein fluid present generates a bright signal in precontrast T1-weighted magnetic resonance images (MRIs) (fig. 15-1) (10,10a). Computerized tomography (CT) scans highlight the frequent presence of peripheral tumoral calcifications. Longstanding craniopharyngiomas can be massive.

Gross Findings. At presentation, the lesions are typically adherent to structures at the base of the brain and frequently indent the floor of the third ventricle. Recurrent tumors are often widely and densely attached to regional structures, particularly the vasculature and the pituitary stalk (fig. 15-2).

Adamantinomatous craniopharyngiomas are almost always cystic and are filled with a turbid, dark brown, "machinery oil" fluid (fig. 15-3) on which float minute, glistening crystals of cholesterol. The characteristic multilaminar construction and notched, "state of Utah," outline of the crystals are evident under polarized light (fig. 15-4).

The presence of gray-yellow flecks of "wet keratin," described below, is a diagnostic finding seen readily through the operating microscope. Recurrent lesions often consist in part of fibrous tissue and grumous debris.

Microscopic Findings. The prominent epithelial lobules, when grouped in a multinodular clover leaf configuration, are both distinctive and diagnostic (fig. 15-5) (11). Cells at the periphery are palisaded, whereas those that are internal are loose textured and readily dehisce to form the so-called stellate reticulum. Progression of the lesion, in concert with degenerative changes about the blood vessels, produces cystic spaces filled with fluid or amorphous debris. The neoplastic cells surrounding the cysts often assume the same palisaded alignment as those on the periphery.

A diagnostic finding is wet keratin, which consists of plump, eosinophilic, keratinized cells whose necrobiotic nature is apparent from their

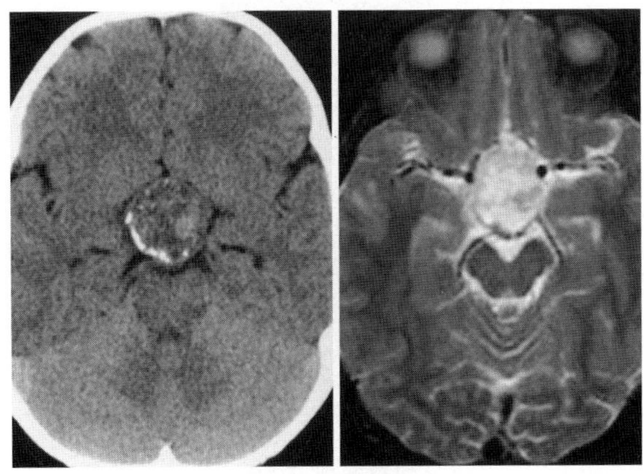

Figure 15-1

ADAMANTINOMATOUS CRANIOPHARYNGIOMA

As well seen by computerized tomography (CT), the lesion often has a peripheral rim of calcification (left). By magnetic resonance imaging (MRI), the suprasellar mass is bright in T2-weighted images (right).

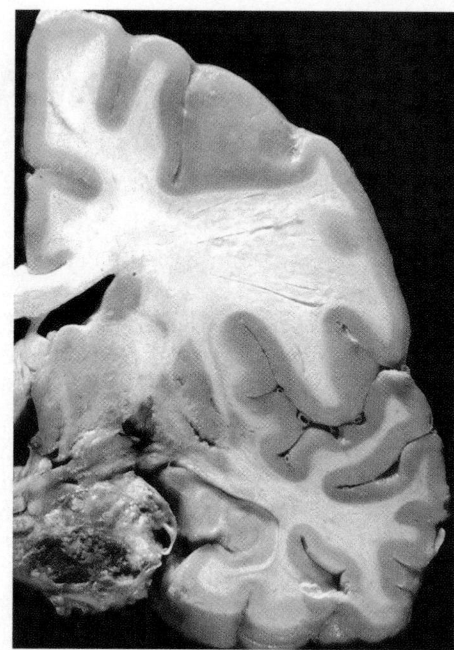

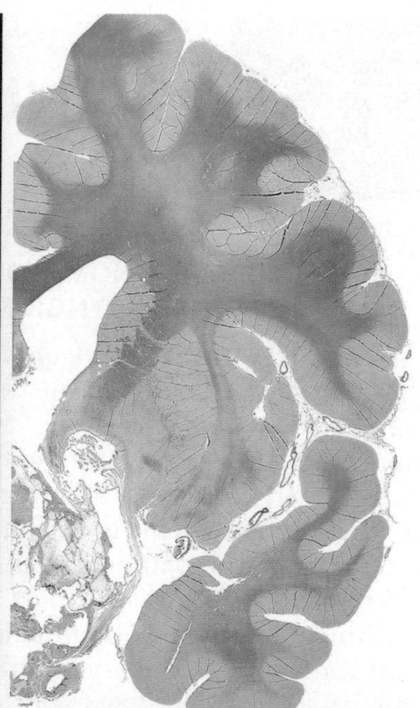

Figure 15-2

ADAMANTINOMATOUS CRANIOPHARYNGIOMA

Left: Grummous debris, cysts, fibrosis, and xanthogranulomatous change contribute substantially to the mass of this recurrent adamantinomatous craniopharyngioma.

Right: A suprasellar location and multicystic architecture are typical features of adamantinomatous craniopharyngioma. Anatomic obstacles to total resection of recurrent craniopharyngiomas are obvious.

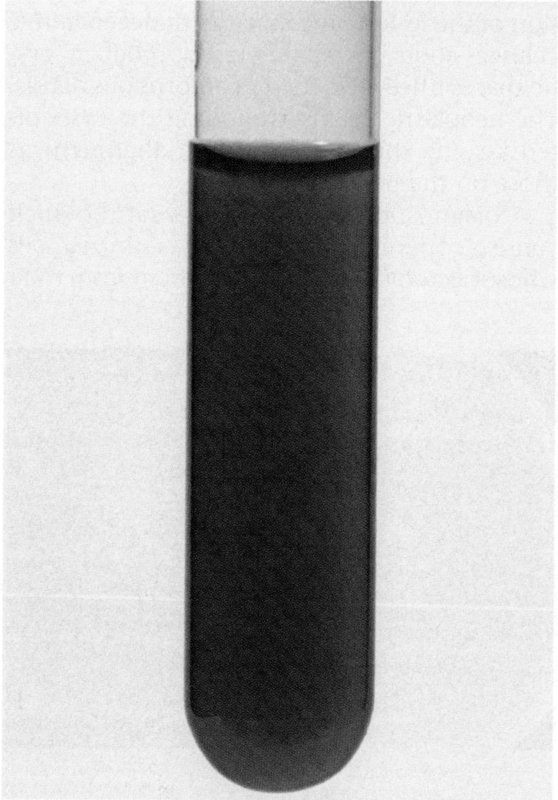

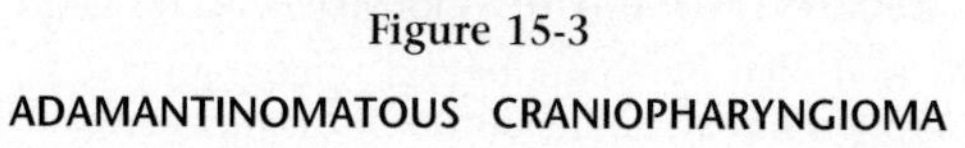

Figure 15-3

ADAMANTINOMATOUS CRANIOPHARYNGIOMA

The lesion typically contains yellow-brown cyst fluid likened to "machinery oil."

Figure 15-4

ADAMANTINOMATOUS CRANIOPHARYNGIOMA

As seen with polarized light, cholesterol crystals in craniopharyngioma cyst fluid are orthogonal, notched, and incrementally lighter with increasing lamination.

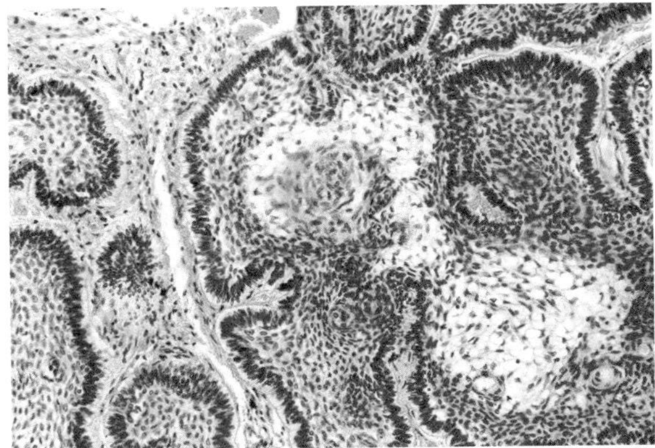

Figure 15-5

ADAMANTINOMATOUS CRANIOPHARYNGIOMA

Spaces within craniopharyngiomas are lined by palisaded cells. Degeneration in the intervening tissue creates the loose component known as stellate reticulum.

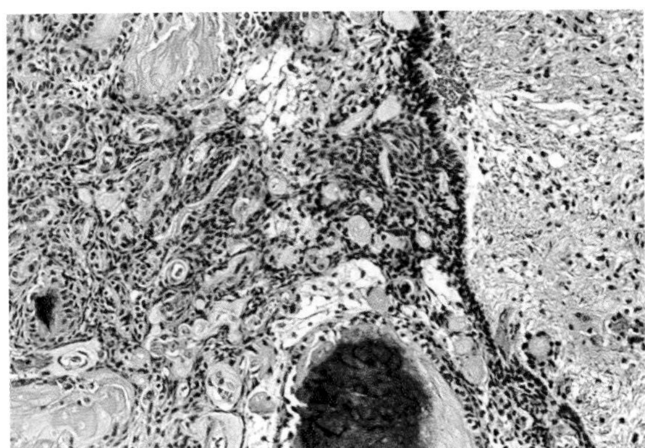

Figure 15-6

ADAMANTINOMATOUS CRANIOPHARYNGIOMA

Some craniopharyngiomas are compact, but retain the characteristic peripheral palisading. Note the necrobiotic nodules, two calcified, of "wet keratin."

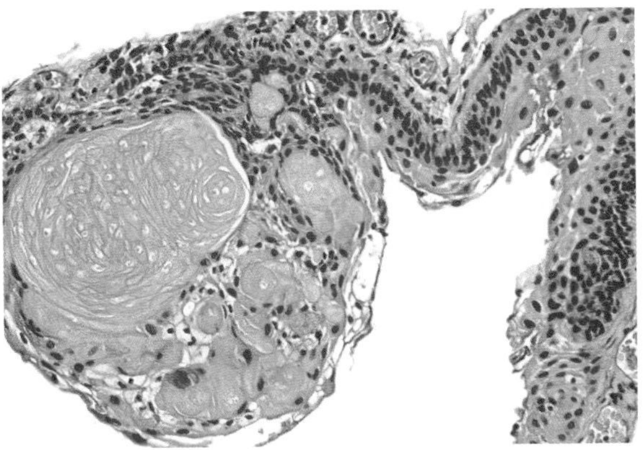

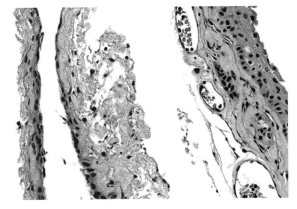

Figure 15-7

ADAMANTINOMATOUS CRANIOPHARYNGIOMA

With progressive cystic change, the epithelium becomes attenuated and loses its specific features. It is thin in the left figure but retains peripheral palisading and, more importantly, a nodule of wet keratin. The epithelium elsewhere in the same lesion varies from simple to stratified squamous (right).

ghost nuclei (figs. 15-6, 15-7, left). Nodules of such "shadow" cells frequently undergo dystrophic calcification.

In macrocystic regions, the epithelium often becomes flattened and may, due to cell stratification, approximate the histologic appearance of epidermoid cyst or papillary craniopharyngioma (fig. 15-7, see fig. 16-25). Such areas generally retain some peripheral palisading, although even this helpful diagnostic feature can be lost in attenuated regions.

In addition to the epithelial components, there is often extensive fibrosis, chronic inflammation, cholesterol clefts, and cystic change. These secondary changes may predominate, particularly in recurrent tumors (fig. 15-8). Often, there is extensive calcification. Thus, in recurrent or irradiated tumors, much of the mass may appear degenerative, with only small remnants of diagnostic epithelium. The finding of wet keratin is diagnostic in such instances, even in the absence of viable epithelium.

Craniopharyngiomas are locally invasive, with tongues of tumor projecting into adjacent brain. Resenting this intrusion, the brain tissue responds with a dense piloid gliosis (fig. 15-9).

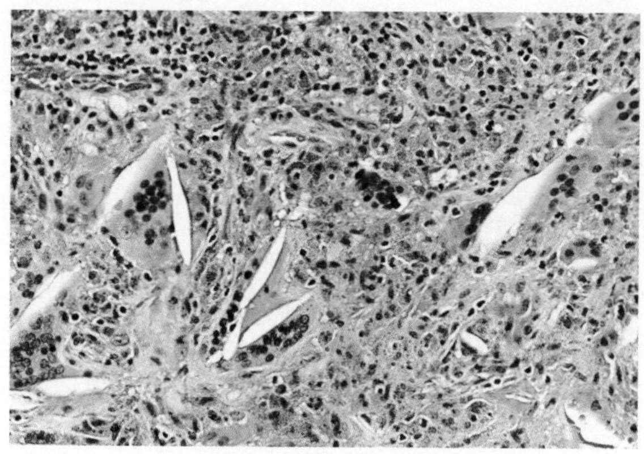

Figure 15-8

ADAMANTINOMATOUS CRANIOPHARYNGIOMA

A prominent xanthogranulomatous reaction is common in recurrent lesions.

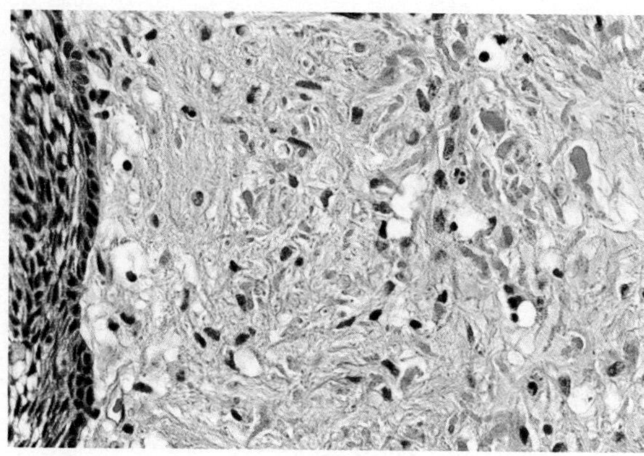

Figure 15-9

ADAMANTINOMATOUS CRANIOPHARYNGIOMA

Craniopharyngiomas elicit a Rosenthal fiber–rich gliosis that should not be interpreted as pilocytic astrocytoma. Unlike the glioma, there are no microcysts.

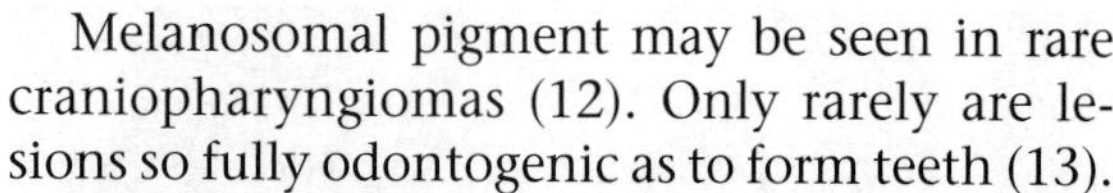

Melanosomal pigment may be seen in rare craniopharyngiomas (12). Only rarely are lesions so fully odontogenic as to form teeth (13).

Immunohistochemical Findings. The neoplasms are immunoreactive for multiple cytokeratins, and consistently for CK7 (14–16). One study recorded negative reactions for CK8 and CK20 (16). Shadow cells appear to be immunoreactive for antibodies to human hair keratin (14).

Ultrastructural Findings. Joined by well-formed desmosomes, the epithelial cells contain bundles of tonofilaments. Basal lamina covers the epithelial cells that abut the stroma, but is absent in those situated internally or lining microcysts. The latter cells have microvilli (8,17,18).

Cytologic Findings. Sheets of cohesive epithelial cells (fig. 15-10) and nodules of wet keratin (fig. 15-11) are characteristic findings (19).

Molecular Findings. There are few, if any, reported chromosomal imbalances as studied by comparative genomic hybridization (20). Mutations in the tumor suppressor gene *β-catenin* have been reported (21)

Differential Diagnosis. Findings that help exclude papillary craniopharyngioma from the differential diagnosis of adamantinomatous craniopharyngioma are well-defined palisading of peripheral epithelial cells, nodules of wet keratin, calcification, and, to a lesser extent, widespread fibrosis. We are not aware of a xanthogranulomatous reaction associated with papillary craniopharyngioma. Nuclear β-catenin staining in the adamantinomatous lesion also helps distinguish it from papillary craniopharyngioma (10a).

Large cystic craniopharyngiomas with attenuated epithelium should not be confused with epidermoid cysts. In most cases, this is not an issue given the architectural complexity of craniopharyngioma. The latter also lack keratohyaline granules and uniform surface maturation to thin anucleate squames. The epidermoid cyst, on the other hand, lacks palisaded cells, wet keratin, calcification, and fibrodegenerative changes.

Specimens obtained from the immediate periphery of a craniopharyngioma can consist in large part of Rosenthal fiber–rich tissue that simulates pilocytic astrocytoma (fig. 15-9). This paucicellular, compact, densely piloid tissue, however, lacks the cellularity and microcysts seen in most pilocytic neoplasms. In any case, the specimen usually includes at least some distinctive palisaded epithelium and/or comes with a prior history of craniopharyngioma.

Intraventricular and cystic lesions are given in Appendices C and D, respectively.

Treatment and Prognosis. Ideally, the treatment of adamantinomatous craniopharyngioma is surgery. Radiotherapy is reserved for incompletely excised or recurrent lesions. The degree to which these tumors are amenable to total excision and the vigor with which this should be pursued remain controversial (22,23). Most recurrences are seen within 5 years of initial surgery.

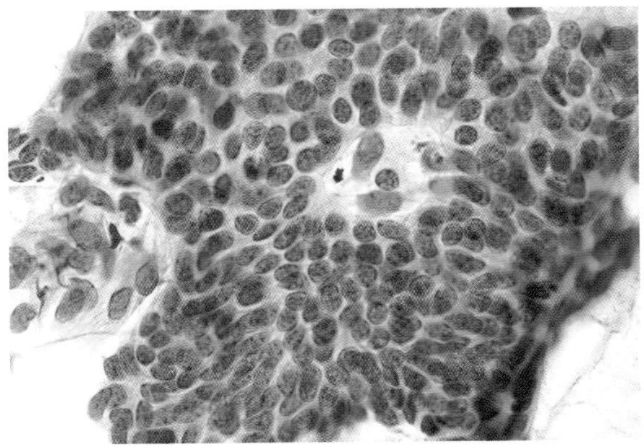

Figure 15-10

ADAMANTINOMATOUS CRANIOPHARYNGIOMA: SMEAR PREPARATION

Sheets of cells establish the epithelial nature of the lesion.

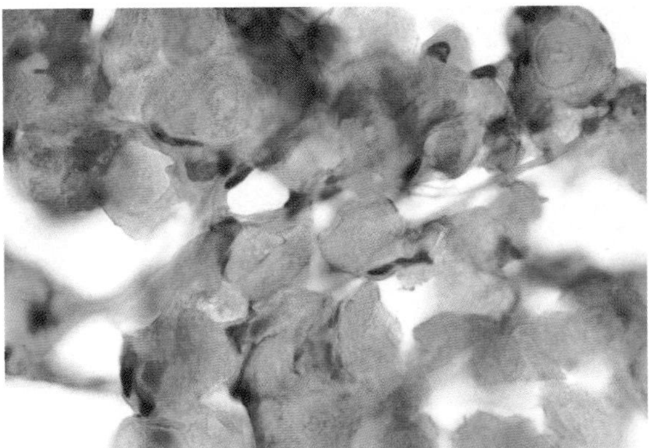

Figure 15-11

ADAMANTINOMATOUS CRANIOPHARYNGIOMA: SMEAR PREPARATION

Plump anucleate squamous epithelial cells (wet keratin) are characteristic of adamantinomatous lesions.

Possibly as the result of radiation, malignant transformation to squamous cell carcinoma has been reported (24,25). Ectopic recurrence, either via cerebrospinal fluid dissemination or by implantation along the operative approach, is rare (26,27,27a).

PAPILLARY CRANIOPHARYNGIOMA

Definition. *Papillary craniopharyngioma* is a very well-differentiated pseudopapillary epithelial neoplasm of the sellar and suprasellar regions.

General Features. While the uncommon papillary craniopharyngioma is classified as a variant of craniopharyngioma, it may well be a distinct entity, as suggested by the longer but more descriptive designation *suprasellar papillary squamous epithelioma* (28). The origin of the lesion is unknown although the presence of occasional ciliated and/or goblet cells in some cases, and a possible transition from Rathke cleft cyst with extensive squamous metaplasia in others, suggest a derivation from the cyst (29–31). The terms *ciliated craniopharyngioma* and *ciliated and goblet cell craniopharyngioma* have been applied to such lesions (29–31). Similarities in cytokeratin expression have been noted between metaplastic Rathke cleft cyst and papillary craniopharyngioma (32).

Clinical Features. In contrast to the classic adamantinomatous lesion, papillary craniopharyngioma is usually encountered in adults, who present with effects upon the visual system, cerebrospinal fluid flow, mental capacity, and personality (28,33,34). (See Appendix D for a list of frequently cystic tumors.) The diencephalic syndrome, with its pronounced weight loss, can also evolve (35).

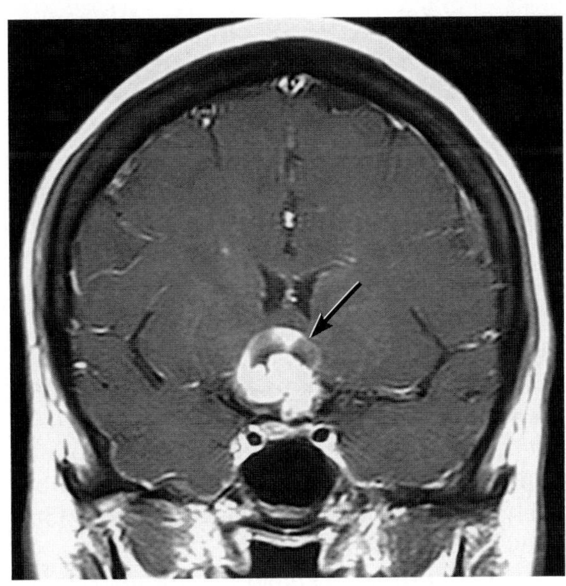

Figure 15-12

PAPILLARY CRANIOPHARYNGIOMA

Papillary craniopharyngiomas (arrow) can be either cystic or solid, but are more likely the latter. This cystic lesion produced headaches, memory loss, and confusion in a 51-year-old woman. Note the papillary profile of the solid component.

Radiologic Findings. The lesions are usually suprasellar rather than intrasellar, and many tumors are entirely within the third ventricle

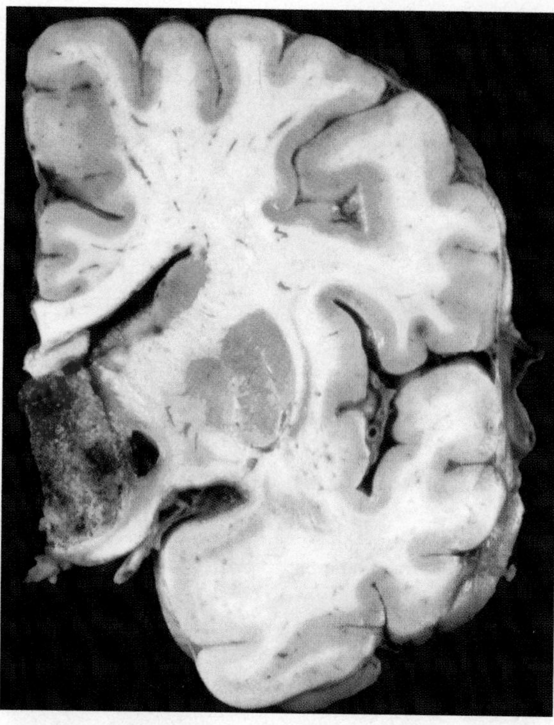

Figure 15-13

PAPILLARY CRANIOPHARYNGIOMA

As is typical, this papillary lesion lies within the third ventricle.

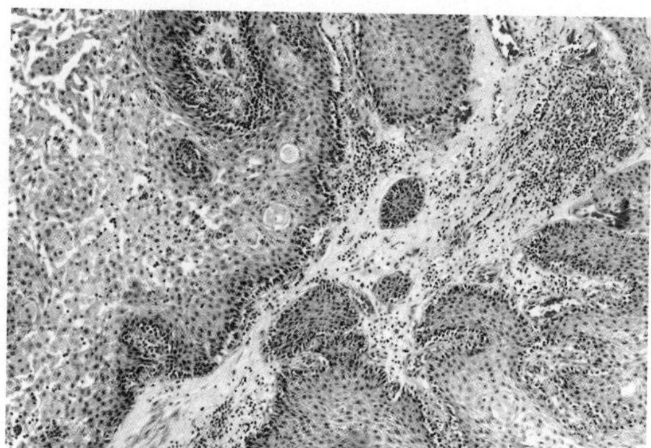

Figure 15-14

PAPILLARY CRANIOPHARYNGIOMA

Papillary craniopharyngioma consists of sheets of well-differentiated squamous epithelium associated with fibrous stroma.

(fig. 15-12). Thus, the profile of the sella may be unaffected. The lesions are variably solid and cystic, and the cyst may be large relative to the size of the sometimes small papillary nodule. Usually, however, papillary tumors are more likely to be solid than adamantinomatous (36). They are not calcified (36a).

Gross Findings. Papillary craniopharyngiomas are encapsulated solid masses without the cholesterol-rich, machinery oil-like content of adamantinomatous craniopharyngiomas (fig. 15-13). In some instances, their smooth surface can be separated readily from the surrounding brain.

Microscopic Findings. Papillary craniopharyngiomas are composed of solid sheets of remarkably well-differentiated squamous epithelial cells interrupted by prominent cores of fibrovascular stroma (fig. 15-14). The cellular sheets typically dehisce, either naturally or during specimen processing, to create prominent pseudopapillae. Small epithelial whorls may be present (fig. 15-15). The single basal layer of somewhat palisaded cells is not as prominent in the adamantinomatous lesion. Stellate reticulum is absent. Papillary craniopharyngiomas may focally have goblet or ciliated cells, and it can be difficult in some cases to distinguish such tumors from Rathke cleft cyst with abundant squamous metaplasia (29–31,34). In our experience, calcification is absent in papillary craniopharyngioma.

Immunohistochemical Findings. The tumor cells are immunoreactive for cytokeratins and epithelial membrane antigen, with consistent staining for CK7 in the superficial, nonbasal layer (37), but not in the basal layer (32,37,38). One study found that, unlike Rathke cleft cysts, the papillary lesions were negative for CK8 and CK20 (38)

MIB-1 labeling indices vary considerably in the papillary lesion, in one small study ranging from 0.4 to an exceptional 32.5 percent (39). Higher rates have been associated with a greater likelihood of recurrence.

Ultrastructural Findings. The cells maintain a high degree of squamous differentiation, since they are interconnected by desmosomes and trussed internally with bundles of tonofilaments (28). Surface microvillous-like projections are common.

Cytologic Findings. The monomorphous, cohesive, overtly epithelial cells are readily apparent in smear preparations (fig. 15-16).

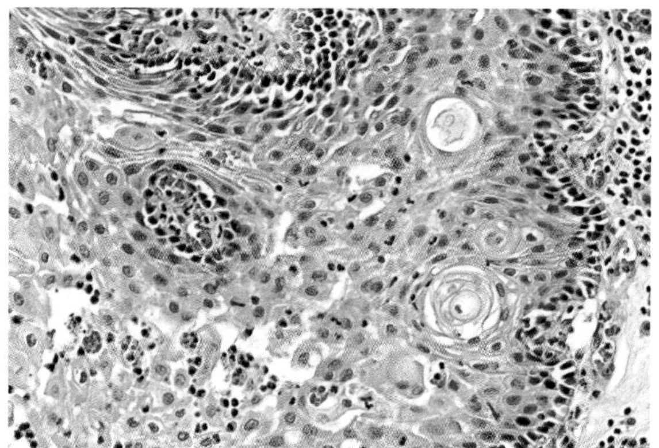

Figure 15-15

PAPILLARY CRANIOPHARYNGIOMA

Clefts within the solid lesion, here surrounded by apoptotic cells, create the pseudopapillary architecture. Small whorls are often present.

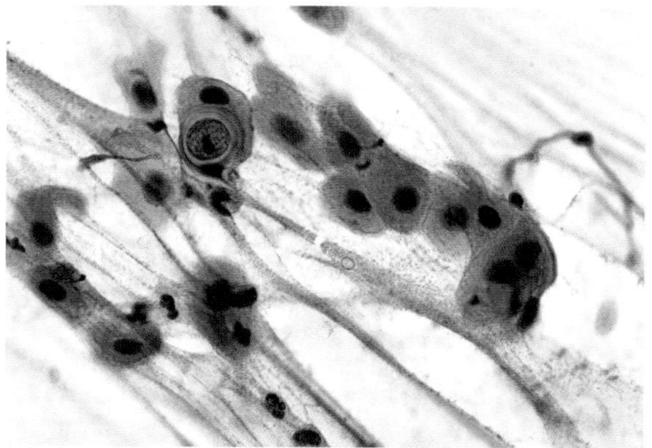

Figure 15-16

PAPILLARY CRANIOPHARYNGIOMA: SMEAR PREPARATION

The squamous quality of the lesion is apparent in cytologic preparations where the plate-like cytoplasm and discrete cell borders are clearly seen. Note the small whorl at the top left.

Differential Diagnosis. The principal differential diagnostic consideration is adamantinomatous craniopharyngioma. The distinction is elementary in a large specimen from a papillary lesion with the classic squamous features, even if vague peripheral palisading is present. The features so characteristic of the adamantinomatous lesion are absent: no well-formed keratin pearls, stellate reticulum, nodules of wet keratin, xanthogranulomatous reaction, cholesterol accumulation, calcification, or fibrosis is seen. On the other hand, small fragments from the cyst wall may be difficult to assess.

The high degree of squamous differentiation present in papillary tumors also prompts consideration of epidermoid and dermoid cysts. Both, however, consist of highly organized squamous cells featuring keratohyaline granules and generating anucleate squames. The papillary lesion shows little squamous maturation, no anucleate squames, and an obvious lack of cutaneous adnexa.

A more difficult issue, both practically and conceptually, is the differentiation from Rathke cleft cyst with extensive squamous metaplasia (see fig. 16-11). In such lesions, there is ciliated epithelium and squamous metaplasia, but also tissue identical to papillary craniopharyngioma (30). The nosology and biologic behavior of such lesions is unclear (29–31).

Intraventricular and cystic lesions are given in Appendices C and D, respectively.

Treatment and Prognosis. As in the case for adamantinomatous craniopharyngioma, surgical excision is the treatment of choice, and, in one series, was more readily and safely accomplished than for the adamantinomatous tumors (33). There were no recurrences in 12 patients with papillary lesions followed for a mean of 7.5 years. Yet another report noted clinical and radiologic differences between papillary and adamantinomatous types, but found no significant differences in resectability or overall survival (34). Meningeal seeding is exceptional (40).

REFERENCES

Adamantinomatous Craniopharyngioma

1. Li TJ, Yu SF. Clinicopathologic spectrum of the so-called calcifying odontogenic cysts: a study of 21 intraosseous cases with reconsideration of the terminology and classification. Am J Surg Pathol 2003;27:372–84.
2. Chen CJ. Suprasellar and infrasellar craniopharyngioma with a persistent craniopharyngeal canal: case report and review of the literature. Neuroradiology 2001;43:760–2.
3. Cooper P, Ransohoff J. Craniopharyngioma originating in the sphenoid bone. J Neurosurg 1972;36:102–6.
4. Link MJ, Driscoll CL, Giannini C. Isolated, giant cerebellopontine angle craniopharyngioma in a patient with Gardner syndrome: case report. Neurosurgery 2002;51:221–6.
5. Solarski A, Panke ES, Panke TW. Craniopharyngioma in the pineal gland. Arch Pathol Lab Med 1978;102:490–1.
6. Bashir E, Lewis P, Edwards R. Posterior fossa craniopharyngioma. Brit J Neurosurg 1996;10: 613–5.
7. Gokalp HZ, Egemen N, Ildan F, Bacaci K. Craniopharyngioma of the posterior fossa. Neurosurgery 1991;29:446–8.
8. Kulkarni V, Daniel RT, Pranatartiharan R. Spontaneous intraventricular rupture of craniopharyngioma cyst. Surg Neurol 2000;54:249–53.
9. Satoh H, Uozumi T, Arita K, et al. Spontaneous rupture of craniopharyngioma cysts. A report of five cases and review of the literature. Surg Neurol 1993;40:414–9.
10. Sartoretti-Schefer S, Wichman W, Aguzzi A, Valavanis A. MR differentiation of adamantinomatous and squamous-papillary craniopharyngiomas. AJNR Am J Neuroradiol 1997;18:77–87.

10a. Hofmann BM, Kreutzer J, Saeger W, et al. Nuclear beta-catenin accumulation as reliable marker for the differentiation between cystic craniopharyngiomas and rathke cleft cells: a clinico-pathologic approach. Am J Surg Pathol 2006:30:1595–603.

11. Petito CK, DeGirolami U, Earle KM. Craniopharyngiomas: a clinical and pathological review. Cancer 1976;37:1944–52.
12. Harris BT, Horoupian DS, Tse V, Herrick MK. Melanotic craniopharyngioma: a report of two cases. Acta Neuropathol (Berl) 1999;98:433–6.
13. Alvarez-Garijo J, Froufe A, Taboada D, Vila M. Successful surgical treatment of an odontogenic ossified craniopharyngioma. Case report. J Neurosurg 1981;55:832–5.
14. Tateyama H, Tada T, Okabe M, Takahashi E, Eimoto T. Different keratin profiles in craniopharyngioma subtypes and ameloblastomas. Pathol Res Pract 2001;197:735–42.
15. Kurosaki M, Saeger W, Ludecke DK. Immunohistochemical localisation of cytokeratins in craniopharyngioma. Acta Neurochir (Wien) 2001;143:147–51.
16. Xin W, Rubin MA, McKeever PE. Differential expression of cytokeratins 8 and 20 distinguishes craniopharyngioma from rathke cleft cyst. Arch Pathol Lab Med 2002;126:1174–8.
17. Ghatak NR, Hirano A, Zimmerman HM. Ultrastructure of a craniopharyngioma. Cancer 1971;27:1465–75.
18. Vilches J, Lopez A, Martinez MC, Gomez J, Barbera J. Scanning and transmission electron microscopy of a craniopharyngioma: x-ray microanalytical study of the intratumoral mineralized deposits. Ultrastruct Pathol 1981;2:343–56.
19. Smith AR, Elsheikh TM, Silverman JF. Intraoperative cytologic diagnosis of suprasellar and sellar cystic lesions. Diagn Cytopathol 1999;20: 137–47.
20. Rickert CH, Paulus W. Lack of chromosomal imbalances in adamantinomatous and papillary craniopharyngiomas. J Neurol Neurosurg Psychiatry 2003;74:260–1.
21. Sekine S, Shibata T, Kokubu A, et al. Craniopharyngiomas of adamantinomatous type harbor beta-catenin gene mutations. Am J Pathol 2002;161:1997–2001.
22. Van Effenterre R, Boch AL. Craniopharyngioma in adults and children: a study of 122 surgical cases. J Neurosurg 2002;97:3–11.
23. Rutka JT. Craniopharyngioma. J Neurosurg 2002;97:1–2.
24. Kristopaitis T, Thomas C, Petruzzelli GJ, Lee JM. Malignant craniopharyngioma. Arch Pathol Lab Med 2000;124:1356–60.
25. Nelson GA, Bastian FO, Schlitt M, White RL. Malignant transformation in craniopharyngioma. Neurosurgery 1988;22:427–9.
26. Liu JM, Garonzik IM, Eberhart CG, Sampath P, Brem H. Ectopic recurrence of craniopharyngioma after an interhemispheric transcallosal approach: case report. Neurosurgery 2002;50: 639–44; discussion 644–5.
27. Novegno F, Di Rocco F, Colosimo Jr C, Lauriola L, Caldarelli M. Ectopic recurrences of craniopharyngioma. Childs Nerv Syst 2002;18:468–73.

27a. Bianco Ade M, Madeira LV, Rosemberg S, Shibata MK. Cortical seeding of a craniopharyngioma after craniotomy: case report. Surg Neurol 2006;66:437–40.

Papillary Craniopharyngioma

28. Giangaspero F, Burger PC, Osborne DR, Stein RB. Suprasellar papillary squamous epithelioma ("papillary craniopharyngioma"). Am J Surg Pathol 1984;8:57–64.
29. Ikeda H, Yoshimoto T. Clinicopathological study of Rathke's cleft cysts. Clin Neuropathol 2002;21:82–91.
30. Matsushima T, Fukui M, Ohta M, Yamakawa Y, Takaki T, Okano H. Ciliated and goblet cells in craniopharyngioma. Light and electron microscopic studies at surgery and autopsy. Acta Neuropathol (Berl) 1980;50:199–205.
31. Oka H, Kawano N, Yagashita S, Kobayashi I, Saegusa H, Kim K. Cliated craniopharyngioma indicates histogenetic relationship to Rathke cleft cyst epithelium. Clin Neuropathol 1997;16:103–6.
32. Tateyama H, Tada T, Okabe M, Takahashi E, Eimoto T. Different keratin profiles in craniopharyngioma subtypes and ameloblastomas. Pathol Res Pract 2001;197:735–42.
33. Adamson TE, Wiestler OD, Kleihues P, Yasargil MG. Correlation of clinical and pathological features in surgically treated craniopharyngiomas. J Neurosurg1990;73:12–7.
34. Crotty TB, Scheithauer BW, Young WF Jr, et al. Papillary craniopharyngioma: a clinicopathological study of 48 cases. J Neurosurg 1995;83: 206–14.
35. Miyoshi Y, Yunoki M, Yano A, Nishimoto K. Diencephalic syndrome of emaciation in an adult associated with a third ventricle intrinsic craniopharyngioma: case report. Neurosurgery 2003;52:224–7.
36. Sartoretti-Schefer S, Wichman W, Aguzzi A, Valavanis A. MR differentiation of adamantinomatous and squamous-papillary craniopharyngiomas. AJNR Am J Neuroradiol 1997;18:77–87.

36a. Hofmann BM, Kreutzer J, Saeger W, et al. Nuclear beta-catenin accumulation as reliable marker for the differentiation between cystic craniopharyngiomas and rathke cleft cells: a clinico-pathologic approach. Am J Surg Pathol 2006:30:1595–603.

37. Kurosaki M, Saeger W, Ludecke DK. Immunohistochemical localisation of cytokeratins in craniopharyngioma. Acta Neurochir (Wien) 2001;143:147–51.
38. Xin W, Rubin MA, McKeever PE. Differential expression of cytokeratins 8 and 20 distinguishes craniopharyngioma from rathke cleft cyst. Arch Pathol Lab Med 2002;126:1174–8.
39. Nishi T, Kuratsu J, Takeshima H, Saito Y, Kochi M, Ushio Y. Prognostic significance of the MIB-1 labeling index for patient with craniopharyngioma. Int J Mol Med 1999;3:157–61.
40. Elmaci L, Kurtkaya-Yapicier O, Ekinci G, et al. Metastatic papillary craniopharyngioma: case study and study of tumor angiogenesis. Neurooncology 2002;4:123–8.

16 BENIGN CYSTIC LESIONS

CYSTS LINED BY COLUMNAR EPITHELIUM

Colloid Cyst of the Third Ventricle

Definition. *Colloid cyst* is a mucus-filled, epithelial-lined cyst in the anterosuperior third ventricle.

General Features. Although colloid cysts were assumed to be neuroectodermal and derived from paraphysial remnants in the region of the third ventricle, evidence now favors an endodermal or ectodermal origin (1–3). Short of their location, colloid cysts are identical to bronchial epithelium and resemble both Rathke cleft and enterogenous cysts. Unlike both bronchial mucosa and the latter two lesions, colloid cysts are not prone to squamous metaplasia. Proponents of the endodermal origin theory need to explain the lesion's ectopic third ventricle location. An endodermal designation is also inconsistent with the ectodermal (stomodeum) origin of the Rathke pouch.

Clinical Features. Intermittent obstruction of cerebrospinal fluid (CSF) flow, a natural consequence of the cyst's position near the foramen of Monro, produces classic symptoms including headaches (the most common expression), sudden transient paralysis of the lower extremities ("drop attacks"), incontinence, personality changes, and, on occasion, dementia (4–6). Young to middle-aged adults are primarily affected; children rarely so (7,8).

Radiologic Findings. In view of the stereotypic intraventricular location, smooth-surfaced contours, cystic nature, and signal characteristics, colloid cysts are radiologically diagnosed with confidence. See Appendix C for a list of intraventricular tumors, and Appendix D for a summary of intracranial and intraspinal cystic lesions. The colloid cyst's mucoid protein-rich contents often have an intrinsically bright (white) signal on nonenhanced T1-weighted magnetic resonance images (MRIs) (fig. 16-1) (9). The closely related, also midline, Rathke cleft cyst has similar imaging characteristics, but is intrasellar and suprasellar, not intraventricular.

Gross Findings. The thin-walled cyst is impacted in the anterosuperior third ventricle in the region of the fornices and the foramina of Monro (fig. 16-2). It is generally loosely attached to these structures, as well as to the choroid plexus. Colloid cysts contain turbid, tenacious mucus that solidifies upon formalin fixation.

Microscopic Findings. The lining epithelium is a simple layer of columnar cells, both goblet and ciliated (fig. 16-3), which, through "pressure atrophy" can become markedly flattened (fig. 16-4). The cyst contents are largely amorphous, but

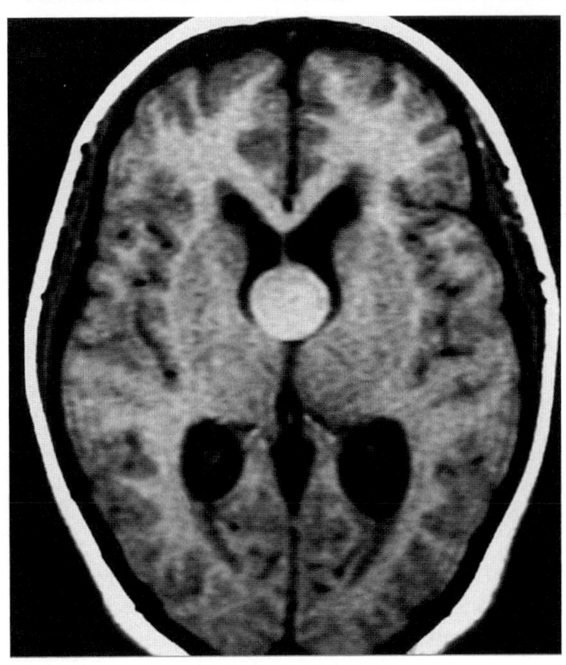

Figure 16-1

COLLOID CYST

Colloid cysts are spherical midline masses situated near the foramina of Monro. Because of their mucoid, protein-rich contents, nearly all are intrinsically bright in precontrast T1-weighted magnetic resonance images (MRIs). (Fig. 16-1 from Fascicle 10, 3rd Series.)

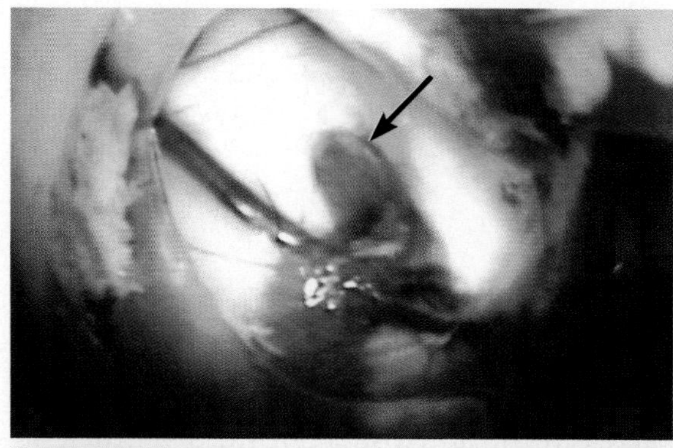

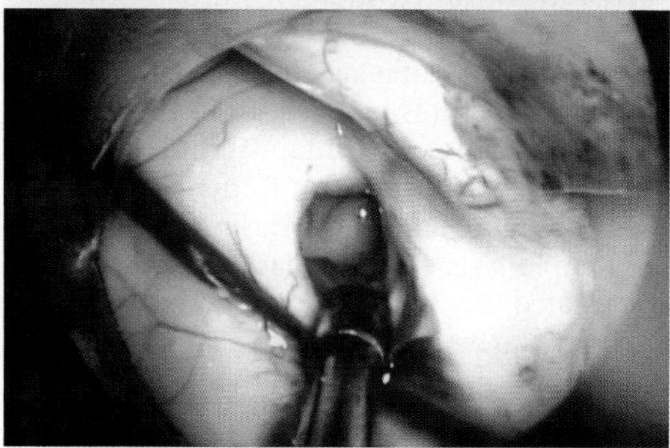

Figure 16-2

COLLOID CYST

As seen through the endoscope, the smooth-surfaced lesion (arrow) bulges through the foramen of Monro (top), an aperture that was opened subsequently by removal of the mass (bottom). (Courtesy of Dr. HD Jho, Pittsburgh, PA.)

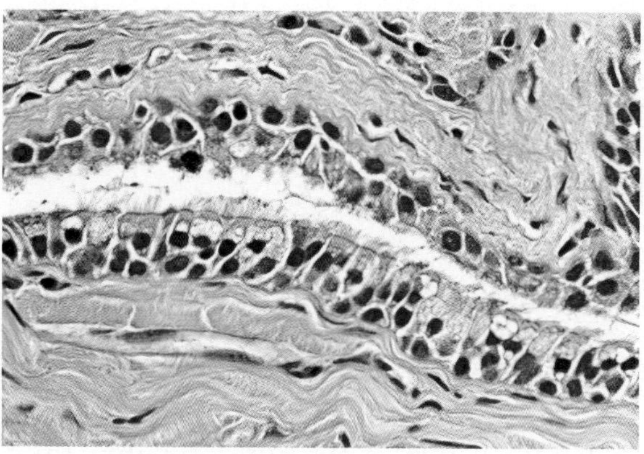

Figure 16-3

COLLOID CYST

In its native state, without pressure atrophy, the ciliated epithelium is tall and pseudostratified.

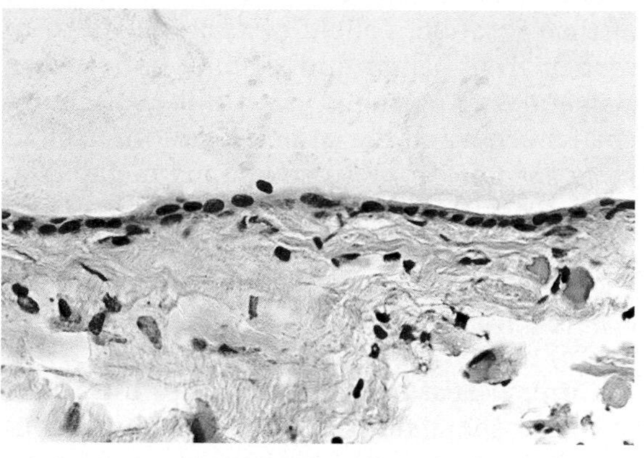

Figure 16-4

COLLOID CYST

Chronic intralesional pressure produces a low cuboidal, or even simple squamous, epithelium.

scattered cell ghosts and masses of filamentous material representing degenerate nucleoprotein may be present (fig. 16-5) (10). The latter resemble organisms of the *Actinomyces* group. A xanthogranulomatous reaction may replace the degenerating epithelium and fill the cyst in longstanding lesions (fig. 16-6) (11,12), accounting for the so-called xanthogranuloma of the third ventricle.

Immunohistochemical Findings. The epithelium of colloid cysts exhibits a distinctive pattern of immunoreactivity, with individual cells positive for cytokeratins and epithelial membrane antigen (EMA) (2,7,13,14). Scattered cells may be reactive for Clara cell–specific antigens (2).

Ultrastructural Findings. The architecture of colloid cysts has been compared to bronchial epithelium, and as many as six cell types have been described (1,3,7,15). These include: 1) ciliated cells, 2) nonciliated cells with surface microvilli, 3) goblet cells, 4) basal cells, 5) occasional electron-lucent Kulchitsky-like cells containing neurosecretory granules, and 6) small cells without specific features of differentiation. The epithelial cells are seated squarely upon a well-formed basal lamina, covered with a finely granular surface coating, and interconnected by desmosomes and apical junctional complexes. As vividly depicted by scanning electron microscopy, the cyst lining is formed both of ciliated cells and nonciliated cells with microvillous and bleb-like apical projections (15).

Cytologic Findings. Abundant mucoid material, macrophages, and epithelial cells are

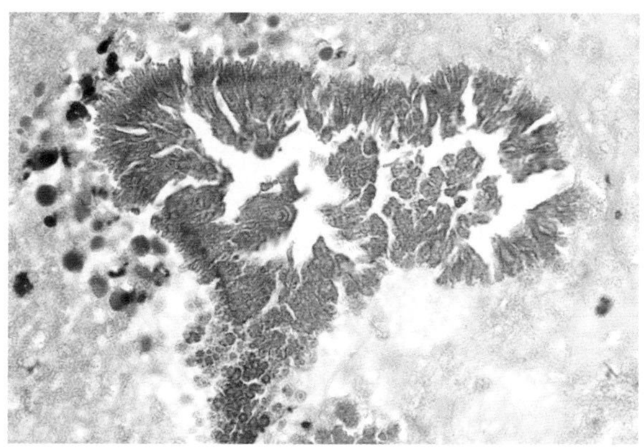

Figure 16-5

COLLOID CYST

Polymerized nucleoprotein within the cyst contents closely resembles infectious organisms.

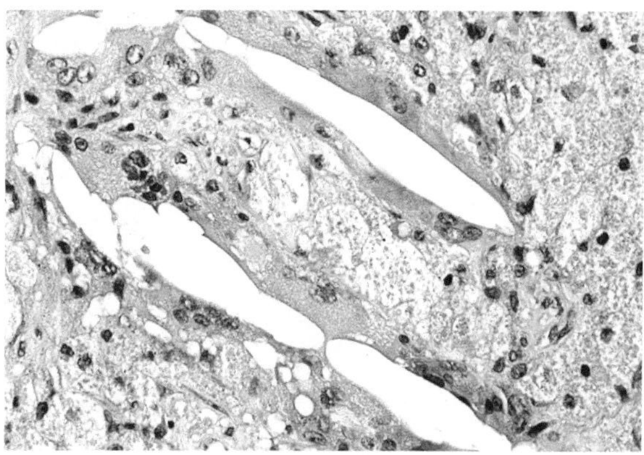

Figure 16-6

COLLOID CYST

A granulomatous reaction replaces the cyst in some instances.

present. The epithelial cells are either individual or fragments of glandular epithelium (16).

Differential Diagnosis. The macroscopic and radiographic characteristics of colloid cysts are so stereotypic that with an adequate specimen little consideration need be given to other entities. Although only scant epithelium may be observed in small specimens, its identification is usually sufficient for a diagnosis. Fragments of normal choroid plexus that are often attached to the outer surface of the cyst should not be interpreted as neoplastic (fig. 16-7). The "cobblestoned" epithelium of normal choroid plexus is neither ciliated, mucin-producing, nor subject to pressure atrophy.

Treatment and Prognosis. Excision is curative. The role of endoscopic resection and the need to remove the entire lesion remain matters of dispute (4,17), especially since many small colloid cysts enlarge little if at all over prolonged periods (18).

Rathke Cleft Cyst

Definition. *Rathke cleft cyst* is an intrasellar or suprasellar mucus-containing cyst lined mainly by columnar ciliated and goblet cells as well as occasional pituitary hormone–producing cells.

General Features. In embryologic terms, the adenohypophysis is derived from proliferating cells lining the cranial termination of the hypophyseal duct (19). As the duct regresses, its distal lumen remains a slit-like space that, within the substance of the pituitary, involutes

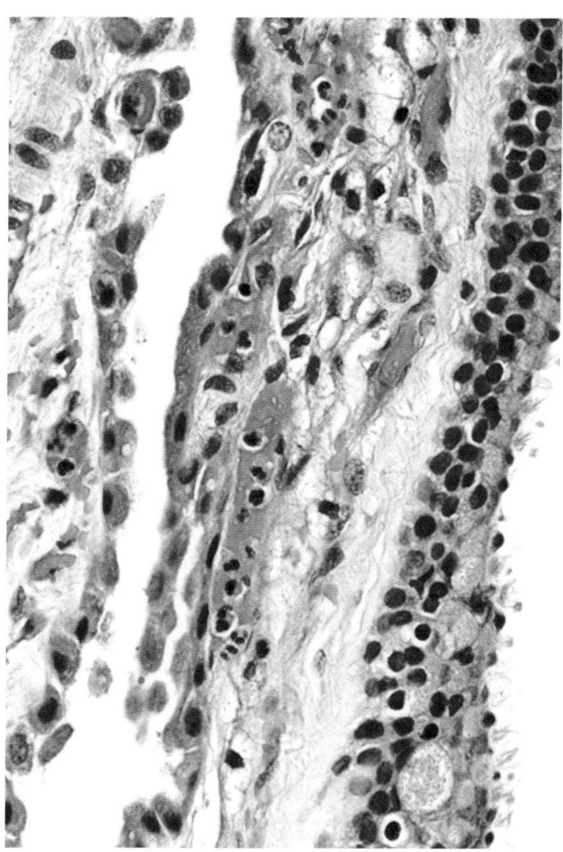

Figure 16-7

COLLOID CYST

Fragments of normal choroid plexus often accompany surgically excised colloid cysts. "Cobblestoned" plexus epithelium at the left is readily distinguished from the ciliated lining of the cyst on the right.

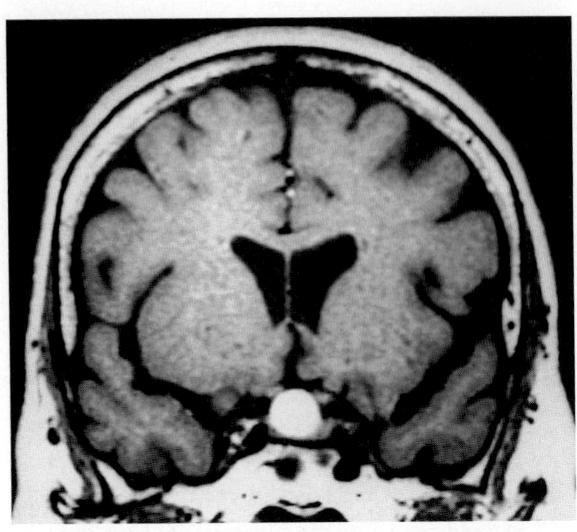

Figure 16-8

RATHKE CLEFT CYST

Rathke cleft cysts generate variable signal characteristics on MRI. Reflecting its high protein concentration, this example is bright in a precontrast T1-weighted image. (Fig. 16-7 from Fascicle 10, 3rd Series.)

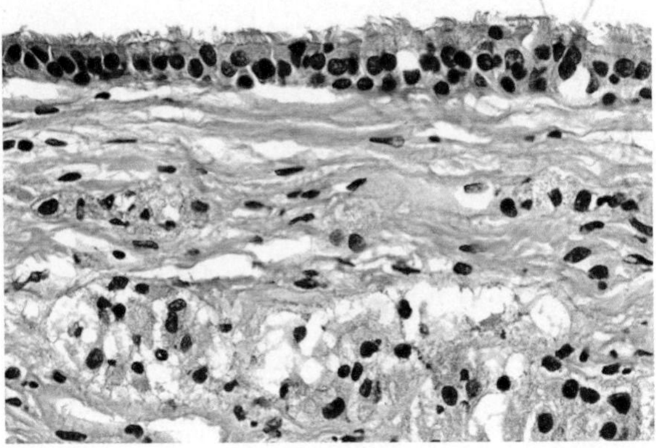

Figure 16-9

RATHKE CLEFT CYST WITH SQUAMOUS METAPLASIA

The classic epithelium is pseudostratified and ciliated.

to form a series of microcysts. Situated at the interface of the anterior and posterior pituitary lobes, these small hollow structures persist throughout life as Rathke cleft remnants. On occasion, their continued and symptomatic enlargement leads to the formation of a Rathke cleft cyst.

Rare pituitary adenomas are accompanied by cysts with ciliated columnar epithelium identical in appearance to that of a Rathke cleft cyst (20–23). In some such cases, transitions have been reported between the cells of the cyst and those of the adenoma, thus suggesting that these are true transitional tumors reflecting the primordial potential of the hypophyseal duct to produce both pituitary parenchyma and Rathke epithelium. Known as a "transitional cell tumor" of the pituitary, it remains a controversial entity.

Clinical Features. Regardless of size, Rathke cleft cysts are frequent incidental postmortem findings. Larger lesions, usually measuring 1 cm or more, are potentially symptomatic in the form of headaches, visual disturbances, or hypothalamic/pituitary dysfunction (24–27). Adults are primarily affected.

Radiologic Findings. Rathke cleft cysts that produce symptoms are all partly intrasellar, but most extend into the suprasellar region. They vary considerably in their MRI appearance in accord with the lining and character of the cyst contents (26–29). Those with a simple epithelium and clear fluid have MRI signal characteristics of CSF; more mucus-filled cysts produce heterogeneous signal characteristics. Some, as a presumed consequence of high protein content, have a hyperintense (white) signal on precontrast T1-weighted images (fig. 16-8). In contrast to adamantinomatous craniopharyngioma, Rathke cleft cysts are not calcified.

Gross Findings. Rathke cleft cysts are thin-walled structures with a content that varies considerably in viscosity. In fragmented specimens, the cyst remnants may be seen as inconspicuous strips adherent to fragments of pituitary tissue. Other specimens consist largely, if not entirely, of mucin.

Microscopic Findings. Captured in its native state, the epithelium of Rathke cleft cysts is well differentiated, columnar, and composed mainly of ciliated, goblet, and nondescript cells. Endocrine cells of anterior lobe type may be present (fig. 16-9). In a fragmented surgical specimen, bits of cyst lining adherent to pituitary tissue may be overlooked easily, and, in small fragments, difficult to distinguish from minute normal intermediate lobe cysts (fig. 16-10). Unlike Rathke cleft cysts, the latter remain small, are asymptomatic, and lack ciliated as well as goblet cells. A thick collagenous wall is more consistent with a Rathke cleft cyst than a fragment of an intermediate lobe cyst.

Frequently, the typical histologic appearance of Rathke cleft cyst is altered by the presence of

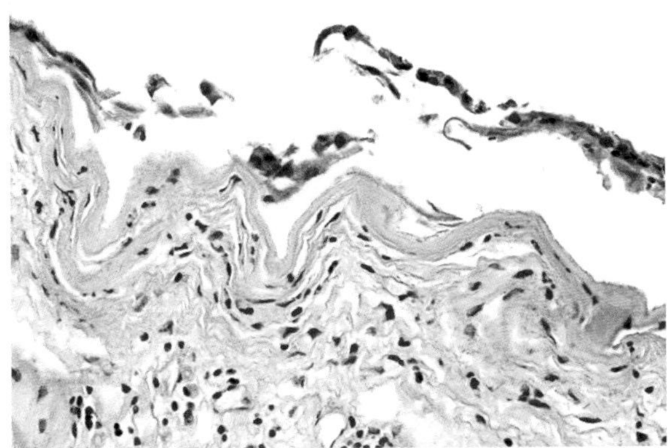

Figure 16-10

RATHKE CLEFT CYST

Only small distorted fragments of epithelium may be present in surgical specimens.

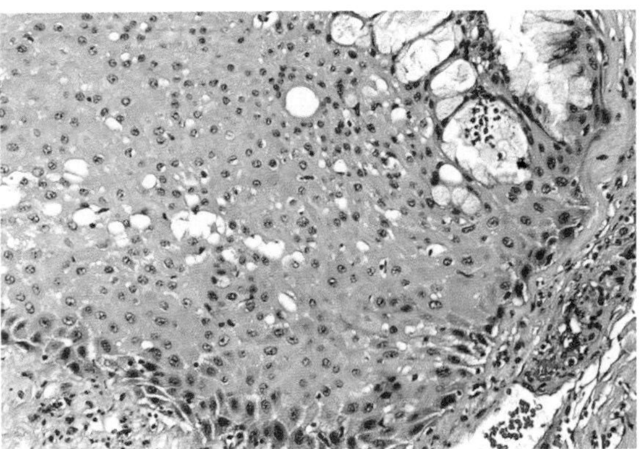

Figure 16-11

RATHKE CLEFT CYST

Squamous metaplasia can be prominent in Rathke cleft cysts, as here where a thick metaplastic layer lifts off the overlying ciliated and goblet cells.

squamous metaplasia (30,31). The latter appears to originate in reserve cells and to proliferate beneath the cyst's surface epithelium (fig. 16-11). In some instances, the metaplastic component predominates as the superficial columnar element sheds or even becomes undetectable.

Like its close relative, the colloid cyst of the third ventricle, xanthogranulomatous degeneration may dominate or entirely obscure the epithelial nature of the lesion (32). Chronic inflammation is not uncommon in the surrounding pituitary (30,31).

Immunohistochemical Findings. The columnar cells are variably positive for cytokeratins and EMA (33,34), as are metaplastic squamous cells. Unlike craniopharyngiomas, both adamantinomatous and papillary, Rathke cleft cysts appear to be immunoreactive for CK8 and CK20 (35). Scattered endocrine cells, often containing one or several pituitary hormones, are positive for chromogranin (15). Reactivity for carcinoembryonic antigen (CEA) has been reported (34).

Ultrastructural Findings. The exquisitely differentiated ciliated and goblet cells are joined by desmosomes and apical junctional complexes. A deeper layer of undifferentiated basal cells is also present (fig. 16-12). Squamous metaplasia adds cells rich in tonofilament bundles and extensively interconnected by well-formed desmosomes.

Cytologic Findings. Both ciliated and mucin-containing cells are readily seen in cytologic preparations (36).

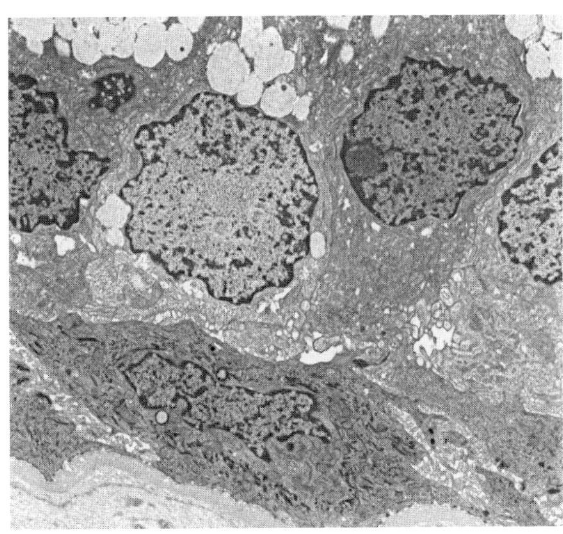

Figure 16-12

RATHKE CLEFT CYST

The cellular anlage of squamous metaplasia is a dark, flattened, horizontally aligned "reserve cell" at the bottom of this electron micrograph. Mucus-producing cells rest on top.

Differential Diagnosis. The diagnosis is self-evident when well-differentiated columnar epithelium and mucin are found in a specimen from a sizable and discrete intrasellar or suprasellar cyst. Small fragments of ordinary columnar epithelium may be difficult to interpret since they could be either from a Rathke cleft cyst or from a small, incidentally biopsied, intermediate lobe

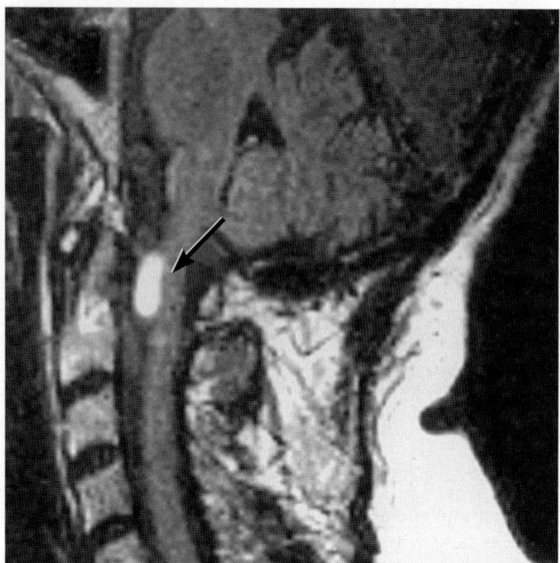

Figure 16-13

ENDODERMAL CYST

As is the case for the related colloid and Rathke cleft cysts, endodermal lesions are often bright in precontrast T1-weighted images. This lesion (arrow), anterior to the spinal cord, produced neck pain in a 52-year-old woman.

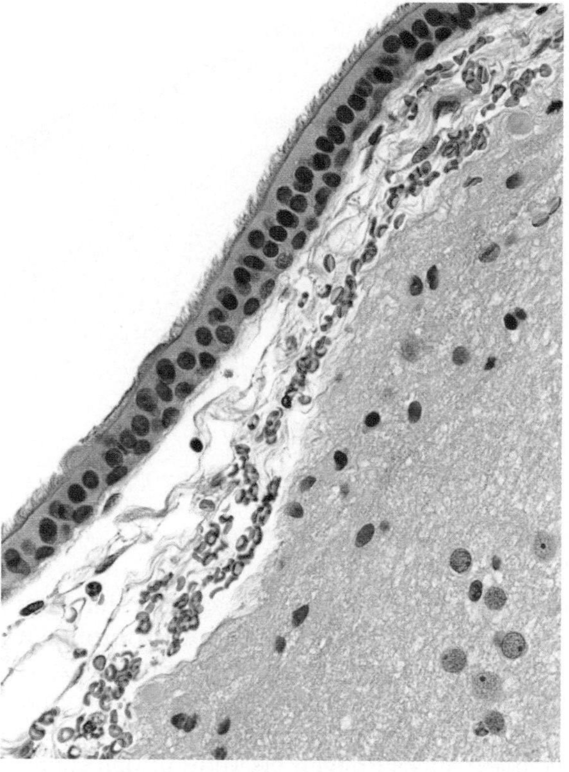

Figure 16-14

ENDODERMAL CYST

Extremely well differentiated and ciliated, endodermal cysts are characteristically located in the subarachnoid space.

cyst. Squamous metaplasia can confuse the issue, although not seriously if it is clearly seen as a secondary or metaplastic change underlying and lifting columnar cells. In certain cases, however, if the columnar epithelium is scant, the metaplasia predominates. In such cases, the metaplastic lining closely mimics that of papillary craniopharyngioma. The possible relationship of such lesions to papillary craniopharyngioma is discussed in chapter 15 (24).

Treatment and Prognosis. Rathke cleft cysts are benign and are cured by simple excision. Although partial removal and drainage may be sufficient (26,37), recurrence is possible (25).

Endodermal (Enterogenous) Cyst

Definition. The *endodermal cyst* is a cyst lined by columnar epithelium of presumed endodermal derivation.

General Features. On the basis of location and histologic features, cysts lined by a simple columnar, ciliated, or goblet cell epithelium have been variably designated as endodermal (38–41), enterogenous (42–47), neurenteric (48–62), enteric (63), enterogenic (64), foregut (65), respiratory (66), bronchogenic (67,68), epithelial (69), and teratomatous (70,71). The term endodermal encompasses these geographically varied lesions that presumably arise from misplaced epithelium of the nasopharynx, respiratory tree, or intestinal tract. Endodermal lesions occur intracranially and throughout the spinal axis.

Clinical Features. Endodermal cysts present at all ages; the posterior fossa and spinal intradural-extramedullary compartment are favored. Some intraspinal lesions are accompanied by osseous abnormalities of the spine anterior to the lesion, and may even connect with a cyst or with normal structures within the thorax or abdomen.

Radiologic Findings. Like the histologically similar Rathke cleft cyst, the contents of endodermal cysts vary in signal characteristics (fig. 16-13) (39).

Gross Findings. Most endodermal cysts are simple, thin-walled sacs filled with gray-white viscous or mucoid material. With few exceptions, their spinal location is intradural and

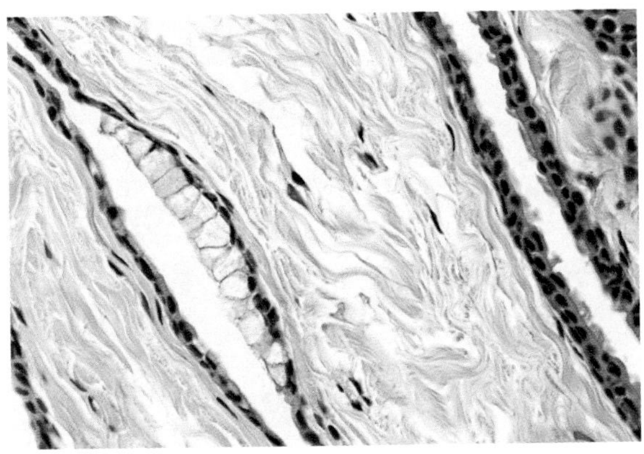

Figure 16-15

ENDODERMAL CYST

Variations in the epithelium include regions that are heavily mucin-producing or more cuboidal than columnar.

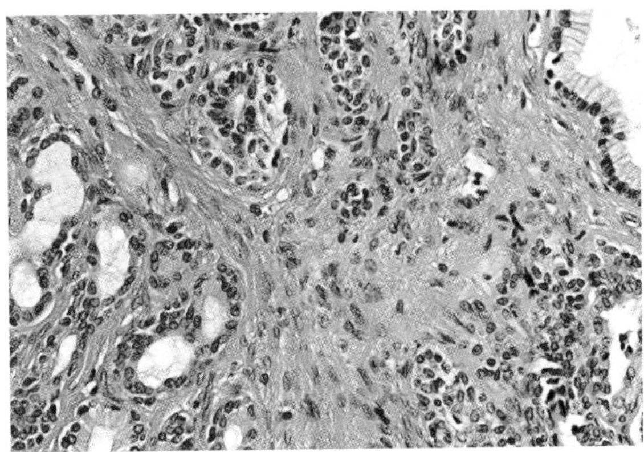

Figure 16-16

ENDODERMAL CYST

Spinal endodermal lesions can be more complex than the classic simple epithelium illustrated in figure 16-14. Well-formed glands are present in this example.

extramedullary. Only rare lesions are intramedullary (59). In other exceptional cases, the cyst lies posterior rather than anterior to the spinal cord (53).

Microscopic Findings. The most common form of endodermal cyst is a simple epithelium resting on a delicate fibrovascular capsule (figs. 16-14, 16-15). Although originally columnar, the epithelium may be converted to a low cuboidal state by chronic pressure. Ciliation is variable. Other cells contain globules of mucus or take the form of goblet cells. Squamous metaplasia is occasionally encountered, as it is in the related Rathke cleft cyst (72). Xanthogranulomatous change can be superimposed (61).

Endodermal cysts are histologically more intriguing when they are more complex (fig. 16-16), especially when they replicate either respiratory or gastrointestinal tract. Bronchogenic lesions in the cervical region may exhibit cartilaginous plates surrounding miniature "bronchi" with pseudostratified, ciliated, and goblet cell epithelium (67,68). Lesions reproducing the gastrointestinal tract often are composed not only of gastric or intestinal mucosa, but also feature a well-defined muscularis mucosa (43,45,51,64,73). Even more complicated posterior lesions may be associated with an overt developmental abnormality, the "split notochord syndrome" (74).

Immunohistochemical Findings. The epithelial cells are typically reactive for EMA, cytokeratins, and CEA (38,50,62,63,69,72,75). Some cells may be reactive for S-100 protein.

Ultrastructural Findings. Ultrastructural studies disclose an epithelium with ciliated cells, goblet cells, and nondescript columnar cells. In bronchogenic-type cysts, occasional neuroendocrine (Kulchitsky) cells as well as cartilage are seen. The ultrastructural features of most endodermal cysts are essentially those of colloid cysts of the third ventricle (67,69). Both ciliated and nonciliated cells are present (56). A surface coating of granulofibrillar material (glycocalyx), a typical feature of endoderm, may also be seen (56,57,67,72). There may be fine "claw-like" projections from the tips of cilia, as in other types of ciliated epithelia (57).

Differential Diagnosis. The presence of well-differentiated epithelium that abuts brain or spinal cord parenchyma understandably prompts consideration of a neuroectodermal or glioependymal cyst. Most lesions with abundant cilia or mucus-containing cells can be relegated to the endodermal category on the basis of their microscopic appearance alone. While immunoreactivity for EMA may be seen in ependyma, and cytokeratins in choroid plexus epithelium, reactivity for both is indicative of an endodermal cyst. Immunoreactivity for CEA is highly supportive of endodermal cyst. Glioependymal cysts stain for S-100 protein, and focally for glial fibrillary acidic protein (GFAP); choroid plexus epithelium is immunopositive for vimentin S-100 protein, transthyretin, and synaptophysin.

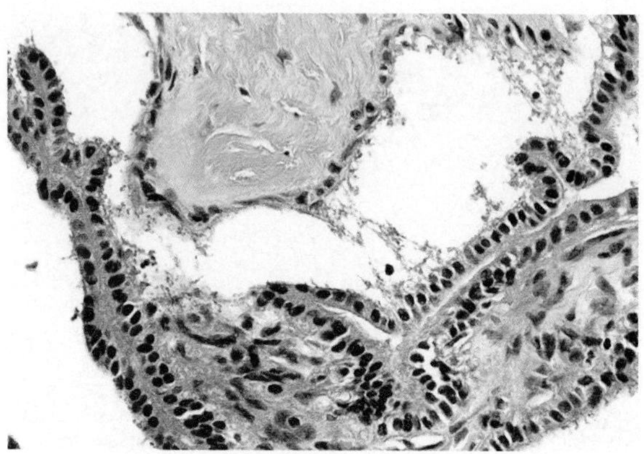

Figure 16-17

ENDOLYMPHATIC SAC TUMOR RESEMBLING ENDODERMAL CYST

The epithelium of endolymphatic sac tumor is more redundant and papillary than that of endodermal cyst.

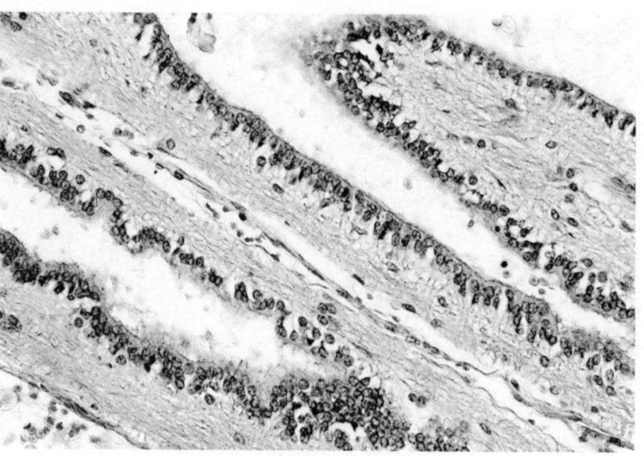

Figure 16-18

EPENDYMAL (GLIOEPENDYMAL) CYST

In contrast to endodermal cysts, the epithelium of ependymal cysts rests on a glial stroma.

At the cerebellopontine angle papillary endolymphatic sac tumor should be considered (fig. 16-17), although the latter arises within, and destroys, bone. It also has a more complicated papillary architecture than the smooth, linear epithelium of endodermal cyst.

Treatment and Prognosis. These slowly growing malformative lesions are cured by simple excision. Nonetheless, recurrence is possible (39,50,56). Rare endodermal cysts disseminate widely within the subarachnoid space (44). Adenocarcinomatous transformation is rare (42).

Ependymal (Glioependymal) Cyst

The increasing perception that simple ciliated or goblet cell-containing intracranial or intraspinal cysts remote from the ventricle are endodermal rather than ependymal has tightened the diagnostic criteria for the latter. An endodermal designation is also appropriate if the epithelium demonstrates squamous epithelium or, ultrastructurally, a surface glycocalyx. The diagnosis of *ependymal cyst* is appropriate for benign, intraparenchymal and often paraventricular cysts in which the lining lacks cilia and exhibits S-100 protein- and occasionally GFAP-reactive cells (fig. 16-18) (76–79). Such cells rest upon native astroglia without an intervening basal lamina. An occasional example is leptomeningeal in location (80). Ependymal cysts are usually intracranial, but spinal (intramedullary) lesions have been reported (81). Purely intraventricular lesions have been reported as well (82).

EPIDERMOID AND DERMOID CYSTS

Definition. *Epidermoid* and *dermoid cysts* are lined by benign keratinizing squamous epithelium. Cutaneous adnexa are present in dermoid cysts.

Clinical Features. Epidermoid cysts occur throughout the neuraxis, but most are intracranial, often at the cerebellopontine angle (83–86). Intraspinal examples are less common (87). Because the lesions tend to surround and envelope, rather than displace, regional structures, symptoms typically occur late in the course when the cyst has already attained considerable size and extent. Epidermoid cysts may also arise in the cranial diploe where they present as a lytic defect.

Although most epidermoid cysts have a presumed developmental origin, the transfer of cutaneous epithelium to the subarachnoid space during diagnostic procedures such as lumbar puncture has been implicated in the pathogenesis of rare intraspinal and intracranial examples (88).

Dermoid cysts are more restricted than epidermoid cysts in terms of patient age and location. Most present in childhood and lie in the midline, being related to a fontanelle (89), the fourth ventricle, or the spinal canal (90). A sinus tract from the cutaneous surface provides a potential portal of entry for infectious organisms in some cases (91–95).

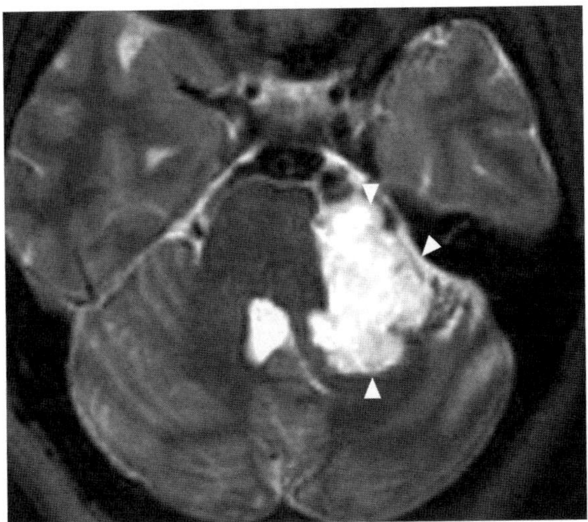

Figure 16-19

EPIDERMOID CYST

The cerebellopontine angle is a favored site for epidermoid cysts, as with this T2-bright mass (arrowheads) that compresses the pons and cerebellum. The lesion often surrounds regional cranial nerves and blood vessels making total excision difficult, if not impossible.

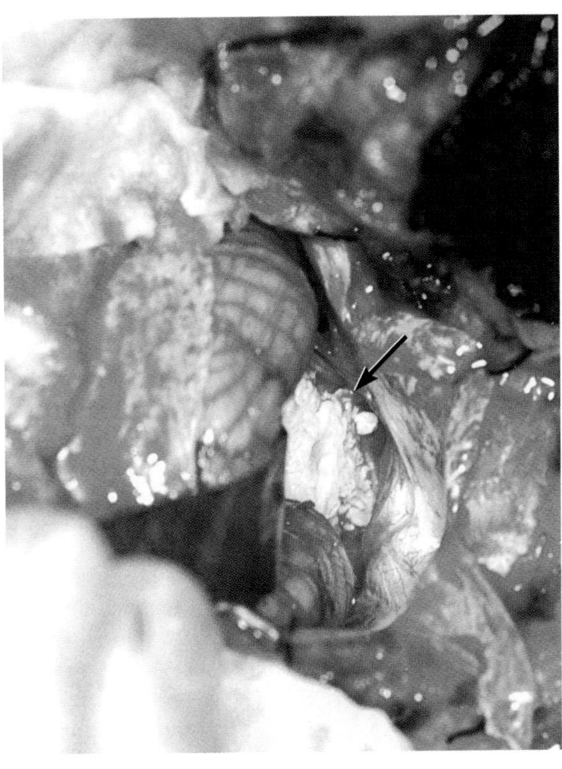

Figure 16-20

EPIDERMOID CYST

Exposed at surgery at the cerebellopontine angle, this typically extra-axial lesion (arrow) has the characteristic pearly sheen of an epidermoid cyst.

Radiologic Findings. Due to their discreteness, extra-axial position, and neuroradiologic signal characteristics of keratinaceous debris, epidermoid cysts are often diagnosed confidently before surgery. Depending upon the amount of lipid in their contents, the lesions exhibit a variable signal on MRI, appearing as either bright or white (when lipid content is high) in T1-weighted images, or black (if lipid content is low) (fig. 16-19) (83,84,96,97). Diffusion MRIs are useful in distinguishing keratinaceous debris, with its restricted diffusion, from mobile water molecules in the fluid of an arachnoid cyst (98,99).

The hair and sebaceous content of a dermoid cyst give it a heterogeneous MRI signal, but the abundant lipid produces high signal intensity in T1-weighted images (100–102). Rupture of both dermoid and epidermoid cysts can produce chemical meningitis (102–104).

Gross Findings. The uniloculate, thin-walled epidermoid cyst has an unmistakable pearly sheen due to its thin wall and content of compact flaky keratin (figs. 16-20, 16-21). The connective tissue–rich wall of the dermoid cyst is thicker, and its greasy content is often matted with hair.

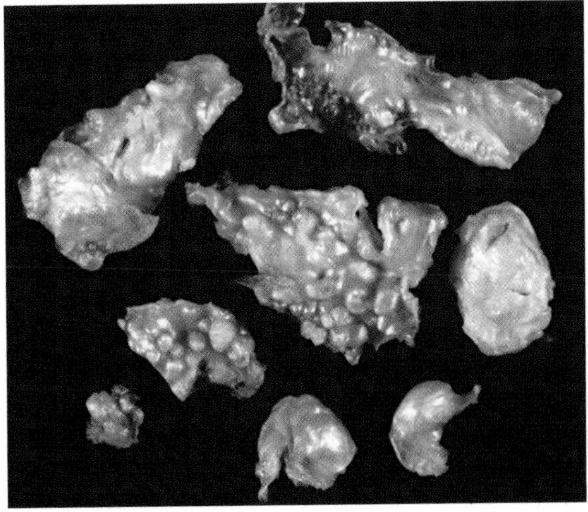

Figure 16-21

EPIDERMOID CYST

Even in fragmented surgical specimens, epidermoid cysts have a remarkable resemblance to the interior of an abalone shell.

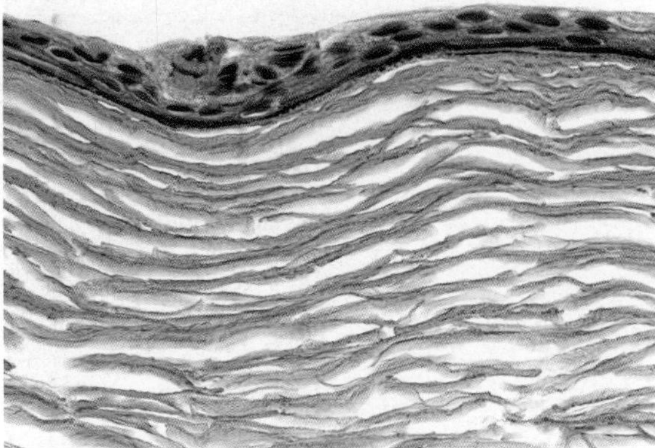

Figure 16-22

EPIDERMOID CYST

Anucleate squames make up the overwhelming majority of the mass. The keratohyaline granules seen here are absent in craniopharyngiomas and central nervous system (CNS) cysts with squamous metaplasia.

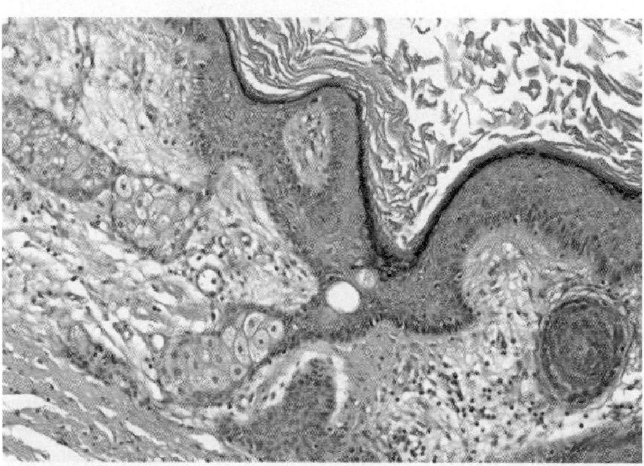

Figure 16-23

DERMOID CYST

The presence of adnexae clearly differentiates this lesion from epidermoid cyst.

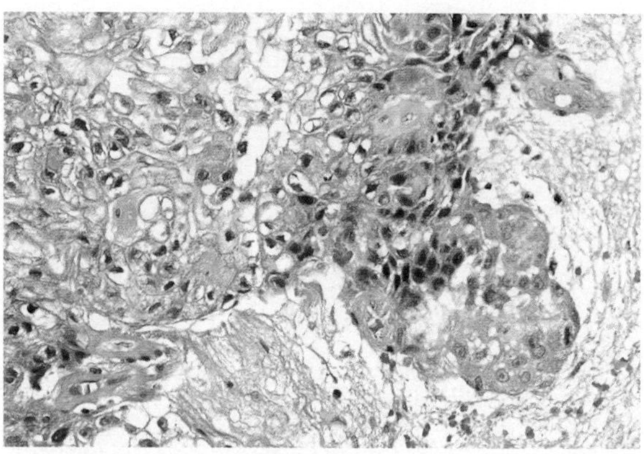

Figure 16-24

EPIDERMOID CYST WITH ANAPLASTIC TRANSFORMATION

Rare epidermoid cysts evolve into invasive carcinoma.

Microscopic Findings. The overwhelming bulk of an epidermoid cyst consists of layered anucleate squames produced by a thin, well-differentiated squamous epithelium, often with keratohyaline granules (fig. 16-22). The proliferating epithelium, often just several cells thick, may not be represented in a small biopsy. Identification of flaky "dry" keratin alone is essentially diagnostic.

The irregularly thickened wall of the dermoid cyst is essentially dermis containing adnexal structures, such as hair follicles, sebaceous glands, and supportive fibroadipose tissue (fig. 16-23). "Shadow cells," indicative of hair matrix differentiation, have been reported (105). Mature bone is present in an occasional case. A foreign body reaction to leaked cyst content is frequently present.

Although epidermoid and dermoid cysts are benign, rare transformation to squamous carcinoma reportedly occurs in both (fig. 16-24) (106–110).

Immunohistochemical Findings. The lining epithelium of both epidermoid and dermoid cysts is immunoreactive for cytokeratins and EMA. The immunophenotypes of the adnexa in dermoid cysts are similar to those of their cutaneous counterparts.

Differential Diagnosis. In most locations, epidermoid cysts are so distinctive that no troublesome diagnostic issues arise. In the suprasellar region, the practical distinction between this cyst and craniopharyngioma, both adamantinomatous and papillary, must be addressed. As is discussed in chapter 15, this is infrequently a practical problem since the delicate epithelium of epidermoid cysts is readily distinguished from the thicker, complex, and palisaded epithelium of adamantinomatous craniopharyngioma. Secondary inflammatory reaction, cholesterol clefts, calcification, and "motor oil" content are lacking in epidermoid cysts. In attenuated cystic areas of craniopharyngiomas, the distinction is not as elementary but readily made by the presence of "wet

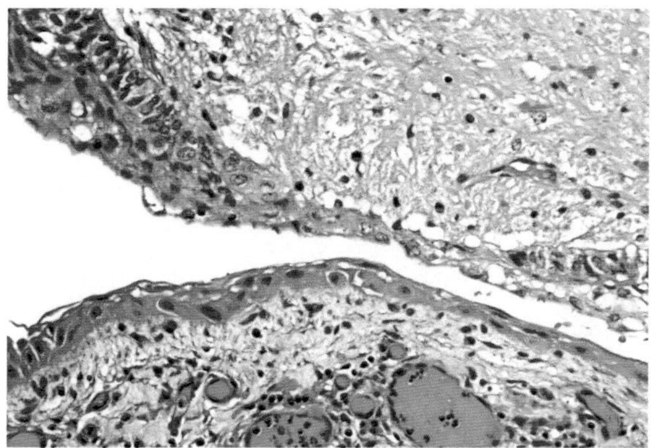

Figure 16-25

CRANIOPHARYNGIOMA RESEMBLING EPIDERMOID CYST

Although craniopharyngiomas do not have surface keratinization, the sloughing of thin surface cells simulates the anuclear squames of the epidermoid lesion. Peripheral palisading is retained in this case.

keratin" and lack of keratohyaline bodies and anucleate squames (fig. 16-25).

The epithelium of the papillary craniopharyngioma is well differentiated and squamous, but is generally thick and disposed in crude papillae with fibrovascular cores. Despite its high degree of differentiation, papillary craniopharyngioma also lacks keratohyaline granules and anucleate squames. Keeping these distinctions in mind, epidermoid cysts can be distinguished from the occasionally attenuated epithelium of either craniopharyngioma variant.

An epidermoid cyst can resemble either endodermal cyst or Rathke cleft cyst with extensive squamous metaplasia. The latter generally shows persistence of surface ciliated or mucin-producing epithelium and a lack of both the keratohyaline granules and anucleate squames so typical of epidermoid cyst, as well as the cutaneous adnexa of the dermoid cyst.

Dermoid cysts must be distinguished from well-differentiated teratomas, such as those arising in the sellar and pineal regions. Dermoid cysts lack endodermal components or highly specialized mesenchyme such as muscle.

Treatment and Prognosis. Some epidermoid and dermoid cysts can be totally excised, but many of the epidermoid variety surround vital structures in the subarachnoid space, such as cranial nerves and major blood vessels. Total resection is then precluded and recurrences become possible (83,84,86).

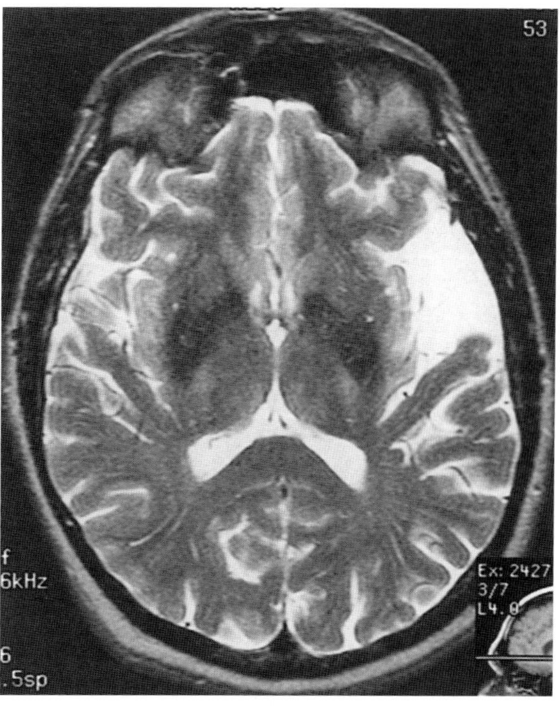

Figure 16-26

ARACHNOID CYST

The bright expansion in the subarachnoid space seen in T2-weighed MRI is typical of arachnoid cyst. The region of the sylvian fissure is the most common supratentorial intracranial site.

ARACHNOID CYST

Definition. *Arachnoid cyst* is a loculated collection of CSF within a non-neoplastic reduplication of the arachnoidal membrane.

Clinical Features. Arachnoid cysts arise within both the cranial (figs. 16-26, 16-27) and spinal leptomeninges (fig. 16-28) (111). Favored intracranial sites include the subarachnoid space overlying the temporal lobe (112–115) and the cerebellopontine angle (116). Large intracranial cysts expand at the expense of the frontal and temporal lobes, and are discovered either as a symptomatic mass or as an incidental neuroradiologic or postmortem finding. Curiously, many sizable arachnoid cysts are not associated with herniation or with displacement of midline structures. This is the basis of a controversy over whether the basic lesion is a cyst displacing the brain, or a regional zone of brain agenesis in

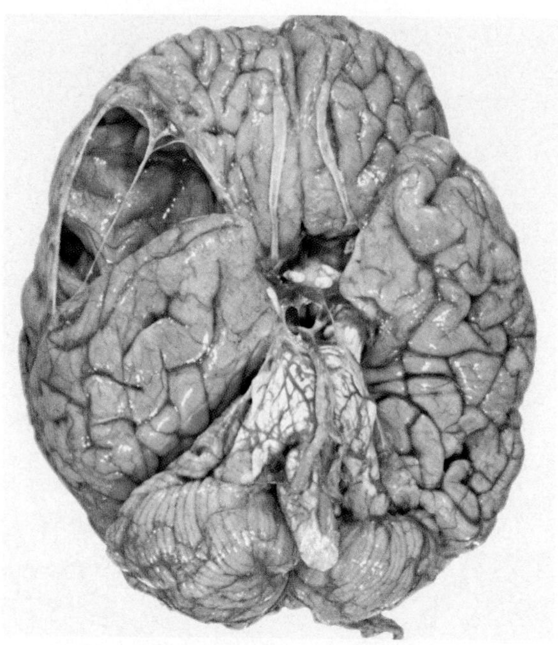

Figure 16-27

ARACHNOID CYST

A readily punctured, delicate membrane circumscribes the loculated fluid of an arachnoid cyst. This lesion opens the sylvian fissure to expose the insular cortex.

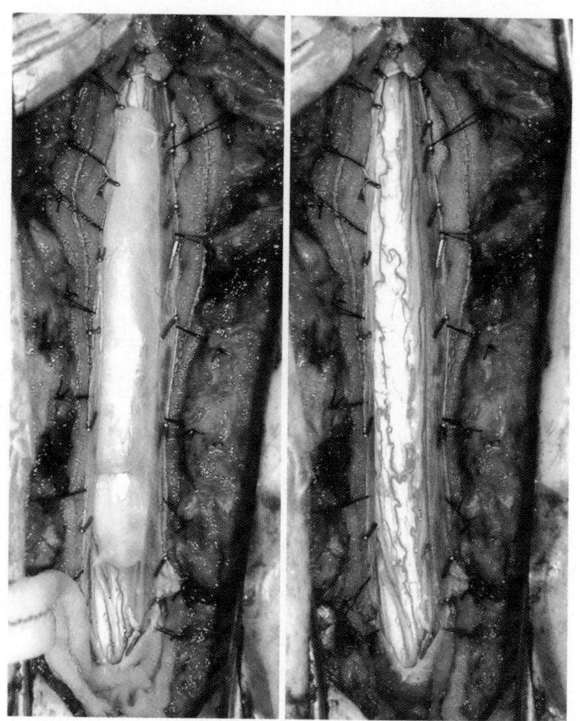

Figure 16-28

ARACHNOID CYST

This large, intradural, extramedullary lesion produced intractable interscapular pain in a 22-year-old woman. At surgery, the translucent lesion is exposed both before (left) and after (right) excision. A ball valve mechanism allowing entry of cerebrospinal fluid into the cyst was apparent intraoperatively. (Courtesy of Dr. Ziya Gokaslan, Baltimore, MD.)

which the defect is passively filled by compensatory expansion of adjacent subarachnoid space. Similar weights and volumes of the indented and nonindented hemispheres suggest that there is no agenesis (113,117).

Radiologic Findings. MRI scans readily identify the CSF content of arachnoid cysts, distinguishing this from the solid, lipid-rich, keratinous or sebaceous content of epidermoid and dermoid cysts.

Gross Findings. The watery content of an arachnoid cyst is readily seen through its delicate translucent wall (figs. 16-27, 16-28). When the wall is punctured, the cyst content escapes to expose the underlying flattened brain. What remains of the cyst wall is an unimpressive delicate membrane.

Microscopic Findings. The wall of an arachnoid cyst consists of a delicate fibrous connective tissue membrane in which meningothelial cells are present, either diffusely or focally (fig. 16-29) (118,119). Lacking epithelium, these cysts differ from other craniospinal cysts that are commonly received as surgical specimens.

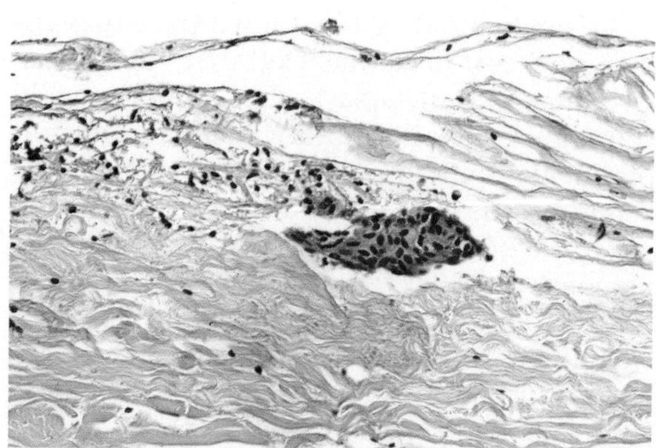

Figure 16-29

ARACHNOID CYST

This lesion, seen macroscopically in figure 16-28, is composed of fibrous tissue containing nests of meningothelial cells.

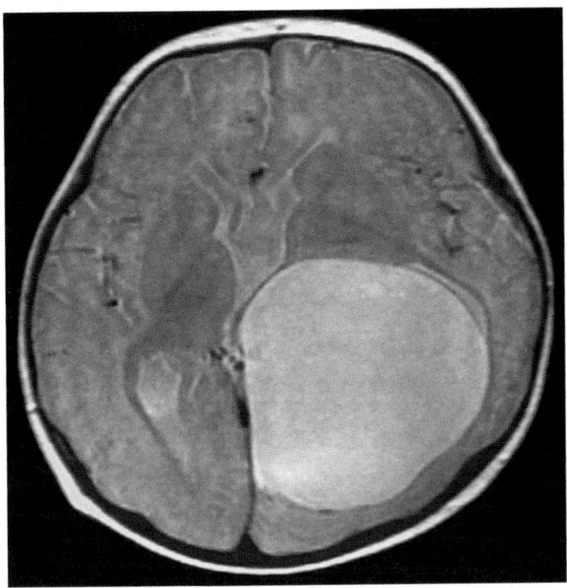

Figure 16-30

CHOROID PLEXUS CYST

As seen in a proton density image, a large cyst fills the atrium and occipital horn of a lateral ventricle of a 5-month-old boy with seizures.

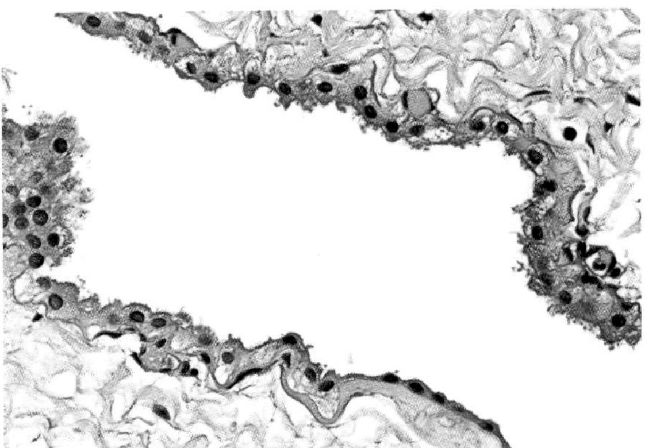

Figure 16-31

CHOROID PLEXUS CYST

The cyst illustrated in figure 16-30 has the cobblestone profile typical of the choroid plexus.

Immunohistochemical Findings. The meningothelial nature of the flattened lining cells is readily established by immunoreactivity for EMA. As expected, the cysts are negative for markers found in epithelial-lined intracranial and intraspinal cysts, such as keratin, CEA, transthyretin, synaptophysin, and GFAP (120).

Treatment and Prognosis. Symptoms are relieved by simple marsupialization of the cyst and release of its trapped fluid. Alternatively, the cyst may be shunted.

OTHER CYSTS

Choroid Plexus Cyst

Small cuboidal or columnar epithelium-lined cysts of the choroid plexus are occasionally encountered as incidental postmortem findings. Symptomatic lesions are rare (121–123). The lining varies from connective tissue alone to columnar epithelium (figs. 16-30, 16-31). The choroid plexus-type epithelium is understandably positive for vimentin, keratin, S-100 protein, transthyretin (122), and synaptophysin.

A presumably distinct form of choroid plexus cyst is commonly encountered in utero as a hypoechoic focus that usually resolves spontaneously by birth (124–127). Histopathologic studies of these common lesions are few, but reports document the presence of a loose stroma surrounding a cyst without an epithelial lining (128), technically removing these lesions from the true cyst category. Markedly flattened cells can be found around the cyst, but it is not clear whether these represent compressed epithelium or simply fibroblasts. A small percentage of these lesions are associated with trisomy 18 and, in some cases, congenital malformations (124–127).

Nerve Root (Tarlov) Cyst

Non-neoplastic leptomeningeal diverticula or duplications that occur primarily around nerve roots of the lower spinal cord are variously termed *nerve root cyst, perineural cyst,* or *Tarlov cyst* (129–134). Most are of no clinical significance and represent incidental radiologic or surgical findings; only a few produce pain (fig. 16-32) (129,134–136). The cysts consist primarily of an arachnoidal membrane. Peripheral nerve fibers and, in some cases, ganglion cells may be present (fig. 16-33).

Synovial Cyst of the Spine

Synovial cysts, sometimes referred to as *ganglion cysts,* are common extradural lesions. They occur particularly at the L4-L5 level where, usually affecting facet joints, they often accompany degenerative joint disease (137–143). Such cysts can also occur in the cervical region. Rare

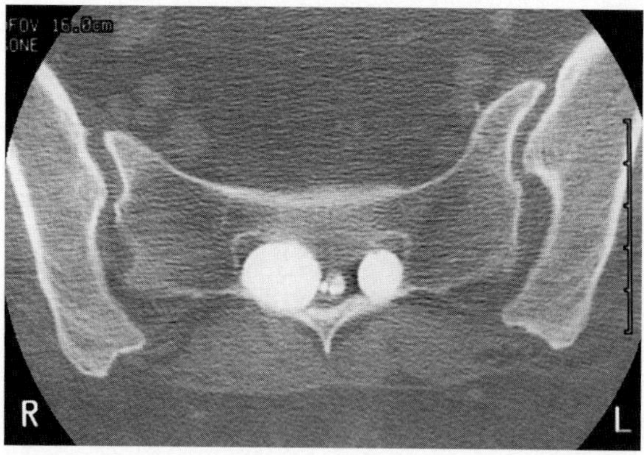

Figure 16-32

NERVE ROOT (TARLOV) CYST

As seen during computerized tomography (CT) myelography, nerve root cysts are spherical lesions related to distal sacral nerve roots.

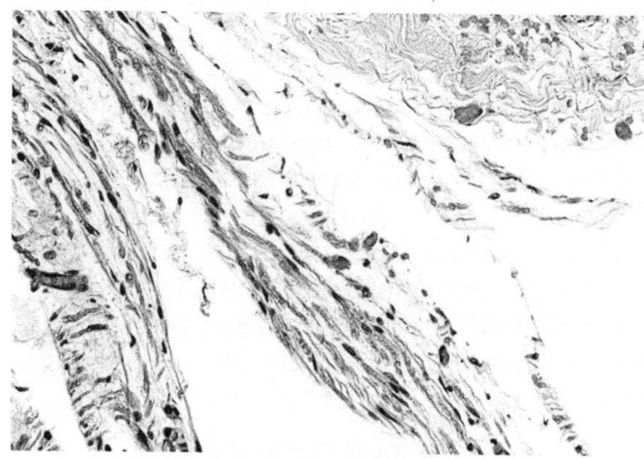

Figure 16-33

NERVE ROOT (TARLOV) CYST

Delicate connective tissue and nerve fibers comprise the rare surgical specimen.

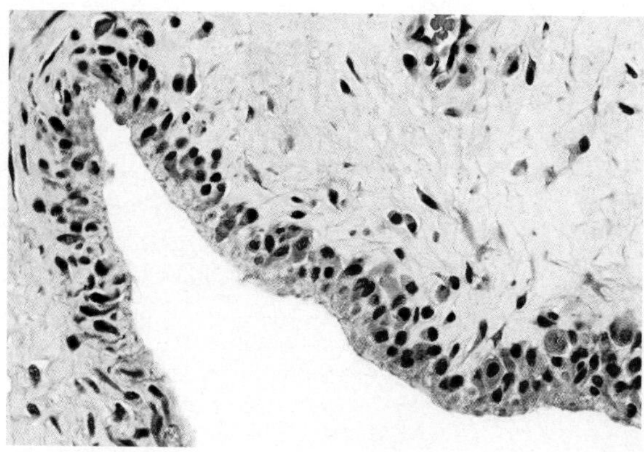

Figure 16-34

SYNOVIAL CYST

A cuboidal epithelium of synovial quality lines the cyst.

examples are not joint associated, such as cysts affecting the transverse ligament near the odontoid process, or the ligamentum flavum. Most lesions are similar in their clinical presentation to the far more common herniated disc. Radiologically, they have a low density center and a denser fibrotic and sometimes calcified rim.

The cyst may be limited to the facet joint or secondarily attached to nearby dura. Its fibrous tissue capsule surrounds a content of gelatinous fluid. Microscopically, the lining may consist either of nondescript connective tissue like that of ordinary ganglion cysts or actual synovium (fig. 16-34). Excision is curative.

Miscellaneous Cysts

Uncommon cysts that defy ready classification include: 1) intraparenchymal examples without an epithelial lining (144,145), 2) choroidal fissure cysts in the medial temporal lobe (146), and 3) the spinal epidural cyst that retains an attachment to the spinal meninges (147–151). The latter consists largely of dense fibrous connective tissue, in some cases of dural quality. A layer of arachnoidal tissue with focal nests of meningothelial cells is seen in other cases.

REFERENCES

Colloid Cyst

1. Ho KL, Garcia JH. Colloid cysts of the third ventricle: ultrastructural features are compatible with endodermal derivation. Acta Neuropathol (Berl) 1992;83:605–12.
2. Lach B, Scheithauer BW, Gregor A, Wick MR. Colloid cyst of the third ventricle. A comparative immunohistochemical study of neuraxis cysts and choroid plexus epithelium. J Neurosurg 1993;78:101–11.
3. Lach B, Scheithauer BW. Colloid cyst of the third ventricle: a comparative ultrastructural study of neuraxis cysts and choroid plexus epithelium. Ultrastruct Pathol 1992;16:331–49.
4. Hellwig D, Bauer BL, Schulte M, Gatscher S, Riegel T, Bertalanffy H. Neuroendoscopic treatment for colloid cysts of the third ventricle: the experience of a decade. Neurosurgery 2003;52: 525–33; discussion 532–3.
5. Hernesniemi J, Leivo S. Management outcome in third ventricular colloid cysts in a defined population: a series of 40 patients treated mainly by transcallosal microsurgery. Surg Neurol 1996;45:2–14.
6. Mathiesen T, Grane P, Lindgren L, Lindquist C. Third ventricle colloid cysts: a consecutive 12-year series. J Neurosurg 1997;86:5–12.
7. Macaulay RJ, Felix I, Jay V, Becker LE. Histological and ultrastructural analysis of six colloid cysts in children. Acta Neuropathol (Berl) 1997;93:271–6.
8. Macdonald RL, Humphreys RP, Rutka JT, Kestle JR. Colloid cysts in children. Pediatr Neurosurg 1994;20:169–77.
9. Maeder PP, Holtas SL, Basibuyuk LN, Salford LG, Tapper UA, Brun A. Colloid cysts of the third ventricle: correlation of MR and CT findings with histology and chemical analysis. AJNR Am J Neuroradiol 1990;11:575–81.
10. Powers JM, Dodds HM. Primary actinomycoma of the third ventricle—the colloid cyst. A histochemical and ultrastructural study. Acta Neuropathol (Berl) 1977;37:21–6.
11. Paulus W, Honegger J, Keyvani K, Fahlbusch R. Xanthogranuloma of the sellar region: a clinicopathological entity different from adamantinomatous craniopharyngioma. Acta Neuropathol (Berl) 1999;97:377–82.
12. Tatter SB, Ogilvy CS, Golden JA, Ojemann RG, Louis DN. Third ventricular xanthogranulomas clinically and radiologically mimicking colloid cysts. Report of two cases. J Neurosurg 1994;81: 605–9.
13. Inoue T, Matsushima T, Fukui M, Iwaki T, Takeshita I, Kuromatsu C. Immunohistochemical study of intracranial cysts. Neurosurgery 1988;23:576–81.
14. Kondziolka D, Bilbao JM. An immunohistochemical study of neuroepithelial (colloid) cysts. J Neurosurg 1989;71:91–7.
15. Leech R, Freeman T, Johnson R. Colloid cyst of the third ventricle. A scanning and transmission electron microscopic study. J Neurosurg 1982;57:108–13.
16. Parwani AV, Fatani IY, Burger PC, Erozan YS, Ali SZ. Colloid cyst of the third ventricle: cytomorphologic features on stereotactic fine-needle aspiration. Diagn Cytopathol 2002;27:27–31.
17. Mathiesen T, Grane P, Lindquist C, von Holst H. High recurrence rate following aspiration of colloid cysts in the third ventricle. J Neurosurg 1993;78:748–52.
18. Pollock BE, Huston J 3rd. Natural history of asymptomatic colloid cysts of the third ventricle. J Neurosurg 1999;91:364–9.

Rathke Cleft Cyst

19. Ikeda H, Suzuki J, Sasano N, Niizuma H. The development and morphogenesis of the human pituitary gland. Anat Embryol (Berl) 1988;178: 327–36.
20. Ikeda H, Yoshimoto T, Katakura R. A case of Rathke's cleft cyst within a pituitary adenoma presenting with acromegaly—do "transitional cell tumors of the pituitary gland" really exist? Acta Neuropathol (Berl) 1992;83:211–5.
21. Kepes JJ. Transitional cell tumor of the pituitary gland developing from a Rathke's cleft cyst. Cancer 1978;41:337–43.
22. Nishio S, Mizuno J, Barrow DL, Takei Y, Tindall GT. Pituitary tumors composed of adenohypophysial adenoma and Rathke's cleft cyst elements: a clinicopathological study. Neurosurgery 1987;21(3):371–7.
23. Swanson SE, Chandler WF, Latack J, Zis K. Symptomatic Rathke's cleft cyst with pituitary adenoma: case report. Neurosurgery 1985;17:657–9.
24. Ikeda H, Yoshimoto T. Clinicopathological study of Rathke's cleft cysts. Clin Neuropathol 2002;21:82–91.
25. Kasperbauer JL, Orvidas LJ, Atkinson JL, Abboud CF. Rathke cleft cyst: diagnostic and therapeutic considerations. Laryngoscope 2002;112:1836–9.

26. Ross DA, Norman D, Wilson CB. Radiologic characteristics and results of surgical management of Rathke's cysts in 43 patients. Neurosurgery 1992;30:173–9.
27. Sade B, Albrecht S, Assimakopoulos P, Vezina JL, Mohr G. Management of Rathke's cleft cysts. Surg Neurol 2005;63(5):459–66; discussion 466.
28. Kucharczyk W, Peck WW, Kelly WM, Norman D, Newton TH. Rathke cleft cysts: CT, MR imaging, and pathologic features. Radiology 1987;165(2):491–5.
29. Sumida M, Uozumi T, Mukada K, Arita K, Kurisu K, Eguchi K. Rathke cleft cysts: correlation of enhanced MR and surgical findings. AJNR Am J Neuroradiol 1994;15:525–32.
30. Kleinschmidt-DeMasters BK, Lillehei KO, Stears JC. The pathologic, surgical, and MR spectrum of Rathke cleft cysts. Surg Neurol 1995;44:19–26; discussion 26–7.
31. Hama S, Arita K, Nishisaka T, et al. Changes in the epithelium of Rathke cleft cyst associated with inflammation. J Neurosurg 2002;96:209–16.
32. Wolfsohn AL, Lach B, Benoit BG. Suprasellar xanthomatous Rathke's cleft cyst. Surg Neurol 1992;38:106–9.
33. Ikeda H, Yoshimoto T, Suzuki J. Immunohistochemical study of Rathke's cleft cyst. Acta Neuropathol (Berl) 1988;77:33–8.
34. Inoue T, Matsushima T, Fukui M, Iwaki T, Takeshita I, Kuromatsu C. Immunohistochemical study of intracranial cysts. Neurosurgery 1988;23:576–81.
35. Xin W, Rubin MA, McKeever PE. Differential expression of cytokeratins 8 and 20 distinguishes craniopharyngioma from rathke cleft cyst. Arch Pathol Lab Med 2002;126:1174–8.
36. Smith AR, Elsheikh TM, Silverman JF. Intraoperative cytologic diagnosis of suprasellar and sellar cystic lesions. Diagn Cytopathol 1999;20:137–47.
37. Midha R, Jay V, Smyth HS. Transsphenoidal management of Rathke's cleft cysts. A clinicopathological review of 10 cases. Surg Neurol 1991;35:446–54.

Endodermal (Enterogenous) Cyst

38. Akaishi K, Hongo K, Ito M, Tanaka Y, Tada T, Kobayashi S. Endodermal cyst in the cerebellopontine angle with immunohistochemical reactivity for CA19-9. Clin Neuropathol 2000; 19:296–9.
39. Bejjani GK, Wright DC, Schessel D, Sekhar LN. Endodermal cysts of the posterior fossa. Report of three cases and review of the literature. J Neurosurg 1998;89:326–35.
40. Cheng JS, Cusick JF, Ho KC, Ulmer JL. Lateral supratentorial endodermal cyst: case report and review of literature. Neurosurgery 2002;51:493–9.
41. Mackenzie IR, Gilbert JJ. Cysts of the neuraxis of endodermal origin. J Neurol Neurosurg Psychiatry 1991;54:572–5.
42. Ho LC, Olivi A, Cho CH, Burger PC, Simeone F, Tihan T. Well-differentiated papillary adenocarcinoma arising in a supratentorial enterogenous cyst: case report. Neurosurgery 1998;43:1474–7.
43. Millis RR, Holmes AE. Enterogenous cyst of the spinal cord with associated intestinal reduplication, vertebral anomalies, and a dorsal dermal sinus. Case report. J Neurosurg 1973;38:73–7.
44. Perry A, Scheithauer BW, Zaias BW, Minassian HV. Aggressive enterogenous cyst with extensive craniospinal spread: case report. Neurosurgery 1999;44:401–5.
45. Rhaney K, Barclay GP. Enterogenous cysts and congenital diverticula of the alimentary canal with abnormalities of the vertebral column and spinal cord. J Pathol Bacteriol 1959;77:457–71.
46. Umezu H, Aiba T, Unakami M. Enterogenous cyst of the cerebellopontine angle cistern: case report. Neurosurgery 1991;28(3):462–6.
47. Christov C, Chretien F, Brugieres P, Djindjian M. Giant supratentorial enterogenous cyst: report of a case, literature review, and discussion of pathogenesis. Neurosurgery 2004;54(3):759–63.
48. Birch BD, McCormick PC. High cervical split cord malformation and neurenteric cyst associated with congenital mirror movements: case report. Neurosurgery 1996;38:813–6.
49. Devkota UP, Lam JM, Ng H, Poon WS. An anterior intradural neurenteric cyst of the cervical spine: complete excision through central corpectomy approach—case report. Neurosurgery 1994;35(6):1150–3; discussion 1153–4.
50. Eynon-Lewis NJ, Kitchen N, Scaravilli F, Brookes GB. Neurenteric cyst of the cerebellopontine angle: case report. Neurosurgery 1998;42:655–8.
51. Harris CP, Dias MS, Brockmeyer DL, Townsend JJ, Willis BK, Apfelbaum RI. Neurenteric cysts of the posterior fossa: recognition, management, and embryogenesis. Neurosurgery 1991;29: 893–8.
52. Kim CY, Wang KC, Choe G, et al. Neurenteric cyst: its various presentations. Childs Nerv Syst 1999;15:333–41.
53. Macdonald RL, Schwartz ML, Lewis AJ. Neurenteric cyst located dorsal to the cervical spine: case report. Neurosurgery 1991;28:583–8.
54. Mendel E, Lese GB, Gonzalez-Gomez I, Nelson MD, Raffel C. Isolated lumbosacral neurenteric cyst with partial sacral agenesis: case report. Neurosurgery 1994;35:1159–62; discussion 1162–3.

55. Menezes AH, Ryken TC. Craniocervical intradural neurenteric cysts. Pediatr Neurosurg 1995; 22:88–95.
56. Morita Y, Kinoshita K, Wakisaka S, Makihara S. Fine surface structure of an intraspinal neurenteric cyst: a scanning and transmission electron microscopy study. Neurosurgery 1990;27:829–33.
57. Morita Y, Kinoshita K, Wakisaka S, Makihara S. Claws of cilia: further observation of ciliated epithelium in neurenteric cyst. Virchows Arch A Pathol Anat Histopathol 1991;418:263–5.
58. Paleologos TS, Thom M, Thomas DG. Spinal neurenteric cysts without associated malformations. Are they the same as those presenting in spinal dysraphism? Br J Neurosurg 2000;14:185–94.
59. Paolini S, Ciappetta P, Domenicucci M, Guiducci A. Intramedullary neurenteric cyst with a false mural nodule: case report. Neurosurgery 2003; 52:243–6.
60. Prasad VS, Reddy DR, Murty JM. Cervico-thoracic neurenteric cyst: clinicoradiological correlation with embryogenesis. Childs Nerv Syst 1996;12:48–51.
61. Shin JH, Byun BJ, Kim DW, Choi DL. Neurenteric cyst in the cerebellopontine angle with xanthogranulomatous changes: serial MR findings with pathologic correlation. AJNR Am J Neuroradiol 2002;23:663–5.
62. Whiting DM, Chou SM, Lanzieri CF, Kalfas IH, Hardy RW. Cervical neurenteric cyst associated with Klippel-Feil syndrome: a case report and review of the literature. Clin Neuropathol 1991;10:285–90.
63. Lach B, Scheithauer BW, Gregor A, Wick MR. Colloid cyst of the third ventricle. A comparative immunohistochemical study of neuraxis cysts and choroid plexus epithelium. J Neurosurg 1993;78:101–11.
64. Piramoon AN, Abbassioun K. Mediastinal enterogenic cyst with spinal cord compression. J Pediatr Surg 1974;9:543–5.
65. Dorsey JF, Tabrisky J. Intraspinal and mediastinal foregut cyst compressing the spinal cord. Report of a case. J Neurosurg 1966;24:562–7.
66. Schelper RL, Kagan-Hallet KS, Huntington HW. Brainstem subarachnoid respiratory epithelial cysts: report of two cases and review of the literature. Hum Pathol 1986;17:417–22.
67. Ho KL, Tiel R. Intraspinal bronchogenic cyst: ultrastructural study of the lining epithelium. Acta Neuropathol (Berl) 1989;78:513–20.
68. Yamashita J, Maloney AF, Harris P. Intradural spinal bronchiogenic cyst. Case report. J Neurosurg 1973;39:240–5.
69. Ho KL, Chason JL. Subarachnoid epithelial cyst of the cerebellum. Immunohistochemical and ultrastructural studies. Acta Neuropathol (Berl) 1989;78:220–4.
70. Hoefnagel D, Benirschke K, Duarte J. Teratomatous cysts within the vertebral canal. Observations on the occurrence of sex chromatin. J Neurol Neurosurg Psychiatry 1962;25:159–164.
71. Rosenbaum TJ, Soule EH, Onofrio BM. Teratomatous cyst of the spinal canal. Case report. J Neurosurg 1978;49:292–7.
72. Del Bigio MR, Jay V, Drake JM. Prepontine cyst lined by respiratory epithelium with squamous metaplasia: immunohistochemical and ultrastructural study. Acta Neuropathol (Berl) 1992;83:564–8.
73. Knight G, Griffiths T, Williams I. Gastrocystoma of the spinal cord. Br J Surg 1955;42:635–8.
74. Bentley JF, Smith JR. Developmental posterior enteric remnants and spinal malformations: the split notochord syndrome. Arch Dis Child 1960;35:76–86.
75. Inoue T, Matsushima T, Fukui M, Iwaki T, Takeshita I, Kuromatsu C. Immunohistochemical study of intracranial cysts. Neurosurgery 1988;23:576–81.

Ependymal (Glioependymal) Cyst

76. Boockvar JA, Shafa R, Forman MS, O'Rourke DM. Symptomatic lateral ventricular ependymal cysts: criteria for distinguishing these rare cysts from other symptomatic cysts of the ventricles: case report. Neurosurgery 2000;46:1229–32; discussion 1232–3.
77. Bouch DC, Mitchell I, Maloney AF. Ependymal lined paraventricular cerebral cysts; a report of three cases. J Neurol Neurosurg Psychiatry 1973;36:611–7.
78. Gherardi R, Lacombe MJ, Poirier J, Roucayrol AM, Wechsler J. Asymptomatic encephalic intraparenchymatous neuroepithelial cysts. Acta Neuropathol (Berl) 1984;63:264–8.
79. Robertson DP, Kirkpatrick JB, Harper RL, Mawad ME. Spinal intramedullary ependymal cyst. Report of three cases. J Neurosurg 1991;75:312–6.
80. Ho KL, Chason JL. A glioependymal cyst of the cerebellopontine angle. Immunohistochemical and ultrastructural studies. Acta Neuropathol (Berl) 1987;74:382–8.
81. Iwahashi H, Kawai S, Watabe Y, et al. Spinal intramedullary ependymal cyst: a case report. Surg Neurol 1999;52:357–61.
82. Pant B, Uozumi T, Hirohata T, et al. Endoscopic resection of intraventricular ependymal cyst presenting with psychosis. Surg Neurol 1996;46:573–8.

Epidermoid and Dermoid Cysts

83. Mohanty A, Venkatrama SK, Rao BR, Chandramouli BA, Jayakumar PN, Das BS. Experience with cerebellopontine angle epidermoids. Neurosurgery 1997;40:24–30.
84. Talacchi A, Sala F, Alessandrini F, Turazzi S, Bricolo A. Assessment and surgical management of posterior fossa epidermoid tumors: report of 28 cases. Neurosurgery 1998;42:242–51; discussion 251–2.
85. Vinchon M, Pertuzon B, Lejeune JP, Assaker R, Pruvo JP, Christiaens JL. Intradural epidermoid cysts of the cerebellopontine angle: diagnosis and surgery. Neurosurgery 1995;36:52–7.
86. Yamakawa K, Shitara N, Genka S, Manaka S, Takakura K. Clinical course and surgical prognosis of 33 cases of intracranial epidermoid tumors. Neurosurgery 1989;24:568–73.
87. Roux A, Mercier C, Larbrisseau A, Dube LJ, Dupuis C, Del Carpio R. Intramedullary epidermoid cysts of the spinal cord. Case report. J Neurosurg 1992;76:528–33.
88. Boyd HR. Iatrogenic intraspinal epidermoid. Report of a case. J Neurosurg 1966;24:105–7.
89. de Carvalho GT, Fagundes-Pereyra WJ, Marques JA, Dantas FL, de Sousa AA. Congenital inclusion cysts of the anterior fontanelle. Surg Neurol 2001;56:400–5.
90. Lunardi P, Missori P, Gagliardi FM, Fortuna A. Dermoid cysts of the posterior cranial fossa in children. Report of nine cases and review of the literature. Surg Neurol 1990;34:39–42.
91. Akhaddar A, Jiddane M, Chakir N, El Hassani R, Moustarchid B, Bellakhdar F. Cerebellar abscesses secondary to occipital dermoid cyst with dermal sinus: case report. Surg Neurol 2002;58:266–70.
92. Goffin J, Plets C, Van Calenbergh F, et al. Posterior fossa dermoid cyst associated with dermal fistula: report of 2 cases and review of the literature. Childs Nerv Syst 1993;9:179–81.
93. Higashi S, Takinami K, Yamashita J. Occipital dermal sinus associated with dermoid cyst in the fourth ventricle. AJNR Am J Neuroradiol 1995;16(Suppl 4):945–8.
94. Logue V, Till K. Posterior fossa dermoid cysts with special reference to intracranial infection. J Neurol Neurosurg Psychiat 1952;14:1–12.
95. Zerris VA, Annino D, Heilman CB. Nasofrontal dermoid sinus cyst: report of two cases. Neurosurgery 2002;51:811–4.
96. Horowitz BL, Chari MV, James R, Bryan RN. MR of intracranial epidermoid tumors: correlation of in vivo imaging with in vitro 13C spectroscopy. AJNR Am J Neuroradiol 1990;11:299–302.
97. Ikushima I, Korogi Y, Hirai T, et al. MR of epidermoids with a variety of pulse sequences. AJNR Am J Neuroradiol 1997;18:1359–63.
98. Chen S, Ikawa F, Kurisu K, Arita K, Takaba J, Kanou Y. Quantitative MR evaluation of intracranial epidermoid tumors by fast fluid-attenuated inversion recovery imaging and echo-planar diffusion-weighted imaging. AJNR Am J Neuroradiol 2001;22:1089–96.
99. Dutt SN, Mirza S, Chavda SV, Irving RM. Radiologic differentiation of intracranial epidermoids from arachnoid cysts. Otol Neurotol 2002;23:84–92.
100. Barkovich AJ, Edwards M, Cogen PH. MR evaluation of spinal dermal sinus tracts in children. AJNR Am J Neuroradiol 1991;12:123–9.
101. Graham DV, Tampieri D, Villemure JG. Intramedullary dermoid tumor diagnosed with the assistance of magnetic resonance imaging. Neurosurgery 1988;23:765–7.
102. Smith AS, Benson JE, Blaser SI, Mizushima A, Tarr RW, Bellon EM. Diagnosis of ruptured intracranial dermoid cyst: value MR over CT. AJNR Am J Neuroradiol 1991;12:175–80.
103. Lunardi P, Missori P, Rizzo A, Gagliardi FM. Chemical meningitis in ruptured intracranial dermoid. Case report and review of the literature. Surg Neurol 989;32:449–52.
104. Stendel R, Pietila TA, Lehmann K, Kurth R, Suess O, Brock M. Ruptured intracranial dermoid cysts. Surg Neurol 2002;57(6):391–8.
105. Hitchcock MG, Ellington KS, Friedman AH, Provenzaie JM, McLendon RE. Shadow cells in an intracranial dermoid cyst. Arch Pathol Lab Med 1995;119:371–3.
106. Asahi T, Kurimoto M, Endo S, Monma F, Ohi M, Takami M. Malignant transformation of cerebello-pontine angle epidermoid. J Clin Neurosci 2001;8:572–4.
107. Gluszcz A. A cancer arising in a dermoid of the brain. A case report. J Neuropathol Exp Neurol 1962;21:383–7.
108. Goldman SA, Gandy SE. Squamous cell carcinoma as a late complication of intracerebroventricular epidermoid cyst. Case report. J Neurosurg 1987;66:618–20.
109. Knorr JR, Ragland RL, Smith TW, Davidson RI, Keller JD. Squamous carcinoma arising in a cerebellopontine angle epidermoid: CT and MR findings. AJNR Am J Neuroradiol 1991;12:1182–4.
110. Lewis AJ, Cooper PW, Kassel EE, Schwartz ML. Squamous cell carcinoma arising in a suprasellar epidermoid cyst. Case report. J Neurosurg 1983;59:538–41.

Arachnoid Cyst

111. Wang MY, Levi AD, Green BA. Intradural spinal arachnoid cysts in adults. Surg Neurol 2003;60(1):49–55; discussion 55–6.
112. Go KG, Houthoff HJ, Blaauw EH, Havinga P, Hartsuiker J. Arachnoid cysts of the sylvian fissure. Evidence of fluid secretion. J Neurosurg 1984;60:803–13.
113. Passero S, Filosomi G, Cioni R, Venturi C, Volpini B. Arachnoid cysts of the middle cranial fossa: a clinical, radiological and follow-up study. Acta Neurol Scand 1990;82:94–100.
114. Wester K. Gender distribution and sidedness of middle fossa arachnoid cysts: a review of cases diagnosed with computed imaging. Neurosurgery 1992;31:940–4.
115. Wester K. Peculiarities of intracranial arachnoid cysts: location, sidedness, and sex distribution in 126 consecutive patients. Neurosurgery 1999;45:775–9.
116. Jallo GI, Woo HH, Meshki C, Epstein FJ, Wisoff JH. Arachnoid cysts of the cerebellopontine angle: diagnosis and surgery. Neurosurgery 1997;40:31–8.
117. Shaw CM. "Arachnoid cysts" of the sylvian fissure versus "temporal lobe agenesis" syndrome. Ann Neurol 1979;5:483–5.
118. Krawchenko J, Collins GH. Pathology of an arachnoid cyst. Case report. J Neurosurg 1979;50:224–8.
119. Starkman SP, Brown TC, Linell EA. Cerebral arachnoid cysts. J Neuropathol Exp Neurol 1958;17:484–500.
120. Inoue T, Matsushima T, Fukui M, Iwaki T, Takeshita I, Kuromatsu C. Immunohistochemical study of intracranial cysts. Neurosurgery 1988;23:576–81.

Choroid Plexus Cyst

121. Giorgi C. Symptomatic cyst of the choroid plexus of the lateral ventricle. Neurosurgery 1979;5(Pt 1):53–6.
122. Inoue T, Matsushima T, Fukui M, Matsubara T, Kitamoto T. Choroidal epithelial cyst of the cerebral hemisphere. An immunohistochemical study. Surg Neurol 1987;28:119–22.
123. Odake G, Tenjin H, Murakami N. Cyst of the choroid plexus in the lateral ventricle: case report and review of the literature. Neurosurgery 1990;27:470–6.
124. Chitkara U, Cogswell C, Norton K, Wilkins IA, Mehalek K, Berkowitz RL. Choroid plexus cysts in the fetus: a benign anatomic variant or pathologic entity? Report of 41 cases and review of the literature. Obstet Gynecol 1988;72:185–9.
125. DeRoo TR, Harris RD, Sargent SK, Denholm TA, Crow HC. Fetal choroid plexus cysts: prevalence, clinical significance, and sonographic appearance. AJR Am J Roentgenol 1988;151:1179–81.
126. Oettinger M, Odeh M, Korenblum R, Markovits J. Antenatal diagnosis of choroid plexus cyst: suggested management. Obstet Gynecol Surv 1993;48:635–9.
127. Reinsch RC. Choroid plexus cysts—association with trisomy: prospective review of 16,059 patients. Am J Obstet Gynecol 1997;176:1381–3.
128. Farhood AI, Morris JH, Bieber FR. Transient cysts of the fetal choroid plexus: morphology and histogenesis. Am J Med Genet 1987;27:977–82.

Nerve Root (Tarlov) Cyst

129. Caspar W, Papavero L, Nabhan A, Loew C, Ahlhelm F. Microsurgical excision of symptomatic sacral perineurial cysts: a study of 15 cases. Surg Neurol 2003;59:101–6.
130. Mummaneni PV, Pitts LH, McCormack BM, Corroo JM, Weinstein PR. Microsurgical treatment of symptomatic sacral Tarlov cysts. Neurosurgery 2000;47:74–9.
131. Nathan H, Rosner S. Multiple meningeal diverticula and cysts associated with duplications of the sheaths of spinal nerve posterior roots. J Neurosurg 1977;47:68–72.
132. Siqueira E, Schaffer L, Kranzler L, Gan J. CT characteristics of sacral perineural cysts. Report of two cases. J Neurosurg 1984;61:596–8.
133. Tarlov IM. Spinal perineurial and meningeal cysts. J Neurol Neurosurg Psychiatry 1970;33:833–43.
134. Voyadzis JM, Bhargava P, Henderson FC. Tarlov cysts: a study of 10 cases with review of the literature. J Neurosurg 2001;95(Suppl):25–32.
135. Paulsen RD, Call GA, Murtagh FR. Prevalence and percutaneous drainage of cysts of the sacral nerve root sheath (Tarlov cysts). AJNR Am J Neuroradiol 1994;15:293–9.
136. Van de Kelft E, Van Vyve M. Chronic perineal pain related to sacral meningeal cysts. Neurosurgery 1991;29:223–6.

Synovial Cyst of the Spine

137. Gadgil AA, Eisenstein SM, Darby A, Cassar Pullicino V. Bilateral symptomatic synovial cysts of the lumbar spine caused by calcium pyrophosphate deposition disease: a case report. Spine 2002;27:E428–31.
138. Gorey MT, Hyman RA, Black KS, Scuderi DM, Cinnamon J, Kim KS. Lumbar synovial cysts eroding bone. AJNR Am J Neuroradiol 1992;13:161–3.

139. Hemminghytt S, Daniels DL, Williams AL, Haughton VM. Intraspinal synovial cysts: natural history and diagnosis by CT. Radiology 1982;145:375–6.
140. Kao CC, Uihlein A, Bickel WH, Soule EH. Lumbar intraspinal extradural ganglion cyst. J Neurosurg 1968;29:168–72.
141. Phuong LK, Atkinson JL, Thielen KR. Far lateral extraforaminal lumbar synovial cyst: report of two cases. Neurosurgery 2002;51:505–8.
142. Sandhu FA, Santiago P, Fessler RG, Palmer S. Minimally invasive surgical treatment of lumbar synovial cysts. Neurosurgery 2004;54:107–11; discussion 111–2.
143. Tillich M, Trummer M, Lindbichler F, Flaschka G. Symptomatic intraspinal synovial cysts of the lumbar spine: correlation of MR and surgical findings. Neuroradiology 2001;43:1070–5.

Miscellaneous Cysts

144. Nakasu Y, Handa J, Watanabe K. Progressive neurological deficits with benign intracerebral cysts. Report of two cases. J Neurosurg 1986;65:706–9.
145. Wilkins RH, Burger PC. Benign intraparenchymal brain cysts without an epithelial lining. J Neurosurg 1988;68:378–82.
146. Sherman JL, Camponovo E, Citrin CM. MR imaging of CSF-like choroidal fissure and parenchymal cysts of the brain. AJNR Am J Neuroradiol 1990;11:939–45.
147. Bergland RM. Congenital intraspinal extradural cyst. Report of three cases in one family. J Neurosurg 1968;28:495–9.
148. Hamburger CH, Buttner A, Weis S. Dural cysts in the cervical region. Report of three cases and review of the literature. J Neurosurg 1998;89:310–3.
149. Nugent GR, Odom GL, Woodhall B. Spinal extradural cysts. Neurology 1959;9:397–406.
150. Papo I, Longhi G, Caruselli G. Giant spinal extradural cyst. Surg Neurol 1977;8:350–2.
151. Uemura K, Yoshizawa T, Matsumura A, Asakawa H, Nakamagoe K, Nose T. Spinal extradural meningeal cyst. Case report. J Neurosurg 1996;85:354–6.

17 TUMOR-LIKE LESIONS OF MALDEVELOPMENTAL OR UNCERTAIN ORIGIN

HYPOTHALAMIC HAMARTOMA

Definition. *Hypothalamic hamartoma* is a malformative mass of neurons and glia in the region of the hypothalamus.

Clinical Features. These generally small, solitary, non-neoplastic lesions are typically incidental radiologic or postmortem findings. Manifestations, if any, include endocrine abnormalities, such as precocious puberty, which are due in some cases to lesional overproduction of gonadotropin-releasing hormone (1–3) or corticotropin-releasing hormone (1,4). In most instances, however, no clinical evidence of endocrine dysfunction is apparent. There may be disorders of vegetative function and psychotic disease. Some pediatric patients convulse periodically into highly distinctive episodes of laughter known as gelastic seizures (5,6).

Radiologic Findings. Hypothalamic hamartomas vary in signal intensity with magnetic resonance imaging (MRI), but most are isointense to cortical gray matter in precontrast T1-weighted images (fig. 17-1) (7). They are not contrast enhancing and do not show interval enlargement.

Gross Findings. The lesions are smooth-surfaced, homogeneous, gray nodules intimately related to the floor of the third ventricle (fig. 17-2). Some have only a vascular connection, and a few are independent of the overlying brain. Most are a centimeter or less in diameter, but a few are substantially larger and can intrude upon the third ventricle.

Microscopic Findings. Hypothalamic hamartomas are composed of fully differentiated neuroglial tissue that, despite its location and minor architectural abnormalities, closely resembles that of normal hypothalamus. The lesions are covered by a thin glial layer analogous to the superficial or molecular layer of the cerebral cortex. The bulk of the nodule is composed of neurons and supportive glia (fig. 17-3) (8). Neurons lie singly or in clusters within fields of unmyelinated axons which, when disposed

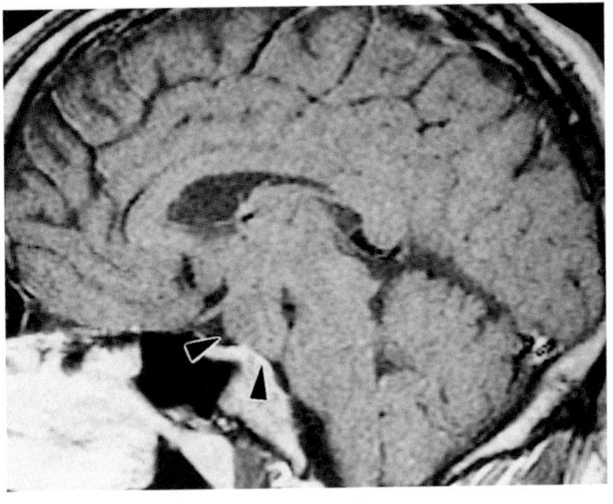

Figure 17-1

HYPOTHALAMIC HAMARTOMA

As is typical, this large hypothalamic hamartoma (arrowheads) is attached to the floor of the third ventricle. (Fig. 17-1 from Fascicle 10, 3rd Series.)

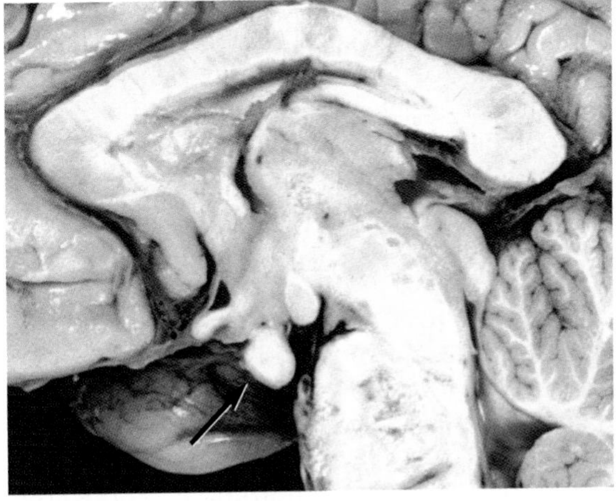

Figure 17-2

HYPOTHALAMIC HAMARTOMA

Smooth surfaced hypothalamic hamartoma (arrow) is often an incidental autopsy finding.

in parallel, closely resemble normal fiber tracts. Glial cells are usually normal in number, but astrocytic hypertrophy (gliosis) may be present. No mitoses are expected.

Immunohistochemical Findings. In individual lesions, a variety of hypothalamic hormones and neurotransmitters have been demonstrated by immunohistochemical methods. These include releasing hormones for growth hormone, corticotropin, thyrotropic hormone, and gonadotropins, and the neurotransmitter somatostatin (9).

Ultrastructural Findings. The mass is composed largely of unmyelinated axons, but scattered neuronal cell bodies and glia are seen. Not surprisingly, electron-dense secretory granules that correspond to hypothalamic-releasing hormones have been demonstrated (10).

Differential Diagnosis. Hypothalamic hamartomas mimic normal gray matter to such an extent that the principal diagnostic challenge is to exclude normal brain tissue, a distinction that can be difficult in small fragmented specimens. Histologically, the distinction is based upon the disorganization of the tissue, frequent clustering of ganglion cells, and astrocytic hypertrophy present in some cases. Awareness of neuroradiologic findings and a discussion with the surgeon regarding the site of the biopsy are obviously critical.

Distinguishing hypothalamic hamartoma from a ganglion cell tumor is a problem in theory, but rarely one in practice. Hypothalamic hamartoma consists of normal-appearing, albeit ectopic, gray matter, whereas ganglion cell tumors are composed of disordered proliferations of abnormal ganglion cells within a considerably more cellular glial stroma. Gangliogliomas typically also feature cytologic pleomorphism, sclerotic vessels, desmoplasia, and chronic inflammation. Central nervous system (CNS) tumors with ganglion cells are given in Appendix G.

Treatment and Prognosis. Hypothalamic hamartomas are biologically indolent and, if resected, are "cured" by surgery alone. This may not be necessary in asymptomatic lesions, but resection does alleviate precocious puberty (11). One approach utilizes stereotactic radiosurgery (12). Gonadotrophin-releasing factor analogs are also employed (3). Gelastic epilepsy may be refractory to treatment.

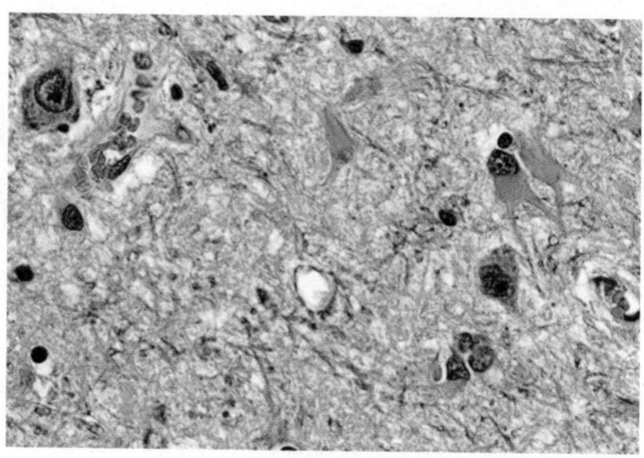

Figure 17-3

HYPOTHALAMIC HAMARTOMA

Hypothalamic hamartomas are well-formed masses of gray matter with a cell density similar to that of normal hypothalamus.

NASAL CEREBRAL HETEROTOPIA ("NASAL GLIOMA")

Definition. *Nasal cerebral heterotopias*, whether extranasal or intranasal, are masses of mature, displaced CNS tissue connected with the CNS either directly or by a simple fibrous tract. When directly connected and better organized, the lesion justifiably can be considered a form of encephalocele. In either case, the process is clearly malformative, despite the popular misnomer *nasal glioma* (13,14).

Clinical Features. Extranasal lesions often present as a visible protuberance near the glabella, whereas intranasal examples produce a nasal mass with resulting "stuffiness." If large, obstruction results (15).

Radiologic Findings. Neuroimaging clarifies the relationship of the mass to the CNS, whether contiguous or independent (13). The epicenter of the lesion is usually slightly off the midline.

Microscopic Findings. Histologically, these lesions are similar to related congenital malformations such as encephaloceles since they consist of islands of neuroglial tissue enmeshed in a matrix of dense connective tissue (figs. 17-4, 17-5). Neurons may be more prominent in some (16). Any question about the nature of this somewhat pale-staining, neural-appearing tissue is resolved by immunostains for glial fibrillary acidic protein (GFAP) and S-100 protein (14, 17), as well as neurofilament protein. With the use of the Masson trichrome stain, the gray

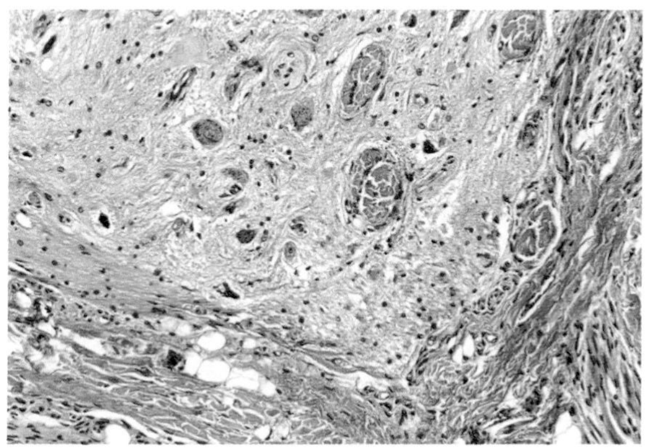

Figure 17-4

NASAL CEREBRAL HETEROTOPIA

The lesion consists of ectopic central nervous system (CNS) tissue, in this case with prominent ganglion cells.

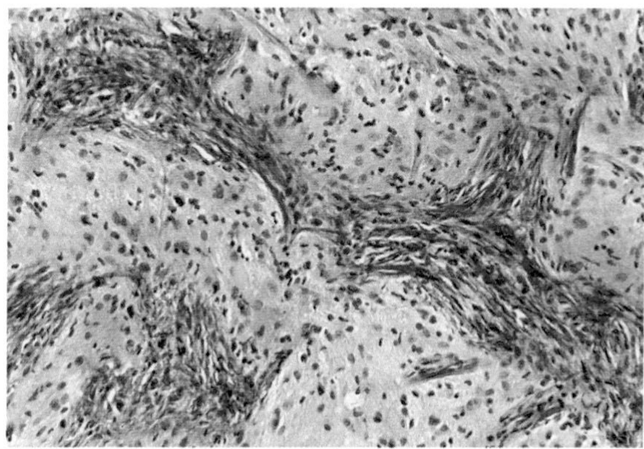

Figure 17-5

NASAL CEREBRAL HETEROTOPIA

The apposition of dense connective tissue with malformed CNS elements is typical of nasal cerebral heterotopia (Masson trichrome stain).

neuroglial component contrasts with the bright blue surrounding fibrous connective tissue (fig. 17-5). The intimate mixture of collagenous tissue and CNS tissue is typical of malformations such as encephaloceles at any site.

Treatment and Prognosis. Nasal neuroglial heterotopias are cured by simple excision.

MENINGIOANGIOMATOSIS

Definition. *Meningioangiomatosis* is an intracortical perivascular proliferation of meningothelial cells and fibroblasts.

General Features. While the lesion has been considered a reactive or hamartomatous process, the molecular findings described below and the association with an overlying meningioma in rare cases, suggest that some lesions could be neoplasms (18–20).

Clinical Features. Meningioangiomatosis is an uncommon entity that occurs either sporadically (21–23) or, less commonly, in patients with neurofibromatosis 2 (NF2), in which case multifocality can be seen (see fig. 18-14) (24). In accord with the cortical localization, seizures are the principal symptom (25). A minority of lesions are associated with an overlying calcifying pseudoneoplasm of the neuraxis.

Radiologic Findings. The intracortical, rather plaque-like lesions have a low (dark) signal on T2-weighted MRIs (fig. 17-6) (25). Only an occasional example is contrast enhancing. Calcification is seen well by computerized tomography (CT).

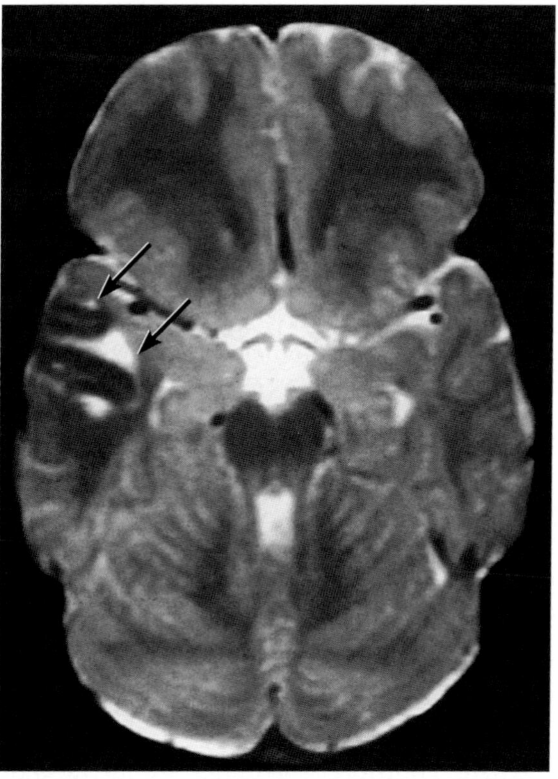

Figure 17-6

MENINGIOANGIOMATOSIS

As seen here in this T2-weighted image, collagen and calcification can combine to produce a black signal within the affected cortex (arrows). A bright signal (edema) is present in subjacent white matter.

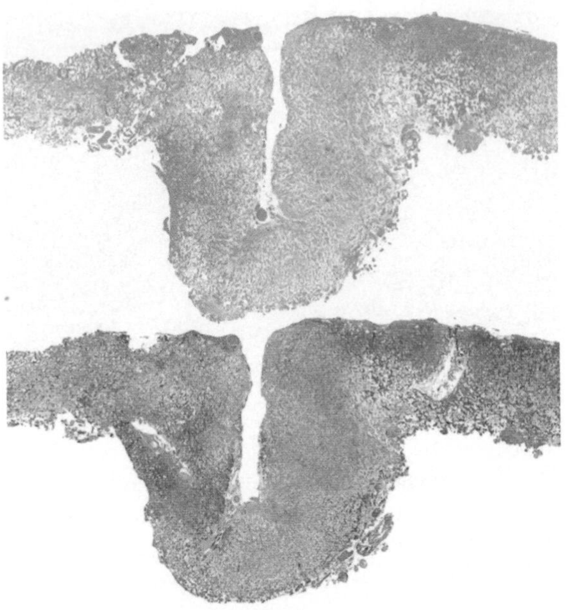

Figure 17-7

MENINGIOANGIOMATOSIS

Paired histologic sections stained with hematoxylin and eosin (H&E) (top) and for connective tissue (bottom) illustrate the degree of collagenization present in advanced cases.

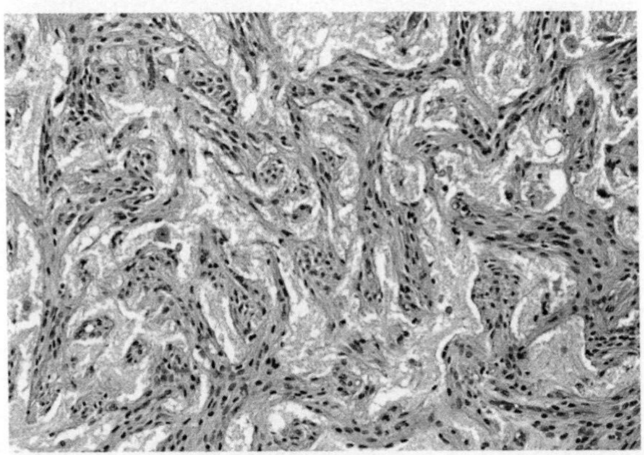

Figure 17-8

MENINGIOANGIOMATOSIS

The perivascular cells in some lesions have typical meningiothelial features.

Gross Findings. The dense lesion thickens and hardens the affected cortex over a sharply defined plaque-like area. Serpentine vessels on the leptomeningeal surface often create the impression of a vascular malformation. Fibrotic and heavily calcified examples can resist and even dull a scalpel (fig. 17-7).

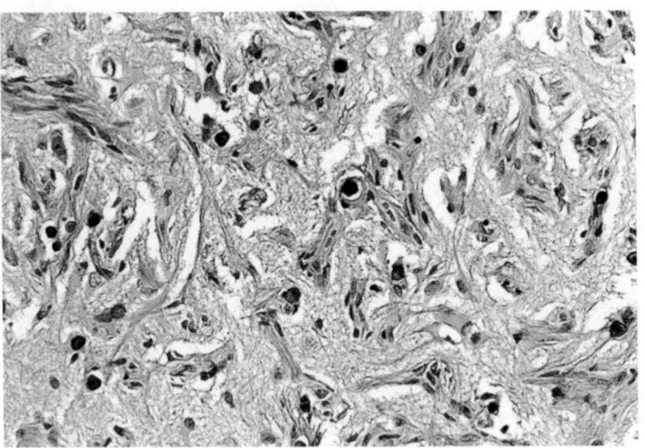

Figure 17-9

MENINGIOANGIOMATOSIS

There is only a thin perivascular layer of fibroblast-like cells in some examples. Scattered calcospherites are common.

Microscopic Findings. The microscopic appearance varies from lesion to lesion. Some are proliferations of elongated, perivascular, fibroblast-like cells, whereas other lesions have more obvious meningothelial features (figs. 17-8, 17-9), sometimes even forming nodules or lobules similar to minute meningiomas. Psammoma bodies are present in varying numbers in either case.

As the ratio of stroma to brain parenchyma increases, the intrinsic cortical architecture is obscured proportionally by downgrowth of the lesion. In advanced cases, the perivascular connective tissue element dominates, leaving only small islands of distorted parenchyma (fig. 17-10). Neurons within or immediately around the lesion may demonstrate neurofibrillary tangles or granulovacuolar degeneration (fig. 17-11) (26–29). Osseous metaplasia may be present in advanced cases.

Immunohistochemical Findings. In some examples, perivascular cells are variably positive for epithelial membrane antigen (EMA), but this reactivity may be scant or absent in tissue that is more spindled and fibroblastic. The MIB-1 index is less than 1 percent (30).

Ultrastructural Findings. The meningothelial nature of the cortical perivascular proliferation is evidenced by cellular interdigitation, scattered cytoplasmic intermediate filaments, and well-formed desmosomes. Paired helical filaments (neurofibrillary change) may fill the cytoplasm of trapped or adjacent neurons (26,27).

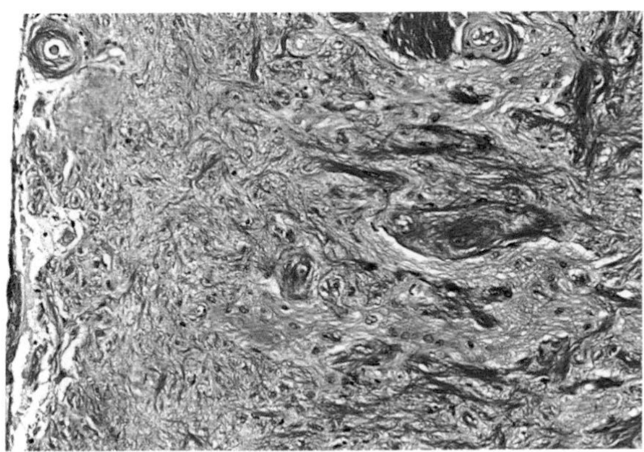

Figure 17-10

MENINGIOANGIOMATOSIS

The intracortical element can be densely collagenized.

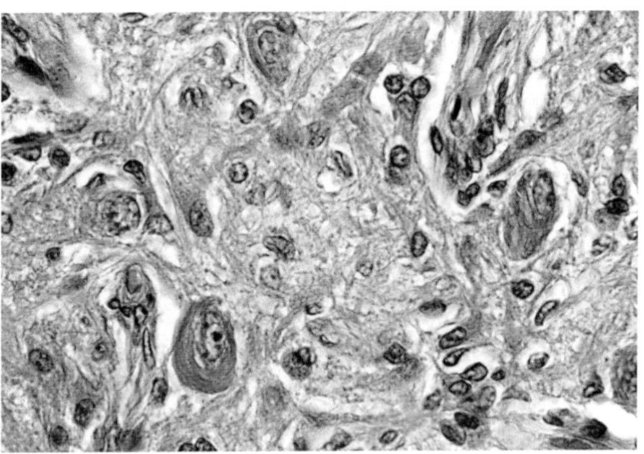

Figure 17-11

MENINGIOANGIOMATOSIS

Ganglion cells trapped within the lesion are susceptible to neurofibrillary change.

Molecular Findings. In complex cases featuring both cortical meningioangiomatosis and an accompanying meningioma, deletion of the *NF2* gene region in both lesions suggests a relationship between the two processes, although which lesion is primary is unclear (20). One report of a lesion in a non-NF2 patient found loss of heterozygosity on microsatellite markers flanking the *NF2* gene (31). No overt meningioma was present, but there were micronodules of meningothelial cells in the adjacent leptomeninges.

Differential Diagnosis. When the meningothelial/fibroblastic character of the intracortical lesion is readily apparent, the diagnosis depends only on recognition of the entity. In small fragmented specimens, particularly the more collagenized examples, the diagnosis is more challenging and one must consider the possibility of another slowly growing and often calcified epileptogenic mass, ganglion cell tumor (see Appendix G for this and other lesions with ganglion cells). In contrast to meningioangiomatosis, ganglion cell tumors other than desmoplastic infantile ganglioglioma are generally less collagenized, contain clearly abnormal neurons, often have perivascular lymphocytes, and do not demonstrate intracortical meningothelial cells. They are also often cystic mass lesions.

An invasive meningioma also becomes suspect if meningothelial cells are prominent. In prototypical forms, the two are distinct lesions. One is an extra-axial meningothelial mass, the other an intracortical, calcified, plaque-like lesion of more fibroblastic character. As discussed above, there is overlap in rare cases where both lesions are present in the same specimen, and the histologic features of the cortical component may be intermediate between those of the two classic lesions.

A vascular malformation may be entertained when leptomeningeal vascularity is prominent, but abnormal intracortical vessels in meningioangiomatosis are evenly spaced and interrupted by islands of cerebral cortex.

Treatment and Prognosis. In most cases, meningioangiomatosis is cured by surgical resection.

CALCIFYING PSEUDONEOPLASM OF THE NEURAXIS

Definition. *Calcifying pseudoneoplasm of the neuraxis* is a non-neoplastic calcified mass of unknown etiology involving either the meninges or paraspinous tissue.

Clinical Features. The lesion may occur in close association with meningioangiomatosis and in patients with NF2 (32). Most are sporadic and arise independently of any recognized predisposing factor (33). Symptoms of mass effect or nerve compression are often present for years. It is unclear whether all reported cases represent the same entity since some occur extra-axially and overlap with calcifying pseudotumors of other types. Despite its unknown etiology, the lesion described here is clearly a unique clinicopathologic entity.

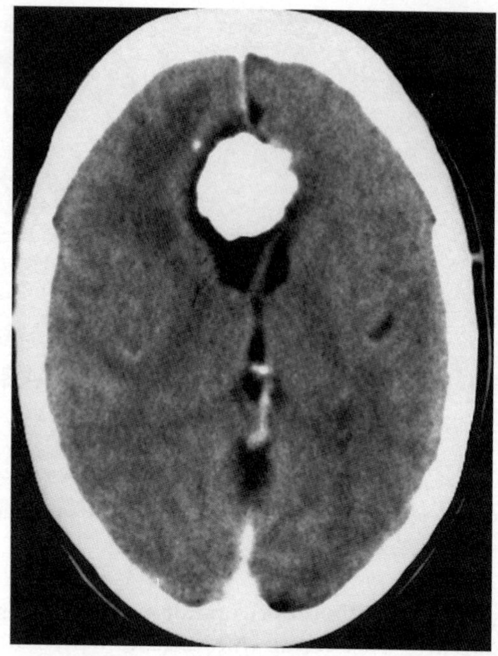

Figure 17-12

CALCIFYING PSEUDONEOPLASM OF THE NEURAXIS

The lesions are well circumscribed, usually leptomeningeal, and densely calcified, as seen here in a computerized tomography (CT) scan. (Fig. 17-11 from Fascicle 10, 3rd Series.)

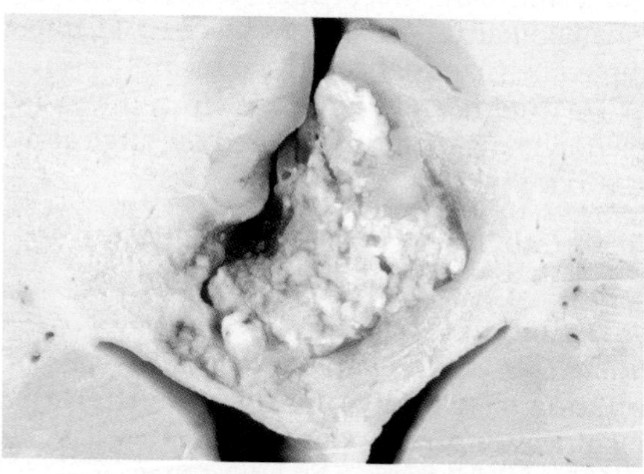

Figure 17-13

CALCIFYING PSEUDONEOPLASM OF THE NEURAXIS

When present, the meningeal component is characteristically gritty and grummous.

Calcified intracranial and intraspinal lesions are listed in Appendix F.

Radiologic Findings. The lesions are discrete, usually superficial, either leptomeningeal or extradural, and densely calcified (fig. 17-12).

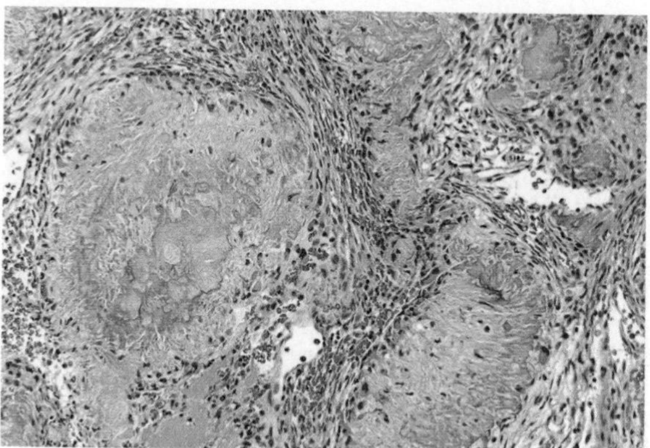

Figure 17-14

CALCIFYING PSEUDONEOPLASM OF THE NEURAXIS

The lesion is a multilobular mass of granular and fibrillar material.

Gross Findings. The well-circumscribed, rock-hard, somewhat granular mass often measures several centimeters and typically lies in the meninges (fig. 17-13). When positioned within a sulcus, a lesion may be partially buried within the substance of the brain. Spinal lesions appear to arise in paraspinous tissue and to secondarily involve bone.

Microscopic Findings. The distinctive, but difficult to describe mass has a curious microscopic appearance, superficially resembling a cross between gouty tophus, tumoral calcinosis, rheumatoid nodule, and osteoma (figs. 17-14, 17-15). It is not surprising that it has been designated variably as *fibroosseous lesion* (34), *unusual fibroosseous lesion* (35,36), and *calcifying pseudotumor of the neuraxis* (33). In spite of its largely amorphous quality, the lesion has distinctive and diagnostic features, including a radial fibrillarity in the outer rind of the calcified mass produced by "rope-like" basophilic material, and a delicate and inconstant covering of somewhat epithelioid cells and occasional giant cells (fig. 17-16). Metaplasia to mature bone and adipose tissue may be present (fig. 17-15).

Immunohistochemical Findings. The superficial rind of radiating cells is variably but often convincingly positive for EMA (fig. 17-17).

Treatment and Prognosis. The lesions are often cured by excision or even enucleation, but significant morbidity and mortality are associated with large skull base examples that

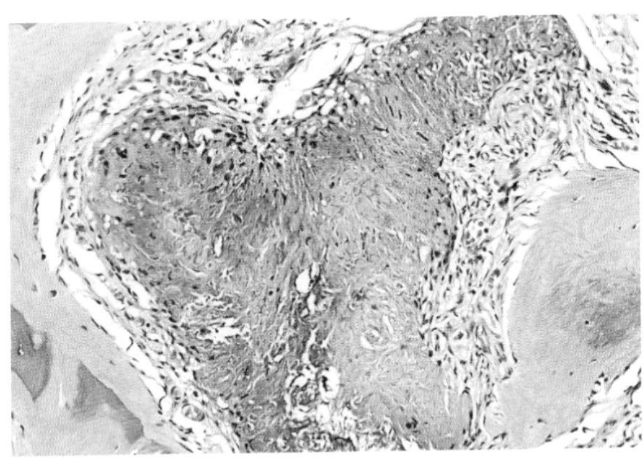

Figure 17-15

CALCIFYING PSEUDONEOPLASM OF THE NEURAXIS

Mature bone is often present. The superficial layer of cells surrounds the basophilic fibrillar core.

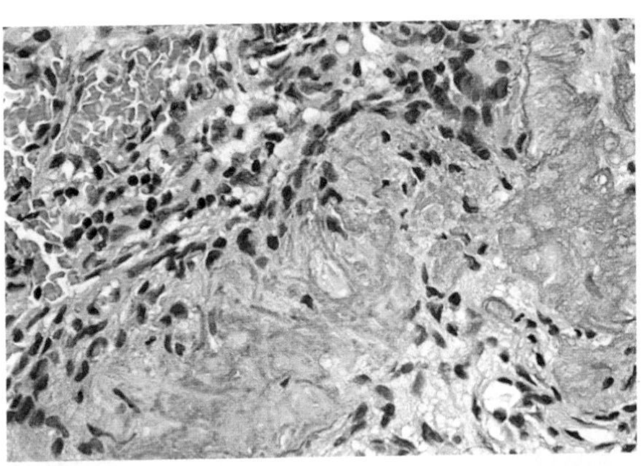

Figure 17-16

CALCIFYING PSEUDONEOPLASM OF THE NEURAXIS

A superficial layer of small polygonal cells is prominent in some cases, as here, but not in others.

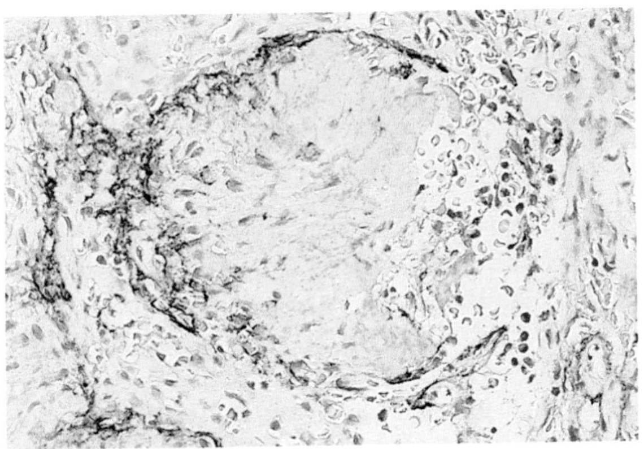

Figure 17-17

CALCIFYING PSEUDONEOPLASM OF THE NEURAXIS

The rind of superficial cells surrounding the amorphous calcification may be immunoreactive for epithelial membrane antigen (EMA).

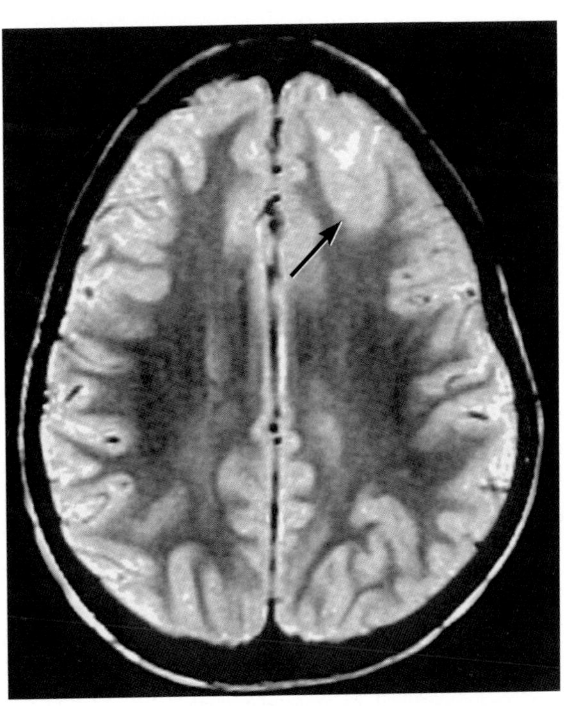

Figure 17-18

CORTICAL DYSPLASIA

A zone of thickened cerebral cortex (arrow) is typical of cortical dysplasia.

compromise vessels and nerves. Some are only incidental autopsy findings (36).

CORTICAL DYSPLASIA

Definition. *Cortical dysplasia* is a focal or diffuse epileptogenic abnormality of the cortex and underlying white matter.

Clinical Features. Children and young adults are most often affected.

Radiologic Findings. The thickened and noncontrast-enhancing cortical expansion is underlaid by an abnormal (bright) signal in T2-weighted images (fig. 17-18) (37).

Gross Findings. The normally sharp gray-white junction is effaced beneath the thickened cortex (fig. 17-19).

Microscopic Findings. The affected region contains large cells with glassy cytoplasm (figs.

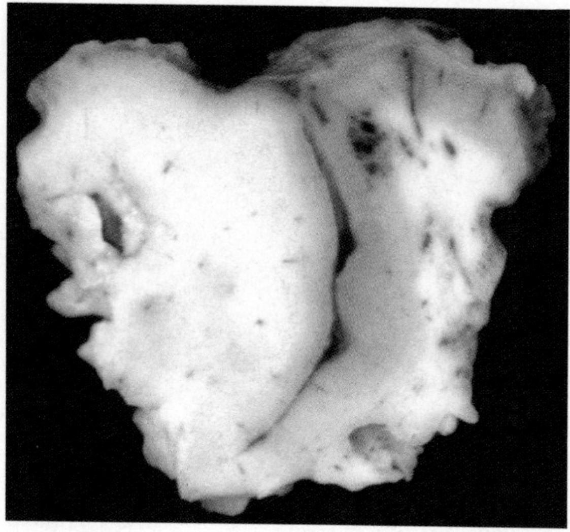

Figure 17-19

CORTICAL DYSPLASIA

Because of subcortical hypomyelination, resected lesions show blurring of the gray-white junction. (Courtesy of Dr. Caterina Giannini, Rochester, MN.)

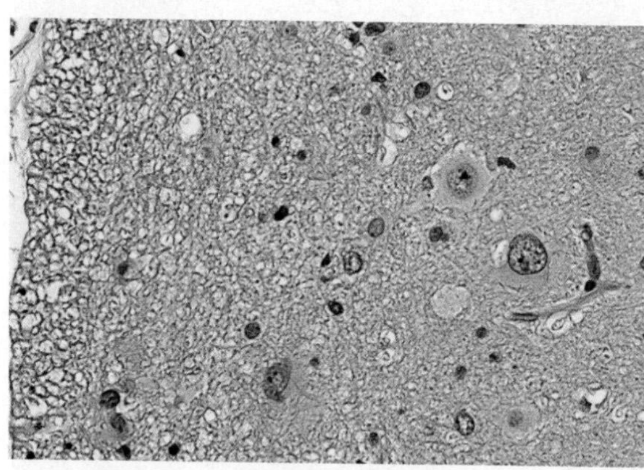

Figure 17-20

CORTICAL DYSPLASIA

Intracortical pleomorphic cells are easily recognized in the most superficial, or molecular, layer.

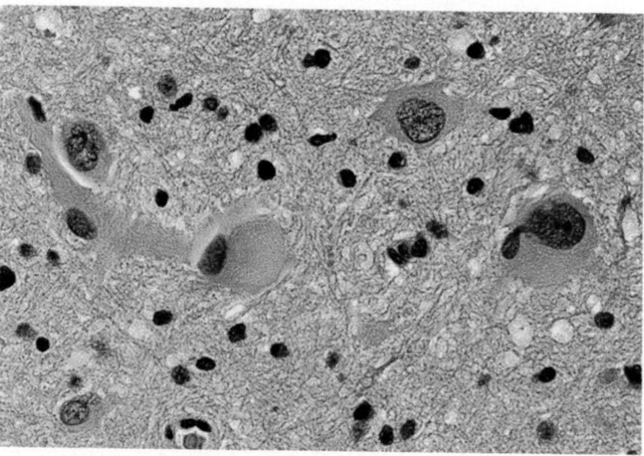

Figure 17-21

CORTICAL DYSPLASIA

Within the cortex are abnormal large cells that appear either neuronal or astrocytic. Neurons are identified by their large nuclei and cytoplasmic Nissl substance. Astrocytes, with glassy cytoplasm, have smaller, more condensed nuclei. It is not always possible, however, to assign such large cells to the neuronal or astrocytic category based on H&E staining alone.

17-20, 17-21) (38–40), of which some are obviously neurons in light of prominent nucleoli and the presence of irregularly clumped Nissl substance (fig. 17-22). These giant neurons are several times larger than their normal counterparts in unaffected cortex. Although their apical dendrites point to the cortical surface, the cells are irregularly distributed, creating considerable disarray of cortical architectonics. Many large cells with glassy cytoplasm and eccentric nuclei ("balloon cells") are astrocytic. Large cells are often situated in the finely fibrillar, paucicellular molecular layer where cells with glassy cytoplasm are not normally present.

In addition to abnormal neurons, other large cells with glassy cytoplasm and prominent processes appear astrocytic. These are found throughout the cortex but are also especially prominent in the molecular layer.

The same abnormal neurons and astrocytes may be even more conspicuous in underlying white matter. Here they are accompanied by gliosis and loss of myelin as the correlate of the increased signal intensity in T2-weighted MRIs (fig. 17-23). As in the cortical tubers of tuberous sclerosis, which cortical dysplasia most closely resembles, it is often difficult to determine whether certain enlarged cells are astrocytes or neurons.

Immunohistochemical Findings. Dysmorphic ganglion cells may be positive for neuronal markers such as neurofilament protein or NeuN, whereas astrocytes are reactive for GFAP and S-100 protein. Some balloon cells are immunonegative for both.

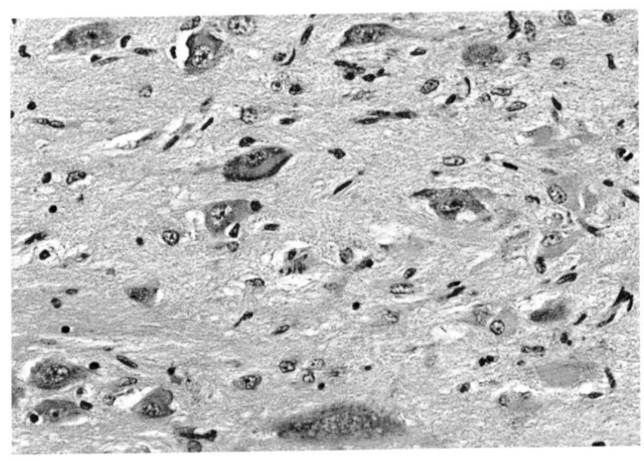

Figure 17-22

CORTICAL DYSPLASIA

Giant cells with Nissl substance are obviously both neuronal and abnormal. Smaller, but abnormal, astrocytes are present as well.

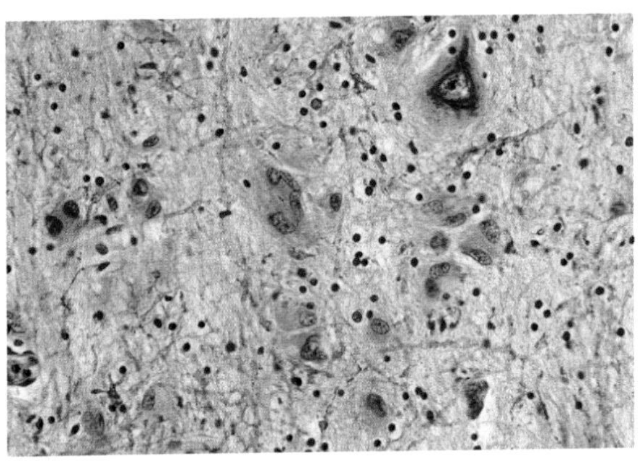

Figure 17-23

CORTICAL DYSPLASIA

The white matter below the cortical component is hypomyelinated and contains abnormal large astrocytes and neurons (H&E/Luxol fast blue stain).

Differential Diagnosis. This malformative process is to be distinguished primarily from a cortical tuber. In contrast to the tuber, the gigantic cells described above are more obviously glial or neuronal. Clinical and neuroimaging features so characteristic of tuberous sclerosis are lacking.

The distinction from ganglioglioma is discussed with that tumor. CNS lesions with ganglion cells are summarized in Appendix G.

Treatment and Prognosis. Treatment is surgical. The non-neoplastic lesions neither enlarge nor recur. Seizure activity may not be cured by excision.

OTHER MALFORMATIVE LESIONS

Other malformative lesions include: *intracranial ectopic cerebellar tissue* (41) and *hamartomatous/ectopic CNS tissue* in the intraspinal (42–44) and intracranial compartments (45,46). *Ectopic adrenal tissue,* consisting of cortex and medulla, is a rare autopsy finding in the CNS (47). In contrast, *ectopic adrenal cortical adenomas* in the CNS tend to be symptomatic and to involve a spinal nerve root (48).

Intracranial and intraspinal lipomas are discussed in chapter 10. Cranial nerve hamartomatous lesions of adipose tissue and/or muscle (lipoma, rhabdomyoma, hamartoma, choristoma, and ectomesenchymoma) are discussed in chapter 13.

REFERENCES

Hypothalamic Hamartoma

1. Nishio S, Fujiwara S, Aiko Y, Takeshita I, Fukui M. Hypothalamic hamartoma. Report of two cases. J Neurosurg 1989;70:640–5.
2. Sato M, Ushio Y, Arita N, Mogami H. Hypothalamic hamartoma: report of two cases. Neurosurgery 1985;16:198–206.
3. Stewart L, Steinbok P, Daaboul J. Role of surgical resection in the treatment of hypothalamic hamartomas causing precocious puberty. Report of six cases. J Neurosurg 1998;88:340–5.
4. Voyadzis JM, Guttman-Bauman I, Santi M, Cogen P. Hypothalamic hamartoma secreting corticotropin-releasing hormone. Case report. J Neurosurg Spine 2004;100:212–6.
5. Kuzniecky R, Guthrie B, Mountz J, et al. Intrinsic epileptogenesis of hypothalamic hamartomas in gelastic epilepsy. Ann Neurol 1997;42:60–7.
6. Valdueza JM, Cristante L, Dammann O, et al. Hypothalamic hamartomas: with special reference to gelastic epilepsy and surgery. Neurosurgery 1994;34:949–58.
7. Boyko OB, Curnes JT, Oakes WJ, Burger PC. Hamartomas of the tuber cinereum: CT, MR, and pathologic findings. AJNR Am J Neuroradiol 1991;12:309–14.
8. Sherwin RP, Grassi JE, Sommers SC. Hamartomatous malformation of the posterolateral hypothalamus. Lab Invest 1962;2:89–97.
9. Asa SL, Scheithauer BW, Bilbao JM, et al. A case for hypothalamic acromegaly: a clinicopathological study of six patients with hypothalamic gangliocytomas producing growth hormone-releasing factor. J Clin Endocrinol Metab 1984;58:796–803.
10. Vaquero J, Carrillo R, Oya S, Martinez R, Lopez R. Precocious puberty and hypothalamic hamartoma. Report on a new case with ultrastructural data. Acta Neurochir (Wien) 1985;74:129–33.
11. Albright AL, Lee PA. Neurosurgical treatment of hypothalamic hamartomas causing precocious puberty. J Neurosurg 1993;78:77–82.
12. Regis J, Bartolomei F, de Toffol B, et al. Gamma knife surgery for epilepsy related to hypothalamic hamartomas. Neurosurgery 2000;47:1343–51; discussion 1351–2.

Nasal Cerebral Heterotopia ("Nasal Glioma")

13. Barkovich AJ, Vandermarck P, Edwards MS, Cogen PH. Congenital nasal masses: CT and MR imaging features in 16 cases. AJNR Am J Neuroradiol 1991;12:105–16.
14. Patterson K, Kapur S, Chandra RS. "Nasal gliomas" and related brain heterotopias: a pathologist's perspective. Pediatr Pathol 1986; 5:353–62.
15. Choudhury AR, Bandey SA, Haleem A, Sharif H. Glial heterotopias of the nose: a report of two cases. Childs Nerv Syst 1996;12:43–7.
16. Mirra SS, Pearl GS, Hoffman JC, Campbell WG Jr. Nasal 'glioma' with prominent neuronal component: report of a case. Arch Pathol Lab Med 1981;105:540–1.
17. Kindblom LG, Angervall L, Haglid K. An immunohistochemical analysis of S-100 protein and glial fibrillary acidic protein in nasal glioma. Acta Pathol Microbiol Immunol Scand [A] 1984;92:387–9.

Meningioangiomatosis

18. Blumenthal D, Berho M, Bloomfield S, Schochet SS Jr, Kaufman HH. Childhood meningioma associated with meningio-angiomatosis. Case report. J Neurosurg 1993;78:287–9.
19. Giangaspero F, Guiducci A, Lenz FA, Mastronardi L, Burger PC. Meningioma with meningioangiomatosis: a condition mimicking invasive meningiomas in children and young adults: report of two cases and review of the literature. Am J Surg Pathol 1999;23:872–5.
20. Sinkre P, Perry A, Cai D, et al. Deletion of the NF2 region in both meningioma and juxtaposed meningioangiomatosis: case report supporting a neoplastic relationship. Pediatr Dev Pathol 2001;4:568–72.
21. Harada K, Inagawa T, Nagasako R. A case of meningioangiomatosis without von Recklinghausen's disease. Report of a case and review of 13 cases. Childs Nerv Syst 1994;10:126–30.
22. Ogilvy CS, Chapman PH, Gray M, de la Monte SM. Meningioangiomatosis in a patient without von Recklinghausen's disease. Case report. J Neurosurg 1989;70:483–5.
23. Sakaki S, Nakagawa K, Nakamura K, Takeda S. Meningioangiomatosis not associated with von Recklinghausen's disease. Neurosurgery 1987; 20:797–801.
24. Stemmer-Rachamimov AO, Horgan MA, Taratuto AL, et al. Meningioangiomatosis is associated with neurofibromatosis 2 but not with somatic alterations of the NF2 gene. J Neuropathol Exp Neurol 1997;56:485–9.

25. Wiebe S, Munoz DG, Smith S, Lee DH. Meningioangiomatosis. A comprehensive analysis of clinical and laboratory features. Brain 1999; 122(Pt 4):709–26.
26. Goates JJ, Dickson DW, Horoupian DS. Meningioangiomatosis: an immunocytochemical study. Acta Neuropathol (Berl) 1991;82:527–32.
27. Halper J, Scheithauer BW, Okazaki H, Laws ER, Jr. Meningio-angiomatosis: a report of six cases with special reference to the occurrence of neurofibrillary tangles. J Neuropathol Exp Neurol 1986;45:426–46.
28. Liu SS, Johnson PC, Sonntag VK. Meningioangiomatosis: a case report. Surg Neurol 1989; 31:376–80.
29. Mokhtari K, Uchihara T, Clemenceau S, Baulac M, Duyckaerts C, Hauw JJ. Atypical neuronal inclusion bodies in meningioangiomatosis. Acta Neuropathol (Berl) 1998;96:91–6.
30. Prayson RA. Meningioangiomatosis. A clinicopathologic study including MIB1 immunoreactivity. Arch Pathol Lab Med 1995;119:1061–4.
31. Takeshima Y, Amatya VJ, Nakayori F, Nakano T, Sugiyama K, Inai K. Meningioangiomatosis occurring in a young male without neurofibromatosis: with special reference to its histogenesis and loss of heterozygosity in the NF2 gene region. Am J Surg Pathol 2002;26:125–9.

Calcifying Pseudoneoplasm of the Neuraxis

32. Halper J, Scheithauer BW, Okazaki H, Laws ER, Jr. Meningio-angiomatosis: a report of six cases with special reference to the occurrence of neurofibrillary tangles. J Neuropathol Exp Neurol 1986;45:426–46.
33. Bertoni F, Unni KK, Dahlin DC, Beabout JW, Onofrio BM. Calcifying pseudoneoplasms of the neural axis. J Neurosurg 1990;72:42–8.
34. Garen PD, Powers JM, King JS, Perot PL Jr. Intracranial fibro-osseous lesion. Case report. J Neurosurg 1989;70:475–7.
35. Jun C, Burdick B. An unusual fibro-osseous lesion of the brain: case report. J Neurosurg 1984; 60:1308–11.
36. Rhodes RH, Davis RL. An unusual fibro-osseous component in intracranial lesions. Hum Pathol 1978;9:309–19.

Cortical Dysplasia

37. Kuzniecky R, Garcia JH, Faught E, Morawetz RB. Cortical dysplasia in temporal lobe epilepsy: magnetic resonance imaging correlations. Ann Neurol 1991;29:293–8.
38. Mischel PS, Nguyen LP, Vinters HV. Cerebral cortical dysplasia associated with pediatric epilepsy. Review of neuropathologic features and proposal for a grading system. J Neuropathol Exp Neurol 1995;54:137–53.
39. Prayson RA, Estes ML. Cortical dysplasia: a histopathologic study of 52 cases of partial lobectomy in patients with epilepsy. Hum Pathol 1995;26:493–500.
40. Taylor DC, Falconer MA, Bruton CJ, Corsellis JA. Focal dysplasia of the cerebral cortex in epilepsy. J Neurol Neurosurg Psychiatry 1971;34: 369–87.

Other Malformative Lesions

41. Chang AH, Kaufmann WE, Brat DJ. Ectopic cerebellum presenting as a suprasellar mass in infancy: implications for cerebellar development. Pediatr Dev Pathol 2001;4:89–93.
42. Brownlee RD, Clark AW, Sevick RJ, Myles ST. Symptomatic hamartoma of the spinal cord associated with neurofibromatosis type 1. Case report. J Neurosurg 1998;88:1099–103.
43. Castillo M, Smith MM, Armao D. Midline spinal cord hamartomas: MR imaging features of two patients. AJNR Am J Neuroradiol 1999;20: 1169–71.
44. Riley K, Palmer CA, Oser AB, Paramore CG. Spinal cord hamartoma: case report. Neurosurgery 1999;44:1125–8.
45. Delalande O, Rodriguez D, Chiron C, Fohlen M. Successful surgical relief of seizures associated with hamartoma of the floor of the fourth ventricle in children: report of two cases. Neurosurgery 2001;49:726–31.
46. Gyure KA, Morrison AL, Jones RV. Intracranial extracerebral neuroglial heterotopia: a case report and review of the literature. Ann Diagn Pathol 1999;3:182–6.
47. Wiener MF, Dallgaard SA. Intracranial adrenal gland; a case report. AMA Arch Pathol 1959;67: 228–33.
48. Mitchell A, Scheithauer BW, Sasano H, Hubbard EW, Ebersold MJ. Symptomatic intradural adrenal adenoma of the spinal nerve root: report of two cases. Neurosurgery 1993;32:658–61; discussion 661–2.

18 DYSGENETIC DISEASES

NEUROFIBROMATOSIS 1 (VON RECKLINGHAUSEN'S DISEASE)

Neurofibromatosis 1 (NF1), also known as *von Recklinghausen's disease*, is inherited in an autosomal dominant fashion from a locus at chromosome 17q11.2, in a large gene that encodes neurofibromin (1,2). Penetrance is high but expression is variable; a high rate of spontaneous mutation adds to the genetic pool of the disease, making it one of the most common Mendelian disorders.

Diagnostic criteria are two or more of the following: 1) six or more café-au-lait spots (the diameter of the largest being more than 5 mm in prepubertal patients and greater than 15 mm in others; 2) two or more neurofibromas of any type, or one plexiform example; 3) axillary or inguinal freckling; 4) optic pathway glioma; 5) two or more Lisch nodules; 6) a distinctive osseous lesion, such as sphenoid dysplasia or thinning of the long bone cortex (with or without pseudoarthrosis); and 7) a first-degree relative with NF1 by the above criteria (3). Table 18-1 summarizes features of NF1, and compares them with those of neurofibromatosis 2 (NF2).

The presence of *neurofibromas*, often multiple, of the intraspinal nerve roots, is the common intradural extramedullary expression of NF1 (fig. 18-1; see figs. 13-27–13-29). Nerve root tumors associated with NF2 are *schwannomas* (2,4,5). Of the peripheral neurofibromas, including those of spinal nerve roots or paraspinal nerves, the plexiform type is best known. It and massive soft tissue neurofibromas are considered pathognomonic of the NF1 syndrome. Plexiform tumors produce unmistakable ropey enlargement of large nerve roots, trunks, and their branches, and may also affect small nerves such as those of the skin. Separate status is given a segmental variant of neurofibromatosis, due to a somatic mutation, in which neurofibromas are confined to one extremity or body region (6–9). A small proportion of peripheral neurofibromas, particularly those of plexiform type, undergo malignant transformation. Occurring in approximately 2 percent of patients with NF1, *malignant peripheral nerve sheath tumors* (MPNSTs) are often high-grade aggressive neoplasms.

Gliomas associated with NF1 are usually *pilocytic astrocytomas*, most classically bilateral tumors of the optic nerves (fig. 18-2) (10–13), although this same neoplasm may also affect cerebral hemispheres, thalamus, hypothalamus, brain stem, cerebellum, or spinal cord (14). Pilocytic astrocytomas of the optic nerve are often more indolent than sporadic equivalents in non-NF1 patients (15). In both groups, but

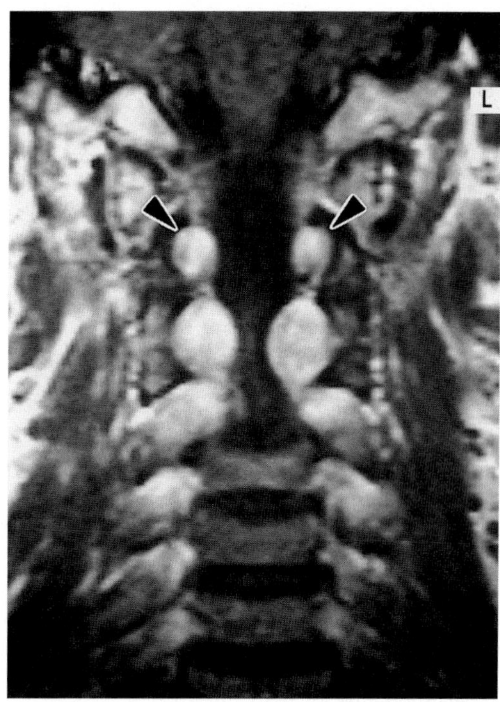

Figure 18-1

NEUROFIBROMATOSIS 1: MULTIPLE SPINAL NEUROFIBROMAS

Multiple, and difficult to control, neurofibromas of spinal nerve roots are common in patients with neurofibromatosis 1 (NF1). Beginning at the C2-C3 level, multiple, white, contrast-enhancing lesions impinge on the spinal cord of a 10-year-old boy. (Fig. 18-1 from Fascicle 10, 3rd Series.)

Table 18-1

COMPARATIVE FEATURES OF NEUROFIBROMATOSIS 1 AND 2

Features	NF1 (Peripheral Form; von Recklinghausen's Disease)	NF2 (Central Form)
Incidence	1/3,000	1/40,000
Prevalence	60/100,000	0.01/100,000
Inheritance	Autosomal dominant	Autosomal dominant
Sporadic occurrence	50%	50%
Chromosome location	17q11.2	22q12
Encoded protein	Neurofibromin	Merlin (schwannomin)
Café-au-lait spots	Often multiple and large	Small, rarely more than 6
Cutaneous neurofibromas	Most patients	Rare
Cutaneous schwannomas	Not associated	70%
Multiple Lisch nodules	Very common	Not associated
Cataracts	Not associated	60 to 80%
Skeletal malformations	Common	Not associated
Astrocytomas (optic, cerebellar, cerebral)	Moderate incidence	Infrequent
Pheochromocytoma	Occasionally present	Not associated
Malignant peripheral nerve sheath tumor	Approximately 2%	Not associated
Intellectual impairment	Associated	Not associated
Vestibular schwannoma	Not associated	Most cases (usually bilateral)
Meningioma	Uncommon	Common
Spinal cord ependymoma	Not associated	Common
Meningioangiomatosis	Not associated	Occasional
Schwannosis	Not associated	Common
Glial hamartomas	Occasional	Very common
Syringomyelia	Not associated	Associated
Posterior subcapsular cataracts	Not associated	Common
Ganglioneuroma	Occasional	Not associated
Gastrointestinal autonomic nerve tumor	Occasional	Not associated
Paraganglioma, including duodenal gangliocytic variant	Occasional	Not associated
Foregut carcinoid tumor, including duodenal calcifying somatostatinoma	Occasional	Not associated
Juvenile xanthogranuloma	Occasional	Not associated
Rhabdomyosarcoma	Occasional	Not associated
Juvenile leukemia (CML)[a]	Occasional	Not associated

[a]CML = chronic myelogenous leukemia.

especially in NF1 patients, spontaneous regression occasionally occurs (16,17). Brain stem astrocytomas include classic contrast-enhancing pilocytic lesions, but also masses that are radiologically "diffuse" and not contrast enhancing. These thus resemble the classic diffuse astrocytoma of the brain stem, but are generally indolent (18,19). Less common central nervous system (CNS) tumors are truly diffuse astrocytomas of varying histologic grade (20–23). A "pleomorphic astrocytoma with continual malignant progression" has been described (24).

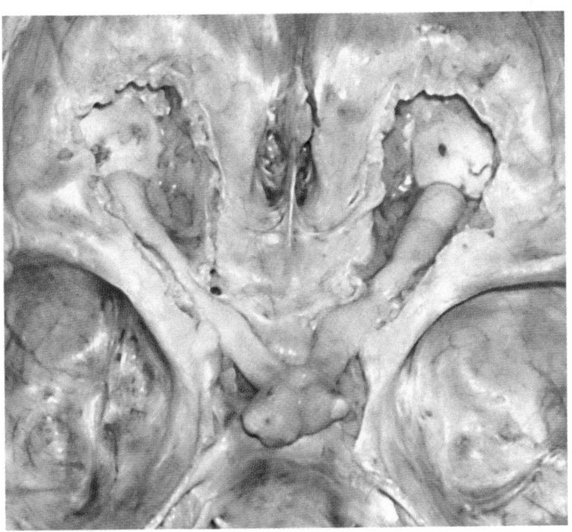

Figure 18-2

NEUROFIBROMATOSIS 1:
BILATERAL OPTIC NERVE ASTROCYTOMAS

Bilateral pilocytic astrocytomas of the optic nerve are almost pathognomonic of NF1.

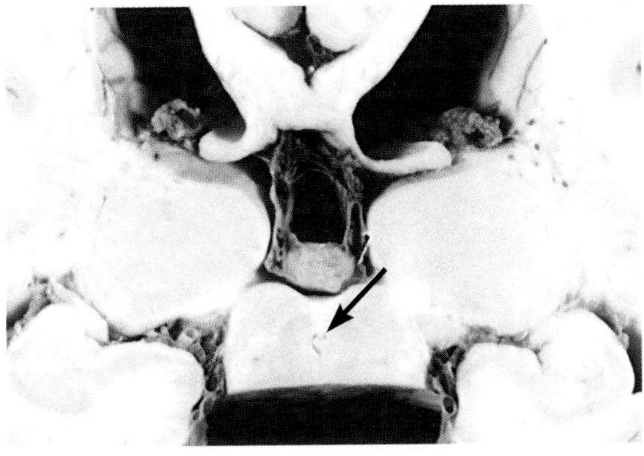

Figure 18-3

NEUROFIBROMATOSIS 1: PERIAQUEDUCTAL GLIOMA

The wall of the cerebral aqueduct is sometimes the site of a glial overgrowth (arrow) that, while small, can produce hydrocephalus.

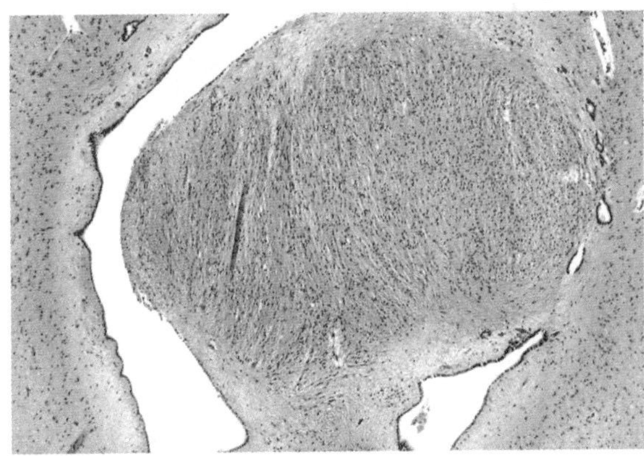

Figure 18-4

NEUROFIBROMATOSIS 1: PERIAQUEDUCTAL GLIOMA

The lesion illustrated in figure 18-3 is composed of small, elongated cells that create a fibrillar background.

An unusual CNS lesion is a minute subependymal excrescence that is usually asymptomatic except in the narrow confines of the Sylvian aqueduct. Here it can obstruct the flow of cerebrospinal fluid and produce hydrocephalus (figs. 18-3, 18-4) (25).

Magnetic resonance imaging (MRI) of NF1 patients often demonstrates foci of high signal intensity in T2-weighted images ("unidentified bright objects") of the deep gray matter (basal ganglia and thalamus), cerebellum, and brain stem (fig. 18-5) (26,27). Lesions of the white matter are also common (28). The significance of these sometimes transient foci is unclear. One correlative histologic study found only "spongiosis" in tissue from the gray matter lesion (26).

Lisch nodules are multifocal pigmented iris hamartomas, the incidence of which increases with age. In one series of patients with NF1 they appeared in all patients by age 21 years (29).

NEUROFIBROMATOSIS 2 (CENTRAL NEUROFIBROMATOSIS)

Far less common than NF1, *neurofibromatosis 2* (NF2) is an autosomal dominant disorder linked to the *NF2* gene situated at chromosome 22q12 (30–32). Penetrance is 100 percent by age 60. The same gene also is implicated in the development of sporadic schwannomas, meningiomas, and spinal ependymomas. Merlin, or schwannomin, is its product. The disease usually presents within the second and third decades, often with deafness (33).

The criteria for the disease continue to evolve. The 2002 criteria of the National Neurofibromatosis Foundation for the diagnosis of NF2 are: a patient with bilateral vestibular schwannomas or an individual with a first-degree family relative with NF2 and any two of the following: meningioma, schwannoma, glioma, juvenile lens opacity (posterior subcapsular cataract or

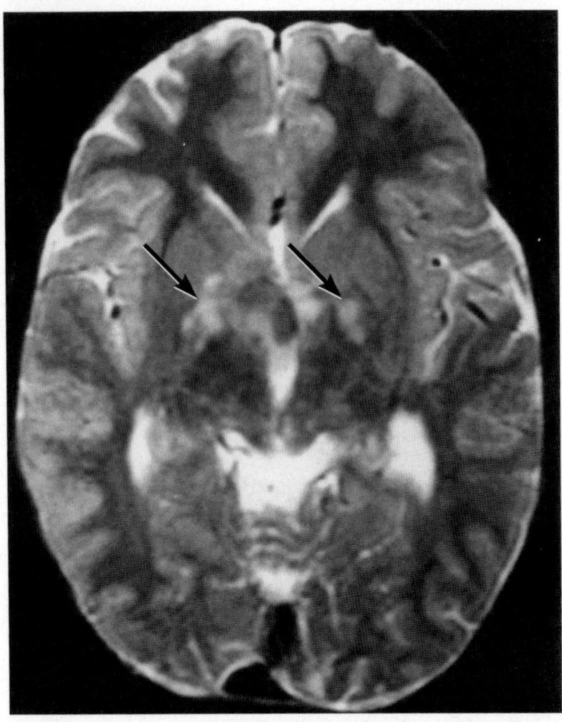

Figure 18-5

NEUROFIBROMATOSIS 1:
"UNIDENTIFIED BRIGHT OBJECTS"

Multiple, somewhat fluffy areas of high signal intensity in T2-weighted images (arrows) are common in the basal ganglia, cerebellum, and brain stem of NF1 patients.

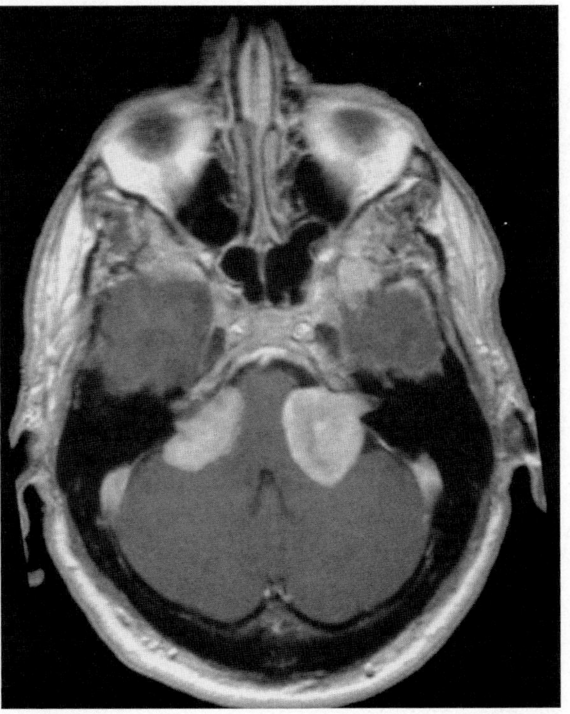

Figure 18-6

NEUROFIBROMATOSIS 2:
BILATERAL VESTIBULAR SCHWANNOMAS

Bilateral vestibular schwannomas are diagnostic of neurofibromatosis 2 (NF2).

cortical cataract) (34). Table 18-1 compares the features of NF2 and NF1.

Intracranially, NF2 is noted for schwannomas and meningiomas, with bilateral vestibular schwannomas being diagnostic of the condition (fig. 18-6). Most of the neoplasms appear after puberty and metachronously in terms of symptom onset (30,33,35,36). Schwannomas also affect other cranial nerves, particularly the trigeminal nerve. Cutaneous schwannomas are seen in half of patients with NF2. Since meningiomas can be innumerable (fig. 18-7), it is not surprising that masses composed of both schwannoma and meningioma can occur at the cerebellopontine angle (figs. 18-8–18-10) (37–39). Meningiomas in the setting of NF2 may be more aggressive than sporadic variants in terms of the frequency of brain invasion (40). Radiologically, intracranial calcifications are common in the choroid plexus, cerebral cortex (see fig. 18-14), and cerebellum (41).

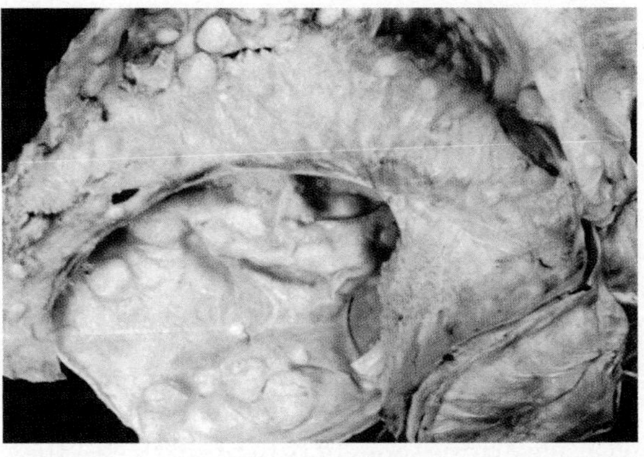

Figure 18-7

NEUROFIBROMATOSIS 2: MENINGIOMATOSIS

Multiple meningiomas are common in the advanced stages of NF2.

Intraspinally, schwannomas usually affect multiple dorsal nerve roots (fig. 18-11) (42,43). In contrast to NF1, where nerve root tumors

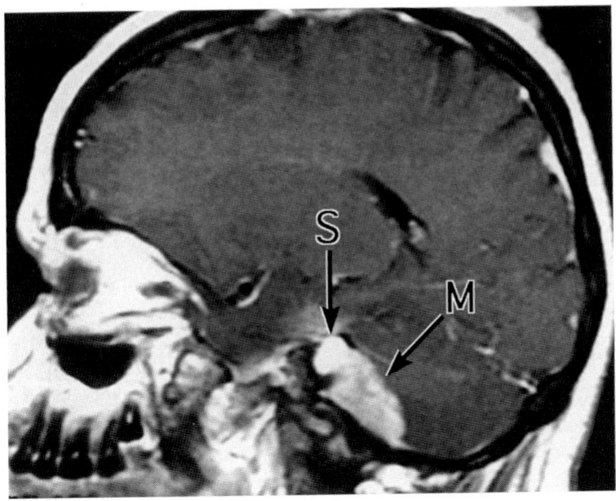

Figure 18-8

NEUROFIBROMATOSIS 2: CEREBELLOPONTINE ANGLE "COLLISION TUMOR" (SCHWANNOMA AND MENINGIOMA)

The two common intracranial tumors in patients with NF2, schwannoma (S) and meningioma (M), can coexist at the cerebellopontine angle. Multiple, small convexity meningiomas are also present.

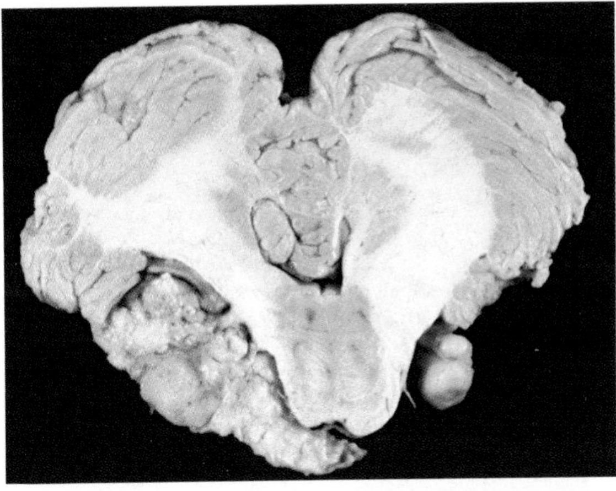

Figure 18-9

NEUROFIBROMATOSIS 2: CEREBELLOPONTINE ANGLE "COLLISION TUMOR" (SCHWANNOMA AND MENINGIOMA)

Meningiomas can collide with schwannomas at the cerebellopontine angle in patients with NF2. The schwannomas here are the bilateral, globular, smooth-surfaced masses. The one on the left is encased by a plaque-like meningioma.

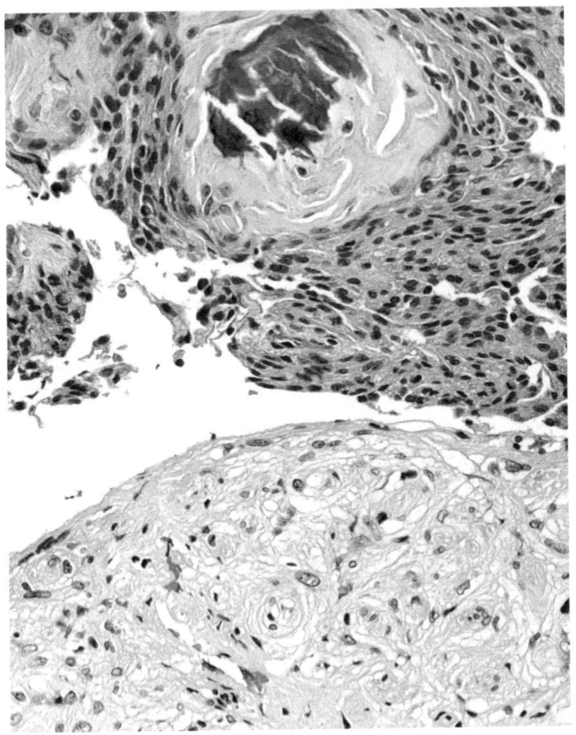

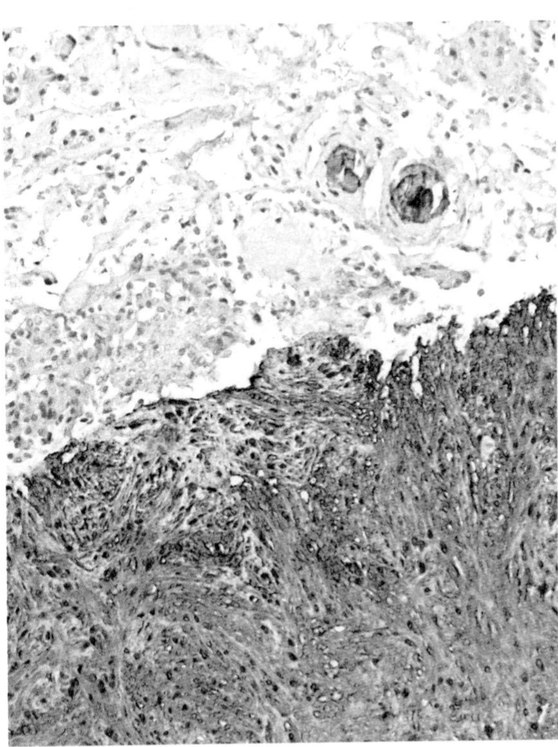

Figure 18-10

NEUROFIBROMATOSIS 2: CEREBELLOPONTINE ANGLE "COLLISION TUMOR" (SCHWANNOMA AND MENINGIOMA)

Left: The mass seen radiologically in figure 18-8 is composed of both meningioma (top) and schwannoma (bottom). Right: In contrast to the meningioma, the schwannoma is strongly immunoreactive for S-100 protein.

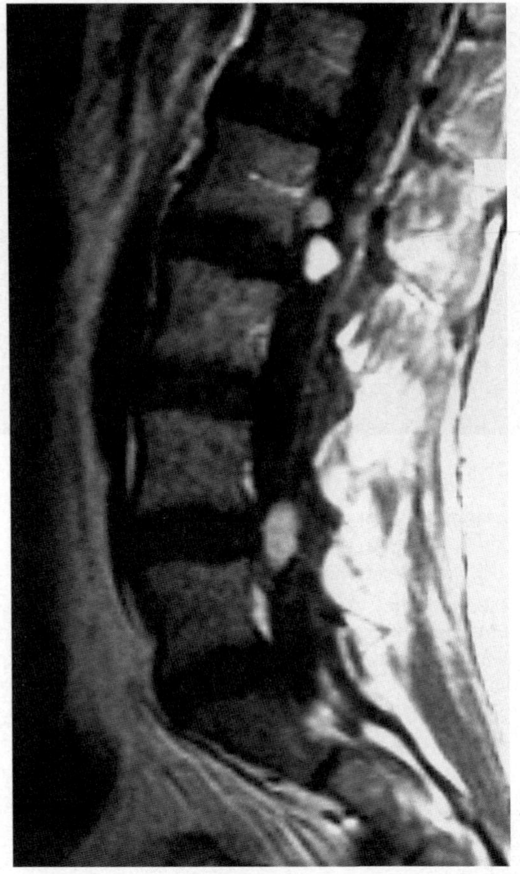

Figure 18-11

NEUROFIBROMATOSIS 2:
MULTIPLE SPINAL SCHWANNOMAS

As seen in this contrast-enhanced magnetic resonance image (MRI), bright intradural-extramedullary schwannomas are common in patients with NF2. (Fig. 18-4 from Fascicle 10, 3rd Series.)

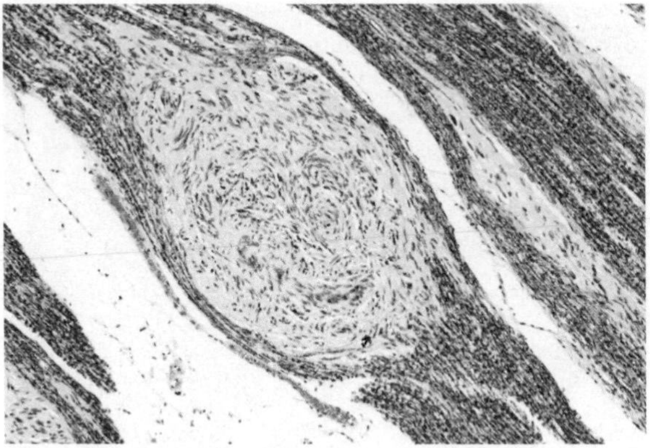

Figure 18-12

NEUROFIBROMATOSIS 2:
SCHWANNOSIS OF SPINAL NERVE ROOT

Focal proliferations of Schwann cells may be encountered within spinal nerve roots in patients with NF2.

are typically neurofibromas (see fig. 18-1), nerve sheath neoplasms in NF2 are schwannomas.

Completing the intraspinal spectrum of this syndrome are multiple meningiomas and ependymomas (44). Associated vascular abnormalities include aneurysms and arteriovenous fistulae (45).

Given the genetic abnormality in NF2, it is not surprising that microscopic study of the CNS expands the spectrum of lesions to include such dysplastic proliferations as: 1) Schwann cell nodules in spinal nerve roots (schwannomatosis) (fig. 18-12); 2) aberrant clusters of ependymal cells in the spinal cord; and 3) minute clusters of pleomorphic amitotic cells resembling glial heterotopias within cerebral gray matter (fig. 18-13, left) (44). The eosinophilic cytoplasm of the amitotic cells and their immunoreactivity for S-100 protein suggest an astrocytic derivation, despite lack of staining for glial fibrillary acidic protein (GFAP) (fig. 18-13, right) (46).

Meningioangiomatosis, either solitary or multiple (fig. 18-14), may be associated with NF 2. Meningioangiomatosis is discussed in chapter 17.

The disease is progressive and debilitating as the result of both intracranial and spinal neoplasms; multiple operations are usually required. In one study, almost 40 percent of patients died before the age of 50 years (33).

VON HIPPEL-LINDAU DISEASE

In its consummate form, *Von Hippel-Lindau* (VHL) *disease* consists of multiple hemangioblastomas of the CNS (47–52), including the cerebellum (fig. 18-15) (53), optic nerve and retina (figs. 18-16, 18-17), spinal cord (fig. 18-18), spinal nerve roots (fig. 18-19) (54,55), and brain stem. Brain stem lesions most often arise in the medulla and project into the fourth ventricle. Rare hemangioblastomas in VHL disease affect the supratentorial leptomeninges (56) or a peripheral nerve (57).

Other lesions associated with VHL disease include renal cell carcinoma; pheochromocytoma; cysts or cystadenomas of the kidneys, liver, pancreas, or epididymis; and well-differentiated but locally aggressive papillary tumors

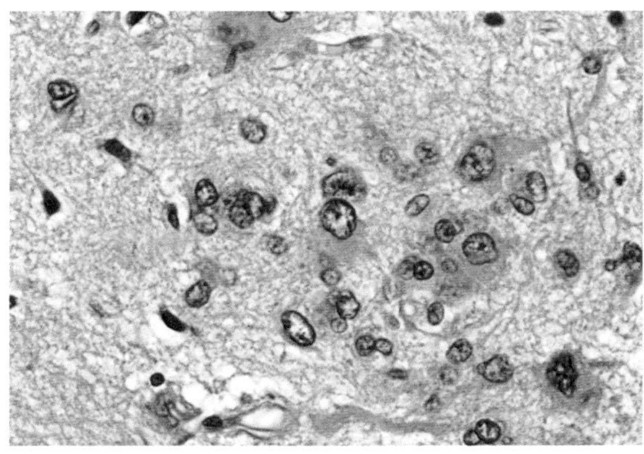

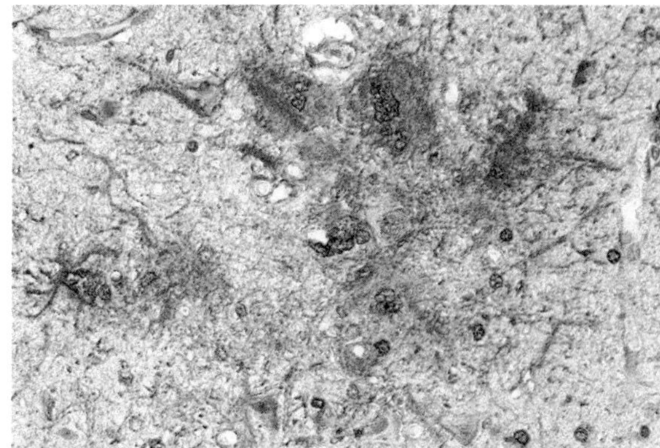

Figure 18-13

NEUROFIBROMATOSIS 2: CORTICAL MICROHAMARTOMAS

Left: Small collections of pleomorphic cells are common in the gray matter of NF2 patients. Right: The abnormal cells are immunoreactive for S-100 protein.

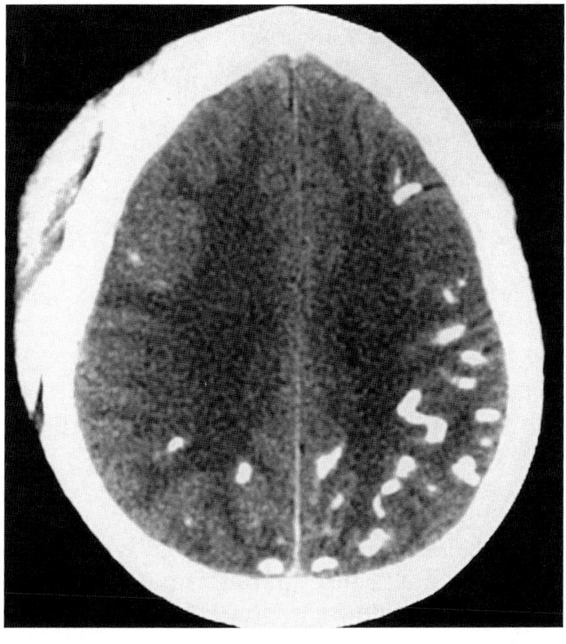

Figure 18-14

NEUROFIBROMATOSIS 2: MENINGIOANGIOMATOSIS (MULTIPLE)

As seen by computerized tomography (CT), multiple, densely calcified, curvilinear profiles are radiologic evidence of multifocal meningioangiomatosis.

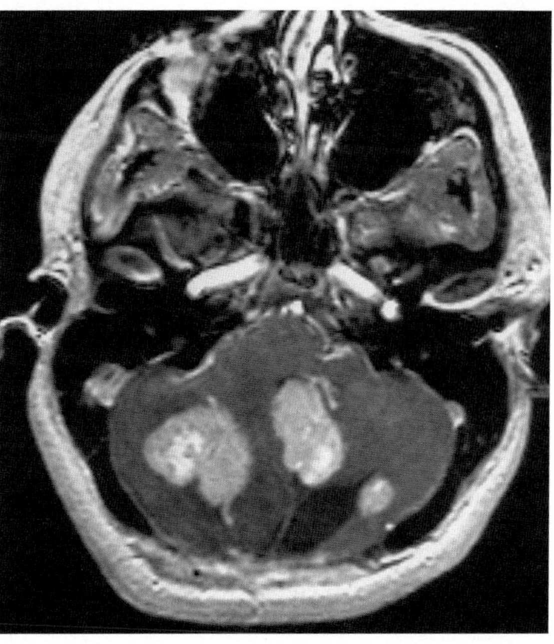

Figure 18-15

VON HIPPEL-LINDAU DISEASE: MULTIPLE CEREBELLAR HEMANGIOBLASTOMAS

Multiple hemangioblastomas are a common expression of this genetic disease. Bilateral enucleation was required for the retinal hemangioblastomas in this 35-year-old woman. (Fig. 18-8 from Fascicle 10, 3rd Series.)

of the inner ear/temporal bone (see figs. 3-351, 16-17) (50,58–60). VHL disease is inherited in an autosomal dominant fashion (61,62). The defective gene resides on the short arm of chromosome 3 (63–65).

Hemangioblastomas, histologically identical at all sites and in both syndrome-associated and sporadic lesions, are discussed in chapter 7.

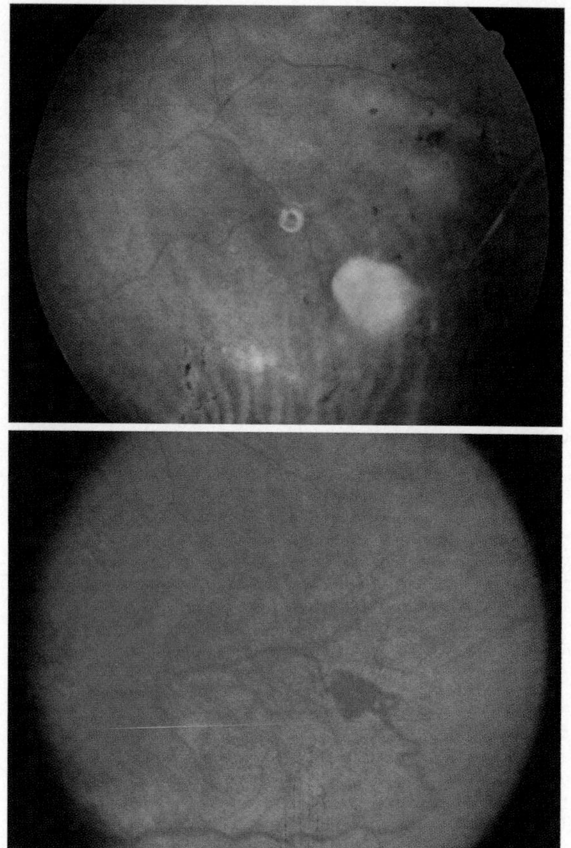

Figure 18-16

VON HIPPEL-LINDAU DISEASE: RETINAL HEMANGIOBLASTOMA

As seen in this fundus photograph (top) and angiogram (bottom), ocular hemangioblastomas are a common component in von Hippel-Lindau (VHL) disease. (Courtesy of Dr. Stephen C. Pollack, Durham, NC.)

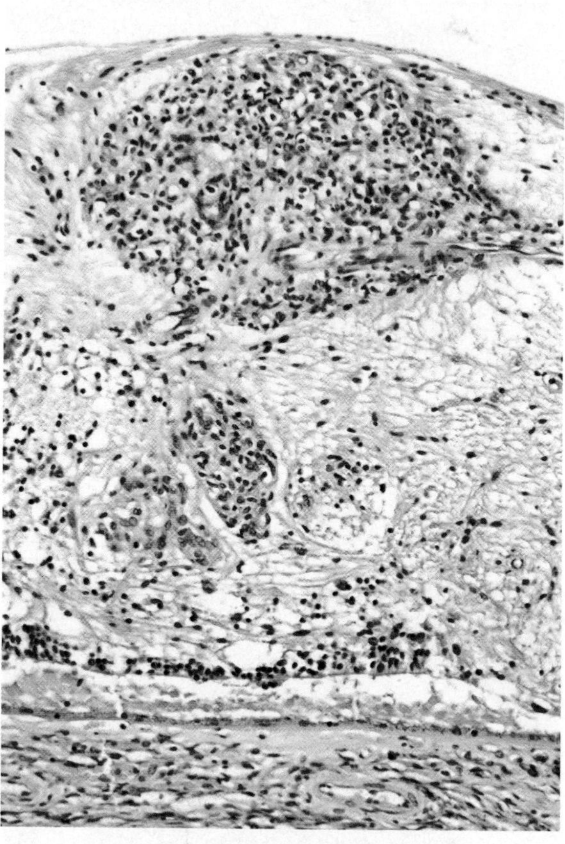

Figure 18-17

VON HIPPEL-LINDAU DISEASE: RETINAL HEMANGIOBLASTOMA

This retinal hemangioblastoma is homologous histologically to its intracranial and intraspinal counterparts. (Courtesy of Dr. Robert Folberg, Iowa City, IA.)

TUBEROUS SCLEROSIS COMPLEX

Tuberous sclerosis complex is characterized by a combination of cutaneous lesions, epilepsy, and associated mental deficiency. Major features of the disease include facial angiofibromas or forehead plaques (fig. 18-20), nontraumatic ungual or periungual fibromas, three or more hypomelanotic macules, shagreen patches, multiple retinal nodular hamartomas, cortical tubers (figs. 18-21–18-23), subependymal giant cell astrocytomas (fig. 18-24, left; see fig. 3-170), subependymal nodules (fig. 18-24, right), cardiac rhabdomyomas, lymphngiomatosis, and renal angiomyolipomas (66–69).

Minor features of the disease include multiple, randomly distributed pits in the dental enamel, hamartomatous rectal polyps, bone cysts, cerebral white matter radial migration lines, gingival fibromas, nonrenal hamartomas, retinal achromatic patches, "confetti" skin lesions, and multiple renal cysts.

A diagnosis of definite tuberous sclerosis requires either two major features or one major plus two minor features. Probable tuberous sclerosis is indicated by one major plus one minor feature, and possible tuberous sclerosis by either one major feature or two or more minor features (66–69).

The responsible genes, *TSC1* and *TSC2,* on chromosomes 9q34 and 16p13, respectively (70–73), are involved in disease expression with equal frequency. Autosomal dominant inheritance is

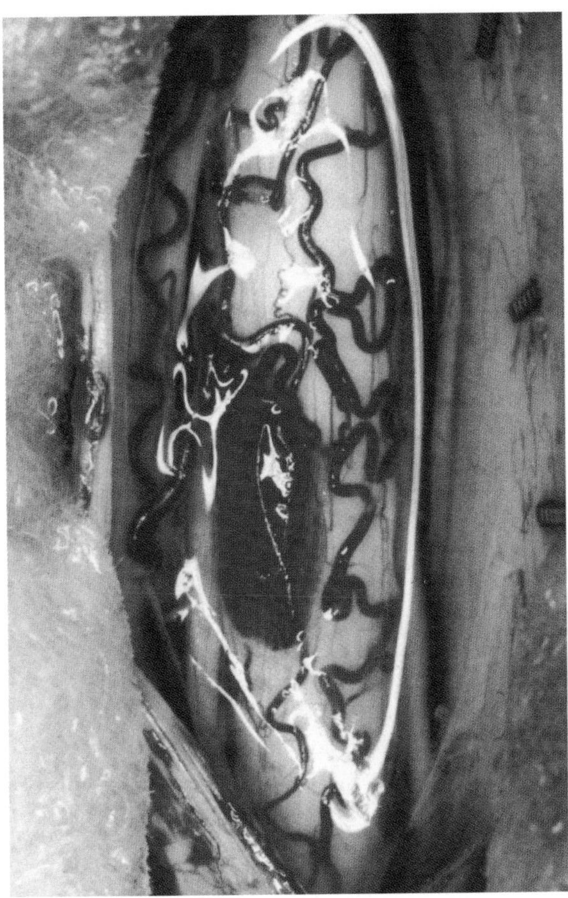

Figure 18-18

VON HIPPEL-LINDAU DISEASE: SPINAL HEMANGIOBLASTOMA

Spinal hemangioblastomas are frequent in patients with VHL disease. Note the characteristic superficial/posterior location of this small lesion at C5-C6. The incidental tumor was detected during workup of a larger, symptomatic hemangioblastoma at C2 in a 29-year-old woman. A renal mass was also present. (Courtesy of Dr. Ziya Gokaslan, Baltimore, MD.)

established in only one third of patients; the remainder presumably have new germline mutations. As with NF1, cases with few manifestations, so-called formes fruste, are common.

Subependymal giant cell astrocytoma (SEGA), a major feature of tuberous sclerosis, is essentially a larger, but also benign, version of the subependymal hamartoma that is restricted to the region of the foramen of Monro. Both the frequently calcified SEGA (fig. 18-24, left) and the *subependymal nodule* are readily evident on computerized tomography (CT) and MRI. *Cortical*

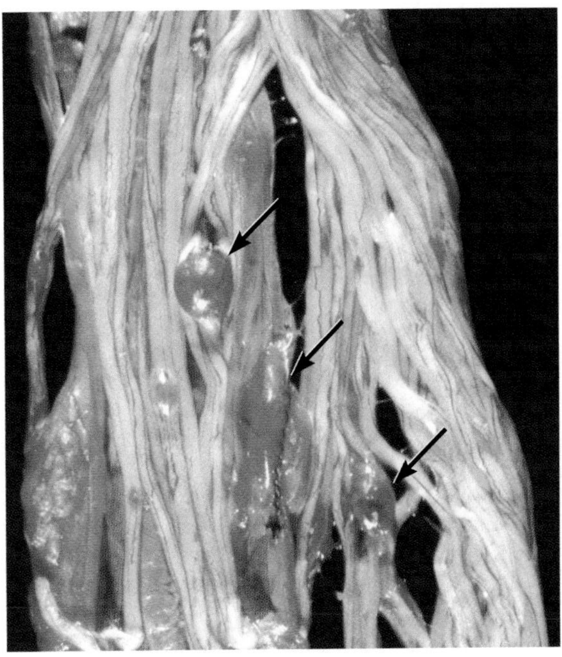

Figure 18-19

VON HIPPEL-LINDAU DISEASE: MULTIPLE SPINAL NERVE ROOT HEMANGIOBLASTOMAS

Whether by subarachnoid spread or multifocal origin, multiple hemangioblastomas (arrows) may stud the nerve roots in patients with VHL disease.

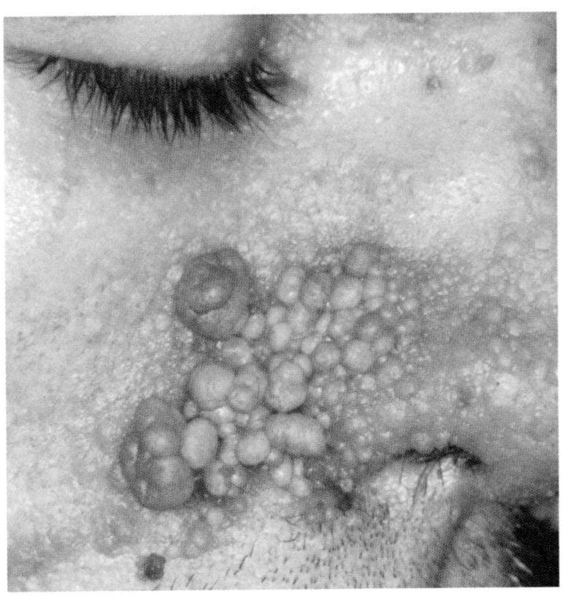

Figure 18-20

TUBEROUS SCLEROSIS

Adenoma sebaceum classically presents as multiple excrescences along the nasolabial fold.

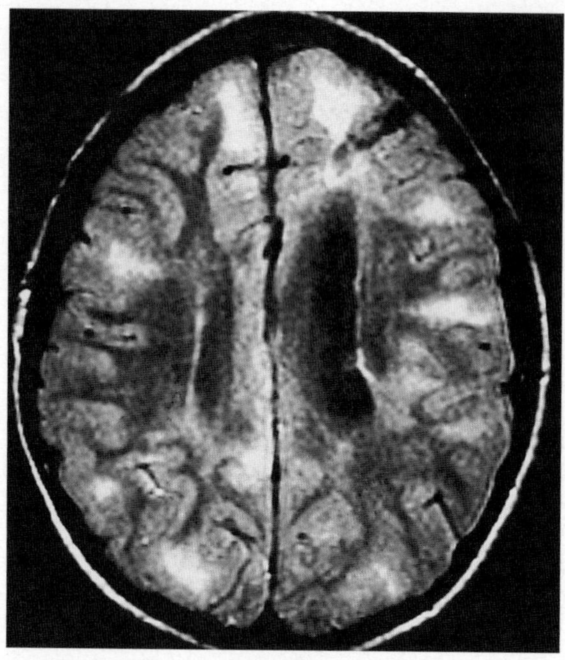

Figure 18-21

TUBEROUS SCLEROSIS

Tubers are apparent on MRI as thickenings of the cerebral cortex associated with wedge-shaped, high signal intensity regions that extend to the ventricular system. (Fig. 18-12 from Fascicle 10, 3rd Series.)

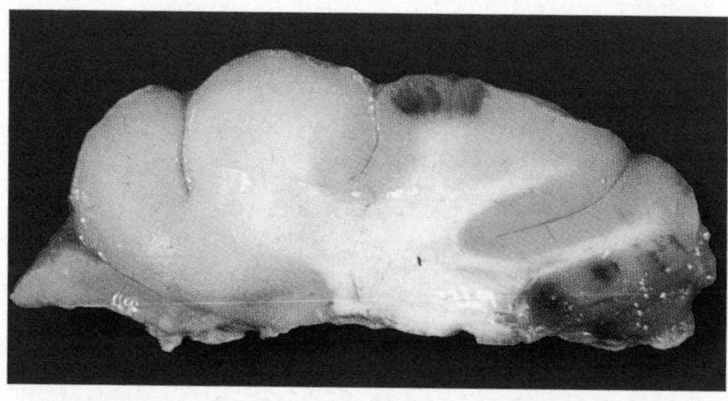

Figure 18-22

TUBEROUS SCLEROSIS: TUBER

Removed for seizure control, the tuber thickens the cerebral cortex and obscures the gray-white junction. (Courtesy of Dr. Caterina Giannini, Rochester, MN.)

tubers (discussed below) are particularly well visualized and appear as expanded regions of high signal intensity in the cerebral cortex in proton-density, T2-weighted, or fluid-attenuated inversion recovery (FLAIR) images (fig. 18-21). A subcortical zone of hypomyelination and gliosis, as well as a line of glial heterotopias radiating toward the underlying ventricle, may be evident (74–76).

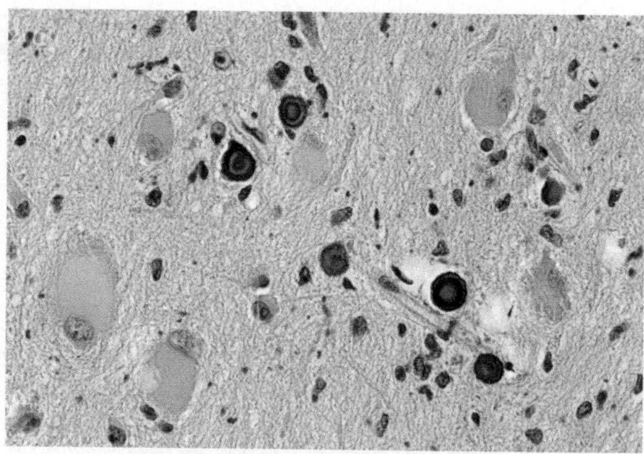

Figure 18-23

TUBEROUS SCLEROSIS: TUBER

Large glassy cells with features intermediate between ganglion cells and astrocytes populate the tuber. Calcification is common.

Although most tubers are found in the cerebral cortex, occasional examples lie within cerebellar folia. Classic tubers are firm, pale, and somewhat raised (fig. 18-22). They vary in size and extent, and many are macroscopically subtle both externally and on cut section. Some are several centimeters across and involve an entire gyrus, while others are evident only microscopically. Fully developed tubers are often umbilicated as the result of intense superficial cortical gliosis. Chronic lesions may become calcified or associated with subcortical cystic degeneration. On cross section, the expanded cortex is ill-defined with respect to the underlying white matter (fig. 18-22).

Microscopically, tubers consist of cortical regions in which neurons are few and cytologically abnormal (fig. 18-23). Cortical lamination is deranged, and extensive gliosis is typical of fully developed examples. Reactive astrocytes often lie concentrated about vessels and throughout the superficial cortex, a feature underlying the central umbilication of tubers. Often clustered, these reactive cells vary from small and multipolar to typical stellate astrocytes. Thus, they differ from the more conspicuous, large globular cells with ample pink cytoplasm and eccentric nuclei that are so characteristic, although not specific, of tubers. Particularly in early tubers, such large globular cells

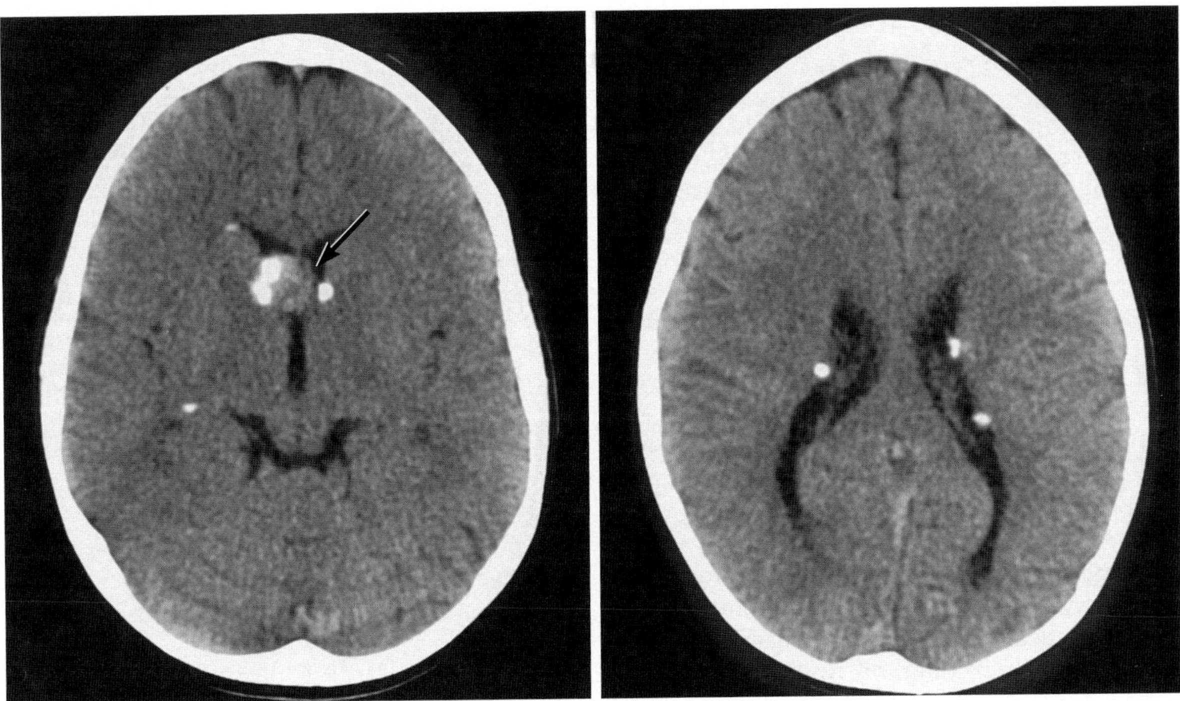

Figure 18-24

TUBEROUS SCLEROSIS: SUBEPENDYMAL GIANT CELL ASTROCYTOMA AND SUBEPENDYMAL HAMARTOMAS

Calcifications of subependymal giant cell astrocytoma (arrow, left) and subependymal hamartomas (right) are best seen by CT.

may be mistaken for bizarre neurons. Similar globular cells often lie clustered about vessels within cerebral white matter, a site at which they align in a radial manner from subcortical white matter to the deep subependymal region.

Histologically, the subependymal nodules are often densely calcified and similar to SEGAs, their symptomatic counterparts near the foramen of Monro. Both lesions have spindle to epithelioid, astrocytic-appearing cells, although pattern variation is most apparent in SEGAs.

Many of the large cells observed in tubers as well as in white matter are weakly periodic acid-Schiff (PAS) positive but, despite their astrocytic appearance, they are not always reactive for GFAP and S-100 protein (77). Ultrastructurally, the cells vary considerably in their content of intermediate filaments but many have an astrocytic appearance (78) consistent with their reported immunoreactivity for alphaB-crystallin (79). Other studies have shown positive staining for neurofilament proteins and an oligodendroglial marker, galactocerebroside (70,80). Collectively, these cells may exhibit structural proteins common to glial cells as well as neurons. Retinal astrocytic hamartomas also resemble subependymal nodules and SEGAs.

Tuberous sclerosis has, overall, a life-shortening effect, with death due variably to renal failure, brain tumors, status epilepticus, and bronchopneumonia, among other causes (81).

COWDEN'S SYNDROME

Cowden's syndrome, or *multiple hamartoma-neoplasia syndrome,* is an autosomal dominant disorder consisting of multiple cutaneous trichilemmomas, oral mucosal papillomas, colonic and small intestinal polyps of varying type (often associated with breast carcinomas), and thyroid abnormalities (goiter, adenomas, and, rarely, carcinomas) (82–85). The relevance of Cowden's syndrome to tumors of the CNS is based on its association with Lhermitte-Duclos disease, a highly distinctive cerebellar lesion discussed in chapter 4. Cowden's syndrome and Lhermitte-Duclos disease result from abnormalities in the tumor suppressor gene, *PTEN.*

GORLIN'S SYNDROME

Also known as *nevoid basal cell carcinoma syndrome, Gorlin's syndrome* features multiple basal cell carcinomas of skin, keratocysts of the jaw, and skeletal abnormalities (86). Lesions of the CNS include nodular/desmoplastic medulloblastomas (see chapter 5) (87), and lamellar calcifications of the falx. Gorlin's syndrome is inherited as an autosomal dominant condition, the genetic basis of which is the *PTCH* tumor suppressor gene.

TURCOT'S SYNDROME

Turcot's syndrome is the coincident appearance of colon cancer and a brain neoplasm. Two clinically and molecularly distinct types are included. One, due to an abnormality in DNA mismatch repair genes, consists of hereditary nonpolyposis colon cancer (HNPCC) and a glial neoplasm, usually glioblastoma (fig. 18-25) (88, 89). As a result of mutation in the adenopolyposis gene, the second Turcot variant combines colon cancer with medulloblastoma (88,89).

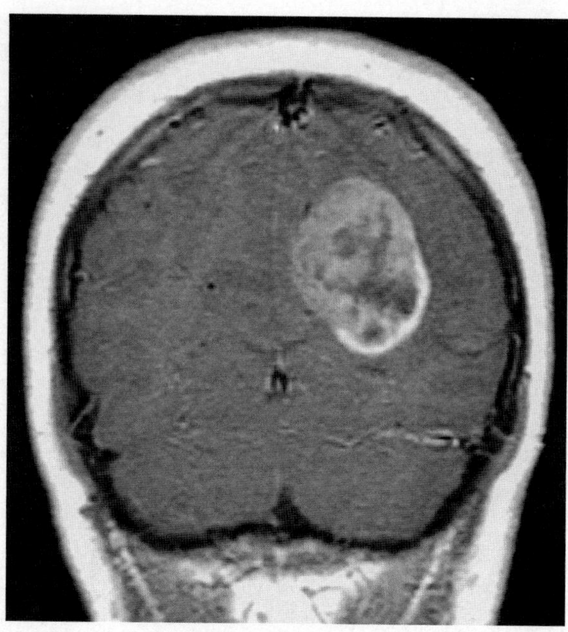

Figure 18-25

TURCOT'S SYNDROME: GLIOBLASTOMA

Only a rare glioblastoma can be explained by the genetic predisposition of Turcot's syndrome.

OTHER DYSGENETIC DISORDERS

The CNS vascular lesions in mesencephalo-oculo-facial angiomatosis (Wyburn-Mason syndrome) and hereditary hemorrhagic telangiectasia (Rendu-Osler-Weber syndrome) are discussed in chapter 10. The glial and chondroid tumors in Ollier's disease and Maffucci's syndrome are described in chapters 3 and 10, respectively. Choroid plexus tumors arising in the context of Aicardi's syndrome and Li-Fraumeni syndrome are considered briefly in chapter 3. Germline mutations in the *INI1* (*hSNF5*) gene and its relation to atypical teratoid/rhabdoid tumor are discussed in chapter 5. Germline mutations in *p53* appear responsible for occasional gliomas (90–93) and medulloblastomas (94). Carney's complex, including spotty skin pigmentation, myxomas (particularly in the heart), multiple endocrine neoplasms, and psammomatous melanotic schwannoma, is cited in the context of the peripheral nerve sheath neoplasms in chapter 13. Trilateral retinoblastoma is discussed in chapter 5. The relationship of some intracranial germ cell tumors to Kleinfelter's syndrome and Down's syndrome is described in chapter 8. The association of Rubinstein-Taybi syndrome and medulloblastoma is discussed briefly in chapter 5.

REFERENCES

Neurofibromatosis 1

1. Ars E, Kruyer H, Morell M, et al. Recurrent mutations in the NF1 gene are common among neurofibromatosis type 1 patients. J Med Genet 2003;40:e82.
2. Mulvihill JJ, Parry DM, Sherman JL, Pikus A, Kaiser-Kupfer MI, Eldridge R. NIH conference. Neurofibromatosis 1 (Recklinghausen disease) and neurofibromatosis 2 (bilateral acoustic neurofibromatosis). An update. Ann Intern Med 1990;113:39–52.
3. Friedman JM. Neurofibromatosis 1: clinical manifestations and diagnostic criteria. J Child Neurol 2002;17:548–54; discussion 571–2, 646–51.
4. Egelhoff JC, Bates DJ, Ross JS, Rothner AD, Cohen BH. Spinal MR findings in neurofibromatosis types 1 and 2. AJNR Am J Neuroradiol 1992;13:1071–7.
5. Halliday AL, Sobel RA, Martuza RL. Benign spinal nerve sheath tumors: their occurrence sporadically and in neurofibromatosis types 1 and 2. J Neurosurg 1991;74:248–53.
6. Calzavara PG, Carlino A, Anzola GP, Pasolini MP. Segmental neurofibromatosis. Case report and review of the literature. Neurofibromatosis 1988;1:318–22.
7. Listernick R, Mancini AJ, Charrow J. Segmental neurofibromatosis in childhood. Am J Med Genet 2003;121A:132–5.
8. Schwarz J, Belzberg AJ. Malignant peripheral nerve sheath tumors in the setting of segmental neurofibromatosis. Case report. J Neurosurg 2000;92:342–6.
9. Tinschert S, Naumann I, Stegmann E, et al. Segmental neurofibromatosis is caused by somatic mutation of the neurofibromatosis type 1 (NF1) gene. Eur J Hum Genet 2000;8:455–9.
10. Czyzyk E, Jozwiak S, Roszkowski M, Schwartz RA. Optic pathway gliomas in children with and without neurofibromatosis 1. J Child Neurol 2003;18:471–8.
11. Guillamo JS, Creange A, Kalifa C, et al. Prognostic factors of CNS tumours in neurofibromatosis 1 (NF1): a retrospective study of 104 patients. Brain 2003;126(Pt 1):152–60.
12. Gutmann DH, Hedrick NM, Li J, Nagarajan R, Perry A, Watson MA. Comparative gene expression profile analysis of neurofibromatosis 1-associated and sporadic pilocytic astrocytomas. Cancer Res 2002;62:2085–91.
13. Rosser T, Packer RJ. Intracranial neoplasms in children with neurofibromatosis 1. J Child Neurol 2002;17:630–7; discussion 646–51.
14. Tekkok IH, Akpinar G, Gungen Y, Sav A. Low-grade astrocytoma of the spinal cord associated with neurofibromatosis type-1: report of a case with poor correlation between histopathology and prognosis. Br J Neurosurg 2003;17:274–7.
15. Chateil JF, Soussotte C, Pedespan JM, Brun M, Le Manh C, Diard F. MRI and clinical differences between optic pathway tumours in children with and without neurofibromatosis. Br J Radiol 2001;74:24–31.
16. Leisti EL, Pyhtinen J, Poyhonen M. Spontaneous decrease of a pilocytic astrocytoma in neurofibromatosis type 1. AJNR Am J Neuroradiol 1996;17:1691–4.
17. Perilongo G, Moras P, Carollo C, et al. Spontaneous partial regression of low-grade glioma in children with neurofibromatosis-1: a real possibility. J Child Neurol 1999;14:352–6.
18. Molloy PT, Bilaniuk LT, Vaughan SN, et al. Brainstem tumors in patients with neurofibromatosis type 1: a distinct clinical entity. Neurology 1995;45:1897–902.
19. Pollack IF, Shultz B, Mulvihill JJ. The management of brainstem gliomas in patients with neurofibromatosis 1. Neurology 1996;46:1652–60.
20. Blatt J, Jaffe R, Deutsch M, Adkins JC. Neurofibromatosis and childhood tumors. Cancer 1986;57:1225–9.
21. Carella A, Medicamento N. Malignant evolution of presumed benign lesions in the brain in neurofibromatosis: case report. Neuroradiology 1997;39:639–41.
22. Gutmann DH, Rasmussen SA, Wolkenstein P, et al. Gliomas presenting after age 10 in individuals with neurofibromatosis type 1 (NF1). Neurology 2002;59:759–61.
23. Gutmann DH, James CD, Poyhonen M, et al. Molecular analysis of astrocytomas presenting after age 10 in individuals with NF1. Neurology 2003;61:1397–400.
24. Yokoo H, Kamiya M, Sasaki A, Hirato J, Nakazato Y, Kurachi H. Neurofibromatosis type 1-associated unusual pleomorphic astrocytoma displaying continual malignant progression. Pathol Int 2001;51:570–7.
25. Rubinstein LJ. The malformative central nervous system lesions in the central and peripheral forms of neurofibromatosis. A neuropathological study of 22 cases. Ann N Y Acad Sci 1986; 486:14–29.

26. DiPaolo DP, Zimmerman RA, Rorke LB, Zackai EH, Bilaniuk LT, Yachnis AT. Neurofibromatosis type 1: pathologic substrate of high-signal-intensity foci in the brain. Radiology 1995;195:721–4.
27. Szudek J, Friedman JM. Unidentified bright objects associated with features of neurofibromatosis 1. Pediatr Neurol 2002;27:123–7.
28. Sevick RJ, Barkovich AJ, Edwards MS, Koch T, Berg B, Lempert T. Evolution of white matter lesions in neurofibromatosis type 1: MR findings. AJR Am J Roentgenol 1992;159:171–5.
29. Lubs ML, Bauer MS, Formas ME, Djokic B. Lisch nodules in neurofibromatosis type 1. N Engl J Med 1991;324:1264–6.

Neurofibromatosis 2 (Central Neurofibromatosis)

30. Mulvihill JJ, Parry DM, Sherman JL, Pikus A, Kaiser-Kupfer MI, Eldridge R. NIH conference. Neurofibromatosis 1 (Recklinghausen disease) and neurofibromatosis 2 (bilateral acoustic neurofibromatosis). An update. Ann Intern Med 1990;113:39–52.
31. Pollack IF, Mulvihill JJ. Neurofibromatosis 1 and 2. Brain Pathol 1997;7:823–36.
32. Seizinger BR, Rouleau G, Ozelius LJ, et al. Common pathogenetic mechanism for three tumor types in bilateral acoustic neurofibromatosis. Science 1987;236:317–9.
33. Evans DG, Huson SM, Donnai D, et al. A clinical study of type 2 neurofibromatosis. Q J Med 1992;84:603–18.
34. Baser ME, Friedman JM, Wallace AJ, Ramsden RT, Joe H, Evans DG. Evaluation of clinical diagnostic criteria for neurofibromatosis 2. Neurology 2002;59:1759–65.
35. Kanter WR, Eldridge R, Fabricant R, Allen JC, Koerber T. Central neurofibromatosis with bilateral acoustic neuroma: genetic, clinical and biochemical distinctions from peripheral neurofibromatosis. Neurology 1980;30:851–9.
36. Martuza RL, Ojemann RG. Bilateral acoustic neuromas: clinical aspects, pathogenesis, and treatment. Neurosurgery 1982;10:1–12.
37. Geddes JF, Sutcliffe JC, King TT. Mixed cranial nerve tumors in neurofibromatosis type 2. Clin Neuropathol 1995;14:310–3.
38. Ludemann W, Stan AC, Tatagiba M, Samii M. Sporadic unilateral vestibular schwannoma with islets of meningioma: case report. Neurosurgery 2000;47:451–4.
39. Sobel RA. Vestibular (acoustic) schwannomas: histologic features in neurofibromatosis 2 and in unilateral cases. J Neuropathol Exp Neurol 1993;52:106–13.
40. Perry A, Giannini C, Raghavan R, et al. Aggressive phenotypic and genotypic features in pediatric and NF2-associated meningiomas: a clinicopathologic study of 53 cases. J Neuropathol Exp Neurol 2001;60:994–1003.
41. Mayfrank L, Mohadjer M, Wullich B. Intracranial calcified deposits in neurofibromatosis type 2. A CT study of 11 cases. Neuroradiology 1990; 32:33–7.
42. Egelhoff JC, Bates DJ, Ross JS, Rothner AD, Cohen BH. Spinal MR findings in neurofibromatosis types 1 and 2. AJNR Am J Neuroradiol 1992;13:1071–7.
43. Halliday AL, Sobel RA, Martuza RL. Benign spinal nerve sheath tumors: their occurrence sporadically and in neurofibromatosis types 1 and 2. J Neurosurg 1991;74:248–53.
44. Rubinstein LJ. The malformative central nervous system lesions in the central and peripheral forms of neurofibromatosis. A neuropathological study of 22 cases. Ann N Y Acad Sci 1986;486:14–29.
45. Schievink WI, Piepgras DG. Cervical vertebral artery aneurysms and arteriovenous fistulae in neurofibromatosis type 1: case reports. Neurosurgery 1991;29:760–5.
46. Wiestler OD, von Siebenthal K, Schmitt HP, Feiden W, Kleihues P. Distribution and immunoreactivity of cerebral micro-hamartomas in bilateral acoustic neurofibromatosis (neurofibromatosis 2). Acta Neuropathol (Berl) 1989;79:137–43.

Von Hippel-Lindau Disease

47. Conway JE, Chou D, Clatterbuck RE, Brem H, Long DM, Rigamonti D. Hemangioblastomas of the central nervous system in von Hippel-Lindau syndrome and sporadic disease. Neurosurgery 2001;48:55–62; discussion 62–3.
48. Couch V, Lindor NM, Karnes PS, Michels VV. von Hippel-Lindau disease. Mayo Clin Proc 2000;75:265–72.
49. Dollfus H, Massin P, Taupin P, et al. Retinal hemangioblastoma in von Hippel-Lindau disease: a clinical and molecular study. Invest Ophthalmol Vis Sci 2002;43:3067–74.
50. Kerr DJ, Scheithauer BW, Miller GM, Ebersold MJ, McPhee TJ. Hemangioblastoma of the optic nerve: case report. Neurosurgery 1995;36: 573–80; discussion 580–1.
51. Neumann HP, Eggert HR, Scheremet R, et al. Central nervous system lesions in von Hippel-Lindau syndrome. J Neurol Neurosurg Psychiatry 1992;55:898–901.
52. Neumann HP, Eggert HR, Weigel K, Friedburg H, Wiestler OD, Schollmeyer P. Hemangioblastomas of the central nervous system. A 10-year study with special reference to von Hippel-Lindau syndrome. J Neurosurg 1989;70:24–30.

53. Slater A, Moore NR, Huson SM. The natural history of cerebellar hemangioblastomas in von Hippel-Lindau disease. AJNR Am J Neuroradiol 2003;24:1570–4.
54. Ismail SM, Cole G. Von Hippel-Lindau syndrome with microscopic hemangioblastomas of the spinal nerve roots. Case report. J Neurosurg 1984;60:1279–81.
55. Lonser RR, Wait SD, Butman JA, et al. Surgical management of lumbosacral nerve root hemangioblastomas in von Hippel-Lindau syndrome. J Neurosurg 2003;99(Suppl 1):64–9.
56. Ishwar S, Taniguchi RM, Vogel FS. Multiple supratentorial hemangioblastomas. Case study and ultrastructural characteristics. J Neurosurg 1971;35:396–405.
57. Giannini C, Scheithauer BW, Hellbusch LC, et al. Peripheral nerve hemangioblastoma. Mod Pathol 1998;11:999–1004.
58. Hamazaki S, Yoshida M, Yao M, et al. Mutation of von Hippel-Lindau tumor suppressor gene in a sporadic endolymphatic sac tumor. Hum Pathol 2001;32:1272–6.
59. Horiguchi H, Sano T, Toi H, Kageji T, Hirokawa M, Nagahiro S. Endolymphatic sac tumor associated with a von Hippel-Lindau disease patient: an immunohistochemical study. Mod Pathol 2001;14:727–32.
60. Richard S, David P, Marsot-Dupuch K, Giraud S, Beroud C, Resche F. Central nervous system hemangioblastomas, endolymphatic sac tumors, and von Hippel-Lindau disease. Neurosurg Rev 2000;23:1–24.
61. Maher ER, Yates JR, Harries R, et al. Clinical features and natural history of von Hippel-Lindau disease. Q J Med 1990;77:1151–63.
62. Melmon KL, Rosen SW. Lindau's disease. Review of the literature and study of a large kindred. Am J Med 1964;36:595–617.
63. Latif F, Tory K, Gnarra J, et al. Identification of the von Hippel-Lindau disease tumor suppressor gene. Science 1993;260:1317–20.
64. Seizinger BR, Rouleau GA, Ozelius LJ, et al. Von Hippel-Lindau disease maps to the region of chromosome 3 associated with renal cell carcinoma. Nature 1988;332:268–9.
65. Vortmeyer AO, Huang SC, Pack SD, et al. Somatic point mutation of the wild-type allele detected in tumors of patients with VHL germline deletion. Oncogene 2002;21:1167–70.

Tuberous Sclerosis Complex

66. Jozwiak S, Schwartz RA, Janniger CK, Bielicka-Cymerman J. Usefulness of diagnostic criteria of tuberous sclerosis complex in pediatric patients. J Child Neurol 2000;15:652–9.
67. Nagib MG, Haines SJ, Erickson DL, Mastri AR. Tuberous sclerosis: a review for the neurosurgeon. Neurosurgery 1984;14:93–8.
68. Roach ES, Gomez MR, Northrup H. Tuberous sclerosis complex consensus conference: revised clinical diagnostic criteria. J Child Neurol 1998;13:624–8.
69. Sparagana SP, Roach ES. Tuberous sclerosis complex. Curr Opin Neurol 2000;13:115–9.
70. consortium Ects. Identification and characterization of the tuberous sclerosis gene on chromosome 16. The European Chromosome 16 Tuberous Sclerosis Consortium. Cell 1993;75: 1305–15.
71. Johnson MW, Emelin JK, Park SH, Vinters HV. Co-localization of TSC1 and TSC2 gene products in tubers of patients with tuberous sclerosis. Brain Pathol 1999;9:45–54.
72. Kwiatkowska J, Jozwiak S, Hall F, et al. Comprehensive mutational analysis of the TSC1 gene: observations on frequency of mutation, associated features, and nonpenetrance. Ann Hum Genet 1998;62 (Pt 4):277–85.
73. Menchine M, Emelin JK, Mischel PS, et al. Tissue and cell-type specific expression of the tuberous sclerosis gene, TSC2, in human tissues. Mod Pathol 1996;9:1071–80.
74. Iwasaki S, Nakagawa H, Kichikawa K, et al. MR and CT of tuberous sclerosis: linear abnormalities in the cerebral white matter. AJNR Am J Neuroradiol 1990;11:1029–34.
75. Nixon JR, Miller GM, Okazaki H, Gomez MR. Cerebral tuberous sclerosis: postmortem magnetic resonance imaging and pathologic anatomy. Mayo Clin Proc 1989;64:305–11.
76. Nixon JR, Houser OW, Gomez MR, Okazaki H. Cerebral tuberous sclerosis: MR imaging. Radiology 1989;170(Pt 1):869–73.
77. Scheithauer BW. The neuropathology of tuberous sclerosis. J Dermatol 1992;19:897–903.
78. Trombley IK, Mirra SS. Ultrastructure of tuberous sclerosis: cortical tuber and subependymal tumor. Ann Neurol 1981;9:174–81.
79. Iwaki T, Wisniewski T, Iwaki A, et al. Accumulation of alpha B-crystallin in central nervous system glia and neurons in pathologic conditions. Am J Pathol 1992;140:345–56.
80. Chou TM, Chou SM. Tuberous sclerosis in the premature infant: a report of a case with immunohistochemistry on the CNS. Clin Neuropathol 1989;8:45–52.
81. Shepherd CW, Gomez MR, Lie JT, Crowson CS. Causes of death in patients with tuberous sclerosis. Mayo Clin Proc 1991;66:792–6.

Cowden's Syndrome

82. Padberg GW, Schot JD, Vielvoye GJ, Bots GT, de Beer FC. Lhermitte-Duclos disease and Cowden disease: a single phakomatosis. Ann Neurol 1991;29:517–23.
83. Albrecht S, Haber RM, Goodman JC, Duvic M. Cowden syndrome and Lhermitte-Duclos disease. Cancer 1992;70:869–76.
84. Derrey S, Proust F, Debono B, et al. Association between Cowden syndrome and Lhermitte-Duclos disease: report of two cases and review of the literature. Surg Neurol 2004;61:447–54.
85. Abel TW, Baker SJ, Fraser MM, et al. Lhermitte-Duclos disease: a report of 31 cases with immunohistochemical analysis of the PTEN/AKT/mTOR pathway. J Neuropathol Exp Neurol 2005;64:341–9.

Gorlin's Syndrome

86. Kimonis VE, Goldstein AM, Pastakia B, et al. Clinical manifestations in 105 persons with nevoid basal cell carcinoma syndrome. Am J Med Genet 1997;69:299–308.
87. Cowan R, Hoban P, Kelsey A, Birch JM, Gattamaneni R, Evans DG. The gene for the naevoid basal cell carcinoma syndrome acts as a tumour-suppressor gene in medulloblastoma. Br J Cancer 1997;76:141–5.

Turcot's Syndrome

88. Hamilton SR, Liu B, Parsons RE, et al. The molecular basis of Turcot's syndrome. N Engl J Med 1995;332:839–47.
89. Qualman SJ, Bowen J, Erdman SH. Molecular basis of the brain tumor-polyposis (Turcot) syndrome. Pediatr Dev Pathol 2003;6:574–6.

Other Dysgenetic Disorders

90. Bogler O, Huang HJ, Kleihues P, Cavenee WK. The p53 gene and its role in human brain tumors. Glia 1995;15:308–27.
91. Kleihues P, Schauble B, zur Hausen A, Esteve J, Ohgaki H. Tumors associated with p53 germline mutations: a synopsis of 91 families. Am J Pathol 1997;150:1–13.
92. Li YJ, Sanson M, Hoang-Xuan K, et al. Incidence of germ-line p53 mutations in patients with gliomas. Int J Cancer 1995;64:383–7.
93. Vital A, Bringuier PP, Huang H, et al. Astrocytomas and choroid plexus tumors in two families with identical p53 germline mutations. J Neuropathol Exp Neurol 1998;57:1061–9.
94. Taylor MD, Mainprize TG, Rutka JT. Molecular insight into medulloblastoma and central nervous system primitive neuroectodermal tumor biology from hereditary syndromes: a review. Neurosurgery 2000;47:888–901.

19 REACTIVE AND INFLAMMATORY MASSES

MACROPHAGE-RICH LESIONS

Demyelinating Disease

Clinical Features. Classic multicentric *demyelinating disease,* or *multiple sclerosis,* is usually recognized on the basis of its clinical and radiographic presentation, without the need for histologic confirmation. The unusual demyelinating lesions approached surgically are often solitary, and sometimes sentinel, expressions of classic multiple sclerosis, or are what later prove to be permanently isolated lesions of another demyelinating disease variant (1). Multiple lesions in some cases, however, prompt such serious consideration of a glioma that a biopsy is performed (2). Like demyelinating lesions in general, those that mimic neoplasms generally affect young adults, but presentation in older individuals in no way excludes this class of disease, although the possibility of a steroid-treated lymphoma must be considered. Metastatic carcinoma may be suspected preoperatively in a rare form of multifocal demyelinating disease that occurs during chemotherapy for colonic carcinoma (3,4).

The clinical, radiologic, and pathologic features of demyelinating diseases are summarized in Table 19-1.

Radiologic Findings. Most demyelinating plaques that are approached surgically lie either adjacent to the lateral ventricles, immediately beneath the cerebral cortex, or within the spinal cord. While plaques are usually small, 1 to 2 cm, massive lesions can be encountered in the Marburg variant of demyelinating disease (5). Like most other pathologic processes, plaques are intrinsically bright in T2-weighted magnetic resonance images (MRIs) or fluid-attenuated inversion recovery (FLAIR) images, including the perilesional zone of edema that is prominent in some cases (6,7). Demyelinating plaques often contrast enhance in a pattern that is highly suggestive of the disease. Unlike neoplasms that enhance in a "rim-like" or circumferential pattern, demyelinating foci do so incompletely in the form of an "open ring" or "horseshoe" (fig. 19-1) (6,8–10). The opening or missing segment faces inwardly in the case of subependymal lesions and outwardly with subcortical plaques. Spinal lesions are nonspecific and contrast enhancing.

Microscopic Findings. The outpouring of macrophages that accompanies demyelination results in a hypercellularity that can closely resemble glioma. This is particularly true at low-power magnification in both frozen and permanent sections (figs. 19-2, 19-3) (5,7–9,11–13).

Table 19-1

CLINICAL, RADIOLOGIC, AND PATHOLOGIC FEATURES OF BIOPSIED DEMYELINATING DISEASE

Clinical
Often in third and fourth decades, but any age group potentially affected
Usually solitary but sometimes multiple
Almost always suspected as a neoplasm; suspicions, if any, about the possibility of demyelinating disease may not be relayed to the pathologist
Radiologic
Any location, but examples approached surgically usually are subependymal, subcortical, or spinal intramedullary
Contrast enhancement in intracranial lesions often assumes an "open ring" or "horseshoe" pattern, the open end facing inwardly in subendymal plaques and outwardly in subcortical lesions
Considerable cerebral edema in some cases
Pathologic
White matter localization
Discrete border
Macrophage infiltrate
Perivascular lymphocytic inflammation
Preserved axons
Reactive astrocytes
Creutzfeldt cells

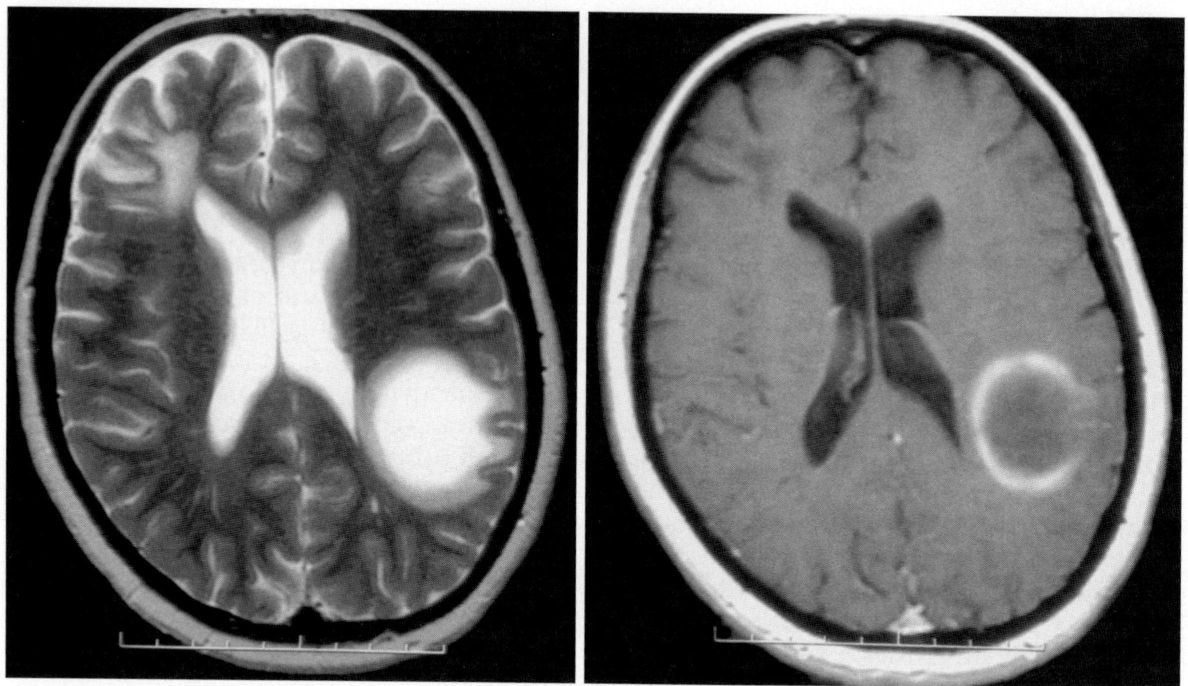

Figure 19-1

DEMYELINATING DISEASE

As is characteristic of demyelinating lesions, the rim of enhancement is open at one end, in contrast to malignant glioma in which it is circumferential. The patient was a 36-year-old woman with a recent history of aphasia. T2-weighted imaging disclosed a hyperintense, expansile lesion in the left parietal lobe. The possibility of demyelinating disease was entertained, but, faced with clinical progression, a needle biopsy was performed to rule out a glioma.

Scattered mitoses and large astrocytes add to the illusion. Recognition of demyelinating lesions thus may not be intuitive, but is facilitated by the checklist approach in Appendix O that requires the pathologist to exclude the entity before considering infiltrating gliomas. Macrophages abound (fig. 19-2C) and axons are preserved (fig. 19-2D).

As enumerated in Table 19-1, there are seven cardinal histologic features of demyelinating disease.

White Matter Localization. Since the disease targets myelin, white matter is principally affected (fig. 19-2A,B). Nonetheless, intracortical myelinated axons can be bared as well.

Discrete Border. Demyelinating plaques are often remarkably well defined even in hematoxylin and eosin (H&E)-stained sections (fig. 19-2A; see Appendix L, fig. 1E), but are especially so after staining for myelin with such methods as Luxol fast blue (fig. 19-2B) or immunohistochemistry for macrophages (fig. 19-2C). Some variants of demyelinating disease are not as well circumscribed at the microscopic level, however. As is evident only in large specimens, such as those studied postmortem, demyelinating plaques are often strikingly perivenous.

Macrophage Infiltrate. High-power magnification brings out the round, uniform, cytologically bland nuclei; vacuolated cytoplasm; and distinct cell borders of the macrophages (figs. 19-2A, 19-3). Once even a few macrophages are identified, it becomes apparent that they are the principal cause of the hypercellularity. The ease with which these cells are recognized varies from case to case, however. Their identity is obvious in some lesions (figs. 19-2A, 19-3A), but not in others, where they may be misinterpreted as cells of an infiltrative glioma, either oligodendroglial, astrocytic, or "mixed" (fig. 19-3B,C,D). As the lipid-laden phagocytes return to the vasculature, they congregate in the perithelial region, where they appear to hesitate before reentering the blood stream. This distinctive

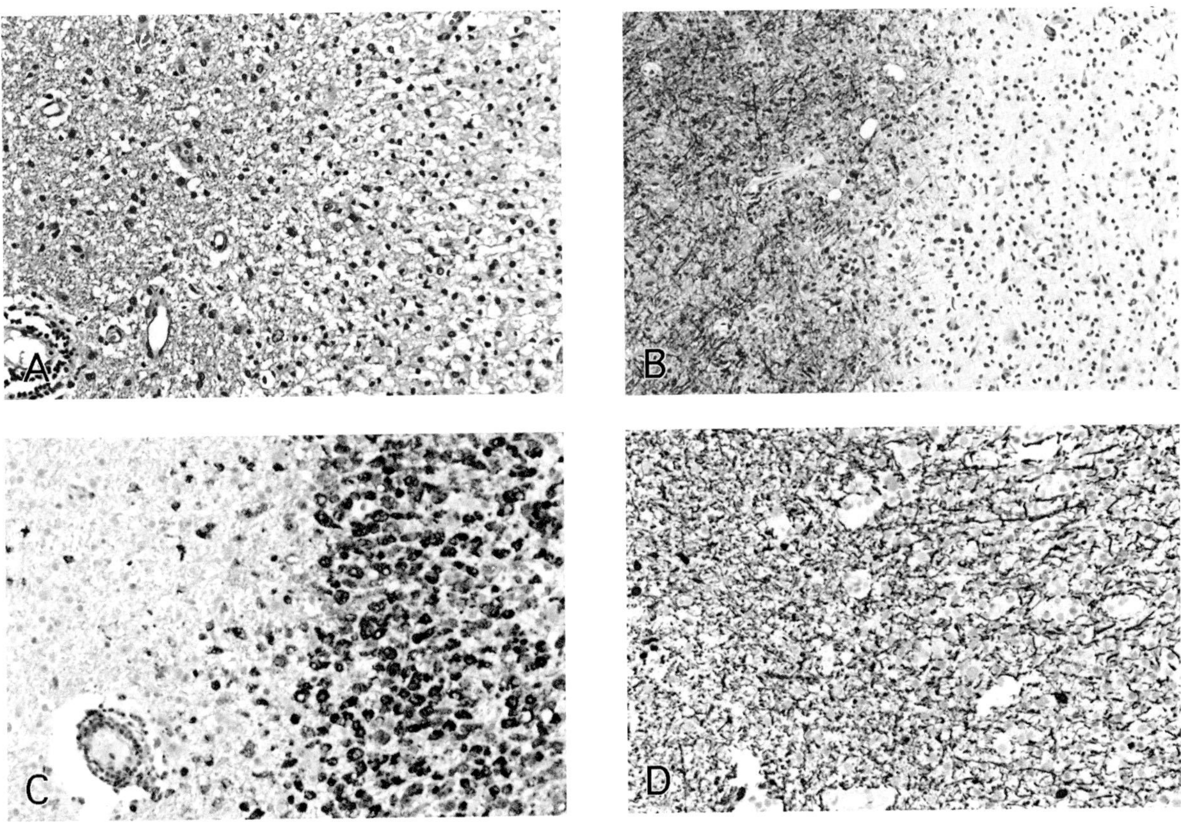

Figure 19-2

CLASSIC PATHOLOGIC FEATURES OF DEMYELINATING DISEASE

A: Demyelinating plaques are localized to white matter, sharply defined, and often associated with an intralesional or perilesional lymphocytic infiltrate.

B: Staining for myelin can be used to demonstrate a plaque's remarkably discrete border (hematoxylin and eosin[H&E]/ Luxol fast blue stain).

C: The sharply defined macrophage-rich lesions are clearly seen after immunostaining for CD68.

D: As is required for the diagnosis, axons are largely, if not completely, preserved. As a consequence, plaques are less obvious in sections stained for axons (here with antibodies to neurofilament protein) than in those stained for macrophages or myelin.

perivascular huddle of monomorphous, vacuolated, periodic acid–Schiff (PAS)-positive cells is typical of both demyelinating and ischemic lesions, but not of neoplasia.

Occasional plaques are only subtly demyelinated ("shadow plaques"); they are more obvious on MRI than in histologic sections. Knowledge of the neuroradiologic appearance of the lesion becomes essential in such cases in which a definitive diagnosis of demyelination may not be possible.

Perivascular Inflammation. Cuffs of lymphocytes around vessels (fig. 19-4) are a common feature of demyelinating disease, although they are neither disease-specific nor present in all cases. These cells alert the pathologist to the possibility of demyelinating disease, certain other non-neoplastic lesions, or low-grade lesions such as ganglion cell tumors.

Preserved Axons. On balance, axis cylinders are spared by demyelinating disease, although some fragmentation and axonal loss occur in active plaques. Axons can be demonstrated with immunohistochemistry for neurofilament protein (fig. 19-2D) or histochemical methods such as that of Bielschowsky. Surviving axons may be less precisely aligned than those in the adjacent normal white matter and can show reactive swellings or "spheroids." Coagulative tissue necrosis, with destruction of axons as well

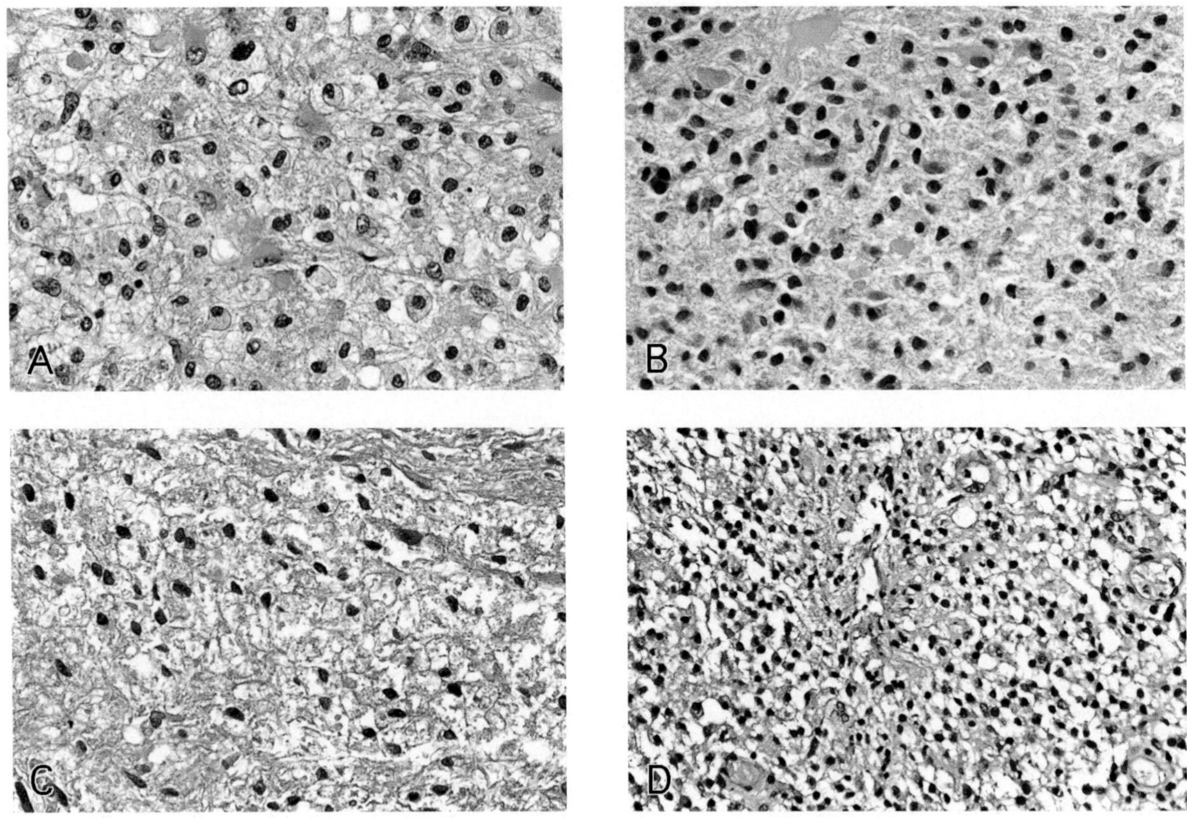

Figure 19-3

HISTOLOGIC VARIATIONS IN DEMYELINATING DISEASE

A: The macrophage nature of the cells is readily apparent in this case given their distinct cell borders, finely granular or foamy cytoplasm, and uniform bland nuclei. Reactive astrocytes with multipolar processes should not be interpreted as neoplastic.

B: Discrete cell borders, hallmarks of macrophages, can be very subtle findings.

C: Demyelinating disease plaques of only moderate cellularity can resemble a grade II infiltrating glioma. The distinct cell borders, here only focally apparent, are a helpful distinguishing feature.

D: Although almost every cell in this field is a macrophage, few have classic features. Cellular lesions such as this are easily misinterpreted as gliomas.

as myelin, usually is limited to rapidly evolving plaques, but can be encountered at sites such as optic nerve or spinal cord where tissue is restrained from swelling.

Reactive Astrocytes. Lying among the macrophages, reactive astrocytes have prominent pink cytoplasm and a lush, stellate array of tapering processes (figs. 19-5, 19-6). There is little actual proliferation of these hypertrophic cells, which are more evenly distributed and more richly endowed with processes than those of a gemistocytic astrocytoma.

Creutzfeldt Cells. These unusual cells are astrocytes featuring prominent glassy cytoplasm and multiple irregular micronuclei of varying size and shape (fig. 19-7). The nuclei have the chromatin density of normal cells, and not the hyperchromasia of karyorrhectic fragments. Creutzfeldt cells are not specific to demyelinating disease, or even to reactive lesions, but are a frequent and useful feature pointing to the possibility of these diseases. What may be precursors of Creutzfeldt cells are large astrocytes exhibiting chromosomes that appear to explode centripetally in a starburst fashion, unaccompanied by subsequent cell division.

Frozen Section. The diagnosis of demyelinating disease may be difficult based on frozen

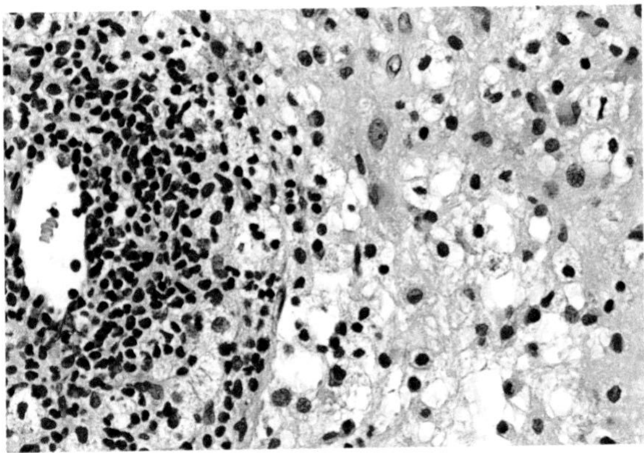

Figure 19-4

DEMYELINATING DISEASE

Perivascular lymphocytic infiltrates should prompt consideration of demyelinating disease. The foamy cells in this case have the classic features of macrophages.

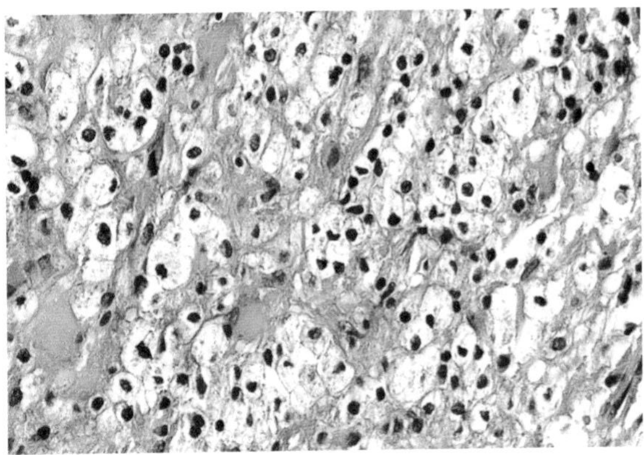

Figure 19-5

DEMYELINATING DISEASE

Large reactive astrocytes may give the false impression of astrocytoma or oligoastrocytoma.

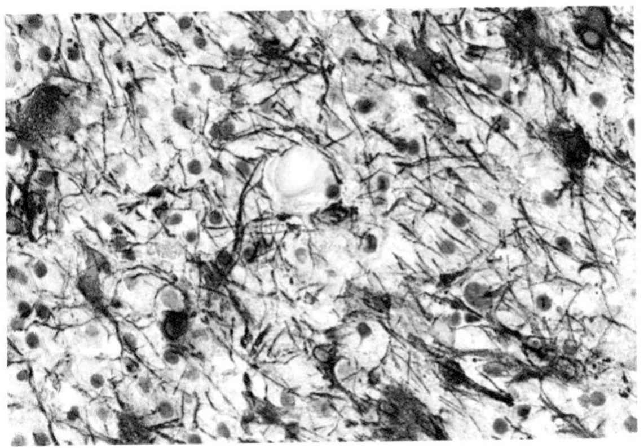

Figure 19-6

DEMYELINATING DISEASE

Large, glial fibrillary acidic protein (GFAP)-positive reactive astrocytes may divert attention from macrophages and abet in the misdiagnosis of glioma.

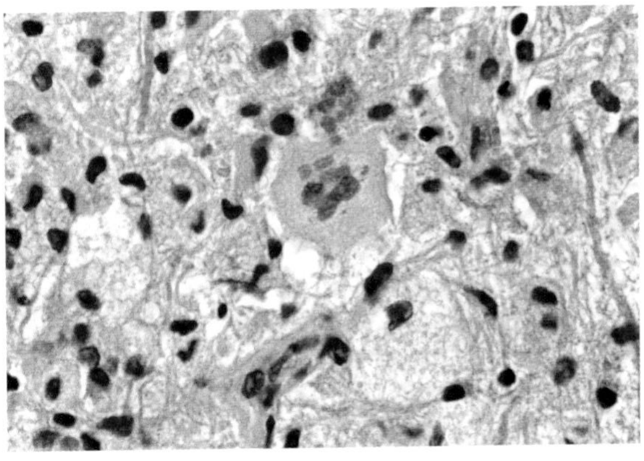

Figure 19-7

DEMYELINATING DISEASE

Reactive astrocytes with multiple, irregular micronuclei (Creutzfeldt cells) suggest the possibility of demyelinating or other non-neoplastic lesions.

section specimens, since the sharp cell borders of the macrophages are often obscured. When sampled, however, the plaque's discrete border is readily apparent, and an inflammatory lymphocytic infiltrate, albeit scant or absent in some cases, is an additional feature pointing to the possibility of the disease (fig. 19-8).

Immunohistochemical Findings. The macrophages comprising much of the cellularity of a plaque are strongly immunoreactive for the histiocytic marker CD68 (fig. 19-2C). Preserved axons can be demonstrated with antibodies to neurofilament protein (fig. 19-2D). The cytoplasm and the symmetric array of processes emanating from reactive astrocytes are impressively positive for glial fibrillary acidic protein (GFAP). As seen in smear or tissue sections, these lush processes are more consistent with gliosis than glioma (fig. 19-6). Perivascular lymphocytes are polyclonal and principally T-cell type (14).

Ultrastructural Findings. Macrophages are identified by their irregular, sometimes filopodia-bearing cell borders without junctional contacts. The cytoplasmic content consists largely

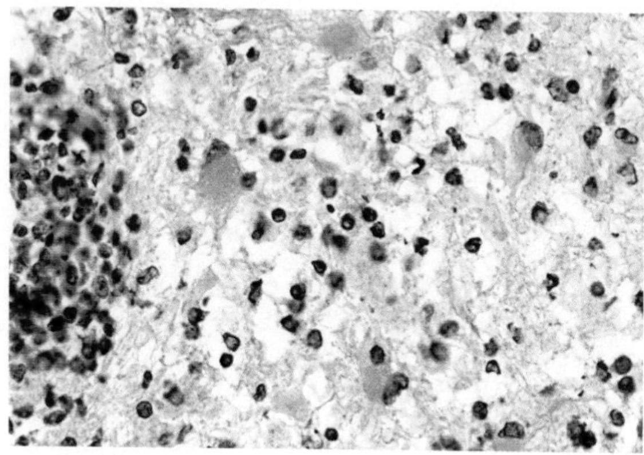

Figure 19-8

DEMYELINATING DISEASE: FROZEN SECTION

Demyelinating lesions can be readily misinterpreted as glioma since the cytologic features of the macrophages are obscured in the freezing process. Although nonspecific, the perivascular lymphocytic infiltrate should raise the possibility of demyelinating disease.

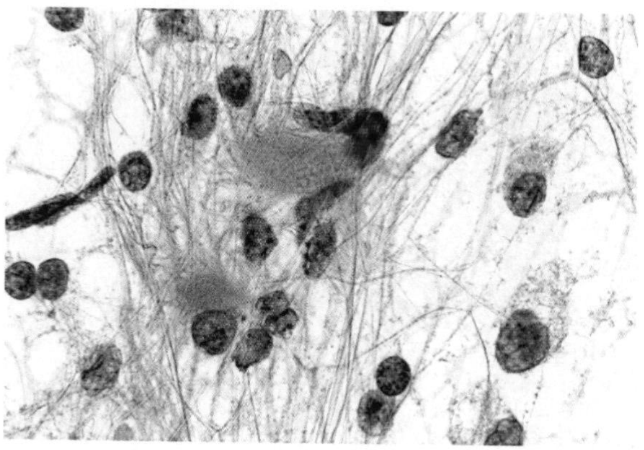

Figure 19-9

DEMYELINATING DISEASE: SMEAR PREPARATION

The combination of macrophages and large reactive astrocytes is typical of demyelinating disease.

of phagocytized myelin debris. Oligodendrocytes are sparse. Remyelination may be in progress as evidenced by unusually thin periaxonal myelin sheaths.

Cytologic Findings. The essential features of macrophages are their plump appearance with round benign-appearing nuclei, granular cytoplasmic contents, and distinct cell borders (fig. 19-9) (15). Reactive astrocytes, with their luxurious array of processes, are often prominent.

Differential Diagnosis. A diagnosis of demyelinating lesion should be considered for any surgical specimen for which a diagnosis of infiltrating glioma is entertained. One must be thus vigilant for the presence of macrophages and liberal in the use of antibodies for their detection. Appendix L summarizes reactive and non-neoplastic entities, such as demyelinating disease and other lesions, that can be confused with a neoplasm, and Appendix M lists the histologic features that suggest a low-grade or non-neoplastic lesion, such as the presence of macrophages or an oligodendroglioma-like appearance.

Non-neoplastic macrophage-rich lesions include cerebral infarct and progressive multifocal leukoencephalopathy. In contrast to demyelinating disease, infarcts have a prominent effect on the cortex, and include ischemic ("red-dead") neurons and hypertrophic capillaries in the acute phase. In principle, axons are preserved in demyelinating disease but lost in infarcts.

In the immunosuppressed patient, the presence of macrophage-rich areas of myelin loss should prompt consideration of progressive multifocal leukoencephalopathy (PML). Due to infection by the opportunistic JC papova virus, the key morphologic finding is large, hyperchromatic, infected oligodendroglia with ground-glass nuclear inclusions that are best seen at the edges of the lesion. As in the plaques of conventional demyelinating disease, those of PML contain reactive astrocytes with striking cytologic atypia. Such cells can be easily misinterpreted as evidence of a glioma.

Given the uniformity, nuclear roundness, and cytoplasmic vacuolization of the macrophages in demyelinating disease, oligodendroglioma intrudes into the differential diagnosis. When reactive astrocytes are particularly prominent, a diagnosis of astrocytoma or of oligoastrocytoma ("mixed glioma") may be considered as well. In contrast to neoplastic oligodendrocytes, however, macrophages have normochromatic nuclei, vacuolated rather than clear cytoplasm (figs. 19-10, 19-11), and immunoreactivity to CD68. The relatively sharp border of plaques in demyelinating disease is also a helpful feature, although circumscription may be seen on the deep aspect of some oligodendrogliomas. Given the stellate configuration and uniform distribution of reactive astrocytes, a diagnosis

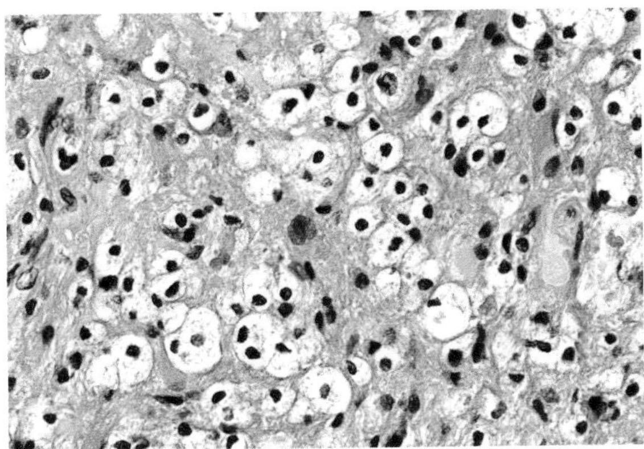

Figure 19-10

DEMYELINATING DISEASE

Foamy or finely granular, not clear, cytoplasm is typical of macrophages. Without this recognition, the lesion can be misinterpreted readily as a glioma.

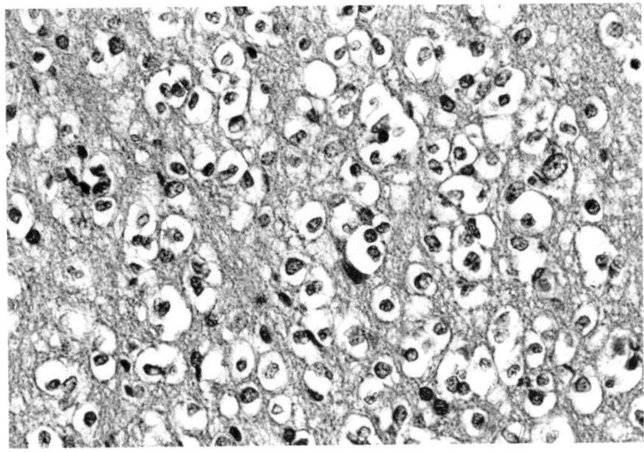

Figure 19-11

OLIGODENDROGLIOMA RESEMBLING DEMYELINATING DISEASE

The water clear cytoplasm of oligodendrocytes differs from the foamy or finely granular cytoplasm of the macrophages shown in figure 19-10.

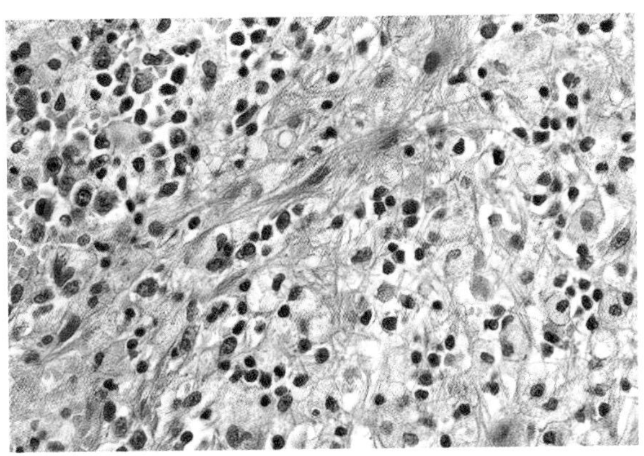

Figure 19-12

PRIMARY CENTRAL NERVOUS SYSTEM LYMPHOMA (PCNSL) TREATED WITH CORTICOSTEROIDS RESEMBLING DEMYELINATING DISEASE

Corticosteroid therapy can convert a PCNSL into a sea of macrophages that resembles demyelinating disease. While residual tumor cells are present in this case (upper left), there may be none in other cases.

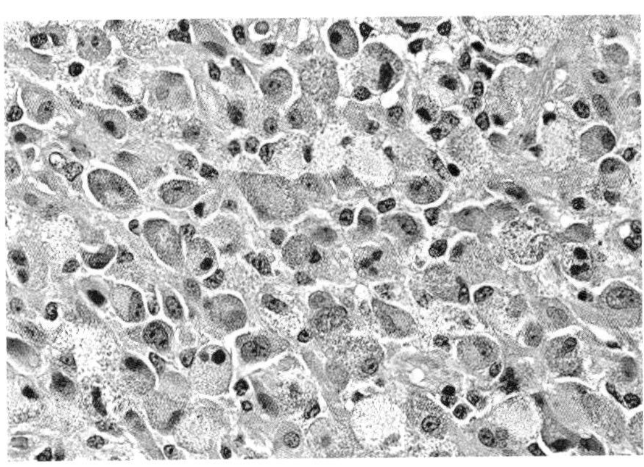

Figure 19-13

GRANULAR CELL ASTROCYTOMA RESEMBLING DEMYELINATING DISEASE

While the cells of granular cell astrocytoma are somewhat larger and their cytoplasm is more coarsely granular, the distinction between the two entities is not always easy.

of astrocytoma can usually be excluded. Reference to neuroimages can be extremely helpful, since a contrast-enhancing lesion, as is often seen in demyelinating disease, is inconsistent with a diagnosis of grade II astrocytoma, oligodendroglioma, or oligoastrocytoma.

Preoperative corticosteroid therapy can deplete a primary central nervous system (CNS) lymphoma of its tumor cells, and leave a mass lesion consisting largely, if not entirely, of macrophages (fig. 19-12). The often older age of the patient with lymphoma is a clue, as is the neoplasm's radiologic profile, i.e., a solidly enhancing mass without the open ring pattern so characteristic of demyelinating disease.

Granular cell astrocytoma (fig. 19-13) also enters into the differential diagnosis since this macrophage-like proliferation may be CD68 positive. The cells of such astrocytomas show obvious nuclear atypia and cytoplasm that is more

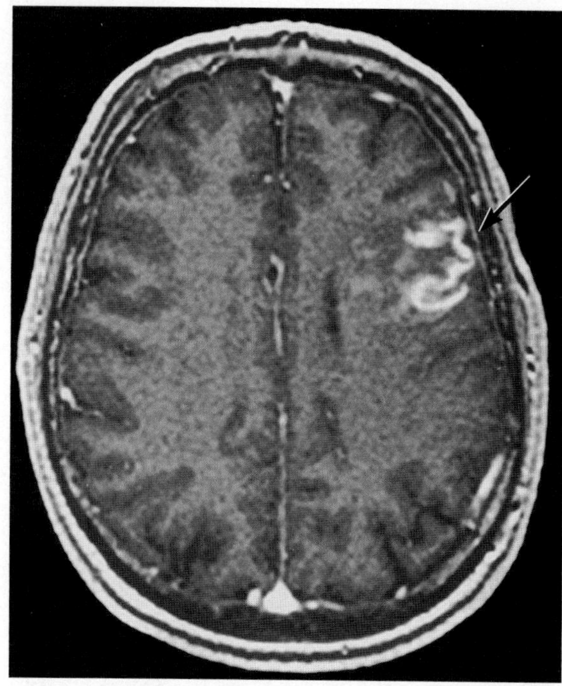

Figure 19-14

CEREBRAL INFARCT

This 45-year-old man presented with right-sided weakness and difficulty with speech. While the gyriform pattern of enhancement (arrow) suggested an infarct, a biopsy was performed to exclude an inflammatory or neoplastic process.

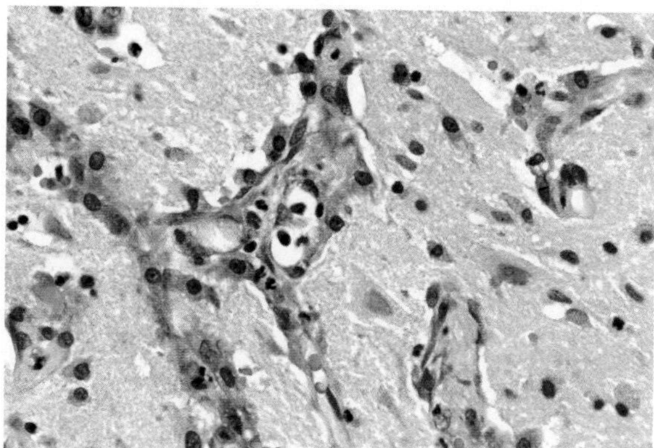

Figure 19-15

CEREBRAL INFARCT

Acute cerebral infarcts are recognized by changes in both neurons and small vessels. Karyolytic "red-dead" neurons are reliable evidence of recent neuronal injury; cells of the capillaries are hypertrophic and, to a lesser extent, hyperplastic.

abundant and coarsely granular rather than finely granular or foamy. The distinction is not always easy, however, as is discussed in chapter 3.

In some cases, a lesion composed largely of macrophages cannot be placed in any of the above categories and must be diagnosed descriptively as a "macrophage infiltrate" or "macrophage-rich lesion."

Treatment and Prognosis. There is great variability in the course of demyelinating disease; some lesions remain unifocal but others are joined later by plaques at other sites. It remains to be established whether the rare glioma that develops in the setting of demyelinating disease is pathogenetically related (16).

Cerebral Infarct

Clinical Features. With the exception of rare instances in which reductive surgery is undertaken to interdict herniation, infarcts are generally biopsied only inadvertently in anticipation of a malignant glioma. Large superficial infarcts may be approached by craniotomy, whereas small, deep-seated lesions in the basal ganglia or thalamus are usually sampled by stereotactic biopsy. Restricted diffusion, as seen on MRI, permits the identification of most infarcts, and has lessened the number approached surgically as glioma suspects.

Radiologic Findings. Large hemispheric infarcts are characteristically wedge shaped and confined to the distribution of a single cerebral blood vessel. Infarcts of the basal ganglia or thalamus, in contrast, are generally round in the axial plane and, lacking the wedge-shape configuration, are more likely to be glioma suspects.

Infarcts contrast enhance after approximately 1 week. Prominent gyriform enhancement is a helpful diagnostic feature of cortical lesions (fig. 19-14).

Gross Findings. Acute arterial cerebral infarcts are soft expansile masses; the gray-white junction is blurred. Venous infarcts are hemorrhagic.

Microscopic Findings. In the acute phase (first 24 hours), before macrophages make their appearance, the lesion is evidenced by changes in the cortex (17,18). Here, the neuropil appears hypereosinophilic and somewhat granular. In addition, large neurons become intensely eosinophilic (red-dead neurons) while their nuclei undergo karyolysis (fig. 19-15). Thereafter, capillaries become reactive with endothelial cells that

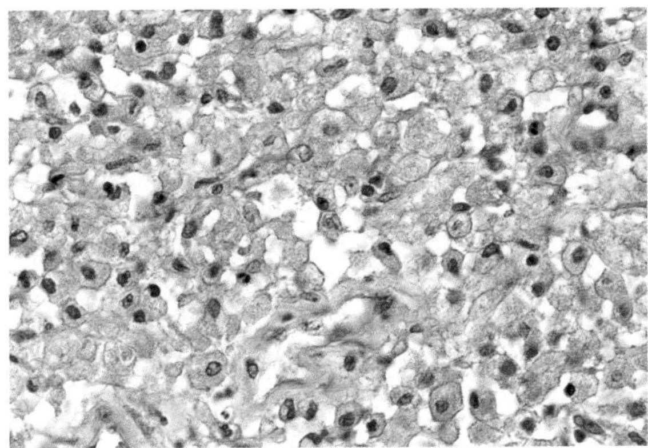

Figure 19-16

CEREBRAL INFARCT

Mature infarcts are macrophage-rich.

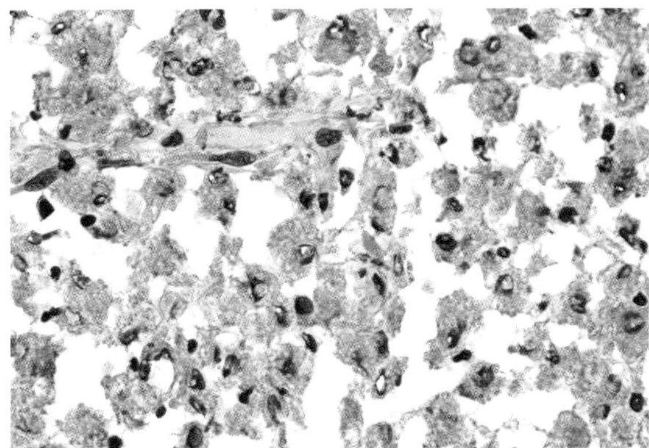

Figure 19-17

CEREBRAL INFARCT: FROZEN SECTION

Macrophages may be misinterpreted as tumor cells unless the possibility of an infarct is considered. The radiologic features of this lesion are illustrated in figure 19-14.

are hypertrophic, cytologically atypical, and sometimes even mitotically active. Later still, infarcts become progressively macrophage rich, producing a hypercellularity resembling that of demyelinating disease or glioma (fig. 19-16).

The cause of infarcts generally eludes histologic study of surgical specimens, but the presence of a thrombosed artery or vein resolves the issue in some cases. An important cause of hemorrhage and infarction in older patients, amyloid angiopathy, must be considered in any patient 60 years of age and older, especially in the face of multiple, superficial hemorrhages. Independent of systemic amyloidosis, this common disease is restricted to subarachnoid and cortical vessels. The diagnosis is confirmed with Congo red or beta-amyloid stains.

Frozen Section. As in demyelinating disease, an abundance of macrophages may be mistaken for glial neoplasia (fig. 19-17).

Immunohistochemical Findings. Macrophages are strongly positive for macrophage markers such as CD68.

Cytologic Findings. The cytologic features of macrophages include crisp cell membranes, cytoplasm filled with granular or foamy debris, and round, cytologically benign nuclei (fig. 19-18).

Differential Diagnosis. The hypercellularity and vascular changes that accompany early infarction may, on first inspection, suggest an infiltrating glioma. Oligodendroglioma is the principal suspect given the round nuclei and generally pale cytoplasm of the macrophages.

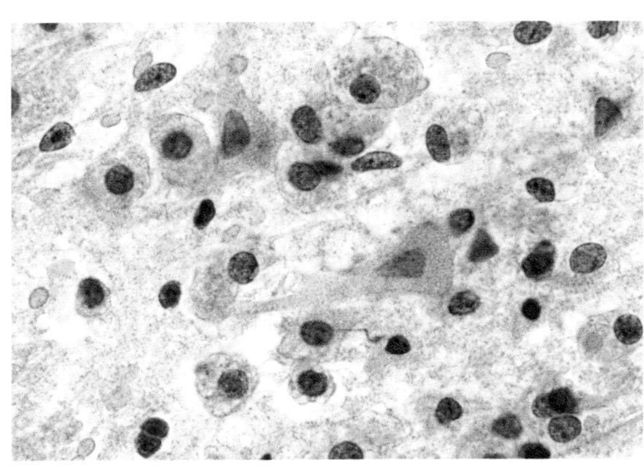

Figure 19-18

CEREBRAL INFARCT: SMEAR PREPARATION

As with other macrophage-rich lesions, companion smear preparations minimize the chance that phagocytes are misinterpreted as tumor cells in tissue sections. Macrophages, with their discrete cell borders and granular/foamy cytoplasm, are clearly seen. Note the two ischemic red-dead neurons.

A misdiagnosis is most likely in biopsies from regions of subtotal necrosis in which such cells are intermixed with viable tissue. The cytoplasmic pallor of infarct-associated macrophages is not the consequence of fluid imbibition during autolysis as is the case in oligodendroglioma, but the result of phagocytized cellular debris that appears more granular or foamy rather than water clear. Immunostains for macrophage markers readily make the distinction.

Appendices L and M include information useful for the diagnosis for cerebral infarcts.

Given its cellularity, accompanying vascular changes, and necrosis, glioblastoma may also be a consideration. This diagnosis is excluded cytologically on the basis of the uniform nuclei, foamy cytoplasm, and discrete cell borders of the macrophages. Telltale clustering of macrophages around blood vessels is another helpful feature.

Infarcts may also resemble granular cell astrocytomas (see figs. 3-96, 19-13), which consist of neoplastic astrocytes with cytologic features reminiscent of macrophages, i.e., discrete cell borders and granular cytoplasm. The neoplastic cells are typically larger than macrophages and their cytoplasm is granular, not foamy. In large specimens, transition to obvious glioma may be seen. Some tumor cells may be positive for GFAP, but this is often weak and scant. Macrophage markers may be positive, albeit not with the intensity of certified macrophages.

Infarcts must be distinguished from demyelinating disease, discussed in the previous section.

Acute and subacute herpes simplex encephalitis has an ischemic component with changes such as red-dead neurons and a macrophage response. Distinguishing features are intranuclear inclusions, at least some perivascular inflammation, and immunoreactivity for the virus.

Treatment and Prognosis. Infarcts resolve within weeks to months and, depending upon their size, become gliotic, rarefied, and even macroscopically cystic. They eventually evolve into contracted foci readily distinguishable from neoplasms by radiologic methods.

Progressive Multifocal Leukoencephalopathy

Definition. *Progressive multifocal leukoencephalopathy* (PML) is a demyelinating opportunistic infection of the CNS caused by the JC papova virus.

Clinical Features. Almost always occurring in the context of immunodeficiency, albeit a state that is not always clinically evident at the time of biopsy, patients typically present with focal neurological deficits. Acquired immunodeficiency syndrome (AIDS) is now the principal predisposing factor (19,20). Polymerase chain reaction (PCR)-based detection of JC virus DNA in cerebrospinal fluid not only supplements, but is replacing, the need for a tissue diagnosis (20–23).

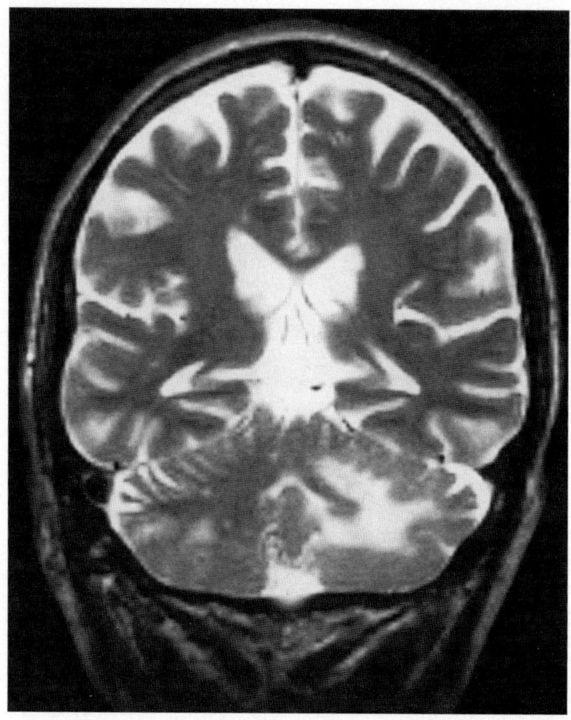

Figure 19-19

PROGRESSIVE MULTIFOCAL LEUKOENCEPHALOPATHY

In the setting of immunocompromise, a bright, white matter lesion with little mass effect on T2-weighted images, as here in the cerebellum, is highly suspicious of progressive multifocal leukoencephalopathy (PML).

Radiologic Findings. Although PML is a multifocal process, the disease often makes its initial appearance as a solitary lesion, with little or no mass effect or contrast enhancement. Such a lesion, especially one in the cerebellum of an AIDS patient, is highly suspicious for PML (fig. 19-19).

Gross Findings. As seen postmortem, the brain contains a myriad of minute, somewhat gelatinous, demyelinated foci that are often associated with one or more large symptomatic lesions in the centrum semiovale (fig. 19-20).

Microscopic Findings. In biopsy specimens, PML appears as multiple, small, macrophage-rich foci in which a minority of oligodendroglial cells have large, round, hyperchromatic nuclei with ground-glass alterations (figs. 19-21–19-23) (24). Such cells are especially prominent at the periphery of the plaque. In most instances, there is little if any chronic inflammation. The foci

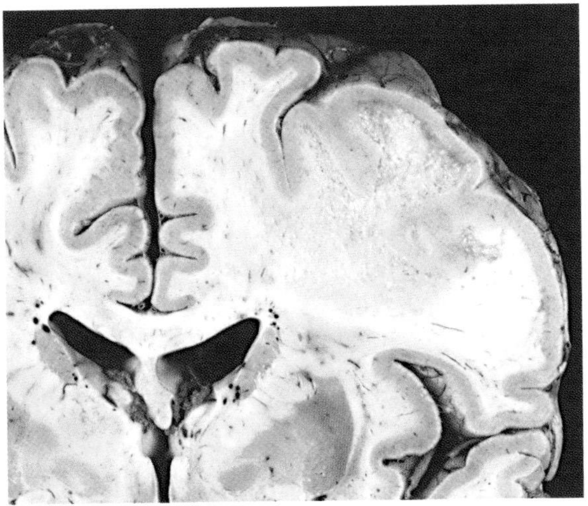

Figure 19-20

PROGRESSIVE MULTIFOCAL LEUKOENCEPHALOPATHY

A large, somewhat gelatinous and contracted lesion in the white matter is typical of PML. Innumerable demyelinated foci are seen along the gray-white junction.

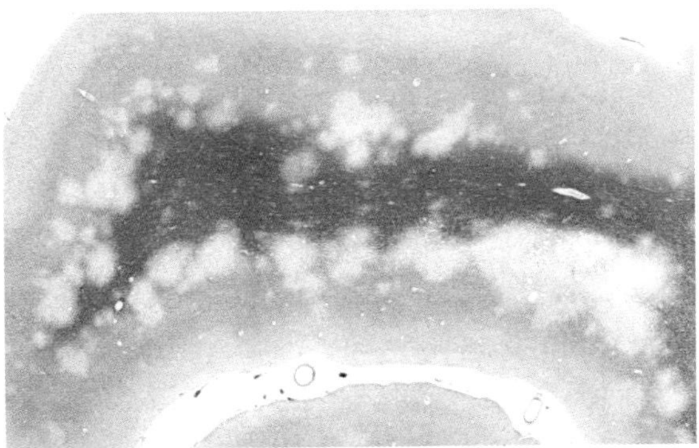

Figure 19-21

PROGRESSIVE MULTIFOCAL LEUKOENCEPHALOPATHY

Multitudinous lesions along the gray-white junction are typical of PML (H&E/Luxol fast blue stain).

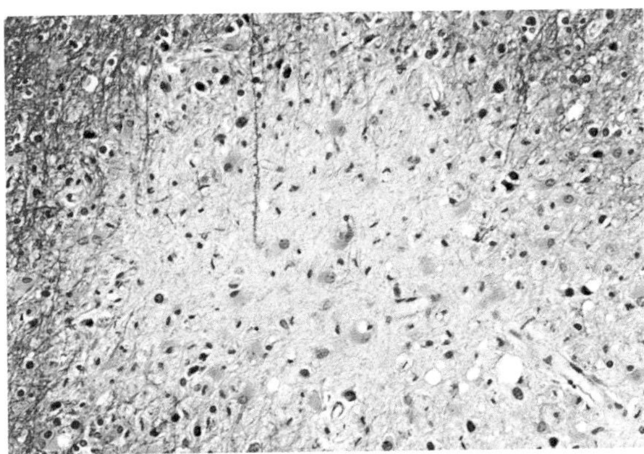

Figure 19-22

PROGRESSIVE MULTIFOCAL LEUKOENCEPHALOPATHY

Staining with H&E/Luxol fast blue emphasizes the lesion's lytic effect on myelin. Large, round, dark nuclei of infected oligodendroglia are at the edge of the lesion.

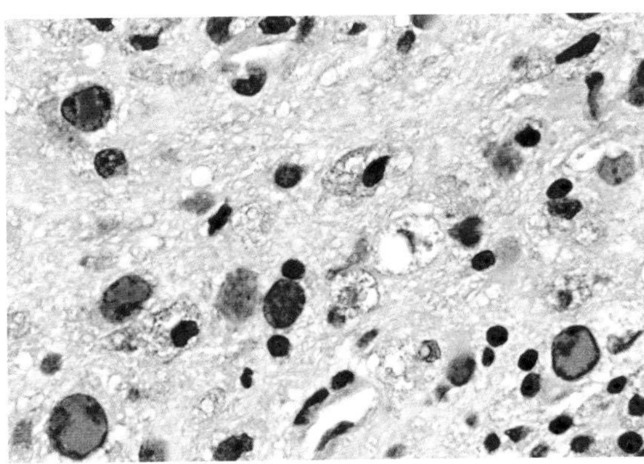

Figure 19-23

PROGRESSIVE MULTIFOCAL LEUKOENCEPHALOPATHY

Dark nuclei with a ground-glass appearance are typical of oligodendrocytes infected by papova virus.

are much more obvious in large specimens stained for myelin than in the typically small, stereotactically-obtained fragments stained only by the H&E method.

While the hyperchromatic oligodendrocytes of PML can themselves be misinterpreted as neoplastic, a misdiagnosis of glioma is even more likely when bizarre reactive astrocytes are present as well. Although typical of the condition, the number and degree of atypia of these cells vary from case to case, being prominent and cytologically alarming in some cases (fig. 19-24) but inconspicuous in others. Unlike infected oligodendrocytes, bizarre astrocytes are not concentrated at the edge of the plaques.

Inflammation is limited in the classic case, but a lymphoplasmacytic response may be present in patients with human immunodeficiency virus (HIV) and a high CD4 count as the result of highly active retroviral therapy (25,26). Such lesions are more likely to be radiologically contrast enhancing, and contain fewer inclusion-bearing cells and atypical astrocytes. Immunohistochemistry may be necessary to find inclusions.

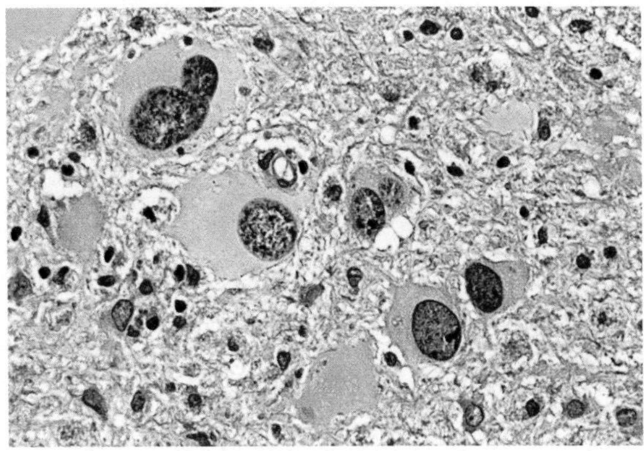

Figure 19-24

PROGRESSIVE MULTIFOCAL LEUKOENCEPHALOPATHY

Nuclear hyperchromasia and pleomorphism of astrocytes can easily be misinterpreted as evidence of an astrocytoma.

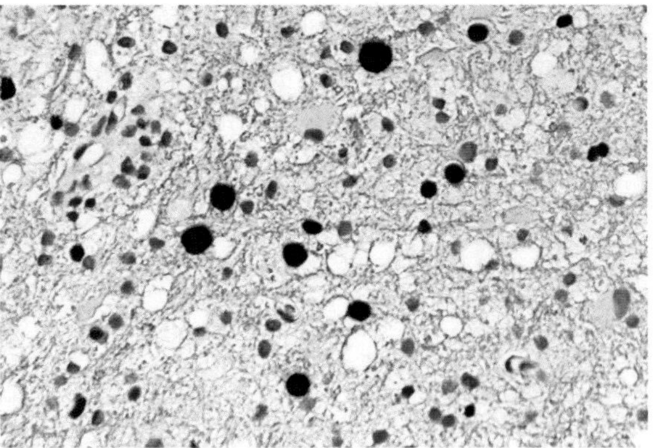

Figure 19-25

PROGRESSIVE MULTIFOCAL LEUKOENCEPHALOPATHY

The large, dark nuclei of virally infected oligodendrocytes react with antibodies against JC virus.

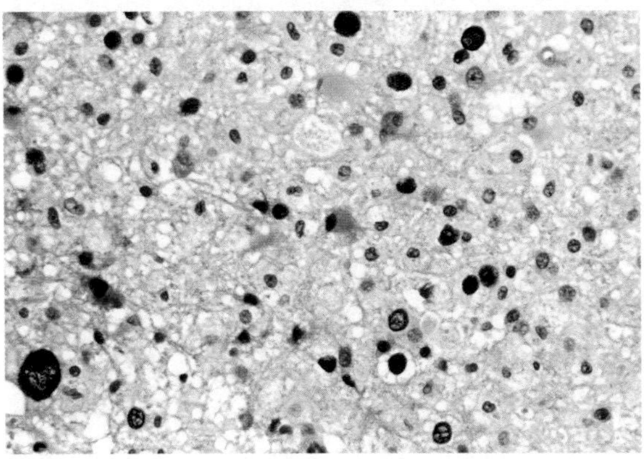

Figure 19-26

PROGRESSIVE MULTIFOCAL LEUKOENCEPHALOPATHY

Virally infected cells are immunoreactive for p53.

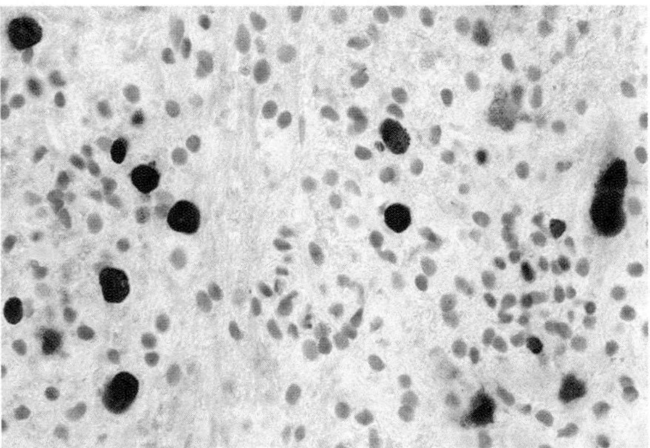

Figure 19-27

PROGRESSIVE MULTIFOCAL LEUKOENCEPHALOPATHY

The high MIB-1 labeling index of infected oligodendrocytes and large astrocytes should not be interpreted as evidence of a neoplasm.

Immunohistochemical Findings. The large oligodendrocytes that have a cytopathic effect are immunoreactive for JC virus (fig. 19-25) and for p53 (fig. 19-26) (27). Immunostaining, as well as in situ hybridization for the JC virus, helps confirm the diagnosis in small specimens. It appears that the JC virus' T antigen (28) binds and stabilizes wild type p53 that accumulates in PML lesions (29). Unlike infected oligodendrocytes, bizarre reactive astrocytes are usually negative, or only weakly immunoreactive, for JC virus, but often strongly positive for p53. They are nonetheless JC positive by in situ hybridization (30,31). Misleadingly, infected oligodendrocytes and astrocytes have a high index with both MIB-1 (fig. 19-27) and proliferating cell nuclear antigen (PCNA) (27).

Ultrastructural Findings. The nuclei of infected oligodendroglia contain small, faintly polygonal virions either packed in a crystalline array or as both spheres and filamentous forms (32–34).

Cytologic Findings. The combination of macrophages, large round nuclei with ground-glass inclusions, and atypical astrocytes is distinctive and highly suggestive of PML (fig. 19-28).

Differential Diagnosis. Distinguishing PML from a glioma requires awareness of the entity and recognition of large oligodendroglial nuclei

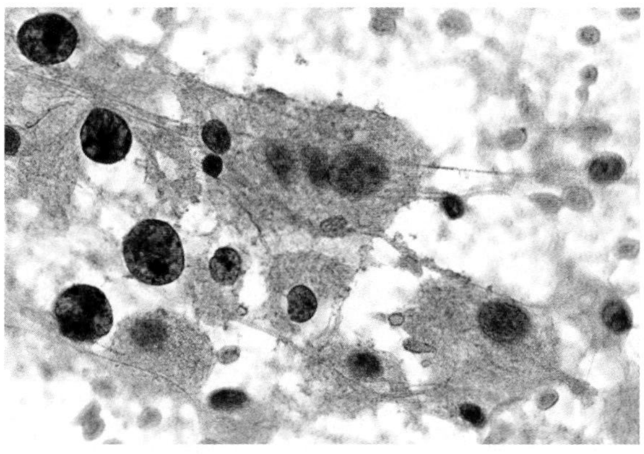

Figure 19-28

PROGRESSIVE MULTIFOCAL LEUKOENCEPHALOPATHY: SMEAR PREPARATION

The mixture of macrophages and oligodendrocytes with viral cytopathic effect is characteristic of PML. Note the multinucleated astrocyte (Creutzfeldt cell) that is common in reactive, macrophage-rich lesions.

with a viral cytopathic effect. The broader differential diagnosis of demyelinating disease, discussed with that entity, applies as well.

Treatment and Prognosis. The progressive disease is almost always fatal. It is not clear whether the oncogenic potential of papova virus is expressed in PML patients (35).

LANGERHANS CELL HISTIOCYTOSIS

Definition. *Langerhans cell histiocytosis* (LCH) is a unifocal or multifocal proliferation of Langerhans histiocytes.

General Features. Historically, LCH, or *histiocytosis X*, was divided into *eosinophilic granuloma*, *Hand-Schüller-Christian disease*, and *Letterer-Siwe disease*. In view of considerable clinical overlap, the spectrum is now characterized as LCH of unifocal, multifocal, or disseminated type. The principal cells exhibit both immunohistochemical and ultrastructural features of Langerhans cells, thus supporting the unifying concept that these clinically dissimilar disorders are related proliferations of a unique cell common to many tissues.

The pathogenesis of LCH is unknown. While it was generally assumed to be an inflammatory rather than a neoplastic process, disseminated disease in the very young can behave in a malignant fashion. Furthermore, clonality has been demonstrated (36).

Most cases of LCH that directly affect the CNS do so by extension from osseous foci, particularly in the skull. This is typically the case in *multifocal eosinophilic granuloma* in which basilar involvement (Hand-Schüller-Christian disease) interferes with hypothalamic function and causes endocrine deficiency states. Children and young adults are most often affected, but no age group is spared. Unifocal or multifocal lesions have been described as occurring within the hypothalamus (37–40), infundibulum (38,39, 41,42), optic chiasm (38,43), choroid plexus (44), cerebral hemispheres (38,45–49), cerebellum (45, 50), or brain stem (45,47,50). The hypothalamus and infundibulum are favored sites of involvement (fig. 19-29). Whether all of these represent infiltrates of Langerhans cells is unclear.

Gross Findings. The frequently yellow lesions vary from localized nodules attached to the dura to granular intraparenchymal infiltrates. Hypothalamic lesions may be discrete or ill-defined processes that also infiltrate the meninges.

Microscopic Findings. In its consummate form, LCH consists of a mixed infiltrate of Langerhans cells, lymphocytes, plasma cells, and focal aggregates of eosinophils (fig. 19-30). Langerhans cells have large, indented or convoluted nuclei. There may also be a conspicuous xanthomatous reaction replete with foamy histiocytes and Touton giant cells. As in chronic osseous foci of LCH, abundant deposition of collagen and reticulin is often present. Striking and vigorous gliosis is elicited in areas of brain involvement.

Immunohistochemical Findings. Langerhans cells are conspicuously positive for S-100 protein, vimentin, and CD1a (fig. 19-31) (37,40). On occasion, S-100 protein staining is minor; in such cases the more specific CD1a stain is particularly valuable. MAC387 has been suggested as a means of distinguishing LCH (immunonegative) from immunoreactive macrophage-derived lesions such as xanthogranulomas (51).

Ultrastructural Findings. Multilobation and complex folding of the nuclear membrane of the Langerhans cells are seen well at the ultrastructural level. Diagnostic of the condition are Birbeck granules, distinctive pentalaminar organelles with cross striations and a frequent expansion resembling the head of a tennis racket (fig. 19-32).

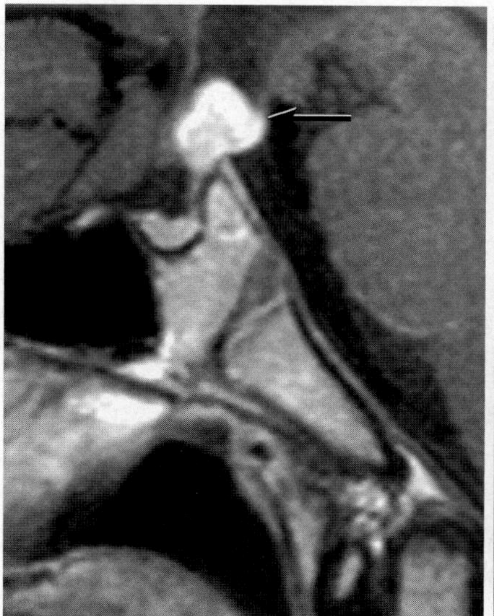

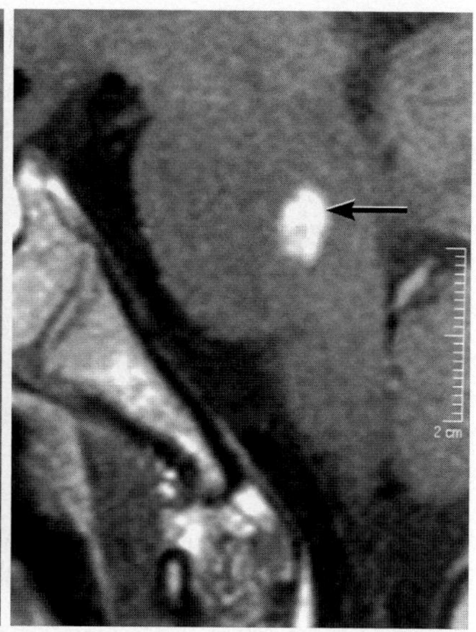

Figure 19-29

LANGERHANS CELL HISTIOCYTOSIS

This 3-year-old boy presented with diabetes insipidus and lesions in the hypothalamus and brain stem (arrows). A lytic lesion was present in one femur and a rib. A CNS component such as this may consist only of macrophages and gliosis.

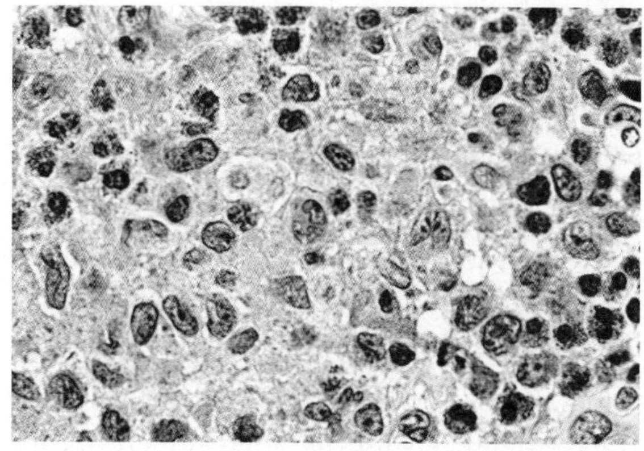

Figure 19-30

LANGERHANS CELL HISTIOCYTOSIS

A mixture of Langerhans cells with clefted nuclei and aggregates of eosinophils is classic.

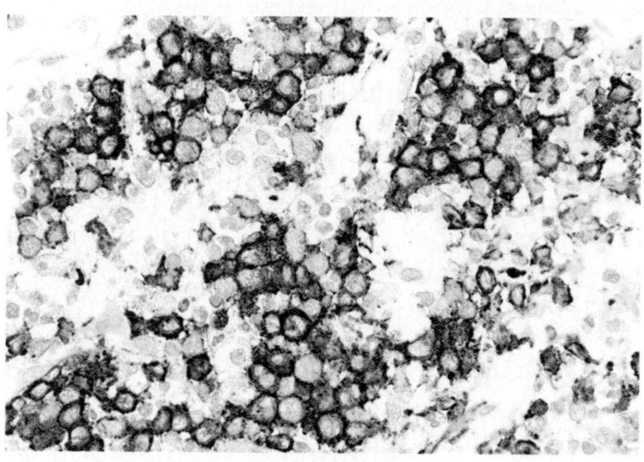

Figure 19-31

LANGERHANS CELL HISTIOCYTOSIS

Surface immunoreactivity for CD1a is characteristic.

Cytologic Findings. The clefted, convoluted nuclei are prominent in cytologic preparations (fig. 19-33). Eosinophils in this venue can be recognized by their bilobed nuclei, a helpful feature since they are often not as obviously granulated or eosinophilic as in histologic sections.

Differential Diagnosis. Given the often small size of hypothalamic/infundibular specimens, the diagnosis may be difficult in routine histologic sections. Immunohistochemistry and electron microscopy aid in distinguishing LCH from large cell lymphoma and nonspecific inflammatory reactions.

Sarcoidosis, a process also favoring the suprasellar region, has granulomas that are lacking in LCH. The identification of S-100- and CD1a-immunoreactive Langerhans cells and, if necessary, the ultrastructural demonstration of Birbeck granules, confirms the diagnosis.

Germinoma is another lymphocyte-rich lesion favoring the suprasellar region that must be distinguished from LCH. The germ cell tumor lacks nuclear folding, has prominent nucleoli, is PAS positive due to the high glycogen content, and is immunoreactive for c-kit (CD117), OCT4, and placental alkaline phosphatase (PLAP).

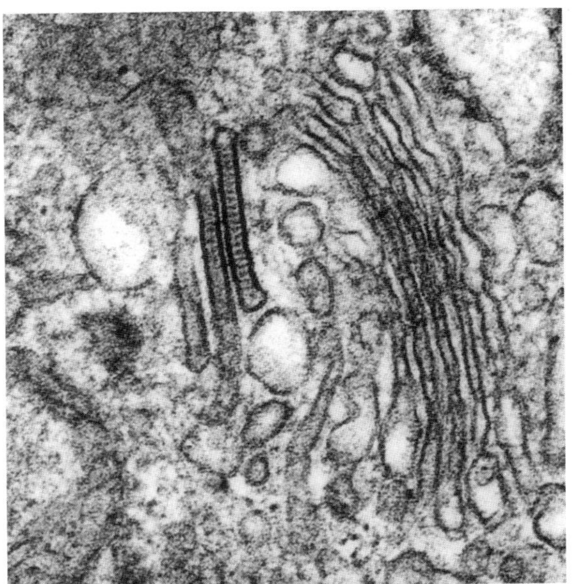

Figure 19-32

LANGERHANS CELL HISTIOCYTOSIS

Cross-striated Birbeck granules are pathognomonic of Langerhans cells.

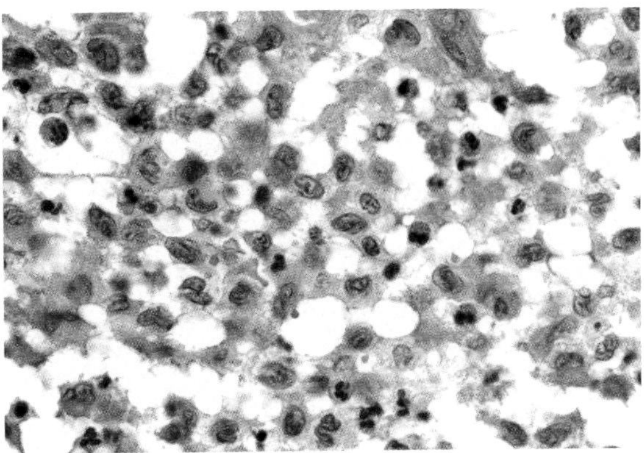

Figure 19-33

LANGERHANS CELL HISTIOCYTOSIS: SMEAR PREPARATION

The nuclear convolutions and clefts characteristic of Langerhans cells are seen well in cytologic preparations.

The prominent gliosis elicited by parenchymal infiltrates of LCH must not be interpreted as either pilocytic or diffuse astrocytoma. The distinction between LCH and sinus histiocytosis with massive lymphadenopathy (Rosai-Dorfman disease) is discussed elsewhere in this chapter.

Treatment and Prognosis. Solitary and even multifocal LCH generally responds to low-dose radiotherapy, although intraparenchymal lesions may be less responsive to treatment. It is not clear whether this treatment is always necessary. Chemotherapy is now an option. Despite significant morbidity in patients with multifocal examples, a fatal outcome is uncommon.

LYMPHOPLASMACYTE-RICH MASS LESIONS

Plasma Cell Granuloma

Definition. *Plasma cell granuloma* is a nonneoplastic chronic inflammatory mass.

Clinical Features. These dura-based meningioma-like masses with a prominent component of lymphocytes and plasma cells occur at all levels of the neuraxis. Adults are primarily affected.

Radiologic Findings. The contrast-enhancing masses are almost always dura-based, and thus usually meningioma suspects (fig. 19-34) (52–56). Less commonly, they are intraparenchymal (57) or arise within the stroma of the choroid plexus (58,59).

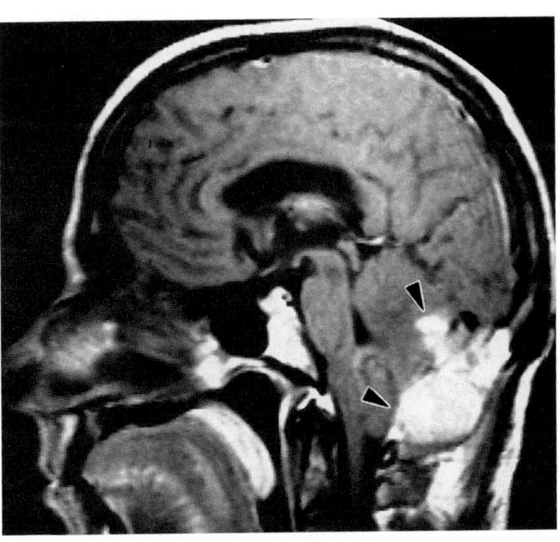

Figure 19-34

PLASMA CELL GRANULOMA

Intracranial plasma cell granulomas (arrowheads) are typically dura-based masses that resemble meningiomas. (Fig. 19-32 from Fascicle 10, 3rd Series.)

Gross Findings. Reflecting the extent of collagenization, the texture and color of the lesions range from soft and gray to firm and white.

Microscopic Findings. Plasma cell granulomas are noted for their mature plasma cells,

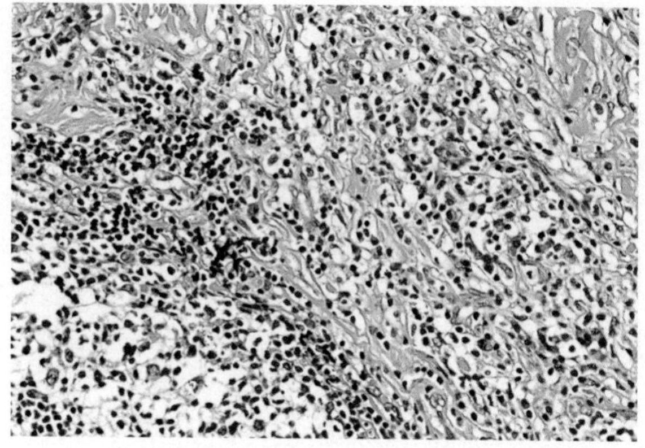

Figure 19-35

PLASMA CELL GRANULOMA

The lesions are composed in varying proportion of chronic inflammatory cells and collagen. Germinal centers are present in some cases.

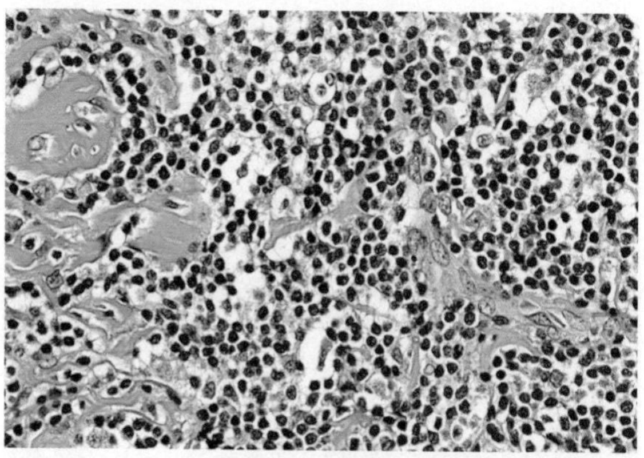

Figure 19-36

PLASMA CELL GRANULOMA

Some inflammatory pseudotumors are masses of mature lymphocytes.

which are often admixed with plasmacytoid and typical lymphocytes, and in some cases macrophages, which may be lipid-laden. Germinal centers are present in some cases (figs. 19-35, 19-36). Russell bodies are often found. A fibroblastic reaction varies considerably in extent and degree. A storiform pattern is observed in some. Dense bands of collagen are common. In cases designated *hyalinizing plasmacytic granulomatosis*, multinucleate giant cells surround hypereosinophilic bands of necrotic collagen (fig. 19-37) (60).

Immunohistochemical Findings. These are mixed T- and B-cell lesions. Immunoglobulin production is polyclonal.

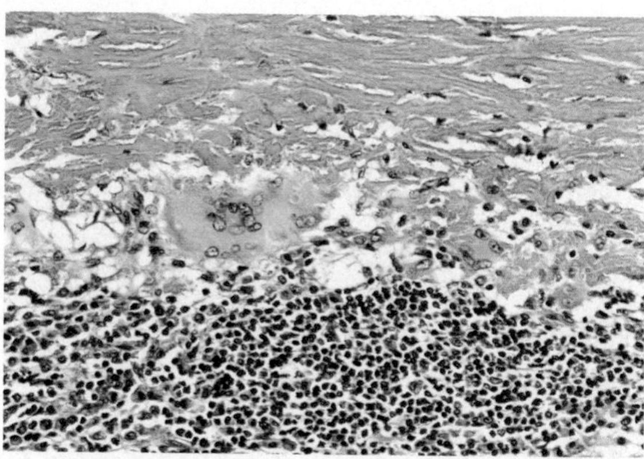

Figure 19-37

HYALINIZING PLASMACYTIC GRANULOMATOSIS

One variant of inflammatory pseudotumor has bands of necrotic, brightly eosinophilic collagen surrounded by multinucleated giant cells.

Differential Diagnosis. As far as possible, infectious lesions must be excluded by stains and immunostains for organisms. Lymphoplasmacyte-rich meningiomas, also diagnostic contenders, are discussed in chapter 9.

The distinction of plasma cell granuloma from plasmacytoma is usually apparent on H&E-stained sections. Plasmacytoma is a cytologically monotonous proliferation without the mixed inflammatory appearance of plasma cell granuloma. Nonetheless, immunohistochemical demonstration of polyclonality may be required. It may be difficult also to distinguish plasma cell granuloma from dura-based, well-differentiated mucosa-associated lymphatic tissue (MALT) lymphomas (see chapter 12) without immunohistochemical and/or molecular assessment.

Rosai-Dorfman disease has its aggregates of large, S-100 protein–positive histiocytes, some showing emperipolesis. This disorder is discussed immediately below. Although rare, the lesion in Castleman's disease is rather distinctive, the most common form featuring hyalinizing germinal centers and hyperplastic vessels.

Treatment and Prognosis. Too few cases have been studied to permit generalization regarding the optimum treatment of patients with plasma cell granuloma. A number of patients have undergone postoperative radiation therapy with resolution of the lesion. On the other hand, many do well without progression or recurrence of disease after resection alone. The

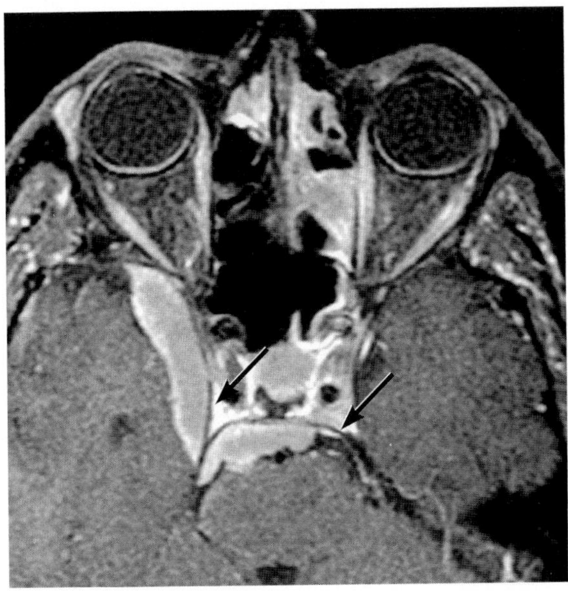

Figure 19-38

EXTRANODAL SINUS HISTIOCYTOSIS WITH MASSIVE LYMPHADENOPATHY (ROSAI-DORFMAN DISEASE)

Rosai-Dorfman disease mimics meningioma both radiologically and macroscopically as in this 69-year-old man with progressive loss of vision. Magnetic resonance imaging (MRI) showed multiple lesions spanning the dura, the dark lamina at the arrows.

Figure 19-39

EXTRANODAL SINUS HISTIOCYTOSIS WITH MASSIVE LYMPHADENOPATHY (ROSAI-DORFMAN DISEASE)

Rosai-Dorfman disease may be simultaneously both intradural (right) and extradural (left). The dura is the eosinophilic lamina coursing vertically through the specimen.

role of radiotherapy is thus debatable, particularly in cases in which in macroscopic or radiographic terms total resection has been achieved.

Extranodal Sinus Histiocytosis with Massive Lymphadenopathy (Rosai-Dorfman Disease)

This uncommon non-neoplastic disorder usually presents as massive cervical lymphadenopathy, with or without concurrent or subsequent involvement of other sites, nodal or extranodal (61). An elevated erythrocyte sedimentation rate and polyclonal gammopathy are common. Of neurosurgical relevance are sometimes multiple, dura-based, intracranial (61–64) or intraspinal masses (fig. 19-38) (62,65,66). Only rarely is the lesion intraparenchymal (67). Meningiomas are almost always suspected preoperatively. Macroscopically, the indurated lesions may be simultaneously intradural and extradural (fig. 19-39). Only occasionally does clinically apparent nodal involvement accompany CNS symptoms.

Histologically, the infiltrates are composed of lymphocytes, plasma cells, and foamy histiocytes that are S-100 protein immunopositive and include multinucleate cells (figs. 19-40–19-43) (68,69). The multinucleate cells often engage in emperipolesis (figs. 19-41–19-43).

Although morphologic similarities exist between LCH and Rosai-Dorfman disease, the former is a more polymorphic infiltrate and includes smaller, CD1a-positive histiocytes with clefted nuclei as well as frequent eosinophils. Birbeck granules are not found in Rosai-Dorfman disease.

Although the meningeal lesions may persist, and in some cases enlarge, the disorder is rarely progressive or lethal (62).

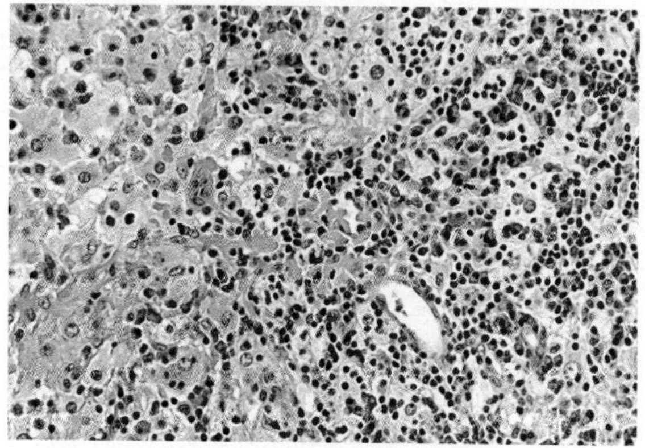

Figure 19-40

EXTRANODAL SINUS HISTIOCYTOSIS WITH MASSIVE LYMPHADENOPATHY (ROSAI-DORFMAN DISEASE)

Histiocytes, individually or in large aggregates, interrupt the dense lymphoplasmacytic infiltrates.

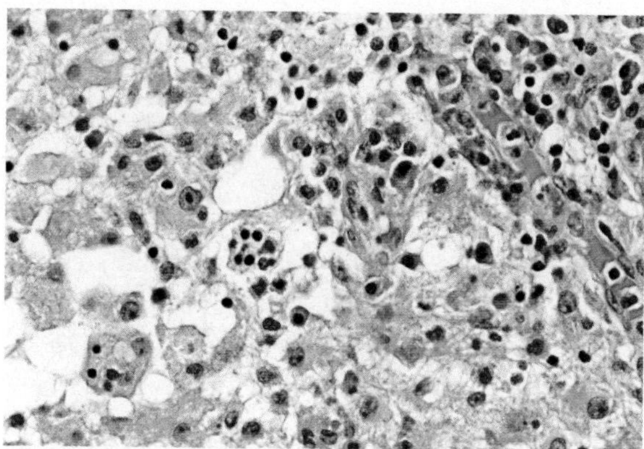

Figure 19-41

EXTRANODAL SINUS HISTIOCYTOSIS WITH MASSIVE LYMPHADENOPATHY (ROSAI-DORFMAN DISEASE)

Large histiocytes with intracytoplasmic lymphocytes (emperipolesis) are characteristic, but sometimes focal, findings.

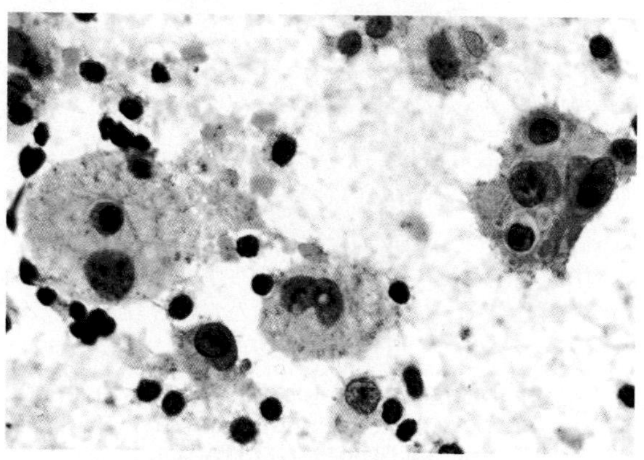

Figure 19-42

EXTRANODAL SINUS HISTIOCYTOSIS WITH MASSIVE LYMPHADENOPATHY (ROSAI-DORFMAN DISEASE): SMEAR PREPARATION

Emperipolesis is well visualized in smear preparations.

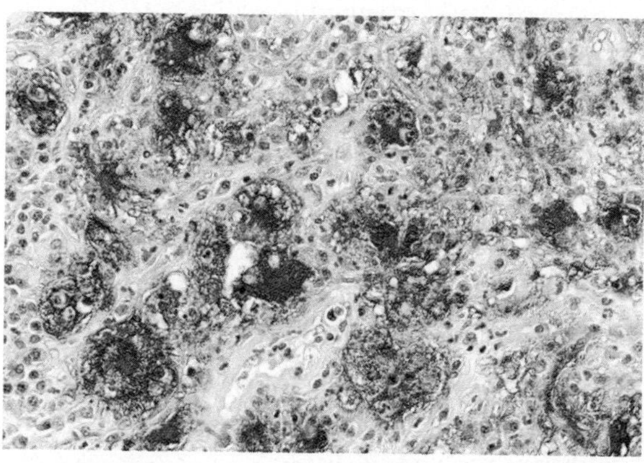

Figure 19-43

EXTRANODAL SINUS HISTIOCYTOSIS WITH MASSIVE LYMPHADENOPATHY (ROSAI-DORFMAN DISEASE)

Large histiocytes are S-100 protein positive.

Castleman's Disease

This uncommon reactive lesion may be systemic or limited to lymph nodes, most often those of the mediastinum. Two overlapping variants, hyaline-vascular and plasma cell types, are recognized. The former is noted for germinal centers with hyalinization and prominent vessels; the latter for numerous plasma cells, often with accompanying Russell bodies, within interfollicular areas (70). The plasma cell variant may be accompanied by anemia, hypogammaglobulinemia, and an elevated erythrocyte sedimentation rate. Histologically, intracranial or intraspinal involvement is unusual, but the lesion has been reported to form meningioma-like masses (71–73).

Although most extracranial lesions are patently benign, too few cases are on record to permit conclusions concerning their biologic behavior.

Sarcoidosis

As it affects the CNS, *sarcoidosis* takes a number of forms, including multiple small leptomeningeal plaques, a meningocerebral focus in the hypothalamus, and solitary/multiple dura-based,

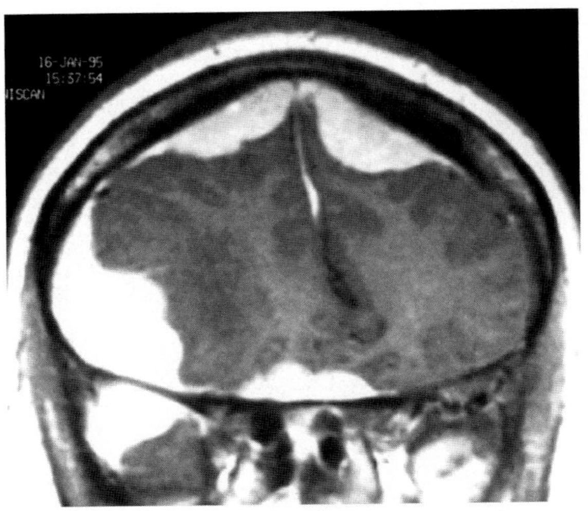

Figure 19-44

SARCOIDOSIS

Dura-based masses are sometimes composed of granulomatous inflammation histologically consistent with sarcoidosis.

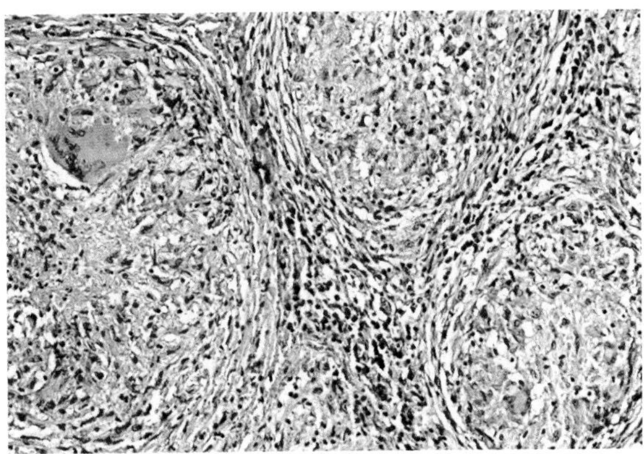

Figure 19-45

SARCOIDOSIS

The presence of multiple compact granulomas, sometimes with scant or focal necrosis, is a cardinal feature of the lesion.

meningioma-like masses (fig. 19-44) (74–79). In the usual scenario, i.e., patients with no evidence of extracranial disease, the diagnosis is obviously problematic and may well be no more than "consistent with sarcoidosis." "Pathogen-free granulomatous disease" is a descriptive and noncommittal designation for lesions that include cases of sarcoidosis as well as other ostensibly noninfectious granulomatous lesions (80).

Histologically, the principal features are multiple, small, discrete granulomas, often perivascular, that have a tendency for gradual fibrosis (fig. 19-45). Focal necrosis within the granulomas is not rare. Sizable geographic zones of necrosis are not expected, but can occur rarely (81).

In the suprasellar region, sarcoidosis must be distinguished from the granulomatous inflammation that accompanies some germinomas (82), as well as from giant cell granuloma of the anterior pituitary, a rare noninfectious process that may ascend from the sella (83).

XANTHOMATOUS LESIONS

A disparate group of CNS lesions is distinctive for cells with abundant intracytoplasmic lipid. Some lesions are diffuse and "metabolic" or "degenerative" in character, whereas others are tumefactive, yet non-neoplastic. Some may even be neoplasms, although it is sometimes unclear whether a lesion is reactive or neoplastic. Langerhans cell histiocytosis is considered separately above.

Xanthogranuloma and Xanthoma of the Choroid Plexus

The stroma, and to a lesser extent the epithelium, of the choroid plexus is subject to xanthic change, typically in the glomus within the trigone, or atrium, of the lateral ventricle. The terms *xanthogranuloma* and *xanthoma* are often used interchangeably (84–89), but one clinicopathologic study suggested that they are two distinct entities (90). Xanthogranulomas, which occasionally present with sign of hydrocephalus, are almost always bilateral, and create a complex neuroradiologic image because of their content of lipid, hemosiderin, and fibrous tissue (fig. 19-46). Macroscopically, they are nodular, partly cystic, and gray-yellow. The tissue reaction is identical to that seen with cholesterol granulomas at other sites, i.e., xanthoma cells, cholesterol clefts, giant cells, hemosiderin, fibrosis, and occasional calcium deposits (fig. 19-47). Trapped choroid plexus epithelium is commonly present.

Xanthomas are asymptomatic, occur more frequently in women, are often associated with hypercholesterolemia, can be either unilateral or bilateral, and consist solely of lipid-engorged histiocytes (fig. 19-48).

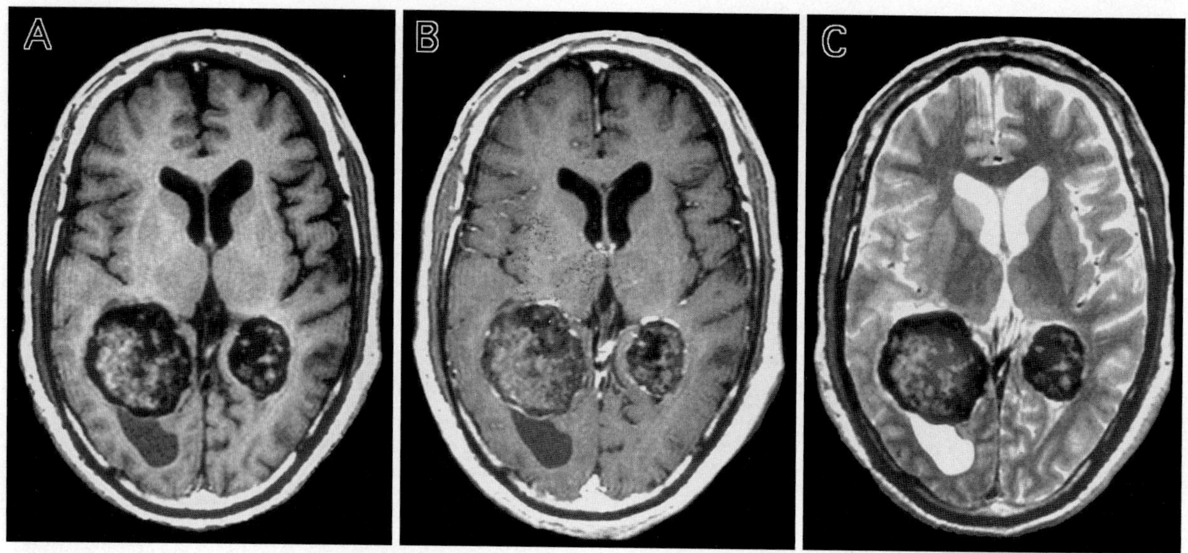

Figure 19-46

XANTHOGRANULOMA OF THE CHOROID PLEXUS

Bilateral masses in the glomus of the choroid plexus are typical of xanthogranulomas. In precontrast (A) and postcontrast (B) T1-weighted images there are mixed signal characteristics reflecting the lesion's content of lipid, hemosiderin, and fibrous tissue. The latter two components generate dark signal on T2-weighted images (C).

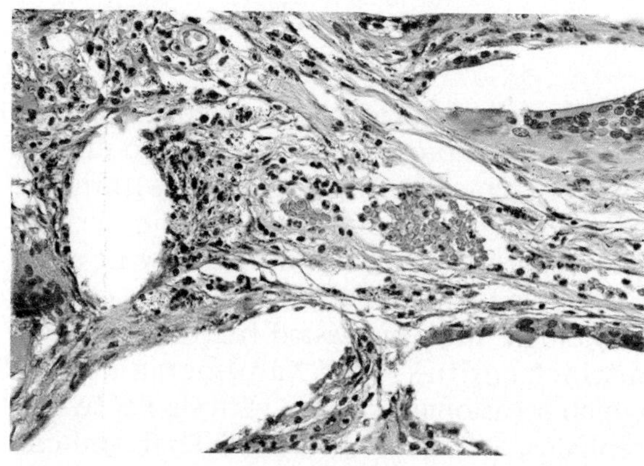

Figure 19-47

XANTHOGRANULOMA OF THE CHOROID PLEXUS

With their cholesterol clefts, foreign body giant cells, hemosiderin, and fibrosis, xanthogranulomas of the choroid plexus are identical to those occurring at other body sites.

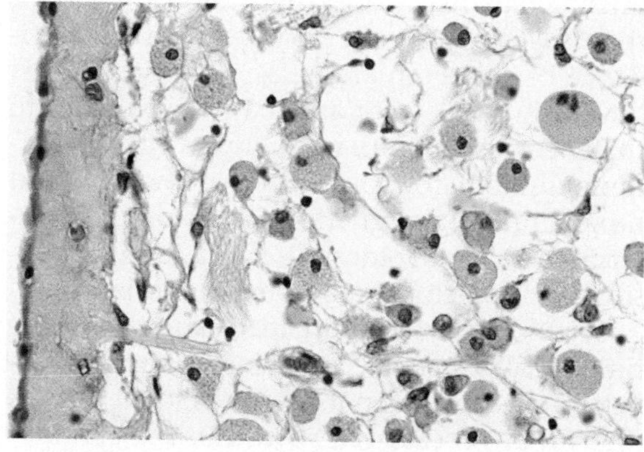

Figure 19-48

XANTHOMA OF THE CHOROID PLEXUS

Composed simply of foamy histiocytes, xanthomas are histologically less complex than xanthogranulomas.

Xanthomas Associated with Hyperlipidemia

Xanthomatous changes in patients with congenital or acquired hyperlipidemia and hyperlipoproteinemia may present as intracranial masses affecting bone and dura. Those in the choroid plexus are discussed above. Histologically, the lesions contain foamy histiocytes and extracellular cholesterol clefts. Superficial cutaneous xanthomas are a frequent associated systemic finding (91–93).

Xanthomas Associated with Systemic Disorders

On rare occasion, intracranial or intraspinal masses appear as part of a systemic disorder. These include *xanthoma disseminatum* (94–97), *juvenile xanthogranuloma* (98), and *xanthogranulomatosis* (99).

Xanthomas in Weber-Christian Disease

Rare dura-based xanthomatous foci may be seen in patients with Weber-Christian disease, a relapsing, nodular, nonsuppurative panniculitis with a broad spectrum of clinical expressions. Grossly, the process resembles the yellow fibrofatty lesions seen at visceral sites of involvement. Histologically, these inflammatory lesions consist, in varying proportions, of histiocytes, giant cells, lymphoplasmacytic infiltrates, fibrosis, and xanthomatous change (100,101).

Xanthic Lesions Unassociated with Systemic Disease

These rare processes include *xanthoma* (102), *juvenile xanthogranuloma* (98,103–105), *xanthogranuloma* (104,106–108), and *fibrous xanthoma* (109–111). Other than cutaneous involvement in some cases, they occur in the absence of a recognizable predisposing disorder. Heterogeneous in nature, these dural or parenchymal lesions share common histologic features including a variable content of xanthomatous cells, fibrous tissue, and chronic inflammation.

Xanthogranulomatous Reaction to Benign Epithelial Cysts

Benign cysts lined by columnar epithelium sometimes are associated with a pronounced xanthogranulomatous reaction. This has been described in colloid cyst (see fig. 16-6), endodermal cyst, and Rathke's cleft cyst as well as craniopharyngioma (see fig. 15-8). In order to avoid a misleadingly simple diagnosis of "xanthogranuloma," careful specimen sampling is in order to find the often sparse residual epithelium.

Erdheim-Chester Disease

This rare multisystem histiocytosis may affect the CNS, most often the cerebellum, spinal cord, pituitary gland, meninges, or orbit (112–115). At least one case has mimicked a brain tumor (115a). Systemic involvement of bone, viscera, and adipose tissue may not be evident. Patients with sellar lesions usually present with diabetes insipidus; the hypothalamus is not involved (114,115). Histologically, the lesions consist of lipid-laden histiocytes and Touton-type multinucleate giant cells, both immunopositive for CD68. Lymphocytes, scant eosinophils, and fibrosis may also be seen. Stains for S-100 protein and CD1a are negative

COLLAGEN VASCULAR DISEASES

Rheumatoid Disease

On rare occasion, *rheumatoid disease* presents as tumefactive, fibroinflammatory meningeal lesions with necrotic centers that are surrounded by epithelioid and scattered multinucleate histiocytes (116–120). Such lesions simulate classic rheumatoid nodules, although radiating epithelioid cells may not be as prominent a feature. Tuberculosis and other necrotizing granulomatous infections should be excluded given the lesion's granulomatous quality.

Wegener's Granulomatosis

Only rarely is the CNS affected by *Wegener's granulomatosis*. A diffuse, dura-based lesion, it extends from a paranasal sinus in some cases. Histologically, it is characterized by necrotizing granulomatous inflammation, geographic necrosis, and small vessel vasculitis (121–123). Vascular occlusion is common.

Systemic Lupus Erythematosus

Cerebral symptoms are common in patients with *systemic lupus erythematosus* (SLE). Cerebral infarction, rather than an inflammatory mass, is the usual consequence (124).

Sjögren's Syndrome

Sjögren's syndrome may occasionally be associated with a steroid-responsive dementia due to lymphocytic microperivasculitis (125,126).

RADIONECROSIS

The complication of craniospinal irradiation relevant to this volume is *delayed radionecrosis*, particularly when it occurs as a rapidly evolving, edema-generating mass that requires surgical decompression or prompts biopsy to exclude a malignant glioma (127–129). Yet another undesirable effect of irradiation, *leukoencephalopathy*, is a diffuse degeneration of the white matter with resultant hydrocephalus ex vacuo. This presents clinically as a progressive, often disabling, loss of intellectual function. Its anatomic substrate is unclear, and the extent of cortical injury remains to be established (127, 130,131). This discussion concentrates on delayed radionecrosis.

Clinical Features. In its classic and indisputable form, delayed radionecrosis occurs after treatment of extracranial malignancies of the head and neck. Its distinction from "radiation effect" in and around the bed of an irradiated brain tumor is discussed below in the section, Differential Diagnosis, and in the section on glioblastoma in chapter 3.

The occurrence of radionecrosis is dose and rate of administration dependent, and generally appears after doses of 50 Gy or more (132). The effect is not immediate: there is a latency period of 6 months to 2 years in most cases. Not surprisingly, the process occurs earlier after stereotactic radiosurgery (133).

Radiologic Findings. Neuroimaging typically shows an aggressive-appearing, contrast-enhancing, edema-generating mass that closely mimics a malignant glioma (fig. 19-49). This is a gross misrepresentation since these lesions are usually hypometabolic, or "cold," by positron emission tomography (PET), whereas recurrent high-grade neoplasms are hypermetabolic or "hot." MRI spectroscopy may be helpful in distinguishing radionecrosis from neoplasia (134). The pattern of contrast enhancement in radionecrosis suggests the diagnosis, since the enhancing rim is thinner, more uniform, and more aligned to the gray-white junction than that of glioblastoma.

Undergoing dystrophic calcification, chronic foci of radionecrosis evolve into dense deposits in the deep white matter, or small foci along the gray-white junction. Surgery is not considered at this late contractile stage.

Gross Findings. In the acute lesion, overlying gyri are swollen but otherwise unremarkable, but the underlying necrotic white matter has a granular quality and is frequently yellow-gray and punctuated by petechiae (fig. 19-50). Cystic change and large chalky areas of calcification appear with time (see figs. 19-56, 19-57).

Microscopic Findings. Radionecrosis is a distinctive form of coagulative necrosis in which multiple, rather discrete, foci coalesce into a large central region of necrosis. White matter is primarily affected, but deep cortical laminae are also involved (figs. 19-51–19-53) (127,135). Radionecrosis is an unusual form of CNS tissue destruction since it lacks the brisk macrophage response so characteristic of infarction and demyelination.

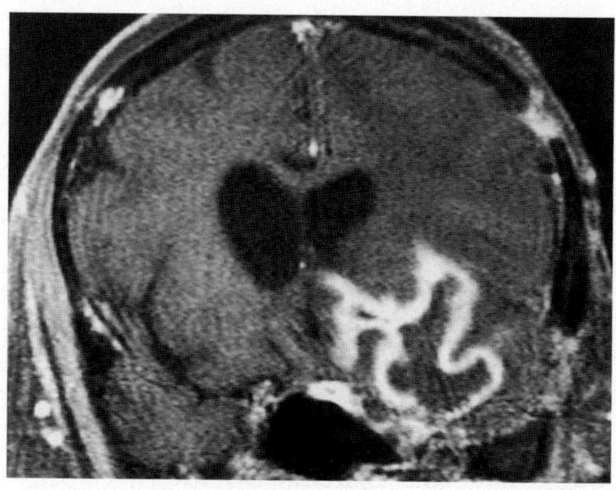

Figure 19-49

RADIONECROSIS

The uniform thickness of the contrast-enhancing rim and its conformity to the gray-white junction suggest radionecrosis. This lesion appeared 14 months after irradiation for ameloblastoma of the orbital apex.

Vascular damage, a prominent feature, has led many to suggest an ischemic basis for chronic delayed radiation injury. In active lesions, fibrinoid necrosis, as evidenced by amorphous thickening and eosinophilia of small blood vessels, is often widespread and accompanied by thrombosis (fig. 19-53). Pools of fibrin may surround vessels. Although not generally prominent, microvascular proliferation, even the glomeruloid form, may be present (fig. 19-54). Chronic changes include recanalization and mural hyalinization, in addition to telangiectases. Granulation tissue may be present; endothelial cells may become large and atypical. Reactive astrocytes are often abundant, particularly surrounding areas of frank necrosis. While these may have a degree of "irradiation atypia," bizarre forms are not expected (fig. 19-55).

The relative paucity of histiocytes may relate to the vascular injury that presumably retards their influx. Failing to undergo timely resolution, some areas remain as granular areas of coagulative necrosis prone to dystrophic calcification (figs. 19-56, 19-57).

The changes seen after stereotactic radiosurgery, such as "gamma knife" therapy, are similar, but evolve more quickly and are more severe, ranging from total parenchymal and vascular necrosis to the more selective white matter insult of ordinary radionecrosis (133).

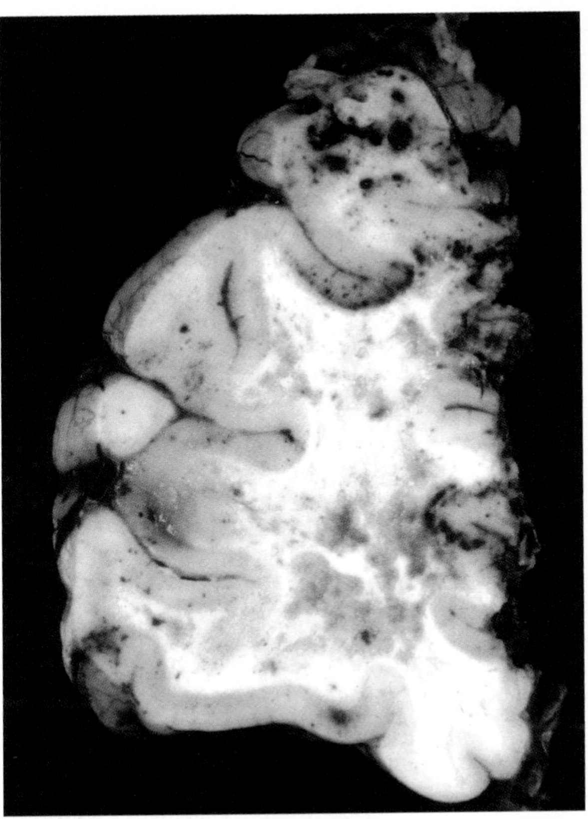

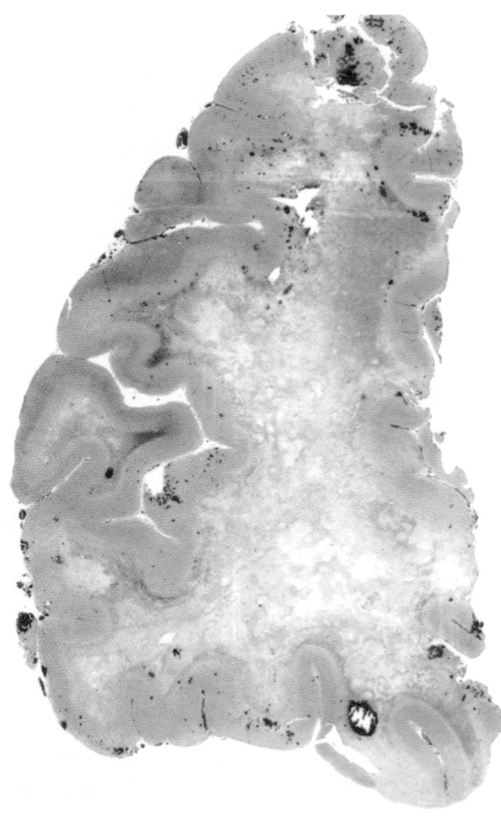

Figure 19-50

RADIONECROSIS

Left: Multiple coalescing lesions comprise delayed radionecrosis in its acute stage. Petechiae are common.

Right: This myelin-stained, whole mount section illustrates the mosaic pattern of white matter disease in delayed radionecrosis (H&E/Luxol fast blue stain).

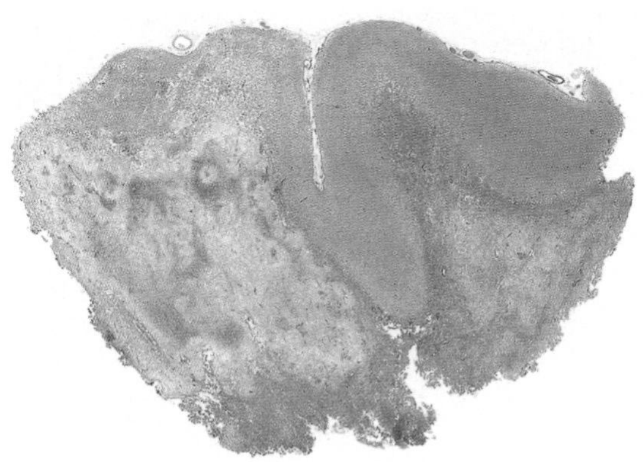

Figure 19-51

RADIONECROSIS

As seen in this whole mount section of the mass illustrated radiologically in figure 19-49, the lesion is composed of confluent, pale foci of white matter necrosis. Encroachment upon deeper cortical lamina is common.

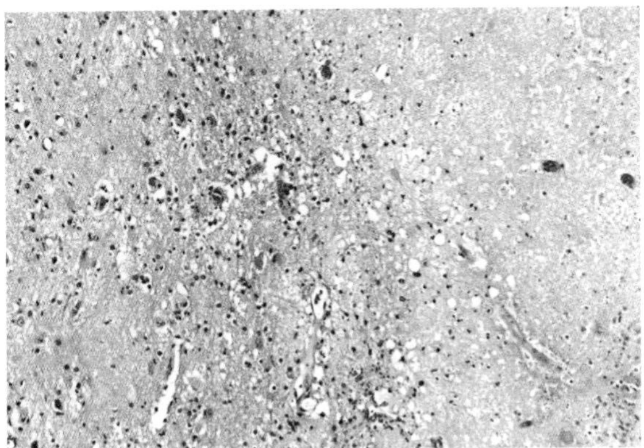

Figure 19-52

RADIONECROSIS

Radionecrosis typically has multiple, discrete and coalescing foci of necrosis that affect both parenchyma and vasculature. Unlike other forms of necrosis, macrophages are few.

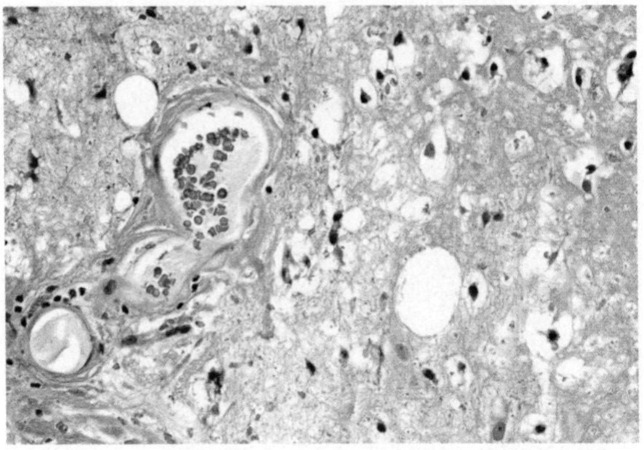

Figure 19-53

RADIONECROSIS

Fibrinoid vascular necrosis is a distinctive feature. Partial tissue necrosis is also common (right).

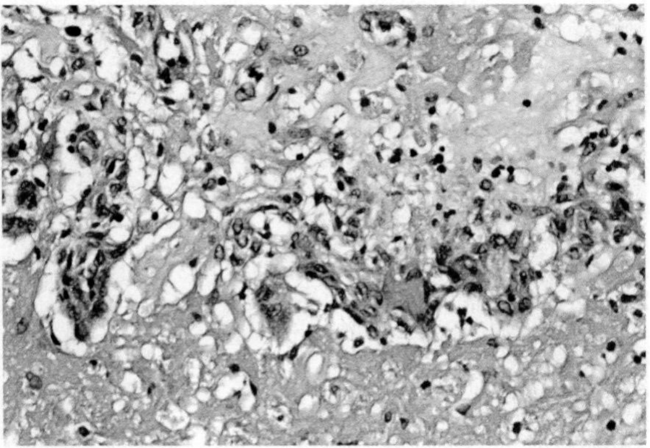

Figure 19-54

RADIONECROSIS

Glomeruloid vascular proliferation, usually to a restrained degree, is present about necrotic foci in some cases.

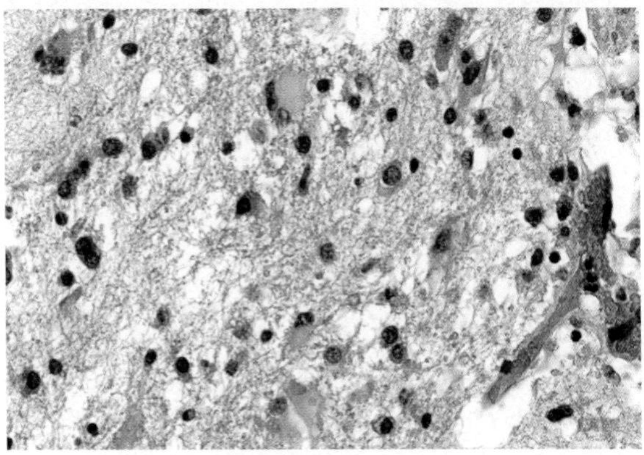

Figure 19-55

RADIONECROSIS

Prominent astrogliosis, as seen here, is common, but marked cytologic atypia ("radiation atypia") is not.

The high concentration of drugs delivered locally by implanted polymer-wafers also produces coagulative necrosis, of both the parenchyma and vessels, similar to radionecrosis.

Differential Diagnosis. In a large specimen, radionecrosis is a distinctive and readily recognized process in which the prominent astrocytes are quickly identified as reactive rather than neoplastic. This simple scenario contrasts with that produced by a previously radiated, necrosis-prone malignant glioma in which two practical problems arise: 1) distinguishing spontaneous tumor necrosis from radiation-induced tumor necrosis and 2) differentiating radiation-induced

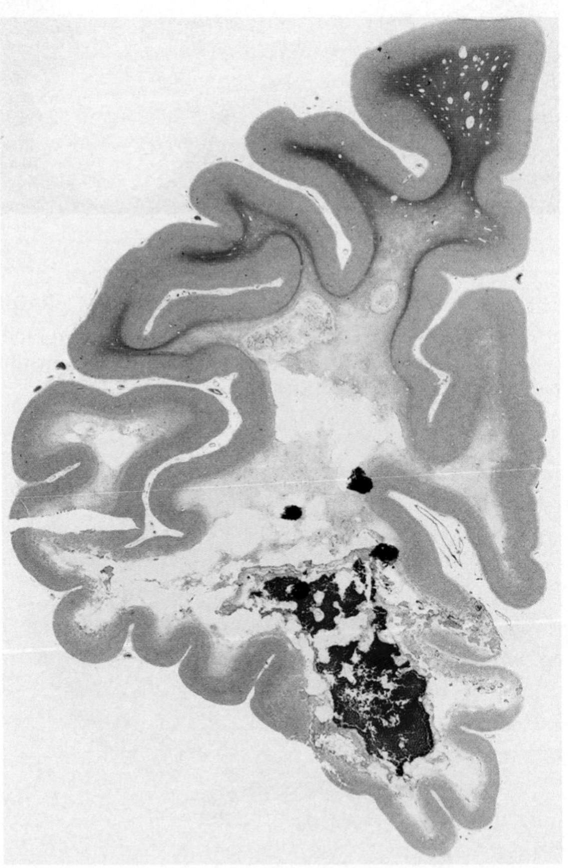

Figure 19-56

RADIONECROSIS

With time, radionecrosis contracts to form rarified, if not partially cystic, tissue that sometimes contains a densely calcified coagulum.

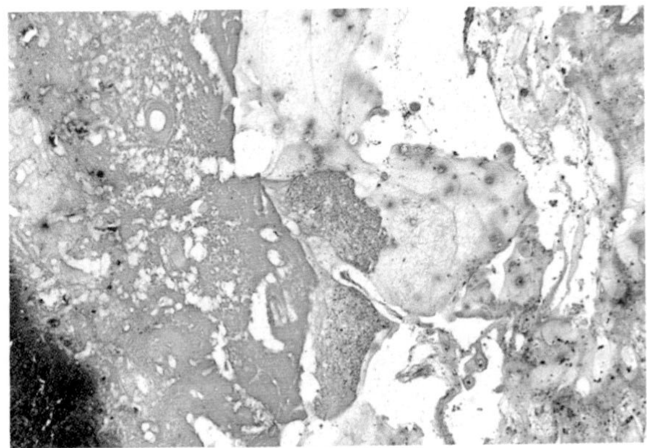

Figure 19-57

RADIONECROSIS

The chronic lesion contains amorphous, often densely calcific, unphotogenic deposits.

tumor necrosis from clinically detrimental radionecrosis in peritumoral parenchyma. These issues are discussed in chapter 3. In brief, all these alterations may be present in specimens from the same patient. Without adequate tissue sampling, familiarity with the current radiographic images, and correlation with clinical as well as radiotherapeutic data, it may be impossible to determine which component is responsible for a given patient's clinical or radiographic deterioration. This is particularly true when only small specimens are provided.

From a purely histologic perspective, the coagulative, as opposed to macrophage-rich, acute radionecrosis most closely resembles toxoplasmosis. The latter may be inflamed, although often only slightly, and contains organisms, albeit ones so few that immunohistochemistry may be required for detection.

Treatment and Prognosis. Simple delayed radionecrosis is generally a self-limited process, but may require surgical decompression, sometimes more than once. In the absence of a life-threatening mass effect, the lesion evolves into a cystic or densely calcified focus. In the presence of a coexistent glioma, the latter generally determines the prognosis.

PYOGENIC ABSCESS

Pyogenic abscesses may be surgically approached either for the purpose of drainage or in anticipation of a malignant glioma. Radiologically, the expansile, edema-inciting lesion resembles glioblastoma except for two distinguishing features, both of which relate to the capsule: 1) the enhancing rim of the abscess is thinner and more uniform than the shaggy ring of a glioblastoma (fig. 19-58) and 2) as it evolves, an abscess capsule acquires a dark signal in T2-weighted MRIs in proportion to the degree of collagenization. In contrast, the cellular rim of a glioblastoma is white, or at least is not dark. Diffusion-weighted imaging (DWI) is also useful in the differential since the contents are bright in abscess, demonstrating restricted diffusion, whereas the necrotic center of a glioblastoma is dark.

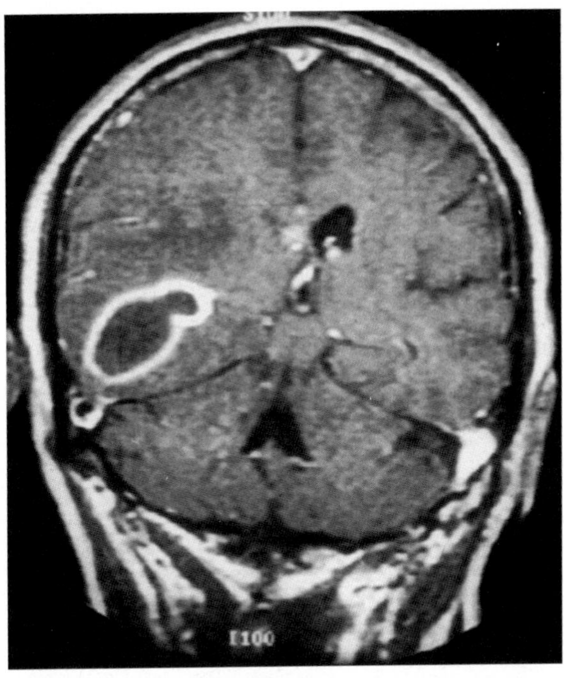

Figure 19-58

PYOGENIC ABSCESS

Unlike the irregular, shaggy rim of a glioblastoma, the contrast-enhancing capsule of an abscess is uniform in width. Note the "daughter abscess" that, as is typical, is forming from the deep or medial aspect of the parent lesion.

Histologically, abscesses begin as ill-defined areas of acutely inflamed parenchyma, termed cerebritis. The established abscess has, outwardly to inwardly, a peripheral zone of edema and chronic parenchymal inflammation, a transition zone of migrating fibroblasts and granulation tissue formation, and central purulent contents (figs. 19-59–19-63). The last may comprise only a small portion of the specimen.

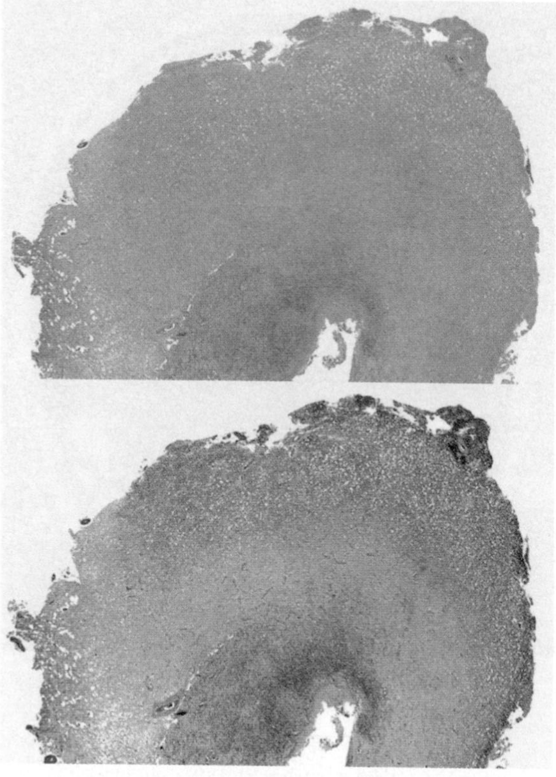

Figure 19-59

PYOGENIC ABSCESS

Paired sections stained for H&E (top) and trichrome (bottom) illustrate the zone of granulation tissue that appears during encapsulation.

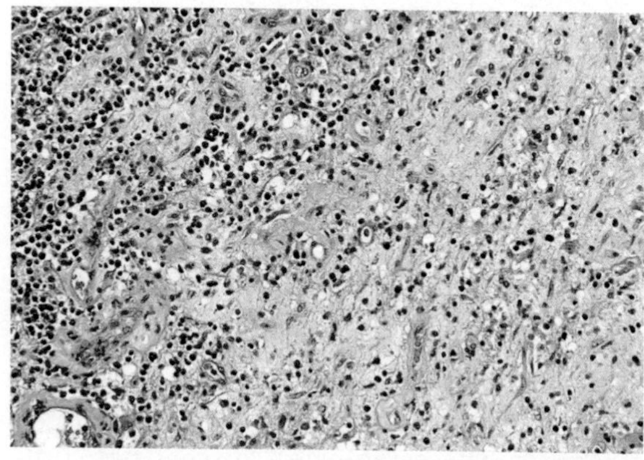

Figure 19-60

PYOGENIC ABSCESS

As seen at the left of the illustration, peripheral tissue is rich in lymphocytes and plasma cells. To the right of this lamina is a zone of active granulation tissue with migrating fibroblasts.

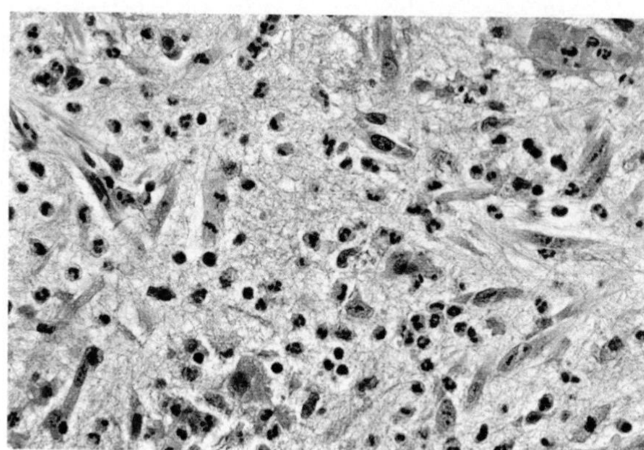

Figure 19-61

PYOGENIC ABSCESS

Mitotically active, migrating fibroblasts ultimately create a fibrous capsule.

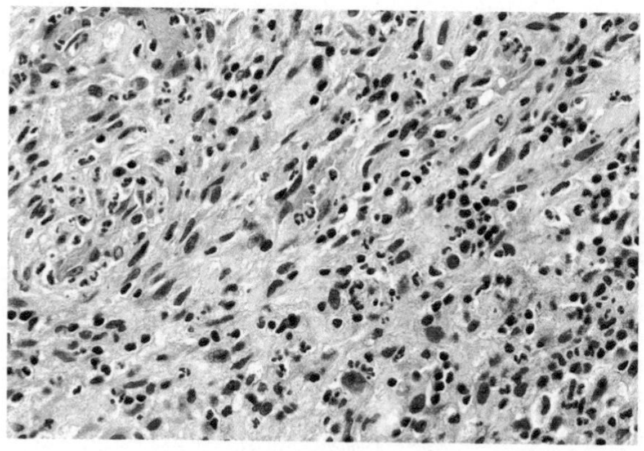

Figure 19-62

PYOGENIC ABSCESS

High cellularity, a degree of cytologic atypia, and a population of fusiform vascular cells create the impression of a neoplasm.

The histologic differential focuses on glioblastoma, particularly the small subset with abundant acute inflammation (fig. 19-64; see fig. 3-67). This can present a challenge, given the cellularity and cytologic atypia of activated fibroblasts in granulation tissue about an abscess. Multiple sections may be necessary to find unequivocally neoplastic tissue, but a firm diagnosis may be elusive and established either by a second biopsy or evolution of clinical events.

Other non-neoplastic lesions that can simulate malignant gliomas are given in Appendix L.

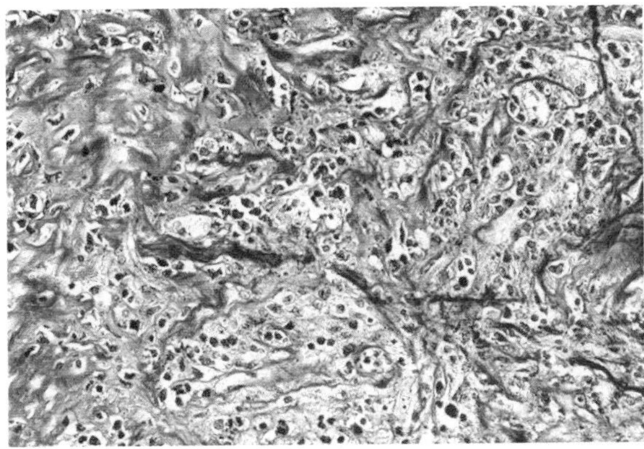

Figure 19-63

PYOGENIC ABSCESS

The mature capsule becomes densely collagenized (Masson trichrome stain).

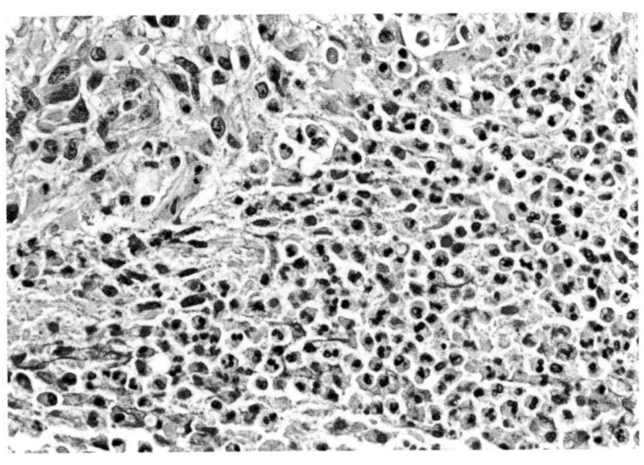

Figure 19-64

ACUTELY INFLAMED GLIOBLASTOMA RESEMBLING PYOGENIC ABSCESS

Glioblastoma with an acute inflammatory component can be difficult to distinguish from abscess. Tumor cells are present at the top left, but only in small numbers.

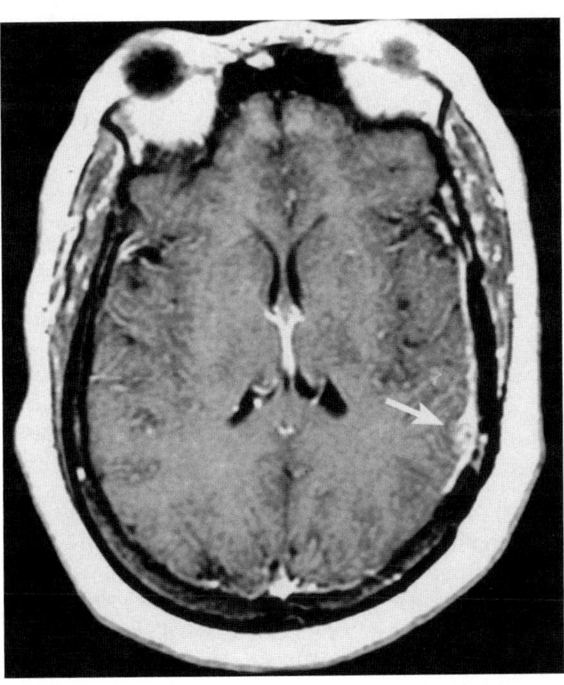

Figure 19-65

TEXTILOMA (GOSSYPIBOMA)

Surgery for a contrast-enhancing lesion (left, arrow) obtained a dura-attached mass assumed preoperatively and macroscopically to be recurrent meningioma (right).

TEXTILOMA (GOSSYPIBOMA)

Used for hemostasis in the operative field, resorbable substances such as Gelfoam®, Surgicell®, and Avitene® are commonly encountered histologically in surgical specimens (136). Less often, unresorbable cotton fibers or intact cotton balls used at the time of the original operation are found incidentally at reoperation or at

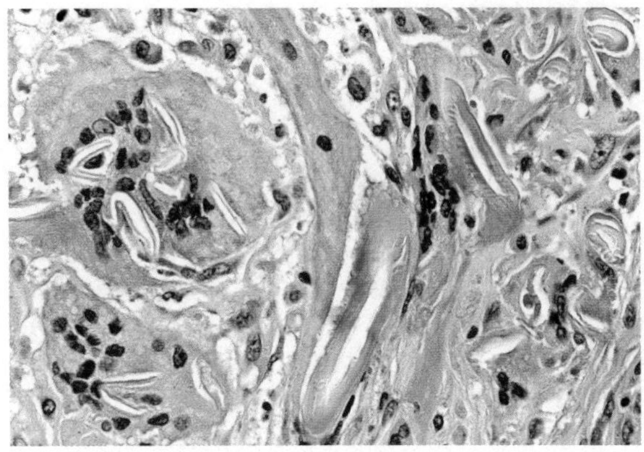

Figure 19-66

TEXTILOMA (GOSSYPIBOMA)

Histologically, the lesion in figure 19-65 was a mass of characteristically hollow cotton fibers that elicited a granulomatous response. No meningioma was identified.

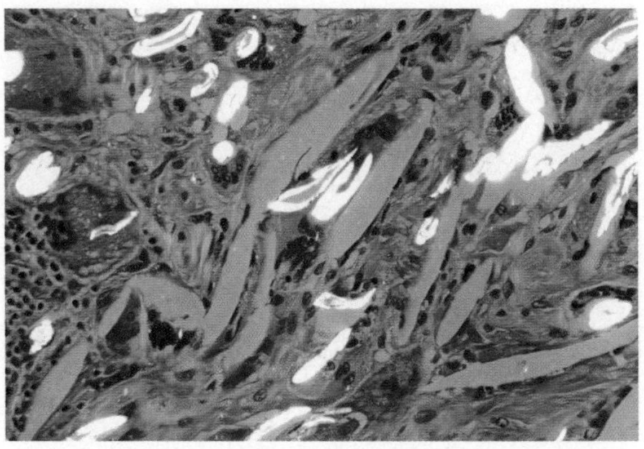

Figure 19-67

TEXTILOMA (GOSSYPIBOMA)

Cotton fibers are intensely birefringent when viewed with polarized light.

autopsy. Rarely, this form of *textiloma*, or *gossypiboma*, is large enough in itself, or in combination with an inflammatory response, to suggest tumor recurrence (fig. 19-65). Cotton fibers, which elicit a vigorous granulomatous response (fig. 19-66), are brilliantly birefringent (fig. 19-67).

REFERENCES

Demyelinating Disease

1. Annesley-Williams D, Farrell MA, Staunton H, Brett FM. Acute demyelination, neuropathological diagnosis, and clinical evolution. J Neuropathol Exp Neurol 2000;59:477–89.
2. Di Patre PL, Castillo V, Delavelle J, Vuillemoz S, Picard F, Landis T. "Tumor-mimicking" multiple sclerosis. Clin Neuropathol 2003;22:235–9.
3. Hook CC, Kimmel DW, Kvols LK, et al. Multifocal inflammatory leukoencephalopathy with 5-fluorouracil and levamisole. Ann Neurol 1992;31:262–7.
4. Savarese DM, Gordon J, Smith TW, et al. Cerebral demyelination syndrome in a patient treated with 5-fluorouracil and levamisole. The use of thallium SPECT imaging to assist in noninvasive diagnosis—a case report. Cancer 1996;77:387–94.
5. Kepes JJ. Large focal tumor-like demyelinating lesions of the brain: intermediate entity between multiple sclerosis and acute disseminated encephalomyelitis? A study of 31 patients. Ann Neurol 1993;33:18–27.
6. Masdeu JC, Quinto C, Olivera C, Tenner M, Leslie D, Visintainer P. Open-ring imaging sign: highly specific for atypical brain demyelination. Neurology 2000;54:1427–33.
7. Nesbit GM, Forbes GS, Scheithauer BW, Okazaki H, Rodriguez M. Multiple sclerosis: histopathologic and MR and/or CT correlation in 37 cases at biopsy and three cases at autopsy. Radiology 1991;180:467–74.
8. Burger PC, Nelson JS, Boyko OB. Diagnostic synergy in radiology and surgical neuropathology: radiographic findings of specific pathologic entities. Arch Pathol Lab Med 1998;122:620–32.
9. Burger PC, Nelson JS, Boyko OB. Diagnostic synergy in radiology and surgical neuropathology: neuroimaging techniques and general interpretive guidelines. Arch Pathol Lab Med 1998;122: 609–19.
10. Passe T, Beauchamp N, Burger P. Neuroimaging in the identification of low-grade and nonneoplastic CNS lesions. The Neurologist 1999;5: 293–9.
11. Hunter SB, Ballinger WE Jr, Rubin JJ. Multiple sclerosis mimicking primary brain tumor. Arch Pathol Lab Med 1987;111:464–8.

12. Sugita Y, Terasaki M, Shigemori M, Sakata K, Morimatsu M. Acute focal demyelinating disease simulating brain tumors: histopathologic guidelines for an accurate diagnosis. Neuropathology 2001;21:25–31.
13. Zagzag D, Miller DC, Kleinman GM, Abati A, Donnenfeld H, Budzilovich GN. Demyelinating disease versus tumor in surgical neuropathology. Clues to a correct pathological diagnosis. Am J Surg Pathol 1993;17:537–45.
14. Estes ML, Rudick RA, Barnett GH, Ransohoff RM. Stereotactic biopsy of an active multiple sclerosis lesion. Immunocytochemical analysis and neuropathologic correlation with magnetic resonance imaging. Arch Neurol 1990;47:1299–303.
15. Raisanen J, Goodman HS, Ghougassian DF, Harper CG. Role of cytology in the intraoperative diagnosis of central demyelinating disease. Acta Cytol 1998;42:907–12.
16. Malmgren RM, Detels R, Verity MA. Co-occurrence of multiple sclerosis and glioma—case report and neuropathologic and epidemiologic review. Clin Neuropathol 1984;3:1–9.

Cerebral Infarct

17. Chuaqui R, Tapia J. Histologic assessment of the age of recent brain infarcts in man. J Neuropathol Exp Neurol 1993;52:481–9.
18. Mena H, Cadavid D, Rushing EJ. Human cerebral infarct: a proposed histopathologic classification based on 137 cases. Acta Neuropathol (Berl) 2004;108:524–30.

Progressive Multifocal Leukoencephalopathy

19. Hair LS, Nuovo G, Powers JM, Sisti MB, Britton CB, Miller JR. Progressive multifocal leukoencephalopathy in patients with human immunodeficiency virus. Hum Pathol 1992;23:663–7.
20. Fong IW, Toma E. The natural history of progressive multifocal leukoencephalopathy in patients with AIDS. Canadian PML Study Group. Clin Infect Dis 1995;20:1305–10.
21. Bogdanovic G, Priftakis P, Hammarin AL, et al. Detection of JC virus in cerebrospinal fluid (CSF) samples from patients with progressive multifocal leukoencephalopathy but not in CSF samples from patients with herpes simplex encephalitis, enteroviral meningitis, or multiple sclerosis. J Clin Microbiol 1998;36:1137–8.
22. Giri JA, Gregoresky J, Silguero P, Garcia Messina O, Planes N. Polyoma virus JC DNA detection by polymerase chain reaction in CSF of HIV infected patients with suspected progressive multifocal leukoencephalopathy. Am Clin Lab 2001;20:33–5.
23. McGuire D, Barhite S, Hollander H, Miles M. JC virus DNA in cerebrospinal fluid of human immunodeficiency virus-infected patients: predictive value for progressive multifocal leukoencephalopathy. Ann Neurol 1995;37:395–9.
24. Richardson EP Jr. Progressive multifocal leukoencephalopathy. N Engl J Med 1961;265:815–23.
25. Hoffmann C, Horst HA, Albrecht H, Schlote W. Progressive multifocal leucoencephalopathy with unusual inflammatory response during antiretroviral treatment. J Neurol Neurosurg Psychiatry 2003;74:1142–4.
26. Di Giambenedetto S, Vago G, Pompucci A, et al. Fatal inflammatory AIDS-associated PML with high CD4 counts on HAART: a new clinical entity? Neurology 2004;63:2452–3.
27. Lammie GA, Beckett A, Courtney R, Scaravilli F. An immunohistochemical study of p53 and proliferating cell nuclear antigen expression in progressive multifocal leukoencephalopathy. Acta Neuropathol (Berl) 1994;88(5):465–71.
28. Greenlee JE, Keeney PM. Immunoenzymatic labelling of JC papovavirus T antigen in brains of patients with progressive multifocal leukoencephalopathy. Acta Neuropathol (Berl) 1986;71: 150–3.
29. Ariza A, von Uexkull-Guldeband C, Mate JL, et al. Accumulation of wild-type p53 protein in progressive multifocal leukoencephalopathy: a flow of cytometry and DNA sequencing study. J Neuropathol Exp Neurol 1996;55:144–9.
30. Hulette CM, Downey BT, Burger PC. Progressive multifocal leukoencephalopathy. Diagnosis by in situ hybridization with a biotinylated JC virus DNA probe using an automated Histomatic Code-On slide stainer. Am J Surg Pathol 1991;15:791–7.
31. Schmidbauer M, Budka H, Shah KV. Progressive multifocal leukoencephalopathy (PML) in AIDS and in the pre-AIDS era. A neuropathological comparison using immunocytochemistry and in situ DNA hybridization for virus detection. Acta Neuropathol (Berl) 1990;80:375–80.
32. Woodhouse MA, Dayan AD, Burston J, et al. Progressive multifocal leukoencephalopathy: electron microscope study of four cases. Brain 1967;90:863–70.
33. Padgett BL, Walker DL, ZuRhein GM, Hodach AE, Chou SM. JC papovavirus in progressive multifocal leukoencephalopathy. J Infect Dis 1976;133:686–90.
34. ZuRhein G. Particles resembling papova viruses in human cerebral demyelinating disease. Science 1965;148:1477–9.

35. Shintaku M, Matsumoto R, Sawa H, Nagashima K. Infection with JC virus and possible dysplastic ganglion-like transformation of the cerebral cortical neurons in a case of progressive multifocal leukoencephalopathy. J Neuropathol Exp Neurol 2000;59:921–9.

Langerhans Cell Histiocytosis

36. Willman CL, Busque L, Griffith BB, et al. Langerhans'-cell histiocytosis (histiocytosis X)—a clonal proliferative disease. N Engl J Med 1994; 331:154–60.
37. Czech T, Mazal PR, Schima W. Resection of a Langerhans cell histiocytosis granuloma of the hypothalamus: case report. Br J Neurosurg 1999;13:196–200.
38. Grois NG, Favara BE, Mostbeck GH, Prayer D. Central nervous system disease in Langerhans cell histiocytosis. Hematol Oncol Clin North Am 1998;12:287–305.
39. Maghnie M, Arico M, Villa A, Genovese E, Beluffi G, Severi F. MR of the hypothalamic-pituitary axis in Langerhans cell histiocytosis. AJNR Am J Neuroradiol 1992;13:1365–71.
40. Mazal PR, Hainfellner JA, Preiser J, et al. Langerhans cell histiocytosis of the hypothalamus: diagnostic value of immunohistochemistry. Clin Neuropathol 1996;15:87–91.
41. Asano T, Goto Y, Kida S, Ohno K, Hirakawa K. Isolated histiocytosis X of the pituitary stalk. J Neuroradiol 1999;26:277–80.
42. Tien RD, Newton TH, McDermott MW, Dillon WP, Kucharczyk J. Thickened pituitary stalk on MR images in patients with diabetes insipidus and Langerhans cell histiocytosis. AJNR Am J Neuroradiol 1990;11:703–8.
43. Smolik EA, Devecerski M, Nelson JS, Smith KR, Jr. Histiocytosis X in the optic chiasm of an adult with hypopituitarism. Case report. J Neurosurg 1968;29:290–5.
44. Morello A, Campesi G, Bettinazzi N, Albeggiani A. Neoplastiform xanthomatous granulomas of choroid plexus in a child affected by Hand-Schuller-Christian disease. Case report. J Neurosurg 1967;26:536–41.
45. Barthez MA, Araujo E, Donadieu J. Langerhans cell histiocytosis and the central nervous system in childhood: evolution and prognostic factors. Results of a collaborative study. J Child Neurol 2000;15:150–6.
46. Bergmann M, Yuan Y, Bruck W, Palm KV, Rohkamm R. Solitary Langerhans cell histiocytosis lesion of the parieto-occipital lobe: a case report and review of the literature. Clin Neurol Neurosurg 1997;99:50–5.
47. Breidahl WH, Ives FJ, Khangure MS. Cerebral and brain stem Langerhans cell histiocytosis. Neuroradiology 1993;35:349–51.
48. Montine TJ, Hollensead SC, Ellis WG, Martin JS, Moffat EJ, Burger PC. Solitary eosinophilic granuloma of the temporal lobe: a case report and long-term follow-up of previously reported cases. Clin Neuropathol 1994;13:225–8.
49. Moscinski LC, Kleinschmidt-DeMasters BK. Primary eosinophilic granuloma of frontal lobe. Diagnostic use of S-100 protein. Cancer 1985;56: 284–8.
50. Gavriel H, Shuper A, Kornreich L, Goshen Y, Yaniv I, Cohen IJ. Diffuse intrinsic brainstem disease with neurologic deterioration: not what it seemed. Med Pediatr Oncol 2000;34:213–4.
51. Fartasch M, Vigneswaran N, Diepgen TL, Hornstein OP. Immunohistochemical and ultrastructural study of histiocytosis X and non-X histiocytoses. J Am Acad Dermatol 1990;23(Pt 1):885–92.

Plasma Cell Granuloma

52. Breidahl WH, Robbins PD, Ives FJ, Wong G. Intracranial plasma cell granuloma. Neuroradiology 1996;38(Suppl 1):S86–9.
53. Gangemi M, Maiuri F, Giamundo A, Donati P, De Chiara A. Intracranial plasma cell granuloma. Neurosurgery 1989;24:591–5.
54. Johnson MD, Powell SZ, Boyer PJ, Weil RJ, Moots PL. Dural lesions mimicking meningiomas. Hum Pathol 2002;33:1211–26.
55. Le Marc'hadour F, Fransen P, Labat-Moleur F, Passagia JG, Pasquier B. Intracranial plasma cell granuloma: a report of four cases. Surg Neurol 1994;42:481–8.
56. Sitton JE, Harkin JC, Gerber MA. Intracranial inflammatory pseudotumor. Clin Neuropathol 1992;11:36–40.
57. Gochman GA, Duffy K, Crandall PH, Vinters HV. Plasma cell granuloma of the brain. Surg Neurol 1990;33:347–52.
58. Bramwit M, Kalina P, Rustia-Villa M. Inflammatory pseudotumor of the choroid plexus. AJNR Am J Neuroradiol 1997;18:1307–9.
59. Pimentel J, Costa A, Tavora L. Inflammatory pseudotumor of the choroid plexus. Case report. J Neurosurg 1993;79:939–42.
60. Nazek M, Mandybur TI, Sawaya R. Hyalinizing plasmacytic granulomatosis of the falx. Am J Surg Pathol 1988;12:308–13.

Extranodal Sinus Histiocytosis with Massive Lymphadenopathy (Rosai-Dorfman Disease)

61. Foucar E, Rosai J, Dorfman R. Sinus histiocytosis with massive lymphadenopathy (Rosai-Dorfman disease): review of the entity. Semin Diagn Pathol 1990;7:19–73.
62. Andriko JA, Morrison A, Colegial CH, Davis BJ, Jones RV. Rosai-Dorfman disease isolated to the central nervous system: a report of 11 cases. Mod Pathol 2001;14:172–8.
63. Kim M, Provias J, Bernstein M. Rosai-Dorfman disease mimicking multiple meningioma: case report. Neurosurgery 1995;36:1185–7.
64. Udono H, Fukuyama K, Okamoto H, Tabuchi K. Rosai-Dorfman disease presenting multiple intracranial lesions with unique findings on magnetic resonance imaging. Case report. J Neurosurg 1999;91:335–9.
65. Foucar E, Rosai J, Dorfman RF, Brynes RK. The neurologic manifestations of sinus histiocytosis with massive lymphadenopathy. Neurology 1982;32:365–72.
66. Osenbach RK. Isolated extranodal sinus histiocytosis presenting as an intramedullary spinal cord tumor with paraplegia. Case report. J Neurosurg 1996;85:692–6.
67. Juric G, Jakic-Razumovic J, Rotim K, Zarkovic K. Extranodal sinus histiocytosis (Rosai-Dorfman disease) of the brain parenchyma. Acta Neurochir (Wien) 2003;145:145–9.
68. Eisen RN, Buckley PJ, Rosai J. Immunophenotypic characterization of sinus histiocytosis with massive lymphadenopathy (Rosai-Dorfman disease). Semin Diagn Pathol 1990;7:74–82.
69. Lopez P, Estes ML. Immunohistochemical characterization of the histiocytes in sinus histiocytosis with massive lymphadenopathy: analysis of an extranodal case. Hum Pathol 1989;20:711–5.

Castleman's Disease

70. Keller AR, Hochholzer L, Castleman B. Hyaline-vascular and plasma-cell types of giant lymph node hyperplasia of the mediastinum and other locations. Cancer 1972;29:670–83.
71. Gianaris PG, Leestma JE, Cerullo LJ, Butler A. Castleman's disease manifesting in the central nervous system: case report with immunological studies. Neurosurgery 1989;24:608–13.
72. Kachur E, Ang LC, Megyesi JF. Castleman's disease and spinal cord compression: case report. Neurosurgery 2002;50:399–402; discussion 402–3.
73. Severson GS, Harrington DS, Weisenburger DD, et al. Castleman's disease of the leptomeninges. Report of three cases. J Neurosurg 1988;69:283–26.

Sarcoidosis

74. Clark WC, Acker JD, Dohan FC Jr, Robertson JH. Presentation of central nervous system sarcoidosis as intracranial tumors. J Neurosurg 1985;63:851–6.
75. Jackson RJ, Goodman JC, Huston DP, Harper RL. Parafalcine and bilateral convexity neurosarcoidosis mimicking meningioma: case report and review of the literature. Neurosurgery 1998;42:635–8.
76. Ranoux D, Devaux B, Lamy C, Mear JY, Roux FX, Mas JL. Meningeal sarcoidosis, pseudo-meningioma, and pachymeningitis of the convexity. J Neurol Neurosurg Psychiatry 1992;55:300–3.
77. Stern BJ, Krumholz A, Johns C, Scott P, Nissim J. Sarcoidosis and its neurological manifestations. Arch Neurol 1985;42:909–17.
78. Tobias S, Prayson RA, Lee JH. Necrotizing neurosarcoidosis of the cranial base resembling an en plaque sphenoid wing meningioma: case report. Neurosurgery 2002;51:1290–4.
79. Quinones-Hinojosa A, Chang EF, Khan SA, McDermott MW. Isolated trigeminal nerve sarcoid granuloma mimicking trigeminal schwannoma: case report. Neurosurgery 2003;52:700–5.
80. Thomas G, Murphy S, Staunton H, O'Neill S, Farrell MA, Brett FM. Pathogen-free granulomatous diseases of the central nervous system. Hum Pathol 1998;29:110–5.
81. Strickland-Marmol LB, Fessler RG, Rojiani AM. Necrotizing sarcoid granulomatosis mimicking an intracranial neoplasm: clinicopathologic features and review of the literature. Mod Pathol 2000;13:909–13.
82. Kraichoke S, Cosgrove M, Chandrasoma PT. Granulomatous inflammation in pineal germinoma. A cause of diagnostic failure at stereotaxic brain biopsy. Am J Surg Pathol 1988;12:655–60.
83. Siqueira E, Tsung JS, Al-Kawi MZ, Woodhouse N. Case report: idiopathic giant cell granuloma of the hypophysis: an unusual cause of panhypopituitarism. Surg Neurol 1989;32:68–71.

Xanthomatous Lesions

84. Ayers W, Haymaker W. Xanthoma and cholesterol granuloma of the choroid plexus. Report of the pathological aspects in 29 cases. J Neuropathol Exp Neurol 1962;19:280–95.
85. Bruck W, Sander U, Blanckenberg P, Friede RL. Symptomatic xanthogranuloma of choroid plexus with unilateral hydrocephalus. Case report. J Neurosurg 1991;75:324–7.
86. Kadota T, Mihara N, Tsuji N, Ishiguro S, Nakagawa H, Kuroda C. MR of xanthogranuloma of the choroid plexus. AJNR Am J Neuroradiol 1996;17:1595–7.

87. Mendez-Martinez OE, Luzardo-Small GD, Cardozo-Duran JJ. Symptomatic bilateral xanthogranulomas of choroid plexus in a child. Br J Neurosurg 2000;14:62–4.
88. Shuangshoti S, Netsky MG. Xanthogranuloma (xanthoma) of choroid plexus. The origin of foamy (xanthoma) cells. Am J Pathol 1966;48:503–33.
89. Wolf A, Cowen D, Graham S. Xanthomas of the choroid plexus in man. J Neuropathol Exp Neurol 1950;9:286–97.
90. Muenchau A, Laas R. Xanthogranuloma and xanthoma of the choroid plexus: evidence for different etiology and pathogenesis. Clin Neuropathol 1997;16:72–6.
91. Akazawa S, Ikeda Y, Toyama K, Miyake S, Takamori M, Nagataki S. Familial type IIa hyperlipoproteinemia associated with a huge intracranial xanthoma. Arch Neurol 1984;41:793–4.
92. Bonhomme GR, Loevner LA, Yen DM, Deems DA, Bigelow DC, Mirza N. Extensive intracranial xanthoma associated with type II hyperlipidemia. AJNR Am J Neuroradiol 2000;21:353–5.
93. Okabe H, Ishizawa M, Matsumoto K, et al. Immunohistochemical analysis of spinal intradural xanthomatosis developed in a patient with phytosterolemia. Acta Neuropathol (Berl) 1992;83:554–8.
94. Chepuri NB, Challa VR. Xanthoma disseminatum: a rare intracranial mass. AJNR Am J Neuroradiol 2003;24:105–8.
95. Hammond RR, Mackenzie IR. Xanthoma disseminatum with massive intracranial involvement. Clin Neuropathol 1995;14:314–21.
96. Giller RH, Folberg R, Keech RV, Piette WW, Sato Y. Xanthoma disseminatum. An unusual histiocytosis syndrome. Am J Pediatr Hematol Oncol 1988;10:252–7.
97. Knobler RM, Neumann RA, Gebhart W, Radaskiewicz T, Ferenci P, Widhalm K. Xanthoma disseminatum with progressive involvement of the central nervous and hepatobiliary systems. J Am Acad Dermatol 1990;23(Pt 2):341–6.
98. Flach DB, Winkelmann RK. Juvenile xanthogranuloma with central nervous system lesions. J Am Acad Dermatol 1986;14:405–11.
99. Miyachi S, Kobayashi T, Takahashi T, Saito K, Hashizume Y, Sugita K. An intracranial mass lesion in systemic xanthogranulomatosis: case report. Neurosurgery 1990;27:822–6.
100. Mangiardi JR, Rappaport ZH, Ransohoff J. Systemic Weber-Christian disease presenting as an intracranial mass lesion. Case report. J Neurosurg 1980;52:134–7.
101. Pick P, Jean E, Horoupian D, Factor S. Xanthogranuloma of the dura in systemic Weber-Christian disease. Neurology 1983;33:1067–70.
102. Kimura H, Oka K, Nakayama Y, Tomonaga M. Xanthoma in Meckel's cave. A case report. Surg Neurol 1991;35:317–20.
103. Ernemann U, Skalej M, Hermisson M, Platten M, Jaffe R, Voigt K. Primary cerebral non-Langerhans cell histiocytosis: MRI and differential diagnosis. Neuroradiology 2002;44:759–63.
104. Okubo T, Okabe H, Kato G. Juvenile xanthogranuloma with cutaneous and cerebral manifestations in a young infant. Acta Neuropathol (Berl) 1995;90:87–92.
105. Paulus W, Kirchner T, Michaela M, et al. Histiocytic tumor of Meckel's cave. An intracranial equivalent of juvenile xanthogranuloma of the skin. Am J Surg Pathol 1992;16:76–83.
106. Bostrom J, Janssen G, Messing-Junger M, et al. Multiple intracranial juvenile xanthogranulomas. Case report. J Neurosurg 2000;93:335–41.
107. Iwasaki Y, Hida K, Nagashima K. Cauda equina xanthogranulomatosis. Br J Neurosurg 2001;15: 72–3.
108. Lesniak MS, Viglione MP, Weingart J. Multicentric parenchymal xanthogranuloma in a child: case report and review of the literature. Neurosurgery 2002;51:1493–8.
109. Carrillo R, Ricoy JR, Herrero-Vallejo J, Bravo G. Fibrous xanthomas of the brain. Report of two cases. Clin Neurol Neurosurg 1975;78:34–40.
110. Kamiryo T, Abiko S, Orita T, Aoki H, Watanabe Y, Hiraoka K. Bilateral intracranial fibrous xanthoma. Surg Neurol 1988;29:27–31.
111. Kepes JJ, Kepes M, Slowik F. Fibrous xanthomas and xanthosarcomas of the meninges and the brain. Acta Neuropathol (Berl) 1973;23: 187–99.
112. Babu RP, Lansen TA, Chadburn A, Kasoff SS. Erdheim-Chester disease of the central nervous system. Report of two cases. J Neurosurg 1997;86:888–92.
113. Wright RA, Hermann RC, Parisi JE. Neurological manifestations of Erdheim-Chester disease. J Neurol Neurosurg Psychiatry 1999;66:72–5.
114. Adle-Biassette H, Chetritt J, Bergemer-Fouquet AM, Wechsler J, Mussini JM, Gray F. Pathology of the central nervous system in Chester-Erdheim disease: report of three cases. J Neuropathol Exp Neurol 1997;56:1207–16.
115. Oweity T, Scheithauer BW, Ching HS, Lei C, Wong KP. Multiple system Erdheim-Chester disease with massive hypothalamic-sellar involvement and hypopituitarism. J Neurosurg 2002;96:344–51.

115a. Veyssier-Belot C, Cacoub P, Caparros-Lefebvre D, et al. Erdheim-Chester disease. Clinical and radiologic characteristics of 59 cases. Medicine (Baltimore) 1996;75:157–69.

Collagen Vascular Diseases

116. Bathon JM, Moreland LW, DiBartolomeo AG. Inflammatory central nervous system involvement in rheumatoid arthritis. Semin Arthritis Rheum 1989;18:258–66.
117. Jackson CG, Chess RL, Ward JR. A case of rheumatoid nodule formation within the central nervous system and review of the literature. J Rheumatol 1984;11:237–40.
118. Johnson MD, Powell SZ, Boyer PJ, Weil RJ, Moots PL. Dural lesions mimicking meningiomas. Hum Pathol 2002;33:1211–26.
119. Karam NE, Roger L, Hankins LL, Reveille JD. Rheumatoid nodulosis of the meninges. J Rheumatol 1994;21:1960–3.
120. Schachenmayr W, Friede RL. Dural involvement in rheumatoid arthritis. Acta Neuropathol (Berl) 1978;42:65–6.
121. Albayram S, Kizilkilie O, Adaletli I, Erdogan N, Kocer N, Islak C. MR imaging findings of spinal dural involvement with Wegener granulomatosis. AJNR Am J Neuroradiol 2002;23:1603–6.
122. Jinnah HA, Dixon A, Brat DJ, Hellmann DB. Chronic meningitis with cranial neuropathies in Wegener's granulomatosis. Case report and review of the literature. Arthritis Rheum 1997; 40:573–7.
123. Murphy JM, Gomez-Anson B, Gillard JH, et al. Wegener granulomatosis: MR imaging findings in brain and meninges. Radiology 1999;213: 794–9.
124. Jennekens FG, Kater L. The central nervous system in systemic lupus erythematosus. Part 1. Clinical syndromes: a literature investigation. Rheumatology (Oxford) 2002;41:605–18.
125. Caselli RJ, Scheithauer BW, Bowles CA, et al. The treatable dementia of Sjogren's syndrome. Ann Neurol 1991;30:98–101.
126. Ferreiro JE, Robalino BD, Saldana MJ. Primary Sjogren's syndrome with diffuse cerebral vasculitis and lymphocytic interstitial pneumonitis. Am J Med 1987;82:1227–32.

Radionecrosis

127. Burger P, Boyko O. The pathology of central nervous system radiation injury. In: Gutin P, Leibel S, Sheline G, eds. Radiation injury to the central nervous system. New York: Raven Press; 1991:191–208.
128. Morris JG, Grattan-Smith P, Panegyres PK, O'Neill P, Soo YS, Langlands AO. Delayed cerebral radiation necrosis. Q J Med 1994;87:119–29.
129. Nelson DR, Yuh WT, Wen BC, Ryals TJ, Cornell SH. Cerebral necrosis simulating an intraparenchymal tumor. AJNR Am J Neuroradiol 1990;11:211–2.
130. Ball WS Jr, Prenger EC, Ballard ET. Neurotoxicity of radio/chemotherapy in children: pathologic and MR correlation. AJNR Am J Neuroradiol 1992;13:761–76.
131. Vigliani MC, Duyckaerts C, Hauw JJ, Poisson M, Magdelenat H, Delattre JY. Dementia following treatment of brain tumors with radiotherapy administered alone or in combination with nitrosourea-based chemotherapy: a clinical and pathological study. J Neurooncol 1999;41:137–49.
132. Marks JE, Wong J. The risk of cerebral radionecrosis in relation to dose, time and fractionation. A follow-up study. Prog Exp Tumor Res 1985;29:210–8.
133. Uematsu Y, Fujita K, Tanaka Y, et al. Gamma knife radiosurgery for neuroepithelial tumors: radiological and histological changes. Neuropathology 2001;21:298–306.
134. Rock JP, Hearshen D, Scarpace L, et al. Correlations between magnetic resonance spectroscopy and image-guided histopathology, with special attention to radiation necrosis. Neurosurgery 2002;51:912–20.
135. Burger PC, Mahley MS Jr, Dudka L, Vogel FS. The morphologic effects of radiation administered therapeutically for intracranial gliomas: a postmortem study of 25 cases. Cancer 1979;44:1256–72.

Textiloma (Gossypiboma)

136. Ribalta T, McCutcheon IE, Neto AG, et al. Textiloma (gossypiboma) mimicking recurrent intracranial tumor. Arch Pathol Lab Med 2004; 128:749–58.

20 METASTATIC AND SECONDARY NEOPLASMS

Metastatic and secondary neoplasms affect the central nervous system (CNS) either by bloodborne dissemination from a distant site or by direct local extension.

CLINICAL AND RADIOLOGIC FEATURES

Metastatic neoplasms are common intracranial masses encountered both at surgery and autopsy. Adults are primarily affected. Although metastatic disease is often entertained in the differential diagnosis of a poorly differentiated malignant brain tumor in children, it is most unusual for systemic malignancy to present as metastatic disease in this age group (1).

Selected references give clinical accounts of metastases from carcinomas of lung (2,3), breast (4,5), kidney (6–10), bladder (11,12), prostate gland (13,14), ovary (15,16), liver (17), esophagus (18), placenta (19,20), islets of Langerhans (21), thyroid gland (22), parathyroid gland (23), visceral neuroendocrine cells (carcinoid tumor) (24), and from tumors of melanocytes (25–28).

Metastases of sarcomas are rare, and occur in proportion to the frequency of the primary tumor. All major types have been represented (29–35).

Metastases to the Parenchyma of the Brain and Spinal Cord

Metastases to CNS parenchyma are almost always discrete, contrast enhancing, and superficial, with an epicenter near the gray-white junction. A zone of cerebral edema, large and commensurate with a rapidly growing mass, typically surrounds the lesion (fig. 20-1). Tumors such as malignant melanoma that frequently disseminate to the CNS often produce multiple lesions, whereas solitary deposits are more likely from tumors not prone to CNS metastasis such as adenocarcinoma of the gastrointestinal tract.

Symptoms of a brain metastasis as first evidence of a malignancy are especially likely with carcinoma of the lung (3). In contrast, late and often a solitary metastasis is classically associated with a primary in the kidney (renal cell carcinoma) or the eye (melanoma).

Metastatic carcinoma to the brain usually affects the cerebral hemispheres (fig. 20-1) and/or the cerebellum (fig. 20-2). Although any cerebral lobe or portion thereof may be involved, the distribution of metastases is not random. As a presumed consequence of circulatory dynamics, "watershed areas" between the anterior, middle, and posterior cerebral arteries are favored sites of secondary deposits (36). The pineal region (37), brain stem (13,38), and choroid plexus (39–41) are uncommon loci. Similarly uncommon is diffuse spread in the subependymal region (42).

Metastases to the parenchyma of the spinal cord are generally found only in the terminal stages of widely metastatic carcinoma from the lung or skin (melanoma). Presentation of a spinal cord metastasis as the first evidence of malignancy is rare (43–45).

Tumors Metastatic to the Meninges and Subarachnoid Space

Dissemination within intracranial and/or intraspinal leptomeninges is a frequent, often late, consequence of CNS metastases (46,47). In other cases, it is the result of perineural extension of squamous cancers of the head and neck. A special variant of meningeal cancer is a diffuse leptomeningeal proliferation known as *carcinomatous meningitis* or, more correctly, *meningeal carcinomatosis* (fig. 20-3) (48–51). By definition, this process is limited largely to the subarachnoid space. Access to the meninges may be attained from metastases to the spine or skull (50,51), and of course, from small parenchymal foci of metastatic carcinoma (46). Gastric carcinoma, the once classic primary source, has been largely supplanted by adenocarcinomas from lung and breast, as well as melanoma of skin (52). Mental "clouding," headaches, and progressive cranial nerve deficits are common clinical expressions. The disorder is suggested when

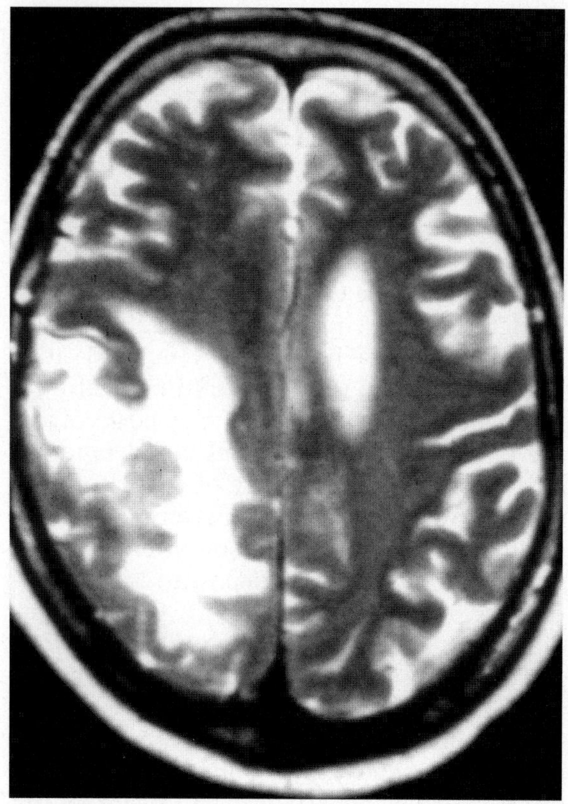

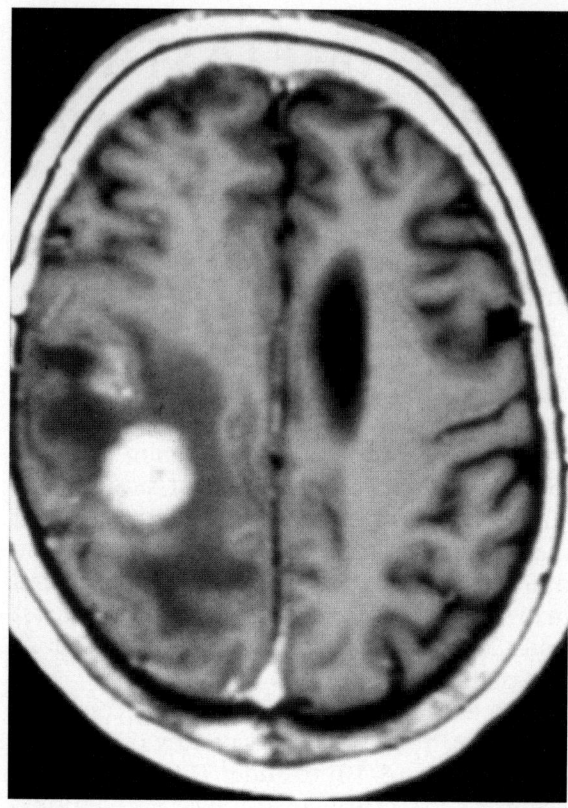

Figure 20-1

METASTATIC CARCINOMA

A discrete, superficial lesion with abundant peritumoral edema (left) and contrast enhancement (right) is a prime suspect for a metastasis.

neuroimaging studies show leptomeningeal thickening and contrast enhancement. Cytologic examination of cerebrospinal fluid is often confirmatory. When undertaken, a meningeal biopsy must include a generous sample of the leptomeninges. Dural samples alone are typically nondiagnostic.

Tumors Metastatic to the Skull and Cranial Dura

Most cranial metastases are irregular lytic lesions that, with inward growth, may compress the underlying brain (fig. 20-4). Less common are globular, dura-based masses with little, if any, detectable calvarial component (53). Extensive osteoblastic metastases, particularly at the skull base, are classically derived from carcinomas of the prostate gland. It is of note that prostatic carcinoma may affect the dura (53), but only rarely involves brain parenchyma. Neuroblastomas in children may undergo extensive metastasis to the skull.

Metastases to the Pituitary Gland and Sellar Region

Metastatic lesions in and about the sella turcica often originate in breast (54) or lung. Secondary involvement of the adenohypophysis may fill its sinusoids. Direct metastasis to the posterior lobe, the only portion of the pituitary gland with a direct arterial supply, may bring about its replacement, with associated diabetes insipidus. Carcinomas of the head and neck, typically adenoid cystic carcinoma, may extend along neurovascular structures to present in the cavernous sinus at a time when the presence of the primary is unsuspected (fig. 20-5; see fig. 20-15) (55,56).

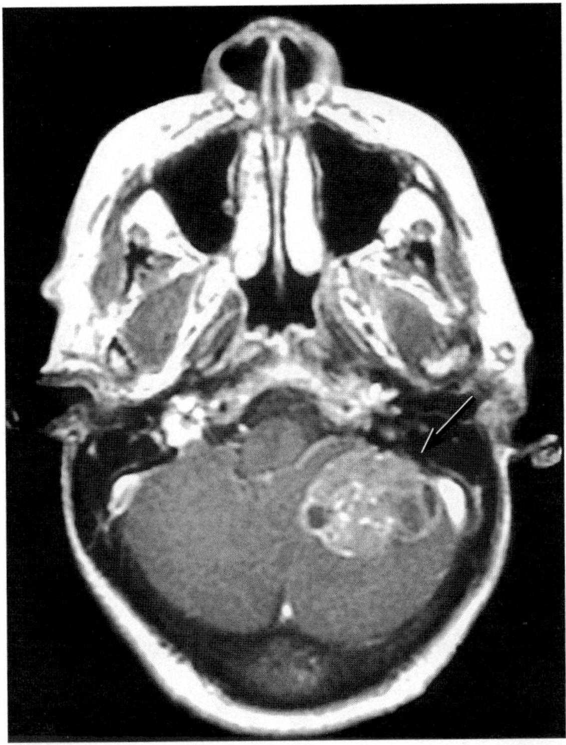

Figure 20-2

METASTATIC CARCINOMA

Discrete, expansile, edema-generating lesions in the cerebellum of an adult are highly suspicious for metastatic carcinoma. Ataxia was the initial symptom of this large lesion (arrow) in a 66-year-old woman with an unbiopsied tumor in the lung. The histochemical and immunohistochemical features of the cerebral lesion are illustrated in figure 20-22.

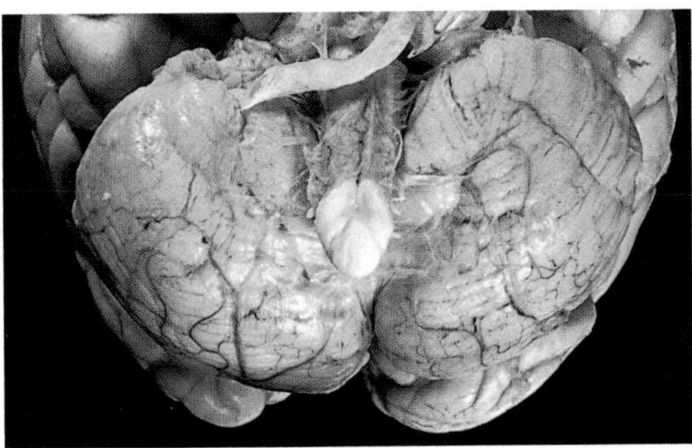

Figure 20-3

MENINGEAL CARCINOMATOSIS

Only uncommonly are central nervous system (CNS) metastases limited to the leptomeninges. The primary in this case was the stomach. The 66-year-old woman had disturbed gait, slurred speech, and headaches.

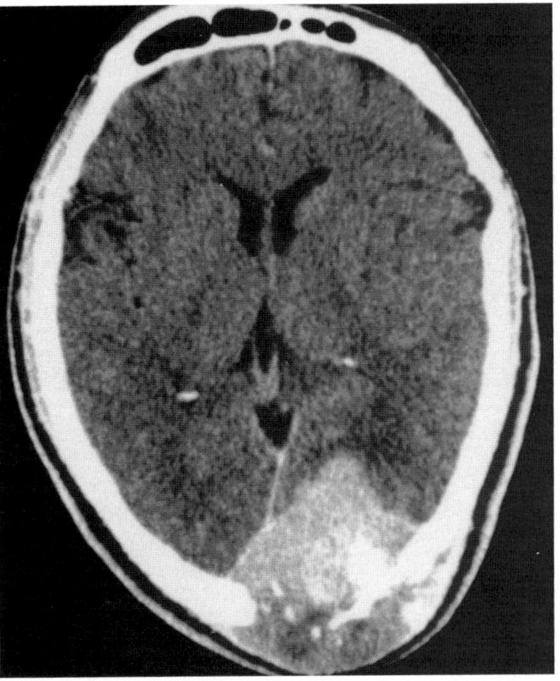

Figure 20-4

METASTATIC CARCINOMA TO THE SKULL

Most metastases to the cranium are lytic masses that can encroach on the underlying brain. A postoperative workup found multiple lesions in the liver, but no primary source, in this 43-year-old man who presented with slurred speech and blurred vision.

Metastases to the Spinal Column and Epidural Space

These frequent causes of spinal cord compression typically originate in breast, lung, or prostate gland (fig. 20-6) (57–59). Particularly with lung cancer, the primary may be unsuspected at the time the metastasis is discovered (58). Metastases from prostate carcinoma, usually osteoblastic, typically "creep" rostrally and laterally from lumbosacral levels to simultaneously involve the base of the skull, lateral bony pelvis, and ribs. Unlike the destructive lytic effects with resultant vertebral collapse that are the result of most vertebral metastases, prostate carcinoma usually is osteoblastic, and back pain is the principal symptom.

PATHOLOGIC FINDINGS

Gross Findings. Metastases to the parenchyma are well circumscribed (fig. 20-7) and surrounded by a zone of edematous white matter

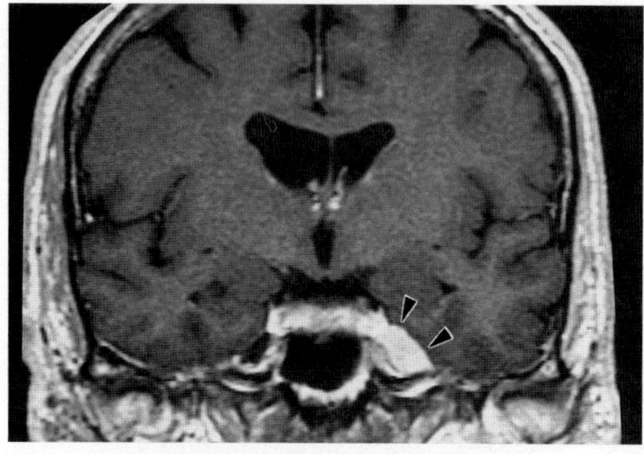

Figure 20-5

ADENOID CYSTIC CARCINOMA INVOLVING THE CAVERNOUS SINUS

Nasopharyngeal adenoid cystic carcinoma can track along nerves to present as a mass in the cavernous sinus (arrowheads). (Fig. 20-1 from Fascicle 10, 3rd Series.)

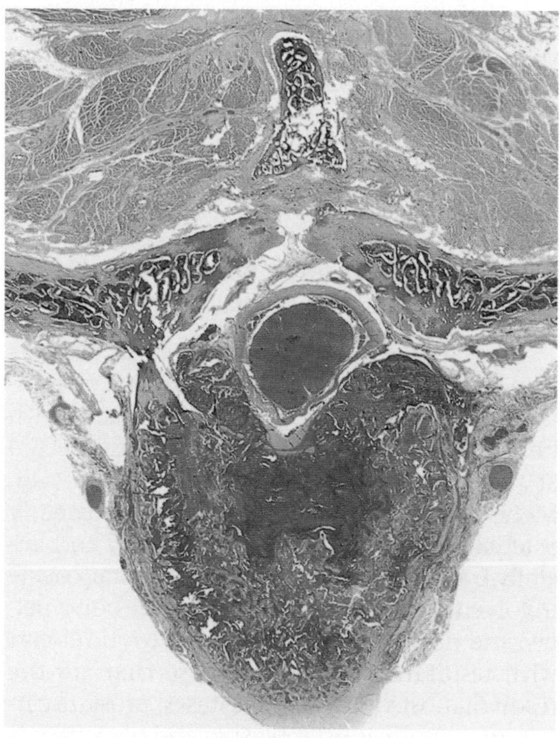

Figure 20-6

METASTATIC CARCINOMA TO THE SPINE AND EPIDURAL SPACE

Extending from a vertebral body, metastatic prostate carcinoma can enter the epidural space and compress the spinal cord.

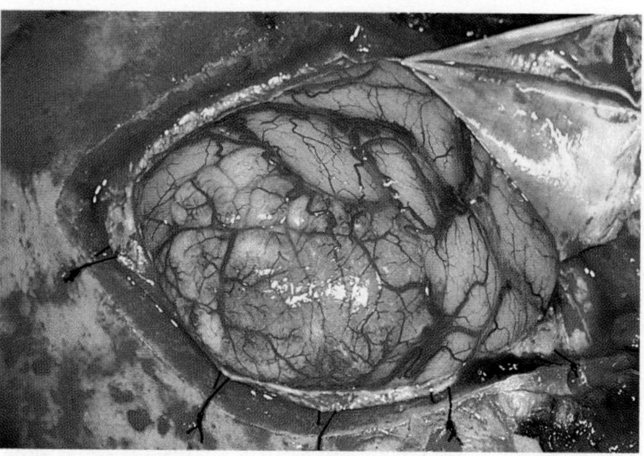

Figure 20-7

METASTATIC CARCINOMA

As in this adenocarcinoma from the lung, cerebral metastases are typically intracortical and discrete.

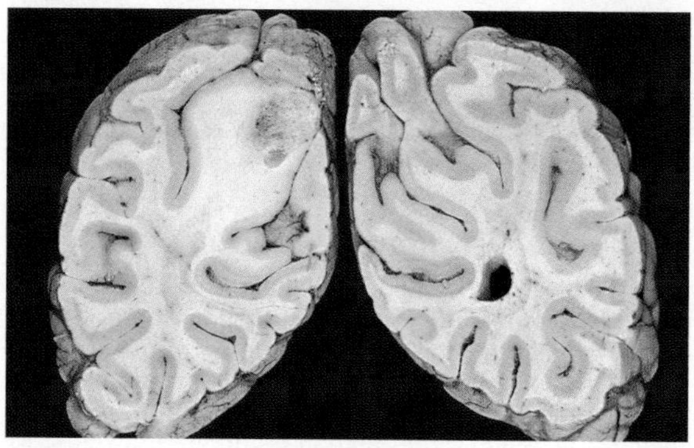

Figure 20-8

METASTATIC CARCINOMA

Cortical metastases elicit a prominent zone of vasogenic edema in subjacent white matter.

(fig. 20-8), as evidenced by a large peritumoral area of increased T2-weighted signal on magnetic resonance imaging (MRI). Extensive hemorrhage most often accompanies metastatic melanoma, choriocarcinoma, and, due primarily to its high incidence, lung carcinoma (60).

Meningeal carcinomatosis opacifies the leptomeninges and can thicken cranial and/or spinal nerve roots (fig. 20-3).

Metastases to the spine are often osteolytic, fleshy, and focally necrotic. Accompanying vertebral collapse is common except with prostate cancer. In contrast, vertebrae hosting prostate cancer are often dense and even eburnated.

Figure 20-9

METASTATIC CARCINOMA

The affinity of metastases for cerebral cortex near the gray-white junction is readily apparent in this example of renal cell carcinoma (hematoxylin and eosin [H&E]/Luxol fast blue stain).

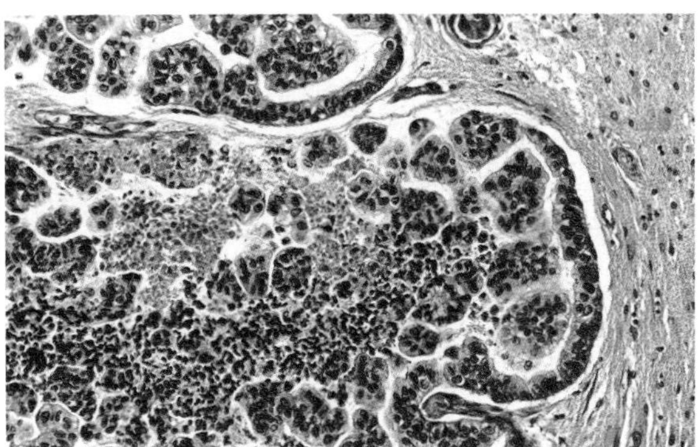

Figure 20-10

METASTATIC CARCINOMA

As in this well-differentiated adenocarcinoma, most metastases have a sharp interface with surrounding brain.

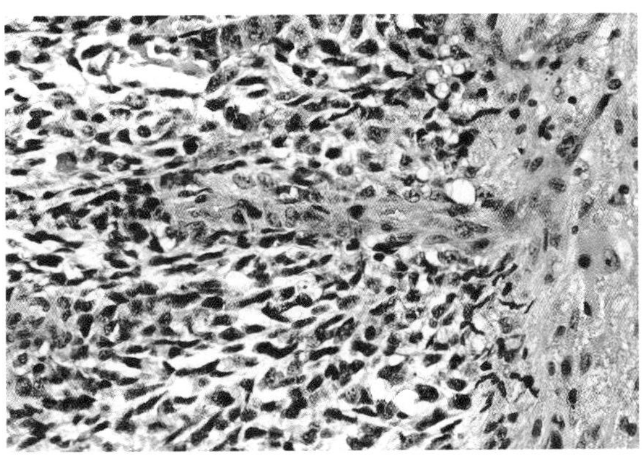

Figure 20-11

METASTATIC SARCOMA

Metastatic sarcomas are usually discrete masses. This undifferentiated lesion arose in the soft tissues of the thigh.

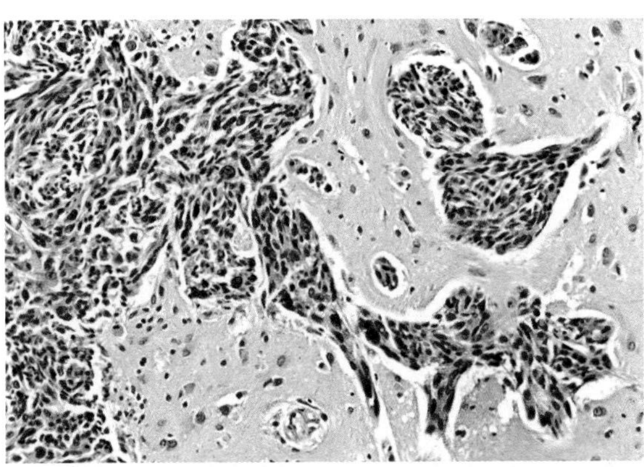

Figure 20-12

METASTATIC CARCINOMA

Some metastases, such as this malignant melanoma, extend along small intracortical vessels.

Microscopic Findings. Parenchymal metastases are usually discrete lesions that largely displace rather than diffusely infiltrate brain parenchyma (figs. 20-9–20-11). They are thus architecturally solid, containing few if any overrun neurons or glia, although small cell carcinomas of lung and carcinoma of the breast may show a degree of infiltrative, often perivascular, growth (fig. 20-12) (61). In contrast, diffuse astrocytomas of all grade and primary CNS lymphomas are frankly permeative tumors that overrun normal structures. As reflected in terms of induration, some metastases contain abundant connective tissue, often as coarse septa (fig. 20-13).

The neoplastic cells in meningeal carcinomatosis proliferate freely within the subarachnoid space and often extend inwardly along the perivascular (Virchow-Robin) spaces of the cerebral cortex (fig. 20-14). In addition, they produce either diffuse or nodular enlargement of cranial and spinal nerves caught up in the process. Differentiation, if any, is usually limited to mucus production by signet ring cells. Other epithelial growth patterns, such squamous differentiation, are uncommon.

Certain head and neck cancers, especially adenoid cystic carcinoma, are freely invasive of cranial nerves (fig. 20-15).

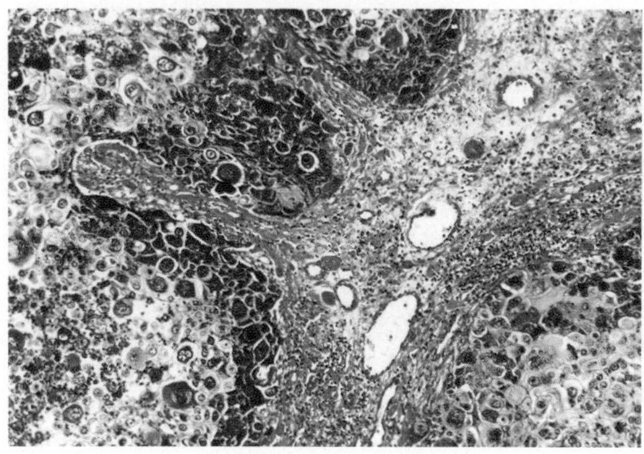

Figure 20-13

METASTATIC CARCINOMA

Lobular architecture, fibrous septa, and epithelial cytologic features are common features of metastases (Masson trichrome stain).

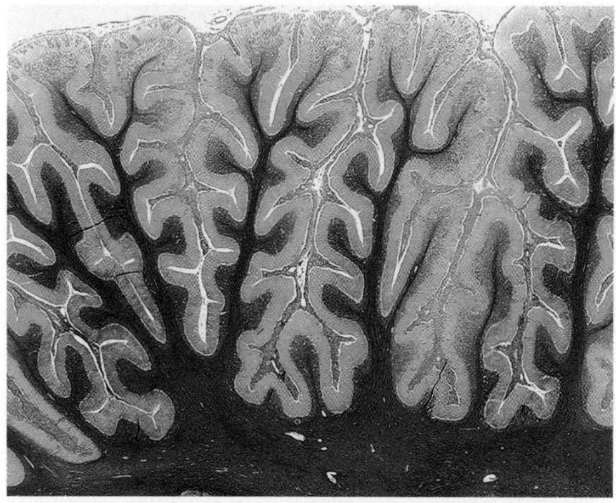

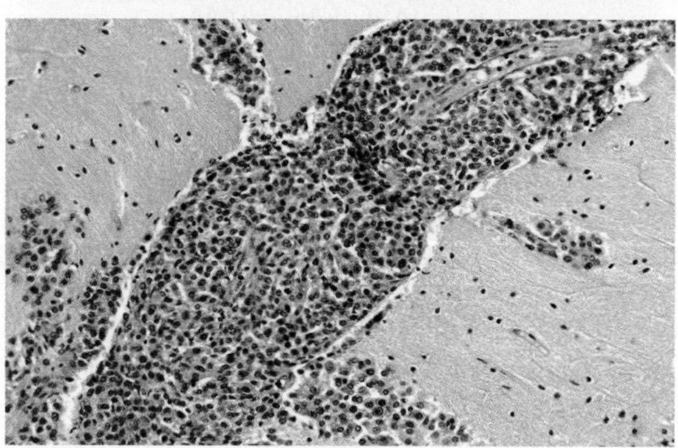

Figure 20-14

MENINGEAL CARCINOMATOSIS

A whole mount histologic section (top) and a photomicrograph (bottom) illustrate metastatic breast cancer filling the cerebellar subarachnoid space. Extension along perivascular spaces is common.

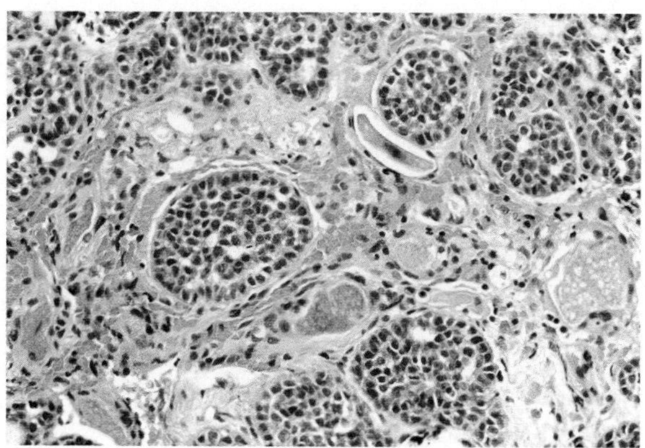

Figure 20-15

ADENOID CYSTIC CARCINOMA INVOLVING THE TRIGEMINAL GANGLION

Adenoid cystic carcinomas have a curious affinity for peripheral nerves. Extending along the fifth cranial nerve, this example has reached the gasserian ganglion.

Metastases often undergo extensive necrosis, often with viable tissue limited to a narrow rim at the periphery of the mass or to perivascular cuffs (fig. 20-16). While uniform perivascular collars of preserved tumor cells are occasionally present in necrotic malignant gliomas, this pattern of sparing is far more characteristic of metastases.

Aside from its presence in some renal cell carcinomas (fig. 20-17) and small cell carcinomas, glomeruloid vascular proliferation, so commonly seen in glioblastomas, is infrequent in metastases.

Histologic features of specific CNS metastases vary considerably, as is illustrated in metastases from lung (figs. 20-18, 20-19; see fig. 20-22), kidney (figs. 20-17, 20-20), ovary (fig. 20-21), breast (fig. 20-23), skin (melanoma) (fig. 20-24), and prostate gland (fig. 20-25).

Frozen Section. The general architectural features of microscopic circumscription, with or without lobularity, and monomorphous epithelial cytologic features, remain diagnostic cornerstones of metastases (figs. 20-18, 20-19). The presence of glandular differentiation, melanin, or mucus greatly simplifies the task.

Metastases often engender marked subacute, gemistocytic-appearing gliosis.

Findings in Needle Biopsy Specimens. Stereotactic needle or trochar specimens, when representative, generally pose no problem in the diagnosis of a metastasis and its distinction from

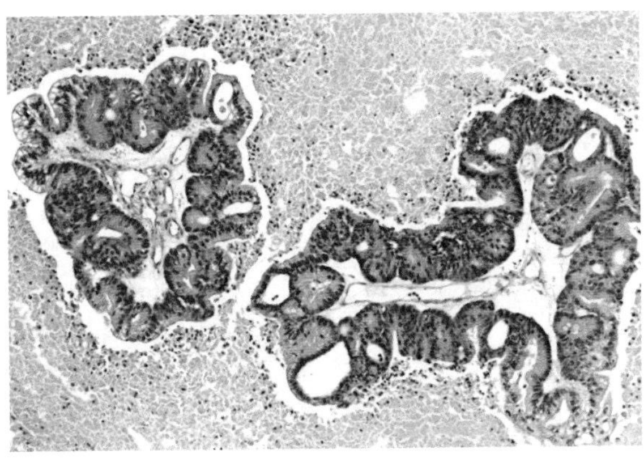

Figures 20-16

METASTATIC CARCINOMA

Necrosis, often extensive in metastases, frequently spares vessels and an isometric collar of perivascular tumor cells.

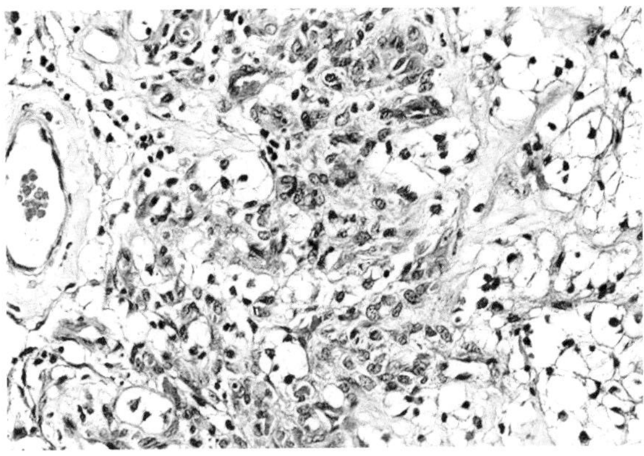

Figure 20-17

METASTATIC RENAL CELL CARCINOMA

Except for renal cell carcinoma, vascular proliferation is uncommon in metastases.

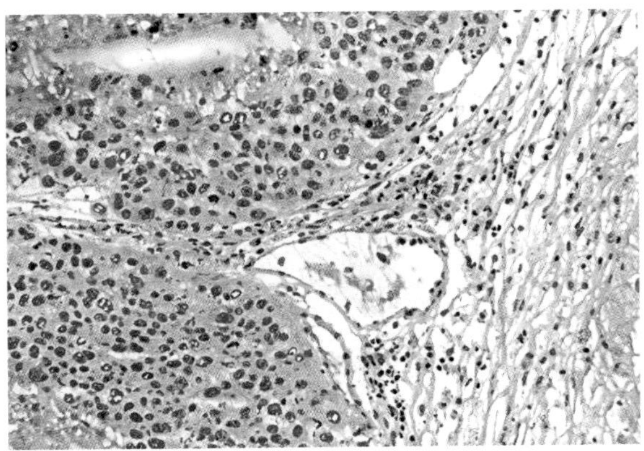

Figure 20-18

METASTATIC CARCINOMA: FROZEN SECTION

A well-circumscribed mass with epithelial features is typical of metastatic carcinoma. Necrosis is frequent.

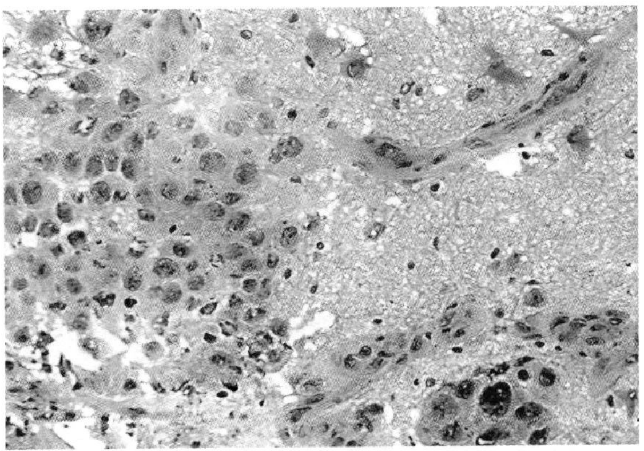

Figure 20-19

METASTATIC CARCINOMA: FROZEN SECTION

The findings of circumscription and large epithelioid cells suggest the diagnosis of metastatic carcinoma. Although the large cells resemble those of lymphoma, they are cohesive and do not freely invade the brain.

a glial primary. Given the firm texture of some metastases, however, flexible biopsy needles may be deflected from the tumor into the surrounding brain where they obtain only edematous and gliotic tissue superficially resembling glioma. Correlation of microscopic findings with radiographic images quickly underscores the discordance between the marginally abnormal histologic picture and the unexplained radiologic profile of an expansile, edema-inciting, contrast-enhancing mass.

Immunohistochemical Findings. In general, metastatic carcinomas share the immunoreactivity of the parent neoplasm, including positivity for various cytokeratins and for epithelial membrane antigen (EMA) (fig. 20-20, right). Immunoreactivity for cytokeratins is understandably common in carcinomas, but its utility in the differential diagnosis is compromised by the frequent reactivity in neoplastic and especially non-neoplastic astrocytes (62,63). This is particularly true of broad-spectrum keratins such as AE1/AE3, and far less so with CAM5.2 (64,65). Unlike gliomas, many metastatic carcinomas are immunoreactive for CK7, CK20, BerEP4, and HEA125 (fig. 20-21, right) (64,66). One study found that CK7 was a reliable marker for adenocarcinomas

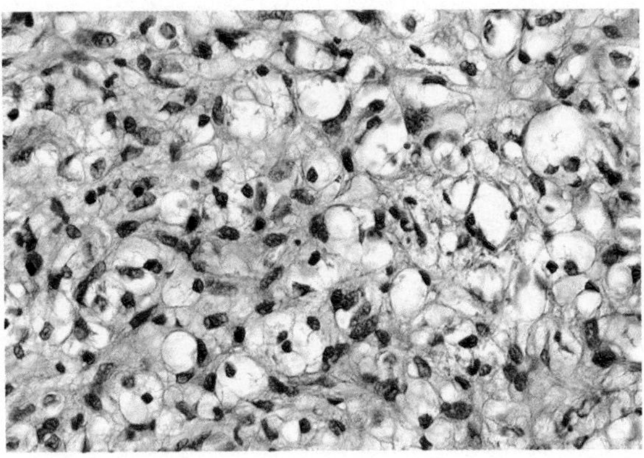

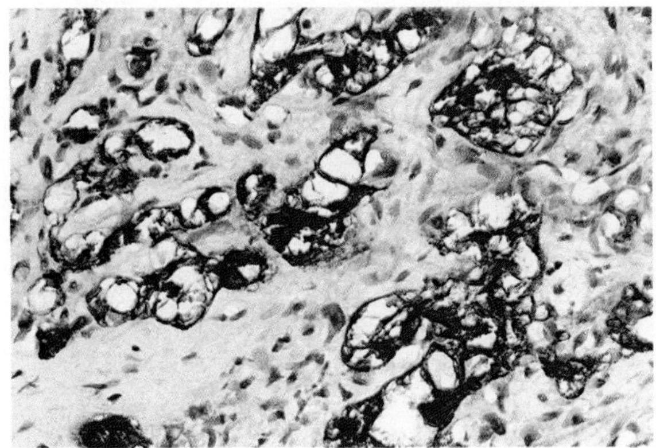

Figure 20-20

METASTATIC RENAL CELL CARCINOMA

Left: The compact, nonglandular appearance of renal cell carcinoma may not bring a kidney primary, or even an epithelial neoplasm, to mind.

Right: Surface staining for epithelial membrane antigen was used to identify the epithelial nature of this spinal epidural lesion. A remote history of nephrectomy for renal cell carcinoma was obtained, and hepatic metastases were documented by computerized tomography (CT).

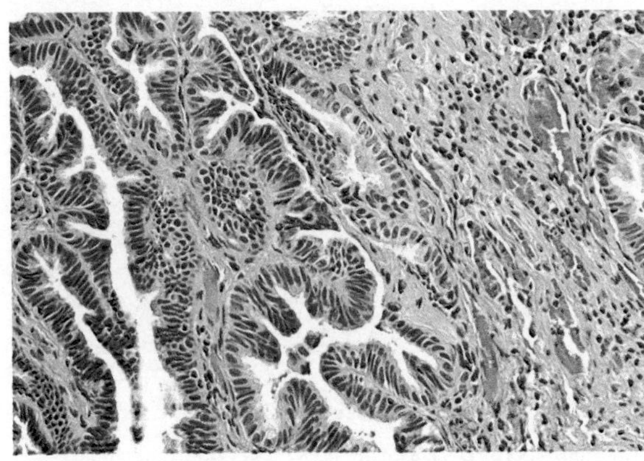

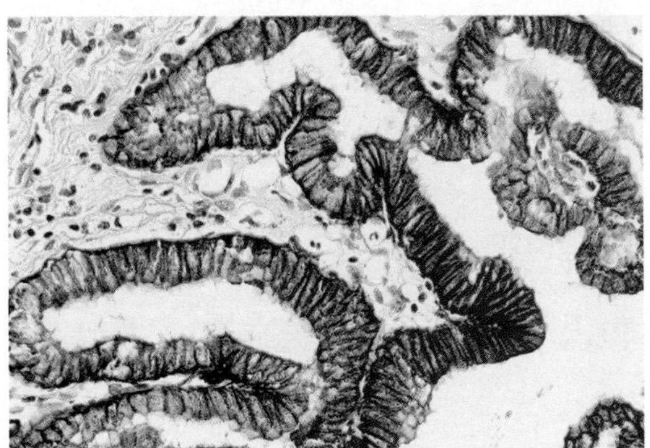

Figure 20-21

METASTATIC OVARIAN CARCINOMA

Left: Papillary lesions, such as this ovarian carcinoma, need to be distinguished from choroid plexus carcinoma.

Right: Immunoreactivity for BerEP4 is helpful since it is common in metastatic carcinomas and alien to choroid plexus neoplasms.

of the lung and breast, whereas CK20 helped identify carcinomas from the gastrointestinal tract (65). In the same study, CDFP-15 (gross cystic disease fluid protein-15) and estrogen receptors were helpful in distinguishing metastases from breast and lung.

Small cell lung carcinomas are often immunoreactive for cytokeratins and EMA, less frequently for synaptophysin, and still less often for chromogranin. Nuclear staining for thyroid transcription factor-1 (TTF1) is a useful marker for cancer of the lung and thyroid gland (fig. 20-22, right) (67).

Renal cell carcinomas are immunoreactive for EMA, RCC, and CD10 (fig. 20-20, right), and may be negative for cytokeratins. Breast carcinomas are often immunoreactive for estrogen receptors (fig. 20-23, right) and amplified for Her2Neu, seen best by fluorescence in situ hybridization. Malignant melanomas usually react variably for S-100 protein, HMB45, Melan-A (fig. 20-24, right), tyrosinase, and microphthalmia transcription factor

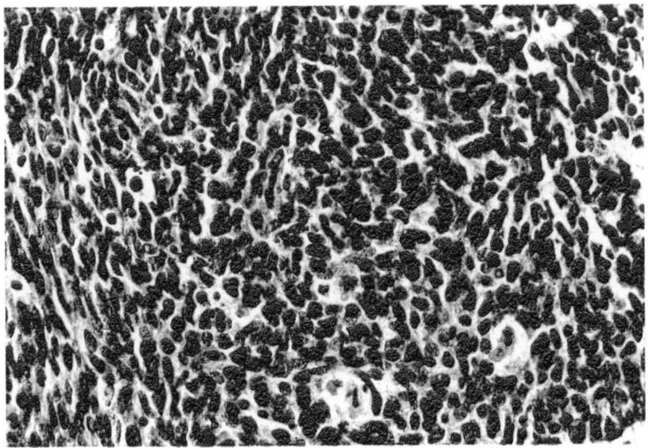

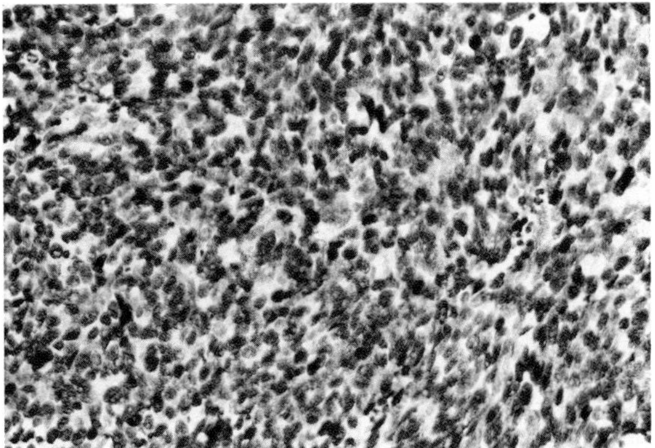

Figure 20-22

METASTATIC CARCINOMA FROM THE LUNG

Left: Small cell lesions, such as this metastatic lung cancer in the cerebellum, must be distinguished from medulloblastoma, even in adults.

Right: Positive nuclear staining for thyroid transcription factor 1 strongly supports a pulmonary origin. The cerebellar lesion was also reactive focally for cytokeratins and chromogranin. The neoplasm's radiologic features are illustrated in figure 20-2.

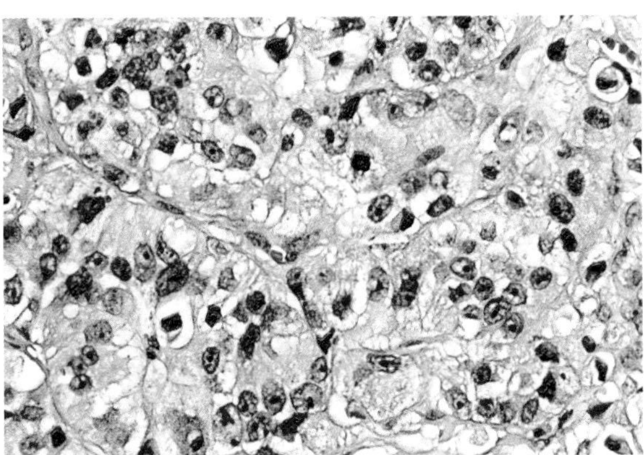

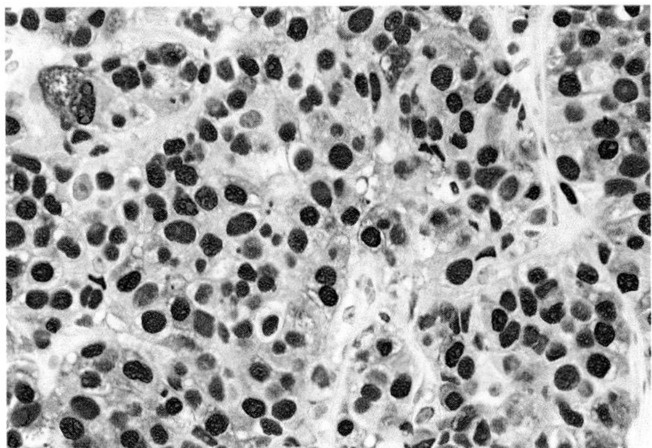

Figure 20-23

METASTATIC BREAST CARCINOMA

Left: This adenocarcinoma has architectural and cytologic features consistent with a breast primary.

Right: Nuclear staining for estrogen receptors helped confirm a mammary origin.

(68). Prostate carcinomas are positive for cytokeratins, EMA, prostate-specific antigen (fig. 20-25, right), and prostatic acid phosphatase. Thyroid cancers are immunoreactive for cytokeratins, EMA, thyroglobin, and TTF1. Hepatocellular carcinomas are often keratin and EMA immunonegative, but can be stained for polyclonal carcinoembryonic antigen (CEA) with which canaliculi are highlighted, and for Hepar A, which targets parenchymal cells. In situ hybridization for albumin also serves to identify this tumor.

An algorithm using the results of the above stains and other methods is successful in identifying the primary source of a metastasis of unknown origin in most cases (69).

Ultrastructural Findings. Cells of squamous cell carcinomas are connected by desmosomes that are in turn joined to bundles of tonofilaments that fan out into the adjacent cytoplasm. Tonofilament bundles may also lie free in the cytoplasm. Adenocarcinomas have intracellular and intercellular microvillous-lined acini and

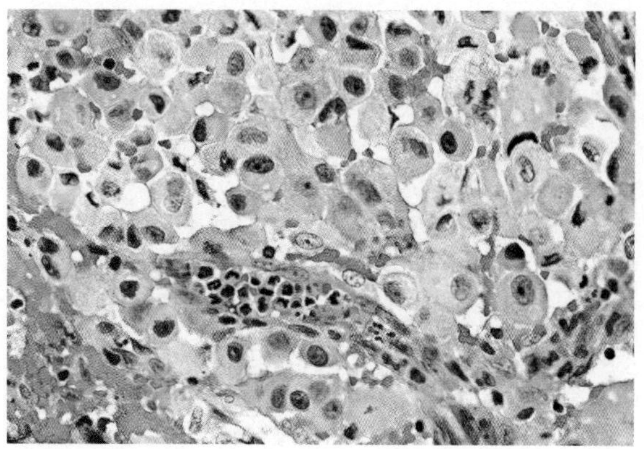
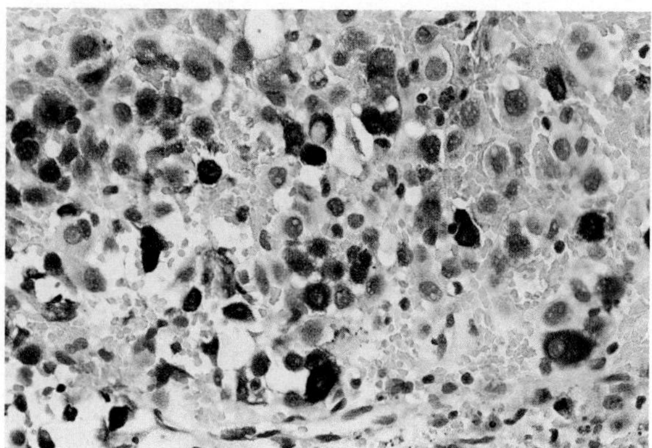

Figure 20-24

METASTATIC MALIGNANT MELANOMA

Left: While melanoma is notoriously polyphenotypic, large, rather discrete cells with distinct borders and prominent nucleoli are common.

Right: Immunoreactivity for Melan-A was instrumental in determining the melanocytic nature of this amelanotic lesion.

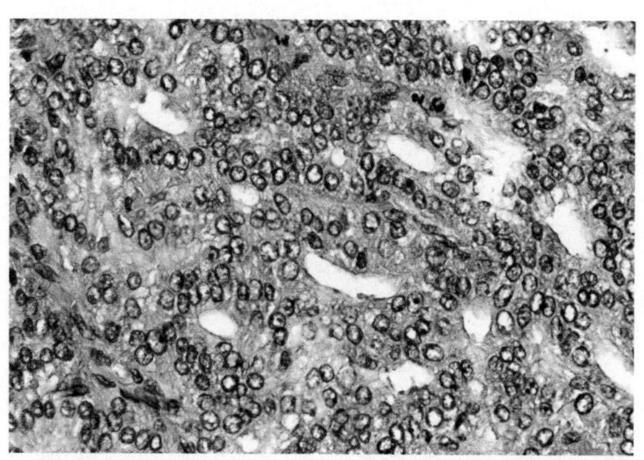
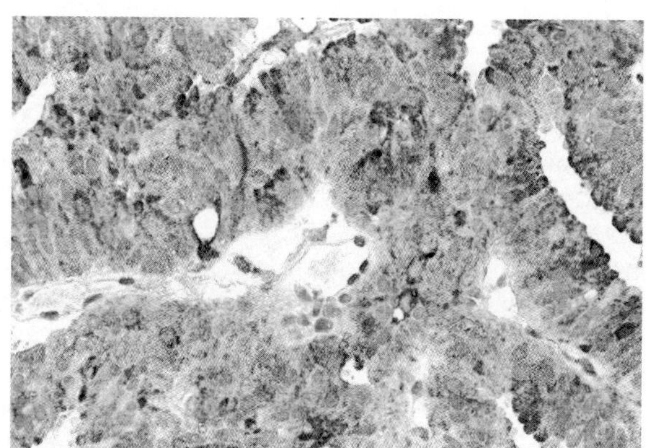

Figure 20-25

METASTATIC PROSTATE CARCINOMA

Left: Although it frequently reaches the spine and the skull, prostate cancer rarely presents as an intraparenchymal central nervous system (CNS) deposit.

Right: Immunohistochemistry for prostate-specific antigen helped confirm the nature of this lesion.

cytoplasmic and intraluminal mucus. Those of the lung are noted for microvilli, Clara granules, "surfactant figures," and features of squamous differentiation (70). Some small cell carcinomas contain occasional neurosecretory granules. Metastatic melanoma has melanosomes in various stages of development, although often in small number. In contrast to carcinomas, intercellular junctions are lacking.

Cytologic Findings. Smears of epithelial neoplasms have sheets or clumps that are more obviously epithelial, or even overtly glandular or secretory (fig. 20-26). Typically, there is cellular monomorphism and cohesion, as well as nuclear molding. Chromatin is coarse and nucleoli are often prominent. Mucus, either in cytoplasmic lumens or between cells, may be present in well-differentiated adenocarcinomas. Skeins of refractile filaments are a feature of squamous carcinoma. Cytoplasmic clearing is a characteristic of renal cell carcinoma. Colon carcinoma metastases typically have columnar cells with elongated nuclei. Melanoma cells can be spindle or epithelioid in configuration, show little cohesion,

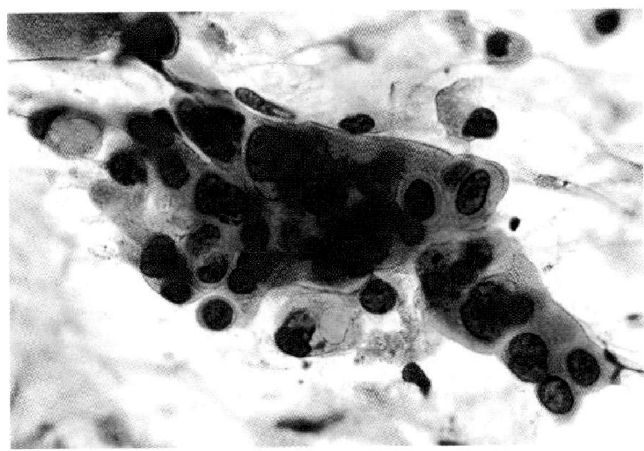

Figure 20-26

METASTATIC ADENOCARCINOMA: SMEAR PREPARATION

Tissue fragments of epithelial cells and signet ring cells are diagnostic of metastatic adenocarcinoma, in this case from the ovary. Note the mucinous background.

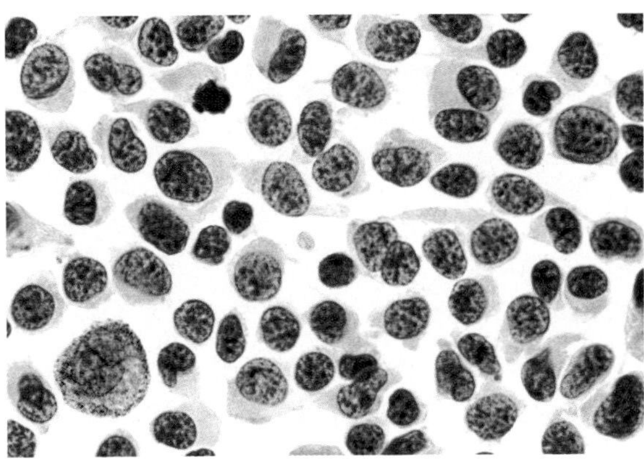

Figure 20-27

METASTATIC MELANOMA: SMEAR PREPARATION

As is typical, melanomas smear out as individual cells. Their cytoplasm varies in quantity and content of melanosomes.

have delicate chromatin but prominent nucleoli, and have highly variable cytoplasmic pigmentation (fig. 20-27). Metastatic sarcomas yield tissue fragments with little tendency to dissociate. High-grade tumors prone to necrosis are often littered with cellular debris, causing the background to be "dirty." Histiocytes and reactive astrocytes are a common accompaniment.

DIFFERENTIAL DIAGNOSIS

At high magnification, glioblastomas with epithelioid (figs. 20-28, 20-29) or rhabdoid (fig. 20-30) cytologic features can closely resemble metastatic carcinoma. Distinguishing features are the more classic features of astrocytoma elsewhere in the lesion, including elongated or plump cells with bright pink cytoplasm and fibrillarity, glial fibrillary acid protein (GFAP) positivity, and an infiltrating quality that can be confirmed by neurofilament protein immunostaining. Especially problematic are epithelial structures in glioblastomas and gliosarcomas (see chapter 3 and figs. 3-60–3-62).

While anaplastic oligodendrogliomas are infiltrating neoplasms, they can have a compact, sheet-like architecture, which, when coupled with epithelioid cytologic features, suggests carcinoma (fig. 20-31, see fig. 3-220). Infiltrative margins, classic features of grade II or III oligodendroglioma elsewhere in the specimen, and S-100 protein or GFAP positivity help resolve the issue.

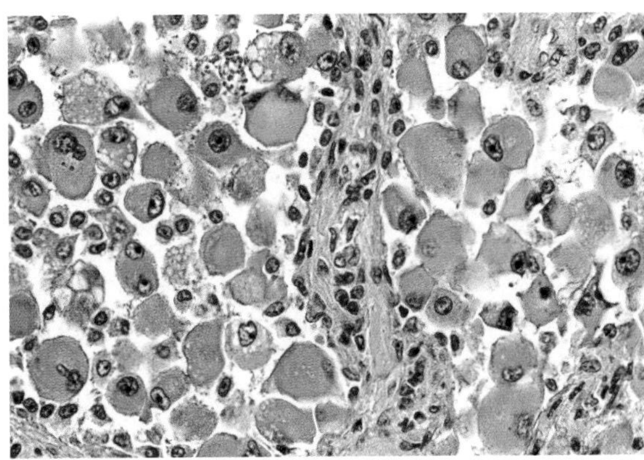

Figure 20-28

GLIOBLASTOMA RESEMBLING METASTATIC CARCINOMA

Lobules of cells with discrete borders and large nuclei create the appearance of an epithelial malignancy.

Primary CNS lymphomas (PCNSLs) can also have densely cellular regions composed of back-to-back, vaguely epithelioid large lymphocytes that mimic metastatic carcinoma (fig. 20-32). PCNSLs are, however, infiltrative tumors with a distinctive angiocentricity. Apoptosis is the common pathway of cell death. Only in the context of immunosuppression is coagulation necrosis an extensive feature. Lymphoid markers are both sensitive and specific in distinguishing lymphomas from metastases. A special rare variant of large cell lymphoma, anaplastic, or formerly,

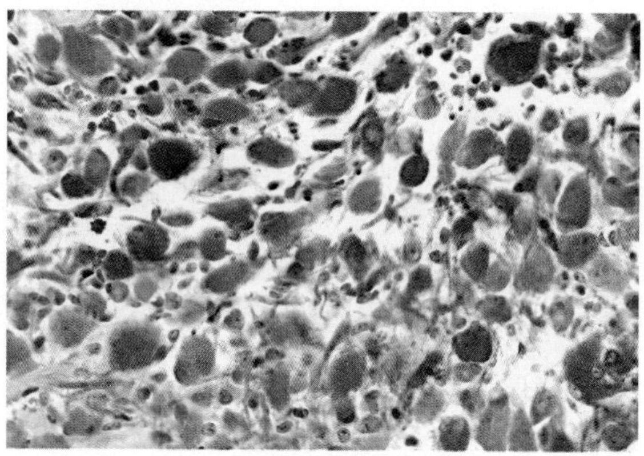

Figure 20-29

GLIOBLASTOMA RESEMBLING METASTATIC CARCINOMA

Glioblastomas can be immunoreactive for epithelial markers, as in this case stained with broad-spectrum antibodies to cytokeratins. Cell processes are unusual in a carcinoma.

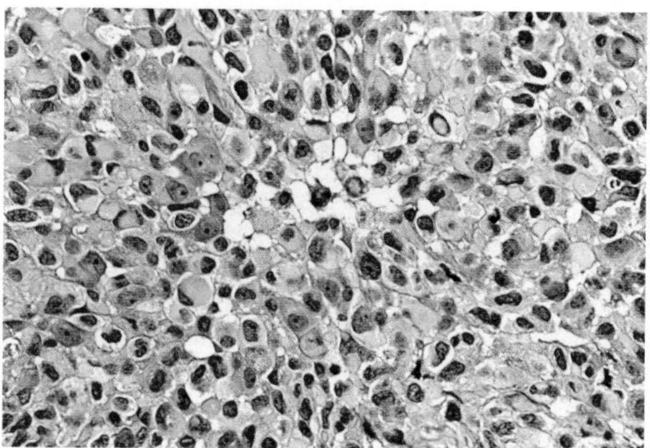

Figure 20-30

RHABDOID GLIOBLASTOMA RESEMBLING METASTATIC CARCINOMA

Distinct cell borders, round nuclei, and prominent nucleoli create the impression of metastatic carcinoma in glioblastoma with rhabdoid features.

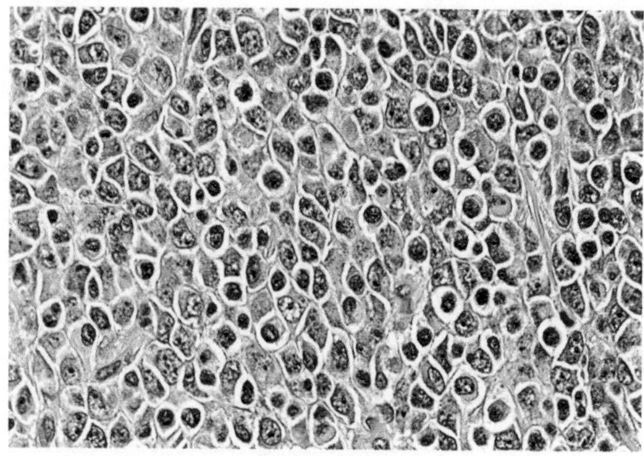

Figure 20-31

ANAPLASTIC OLIGODENDROGLIOMA RESEMBLING METASTATIC CARCINOMA

An anaplastic (grade III) oligodendroglioma with back-to-back cells with large round nuclei can mimic metastatic carcinoma. Elsewhere, lesions such as this show their capacity for infiltration and, at least focally, usually have features of classic oligodendroglioma.

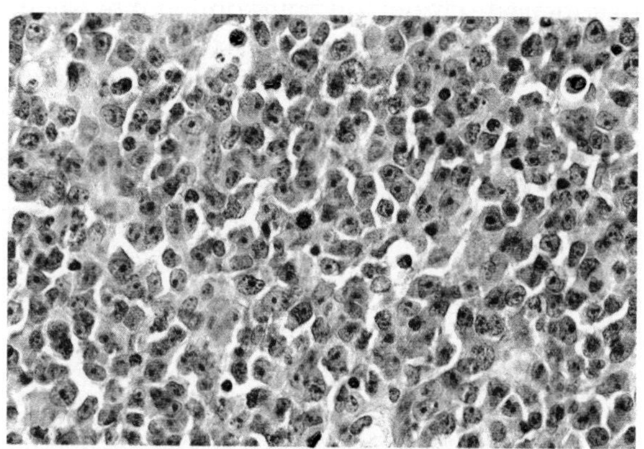

Figure 20-32

PRIMARY CNS MALIGNANT LYMPHOMA RESEMBLING METASTATIC CARCINOMA

The "epithelioid" features can closely mimic metastatic carcinoma.

Ki-1, lymphoma can mimic metastatic carcinoma both cytologically and immunohistochemically given its immunoreactivity for EMA.

Differentiating metastatic renal cell carcinoma to the cerebellum from hemangioblastoma can be challenging, particularly since both are components of von Hippel-Lindau disease (fig. 20-33). Even metastases of renal cell cancer to hemangioblastoma have been recorded (71,72). The findings in occasional hemangioblastomas of scattered mitoses and a surprisingly high MIB-1 rate do nothing to facilitate the distinction. The deciding factor is usually immunohistochemistry, particularly for EMA and CD10, both positive in renal cell carcinoma but not hemangioblastoma (see fig. 20-20, right). As is discussed in chapter 7, hemangioblastomas may be immunoreactive for inhibin. The exclusion of renal cell carcinoma can be even more

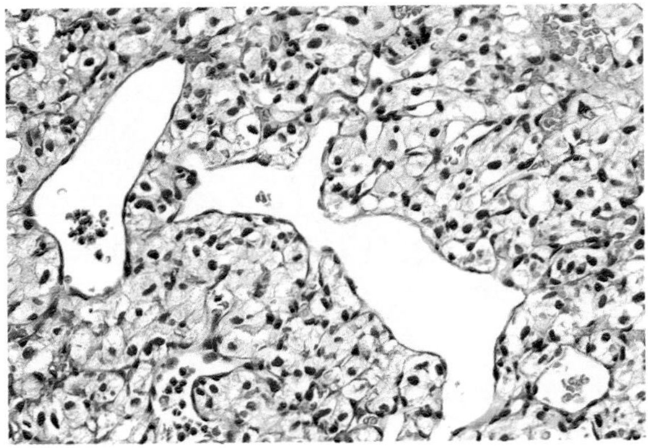

Figure 20-33

HEMANGIOBLASTOMA RESEMBLING METASTATIC CLEAR CELL CARCINOMA

Hemangioblastomas can closely resemble renal cell carcinomas.

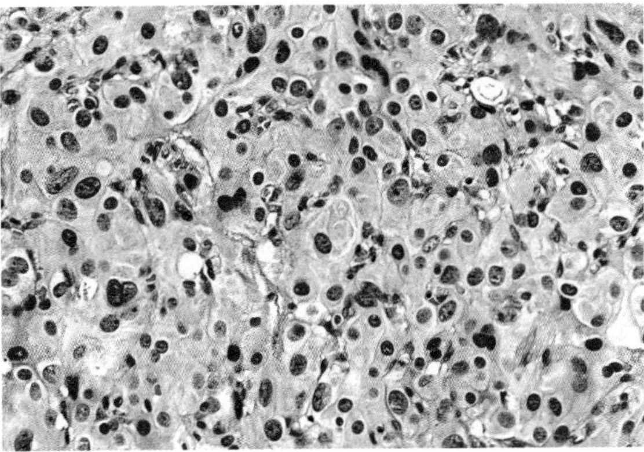

Figure 20-34

WELL-DIFFERENTIATED (GRADE I) MENINGIOMA RESEMBLING METASTATIC CLEAR CELL CARCINOMA

Nuclear pleomorphism and well-defined cell borders combine to simulate carcinoma. Thick-walled vessels, benign nuclear cytologic features, and absence of anaplasia help distinguish the two lesions.

conclusive given the hemangioblastoma's reactivity for both inhibin and neuron-specific enolase (NSE). Radiologic scanning of the kidneys may be warranted in equivocal cases.

Small cell carcinomas metastatic to the cerebellum of an adult must be distinguished from medulloblastoma, a sometimes difficult, if not impossible, task since both exhibit neural or neuroendocrine features. A nodular architecture with "pale islands" simplifies the procedure, but many medulloblastomas are not so obliging. Immunohistochemical mainstays are epithelial markers such as cytokeratins and EMA, as well as TTF1 (see fig. 20-22). A chest computerized tomography (CT) scan may be helpful in detecting a primary.

From the narrow perspective of high magnification, particularly at the time of frozen section, meningiomas (fig. 20-34), particularly those with epithelial features such as the chordoid or rhabdoid variant, may mimic metastatic carcinoma. Distinguishing features of low-grade meningiomas are hyalinized vessels, features of classic meningioma elsewhere in the specimen, cytologic features such as nuclear-cytoplasmic inclusions, and lack of anaplasia. Anaplastic (grade III) meningiomas often retain sufficient meningothelial features to readily establish the diagnosis. Only an occasional meningioma is fully anaplastic (fig. 20-35) or qualifies for the term "sarcomatoid." While meningiomas are EMA positive, they are variably positive for cytokeratins as discussed in chapter 9.

Metastatic papillary carcinomas prompt consideration of choroid plexus papilloma or carcinoma. The latter is rare in adults and usually restricted to the fourth ventricle, a most unusual site for a metastasis. Nevertheless, there are well-differentiated papillary lesions that, despite their presentation like a CNS primary (fig. 20-36), are found after a more detailed workup to be associated with an extracerebral mass, usually in the lung. Immunohistochemically, choroid plexus tumors are immunoreactive for cytokeratins, vimentin, S-100 protein, and synaptophysin. Papillomas often feature some GFAP reactivity. EMA staining, if present, is minor. BerEP4 stains many metastatic papillary carcinomas (see fig. 20-21, right), but not primary choroid plexus neoplasms (66). Metastases may be positive for CEA, although the same may be true of occasional choroid plexus carcinomas.

TREATMENT AND PROGNOSIS

Many clinical variables determine the outlook for patients with metastatic disease. Aside from some patients with resectable solitary metastases, as from renal cell carcinoma, or with early osseous lesions of prostate cancer, most patients with metastatic CNS disease die within several months.

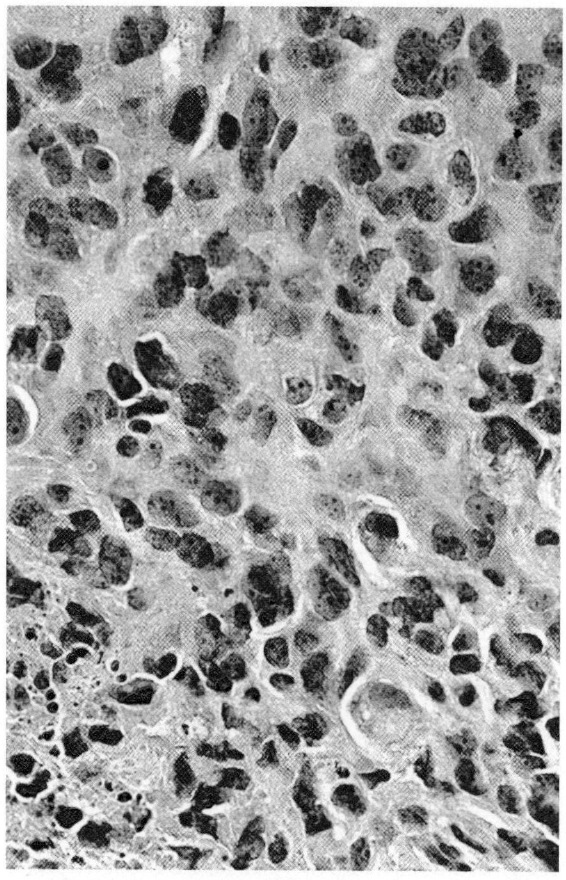

Figure 20-35

ANAPLASTIC MENINGIOMA RESEMBLING METASTATIC CELL CARCINOMA

Only rare meningiomas are anaplastic enough to simulate carcinoma.

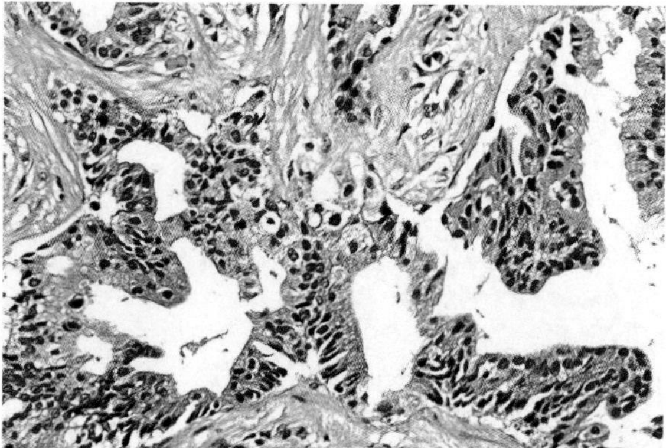

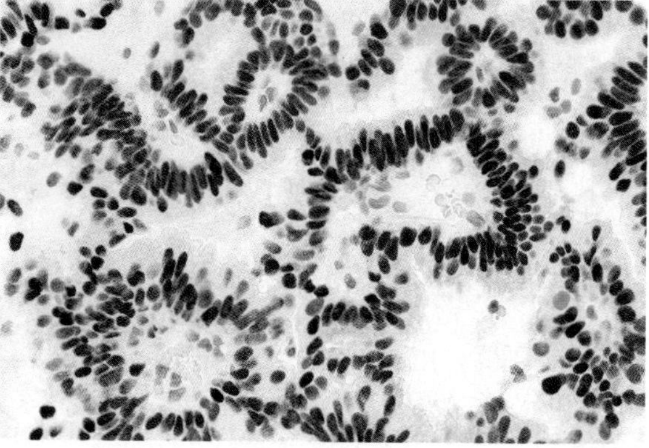

Figure 20-36

METASTATIC PAPILLARY CARCINOMA RESEMBLING CHOROID PLEXUS CARCINOMA

Although this papillary neoplasm (top) was felt preoperatively to be a primary brain tumor, nuclear immunostaining for thyroid transcription factor-1 suggested a pulmonary or thyroid source (bottom). A needle biopsy obtained identical tissue from a small pulmonary mass.

REFERENCES

1. Curless RG, Toledano SR, Ragheb J, Cleveland WW, Falcone S. Hematogenous brain metastasis in children. Pediatr Neurol 2002;26:219–21.
2. Hirsch FR, Paulson OB, Hansen HH, Vraa-Jensen J. Intracranial metastases in small cell carcinoma of the lung: correlation of clinical and autopsy findings. Cancer 1982;50:2433–7.
3. Trillet V, Catajar JF, Croisile B, et al. Cerebral metastases as first symptom of bronchogenic carcinoma. A prospective study of 37 cases. Cancer 1991;67:2935–40.
4. Boogerd W, Vos VW, Hart AA, Baris G. Brain metastases in breast cancer; natural history, prognostic factors and outcome. J Neurooncol 1993;15:165–74.
5. Tsukada Y, Fouad A, Pickren JW, Lane WW. Central nervous system metastasis from breast carcinoma. Autopsy study. Cancer 1983;52:2349–54.
6. Ammirati M, Samii M, Skaf G, Sephernia A. Solitary brain metastasis 13 years after removal of renal adenocarcinoma. J Neurooncol 1993;15:87–90.

7. Franke FE, Altmannsberger M, Schachenmayr W. Metastasis of renal carcinoma colliding with glioblastoma. Carcinoma to glioma: an event only rarely detected. Acta Neuropathol (Berl) 1990;80:448–52.
8. Radley MG, McDonald JV, Pilcher WH, Wilbur DC. Late solitary cerebral metastases from renal cell carcinoma: report of two cases. Surg Neurol 1993;39:230–4.
9. Seaman EK, Ross S, Sawczuk IS. High incidence of asymptomatic brain lesions in metastatic renal cell carcinoma. J Neurooncol 1995;23:253–6.
10. Roser F, Rosahl SK, Samii M. Single cerebral metastasis 3 and 19 years after primary renal cell carcinoma: case report and review of the literature. J Neurol Neurosurg Psychiatry 2002;72:257–8.
11. Davis RP, Spigelman MK, Zappulla RA, Sacher M, Strauchen JA. Isolated central nervous system metastasis from transitional cell carcinoma of the bladder: report of a case and review of the literature. Neurosurgery 1986;18:622–4.
12. Vinchon M, Ruchoux MM, Sueur JP, Assaker R, Christiaens JL. Solitary brain metastasis as only recurrence of a carcinoma of the bladder. Clin Neuropathol 1994;13:338–40.
13. Gupta A, Baidas S, Cumberlin RK. Brain stem metastasis as the only site of spread in prostate carcinoma. A case report. Cancer 1994;74:2516–9.
14. McCutcheon IE, Eng DY, Logothetis CJ. Brain metastasis from prostate carcinoma: antemortem recognition and outcome after treatment. Cancer 1999;86:2301–11.
15. Dauplat J, Nieberg RK, Hacker NF. Central nervous system metastases in epithelial ovarian carcinoma. Cancer 1987;60:2559–62.
16. Stein M, Steiner M, Klein B, et al. Involvement of the central nervous system by ovarian carcinoma. Cancer 1986;58:2066–9.
17. McIver JI, Scheithauer BW, Rydberg CH, Atkinson JL. Metastatic hepatocellular carcinoma presenting as epidural hematoma: case report. Neurosurgery 2001;49:447–9.
18. Gabrielsen TO, Eldevik OP, Orringer MB, Marshall BL. Esophageal carcinoma metastatic to the brain: clinical value and cost-effectiveness of routine enhanced head CT before esophagectomy. AJNR Am J Neuroradiol 1995;16:1915–21.
19. Athanassiou A, Begent RH, Newlands ES, Parker D, Rustin GJ, Bagshawe KD. Central nervous system metastases of choriocarcinoma. 23 years' experience at Charing Cross Hospital. Cancer 1983;52:1728–35.
20. Ishizuka T, Tomoda Y, Kaseki S, Goto S, Hara T, Kobayashi T. Intracranial metastasis of choriocarcinoma. A clinicopathologic study. Cancer 1983;52:1896–903.
21. Sabo RA, Kalyan-Raman UP. Multiple intracerebral metastases from an islet cell carcinoma of the pancreas: case report. Neurosurgery 1995;37:326–8.
22. Aguiar PH, Agner C, Tavares FR, Yamaguchi N. Unusual brain metastases from papillary thyroid carcinoma: case report. Neurosurgery 2001;49:1008–13.
23. Tyler D 3rd, Mandybur G, Dhillon G, Fratkin J. Intracranial metastatic parathyroid carcinoma: case report. Neurosurgery 2001;48:937–40.
24. Nida TY, Hall WA, Glantz M, Brent CH. Metastatic carcinoid tumor to the orbit and brain. Neurosurgery 1992;31:949–52.
25. Atlas SW, Grossman RI, Gomori JM, et al. MR imaging of intracranial metastatic melanoma. J Comput Assist Tomog 1987;11:577–82.
26. Sampson JH, Carter JH Jr, Friedman AH, Seigler HF. Demographics, prognosis, and therapy in 702 patients with brain metastases from malignant melanoma. J Neurosurg 1998;88:11–20.
27. Wronski M, Arbit E. Surgical treatment of brain metastases from melanoma: a retrospective study of 91 patients. J Neurosurg 2000;93:9–18.
28. Zacest AC, Besser M, Stevens G, Thompson JF, McCarthy WH, Culjak G. Surgical management of cerebral metastases from melanoma: outcome in 147 patients treated at a single institution over two decades. J Neurosurg 2002;96:552–8.
29. Baram TZ, van Tassel P, Jaffe NA. Brain metastases in osteosarcoma: incidence, clinical and neuroradiological findings and management options. J Neurooncol 1988;6:47–52.
30. Espat NJ, Bilsky M, Lewis JJ, Leung D, Brennan MF. Soft tissue sarcoma brain metastases. Prevalence in a cohort of 3829 patients. Cancer 2002;94:2706–11.
31. Feeney JJ, Popek EJ, Bergman WC. Leiomyosarcoma metastatic to the brain: case report and literature review. Neurosurgery 1985;16:398–401.
32. Jeffery DR, Ford CC. Fibrosarcomatous metastasis to the central nervous system with overt hemorrhage: case report and review of the literature. J Neurol Sci 1995;134:207–9.
33. Lewis AJ. Sarcoma metastatic to the brain. Cancer 1988;61:593–601.
34. Salvati M, Cervoni L, Caruso R, Gagliardi FM, Delfini R. Sarcoma metastatic to the brain: a series of 15 cases. Surg Neurol 1998;49:441–4.
35. Yu L, Craver R, Baliga M, et al. Isolated CNS involvement in Ewing's sarcoma. Med Pediatr Oncol 1990;18:354–8.

36. Delattre JY, Krol G, Thaler HT, Posner JB. Distribution of brain metastases. Arch Neurol 1988; 45:741–4.
37. Weber P, Shepard KV, Vijayakumar S. Metastases to pineal gland. Cancer 1989;63:164–5.
38. Weiss HD, Richardson EP Jr. Solitary brainstem metastasis. Neurology 1978;28:562–6.
39. Al-Anazi A, Shannon P, Guha A. Solitary metastasis to the choroid plexus. Case illustration. J Neurosurg 2000;92:506.
40. Kohno M, Matsutani M, Sasaki T, Takakura K. Solitary metastasis to the choroid plexus of the lateral ventricle. Report of three cases and a review of the literature. J Neurooncol 1996;27:47–52.
41. Qasho R, Tommaso V, Rocchi G, Simi U, Delfini R. Choroid plexus metastasis from carcinoma of the bladder: case report and review of the literature. J Neurooncol 1999;45:237–40.
42. Vannier A, Gray F, Gherardi R, Marsault C, Degos JD, Poirier J. Diffuse subependymal periventricular metastases. Report of three cases. Cancer 1986;58:2720–5.
43. Costigan DA, Winkelman MD. Intramedullary spinal cord metastasis. A clinicopathological study of 13 cases. J Neurosurg 1985;62:227–33.
44. Miller DJ, McCutcheon IE. Hemangioblastomas and other uncommon intramedullary tumors. J Neurooncol 2000;47:253–70.
45. Taniura S, Tatebayashi K, Watanabe K, Watanabe T. Intramedullary spinal cord metastasis from gastric cancer. Case report. J Neurosurg 2000;93(Suppl 1):145–7.
46. Olson ME, Chernik NL, Posner JB. Infiltration of the leptomeninges by systemic cancer. A clinical and pathologic study. Arch Neurol 1974;30: 122–37.
47. Wasserstrom WR, Glass JP, Posner JB. Diagnosis and treatment of leptomeningeal metastases from solid tumors: experience with 90 patients. Cancer 1982;49:759–72.
48. Abdo AA, Coderre S, Bridges RJ. Leptomeningeal carcinomatosis secondary to gastroesophageal adenocarcinoma: a case report and review of the literature. Can J Gastroenterol 2002;16:807–11; quiz 831–2.
49. Gasecki AP, Bashir RM, Foley J. Leptomeningeal carcinomatosis: a report of 3 cases and review of the literature. Eur Neurol 1992;32:74–8.
50. Gonzalez-Vitale JC, Garcia-Bunuel R. Meningeal carcinomatosis. Cancer 1976;37:2906–11.
51. Kokkoris CP. Leptomeningeal carcinomatosis. How does cancer reach the pia-arachnoid? Cancer 1983;51:154–60.
52. Lisenko Y, Kumar AJ, Yao J, Ajani J, Ho L. Leptomeningeal carcinomatosis originating from gastric cancer: report of eight cases and review of the literature. Am J Clin Oncol 2003;26:165–70.
53. Kleinschmidt-DeMasters BK. Dural metastases. A retrospective surgical and autopsy series. Arch Pathol Lab Med 2001;125:880–7.
54. Marin F, Kovacs KT, Scheithauer BW, Young WF Jr. The pituitary gland in patients with breast carcinoma: a histologic and immunocytochemical study of 125 cases. Mayo Clin Proc 1992;67:949–56.
55. Alleyne CH, Bakay RA, Costigan D, Thomas B, Joseph GJ. Intracranial adenoid cystic carcinoma: case report and review of the literature. Surg Neurol 1996;45:265–71.
56. Gormley WB, Sekhar LN, Wright DC, et al. Management and long-term outcome of adenoid cystic carcinoma with intracranial extension: a neurosurgical perspective. Neurosurgery 1996;38:1105–12; discussion 1112–3.
57. Chamberlain MC, Kormanik PA. Epidural spinal cord compression: a single institution's retrospective experience. Neuro-oncol 1999;1:120–3.
58. Schiff D, O'Neill BP, Suman VJ. Spinal epidural metastasis as the initial manifestation of malignancy: clinical features and diagnostic approach. Neurology 1997;49:452–6.
59. Stark RJ, Henson RA, Evans SJ. Spinal metastases. A retrospective survey from a general hospital. Brain 1982;105(Pt 1):189–213.
60. Yuguang L, Meng L, Shugan Z, et al. Intracranial tumoural haemorrhage—a report of 58 cases. J Clin Neurosci 2002;9:637–9.
61. Madow L, Alpers BJ. Encephalitic form of metastatic carcinoma. AMA Arch Neurol Psychiat 1951;65:161–71.
62. Cosgrove M, Fitzgibbons PL, Sherrod A, Chandrasoma PT, Martin SE. Intermediate filament expression in astrocytic neoplasms. Am J Surg Pathol 1989;13:141–5.
63. Ng HK, Lo ST. Cytokeratin immunoreactivity in gliomas. Histopathology 1989;14:359–68.
64. Oh D, Prayson RA. Evaluation of epithelial and keratin markers in glioblastoma multiforme: an immunohistochemical study. Arch Pathol Lab Med 1999;123:917–20.
65. Perry A, Parisi JE, Kurtin PJ. Metastatic adenocarcinoma to the brain: an immunohistochemical approach. Hum Pathol 1997;28:938–43.
66. Gottschalk J, Jautzke G, Paulus W, Goebel S, Cervos-Navarro J. The use of immunomorphology to differentiate choroid plexus tumors from metastatic carcinomas. Cancer 1993;72:1343–9.
67. Oliveira AM, Tazelaar HD, Myers JL, Erickson LA, Lloyd RV. Thyroid transcription factor-1 distinguishes metastatic pulmonary from well-differentiated neuroendocrine tumors of other sites. Am J Surg Pathol 2001;25:815–9.

68. Miettinen M, Fernandez M, Franssila K, Gatalica Z, Lasota J, Sarlomo-Rikala M. Microphthalmia transcription factor in the immunohistochemical diagnosis of metastatic melanoma: comparison with four other melanoma markers. Am J Surg Pathol 2001;25:205–11.
69. DeYoung BR, Wick MR. Immunohistologic evaluation of metastatic carcinomas of unknown origin: an algorithmic approach. Semin Diagn Pathol 2000;17:184–93.
70. Mrak RE. Origins of adenocarcinomas presenting as intracranial metastases. An ultrastructural study. Arch Pathol Lab Med 1993;117:1165–9.
71. Hamazaki S, Nakashima H, Matsumoto K, Taguchi K, Okada S. Metastasis of renal cell carcinoma to central nervous system hemangioblastoma in two patients with von Hippel-Lindau disease. Pathol Int 2001;51:948–53.
72. Mottolese C, Stan H, Giordano F, Frappaz D, Alexei D, Streichenberger N. Metastasis of clear-cell renal carcinoma to cerebellar hemangioblastoma in von Hippel Lindau disease: rare or not investigated? Acta Neurochir (Wien) 2001;143:1059–63.

APPENDICES

APPENDIX A

CENTRAL NERVOUS SYSTEM NEOPLASMS BY DEGREE OF CIRCUMSCRIPTION

Infiltrating
- Diffuse astrocytomas (grades II, III, and IV)
- Granular cell astrocytoma
- Gliomatosis cerebri
- Protoplasmic astrocytoma
- Pleomorphic xanthoastrocytoma (cerebral cortical component)
- Oligodendrogliomas (grades II and III)
- Choroid plexus carcinoma (some cases)
- Embryonal tumors
- Pineoblastoma
- Anaplastic meningioma
- Malignant mesenchymal tumors
- Primary central nervous system (CNS) lymphoma
- Microgliomatosis

Well-Circumscribed[a]
- Astrocytomas
  - Gliosarcoma (mesenchymal component)
  - Pilocytic astrocytoma
  - Pleomorphic xanthoastrocytoma (leptomeningeal component)
  - Subependymal giant cell astrocytoma (tuberous sclerosis)
  - Granular cell tumor of the infundibulum
  - Pituicytoma
- Ependymoma
- Subependymoma
- Astroblastoma
- Chordoid glioma of the third ventricle
- Choroid plexus papilloma
- Neuronal and glioneuronal tumors
  - Ganglion cell tumors, including desmoplastic infantile ganglioglioma
  - Neurocytic tumors
- Hemangioblastoma
- Pineocytoma
- Well-differentiated meningioma
- Paraganglioma
- Schwannoma
- Melanocytoma (some cases)
- Craniopharyngioma
- Benign cystic lesions

[a]The designation well-circumscribed refers to the degree of circumscription at the macroscopic level. Most astrocytomas, even well-circumscribed types such as pilocytic astrocytoma, are to some extent infiltrative at the microscopic level. The infiltrative properties of some lesions, e.g., extensively nodular medulloblastomas, are not clear.

APPENDIX B

INTRASPINAL LESIONS BY ANATOMIC LOCATION

Extradural
- Herniated nucleus pulposus
- Metastatic neoplasm
- Lymphoma
- Myeloma
- Bacterial abscess
- Tuberculoma
- Primary spine and soft tissue tumors
- Calcific pseudoneoplasm of the neuraxis
- Vascular malformations

Intradural
- Extramedullary
  - Schwannoma
  - Meningioma
  - Hemangiopericytoma
  - Solitary fibrous tumor
  - Mesenchymal chondrosarcoma
  - Melanocytic neoplasm
  - Meningeal gliomatosis
  - Non-neoplastic cysts (endodermal; arachnoid, including Tarlov cyst of nerve root)
  - Meningeal carcinomatosis (subarachnoid)
  - Drop metastasis (subarachnoid)
  - Inflammatory/infectious lesions, e.g., sarcoidosis, tuberculosis
  - Vascular malformations
  - Calcific pseudoneoplasm of the neuraxis
- Intramedullary
  - Ependymoma
    - Cellular
    - Myxopapillary (filum terminale)
    - Giant cell
    - Tanycytic
  - Subependymoma
  - Astrocytoma
    - Pilocytic
    - Diffuse
  - Hemangioblastoma
  - Paraganglioma (filum terminale)
  - Ganglion cell tumor
  - Schwannoma (rare)
  - Oligodendroglioma (rare)
  - Inflammatory/infectious lesions, e.g., demyelinating disease, sarcoidosis, tuberculosis
  - Metastatic neoplasm (rare)
  - Vascular malformations

APPENDIX C

INTRAVENTRICULAR TUMORS

Ependymoma (fourth, lateral, and third)

Subependymoma (lateral and fourth)

Subependymal giant cell astrocytoma (tuberous sclerosis) (lateral)

Chordoid glioma (third)

Choroid plexus tumors (lateral and third - children; fourth - adults)

Pilocytic astrocytoma (third, fourth, and lateral)

Central neurocytoma (lateral and with extension into the third in some cases)

Pineal parenchymal tumors (posterior third)

Germ cell tumors (posterior third)

Craniopharyngioma, especially papillary variant (third)

Colloid cyst (third)

Meningioma (uncommon; lateral, third, or fourth)

Choroid plexus xanthogranuloma and xanthoma (lateral - trigone)

Papillary tumor of the pineal region (posterior third)

Dysembryoplastic neuroepithelial tumor of septum pellucidum (lateral)

APPENDIX D

FREQUENTLY CYSTIC LESIONS

Pilocytic astrocytoma

Pleomorphic xanthoastrocytoma

Ganglioglioma

Desmoplastic infantile ganglioglioma

Extraventricular neurocytic neoplasm

Ependymoma (supratentorial and spinal)

Astroblastoma

Hemangioblastoma

Meningioma (occasional)

Craniopharyngioma

Schwannoma (larger lesions)

Teratoma

Cysts

- Colloid
- Rathke cleft
- Endodermal
- Ependymal ("glioependymal")
- Epidermoid and dermoid
- Arachnoid
- Choroid plexus
- Nerve root (Tarlov)
- Spinal synovial
- Pineal

APPENDIX E

OLIGODENDROGLIOMA-LIKE LESIONS AND NORMAL TISSUES WITH OLIGODENDROGLIOMA-LIKE FEATURES

Normal brain with:

- Artifactual perinuclear halos about cortical neurons
- Apparent increase in normal oligodendroglia (especially about chronic lesions)

Macrophage-rich lesions

- Demyelinating disease
- Progressive multifocal leukoencephalopathy
- Cerebral infarct
- Primary CNS lymphoma after corticosteroid treatment

Diffuse astrocytomas

Pilocytic astrocytoma

Clear cell ependymoma

Dysembryoplastic neuroepithelial tumor

Neurocytic neoplasms

- Central
- Extraventricular

Rosette-forming glioneuronal tumor of the fourth ventricle

APPENDIX F

FREQUENTLY CALCIFIED TUMORS

Oligodendroglioma (highly suggestive when gyriform)

Meningioma (especially spinal and olfactory groove)

Neurocytic neoplasms

Subependymal giant cell astrocytoma and hamartomas (tuberous sclerosis)

Ependymoma (intracranial)

Subependymoma (especially fourth ventricular)

Ganglion cell tumors

Pilocytic astrocytoma

Choroid plexus papilloma (fourth ventricular)

Pineal cyst (attached normal pineal)

Craniopharyngioma (adamantinomatous)

Meningioangiomatosis

Mature teratomas (some)

Calcifying pseudoneoplasm of the neuraxis

Vascular malformations

APPENDIX G

TUMORS WITH GANGLION CELLS OR SIMILAR CELLS

Infiltrating gliomas with trapped neurons

Ganglion cell tumors (gangliocytoma, ganglioglioma)

Desmoplastic infantile ganglioglioma

Dysplastic gangliocytoma (Lhermitte-Duclos disease)

Hypothalamic hamartoma

Medulloblastoma (uncommon)

Ganglioneurocytoma

Ganglioneuroblastoma

Pleomorphic xanthoastrocytoma (occasional)

Pilocytic astrocytoma (occasional)

Paraganglioma (in some cases)

Pituitary adenoma with ganglion cell metaplasia

Cortical dysplasia

Cortical tuber

Subependymal giant cell astrocytoma (tuberous sclerosis)

Nasal cerebral heterotopia (“nasal glioma”)

Meningioangiomatosis (trapped cortical neurons)

Lipoma of cranial nerve VIII

APPENDIX H

MELANOTIC/PIGMENTED LESIONS

Melanocytic neoplasms (melanocytoma, melanocytic tumor of intermediate differentiation, melanoma, neurocutaneous melanosis)

Melanotic schwannoma

Melanotic medulloblastoma

Pineal anlage tumor

Melanotic neuroectodermal tumor of infancy

Teratoma

Miscellaneous lesions[a]

- Choroid plexus neoplasms
- Ependymoma
- Ganglion cell tumor
- Astrocytoma
- Neurocytoma
- Pleomorphic xanthoastrocytoma

---

[a]All only rarely pigmented, and in some cases with lipofuscin rather than melanosome-associated pigment.

APPENDIX I

CENTRAL NERVOUS SYSTEM TUMORS WITH MYOID DIFFERENTIATION

Rhabdomyosarcoma

Medullomyoblastoma

Small cell embryonal tumor (peripheral neuroendocrine tumor [PNET]) with muscle differentiation

Leiomyoma and leiomyosarcoma

Neuromuscular hamartoma of peripheral (cranial) nerve

Teratoma

Rhabdomyoma (eighth cranial nerve)

Atypical teratoid tumor (immunohistochemical finding only, some cases)

Pineal anlage tumor

APPENDIX J

CENTRAL NERVOUS SYSTEM TUMORS WITH RIBBONING OR PALISADES

Glioblastoma

Oligodendroglioma

Pilocytic astrocytoma

Medulloblastoma

Neuroblastoma

Polar spongioblastoma

Paraganglioma (uncommon)

Malignant neoplasm following irradiation of acute lymphocytic leukemia

APPENDIX K

CENTRAL NERVOUS SYSTEM TUMORS POTENTIALLY OVERGRADED AS AGGRESSIVE OR ANAPLASTIC NEOPLASMS

Pilocytic Astrocytoma is overgraded when mistaken for diffuse astrocytoma of any grade. This error is especially likely with lesions in the cerebral hemispheres and spinal cord where pilocytic astrocytomas are incorrectly assumed to be rare. Pilocytic astrocytoma should always be included in the differential diagnosis of a discrete, contrast-enhancing, cystic lesion. Vascular proliferation is common.

Ganglion Cell Tumors may be overgraded as diffuse astrocytoma if the sometimes inconspicuous neural component is overlooked. The pleomorphism of both neurons and glia can be misinterpreted as evidence of anaplasia, particularly at the time of frozen section. Vascular proliferation is frequent in the cyst wall, and in some cases within the tumor parenchyma.

Pleomorphic Xanthoastrocytoma (PXA) may, given its pleomorphism, be mistaken for a malignant glioma. This error is avoided by awareness of the clinical (young age, association with seizures) and radiologic (superficial enhancing nodule with underlying cyst) features of classic PXA, and the disparity between extensive cytologic atypia and the absence of mitotic figures and necrosis.

Hemangioblastoma can closely resemble diffuse astrocytoma in frozen sections, whereas the non-neoplastic cyst wall can simulate pilocytic astrocytoma. The lesion can resemble clear cell carcinoma in permanent sections. Awareness of the tumor's radiologic features and its occurrence at specific sites (medulla and spinal cord) minimizes the likelihood of diagnostic confusion.

Neurocytic Neoplasms are overgraded if mistaken for oligodendroglioma, ependymoma, or neuroblastoma. The central neurocytoma is readily recognized in the context of its clinical and radiographic features (a large intraventricular, often calcified, midline mass with no apparent intraparenchymal component). Extraventricular neurocytomas also have radiologic clues in the form of the lesion's discreteness, contrast enhancement, and frequent cyst formation and calcification).

Dysembryoplastic Neuroepithelial Tumor (DNT) may be similar, if not identical, to oligodendroglioma, particularly in small fragmented specimens. The presence of occasional trapped ganglion cells may suggest a ganglion cell tumor. Recognition is aided by awareness of the DNT's stereotypic clinical (young patient, history of partial complex seizures), radiologic (intracortical location), and histologic (often mucoid, loose-textured nodules, "floating neurons") features.

Desmoplastic Infantile Ganglioglioma may be misinterpreted as a meningeal sarcoma, malignant glioma, or embryonal tumor depending upon the component sampled. The lesion is recognized by its presentation in infancy, radiologic profile (large superficial solid tumor mass, underlying cyst), and histologic findings that include a prominent desmoplastic component.

APPENDIX L

NON-NEOPLASTIC LESIONS POTENTIALLY MISINTERPRETED AS NEOPLASMS

Reactive Gliosis of either fibrillary or piloid type can be misinterpreted as a glioma if attention is focused on cytoplasmic features and not on cell density and distribution. Unlike well-differentiated diffuse astrocytoma, gliosis of the fibrillary type has an even cell distribution and little or no hypercellularity. In addition, the cells of active fibrillary gliosis are typically stellate, with long, tapering and circumferentially radiating processes. There is thus cellular hypertrophy but little hyperplasia. Multinucleated astrocytes (Creutzfeldt cells) are common in subacute lesions such as demyelinating disease.

Piloid Gliosis, a common finding in the environs of chronic mass lesions, typically affects the hypothalamus, cerebellum, and spinal cord. Microscopically, it consists of elongate, densely fibrillar cells accompanied by Rosenthal fibers, but lacks the spongy microcystic component so common in pilocytic astrocytoma.

Demyelinating Disease, a macrophage-rich lesion, may be misinterpreted as glioma due to its relatively high cellularity. The large reactive, occasionally atypical, astrocytes lend additional overtones of neoplasia. The diagnosis depends on vigilance enforced by the approach presented in *Appendix O* that demands that reactive lesions be excluded before the diagnosis of a neoplasm is entertained. Heightened suspicion is appropriate in a young patient presenting with one or more white matter lesions, especially when the latter are paraventricular or subcortical. On magnetic resonance imaging (MRI), the "open ring" pattern of enhancement is highly suggestive of the entity. Perivascular inflammation, although sometimes present in gliomas, also points to the possibility of demyelinating disease. Smear preparations aid in the recognition of macrophages and reactive astrocytes. Multinucleated Creutzfeldt cells are usually present. Immunohistochemical stains for macrophage markers help to confirm the diagnosis.

Progressive Multifocal Leukoencephalopathy may be misconstrued as a malignant glioma because of the often considerable pleomorphism and hyperchromasia of oligodendroglia and astrocytes within this demyelinating lesion. Identification depends on recognizing the macrophage-rich background and virus-infected oligodendroglia that are most abundant at the edge of the lesion.

Cerebral Infarct may be interpreted as a glioma, particularly a malignant variety, because of the high cellularity and hypertrophic, and sometimes even hyperplastic, vascular changes. As in demyelinating disease, macrophages dominate the subacute lesion.

Pineal Cyst may be mistaken for either: 1) a glioma, if, out of the radiologic of context, attention is focused on the gliotic wall surrounding the cyst cavity or 2) a pineocytoma, if consideration is directed toward the attached normal gland. Recognition is aided by the radiologic finding of a largely thin-walled pineal cyst, and by familiarity with the histologic features of normal pineal parenchyma and how closely these features can resemble those of a neoplasm.

Meningioangiomatosis may be misinterpreted as: 1) meningioma, if meningothelial cells are prominent; 2) glioma, because of the overall hypercellularity; or 3) ganglion cell neoplasm, because of trapped preexisting neurons. The diagnosis depends largely on the recognition of the entity as an intracortical plaque-like lesion with a perivascular proliferation of meningothelial cells and fibroblasts.

Hypothalamic Hamartoma may be misinterpreted as a ganglion cell tumor if attention is focused only on the neurons. Identification is simplified by: 1) cognizance of radiologic features; 2) degree of cellularity similar to normal gray matter; and 3) relative degree of organization of mature neurons and glia.

Dysplastic Gangliocytoma (Lhermitte-Duclos Disease), a cerebellar lesion, may be misinterpreted as a glioma or a ganglion cell tumor because of its abnormal, large and closely packed cells. The lesion is recognized by its characteristic features on MRI, and laminar architecture in histologic sections.

APPENDIX M

HISTOPATHOLOGIC FEATURES SUGGESTING LOW-GRADE OR NON-NEOPLASTIC LESION

Cystic Architecture showing large fluid-filled cavities is typical of well-circumscribed grades I and II lesions, as summarized in Appendix D. They are often surrounded by a layer of proliferating vessels as is illustrated in a ganglion cell tumor in figure A-1A.

Leptomeningeal Localization is usually present in grade I or II tumors such as pleomorphic xanthoastrocytoma (fig. A-1B,C), ganglion cells tumors (including desmoplastic infantile ganglioglioma), and some pilocytic astrocytomas. In some cases, the principal clue to the superficial location of the lesion comes from the presence within the mass of large muscularized arteries of the sort not found in the CNS parenchyma (fig. A-1C).

Sharp Borders can be present in high-grade lesions, even glioblastoma, but are more characteristic of low-grade noninfiltrating processes, summarized in Appendix A, and illustrated here in an intraparenchymal neurocytoma (fig. A-1D). An important non-neoplastic lesion, demyelinating disease, also has sharp borders (fig. A-1E).

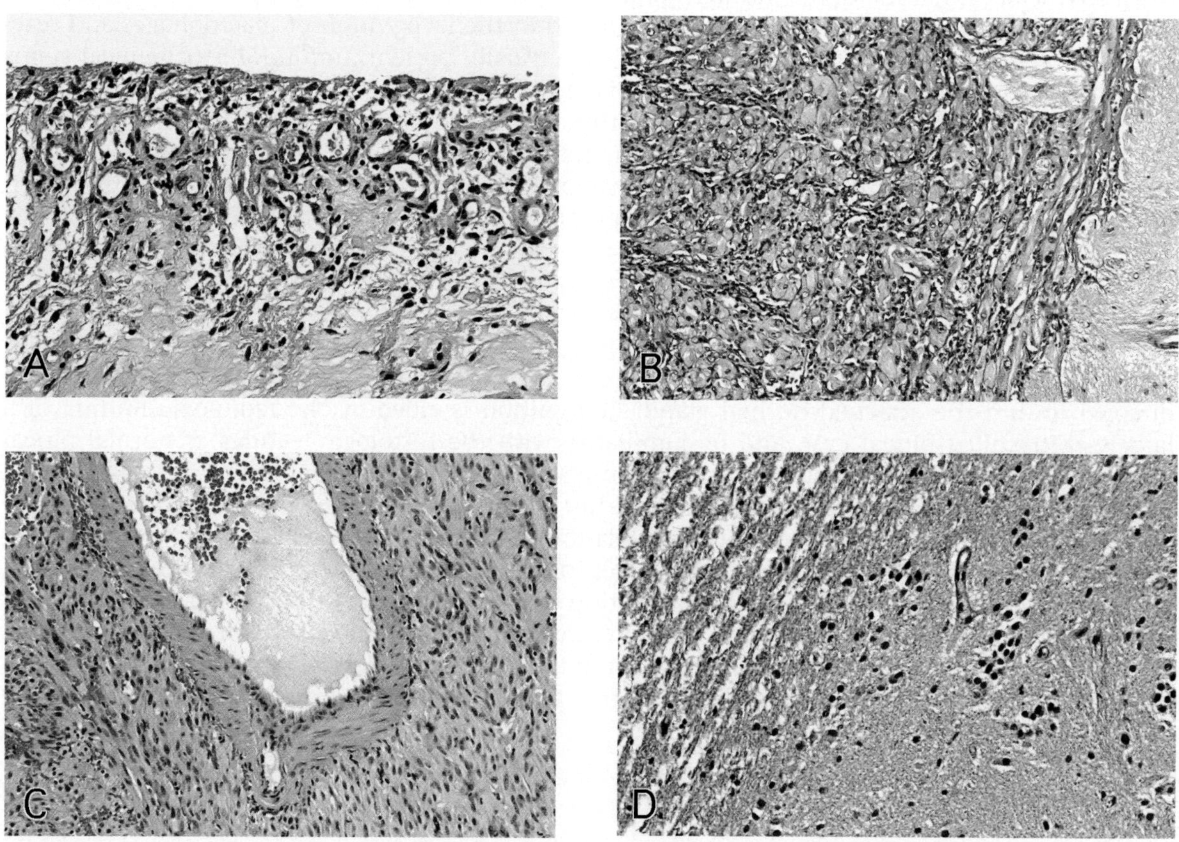

Figure A-1

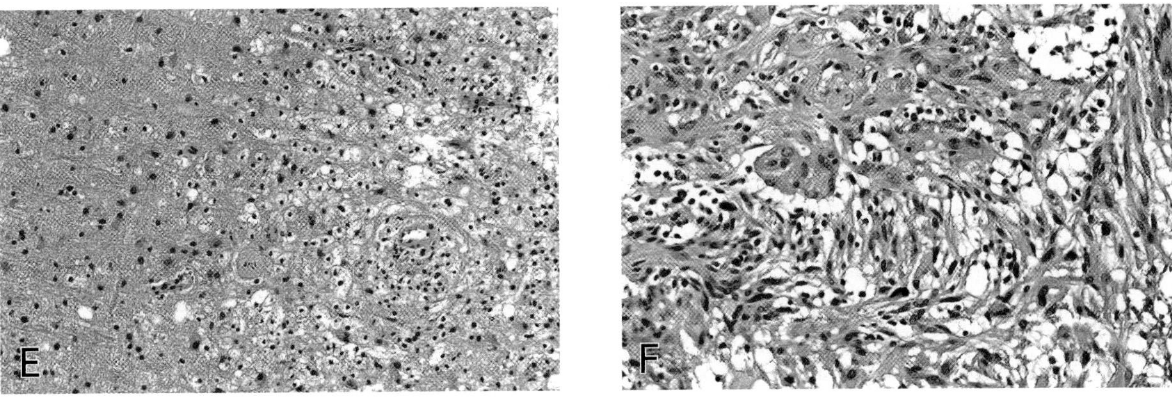

Figure A-1 (Continued)

Compact, Noninfiltrative Architecture with few if any preexisting normal parenchymal elements is usually a feature of the well-circumscribed lesions enumerated in Appendix A, and illustrated in a pilocytic astrocytoma (fig. A-1F). At first glance, lesions such as this can resemble glioblastoma.

Sclerotic, Hyalinized Vessels are common in long-standing lesions such as pilocytic astrocytoma (fig. A-2A), ganglion cell tumors, and some neurocytic neoplasms.

Chronic Inflammatory Cells are present in some infiltrating gliomas, especially gemistocytic astrocytoma, but are more common in ganglion cell tumors (fig. A-2B) and pleomorphic xanthoastrocytomas. These cells always are an indication of the possibility of an inflammatory state such as demyelinating disease or a low-grade neoplasm.

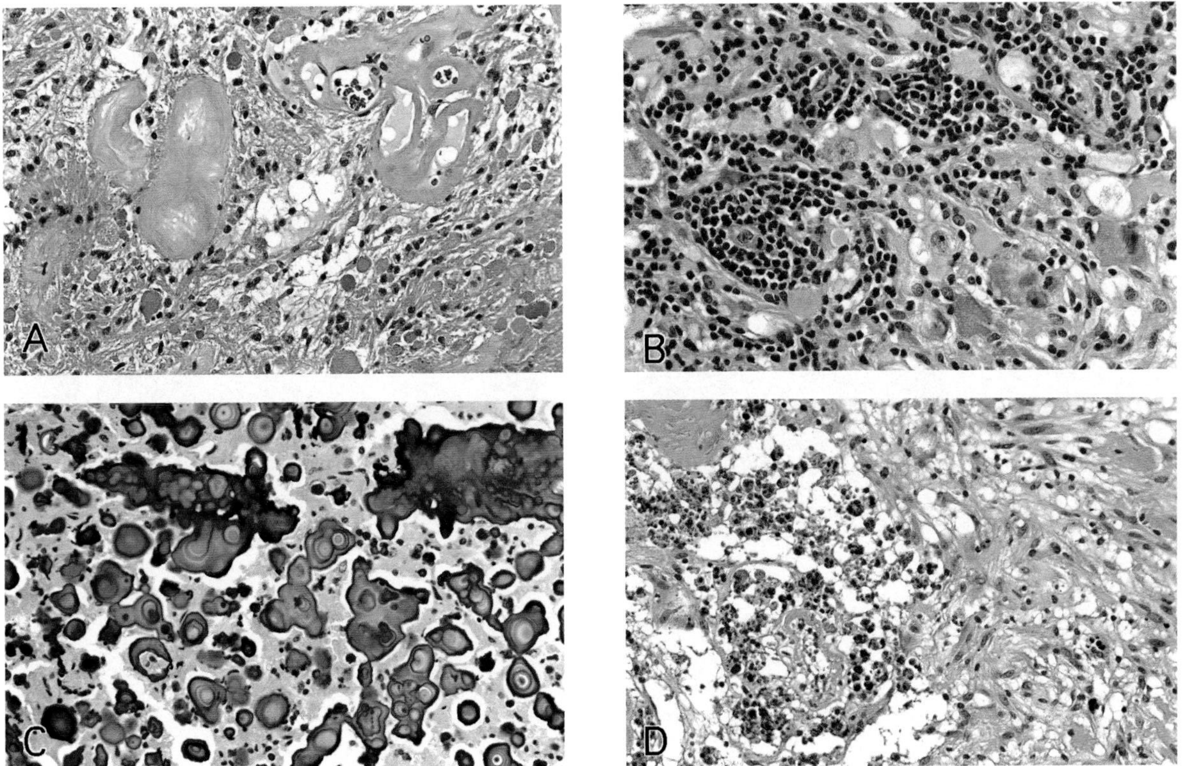

Figure A-2

Dense Calcification, while present in some infiltrating and even anaplastic neoplasms, is more often a feature of chronic, well-circumscribed tumors such as neurocytic neoplasms (fig. A-2C) and subependymoma. Calcified CNS neoplasms are summarized in Appendix F.

Abundant Hemosiderin is considerably more common in long-standing, and hence low-grade, lesions such as pilocytic astrocytoma (fig. A-2D) and vascular malformations such as cavernous angioma.

Eosinophilic Granular Bodies are common in pleomorphic xanthoastrocytoma (fig. A-3A), ganglion cell tumors, and pilocytic astrocytoma (see fig. A-2A).

Rosenthal Fibers are a hallmark of pilocytic astrocytoma (fig. A-3B) and piloid gliosis. Piloid gliosis is common in: 1) the hypothalamic region about craniopharyngiomas; 2) the cerebellum about hemangioblastomas; and 3) the spinal cord about any chronic mass, such as hemangioblastoma or ependymoma. Only rarely do Rosenthal fibers occur in diffuse astrocytomas.

Oligodendroglioma-Like Cells (OLCs) (Including Macrophages), or small clear cells with round nuclei, are found in many CNS neoplastic (often low-grade) and non-neoplastic entities. Figure A-3C and D illustrate OLCs in an extraventricular neurocytoma and macrophage-rich lesions such as demyelinating disease, respectively.

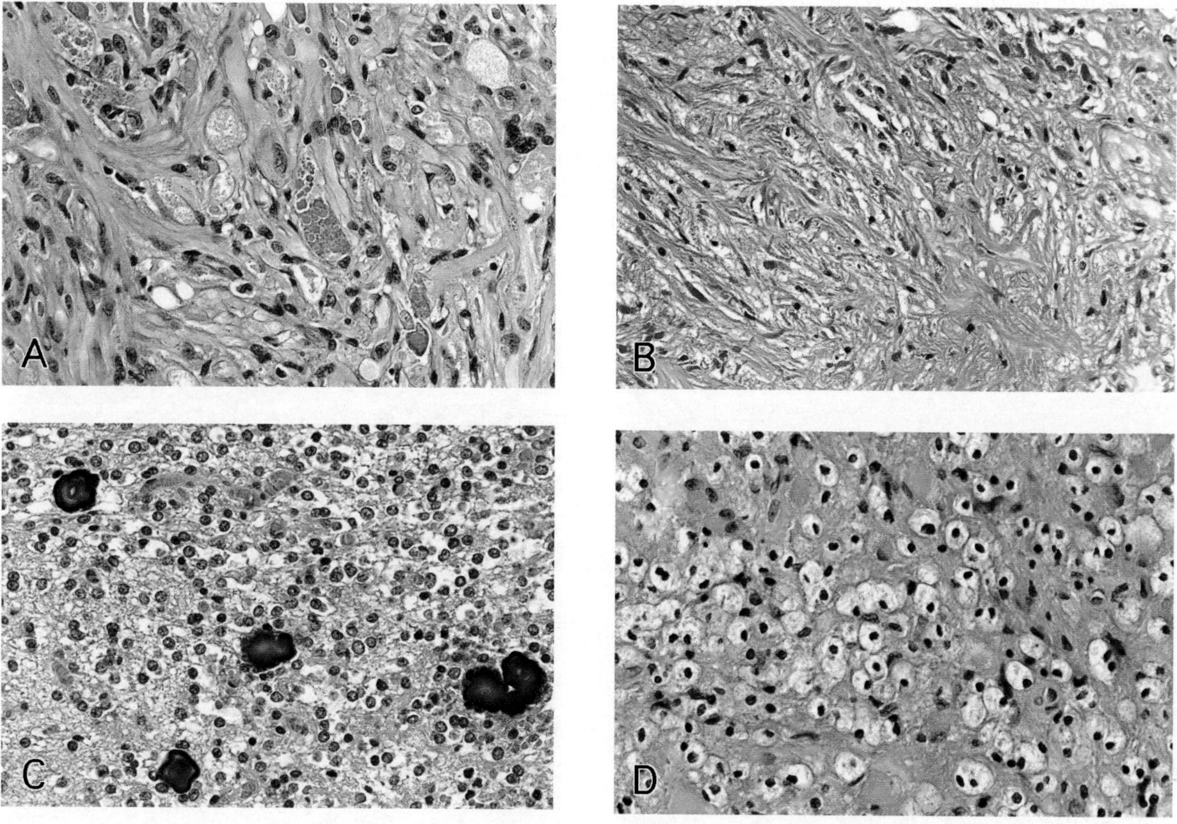

Figure A-3

APPENDIX N

SUSPECT DIAGNOSES IN SURGICAL NEUROPATHOLOGY

Malignant tumor (especially in young adults) with either a long clinical history (usually seizures) or scalloping of the inner table of the skull.

Well-differentiated (grade II) diffuse astrocytoma or oligodendroglioma in the face of a contrast-enhancing lesion on MRI or computerized tomography (CT).

Diffuse astrocytoma or oligodendroglioma of any grade when faced with a cystic lesion, especially one with a mural nodule.

Malignant neoplasm in the absence of neuroradiologic evidence of mass effect and/or peritumoral edema.

Glioma in the setting of a macrophage-rich lesion, especially one that is multifocal.

Infiltrating glioma, especially high grade, for a tumor with Rosenthal fibers or eosinophilic granular bodies.

Glioma or pineocytoma in the face of a cystic lesion in the pineal gland.

APPENDIX O

ALGORITHM FOR ANALYSIS OF CENTRAL NERVOUS SYSTEM SURGICAL SPECIMENS

Is the tissue abnormal?

Is the tissue neoplastic? (exclude reactive lesions such as demyelinating disease)

What is the tumor type? (exclude grades I and II neoplasms such as pilocytic astrocytoma, neurocytic neoplasms, and pleomorphic xanthoastrocytoma)

What is the tumor grade? (be cautious in assigning grade III or IV)

Is the pathologic diagnosis concordant with clinical and radiologic findings?

Index*

*In a series of numbers, those in boldface indicate the main discussion of the entity.

D

E

F

G

H

I

L

M

N

O

P

R

S

T

V

W

X

Z